MOLECULAR REARRANGEMENTS IN ORGANIC SYNTHESIS

MOLECULAR REARRANGEMENTS IN ORGANIC SYNTHESIS

Edited by

CHRISTIAN M. ROJAS
Barnard College, New York, NY, USA

WILEY

Published by John Wiley & Sons, Inc., Hoboken, New Jersey Published simultaneously in Canada

Library of Congress Cataloging-in-Publication Data:

Molecular rearrangements in organic synthesis / Edited by Christian M. Rojas.
pages cm
Includes bibliographical references and index.
ISBN 978-1-118-34796-6 (cloth)
1. Rearrangements (Chemistry) 2. Organic compounds–Synthesis. I. Rojas, Christian Miguel, editor.
QD281.R35M65 2015
547′.2–dc23

2015013578

Typeset in 10/12pt Times by SPi Global, Chennai, India.

10 9 8 7 6 5 4 3 2 1

1 2016

CONTENTS

LIST OF CONTRIBUTORS

Laura L. Anderson, Department of Chemistry, University of Illinois at Chicago, Chicago, IL, USA

Martin G. Banwell, Research School of Chemistry, Institute of Advanced Studies, The Australian National University, Canberra, Australia

Hongli Bao, Key Laboratory of Coal to Ethylene Glycol and Its Related Technology, Fujian Institute of Research on the Structure of Matter, Chinese Academy of Sciences, Fuzhou, Fujian, China

David J.-Y. D. Bon, Research School of Chemistry, Institute of Advanced Studies, The Australian National University, Canberra, Australia

Noritaka Chida, Department of Applied Chemistry, Keio University, Kohoku-ku, Yokohama, Japan

Yoann Coquerel, Aix Marseille Université, Centrale Marseille, CNRS, iSm2 UMR 7313, Marseille, France

Donald Craig, Department of Chemistry, Imperial College London, South Kensington Campus, London, UK

Laurent El Kaim, Laboratoire de Synthèse Organique, CNRS, Ecole Polytechnique, ENSTA ParisTech – UMR 7652, Universtité Paris-Saclay, Palaiseau, France

Xiaoming Feng, Key Laboratory of Green Chemistry and Technology, Ministry of Education, College of Chemistry, Sichuan University, Chengdu, P.R. China

Diego Gamba-Sánchez, Laboratory of Organic Synthesis, Bio and Organocatalysis, Chemistry Department, Universidad de los Andes, Bogotá, Colombia

Alison Xiang Gao, Department of Chemistry, The Scripps Research Institute, Jupiter, FL, USA

Fabián Garzón-Posse, Laboratory of Organic Synthesis, Bio and Organocatalysis, Chemistry Department, Universidad de los Andes, Bogotá, Colombia

Gunda I. Georg, University of Minnesota, Department of Medicinal Chemistry and Institute for Therapeutics Discovery and Development, Minneapolis, MN, USA

Nicolas Gigant, Institut de Chimie Organique et Analytique, UMR 7311 CNRS, Université d'Orléans, rue de Chartres, Orléans Cedex 2, France

Isabelle Gillaizeau, Institut de Chimie Organique et Analytique, UMR 7311 CNRS, Université d'Orléans, rue de Chartres, Orléans Cedex 2, France

Laurence Grimaud, Département de Chimie, Ecole Normale Supérieure-PSL Research University, Paris, France; CNRS, UMR 8640, PASTEUR, Sorbonne Universités, UPMC Univ Paris 06, Paris, France

Michael Harmata, Department of Chemistry, University of Missouri-Columbia, Columbia, MO, USA

Joanne E. Harvey, School of Chemical and Physical Sciences, Victoria University of Wellington, Wellington, New Zealand

Trinh A. D. Holth, University of Minnesota, Department of Medicinal Chemistry and Institute for Therapeutics Discovery and Development, Minneapolis, MN, USA

Oliver E. Hutt, CSIRO, Industrial Chemistry and Biotechnology, Clayton South, Australia

Minoru Isobe, Department of Chemistry, National Tsing Hua University, Hsinchu, Taiwan, ROC

Lili Lin, Key Laboratory of Green Chemistry and Technology, Ministry of Education, College of Chemistry, Sichuan University, Chengdu, P.R. China

Xiaohua Liu, Key Laboratory of Green Chemistry and Technology, Ministry of Education, College of Chemistry, Sichuan University, Chengdu, P.R. China

Janis Louie, Department of Chemistry, University of Utah, Salt Lake City, UT, USA

Graham K. Murphy, Department of Chemistry, University of Waterloo, Waterloo, ON, Canada

Chatchawan Ploysuk, Department of Chemistry, National Tsing Hua University, Hsinchu, Taiwan, ROC

Jean Rodriguez, Aix Marseille Université, Centrale Marseille, CNRS, iSm2 UMR 7313, Marseille, France

Christian M. Rojas, Department of Chemistry, Barnard College, New York, NY, USA

Michiko Sasaki, Department of Synthetic Organic Chemistry, Institute of Biomedical and Health Sciences, Hiroshima University, Hiroshima, Japan

Scott A. Snyder, Department of Chemistry, The Scripps Research Institute, Jupiter, FL, USA; Department of Chemistry, Columbia University, New York, NY, USA

Kei Takeda, Department of Synthetic Organic Chemistry, Institute of Biomedical and Health Sciences, Hiroshima University, Hiroshima, Japan

Uttam K. Tambar, Department of Biochemistry, The University of Texas Southwestern Medical Center at Dallas, Dallas, TX, USA

Ashish Thakur, Department of Chemistry, University of Utah, Salt Lake City, UT, USA

Stephen B. Thomas, Department of Chemistry, Columbia University, New York, NY, USA

F. G. West, Department of Chemistry, University of Alberta, E3-43 Gunning-Lemieux Chemistry Centre, Edmonton, AB, Canada

Takashi Yamazaki, Division of Applied Chemistry, Institute of Engineering, Tokyo University of Agriculture and Technology, Tokyo, Japan

Lin Zhou, Key Laboratory of Green Chemistry and Technology, Ministry of Education, College of Chemistry, Sichuan University, Chengdu, P.R. China

PREFACE

There is an aesthetically pleasing quality – a Kekuléan snake-seizing-its-tail appeal – to molecular rearrangements. A chemist using the arrow-pushing formalism to outline a rearrangement reaction feels a certain tactile satisfaction in charting the electron flow, an aspect that makes these reactions fun for instructors to teach or for students to work through at the blackboard. Rearrangements often have perfect atom economy, satisfying the frugal: bonds reorganized and nothing wasted. In some cases, a small molecule or other appropriate leaving group is extruded, initiating the rearrangement; sometimes, a catalyst is involved. Whatever the details, molecular rearrangements possess an inherent elegance that makes them especially appealing to students and practitioners of organic chemistry.

But molecular rearrangements do not just satisfy the intellect; they are irreplaceable tools in getting the work of organic chemical synthesis done. Different types of rearrangements enable synthetically useful operations, including ring expansions as in certain Beckmann rearrangements, ring-contracting reactions (e.g., the quasi-Favorskii and Ramberg–Bäcklund rearrangements), and functional group transpositions as occur in Payne and Mislow–Evans rearrangements. Another synthetically essential hallmark of many rearrangements is their stereospecificity, providing, as in the Claisen rearrangement, a means to parlay more readily established stereochemical features of the starting material into product stereochemistry that might otherwise be difficult to access.

How to group different molecular rearrangements? There is an understandable desire to classify and categorize, to find common themes, but there are many possible points of comparison. One could focus on functionality within the starting material, the type of product accessible via the rearrangement, or the reaction mechanism. To a certain extent, however, any categorization scheme is idiosyncratic and imperfect. For example, mechanistic categorization can be difficult for certain reactions, such as the vinylcyclopropane–cyclopentene rearrangement, which may have mechanistic variants, including radical, ionic, or metal-catalyzed versions. In this book, rearrangements are sorted mainly by changes in connectivity during the reaction: 1,2-migrations, 1,3-transpositions, or ipso rearrangements. However, sigmatropic rearrangements of differing order, [3,3] versus [2,3], are grouped together as a mechanistic category so as to hew to tradition and not confound the reader.

The purpose of this book is to provide readers with a clear and interesting point of departure for thought and further investigation. There is a clear focus on synthetic utility, analyzing how rearrangement reactions can meet challenges in the synthesis of complex molecular structures, including biologically active natural products. Importantly, the chapters are not intended as comprehensive reviews, and each is rendered in a distinctive voice. Some chapters take a broad outlook, while others concentrate on the authors' own contributions. Some chapters are more limited to the recent literature, while others provide historical context and discuss seminal studies in addition to current examples. By highlighting key examples and describing recent progress in the field, this book aims to help stimulate creative rearrangement-mediated solutions to contemporary challenges in synthetic organic chemistry.

In the realization of this volume, I am indebted foremost to the chapter authors for their willingness to participate in the project and their insightful, thorough treatment of the rearrangement topics. I am also grateful to Jonathan Rose, my editor at Wiley, for his patient guidance in completion of the book. Finally, I thank Ms. Jenny Lam for her invaluable assistance in compiling the manuscript components.

Christian M. Rojas
Barnard College
New York, NY
March 2015

PART I

1,2-MIGRATIONS

1

PINACOL AND SEMIPINACOL REARRANGEMENTS IN TOTAL SYNTHESIS

Alison Xiang Gao

Department of Chemistry, The Scripps Research Institute, Jupiter, FL, USA

Stephen B. Thomas

Department of Chemistry, Columbia University, NY, USA

Scott A. Snyder

Department of Chemistry, The Scripps Research Institute, Jupiter, FL, USA; Department of Chemistry, Columbia University, NY, USA

1.1 INTRODUCTION

Among the array of reactions available to alter molecular complexity, pinacol and semipinacol rearrangements have a particularly long history, constituting among the very first (if not the first) rearrangement reactions discovered by synthetic chemists.[1] However, despite being known for over a century and a half, their use in complex natural product synthesis has only recently come of age. Indeed, with a clearer understanding of the factors governing their regio- and stereoselectivity, as well as more powerful variants (including asymmetric) that can induce the rearrangement under mild conditions, these processes have a number of specific, but highly valuable, applications whose wealth is beginning to be tapped with ever greater frequency. This chapter seeks to provide a sense of the current state of the art of both pinacol and semipinacol processes, discussing each separately under the rubric of recent applications.

Molecular Rearrangements in Organic Synthesis, First Edition. Edited by Christian M. Rojas.

1.2 PINACOL REACTION

1.2.1 Background and Introduction

We begin with the pinacol rearrangement, a reaction process whose name derives from the starting material used in the earliest known example of the transformation. That event, the exposure of pinacol (**1**, 2,3-dimethylbutane-2,3-diol; Scheme 1.1) to sulfuric acid, produced pinacolone (**2**, 3,3-dimethylbutane-2-one). Although this reaction was first performed by Fittig in 1860,[2] it was not until the early 1870s that the actual structure of the product was confirmed by Butlerov[3]; this lapse not only reflects the challenges of determining structure in that era but also the fact that rearrangements were effectively unknown. In fact, a contemporary publication by Kekulé (his seminal paper on representing organic structures) included rules which suggested that carbon skeletal rearrangements could not occur.[4] In any event, by the end of the 19th century, the overall process depicted for the conversion of **1** to **2** was clear in terms of starting material and product. As additional substrates proved amenable to the process, the term pinacol rearrangement has since been used more broadly to define the conversion of any acyclic or cyclic vicinal diol into an aldehyde or ketone under acidic (proton or Lewis) conditions.[5]

Critically, as with many other skeletal rearrangements, there are subtleties that define both its mechanism as well as the products that can be generated from a given substrate under a specific set of reaction conditions. One early clue to that complexity derived from the observation by Danilov in 1917 that a single diol substrate (**4**) could yield different carbonyl-containing products (**3** or **5**) based solely on the strength of the acid deployed (Scheme 1.1b).[6] Although these outcomes reflect kinetic control, that analysis, in and of itself, is insufficient given that separate study has shown that both **4** and **5** can interconvert, suggesting the prospect of reversibility as part of the pinacol rearrangement process itself. Indeed, that concept was beautifully illustrated by Fry in a subsequent series of elegant ^{14}C-labeling studies which showed the transposition of carbon atoms of an array of pinacol "products" exposed to strongly acidic conditions (Scheme 1.1d).[7] A second critical observation resides in the fact that a number of pinacol rearrangements produce epoxides in addition to the standard carbonyl-containing adduct. In some cases, these materials can be induced to rearrange to standard pinacol products, while in others, they are inert to the reaction conditions. Finally, there is a wide range of contexts, employing both protic acids and Lewis acids under varying reaction conditions, that can induce the rearrangement to occur for most individual substrates (with expected variations in yield).

Collectively, these findings indicate that very few mechanistic conclusions can be drawn in general terms for all diols under all conditions. However, what is reasonable to presume, and/or consider, is participation of a substrate along the generalized mechanistic process shown in Scheme 1.1 while concurrently taking into account what is reasonable to occur under a given set of conditions. For instance, in strongly acidic aqueous media, diol substrate **6** is likely in equilibrium with epoxide **8**, with the more stabilized cation (**7**) being both the connecting intermediate and the active species for rearrangement (Scheme 1.1c). In this mechanistic paradigm, either **6** or

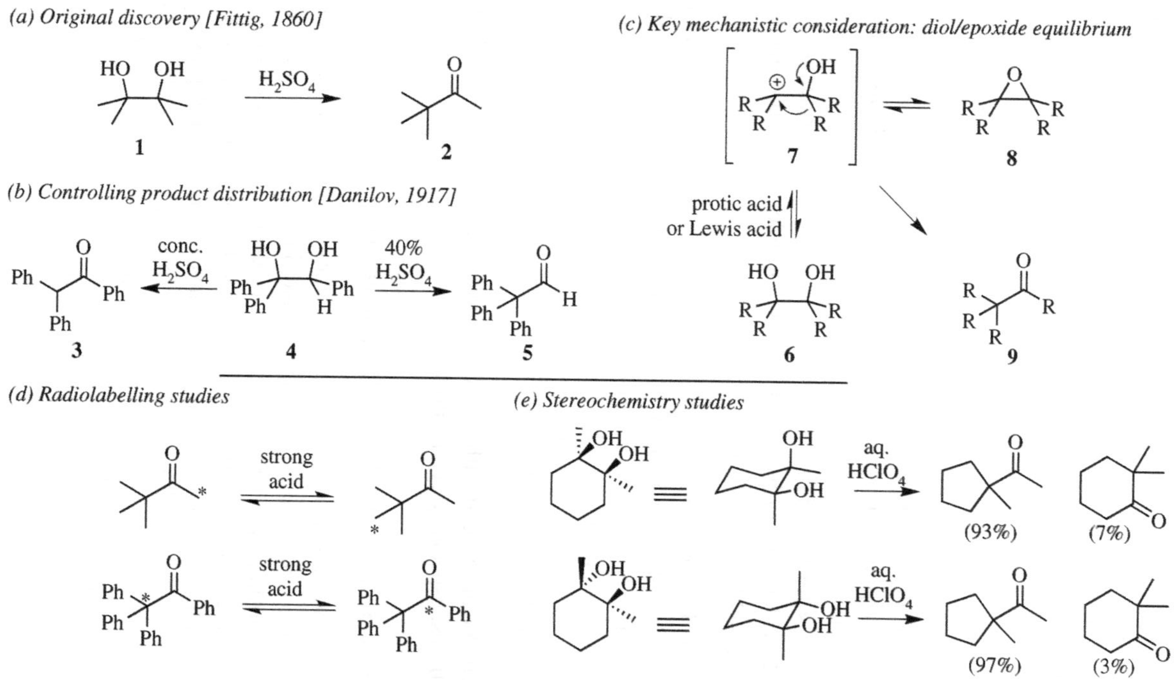

Scheme 1.1 The pinacol rearrangement: discovery and key considerations.

8 could be viewed as a reasonable starting material for the pinacol rearrangement, with the two substrates being effectively equivalent from a product-determining perspective. Under milder conditions, particularly as promoted by Lewis acids and/or when a good nucleophile is present, alternate pathways may proceed from **6**, **7**, and/or **8** to afford isolable intermediates and/or side products in addition to the desired pinacol adduct.

To put these thoughts, and the dozens of successful examples of the process, in more specific terms, the following generalizations can be made about pinacol rearrangements:

- Virtually any cyclic or acyclic vicinal diol can undergo the rearrangement, with aldehydes or ketones formed based solely on the substitution pattern of the diol.
- The reaction occurs in an exclusively intramolecular fashion. As such, symmetrical diols will yield a single product, while unsymmetrical diols may lead to product mixtures.
- The product is formed via the more stable carbocation intermediate, with the final product determined by the migratory aptitude of the substituents at the neighboring alcohol-bearing carbon.

What the pinacol rearrangement provides from a strategic perspective is the ability to generate carbonyl compounds with a high degree of substitution at the alpha position (particularly tertiary and quaternary systems), as well as to effect ring contraction and/or expansion with a high degree of regiocontrol in appropriate systems. Few other, if any, methods provide access to such products as readily.

Equally important is the following additional observation:

- The reaction can proceed with a high degree of stereoselectivity with appropriate substrates, especially cyclic diols.

1.2.2 Stereochemistry of the Pinacol Rearrangement

The alignment of orbitals as part of the bond migration itself ensures that stereochemical information encoded in the starting material can be expressed with high fidelity in the rearrangement if the substrate is designed appropriately. Again, however, it is critical to note that substrate-specific subtleties can also play a role. One representative example along these lines rounds out the presentation in Scheme 1.1. In this work by Bunton and Carr, exposure of two different diastereomers of 1,2-dimethyl-1,2-cyclohexanediol to aqueous $HClO_4$ at 60 °C afforded nearly indistinguishable distributions of two products, favoring the expected ring-contracted adduct (Scheme 1.1e).[8] This outcome suggests the intermediacy of the same carbocation intermediate. However, because the final product distributions are not exactly identical, there must be a slight stereochemical memory effect that contributes to (but clearly does not dominate) the reaction process. Intriguingly, even larger differences are found with analogous five-membered systems. This key component of effecting stereocontrol will be a critical point of discussion in the case studies that follow in later sections.

1.2.3 Preparation of Substrates for the Pinacol Rearrangement

The diol substrates needed for the pinacol rearrangement are readily constructed through a variety of procedures as denoted in Scheme 1.2. Though the overall diversity of methods is not as numerous as for some functional groups, the potential advantage, at least from a strategic perspective, is that there are only a select number of choices, streamlining synthetic planning. Critically, if the stereospecific, or chiral, preparation of these materials is necessary from achiral precursors, then only dihydroxylation, epoxidation, or α-functionalization of carbonyls are reasonable approaches.

1.2.4 Applications of the Pinacol Reaction in Complex Molecule Synthesis

1.2.4.1 Two Key Case Studies To see the general precepts of the preceding sections in action, we begin our discussion of specific examples with an elegant application of the pinacol rearrangement as part of a synthesis of the originally proposed structure of diazonamide A (**13**, Scheme 1.3), a marine natural product with antitumor activities originally harvested from the colonial ascidian *Diazona angulata*. This work, published in 2000 by the Harran group, highlights a number of the critical features in substrate design needed to render a pinacol rearrangement having high levels of regio- and stereocontrol, especially in a complex context.[9] Seeking to forge the all-carbon quaternary center linking the two 12-membered rings of the target, the Harran team designed 13-membered intermediate **10** in hopes that it could undergo a regio- and stereospecific pinacol rearrangement (and ring contraction) by way of a bridging phenonium intermediate that would communicate the original stereochemistry to forge the needed isomer of the quaternary center. Pleasingly, exposure of a 5:1 mixture of diols (favoring **10** as drawn and whose structure was confirmed by X-ray) to *p*-TsOH at elevated temperature afforded compound **12** in 54% overall yield following a terminal protection of the free amine unveiled during the acid-initiated rearrangement. Complete stereochemical fidelity was achieved in this process, with **12** arising only from the drawn epimer of **10**; the alternate, minor stereochemical diol isomer gave the alternate chirality at the newly formed quaternary carbon. Apart from providing the means to ultimately

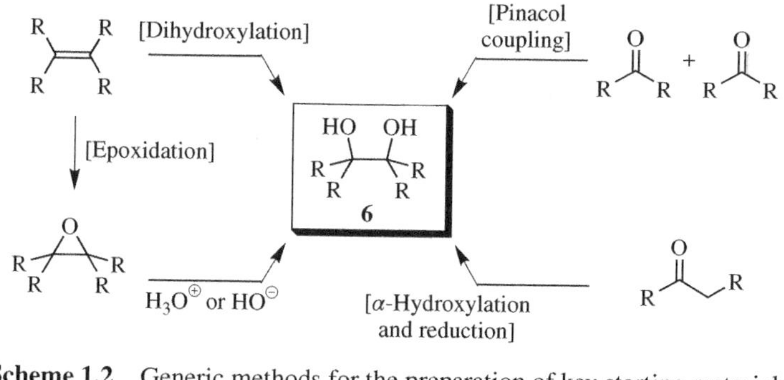

Scheme 1.2 Generic methods for the preparation of key starting materials.

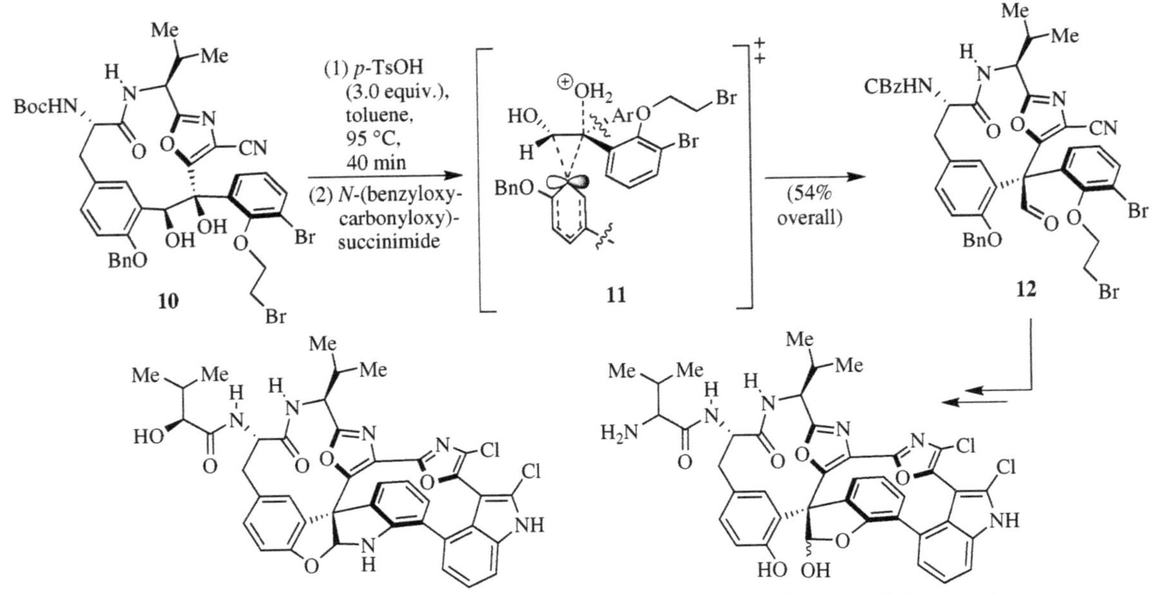

14: revised structure of diazonamide A

13: originally proposed structure of diazonamide A

Scheme 1.3 Use of a pinacol rearrangement as part of the total synthesis of the original structure of diazonamide A (**13**).[9a]

determine that the originally assigned structure of diazonamide A was incorrect (the Harran group revised it correctly to structure **14**), this work highlights how positional carbocation control, migrating group control, and stereocontrol can be merged together to accomplish a beautiful ring contraction, leading to a highly strained final product.

A more recent example which similarly displays key components of regio- and stereocontrol in a pinacol rearrangement process, with some additional critical twists, derives from the Suzuki group's approach to seragakinone A and BE-43472A (**19** and **20**, Scheme 1.4).[10] In their pursuit of these, and several other related polyketide targets, the Suzuki team required a general and effective method for establishing quaternary stereogenic centers at the angular position in these polycycles, a challenge that could not be addressed successfully by more conventional approaches such as enolate alkylation or nucleophilic displacement. Their more indirect solution utilizing the pinacol rearrangement is shown in Scheme 1.4, an approach that hinged upon the ability to access the requisite diol in stereodefined form (either **15** or **17**) and achieve regiospecific carbocation formation at the desired migration terminus with a subsequent stereospecific 1,2-shift (to generate **16** and **18**). The regiochemical requirement was the critical challenge, and its solution was the installation of the isoxazole ring within the substrates that afforded the stabilization necessary to render cation formation at the requisite site preferable over its mutually reinforcing tertiary allylic and benzylic alternative. As shown in the lower part of Scheme 1.4, the induced stabilization is so significant that the formation of **21** from generalized starting material **22** is effectively prevented, allowing for the desired 1,2-suprafacial shift to occur, with the cyclic constraints of the system driving stereoselectivity. Worth noting is that varying degrees of enantioselectivity were observed based on the migratory aptitude of the alkyl group, with better migrating groups affording enhanced enantioselectivities and highlighting the critical contribution of this component in the final, successful synthesis design.

1.2.4.2 Stereocontrol with Acyclic Diols In the absence of a preexisting cyclic system, achieving similar levels of stereocontrol with chiral diols can be difficult. An instructive example along these lines derives from the Pettit group in work targeting the antitumor natural product hydroxyphenstatin (**29**, Scheme 1.5).[11a] In this case, enantiomerically pure diol **27** was obtained through a Sharpless asymmetric dihydroxylation of the precursor olefin. Subsequent exposure to $BF_3 \cdot OEt_2$ in tetrahydrofuran (THF) at ambient temperature for 1 h afforded aldehyde **28** in 68% yield, but as a racemate (as determined by optical rotation experiments and X-ray crystallography). That absence of chiral control could be the result of racemization of the product given that the aldehyde is bis-benzylic, though it is worth noting that in other cases involving such aldehydes, that process has not been observed[11b]; alternatively, a stepwise mechanism through the intermediacy of two possible carbocationic intermediates could also be the cause for the loss of chiral information. Although in this case that outcome was fine in regard to the final benzophenone target, where stereochemistry was of no consequence, the fact that the chiral diol led to racemic product

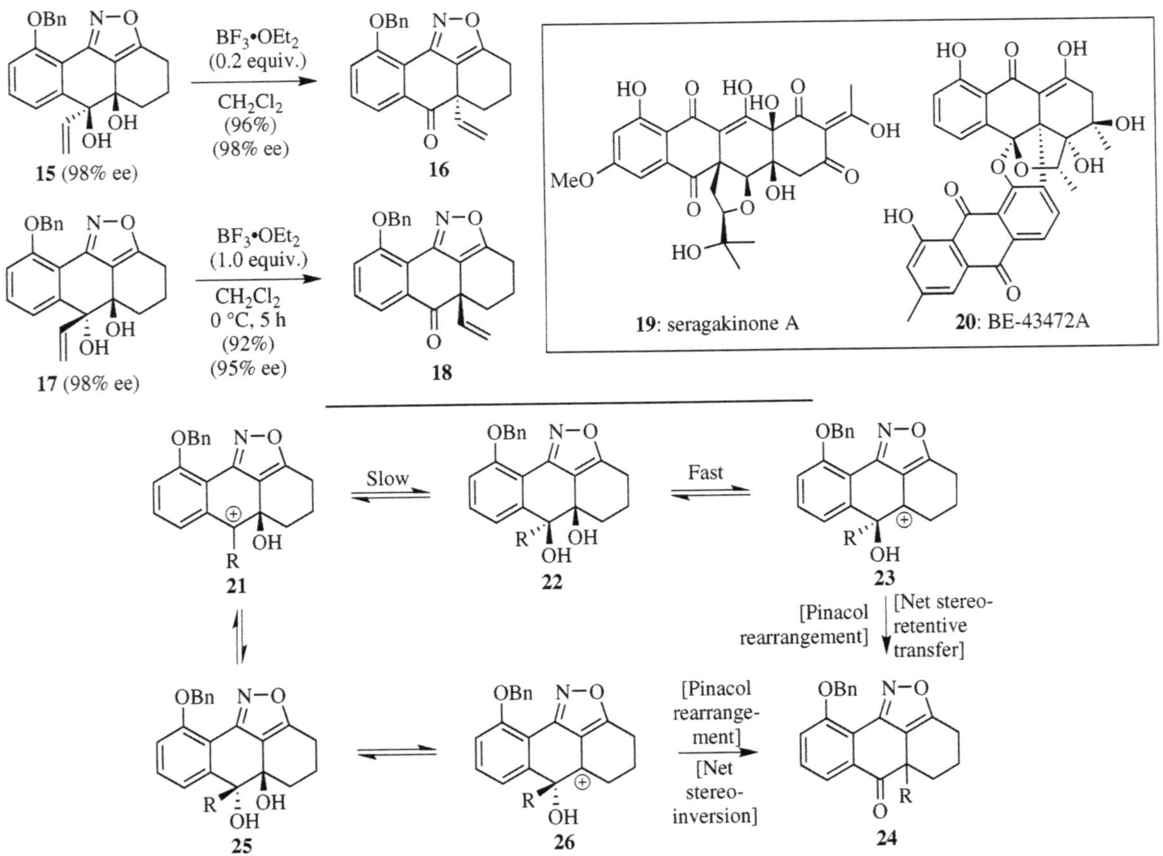

Scheme 1.4 An isoxazole-directed pinacol rearrangement as part of total syntheses of seragakinone A and BE-43472A (**19** and **20**).[10]

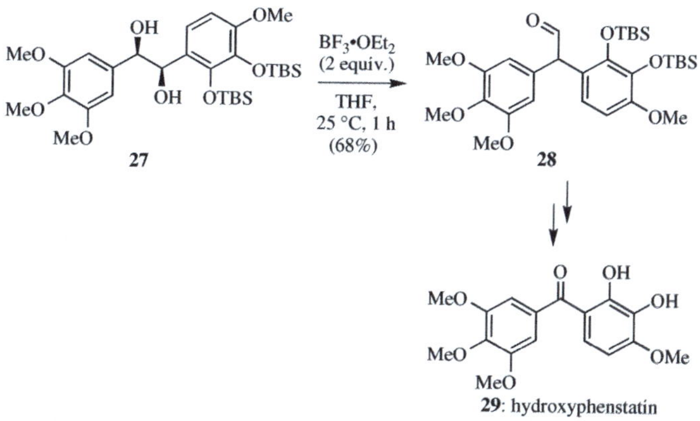

Scheme 1.5 Use of a pinacol rearrangement as part of the total synthesis of hydroxyphenstatin (**29**).[11a]

in this pinacol rearrangement highlights that in acyclic cases, absolute stereocontrol can be very difficult (if not sometimes impossible) to achieve.

In fact, there were no effective solutions to this general challenge with acyclic diols prior to 2010 when Antilla and coworkers provided an example of an asymmetric pinacol rearrangement involving racemic substrates of general flavor **30** (Scheme 1.6a, R = aryl).[12] These starting materials were very carefully designed, noting that it was expected that upon complexation with a chiral phosphoric acid (to afford **32**), acid-mediated dehydration could lead to iminium species **33**, a reactive intermediate previously shown to be compatible to chiral phosphoric acids for asymmetric nucleophile addition. Here, that nucleophile would be the internal migrating group in the pinacol rearrangement to forge chiral **31**. A variety of BINOL-derived phosphoric acids worked well in the process, with backbone variation directly correlating with enhancements in enantiospecificity. Chiral selectivity is postulated overall to arise via favorable electrostatic and hydrogen bonding interactions which constrain the 1,2-aryl shift to proceed with stereocontrol.

Inspired by this unique precedent, members of our group used the concept of a chiral phosphoric acid initiator to convert the diol diastereomers of **34** into the quaternary carbon of aldehyde **36**, a key precursor toward a total synthesis of the resveratrol dimer hopeanol (**37**, Scheme 1.6b).[13] The key ideas here hinge upon the ionization ability of the bis-benzylic tertiary alcohol versus the secondary benzylic alcohol affording requisite regiocontrol, while the chiral phosphoric acid could potentially impart some exogenous stereocontrol to enhance throughput to the desired diastereomer.

In initial probes of the process, traditional acid sources such as *p*-TsOH did indeed furnish pinacol rearrangement product **36** in moderate yield and diastereoselectivity along with a small amount of an epoxide side-product (which could not be converted to pinacol-rearranged material). However, the critical observation was that one diastereomer of the starting material reacted more quickly and under milder conditions than the other, suggesting that this diastereomer had a more favorable

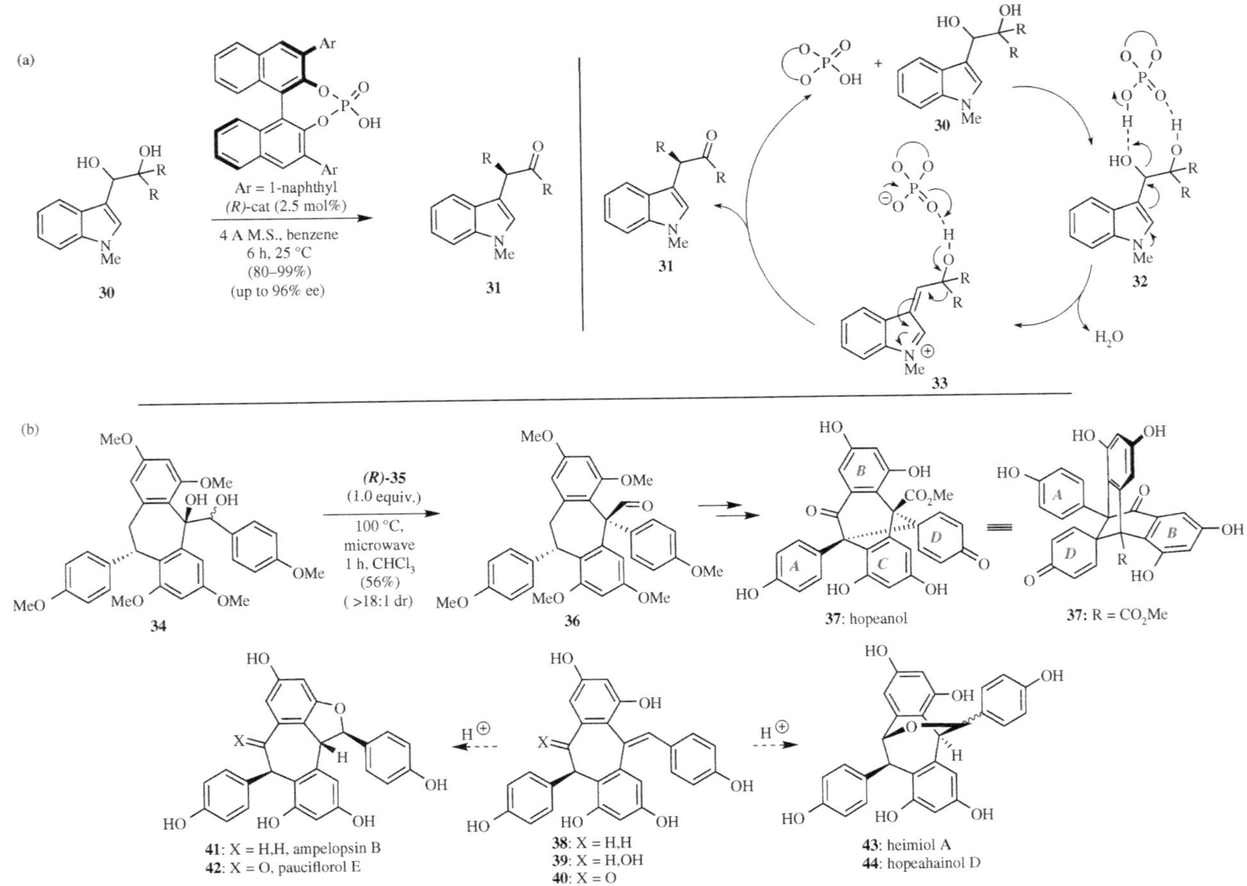

Scheme 1.6 Development of an asymmetric pinacol rearrangement by Antilla and coworkers[12] and related application to a total synthesis of hopeanol (**37**).[13]

stereochemical alignment than the other with facilitated transposition of the migrating aryl group due to advantageous orbital alignment with the departing alcohol. Intriguingly, when chiral phosphoric acids of BINOL flavor were instead employed to induce the rearrangement, the overall diastereoselectivity and yield were dramatically enhanced. Thus far, there is no evidence that the chiral phosphoric acids overturned the initial ratio of diastereomers of **34** that entered the reaction; however, it did indeed promote a rapid and clean reaction of the more reactive isomer such that it funneled toward desired product, while the less reactive isomer remained largely unaffected.

Thus, it would seem that the chiral phosphoric acid's reinforcement (and potential acceleration) of this preferential reactivity profile ensured that virtually none of the undesired diastereomer of **36** was formed under the reaction conditions. It seems appropriate, therefore, in an empirical sense to conclude that this increase in selectivity can be attributed to specific interactions between the diol moiety of the substrate and the chiral acid. Significantly, this notion may be applicable to other systems where such opportunities may exist. And, as a side note, the pinacol reaction process here may have biogenetic relevance as well, considering the structures of related natural products that are presumed to arise via acid-catalyzed processes from various oxidized forms of **38–40**. In this case, starting diol **34** is the fully oxidized version of these substrates (**38–40**) in terms of the double bond.

1.2.4.3 ***Pinacol Rearrangements in a Cascade Process*** We conclude this section with two final examples, each of which highlights two additional, and critical, components of pinacol rearrangement chemistry. What is emphasized here is the power of pinacol rearrangements when coupled with additional transformations in a cascade, or domino, set of processes where those events are promoted through a different intermediate other than the traditional carbocation generated directly by ionization of a diol. The first is slightly older work than the examples already described, and issued from the Overman group in 2003 where that key initiating event is a Prins cyclization.[14] As shown in a generic format (based on the protecting group used), the sequence afforded expedient and highly stereoselective constructions of the THF ring systems at the heart of several *Laurencia* sesquiterpenes (**50–52**, Scheme 1.7) through the mechanistic process delineated. Critically, enantiopure starting material (as in **45**) translated smoothly into enantiopure **47**, a material amenable for elaboration to several targets in the family.

The second example is more recent and was a critical part of a 14-step, gram-scale synthesis of ingenol (**54**, Scheme 1.8) by the Baran research group.[15] This case illustrates that the starting material for a successful pinacol rearrangement does not necessarily have to be a vicinal diol. Inspired by the structural similarities of this natural product with the related antitumor compound phorbol (**53**), these researchers postulated that perhaps the phorbol skeleton, in the form of a less oxidized intermediate such as **55**, could undergo a pinacol rearrangement using a stabilized vinylic carbocation to create the ingenol core. Only ring expansion to a seven-membered ring, not the nine-membered alternative, was expected even though the cycloheptanone is intramolecularly transfused onto the established scaffold. In practice, Baran and coworkers found that the TMS-protected vinylic alcohol **56**, when

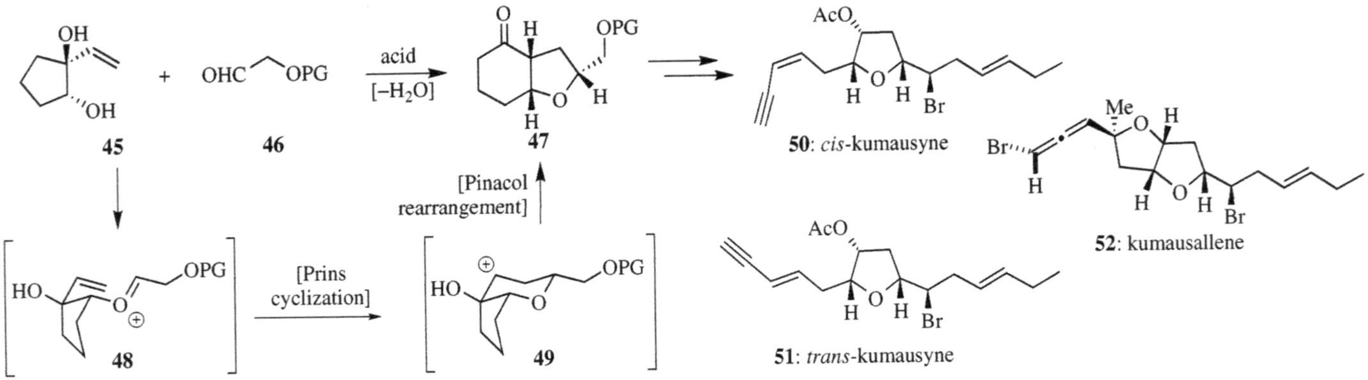

Scheme 1.7 Development of a Prins–pinacol rearrangement strategy for *Laurencia* sesquiterpenes (**50–52**).[14]

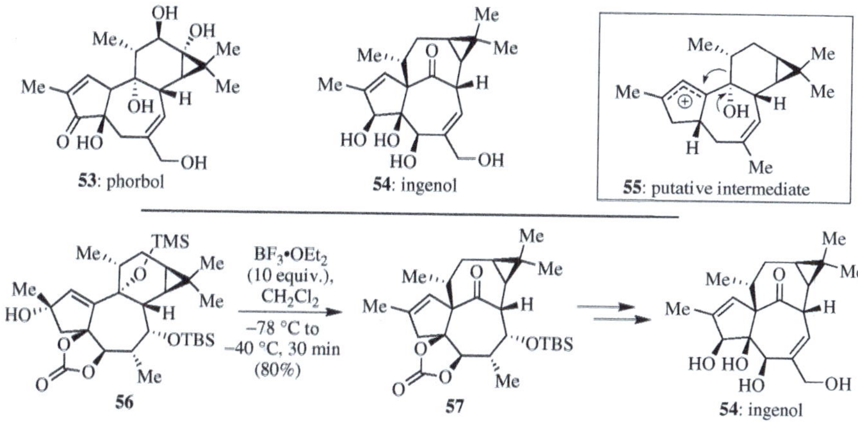

Scheme 1.8 Use of a vinylogous pinacol rearrangement as part of the total synthesis of ingenol (**54**).[15]

exposed to $BF_3{\cdot}OEt_2$, rearranged to the desired framework in 80% yield. Intriguingly, use of the unprotected alcohol variant of **56** and a variety of other alternatives failed to deliver the needed pinacol-rearranged target. Although formally this process can only be called a pinacol-type rearrangement, or more accurately a semipinacol rearrangement since the starting material is not explicitly a vicinal diol, it highlights additional variations of the process of high value in complex contexts, adds to the substrate scope and versatility of the pinacol rearrangement, and serves as a wonderful introduction to the seemingly more predictable, and controllable, rearrangement processes which will consume the remaining pages of this chapter.

1.3 SEMIPINACOL REARRANGEMENT

1.3.1 Background and Introduction

Semipinacol rearrangement events were first defined as a special type of pinacol rearrangement by Tiffeneau in 1923.[16] In Tiffeneau's original conception, these reactions involved migration toward the secondary carbon center on a tertiary/secondary diol as shown in Scheme 1.9 (**58** → **59**), the reverse regiochemistry of the typical pinacol rearrangement. Currently, however, this definition no longer applies, with the term semipinacol rearrangement referring to any process reminiscent of a pinacol rearrangement that utilizes a nondiol-based starting material.[14,17] Hence, the Baran ingenol example shown in Scheme 1.8 is technically a semipinacol rearrangement since it is a vinylogous diol, not a 1,2-diol, which served as the starting material. More formally, the key defining feature of a semipinacol rearrangement with these alternate starting materials is:

- 1,2-Migration of a C—C or C—H bond that is centered on the oxygen-bearing carbon and that occurs toward a vicinal electrophilic carbon center, generating a carbonyl group at the end of the process.

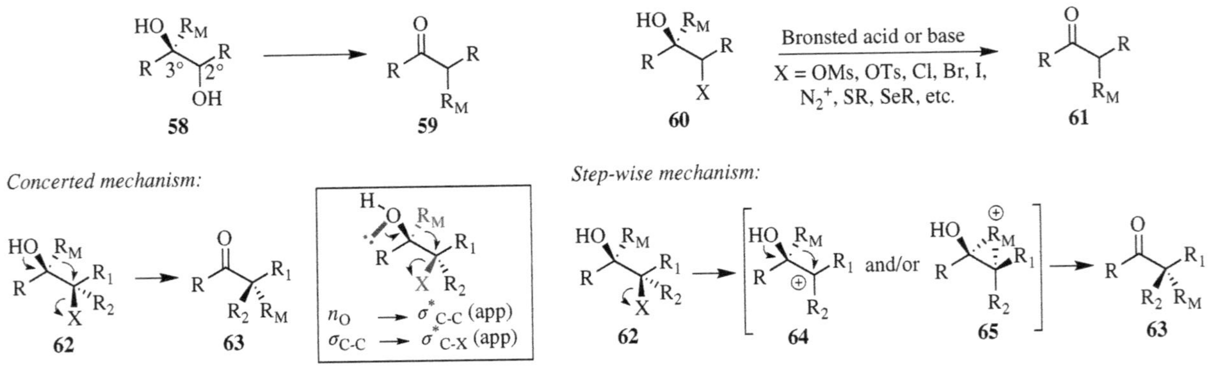

Scheme 1.9 General features and possible mechanisms of the semipinacol rearrangement.

This concept is represented by the generalized conversion of **60** into **61** (Scheme 1.9), highlighting the large variety of species that can serve as the departing group in the process. Among these, sulfonates, halides, N_2, thiolates, and selenolates are the most common.

1.3.2 Mechanism of the Semipinacol Rearrangement

Once activated by an appropriate Lewis acid, metal species, or even base (as we will see shortly in the examples that follow), the semipinacol rearrangement readily proceeds in a concerted manner. Stereoelectronically, the most favorable orbital alignment for the process is *anti*periplanar (as highlighted schematically). That concertedness translates into high stereospecificity, irrespective of whether the semipinacol rearrangement leads to ring expansion, ring contraction, and/or carbonyl homologation. A less common but possible mechanistic alternative is the stepwise process shown in Scheme 1.9 invoking the intermediacy of carbocation **64** or its migrating group-stabilized variant **65**. This pathway, too, can afford stereospecificity but can also account for possible erosion in stereocontrol. Critical, though, is that these semipinacol processes have far more mechanistic harmony than their pinacol counterparts, affording greater assurance of selective and predictable product formation, as we will see.

1.3.3 Selected Variants of Semipinacol Rearrangements

1.3.3.1 Electrophilic Activation of Allylic Alcohols Outside of the specific manifold illustrated in Scheme 1.9, the semipinacol rearrangement is also broad enough in scope to include a number of substrate types in addition to 1,2-difunctionalized systems. One of the most common variants involves electrophilic activation of the C=C double bond within allylic alcohols and their derivatives. Scheme 1.10 shows this overall process and indicates which electrophiles typically promote the process; activation conditions are dependent on the intra- or intermolecular nature of the rearrangement event. The intramolecular variant, achievable via oxocarbenium, thiocarbenium, and iminium ions, is also known as the Prins–pinacol rearrangement, an example of which was shown in Scheme 1.7. Critical to note, however, is that with these substrates, following activation of the double bond, the migrating group undergoes a 1,2-shift to the electrophilic carbon center, driven exclusively by ring opening of the cyclic cationic intermediate; the migratory aptitude of the shifting group is not a dominant factor. As before, however, the robust stereoselectivity of rearrangements

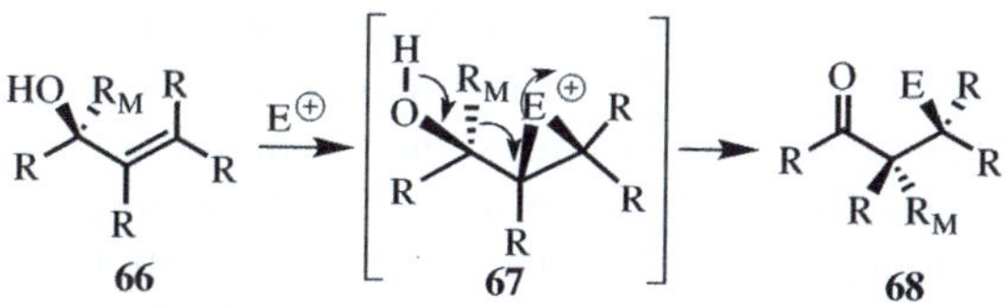

Electrophiles for intramolecular cases: oxocarbeniums, thiocarbeniums, iminiums
Electrophiles for intermolecular cases: haloniums, seleniums, proton, Lewis acids

Scheme 1.10 Semipinacol rearrangements via electrophile activation of allylic alcohols.

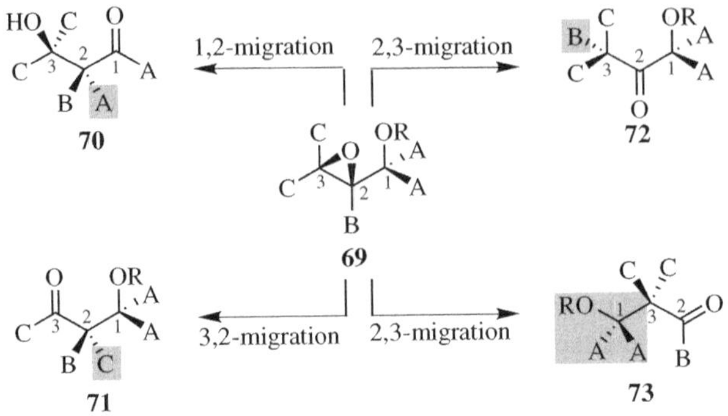

Scheme 1.11 Possible rearrangements of hydroxyepoxide derivatives in semipinacol processes; migrating group is highlighted in each product.

with these starting materials relies on the *anti*periplanar orientation of the C—R_M and C—E bonds (cf. **67**).

1.3.3.2 Epoxy Alcohol Rearrangements Similarly, epoxy alcohols and their derivatives can also undergo semipinacol rearrangements. The difference here from the examples in Scheme 1.10 just denoted is the ability to isolate the starting material (epoxide **69** in Scheme 1.11 vs. transient intermediate **67** in Scheme 1.10). Because both the C-2 and C-3 positions are highly electrophilic (in accordance with the numbering shown for the generalized 2,3-epoxy alcohol **69** within Scheme 1.11), a range of migrations are possible, including 1,2-, 2,3-, and 3,2-shifts to produce **70–73**. As a result, application of semipinacol rearrangements with these materials allows direct access to many synthetically useful functionalities such as *β*-halo and *β*-amino ketones as well as aldol-type products. And, with careful design, the resulting electrophilic carbon centers in the product can also be used for further chemistry, including tandem reactions. The mechanism for these transformations is analogous to the rearrangement of allylic alcohols, where the migrating group generally travels *anti* to the epoxide, accounting for the excellent stereospecificity observed in most cases. Of particular note, aldol-type products bearing a stereogenic quaternary carbon at C-1 can be generated from the rearrangement with excellent diastereoselectivity; this outcome remains a challenge using classical aldol strategies, highlighting a key use for semipinacol processes within a variety of synthetic contexts.

1.3.4 Key Features of the Semipinacol Rearrangement

To summarize this introduction to semipinacol processes, this variant of the rearrangement has several distinguishing features compared to its pinacol cousin that make it particularly well suited to applications in synthesis. First, semipinacol reactions typically proceed under much milder conditions. Whereas the classical pinacol rearrangement usually requires a strong acid to form the reactive carbocation, less drastic conditions, including basic ones, work in the semipinacol manifold, thereby greatly enhancing the overall array of functional and protecting groups compatible with the process. Second, excellent stereochemical control can be achieved for both cyclic and acyclic systems because of the generally concerted semipinacol mechanism, one where the migrating group is *anti*periplanar to the leaving group, reliably leading to inversion of configuration. This remarkable feature allows for the generation of highly substituted contiguous stereogenic centers and the diastereospecific construction of complex structural motifs such as spirocycles and sterically hindered fused ring junctions. Third, migratory aptitude is not the dominant factor controlling the product in a semipinacol rearrangement. Instead, factors such as release of ring strain or the stereoelectronic preference for *anti*periplanar alignment play a more important role in determining which bond will migrate. Lastly, given the versatility of the precursors to the semipinacol rearrangement, it is unsurprising that a variety of methods exist to prepare these substrates enantioselectively, including the Sharpless asymmetric epoxidation of allylic alcohols. For these reasons, the semipinacol rearrangement has proven far more applicable to challenging synthetic endeavors, and the following case studies provide specific examples of the generic strategies outlined previously. We have chosen these select applications to highlight the diversity of substrates, conditions, and complex high-value products accessible through semipinacol rearrangement processes.

1.3.5 Examples of Semipinacol Rearrangements in Total Synthesis

We begin these case studies with a total synthesis of the macrolide natural product protomycinolide IV (**80**, Scheme 1.12) as accomplished by Tsuchihashi and coworkers in 1985.[18] This work, although older than many of the examples we will present in this chapter, is significant in that it highlights how very mild conditions, namely, exposure to trialkylaluminum species, can be extremely effective in imposing a concerted semipinacol rearrangement as a result of the electron-deficient metal center coordinating oxygen atoms on both the alcohol and a vicinal sulfonate leaving group. Indeed, as shown, exposure of **74** to $AlEt_3$ in CH_2Cl_2 at −78 °C effected the desired event leading to **76** in 80% yield, with complete stereocontrol (i.e., conservation) of the alkene stereochemistry in the vinyl migration achieved via the intermediacy of **75**, leading to the final inversion of configuration at the sulfonate-bearing center. Here, aluminum activation facilitated the ability of the sulfonate to serve as a leaving group, with the TMS group on the alkene increasing the ability of the hydroxylated carbon to electronically support δ^+ and facilitating the final migration. Similar chemistry was also used to prepare chiral lactone **78**, affording facile and rapid access to the needed pieces to complete an effective total synthesis of the target molecule (**80**).

A more recent, but similar, deployment of such mild conditions for effecting a semipinacol rearrangement can be found in the opening steps of the Corey and Kingsbury synthesis of isoeudunol (**86**) and related terpenes such as β-araneosene (**87**, Scheme 1.13).[19] Here, in an effort to construct the transfused 5,11-membered ring system found in these and a number of other targets, these researchers began by effecting a ring expansion of cyclopropyl carbinol **81** (prepared from a Kulinkovich cyclopropanation), using $AlMe_3$ to induce a concerted rearrangement. As shown with the

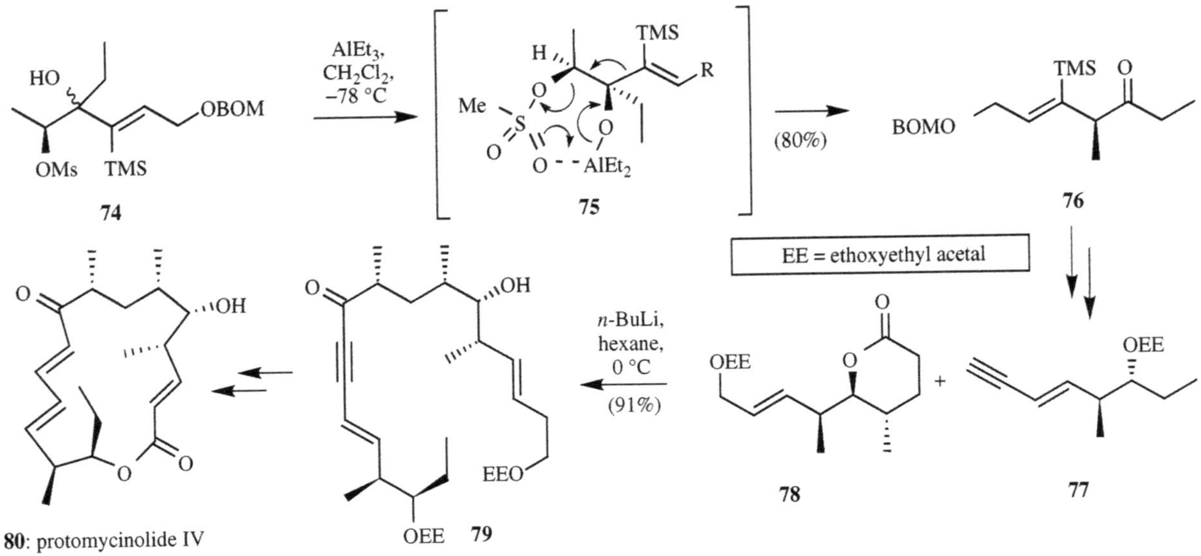

Scheme 1.12 Development of a mesylate-assisted semipinacol rearrangement for protomycinolide IV (**80**).[18a]

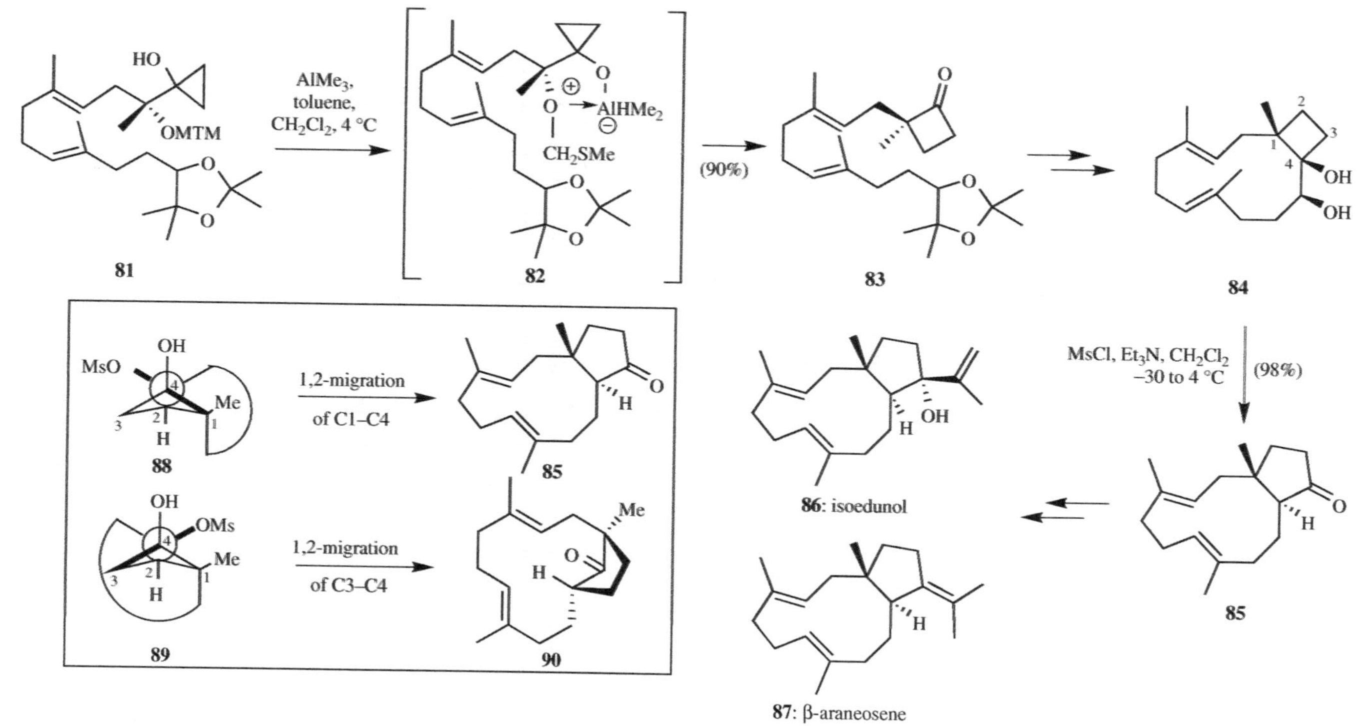

Scheme 1.13 Application of a ring-expanding semipinacol rearrangement to the total synthesis of isoedunol (**86**) and β-araneosene (**87**).[19]

structure of intermediate **82**, the Lewis acid played a bis-coordinating role in ensuring that the desired cyclobutanone could arise with high stereospecificity. If Brønsted acids such as catalytic PPTS were used instead to initiate the process (species without such a bis-coordination capability), significant loss of enantiopurity was observed due to partial racemization of the lone chiral center, presumably via a nonconcerted semipinacol rearrangement.

However, this event would not be the only semipinacol rearrangement of the sequence. Indeed, once **83** was elaborated to diol **84** (using a SmI_2-mediated pinacol coupling to create this key motif), a second semipinacol variant, this time promoted under basic conditions by the formation of a secondary mesylate, effected another ring expansion. As long as the diol was cis, the desired migration to **85** occurred smoothly through transition state **88** with migration of C1—C4 bond; however, for the trans-diol variant of **84** (drawn here as mesylate derivative **89**), migration occurred instead through the C3—C4 bond to afford the undesired bridged 5,12-bicyclopentanone (i.e., **90**). The conformational rigidity of the adjoining 12-membered ring is presumed to play a critical role in this process, making extensions of this exact reaction to other systems difficult to predict. Nevertheless, the more global lesson of this work is that semipinacol chemistry can be readily used to effect a number of critical ring expanding bond constructions with high stereocontrol under very mild conditions.

As one final example of a mild semipinacol rearrangement using diol derivatives as starting materials, and to highlight that stepwise versions can also be quite effective when properly deployed, we present Rawal's synthesis of *epi*-ent-elisabethin A (**94**, Scheme 1.14).[20] In this example, the displacement of a mesylate group formed from substrate **91** was induced by the *ortho*- and *para*-disposed electron-donating OMe groups on the aromatic ring upon heating in MeOH in the presence of $CaCO_3$ at 50 °C, forming a bridged quinone methide intermediate (**92**). The culmination of this stepwise semipinacol rearrangement involved 1,2-aryl migration from the ketal, driven by rearomatization, and afforded ester **93** in 72% yield. Further elaboration featuring an *endo*selective intramolecular Diels–Alder reaction and oxidative cyclization then furnished the target molecule. This synthesis strategy highlights an extremely clever and nonobvious use of a semipinacol process, in that few direct traces of the original substrate are present in the final product except for the critical chiral centers established with the formation of **93**.

Outside of diol derivatives, a number of other substrate types with alternative leaving groups work well as semipinacol substrates. Next, we present an example using a

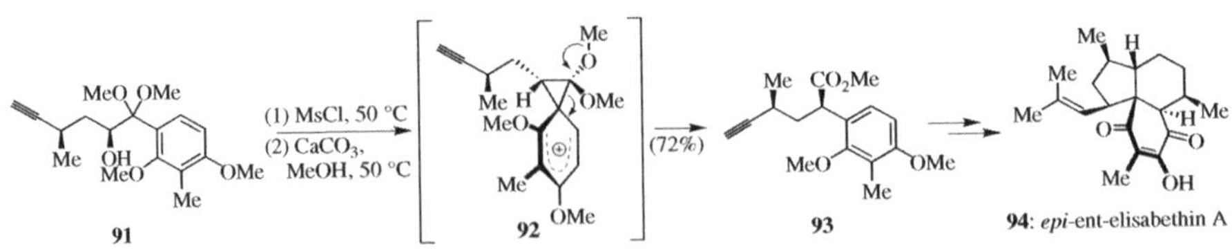

Scheme 1.14 Use of a base-mediated semipinacol rearrangement as part of the total synthesis of *epi*-ent-elisabethin A (**94**).[20]

halide leaving group, where the generation of a halohydrin intermediate is critical for effecting the semipinacol rearrangement step. In these events, the halogen is typically removed (and thus the semipinacol process is initiated) by the addition of a metal salt which has a high affinity for a halide (such as Ag(I)). Classically, these events are also known as "quasi-Favorskii" rearrangements because either steric hindrance or the absence of an appropriate α-proton prevents the formation of a cyclopropanone through a standard Favorskii process, instead allowing for migration of a C—C bond through a semipinacol event. One recent use of this chemistry derives from the Harmata group's synthesis of the core of tricycloclavulone (**100**, Scheme 1.15).[21] Here, after forming **95** through a [4+3] cycloaddition, treatment with vinyl lithium species **96** in THF at −78 °C effected the formation of halohydrin anion **97**. Despite the inability to generate a cyclopropanone from **95** through a Favorskii-type process due to the rigidity of the system, the semipinacol alternative proceeded instead to deliver **98** in 90% yield upon warming to −30 °C. As expected, this ring-contracted product reflects a stereospecific transfer in which the migrating C—C bond is *anti*periplanar to the leaving group in the more favorable chair conformation (as drawn within **97**). The resultant enone (**98**) was elaborated into the tricycloclavulone core (**99**) through a series of steps including a ring-closing metathesis reaction.

Halogen atoms, besides serving as leaving groups in their halide form, can also activate substrates for semipinacol rearrangements in their halonium form, affording β-halocarbonyl products from allylic alcohol starting materials. One recent and highly illustrative application of such a process to generate a chlorinated motif as found directly in a natural product derives from the Wood group's total synthesis of welwitindolinone A isonitrile (**104**, Scheme 1.16).[22] Here, exposure of allylic alcohol **101** to NaOCl and $CeCl_3 \cdot 7H_2O$ effected chloronium generation *in situ* to form intermediate **102** via electrophile addition on the concave face. Critical to controlling this addition from the seemingly more hindered face of the substrate was the large TIPS-protected alcohol, which is presumed to reside in a pseudoaxial position as drawn within **102** and thus blocks the less hindered convex face. Stereospecific alkyl migration *anti* to the chloronium ion via a concerted mechanism afforded **103** as a single stereoisomer in which both the C-12 quaternary carbon and the adjoining C-13 stereocenter were fashioned as needed for the final target.

An even wider array of substrates can be induced to participate in semipinacol processes, and often additional cascades, if metals are used to initiate the rearrangement. As one recent example, Nemoto, Ihara, and coworkers developed a palladium-promoted cascade involving semipinacol ring expansion/intramolecular insertion of an isopropenylcyclobutanol as part of an asymmetric total synthesis of (+)-equilenin (**110**, Scheme 1.17).[23] As shown, coordination of the palladium onto the isopropenyl group of **105** activated its double bond in the form of complex **106**, generating the electrophilic carbon center necessary to induce carbonyl bond formation and concomitant alkyl migration to **107**. The subsequent olefin insertion and β-hydride elimination then afforded the complete steroid core, favoring the trans-isomer (**108**) needed for the target molecule. Of note, the diastereoselectivity of this ring expansion process was governed by the conformation of the isopropenyl group in the cascade process (**111** vs. **112**). In nonpolar or noncoordinating solvents,

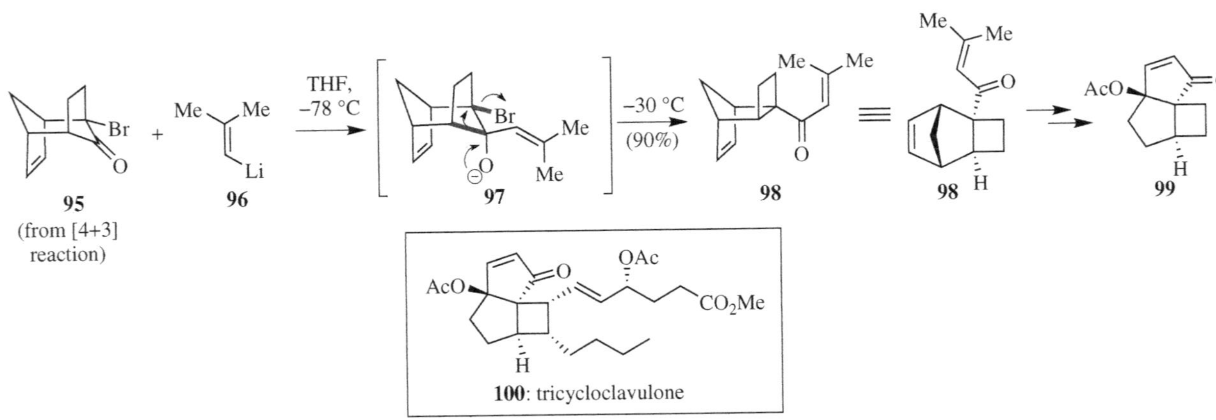

Scheme 1.15 Development of a [4+3]-cycloaddition-Favorskii sequence and its application in the synthesis of tricycloclavulone core (**99**).[21]

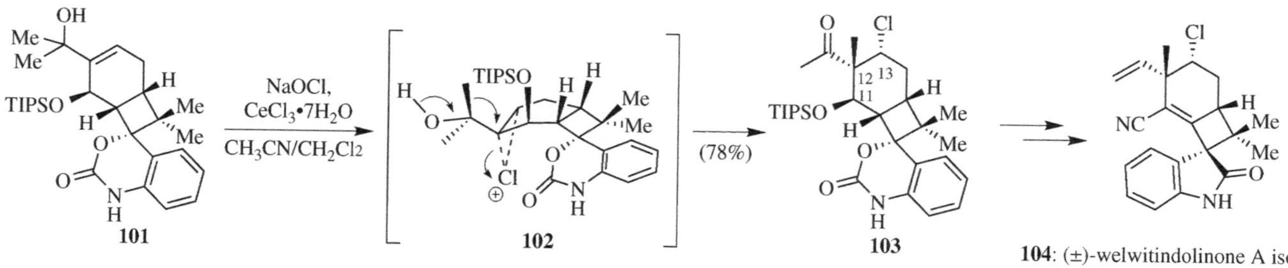

104: (±)-welwitindolinone A isonitrile

Scheme 1.16 Use of a chloronium ion-mediated semipinacol rearrangement in the total synthesis of (±)-welwitindolinone A isonitrile (**104**).[22a]

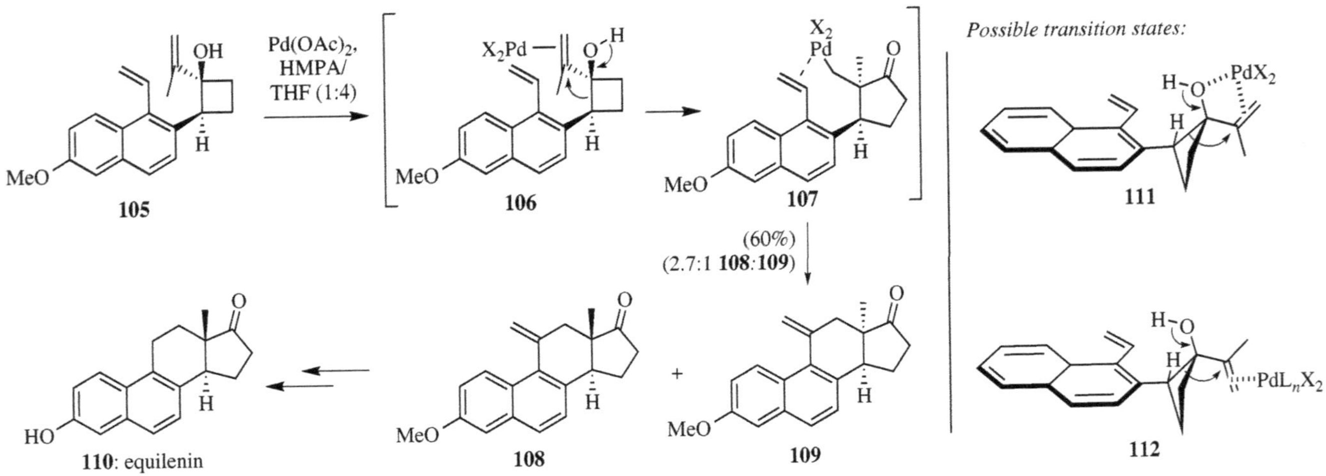

Scheme 1.17 Semipinacol rearrangement as part of the asymmetric total synthesis of (+)-equilenin (**110**).[23]

ring expansion gave predominantly the *cis*-product (**109**) via transition state **111**, where the hydroxy group occupies a coordination site on palladium. In polar or coordinating solvents such as HMPA, the rearrangement could proceed instead through transition state **112** with only palladium bound to the olefin, preferentially affording **108**. On a more general level, this ring expansion process is driven by release of ring strain, and though an extra carbon atom not needed for equilenin (**110**) was installed as a result of the requirements of completing the reaction cascade, that atom could be easily cleaved making its presence more than worthwhile for what the tandem semipinacol process was able to accomplish overall.

Another example of the synthetic value of transition metals in initiating cascades that include semipinacol processes derives from the Toste group. In this case, shown in Scheme 1.18, a gold(I)-catalyzed cycloisomerization/semipinacol ring expansion was followed by a similar shift as in **106** → **107** to assemble the entire polycyclic core of the target.[24] Just as in the Corey work described in Scheme 1.13, the cyclopropanol framework was prepared by the Kulinkovich reaction, and after installation of the alkyne by a Seyferth–Gilbert homologation using the Ohira–Bestmann reagent to afford **113**, the stage was set for the critical operation. In the event, exposure of this substrate to the complex formed from $AuPPh_3Cl$ and $AgSbF_6$ generated the Au(I)-activated cationic intermediate **115**, which rearranged to give cyclobutanone **116** as a single diastereomer in 87% yield. The high stereoselectivity of this process can be understood by considering the drawn intermediates (**114** and **115**) presumed to account for the observed product. The only other alternative (not drawn) would be of much higher energy since it would possess a *trans*-diquinane motif, one that would have led instead to a high-energy *trans*-cyclobutanone if it was product determining. Overall, this powerful strategy allowed for rapid construction of the angular triquinane ring system with the methyl-bearing stereocenter; after elaboration of **116** into **117**, a palladium-catalyzed ring expansion effected in the presence of DDQ or benzoquinone afforded a 4:1 mixture of products favoring migration of the more substituted C—C bond. Oxidation state adjustments and removal of the

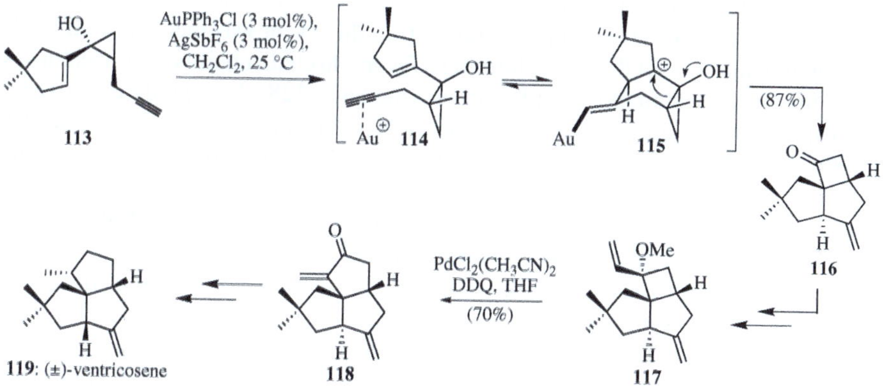

Scheme 1.18 Use of a double ring expansion strategy in the total synthesis of (±)-ventricosene (**119**).[24]

heteroatom critical to both rearrangements then completed the total synthesis of ventricosene (**119**).

Finally, we will round out our presentation of semipinacol rearrangements with a number of examples using 2,3-epoxy alcohols and their derivatives as substrates. We begin with two different approaches to ingenol (**54**), a target mentioned earlier in the context of Scheme 1.8 as likely arising in nature from a pinacol-like rearrangement of an appropriate terpene precursor. Here, we show two different approaches, one from the Cha group and the other from Tanino and Kuwajima, which used very different epoxy alcohols to accomplish the same general conversion, affording a highly strained transfused architecture. In the Cha case, treatment of **120** with $AlMe_3$ at −78 °C activated the epoxide and induced the alkyl shift to give the carbocyclic core of **122** in 82% yield.[25] Of note, the stereochemistry of the alcohol group at C-4 is crucial in ensuring the migrating group is the requisite C9—C11 bond. As modeled in the presumed transition state with the alcohol group properly in place (**121**), the C9—C11 bond is *anti*periplanar to the epoxide C10—O bond, providing the orbital alignment required for the desired rearrangement to **122**. On the other hand, with the alternate C4-alcohol stereochemistry (not shown), the C8—C9 bond can migrate as well. In the Tanino and Kuwajima work, a different epoxy alcohol (**123**) underwent stereospecific 1,2-migration upon treatment with $AlMe_3$ to afford a similar framework (**125**) in 76% yield.[26] Globally, these powerful approaches allow for facile ring size manipulation to access fused 5,7,7-ring systems without the loss of any stereochemical information. They highlight, in combination with the Baran case presented earlier, the variety of pinacol and semipinacol processes that can be deployed creatively to solve the challenging synthetic problem of the ingenol core, showing the true versatility of these rearrangements (Scheme 1.19).

Next, we consider a creative semipinacol approach leading to a total synthesis of stemonamine (**131**, Scheme 1.20) from Tu and coworkers.[27] In this case, a tandem semipinacol rearrangement/Schmidt reaction was deployed to construct the functionalized aza-quaternary carbon center of the target, using $TiCl_4$ to activate the two oxygen groups of substrate **126**. The cascade began with the semipinacol rearrangement in which alkyl group migrated to the adjacent quaternary carbon to open up the epoxide. The carbonyl group generated from this event was then attacked by the nucleophilic nitrogen of the azide, forming a favorable chair, chair hetero-decalin system. This event presumably afforded an intermediate (**128**) that readily underwent a stereospecific 1,2-migration, driven by loss of N_2, to furnish 5,7-bicyclic lactam **129** in 68% yield. The rapid access to this bicyclic system underscores the power of the approach, with subsequent operations including ozonolysis of the pendant alkene, alcohol oxidation, and aldol condensation yielding a tricyclic core (**130**) that was then elaborated smoothly into stemonamine (**131**).

As our final case study, we present the use of a semipinacol rearrangement strategy to generate the spirocyclic core of the potent antitumor natural product fredericamycin A (**135**, Scheme 1.21) using acrylated epoxy alcohols.[28] Following extensive model studies to probe the stereoselectivity of the general process, the Kita group was able to establish that treatment of *cis*-epoxy acylates **136** with $BF_3{\cdot}OEt_2$ led to

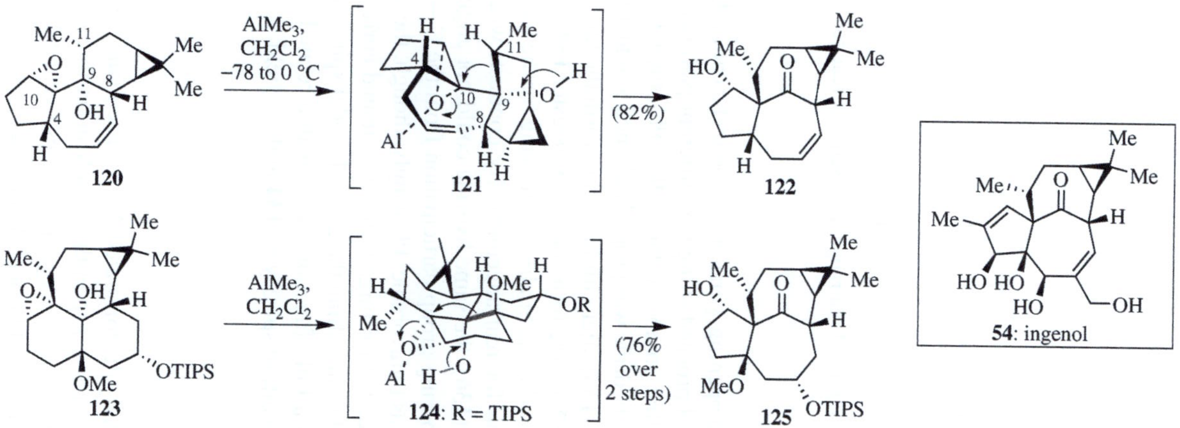

Scheme 1.19 Use of a semipinacol rearrangement to access unique core in the total synthesis of ingenol (**54**).[25,26]

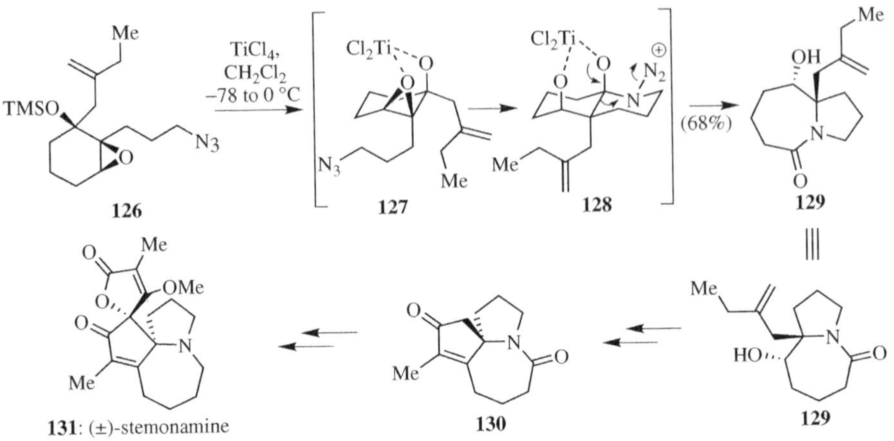

Scheme 1.20 A tandem semipinacol rearrangement/Schmidt reaction in the total synthesis of (±)-stemonamine (**131**).[27]

regioselective epoxide ring opening, generating a benzylic cation **137**. From carbocation **137**, the reaction could proceed along two different pathways dependent on the nature of group R. When that group was large (R = Ph), the intermediate adopted a conformation such that the hydride was properly aligned to undergo 1,2-shift (path a) to give the enone product (**138**) after elimination to form the tetrasubstituted alkene. When R was smaller (R = Me), the formation of orthoester **139** was the predominant process due to the absence of steric hindrance allowing for the methyl group to adopt the requisite conformation for rearrangement. By contrast, when the *trans*-epoxy acetate was used instead (**136**), the same hydride could not align with the adjacent empty orbital, and neighboring group participation from the C—O bond leading to the five-membered ring was unfavorable. Pleasingly, though, the result was that only the desired C—C bond needed for migration was oriented correctly with respect to the empty *p*-orbital, thereby furnishing the desired model spirocycle (**141**). Application of this knowledge to a fully functional case with a related *trans*-epoxy acylate (**132**) worked smoothly to afford spirocycle **134** in 94% yield.

1.4 CONCLUSION

As this chapter has hopefully demonstrated, both pinacol and semipinacol processes have undergone a recent renaissance in their application to complex molecule synthesis, affording a number of highly elegant solutions to challenging problems. And, when combined in tandem with additional chemical events, the rearrangements reach an even greater level of overall power. Further investigation will certainly uncover additional areas of value for these processes in accessing complex molecules. Of special note is that a number of recent explorations into novel, asymmetric variants of pinacol and semipinacol processes have provided and should continue to provide

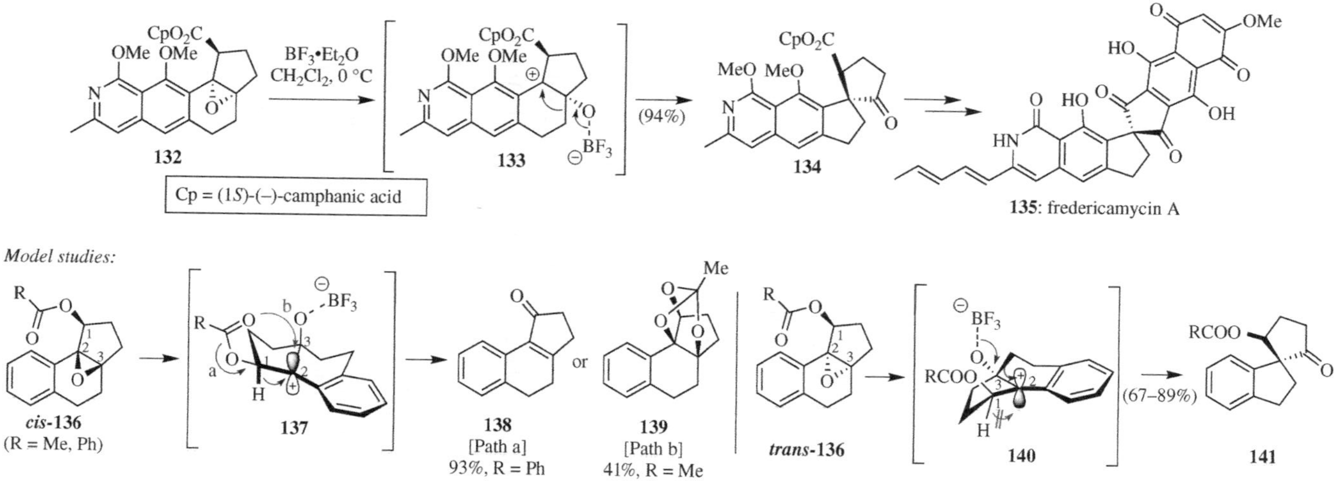

Scheme 1.21 A semipinacol rearrangement strategy as part of the total synthesis of fredericamycin A (**135**).[28a]

greatly enhanced tools for enantiospecific synthesis.[29] Thus, though these general reactions have been known for many decades, the potential for additional discoveries remains high. We hope this compilation will serve as inspiration for some of those future creative applications.

REFERENCES

1. Rickborn, B. "The pinacol rearrangement" In *Comprehensive Organic Synthesis*; Trost, B. M.; Fleming, I., Eds.; Pergamon: Oxford, UK, **1991**, Vol. 3, pp. 721–732.
2. Fittig, R. *Justus Liebigs Ann. Chem.* **1860**, *114*, 54–63.
3. Butlerov, A. *Justus Liebigs Ann. Chem.* **1873**, *170*, 151–162.
4. Kekulé, A. *Justus Liebigs Ann. Chem.* **1858**, *106*, 129–159.
5. Kurti, L.; Czako, B. *Strategic Applications of Named Reactions in Organic Synthesis*; Elsevier Academic Press: Burlington, MA, **2005**, pp. 350–351.
6. Danilov, S. *J. Russ. Phys. Chem. Soc.* **1917**, *49*, 282–289.
7. (a) Rothrock, T. S.; Fry, A. *J. Am. Chem. Soc.* **1958**, *80*, 4349–4354. (b) Fry, A.; Carrick, W. L.; Adams, C. J. *J. Am. Chem. Soc.* **1958**, *80*, 4743–4744.
8. Bunton, C. A.; Carr, M. D. *J. Chem. Soc.* **1963**, 5854–5861 and 5861–5867.
9. (a) Chen, X.; Esser, L.; Harran, P. G. *Angew. Chem. Int. Ed.* **2000**, *39*, 937–940. (b) Li, J.; Jeong, S.; Esser, L.; Harran, P. G. *Angew. Chem. Int. Ed.* **2001**, *40*, 4765–4770.
10. Suzuki, K.; Takikawa, H.; Hachisu, Y.; Bode, J. W. *Angew. Chem. Int. Ed.* **2007**, *46*, 3252–3254.
11. (a) Pettit, G. R.; Lippert, J. W.; Herald, D. L. *J. Org. Chem.* **2000**, *65*, 7438–7444. (b) Snyder, S. A.; Gollner, A.; Chiriac, M. I. *Nature* **2011**, *474*, 461–466.
12. Liang, T.; Zhang, Z.; Antilla, J. C. *Angew. Chem. Int. Ed.* **2010**, *49*, 9734–9736.
13. Snyder, S. A.; Thomas, S. B.; Mayer, A. C.; Breazzano, S. P. *Angew. Chem. Int. Ed.* **2012**, *51*, 4080–4084.
14. Overman, L. E.; Pennington, L. D. *J. Org. Chem.* **2003**, *68*, 7143–7157.
15. Jørgensen, L.; McKerrall, S. J.; Kuttruff, C. A.; Ungeheuer, F.; Felding, J.; Baran, P. S. *Science* **2013**, *341*, 878–882.
16. Coveney, D. J. "The semipinacol and other rearrangements" In Comprehensive Organic Synthesis; Trost, B. M.; Fleming, I., Eds.; Pergamon: Oxford, UK, **1991**, Vol. 3, pp. 777–801.
17. For reviews, see: (a) Overman, L. E. *Acc. Chem. Res.* **1992**, *25*, 352–359. (b) Kurti, L., Czako, B. Strategic Applications of Named Reactions in Organic Synthesis; Elsevier Academic Press: Burlington, MA, **2005**, pp. 366–367. (c) Song, Z. L.; Fan, C. A.; Tu, Y. Q. *Chem. Rev.* **2011**, *111*, 7523–7556.
18. (a) Suzuki, K.; Tomooka, K.; Matsumoto, T.; Katayama, E.; Tsuchihashi, G. *Tetrahedron Lett.* **1985**, *26*, 3711–3714. (b) Suzuki, K.; Tomooka, K.; Katayama, E.; Matsumoto, T.; Tsuchihashi, G. *J. Am. Chem. Soc.* **1986**, *108*, 5221–5229.
19. Kingsbury, J. S.; Corey, E. J. *J. Am. Chem. Soc.* **2005**, *127*, 13813–13815.
20. Waizumi, N.; Stankovic, A. R.; Rawal, V. H. *J. Am. Chem. Soc.* **2003**, *125*, 13022–13023.
21. Harmata, M.; Wacharasindhu, S. *Org. Lett.* **2005**, *7*, 2563–2565.

22. (a) Reisman, S. E.; Ready, J. M.; Hasuoka, A.; Smith, C. J.; Wood, J. L. *J. Am. Chem. Soc.* **2006**, *128*, 1448–1449. (b) Reisman, S. E.; Ready, J. M.; Weiss, M. M.; Hasuoka, A.; Hirata, M.; Tamaki, K.; Ovaska, T. V.; Smith, C. J.; Wood, J. L. *J. Am. Chem. Soc.* **2008**, *130*, 2087–2100.

23. Yoshida, M.; Ismail, M. A. H.; Nemoto, H.; Ihara, M. *J. Chem. Soc., Perkin Trans. 1* **2000**, *16*, 2629–2635.

24. Sethofer, S. G.; Staben, S. T.; Hung, O. Y.; Toste, F. D. *Org. Lett.* **2008**, *10*, 4315–4318.

25. Epstein, O. L.; Cha, J. K. *Angew. Chem. Int. Ed.* **2005**, *44*, 121–123.

26. Tanino, K.; Onuki, K.; Asano, K.; Miyashita, M.; Nakamura, T.; Takahashi, Y.; Kuwajima, I. *J. Am. Chem. Soc.* **2003**, *125*, 1498–1500.

27. Zhao, Y. M.; Gu, P. M.; Tu, Y. Q.; Fan, C. A.; Zhang, Q. W. *Org. Lett.* **2008**, *10*, 1763–1766.

28. (a) Kita, Y.; Higuchi, K.; Yoshida, Y.; Iio, K.; Kitagaki, S.; Akai, S.; Fujioka, H. *Angew. Chem. Int. Ed.* **1999**, *38*, 683–686. (b) Kita, Y.; Higuchi, K.; Yoshida, Y.; Iio, K.; Kitagaki, S.; Ueda, K.; Akai, S.; Fujioka, H. *J. Am. Chem. Soc.* **2001**, *123*, 3214–3222.

29. (a) For a review, see: Hennecke, U. *Chem. Asian. J.* **2012**, *7*, 456–465. For selected examples, see: (b) Chen, Z.-M.; Zhang, Q.-W.; Chen, Z.-H.; Li, H.; Tu, Y.-Q.; Zhang, F.-M.; Tian, J.-M. *J. Am. Chem. Soc.* **2011**, *133*, 8818–8821. (c) Li, H.; Zhang, F.-M.; Tu, Y.-Q.; Zhang, Q.-W.; Chen, Z.-M.; Chen, Z.-H.; Li, J. *Chem. Sci.* **2011**, *2*, 1839–1841.

2

BAEYER–VILLIGER (BV) OXIDATION/REARRANGEMENT IN ORGANIC SYNTHESIS

LIN ZHOU, LILI LIN, XIAOHUA LIU, AND XIAOMING FENG
Key Laboratory of Green Chemistry and Technology, Ministry of Education, College of Chemistry, Sichuan University, Chengdu, P.R. China

2.1 INTRODUCTION

In 1899, Baeyer and Villiger found that menthone **1a** and tetrahydrocarvone **1b** could be oxidized to the corresponding lactones (Scheme 2.1) in a mixture of sodium persulfate and concentrated sulfuric acid (Caro's acid).[1] The persulfuric acid was subsequently replaced by an organic peracid. Later, the transformation of ketones into esters and cyclic ketones into lactones or hydroxy acids by peroxy acids became known as the Baeyer–Villiger (BV) oxidation/rearrangement.[2,3] Besides ketones and cyclic ketones, aldehydes and α-diketones can also be oxidized, providing esters and anhydrides, respectively (Scheme 2.2).

2.2 MECHANISM

The mechanism of the BV oxidation/rearrangement was first studied by Criegee in 1948[4] and was clarified by Doering and Dorfman in 1953.[5] The generally accepted mechanism involves a simple two-step process. As shown in Scheme 2.3, the peroxy acid initially adds as nucleophile at the polarized carbonyl carbon to form an intermediate species **3** known as the Criegee adduct (step **a**). Then, one of the alkyl groups R^M migrates onto the peroxygen with the concomitant release of the carboxylate anion

Molecular Rearrangements in Organic Synthesis, First Edition. Edited by Christian M. Rojas.

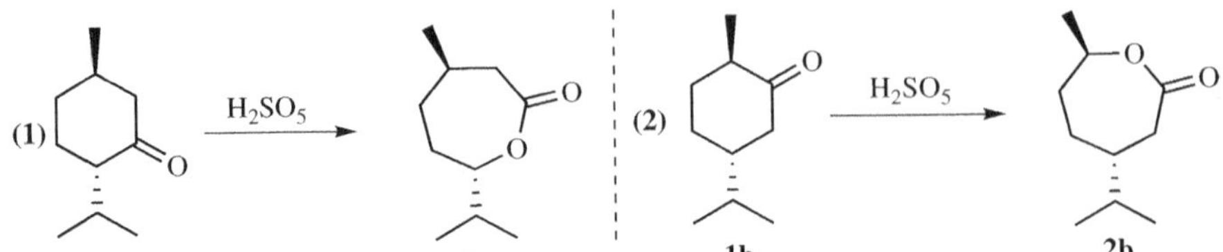

Scheme 2.1 Oxidation of menthone **1a** and tetrahydrocarvone **1b** with Caro's acid.

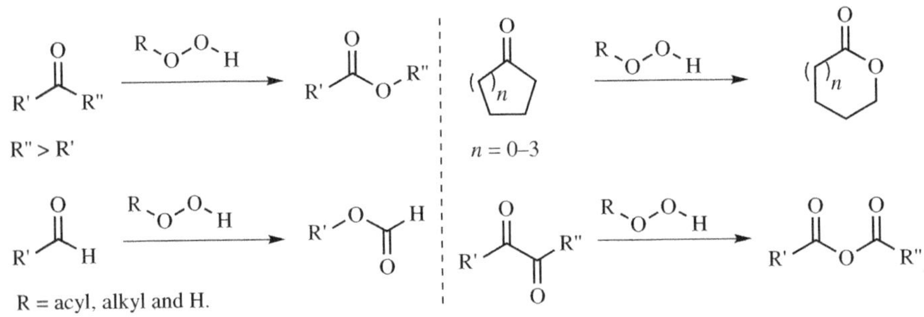

Scheme 2.2 The BV oxidation/rearrangement of carbonyl compounds.

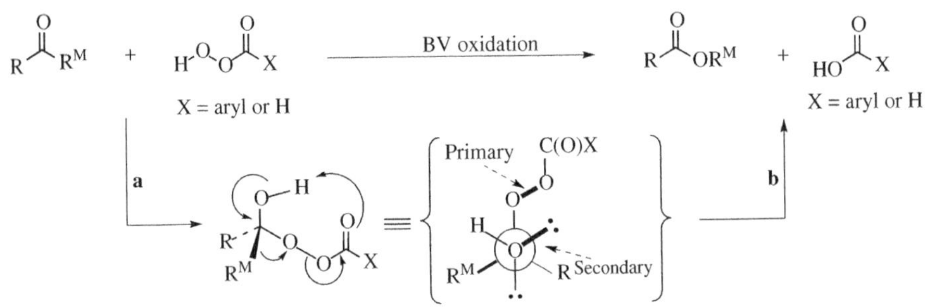

Scheme 2.3 The mechanism of the BV oxidation of ketones. R^M denotes the migrating group.

(step **b**). The migration step was suggested for the first time by Berson and Suzuki in 1959.[6] It was believed to account for the selectivity of the reaction and is also rate determining in the vast majority of cases.

Generally, the migration abilities of alkyl groups increase proportionately with their partial cation stabilization. As a result, electron-donating and bulky substituents in the migrating group R^M strongly facilitate the rearrangement. This is consistent with the well-known migratory aptitude in the BV reaction: tertiary alkyl > cyclohexyl > secondary alkyl > benzyl > phenyl > primary alkyl > cyclopentyl ≫ methyl.

Additionally, it was found that there were stereoelectronic effects in the Criegee intermediate **3** influencing the rearrangement step (Scheme 2.3).[7] The primary stereoelectronic effect is that the migrating group R^M needs to be antiperiplanar to the O—O bond of the leaving group, while the secondary stereoelectronic effect is that R^M is antiperiplanar to a lone pair of the hydroxyl group. In other words, proper alignment is required for the rearrangement step.[8] In some cases, the primary stereoelectronic effect is more important than migratory aptitude in determining the course of the rearrangement.[9]

In terms of stereochemistry, the BV oxidation/rearrangement is stereoselective, and the migrating group retains its configuration in the corresponding ester product. Synthetic utility is also predictable since in the very great majority of cases, the BV oxidation tolerates the presence of many functional groups, such as C═C double bonds, C≡C triple bonds, and hydroxyl groups.

The hydroxyl proton in the intermediate **3** can migrate intramolecularly only with acylperoxo-type oxidants. Hence, acylperoxo-type oxidants are more effective than alkylperoxo oxidants. Additionally, it was observed that the presence of electron-withdrawing substituents on the peroxy acid increased the reaction rate because they made the conjugate base a better leaving group. Therefore, the reactivity of oxidants follows this order: CF_3CO_3H > monopermaleic acid > monoperphthalic acid > 3,5-dinitroperbenzoic acid > *p*-nitroperbenzoic acid > *m*-chloroperoxobenzoic acid (*m*-CPBA) ~ performic acid > perbenzoic acid > peracetic acid ≫ H_2O_2 > *t*-BuOOH. BV oxidations using oxidants with low activity generally require catalysts. It should be noted that the two steps in the mechanism have similar activation energies in many cases. Hence, both steps may need to be promoted by the catalysts. In general, however, and with some exceptions, the rearrangement step is more important in determining the properties of the overall reaction.[4]

2.3 SYNTHETIC APPLICATIONS

2.3.1 Oxidation of Linear Ketones to Esters

α-Methyl-α'-multisubstituted alkyl ketones are important in organic synthesis. Because the methyl substituent is known to be a bad migrating group, the multisubstituted alkyl group migrates preferentially in the rearrangement step, generating multisubstituted alkyl acetate esters with the absolute configuration of the migrating chiral carbon atoms retained. The resulting esters are not only structural units of many useful natural products but can also be further transformed to alcohols and carboxylic acids.

The optically active tropane alkaloid Bao Gong Teng A, isolated from the Chinese herb *Erycibe obtusifolia* Benth,[10] is an azabicyclo-[3.2.1]octane compound substituted with an acetate ester and a hydroxyl group. By a regio- and stereocontrolled [5+2] cycloaddition of $TpMo(CO)_2(\eta^3$-pyridinyl) scaffold **4** with methyl vinyl ketone **5**, the azabicyclo-[3.2.1]octane framework **6** bearing a ketone group could be prepared in one step.[11] The transformation of ketone to ester was through a BV oxidation of *N*-Cbz-(−)-Bao Gong Teng A **7** with *m*-CPBA as the oxidant. The multisubstituted

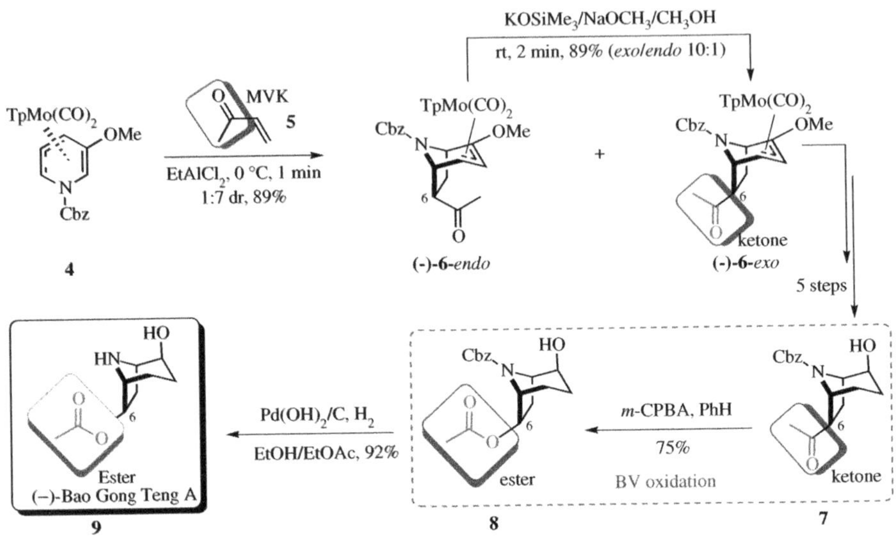

Scheme 2.4 The total synthesis of (−)-Bao Gong Teng A **9**.

alkyl group of **7** migrated preferentially, and the desired acetate ester **8** was obtained. During the process, the hydroxyl group was unaffected and the configuration at C6 was completely retained. Finally, the total synthesis was completed by removal of the Cbz protecting group (Scheme 2.4).[12]

The BV oxidation at an acetyl carbonyl group can occur selectively in the presence of an α,β-unsaturated cyclohexenone under suitable conditions. In the synthesis of *ent*-peribysin E from (*S*)-carvone developed by Danishefsky and coworkers,[13] compound **12** was a key intermediate, and the ester was ultimately transformed to the required C2—OH group. After constructing the bicyclic ring system **10**, the isopropenyl group was converted into the corresponding ketone **11** by a concurrent Johnson–Lemieux oxidation.[14] Following a BV oxidation of **11** with *m*-CPBA as oxidant in CH_2Cl_2 at 0 °C to room temperature, the secondary acetate at C2 with the desired relative configuration for the total synthesis was in hand. Notably, the unsaturated ketone remained unchanged during the oxidation process. After hydrolysis, the desired OH group could be revealed (Scheme 2.5).

In Marchand–Brynaert group's synthesis of carbapenems,[15] β-lactam antibiotics **16**,[16] a common key intermediate is acetoxyazetidinone **15** (Scheme 2.6) containing three of the four stereocenters of the final compounds. For introducing the acetoxy group, the BV oxidation method was attractive for its relatively mild conditions and avoidance of toxic metals. However, the methyl azetidinyl ketone (**17**, R = Me) was difficult to obtain because one of the necessary starting materials was not commercially available. So, other substituted azetidinyl ketones **17** were synthesized. A key question was what substitution would be suitable to obtain azetidinyl group-migrated products **14** versus the regioisomeric esters **18**. The azetidinyl function is a good migrating group in electron-deficient BV rearrangements

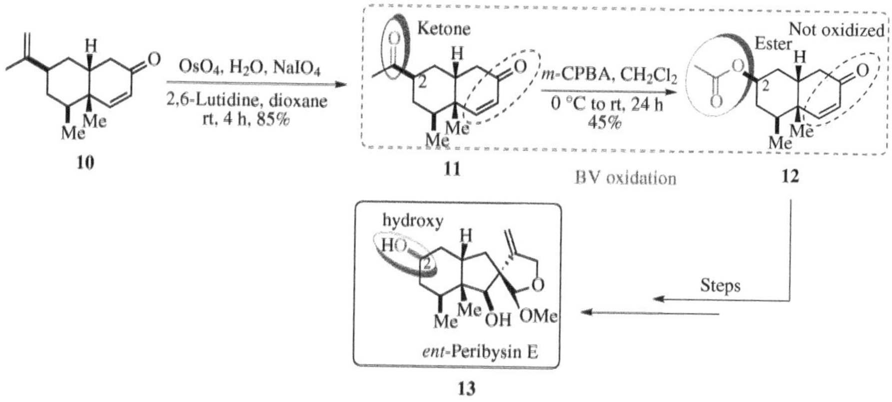

Scheme 2.5 The total synthesis of *ent*-peribysin E **13**.

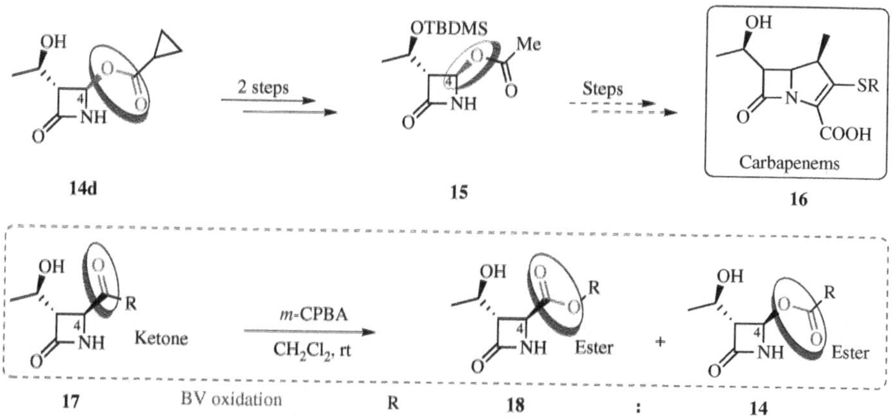

R	18	:	14
a: Ph	0		100
b: *t*-Bu	100		0
c: *i*-Pr	48		52
d: *c*-Pr	0		100

Scheme 2.6 The total synthesis of carbapenems **16**.

since it can stabilize a partial positive charge created at C4. For comparison to other good migrating groups, the migrating ability of the azetidinyl group versus phenyl, secondary alkyl, and tertiary alkyl groups in the BV reaction was studied. The experimentally determined order of migration was *t*-Bu > *i*-Pr = azetidinyl > *c*-Pr = Ph. Thus, the cyclopropyl group was chosen for application in the total synthesis. The particular lack of migratory aptitude of the cyclopropyl group compared to *t*-Bu and *i*-Pr groups may result from the electronic properties of the exocyclic C—C bond of the cyclopropane ring (orbitals with sp^2 character) and the reduced number of C—H

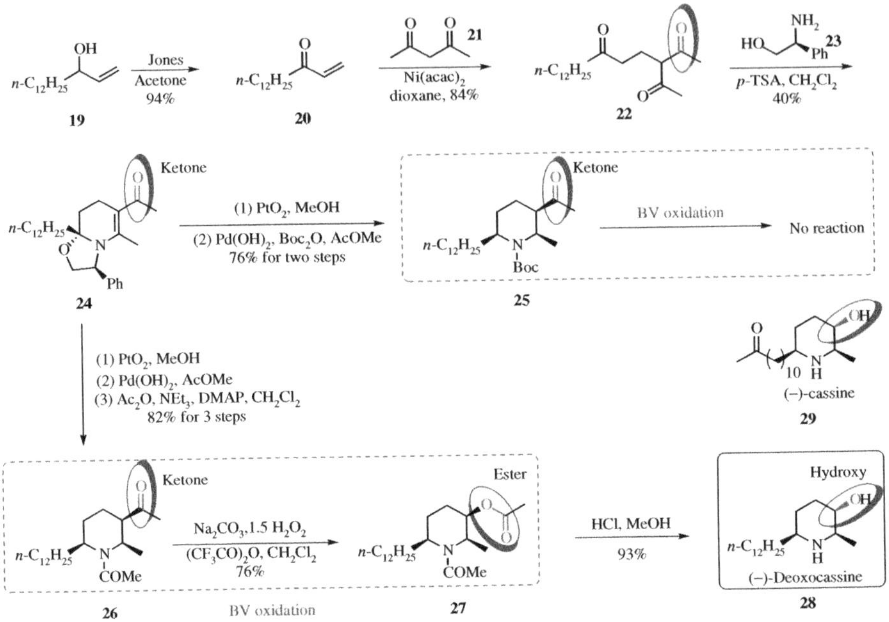

Scheme 2.7 The total synthesis of the (−)-deoxocassine **28**.

bonds (four bonds) involved in the hyperconjugative stabilization of the δ^+ charge, while in *t*-Bu and *i*-Pr, nine and six bonds are involved, respectively.

Interestingly, the nature of a remote N-protecting group can affect the BV oxidation of ketones attached to the N-heterocycle. In their total synthesis of the hydroxypiperidine (−)-deoxocassine **28**, a simple 3-hydroxy analogue of piperidine alkaloid (−)-cassine **29**, Vanucci-Bacqué and coworkers used *N*-acetyl piperidine **27** as the key intermediate (Scheme 2.7).[17] This was constructed from the allylic alcohol **19**, acetylacetone **21**, and (*S*)-phenylglycinol **23**. Unfortunately, the BV oxidation of Boc-protected piperidine **25** did not occur under various experimental conditions, though the reason for this failure was unclear. But with an *N*-acetyl protecting group,[18] the BV oxidation of **26** occurred successfully using sodium percarbonate and trifluoroacetic anhydride, affording ester **27** as a single isomer. Subsequent hydrolysis under acidic conditions gave the corresponding hydroxypiperidine **28** in high yield.

Though aryl substituents are known to be bad migrating groups, their migrating ability is higher than that of a primary alkyl group. Huang's and Jiang's group reported that compound **33**, which can be constructed by the asymmetric vinylogous aldol reaction of the carbonyl-activated allyl nucleophile **30** with isatin **31**, serves as the common intermediate for the divergent synthesis of 3-hydroxy-2-oxindole derivatives (Scheme 2.8).[19] The common-intermediate strategy to multiple products through divergent synthesis is different from the traditional "single target" approach in total synthesis. By combining a sequence of reduction and BV oxidation

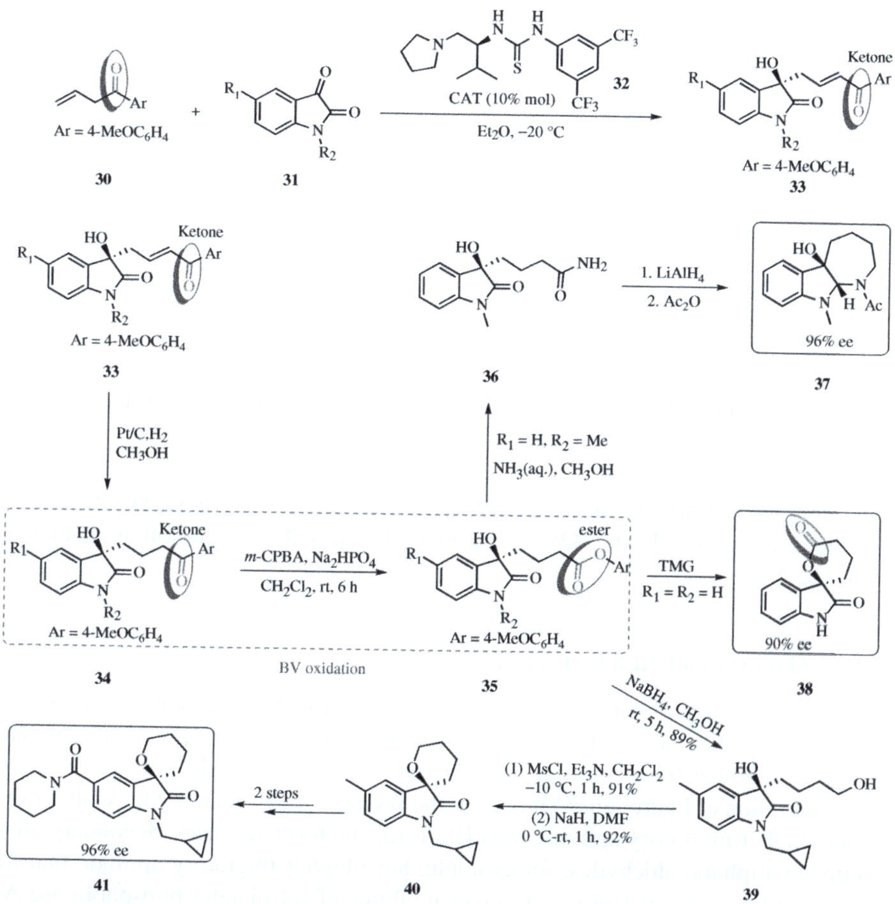

Scheme 2.8 The total synthesis of the 3-hydroxy-2-oxindole derivatives.

with *m*-CPBA, the desired aryl-migrated esters **35** could be obtained. To favor the aryl migration, an electron-donating methoxy group was introduced in the para-position of the benzene ring. Subsequently, through treatment of **35** with 1,1,3,3-tetramethylguanidine (TMG), spirolactone **38** was conveniently obtained in 90% ee. Alternatively, by reaction in a solution of ammonia in methanol, compound **35** could be converted to amide **36**. Treatment of **36** with $LiAlH_4$ followed by acetic anhydride afforded the hexahydroazepino[2,3-*b*]indole **37** without compromising enantiopurity. Finally, through a sequence of reduction with $NaBH_4$, reaction with MsCl, and treatment with NaH, **35** could be converted to spiroether **40**, which could be further converted in two steps into the CB2 agonist **41** with 96% ee.

The migratory aptitude of a phenyl substituent is also greater than that of the epoxide functional group. In the BV oxidation step during Coltart group's synthesis of the important antimalarial drug mefloquine hydrochloride,[20] the phenyl substituent

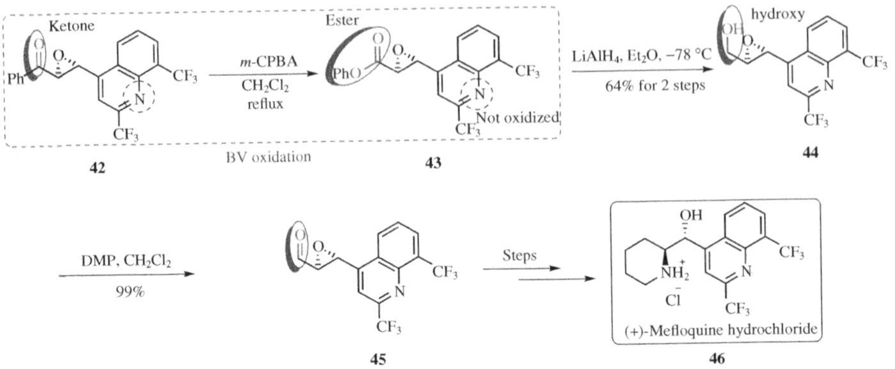

Scheme 2.9 The total synthesis of (+)-mefloquine hydrochloride **46**.

migrated upon treatment of ketone **42** with *m*-CPBA in refluxing CH_2Cl_2, affording the desired phenyl ester **43** while leaving the epoxide and quinoline unchanged (Scheme 2.9).

2.3.2 Oxidation of Aldehydes to Esters

The aldehyde is arguably the most versatile carbonyl functional group and more reactive than any other carbonyl groups toward a plethora of nucleophilic reactions. The BV oxidation can be a common way to transform an aldehyde into a carboxylic acid (migration of hydrogen);[21] however, the BV oxidation of aldehydes with groups having higher migratory aptitude than H affords the formate esters. Secondary alkyl groups in aliphatic aldehydes, for example, have higher migratory aptitude than H. In the synthesis of cinatrin C_1, a potent inhibitor of rat platelet phospholipase A2 (PLA2),[22] reported by Hatakeyama and coworkers (Scheme 2.10),[23] aldehyde **48** was formed from primary alcohol **47** via Dess–Martin oxidation. Then the aldehyde **48** was subjected to BV oxidation using *m*-CPBA as oxidant and the secondary alkyl group transferred, generating formate ester **49**. Methanolytic removal of the formyl group of **49** followed by TPAP oxidation of the lactol led to the formation of γ-lactone **50**, which could be further converted to the target molecule (−)-cinatrin C_1 **51** in several steps.

2.3.3 Oxidation of Cyclic Ketones to Lactones

BV oxidations of the four-, five-, six-, and seven-membered-ring ketones can provide the corresponding γ-, δ-, ε-, and ζ-lactones, respectively. These are not only important structural cores of numerous natural products but are also key intermediates in many organic syntheses. Reactivity in the BV oxidation decreases gradually along with the expansion of the ring size (from 4 to 7), implying that ring strain within the smaller-ring ketones favors the BV oxidation.

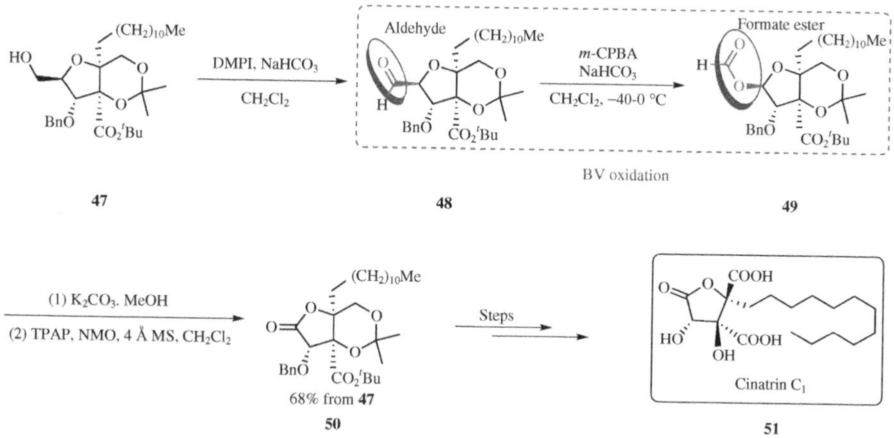

Scheme 2.10 The total synthesis of the (−)-cinatrin C_1 **51**.

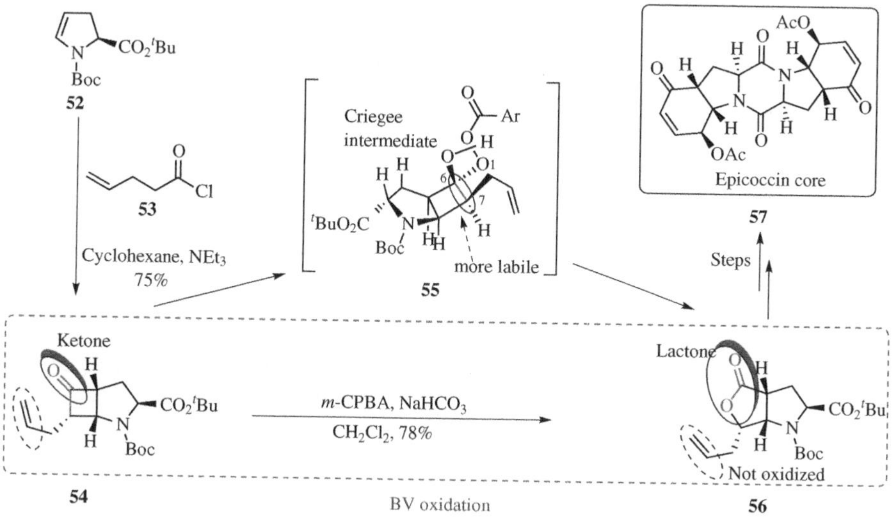

Scheme 2.11 The total synthesis of the epicoccin core **57**.

In Bräse group's synthesis of precursor **56** for the epicoccin core **57**,[24] the bicyclic γ-lactone **56** was an important intermediate. The bicyclic cyclobutanone **54** was constructed through a [2+2] cycloaddition of enecarbamate **52** with an allylketene, generated *in situ* from the corresponding pent-4-enoyl chloride **53** and triethylamine. Subsequently, a completely regioselective BV oxidation of the cyclobutanone **54** with *m*-CPBA as oxidant in the presence of $NaHCO_3$ successfully converted **54** into lactone **56**. Notably, the terminal double bond was not oxidized under these conditions (Scheme 2.11). The high regioselectivity of this BV oxidation could be due to localized strain in the Criegee cyclobutane intermediate **55**. In accord with literature precedent,[25] the steric interactions among the allyl, hydroxyl, and pyrrolidine ring in the all-cis arrangement with in the endo alkyl framework of the substrate **54** probably makes the C6—C7 bond more labile. This promotes the O1 insertion at this position, affording the desired product **56**.

The BV oxidation of cyclopentanones can provide one-step construction of δ-lactones, which are important frameworks existing in various natural products. For example, (+)-tanikolide **63**, a 6,6-disubstituted δ-lactone, is a γ-lactone metabolite of the marine cyanobacterium *Lyngbya majuscula* and shows antifungal activity.[26] In the concise and enantioselective total synthesis described by Deng's group (Scheme 2.12),[27] the enantioselective conjugate addition of β-ketoester **58** to acrolein **59** with 9′-phenanthryl substituted quinuclidine QD as catalyst smoothly provided the desired product **60** in quantitative yield with virtually perfect enantioselectivity (>99% ee). Following several transformations, the keto alcohol **62** with the two aliphatic side chains necessary for tanikolide was obtained. At last, a BV oxidation of **62** with *m*-CPBA completed the total synthesis of (+)-tanikolide **63**. Importantly, the hydroxyl group was unaffected, and the absolute configuration at the migrating chiral center was unchanged.

Though the BV oxidation of cyclopentanones to δ-lactones is usually realized easily, there are cases when reaction conditions need to be carefully screened. In Lei's synthesis of psilostachyin C **66** (Scheme 2.13),[28] isolated from *Ambrosia psilostachya* DC,[29] the last step was the BV oxidation of cyclopentanone **65**. However, the oxidation turned out to be a great challenge. The commonly used BV oxidant such as *m*-CPBA, CH_3CO_3H, $PhSeO_2H/H_2O_2$, $TFAA/H_2O_2$, $H_2O_2/NaOH$, and Ph_3COOLi led to either poor reactivity or excessive side reactions. After extensive screening

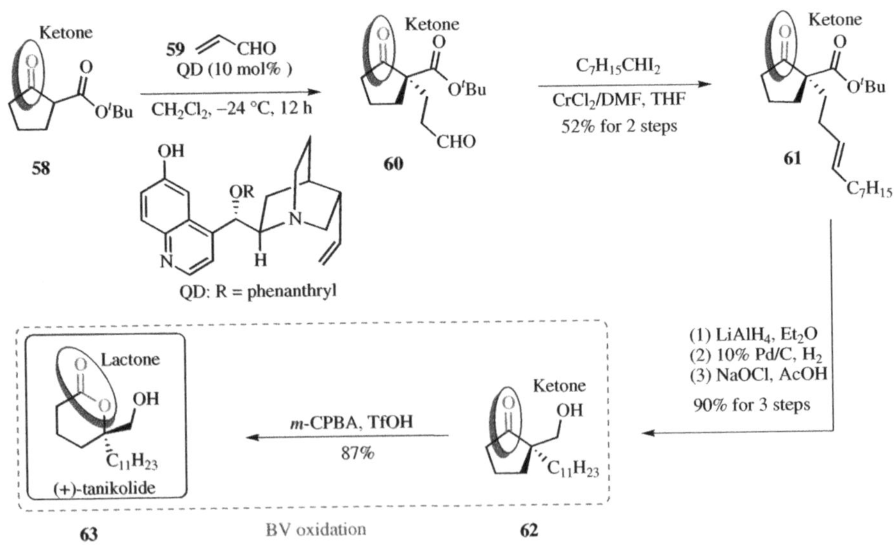

Scheme 2.12 The total synthesis of (+)-tanikolide **63**.

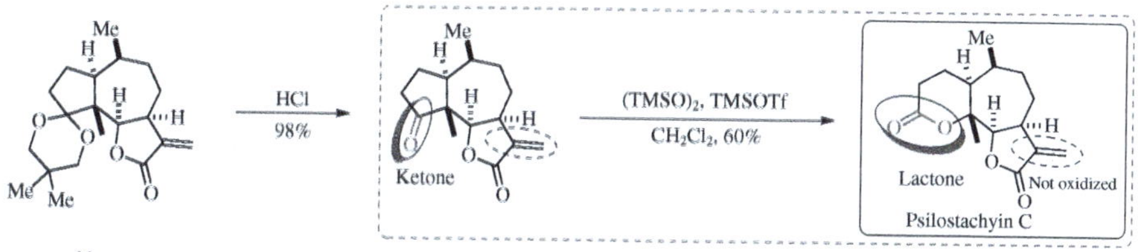

64 65 66

Scheme 2.13 The total synthesis of the psilostachyin C **66**.

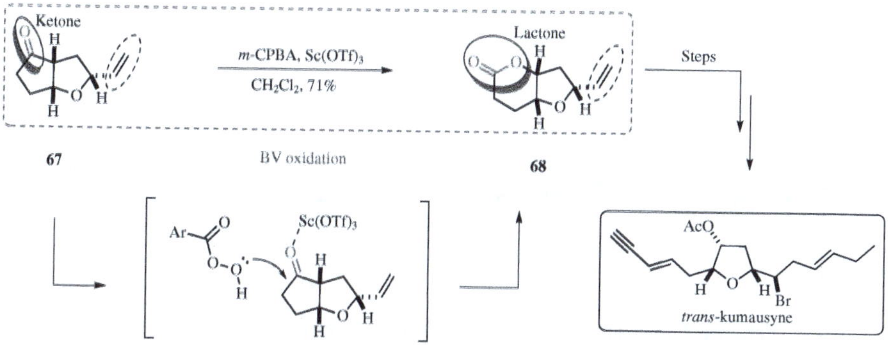

67 68 69

Scheme 2.14 The total synthesis of (±)-*trans*-kumausyne **69**.

of reaction conditions, bis(trimethylsilyl)-peroxide $(TMSO)_2$ along with TMSOTf was identified as an effective oxidation system, allowing for completion of the total synthesis of racemic psilostachyin C **66**, in which the C=C double bond was not affected.

Though usually tolerated in BV oxidation conditions, as in the example, a C=C double bond has the potential to undergo competitive olefin epoxidation. The presence of a suitable Lewis acid can help resolve the problem. In the Phillips group's synthesis of (±)-*trans*-kumausyne **69** (Scheme 2.14),[30] when performing the BV oxidation of **67** with *m*-CPBA alone, mixtures of products derived from epoxidation and BV oxidation were produced. On the other hand, when ketone **67** was subjected to *m*-CPBA in the presence of scandium triflate, which could activate the C=O carbonyl group and promote the BV oxidation, the bicyclic lactone **68** was produced in 71% yield.

As another example of a selective BV reaction, under suitable oxidation conditions, a cyclopentanone could be oxidized in the presence of an α,β-unsaturated cyclohexenone. In the Williams group's total synthesis of the acetylcholinesterase inhibiting fawcettimine alkaloids,[31] including fawcettimine **76**, fawcettidine **77**, lycoflexine **78**, and lycoposerramine B **79**,[32] a stereoselective Diels–Alder reaction of enone **71** with diene **70** provided an efficient and unified strategy to construct *cis*-fused 6,5-carbocycles such as **72** with one all-carbon quaternary center. In a

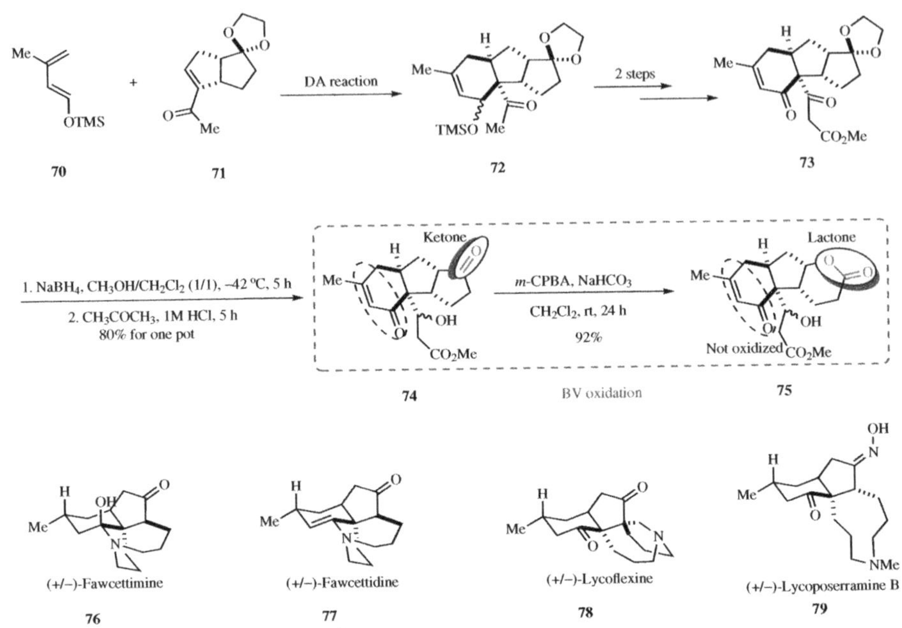

Scheme 2.15 The total synthesis of the fawcettimine class alkaloids.

key BV oxidation step, tricyclic keto-enone **74** provided the target intermediate **75** in high yield with *m*-CPBA as the oxidant in the presence of $NaHCO_3$. Although compound **74** is a keto-enone, the BV oxidation was found to be highly selective, and only the desired lactone **75** was isolated. It may be that the conjugation lowers the electrophilicity of the enone carbonyl group compared to the cyclopentanone, so that the nucleophilic peroxide reagent adds to the unconjugated carbonyl site, affording lactone product **75** (Scheme 2.15).

Under different conditions, meanwhile, an α,β-unsaturated cyclohexenone can be oxidized to a seven-membered enol lactone. The vinyl group of a primary alkyl- or secondary alkyl-substituted α,β-unsaturated ketone showed preferential migration in the BV oxidation.[33] In the Tokuyama group's synthesis of (−)-acetylaranotin **86** (Scheme 2.16),[34] the proline-fused dihydrooxepine ring was formed efficiently by an unusual vinylogous Rubottom oxidation followed by the regioselective BV oxidation. Starting from L-Cbz-tyrosine **80**, the intermediate **84** was synthesized efficiently in several steps including the vinylogous Rubottom transformation of **81–82**. In constructing the dihydrooxepine skeleton **85**, a BV oxidation of enol lactone **84** worked well since the sp^2-hybridized carbon of the α,β-unsaturated cyclic ketone showed a higher migratory aptitude than the α' site.[33,35] After examining a variety of oxidants, it was found that a combination of TFAA/UHP gave the best results, which provided enol lactone **85** exclusively.

Additionally, the regioselective BV oxidation of primary alkyl-substituted α,β-unsaturated ketones[35] to prepare seven-membered lactones was also reported

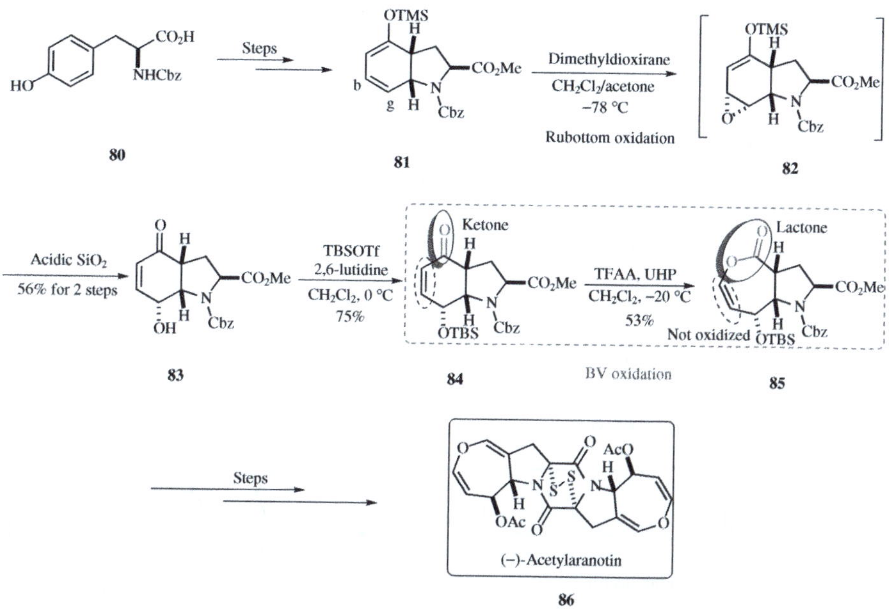

Scheme 2.16 The total synthesis of the (−)-acetylaranotin **86**.

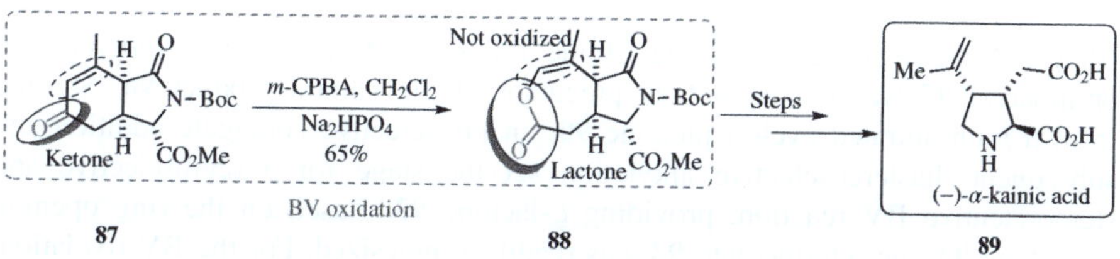

Scheme 2.17 The total synthesis of the (−)-α-kainic acid **89**.

by Schultz and Pettus using TFAA/UHP systems. The true oxidant under these conditions is pertrifluoroacetic acid, generated *in situ* by reaction of trifluoroacetic anhydride with urea–hydrogen peroxide in dichloromethane in the presence of sodium hydrogen phosphate. However, in the Jung group's synthesis of (−)-α-kainic acid (Scheme 2.17),[36] isolated from the marine alga *Digenea simplex*,[37] the ring opening of cyclohexenone **87** using the BV oxidation resulted in enol lactone **88** which was unstable upon prolonged reaction times, resulting in a low yield of the product under Schultz's reaction conditions. Alternatively, *m*-CPBA proved to be an excellent oxidation reagent. Buffered with Na_2HPO_4 in CH_2Cl_2, this BV oxidation system gave the desired product **88** in moderate yield and regioselectivity, with only a trace amount of the corresponding epoxide byproduct.

Cycloheptanones can also undergo BV oxidation, providing eight-membered ζ-lactones. In Minnaard group's synthesis of phthiocerol dimycocerosate A

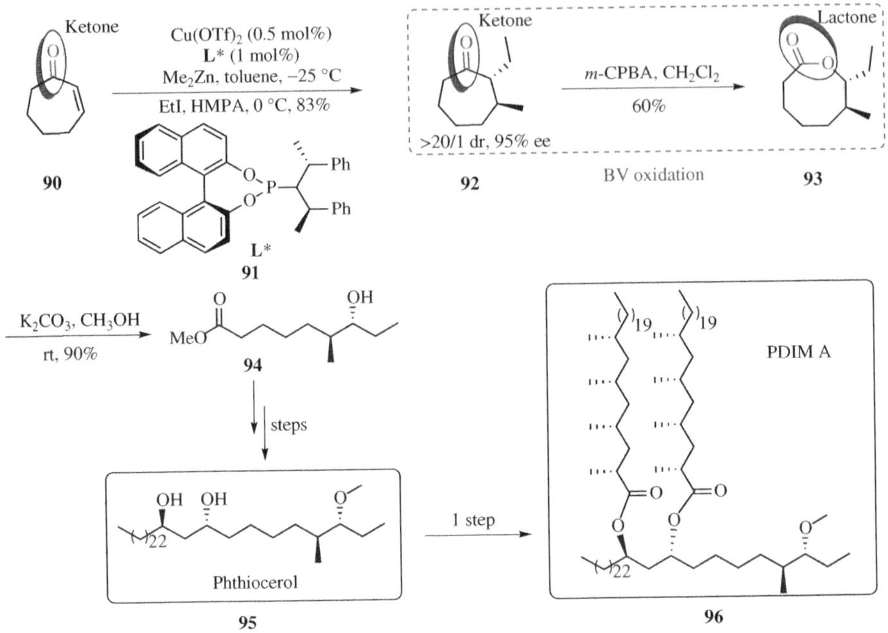

Scheme 2.18 The total synthesis of PDIM A.

(PDIM A) (Scheme 2.18),[38] one of the lipids of *Mycobacterium tuberculosis*, phthiocerol **95** was the immediate precursor of the natural product **96**. Starting from α,β-unsaturated cycloheptenone **90**, enantioselective conjugate addition and subsequent diastereoselective alkylation set the stage for a regioselective and stereoretentive BV reaction, providing ζ-lactone **93**. Based on the ring opening of lactone **93**, the intermediate **94** was readily synthesized. For the BV oxidation, despite extensive experimentation, the best results, achieved using *m*-CPBA, still gave moderate efficiency (60% yield). Undoubtedly, this outcome is due to the increase in ring strain incurred on progressing from a seven- to an eight-membered ring during the course of the C-to-O shift in the reaction. Versatile solutions to this problem are scarce.

2.3.4 Asymmetric BV Oxidation/Rearrangement

Given that oxidation of cyclic ketones to lactones by the BV oxidation/rearrangement is such an important synthetic tool in organic chemistry, catalytic asymmetric variants have also been developed to obtain enantioenriched chiral lactones via desymmetrization and kinetic resolution of prochiral or racemic ketones (Scheme 2.19), respectively. Though not widely used in total synthesis yet, these approaches are highly attractive for their advantages in terms of chiral amplification and obtaining optically pure compounds.

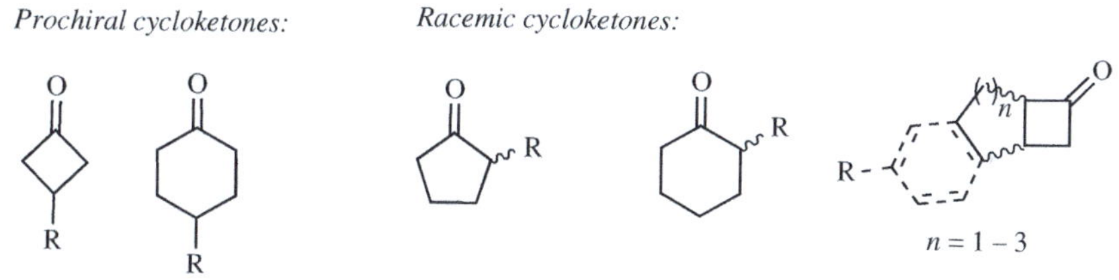

Scheme 2.19 Prochiral and racemic ketones in the asymmetric BV oxidation.

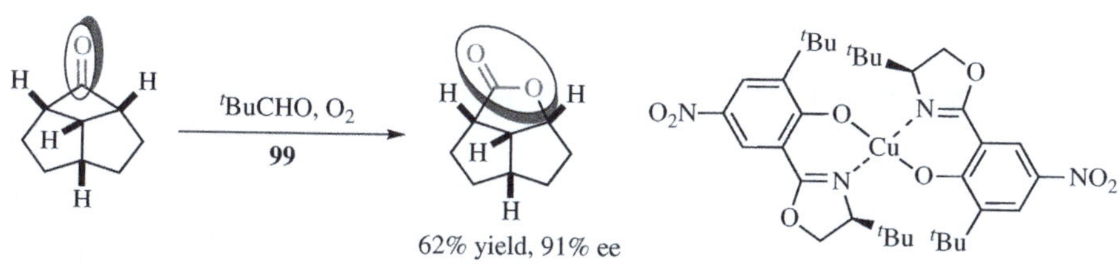

97 98 99

Scheme 2.20 Desymmetrization of tricyclic ketone **97**.

2.3.4.1 ***Desymmetrization of Prochiral Cyclic Ketones*** Because ring-strained cyclobutanones are more reactive in BV oxidations than other cyclic ketones, they have been well studied in the catalytic asymmetric BV oxidation. The first example of desymmetrization of prochiral ketones via the asymmetric BV oxidation was reported by Bolm *et al.* in 1997 with a chiral copper complex as catalyst.[39] In the Bolm group's study, a tricyclic ketone **97** afforded the corresponding lactone **98** with an asymmetric induction of up to 91% ee (Scheme 2.20). Since then, impressive results have been achieved for the desymmetrization of prochiral 3-substituted cyclobutanones **100** with chiral metal complexes or organocatalysts.[40] The BV oxidation of prochiral cyclobutanones **100** with hydrogen peroxide as the oxidant was successfully realized by Ding's group using a chiral Brønsted acid catalyst **102**. This approach afforded various aryl γ-lactones **101** in 91–99% yield with 82–93% ee and benzyl-substituted γ-lactones **101** in 99% yield and 55–58% ee (Scheme 2.21).[40h,i] Feng's group has also developed an efficient chiral *N,N′*-dioxide–Sc^{III} complex catalyst for the asymmetric BV oxidation of prochiral cyclobutanones **100**.[41] With *m*-CPBA as oxidant, the desired aryl-substituted γ-lactones **101** were obtained in 71–84% yield with 87–91% ee (Scheme 2.21).

For the BV oxidation of 4-alkylcyclohexanones **103** to afford chiral ε-lactones **104**, biocatalysts have shown excellent enantiocontrol.[42] For example, baker's yeast and engineered *Escherichia coli* could convert the prochiral ketones **103** to the corresponding desymmetrized lactones **104** (Scheme 2.22),[42a,b] in which side reactions of the hydrolysis of the lactone and subsequent oxidation were minimized.

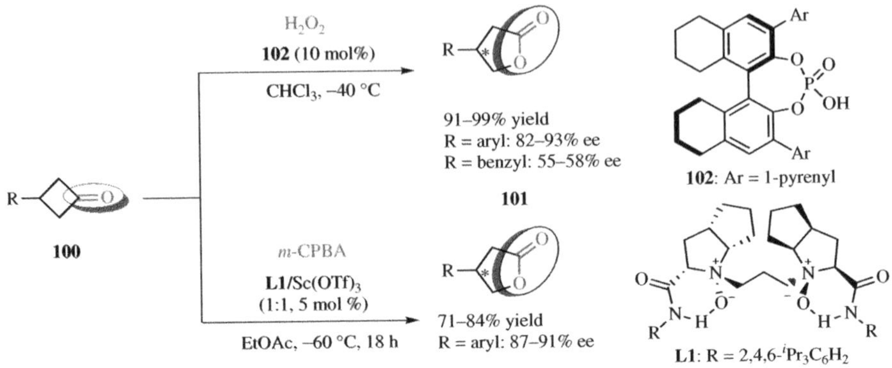

Scheme 2.21 Desymmetrization of prochiral cyclobutanones **100**.

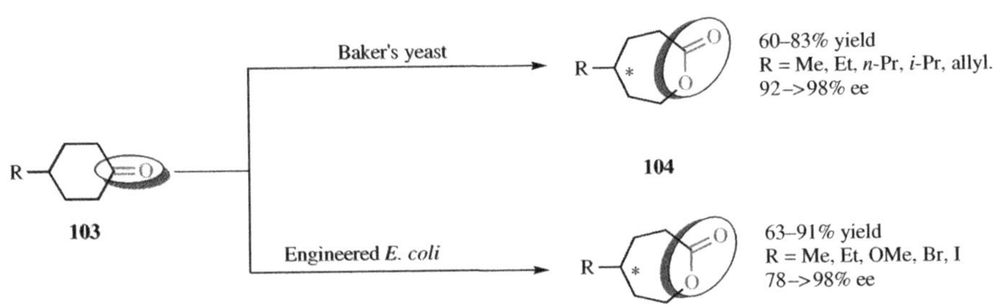

Scheme 2.22 Desymmetrization of prochiral cyclohexanones **103** mediated by biocatalysts.

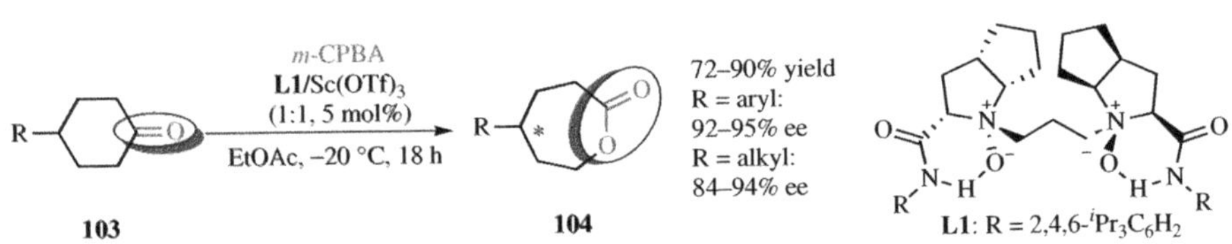

Scheme 2.23 Desymmetrization of prochiral cyclohexanones **103** catalyzed by a chiral *N,N′*-dioxide–Sc^{III} complex.

On the other hand, there are fewer artificial catalysts that have been explored in the desymmetrization of prochiral cyclohexanones **103**.[41,43] To date, the only nonenzymatic catalytic system for this reaction has been developed by Feng's group. The *N,N′*-dioxide **L1** coordinated with Sc^{III} was found to be highly efficient, affording various aryl- and alkyl-substituted ε-lactones **104** in 72–90% yield and 84–95% ee (Scheme 2.23).[41]

2.3.4.2 *Kinetic Resolution of Racemic Ketones* There are three types of kinetic resolution for racemic ketones:

1. Dynamic kinetic resolution (DKR), in which the racemic ketone is converted into a single optically pure lactone.
2. Classical kinetic resolution (KR), in which one of the enantiomers of the ketone is converted into a different optically pure lactone, and the other ketone enantiomer recovered.
3. Parallel kinetic resolution (PKR), in which each of the two ketone enantiomers is converted into two types of optically pure lactone. PKR can be especially suitable for bicyclobutanones.

DKR of racemic ketones via the BV oxidation is very rare. Notably, the CHMO-containing (CHMO = cyclohexanone monooxygenase) recombinant *E. coli* sp. was found to be highly efficient in the DKR of 2-substituted cyclopentanone *rac*-**105** at pH 9. The ketone substrate **105** underwent facile racemization via keto–enol tautomerism, and the lactone product (*R*)-**106** was isolated in 85% yield and 96% ee (Scheme 2.24).[44]

The first artificial catalysts for the BV oxidation of racemic cyclic ketones via classical kinetic resolutions[45] were independently and almost simultaneously reported by Bolm *et al.*[45a] and Strukul and coworkers.[45b] These studies produced chiral lactones in moderate ee using chiral Cu complex **99** (Scheme 2.25) and diphosphane/Pt complex **111** (Scheme 2.26) as catalyst, respectively. After a nearly blank near 20-year hiatus in this field, Feng's group successfully developed a highly efficient catalytic asymmetric *N,N'*-dioxide–Sc^{III} complex catalyst system for the kinetic resolution of racemic 2-arylcyclohexanones **107** (Scheme 2.27).[41] Unlike with other systems described previously, the kinetic resolution of racemic 2-arylcyclohexanones **107** was realized via an abnormal BV oxidation, affording the ε-lactones **112** with a reversal of migratory aptitude. Both the desired lactones **112** and the recovered ketones **107** were obtained with high ee.

Asymmetric BV oxidations can also be used in PKR, where the two different enantiomers of a racemic cyclic ketone lead selectively to the two regioisomeric lactones. For example, in 2004, the BV oxidation of racemic bicyclobutanones was successfully realized by the Katsuki group using Zr(salen) complex **116**.[46] In this study, the authors demonstrated that Zr(salen) complex **116** exhibits asymmetric catalytic activity in the BV oxidation of racemic bicyclo[3.2.0]alkan-5-ones **113**,

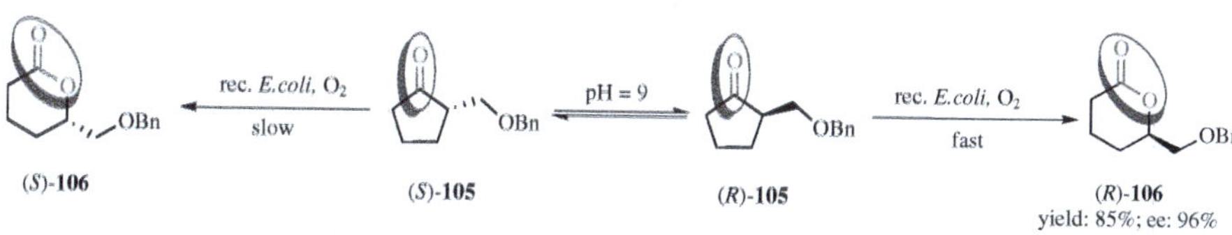

Scheme 2.24 Dynamic kinetic resolution of ketone **105** catalyzed by CHMO-containing rec. *E. coli*.

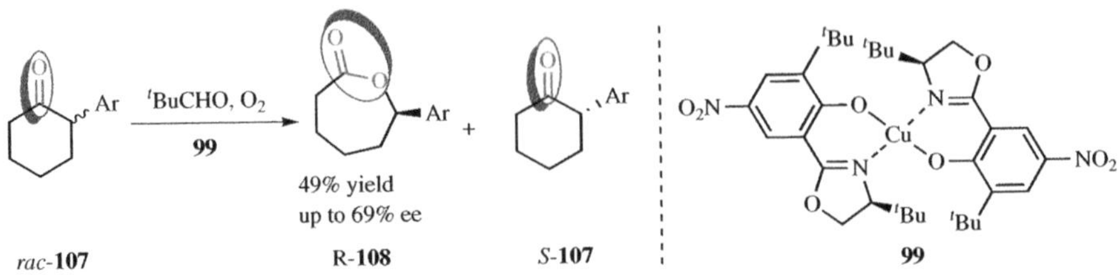

Scheme 2.25 Kinetic resolution of cyclic ketones **107** catalyzed by a chiral Cu complex **99**.

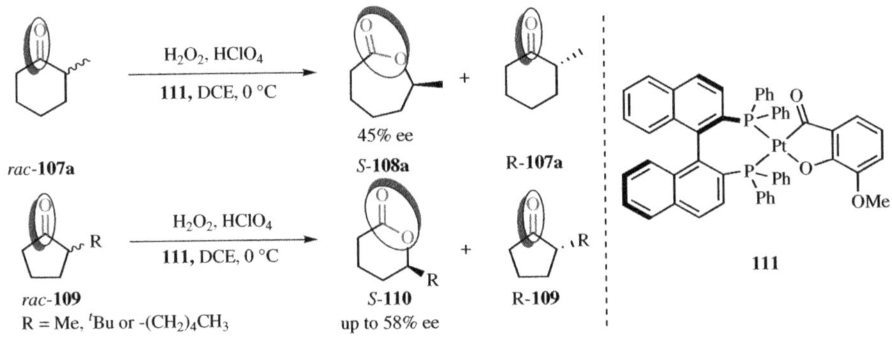

Scheme 2.26 Kinetic resolution of cyclic ketones catalyzed by a chiral Pt complex **111**.

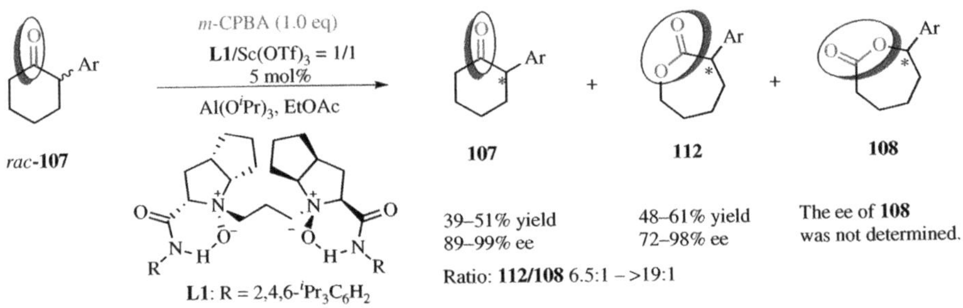

Scheme 2.27 Kinetic resolution of cyclic ketones **107** catalyzed by a chiral *N,N′*-dioxide–Sc^{III} complex.

in which one enantiomer of **113** led to the normal lactone (NL) **114** and the other enantiomer led to the abnormal lactone (AL) **115**, both with excellent ee values (Scheme 2.28).

Additionally, if one enantiomer of racemic bicyclobutanones can undergo both normal and abnormal BV oxidation, one regioisomer of lactones will be obtained in lower ee. For example, a chiral copper catalyst system developed by Bolm *et al.* was extended to the asymmetric BV oxidation of racemic substituted bicyclobutanones

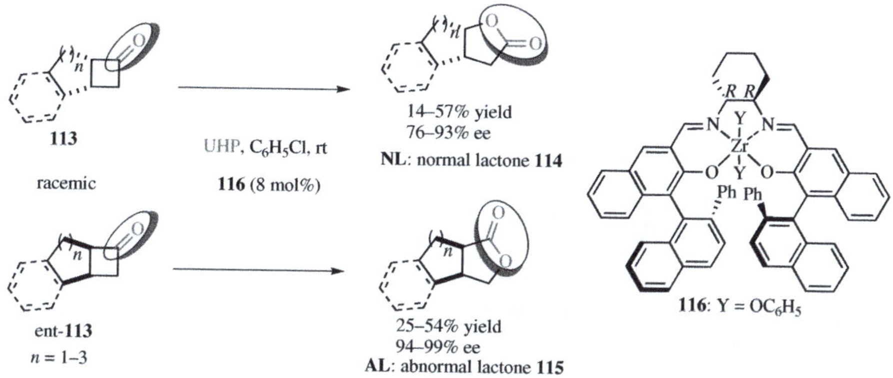

Scheme 2.28 Parallel kinetic resolution of racemic bicyclic cyclobutanones catalyzed by chiral Zr complex **116**.

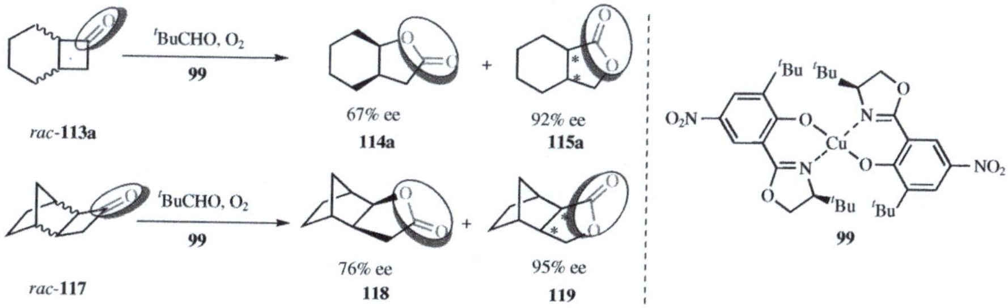

Scheme 2.29 Chiral Cu complex **99** catalyzed asymmetric BV oxidation of racemic bicyclic cyclobutanones.

113a and **117**, affording the corresponding enantioenriched chiral lactones **115a** and **119** with high ee but **114a** and **118** with moderate ee (Scheme 2.29).[47a] Since then, further impressive results have been achieved for the BV oxidation of racemic bicyclobutanones with chiral metal complexes or organocatalysts.[47b–e] For example, Ding's group has successfully developed a highly efficient chiral Brønsted acid catalyst **102** for the asymmetric BV oxidation of racemic bicyclobutanones **120**,[48] affording a range of chiral lactones with the NL products **121** as the major regioisomer in moderate to high NL/AL ratios (from 2.1/1 to 11.5/1), whereas the minor abnormal isomers **122** could be obtained with good to excellent ee values (Scheme 2.30).

2.3.5 Oxidation of α-Diketones to Anhydrides

Submitting α-diketones to the BV oxidation affords anhydrides. In Rigby *et al.*'s synthesis of 9-*epi*-pentalenic acid **130**,[49] isolated from the fermentation broth of *Streptomyces griseochromogens*,[50] anhydride **129** was a key intermediate. The

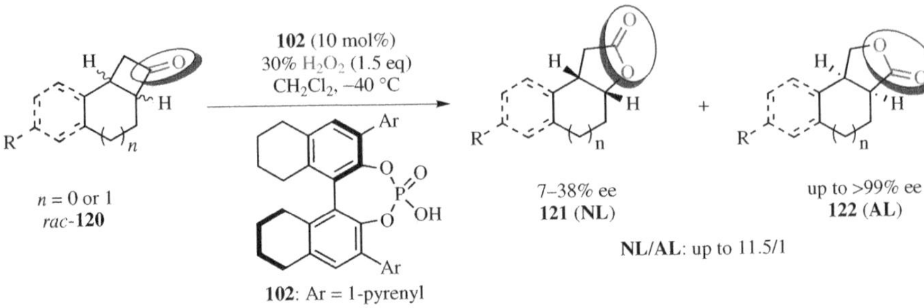

Scheme 2.30 Chiral Brønsted acid **102** catalyzed asymmetric BV oxidation of racemic bicyclic cyclobutanones **120**.

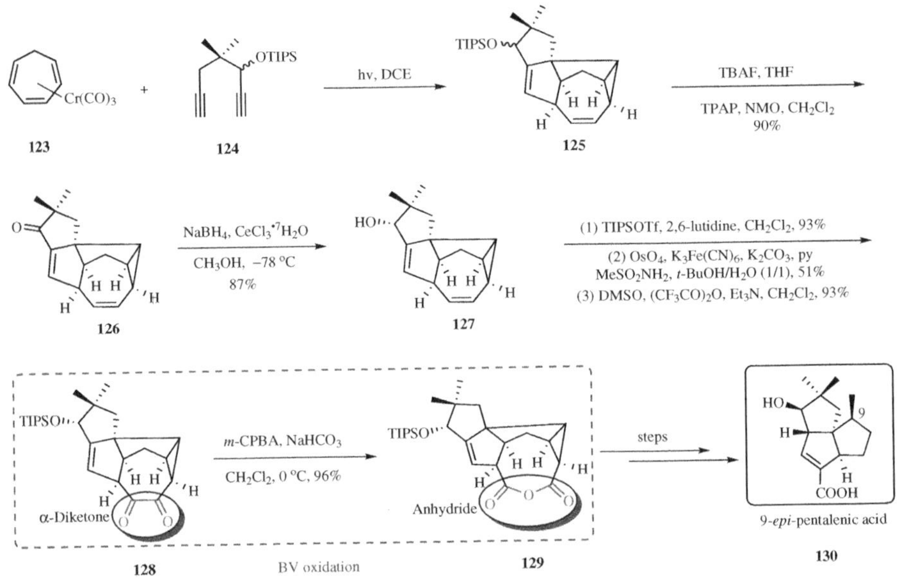

Scheme 2.31 The total synthesis of 9-*epi*-pentalenic acid **130**.

regioselective higher-order three-component cycloaddition of tethered diyne **124** with chromium(0)-complexed cyclic triene **123** generated five rings and six stereogenic centers of product **125** in one step. The cycloaddition product **125** was then successfully advanced to intermediate **128** in several steps. Treatment of α-diketone **128** with 1.3 equiv. *m*-CPBA, buffered with an excess of $NaHCO_3$ at 0 °C, provided the cyclic anhydride **129** in excellent yield. Notably, no trace of epoxide formation was observed at the trisubstituted alkene. The formation of eight-membered ring anhydrides by the BV reaction of cyclic seven-membered vicinal diones is very rare in the literature. This outcome is all the more significant in Rigby's synthesis because it left a potentially reactive olefin untouched during the reaction (Scheme 2.31).

2.4 SUMMARY AND OUTLOOK

In conclusion, extensive efforts have been devoted to developing BV oxidation/rearrangement processes for organic synthesis since the discovery of the reaction over 110 years ago. Its high regioselectivity, functional group tolerance, numerous oxidant options, and mild reaction conditions make the BV oxidation/rearrangement a very attractive synthetic method. From the perspective of green synthesis, meanwhile, there is still room for improvement in the application of the BV oxidation/rearrangement. For example, H_2O_2 is an environmentally friendly oxidant, but it is less applied in organic synthesis due to its low reactivity and presence as an aqueous solution. Additionally, green solvents would be desirable because the typical BV solvent, CH_2Cl_2, is far less than idea from an environmental perspective. Development of new catalysts for the BV oxidation/rearrangement will certainly help resolve these problems.

REFERENCES

1. Baeyer, A.; Villiger, V. *Ber. Dtsch. Chem. Ges.* **1899**, *32*, 3625–3633.
2. For selected reviews, see: (a) Krow, G. R. *Org. React.* **1993**, *43*, 251–798. (b) Renz, M.; Meunier, B. *Eur. J. Org. Chem.* **1999**, 737–750. (c) Strukul, G. *Angew. Chem. Int. Ed.* **1998**, *37*, 1198–1209. (d) ten Brink, G.-J.; Arends, I. W. C. E.; Sheldon, R. A. *Chem. Rev.* **2004**, *104*, 4105–4123.
3. (a) Bolm, C.; Beckmann, O. In Comprehensive Asymmetric Catalysis; Jacobsen, E. N., Pfaltz, A., Yamamoto, H., Eds.; Springer: Berlin, **1999**; Vol. 2, pp. 803–809. (b) Bolm, C. In Advances in Catalytic Processes; Doyle, M. P., Ed.; JAI Press: Greenwich, CT, **1997**; Vol. 2, pp. 43–68.
4. Criegee, R. *Justus Liebigs Ann. Chem.* **1948**, *560*, 127–135.
5. Doering, W. von. E.; Dorfman, E. *J. Am. Chem. Soc.* **1953**, *75*, 5595–5598.
6. Berson, J. A.; Suzuki, S. *J. Am. Chem. Soc.* **1959**, *81*, 4088–4094.
7. (a) Cádenas, R.; Reyes, L.; Langúnez-Otero, J.; Cetina, R. J. *Mol. Struct.: THEOCHEM* **2000**, *497*, 211–225. (b) For a study on intra- and intermolecular isotope effects, see: Singleton, D. A.; Szymanski, M. J. *J. Am. Chem. Soc.* **1999**, *121*, 9455–9456. (c) Hawthorne, M. F.; Emmons, W. D. *J. Am. Chem. Soc.* **1958**, *80*, 6398–6404. (d) Okuno, Y. *Chem. Eur. J.* **1997**, *3*, 212–218.
8. (a) Noyori, R.; Sato, T.; Kobayashi, H. *Tetrahedron Lett.* **1980**, *21*, 2569–2572. (b) Noyori, R.; Kobayashi, H.; Sato, T. *Tetrahedron Lett.* **1980**, *21*, 2573–2576. (c) Chandrasekhar, S.; Roy, C. D. *Tetrahedron Lett.* **1987**, *28*, 6371–6372. (d) Chandrasekhar, S.; Roy, C. D. *J. Chem. Soc., Perkin Trans. 2* **1994**, 2141–2143. (e) Goodman, R. M.; Kishi, Y. *J. Am. Chem. Soc.* **1998**, *120*, 9392–9393. (f) Krasutsky, P. A.; Kolomitsyn, I. V.; Kiprof, P.; Carlson, R. M.; Fokin, A. A. *J. Org. Chem.* **2000**, *65*, 3926–3933.
9. Crudden, C. M.; Chen, A. C.; Calhoun, L. A. *Angew. Chem. Int. Ed.* **2000**, *39*, 2851–2855.
10. Yao, T.-R.; Chen, Z.-N. *Yaoxue Xuebao* **1979**, *12*, 731–735; *Chem. Abstr.* **1980**, *93*, 101406n.
11. Malinakova, H. C.; Liebeskind, L. S. *Org. Lett.* **2000**, *2*, 3909–3911.
12. Zhang, Y.; Liebeskind, L. S. *J. Am. Chem. Soc.* **2006**, *128*, 465–472.

13. (a) Angeles, A. R.; Dorn, D. C.; Kou, C. A.; Moore, M. A. S.; Danishefsky, S. J. *Angew. Chem. Int. Ed.* **2007**, *46*, 1451–1454. (b) Angeles, A. R.; Waters, S. P.; Danishefsky, S. J. *J. Am. Chem. Soc.* **2008**, *130*, 13765–13770.
14. Pappo, R.; Allen, Jr., D. S.; Lemieux, R. U.; Johnson, W. S. *J. Org. Chem.* **1956**, *21*, 478–479.
15. Laurent, M.; Cérésiat, M.; Marchand-Brynaert, J. *J. Org. Chem.* **2004**, *69*, 3194–3197.
16. Walsh, C. *Nature* **2000**, *406*, 775–781.
17. Noël, R.; Vanucci-Bacqué, C.; Fargeau-Bellassoued, M.-C.; Lhommet, G. *Eur. J. Org. Chem.* **2007**, 476–486.
18. Ma, D.; Ma, N. *Tetrahedron Lett.* **2003**, *44*, 3963–3965.
19. Zhu, B.; Zhang, W.; Lee, R.; Han, Z.; Yang, W.; Tan, D.; Huang, K.-W.; Jiang, Z. *Angew. Chem. Int. Ed.* **2013**, *52*, 6666–6670.
20. Knight, J. D.; Sauer, S. J.; Coltart, D. M. *Org. Lett.* **2011**, *13*, 3118–3121.
21. Murray, A. T.; Matton, P.; Fairhurst, N. W. G.; John, M. P.; Carbery, D. R. *Org. Lett.* **2012**, *14*, 3656–3659.
22. (a) Murakami, M.; Taketomi, Y.; Miki, Y.; Sato, H.; Hirabayashi, T.; Yamamoto, K. *Prog. Lipid Res.* **2011**, *50*, 152–192. (b) Dennis, E. A.; Cao, J.; Hsu, Y.-H.; Magrioti, V.; Kokotos, G. *Chem. Rev.* **2011**, *111*, 6130–6185.
23. Urabe, F.; Nagashima, S.; Takahashi, K.; Ishihara, J.; Hatakeyama, S. *J. Org. Chem.* **2013**, *78*, 3847–3857.
24. Gross, U.; Nieger, M.; Bräse, S. *Org. Lett.* **2009**, *11*, 4740–4742.
25. de Faria, A. R.; Salvador, E. L.; Correia, C. R. D. *J. Org. Chem.* **2002**, *67*, 3651–3661.
26. Singh, I. P.; Milligan, K. E.; Gerwick, W. H. *J. Nat. Prod.* **1999**, *62*, 1333–1335.
27. Wu, F.; Hong, R.; Khan, J.; Liu, X.; Deng, L. *Angew. Chem. Int. Ed.* **2006**, *45*, 4301–4305.
28. Li, C.; Tu, S.; Wen, S.; Li, S.; Chang, J.; Shao, F.; Lei, X. *J. Org. Chem.* **2011**, *76*, 3566–3570.
29. Kagan, H. B.; Miller, H. E.; Renold, W.; Lakshmikantham, M. V.; Tether, L. R.; Herz, W.; Mabry, T. J. *J. Org. Chem.* **1966**, *31*, 1629–1632.
30. Chandler, C. L.; Phillips, A. J. *Org. Lett.* **2005**, *7*, 3493–3495.
31. For reviews of *Lycopodium* alkaloids, see: (a) Ayer, W. A. *Nat. Prod. Rep.* **1991**, *8*, 455–463. (b) Ma, X.; Gang, D. R. *Nat. Prod. Rep.* **2004**, *21*, 752–772. (c) Hirasawa, Y.; Kobayashi, J.; Morita, H. *Heterocycles* **2009**, *77*, 679–729.
32. Pan, G.; Williams, R. M. *J. Org. Chem.* **2012**, *77*, 4801–4811.
33. Krafft, G. A.; Katzenellenbogen, J. A. *J. Am. Chem. Soc.* **1981**, *103*, 5459–5466.
34. Fujiwara, H.; Kurogi, T.; Okaya, S.; Okano, K.; Tokuyama, H. *Angew. Chem. Int. Ed.* **2012**, *51*, 13062–13065.
35. Schultz, A. G.; Pettus, L. *J. Org. Chem.* **1997**, *62*, 6855–6861.
36. Jung, Y. C.; Yoon, C. H.; Turos, E.; Yoo, K. S.; Jung, K. W. *J. Org. Chem.* **2007**, *72*, 10114–10122.
37. Murakami, S.; Takemoto, T.; Shimizu, Z. *J. Pharm. Soc. Jpn.* **1953**, *73*, 1026–1028.
38. (a) Casas-Arce, E.; ter Horst, B.; Feringa, B. L.; Minnaard, A. J. *Chem. Eur. J.* **2008**, *14*, 4157–4159. (b) Barroso, S.; Geerdink, D.; ter Horst, B.; Casas-Arce, E.; Minnaard, A. J. *Eur. J. Org. Chem.* **2013**, 4642–4654.
39. (a) Bolm, C.; Schlingloff, G.; Bienewald, F. *J. Mol. Catal. A: Chem.* **1997**, *117*, 347–350. (b) Bolm, C.; Luong, T. K. K.; Schlingloff, G. *Synlett* **1997**, 1151–1152.

40. (a) Sugimura, T.; Fujiwara, Y.; Tai, A. *Tetrahedron Lett.* **1997**, *38*, 6019–6022. (b) Shinohara, T.; Fujioka, S.; Kotsuki, H. *Heterocycles* **2001**, *55*, 237–242. (c) Bolm, C.; Beckmann, O.; Cosp, A.; Palazzi, C. *Synlett* **2001**, 1461–1463. (d) Uchida, T.; Katsuki, T. *Tetrahedron Lett.* **2001**, *42*, 6911–6914. (e) Murahashi, S.-I.; Ono, S.; Imada, Y. *Angew. Chem. Int. Ed.* **2002**, *41*, 2366–2368. (f) Bolm, C.; Frison, J.-C.; Zhang, Y.; Wulff, W. D. *Synlett* **2004**, 1619–1621. (g) Frison, J.-C.; Palazzi, C.; Bolm, C. *Tetrahedron* **2006**, *62*, 6700–6706. (h) Xu, S.; Wang, Z.; Zhang, X.; Zhang, X.; Ding, K. *Angew. Chem. Int. Ed.* **2008**, *47*, 2840–2843. (i) Xu, S.; Wang, Z.; Li, Y.; Zhang, X.; Wang, H.; Ding, K. *Chem. Eur. J.* **2010**, *16*, 3021–3035.

41. Zhou, L.; Liu, X. H.; Ji, J.; Zhang, Y. H.; Hu, X. L.; Lin, L. L.; Feng, X. M. *J. Am. Chem. Soc.* **2012**, *134*, 17023–17026.

42. (a) Taschner, M. J.; Black, D. J.; Chen, Q.-Z. *Tetrahedron: Asymmetry* **1993**, *4*, 1387–1390. (b) Stewart, J. D.; Reed, K. W.; Martinez, C. A.; Zhu, J.; Chen, G.; Kayser, M. M. *J. Am. Chem. Soc.* **1998**, *120*, 3541–3548. (c) Mihovilovic, M. D.; Chen, G.; Wang, S.; Kyte, B.; Rochon, F.; Kayser, M. M.; Stewart, J. D. *J. Org. Chem.* **2001**, *66*, 733–738. (d) Rial, D. V.; Bianchi, D. A.; Kapitanova, P.; Lengar, A.; van Beilen, J. B.; Mihovilovic, M. D. *Eur. J. Org. Chem.* **2008**, 1203–1213. (e) Reetz, M. T.; Brunner, B.; Schneider, T.; Schulz, F.; Clouthier, C. M.; Kayser, M. M. *Angew. Chem. Int. Ed.* **2004**, *43*, 4075–4078.

43. (a) Paneghetti, C.; Gavagnin, R.; Pinna, F.; Strukul, G. *Organometallics* **1999**, *18*, 5057–5065. (b) Cavarzan, A.; Bianchini, G.; Sgarbossa, P.; Lefort, L.; Gladiali, S.; Scarso, A.; Strukul, G. *Chem. Eur. J.* **2009**, *15*, 7930–7939.

44. Berezina, N.; Alphand, V.; Furstoss, R. *Tetrahedron: Asymmetry* **2002**, *13*, 1953–1955.

45. (a) Bolm, C.; Schlingloff, G.; Weickhardt, K. *Angew. Chem. Int. Ed. Engl.* **1994**, *33*, 1848–1849. (b) Gusso, A.; Baccin, C.; Pinna, F.; Strukul, G. *Organometallics* **1994**, *13*, 3442–3451. (c) Peng, Y. G.; Feng, X. M.; Yu, K. B.; Li, Z.; Jiang, Y. Z.; Yeung, C.-H. *J. Organomet. Chem.* **2001**, *619*, 204–208.

46. Watanabe, A.; Uchida, T.; Irie, R.; Katsuki, T. *Proc. Natl. Acad. Sci. U.S.A.* **2004**, *101*, 5737–5742.

47. (a) Bolm, C.; Schlingloff, G. *J. Chem. Soc., Chem. Commun.* **1995**, 1247–1248. (b) Lopp, M.; Paju, A.; Kanger, T.; Pehk, T. *Tetrahedron Lett.* **1996**, *37*, 7583–7586. (c) Kanger, T.; Kriis, K.; Paju, A.; Pehk, T.; Lopp, M. *Tetrahedron: Asymmetry* **1998**, *9*, 4475–4482. (d) Bolm, C.; Beckmann, O. *Chirality* **2000**, *12*, 523–525. (e) Colladon, M.; Scarso, A.; Strukul, *G. Synlett* **2006**, 3515–3520.

48. Xu, S.; Wang, Z.; Zhang, X.; Ding, K. *Eur. J. Org. Chem.* **2011**, 110–116.

49. Rigby, J. H.; Laxmisha, M. S.; Hudson, A. R.; Heap, C. H.; Heeg, M. J. *J. Org. Chem.* **2004**, *69*, 6751–6760.

50. Seto, H.; Sasaki, T.; Uzawa, J.; Takeuchi, S.; Yonehara, H. *Tetrahedron Lett.* **1978**, *19*, 4411–4412.

3

THE WOLFF REARRANGEMENT: TACTICS, STRATEGIES AND RECENT APPLICATIONS IN ORGANIC SYNTHESIS

YOANN COQUEREL AND JEAN RODRIGUEZ

Aix Marseille Université, Centrale Marseille, CNRS, iSm2 UMR 7313, Marseille, France

3.1 INTRODUCTION

The Wolff rearrangement is the conversion of α-diazocarbonyl derivatives into ketenes. The reaction proceeds by extrusion of nitrogen and a 1,2-shift of the R^2 substituent of the α-diazocarbonyl substrate (Scheme 3.1). The Wolff rearrangement is one of the fundamental reactions of diazo compounds and has found many applications in organic synthesis, drug delivery,[1] and photolithography.[2] This chapter focuses on applications of the Wolff rearrangement in modern synthetic organic chemistry. An excellent and comprehensive review on the Wolff rearrangement appeared in 2002,[3] and some more recent research advances using the Wolff rearrangement have also been reviewed.[4] The Wolff rearrangement can be activated thermally, photochemically, or catalytically in the presence of silver(I), or less often rhodium(II) salts, or silver nanoclusters.[5] In most applications, the ketene product of the Wolff rearrangement is not isolated but used directly in nucleophilic additions or cycloaddition reactions. Significantly, the Wolff rearrangement has contributed to boost the development of ketene chemistry, avoiding the scope-limiting issues of having to use stable ketenes, ketene precursors (e.g., acyl halides) combined with additives, or precursors that produce unwanted or troublesome coproduct(s).

Molecular Rearrangements in Organic Synthesis, First Edition. Edited by Christian M. Rojas.

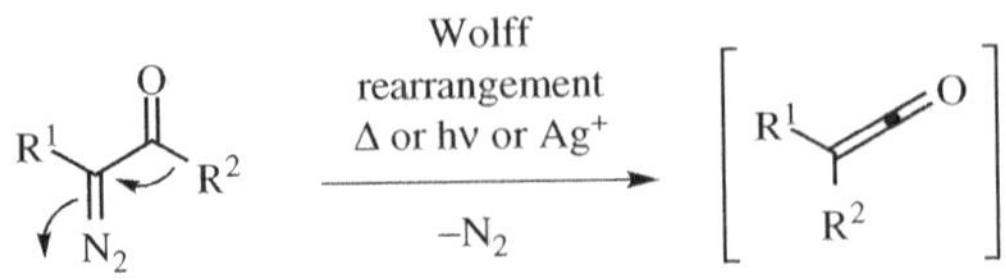

Scheme 3.1 The Wolff rearrangement.

Notably, if the R^1 and R^2 substituents of the diazo compound are different, the ketene product of the Wolff rearrangement is prochiral, opening opportunities for asymmetric reactions.

It was back in 1902 when Ludwig Wolff (1875–1919), a professor of analytical chemistry at the University of Jena, discovered the thermal rearrangement of ethyl 2-diazo-3-oxobutyrate in water to produce ethane-1,1-dicarboxylic acid, a reaction that has since been named after him.[6] Although it is now over 100 years old, the Wolff rearrangement is still alive and well, with prominent applications in modern synthetic organic chemistry. The main applications of the Wolff rearrangement in organic synthesis are the homologation of carboxylic acids (Arndt–Eistert reaction), the one-carbon ring contraction of cyclic α-diazoketones, and, more generally, the *in situ* production of ketenes and α-oxo-ketenes as reactive intermediates under additive-free and coproduct-free conditions (except for the inert N_2 gas).

3.2 TACTICS AND STRATEGIES VIA THE WOLFF REARRANGEMENT

When planning a synthesis, the identification of structural elements amenable to synthesis via the Wolff rearrangement is not an easy task because of the modifications in chain lengths or ring sizes associated with the rearrangement. However, some generic structural patterns of fundamental importance for synthesis (highlighted with gray backgrounds in Schemes 3.2–3.5) are particularly well suited for the use of the title rearrangement in their synthesis. Some contemporaneous applications of the retrosynthetic approaches presented here are detailed in Section 3.5.

3.2.1 Methylene Carboxylic Acid Derivatives (Arndt–Eistert Reaction)

The so-called Arndt–Eistert reaction[7] is the Wolff rearrangement of diazomethyl ketones followed by trapping of the resulting monosubstituted ketenes with

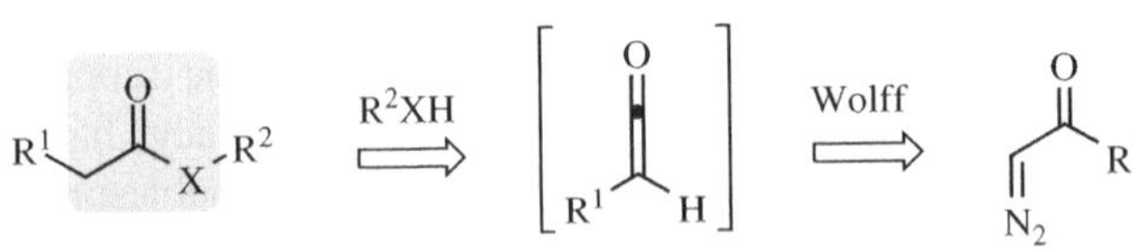

R^1 = alkyl, alkenyl, aryl; X = NR, O, S

Scheme 3.2 The Arndt–Eistert reaction.

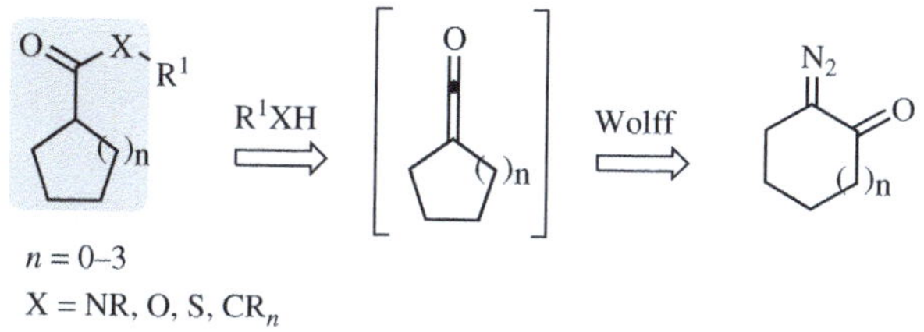

Scheme 3.3 The ring contraction of α-diazoketones.

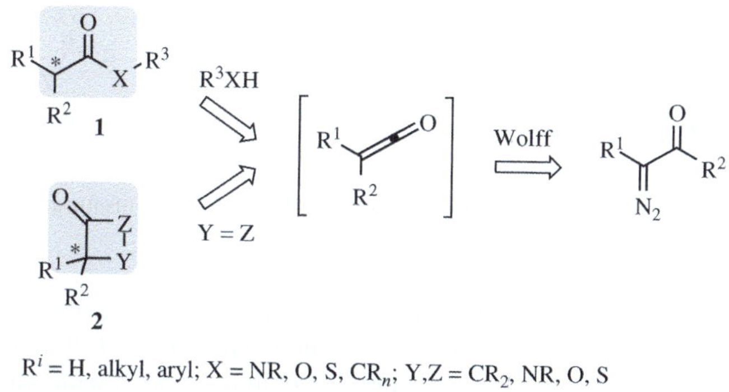

Scheme 3.4 Reactions of ketenes obtained by the Wolff rearrangement.

heteronucleophiles to give the corresponding homologated methylene carboxylic acid derivatives (Scheme 3.2). The reaction is very general, and extensions of the method have also allowed for the homologation of diazomethyl esters and amides ($R^1 = OR$, NR_2) and the use of carbon-based nucleophiles ($X = R_nC$). The rearrangement is generally best performed under silver(I) ion catalysis, but thermal and photochemical activations are also possible.

3.2.2 One-Carbon Ring Contraction of Cyclic α-Diazoketones

The Wolff rearrangement is highly efficient for the one-carbon ring contraction of cyclic α-diazoketones (Scheme 3.3). This strategy has proven useful for the synthesis of strained ring systems such as cyclobutanes ($n = 0$) and bridged polycyclic compounds and for the total synthesis of natural products and other biologically active complex molecules. Of course, in the latter cases, most examples concern the ring contraction of cyclohexanones into cyclopentanecarboxylic derivatives because of the synthetic availability of six-membered rings. The reaction can be activated both under photochemical and thermal conditions.

3.2.3 Reactions of Ketenes and α-Oxo-Ketenes

The Wolff rearrangement is a general reaction that proceeds under streamlined conditions, requiring no additives when performed thermally or photochemically and

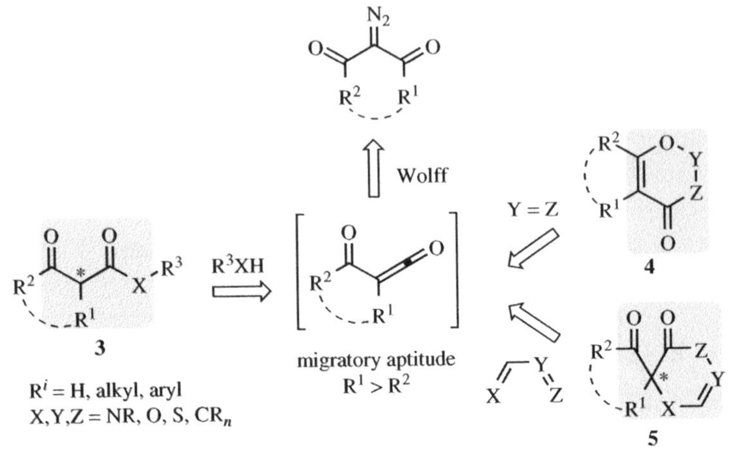

Scheme 3.5 Reactions of α-oxo-ketenes with nucleophiles and unsaturated compounds.

producing no reactive coproducts. It is thus a very convenient source of ketenes and α-oxo-ketenes for the exploration of their reactivity. Ketenes follow two principal types of reactivity: they undergo nucleophilic additions with a wide range of nucleophiles, even with poorly nucleophilic species, to afford carboxylic acid derivatives **1**, and they react with a variety of 2π unsaturated compounds in [2+2] cycloaddition processes to give oxo-four-membered rings **2** (Scheme 3.4). For example, their cycloaddition with imines (Staudinger reaction) is one of the most popular routes to β-lactams.[8] It can be noted that when the R^1 and R^2 substituents of the ketene are different, these reactions lead to chiral products.

In the case of 2-diazo-1,3-diketone and 2-diazo-1,3-ketoester substrates (Scheme 3.5), the Wolff rearrangement produces α-oxo-ketenes, a reaction that Ludwig Wolff himself studied in his early work. The control of the relative migratory aptitude of the R^1 and R^2 groups is a crucial parameter for the synthetic utility of this approach and essentially depends on the nature of the R^i groups, the geometry of the diazo substrate, and the activation method employed (see Section 3.3). Similarly to other ketenes, α-oxo-ketenes readily react with a large array of nucleophiles and produce monosubstituted 2-alkyl (or alkenyl or aryl)-1,3-dicarbonyl compounds **3** (Scheme 3.5). The advantages of this method over traditional alkylation techniques via the enolate of the corresponding 2-methylene-1,3-dicarbonyl derivatives are the suppression of both competitive O-alkylation and double C-alkylation.

Cycloaddition reactions of α-oxo-ketenes, meanwhile, have a different periselectivity than those of simple ketenes.[9] Indeed, α-oxo-ketenes react with 2π unsaturated systems almost exclusively as 1,3-oxadienes in inverse-demand Diels–Alder reactions to produce dihydropyranone derivatives **4** with excellent and predictable regioselectivity (Scheme 3.5).[10] The α-oxo-ketenes are especially prone to undergo [4+2] cycloadditions with heterodienophiles, such as imines or nitriles for the synthesis of 4*H*-1,3-oxazin-4-one derivatives, carbonyl groups for the preparation of

4*H*-1,3-dioxin-4-ones, and electron-rich alkenes such as enamines or enol ethers to give 2,3-dihydropyran-4-ones. The α-oxo-ketenes can also undergo [4+2] cycloadditions with heterocumulenes such as carbodiimides and isocyanates. Examples of [2+2] cycloadditions to the ketene function of α-oxo-ketenes are extremely rare and specific.[11] With the exception of their cyclodimerization, examples of 6π cycloadditions involving the C=C bond of α-oxo-ketenes as the dienophile were until recently unknown. The reactivity of α-oxo-ketenes as dienophiles or dipolarophiles was recently uncovered and has provided synthetically valuable reactions for the straightforward synthesis of spirobicyclic compounds **5**, which will be detailed at the end of this chapter (Scheme 3.5). Because of the intrinsic prochiral nature of α-oxo-ketenes, their reactions generally lead to chiral products.

3.3 MECHANISTIC FEATURES AND SELECTIVITY ISSUES OF THE WOLFF REARRANGEMENT

α-Diazocarbonyl compounds are more stable than simple diazo compounds due to the electron donation from the diazo group to the carbonyl moiety, as shown clearly in the corresponding resonance structures (Scheme 3.6). As a direct consequence, and unless restricted by bulky substituents, α-diazocarbonyls usually adopt a planar geometry with the two possible conformers, *s*-cis and *s*-trans. The electronic interactions between the negatively charged oxygen atom and the positively charged nitrogen atom contribute to the stabilization of the *s*-cis conformation. In the case of acyclic 2-diazo-1,3-dicarbonyl compounds, meanwhile, up to four distinct planar conformations can be adopted.

Mechanistically, the Wolff rearrangement can either proceed by a concerted nitrogen extrusion/1,2-shift or in a stepwise manner via a singlet state α-oxo-carbene intermediate (Scheme 3.6). Photolytic Wolff rearrangements have been extensively

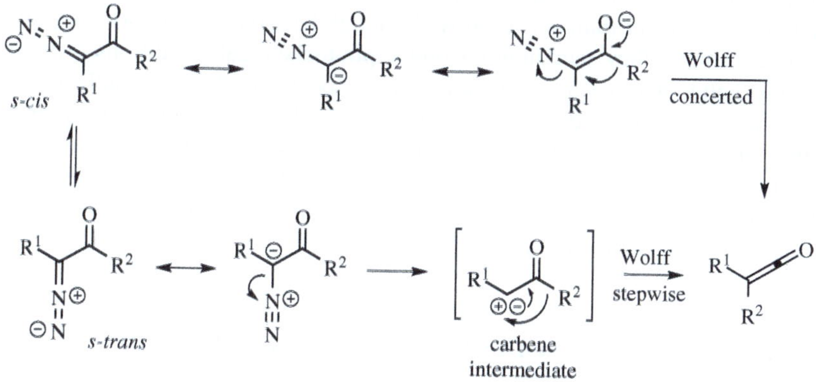

Scheme 3.6 Electronic structure of α-diazocarbonyl compounds and mechanism of the Wolff rearrangement.

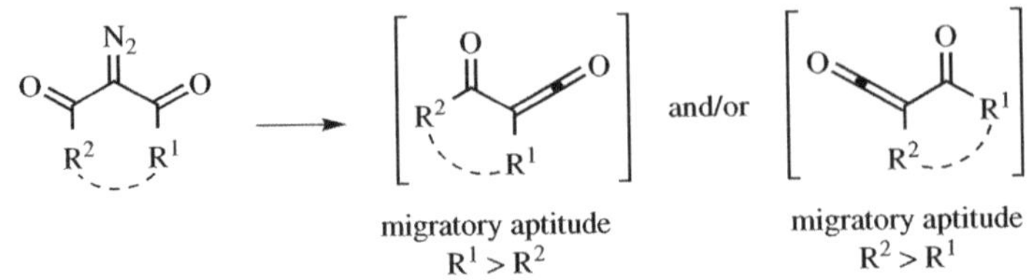

For photochemical reactions: H > alkyl ≥ aryl > SR > OR ≥ NR_2
For thermal reactions: H > aryl ≥ alkyl >> heteroatoms

Scheme 3.7 Migratory aptitudes in the Wolff rearrangement.

studied by ultrafast transient absorption spectroscopy, and it is now clear that both stepwise and concerted mechanisms operate, often competitively.[12,13] The concerted mechanism seems to require an *s*-cis conformation of the α-diazocarbonyl compound, and accordingly, the Wolff rearrangement of cyclic α-diazoketones that are locked in the *s*-cis conformation generally proceed via the concerted mechanism. On the contrary, with acyclic 2-diazo-1,3-ketoesters, the Wolff rearrangement preferentially occurs following the stepwise mechanism. Other factors such as ring strain and carbene stability (singlet–triplet state interconversion) also influence the mechanism and the fate of the reaction.

The Wolff rearrangement of nonsymmetrical 2-diazo-1,3-dicarbonyl compounds can lead to two regioisomers as a function of the relative migratory aptitudes of the carbonyl substituents (Scheme 3.7). Migratory aptitudes essentially depend on the nature of the migrating group, but other factors such as the conformation of the α-diazocarbonyl compound, the activation method, and the actual mechanism at hand (concerted or stepwise) can also influence the preferential migration of one substituent versus the other. The accumulation of experimental data has allowed a qualitative ranking of the migratory aptitudes in the Wolff rearrangement of 2-diazo-1,3-dicarbonyl compounds. For photochemically promoted Wolff rearrangements, hydrogen atoms migrate preferentially over alkyl groups that themselves usually migrate better than aryl groups. Although heteroatoms are not the preferred migrating groups, they can migrate under photochemical conditions with sulfur migrating better than oxygen and nitrogen, respectively. For thermally promoted Wolff rearrangements, hydrogen atoms remain the best migrating groups, and aryl or vinyl substituents generally migrate better than alkyl groups, while heteroatoms normally do not migrate. It should be stressed here that an attractive feature of the Wolff rearrangement is that all these migrations occur with complete retention of the chiral information of the migrating group.

3.4 PREPARATION OF α-DIAZOCARBONYL COMPOUNDS

The two principal and most general synthetic routes to α-diazocarbonyl compounds are (1) acylation of diazoalkanes and (2) diazo-transfer reactions to carbonyl

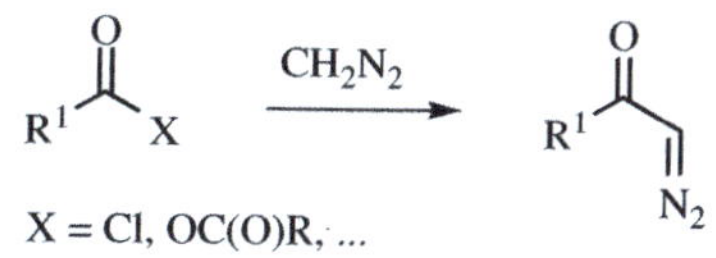

Scheme 3.8 Acylation of diazomethane.

compounds with sulfonyl azide reagents. *Caution:* Diazo compounds and azide reagents are presumed to be toxic and potentially explosive,[14] and appropriate safety measures should be deployed when handling these compounds.

3.4.1 Acylation of Diazoalkanes

The acylation of diazoalkanes is the method of choice to prepare diazomethyl ketones, the substrates of the Arndt–Eistert reaction. Despite the potential hazards associated with the use of diazomethane,[15] its reaction in excess amount (generally 2–6 equiv.) with acyl halides or mixed anhydrides remains the most important synthetic route to these substrates (Scheme 3.8). Alternative acylating agents include acyl mesylates[16] and *in situ* generated acyl phosphonium salts.[17] Peptide coupling reagents can also be used to activate the carboxylic acid derivative toward nucleophilic substitution with diazomethane.[18] An improved synthesis of α-diazoketones from acyl halides requiring only a single equivalent of diazomethane was recently described.[19] In order to avoid the use of diazomethane, the conversion of acyl chlorides into diazomethyl ketones can also be realized with trimethylsilyldiazomethane[20] or *N*-isocyanotriphenyliminophosphorane.[21]

3.4.2 Diazo-Transfer Reactions

The direct diazo transfer from sulfonyl azides to methyl ketones is usually not a practicable process, with the exception of the α-diazotization of ketones with 2,4,6-triisopropylbenzenesulfonyl azide under phase-transfer conditions.[22] However, diazomethyl ketones amenable to the Arndt–Eistert reaction can be easily prepared in two steps by a formylation–deformylation diazo-transfer sequence (Regitz procedure[23]). A related practical method advantageously applies the detrifluoroacetylation of α-trifluoroacetyl ketones (Scheme 3.9a).[24] The preparation of 2-diazo-1,3-dicarbonyl compounds is commonly best performed with *p*-toluenesulfonyl azide (TsN_3) as the diazo-transfer reagent (Scheme 3.9b).[25,26] An issue often associated with the use of a sulfonyl azide diazo-transfer reagent is the coproduction of the corresponding sulfonyl amide, which usually requires a tedious purification step. Several alternative reagents have been developed to bypass this technical problem. Thus, methanesulfonyl azide (MsN_3)[27] and *p*-acetamidobenzenesulfonyl azide (*p*-ABSA)[28] have been found to be efficient complementary reagents, and a polystyrene-supported benzenesulfonyl azide ($PS\text{-}SO_2N_3$)[29] has been proposed as a safer alternative to TsN_3, which also enables

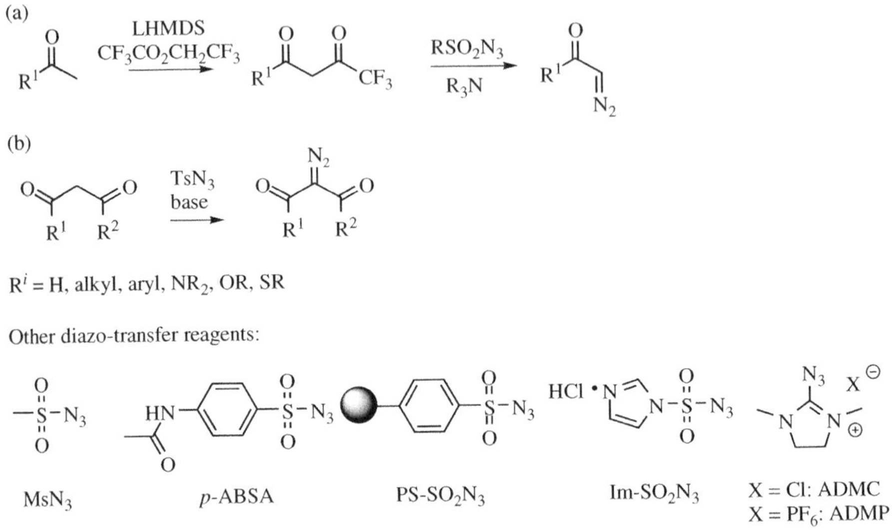

Scheme 3.9 Diazo-transfer reactions.

easy purification by simple filtration of the polystyrene-supported benzenesulfonamide coproduct. Imidazole-1-sulfonyl azide ($Im\text{-}SO_2N_3$) hydrochloride has advantages as an efficient reagent for diazo-transfer reactions to primary amines (compared to the standard trifluoromethanesulfonyl azide reagent), but its efficiency in diazo-transfer reactions to activated methylene compounds is moderate.[30] Recently, 2-azido-1,3-dimethylimidazolinium chloride (ADMC) was reported as an effective diazo-transfer reagent to activated methylene compounds.[31a] This new reagent is not sulfonyl azide-based, avoiding issues linked to the coproduction of sulfonyl amides, but due to its hygroscopic character, the reagent must be prepared *in situ* immediately before use. The corresponding 2-azido-1,3-dimethylimidazolinium hexafluorophosphate (ADMP) was also prepared and found to be less hygroscopic and easier to handle.[31b]

3.4.3 Alkyne Cycloaddition

Strained cyclic alkynes react with nitrous oxide by 1,3-dipolar cycloaddition to give 1,2,3-oxadiazole intermediates that undergo spontaneous ring fragmentation, affording the corresponding α-diazoketones, which are prone to the Wolff rearrangement (Scheme 3.10).[32] Notably, the reaction proceeds with high atom economy. The application of this promising approach to concrete synthetic problems remains to be demonstrated (e.g., regioselectivity of the cycloaddition, scope of alkyne substrates).

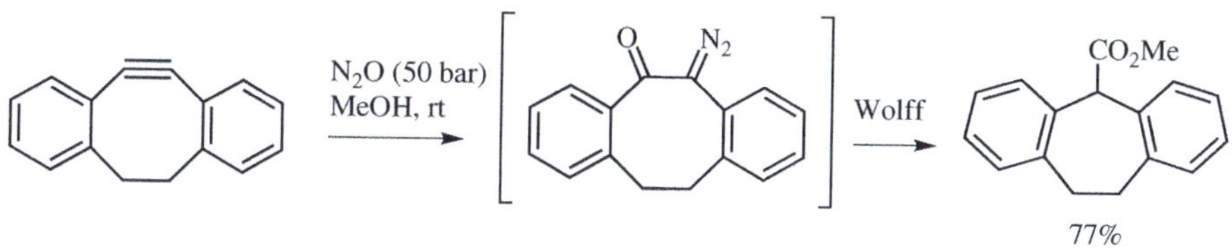

Scheme 3.10 Generation of diazoketones by 1,3-dipolar cycloaddition of nitrous oxide and cyclic alkynes.

3.5 RECENT SYNTHETIC APPLICATIONS OF THE WOLFF REARRANGEMENT

3.5.1 Wolff Rearrangements Leading to Ketene Intermediates

The Wolff rearrangement is an extremely efficient reaction that produces ketenes quantitatively. For example, stable (trialkylsilyl)arylketenes can be obtained by photochemical[33a,b] or rhodium(II)-catalyzed[33c] Wolff rearrangement of α-silyl α-diazo aryl ketones. And because ketenes are exceptional synthetic platforms,[34] the Wolff rearrangement has found a multitude of synthetic applications in organic chemistry. Among these, variations of the Arndt–Eistert reaction remain of tremendous importance. For example, one popular synthetic approach to β-amino acids, compounds of primary importance for their biological properties, relies on the Arndt–Eistert reaction of α-amino acid derivatives (Scheme 3.11).[35] An interesting application of this reaction allowed the synthesis of β-hexapeptide derivatives containing crown ether moieties that adopt helical conformations.[36] In another application, a total synthesis of grassypeptolide A was recently achieved exploiting the Arndt–Eistert reaction of a D-alanine derivative in the early stages of the synthesis for the preparation of the corresponding enantiopure β-amino acid derivative.[37] By analogy to the preparation of β-amino acid derivatives, β-alkoxy acid derivatives can be obtained via the Arndt–Eistert reaction from the α-alkoxy carboxylic acids. This approach was recently taken for the development of a potent HIV-1 nucleotide-competing reverse transcriptase inhibitor.[38]

Seki and Georg recently demonstrated a particularly interesting application of the Arndt–Eistert reaction to the homologation of α-amino acids. They prepared cyclic enaminones **7** from functionalized α-amino acid derivatives using vinylogous amides as intramolecular carbon nucleophiles to trap the intermediate ketene (Scheme 3.12).[39] The diazomethyl ketone precursors **6** were readily prepared from α-amino acids and the corresponding alkynes. On treatment with silver(I) benzoate, they underwent clean Wolff rearrangement to produce the corresponding ketene intermediates that were trapped intramolecularly by the enamine nucleophile in a 6-exo-dig cyclization. It should be noted that examples of C=C bond formation in nucleophilic addition reactions to ketenes are relatively rare. Importantly, complete retention of stereochemistry was observed with chiral, enantiopure diazo substrates.

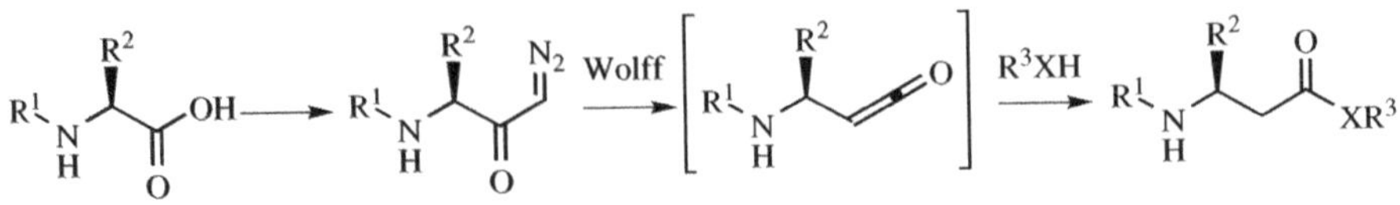

Scheme 3.11 General scheme of the homologation of α-amino acids.

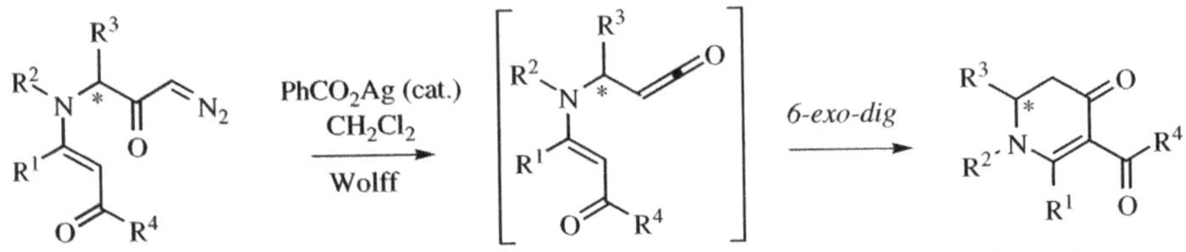

6

7: 76–99%

R^1 = H, CO_2Et; R^2, R^3 = alkyl; R^4 = Me, Ph, OEt

Scheme 3.12 Arndt–Eistert-based synthesis of cyclic enaminones.

In the course of their synthetic approach to aspidosperma alkaloid natural products, Cho and coworkers reported an Arndt–Eistert reaction involving the migration of a chiral trisubstituted carbon center to form the ketene intermediate (Scheme 3.13a).[40] The main difficulty encountered in this approach was *not* the silver(I)-catalyzed Arndt–Eistert reaction that reliably afforded the desired homologated product **9** in 75% yield, but the preparation of the diazomethyl ketone precursor **8** in a sterically crowded environment. Indeed, the corresponding methyl ester was converted into the desired diazo precursor following a three-step hydrolysis–activation–diazotization sequence in only 33% yield. Liang and coworkers recently reported a similar reaction sequence in their work on the total synthesis of echinopine natural products (Scheme 3.13b).[41] In this case, it is not clear which step limits the overall yield, but here again, preparation of the diazo precursor **10** in a sterically demanding environment might be responsible.

Burtoloso and coworkers have studied Arndt–Eistert reactions of α,β-unsaturated diazomethyl ketones, easily prepared by Horner–Wadsworth–Emmons olefination. This approach provided as a convenient source of β,γ-unsaturated carboxylic acid derivatives for the Burtoloso group's synthetic route to polyhydroxylated indolizidine natural products (Scheme 3.14).[42] The main advantage of this strategy for the targeted class of compounds is that it produces unsaturated products (e.g., **11**) in which the double bond is a precursor for the oxygenated positions in the final product. The application in the synthesis of this kind of Wolff rearrangement, which affords a vinyl ketene intermediate, is relatively rare. The reaction was best performed under photochemical conditions, but silver(I)-catalyzed Wolff rearrangement was also possible, though less practical and efficient. As expected, the vinylogous chiral center was not epimerized under either set of conditions, and it is interesting to note that the rearrangement was stereospecific: (*E*)-configured double bonds gave the corresponding (*E*)-vinyl ketenes, while (*Z*)-double bonds gave the (*Z*)-vinyl ketenes.

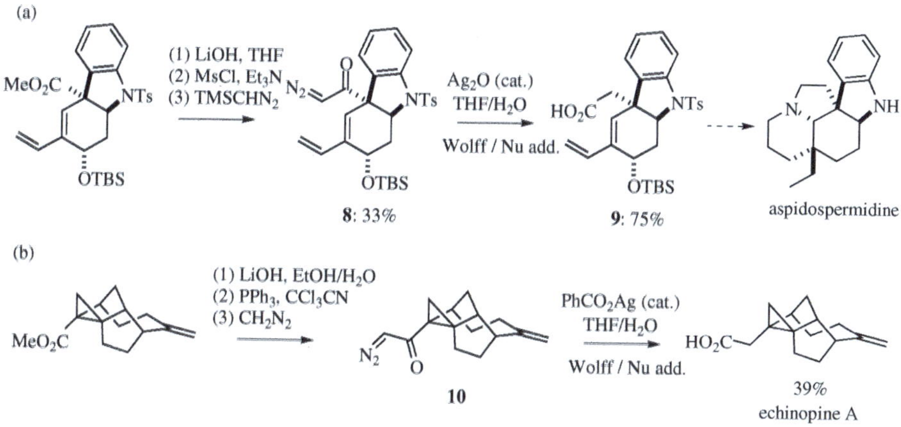

Scheme 3.13 Total syntheses of natural products using Arndt–Eistert reactions in sterically demanding settings.

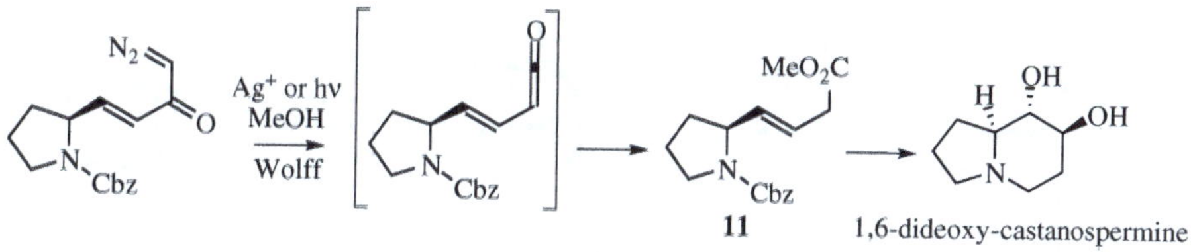

Scheme 3.14 Arndt–Eistert reactions of α,β-unsaturated diazomethyl ketones.

Combining the Wolff rearrangement with other reactions in domino processes is an attractive strategy for the elaboration of complex molecules in a single operation. Wang and coworkers have reported a beautiful example of such a reaction combination in their synthetic approach to substituted pyridines (Scheme 3.15).[43] The reaction between a diazomethyl ketone and an ethyl 2-azido-penta-2,4-dienoate derivative in the presence of triphenylphosphine at 140 °C afforded the corresponding pyridines in good yields. The postulated mechanism for this reaction involves initial and parallel formation of the ketene **13** by Wolff rearrangement of the diazo compound and the formation of a phosphazene **12** from the reaction of the azide with the phosphine (Staudinger–Meyer reaction). These two reactive intermediates combine through an aza-Wittig reaction to afford the corresponding *N*-vinyl ketenimine **14**, which in turn undergoes a 6π electrocyclization and a double-bond isomerization to finally give the pyridine product **15**. Notably, the reaction exploits the reactivity of both diazo and azido compounds, as well as the concept of domino processes, features that will be discussed again later in this chapter.

The combination of the Wolff rearrangement with other types of rearrangements in a synthetic sequence can be particularly effective. One spectacular example has been reported by Stoltz and coworkers, who realized a domino Wolff/Cope sequence

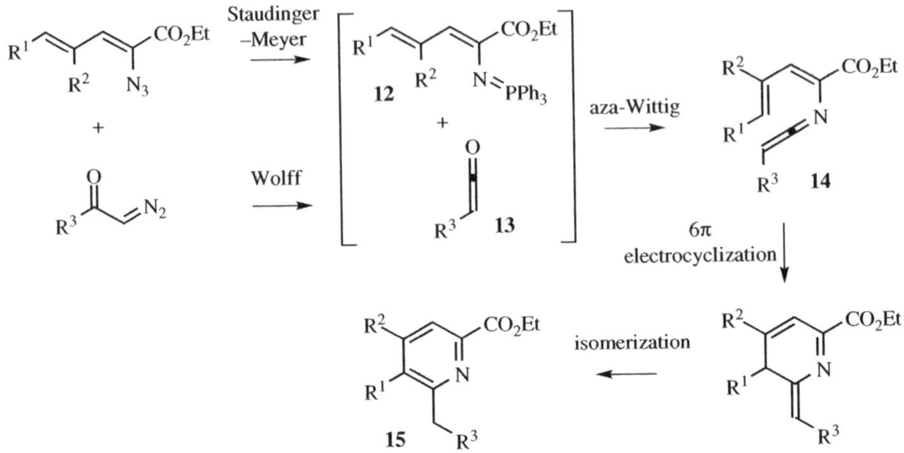

Scheme 3.15 Domino Wolff/aza-Wittig/electrocyclization sequence.

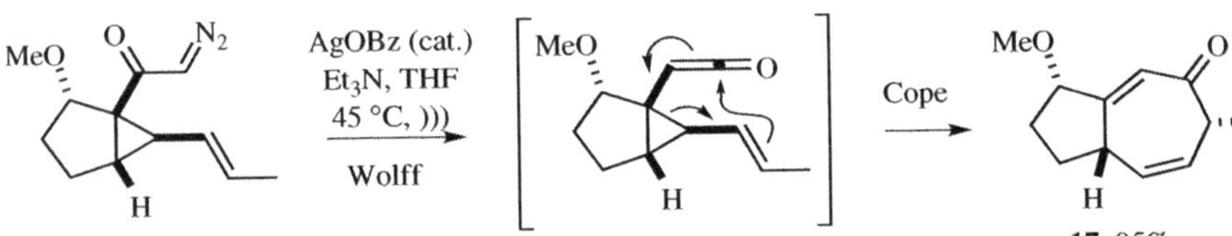

Scheme 3.16 Domino Wolff/Cope rearrangements sequence.

for the stereoselective synthesis of functionalized bicyclo[5.3.0]decane systems, a bicyclic framework ubiquitous in terpenoid natural products (Scheme 3.16).[44] For this work, the Stoltz group prepared the required diazomethyl ketone precursor **16**, which was subjected to a number of known conditions to promote its Wolff rearrangement. In early experiments, only the normal Arndt–Eistert homologated product could be obtained. However, it was found that the desired domino sequence could be triggered under a combination of silver(I) catalysis and sonochemical activation,[45] affording exclusively the desired Wolff/Cope product **17** in excellent yield. The scope and limitations of the reaction have been rationalized by a thorough computational study.[46]

Other than Arndt–Eistert reactions and related transformations, applications of the Wolff rearrangement of α-diazocarbonyl compounds principally concern ring contraction reactions. The method is particularly valuable for the preparation of strained ring systems such as cyclobutane derivatives. For example, the photochemical Wolff rearrangement of 2-diazo-3-oxo-5,10,15,20-tetraphenylchlorins in the presence of nucleophilic and biomimetic substrates (1-butanol, tosylhydrazine, and tetrahydrofurfuryl alcohol) afforded, among other products, the corresponding rare ring-contracted azeteoporphyrinoids in 11–34% yields (Scheme 3.17).[47] These

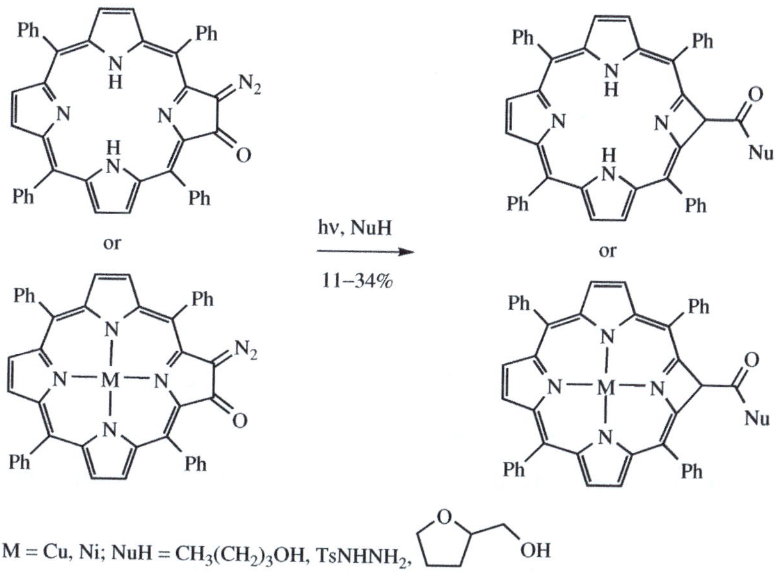

Scheme 3.17 Wolff rearrangement of α-diazo-oxochlorins.

porphyrinoids show intense absorption bands throughout the visible spectral region, indicating that they maintain porphyrinoid aromaticity.

Another notable ring contraction mediated by the Wolff rearrangement was achieved as the last step in preparation of an advanced intermediate for the total synthesis of dibromopalau'amine.[48] Feldman *et al.* prepared the necessary α-diazoketone precursor starting from cyclohexene **18** (Scheme 3.18). Early attempts to introduce the keto group using epoxidation, hydroboration/oxidation, or Wacker-type oxidations were fruitless, but the alcohols **19** could be obtained by oxymercuration techniques, though as a regioisomeric and diastereomeric mixture. The mixture was directly oxidized to afford the corresponding regioisomeric ketones **20**, which was of no consequence since the two products would converge to the same ring-contracted product after the Wolff rearrangement. The installation of the α-diazo group required the activation of the ketones by means of their trifluoroacetyl derivatives, and the diazo-transfer reaction could then be performed with methanesulfonyl azide to afford a mixture of the two α-diazoketones **21** as the main regioisomers. Irradiation of a methanolic solution of this material furnished the desired ring-contracted product **22**, representing the pentacyclic core of dibromopalau'amine, as a 1:1 mixture of the two diastereomeric methyl esters. This work nicely illustrates that the Wolff rearrangement is a very selective and valuable reaction that can be performed even on highly complex and densely functionalized molecules.

Another interesting Wolff-mediated ring contraction applied to the total synthesis of natural products was realized in the course of an approach to the DEF ring system of fredericamycin, a natural product with potent antibiotic and antitumor activities

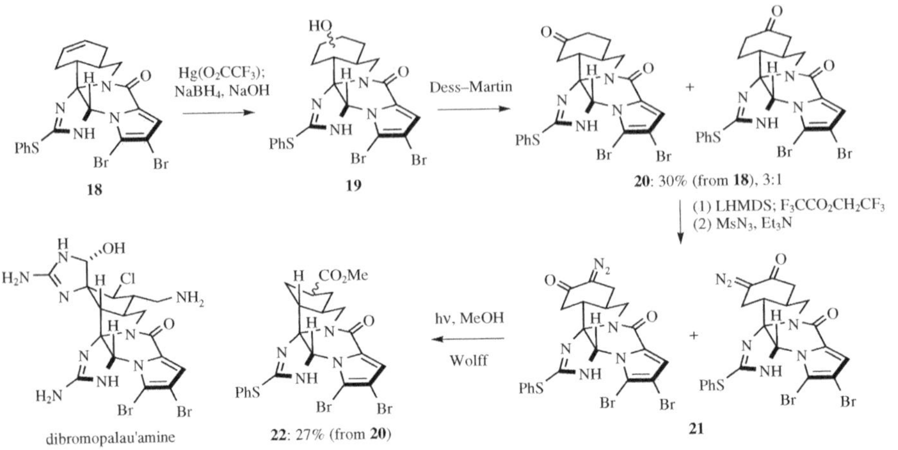

Scheme 3.18 Late-stage Wolff rearrangement in the total synthesis of a complex molecule.

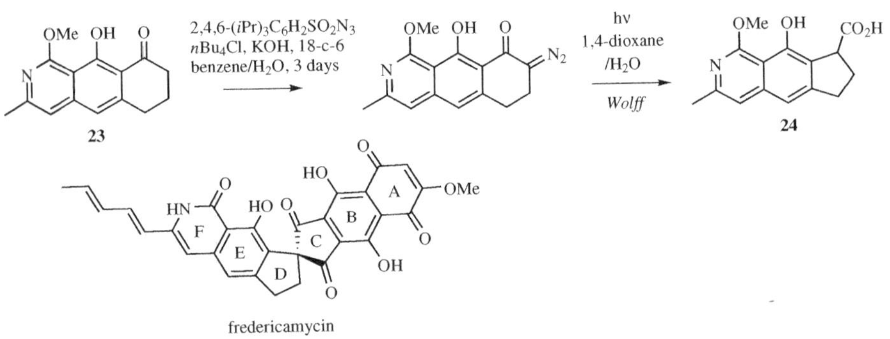

Scheme 3.19 Wolff rearrangement en route to the total synthesis of fredericamycin.

(Scheme 3.19).[49] In this case, the diazo-transfer reaction was performed from the nonactivated but easily enolizable ketone **23** using 2,4,6-triisopropylbenzenesulfonyl azide under phase-transfer conditions and prolonged reaction times.[22] The photochemical Wolff rearrangement then afforded the desired ring-contracted carboxylic acid product **24**.

The [2+2] cycloaddition between ketenes and imines (Staudinger reaction[8]) is a reliable route to β-lactams, a class of compounds with a fundamental role as antibacterial agents in medicinal chemistry and also as versatile synthetic intermediates. Although the Wolff rearrangement is usually not the preferred route for the formation of the ketene intermediate in these reactions, a Wolff-mediated Staudinger reaction was recently described for the preparation of spirooxindole compounds with antibacterial and antifungal activities (Scheme 3.20).[50] Notably, the azetidinone products **25** exhibiting two contiguous tetrasubstituted carbon atoms were obtained in good yields.

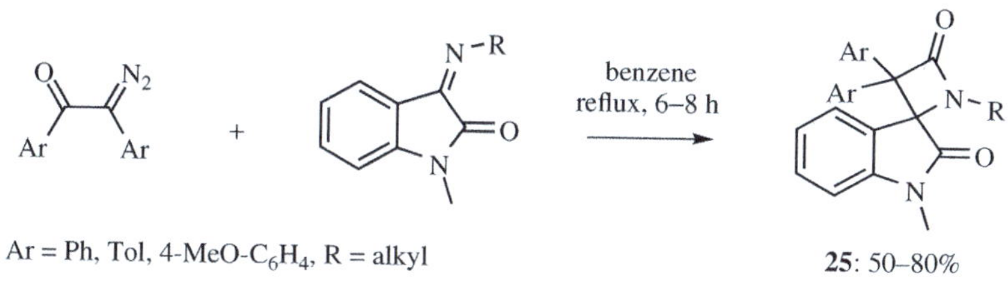

Scheme 3.20 A Wolff-mediated Staudinger reaction.

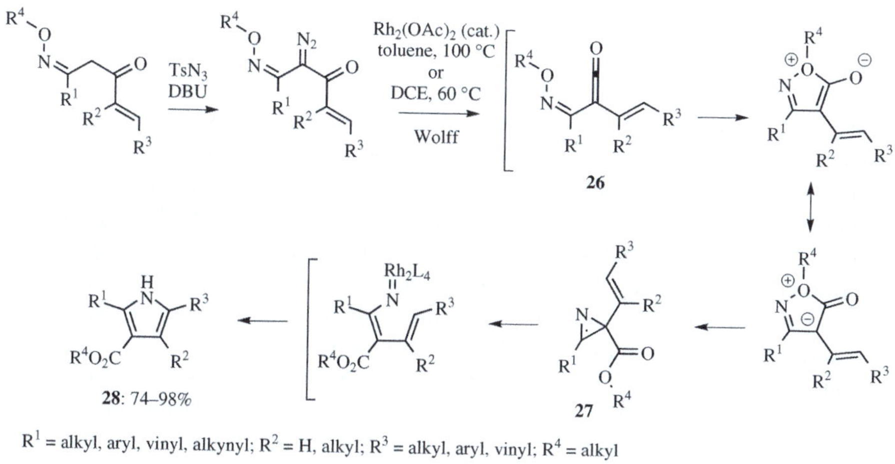

Scheme 3.21 Domino rearrangement route to pyrroles initiated by Wolff rearrangement of α-diazo-β-oximino enones.

Park and coworkers recently disclosed a spectacular application of the Wolff rearrangement as the key step of an unusual domino rearrangement sequence. This creative approach developed a rhodium(II)-catalyzed cascade reaction of α-diazo-β-oximino enones as a direct route to substituted pyrroles (Scheme 3.21).[51] The proposed mechanism of the transformation involves initial formation of a vinyl ketene **26** from the diazo substrate by a Wolff rearrangement. Nucleophilic addition of the oxygen atom of the oxime ether moiety to the ketene leads to the formation of an ylide that undergoes a novel rearrangement to afford a 2-vinyl-2*H*-azirine-2-carboxylic ester intermediate (**27**) that can be isolated if the domino process is interrupted (reaction at lower temperature). Finally, a rhodium–nitrene complex, formed via ring opening of the azirine, leads to a C—H insertion reaction, affording the pyrrole product **28**.[52] The products were obtained in good to excellent yields, and the reaction appears to have a wide scope.

3.5.2 Wolff Rearrangements Leading to α-Oxo-Ketene Intermediates

The Wolff rearrangement of 2-diazo-1,3-dicarbonyl compounds produces α-oxo-ketenes, a class of ketenes with enhanced reactivity.[9] Therefore, only

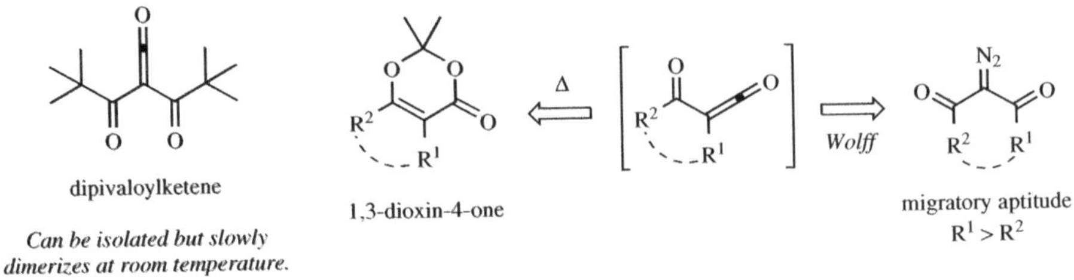

Scheme 3.22 Example of a stable α-oxo-ketene and popular methods for preparing α-oxo-ketenes.

very sterically hindered and/or inductively stabilized α-oxo-ketenes can be isolated (Scheme 3.22), and these have served mainly for structural and spectroscopic studies. Meanwhile, synthetic applications of these reactive intermediates typically involve their formation *in situ*, followed by rapid trapping with species already present in the reaction mixture. By far, the most commonly used method for generating α-oxo-ketenes is thermolysis (or sometimes photolysis) of the corresponding 1,3-dioxin-4-ones, rather than the Wolff rearrangement (Scheme 3.22). However, in the past few years, some interesting and valuable reactions of α-oxo-ketenes generated by Wolff rearrangements have been reported.

One such application of the Wolff rearrangement concerns the synthesis of 1,3-dicarbonyl compounds by the reaction of α-oxo-ketenes with nucleophiles. This reaction has been known for a long time, but its development was hampered by the lack of a practical method for generating the ketenes (long reaction times under UV irradiation or at elevated temperatures were required). With the advent of microwave-assisted organic synthesis in the last decade, the situation has changed because elevated temperature can be reached in minutes or even seconds, most often using sealed reaction vessels under safe and user-friendly conditions. In 2009, we reported that the irradiation of a toluene solution of 2-diazo dimedone (5,5-dimethylcyclohexane-1,3-dione) in the presence of a single equivalent of methanol for less than 3 min produced the corresponding ring-contracted β-ketoester in nearly quantitative yield (Scheme 3.23),[53] which represented a very significant improvement over known conditions for performing this reaction.[54] The scope of the reaction was examined with a variety of 2-diazo-1,3-dicarbonyl compounds and heteronucleophiles, and the method was found to be very general, allowing for the expeditious and high-yielding preparation of a number β-ketoesters, amides, and thioesters under eco-compatible conditions. Importantly, the method was also suitable to prepare products that would hardly be available by other methods, including transesterification and transamidation (Scheme 3.23). Indeed, even poorly nucleophilic alcohols and amines such as phenol and p-toluenesulfonamide could be used, and virtually no loss of efficiency was observed with bulky tertiary alcohols, allowing the straightforward preparation of challenging substrates such as *tert*-butyl 2-oxocyclobutanecarboxylate. The neutral conditions of the reaction also allow the use of functionalized nucleophiles bearing, for example, a methyl ester functional

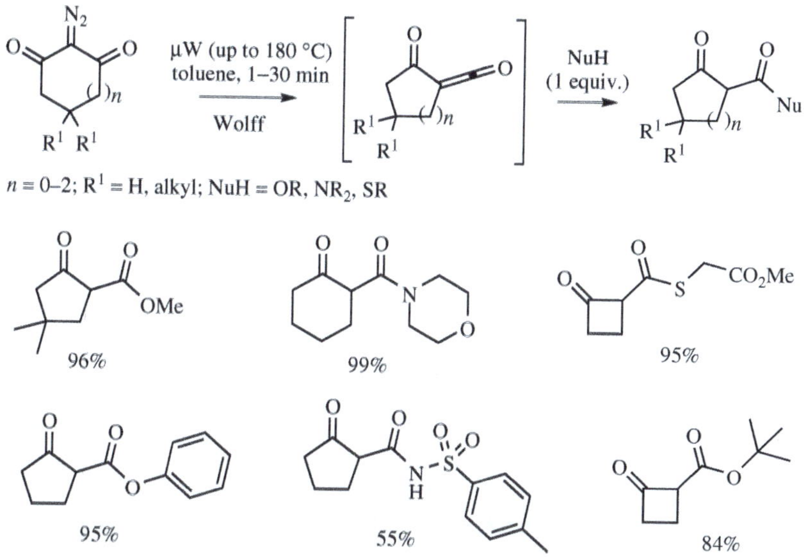

Scheme 3.23 Microwave-assisted Wolff rearrangement for the preparation of 1,3-dicarbonyl compounds.

group. Because of these attractive features, the method has now found many applications in our and other laboratories.[55]

In one notable set of examples, our group described a fully microwave-assisted consecutive reaction involving a Wolff rearrangement/α-oxo-ketene trapping/cross metathesis/Michael addition sequence from cyclic 2-diazo-1,3-diketones, (homo)allylic alcohols or amines, and acrylic derivatives.[56] This approach enabled the synthesis of α-spirolactones and spirolactams **29** in a single operation (Scheme 3.24). In related work, the microwave-assisted Wolff rearrangement of 2-diazo-1,3-diketones was also used in domino benzannulations of α-oxo-ketenes for the preparation of 1,3-dihydro-2*H*-1,5-benzodiazepin-2-ones **30** (Scheme 3.25).[57] In the presence of *o*-phenylenediamine derivatives, the microwave-assisted Wolff rearrangement triggered a domino Wolff rearrangement/nucleophilic addition/intramolecular imination sequence, which resulted in an expeditious synthetic approach to this class of drug-like compounds.

One further application of the microwave-assisted Wolff rearrangement of 2-diazo-1,3-dicarbonyl compounds involves the direct Friedel–Crafts α-ketoacylation of heteroaromatic compounds in the absence of any catalyst or additive and under halogen-free conditions (Scheme 3.26).[58] In contrast to standard Friedel–Crafts reactions that are often complicated by regioselectivity and/or efficiency issues, the microwave-assisted Friedel–Crafts α-ketoacylation via α-oxo-ketenes proceeded cleanly and with high regioselectivity. It should be noted that there are only a few known examples of formation of carbon–carbon bonds by nucleophilic addition to α-oxo-ketenes and also that the possible [4+2] oxa-Diels–Alder cycloaddition products (Scheme 3.5) were not detected.

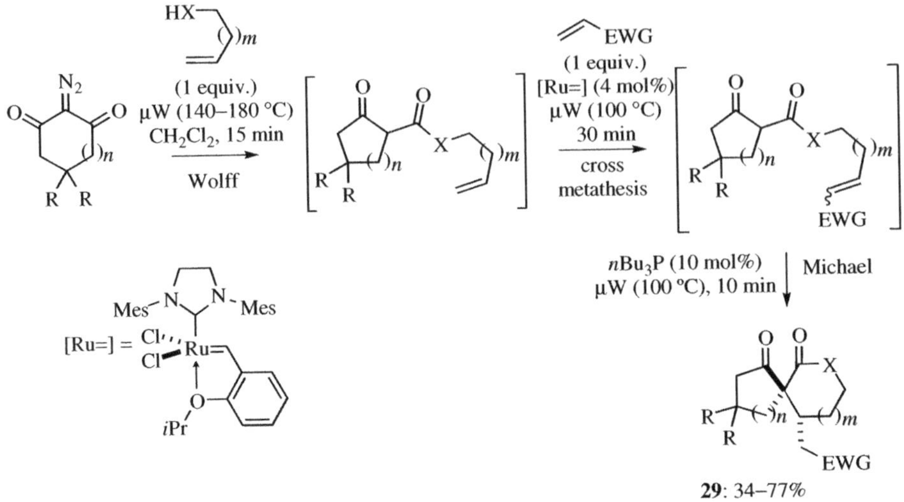

$n = 1,2$; $m = 0,1$; R = H, Me; X = O, NR; EWG = CO_2Me, CN, COMe

Scheme 3.24 Microwave-assisted step- and atom-economical approach to α-spirolactones and spirolactams.

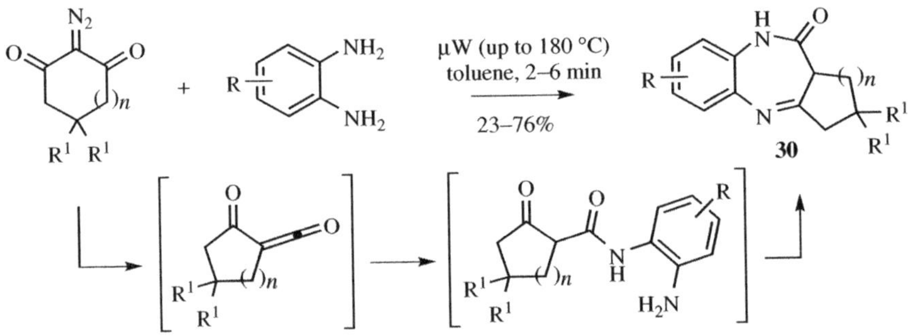

$n = 0–3$; R^1 = H, alkyl; R = H, Cl, Me

Scheme 3.25 Microwave-assisted domino benzannulation of α-oxo-ketenes.

As outlined in Section 3.2.3 (see Scheme 3.5), α-oxo-ketenes are excellent 1,3-oxadienes in inverse-demand Diels–Alder [4+2] cycloadditions, and their reactions with preformed imines and isocyanates lead, respectively, to 1,3-oxazine-4-ones and 1,3-oxazine-2,4-diones, structural units found in some biologically active natural and synthetic products. Based on a literature precedent for the catalyst-free microwave-assisted synthesis of aldimines,[59] we recently demonstrated that a domino three-component synthesis of oxazinones **31** was possible by simple microwave irradiation of a 1:1:1 mixture of a cyclic 2-diazo-1,3-diketone, an aldehyde, and a primary amine in toluene. The result is an efficient imination/Wolff

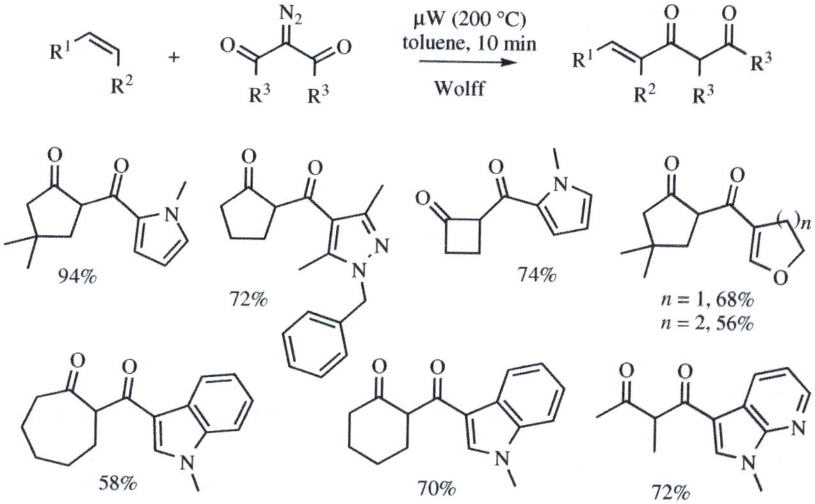

Scheme 3.26 Catalyst- and halogen-free regioselective Friedel–Crafts α-ketoacylations.

rearrangement/oxa-aza-Diels–Alder sequence (Scheme 3.27a).[60] Importantly, these reactions proceed efficiently only if formation of the C═N double bond precedes generation of the α-oxo-ketene by the Wolff rearrangement. This avoids nucleophilic addition of the amine component to the α-oxo-ketene, leading to a β-ketoamide product. As the imination reactions occur at lower temperature than the Wolff rearrangement, an electronically set temperature ramp (20–140 °C over 2 min) efficiently controls the sequential formation of first the imine and then the α-oxo-ketene. When both the aldehyde and amine components were specifically chosen to enable a subsequent intramolecular Diels–Alder reaction, a pentacyclic oxazinone (e.g., **32**) was obtained via an impressive imination/Wolff rearrangement/intermolecular hetero-Diels–Alder/intramolecular Diels–Alder domino sequence (Scheme 3.27b). Remarkably, the reaction allows the stereocontrolled creation of four stereogenic carbon atoms, six chemical bonds, and four rings in a single catalyst-free reaction in relatively good yields, given the increase in molecular complexity.

We anticipated that a similar strategy would be applicable to the synthesis of 1,3-oxazine-2,4-diones, involving the *in situ* formation of isocyanates by thermal Curtius rearrangement of acyl azides under the same conditions of the Wolff rearrangement. Indeed, temperature-regulated microwave irradiation of a 1:1 mixture of a 2-diazo-1,3-diketone and an acyl azide afforded the desired 1,3-oxazine-2,4-dione products **33** (Scheme 3.28).[60] This sequence nicely illustrates the utilization of microwave activation for the selective one-pot thermal rearrangements of azido and diazo compounds.

The extension of the three-component reaction presented in Scheme 3.27 to α,β-unsaturated aldehydes uncovered a new generic type of reactivity of α-oxo-ketenes. Indeed, microwave irradiation of a 1:1:1 mixture of a cyclic

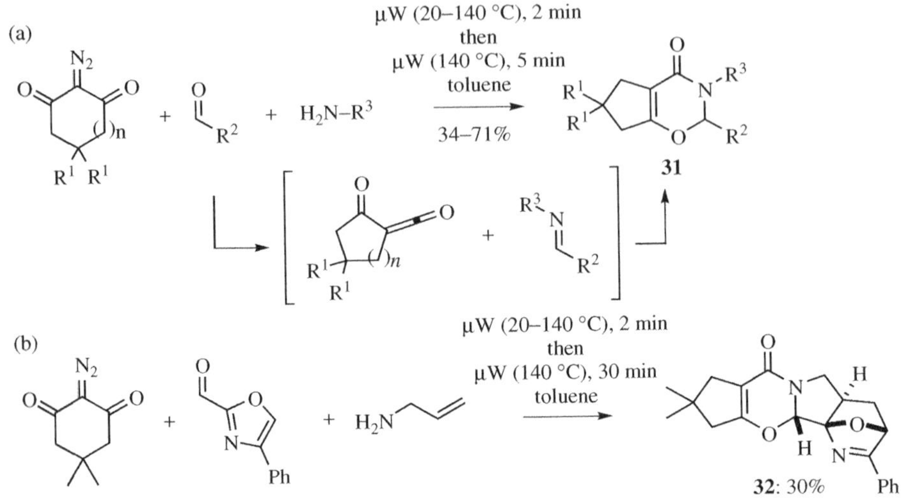

Scheme 3.27 Wolff rearrangement-mediated domino three-component synthesis of 1,3-oxazin-4-ones.

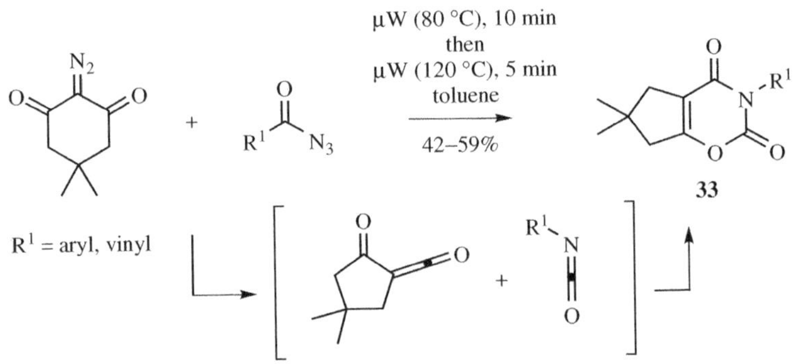

Scheme 3.28 A Wolff/Curtius rearrangement domino approach to 1,3-oxazine-2,4-diones.

2-diazo-1,3-diketone, an α,β-unsaturated aldehyde, and a primary amine in toluene furnished α-spiro-δ-lactam products **36** bearing two contiguous stereogenic centers, as single diastereomers in good yields (Scheme 3.29).[61] These lactam products resulted from a formal aza-Diels–Alder reaction ([2+4] cycloaddition) between the 1-azadiene **35** generated *in situ* and the C=C bond of the α-oxo-ketene intermediate **34**. Overall, this reaction creates four covalent bonds and reveals that α-oxo-ketenes are actually good dienophiles. Based on an experimental and theoretical study, the mechanism of this reaction involves initial nucleophilic addition of the 1-azadiene to the ketene carbon atom to reversibly produce two distinct zwitterionic intermediates. Each of these intermediates can then evolve reversibly by six-electron disrotatory electrocyclization to afford either the oxazinone product **37** or the spirolactam

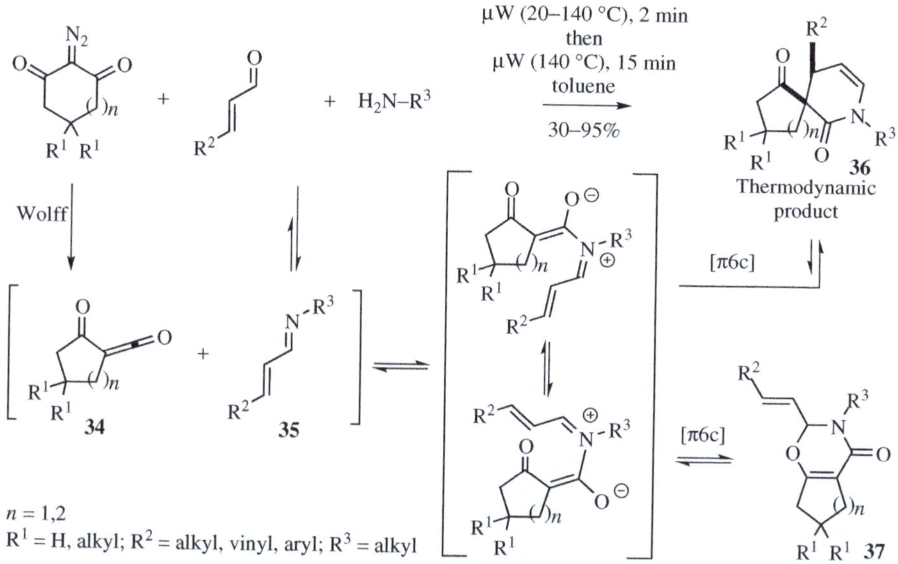

Scheme 3.29 Wolff rearrangement-mediated domino three-component synthesis of α-spiro-δ-lactams.

product **36**. Given the ready reversibility of these steps, the periselectivity of the reaction would be determined by the relative energies of the two possible products (thermodynamic control). In the same study, a consecutive aza-Wittig/Wolff rearrangement/[2+4] cycloaddition sequence was also developed for an analogous reaction with aromatic amines (R^3 = aryl).

Following the same concept, a three-component stereoselective entry to pyrazolidinones, involving a 1,3-dipolar cycloaddition of azomethine imines and α-oxo-ketenes, was also reported. Thus, the microwave irradiation of a 1:1:1 mixture of a 2-diazo-1,3-diketone, an aldehyde or a ketone, and a substituted hydrazine in toluene led to the expected spiropyrazolidin-3-ones **38** (Scheme 3.30).[62] The success of this reaction demonstrates that α-oxo-ketenes are also excellent dipolarophiles and that two adjacent stereocenters can be formed with high selectivity and efficiency. The reaction proceeds by initial formation of a hydrazone **40** from the carbonyl compound and the hydrazine, which upon heating is in equilibrium with the corresponding azomethine imine 1,3-dipole **39**. Wolff rearrangement of the diazo compound, meanwhile, provides the ketene dipolarophile for the cycloaddition, leading to the pyrazolidinone product **38**. When isatin (indoline-2,3-dione) derivatives were used as the carbonyl component of the domino three-component process, the reaction afforded an original class of spirooxindoles, privileged heterocyclic molecules generally endowed with important biological properties. Overall, this new reaction of α-oxo-ketenes allows a straightforward and highly chemo-, regio-, and diastereoselective synthesis of monocyclic, structurally distinct spirobicyclic, and bis-spirotricyclic pyrazolidin-3-ones.

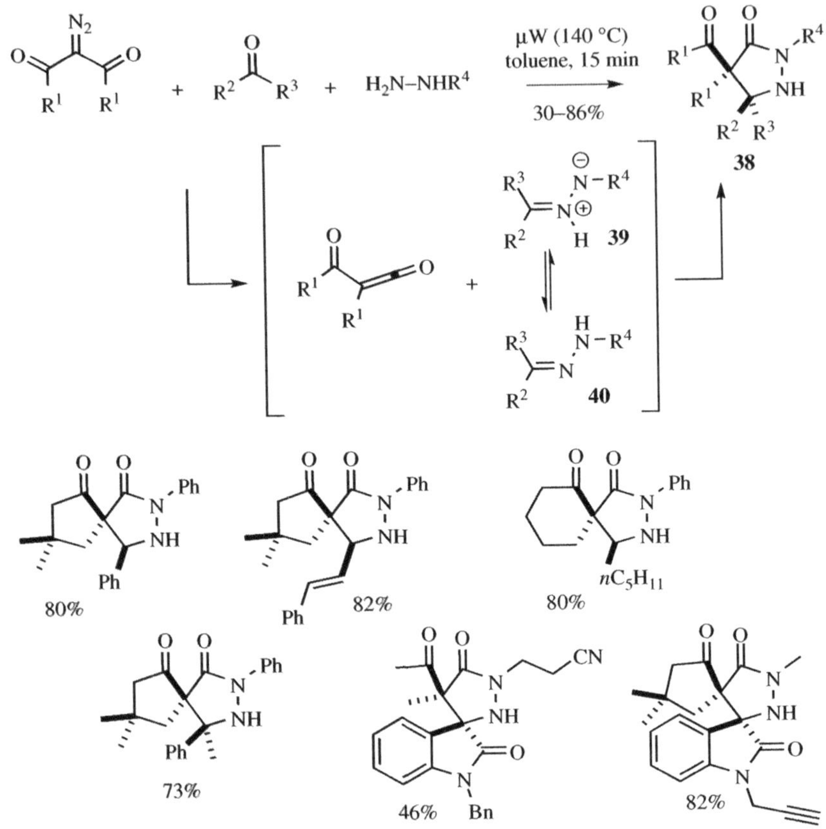

Scheme 3.30 Wolff rearrangement-mediated domino three-component synthesis of pyrazolidinones.

3.6 CONCLUSION AND OUTLOOK

In the past decade, the Wolff rearrangement has steadily continued to play a role in sometimes spectacular applications in organic synthesis. Its mechanism is now correctly understood in intimate detail due to extensive experimental and theoretical studies. However, in his 2002 comprehensive review on the reaction, Kirmse noted that little was known about the mechanism of the catalytically induced Wolff rearrangement.[3] Since then, more and more examples of silver(I)- and rhodium(II)-catalyzed Wolff rearrangements have been reported, but, 12 years later, the mechanistic questions remain almost the same. For example, why do silver(I) ions nicely catalyze the Wolff rearrangement of diazomethyl ketones in the Arndt–Eistert reaction but barely promote the rearrangement of other α-diazo compounds? Similarly, why is rhodium(II) catalysis not generalizable to all types of Wolff rearrangements?

Significantly, the newly discovered reactivity of α-oxo-ketenes as dienophiles and dipolarophiles in cycloaddition reactions complements their known chemistry as 1-oxadienes. This important extension of the Wolff rearrangement opens up new strategic possibilities in retrosynthetic planning and considerably expands the scope of possible synthetic applications. These pathways remain largely to be explored.

Finally, the use of the Wolff rearrangement in combination with other reactions or rearrangements in complex consecutive or domino processes has emerged as a powerful strategy for the rapid increase of molecular complexity. Because ketenes obtained by the Wolff rearrangement are usually prochiral, enantioselective versions of the reactions described herein are highly desirable. This is certainly one of the most important challenges for the development of the Wolff rearrangement in forthcoming years.

REFERENCES

1. Liu, G.; Liu, W.; Dong, C.-M. *Polym. Chem.* **2013**, *4*, 3431–3443.
2. Yang, Y.; Huang, S.; He, H.; Mau, A. W. H.; Dai, L. *J. Am. Chem. Soc.* **1999**, *121*, 10832–10833.
3. Kirmse, W. *Eur. J. Org. Chem.* **2002**, 2193–2256.
4. (a) Zhang, Z.; Wang, J. *Tetrahedron* **2008**, *64*, 6577–6605. (b) Candeias, N. R.; Afonso, C. A. M. *Curr. Org. Chem.* **2009**, *13*, 763–787.
5. (a) Sudrik, S. G.; Chaki, N. K.; Chavan, V. B.; Chavan, S. P.; Chavan, S. P.; Sonawane, H. R.; Vijayamohanan, K. *Chem. Eur. J.* **2006**, *12*, 859–864. (b) Sudrik, S. G.; Sharma, J.; Chavan, V. B.; Chaki, N. K.; Sonawane, H. R.; Vijayamohanan, K. P. *Org. Lett.* **2006**, *8*, 1089–1092.
6. Wolff, L. *Justus Liebigs Ann. Chem.* **1902**, *325*, 129–195.
7. Arndt, F.; Eistert, B. *Ber. Dtsch. Chem. Ges. B* **1935**, *68*, 200–208.
8. (a) Palomo, C.; Aizpurua, J. M.; Ganboa, I.; Oiarbide, M. *Eur. J. Org. Chem.* **1999**, 3223–3235. (b) Singh, G. S. *Tetrahedron* **2003**, *59*, 7631–7649. (c) Brandi, A.; Cicchi, S.; Cordero, F. M. *Chem. Rev.* **2008**, *108*, 3988–4035. (d) Cossío, F. P.; Arrieta, A.; Sierra, M. A. *Acc. Chem. Res.* **2008**, *41*, 925–936. (e) Aranda, M. T.; Perez-Faginas, P.; Gonzalez-Muniz, R. *Curr. Org. Chem.* **2009**, *6*, 325–341.
9. (a) Wentrup, C.; Heilmayer, W.; Kollenz, G. *Synthesis* **1994**, 1219–1248. (b) Kollenz, G.; Ebner, S. In Science of Synthesis: Houben-Weyl Methods of Molecular Transformations; Danheiser, R., Ed.; Georg Thieme Verlag: Stuttgart, Germany, **2006**; Vol. 23, pp. 271–349, Chapter 9. (c) Reber, K. P.; Tilley, S. D.; Sorensen, E. J. *Chem. Soc. Rev.* **2009**, *38*, 3022–3034.
10. Birney, D. M.; Wagenseller, P. E. *J. Am. Chem. Soc.* **1994**, *116*, 6262–6270.
11. (a) Xu, J.; Zhang, Q.; Chen, L.; Chen, H. *J. Chem. Soc., Perkin Trans. 1* **2001**, 2266–2268. (b) Reddy, B. V. S.; Karthik, G.; Rajasekaran, T.; Antony, A.; Sridhar, B. *Tetrahedron Lett.* **2012**, *53*, 2396–2401.
12. Burdzinski, G.; Platz, M. S. *J. Phys. Org. Chem.* **2010**, *23*, 308–314.
13. For recent studies, see: (a) Burdzinski, G.; Kubicki, J.; Sliwa, M.; Réhault, J.; Zhang, Y.; Vyas, S.; Luk, H. L.; Hadad, C. M.; Platz, M. S. *J. Org. Chem.* **2013**, *78*, 2026–2032. (b) Cui, G.; Thiel, W. *Angew. Chem. Int. Ed.* **2013**, *52*, 433–436. (c) Guan, J.; Randall, K. R.;

Schaefer, H. F.; Li, H. *J. Phys. Chem. A* **2013**, *117*, 2152–2159. (d) Li, Q.; Migani, A.; Blancafort, L. *J. Phys. Chem. Lett.* **2012**, *3*, 1056–1061.
14. (a) Black, T. H. *Aldrichim. Acta* **1983**, *16*, 3–10. (b) Hazen, G. G.; Weinstock, L. M.; Connell, R.; Bollinger, F. W. *Synth. Commun.* **1981**, *11*, 947–956.
15. Proctor, L. D.; Warr, A. J. *Org. Process Res. Dev.* **2002**, *6*, 884–892.
16. Nicolau, K. C.; Baran, P. S.; Zhong, Y. L.; Choi, H. S.; Fong, K. C.; He, Y.; Yoon, W. H. *Org. Lett.* **1999**, *1*, 883–886.
17. Cuevas-Yañez, E.; García, M. A.; de la Mora, M. A.; Muchowski, J. M.; Cruz-Almanza, R. *Tetrahedron Lett.* **2003**, *44*, 4815–4817.
18. Siciliano, C.; De Marco, R.; Guidi, L. E.; Spinella, M.; Liguori, A. *J. Org. Chem.* **2012**, *77*, 10575–10582.
19. Pace, V.; Verniest, G.; Sinisterra, J.-V.; Alcántara, A. R.; De Kimpe, N. *J. Org. Chem.* **2010**, *75*, 5760–5763.
20. (a) Aoyama, T.; Shioiri, T. *Tetrahedron Lett.* **1980**, *21*, 4461–4462. (b) Aoyama, T.; Shioiri, T. *Chem. Pharm. Bull.* **1981**, *29*, 3249–3255.
21. (a) Aller, E.; Molina, P.; Lorenzo, Á. *Synlett* **2000**, 526–528. (b) Bio, M. M.; Javadi, G.; Song, Z. J. *Synthesis* **2005**, 19–21.
22. (a) Lombardo, L.; Mander, L. N. *Synthesis* **1980**, 368–369. (b) Uyehara, T.; Takehara, N.; Ueno, M.; Sato, T. *Bull. Chem. Soc. Jpn.* **1995**, *68*, 2687–2694.
23. Regitz, M.; Rüer, J.; Liedhegener, A. In Organic Syntheses; Noland, W. E., Ed.; Wiley: New York, **1988**; Coll. Vol. *VI*, pp. 389–392.
24. (a) Danheiser, R. L.; Miller, R. F.; Brisbois, R. G.; Park, S. Z. *J. Org. Chem.* **1990**, *55*, 1959–1964. (b) Danheiser, R. L.; Miller, R. F.; Brisbois, R. G. *Org. Synth.* **1996**, *73*, 134–143. (c) Danheiser, R. L.; Miller, R. F.; Brisbois, R. G. In Organic Syntheses; Freeman, J. P., Ed.; Wiley: New York, **1998**; Coll. Vol. *IX*, pp. 197–202.
25. (a) Regitz, M. *Tetrahedron Lett.* **1964**, 1403–1407. (b) Regitz, M. *Angew. Chem., Int. Ed. Engl.* **1967**, *6*, 733–749. (c) Regitz, M.; Hocker, J.; Liedhegener, A. Org. Synth. Coll. Vol. V **1973**, 179–183.
26. Presset, M.; Mailhol, D.; Coquerel, Y.; Rodriguez, J. *Synthesis* **2011**, 2549–2552.
27. Taber, D. F.; Ruckle Jr, R. E.; Hennessy, M. J. *J. Org. Chem.* **1986**, *51*, 4077–4078.
28. Baum, J. S.; Shook, D. A.; Davies, H. M. L.; Smith, H. D. *Synth. Commun.* **1987**, *17*, 1709–1716.
29. Green, G. M.; Peet, N. P.; Metz, W. A. *J. Org. Chem.* **2001**, *66*, 2509–2511.
30. Goddard-Borger, E. D.; Stick, R. V. *Org. Lett.* **2007**, *9*, 3797–3800.
31. (a) Kitamura, M.; Tashiro, N.; Okauchi, T. *Synlett* **2009**, 2943–2944. (b) Kitamura, M.; Tashiro, N.; Miyagawa, S.; Okauchi, T. *Synthesis* **2011**, 1037–1044.
32. Banert, K.; Plefka, O. *Angew. Chem. Int. Ed.* **2011**, *50*, 6171–6174.
33. (a) Maas, G.; Brückmann, R. *J. Org. Chem.* **1985**, *50*, 2801–2802. (b) Brückmann, R.; Schneider, K.; Maas, G. *Tetrahedron* **1989**, *45*, 5517–5530. (c) Marsden, S. P.; Pang, W.-K. *J. Chem. Soc., Chem. Commun.* **1999**, 1199–1200.
34. (a) Allen, A. D.; Tidwell, T. T. *Chem. Rev.* **2013**, *113*, 7287–7342. (b) Tidwell, T. T. *Eur. J. Org. Chem.* **2006**, 563–576.
35. (a) Matthews, J. L.; Braun, C.; Guibourdenche, C.; Overhand, M.; Seebach, D. In Enantioselective Synthesis of β-Amino Acids; Juaristi, E., Ed.; Wiley-VCH: New York, **1997**, pp. 105–126. (b) Abele, S.; Seebach, D. *Eur. J. Org. Chem.* **2000**, 1–15.

36. Dutot, L.; Gaucher, A.; Elkassimi, K.; Drapeau, J.; Wakselman, M.; Mazaleyrat, J.-P.; Peggion, C.; Formaggio, F.; Toniolo, C. *Chem. Eur. J.* **2008**, *14*, 3154–3163.

37. (a) Liu, H.; Liu, Y. Q.; Xu, Z. S.; Ye, T. *Chem. Commun.* **2010**, *46*, 7486–7488. (b) Liu, H.; Liu, Y.; Wang, Z.; Xing, X.; Maguire, A. R.; Luesch, H.; Zhang, H.; Xu, Z.; Ye, T. *Chem. Eur. J.* **2013**, *19*, 6774–6784.

38. James, C. A.; DeRoy, P.; Duplessis, M.; Edwards, P. J.; Halmos, T.; Minville, J.; Morency, L.; Morin, S.; Simoneau, B.; Tremblay, M.; Bethell, R.; Cordingley, M.; Duan, J.; Lamorte, L.; Pelletier, A.; Rajotte, D.; Salois, P.; Tremblay, S.; Sturino, C. F. *Bioorg. Med. Chem. Lett.* **2013**, *23*, 2781–2786.

39. (a) Seki, H.; Georg, G. I. *J. Am. Chem. Soc.* **2010**, *32*, 15512–15513. (b) Seki, H.; Georg, G. I. *Org. Lett.* **2011**, *13*, 2147–2149.

40. Tam, N. T.; Jung, E.-J.; Cho, C.-G. *Org. Lett.* **2010**, *12*, 2012–2014.

41. Xu, W.; Wu, S.; Zhou, L.; Liang, G. *Org. Lett.* **2013**, *15*, 1978–1981.

42. (a) Pinho, V. D.; Burtoloso, A. C. B. *Tetrahedron Lett.* **2012**, *53*, 876–878. (b) Bernardim, B.; Pinho, V. D.; Burtoloso, A. C. B. *J. Org. Chem.* **2012**, *77*, 9926–9931.

43. Chen, Z.-B.; Hong, D.; Wang, Y.-G. *J. Org. Chem.* **2009**, *74*, 903–905.

44. Sarpong, R.; Su, J. T.; Stoltz, B. M. *J. Am. Chem. Soc.* **2003**, *125*, 13624–13625.

45. Winum, J.-Y.; Kamal, M.; Leydet, A.; Roque, J.-P.; Montero, J.-L. *Tetrahedron Lett.* **1996**, *37*, 1781–1782.

46. Su, J. T.; Sarpong, R.; Stoltz, B. M.; Goddard III, W. A. *J. Am. Chem. Soc.* **2004**, *126*, 24–25.

47. Köpke, T.; Pink, M.; Zaleski, J. M. *J. Am. Chem. Soc.* **2008**, *130*, 15864–15871.

48. (a) Feldman, K. S.; Nuriye, A. Y. *Org. Lett.* **2010**, *12*, 4532–4535. (b) Feldman, K. S.; Nuriye, A. Y.; Li, J. *J. Org. Chem.* **2011**, *76*, 5042–5060.

49. Braun, M.; Kergoët, G.; Kruska, F.; Frank, W. *Synthesis* **2010**, 2023–2026.

50. Singh, G. S.; Luntha, P. *Eur. J. Med. Chem.* **2009**, *44*, 2265–2269.

51. Jiang, Y.; Chan, W. C.; Park, C.-M. *J. Am. Chem. Soc.* **2012**, *134*, 4104–4107.

52. Padwa, A.; Smolanoff, J.; Tremper, A. *J. Am. Chem. Soc.* **1975**, *97*, 4682–4691.

53. Presset, M.; Coquerel, Y.; Rodriguez, J. *J. Org. Chem.* **2009**, *74*, 415–418.

54. Froborg, J.; Magnusson, G. *J. Am. Chem. Soc.* **1978**, *100*, 6728–6733.

55. (a) Mohanan, K.; Coquerel, Y.; Rodriguez, J. *Org. Lett.* **2012**, *14*, 4686–4689. (b) Mailhol, D.; Sanchez Duque, M. M.; Raimondi, W.; Bonne, D.; Constantieux, T.; Coquerel, Y.; Rodriguez, J. *Adv. Synth. Catal.* **2012**, *354*, 3523–3532. (c) Mailhol, D.; Castillo, J.-C.; Mohanan, K.; Abonia, R.; Coquerel, Y.; Rodriguez, J. *ChemCatChem* **2013**, *5*, 1192–1199. (d) Reeves, C. M.; Eidamshaus, C.; Kim, J.; Stoltz, B. M. *Angew. Chem. Int. Ed.* **2013**, *52*, 6718–6721.

56. (a) Boddaert, T.; Coquerel, Y.; Rodriguez, J. *Adv. Synth. Catal.* **2009**, *351*, 1744–1748 (Erratum: Boddaert, T.; Coquerel, Y.; Rodriguez, J. *Adv. Synth. Catal.* **2009**, *351*, 2541). (b) Boddaert, T.; Coquerel, Y.; Rodriguez, J. *Eur. J. Org. Chem.* **2011**, 5061–5070. (c) Boddaert, T.; Coquerel, Y.; Rodriguez, J. *Chem. Eur. J.* **2011**, *17*, 2048–2051.

57. Castillo, J.-C.; Presset, M.; Abonia, R.; Coquerel, Y.; Rodriguez, J. *Eur. J. Org. Chem.* **2012**, 2338–2345.

58. Mohanan, K.; Presset, M.; Mailhol, D.; Coquerel, Y.; Rodriguez, J. *Chem. Eur. J.* **2012**, *18*, 9217–9220.

59. Paquin, L.; Hamelin, J.; Texier-Boullet, F. *Synthesis* **2006**, 1652–1656.
60. Presset, M.; Coquerel, Y.; Rodriguez, J. *Org. Lett.* **2009**, *11*, 5706–5709.
61. Presset, M.; Coquerel, Y.; Rodriguez, J. *Org. Lett.* **2010**, *12*, 4212–4215.
62. Presset, M.; Mohanan, K.; Hamann, M.; Coquerel, Y.; Rodriguez, J. *Org. Lett.* **2011**, *13*, 4124–4127.

4

ALKYL AND ACYL AZIDE REARRANGEMENTS

Isabelle Gillaizeau and Nicolas Gigant

Institut de Chimie Organique et Analytique, UMR 7311 CNRS, Université d'Orléans, rue de Chartres, Orléans Cedex 2, France

4.1 INTRODUCTION

Alkyl and acyl azides represent a very important class of compounds since they can participate in a wide array of reactions, making them important building blocks for the construction of new carbon–nitrogen bonds. Given the growing ubiquity of nitrogen-containing molecules in natural products and drugs, azide chemistry remains a fundamental task in organic synthesis. This has been driven by the unique reactivity of azides which could rearrange to ultimately give a product of markedly different structure. Investigations started more than one century ago with the Schmidt reaction involving an alkyl azide rearrangement and with the remarkable Curtius reaction with its corresponding acyl azide rearrangement. In both cases, the driving force of the reaction is the loss of simple molecular nitrogen as the by-product. Mechanistically, a reactive intermediate is generated and typically allowed to react with a nucleophile (e.g., water, alcohols, amines). A similar mechanistic pathway is involved in the Lossen reaction and the Hofmann reaction, where an isocyanate intermediate is conventionally engaged in new chemical transformations.

During the last century, the enormous impact of these alkyl and acyl azide rearrangements has of course been abundantly reported in many important papers. The goal of this chapter is to survey recent developments in this field. The discussion will first concentrate on the Schmidt reaction and the Curtius reaction. Later in the

Molecular Rearrangements in Organic Synthesis, First Edition. Edited by Christian M. Rojas.

chapter, related rearrangements including the Lossen reaction and the Hofmann reaction will also be described. A brief historical highlight will be given for each reaction, followed by a general mechanistic description. In addition, key recent developments both in synthetic methodology and in total synthesis will be addressed in each section.

4.2 ALKYL AZIDE REARRANGEMENTS

The alkyl azide rearrangement is now best known under the name of Schmidt reaction.[1] Discovered in 1923 by Schmidt,[2] this type of reaction, as originally reported, involved an acid-catalyzed process between hydrazoic acid and various electrophiles such as carboxylic acids, aldehydes, or ketones, giving, respectively, the corresponding amines, nitriles, and formamides or amides (Scheme 4.1). Mechanistically, after activation of the electrophile under acidic conditions, nucleophilic addition of hydrazoic acid can generate an azidohydrin intermediate **1**. After loss of water, rearrangement by alkyl group migration with concomitant loss of molecular nitrogen provided cationic nitrilium intermediate **2**. Addition of water then gave the desired amide **3**.

Following this seminal work, extensions from hydrazoic acid to the use of alkyl azides in the Schmidt reactions were rare. Although the reaction between methyl azide and acetophenone did not afford the desired amide,[3] Boyer and Hamer were able to conduct a low-yielding rearrangement upon combination of aromatic aldehydes with some alkyl azides (cf. **4**) (Scheme 4.2).[4] At the same time, this group reported an efficient synthesis of oxazolidine derivatives **5** starting from aromatic aldehydes and 2- or 3-hydroxy azides.[5] The better reactivity for this second example can be rationalized by the fact that the addition of azide is intramolecular and occurs after the formation of an oxonium ion **6**. In spite of these efforts, this reaction was almost totally forgotten until the early 1990s when major breakthroughs were described.

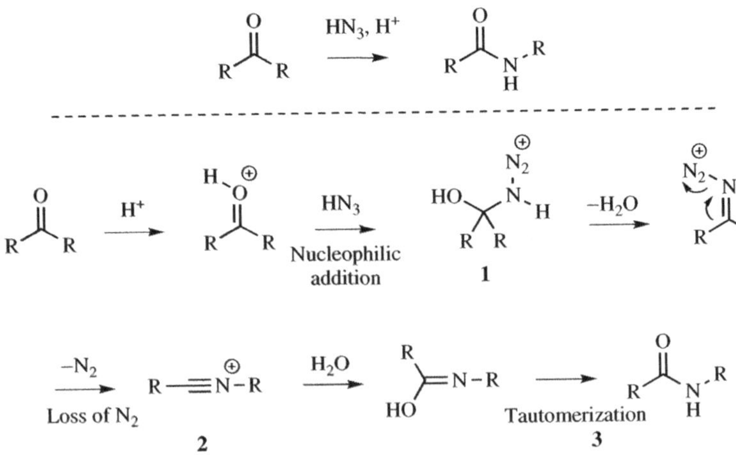

Scheme 4.1 Schmidt reaction.

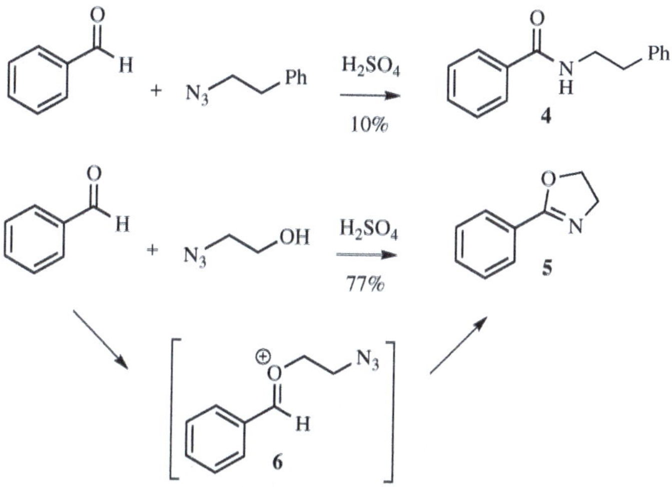

Scheme 4.2 Examples of reactions of alkyl azides and aldehydes.

Inspired by these earlier studies, Aubé and Milligan developed a novel intramolecular reaction of alkyl azides with simple ketones under mild conditions (Scheme 4.3).[6] After treatment of the azido-tethered ketones **7** under acidic conditions, the desired lactams **8** were isolated in good to high yields. The detailed scope and features of this rearrangement have been established: cyclic as well as acyclic ketones are tolerated, four or five atoms between the ketone and the azido group are required, and chirality transfer is complete when starting from enantiopure material. Mechanistically, migration and N_2-loss occur from an azidohydrin intermediate **9**. The results of DFT calculations by Tantillo and coworkers[7] indicated that such a process is concerted, involving N_2 departure and a shift of the alkyl group that is antiperiplanar to the N_2 leaving group. Intermediates **9a–c** will lead to a fused lactam, while intermediate **9d** will lead to a bridged lactam **8**. The roles of steric effects, lone pair cation, and cation–π interactions on the fused/bridged product selectivity have been characterized and quantified theoretically. Experimentally, only fused lactams were isolated; no bridged lactams were observed due to the probable high instability of the amide bond in these structures. It is also important to note that Lewis acids are required when the process involves starting materials bearing electron-withdrawing groups or formation of large rings; Brønsted acids are not strong enough to initiate the reaction when the rate of rearrangement is reduced.

This intramolecular reaction has been well explored and expanded to aldehydes and ketal derivatives; however, its major limitation is the requirement of excess Lewis or Brønsted acid in order to obtain complete conversion (cf. **10**). This lack of reactivity can be explained by strong product inhibition, behavior frequently present for any reaction that transforms a ketone to an amide. However, after extensive screening of conditions, the Aubé group recently reported a catalytic intramolecular Schmidt reaction using either $TiCl_4$ or AcCl in the presence of hexafluoro-2-propanol (Scheme 4.4).[8] The strong hydrogen-bonding ability of this solvent is the key point

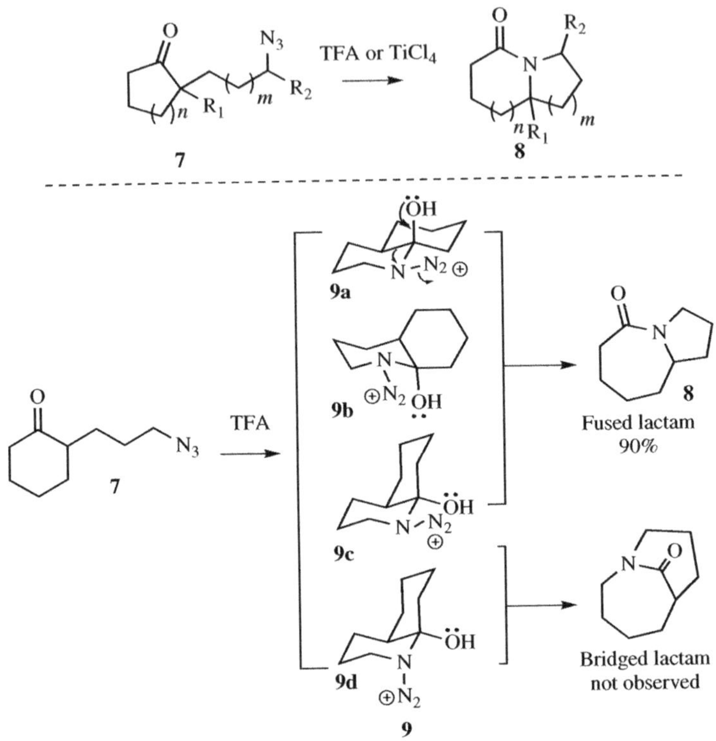

Scheme 4.3 Intramolecular Schmidt reactions of cyclic azidoketones.

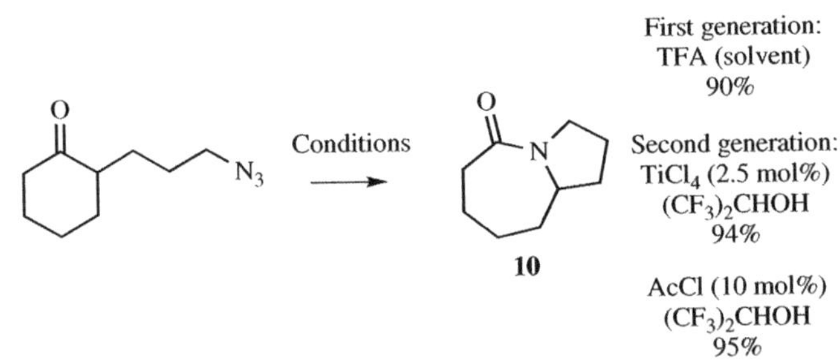

Scheme 4.4 Acid promoted intramolecular Schmidt reaction.

to efficiently perform the rearrangement with a broad substrate scope. Notably, the method using AcCl provides a metal-free catalytic reaction.

The acid-promoted intermolecular Schmidt reaction has been developed, though with moderate success (Scheme 4.5).[9] Indeed, several products can be isolated depending on the alkyl azides used and the nature of the acid. Generally, cyclohexanones and related cyclic ketones can generate the traditional Schmidt adduct **11** in the presence of $TiCl_4$, whereas a Mannich reaction is privileged when benzyl azide

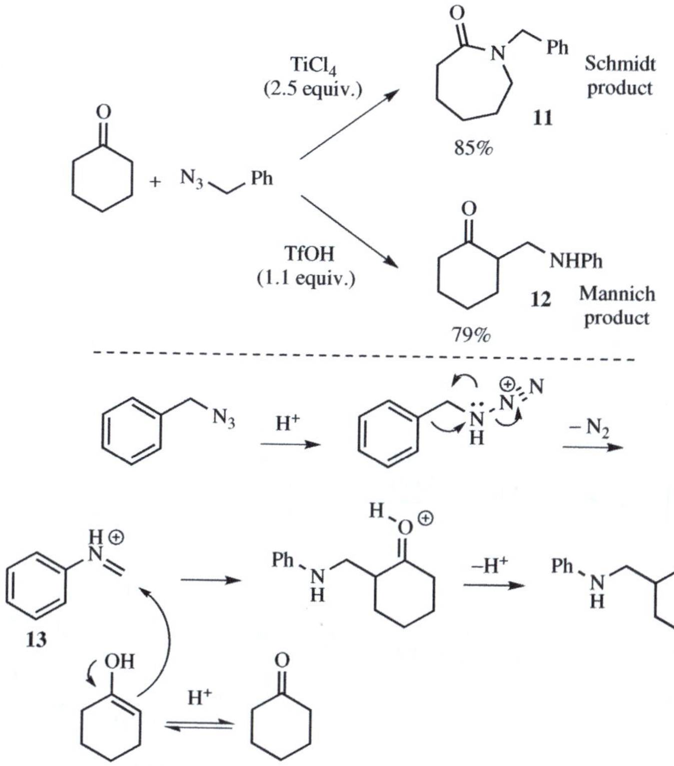

Scheme 4.5 Schmidt versus Mannich pathways in intermolecular reactions.

and triflic acid are involved, giving scaffold **12**. In the latter case, the acid-promoted rearrangement of benzyl azide provides an iminium intermediate **13** which could be trapped by the carbonyl compound via a modified Mannich reaction. The scope of these reactions is limited to relatively simple ketones.

On the basis of Boyer's early findings,[5] it is possible to overcome many of these limitations by using ω-hydroxy azides **14** instead of simple alkyl azides (Scheme 4.6).[10] The reaction is facilitated by the formation of an oxonium ion **15** under acidic conditions, leading to intramolecular rearrangement. Basic workup leads to the final lactam product **16**. The method is quite general; covering a large range of cyclic or acyclic polyfunctionalized ketones as well as hindered hydroxy azides.

Starting from unsymmetrical ketones, two regioisomeric lactams can be formed (Scheme 4.7).[11] Experimentally, when the reaction is performed with methyl, ethyl, or inductively electron-withdrawing groups such as phenyl or methoxy at the α-position of the cyclohexanone (cf. **17–18**), migration of the less bulky group is slightly or totally favored. However, bulkier alkyl groups linked at the same position gave the opposite migration regioselectivity (cf. **19**). Finally, high

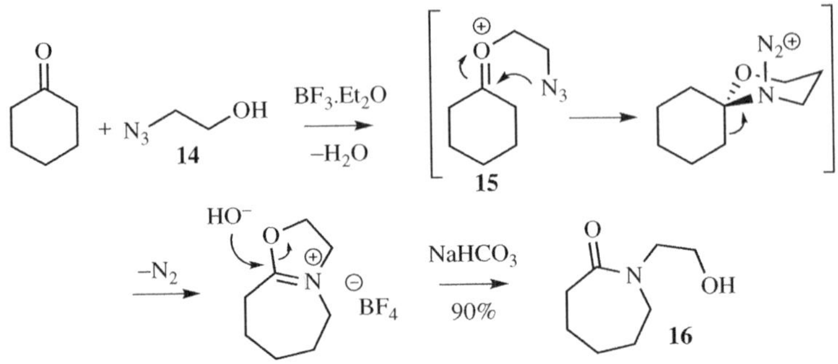

Scheme 4.6 Intermolecular Schmidt reactions using hydroxyalkyl azides.

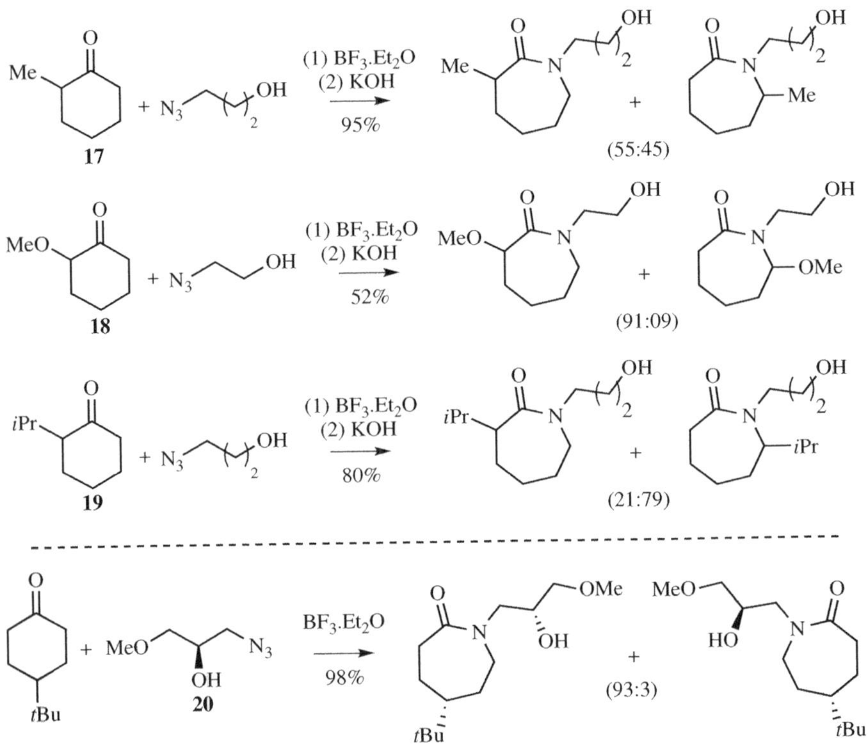

Scheme 4.7 Regiochemical possibilities for the Schmidt reaction of hydroxyalkyl azides with ketones.

diastereoselectivity levels are generally observed when the hydroxyalkyl azide **20** also contains a stereocenter.[12]

Beyond their mechanistic elegance, these acid-promoted rearrangements between aliphatic azides and carbonyl compounds are attractive because they open many opportunities in total synthesis. Among these studies, we highlight the total synthesis of (±)-stenine via an intermolecular Diels–Alder/Schmidt sequence (Scheme 4.8).[13] In this approach, the triene **21** (E:Z=85:15), obtained after 13 linear steps, was treated in the presence of $MeAlCl_2$, furnishing a mixture of three lactams in a good overall yield. Fortunately, the major isomer **22** contained the tricyclic ring system of the natural product. Regarding the mechanism, the two major isomers come from the same Diels–Alder intermediate via an endo transition state. After subsequent nucleophilic addition of the azide to the carbonyl group, the intermediate **23** bearing an equatorial N_2^+ group would provide the desired scaffold in a satisfying yield. After removal of the benzyl group, the fourth ring **24** was established by oxidation of the primary alcohol following by iodolactonization in 80% yield over three steps. With this intermediate in hand, the end of the synthesis proceeded with introduction of two alkyl groups to give the target molecule.

Aldehydes have also been employed using the Schmidt reaction for total synthesis. For instance, Wang and coworkers designed an unprecedented synthesis of (S)-tylophorine **25** via a one-pot intramolecular Schmidt/Bischler–Napieralski/imine reduction cascade sequence (Scheme 4.9).[14] Nevertheless, in comparison to ketones, few examples of alkyl azide rearrangements with aldehydes have been reported due to competition between alkyl migration and hydride migration during the nitrogen extrusion step.[15] In Wang's work, $TiCl_4$ and TFA were recognized as the best promoters for the reaction, and no trace of the undesired lactam was detected.

Use of carbocations to initiate the Schmidt reaction has been the subject of several investigations during the last two decades. The cationic intermediate can be generated from secondary or tertiary alcohols or from olefins. The first example was described by the Pearson group for the synthesis of a bicyclic enamine.[16] However, various drawbacks still remain when carbocations are involved: limited stability of the intermediates, low selectivity, and the need for harsh acidic conditions. The use of alkenes in combination with metals offers the opportunity to overcome some of these limitations. For example, Pearson *et al.* have made a number of contributions in this area, and they have found that $Hg(OTf)_2$ can activate alkyl azide rearrangement with azido-alkenes under mild conditions (Scheme 4.10).[17] Even if yields are moderate, selectivity and functional group tolerance are greatly improved. In this version of the Schmidt reaction, addition of the mercury to the double bond provides a mercurinium ion **26** which is then opened by the azide, giving the aminodiazonium ion **27**. After migration of the cyclic bond and generation of iminium ion intermediate **28**, only one regioisomer **29** was isolated, while an equimolar mixture **30** was obtained under acidic conditions.[18]

As with alkene precursors, there is often a lack of selectivity in Schmidt reactions of carbocations generated from alcohols under acidic reaction conditions. However, the Renaud group reported the first example of an intramolecular Schmidt reaction involving a primary electrophilic carbon center for the synthesis of (−)-indolizidine under nonacidic conditions (Scheme 4.11).[19] After activation of the alcohol **31** via triflation, alkyl migration and extrusion of molecular N_2 occurred smoothly. After subsequent reduction of **32**, the desired structure **33** was obtained in high yield and excellent optical purity (98% ee). This concise approach enabled the synthesis of (−)-indolizidine in seven linear steps and 27% overall yield.

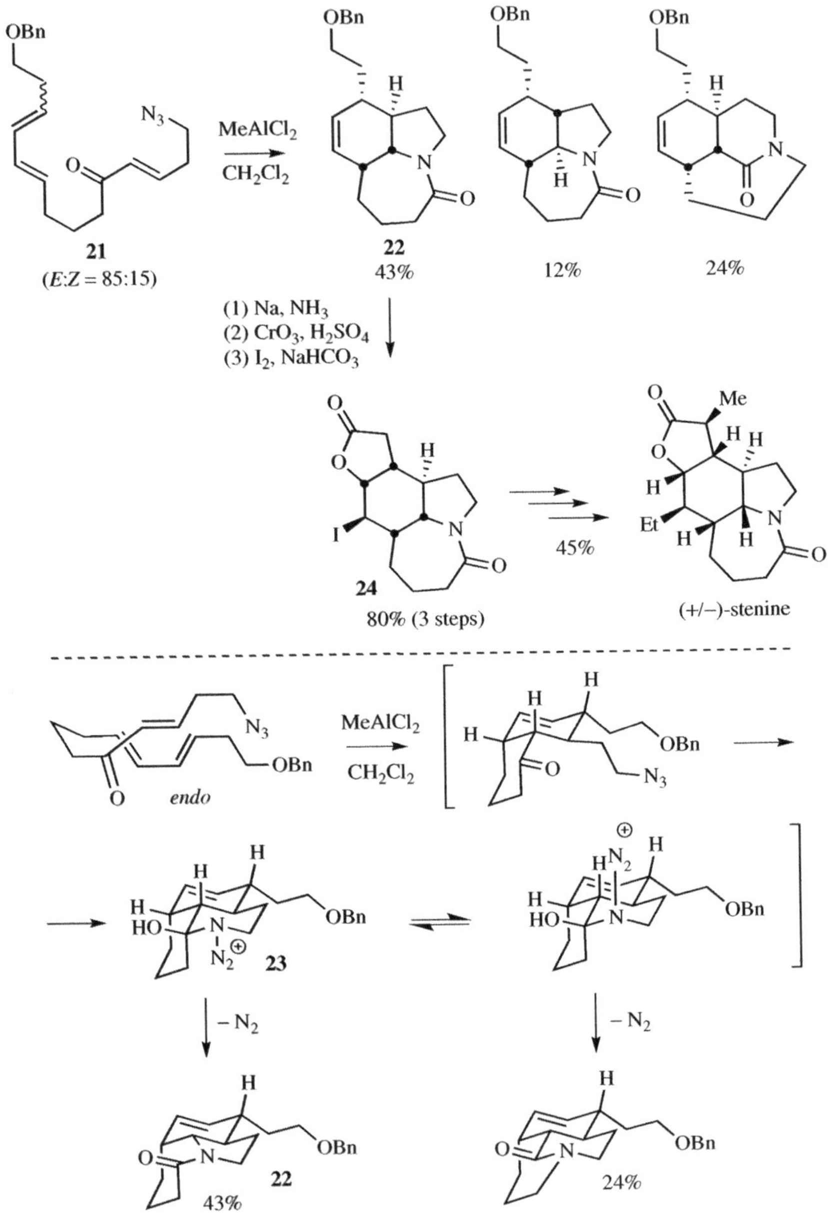

Scheme 4.8 Total synthesis of (±)-stenine via an intermolecular Diels–Alder/Schmidt sequence.

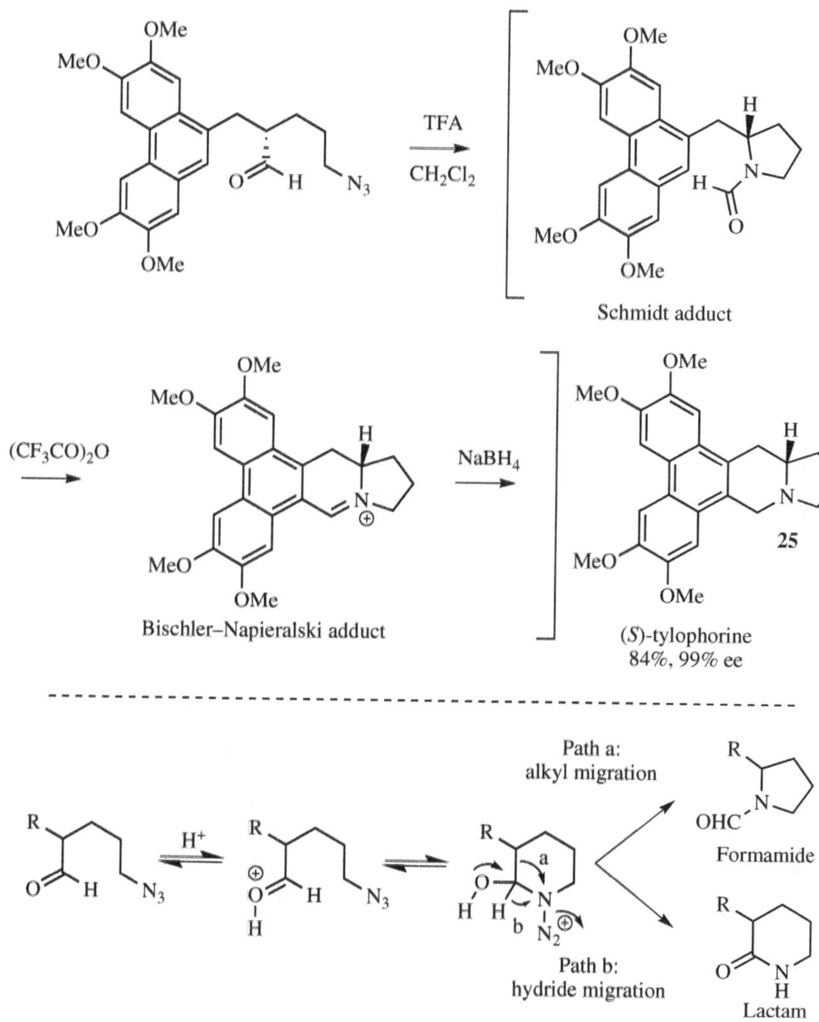

Scheme 4.9 Synthesis of (S)-tylophorine via a one-pot intramolecular Schmidt/Bischler–Napieralski/imine reduction cascade sequence.

Another recent study was accomplished by Gigant and Gillaizeau using electron-rich alkenes and more especially enamides (Scheme 4.12).[20] Inspired by Aubé and coworkers's previous work,[8] the Gillaizeau group studied the benzyl azide-to-iminium rearrangement in the presence of enamides and triflic acid for the direct construction of novel pyrido-fused tetrahydroquinoline scaffolds **34**. From a synthetic point of view, after benzyl azide protonation and formation of an iminium ion intermediate **35** with concomitant loss of molecular nitrogen, the first carbon–carbon bond was created by nucleophilic addition of the enamide. Subsequent cyclization onto the newly formed *N*-sulfonyliminium ion **36** via a

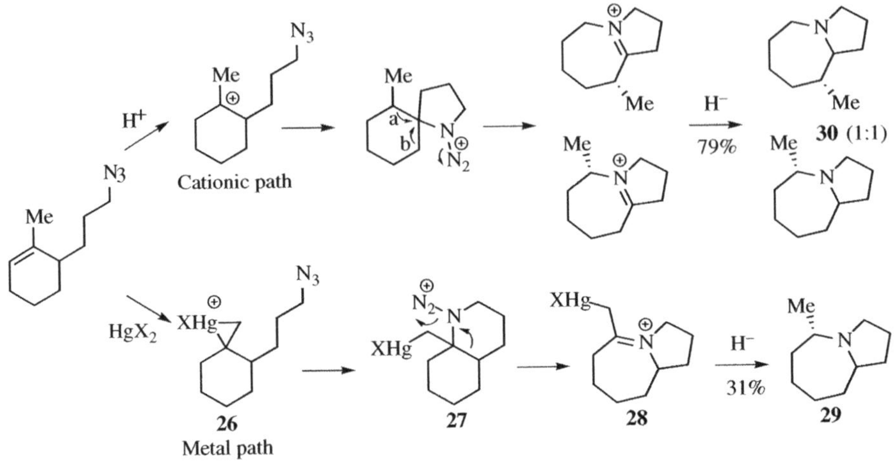

Scheme 4.10 Mechanism of cationic or Hg(II) promoted Schmidt reactions.

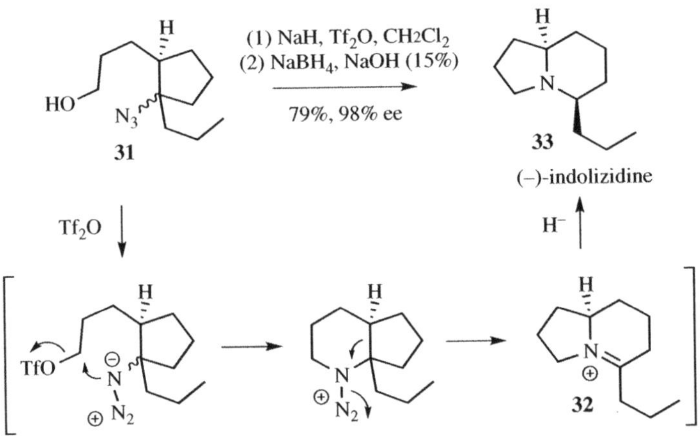

Scheme 4.11 Intramolecular Schmidt reactions of alkyl azides under non acidic conditions.

Pictet–Spengler reaction occurred, providing a second carbon–carbon bond (path a). On the other hand, with C2-substitued enamides, proton elimination is faster than cyclization, and 2,3-difunctionalized enamides **37** were obtained (path b). Numerous of privileged structures were synthesized by application of this versatile strategy.

Toste and coworkers have discovered a gold(I)-catalyzed acetylenic Schmidt reaction for the synthesis of polysubstituted pyrroles **38** (Scheme 4.13).[21] In this reaction, it was suggested that the metal plays two roles. It both activates the alkyne bond (cf. **39**) and also stabilizes the π-intermediate **40** by electron donation. This powerful reaction tolerates both alkyl and aryl groups, and alkyne chemoselectivity was complete starting from a 1,5-enyne.

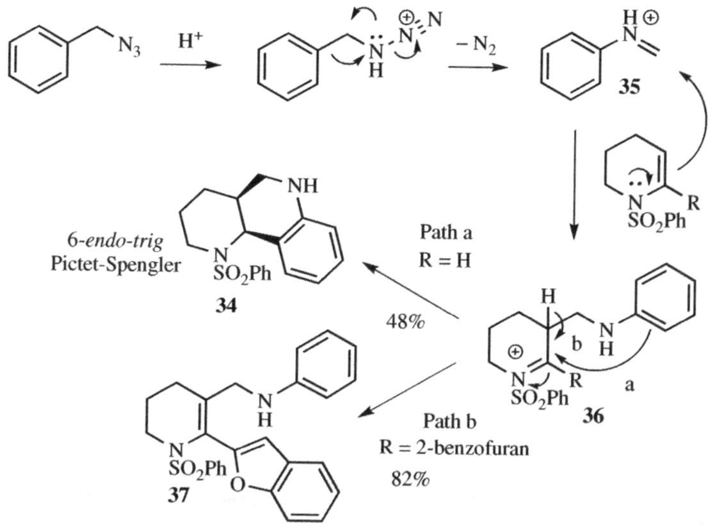

Scheme 4.12 Acid-promoted benzyl-azide-to-iminium rearrangement and intermolecular reaction with electron-rich enamides.

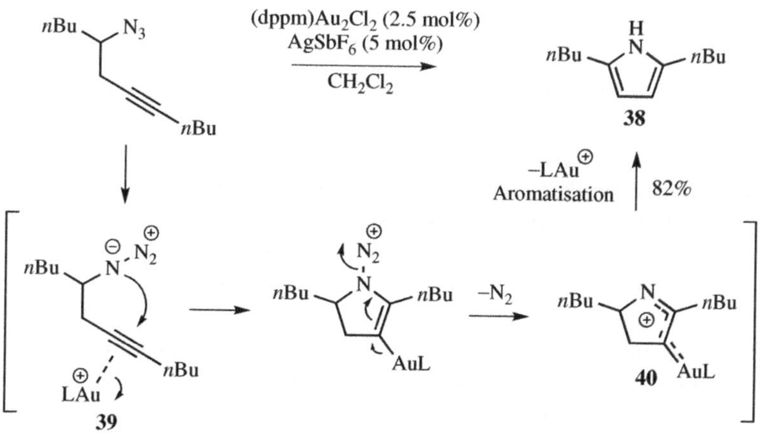

Scheme 4.13 Mechanism of Au(I) promoted acetylenic Schmidt reactions.

Recently, Chen and coworkers[22] reported an efficient new method to construct a series of densely functionalized 4,5-dihydro-1*H*-azepine products **41** from the intermolecular reaction of alkyl azides with propargylic esters. In this approach, sequential reaction of vinyl-gold carbenoids **42** with alkyl azides and formation of vinyl imine intermediates **43** may be involved (Scheme 4.14). Then a subsequent formal [4+3] cycloaddition with another molecule of vinyl-gold carbenoid **42** afford the desired azepine **41**.

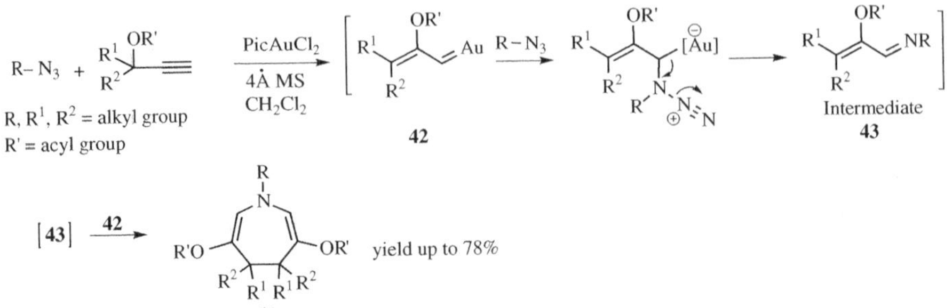

Scheme 4.14 Gold catalyzed synthesis of azepines via an intermolecular Diels–Alder cycloaddition.

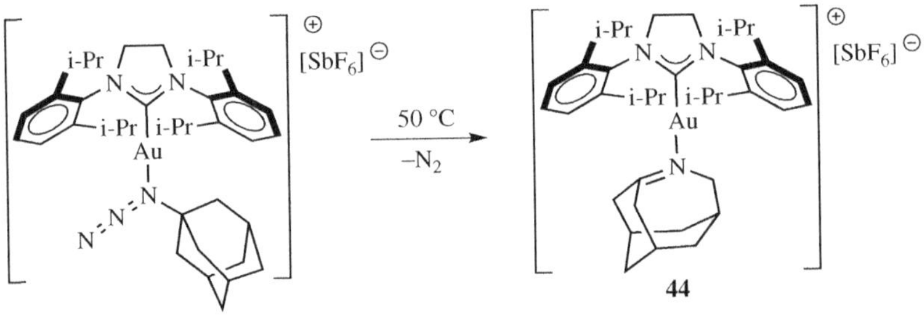

Scheme 4.15 Organoazide complex of Au(I).

Cundari and coworkers have also demonstrated for the first time that it is possible to stabilize organoazide complexes of gold(I) (Scheme 4.15).[23] They have found that the N_2 expulsion from gold-bound 1-azidoadamantane (1-AdNNN), 2-azidoadamantane (2-AdNNN), and azidocyclohexane (CyNNN) leads to rearrangement products featuring endocyclic imine moieties in a seven-membered heterocyclic framework as in 4-azahomoadamant-3-ene, 4-azahomoadamant-4-ene, and 1-azacyclohet-1-ene or exocyclic groups as in cyclohexanimine. All these products of dinitrogen extrusion **44** were isolated as their N-coordinated $[(SIPr)\text{-}Au]^+$ adducts. A computational study was performed to understand the observed structures of gold-coordinated 1-AdNNN and 2-AdNNN and their nitrogen elimination pathways. The calculations indicated that the conversion of the organoazide complex to the imine is a concerted process without a nitrene/nitrenoid intermediate. In addition, kinetic studies of $[(SIPr)AuN(2\text{-}Ad)NN][SbF_6]$ from 30 to 50 °C indicate that nitrogen elimination is a first-order process.

Finally, both epoxides and α,β-unsaturated ketones are also suitable substrates for the Schmidt reaction. Epoxide-based Schmidt reaction is hardly disclosed.[24] As an example, aryl epoxyazides **45** undergo efficient electron-deficient reaction cascades mediated by Lewis acids, leading to regiospecific amination of the aromatic ring (Scheme 4.16). An acid-induced opening of the epoxide **45** by azide attack at

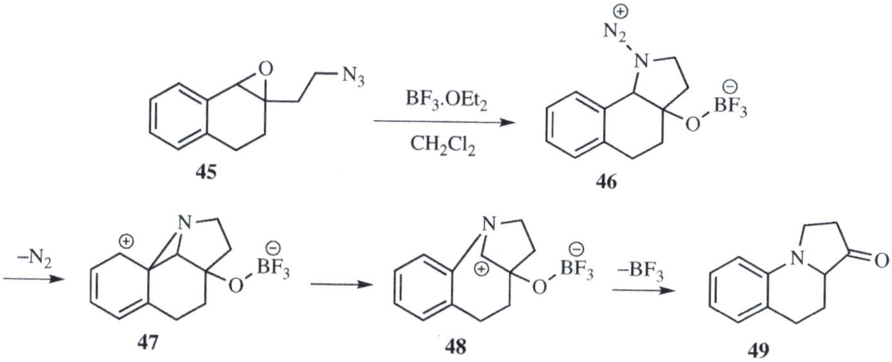

Scheme 4.16 Intramolecular reactions of alkyl azides with epoxides.

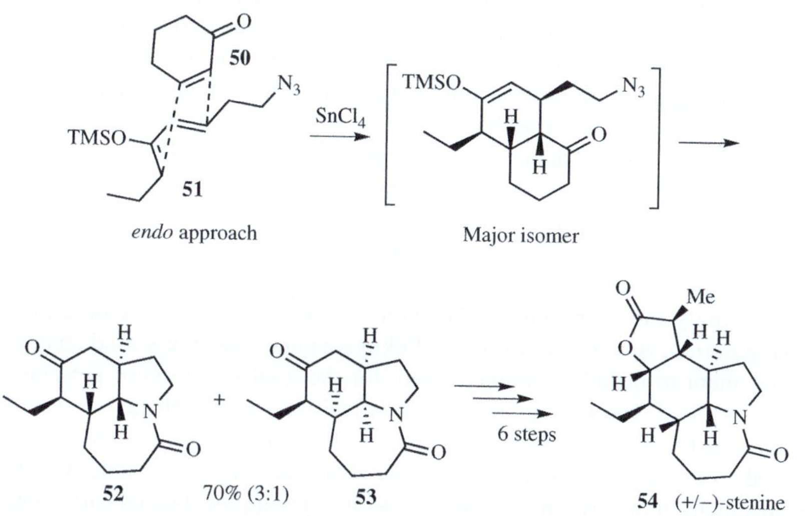

Scheme 4.17 Intermolecular Diels–Alder/Schmidt reaction sequence with enones.

the benzylic position gave the intermediate **46**; then subsequent nitrogen extrusion furnished carbocation **47** following by cleavage of the aziridine (cf. **48**). The tricyclic scaffold **49** is finally isolated after regiospecific ring contraction with assistance from the oxygen atom.

Otherwise, the use of α,β-unsaturated ketones has been demonstrated in several total synthesis efforts including in an improved total synthesis of (±)-stenine reported by Aubé's group (Scheme 4.17).[25] Simple commercially available cyclohexenone **50** reacted with diene **51** in an intermolecular Diels–Alder/Schmidt reaction sequence. $SnCl_4$ was found to be the best Lewis acid in this reaction, giving Diels–Alder/Schmidt adducts **52** and **53**, isolated in excellent global yield (70%)

and with an exo–endo ratio (3:1), and that favored the stereochemistry needed for the natural product in the major lactam **54**. This selectivity can be explained by an exo addition during the Diels–Alder cycloaddition; the endo alternative encounters significant steric encumbrance between one of the protons with the incoming nucleophilic silyl enol ether. By application of the Diels–Alder/Schmidt sequence, the synthesis was accomplished in only nine steps and 14% overall yield.

4.3 ACYL AZIDE REARRANGEMENTS

The Curtius rearrangement is a general reaction that involves an acyl azide rearrangement for the preparation of isocyanates and their related compounds. Thanks to the rich reactivity of isocyanates, these intermediates can be converted to a variety of related compounds, often *in situ* (Scheme 4.18).[26] Discovered by Curtius,[27] this reaction has been widely explored for the last century for its extensive synthetic utility. As a result, both the preparation of the acyl azides without the need for isolation due to their instability and the subsequent reactions of the derived isocyanates have been well developed. From a mechanistic point of view, it is well established that the C—N bond migration occurs concomitantly with the loss of molecular nitrogen. The resulting isocyanates can next be transformed into primary amines by hydrolysis or can react with various nucleophiles, giving a range of N-acyl derivatives. From a synthetic point of view, this reaction offers a way to easily convert a carboxyl group into nitrogen-containing functionality. Alky-, vinyl-, and arylacyl azides are all tolerated, including within the complex molecular frameworks required for natural product total synthesis.

Interestingly, the migration of sp^3 carbons is stereospecific with complete retention of configuration. As a recent example, Fukuyama and coworkers used an acyl azide rearrangement to install an amine moiety for the total synthesis of gelsemoxonine (Scheme 4.19).[28] The Curtius reaction was fruitfully accomplished in a three-step sequence starting from a simple carboxylic acid. After formation of the acyl chloride followed by addition of sodium azide, the newly formed acyl azide underwent thermal rearrangement in the presence of benzyl alcohol, giving the desired carbamate **55** in a stereoselective fashion.

In addition, alkene geometry is totally preserved in the Curtius rearrangement. One appealing example was published by the De Brabander group for the total synthesis and structure revision of the marine metabolite palmerolide A

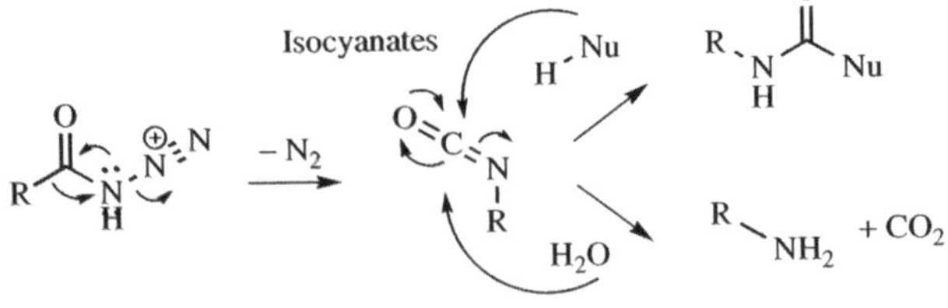

Scheme 4.18 Curtius rearrangement.

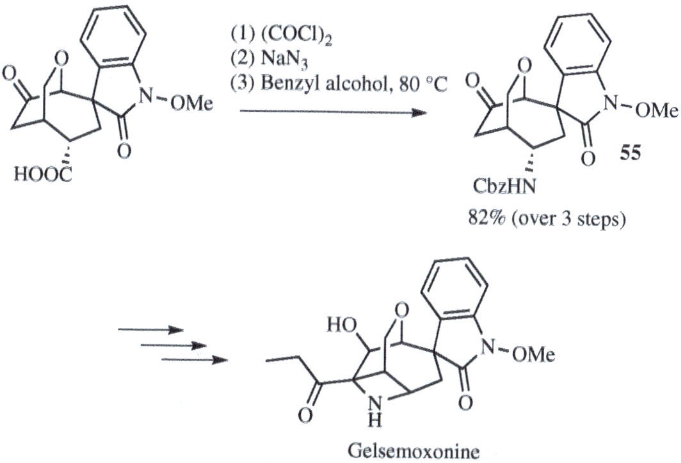

Scheme 4.19 Synthesis of gelsemoxonine via Curtius rearrangement.

(Scheme 4.20).[29] The key acyl azide having the (*E*) configuration was obtained from carboxylic acid **56** using diphenylphosphoryl azide (DPPA). Curtius rearrangement occurred upon heating, followed by isocyanate trapping with the appropriate Grignard reagent. The desired highly functionalized scaffold **57** was isolated with complete (*E*) stereocontrol of the newly formed enamide functionality. Subsequent selective removal of protecting groups and functional group manipulation provided the initial proposed structure of the natural product.

While acyl azides are often obtained by activation of carboxylic acids as the acyl chlorides and treatment with NaN_3, the example in Scheme 4.20 used the Shioiri–Ninomiya–Yamada modification.[30] This one-pot reaction typically involves DPPA, a base, and a nucleophile, generally an alcohol. The main advantage of this procedure is that it avoids isolation of the intermediate acyl azide, an important benefit given the instability of this intermediate in many cases. The carboxylic acid scope is quite general; both aliphatic and aromatic substrates are tolerated. In terms of the nucleophile, benzyl alcohols as well as tertiary alcohols are suitable coupling partners.

Another synthetic approach to acyl azides was reported by Weinstock one decade earlier.[31] In this methodology, acyl azides are prepared via carboxylic–carbonic anhydrides. Generally, carboxylic acids are treated with ethyl chloroformate, generating mixed anhydrides which are trapped *in situ* by sodium azide. This simple strategy was applied recently in the synthesis of (±)-spisulosine (Scheme 4.21).[32] After only 5 min in the presence of ethyl chloroformate, the appropriate carboxylic acid **58** was next treated with sodium azide. The intermediate ene-isocyanate **59** was obtained after heating, and this latter is finally hydrolyzed by water. The corresponding methyl ketone **60** was isolated in 40% overall yield.

The Lebel group has developed an efficient process for the Curtius rearrangement which allows the direct conversion of carboxylic acids into carbamates with sodium

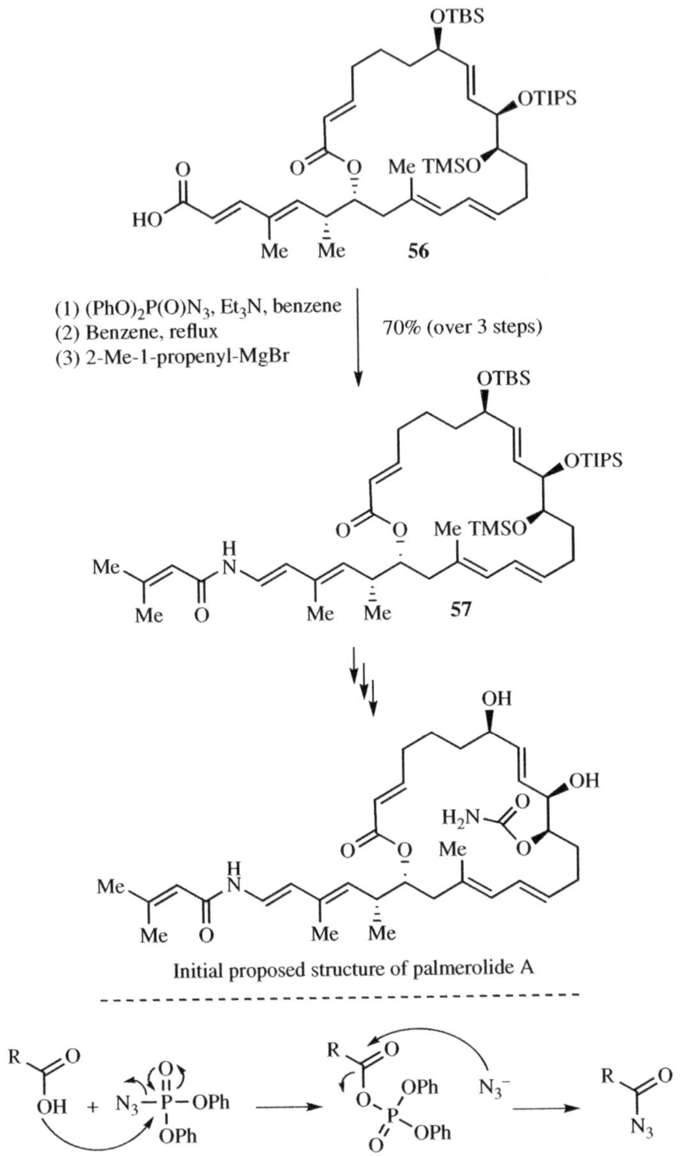

Scheme 4.20 Curtius rearrangement using diphenyl phosphoryl azide.

azide in the presence of tetrabutylammonium bromide and zinc(II) triflate under very mild conditions (cf. **61**) (Scheme 4.22).[33] Mechanistically, the authors suggested a similar pathway to the Shioiri version: *tert*-butyl azidoformate is generated during the process and plays a role analogous to DPPA. The zinc complex is not involved during isocyanate formation, but its presence favors the carbamate formation. Regarding the

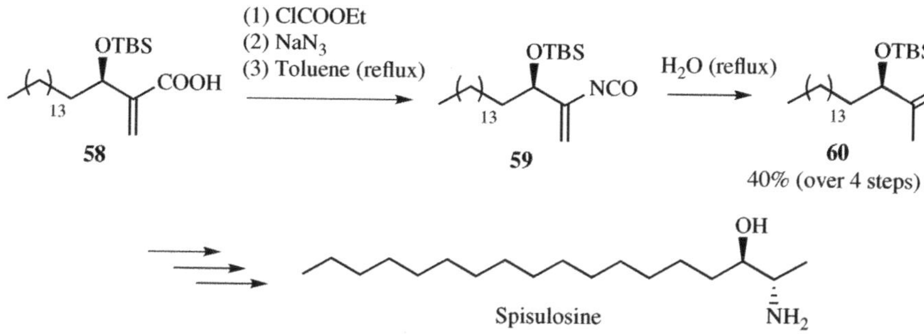

Scheme 4.21 Curtius process via mixed carboxylic–carbonic anhydrides.

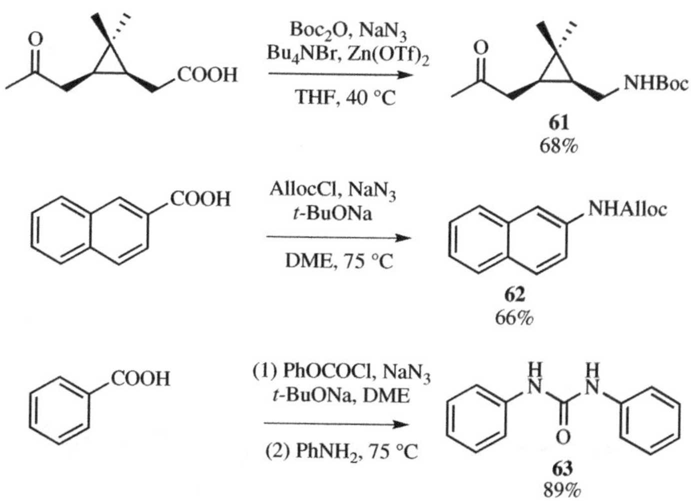

Scheme 4.22 Conversion of carboxylic acids into carbamates and ureas.

scope of this transformation, only aliphatic carboxylic acids are tolerated. However, this limitation was overcome by using chloroformate derivatives in association with a catalytic amount of *tert*-butoxide as a base in DME (cf. **62**).[34] These reaction conditions provide efficient conversions in the presence of a range of functional groups, and the desired carbamates are isolated in good yields. Finally, the use of phenyl chloroformate, which produces a less nucleophilic alcohol compared to allylchloroformate, for instance, enabled the use of amines as nucleophiles for the synthesis of ureas **63**.

To illustrate the high preparative potential of the Curtius rearrangement of acyl azides, we highlight the synthesis of (−)-oseltamivir, achieved using microreactor technology (Scheme 4.23).[35] The key rearrangement step worked efficiently on up to a 10 g scale to afford the desired advanced synthetic intermediate **64** after simple recrystallization. This strategy includes numerous advantages: short reaction time, safe reaction conditions, and low cost.

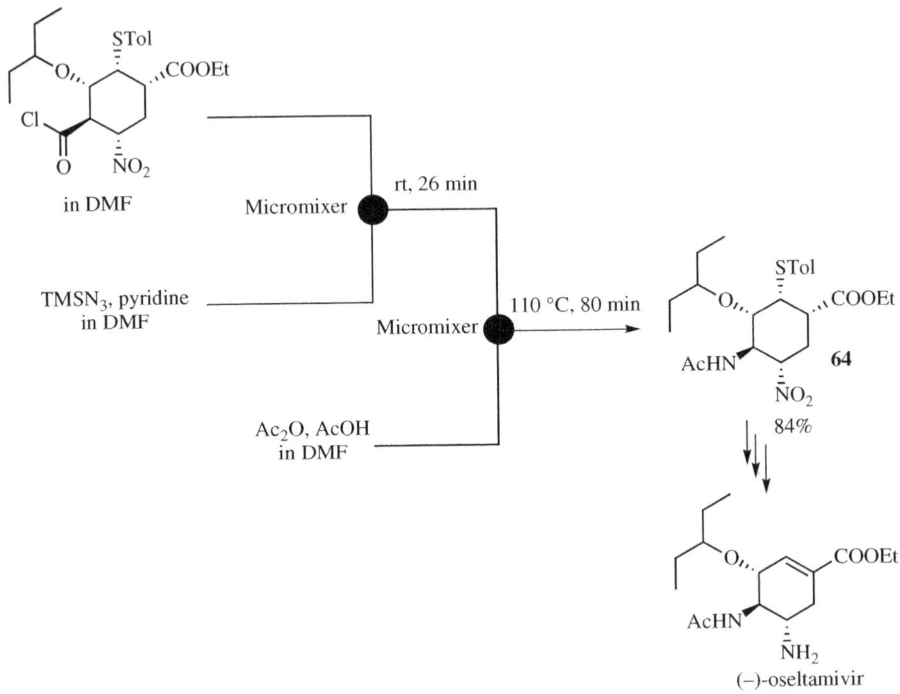

Scheme 4.23 A microflow reaction involving the Curtius rearrangement.

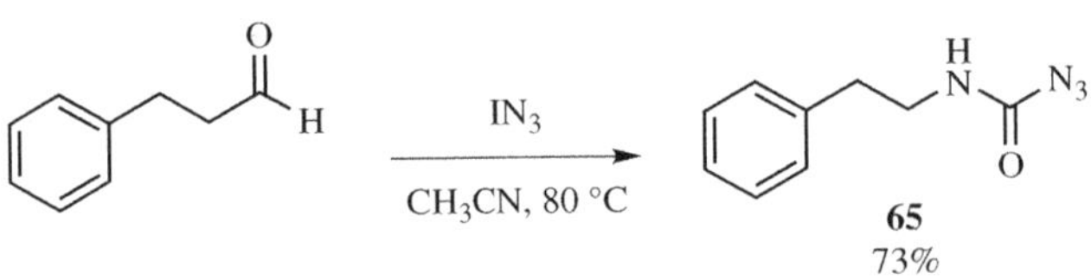

Scheme 4.24 Curtius rearrangement from aldehydes.

Aldehydes can also be precursors for the Curtius acyl azide rearrangement (Scheme 4.24).[36] Aliphatic and aromatic aldehydes can be converted to the corresponding acyl azides by treatment with iodine azide. Subsequent heating gave various carbamoyl azides in high yields. A radical pathway, proceeding by homolysis of the iodine–azide bond, was proposed. The resulting carbamoyl azides **65** are relevant precursors for the synthesis of amines or ureas, for example.

4.4 HOFMANN REARRANGEMENT

The Hofmann rearrangement, commonly referred to as the Hofmann degradation, converts a primary carboxamide to a primary amine (Scheme 4.25).[37] After formation

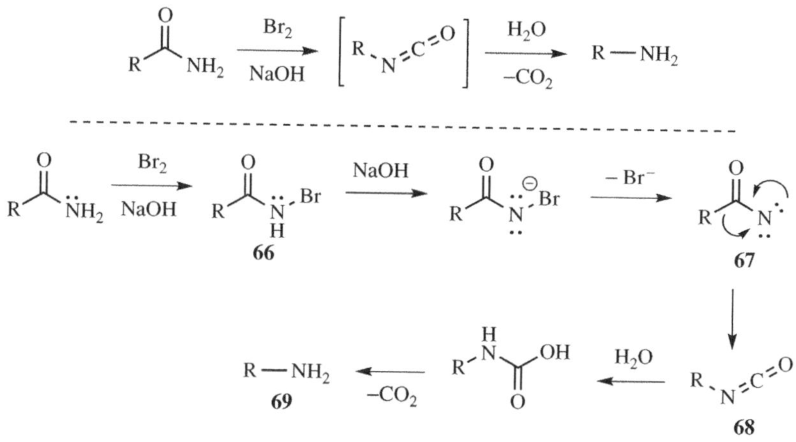

Scheme 4.25 Mechanism of Hofmann rearrangement.

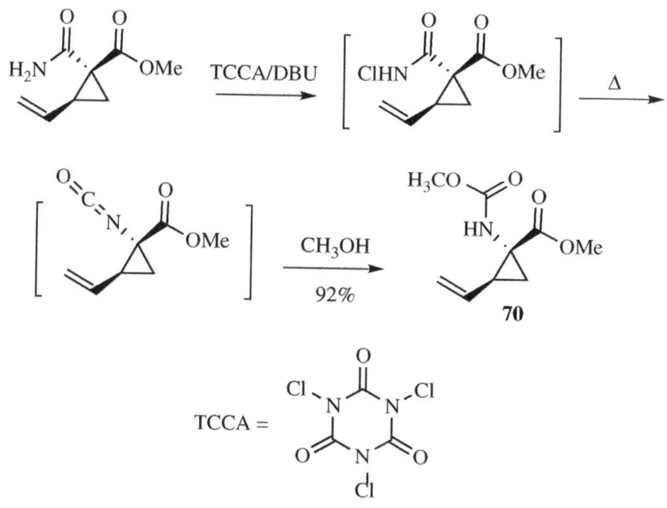

Scheme 4.26 Acid-mediated Hofmann rearrangement.

of an *N*-bromoamide **66** under base-mediated bromination conditions, a nitrene **67** is generated via α-elimination of HBr. Subsequent rearrangement of the nitrene results in the formation of an isocyanate **68** which is hydrolyzed to a primary amine **69** after CO_2 extrusion.

Recently, an efficient and scalable trichloroisocyanuric acid-mediated Hofmann rearrangement was employed to synthesize an unnatural cyclopropane-containing amino acid moiety **70** that is becoming a significant building block in a number of new drug candidates (Scheme 4.26).[38] The reaction proceeded smoothly in the

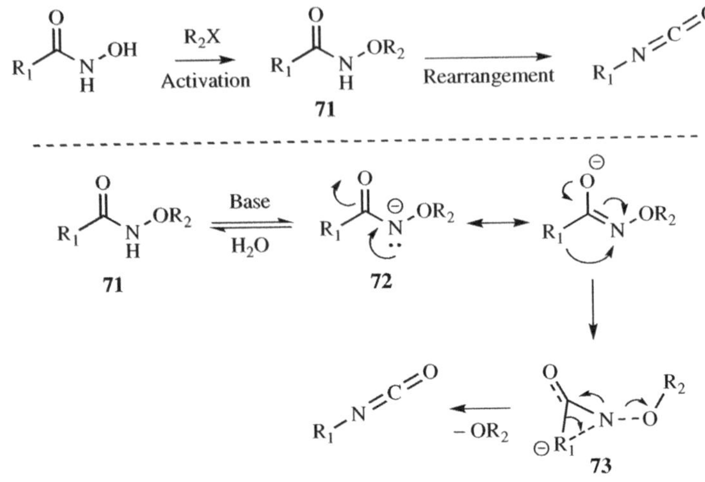

Scheme 4.27 Mechanism of Lossen rearrangement.

presence of large variety of functional groups including olefin, ester, halogen, ether, heteroaromatic, cyclopropyl, and nitro functionality.

4.5 LOSSEN REARRANGEMENT

The Lossen rearrangement, discovered in 1872,[39] is the transformation of a hydroxamic acid derivative to an isocyanate by using an activated hydroxamic acid **71** as the key intermediate (Scheme 4.27).[40] This activation can be achieved by O-acylation, O-arylation, or O-sulfonylation. From a mechanistic point of view, a base or a thermal treatment can initiate the reaction, giving a rearrangement of the corresponding anion **72** via a three-membered ring transition state **73**.

Since its initial discovery and given that isocyanates are versatile synthetic intermediates in organic chemistry, the Lossen rearrangement has been developed extensively. Isocyanates can, in fact, be easily transformed to ureas and amines, utilized in multicomponent reactions such as the Ugi or Passerini reactions, or, more recently, applied as versatile C1 building blocks in palladium catalysis. More than 50 years ago, it was shown that the Lossen rearrangement occurs in a stereospecific fashion.[41] Three of the major challenges associated with the Lossen process in synthesis are (1) the competitive dimerization of activated hydroxamates, (2) the limited availability of hydroxamic acids as starting materials, and (3) the formation of many stoichiometric by-products in the activation/rearrangement sequence, which complicates application to large-scale synthesis.[42]

With this background, recent examples of the Lossen rearrangement in synthesis include methods for the preparation of carbamates and amines via environmentally friendly routes (Scheme 4.28).[43] For the carbamate product **74**, a catalytic procedure

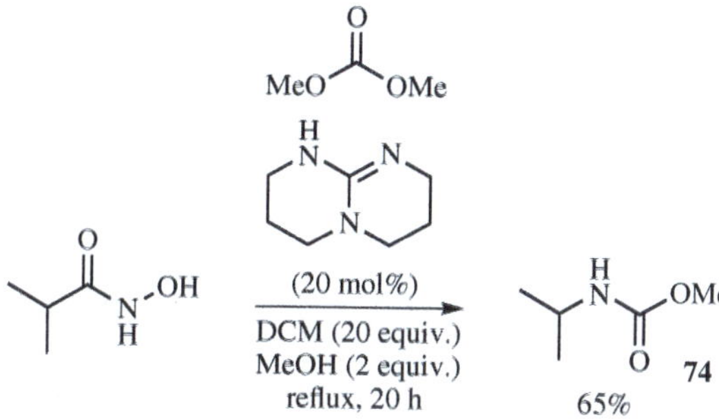

Scheme 4.28 Catalytic Lossen rearrangement.

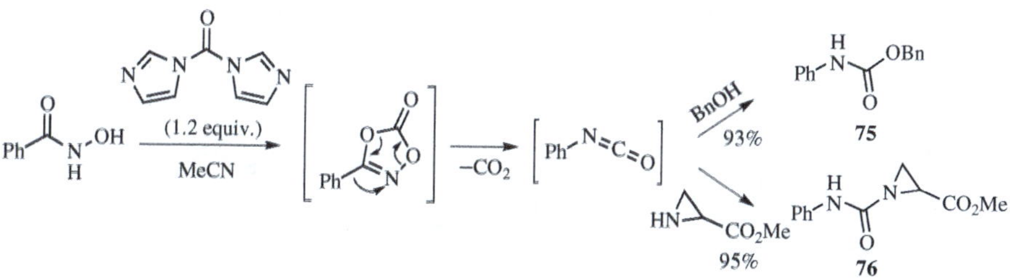

Scheme 4.29 Carbonyldiimidazole-mediated Lossen rearrangement.

by tertiary amine bases was achieved, avoiding large amounts of by-products and offering the possibility for recycling of solvents. Dimethyl carbonate proved to be an efficient *in situ* activator and has the benefit of being a relatively nontoxic reagent. A range of aliphatic methyl carbamates were isolated in good yields under mild conditions.

Large-scale application of the Lossen rearrangement can also be envisaged following the procedure reported by Dubé *et al.* (Scheme 4.29).[44] During this study, carbonyldiimidazole was found to be the best promoter for the reaction. Even if imidazole and carbon dioxide are generated as the by-products, this protocol is simple and avoids the use of traditional Lossen reagents which tend to be either toxic or hazardous. The expected isocyanates were captured by primary amines or alcohols, furnishing the corresponding ureas **75** or carbamates **76** in high yields.

Hydroxamic acids are the most common precursors for the Lossen rearrangement, but related functional groups such as thiohydroxamic acids[45] or trifluoromethane sulfonylimides[46] can also participate in the rearrangement. For these latter conditions, in the presence of thionyl chloride or phosphorus pentachloride, the aza-Lossen rearrangement occurs to generate the corresponding chloroformamidines **77** (Scheme 4.30). Due to their relative instability, these initial products were immediately trapped with morpholine. The author suggested that a five-membered cyclic transition state **78** is involved in the mechanism of the process.

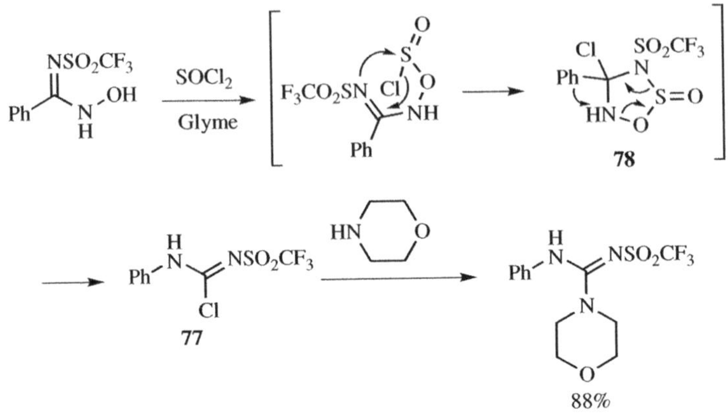

Scheme 4.30 Aza-Lossen rearrangement.

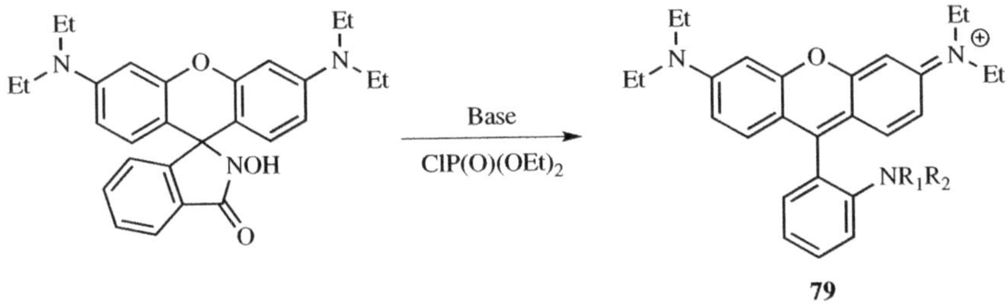

Scheme 4.31 Lossen rearrangement of rhodamine–hydroxamate.

A recent significant application of the Lossen reaction involved the synthesis of a highly fluorescent compound **79** as a nerve agent stimulant (Scheme 4.31).[47] Starting from a nonfluorescent rhodamine–hydroxamate, an efficient rearrangement occurs in the presence of diethyl chlorophosphate under alkaline conditions.

Carbohydrates are also suitable scaffolds for the Lossen rearrangement (Scheme 4.32).[48] The reactions of sugar hydroxamic acids were developed to evaluate the specific degradation of pectins **80** via a carbodiimide activation of galacturonic acid hydroxamates **81**. Treatment with EDC (1-ethyl-3-(3-dimethylaminopropyl) carbodiimide) gave the anticipated isourea **82** which was immediately transformed to an isocyanate before degradation to a primary amine. Beyond this particular application of this elegant method, it is noteworthy that the structure elucidation of pectins remains an important research area having considerable applications in the food industry, for instance.

Bioactive molecules can also be obtained through a Lossen rearrangement as the key step (Scheme 4.33).[49] A 2-*N*-dihydroxybenzamide was subjected to triphenylphosphine and diethyl azodicarboxylate, providing the cyclized adduct **83**

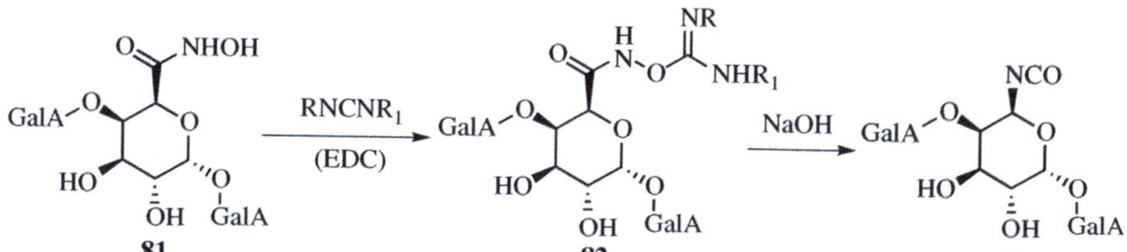

Scheme 4.32 Carbodiimide-mediated Lossen rearrangement of methyl-esterified galacturonic acid residues.

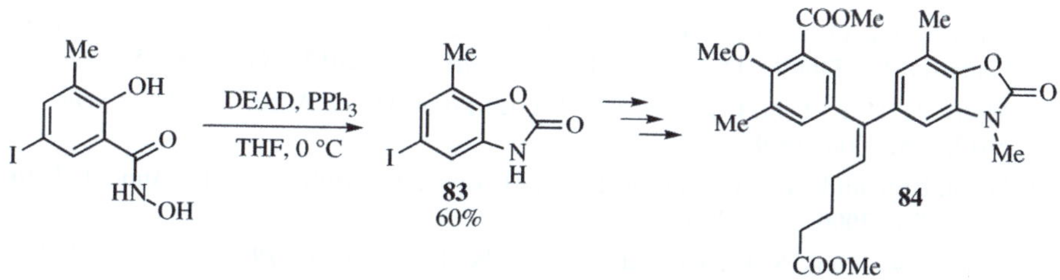

Scheme 4.33 Access to bioactive molecules via Lossen rearrangement.

in good yield. The desired benzoxazolone ring was then subjected to a Sonogashira cross-coupling, and after several additional steps, the target molecule **84**, having promising anti-HIV proprieties, was isolated.

4.6 CONCLUSION

The selected studies presented in this chapter showcase that alkyl and acyl azide rearrangements have become incontestable tools for the creation of new and original nitrogen–carbon bonds. As reflected in the earlier presented examples, these transformations have inspired new synthetic methods with applications across all fields of chemical synthesis, particularly in the total synthesis of natural and bioactive products. Furthermore, in most cases, the choice of appropriate reaction parameters revealed beneficial effects, allowing the use of milder conditions and resulting in an increase on the selectivity of the rearrangements. From a synthetic point of view, the fact that these reactions proceed with simple loss of molecular nitrogen or its equivalent makes these rearrangement processes highly attractive in terms of atom economy. However, most of these chemical processes suffer from the use of stoichiometric promoters. Only few of them are catalyzed, and the development of novel and more efficient catalytic systems is a current challenge. Due to the high impact of these alkyl and acyl azide rearrangements, it is obvious that many new synthetic transformations remain to be discovered in the coming years.

REFERENCES

1. For an excellent review, see: Grecian, S.; Aubé, J. In Organic Azides: Syntheses and Applications; Bräse, S.; Banert, K., Eds.; John Wiley & Sons, Inc., **2010**, pp. 191–237.
2. For seminal reports, see: (a) Schmidt, K. F. Z. *Angew. Chem.* **1923**, *36*, 511–524. (b) Schmidt, K. F. Z. *Ber. Dtsch. Chem. Ges.* **1924**, *57*, 704–706.
3. (a) Briggs, L. H.; De Ath, G. C.; Ellis, S. R. *J. Chem. Soc.* **1942**, 61–63. (b) Smith, P. A. S. *J. Am. Chem. Soc.* **1948**, *70*, 320–323.
4. Boyer, J. H.; Hamer, J. *J. Am. Chem. Soc.* **1955**, *77*, 951–954.
5. Boyer, J. H.; Canter, F. C.; Hamer, J.; Putney, R. K. *J. Am. Chem. Soc.* **1956**, *78*, 325–327.
6. Aubé, J.; Milligan, G. L. *J. Am. Chem. Soc.* **1991**, *113*, 8966–8967.
7. (a) Gutierrez, O.; Aubé, J.; Tantillo, D. J. *J. Org. Chem.* **2012**, *77*, 640–647. (b) Liu, R.; Gutierrez, O.; Tantillo, D. J.; Aubé, J. *J. Am. Chem. Soc.* **2012**, *134*, 6528–6531.
8. Motiwala, H. F.; Fehl, C.; Li, S.-W.; Hirt, E.; Porubsky, P.; Aubé, J. *J. Am. Chem. Soc.* **2013**, *135*, 9000–9009.
9. Desai, P.; Schildknedt, K.; Agrios, K. A.; Mossman, C.; Milligan, G. L.; Aubé, J. *J. Am. Chem. Soc.* **2000**, *122*, 7226–7232.
10. Gracias, V.; Frank, K. E.; Milligan, G. L.; Aubé, J. *Tetrahedron* **1997**, *53*, 16241–16252.
11. Smith, B. T.; Gracias, V.; Aubé, J. *J. Org. Chem.* **2000**, *65*, 3771–3774.
12. For a recent example, see: Ribelin, T.; Katz, C. E.; English, D. G.; Smith, S.; Manukyan, A. K.; Day, V. W.; Neuenswander, B.; Poutsma, J. L.; Aubé, J. *Angew. Chem. Int. Ed.* **2008**, *47*, 6233–6235.
13. Golden, J. E.; Aubé, J. *Angew. Chem. Int. Ed.* **2002**, *41*, 4316–4318.
14. Su, B.; Chen, F.; Wang, Q. *J. Org. Chem.* **2013**, *78*, 2775–2779.
15. Lee, H.-L.; Aubé, J. *Tetrahedron* **2007**, *63*, 9007–9015.
16. Pearson, W. H.; Schkeryantz, J. M. *Tetrahedron Lett.* **1992**, *33*, 5291–5294.
17. Pearson, W. H.; Hutta, D. A.; Fang, W.-K. *J. Org. Chem.* **2000**, *65*, 8326–8332.
18. Pearson, W. H.; Walavalkar, R.; Schkeryantz, J. M.; Fang, W.-K.; Blickensdorf, J. D. *J. Am. Chem. Soc.* **1993**, *115*, 10183–10194.
19. Kapat, A.; Nyfeler, E.; Giuffredi, G. T.; Renaud, P. *J. Am. Chem. Soc.* **2009**, *131*, 17746–17747.
20. Gigant, N.; Gillaizeau, I. *Org. Lett.* **2012**, *14*, 4622–4625.
21. Gorin, D. J.; Davis, N. R.; Toste, F. D. *J. Am. Chem. Soc.* **2005**, *127*, 11260–11261.
22. Liu, H.; Li, X.; Chen, Z.; Hu, W.-X. *J. Org. Chem.* **2012**, *77*, 5184–5190.
23. Dash, C.; Yousufuddin, M.; Cundari, T. R.; Dias, H. V. R. *J. Am. Chem. Soc.* **2013**, *135*, 15479–15488.
24. (a) Reddy, P. G.; Varghese, B.; Baskaran, S. *Org. Lett.* **2003**, *5*, 583–585. (b) Lang, S.; Kennedy, A. R.; Murphy, J. A.; Payne, A. H. *Org. Lett.* **2003**, *5*, 3655–3658.
25. Zeng, Y.; Aubé, J. *J. Am. Chem. Soc.* **2005**, *127*, 15712–15713.
26. For recent reviews, see: (a) Bräse, S.; Gil, C.; Knepper, K.; Zimmermann, V. *Angew. Chem. Int. Ed.* **2005**, *44*, 5188–5240. (b) Rojas, C. M. In Name Reactions for Homologations; Li, J. J.; Ed.; John Wiley & Sons, Inc., **2009**, pp. 136–163.
27. Curtius, T. *Ber. Dtsch. Chem. Ges.* **1890**, *23*, 3023–3033.

28. Shimokawa, J.; Harada, T.; Yokoshima, S.; Fukuyama, T. *J. Am. Chem. Soc.* **2011**, *133*, 17634–17637.
29. Jiang, X., Liu, B.; Lebreton, S.; De Brabander, J. K. *J. Am. Chem. Soc.* **2007**, *129*, 6386–6387.
30. Shioiri, T.; Ninomiya, K.; Yamada, S. *J. Am. Chem. Soc.* **1972**, *94*, 6203–6205.
31. Weinstock, J. *J. Org. Chem.* **1961**, *26*, 3511.
32. Amarante, G. W.; Cavallaro, M.; Coelho, F. *Tetrahedron Lett.* **2010**, *51*, 2597–2599.
33. Lebel, H.; Leogane, O. *Org. Lett.* **2005**, *7*, 4107–4010.
34. Lebel, H.; Leogane, O. *Org. Lett.* **2006**, *8*, 5717–5720.
35. Ishikawa, H.; Bondzic, B. P.; Hayashi, Y. *Eur. J. Org. Chem.* **2011**, 6020–6031.
36. Marinescu, L.; Thinggaard, J.; Thomsen, I. B.; Bols, M. *J. Org. Chem.* **2003**, *68*, 9453–9455.
37. Hofmann, A. W. *Ber. Dtsch. Chem. Ges.* **1881**, *14*, 2725–2736.
38. Crane, Z. D.; Nichols, P. J.; Sammakia, T.; Stengel, P. J. *J. Org. Chem.* **2011**, *76*, 277–280.
39. Lossen, W. *Liebigs Ann. Chem.* **1872**, *175*, 271.
40. Yale, H. L. *Chem. Rev.* **1943**, *33*, 209–256.
41. Bauer, L.; Miarka, S. V. *J. Org. Chem.* **1959**, *24*, 1293–1296.
42. Marzoni, G.; Varney, M. D. *Org. Process Res. Dev.* **1997**, *1*, 81–84.
43. Kreye, O.; Wald, S.; Meier, M. A. R. *Adv. Synth. Catal.* **2013**, *355*, 81–86.
44. Dubé, P.; Nathel, N. F. F.; Vetelino, M.; Couturier, M.; Abossafy, C. L.; Pichette, S.; Jorgensen, M. L.; Hardink, M. *Org. Lett.* **2009**, *11*, 5622–5625.
45. Walter, W.; Schaumann, E. *Synthesis* **1971**, *3*, 111–130.
46. Yagupolskii, L. M.; Shelyazhenko, S. V.; Maletina, I. I.; Sokolenko, L. V.; Chernega, A. N.; Rusanov, E. B.; Tsymbal, I. F. *J. Fluorine Chem.* **2007**, *128*, 515–523.
47. Han, S.; Xue, Z.; Wang, Z.; Wen, T. B. *Chem. Commun.* **2010**, *46*, 8413–8415.
48. Needs, P. W.; Rigby, N. M.; Ring, S. G.; MacDougall, A. J. *Carbohydrate Res.* **2001**, *333*, 47–58.
49. Deng, B.-L.; Cullen, M. D.; Zhou, Z.; Hartman, T. L.; Buckheit, R. W.; Pannecouque, C.; De Clercq, E.; Fanwick, P. E.; Cushman, M. *Bioorg. Med. Chem.* **2006**, *14*, 2366–2374.

5

BECKMANN REARRANGEMENTS AND FRAGMENTATIONS IN ORGANIC SYNTHESIS

Trinh A. D. Holth

University of Minnesota, Department of Medicinal Chemistry and Institute for Therapeutics Discovery and Development, Minneapolis, MN, USA

Oliver E. Hutt

CSIRO, Industrial Chemistry and Biotechnology, Clayton South, Australia

Gunda I. Georg

University of Minnesota, Department of Medicinal Chemistry and Institute for Therapeutics Discovery and Development, Minneapolis, MN, USA

5.1 INTRODUCTION

Since its discovery in 1886 by Beckmann,[1] the eponymous rearrangement reaction has served as a useful synthetic tool to incorporate a nitrogen atom into a molecule through the modification of a ketone or an aldehyde. As such, this rearrangement remains an important method to convert oximes in open chain and cyclic systems to the corresponding amides and lactams. In specific cases, a Beckmann fragmentation can take place to form a nitrile group. Due to their versatility and wide synthetic scope, the Beckmann reactions have been extensively reviewed.[2] Given these previous reviews, this chapter will only briefly cover the different transformations, provide some important historical examples of the utility of Beckmann reactions in the total synthesis of natural products, and then examine novel strategic applications and their

Molecular Rearrangements in Organic Synthesis, First Edition. Edited by Christian M. Rojas.

synthetic utility with a focus on those examples from the past decade until February 2014.

5.1.1 Beckmann Rearrangement

Of the possible Beckmann transformations (Scheme 5.1), the Beckmann rearrangement is the best known and, in general, involves the conversion of a ketone or aldehyde **1** to an oxime **2** (Scheme 5.1). The oxime **2** is then converted to an amide **3** under acidic and dehydrating conditions, whereby the carbon bond anti to the oxime hydroxyl group migrates to the nitrogen. Beckmann rearrangement of aldoximes (where R^1 or $R^2 = H$) produces primary amides via hydrogen migration in a non-stereospecific manner, while acyclic ketoximes produce secondary amides, and cyclic ketoximes yield lactams.

While the classic Beckmann reaction requires a strong acid to promote the rearrangement, recent efforts to use milder, greener, and nonacidic conditions for this transformation have led to the use of other reagents[2b,3] including Lewis acids, solid catalysts, ionic liquids,[4] and organocatalysts.[5] Oxime activation begins by transforming the *N*-hydroxyl group of the oxime into a better leaving group through reaction with an acid, esterification, or etherification. As shown in Scheme 5.2, the hydroxyl group of oxime **2** is converted into a better leaving group with an electrophile, subsequently forming oxonium intermediate **4**. The Beckmann rearrangement proceeds via the departure of the leaving group and migration of the substituent situated anti to the nitrogen—oxygen (N—O) bond of **4** to the nitrogen, thereby producing nitrilium ion **5**.[2c] This process occurs to maximize the antiperiplanar electron delocalization of the carbon–carbon (C—C) σ orbital to the antibonding σ^* orbital of the N—O bond.[6] A computational study provided evidence that bond migration and N—O bond scission are likely to occur simultaneously.[7] The nitrilium ion **5** reacts with a nucleophile such as water to form intermediate **6** that is deprotonated to furnish imidic acid **7**, which tautomerizes to amide **8**.

The oxime can exist in two distinct isomeric forms: the (*E*)- or anti-isomer and the (*Z*)- or syn-isomer (Scheme 5.3). The (*E*)-isomer **9** will rearrange to give the *N*-isopropylamide, **10** while the (*Z*)-isomer **9** will yield the *N*-methylamide **11** under normal Beckmann reaction conditions with aluminum oxide.[8] In some reactions,

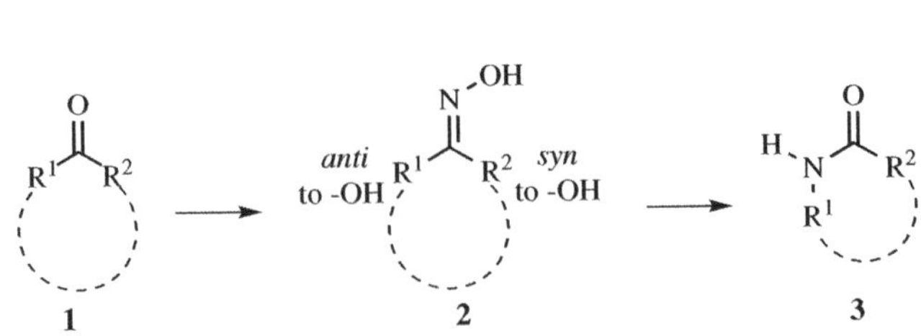

R^1 or R^2 = H, alkyl, or aryl
$R^1 \neq R^2$ when R^1 = H

Scheme 5.1 General mechanism for the Beckmann rearrangement.

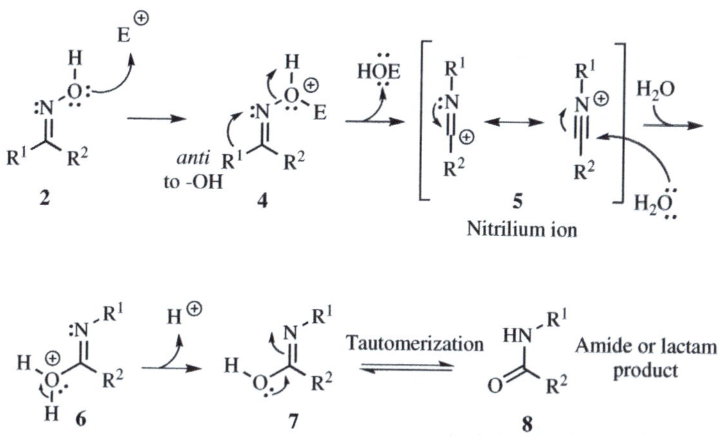

Scheme 5.2 Beckmann rearrangement mechanism.

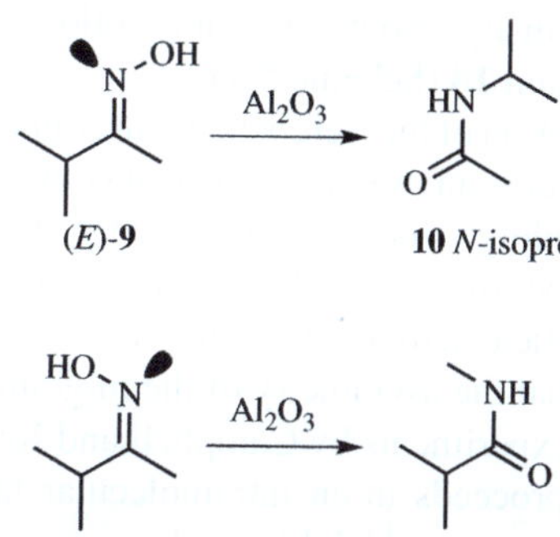

Scheme 5.3 Beckmann rearrangement of geometric oxime isomers.

especially those carried out under acidic conditions and in protic solvents, interconversion of the two isomers can take place, which accounts for non-stereospecific conversions that can be observed. When both alkyl groups can migrate, the amide formed in higher proportion is derived from the migration of the more bulky group to impart some regioselectivity.[2b]

In general, while the carbon bond anti to the leaving group on the nitrogen atom migrates, there are exceptions. In a recent study by Ohwada *et al.*, it was found (Scheme 5.4) that the presence of a nucleophilic halogen, specifically bromine or chlorine in the peri-position as in **12**, afforded the unexpected regioisomeric lactam **15**.[9] This syn-migration is due to a nucleophilic monovalent halogen atom participating in the formation of imino-halonium cation **13**. Transition state calculations showed that the syn-migration pathway resembles an S_N2 reaction and that the halogen atom can stabilize the imino cation by electron donation. In contrast, if a hydrogen

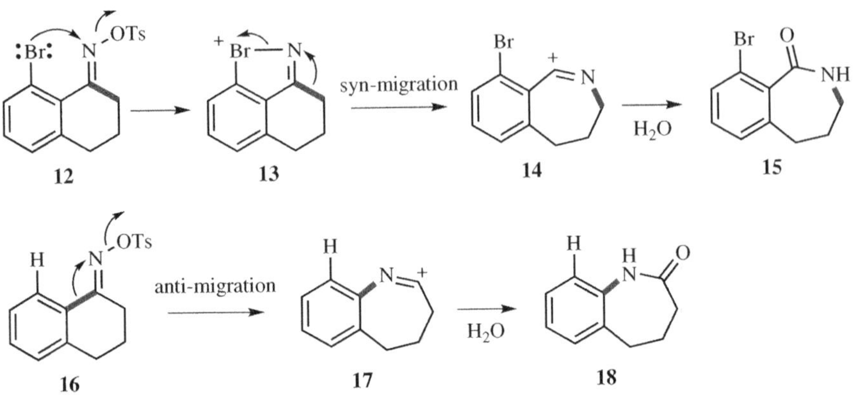

Scheme 5.4 Anti- versus syn-migration of tetralone oximes.

is in the *peri*-position, the normal Beckmann reaction takes place and anti migration of oxime tosylate **16** results in aryl group migration, followed by addition of water to intermediate **17** to form lactam **18** (Scheme 5.4).

In their classical investigation of the stereochemical course of the Beckmann transformation, Kenyon and Young analyzed the rearrangement of an optically active ketoxime into its corresponding amide (Scheme 5.5).[10] They treated the dextrorotatory oxime **19** with phosphorus pentachloride to afford the dextrorotatory amide **20**.[11] They compared the optical activity of the amide product with its corresponding DL-amide and determined that the asymmetry of the migrating group was preserved. This work and subsequent experiments by Campbell and Kenyon confirmed that the Beckmann rearrangement proceeds in an intramolecular fashion with retention of configuration of the migrating group.[12] Additional experiments by Hill *et al.* demonstrated retention of stereochemistry of the Beckmann transformation on a system with a trisubstituted stereogenic carbon (Scheme 5.5).[13] *cis*-Decalin oxime **21** was treated with *p*-toluenesulfonyl chloride in pyridine to yield *cis*-amide **22** in high yield. Reaction with PCl_5 in chloroform, however, led to Beckmann fragmentation (formation of acetonitrile, see also Scheme 5.7).

Vanos and Lambert were the first to provide experimental evidence that the cyclopropenium-activated Beckmann rearrangement is self-propagating rather than an organocatalytic process.[14] Subsequently, using computational investigations (DFT calculations), Eriksson and coworkers demonstrated that in the case of certain organo-mediated Beckmann rearrangements (3,3-dichloro-1,2-diphenylcyclopropene, TsCl, cyanuric chloride, and other reagents), a self-propagating mechanism is energetically favored over other mechanisms (Scheme 5.6).[15] Rather than catalyzing the reaction, the chloride reagent **24** only initiates the reaction with oxime **23** to yield oxime complex **25** that subsequently forms nitrilium ion **26**.[14] Next, another molecule of oxime **23** adds to nitrilium ion **26** to form dimer-like structure **27** that produces the rearrangement product **28** while simultaneously regenerating nitrilium ion **26**.

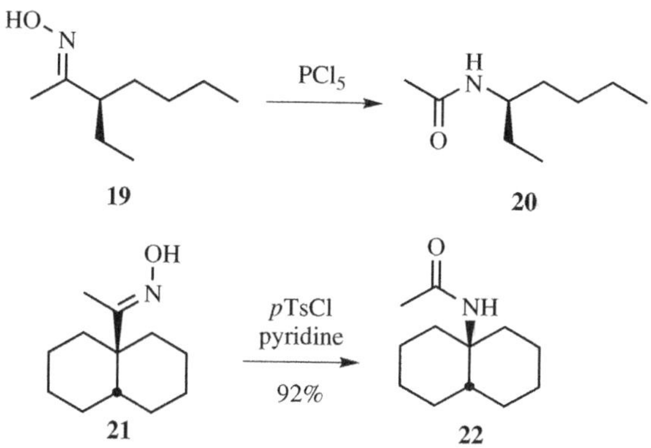

Scheme 5.5 Retention of stereochemistry during Beckmann rearrangement.

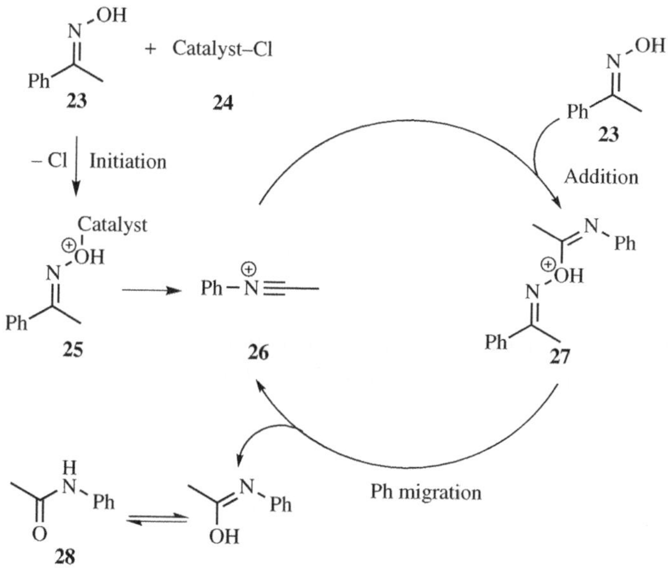

Scheme 5.6 Proposed self-initiation mechanism during Beckmann rearrangement.

5.1.2 Beckmann Fragmentation

First discovered by Wallach in 1889, the Beckmann fragmentation forms a nitrile product instead of an amide (Scheme 5.7). Sometimes called the abnormal or second-order Beckmann rearrangement, or Beckmann fission, this reaction begins like the Beckmann rearrangement with oximes **2** derived from linear or cyclic ketones **1**. The fragmentation occurs when the C—C α bond breaks after shifting to the nitrogen, which results in the formation of carbocation **29** and nitrile **30**

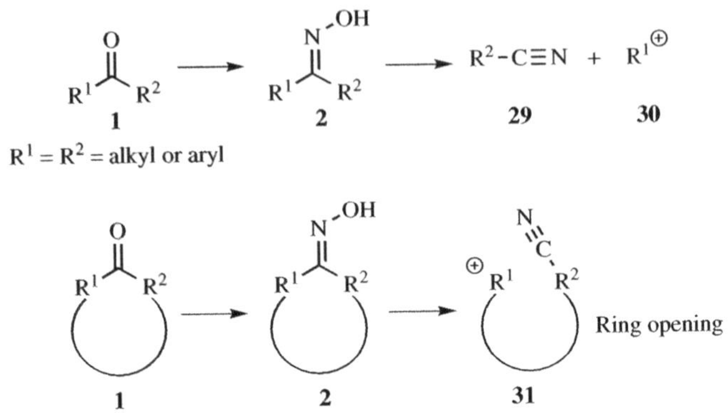

Scheme 5.7 Beckmann fragmentation.

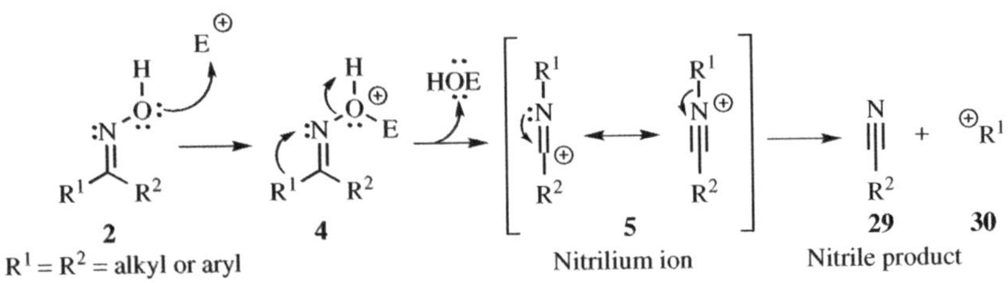

Scheme 5.8 Beckmann fragmentation mechanism.

(Scheme 5.8). In a cyclic system, the fragmentation causes the ring of cyclic oxime **2** to open to generate the noncyclic nitrile product **31**.

Whether Beckmann rearrangement or fragmentation takes place is wholly dependent on the exact reaction conditions employed. For example, reaction of oxime **21** with *p*-toluenesulfonyl chloride in pyridine furnished amide **22** (Scheme 5.5), but when PCl_5 was employed, Beckmann fragmentation took place (see also Schemes 5.23, 5.25–5.27). In general, fragmentation can take place when the hydroxyl group of an oxime is anti to a group that can stabilize a positive charge such as a quaternary carbon or a heteroatom. After activation of oxime **2** by the addition of an electrophile, oxonium ion **4** is formed. The Beckmann rearrangement generates nitrilium ion **5**, which releases carbocation **30** and nitrile **29** (Scheme 5.8).

Neighboring groups with lone pairs such as nitrogen, oxygen, or sulfur that are also able to stabilize a positive charge can assist the Beckmann fragmentation reaction as shown in Scheme 5.9. This type of Beckmann fragmentation was instrumental in determining the structure of the morphine alkaloids. The Beckmann fragmentation provided evidence that a piperidine ring is connected to carbon 13 (cf. **32**, Scheme 5.9) and ruled out a C5-connected azepine ring structure **33**. Thus, upon treatment with thionyl chloride, dihydrocodeine-derived oximes **32** and **33** would afford either aldehyde **34** or ketone **35**, respectively.[16] Formation of aldehyde **34** confirmed that an

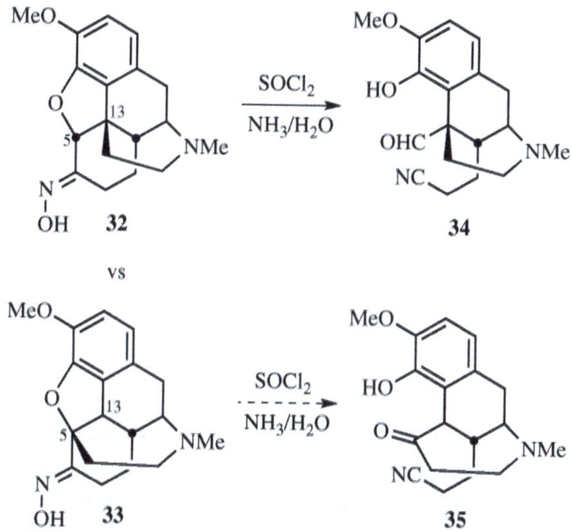

Scheme 5.9 Neighboring group assistance in the determination of morphine alkaloid structure.

N-methyl 6-membered piperidine ring with attachment at the C13 position was the correct structure. The formation of ketone **35** would have confirmed the presence of an *N*-methyl seven-membered azepine ring that is connected to C5.

5.1.3 Photo-Beckmann Rearrangement

In 1963, Mayo and coworkers reported that the direct irradiation of an aldoxime **2**, derived from an aldehyde **1**, formed amide **37** through oxaziridine intermediate **36**, and thus concluded that this photochemical reaction proceeds through an intramolecular oxygen migration (Scheme 5.10).[17] The reaction mechanism was later determined through studies of ^{18}O-labeled oximes.[18] This reaction is called the photochemical Beckmann rearrangement[19] and, in contrast to the typical Beckmann rearrangement, is not accompanied by the corresponding α-fission to produce nitrile products (i.e., fragmentation pathway).

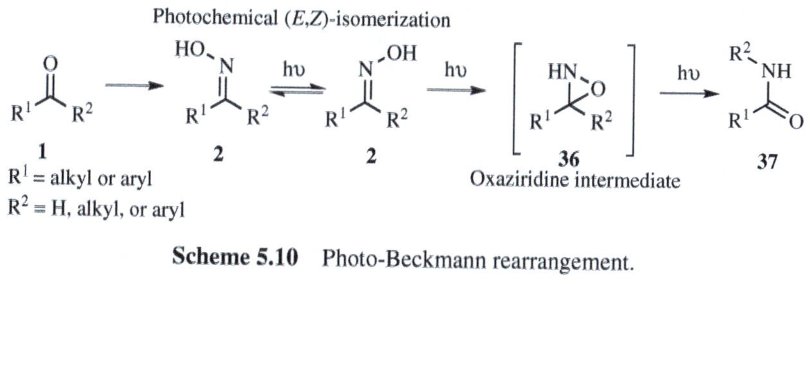

Scheme 5.10 Photo-Beckmann rearrangement.

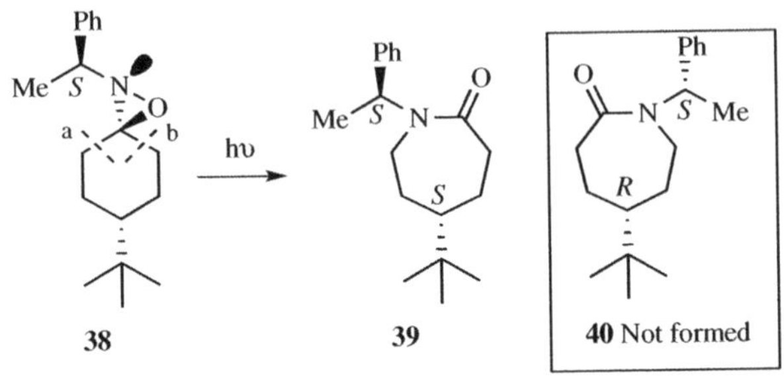

Scheme 5.11 Photo-Beckmann oxaziridine rearrangement mechanism.

The photo-Beckmann reaction provides the regioisomeric amide compared to the amide typically obtained from the traditional acid-catalyzed Beckmann rearrangement in which the carbon bond anti to the hydroxyl group of the oxime migrates. Instead, the photo-Beckmann reaction is under stereoelectronic control, whereby the group anti to the lone pair of the oxaziridine migrates to the nitrogen. While physical evidence to show the formation of oxaziridines has been unsuccessful, empirical evidence suggests that the photo-Beckmann rearrangement involves the photochemical (*E,Z*)-isomerization of oximes, followed by the transformations of the excited singlet (*E*)- and (*Z*)-oximes into oxaziridine intermediates before reorganization of the resulting singlet excited oxaziridines to the amides in a fully concerted manner.[20] The stereoelectronic control was unambiguously verified by photo-Beckmann reaction with oxaziridine **38** in which both bonds adjacent the oxaziridine are chemically equivalent, thus ruling out steric effects that could influence the reaction (Scheme 5.11).[21] Oxaziridine **38** was prepared from the corresponding chiral imine by oxidation with *m*-chloroperoxybenzoic acid. The configurations of oxaziridine **38** and lactam **39** were determined by X-ray analysis and NMR spectroscopy to confirm that the position of the nitrogen lone pair is anti to bond "a." Thus, oxaziridine **38** generated amide **39** by breaking bond "a" and not bond "b," and lactam **40** did not form.

5.2 STRATEGIC PLANNING: A HISTORICAL PERSPECTIVE

Incorporation of the Beckmann reaction during retrosynthetic analysis can be a powerful method for planning carbon–nitrogen bond installation or ring expansion. From an aldehyde or ketone precursor, the Beckmann reaction typically affords an amide functional group or a lactam if the precursor is a cyclic ketone. Lactam formation leads to ring expansion and can therefore be used as a strategy in complex natural product synthesis. In cases where the hydroxyl group of the oxime is anti to a quaternary carbon, nitrile formation can result from a Beckmann fragmentation. If the quaternary carbon is part of a cyclic structure, this causes a ring opening effect. In addition, natural products with ketone functional groups such as camphor, fenchone,

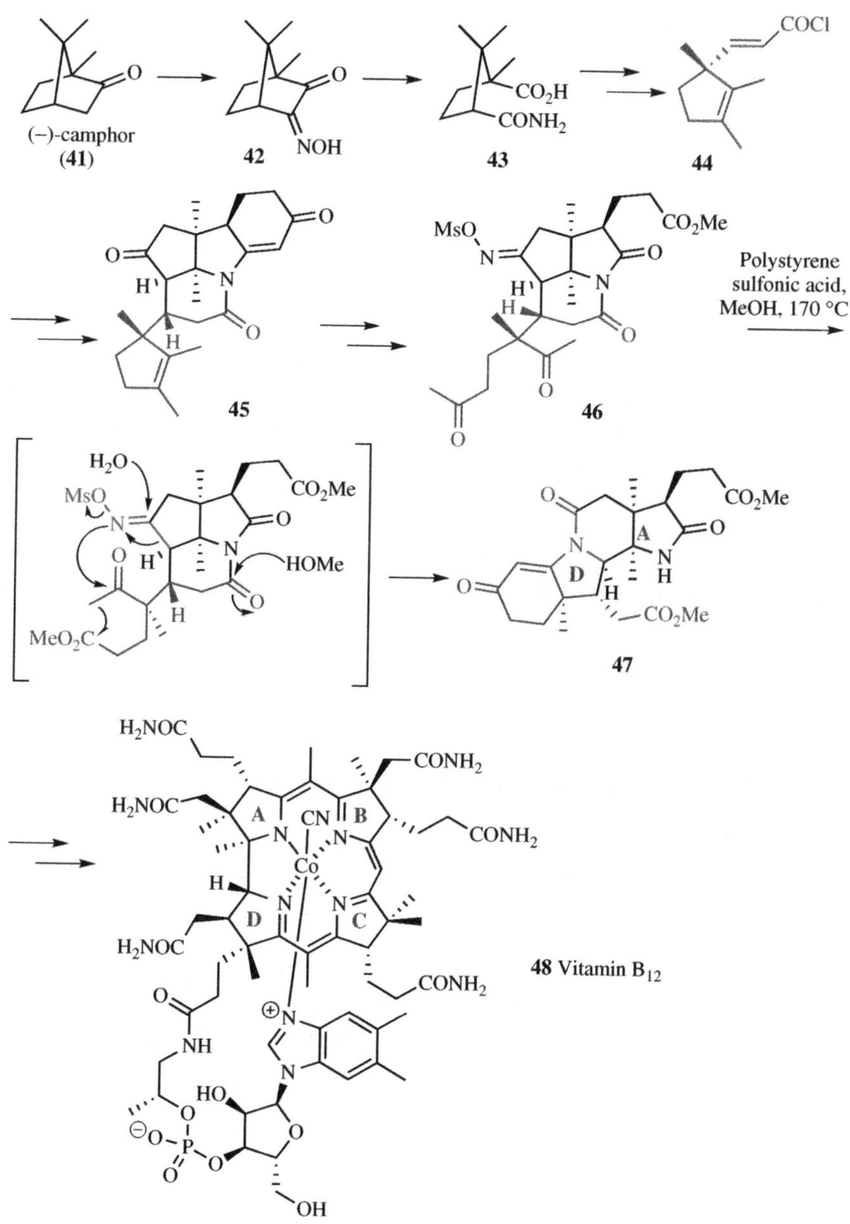

Scheme 5.12 Total synthesis of vitamin B_{12}.

and related derivatives can be converted into chiral nonracemic nitrogen-containing synthetic building blocks by utilizing Beckmann reactions.

In one of the classic examples in natural product synthesis, Woodward, Eschenmoser, and coworkers demonstrated the powerful chemistry of the Beckmann reaction in the synthesis of vitamin B_{12} (**48**, Scheme 5.12).[22] The Beckmann rearrangement had a key role in establishing the D-ring of the complex structure. Acid chloride precursor **44** was constructed from (−)-camphor (**41**)[23] by α-oxidation[24] followed by oxime formation at the less hindered ketone[25] to furnish **42**; the direct α-oximation of **41** with amyl nitrite to form **42** has also been reported.[26] The Beckmann rearrangement of oxime **42** followed by hydrolysis afforded amide **43**,[27] which after additional steps produced key intermediate **44**.[22a] Subsequent transformations led to ketone intermediate **45** that was converted to *O*-methylsulfonyl oxime **46** for a late-stage Beckmann rearrangement. Upon treatment with polystyrene sulfonic acid in methanol heated to 170 °C, the oxime mesylate **46** underwent Beckmann rearrangement and formed the six-membered lactam ring, which then participated in a second transformation to form the D-ring through a dehydrative cyclization to generate the highly functionalized 2,2′-bipyrrolidine ring system **47**. This part of the synthesis provided rings A and D of the vitamin B_{12} structure and reveals the utility of the Beckmann reaction in complex natural product total synthesis.

Another classical example is Stork's 1972 total synthesis of racemic byssochlamic acid (**51**), which constitutes the first application of a Beckmann fragmentation in the total synthesis of a complex natural product (Scheme 5.13). Bicyclic oxime **49** underwent Beckmann fragmentation to generate the nine-membered intermediate **50**, which was converted to (±)-byssochlamic acid (**51**) after ethyl

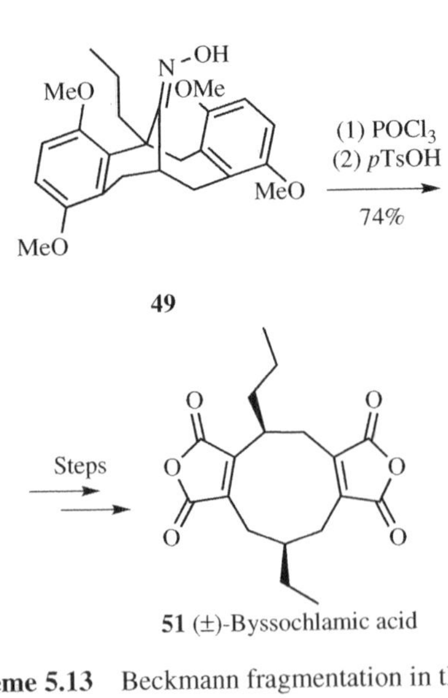

49 **50** **51** (±)-Byssochlamic acid

Scheme 5.13 Beckmann fragmentation in the synthesis of byssochlamic acid.

substituent formation and conversion of the benzene rings to the two maleic anhydrides.[28]

5.3 RECENT APPLICATIONS TOWARD THE SYNTHESIS OF NATURAL PRODUCTS

Natural products are derived from plant, marine, and microbial sources and often possess biological activity. They are a source of new drug products each year and continue to play a significant role in the drug discovery and development process.[29] Herein, we will discuss examples of recent natural product synthesis that utilize the previously discussed Beckmann reactions: rearrangement, fragmentation, and photo-Beckmann.

5.3.1 Toward the Total Synthesis of Pinnaic Acid and Halichlorine

The Beckmann rearrangement reaction was a key step in the total synthesis of two marine natural products, pinnaic acid (**52**) and halichlorine (**53**) (Figure 5.1).[30] Pinnaic acid and halichlorine are of interest because they exhibit anti-inflammatory effects through inhibition of a cytosolic 85 kDa phospholipase ($cPLA_2$) *in vitro*.[31] Halichlorine is also a promising antihypertensive agent due to its inhibition of L-type Ca^{2+} channels in vascular smooth muscle cells.[32]

Both natural products contain an interesting azaspirodecane core structure and have been subjects of multiple total syntheses.[33] A recent synthesis by Arimoto and coworkers featured a regioselective Beckmann rearrangement of oxime **55**, derived from ketone **54**, to install the nitrogen atom of the spirocycle (Scheme 5.14).[30] Initially, a standard two-step Beckmann procedure was employed with oxime formation using hydroxylamine, followed by Beckmann rearrangement to provide the lactams **57** and **58** in 84% yield over two steps but in a 3:1 ratio of lactams **57** and **58**, respectively (Scheme 5.14, conditions "a"). This outcome reflected the modest selectivity in the formation of the (*E*)- and (*Z*)-configured oximes **55** and **56**, which after Beckmann rearrangement furnished amides **57** and **58**, respectively, in a stereospecific manner. Arimoto exploited the bias for (*E*)-oxime formation by employing the sterically hindered hydroxylamine reagent, *O*-mesitylenesulfonylhydroxylamine (MSH),

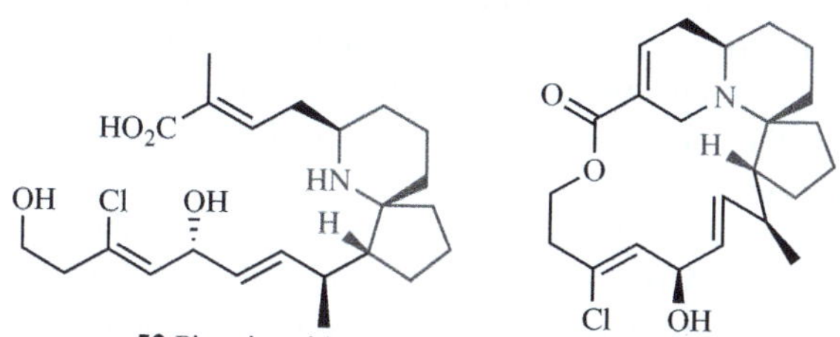

52 Pinnaic acid **53** Halichlorine

FIGURE 5.1 Structures of pinnaic acid and halichlorine

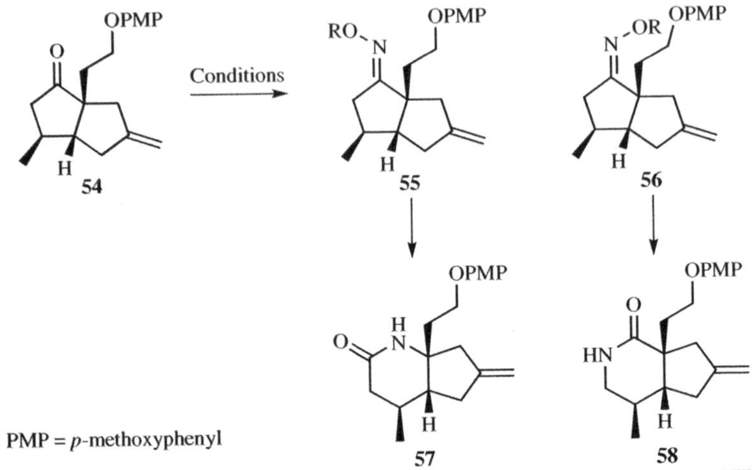

Conditions	Yields
(a) i. $NH_2OH{\cdot}HCl$ (5 equiv.), DABCO (1.1 equiv.), MeOH, rt ii. *p*TsCl (1.2 equiv.), NaOH (aq.), acetone, reflux	84%, over 2 steps (**57**:**58** = 3:1)
(b) MSH (1.5 equiv., solid), CH_2Cl_2, rt; then silica gel	85%, exclusively **57**

DABCO = 1,4-diazabicyclo[2.2.2]octane, MSH = *O*-mesitylenesulfonylhydroxylamine, Ts = toluenesulfonyl

Scheme 5.14 Beckmann reaction optimization.

to convert the hindered ketone **54** into the desired lactam **57** in 85% yield on 5 g scale. This outcome is a consequence of the enhanced selectivity for (*E*)-oxime formation, as the corresponding (*Z*)-oxime does not form due to the steric bulk of the mesitylenesulfonyl moiety. Thus, Arimoto *et al.* were able to efficiently control the regioselectivity of the reaction. The addition of MSH as a solid to the ketone was important to avoid degradation of MSH.

The optimized Beckmann rearrangement was used to install the amine functionality at the quaternary center of the azaspirodecane core **61** of pinnaic acid and halichlorine (Scheme 5.15). Thus, after ketone **54** was converted to the lactam **57**, subsequent reductive cleavage and amine protection gave alcohol **59**. Elaboration of the PMP-protected hydroxyethyl side chain of **59** provided enone **60**, which smoothly underwent a reductive amination to yield the azaspirodecane core **61**. This advanced intermediate was then elaborated into the natural products pinnaic acid (**52**) and halichlorine (**53**).

Related chemistry was reported by Saeki and Toyota for the racemic synthesis of natural product *cis*-195A (**64**), an alkaloid that has been isolated from dendrobatid frog skin extracts (Scheme 5.16).[34] A 1:1 mixture of indanone oxime stereoisomers **62** underwent a regioselective Beckmann reaction to afford lactam **63**. The excellent regiochemistry observed in this reaction must be a result of imine inversion

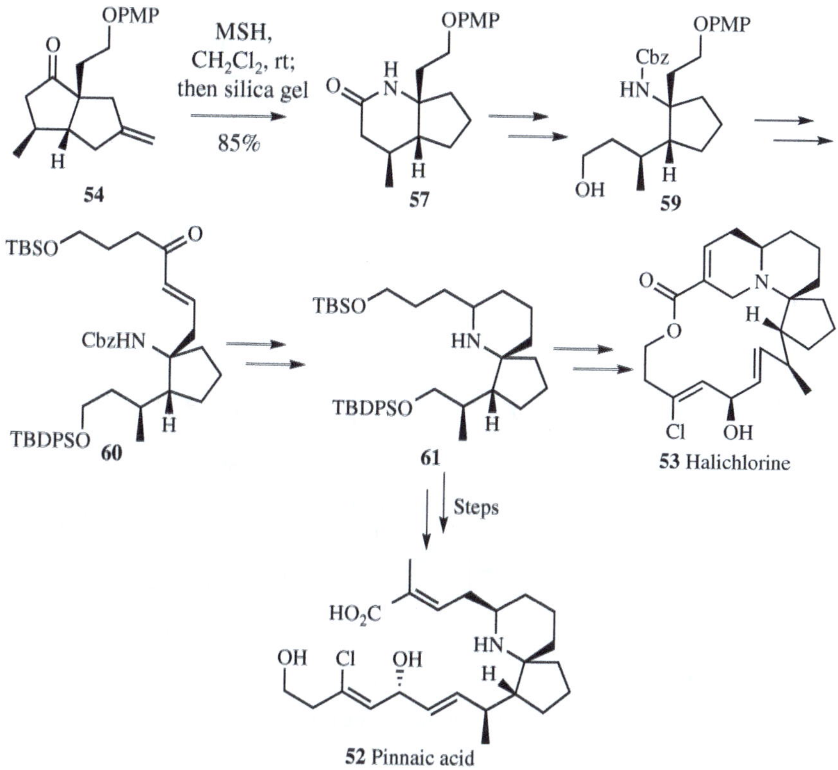

MSH = *O*-mesitylenesulfonylhydroxylamine; Cbz = carboxybenzyl;
PMP = *p*-methoxyphenyl; TBDPS = *tert*-butyldiphenylsilyl

Scheme 5.15 Toward the synthesis of **52** and **53**.

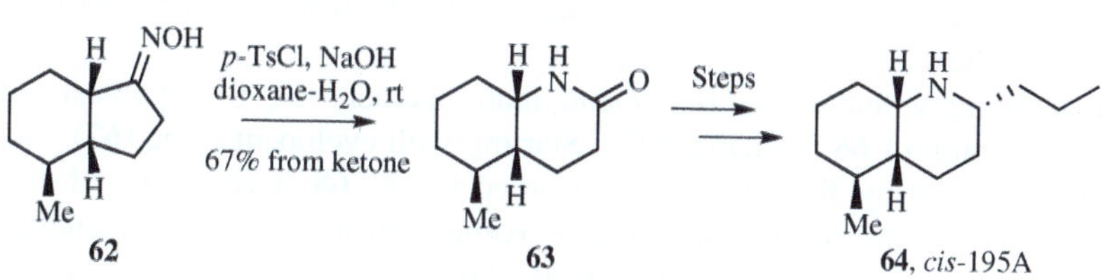

Scheme 5.16 Synthesis of **64**.

under the reaction conditions. The natural product **64** was prepared in three additional steps.

5.3.2 Access to the Core of Mersicarpine

Mersicarpine (**65**, Figure 5.2) is an atypical tetracyclic dihydroindole alkaloid isolated from *Kopsia* plants by Kam *et al.* in 2004.[35] Its structure contains a seven-membered cyclic imine that is fused with both an indoline and a δ-lactam around a tertiary

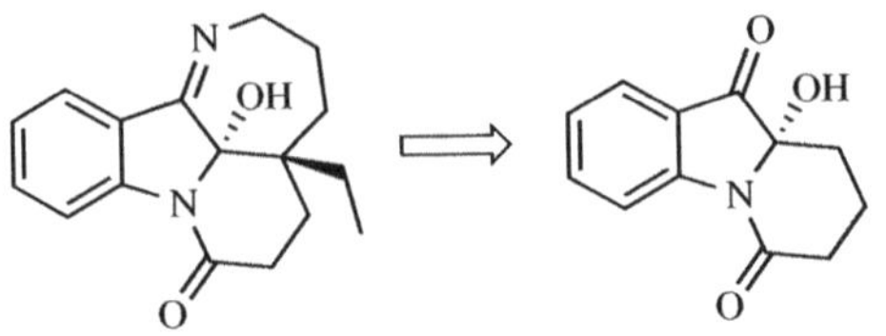

FIGURE 5.2 Mersicarpine and key intermediate **66**

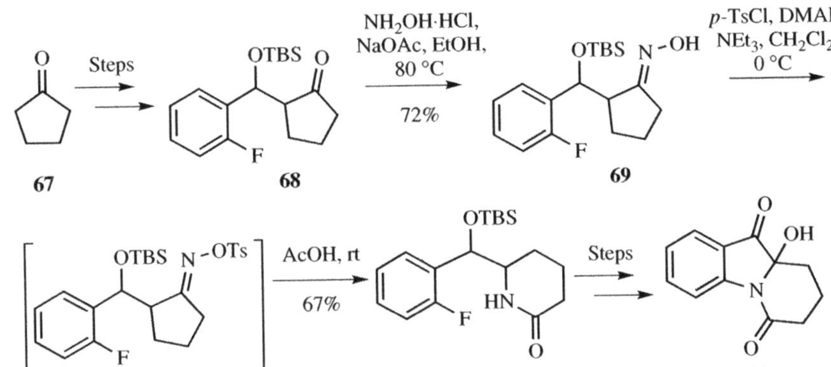

Scheme 5.17 Synthesis of the mersicarpine core structure **66**.

hydroxyl group. Given its unique structural features and the reports of other biologically active indole alkaloids isolated from *Kopsia* plants,[36] mersicarpine (**65**) has been an attractive target for total synthesis.[37] While previous reports all generated the indoline moiety from the indole itself, Li and Liang explored a novel strategy to obtain the dihydroindoline core **66** that could have wider applications in indole alkaloid synthesis.[38]

Their strategy features a Beckmann rearrangement for the synthesis of the δ-lactam portion of **66** (Scheme 5.17). Starting with cyclopentanone (**67**), an aldol reaction followed by TBS protection afforded ketone **68**. Treatment with hydroxylamine gave oxime **69** and subsequent oxime activation with *p*-toluenesulfonyl chloride produced intermediate **70**. The Beckmann rearrangement of intermediate **70** with acetic acid at room temperature proceeded with excellent regioselectivity as a result of its (*E*)-imine geometry to provide δ-lactam **71** in good yield. Deprotection of the hydroxyl group, followed by Dess–Martin oxidation, nucleophilic aromatic substitution, and an enolate autoxidation process generated the mersicarpine core structure **66**.

5.3.3 Formal Synthesis of Gephyrotoxin 287C

Gephyrotoxin is a naturally occurring alkaloid isolated from the skin secretion of a tropical frog (Figure 5.3).[39] Its absolute configuration was first assigned by Daly *et al.*

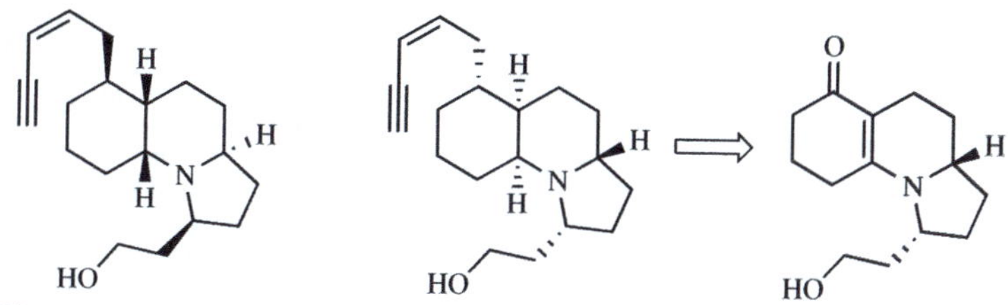

72 (+)-Gephyrotoxin 287C **73** (−)-Gephyrotoxin 287C (natural) **74** Kishi's intermediate

FIGURE 5.3 Structure of gephyrotoxin 287C and Kishi's intermediate **74**

as **72** by X-ray analysis,[39] but Kishi's total synthesis of the natural product established its absolute stereochemistry as **73**.[40] While gephyrotoxin is relatively nontoxic, studies have shown that it displays an array of neurological activities, prompting synthetic efforts[41] including many formal syntheses showcasing abbreviated efforts toward Kishi's intermediate **74**.

Lessard, Spino, and coworkers developed the formation of *N*-heterocycles from cycloalkanones using a photochemical ring contraction reaction coupled with a Beckmann rearrangement ring expansion.[42] Their methodology is an impressive approach to transform disubstituted cycloalkanones into N-heterocycles of the same ring size (Scheme 5.18).[43] The reaction is stereospecific with retention of stereochemistry of the starting material. *cis*-Ketone **75** yielded *cis*-piperidine **78** and *trans*-ketone **75** afforded *trans*-piperidine **78**.[44] The formation of lactam enantiomers *cis*-**76** and *cis*-**77** from the Beckmann rearrangement of *cis*-**75** is inconsequential because upon irradiation of the *N*-chlorolactams, formed by chlorination, and subsequent treatment with methanol and a base, the photochemical ring contraction carbamate *cis*-**78** results. The Beckmann rearrangement of *trans*-**75** yields the lactam intermediate *trans*-**76**/*trans*-**77**, which are identical, and thus furnish the photochemical ring contraction product *trans*-**78**.

This methodology was used for a short and efficient synthesis of Kishi's pyrrolidine intermediate **74** leading to the formal synthesis of the natural product (−)-gephyrotoxin 287C (**73**, Figure 5.3). The initially desired lactam **80** was prepared in one step from bicyclic ketone **79** by direct Beckmann rearrangement with hydroxylamine-*O*-sulfonic acid in formic acid (Scheme 5.19). The lactam **80** was first converted to an *N*-chlorolactam before irradiation with 254 nm UV light to promote the ring contraction, and subsequent steps, similar to those shown in Scheme 5.18, generated benzyl carbamate **81**. Ozonolysis of **81** with reductive workup yielded bis-(2-hydroxyethyl)pyrrolidine **82** that was used to prepare Kishi's intermediate **74**.[40]

Spino and coworkers have also employed an alternative and more efficient two-step approach to amide **80** by reaction of ketone **79** with hydroxylamine to give oxime **83**. Treatment of oxime **83** with *p*-toluenesulfonyl chloride under basic conditions yielded the desired lactam **80** in 55% yield over two steps (Scheme 5.20).

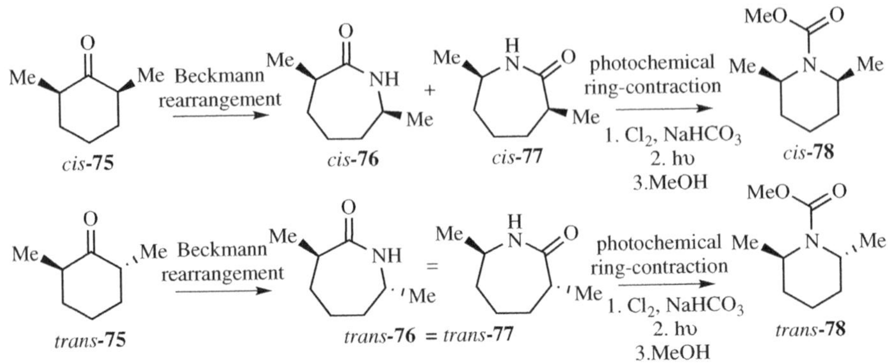

Scheme 5.18 Nitrogen heterocycle formation from Beckmann rearrangement reaction.

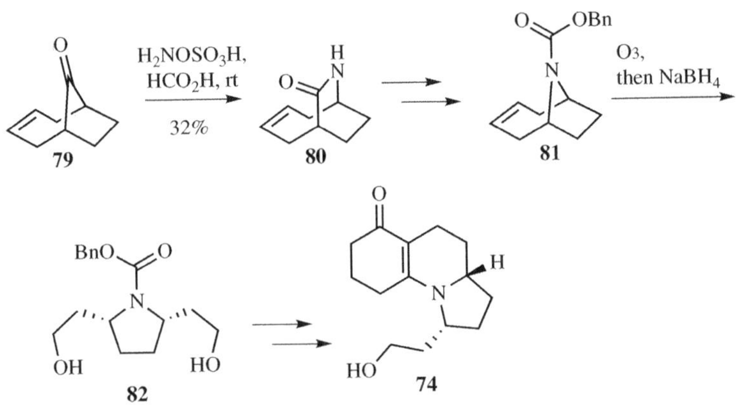

Scheme 5.19 Synthesis of Kishi's intermediate **74**.

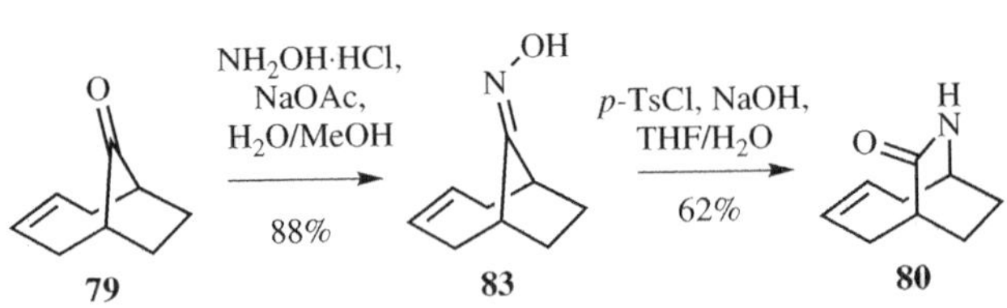

Scheme 5.20 Two-step synthesis of intermediate **80**.

5.3.4 First Asymmetric Total Synthesis of (+)-Sparteine

In 2002, Aubé and coworkers used a photochemical Beckmann rearrangement as a key reaction to install one of the two nitrogens into the tetracyclic structure during the first asymmetric total synthesis of (+)-sparteine (**85**, Figure 5.4).[45] Its

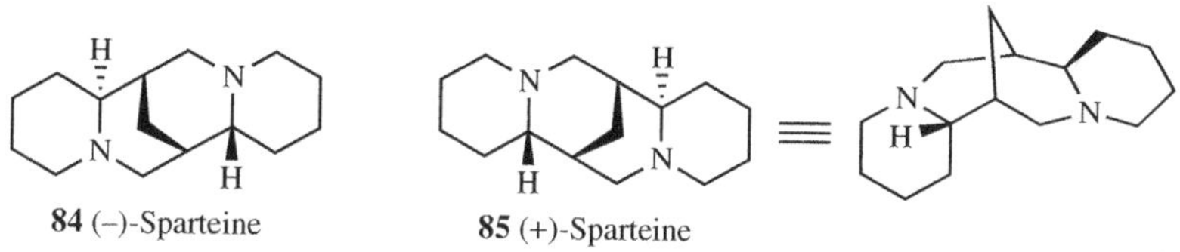

FIGURE 5.4 Structures of sparteines

enantiomer, (−)-sparteine (**84**), is commercially available at reasonable cost because it can be easily isolated from papilionaceous plants, and (−)-sparteine has been a useful chiral diamine in asymmetric synthesis. To prepare the less readily available (+)-sparteine (**85**), Aubé and coworkers initially focused on an intramolecular Schmidt reaction to construct the nitrogen heterocycles. After unsuccessful attempts to effect an intramolecular Schmidt reaction to form the last ring in the multicyclic structure, they turned their attention to the photo-Beckmann rearrangement to complete the synthesis of (+)-sparteine. This is a classic example of the utility of the photo-Beckmann reaction for nitrogen-ring installation.

After installing the first lactam via an intramolecular Schmidt reaction, Aubé and coworkers obtained intermediate **86**, which they elaborated to N,O-bis-protected hydroxylamine **87**. Intermediate **87** was treated with trifluoroacetic acid to remove the Boc protecting groups, and subsequent intramolecular condensation gave the endocyclic nitrone intermediate **88** (Scheme 5.21). Irradiation in benzene at 254 nm then initiated a photo-Beckmann rearrangement of the nitrone to produce oxaziridine intermediate **89**, which subsequently rearranged to lactam **90**.[46] The stereoelectronic effect of the oxaziridine nitrogen promotes the migration of bond antiperiplanar to the nitrogen lone pair (see also Scheme 5.11). Treatment of lactam **90** with lithium aluminum hydride reduced the lactam carbonyl and afforded (+)-sparteine.

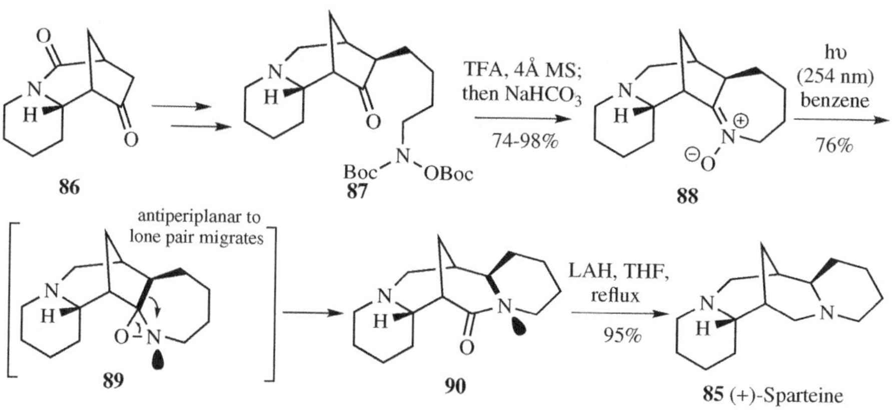

Scheme 5.21 Synthesis of (+)-sparteine (**85**).

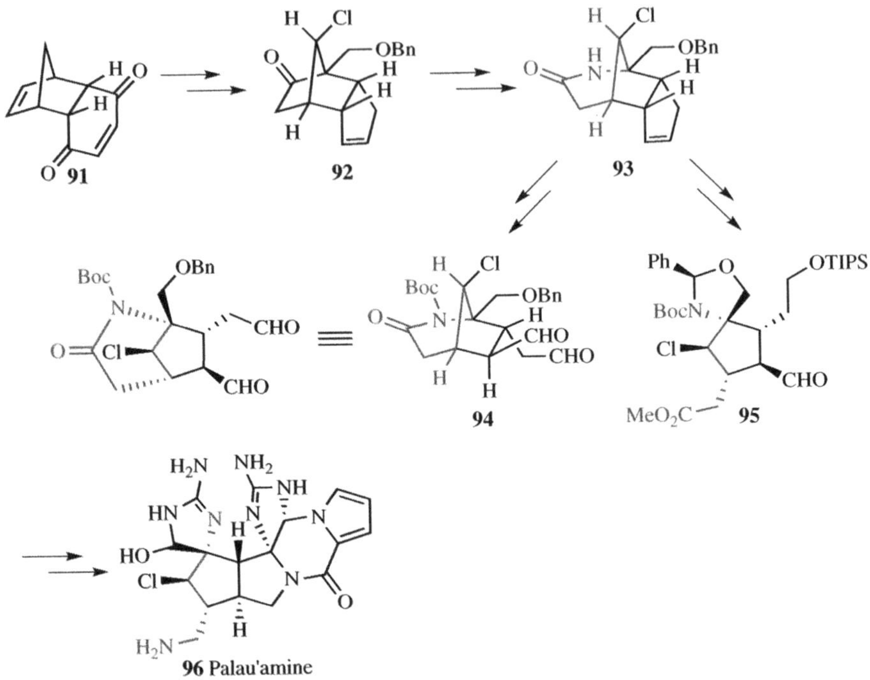

Scheme 5.22 Toward the synthesis of palau'amine (**96**).

5.3.5 Synthesis of the Chloropentane Core of Palau'amine

Gin and coworkers reported the synthesis of the chloropentane scaffold **94** of the cytotoxic and immunosuppressive polycyclic guanidine alkaloid palau'amine (**96**) by employing a Diels–Alder/[3,3]-sigmatropic rearrangement sequence and a Beckmann rearrangement (Scheme 5.22) as key reaction steps.[47] The synthesis started with dihydroquinone **91**, which is readily available from a Diels–Alder reaction between benzoquinone and cyclopentadiene.[48] Dihydroquinone **91** was carried forward in multiple steps, including a [3,3]-sigmatropic rearrangement, producing ketone **92**. After oxime formation, regioselective Beckmann rearrangement provided lactam **93**. From intermediate **93**, the bridged and nonbridged cyclopentanes **94** and **95** were obtained, which are central to the palau'amine structure (**96**).

5.3.6 The Total Synthesis of (−)-Elegansidiol

In their efforts to explore new approaches for the synthesis of oxygenated monocarbocyclic terpenoids, Cao *et al.* utilized the Beckmann fragmentation reaction for the total synthesis of (−)-elegansidiol (**101**, Scheme 5.23).[49] They were able to obtain oxime **98** from the readily available Hajos ketone **97** using known procedures.[50] After different attempts to promote the Beckmann fragmentation, Cao's group found that the Magnus protocol, which utilizes a suspension of phosphorus pentachloride and

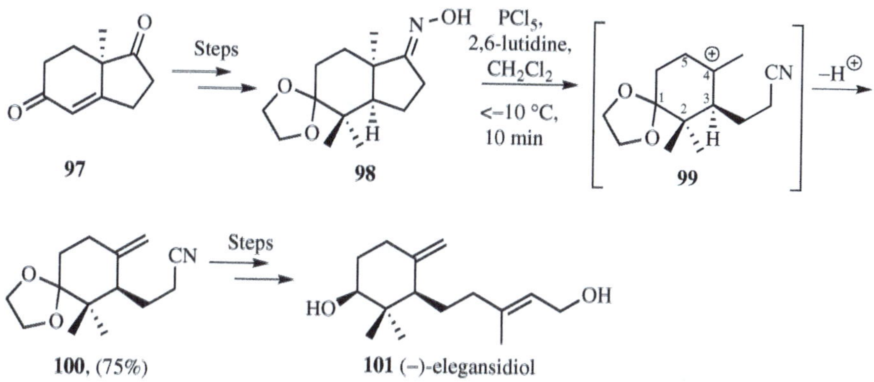

Scheme 5.23 Synthesis of (−)-elegansidiol.

2,6-lutidine[51] in dichloromethane, gave the best regioselectivity for the methylene nitrile product **100** from intermediate **99**. Further elaboration of intermediate **100** afforded the natural product elegansidiol.

5.4 ACCESS TO DIVERSE SCAFFOLDS VIA THE BECKMANN REACTION

5.4.1 Diterpene Hydrocarbons

The following examples show the versatility of the Beckmann reaction in the formation of novel structures by exploring the *ent*-kaurene skeleton of stevioside.

5.4.1.1 Synthesis of* ent-*Kaurene Synthase Inhibitors In their efforts to study the mechanism of terpene synthases, Coates and coworkers synthesized aza derivatives of *ent*-kaurene.[52] The two heterocyclic analogs, 16-aza-trachylobane (**107**) and 16-aza-beyerane (**108**), were synthesized from *ent*-beyeran-16-one (**103**), which was obtained from the natural product stevioside (**102**, Scheme 5.24). These analogs were assessed for their utility as active site probes for various *ent*-diterpene cyclases and as selective inhibitors of gibberellin biosynthesis in plants.

To obtain the aza analogs **107** and **108** from the natural product stevioside (**102**), glycosidic bond hydrolysis was followed by subsequent transformations to afford *ent*-beyeran-16-one (**103**). Oxime formation with hydroxylamine gave **104**, and a subsequent Beckmann fragmentation reaction using tosyl chloride accomplished cleavage of the D-ring of *ent*-beyeranone to produce nitriles **105** and **106**. From nitrile **105**, further reactions yielded aza-trachylobane (**107**) and aza-beyerane (**108**).

5.4.1.2 Skeletally Diverse and Stereochemically Complex Library Templates Our own group has applied a diversity-oriented approach to synthesize multiple templates from steviol and isosteviol, the aglycones derived from the natural product

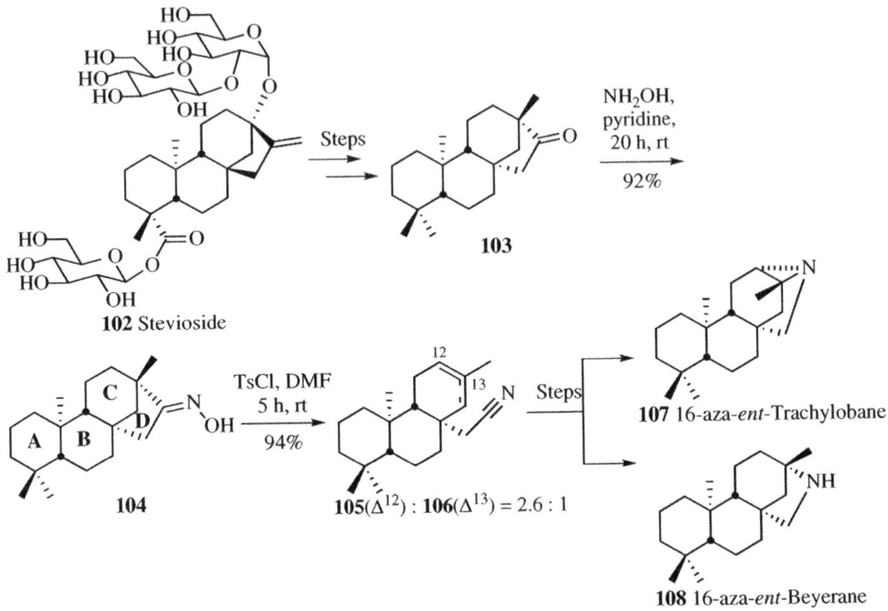

Scheme 5.24 Synthesis of **107** and **108** from stevioside.

stevioside (**102**, Scheme 5.24).[53] Stevioside is a complex natural product that is used as a sweetener and is readily available in kilogram quantities. Therefore, it is an excellent starting material for the synthesis of stereochemically complex templates for library synthesis. Multiple scaffolds were prepared from Beckmann rearrangement, Beckmann fragmentation, and photo-Beckmann reactions. Isosteviol ester **109** was treated with hydroxylamine to form oxime **110** (Scheme 5.25). Upon addition of thionyl chloride in chloroform, products from both the Beckmann fragmentation (nitrile **111**) and Beckmann rearrangement (lactam **112**) formed. Similarly, Wu *et al.* observed the same fragmentation and rearrangement products under a range of acidic conditions with an analogous isosteviol oxime ester.[54]

When the Beckmann reaction was performed with mesylated oxime intermediate **113**, lactam **112**, from the Beckmann rearrangement, was obtained as the sole product (Scheme 5.26), thus shutting down the Beckmann fragmentation pathway.[53] This reaction is similar to what White and coworkers observed during their morphine synthesis. Beckmann reaction of an intermediate brosylated oxime in acetic acid provided the desired lactam, whereas the reaction of the corresponding oxime under acidic conditions did not yield any lactam.[55]

On the other hand, selective Beckmann fragmentation was achieved with this scaffold by reaction of isosteviol oxime **110** with acetic anhydride in acetonitrile, yielding the corresponding acetate. Subsequent treatment with *p*-toluenesulfonic acid produced the Δ^6:Δ^7 olefinic nitriles **111** and **114** in an 8:1 ratio of, respectively (Scheme 5.27).

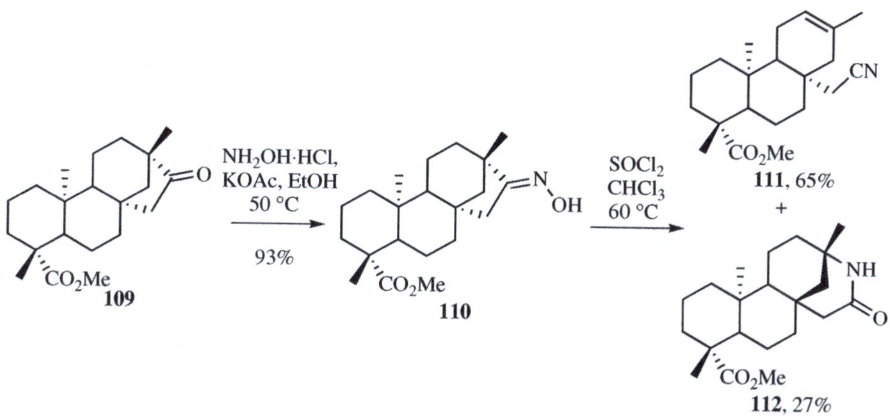

Scheme 5.25 Beckmann rearrangement and fragmentation of oxime **110**.

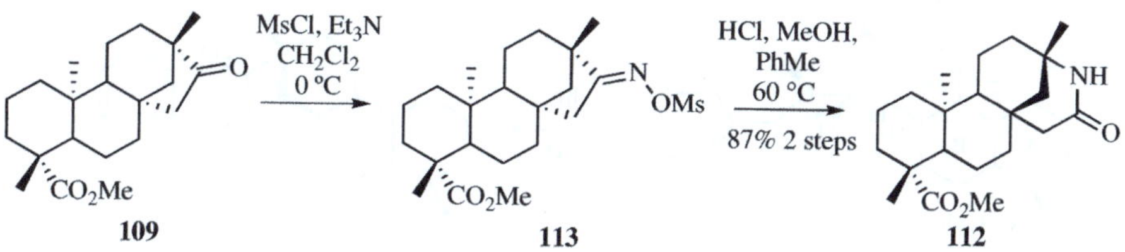

Scheme 5.26 Beckmann rearrangement of mesylate **113** to form lactam **112**.

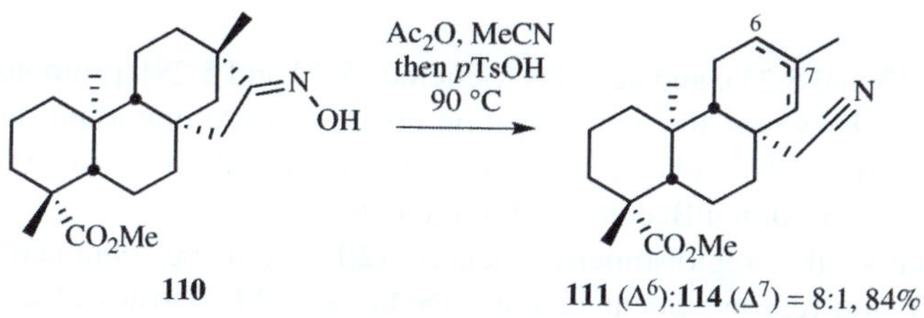

Scheme 5.27 Beckmann fragmentation to generate nitriles **111** and **114**.

The Beckmann fragmentation reaction of the steviol scaffold was also evaluated (Scheme 5.28). Steviol oxime **116** was prepared from steviol ester **115**, and upon reaction with acetic anhydride and *p*-toluenesulfonic acid in acetonitrile at 90 °C, the expected nitrile **117** was generated exclusively in 67% yield.

The experiments shown previously and results from other groups, shown below, suggest that the reaction outcomes are partly determined by the character of the oxime leaving groups. Good leaving groups such as sulfonates promote Beckmann rearrangement (Scheme 5.26; see also Schemes 5.5, 5.12, and 5.17), whereas poorer leaving groups, such as the oxonium ion, tetrachlorooxidophosphorane formed from PCl_5

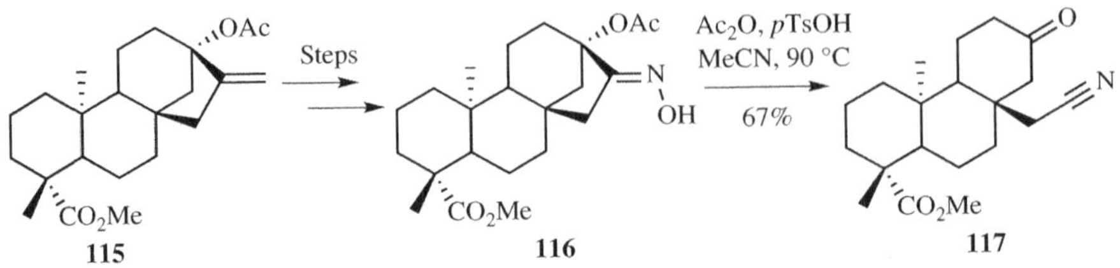

Scheme 5.28 Beckmann fragmentation of steviol oxime **116**.

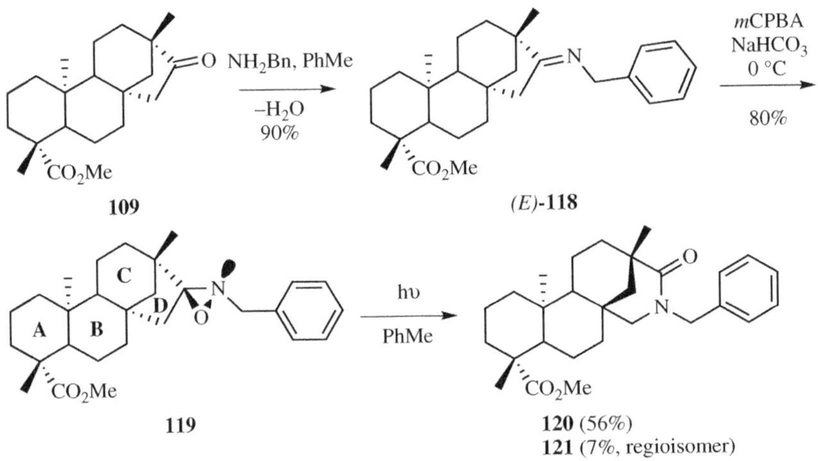

Scheme 5.29 Photo-Beckmann reaction of oxaziridine **119**.

(Schemes 5.13 and 5.23), and acetates (Schemes 5.27 and 5.28), promote Beckmann fragmentation. However, the exact reaction conditions are important. For example, reaction of oxime **104** (Scheme 5.24) with *p*-toluenesulfonic acid in DMF in the absence of base promoted Beckmann fragmentation.

Additionally, the regioisomeric lactam **120** could be obtained through a photo-Beckmann rearrangement reaction (Scheme 5.29). Isosteviol ester **109** was converted to the corresponding oxaziridine **119** through heating in the presence of benzylamine under dehydrating conditions to afford imine **118** as the major diastereoisomer (7:1 ratio of (*E*)- to (*Z*)-imines as determined by NMR), followed by treatment with *m*-chloroperoxybenzoic acid to yield oxaziridine **119**. Notably, the oxidation occurred exclusively at the exo face of the CD ring. Its structure was not independently confirmed but determined from the resulting amide **120**. Photolysis with a 254 nm mercury lamp regioselectively delivered the expected lactam **120** in 56% yield and the minor regioisomer **121** in 7% yield. The stereoelectronic effect of the oxaziridine ring promoted the migration of the bond antiperiplanar to the nitrogen lone pair. Formation of minor isomer **121** resulted from the *exo*-oxaziridine derived from the other imine isomer, (*Z*)-**118**. This work showcases the complementary nature of the Beckmann reactions and how this chemistry was employed for

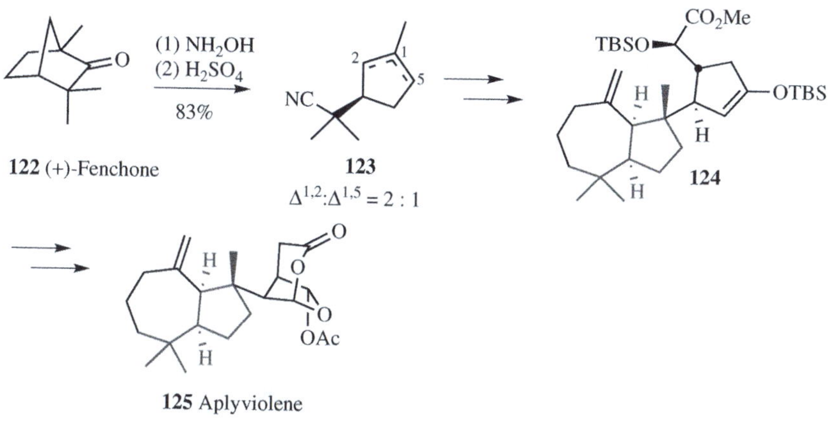

Scheme 5.30 Synthesis of **124**.

developing diverse and stereochemically complex templates for further exploration of diterpenes in library synthesis.

5.4.1.3 ***Toward the Synthesis of (−)-Aplyviolene*** Schnermann and Overman reported the synthesis of diterpene (−)-aplyviolene (**125**), which is a member of the spongian class of diterpene natural products (Scheme 5.30).[56] The synthesis began with a Beckmann fragmentation of (+)-fenchone (**122**) to produce a 2:1 mixture of alkene regioisomers favoring the $\Delta^{1,2}$-isomer **123**. Reduction of the nitrile, Wittig reaction of the corresponding aldehyde, and subsequent steps eventually formed the bicyclic ring system **124**, which the Overman group had previously used to synthesize (−)-aplyviolene (**125**).[57] This synthetic route to **124** improved upon the original total synthesis by requiring five fewer isolated intermediates.

5.4.2 Monoterpene-Related Synthesis

In their efforts to investigate pinene biosynthesis, Coates and coworkers prepared 2-azapinanes, which are aza analogs of the pinyl carbocation intermediates involved in the biosynthetic pathway.[58] The Coates group used both enantiomers of α-pinene as the starting materials, though only the synthesis from the (−)- isomer is shown in Scheme 5.31. α-Pinene (**126**) was converted to ketoacid **127**, which underwent Beckmann rearrangement upon oxime formation to furnish four-membered ring intermediate **128**. Cyclization afforded **129** followed by standard transformations to yield the desired 2-azapinane (**130**).

5.4.3 Furopyran Synthesis

The natural products dysiherbaine (**131**) and neodysiherbaine A (**132**), kainate receptor agonists, and malayamycin A (**133**), an antifungal agent, all contain a furopyran

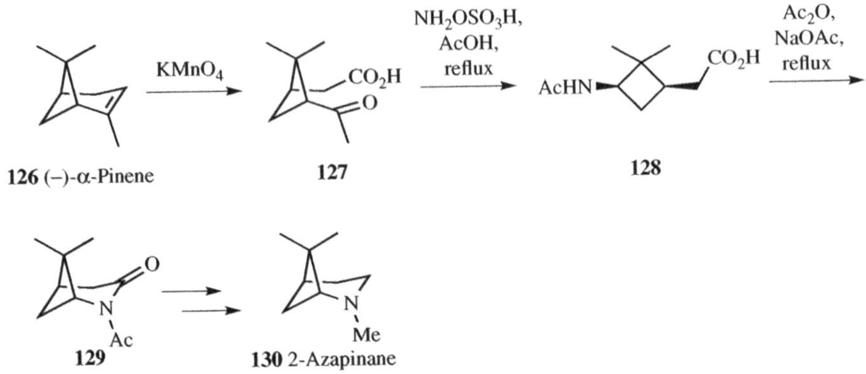

Scheme 5.31 Synthesis of **130**.

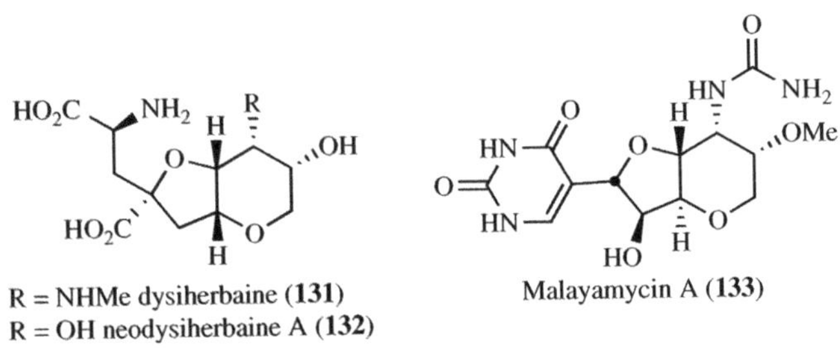

FIGURE 5.5 Furopyran natural products

core structure (Figure 5.5). Krishna reported an approach toward the synthesis of this furopyran core structure that featured a Beckmann fragmentation (Scheme 5.32).[59]

Intermediate **134**, prepared in four steps from readily available starting materials, underwent a [5+2] cycloaddition to furnish multicyclic intermediate **135**. After reduction of the double bond, oxime formation and subsequent Beckmann fragmentation, assisted by the adjacent oxygen, yielded furopyrans **137–139**. The intermediate oxonium ion **136** could be trapped by a hydride, methanol, or water (conditions *a–c*).

5.4.4 Synthesis of a Pyrazine Imide

The synthesis of porphyrinoids containing nonpyrrolic heterocycles is of interest for studying the intrinsic properties of porphoryins. These porphyrin-like molecules possess varying absorption and emission properties and abilities to interact with metals or other analytes.[60] Akhigbe and Brückner converted meso-tetraphenylporphyrine **140** to diketone **141** using osmium tetroxide and DDQ oxidation (Scheme 5.33). Oxime formation followed by a Beckmann rearrangement provided the ring-expanded

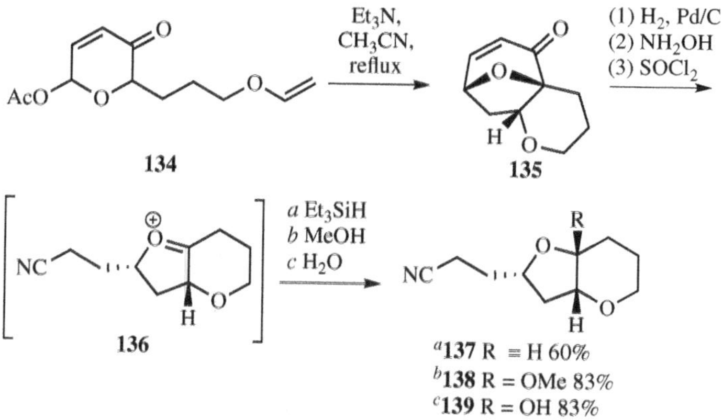

Scheme 5.32 Synthesis of furopyran derivatives.

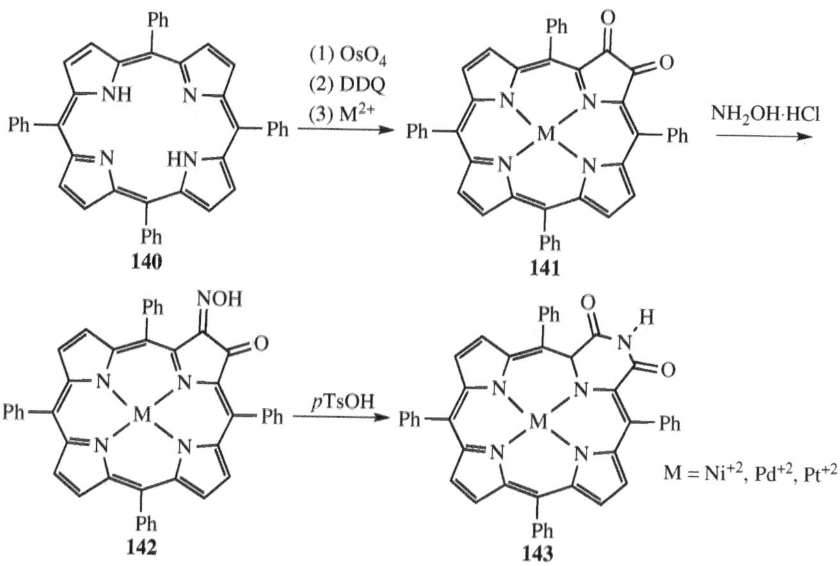

DDQ = 2,3-Dichloro-5,6-dicyano-1,4-benzoquinone

Scheme 5.33 Synthesis of pyrazine imide **143**.

metallopyrazine imide **143**, which showed metallochlorine-like optical properties. The demetallated derivative of **143** has a porphyrin-like optical spectrum.[61] In terms of further elaboration, the imide in **143** provides a functional group that can be derivatized.

5.5 FORMATION OF HETEROCYCLIC SCAFFOLDS

Another aspect of the Beckmann reaction is its ability to form different scaffolds through an intramolecular nucleophile-trapping mechanism. When the oxime contains a nucleophile as one of its substituents, there is the potential for ring closures to take place.[2c] The cyclization involves intramolecular nucleophilic addition to the nitrilium ion generated from the oxime during the Beckmann rearrangement process (Scheme 5.34). When the nucleophile-containing group is anti to the oxime hydroxyl (as in **144**), the Beckmann rearrangement will form ensuing nitrilium **145**, which undergoes a 5-endo-dig cyclization to generate heterocycle **146**.[62] Inversely, when the nucleophile is within the substituent syn to the oxime hydroxyl (cf. **147**), cyclic imine **149** results from a 5-exo-dig cyclization of nitrilium intermediate **148** of oxime **147**. The following sections will showcase recent examples in which the Beckmann rearrangement reaction has been used to produce a variety of heterocyclic and non-heterocyclic scaffolds.

5.5.1 Oxadiazoles

1,3,4-Oxadiazoles[63] and 1,2,4-oxadiazoles[64] are heterocyclic aromatic compounds that appear in many bioactive molecules.[65] Previous methods for the synthesis of 1,2,4-oxadiazoles include the coupling of amidoximes with carboxylic acid derivatives,[66] aerobic C—H oxygenation of amidoximes,[67] or a cyclization of nitrile oxides to nitriles.[68] Telvekar and Takale developed the preparation of 1,2,4-oxadiazoles from substituted diketone derivatives through a Beckmann rearrangement process (Scheme 5.35).[69] When treated with diphosphorus tetraiodide in dichloromethane at room temperature, dioximes **150** formed the Beckmann products, 1,2,4-oxadiazoles **151**, in excellent yields.

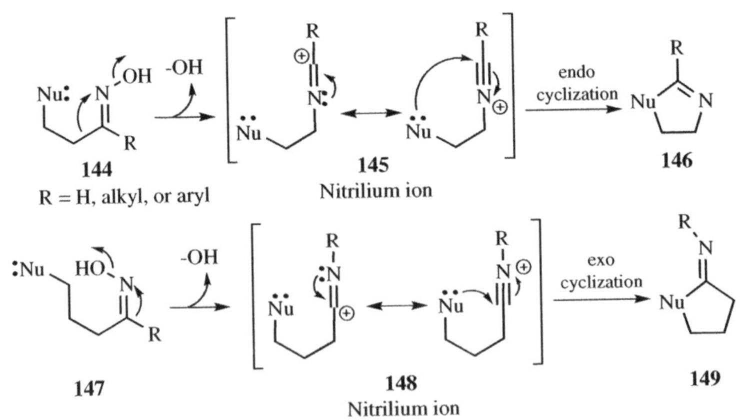

Scheme 5.34 Intramolecular cyclization of oxime.

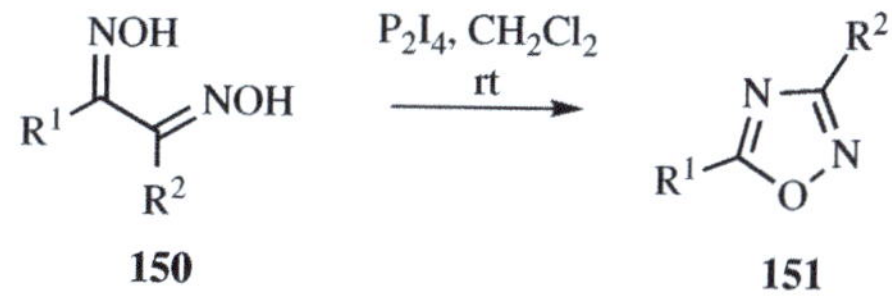

Scheme 5.35 Oxadiazole formation.

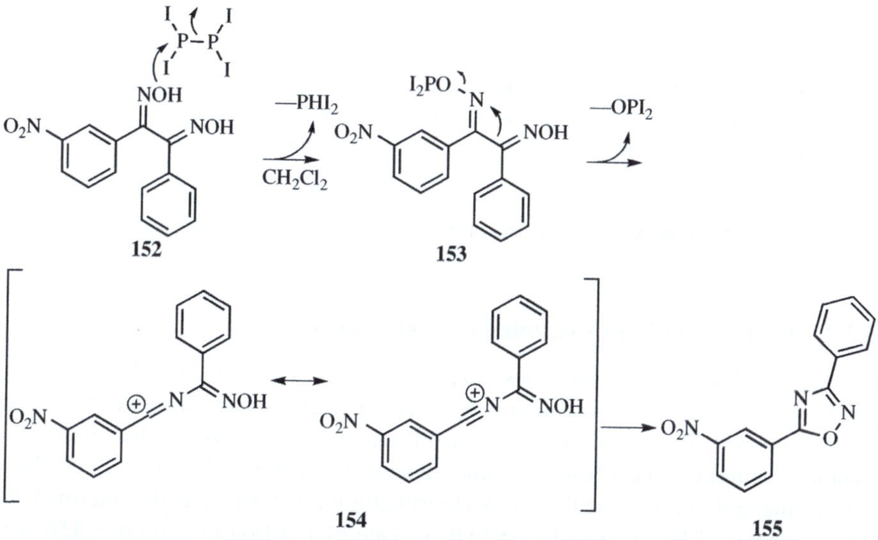

Scheme 5.36 Proposed mechanism for oxadiazole formation.

Mechanistically, the rearrangement occurs at one of the oxime groups followed by cyclization to form the oxadiazole ring (Scheme 5.36). Diphosphorus tetraiodide exhibits high affinity for oxygen and promotes substitution, dehydration, or reduction reactions[70]; the major driving force for the reaction is phosphoryl bond. Upon addition of diphosphorus tetraiodide to dioxime **152**, intermediate **153** is produced. A Beckmann rearrangement releases phosphonic diiodide to generate nitrilium intermediate **154** that cyclizes to form oxadiazole **155**.

The reaction is apparently of limited scope as only five examples were reported. Symmetrical phenyl, 2-furyl and ethyl substitutions on both sides of the dioxime produced the oxadiazole product. Similarly, in one unsymmetrical case, with methyl and phenyl substituents, 3-methyl-5-phenyl-1,2,4-oxadiazole was generated. However, in cases where there are identical substituted benzene rings on both sides of the dioxime (*p*-MeO, *p*-NH_2, *m*-NO_2), a Beckmann fragmentation took place, resulting in the formation of the corresponding arylnitrile products.[69]

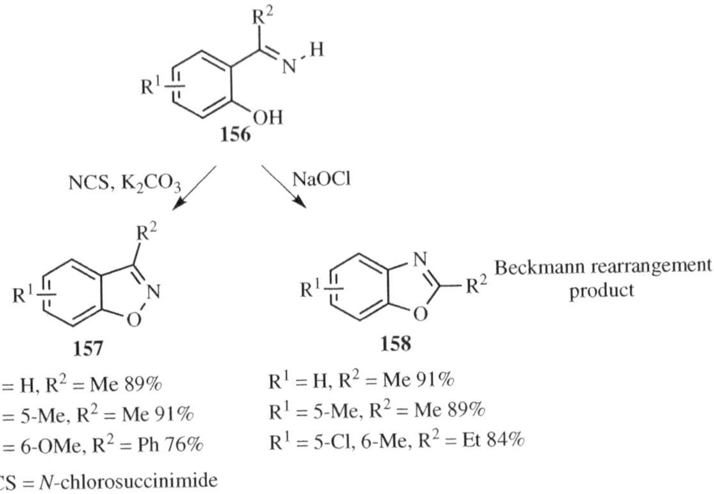

Scheme 5.37 Synthesis of benzisoxazoles and benzoxazoles.

5.5.2 Synthesis of Benzisoxazoles and Benzoxazoles

Benzisoxazoles and benzoxazoles appear as scaffolds in many pharmaceutical products[71], and the benzoxazole moiety is also present in natural products.[72] Chen *et al.* reported a synthesis of benzoxazoles and benzisoxazoles that allows the synthesis of both isomers from the same substrate (Scheme 5.37).[73] In both synthetic routes, the authors propose that an *N*-chloroketimine is formed as the intermediate. The synthesis of benzisoxazoles **157** takes place by reaction of ketimines **156** with NCS in the presence of base, which promotes the addition of the phenol to the intermediate *N*-chloroketimine. Reaction with hypochlorite in the absence of base leads to Beckmann rearrangement to form benzoxazoles **158** via aryl migration and addition of the phenol to the intermediate nitrilium ion.

5.5.3 Isatins

Isatins are indole derivatives with broad use in synthetic dye production and are intermediates in the synthesis of other heterocyclic molecules.[74] They also possess a variety of biological activities.[75] Aksenov *et al.* reported a one-pot synthesis of isatins using ethyl nitroacetate with substituted benzenes such as anisole (**159**) in polyphosphoric acid (Scheme 5.38).[76] The process for isatin formation likely includes a hybrid between the Nef and Vilsmeier reactions of anisole **159** and nitro ester **160** to form the oxime intermediate **161**, which then undergoes a Beckmann rearrangement to give the anilide **162**. Subsequent intramolecular acylation yields isatin **163**.

This chemistry is reported to be the first example of an isatin synthetic method that does not rely on the annelation of a five-membered ring onto aromatic rings with the participation of preexisting aryl substituents. Aksenov and coworkers also applied

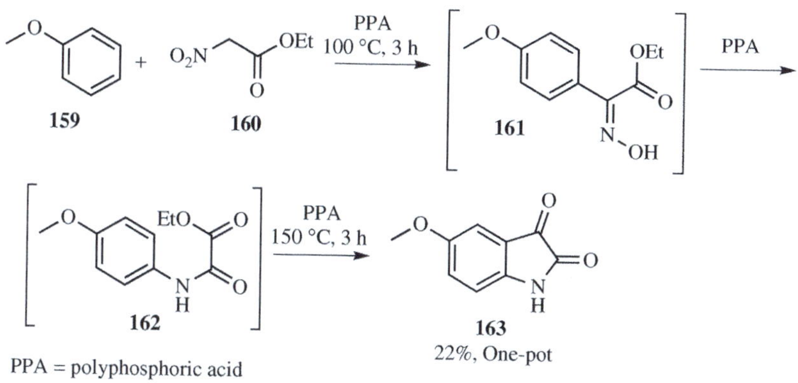

Scheme 5.38 Formation of isatin **163**.

this methodology to methylenedioxybenzene and ethylenedioxybenzene with yields of 27% and 26%, respectively.

5.5.4 *N*-Imidoylbenzotriazoles

It is well known that oximes can be converted to imidates, thioimidates, imidoyl halides and cyanides, and imines through trapping of the intermediate nitrilium ion with the appropriate nucleophiles.[2b] Imidoylbenzotriazoles are useful substitutes for imidoyl chlorides.[77] They are chemically stable and versatile reagents, which can be used to synthesize a variety of heterocycles and diverse compounds with nitrogen-containing functional groups. Deng and coworkers developed a one-pot procedure to obtain *N*-imidoylbenzotriazoles such as **165** in good to excellent yields via a Beckmann rearrangement of ketoximes such as **164** (Scheme 5.39).[78] They examined the scope of the reaction with diarylketoxime, arylalkylketoximes, dialkylketoximes, and cyclohexanone oxime substrates and found that all arylalkylketoximes produced the corresponding benzotriazoles in yields between 91% and 97% in toluene.

In the proposed mechanism, treatment of oxime **2** with sulfonyl chloride in the presence of a base would form the oxime sulfonate intermediate **166**, which would

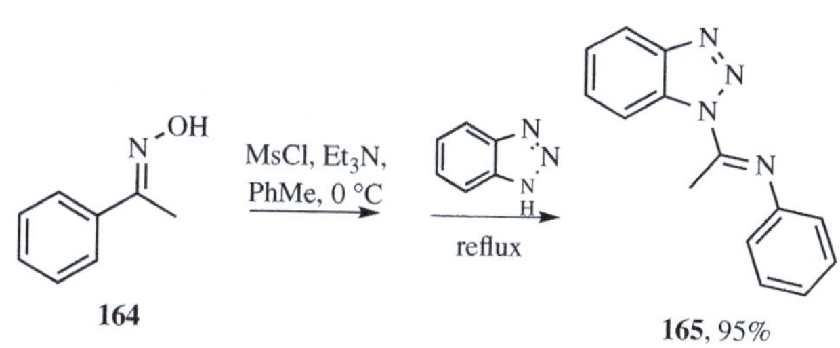

Scheme 5.39 Formation of imidoylbenzotriazole **165**.

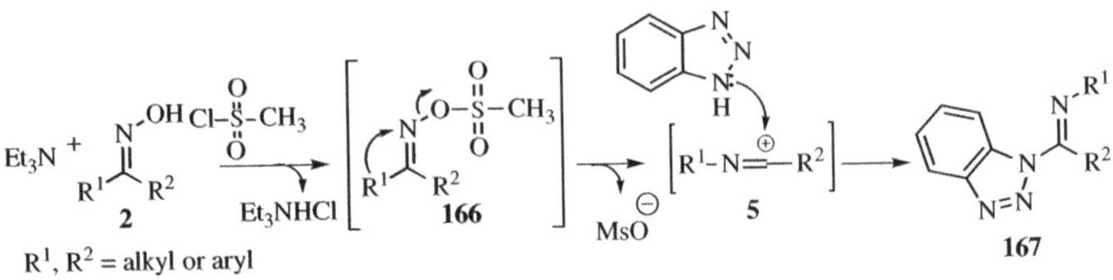

Scheme 5.40 Proposed mechanism for imidoylbenzotriazole formation.

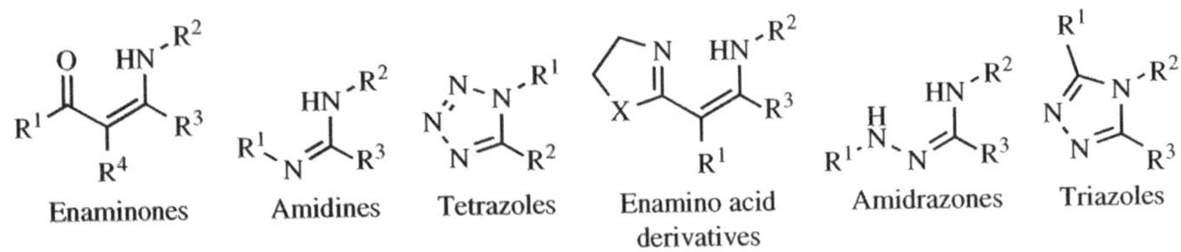

FIGURE 5.6 Heterocycles and functional groups obtained from imidoylbenzotriazoles

undergo Beckmann rearrangement to form nitrilium **5** (Scheme 5.40). Subsequent nucleophilic attack by benzotriazole would furnish imidoylbenzotriazoles **167**. In unsymmetrical cases with aryl and alkyl substitutions, a regioselective migration of the aryl (R^1) group of the (*E*)-oxime takes place.

Imidoylbenzotriazoles serve as attractive intermediates to produce heterocycles and compounds carrying various functional groups including enaminones,[79] amidines,[80] tetrazoles,[81] and triazoles[82] (Figure 5.6).

5.6 SYNTHESIS OF FUNCTIONAL GROUPS

5.6.1 Amidines

Another synthetically useful functional group in organic synthesis accessible via Beckmann rearrangement is the amidine. This structure has become a common synthetic building block for heterocyclic chemistry[83] and natural product synthesis.[84] Known methods to prepare the amidine moiety require transition-metal catalysts[85] or direct use of amines.[86] Ashfeld and coworkers have constructed amidines by coupling oximes, such as arylalkylketoxime **168**, with tosyl azide via a mild, chemoselective Beckmann-like rearrangement under metal-free and amine-free conditions to generate amidine **169** (Scheme 5.41).[87] The researchers expanded the scope of the reaction to dialkylketoximes, diarylketoximes, and cyclic oximes. The formation of the amidine is consistent with a Beckmann-like rearrangement in which the antiperiplanar oxime substituent migrates preferentially. Addition of TBAF to **169** gave dephosphorylated amidine **170**.

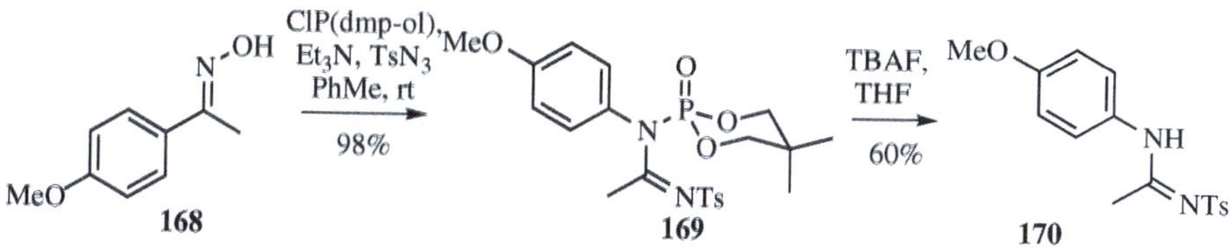

ClP(dmp-ol) = 3,3-dimethylpropanediol chlorophosphite
TBAF = Tetrabutylammonium fluoride

Scheme 5.41 Formation of amidine **170**.

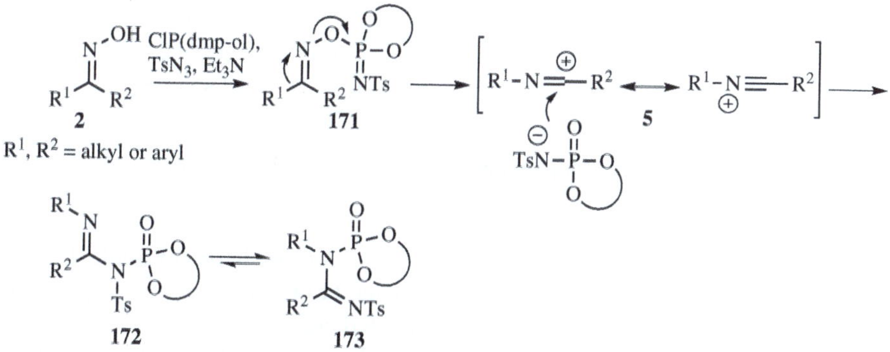

Scheme 5.42 Mechanism for formation of amidines.

Ashfeld and coworkers proposed that the initial phosphinylation of oxime **2** and subsequent Staudinger-like reduction of the tosyl azide forms iminophosphorane **171** (Scheme 5.42). The formation of the phosphoryl bond upon rearrangement is a likely driving force behind the regioselective migration of the *trans*-oxime substituent (R^1 in **171**) to afford the nitrilium ion **5**. The corresponding anionic phosphoramide is released and then adds to the nitrilium ion amidine **17**. A subsequent 1,3-phosphoryl shift leads to the final amidine product **173**.

5.6.2 Thioamides

Certain thioamides exhibit antithyroid properties,[88] have been used in peptide isostere development,[89] and are key synthetic intermediates in heterocyclic chemistry.[90] Li *et al.* reported that Beckmann rearrangement of oximes with phosphorus pentasulfide is a facile and mild reaction to produce secondary thioamides (Scheme 5.43).[91] Previous Beckmann rearrangement routes toward thioamide synthesis required thiophosphoryl chloride, which is a moisture-sensitive and highly odorous reagent,[92] or the use of transition-metal reagents.[93] Deng and coworkers also developed a less malodorous, one-pot procedure for the synthesis of thioamides employing the Beckmann rearrangement.[94] However, some of the yields obtained by Deng and coworkers were

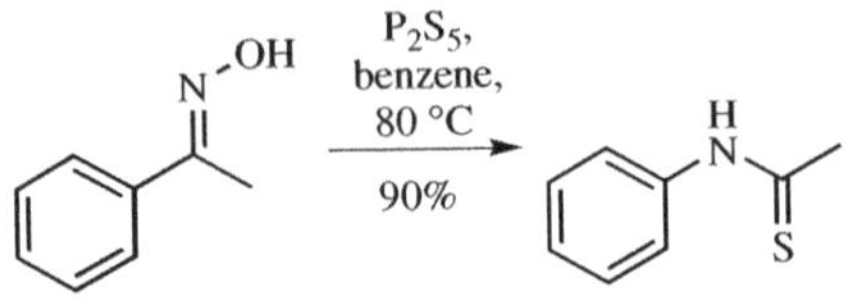

174 175

Scheme 5.43 Formation of thioamide **175**.

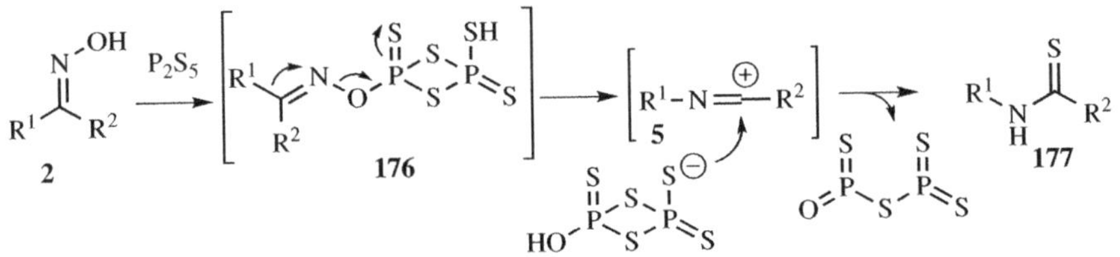

2 176

Scheme 5.44 Proposed mechanism for thioamide formation.

lower when compared to the results reported by Li's group. In one example, Li *et al.* found that upon treatment of acetophenone oxime (**174**) with phosphorus pentasulfide, Beckmann rearrangement yielded *N*-phenylethenethioamide (**175**).

In the mechanism proposed by Li and coworkers, ketoxime **2** reacts with phosphorus pentasulfide to furnish activated ester intermediate **176**, which undergoes the Beckmann rearrangement to deliver the iminocarbocation **5** and the thiolate counteranion (Scheme 5.44). Nucleophilic addition of the thiolate then produces the desired thioamide **177**.

In Li's study, a variety of substrates generated thioamides **177** in notable yields from symmetric and asymmetric oximes (Table 5.1). Electron-donating groups present on the aromatic ring gave higher yields (83–90%), while electron-withdrawing groups on the aromatic ring tended to slow down the reaction and gave lower yields (65–80%). Aliphatic ketoximes were converted to their respective secondary thioamides in good yields (65–83%).

5.6.3 Synthesis of DNA-Photocleaving Agents

DNA-photocleaving agents have gained interest due to their ability upon light activation to induce chemical reactions that cleave DNA. The development of these synthetic organic molecules enables researchers to study DNA properties and design new drugs targeting DNA. One notable fluorophore for DNA cleavage is pyrene. This polycyclic aromatic hydrocarbon acts as a DNA intercalator and can be photochemically excited at longer wavelengths than the DNA absorption band.

Chowdhury *et al.* have developed cell-permeable oxime esters, which can act as intercalators and DNA-cleaving agents upon excitation at wavelengths longer than 350 nm.[95] They monitored the photolysis of (*E*)-pyrene oxime ester **179**, formed

TABLE 5.1 Scope of Thioamide Formation

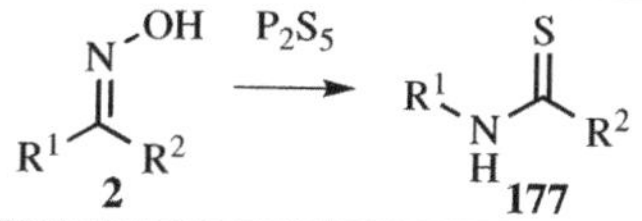

R^1	R^2	Time (h)	Yield (%)
Ph	Ph	0.5	85
4-MeC_6H_4	Me	0.5	90
3-ClC_6H_4	Et	2.5	80
$3\text{-NO}_2C_6H_4$	Me	5	68
4-BrC_6H_4	$MeSCH_2$	1	72
Me	Me	2	83
$(CH_2)_{11}$		2	80

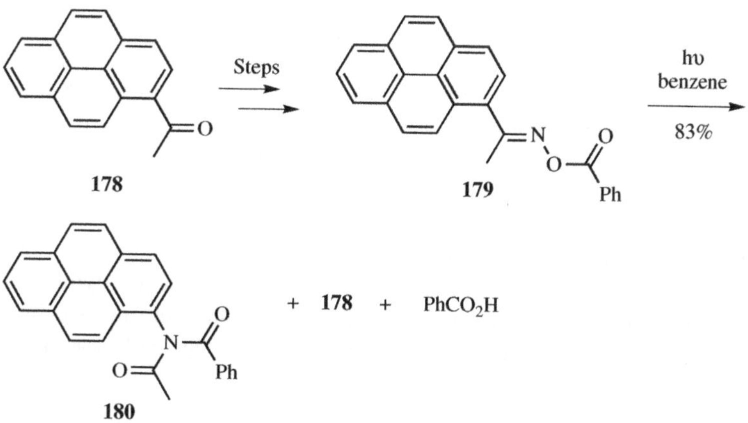

Scheme 5.45 Formation of photocleaving agents.

from pyrene ketone **178**, in an oxygen-free methanol/water solution using a 125 W medium pressure mercury lamp (>350 nm) with a 0.1 M copper(II) sulfate solution as a UV cutoff filter (Scheme 5.45). After 10 min of irradiation, the oxime ester gave photo-Beckmann rearranged benzamide **180** along with acetyl pyrene and benzoic acid as the three major photoproducts. They presumed that the acetyl pyrene and benzoic acid formed from light-induced N—O bond cleavage.

Chowdhury proposed that the formation of the photo-Beckmann rearrangement product **180** from **179** resulted from fission at the oxygen–carboxyl bond to form pyrene iminyloxy radical **181** and benzoyl radical **182** (Scheme 5.46). The radical pair recombines to form oxaziridine intermediate **183**, which generates the *N*-acetyl-*N*-(pyren-6-yl)benzamide product **180**. In addition, the researchers

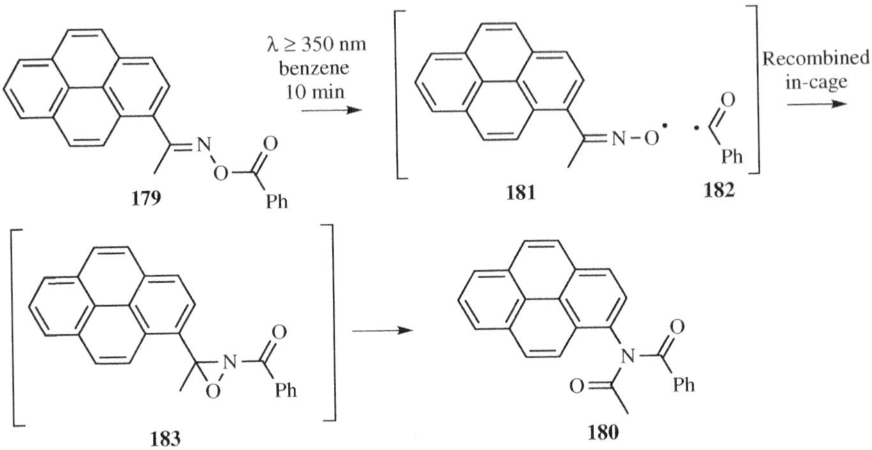

Scheme 5.46 Proposed mechanism for the formation of the benzamide pyrene product.

reported some solvent effects on the photolysis of the pyrene oxime ester conjugate. In benzene, the photo-Beckmann benzamide product was seen exclusively, while both photo-Beckmann rearrangement and fragmentation products **178** and benzoic acid were observed in acetonitrile. Moreover, both methanol and aqueous methanol increased the yield of the photofragmentation products. This work supports the use of (*E*)-pyrene oxime esters as DNA intercalators and DNA-photocleaving agents that form photo-Beckmann rearranged products and N—O bond cleavage upon irradiation. Chowdhury *et al.* tested the oxime esters *in vitro* and found they not only permeated mouse embryonic fibroblasts but also showed greater cytotoxicity upon photoexcitation. Further modifications of the oxime esters are needed in order for DNA-photocleaving agents to exhibit specificity to cancer cells for targeted drug delivery.

5.7 SUMMARY AND OUTLOOK

As shown in this and prior reviews,[2] the Beckmann reaction is a valuable procedure to generate diverse products, generating amides regioselectively from the Beckmann rearrangement and the photo-Beckmann reaction, and nitriles from the Beckmann fragmentation reaction. The Beckmann reaction products can undergo a variety of subsequent transformations toward targeted compounds by functional group conversions. Linear amide reaction products can be hydrolyzed to form an amine and an acid. Hydrolysis of lactams involves ring breaking to generate an amino acid. The amide carbonyl group of acyclic and cyclic amides can be reduced to form acyclic and cyclic amines, respectively. The Beckmann fragmentation involves the breaking a carbon–carbon bond and formation of a nitrile, which can be transformed to other functional groups using standard chemistry. Whether Beckmann rearrangement or fragmentation takes place is determined by the ability of the group anti to the oxime to stabilize a carbocation. However, good leaving groups, such as sulfonates, can promote Beckmann rearrangement in these cases. The Beckmann reaction has been employed in creative ways for natural product syntheses, preparation of biological probes, and for compound libraries derived from the natural product stevioside. In addition, multiple synthetic efforts have expanded the utility of the Beckmann reaction from its ring opening and ring expansion aspects to its utility in generating novel cyclized products. Using the Beckmann reactions can generate diverse nitrogen-containing heterocycles and a number of different functional groups that are useful for other chemical transformations.

A continued challenge for carrying out a Beckmann reaction is the requirement for strong activating reagents and harsh reaction conditions that sometimes require stoichiometric amounts of strong acid and high temperatures. As a result, the yields can be low because of compound decomposition and the regioselectivity of the reaction can be reduced as a result of oxime isomerization. Much effort has therefore been devoted to discovering milder reaction conditions for the Beckmann reaction,[96] including self-propagation reactions (Scheme 5.6). The photo-Beckmann reaction has not yet been explored extensively despite the comparatively mild reaction conditions employed and the excellent regioselectivity of the reaction, which is under stereoelectronic control (Schemes 5.11 and 5.21). For the photo-Beckmann reaction, an imine can be first oxidized diastereoselectively with a peracid to provide the corresponding oxaziridine. This oxidation results in a stable geometry once formed and will generally not undergo isomerization, because of the high energy barrier for nitrogen inversion in oxaziridines. The subsequent photochemical reaction is usually a clean reaction to provide the amide. The amide generated from an (*E*)-oxime will be regioisomeric to the amide generated from the (*E*)-oxime by the regular Beckmann rearrangement. Since the photo-Beckmann reaction has seen limited applications so far, it can be expected that this reaction will be employed more frequently in the future.

REFERENCES

1. Beckmann, E. *Ber. Dtsch. Chem. Ges.* **1886**, *19*, 988–993.
2. (a) Wang, Z. Beckmann rearrangement and Beckmann fragmentation. In *Comprehensive Organic Name Reactions and Reagents*; John Wiley & Sons, Inc., **2010**, pp. 288–295. (b) Pereira, M. M. A.; Santos, P. P. Rearrangements of hydroxylamines, oximes, and hydroxamic acids. In *The chemistry of Hydroxylamines, Oximes and Hydroxamic Acids*; Rappoport, Z.; Liebman, J. F., Eds.; John Wiley & Sons, Inc., **2009**, pp. 343–498. (c) Gawley, R. E. The Beckmann reactions: rearrangements, elimination-additions, fragmentations, and rearrangement-cyclizations. In *Organic Reactions*; Kende, A. S., Ed.; John Wiley & Sons, Inc.: New York, **2004**; Vol. *35*, pp. 1–420. (d) Craig, D. The Beckmann and related reactions. In *Comprehensive Organic Synthesis*; Trost, B. M.; Fleming, I., Eds.; Oxford, **1991**; Vol. *7*, pp. 689–702. (e) Conley, R. T.; Ghosh, S. Abnormal Beckmann rearrangements. In *Mechanisms of Molecular Migration*; Thyagarajan, S., Ed.; Wiley-Interscience: New York, **1971**; Vol. 4, pp. 197–308. (f) Drahl, M. A.; Manpadi, M.; Williams, L. J. *Angew. Chem., Int. Ed.* **2013**, *52*, 11222–11251.

3. Kaur, G.; Rajput, J. K.; Arora, P.; Devi, N. *Tetrahedron Lett.* **2014**, *55*, 1136–1140.
4. Peng, J. J.; Deng, Y. Q. *Tetrahedron Lett.* **2001**, *42*, 403–405.
5. Furuya, Y.; Ishihara, K.; Yamamoto, H. *J. Am. Chem. Soc.* **2005**, *127*, 11240–11241.
6. Yamamoto, Y.; Hasegawa, H.; Yamataka, H. *J. Org. Chem.* **2011**, *76*, 4652–4660.
7. Yamabe, S.; Tsuchida, N.; Yamazaki, S. *J. Org. Chem.* **2005**, *70*, 10638–10644.
8. Clayden, J.; Greeves, N.; Warren, S. *Organic Chemistry*; Oxford University Press: Oxford, **2012**, pp. 958–961.
9. Ohwada, T.; Tani, N.; Sakamaki, Y.; Kabasawa, Y.; Otani, Y.; Kawahata, M.; Yamaguchi, K. *Proc. Natl. Acad. Sci. U. S. A.* **2013**, *110*, 4206–4211.
10. Kenyon, J.; Young, D. P. *J. Chem. Soc.* **1941**, 263–267.
11. Porter, J. R.; Traverse, J. F.; Hoveyda, A. H.; Snapper, M. L. *J. Am. Chem. Soc.* **2001**, *123*, 10409–10410.
12. Campbell, A.; Kenyon, J. *J. Chem. Soc.* **1946**, 25–27.
13. Hill, R. K.; Conley, R. T.; Chortyk, O. T. *J. Am. Chem. Soc.* **1965**, *87*, 5646–5651.
14. Vanos, C. M.; Lambert, T. H. *Chem. Sci.* **2010**, *1*, 705–708.
15. Tian, B. X.; An, N.; Deng, W. P.; Eriksson, L. A. *J. Org. Chem.* **2013**, *78*, 6782–6785.
16. Schöpf, C. *Justus Liebigs Ann. Chem.* **1927**, *452*, 211–267.
17. Amin, J. H.; Mayo, P. D. *Tetrahedron Lett.* **1963**, *4*, 1585–1589.
18. Ogata, Y.; Takagi, K.; Mizuno, K. *J. Org. Chem.* **1982**, *47*, 3684–3687.
19. Izawa, H.; De Mayo, P.; Tabata, T. *Can. J. Chem.* **1969**, *47*, 51–62.
20. Suginome, H. *E,Z*-isomerization and accompanying photoreactions of oximes, oxime ethers, nitrones, hydrazones, imines, azo- and azoxy compounds, and various applications. In *CRC Handbook of Organic Photochemistry and Photobiology*; Horspool, W. M.; Lenci, F., Eds.; CRC Press: Boca Raton, FL, **2004**, pp. 1–55, Chapter 94.
21. Lattes, A.; Oliveros, E.; Riviere, M.; Belzecki, C.; Mostowicz, D.; Abramskj, W.; Piccinnileopardi, C.; Germain, G.; Vanmeerssche, M. *J. Am. Chem. Soc.* **1982**, *104*, 3929–3934.
22. (a) Woodward, R. B. *Pure Appl. Chem.* **1973**, *33*, 145–177. (b) Eschenmoser, A.; Wintner, C. E. *Science* **1977**, *196*, 1410–1426.
23. Woodward, R. B. *Pure Appl. Chem.* **1968**, *17*, 519–547.
24. Burgstahler, A. W.; Abdel-Rahman, M. O. *J. Am. Chem. Soc.* **1963**, *85*, 173–180.
25. Rapoport, H.; Williams, A. R. *J. Am. Chem. Soc.* **1949**, *71*, 1774–1778.
26. Chen, Y. K.; Jeon, S.-J.; Walsh, P. J.; Nugent, W. A. *Org. Synth.* **2005**, *82*, 87–92.
27. Horning, E. C.; Stromberg, V. L.; Lloyd, H. A. *J. Am. Chem. Soc.* **1952**, *74*, 5153–5155.
28. Stork, G.; Tabak, J. M.; Blount, J. F. *J. Am. Chem. Soc.* **1972**, *94*, 4735–4737.
29. Newman, D. J.; Cragg, G. M. *J. Nat. Prod.* **2012**, *75*, 311–335.
30. Xu, S.; Unabara, D.; Uemura, D.; Arimoto, H. *Chem. Asian J.* **2014**, *9*, 367–375.
31. Chou, T.; Kuramoto, M.; Otani, Y.; Shikano, M.; Yazawa, K.; Uemura, D. *Tetrahedron Lett.* **1996**, *37*, 3871–3874.
32. Tsubosaka, Y.; Murata, T.; Kinoshita, K.; Yamada, K.; Uemura, D.; Hori, M.; Ozaki, H. *Eur. J. Pharmacol.* **2010**, *628*, 128–131.

33. (a) Zhang, H. L.; Zhao, G.; Ding, Y.; Wu, B. *J. Org. Chem.* **2005**, *70*, 4954–4961. (b) Christie, H. S.; Heathcock, C. H. *Proc. Natl. Acad. Sci. U. S. A.* **2004**, *101*, 12079–12084. (c) Matsumura, Y.; Aoyagi, S.; Kibayashi, C. *Org. Lett.* **2004**, *6*, 965–968. (d) Liu, D.; Acharya, H. P.; Yu, M.; Wang, J.; Yeh, V. S.; Kang, S.; Chiruta, C.; Jachak, S. M.; Clive, D. L. *J. Org. Chem.* **2009**, *74*, 7417–7428. (e) Trauner, D.; Schwarz, J. B.; Danishefsky, S. J. *Angew. Chem., Int. Ed.* **1999**, *38*, 3542–3545. (f) Xu, S.; Arimoto, H.; Uemura, D. *Angew. Chem., Int. Ed.* **2007**, *46*, 5746–5749.
34. Saeki, M.; Toyota, M. *Tetrahedron Lett.* **2010**, *51*, 4620–4622.
35. Kam, T. S.; Subramaniam, G.; Lim, K. H.; Choo, Y. M. *Tetrahedron Lett.* **2004**, *45*, 5995–5998.
36. Lim, K. H.; Hiraku, O.; Komiyama, K.; Koyano, T.; Hayashi, M.; Kam, T. S. *J. Nat. Prod.* **2007**, *70*, 1302–1307.
37. (a) Zhong, X.; Li, Y.; Han, F. S. *Chem. Eur. J.* **2012**, *18*, 9784–9788. (b) Iwama, Y.; Okano, K.; Sugimoto, K.; Tokuyama, H. *Org. Lett.* **2012**, *14*, 2320–2322. (c) Nakajima, R.; Ogino, T.; Yokoshima, S.; Fukuyama, T. *J. Am. Chem. Soc.* **2010**, *132*, 1236–1237. (d) Biechy, A.; Zard, S. Z. *Org. Lett.* **2009**, *11*, 2800–2803. (e) Magolan, J.; Carson, C. A.; Kerr, M. A. *Org. Lett.* **2008**, *10*, 1437–1440.
38. Li, Z. N.; Liang, G. X. *Tetrahedron Lett.* **2013**, *54*, 242–244.
39. Daly, J. W.; Witkop, B.; Tokuyama, T.; Nishikawa, T.; Karle, I. L. *Helv. Chim. Acta* **1977**, *60*, 1128–1140.
40. Fujimoto, R.; Kishi, Y. *Tetrahedron Lett.* **1981**, *22*, 4197–4198.
41. (a) Hart, D. J.; Kanai, K. *J. Am. Chem. Soc.* **1983**, *105*, 1255–1263. (b) Overman, L. E.; Lesuisse, D.; Hashimoto, M. *J. Am. Chem. Soc.* **1983**, *105*, 5373–5379. (c) Shirokane, K.; Wada, T.; Yoritate, M.; Minamikawa, R.; Takayama, N.; Sato, T.; Chida, N. *Angew. Chem., Int. Ed.* **2014**, *53*, 512–516. (d) Fujimoto, R.; Kishi, Y.; Blount, J. F. *J. Am. Chem. Soc.* **1980**, *102*, 7154–7156.
42. Pichette, S.; Winter, D. K.; Lessard, J.; Spino, C. *J. Org. Chem.* **2013**, *78*, 12532–12544.
43. Drouin, A.; Lessard, J. *Tetrahedron Lett.* **2006**, *47*, 4285–4288.
44. Winter, D. K.; Drouin, A.; Lessard, J.; Spino, C. *J. Org. Chem.* **2010**, *75*, 2610–2618.
45. Smith, B. T.; Wendt, J. A.; Aube, J. *Org. Lett.* **2002**, *4*, 2577–2579.
46. Zeng, Y. B.; Smith, B. T.; Hershberger, J.; Aubé, J. *J. Org. Chem.* **2003**, *68*, 8065–8067.
47. Bultman, M. S.; Ma, J.; Gin, D. Y. *Angew. Chem., Int. Ed.* **2008**, *47*, 6821–6824.
48. Tanaka, H.; Kozuki, Y.; Ogasawara, K. *Tetrahedron Lett.* **2002**, *43*, 4175–4178.
49. Cao, L. Y.; Sun, J. W.; Wang, X. Y.; Zhu, R.; Shi, H. J.; Hu, Y. F. *Tetrahedron* **2007**, *63*, 5036–5041.
50. (a) Arseniyadis, S.; Yashunsky, D. V.; de Freitas, R. P.; Dorado, M. M.; Potier, P.; Toupet, L. *Tetrahedron* **1996**, *52*, 12443–12458. (b) Arseniyadis, S.; Rodriguez, R.; Dorado, M. M.; Alves, R. B.; Ouazzani, J.; Ourisson, G. *Tetrahedron* **1994**, *50*, 8399–8426.
51. Magnus, P.; Rainey, T. *Tetrahedron* **2001**, *57*, 8647–8651.
52. Roy, A.; Roberts, F. G.; Wilderman, P. R.; Zhou, K.; Peters, R. J.; Coates, R. M. *J. Am. Chem. Soc.* **2007**, *129*, 12453–12460.
53. Hutt, O. E.; Doan, T. L.; Georg, G. I. *Org. Lett.* **2013**, *15*, 1602–1605.
54. Wu, Y.; Yang, J. H.; Dai, G. F.; Liu, C. J.; Tian, G. Q.; Ma, W. Y.; Tao, J. C. *Bioorg. Med. Chem.* **2009**, *17*, 1464–1473.
55. White, J. D.; Hrnciar, P.; Stappenbeck, F. *J. Org. Chem.* **1997**, *62*, 5250–5251.

56. Schnermann, M. J.; Overman, L. E. *Angew. Chem., Int. Ed.* **2012**, *51*, 9576–9580.

57. Schnermann, M. J.; Overman, L. E. *J. Am. Chem. Soc.* **2011**, *133*, 16425–16427.

58. Faraldos, J. A.; Kariuki, B. M.; Coates, R. M. *Org. Lett.* **2011**, *13*, 836–839.

59. Krishna, U. M. *Tetrahedron Lett.* **2010**, *51*, 2148–2150.

60. (a) Lash, T. D. Carbaporphyrins and related systems. Synthesis, characterization, reactivity and insights into porphyrinoid aromaticity. In *Handbook of Porphyrin Science*; Kadish, K. M.; Smith, K. M.; Guilard, R., Eds.; World Scientific: Singapore, **2010**; Vol. *16*, pp. 1–329. (b) Xie, Y.; Morimoto, T.; Furuta, H. *Angew. Chem., Int. Ed.* **2006**, *45*, 6907–6910.

61. Akhigbe, J.; Brückner, C. *Eur. J. Org. Chem.* **2013**, 3876–3884.

62. Baldwin, J. E. *J. Chem. Soc., Chem. Commun.* **1976**, 734–736.

63. (a) Dolman, S. J.; Gosselin, F.; O'Shea, P. D.; Davies, I. W. *J. Org. Chem.* **2006**, *71*, 9548–9551. (b) Khan, M. T. H.; Choudhary, M. I.; Khan, K. M.; Rani, M.; Atta-ur-Rahman. *Bioorg. Med. Chem.* **2005**, *13*, 3385–3395.

64. (a) Maftei, C. V.; Fodor, E.; Jones, P. G.; Franz, M. H.; Kelter, G.; Fiebig, H.; Neda, I. *Beilstein J. Org. Chem.* **2013**, *9*, 2202–2215. (b) Oliveira, V. S.; Pimenteira, C.; da Silva-Alves, D. C. B.; Leal, L. L. L.; Neves, R. A. W.; Navarro, D. M. A. F.; Santos, G. K. N.; Dutra, K. A.; dos Anjos, J. V.; Soares, T. A. *Bioorg. Med. Chem.* **2013**, *21*, 6996–7003.

65. Bostrom, J.; Hogner, A.; Llinas, A.; Wellner, E.; Plowright, A. T. *J. Med. Chem.* **2012**, *55*, 1817–1830.

66. Das, B. C.; Tang, X. Y.; Rogler, P.; Evans, T. *Tetrahedron Lett.* **2012**, *53*, 3947–3950.

67. Zhang, F. L.; Wang, Y. F.; Chiba, S. *Org. Biomol. Chem.* **2013**, *11*, 6003–6007.

68. Nishiwaki, N.; Kobiro, K.; Hirao, S.; Sawayama, J.; Saigo, K.; Ise, Y.; Okajima, Y.; Ariga, M. *Org. Biomol. Chem.* **2011**, *9*, 6750–6754.

69. Telvekar, V. N.; Takale, B. S. *Synth. Commun.* **2013**, *43*, 221–227.

70. (a) Suzuki, H.; Fuchita, T.; Iwasa, A.; Mishina, T. *Synthesis* **1978**, 905–908. (b) Krief, A.; Telvekar, V. Diphosphorus tetraiodide. In *e-EROS Encyclopedia of Reagents for Organic Synthesis*; John Wiley & Sons, Inc., **2009**, pp. 1–5.

71. (a) Dubrovskiy, A. V.; Jain, P.; Shi, F.; Lushington, G. H.; Santini, C.; Porubsky, P.; Larock, R. C. *ACS Comb. Sci.* **2013**, *15*, 193–201. (b) Taylor, R. D.; Maccoss, M.; Lawson, A. D. *J. Med. Chem.* **2014**, *57*, 5845–5859. (c) Nieddu, G.; Giacomelli, G. *Tetrahedron* **2013**, *69*, 791–795.

72. Don, M. J.; Shen, C. C.; Lin, Y. L.; Syu, W. J.; Ding, Y. H.; Sun, C. M. *J. Nat. Prod.* **2005**, *68*, 1066–1070.

73. Chen, C. Y.; Andreani, T.; Li, H. M. *Org. Lett.* **2011**, *13*, 6300–6303.

74. da Silva, J. F. M.; Garden, S. J.; Pinto, A. C. *J. Braz. Chem. Soc.* **2001**, *12*, 273–324.

75. Pandeya, S. N.; Smitha, S.; Jyoti, M.; Sridhar, S. K. *Acta Pharm.* **2005**, *55*, 27–46.

76. Aksenov, N. A.; Aksenov, A. V.; Aksenova, I. V.; Smushkevich, Y. I. *Chem. Heterocycl. Compd.* **2013**, *49*, 645–647.

77. (a) Sutharchanadevi, M.; Murugan, R. Eight-membered rings with one nitrogen atom. In *Comprehensive Heterocyclic Chemistry II*; Katritzky, A. R.; Rees, C. W.; Scriven, E. F. V., Eds.; Elsevier: Oxford, **1996**; Vol. 9, pp. 403–428. (b) Katritzky, A. R.; Monteux, D. A.; Tymoshenko, D. O. *Org. Lett.* **1999**, *1*, 577–578.

78. Pi, H. J.; Liu, L. F.; Jiang, S. S.; Du, W. T.; Deng, W. P. *Tetrahedron* **2010**, *66*, 6097–6100.

79. Katritzky, A. R.; Hayden, A. E.; Kirichenko, K.; Pelphrey, P.; Ji, Y. *J. Org. Chem.* **2004**, *69*, 5108–5111.

80. Katritzky, A. R.; Cai, C. M.; Singh, S. K. *J. Org. Chem.* **2006**, *71*, 3375–3380.

81. Katritzky, A. R.; Cai, C. M.; Meher, N. K. *Synthesis* **2007**, 1204–1208.

82. Katritzky, A. R.; Khashab, N. M.; Kirichenko, N.; Singh, A. *J. Org. Chem.* **2006**, *71*, 9051–9056.

83. Yan, S. J.; Tang, Y. J.; Yu, F. C.; Lin, J. *Helv. Chim. Acta* **2011**, *94*, 487–490.

84. Doveston, R. G.; Steendam, R.; Jones, S.; Taylor, R. J. K. *Org. Lett.* **2012**, *14*, 1122–1125.

85. (a) Wang, Y. F.; Zhu, X.; Chiba, S. *J. Am. Chem. Soc.* **2012**, *134*, 3679–3682. (b) DeKorver, K. A.; Johnson, W. L.; Zhang, Y.; Hsung, R. P.; Dai, H. F.; Deng, J.; Lohse, A. G.; Zhang, Y. S. *J. Org. Chem.* **2011**, *76*, 5092–5103.

86. Bae, I.; Han, H.; Chang, S. *J. Am. Chem. Soc.* **2005**, *127*, 2038–2039.

87. Fleury, L. M.; Wilson, E. E.; Vogt, M.; Fan, T. J.; Oliver, A. G.; Ashfeld, B. L. *Angew. Chem., Int. Ed.* **2013**, *52*, 11589–11593.

88. Cooper, D. S. *New Engl. J. Med.* **1984**, *311*, 1353–1362.

89. Choudhary, A.; Raines, R. T. *Chembiochem* **2011**, *12*, 1801–1807.

90. Jagodzinski, T. S. *Chem. Rev.* **2003**, *103*, 197–227.

91. Li, J. S.; Cheng, C.; Zhang, X. R.; Li, Z. W.; Cai, F. F.; Xue, Y.; Liu, W. D. *Chin. J. Chem.* **2012**, *30*, 1687–1689.

92. Pandey, L. K.; Pathak, U.; Mathur, S.; Suryanarayana, M. V. S. *Synthesis* **2012**, *44*, 377–379.

93. Noei, J.; Kosropour, A. R. *Tetrahedron Lett.* **2008**, *49*, 6969–6971.

94. Liu, L. F.; An, N.; Pi, H. J.; Ying, J.; Du, W. T.; Deng, W. P. *Synlett* **2011**, 979–981.

95. Chowdhury, N.; Dutta, S.; Dasgupta, S.; Singh, N. D. P.; Baidya, M.; Ghosh, S. K. *Photochem. Photobiol. Sci.* **2012**, *11*, 1239–1250.

96. (a) De Luca, L.; Giacomelli, G.; Porcheddu, A. *J. Org. Chem.* **2002**, *67*, 6272–6274. (b) Yadav, L. D. S.; Patel, R.; Srivastava, V. P. *Synthesis* **2010**, 1771–1776. (c) Kaur, N.; Sharma, P.; Kishore, D. *J. Chem. Pharm. Res.* **2012**, *4*, 1938–1946.

6

BROOK REARRANGEMENT

Michiko Sasaki and Kei Takeda

Department of Synthetic Organic Chemistry, Institute of Biomedical and Health Sciences, Hiroshima University, Hiroshima, Japan

6.1 INTRODUCTION

Brook rearrangement, discovered by Brook in 1958,[1] is an intramolecular anionic shift of a silyl group from a carbon atom to an oxygen atom (Scheme 6.1). Although this originally only reflected a 1,2-silyl shift (**1** → **2**), all types of 1,*n*-silyl shifts (**3** → **4**) are now commonly referred to as Brook rearrangements.[2] Aza-[3] or thia[4] -variants of the rearrangement involving silyl shifts from carbon to nitrogen or sulfur, respectively, have also been observed. A silyl shift from oxygen to carbon, the reverse process, is called a retro-Brook rearrangement.[5]

Although 1,*n*-Brook rearrangements other than the 1,2-shift have been becoming increasingly important from a synthetic point of view,[2,6] the main focus in this chapter will be on the 1,2-C-to-O silyl shift because it uniquely enables a carbonyl carbon atom in acylsilanes **5** to act as a 1,1-dipole (electrophilic/nucleophilic character) **6** in combination with a nucleophile (Scheme 6.2).

The term "Brook rearrangement" refers to the 1,2-Brook rearrangement throughout this chapter unless otherwise indicated. Also, the use of Brook rearrangements in tandem bond formation strategies up to the year 2000 has been excellently reviewed by Moser.[2e]

In this chapter, we present unique aspects of the rearrangement with a focus on (1) the production of carbanions under mild conditions, (2) the generation of synthetically valuable enol silyl ethers that are often the products, (3) the potential for incorporation of the Brook rearrangement in tandem and sequential anion relay

Molecular Rearrangements in Organic Synthesis, First Edition. Edited by Christian M. Rojas.

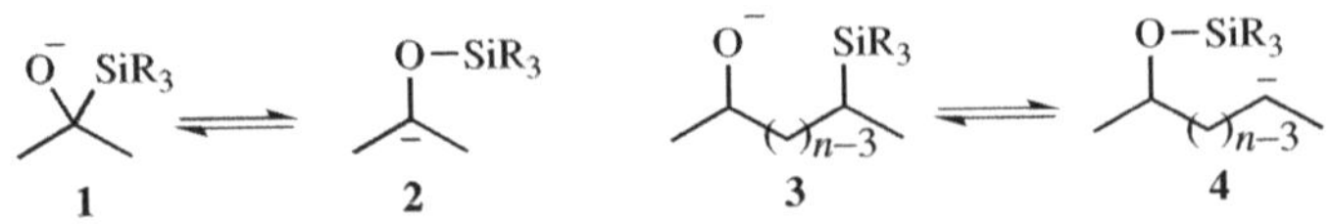

Scheme 6.1 1,2- and 1,*n*-silyl migrations.

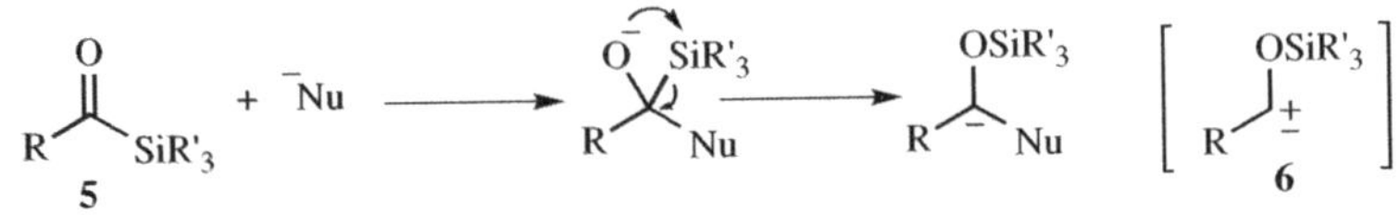

Scheme 6.2 Acylsilanes as 1,1-dipoles.

processes, and (4) the utility of the rearrangement in generating chiral, configurationally stable carbanions.

6.2 MECHANISM[2a,b]

A Brook rearrangement is generally an equilibrium via a pentacoordinate silicon transition state or intermediate, the latter being suggested as the structure in the gas phase by calculations.[7] Kinetic studies have revealed that a considerable negative charge develops in the transition state ($\rho \approx 4$) and that $\Delta S^{\ddagger}$ is greatly negative, being consistent with the cyclic species **9** (Scheme 6.3).[8] Although the driving force of the rearrangement is a strong Si—O bond relative to the Si—C bond, when an α-silyl alkoxide **8** is generated under conditions in which there is no electrophilic quenching agent, for example, addition of an organometallic reagent to acylsilanes, the position of equilibrium is determined by the relative stabilities of an oxyanion **8** and a carbanion **10**.

The stabilities can be altered by several factors including substituents on the carbanionic center, substituents on the silicon atom, the counter cation, and the solvent.[8] When the resulting carbanion **10** can be stabilized by an electron-withdrawing group or intercepted by a process such as elimination, the equilibrium can be shifted toward the carbanion side. Replacement of an alkyl group on silicon by a phenyl group or change of the counter cation from Li^+ to Na^+ or K^+, less coordinating to oxygen, accelerates the rearrangement. However, with electron-donating substituents such as an alkyl group that would destabilize the carbanionic center of **10** or when a metal ion such as Mg^{2+} strongly coordinates to the oxyanion, the rearrangement does not occur and α-silyl alcohols **7** result instead. The solvent can also substantially affect the rate of rearrangement. While in THF, a strongly donating solvent, the rearrangement can proceed at a synthetically viable rate, the rearrangement is strongly retarded in Et_2O and practically does not occur in toluene, leading the α-silyl alcohol. On the other hand, coordinating solvents such as HMPA favor the rearrangement by weakening the coordination of the metal cations to the oxyanion in **8**.

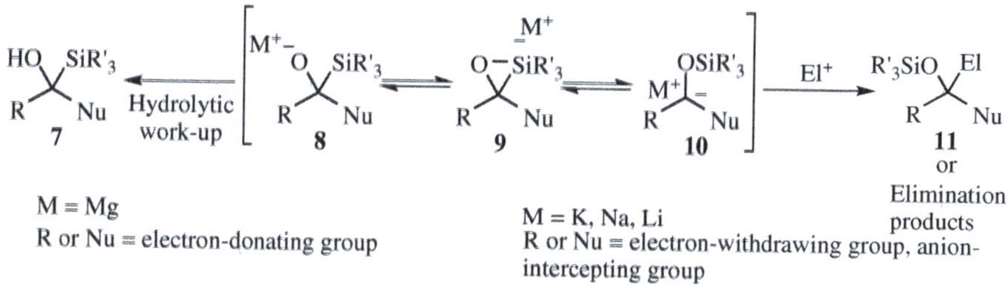

Scheme 6.3 1,2-Silyl migration as an equilibrium process.

Although retention of configuration at the silicon atom is observed when Si-stereogenic substrates are used,[9] the stereochemistry at the carbon atom depends on whether the resulting carbanion is stabilized by an anion-delocalizing group such as phenyl. Thus, the rearrangement proceeds with inversion of configuration when stabilized carbanions intervene[10] but with retention of configuration via unstabilized carbanions.[11]

6.3 METHODS FOR GENERATION OF α-SILYL ALKOXIDES

Several strategies are used to generate an α-silyl alkoxide as shown in Scheme 6.4.

Reactions of acylsilanes with a nucleophile (Scheme 6.4, Eq. 1) are among the most common and versatile methods for generation of an α-silyl alkoxide because various combinations of the two components for facilitating Brook rearrangement are possible.

Although deprotonation of the corresponding α-silyl alcohols (Scheme 6.4, Eq. 2) is the most direct method, this is of limited practical use because (1) alcohols are usually prepared by protonation of the alkoxides obtained by other methods (e.g., Scheme 6.4, Eq. 1), and (2) the incipient carbanion can be rapidly and irreversibly protonated by unrearranged α-silyl alcohol.

Addition of silylmetallic reagents to aldehydes or ketones (Scheme 6.4, Eq. 3) is a reaction that was first employed for generation of a silyl alkoxide by Gillman and Wu,[12] which led to the discovery of the Brook rearrangement. Although this method provides a facile route to the general synthesis of α-silyl alkoxides, the major drawback lies in that organosilyl metal reagents are only readily generated when there is at least one anion-stabilizing group such as phenyl attached to the silicon.[13] However, incorporation of a phenyl group on silicon can lead to a decrease in the stability of the silyl ether Brook product.

Regioselective β-ring opening of α,β-epoxysilanes (Scheme 6.4, Eq. 4) can also serve as a source of α-silyl alkoxides. Although attack of a nucleophile generally occurs at the α-position to produce β-silyl alkoxides, when the silyl group is bulky enough to suppress α-attack, β-ring opening can occur.

A carbanion generated at the γ-position of α,β-epoxysilanes via a process such as deprotonation can cause β-ring opening to provide α-silyl alkoxides (Scheme 6.4,

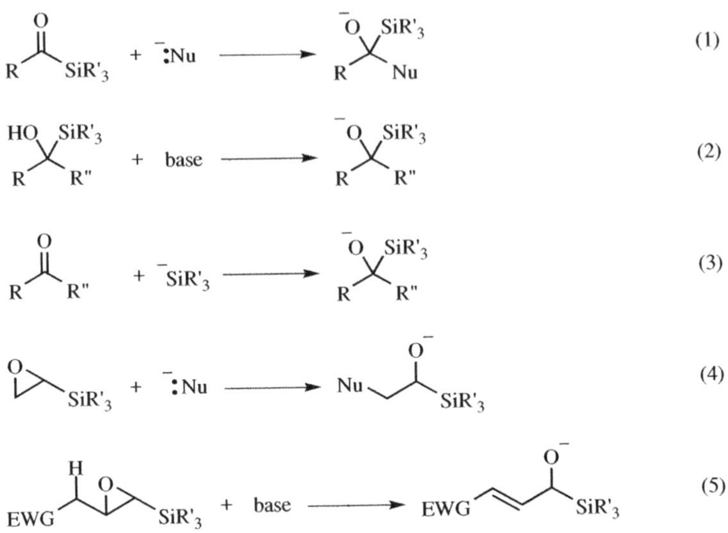

Scheme 6.4 Methods for the generation of α-silyl alkoxides.

Eq. 5). Taking into consideration the ready availability of enantiomerically pure epoxides, this method opens the possibility for using epoxides as a source of chiral carbanions via Brook rearrangement.

6.4 SYNTHETIC REACTIONS USING BROOK REARRANGEMENTS IN THE REACTIONS OF ACYLSILANES WITH NUCLEOPHILES

Brook rearrangements are favorable enough for further synthetic transformation only when the resulting carbanion bears electron acceptors, such as a conjugating group or a leaving group. Consequently, synthetic uses of the rearrangements were rather limited, despite the pioneering studies by Brook, until the seminal work by both Kuwajima's[14] and Reich's[15] groups in the late 1970s to early 1980s. The success in synthetic applications of the Brook rearrangement depends on how one can shift the equilibrium toward the carbanion side (see Scheme 6.3). In this regard, reactions of acylsilanes and organometallic reagents provide various combinations of the two components that allow facile C-to-O silyl migration. This section is subdivided according to the mode of facilitating these rearrangements and begins with methods for preparation of the acylsilane substrates.

6.4.1 Preparation of Acylsilanes

Numerous methods are available for the preparation of acylsilanes and have been adequately covered in reviews.[16,2f] Hence, procedures potentially applicable for a large-scale preparation are briefly described here.

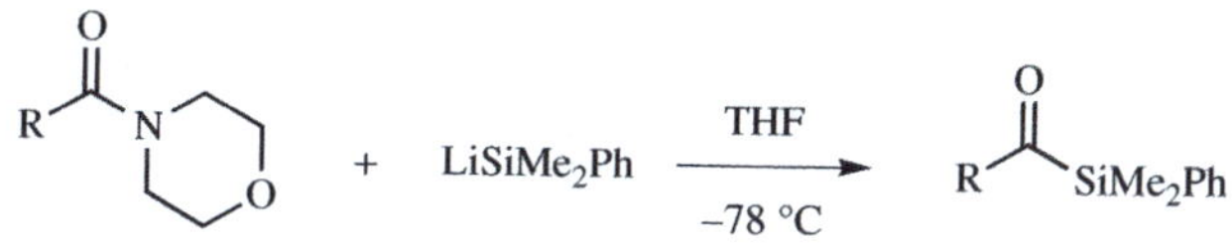

Scheme 6.5 Synthesis of acylsilanes from morpholine amides.

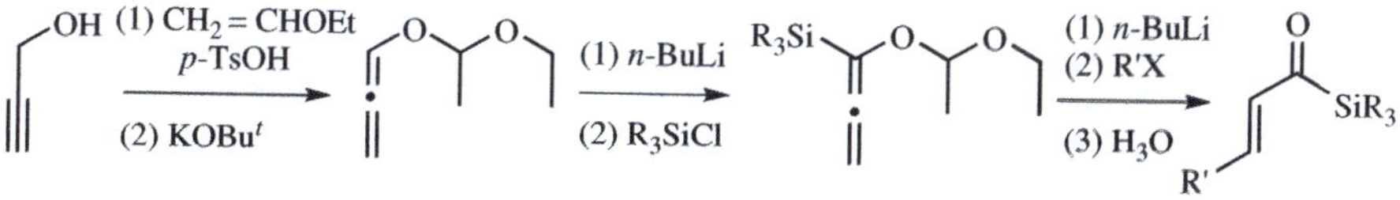

Scheme 6.6 Synthesis of α,β-unsaturated acylsilanes from 1-silyl-1-alkoxyallenes.

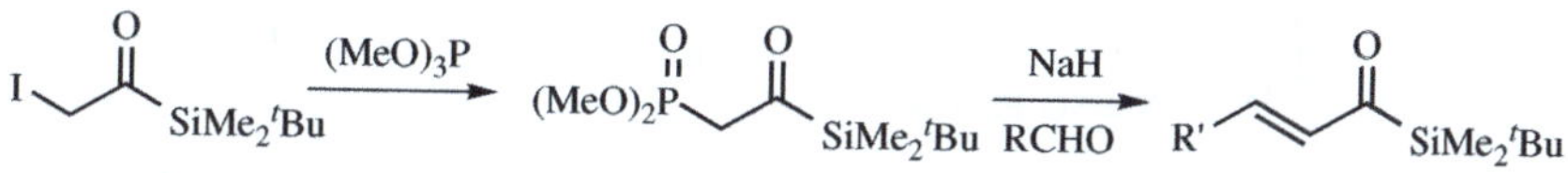

Scheme 6.7 Synthesis of α,β-unsaturated acylsilanes by Horner–Wadsworth–Emmons reaction of α-phosphonoacylsilane.

For alkyl-substituted acylsilanes, conventionally prepared through the use of 1,3-dithianes,[17] Scheidt and coworkers reported a method that uses reactions of morpholine amides with dimethylphenylsilyllithium, affording the corresponding acylsilanes in good yields (Scheme 6.5).[18]

Versatile preparation of conjugated acylsilanes involves hydrolysis of α-silylated allenyl ethers that can be derived from lithiated allenyl ethers.[19] The method is applicable to the synthesis of not only a variety of substituted α,β-unsaturated acylsilanes but also acetylenic acylsilanes (Scheme 6.6).

An alternative method for substituted α,β-unsaturated acylsilanes, especially for β-aryl derivatives, uses Horner–Wadsworth–Emmons reaction of α-phosphonoacylsilanes, which can be derived from α-iodoacylsilanes via Arbuzov reaction (Scheme 6.7).[20]

6.4.2 Reaction of Acylsilanes with Nucleophiles, Either of Which Contains an α-Leaving Group

The dawn of synthetic application of the Brook rearrangement dates to 1967 with Brook's pioneering work on reactions of acylsilanes with diazomethane[21] or Wittig reagents,[22] which can be regarded as nonmetallic nucleophiles containing an α-leaving group. The results of these reactions, particularly with Wittig reagents, showed that the rearrangement could occur with β-elimination to form an enol silyl ether **12** if the generated carbanion had an α-leaving group. This concept became the cornerstone for synthetic application of the Brook rearrangement (Scheme 6.8).

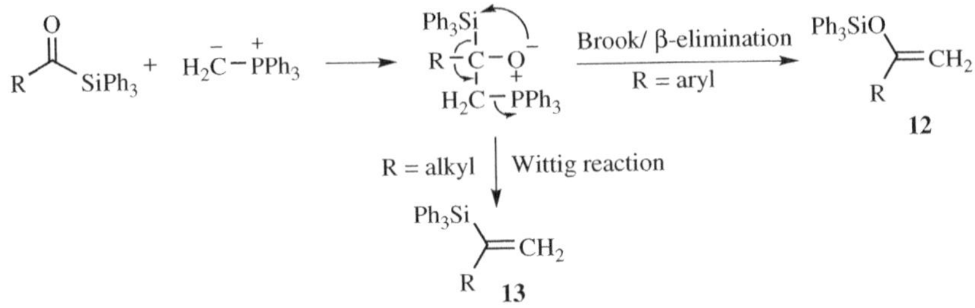

Scheme 6.8 Reactions of acylsilanes with Wittig reagents.

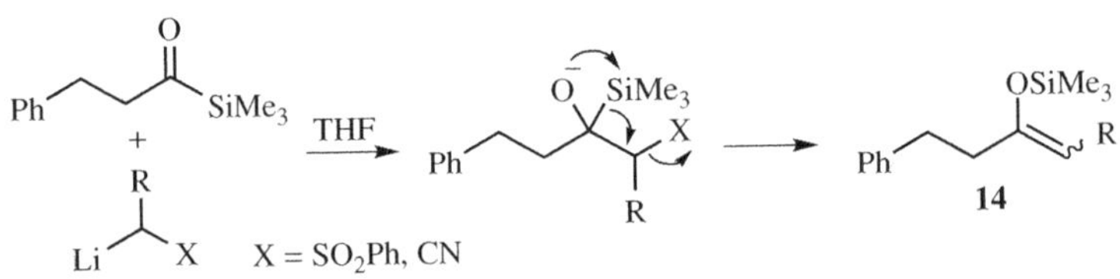

Scheme 6.9 Reactions of acylsilanes with α-lithiated sulfones or nitriles.

In this system, however, the presence of an anion-stabilizing acyl group (R in Scheme 6.8) such as phenyl is essential for success of the rearrangement, and alkyl-substituted acylsilanes afforded normal Wittig reaction products **13**. This side reaction presented an obstacle to applications in organic synthesis for a long time.

A major breakthrough came in 1979 from Reich *et al.*, who circumvented the problem by using organolithium nucleophiles bearing an α-leaving group such as phenylsulfonyl and cyano.[23] Thus, reaction of 3-phenylpropanoylsilane with $LiCH_2X$ ($X = SO_2Ph$, CN) afforded enol silyl ether **14** via Brook rearrangement followed by β-elimination, enabling regioselective formation of enol silyl ethers (Scheme 6.9). In this case, an activating acyl substituent such as aryl is not necessarily required.

Related reactions were reported by Kuwajima *et al.* in 1979, who used 2-lithiofuran as a nucleophile. Here, C—O bond cleavage in the furan shifted the equilibrium toward the carbanion side to give dienol silyl ether derivatives **15** (Scheme 6.10).[24]

An α-bond cleavage driven by relief of ring strain can force the Brook-mediated reaction to completion. Clayden *et al.* found that reaction of benzoyltrimethylsilane with cyclopropyllithium affords phenyl ketone **16** via a Brook rearrangement followed by a 1,5-silyl migration (Scheme 6.11).[25] The moderate yields observed are attributed to competitive formation of an (*E*)-enolate, incapable of 1,5-silyl migration, and insufficient driving force from the release of cyclopropane ring strain alone. However, with change in the leaving atom from carbon to oxygen (i.e., epoxide instead of cyclopropane), relief of ring strain was proved to be an effective driving force for a Brook rearrangement.[26]

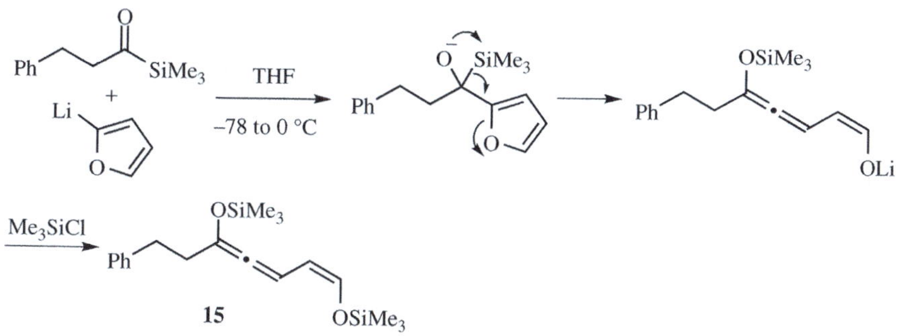

Scheme 6.10 Reactions of acylsilanes with 2-lithiofuran.

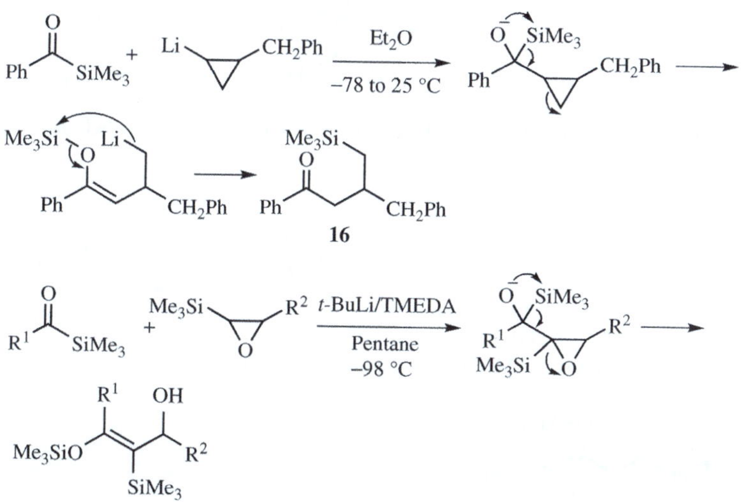

Scheme 6.11 1,2-Silyl migrations driven by relief of ring strain.

Reich *et al.* also reported a similar elimination strategy that involves a combination of acylsilanes bearing an α-leaving group and an alkyllithium.[23a,b] Thus, reaction of α-phenylthioacylsilane **17**, derived from the corresponding lithium enolate, with a variety of nucleophiles proceeded smoothly at lower temperatures to give an enol silyl ether **18** in a stereochemically defined manner (Scheme 6.12).

Similarly, addition of alkyl, vinyl, or aryllithiums to α-halo-α,β-unsaturated acylsilanes generates α-silyl alkoxide **19**, which can collapse to give allenol silyl ethers **20** (Scheme 6.13).[27]

6.4.3 Reaction of Acylsilanes with Nucleophiles, in Which Generated Carbanions Can Be Stabilized by Conjugating Groups

The equilibrium between α-silyl alkoxides and siloxy carbanions can be shifted toward the carbanion side by introduction of a conjugating group into either or both

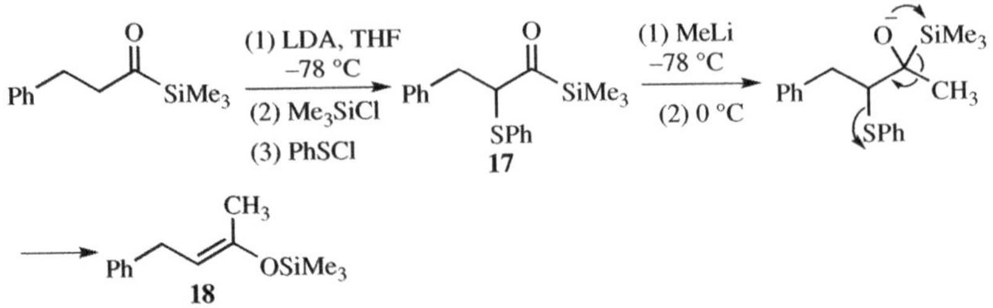

Scheme 6.12 Reactions of α-phenylthioacylsilanes with alkyllithium.

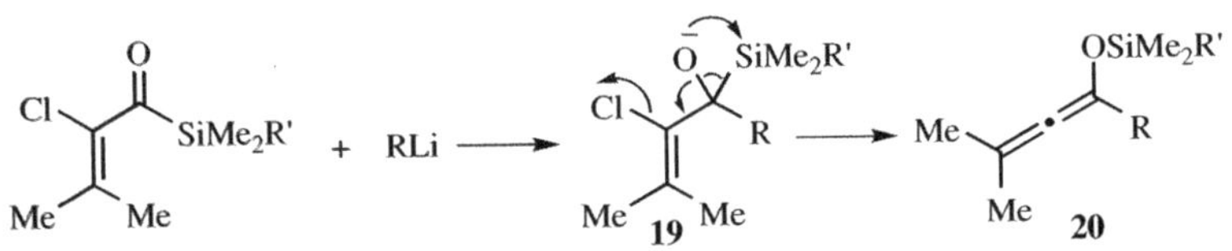

Scheme 6.13 Formation of siloxyallenes from α-halo-α,β-unsaturated acylsilanes and alkyllithiums.

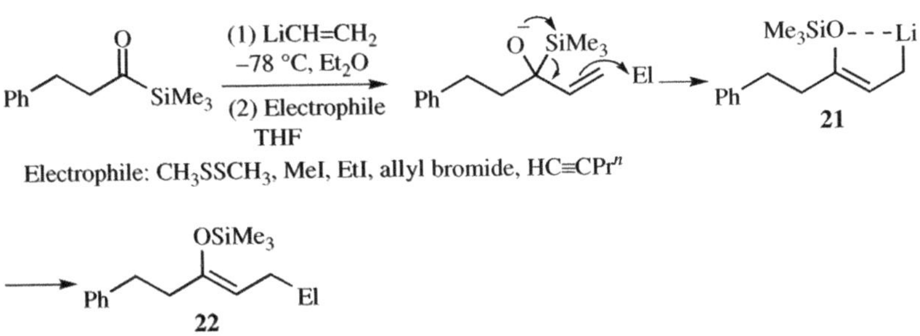

Scheme 6.14 Stereoselective formation of α-substituted enol silyl ethers from acylsilanes and vinyllithium.

the acylsilane and the nucleophile. In 1980, Reich *et al.* reported that treatment of alkyl-substituted acylsilanes with vinyllithium followed by a variety of electrophiles affords α-substituted enol silyl ethers **22** via a siloxy allyllithium intermediate **21** (Scheme 6.14).[28] Similar reactions using phenyllithium give products in which electrophilic quenching occurred at the benzylic position.

Using an alkynyllithium instead of a vinyllithium afforded the siloxyallenyllithium, which can be trapped by an alkylating agent (Scheme 6.15). In this case, the rearrangement did not take place in the absence of an alkylating agent, suggesting that the lithium alkoxide **23** is in equilibrium with a small amount of the allenyl lithium intermediates.

To circumvent this problem, Marek and coworkers utilized transmetalation to an organozinc species from the adduct of alkynyl Grignard addition to acylsilanes.

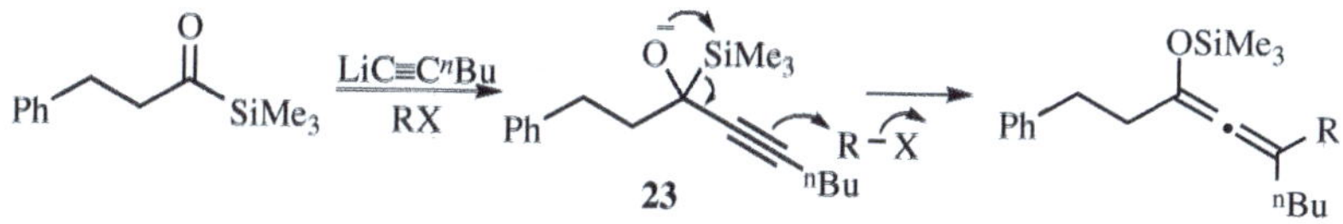

Scheme 6.15 Formation of siloxyallenes from acylsilanes and alkynyllithiums.

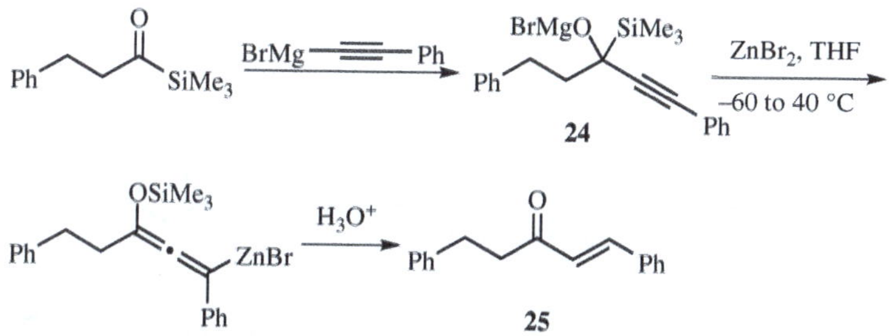

Scheme 6.16 Formation of siloxyallenes via zinc salt-promoted Brook rearrangement.

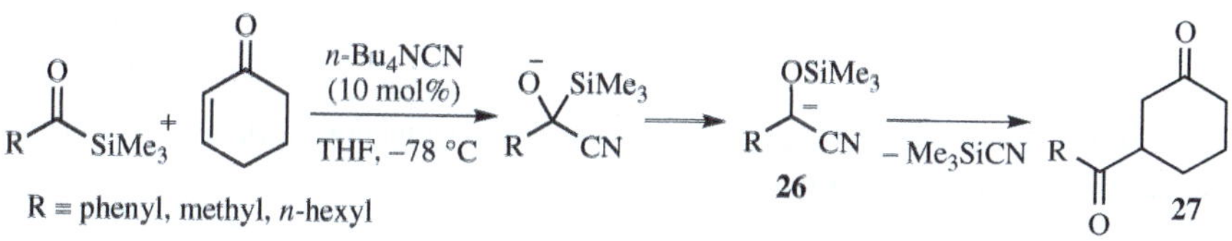

Scheme 6.17 Cyanide ion-catalyzed Michael addition of acylsilanes to α,β-unsaturated carbonyl compounds.

Transmetalation shifts the equilibrium toward the allenyl anion because of the weaker Zn—O bond compared to the Mg—O bond.[29] Thus, when zinc bromide is added to α-silylmagnesium carbinol **24**, derived from the reaction of acylsilane with alkynylmagnesium bromide, enone **25** is obtained after acidic hydrolysis (Scheme 6.16).

Cyanide ion, a nitrogen variant of an alkynyl anion, is expected to facilitate the Brook rearrangement due to the anion-stabilizing capability of the resulting nitrile group. Conventional cyanide ion sources such as KCN, however, are not sufficient because of their low solubility in common organic solvents to generate an appropriate concentration of the α-silyl alkoxide for rearrangement. Degl'Innocenti et al. found that reaction of a catalytic amount of n-Bu_4NCN with acylsilanes affords an anion of O-silyl cyanohydrin **26** that can participate in 1,4-addition to α,β-unsaturated ketones to give 1,4-diketone **27** with liberation of Me_3SiCN (Scheme 6.17).[30,23b] Notably, this process is compatible with alkyl-substituted acylsilanes and represents a nucleophilic acylation that does not require a metalation step.

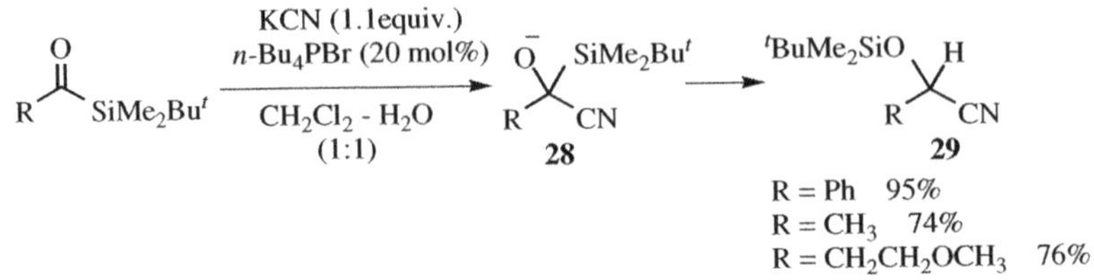

Scheme 6.18 Formation of O-silylcyanohydrins by cyanide ion-initiated Brook rearrangement under phase-transfer catalytic conditions.

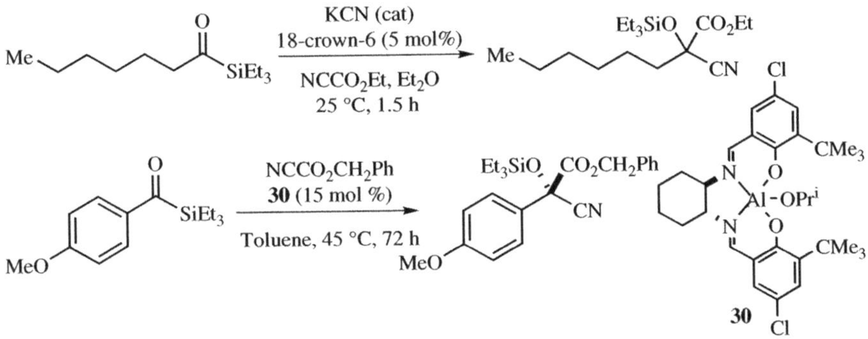

Scheme 6.19 Enantioselective cyanation/Brook rearrangement/C-acylation of acylsilanes catalyzed by chiral metal alkoxides.

Takeda and Ohnishi reported that KCN can be effective for the rearrangement under phase-transfer catalysis conditions.[31] Reaction with acylsilanes, including alkyl-substituted ones, in a CH_2Cl_2-H_2O two-phase system at room temperature proceeds without the quenching of α-silyl alkoxide **28** by water to give *O*-silyl cyanohydrin derivative **29** (Scheme 6.18).

Johnson and coworkers successfully extended this approach to a catalytic system using alkyl cyanoformates, in which cyanide ion is expelled upon C-acylation of the Brook-derived carbanion with the cyanoformate (Scheme 6.19).[32] Furthermore, the Johnson group developed an enantioselective variant by carrying out the reaction in the presence of a chiral aluminum complex **30**.[33]

Anion-stabilizing heteroatom nucleophiles can also facilitate Brook rearrangement upon addition to acyl silanes. A carbanion generated via Brook rearrangement in the adduct from lithium diethylphosphite and acylsilanes can undergo intramolecular Michael addition to an α,β-unsaturated ester, affording four- to six-membered carbocycles **31** as single diastereomers together with uncyclized products (Scheme 6.20).[34]

When the acylsilane carbonyl group is conjugated with a carbon–carbon multiple bond, Brook rearrangement should occur following attack of a nucleophile in the

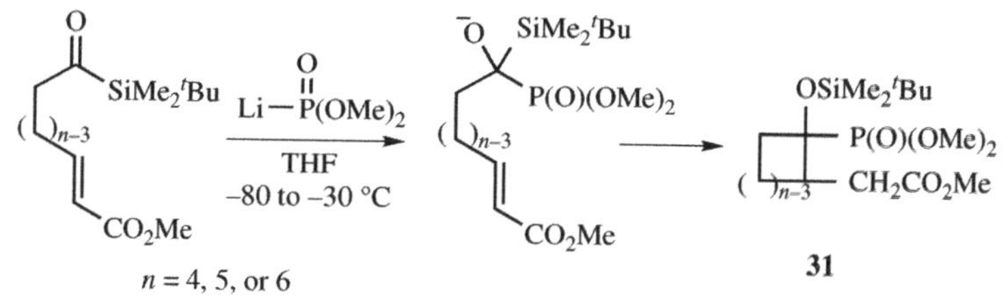

Scheme 6.20 Lithium phosphite-initiated Brook rearrangement/intramolecular Michael additions.

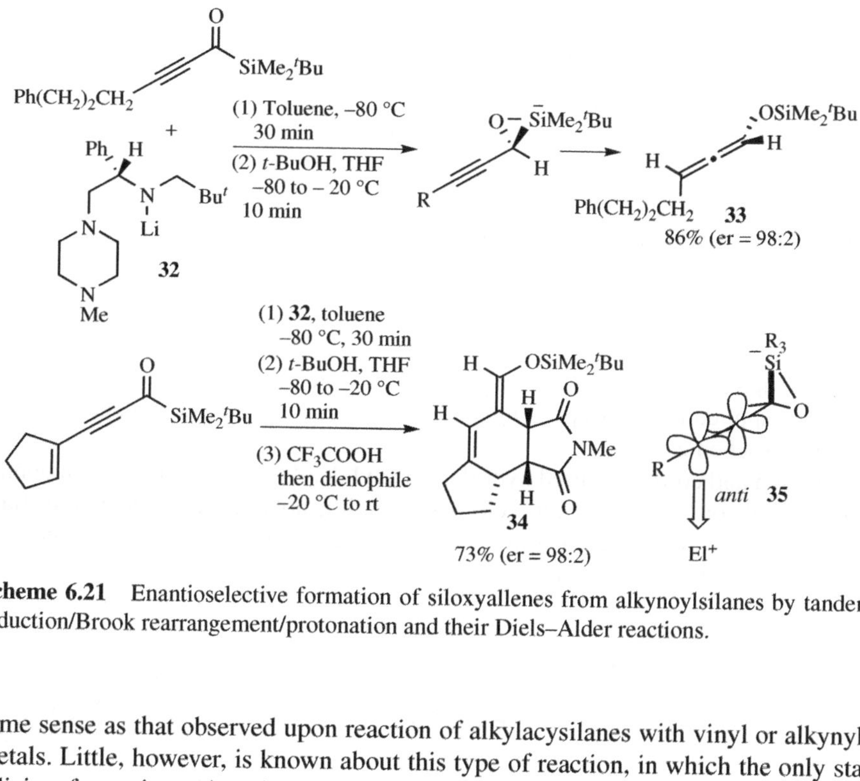

Scheme 6.21 Enantioselective formation of siloxyallenes from alkynoylsilanes by tandem reduction/Brook rearrangement/protonation and their Diels–Alder reactions.

same sense as that observed upon reaction of alkylacysilanes with vinyl or alkynylmetals. Little, however, is known about this type of reaction, in which the only stabilizing factor is conjugation with a carbon–carbon multiple bond. Takeda *et al.* reported that enantioenriched α-silyl propargyl alkoxides, generated from alkynoylsilanes by enantioselective reduction[35] using a chiral lithium amide **32**, undergo Brook rearrangement followed by antimode protonation (**35**) by t-BuOH to give siloxyallene derivatives **33** in an excellent enantioselectivity (Scheme 6.21).[36] This tandem process has been successfully extended to enynoylsilanes to give vinylallenes, which can be trapped by a Diels–Alder reaction to afford **34**.

Acylsilanes whose carbonyl groups are conjugated with a carbon–oxygen double bond are excellent substrates for Brook rearrangement after nucleophilic attack

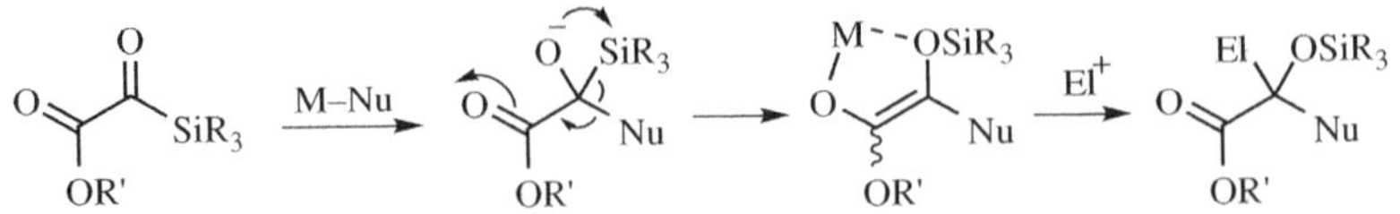

Scheme 6.22 Three-component coupling reactions of silylglyoxylates, nucleophiles, and electrophiles.

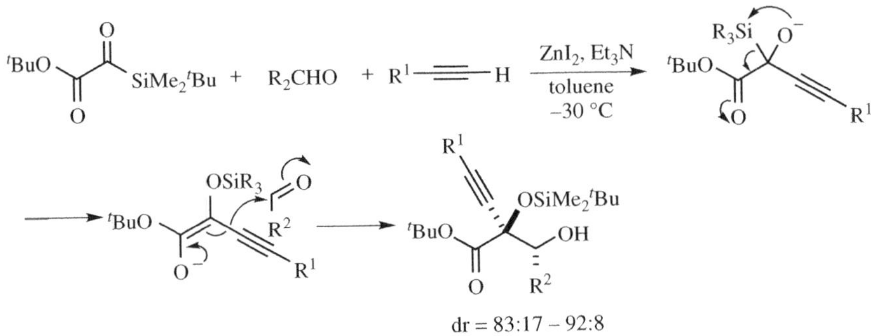

Scheme 6.23 Three-component coupling reactions of silylglyoxylates, acetylides, and aldehydes.

because of better stabilization of the generated carbanion. The resulting enolate can participate in reactions with various electrophiles. Johnson and coworkers successfully applied this concept using silylglyoxylates[37] in three-component coupling reactions (Scheme 6.22).

Johnson and coworkers used hydride,[38] acetylide,[39] or vinyl anion[40] nucleophiles and aldehyde or Michael acceptor electrophiles. It is notable that nucleophilic attack by acetylide ion takes place chemoselectively at the acylsilane carbonyl over the aldehyde carbonyl, likely due to a kinetic preference (Scheme 6.23).

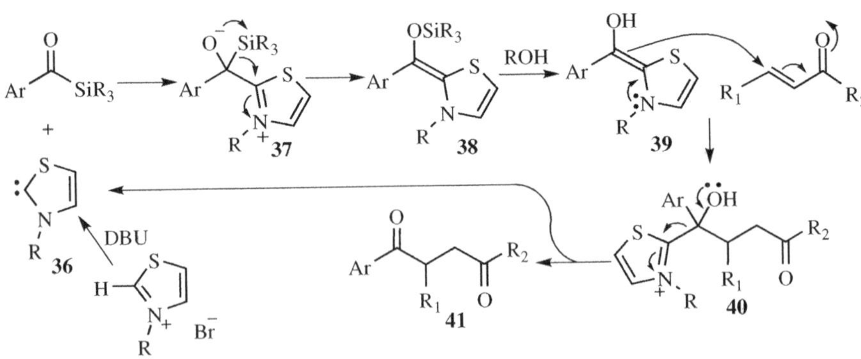

Scheme 6.24 Sila-Stetter reactions between acylsilanes and α,β-unsaturated ketones catalyzed by heterocyclic nucleophilic carbenes.

The nucleophilic and electron-accepting properties of heterocyclic nucleophilic carbenes **36** were also used in combination with the electrophilic/nucleophilic character of acylsilanes via Brook rearrangement, leading to the invention of a sila-Stetter reaction by Scheidt and coworkers (Scheme 6.24).[41] The iminium structure in **37**, generated by addition of the carbene catalyst **36** to the acylsilanes, promotes a Brook rearrangement to afford enol silyl ether **38**. The alcohol additive present in the reaction causes desilylation to produce nucleophilic enaminol **39**, which adds to α,β-unsaturated ketones to give **40**. The formation of aryl ketone expels the carbene catalyst and produces 1,4-diketone **41**.

6.4.4 Use of Ketone Enolates as Nucleophiles Containing an Electron-Accepting Moiety

Enolates, after addition to acylsilanes, can serve as acceptors of a β-carbanion generated by Brook rearrangement due to the electrophilic character of a carbonyl group. This conceptually new approach was first demonstrated in 1993 by Takeda *et al.*[42] Reaction of benzoyltrimethylsilane with enolates of alkyl methyl ketones produces *cis*-1,2-cyclopropanediol derivatives such as **42**, which are formed by internal carbonyl attack of the resulting carbanion (Scheme 6.25). The same reaction was observed with crotonoylsilane to give **43**. Cyclopropanediols were not formed from alkanoylsilanes, indicating that in order for the silyl migration to be efficient, both the presence of a neighboring ketone group and stabilization by the adjacent unsaturation are important.

Introduction of an α-carbanion-stabilizing group on the β-position of acryloylsilanes further enhanced the synthetic potential of the rearrangement in combination with the electron-accepting ability of a carbonyl group.[43] Highly functionalized cyclopentanone derivatives can be obtained by reactions of β-phenylthio or β-trimethylsilyl derivatives of acryloylsilanes with methyl ketone enolates, in which the products are formed via different pathways depending on the α-anion-stabilizing ability of the β-substituent (Scheme 6.26). With the more α-anion-stabilizing phenylthio group (see also: Section 6.5, Table 6.1), the carbanion generated by Brook rearrangement delocalizes to give an intermediate **44** and then cyclizes to **45**. When the β-substituent is a less anion-stabilizing trimethylsilyl group, the carbanion generated by Brook rearrangement attacks internally on the carbonyl group without allylic delocalization to give vinylcyclopropanolates **46**, which undergo oxyanion-accelerated vinylcyclopropane–cyclopentene rearrangement to give **47**. These proposed reaction pathways have been established by mechanistic

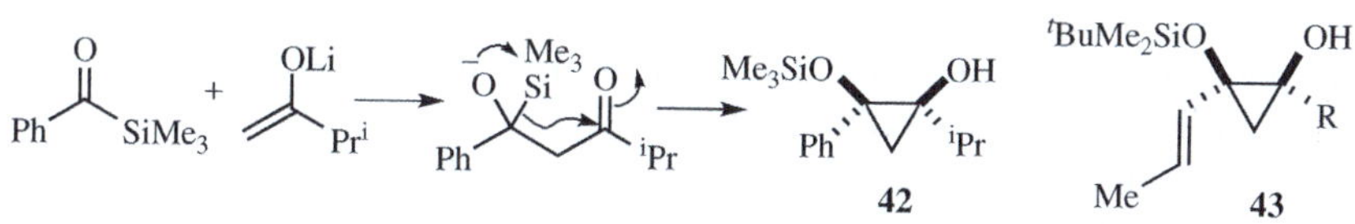

42 **43**

Scheme 6.25 Formation of 1,2-cyclopropanediols from acysilanes with ketone enolates.

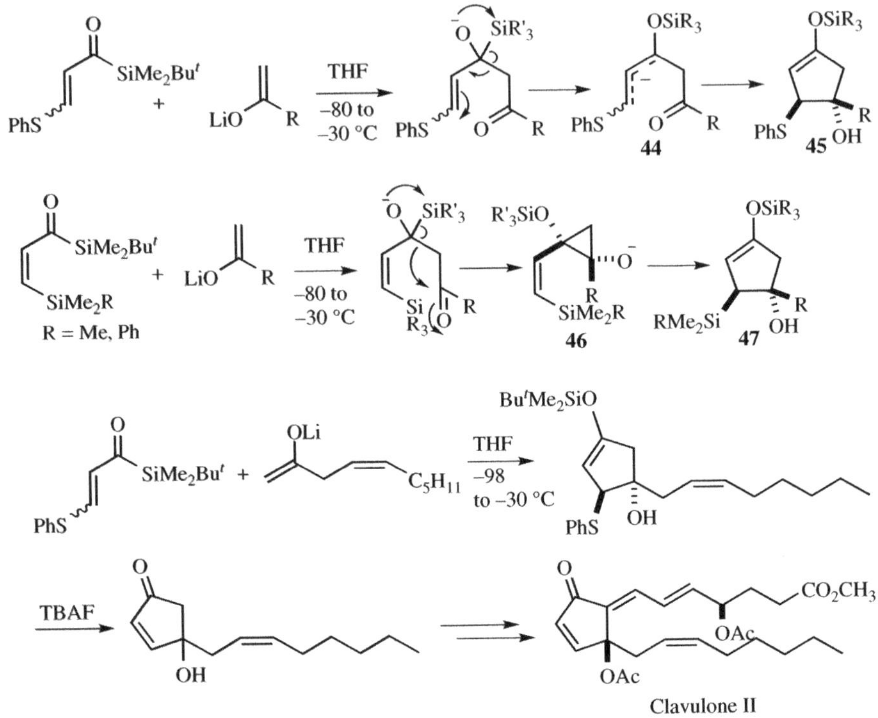

Scheme 6.26 Brook rearrangement-mediated [3+2] annulations.

studies including *in situ* generation of **46** by another route.[44] This method was applied to the synthesis of natural products, including clavulones.[42,43a,45]

Detailed mechanistic analysis of the [3+2] annulation led to the development of a Brook rearrangement-mediated [3+4] annulation (Scheme 6.27).[46,19a,b] Thus, use of an alkenyl methyl ketone enolate **49** in the reaction with (*E*)- or (Z)-**48** resulted in the formation of a 5,6-*cis*- or 5,6-*trans*-substituted derivative **52**, respectively. The observed stereospecificity results from an anionic oxy-Cope rearrangement of 1,2-divinylcyclopropanediol intermediate **51**, generated via Brook rearrangement of the 1,2-adduct **50**.[47] Isolation of vinylcyclopropanol derivatives from the reaction with the lithium enolate of 2′-bromoacetophenone provides strong support for the proposed mechanism. Further support is obtained from the fact that whereas the reaction of 1,2-divinylcyclopropyl acetates **53** with 2 equiv. of MeLi affords cycloheptenones stereospecifically, the reaction of *β*-phenylthioacryloylsilanes, which generates not vinylcyclopropanolates but delocalized allylcarbanions, produces the corresponding [3+2] annulation products. This method has been applied to synthesis of the tricyclic core of allocyathin B_2.[48]

This synthetic strategy could be extended beyond enolates of alkenyl methyl ketones. The use of enolate **54** ($X=CH_2$), derived from 2-cycloheptenone, as the four-carbon unit in [3+4] annulation instead of the enolates of alkenyl methyl

TABLE 6.1 Evaluation of α-carbanion Stabilizing Ability of Substituents Based on the Relative Rates of Brook Rearrangement

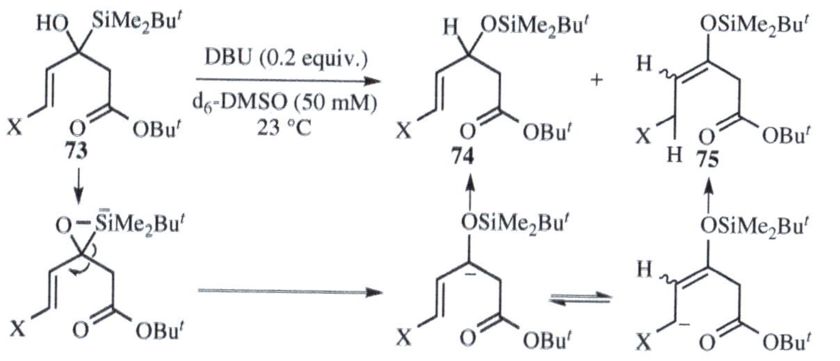

X	product	$t_{1/2}$ (min)
S(O)Ph	**75**	<1.0
SPh	**75**	3.2
Ph	**75**	5.5
$SiMe_3$	**74** + **75** (ca 1:2)	27.5
Cl	**74**	**73**: **74** = 2.2 : 1 (25 h)
Br	**74**	**73**: **74** = 4.5 : 1 (43 h)

ketones produced bicyclo[3.3.2]decenone derivatives **55** (Scheme 6.28).[49] The two-atom internal tether in these products could be oxidatively cleaved after conversion to α-hydroxy ketone **56** to give the *cis*-3,4,8-trisubstituted cyclooctenone enol silyl ethers **57** stereoselectively. This method has also been successfully applied to the construction of oxygen eight-membered heterocycles using enolates of 6-oxacyclohept-2-en-1-one **54** (X = O), affording eight-membered oxygen heterocycles **57** (X = O).[50] The versatility of the annulation has been demonstrated through the formal total synthesis of (+)-prelaureatin, a biogenetic precursor of several members of the laurenan structural subclass.[51]

Another approach to the construction of eight-membered carbocycles using the same concept involves reaction of a vinyllithium with an acylsilane containing an appropriately positioned alkenoyl moiety.[52] Reaction of β-alkenoylsilanes **58** with β-(trimethylsilyl)vinyllithium affords eight-membered carbocycles **61** via a tandem process that involves (1) Brook rearrangement in an adduct **59**, (2) formation of divinylcyclobutanolate **60** via an internal carbonyl attack by the resulting carbanion, and (3) anionic oxy-Cope rearrangement (Scheme 6.29).

Scheidt and coworkers reported the addition of amide enolates to acylsilanes for generation of β-silyloxy homoenolate equivalents **62**, based on the fact[53] that less electrophilic β-carbonyl groups disfavor the formation of cyclopropanolates **63** by internal carbanion attack (Scheme 6.30).[54] Instead, the carbanion generated *in situ* can be trapped by alkyl halides, aldehydes, ketones, and imines. The use of optically active amide enolates delivers β-hydroxy amides with high levels of diastereoselectivity.

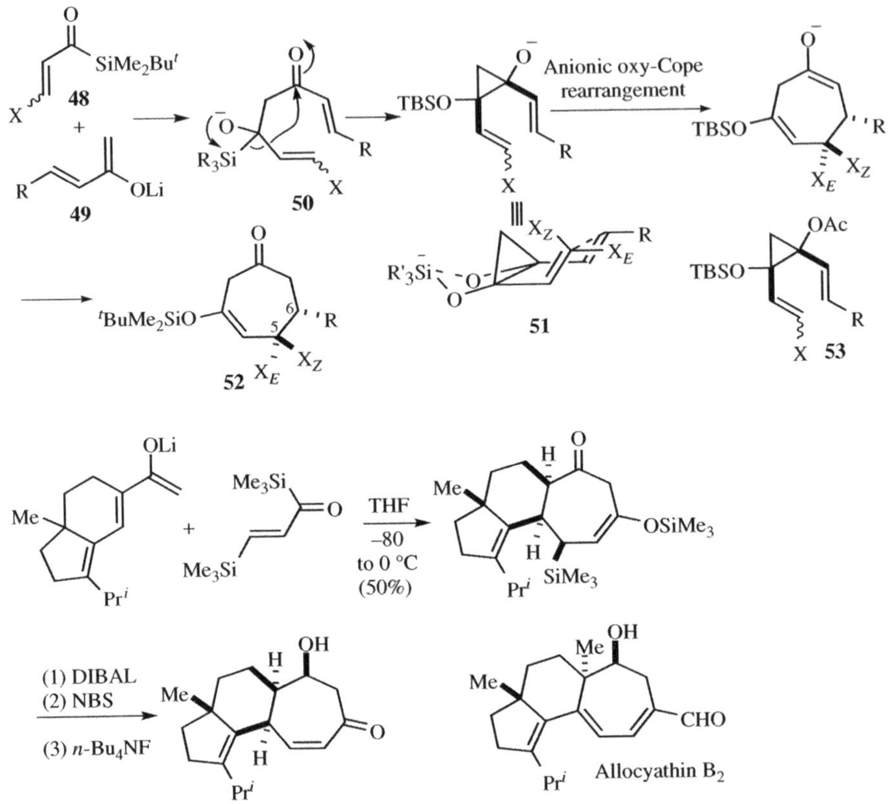

Scheme 6.27 Brook rearrangement-mediated [3+4] annulations.

6.5 SYNTHETIC REACTIONS USING BROOK REARRANGEMENTS TRIGGERED BY DEPROTONATION OF α-SILYL ALCOHOLS

The common approach to α-silyl alcohols involves acylsilane-based processes such as nucleophilic addition of Grignard reagents,[55] reduction,[36,56,16c] and retro-Brook rearrangement[5c,57] in α-lithio silyl ethers.

In his pioneering work[14] on synthetic application of a Brook rearrangement, Kuwajima and coworkers found that α-silyl allylalcohols **64** can be isolated in the reactions of acylsilanes with vinylmagnesium bromide, in contrast to the case of its lithio variant, because of the higher covalent character of an O—Mg bond (Scheme 6.31).[58] A merit for isolating α-silyl alcohols is that further reaction can be controlled depending on whether a catalytic or a stoichiometric amount of a base is used. Thus, with a catalytic amount of n-BuLi, α-siloxyallylanion **66** generated by Brook rearrangement is in equilibrium with an internally chelated species **67**, which undergoes protonation by the starting silyl alcohol, forming enol silyl ether **68** and regenerating **65**. On the other hand, the use of a stoichiometric amount of

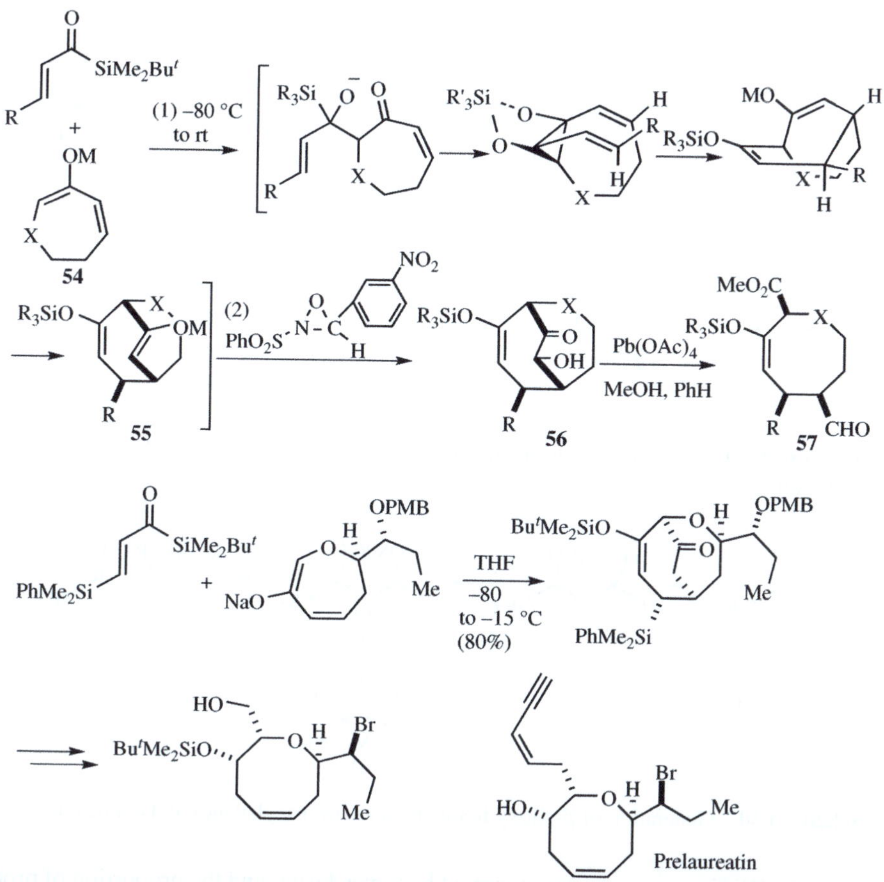

Scheme 6.28 Formation of eight-membered carbocycles and oxygen-heterocycles via Brook rearrangement-mediated [3+4] annulations.

the base enables allylic alkylation of homoenolate equivalent **67**, leading to products **69**. This method is applicable to the synthesis of siloxyallenes via zinc acetylide addition and isolation of the corresponding propargyl α-silyl alcohols (e.g., **70**).[59] This concept was applied to an enantioselective synthesis of siloxyallenes **71** via an alkyne addition using tridentate Schiff base ligand **72** (Scheme 6.32).[60]

Since the rate of a Brook rearrangement depends on the α-anion-stabilizing ability of the substituent on the carbon atom where the carbanion is generated, comparison of rearrangement rates provides a method for semiquantitative evaluation of α-carbanion-stabilizing ability. Takeda *et al.* focused on a DBU-catalyzed Brook rearrangement of α-silyl allylalcohol derivatives **73** (Table 6.1).[53] These α-silyl alcohol can be isolated from reactions of β-substituted acryloylsilanes with an ester enolate, because a Brook rearrangement is suppressed by the lower electrophilicity of an ester carbonyl.[54] As the α-carbanion-stabilizing ability of the substituent X increases, the

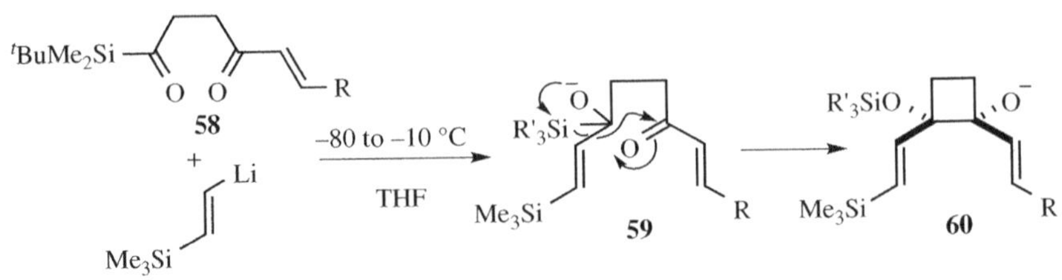

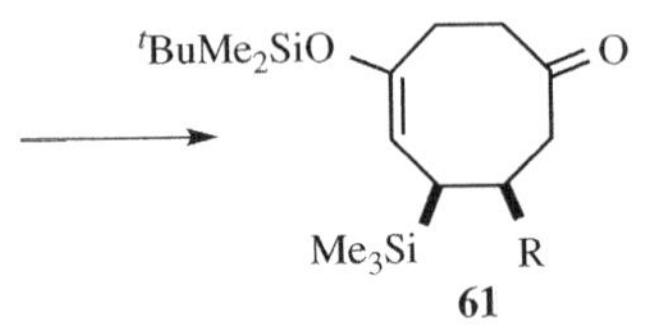

R	Yield (%)
CH_3	45
n-Pr	42
n-C_7H_{15}	45
c-hex	42

Scheme 6.29 Formation of eight-membered carbocycles from *β*-alkenoyl acylsilanes and vinyllithiums.

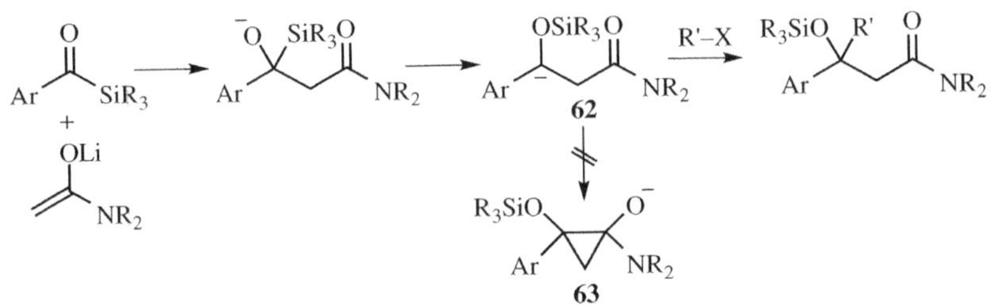

Scheme 6.30 Generation of *β*-siloxy homoenolates from acylsilanes and amide enolates.

rate of the DBU-promoted rearrangement becomes faster, and the proportion of product protonated at the *α*-position of group X increases. The reaction proceeded at a reasonable rate when using 0.2 equiv. of DBU in d_6-DMSO at 23 °C. Under these conditions, the half-lives of the reactions were measured. The *α*-carbanion-stabilizing ability of a phenylthio group is considerably greater than that of a trimethylsilyl group and even greater than that for a phenyl group.

The unique feature of a Brook rearrangement, that is, that carbanions can be generated by deprotonation of an alcoholic proton by a relatively weak base, has been used to evaluate the propensity for racemization of configurationally very labile carbanions such as those next to a cyano or an ester group.[61] Treatment of *α*-silyl alcohols **76** with a catalytic amount of *t*-BuOLi in *t*-BuOH afforded **78**, a Brook rearrangement-mediated S_E2'-type solvolytic protonation product, with ee varying between 0% and 96% (Table 6.2). The ee values observed in **78** reflect the effects of substituents X on the configurational stability of a lithiocarbanion **77**. The results shown in Table 6.2 indicate that the method makes it possible to semiquantitatively evaluate the propensity for racemization of a carbanion next to a conjugative electron-withdrawing group or an anion-stabilizing heteroatom substituent, which has been impossible by other means including the Hoffmann test.[62]

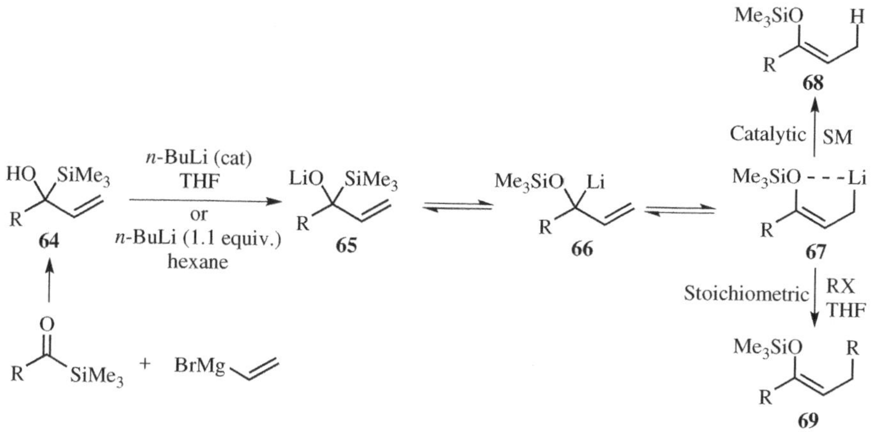

Scheme 6.31 Stoichiometry-dependent product distribution in base-mediated reactions of α-silyl allylalcohols.

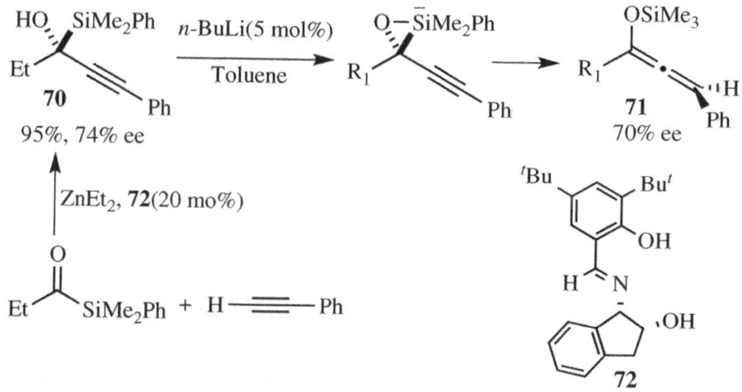

Scheme 6.32 Base-catalyzed rearrangement of propargyl α-silyl alcohols to siloxyallenes.

Scheidt recently reported that α-silyl alkoxides generated from α-silyl silylethers **79** by fluoride-induced desilylation instead of deprotonation of α-silyl alcohol can be trapped by primary alkyl and by allylic and benzylic electrophiles via a Brook rearrangement (Scheme 6.33).[63]

6.6 SYNTHETIC REACTIONS USING BROOK REARRANGEMENTS TRIGGERED BY ADDITION OF SILYLMETALLIC REAGENTS

There are few synthetic reactions in which silylmetallic reagents cause a Brook rearrangement in reactions with carbonyl compounds. Most of them involve the reaction with substrates bearing an α-leaving group such as a siloxy (alkoxy) or

TABLE 6.2 Propensity for Racemization of Lithiocarbanions Next to an Electron-withdrawing Group

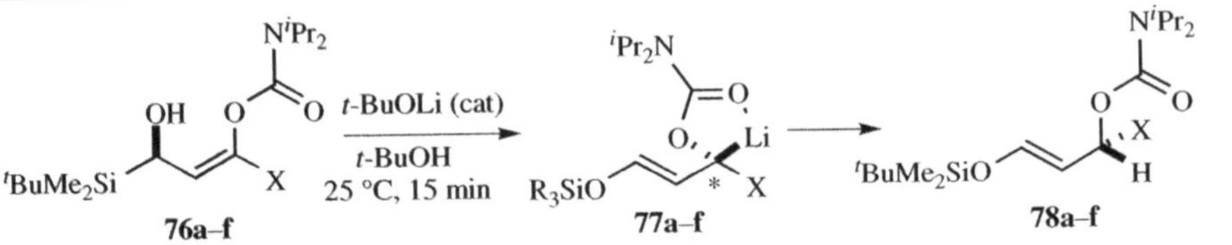

76	X	yield [%]	ee [%]
a	SO_2*p*-Tol	100	97
b	$P(O)(OEt)_2$	97	96
c	CN	91	63
d	$CONMe_2$	85	34
e	CO_2Et	92	32
f	C(O)Ph	57	0

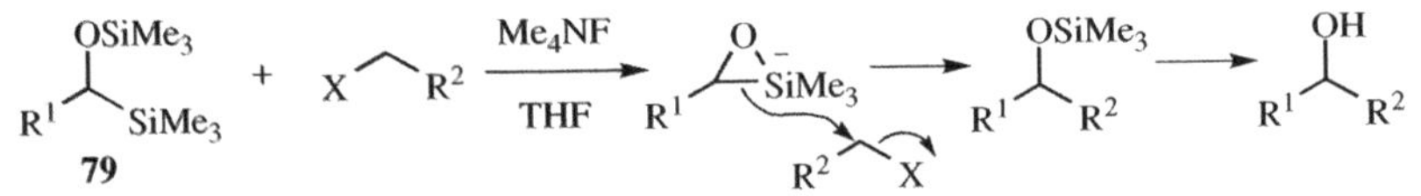

Scheme 6.33 Fluoride-catalyzed generation of α-siloxycarbanion equivalents from α-silyl silylethers and their electrophilic trapping.

alkyl (phenyl)thio group. Since the first introduction of reductive elimination of an oxygen leaving group α to carbonyl via silyl migration by Corey *et al.*[64] in 1980, the mechanism involving Brook rearrangement has been established by Fleming *et al.*,[65] and their synthetic utility has been demonstrated by Hartel and coworkers[66] (Scheme 6.34). Similar results are obtained with substrates possessing a sulfur leaving group.[67,23b]

Reactions of a silyllithium reagent with carboxylic acid derivatives, including esters, nitriles, acid chlorides, and thioamides, have been extensively investigated by Fleming. He found that the reaction of acid chlorides with 2 equiv. of $PhMe_2SiLi$ affords 1,1-disilylalkoxide **80**, in which a Brook rearrangement followed by an alkylation of the resulting α-silyl carbanion **81** occurs to give **82** (Scheme 6.35).[68] Roles of the second silyl group are stabilization of the carbanion and acceleration of the rearrangement by the relief of steric congestion between the two silyl groups. Isolation of 1,1-disilyl alcohols is possible by low-temperature quenching, and the alcohol derived from pivaloyl chloride serves as a hindered carbon base via a Brook rearrangement.

Phenyldimethylsilyllithium reacts with amides in a variety of ways, depending on the stoichiometry, temperature, and structure of the amide.[69] One of the most remarkable reactions is reaction of dimethylamide **83** with $PhMe_2SiLi$, generating a carbene

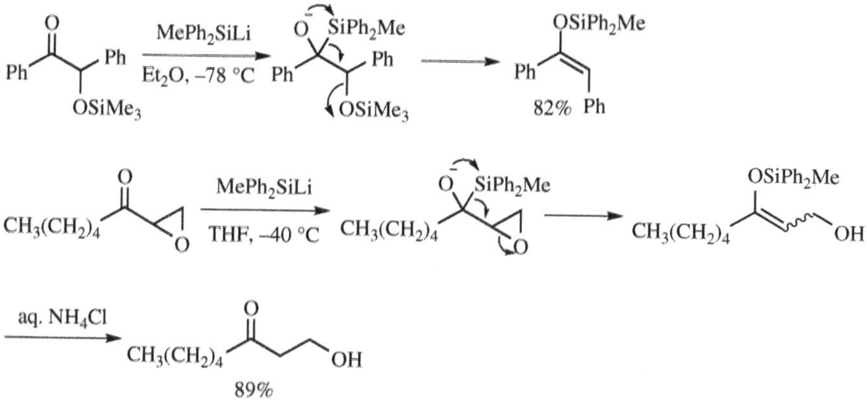

Scheme 6.34 Formation of enol silyl ethers via reductive elimination of an α-oxygen leaving group from ketones.

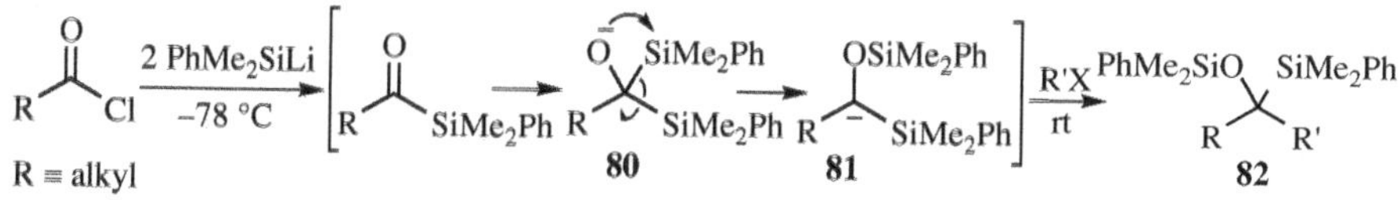

Scheme 6.35 Generation of α-silyl carbanions from acid chlorides and two equivalents of silyllithium reagent.

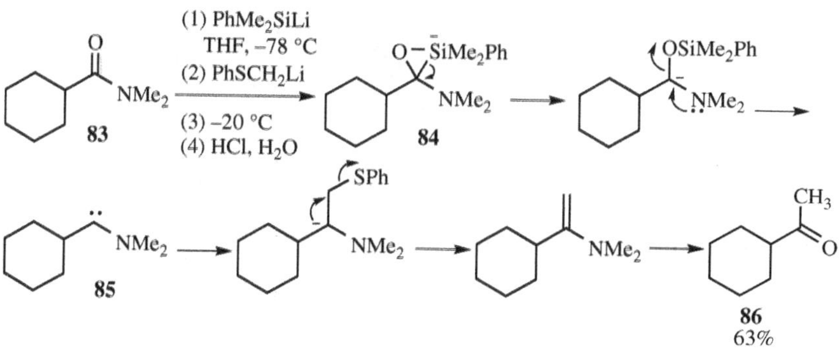

Scheme 6.36 Generation of carbenes from amides and silyllithiums.

intermediate **85** (Scheme 6.36). Thus, treatment of amide **83** with $PhMe_2SiLi$ produces silicate intermediate **84**, in which Brook rearrangement followed by expulsion of a silanoxide ion by the dimethylamino group occurs to give α-amino carbene **85**. The carbene species is trapped by phenylthiomethyllithium to give methyl ketone **86** after β-elimination followed by acidic hydrolysis of the resulting enamine.

6.7 SYNTHETIC REACTIONS USING BROOK REARRANGEMENTS IN α-SILYL ALKOXIDES GENERATED VIA REGIOSELECTIVE β-RING-OPENING OF α,β-EPOXYSILANES BY A NUCLEOPHILE

α,β-Epoxysilanes have been extensively used as versatile building blocks for organic synthesis because of their selectivity in α-ring-opening reactions, where nucleophilic attack occurs regioselectively to produce β-hydroxysilanes. This α-ring-opening process provides a highly stereospecific olefin synthesis when coupled with syn- or anti-elimination, depending on the reaction conditions. When, however, the nucleophile is a heteroatom with leaving group ability and the silyl group is bulky enough to suppress α-attack, β-ring opening occurs to provide an enol silyl ether via Brook rearrangement in the resulting α-oxidosilane, followed by anti-elimination. Thus, Cuadrad and González-Nogal reported that an α-hindered *tert*-butyldiphenylsilylepoxide in the reaction with lithium thiophenoxide gives an enol silyl ether derivative **87** selectively via β-attack and β-ring opening followed by a Brook rearrangement with elimination of a phenylthio group (Scheme 6.37).[70] This is in sharp contrast to the result obtained with the less congested silyl derivatives, which afforded product **88**, derived from α-ring opening followed by a Peterson reaction.

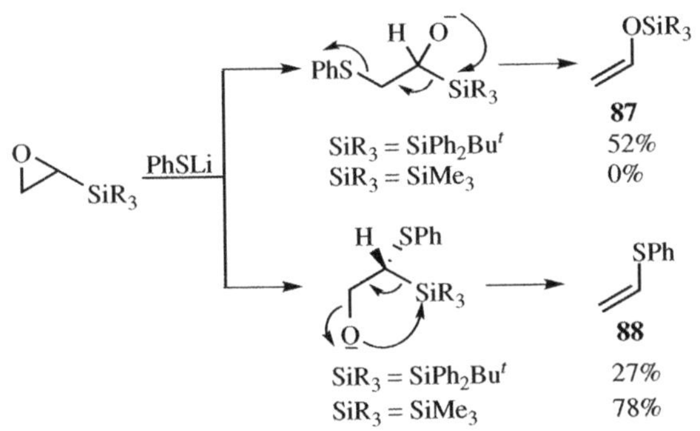

Scheme 6.37 Nucleophilic ring opening of α,β-epoxysilanes.

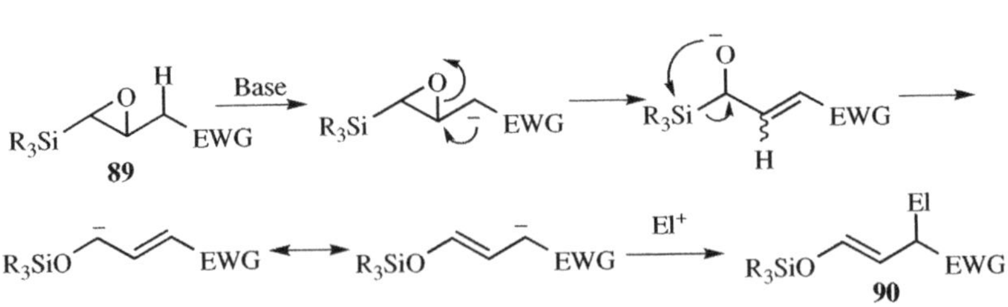

Scheme 6.38 Epoxysilane rearrangement.

6.8 SYNTHETIC REACTIONS USING BROOK REARRANGEMENTS IN α-SILYL ALKOXIDES GENERATED BY A BASE-INDUCED RING OPENING OF α,β-EPOXYSILANES

A process in which a base-induced ring opening of α,β-epoxysilanes occurs in tandem with Brook rearrangement was first observed by Jung and Nichols.[71] A similar but more versatile sequence has been devised by Takeda and coworkers in developing an asymmetric version of the Brook rearrangement-mediated [3+2] annulation (Section 6.4.4). This approach was named the epoxysilane rearrangement by the authors (Scheme 6.38).[72] Key concepts include the idea that deprotonation of an α,β-epoxysilane **89** bearing an anion-stabilizing electron-withdrawing group at the γ-position in the presence of an electrophile provides highly functionalized enol silyl ethers **90** via a tandem process that involves a base-promoted opening of the epoxide, Brook rearrangement, and reaction of the resulting allylic anion with the electrophile.

Initial work was carried out using *O*-silyl cyanohydrins of β-silyl-α,β-epoxyaldehydes **91**. Reaction of **91** with LDA in the presence of an alkylating agent produced α-alkylated cyanohydrins **92** (Scheme 6.39).

After detailed mechanistic studies,[72b] the reaction has been developed into a unique synthetic methodology for preparation of highly functionalized enol silyl ether derivatives. Variants of the epoxysilane-opening concept include (1) Michael-initiated ring closure-type reaction with bisenoate **93** (Scheme 6.40, Eq. 1),[73] (2) Wittig reaction using a phosphonio group instead of a nitrile group (Scheme 6.40, Eq. 2),[74] (3) the use of **94** as an acrolein β-anion equivalent (Scheme 6.40, Eq. 3),[75] and (4) nitrile anion cyclization with epoxysilanes (Scheme 6.40, Eqs 4a and 4b).[76]

Methods other than deprotonation are also applicable to generate a carbanion that can trigger the epoxysilane rearrangement. A carbanion generated by the reaction of acylsilanes **95** with cyanide ion induces epoxysilane rearrangement to give C-acylation products **96** via a twofold Brook rearrangement-mediated tandem sequence (Scheme 6.41).[77] Michael addition of a lithium enolate of α-chloroacetamide to enoate **97** bearing an epoxysilane moiety at the α-position initiates the rearrangement to afford **100** stereoselectively.[78] The cis relationship of the ester and amide groups in cyclopropane **100** can be explained by the formation of chelation structures **99**. A rationale for the exclusive formation of the internal (*Z*)-olefin is that silicate intermediate **98** reacts in an *s-cis* diene conformation and faster than its conformational change.

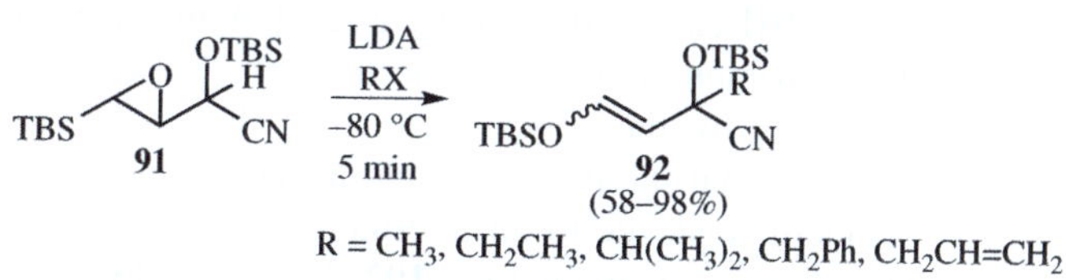

Scheme 6.39 Tandem base-promoted ring opening/Brook rearrangement/allylic alkylation of O-silyl cyanohydrins of β-Silyl-α,β-epoxyaldehydes.

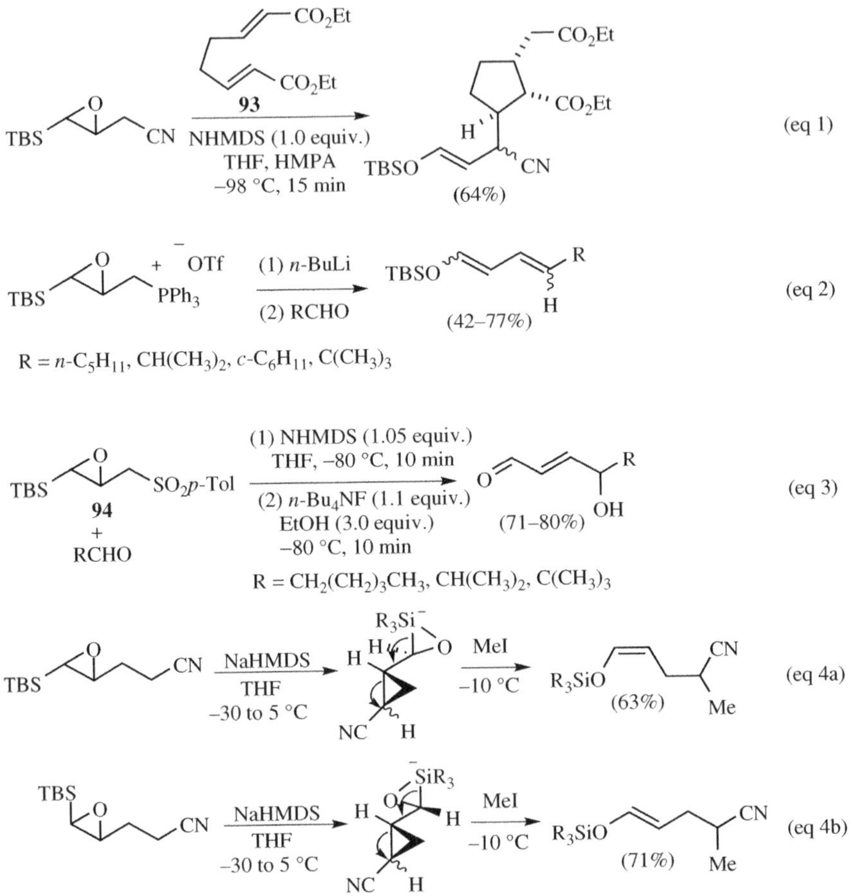

Scheme 6.40 Deprotonation-initiated epoxysilane rearrangements.

An advantageous aspect of the epoxysilane rearrangements is the ready availability of substrates in enantiomerically pure form, providing a source for chiral carbanions.[79] Configurationally, unstable carbanions generated at the position between an aryl group and a vinyl group via epoxysilane rearrangement using enantioenriched substrate **101** can be trapped by a [2,3]-Wittig rearrangement that is stereospecific at the carbanion center (Scheme 6.42).[80] The steric course of the reactions is greatly affected by the solvent, which is due to solvent-dependent differences in the configurational stability of the chiral carbanions.[81]

Extension of the tandem sequence to an intermolecular variant has been made through the use of *O*-carbamoyl cyanohydrins **102** (Scheme 6.43).[82] Although the extent of chirality transfer is moderate, the configurationally extremely labile α-nitrile carbanion can be trapped without complete racemization. This finding, attributable

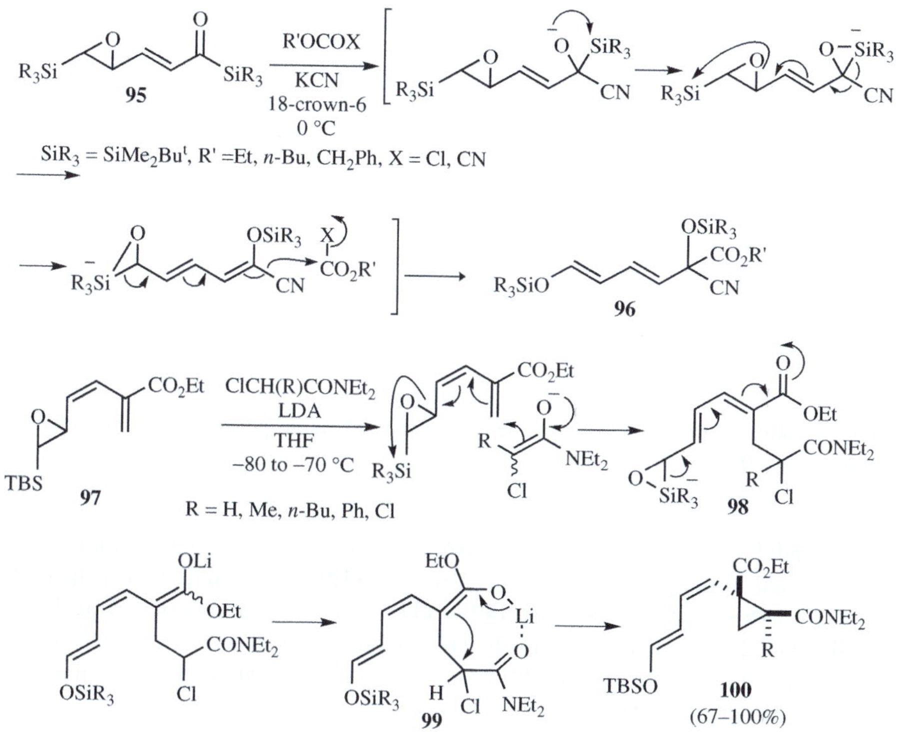

Scheme 6.41 Brook rearrangement- or Michael addition-induced epoxysilane rearrangements.

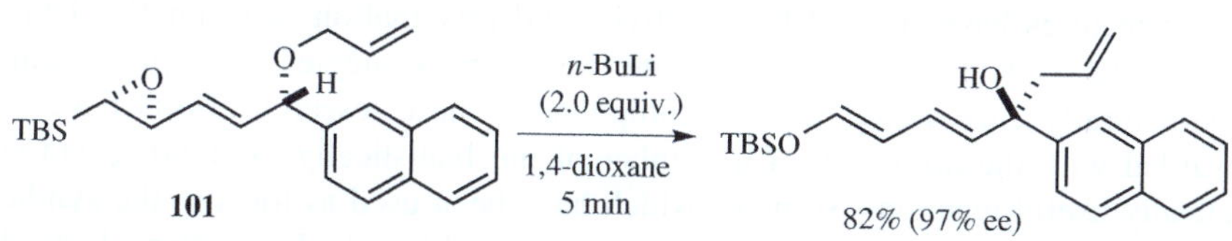

Scheme 6.42 Enantioselective trapping of a carbanion generated by epoxysilane rearrangement by [2,3]-Wittig rearrangement.

to internal chelation, led to the use of a carbamoyl group as a fixing agent for a chiral carbanion (**104** → **105**).[83]

The epoxysilane rearrangement has also been combined with [3+4] annulation for the construction of densely functionalized seven- and eight-membered carbocycles, representing a further extension of a stereocontrolled anion relay.[84] Reactions of δ-silyl-γ,δ-epoxy-α,β-unsaturated acylsilane **106** with an alkenyl methyl ketone enolate afforded highly functionalized cycloheptenone derivative **112** via a tandem process shown in Scheme 6.44. The reactions using an opposite combination of

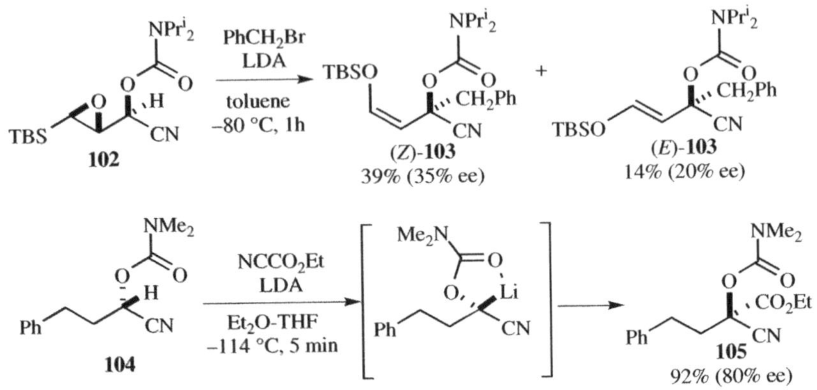

Scheme 6.43 Enantioselective trapping of α-nitrile carbanions generated via chirality transfer from epoxides.

three- and four-carbon unit (**108** + **109**), in which an epoxysilane moiety was incorporated in the four-carbon piece, also give **112** via the same intermediate **107** after 1,4-O-to-O silyl migration (**110** → **111**). Use of enantioenriched acylsilane (−)-**113** and 2-cycloheptenone enolate gave a moderate level (62% ee) of asymmetric induction in the bicyclic ketone **114**, which was transformed to highly functionalized cyclooctene derivative **115**.

6.9 CONCLUSION

Since the discovery of the Brook rearrangement more than half a century ago, its unique features have spurred the creativity and imagination of synthetic chemists who desire to devise new reactions and ingenious strategies for rapid assembly of structurally challenging organic compounds. These research efforts have been rewarded with the discovery of a number of mechanistically fascinating and synthetically useful reactions, some of which have been used as tools in the synthesis of structurally challenging organic compounds. Although the primary focus has so far been "two-dimensional" anion relay of a carbanion generated by Brook rearrangement, it is expected that the manipulation of the "three-dimensional" features of the rearrangement in a catalytic or a noncatalytic manner, including generation of chiral carbanions, will play a pivotal role in the field of asymmetric synthesis.

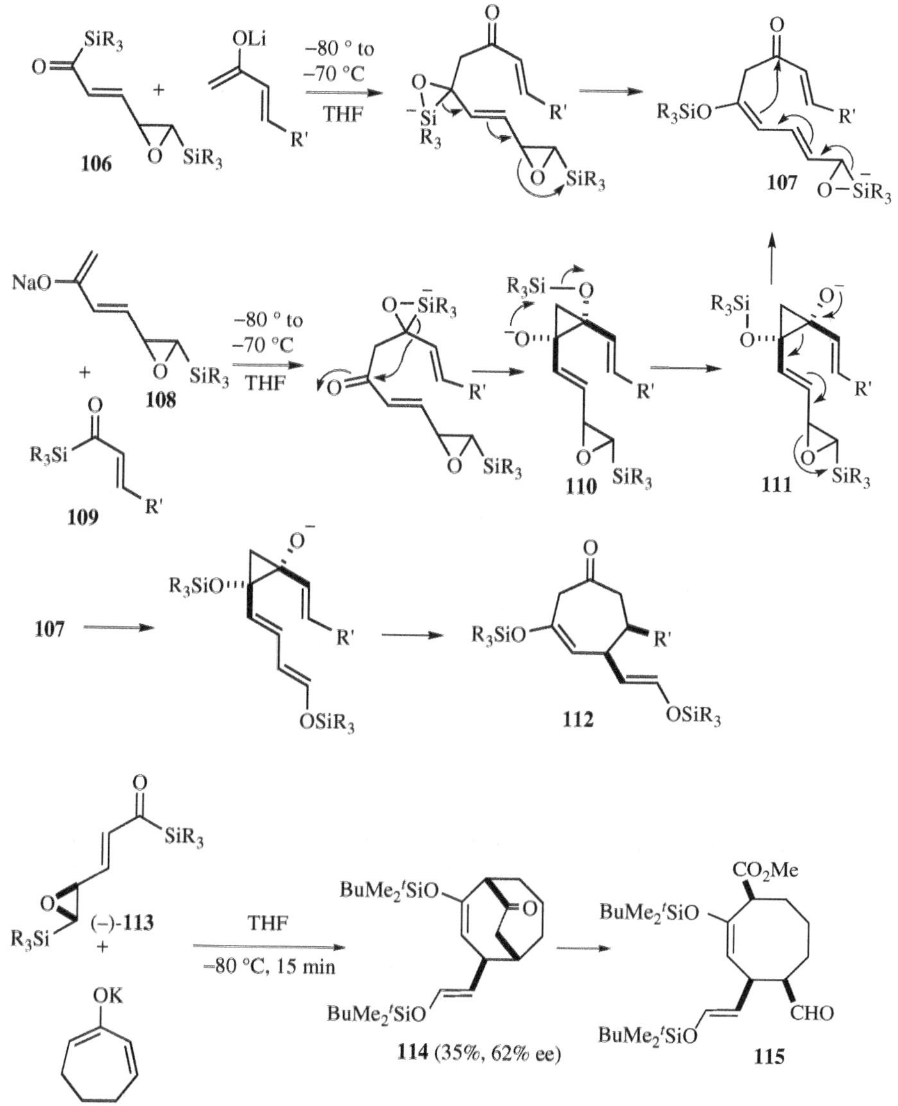

Scheme 6.44 Stereocontrolled formation of seven- and eight-membered carbocycles using a combination of Brook rearrangement-mediated [3+4] annulation and epoxysilane rearrangement.

REFERENCES

1. Brook, A. G. *J. Am. Chem. Soc.* **1958**, *80*, 1886–1889.
2. (a) Brook, A. G. *Acc. Chem. Res.* **1974**, *7*, 77–84. (b) Brook, A. G.; Bassindale, A. R. In *Rearrangements in Ground and Excited States*; de Mayo, P., Ed.; Academic Press: New York, **1980**, pp. 149–221. (c) Brook, M. A. *Silicon in Organic, Organometallic, and Polymer Chemistry*; John Wiley & Sons, Inc., **2000**. (d) Colvin, E. W. *Silicon in Organic Synthesis*; Butterworths: London, **1981**; pp. 30–37. (e) Moser, W. H. *Tetrahedron* **2001**, *57*, 2065–2084. (f) Page, P. C. B.; Klair, S. S.; Rosenthal, S. *Chem. Soc. Rev.* **1990**, *19*, 147–195. (g) Patrocinio, A. F.; Moran, P. J. S. *J. Braz. Chem. Soc.* **2001**, *12*, 7–31. (h) Schaumann, E.; Kirschning, A. *Synlett*, **2007**, 177–190.
3. (a) Brook, A. G.; Duff, J. M. *J. Am. Chem. Soc.* **1974**, *96*, 4692–4693. (b) Honda, T.; Mori, M. *J. Org. Chem.* **1996**, *61*, 1196–1197.
4. (a) Takeda, K.; Sumi, K.; Hagisawa, S. *J. Organomet. Chem.* **2000**, *611*, 449–454. (b) Kawashima, T. *Bull. Chem. Soc. Jpn.* **2003**, *76*, 471–483.
5. (a) Speier, J. L. *J. Am. Chem. Soc.* **1952**, *74*, 1003–1010. (b) West, R.; Lowe, R.; Stewart, H. F.; Wright, A. *J. Am. Chem. Soc.* **1971**, *93*, 282–283. (c) Wright, A.; West, R. *J. Am. Chem. Soc.* **1974**, *96*, 3214–3222. (d) Linderman, R. J.; Ghannam, A. *J. Am. Chem. Soc.* **1990**, *112*, 2392–2398.
6. (a) Smith, A. B. III; Adams, C. M. *Acc. Chem. Res.* **2004**, *37*, 365–377. (b) Smith, A. B., III; Wuest, W. M. *Chem. Commun.* **2008**, 5883–5895.
7. (a) Antoniotti, P.; Tonachini, G. *J. Org. Chem.* **1993**, *58*, 3622–3636. (b) Antoniotti, P.; Canepa, C.; Tonachini, G. *J. Org. Chem.* **1994**, *59*, 3952–3959.
8. Brook, A. G.; LeGrow, G. E.; MacRae, D. M. *Can. J. Chem.* **1967**, *45*, 239–253.
9. Brook, A. G.; Warner, C. M. *Tetrahedron Lett.* **1962**, *3*, 815–819.
10. (a) Brook, A. G.; Pascoe, J. D. *J. Am. Chem. Soc.* **1971**, *93*, 6224–6227. (b) Biernbaum, M. S.; Mosher, H. S. *J. Am. Chem. Soc.* **1971**, *93*, 6221–6223.
11. (a) Hudrlik, P. F.; Hudrlik, A. M.; Kulkarni, A. K. *J. Am. Chem. Soc.* **1982**, *104*, 6809–6811. (b) Wilson, S. R.; Haque, M. S.; Misra, R. N. *J. Org. Chem.* **1982**, *47*, 747–748. (c) Simov, B. P.; Wuggenig, F.; Mereiter, K.; Andres, H.; France, J.; Schnelli, P.; Hammerschmidt, F. *J. Am. Chem. Soc.* **2005**, *127*, 13934–13940. (d) Kapeller, D. C.; Brecker, L.; Hammerschmidt, F. *Chem. Eur. J.* **2007**, *13*, 9582–9588.
12. Gilman, H.; Wu, T. C. *J. Am. Chem. Soc.* **1953**, *75*, 2935–2936.
13. Fleming, I.; Roberts, R. S.; Smith, S. C. *J. Chem. Soc., Perkin Trans. 1* **1998**, 1209–1214.
14. Kuwajima, I. *J. Organometal. Chem.* **1985**, *285*, 137–148.
15. Reich, H. J. *J. Org. Chem.* **2012**, *77*, 5471–5491.
16. (a) Ricci, A.; Degl'Innocenti, A. *Synthesis* **1989**, 647–660. (b) Qi, H.; Curran, D. P. In *Comprehensive Organic Functional Group Transformations*, Katritzky, A. R.; Meth-Cohn, O.; Rees, C. W.; Moody, C. J., Eds.; Pergamon: Oxford, **1995**, pp. 409–431. (c) Cirillo, P. F.; Panek, J. S. *Org. Prep. Proc. Int.* **1992**, *24*, 553–582. (d) Zhang, H.-J.; Priebbenow, D. L.; Bolm, C. *Chem. Soc. Rev.* **2013**, *42*, 8540–8571.
17. (a) Brook, A. G.; Duff, J. M.; Jones, P. F.; Davis, N. R. *J. Am. Chem. Soc.* **1967**, *89*, 431–434. (b) Corey, E. J.; Seebach, D.; Freedman, R. *J. Am. Chem. Soc.* **1967**, *89*, 434–436.
18. Lettan, R. B.: Milgram, B. C.; Scheidt, K. A. *Org. Synth.* **2007**, *84*, 22–31; *Org. Synth.* **2009**, Coll. Vol. 11, 197–204.

19. (a) Takeda, K.; Nakajima, A.; Takeda, M.; Yoshii, E. *Org Synth.* **1999**, *76*, 199–213; *Org. Synth.* **2004**, Coll. Vol. 10, 156. (b) Sasaki, M.; Takeda, K. *Org. Synth.* **2012**, *89*, 267–273, Discussion Addendum. (c) Reich, H. J.; Kelly, M. J.; Olson, R. E.; Holtan, R. C. *Tetrahedron* **1983**, *39*, 949–960.
20. Nowick, J. S.; Danheiser, R. L. *J. Org. Chem.* **1989**, *54*, 2798–2802.
21. Brook, A. G.; Limburg, W. W.; MacRae, D. M.; Fieldhouse, S. A. *J. Am. Chem. Soc.* **1967**, *89*, 704–706.
22. Brook, A. G.; Fieldhouse, S. A. *J. Organomet. Chem.* **1967**, *10*, 235–246.
23. (a) Reich, H. J.; Rusek, J. J.; Olson, R. E. *J. Am. Chem. Soc.* **1979**, *101*, 2225–2227. (b) Reich, H. J.; Holtan, R. C.; Bolm, C. *J. Am. Chem. Soc.* **1990**, *112*, 5609-5617.
24. Kuwajima, I.; Atsumi, K.; Tanaka, T.; Inoue, T. *Chem. Lett.* **1979**, 1239–1240.
25. Clayden, J.; Watson, D. W.; Chambers, M. *Tetrahedron* **2005**, *61*, 3195–3203.
26. Song, Z.; Kui, L.; Sun, X.; Li, L. *Org. Lett.* **2011**, *13*, 1440–1443.
27. Reich, H. J.; Eisenhart, E. K.; Olson, R. E.; Kelly, M. J. *J. Am. Chem. Soc.* **1986**, *108*, 7791–7800.
28. Reich, H. J.; Olson, R. E.; Clark, M. C. *J. Am. Chem. Soc.* **1980**, *102*, 1423–1424.
29. (a) Unger, R.; Cohen, T.; Marek, I. *Org. Lett.* **2005**, *7*, 5313–5316. (b) Unger, R.; Cohen, T.; Marek, I. *Tetrahedron* **2010**, *66*, 4874–4881.
30. Degl'Innocenti, A. D.; Ricci, A.; Mordini, A.; Reginato, G.; Colotta, V. *Gazz. Chim. Ital.* **1987**, *117*, 645–648.
31. Takeda, K.; Ohnishi, Y. *Tetrahedron Lett.* **2000**, *41*, 4169–4172.
32. Linghu, X.; Nicewicz, D. A.; Johnson, J. S. *Org. Lett.* **2002**, *4*, 2957–2960.
33. Nicewicz, D. A.; Yates, C. M.; Johnson, J. S. *J. Org. Chem.* **2004**, *69*, 6548–6555.
34. Takeda, K.; Tanaka, T. *Synlett* **1999**, 705–708.
35. Takeda, K.; Ohnishi, Y.; Koizumi, T. *Org. Lett.* **1999**, *1*, 237–240.
36. Sasaki, M.; Kondo, Y.; Kawahata, M.; Yamaguchi, K. Takeda, K. *Angew. Chem., Int. Ed.* **2011**, *50*, 6375–6378.
37. (a) Boyce, G. R.; Greszler, S. N.; Johnson, J. S.; Linghu, X.; Malinowski, J. T.; Nicewicz, D. A.; Satterfield, A. D.; Schmitt, D. C.; Steward, K. M. *J. Org. Chem.* **2012**, *77*, 4503–4515. (b) Nicewicz, D. A.; Brétéché, G.; Johnson, J. S. *Org. Synth.* **2008**, *85*, 278–286.
38. Linghu, X.; Satterfield, A. D.; Johnson, J. S. *J. Am. Chem. Soc.* **2006**, *128*, 9302–9303.
39. (a) Nicewicz, D. A.; Johnson, J. S. *J. Am. Chem. Soc.* **2005**, *127*, 6170–6171. (b) Boyce, G. R.; Liu, S.; Johnson, J. S. *Org. Lett.* **2012**, *14*, 652–655.
40. Boyce, G. R.; Johnson, J. S. *Angew. Chem., Int. Ed.* **2010**, *49*, 8930–8933.
41. Mattson, A. E.; Bharadwaj, A. W.; Scheidt, K. A. *J. Am. Chem. Soc.* **2004**, *126*, 2314–2315.
42. Takeda, K.; Nakatani, J.; Nakamura, H.; Sako, K.; Yoshii, E.; Yamaguchi, K. *Synlett* **1993**, 841–843.
43. (a) Takeda, K.; Fujisawa, M.; Makino, T.; Yoshii, E.; Yamaguchi, K. *J. Am. Chem. Soc.* **1993**, *115*, 9351–9352. (b) Takeda, K.; Ohtani, Y.; Ando, E.; Fujimoto, K.; Yoshii, E.; Koizumi, T. *Chem. Lett.* **1998**, 1157–1158.
44. Takeda, K.; Yamawaki, K.; Hatakeyama, N. *J. Org. Chem.* **2002**, *67*, 1786–1794.
45. Takeda, K.; Kitagawa, K.; Nakayama, I.; Yoshii, E. *Synlett* **1997**, 255–256.

46. (a) Takeda, K.; Takeda, M.; Nakajima, A.; Yoshii, E. *J. Am. Chem. Soc.* **1995**, *117*, 6400–6401. (b) Takeda, K.; Nakajima, A.; Yoshii, E. *Synlett* **1996**, 753–754. (c) Takeda, K.; Ohtani, Y. *Org. Lett.* **1999**, *1*, 677–680.
47. Takeda, K.; Nakajima, A.; Takeda, M.; Okamoto, Y.; Sato, T.; Yoshii, E.; Koizumi, T.; Shiro, M. *J. Am. Chem. Soc.* **1998**, *120*, 4947–4959.
48. Takeda, K.; Nakane, D.; Takeda, M. *Org. Lett.* **2000**, *2*, 1903–1905.
49. Takeda, K.; Sawada, Y.; Sumi, K. *Org. Lett.* **2002**, *4*, 1031–1033.
50. Sawada, Y.; Sasaki, S.; Takeda, K. *Org. Lett.* **2004**, *6*, 2277–2279.
51. (a) Sasaki, M.; Hashimoto, A.; Tanaka, K.; Kawahata, M.; Yamaguchi, K.; Takeda, K. *Org. Lett.* **2008**, *10*, 1803–1806. (b) Sasaki, M.; Oyamada, K.; Takeda, K. *J. Org. Chem.* **2010**, *75*, 3941–3943.
52. Takeda, K.; Haraguchi, H.; Okamoto, Y. *Org. Lett.* **2003**, *5*, 3705–3707.
53. Takeda, K.; Ubayama, H.; Sano, A.; Yoshii, E.; Koizumi, T. *Tetrahedron Lett.* **1998**, *39*, 5243–5246.
54. (a) Lettan, R. B.; Reynolds, T. E.; Galliford, C. V.; Scheidt, K. A. *J. Am. Chem. Soc.* **2006**, *128*, 15566–15567. (b) Lettan, R. B.; Woodward, C. C.; Scheidt, K. A. *Angew. Chem., Int. Ed.* **2008**, *47*, 2294–2297. (c) Lettan, R. B.; Galliford, C. V.; Woodward, C. C.; Scheidt, K. A. *J. Am. Chem. Soc.* **2009**, *131*, 8805–8814.
55. Kuwajima, I.; Kato, M. *Chem. Commun.* **1979**, 708–709.
56. (a) Buynak, J. D.; Strickland, J. B.; Lamb, G. W.; Khasnis, D.; Modi, S.; Williams, D.; Zhang, H. M. *J. Org. Chem.* **1991**, *56*, 7076–7083. (b) Arai, N.; Suzuki, K.; Sugizaki, S.; Sorimachi, H.; Ohkuma, T. *Angew. Chem., Int. Ed.* **2008**, *47*, 1770–1773. (c) Lou, W.-Y.; Zong, M.-H.; Smith, T. J. *J. Green Chem.* **2006**, *8*, 147–155.
57. Sakaguchi, K.; Fujita, M.; Suzuki, H.; Higashino, M.; Ohfune, Y. *Tetrahedron Lett.* **2000**, *41*, 6589–6592.
58. (a) Kuwajima, I.; Kato, M. *Tetrahedron Lett.* **1980**, *21*, 2745–2748. (b) Kato, M.; Mori, A.; Oshino, H.; Enda, J.; Kobayashi, K.; Kuwajima, I. *J. Am. Chem. Soc.* **1984**, *106*, 1773–1778.
59. Kuwajima, I.; Kato, M. *Tetrahedron Lett.* **1980**, *21*, 623–626.
60. Reynolds, T. E.; Bharadwaj, A. R.; Scheidt, K. A. *J. Am. Chem. Soc.* **2006**, *128*, 15382–15383.
61. Sasaki, M.; Fujiwara, M.; Kotomori, Y.; Kawahata, M.; Yamaguchi, K.; Takeda, K. *Tetrahedron* **2013**, *69*, 5823–5828.
62. Hoffmann, R. W. In *Stereochemical Aspects of Organolithium Compounds, Topics in Stereochemistry*; Gawley, R. E.; Siegel, J., Eds.; Wiley: New York, **2010**; Vol. 26, pp. 165–188.
63. Brekan, J. A.; Chernyak, D.; White, K. L. Scheidt, K. A. *Chem. Sci.* **2012**, *3*, 1205–1210.
64. Corey, E. J.; Tius, M. A.; Das, J. *J. Am. Chem. Soc.* **1980**, *102*, 1742–1744.
65. Fleming, I.; Roberts, R. S.; Smith, S. C. *J. Chem. Soc., Perkin Trans. 1* **1998**, 1215–1228.
66. (a) Robertson, B. D.; Hartel, A. M. *Tetrahedron Lett.* **2008**, *49*, 2088–2090. (b) Reynolds, S. C.; Wengryniuk, S. E.; Hartel, A. M. *Tetrahedron Lett.* **2007**, *48*, 6751–6753.
67. Vedejs, E.; Arnost, M. J.; Eustache, J. M.; Krafft, G. A. *J. Org. Chem.* **1982**, *47*, 4384–4386.
68. Fleming, I.; Lawrence, A. J.; Richardson, R. D.; Surry, D. S.; West, M. C. *Helv. Chim. Acta* **2002**, *85*, 3349–3365.

69. Buswell, M.; Fleming, I.; Ghosh, U.; Mack, S.; Russell, M.; Clark, B. P. *Org. Biomol. Chem.* **2004**, *2*, 3006–3017.

70. Cuadrado, P.; González-Nogal, A. M. *Tetrahedron Lett.* **2000**, *41*, 1111–1114.

71. Jung, M. E.; Nichols, C. J. *J. Org. Chem.* **1996**, *61*, 9065–9067.

72. (a) Takeda, K.; Kawanishi, E.; Sasaki, M.; Takahashi, Y.; Yamaguchi, K. *Org. Lett.* **2002**, *4*, 1511–1514. (b) Sasaki, M.; Kawanishi, E.; Nakai, Y.; Matsumoto, T.; Yamaguchi, K.; Takeda, K. *J. Org. Chem.* **2003**, *68*, 9330–9339. (c) Sasaki, M.; Takeda, K. *J. Synth. Org. Chem., Jpn.* **2006**, *64*, 1148–1158. (d) Sasaki, M.; Ikemoto, H.; Takeda, K. *Heterocycles* **2009**, *78*, 2919–2941.

73. Matsumoto, T.; Masu, H.; Yamaguchi, K.; Takeda, K. *Org. Lett.* **2004**, *6*, 4367–4369.

74. Sasaki, M.; Horai, M.; Takeda, K. *Tetrahedron Lett.* **2006**, *47*, 9271–9273.

75. Sasaki, M.; Takeda, K. *Org. Lett.* **2004**, *6*, 4849–4851.

76. (a) Okugawa, S.; Takeda, K. *Org. Lett.* **2004**, *6*, 2973–2975. (b) Okugawa, S.; Masu, H.; Yamaguchi, K.; Takeda, K. *J. Org. Chem* **2005**, *70*, 10515–10523.

77. (a) Tanaka, K.; Takeda, K. *Tetrahedron Lett.* **2004**, *45*, 7859–7861. (b) Tanaka, K.; Masu, H.; Yamaguchi, K.; Takeda, K. *Tetrahedron Lett.* **2005**, *46*, 6429–6432.

78. Okamoto, N.; Sasaki, M.; Kawahata, M.; Yamaguchi, K.; Takeda, K. *Org. Lett.* **2006**, *8*, 1889–1891.

79. Sasaki, M.; Takeda, K. *Synlett* **2012**, *23*, 2153–2164.

80. Sasaki, M.; Higashi, M.; Masu, H.; Yamaguchi, K.; Takeda, K. *Org. Lett.* **2005**, *7*, 5913–5915.

81. Ikemoto, H.; Sasaki, M.; Takeda, K. *Eur. J. Org. Chem.* **2010**, 6643–6650.

82. Sasaki, M.; Kawanishi, E.; Shirakawa, Y.; Kawahata, M.; Masu, H.; Yamaguchi, K.; Takeda, K. *Eur. J. Org. Chem.* **2008**, 3061–3064.

83. (a) Sasaki, M.; Takegawa, T.; Ikemoto, H.; Kawahata, M.; Yamaguchi, K.; Takeda, K. *Chem. Commun.* **2012**, *48*, 2897–2899. (b) Ikemoto, H.; Sasaki, M.; Kawahata, M.; Yamaguchi, K.; Takeda, K. *Eur. J. Org. Chem.* **2011**, 6553–6557. (c) Sasaki, M.; Takegawa, T.; Sakamoto, K.; Kotomori, Y.; Otani, Y.; Ohwada, T.; Kawahata, M.; Yamaguchi, K.; Takeda, K. *Angew. Chem. Int. Ed.* **2013**, *52*, 12956–12960.

84. Nakai, Y.; Kawahata, M.; Yamaguchi, K.; Takeda, K. *J. Org. Chem.* **2007**, *72*, 1379–1387.

PART II

1,2-MIGRATIONS VIA THREE-MEMBERED RINGS

7

THE QUASI-FAVORSKII REARRANGEMENT

MICHAEL HARMATA
Department of Chemistry, University of Missouri-Columbia, Columbia, MO, USA

7.1 INTRODUCTION

7.1.1 The Favorskii Rearrangement

The Favorskii rearrangement is a base-catalyzed conversion of an α-haloketone into an acid or acid derivative such that the ketone α and α' carbons come to share a bond, and the carbonyl carbon of the ketone becomes the carbonyl carbon of the carboxylic acid derivative (Scheme 7.1).[1] The reaction can be regioselective, and its outcome is dependent not only on structural features of the starting material but on reaction conditions as well.

The overall process is quite interesting mechanistically. Typical conditions consist of treating the starting material with an alkoxide base in a polar solvent, but there are many options available. A possible mechanism for the process is shown in Scheme 7.2. Deprotonation of the ketone **1** results in the enolate **2**. This can either lead directly to cyclopropanone **4** by expulsion of the leaving group or proceed initially to an oxyallylic cation, which then undergoes disrotatory ring closure to the cyclopropanone. Nucleophilic attack on the cyclopropanone affords a tetrahedral intermediate that then fragments. A carbanion such as **6** is often written as an intervening species; this is protonated to give the product(s) observed.

However, when the α-haloketone is treated with base under conditions of isotopic exchange, there are some cases in which no α-H/D exchange takes places, so there can be no equilibrium between **1** and **2**. A comparison between two cases illustrates

Molecular Rearrangements in Organic Synthesis, First Edition. Edited by Christian M. Rojas.

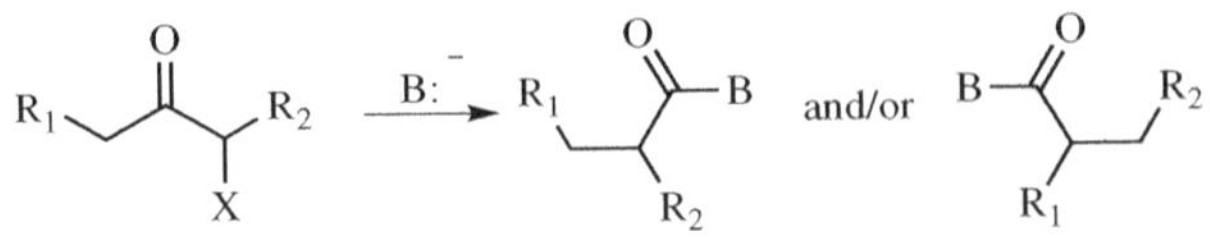

Scheme 7.1 The Favorskii rearrangement.

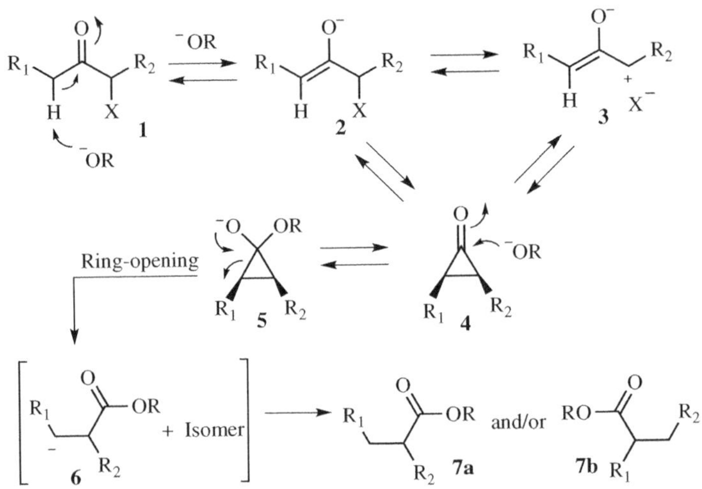

Scheme 7.2 Mechanism of the Favorskii rearrangement.

a great deal. Bordwell *et al.* reported that the Favorskii rearrangement of chloroketones represented by **8** proceeded in sodium methoxide/methanol to afford the corresponding esters **9** in high yield, except when the aryl group in the starting material was substituted with an electron-withdrawing group (Scheme 7.3).[2] When the parent compound (**8**, Ar = phenyl) was treated with NaOMe in MeOD, the product was formed with over 90% deuterium incorporation, meaning that the enolate exchange equilibrium had occurred. Hammett studies gave a ρ value of around −5.0, suggesting the accumulation of carbocation character in the rate-determining step (RDS) of the reaction, that is, loss of chloride. Switching from chloride to bromide increased the rate of reaction approximately 63-fold, as did switching to a more polar reaction medium; the rate of the reaction in 50% aqueous methanol was approximately 200 times that in pure methanol. The reaction of the parent system occurred with an activation energy of 23 kcal/mol and a positive entropy of activation ($\Delta S^{\neq} = +16$ eu). There was no common ion effect.

What species forms in the RDS? The easy answer is that S_N1 loss of chloride from the enolate (enol) produces an oxyallylic (hydroxyallylic) cation that cyclizes in a disrotatory fashion to the corresponding cyclopropanone. The cyclization may be readily reversible, though generally cyclopropanones are considered more stable than oxyallylic cations.[3] However, decades after the Bordwell study, Sorensen and

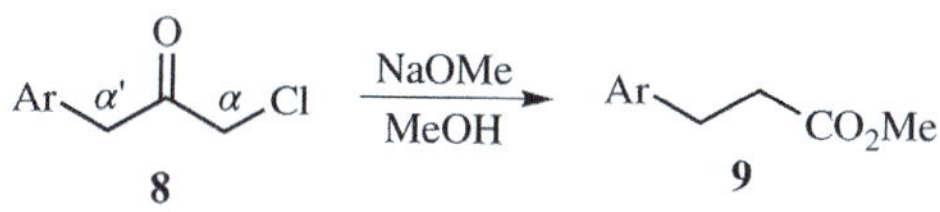

Scheme 7.3

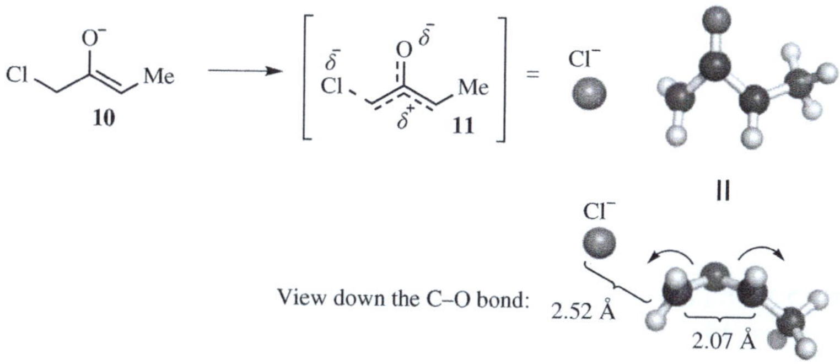

FIGURE 7.1 Favorskii rearrangement of a benzyl ketone.

coworkers examined the chemistry of chloroenolates computationally and found transition states for the direct conversion of such species to cyclopropanones, via both retentive and invertive pathways with respect to the stereochemistry of the carbon bearing the leaving group.[4]

For example, in the conversion of enolate **10** to methylcyclopropanone, two transition states were found. The one corresponding to displacement of the chloride in a more traditional sense, that is, inversion, is shown in Figure 7.1. The structure indicates that the departure of chloride is assisted by the π system and that the carbon backbone is well on its way to undergoing a disrotatory ring closure. This is very reminiscent of what happens in the solvolysis of cyclopropyl bromides, where disrotatory ring-opening occurs so as to assist departure of the leaving group.[5] Interestingly, the energy barrier calculated for this process was ~20 kcal/mol, very close to that observed by Bordwell for the reaction of **8**.

Another transition state for the process is retentive and is shown in Figure 7.2. While about 5 kcal/mol higher in energy than the invertive transition state, this structure indicates the possibility for a stereochemically different ring closure. Carbon–chlorine bond cleavage is more advanced in this structure and carbon–carbon bond formation less advanced compared to the invertive counterpart, though distortion toward a disrotatory ring closure is visually evident. This path looks more like an S_N1 process, where cyclization of an intimate ion pair leads to cyclopropanone without necessarily giving rise to an oxyallylic cation intermediate. Inclusion of solvent models in the calculations did not change the results, though it is not clear how hydrogen bonding might affect the structures of both transition states.

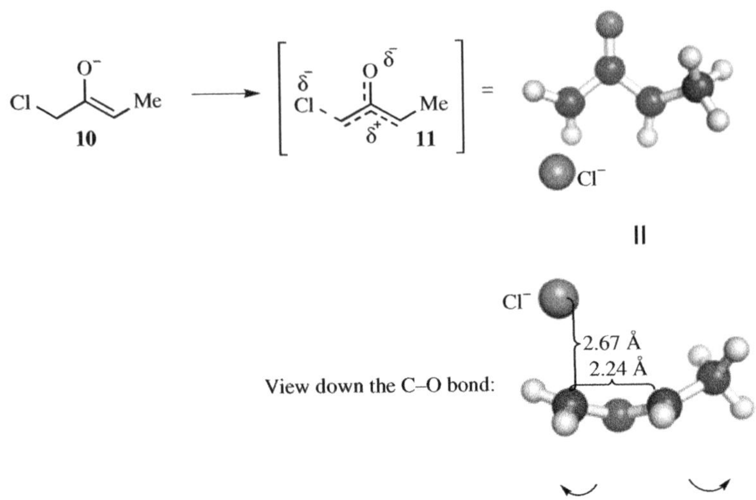

FIGURE 7.2 Transition structure **11** for along path for retentive formation of methylcyclopropanone from **10**.

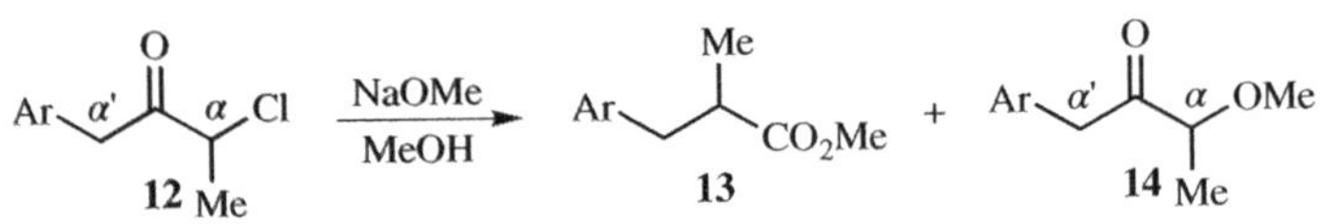

Scheme 7.4 A mechanistically (kinetically) different Favorskii rearrangement.

Exchanging a single hydrogen for a methyl group at the α-position in **8** leads to dramatic changes (Scheme 7.4).[6] With chloroketone **12**, no deuterium exchange is observed.

No difference in rates between chlorides and bromides is observed. The reaction is over 200 times faster than that of **8**, but the ρ value according to Hammett studies is 1.4. The RDS of the reaction is now enolate formation. Presumably cyclopropanone formation occurs by the same mechanism as before, but it is faster relative to the initial deprotonation. Side-product formation (**14**) is more prevalent here, as a function of several variables, including base concentration.

There are more potential complications. Bordwell and Carlson have implicated a protonated version of the oxyallylic cation **3** (see Scheme 7.2) in the formation of side products in the Favorskii rearrangement.[7] Treatment of the stable cyclopropanone **15** under Favorskii reaction conditions with isotope incorporation results in the formation of esters with nearly complete retention of configuration (Scheme 7.5),[8] making the high-energy intermediate **6** in Scheme 7.2 seem less likely and perhaps favoring a mechanism in which ring-opening and protonation occur in concert to directly give the ester products observed in the reaction. This will no doubt be a function of the cyclopropanone hemiacetal structure. Finally, there have been proposals

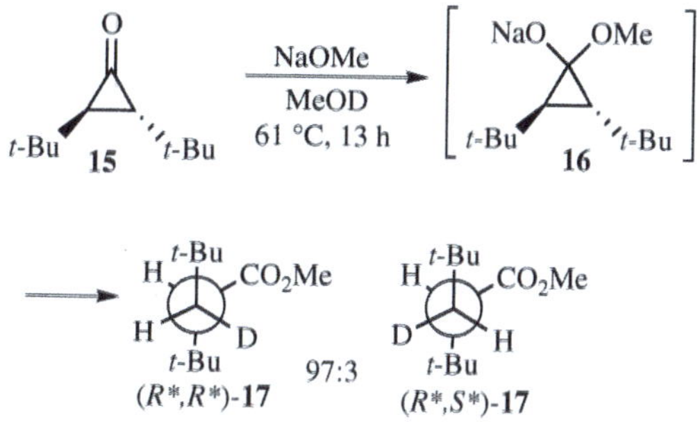

Scheme 7.5 Stereoselective ring-opening (retention) of a cyclopropanone under conditions of the Favorskii rearrangement.

regarding the evolution of the cyclopropanone that include formation of a chloride adduct[9] as well as direct fragmentation to an acylium ion.[10] Deeper consideration of these possibilities will not be attempted here.

It is worth noting that the putative occurrence of an oxyallylic cation (or its protonated form) in the Favorskii process has broad implications beyond the rearrangement itself. For example, oxyallylic cation intermediates are important players in several other reactions, such as the synthetically valuable (4+3)-cycloaddition reaction.[11]

7.1.2 The Quasi-Favorskii Rearrangement

A second mechanism that has been considered for the Favorskii rearrangement is shown in Scheme 7.6.[12] In this case, a nucleophile attacks the carbonyl group to give a tetrahedral intermediate that evolves via a 1,2-shift with concomitant expulsion of the leaving group. This mechanism was rejected in many cases because the evidence accumulated suggested that enolate formation with the intermediacy of a cyclopropanone was the only mechanism that would explain the experimental results.[13] Nevertheless, in cases where enolate formation is quite slow or simply impossible, this mechanism is energetically accessible and leads to rearrangement products.

One early example of this so-called quasi-Favorskii rearrangement is shown in Scheme 7.7. Cope and Synerholm reported that the reaction of **21** with sodium

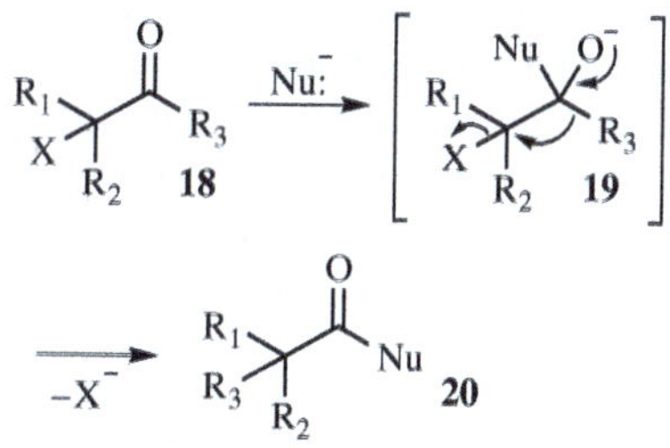

Scheme 7.6 Mechanism of the quasi-Favorskii rearrangement.

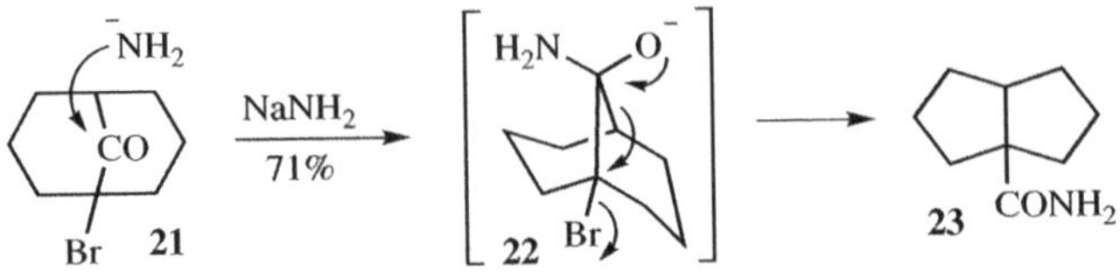

Scheme 7.7 An example of a quasi-Favorskii rearrangement.

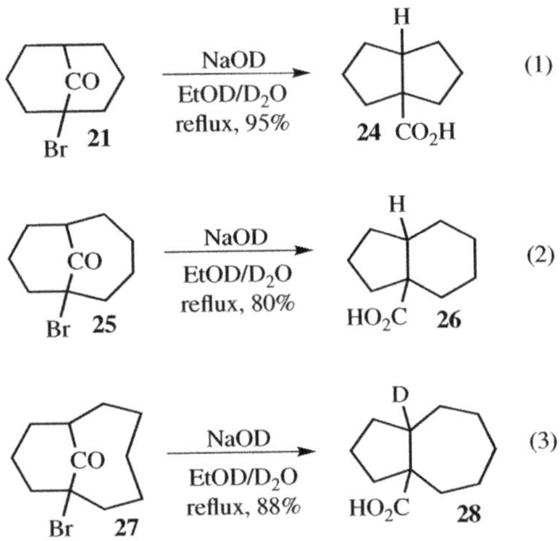

Scheme 7.8 The dividing line between quasi-Favorskii and Favorskii mechanisms in bicyclic bromoketones.

amide afforded the amide **23** in good yield.[14] In this case, Bredt's rule predicts that enolate formation is impossible.[15] Nucleophilic attack at the carbonyl is preferred, and the resulting tetrahedral intermediate undergoes a sigma bond shift and expulsion of bromide to afford the 5,5-fused product. A bit more on mechanism will be discussed later, but what is shown in Scheme 7.7 depicts the essence of the quasi-Favorskii rearrangement.

An elegant study by Warnhoff *et al.* systematically analyzed mechanism as a function of structure in bridged bicyclic ketones related to **21**.[16] Treatment of **21** with NaOD in refluxing EtOD/D_2O afforded the corresponding acid **24** in 95% yield with no incorporation of deuterium (Scheme 7.8, Eq. 1). Had this been a normal Favorskii reaction, one deuterium should have been incorporated. Similarly, when **25** was treated under the same conditions, the ring contraction product **26** contained no deuterium (Scheme 7.8, Eq. 2). However, with **27**, essentially a completely monodeuterated compound was obtained, indicative of a normal Favorskii reaction (Scheme 7.8, Eq. 3).

These results are entirely in line with what one might expect from the feasibility of enolate formation in the bicyclic systems **25** and **27**. However, more recent

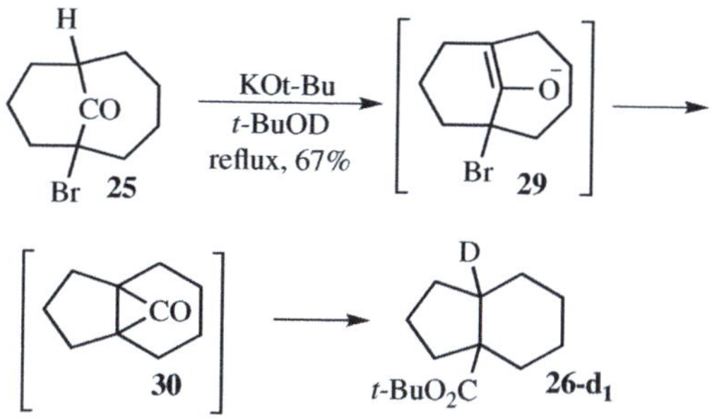

Scheme 7.9 Evidence for the formation of **26** from **25** via a cyclopropanone intermediate.

studies on the question of bridgehead enolates[17] suggest that bridgehead enolates are more accessible than might be commonly appreciated. Indeed, when **25** was treated with *t*-BuOK in *t*-BuOD at reflux, the corresponding ester was formed in good yield (Scheme 7.9). This product was 83% deuterated, which represents complete deuteration when the amounts of OH present in the solvent and isotope effects are taken into consideration. Recovered starting material contained no label. The reaction must have proceeded via a cyclopropanone (**30**), implying enolate formation (**29**). This was rationalized on the basis of increased base strength of *tert*-butoxide relative to hydroxide.

This kind of result has synthetic implications. In the quasi-Favorskii mechanism for the substrates shown in Scheme 7.8, an enantiomerically pure starting material will give an enantiomerically pure product. On the other hand, complete racemization is expected if products are produced via a normal Favorskii mechanism, as the intermediate cyclopropanone (e.g., **30**) would be achiral.

Finally, consideration of the mechanism of the quasi-Favorskii rearrangement quickly reveals that there are a number of related reactions, including the benzylic acid rearrangement and the pinacol rearrangement.[18,19] Indeed, the mechanism of the quasi-Favorskii rearrangement is often referred to as "semibenzylic" or "semipinacol." An exceptional review of the application of semipinacol rearrangements in synthesis has appeared recently.[20] Consequently, the present chapter will focus on processes that proceed via an anionic tetrahedral intermediate that breaks down via expulsion of some leaving group and concomitant bond migration to generate the product observed.

7.2 RETRONS OF THE QUASI-FAVORSKII REARRANGEMENT

Applying any reaction to a synthetic target requires knowledge of the retrons that make the reaction applicable at some point in the synthesis. Partial retrons – those keying elements that resemble, but are not quite the same as the direct product of a particular reaction – are also quite important.

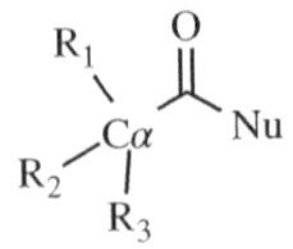

FIGURE 7.3 The quasi-Favorskii rearrangement retron.

The quasi-Favorskii retron is shown in Figure 7.3. From the examples shown thus far, it would seem as though the groups R_1—R_3 would all be carbons and that the Nu group would be an oxygen derived from either water or an alcohol. This is reasonable, but it is by no means the entire story. In fact, almost any atom or group that can be a nucleophile in other organic transformations can participate as a nucleophile in quasi-Favorskii reactions.[21] While intervention of the regular Favorskii process limits the nature of the R_1—R_3 groups, we will show that it is possible to prepare carbonyl compounds that bear an α-hydrogen using a quasi-Favorskii reaction.

Partial retrons are more numerous than retrons for the quasi-Favorskii rearrangement. One can focus on the possible carbonyl product and consider all those groups that could be derived from those carbonyl groups. Furthermore, one of the classic partial retrons for a quasi-Favorskii rearrangement is simply a tertiary or quaternary carbon embedded in a polycyclic carbon framework. These synthetic strategies will be illustrated with examples later in this chapter.

7.3 MECHANISTIC CONSIDERATIONS IN THE QUASI-FAVORSKII REARRANGEMENT

In addition to mechanistic features discussed earlier, there appears to be a stereoelectronic requirement in the quasi-Favorskii rearrangement, as might be expected. For example, Baudry *et al.* reported that the quasi-Favorskii process for ketones *cis*- and *trans*-**31** was stereospecific and occurred with inversion of configuration at the carbon bearing the bromine leaving group (Scheme 7.10).[22] Thus, attack of hydroxide on the carbonyl group of the respective ketones gave intermediates that could adopt conformations in which the migrating group was antiperiplanar to the leaving group, facilitating the migration process.

Other examples of this type of stereoelectronic consideration exist,[23] in line with standard interpretations of such effects in polar organic reactions. But complications can arise in which the paradigm seems to break down, as in the example shown in Scheme 7.11. In this case, it was proposed that neighboring group participation resulted in a stereochemically labile species (**35**), leading to loss of stereochemical control.[24] It is thus likely that rapid isomerization interfered with an otherwise mechanistically well-defined process.

The quasi-Favorskii rearrangement has been used often in the synthesis of unnatural, complex, polycyclic structures. Indeed, this method is uniquely suited for such targets. However, undesired side reactions can occur. An example, discussed here within its mechanistic context, is based on a report by Ueda and coworkers.[25] They

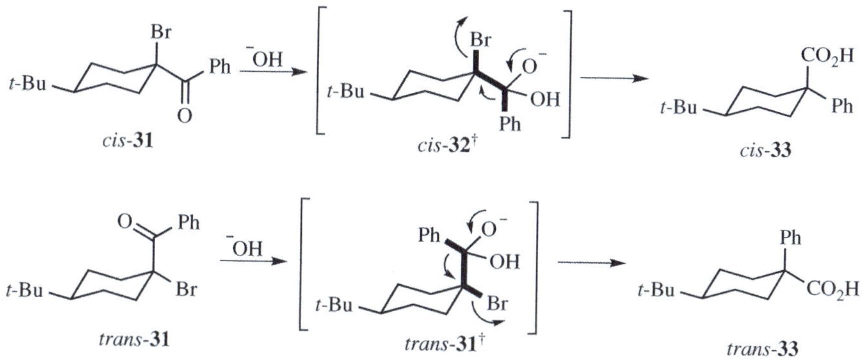

Scheme 7.10 Stereospecificity of the quasi-Favorskii rearrangement.

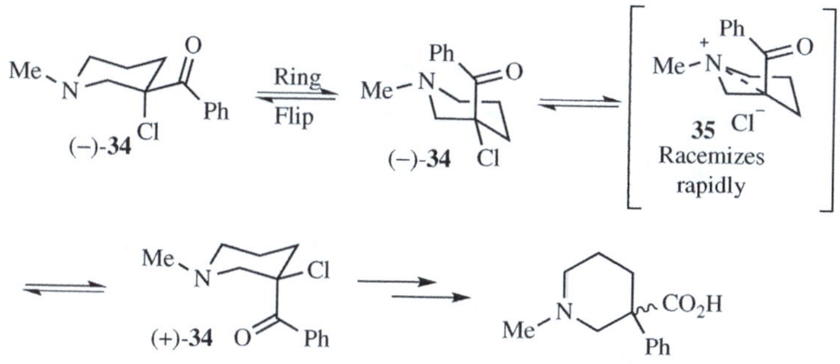

Scheme 7.11 A non-stereoselective quasi-Favorskii reaction.

treated the polycyclic dibrominated diketone **37** with potassium hydroxide in the expectation of obtaining cubane-1,3-dicarboxylic acid **38**. Given what was known about quasi-Favorskii reactions at the time (*vide supra*), the plan made perfect sense. However, exposure of **37** to 5% KOH for 15 min at 80 °C afforded not **38**, but the cyclopropyl lactone **43** (Scheme 7.12). A proposed mechanism for the process began with a Haller–Bauer cleavage, always a potential risk in quasi-Favorskii rearrangements, to give **40**, followed by a Grob fragmentation and ending with a π-assisted departure of a bromide and intramolecular trapping to give **43**. While this example is somewhat extreme, it does highlight the Haller–Bauer process, an important reaction in itself and a potential path to side products in any reaction that involves attack of a nucleophile at a ketone or aldehyde carbonyl group.

7.4 THE PREPARATION OF SUBSTRATES FOR THE QUASI-FAVORSKII REARRANGEMENT

In principle, any reaction that results in the placement of a leaving group next to a carbonyl provides a method to prepare substrates for the quasi-Favorskii rearrangement.

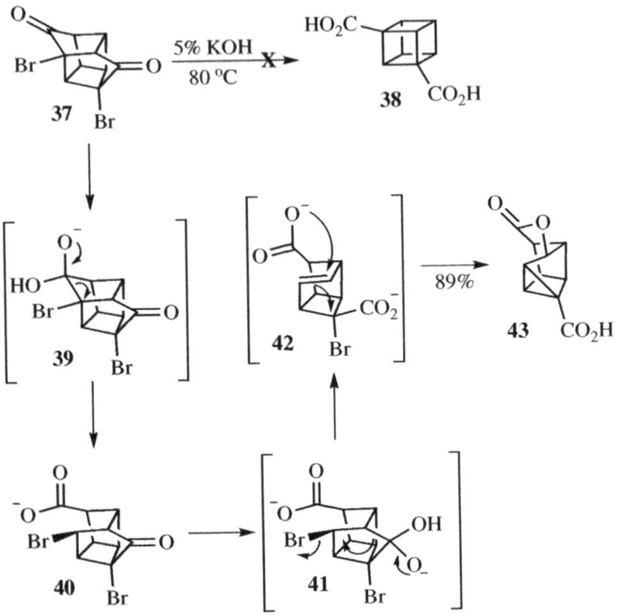

Scheme 7.12 Haller–Bauer and other side reactions in an attempted quasi-Favorskii rearrangement.

Substrates include ketones that might be candidates for a normal Favorskii rearrangement, because even these compounds can undergo a quasi-Favorskii reaction under the right circumstances, as illustrated later in this chapter.

Nevertheless, a majority of quasi-Favorskii rearrangements are performed on bridgehead halides that contain ketones in an appropriate location. The synthesis of these systems can be more challenging, but a number of strategies have been devised.

Certain cycloaddition/cyclization processes produce substrates for the quasi-Favorskii rearrangement with good yield and selectivity. For example, the intramolecular cycloaddition of ketene **45**, available by dehydrohalogenation of the corresponding acid halide **44**, afforded **46** in moderate yield (Scheme 7.13). This compound reacted smoothly with lithium hydroxide under mild conditions to form the carboxylic acid **47**.[26]

Recently, Danishefsky and coworkers reported conditions for the Diels–Alder reaction of 2-bromocycloalkenones with dienes.[27] For example, the cycloaddition of 2-bromocyclobutenone with various dienes took place in good yield at room temperature without the need for a catalyst. The reactions of the corresponding cyclopentenone and cyclohexenone were also reported, but these required more vigorous conditions than **48** for successful cycloaddition. However, the cycloadducts could certainly serve as educts in the quasi-Favorskii process (Scheme 7.14).

It is known that α-halocyclobutanones react via a quasi-Favorskii mechanism to produce ring contraction products.[28,29] Indeed, treatment of **49–51** with sodium hydroxide resulted in the formation of products **52–54**, respectively. The mild

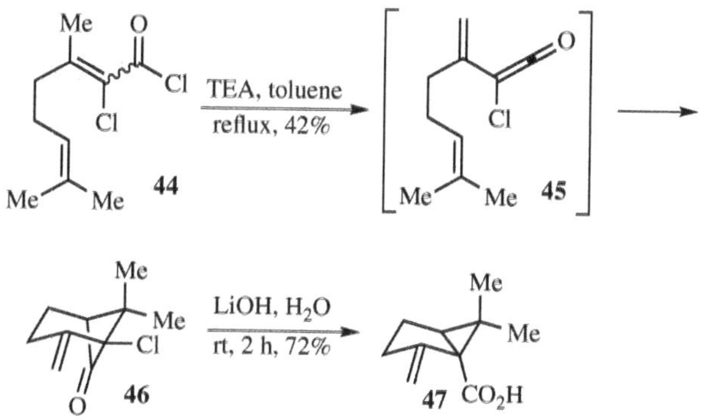

Scheme 7.13 [2+2]-Cycloaddition approach to α-chloroketones.

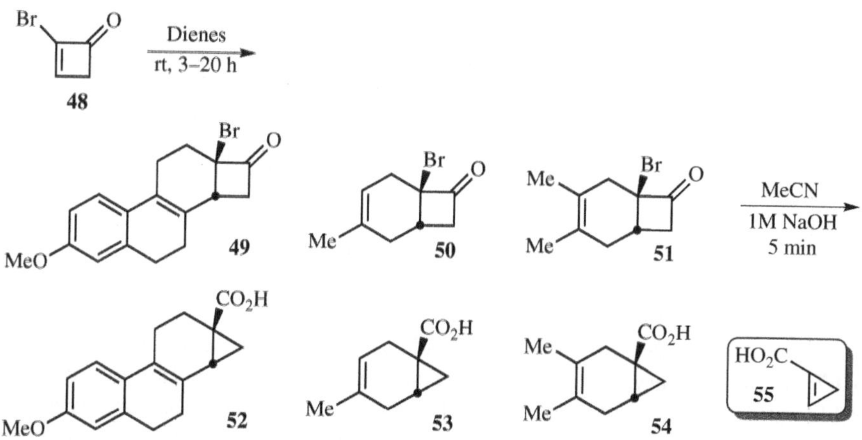

Scheme 7.14 Diels–Alder cycloaddition of 2-bromocyclobutenone (**48**) and quasi-Favorskii rearrangement of the cycloadducts.

reaction conditions are noteworthy, as is the fact that the overall process makes **48** a functional equivalent of **55** as a dienophile in the Diels–Alder reaction.

A single example was given in the same paper on the intramolecular Diels–Alder reaction of a regioisomeric halocyclobutenone. Treatment of the alcohol **56** with $BF_3{\cdot}Et_2O$ resulted in the formation of **57**, which could be observed by NMR and underwent cycloaddition to afford **58** (Scheme 7.15). The crude product was treated with hydroxide to afford the quasi-Favorskii product **59** in 27% overall yield. The particular dienophile/quasi-Favorskii combination used in this case is the functional equivalent of dienophile **60**.

Our group has generated precursors to the quasi-Favorskii rearrangement by using the (4+3)-cycloaddition reaction. For example, treatment of either 2,2-dichlorocyclopentanone (available from **61**) or 2,5-dibromocyclopentanone

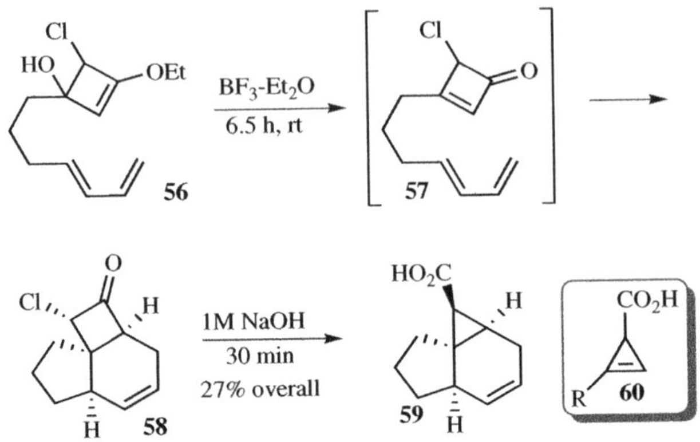

Scheme 7.15 Intramolecular Diels–Alder reaction of a chlorocyclobutenone.

(**65**) with triethylamine (TEA) in the presence of dienes in a polar solvent resulted in the formation of (4+3)-cycloadducts.[30] These were often sensitive and thus characterized as their reduction products as shown in Schemes 7.16 and 7.17.

Both haloketone products and the corresponding alcohols are ideally suited for quasi-Favorskii reactions. For example, the reaction of **69** with dilithiomethylphenyl sulfone produced **70** in 39% overall yield from 2,5-dibromocyclopentanone (**65**) (Scheme 7.18, Eq. 1). When treated with excess of methyllithium, **67** led to the polycyclic cyclobutane **71** as a result of carbonyl addition, quasi-Favorskii rearrangement and further nucleophilic addition of the organometallic to the methyl ketone revealed in the rearrangement step (Scheme 7.18, Eq. 2).

We were able to control the course of addition of certain organometallic reagents to **67**. While primary, unfunctionalized sp^3-hybridized organolithiums could be problematic with respect to carbonyl addition (e.g., side reactions),[31] sp^2-hybridized species were better behaved.[32] Further, we found that in varying the amount of the organometallic reagent and the reaction temperature, we could often steer the outcome of the reaction to simple addition, addition plus quasi-Favorskii rearrangement, or addition/quasi-Favorskii/addition as illustrated in Scheme 7.19. All of these cases demonstrate that the (4+3)-cycloaddition/quasi-Favorskii sequence is a synthetic equivalent for the endo-selective [4+2]-cycloaddition of substituted cyclobutenes.

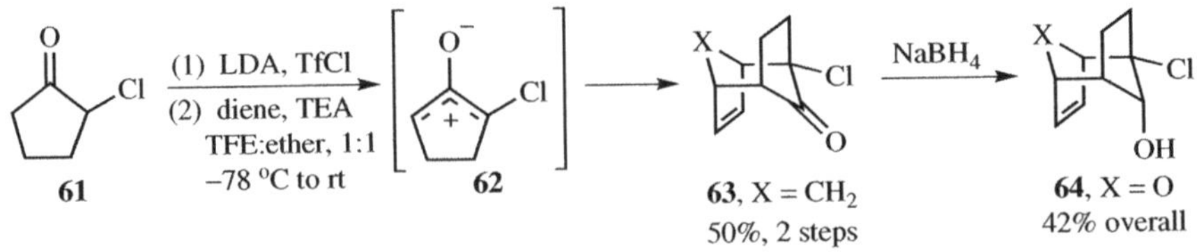

Scheme 7.16 (4+3)-Cycloaddition of a chlorinated cyclopentenyl oxyallylic cation.

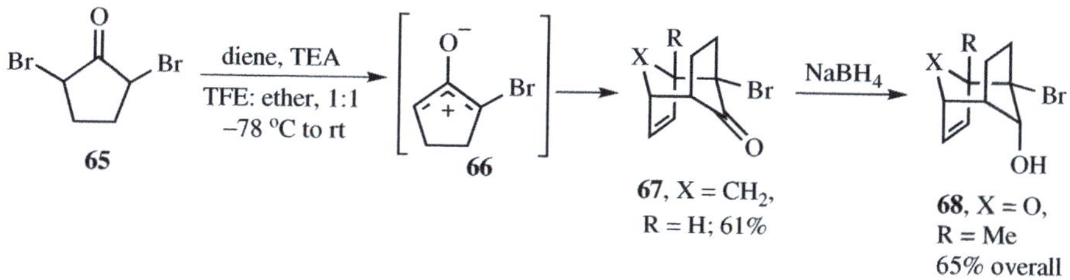

Scheme 7.17 (4+3)-Cycloaddition of a brominated cyclopentenyl oxyallylic cation.

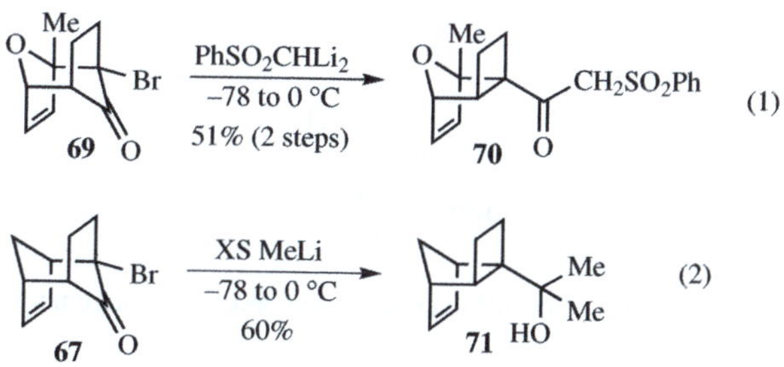

Scheme 7.18 Quasi-Favorskii rearrangements from (4+3)-cycloadducts.

Cyclopentene equivalents for the [4+2]-cycloaddition are also possible.[33] Six-membered ring oxyallylic species undergo the (4+3)-cycloaddition with facility as illustrated in Scheme 7.20. Thus, the reaction of 2,6-dibromocyclohexanone with the furan **77** in the presence of base afforded **78** regioselectively in 66% yield. Reduction of the carbonyl followed by treatment with potassium hydride effected the quasi-Favorskii rearrangement. For isolation and characterization purposes, the immediate rearrangement product, an aldehyde, was reduced to the corresponding alcohol **79**, produced in 60% yield from **78**.

An intramolecular (4+3)-cycloaddition illustrates the potential of the quasi-Favorskii approach in the preparation of complex ring systems.[34] Reaction of readily available alcohol **80** with triflic anhydride afforded the cycloadduct **81** stereoselectively in 65% yield (Scheme 7.21). Treatment of this compound with lithium aluminum hydride (LAH) afforded an essentially quantitative yield of alcohol **82** from a sequence of reduction, quasi-Favorskii rearrangement, and further reduction.

Kraus and Shi used condensation reactions for the assembly of precursors to the quasi-Favorskii rearrangement.[35] Reaction of the bromoketoester **83** with ethylamine and formaldehyde gave **84**, the result of a double Mannich reaction (Scheme 7.22). Under acidic conditions, methyl vinyl ketone combined with **83** to give the annulation product **87**. Each of these products, **84** and **87**, was a good substrate for the quasi-Favorskii rearrangement. Ketone **84** reacted with the enolate of acetylcyclohexene (**85**) to produce **86** in 60% yield. Ketone **87**, meanwhile,

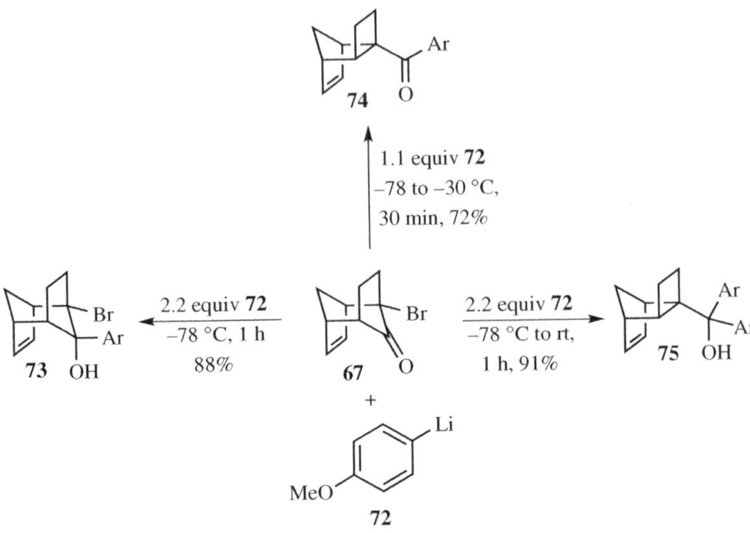

Scheme 7.19 Control over quasi-Favorskii rearrangements.

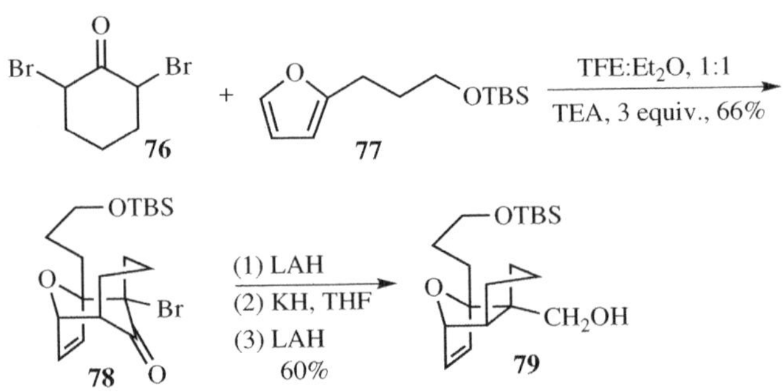

Scheme 7.20 Halogenated cyclohexenyl oxyallylic cations as cyclopentene equivalents via a (4+3)-cycloaddition/quasi-Favorskii reaction sequence.

afforded the ketophosphonate **89** in 70% yield upon reaction with dimethyl lithiomethylphosphonate (**88**). The application of this methodology to synthesis will be discussed in the following text. These examples illustrate that a wide variety of nucleophiles can initiate the quasi-Favorskii rearrangement, an important aspect of this process that allows for diversity in synthesis.[36]

Finally, a common way of introducing a leaving group, generally a halogen, at a bridgehead position is a halogenative (oxidative) decarboxylation. As a single example, Moriarty *et al.* treated **90** with iodobenzene diacetate, iodine and AIBN to give the iodoketone **91** in 92% yield (Scheme 7.23).[37] Reaction of this compound with aqueous sodium hydroxide afforded the quasi-Favorskii rearrangement product **92** in 50% yield.

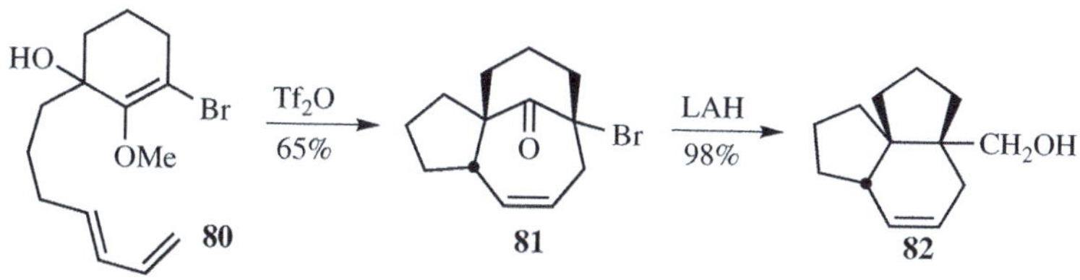

Scheme 7.21 Intramolecular (4+3)-cycloaddition and subsequent quasi-Favorskii rearrangement.

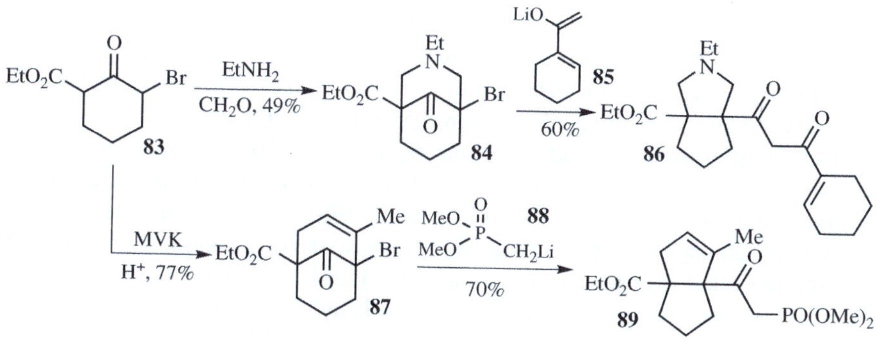

Scheme 7.22 Condensation reactions used to make precursors for the quasi-Favorskii rearrangement.

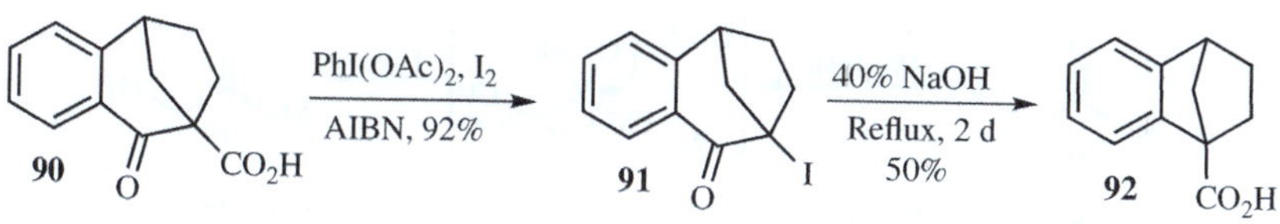

Scheme 7.23 Halogenative (oxidative) decarboxylation as a route to precursors of the quasi-Favorskii rearrangement.

7.5 APPLICATIONS OF THE QUASI-FAVORSKII REARRANGEMENT IN SYNTHESIS

7.5.1 Methodology

The preceding discussion on the synthesis of substrates for the quasi-Favorskii rearrangement clearly contained data on methodology development. This section expands that coverage to include further methodological advances and applications of the quasi-Favorskii rearrangement to target-oriented synthesis.

The addition of organometallics – Grignard reagents in particular – to α-haloketones has a very long history. As early as 1903, Tiffeneau reported that addition of phenylmagnesium chloride to chloroacetone resulted in the formation of phenylacetone after heating (Scheme 7.24).[38] This outcome appears not to have been conceived at the time as proceeding through a quasi-Favorskii mechanism, but eventually it was formulated as such.[39] As related afterward by Tiffeneau himself, this series of discoveries would make for a good chemical history essay.[40]

The quasi-Favorskii reaction has been important in the alkylation of cycloalkanones, cyclohexanone in particular. Bouveault and Chereau showed that 2-chlorohexanone (**96**) reacted with alkyl Grignard reagents to afford 2-alkylcyclohexanones.[41] Bartlett and Rosenwald[42] and Tiffeneau and Tchoubar[39b] developed this chemistry further. Thus, addition of a methyl Grignard reagent to **96** gave **97** as the major diastereomer. Heating the adduct in dry benzene at reflux for 2 h then provided 2-methylcyclohexanone (**99**) in 43% yield (Scheme 7.25). Interestingly, heating the isolated chlorohydrin in 33% NaOH was reported by Bartlett to give only acetylcyclopentane (**101**) in 50% yield. It is thus conceivable that the first reaction takes place via the conformation represented in **97b**, with the methyl group disposed antiperiplanar to the chloride, the departure of which is assisted by the magnesium cation.[43] In the second case, the smaller size of the sodium alkoxide may lead to a preferred conformation in which it is axial (**100**), placing a ring bond antiperiplanar to the equatorial leaving group and leading to the ring-contracted product **101**.

Installing alkyl groups adjacent to ketones is relatively easy based on enolate chemistry but the same is not true of aryl and alkenyl groups. Using quasi-Favorskii methodology presents one solution to this synthetic problem. In fact, because the migratory aptitude of π systems is generally higher than that of alkyl groups, the

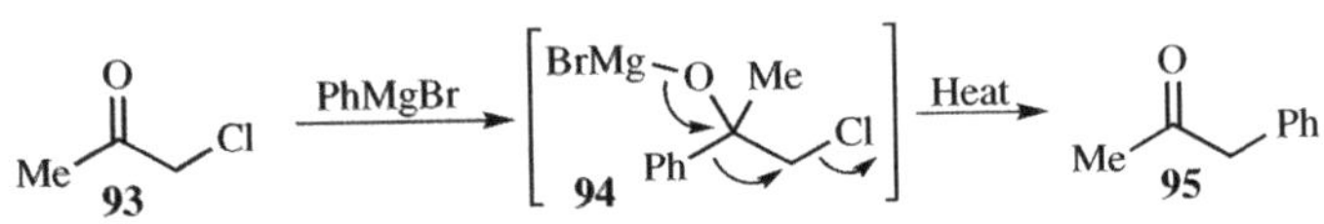

Scheme 7.24 Quasi-Favorskii route to phenylacetone (**95**).

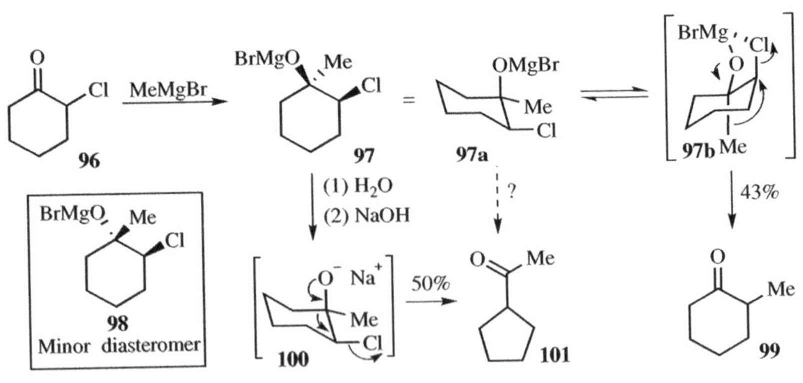

Scheme 7.25 Quasi-Favorskii rearrangement of chlorohydrin alkoxides.

quasi-Favorskii process works better for aryl or alkenyl shifts. In general, the procedure for this reaction involves heating a magnesium alkoxide in noncoordinating solvents at temperatures at least as high as that of boiling benzene. Such a reaction with phenylmagnesium bromide and 2-chlorocyclohexanone has been investigated and applied in some simple syntheses.[44,43] In this process, there are two stereoelectronic conditions for a facile reaction (Figure 7.4). First, the σ bond that is formally migrating must be antiperiplanar to the leaving group. However, the increased ease with which unsaturated systems migrate is no doubt due to π participation, so that the π system must overlap with the antibonding orbital of the carbon-leaving group bond.[45] In the case shown, the plane of the benzene ring must be perpendicular to the plane containing the C—Cl and the C—C (phenyl) bonds. If the π system is sufficiently electron rich, migration can take place under mild conditions. Bachman *et al.* reported that the reaction of *p*-anisylmagnesium bromide with **96** gave the corresponding ketone **102** directly after 15 h at room temperature (Scheme 7.26).[46]

The reaction was expanded in scope by Huang[47] and Hussey and Herr.[48] Figure 7.5 shows some compounds prepared by the process. Interestingly, Hussey and Herr developed both one-pot and two-pot procedures for the process, the latter involving isolation of the intermediate halohydrin and giving yields better than the one-pot alternative.

Pei *et al.* at Merck recently added an innovative twist to this process that resulted in a new indole synthesis.[49] Treatment of *ortho*-aminochloroacetophenones with Grignard or organolithium reagents led to the formation of indoles in very good yields (Scheme 7.27). After nucleophilic addition to the carbonyl group, the electron-rich anilino group migrates in a quasi-Favorskii process to produce ketones **105** that can then cyclize to indoles.

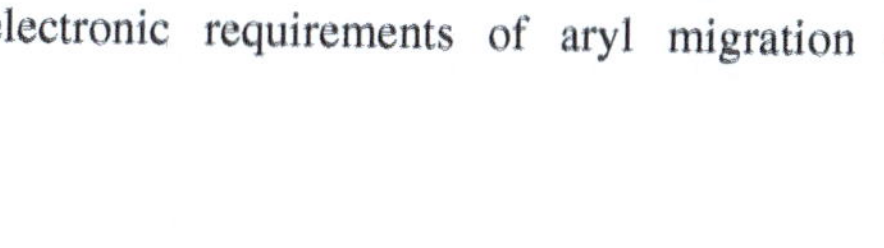

FIGURE 7.4 Stereoelectronic requirements of aryl migration in the quasi-Favorskii rearrangement.

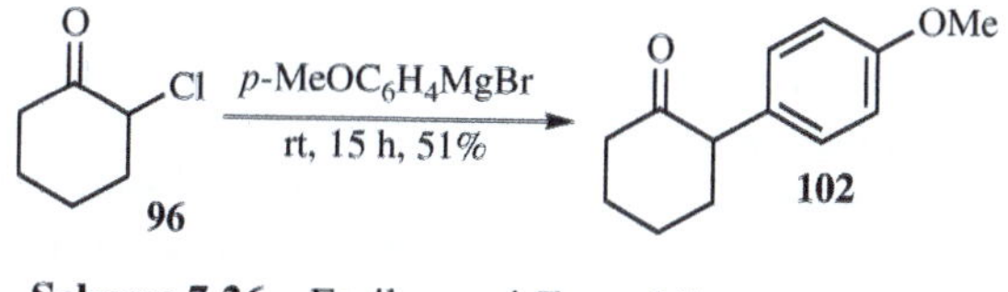

Scheme 7.26 Facile quasi-Favorskii rearrangement.

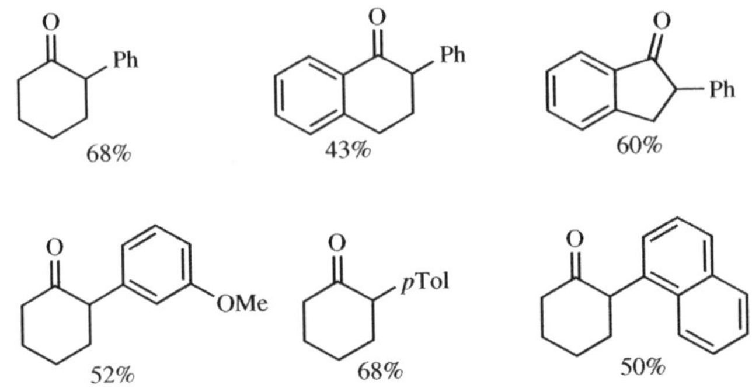

FIGURE 7.5 Some ketones, with yields, produced from the reaction of chloroketones and Grignard reagents.

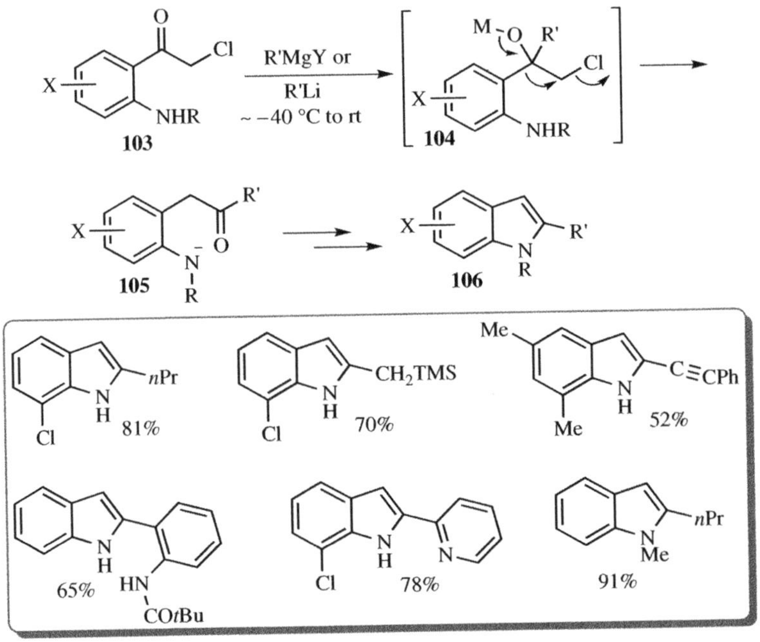

Scheme 7.27 Indole synthesis based on the quasi-Favorskii rearrangement.

Both alkenyl and alkynyl groups can also participate in the quasi-Favorskii rearrangement. Butsugan *et al.* reported that the copper-catalyzed reaction of bromohydrin **107** with Grignard reagents resulted in the formation of products arising from a quasi-Favorskii rearrangement followed by nucleophilic addition to the resulting methyl ketone.[50] An example showing products from two distinct pathways is shown in Scheme 7.28. In this case, the major product **111** is derived from a quasi-Favorskii rearrangement, while the minor product **112** results from direct bromide displacement

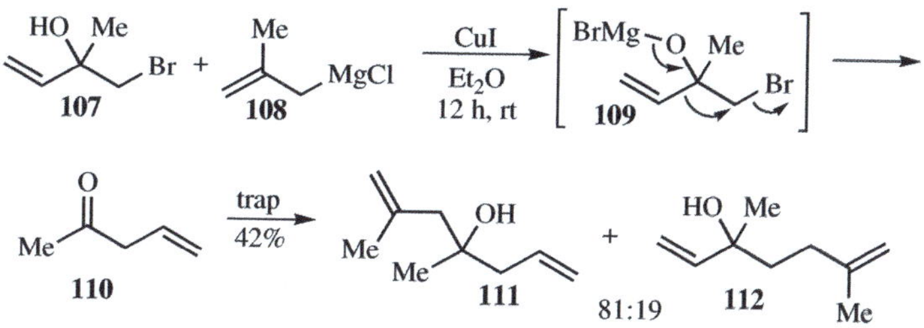

Scheme 7.28 Migration of a vinyl group in a quasi-Favorskii rearrangement.

or from ring closure to the corresponding epoxide followed by ring-opening. Interestingly, this reaction appears rather idiosyncratic, favoring the desired rearrangement addition process with allylic Grignard reagents but only managing a low yield of the quasi-Favorskii rearrangement product with aliphatic Grignard reagents. However, aliphatic organolithium reagents appear to perform better in this process, while phenyllithium and benzyllithium give products based on epoxide formation from the bromohydrin.[51] This chemistry has been applied in more complex form by the Paquette and Barriault groups, aspects of which will be presented later.[52] Finally, certain alkynyl Grignard reagents can also participate in this kind of process,[53] as can cobalt complexes of alkynes.[54]

In the course of developing a synthesis of macrocycles by ring expansion, Wender *et al.* demonstrated that deprotonation of **113** with EtMgBr in a mixture of ether and benzene, followed by heating for 2 min at 75 °C, resulted in the formation of ketone **114** in 66% yield (Scheme 7.29).[55] The facility of this reaction is noteworthy, and further mention of the process will be made later in this chapter.

While magnesium seems to be the metal ion of choice for inducing the quasi-Favorskii rearrangement of halohydrins,[56] there are alternatives. Xue and coworkers reported that treatment of bromohydrins with diethylzinc resulted in quasi-Favorskii rearrangements, often in very high yield.[57] The study focused on aryl migration (Scheme 7.30, Eqs 1 and 2), but by no means exclusively (Scheme 7.30, Eq. 3). Mechanistically, these reactions could be assumed to proceed via zinc alkoxide formation followed by zinc-assisted departure of the leaving group with concomitant 1,2-shift of the migrating group. Also noteworthy among the examples shown is the tolerance of the ester group to the reaction conditions, something that might not be easy to achieve if a Grignard reagent were used to induce the rearrangement.

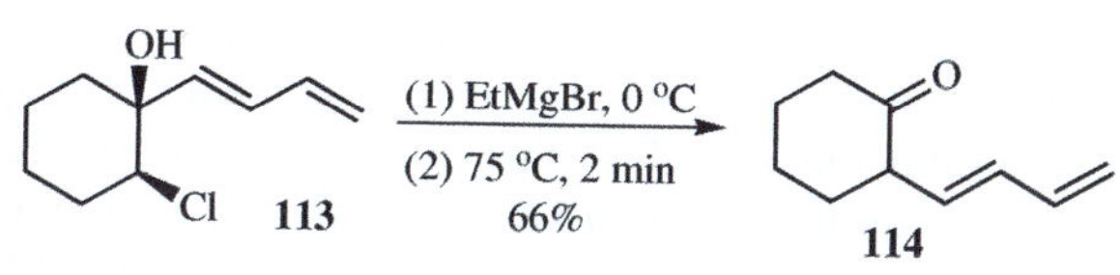

Scheme 7.29 Migration of a butadienyl group in a quasi-Favorskii rearrangement.

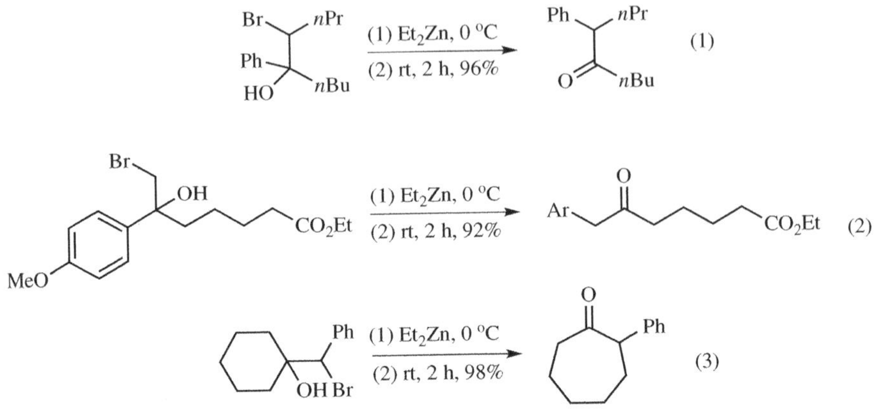

Scheme 7.30 Diethylzinc-mediated quasi-Favorskii rearrangement of tertiary bromohydrins.

The starting materials in Scheme 7.30 are all tertiary alcohols. Zhang and coworkers introduced a new reagent for the conversion of related secondary alcohols to aldehydes via a quasi-Favorskii rearrangement.[58,59] Thus, a number of secondary alcohol bromohydrins afforded rearranged aldehydes when exposed to $EtZnOCOCF_3$ in dichloromethane for several hours at room temperature. The reactions generally showed very high to complete regioselectivity with respect to the migrating group (Ar > H), even when the aryl group bore an electron-withdrawing substituent (Scheme 7.31, Eq. 1). Sterically hindered aldehydes were available using this procedure (Scheme 7.31, Eq. 2). However, attempts to make phenylacetaldehyde and other simple congeners resulted in a dehydrative dimerization (aldol condensation) of the initial product (Scheme 7.31, Eq. 3).

To end this section, we turn to an interesting rearrangement of epoxyalcohols. Covering all semipinacol rearrangements of species of this type is well beyond the scope of this chapter; a recent review must suffice to fill in the many gaps.[18d] In 1986, Magnusson and Bergman, seeking both to study the mechanism of this process and to improve it[60] as a synthetic tool, reported that cyclopentene-1-carboxaldehydes were formed with useful selectivity and yields when 2,3-epoxycyclohexanols were treated with lithium bromide in refluxing 1,2-dimethoxyethane (DME).[61] As examples, Scheme 7.32 shows the conversion of three epoxycyclohexanols to the corresponding cyclopentenecarboxaldehydes. The conversion of **119–120** was performed on large scale and proceeded in 60% isolated yield.

The proposed mechanism of the process includes a quasi-Favorskii rearrangement, as illustrated in Scheme 7.33 for the transformation of **117** into **118**. Attack of bromide on the epoxide is expected to occur in a trans diaxial fashion, in accord with the Fürst–Plattner rule,[62] at the position of the epoxide distal to the electron-withdrawing hydroxyl group. The immediate result of this ring opening would be **121**, which is in rapid conformational equilibrium with **122**. Migration of a σ bond from this latter conformation with concomitant expulsion of bromide would give **123**, which dehydrates to afford the observed product.

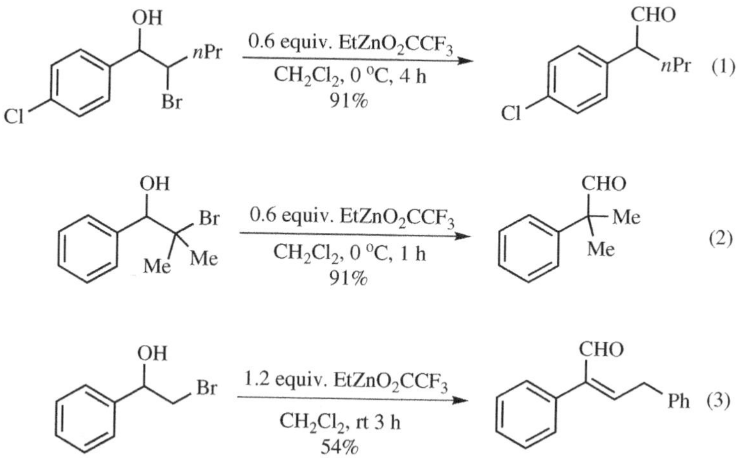

Scheme 7.31 Trifluoroacetoxyethylzinc-mediated quasi-Favorskii rearrangement of secondary bromohydrins.

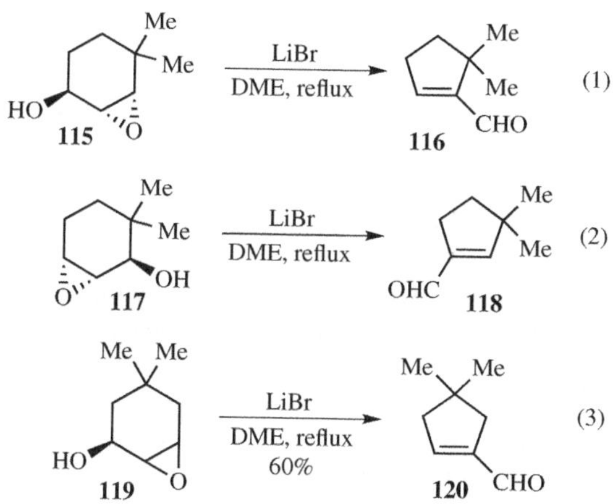

Scheme 7.32 Quasi-Favorskii approach to cyclopentene carboxyaldehydes.

More complex structures can also undergo this rearrangement. For example, Figure 7.6 shows a selection of aldehydes obtained from various epoxyhexopyranosides.[63] Though the instability of the regioisomers of these aldehydes led to lower than expected yields, the method represents a very convenient access to compounds of the type shown.

7.5.2 Natural Products

The extent of application of the quasi-Favorskii rearrangement to total synthesis depends to a certain degree on how one characterizes the many semi-benzylic and

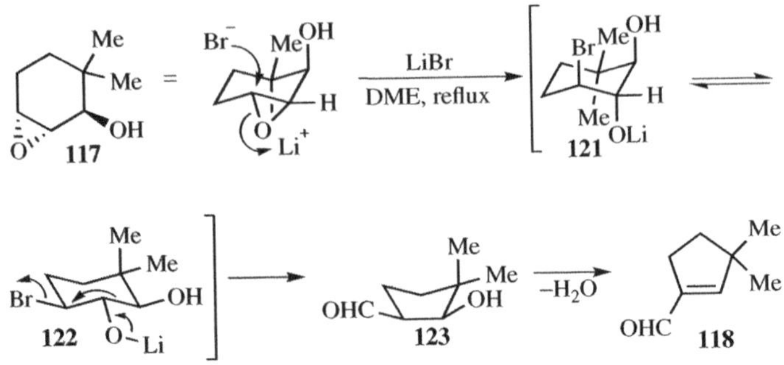

Scheme 7.33 Mechanism for the conversion of **117–123**.

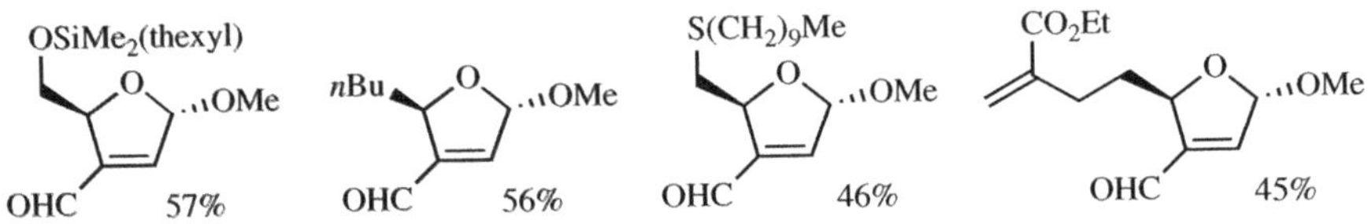

FIGURE 7.6 Selection of sugar-derived aldehydes prepared by a quasi-Favorskii rearrangement.

semipinacol rearrangements that are available. For the purposes of this chapter, the cited examples are those that contain processes in which an alkoxide "pushes" a bond that displaces an adjacent leaving group, though the departure of the leaving group might be assisted by a Lewis acidic metal ion.

7.5.2.1 ***Muscone (Wender)*** The Wender *et al.* synthesis of muscone began with chloroketone **124**.[64] Reaction with the vinyl acetylide **125** afforded alcohol **126**. Reaction of the propargylic alcohol with LAH led to alkyne reduction, quasi-Favorskii rearrangement, and a ketone reduction to afford **127** in 91% yield over two steps. Oxidation to the ketone and a second carbonyl addition/reduction sequence led to the polyene **128** in 45% yield, though starting material was recovered, making the corrected yield somewhat higher. An oxyanionic Cope rearrangement ensued when **128** was treated with potassium hydride at room temperature. It is this process, formally a [5,5]-sigmatropic shift, which forms the core of Wender's macroexpansion methodology. Exhaustive hydrogenation of **129** afforded the target muscone (Scheme 7.34).

The quasi-Favorskii step in the Wender synthesis is especially noteworthy for the mild conditions under which it takes place. A number of substrates were examined as part of the methodology development associated with this synthesis. It appears that rearrangement takes place from the hydroaluminated intermediate and is very rapid even at −20 °C. An interesting twist on the chemistry is shown in Scheme 7.35. Reduction of **131** with LAH followed by quenching with iodine afforded **134** in 81% yield. With DIBAL as reductant, the reaction was still rapid, but the organoaluminum intermediate that rearranged was presumably different from that in the LAH

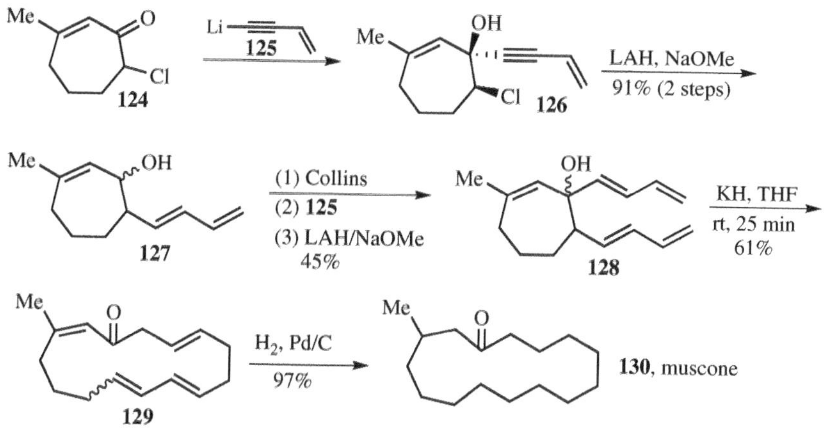

Scheme 7.34 Wender's synthesis of muscone.

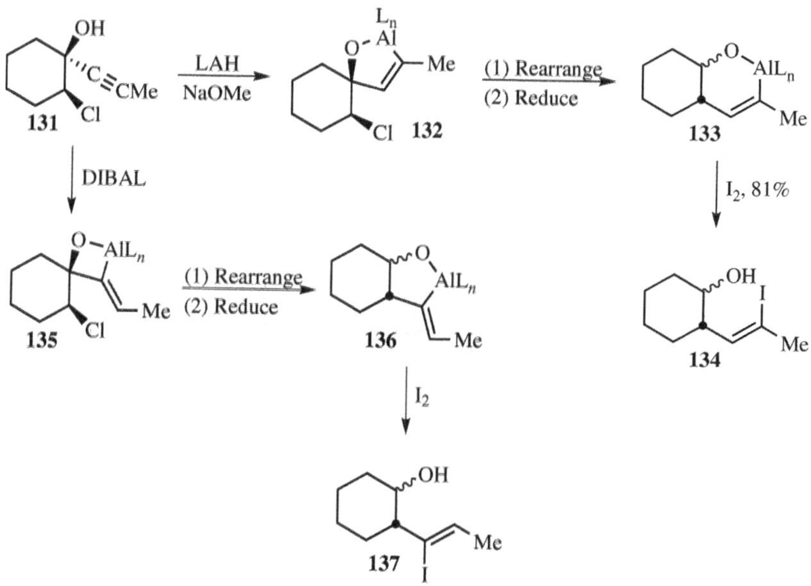

Scheme 7.35 Possible mechanism for reductive rearrangement of propargylic chlorohydrins.

reduction as evidenced by its trapping with iodine leading to **137**. The implication is that vinylmetals have a propensity to migrate beyond that of simple alkenes, and that predilection may be enhanced by the formation of cyclic structures that enforce geometries favorable to the migration process.

7.5.2.2 (−)-(3Z)-Cembrene A (Wender) The synthesis of the title natural product in many ways parallels that of muscone. Wender and Holt's synthesis began

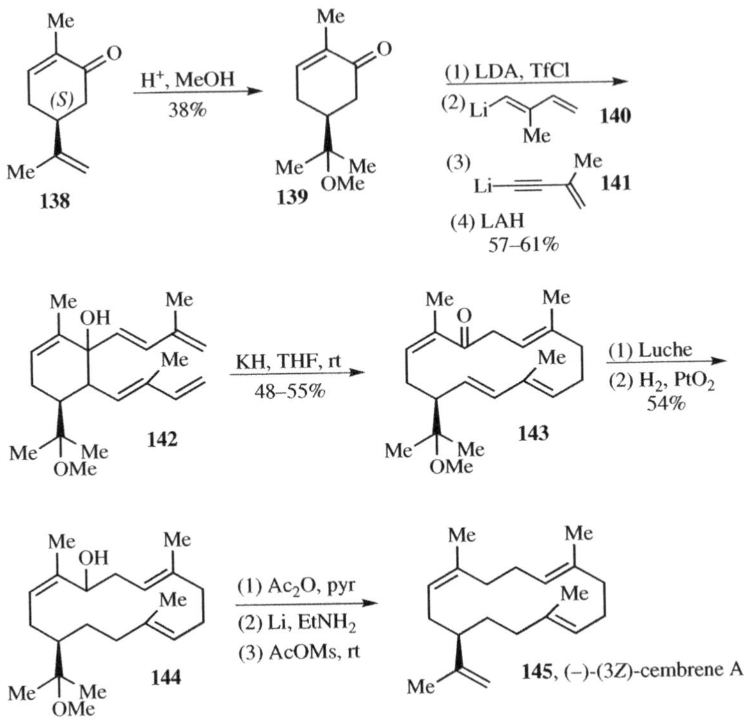

Scheme 7.36 Synthesis of (−)-(3Z)-cembrene A.

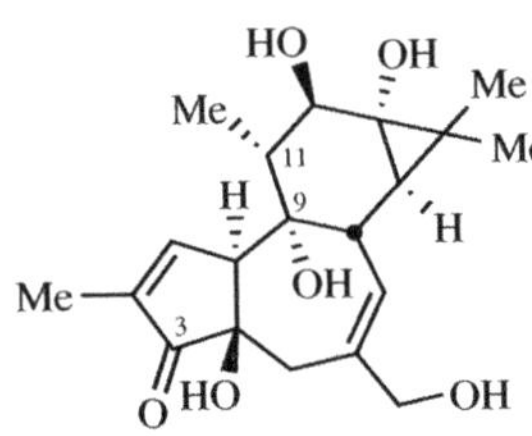

146, phorbol

FIGURE 7.7 Structure of phorbol.

with (*S*)-carvone **138**, which was converted to the methyl ether **139** using simple acid-catalyzed methanolysis of the electron-rich double bond (Scheme 7.36).[65] Chlorination of **139** was followed by a sequence of events including nucleophilic addition of **140** to the ketone, addition of **141** followed by heating and finally LAH reduction of the propargylic alkoxide to form **142**. The last three steps of the sequence were carried out in one pot, demonstrating that lithium alkoxides of chlorohydrins can also undergo the quasi-Favorskii rearrangement.

The next key step in the sequence was the macroexpansion, smoothly executed by treatment of **142** with potassium hydride and 18-crown-6 in THF at room temperature. Reduction of the resultant ketone and selective reduction of the least hindered olefin afforded **144**. An allylic deoxygenation and deprotection of the isopropenyl group with acetyl mesylate gave the target natural product **145**.

7.5.2.3 ***Phorbol Skeleton (Paquette)*** Phorbol esters have been known for many years as tumor promoters and activators of protein kinase C.[66] Their biological activity and the unique terpene scaffold of these esters have stimulated a great deal of interest in phorbol and its congeners as synthetic targets, resulting in recent exciting developments (Figure 7.7).[67]

In 1994, Paquette *et al.* reported a synthesis of the phorbol skeleton using an approach involving a quasi-Favorskii rearrangement.[68] The vinyl bromide **147** (Scheme 7.37) was derived from (+)-carvone. Halogen–metal exchange with *t*-BuLi followed by metal–metal exchange yielded an organocerate for addition to 2-chlorocyclohexanone (**96**). Two separable diastereomers were formed in a 1:1 ratio; only the one leading to the product of interest is shown in the scheme. This compound was treated with vinylmagnesium bromide in THF/benzene, and the mixture was heated to 80 °C for 30 min to give the quasi-Favorskii/trapping product **149**. Anionic oxy-Cope rearrangement of **149** produced the 10-membered ring ketone **150** in 77% yield, stereochemistry being dictated by rearrangement to the convex face of the carene ring system. Oxidation of the double bond in **150** with *m*CPBA was completely stereorandom; only one of the product isomers and its yield are illustrated. Treatment of this isomer (**151**) with LDA effected an intramolecular alkylation in 71% yield, resulting in the formation of **152**, which

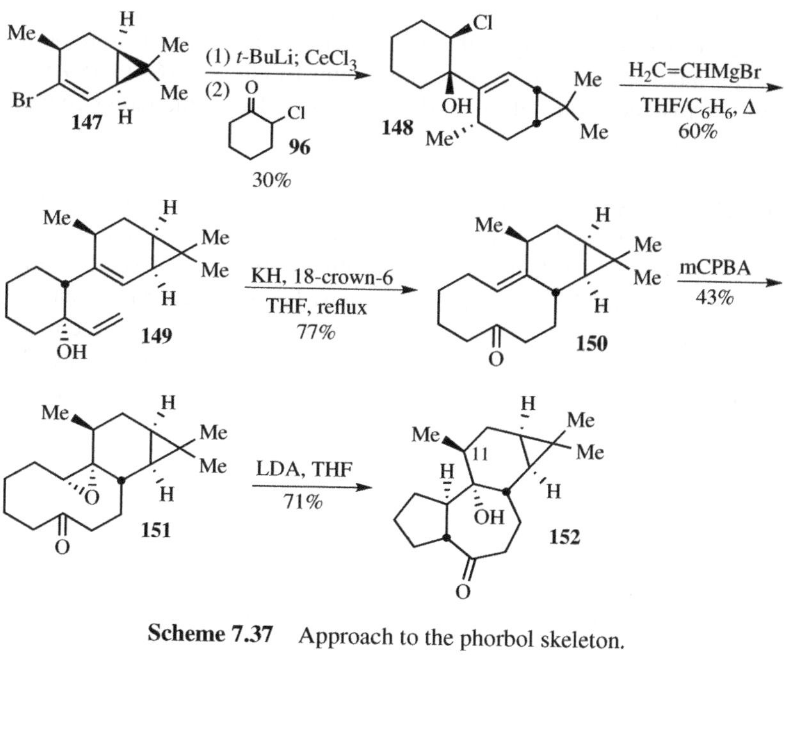

Scheme 7.37 Approach to the phorbol skeleton.

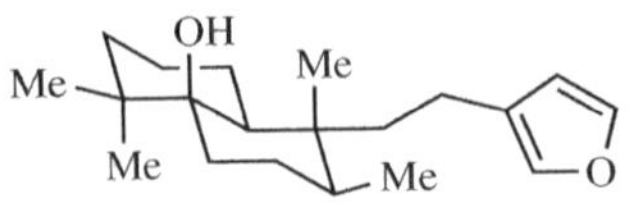

FIGURE 7.8 Structure of ambliol B.

contains the complete carbocyclic skeleton of phorbol (**146**), but is epimeric with **146** at carbon 11.

7.5.2.4 Desdimethyl Ambliol B (Barriault) Barriault *et al.* developed an approach to ambliol B (Figure 7.8) in the course of their studies on a domino oxy-Cope/ene/Claisen sequence.[69] Low-temperature halogen–metal exchange of the bromide **153** followed by trapping with 2-chlorocyclohexanone **96** afforded chlorohydrin **154** in 52% yield (Scheme 7.38). A reaction much like that in Paquette's phorbol approach, but without trapping, took place when **154** was treated with vinylmagnesium bromide in THF/TMEDA at reflux to give ketone **155** in 32% yield. Subsequently, addition of the functionalized vinyllithium **156** afforded key intermediate **157** in 43% yield. Thermolysis of **157** at 200 °C led to the polycycle **161** through a series of intermediates arising from oxy-Cope rearrangement (**158**), ene reaction (**159**), and Claisen rearrangement (**160**).

Wolff–Kishner reduction of **161** followed by a metathesis dimerization process and Raney nickel desulfurization afforded **162** in high yield. The metathesis reaction was performed to circumvent problems associated with the desulfurization of the compound bearing only the terminal vinyl group. In this clever approach, oxidative cleavage gave the same aldehyde that would have been obtained from the terminal alkene. This aldehyde reacted with organolithium **163** to give **164** in 35% yield over two steps. A two-step, site-selective, deoxygenation led to the target compound in reasonable yield.

7.5.2.5 Modhephene (Kraus) The formal synthesis of the propellane sesquiterpene natural product modhephene (**168**) was accomplished by Kraus and Shi using the chemistry shown in Scheme 7.39.[70] As already mentioned (see Scheme 7.22), the reaction of **88** with **87** produced **89** via a quasi-Favorskii rearrangement. Directed hydrogenation followed by intramolecular Horner–Emmons reaction led to **166**. The fact that the olefination proceeded with an ester carbonyl is noteworthy. Addition of methyllithium and hydrolysis afforded the enone **167**, an intermediate in Curran group's synthesis of modhephene (and isomodhephene).[71]

7.5.2.6 Spatol (Harmata) As part of our program exploring the (4+3)-cycloaddition/quasi-Favorskii reaction sequence, we undertook a formal total synthesis of spatol (Figure 7.9), an interesting natural product possessing a unique diepoxide, a 5-4-5 ring-fused carbocyclic core, and antitumor properties.[72]

The readily available dichlorocyclobutanone **170a** (Scheme 7.40) was treated with diazomethane to give the corresponding dichlorocyclopentanone **170b**

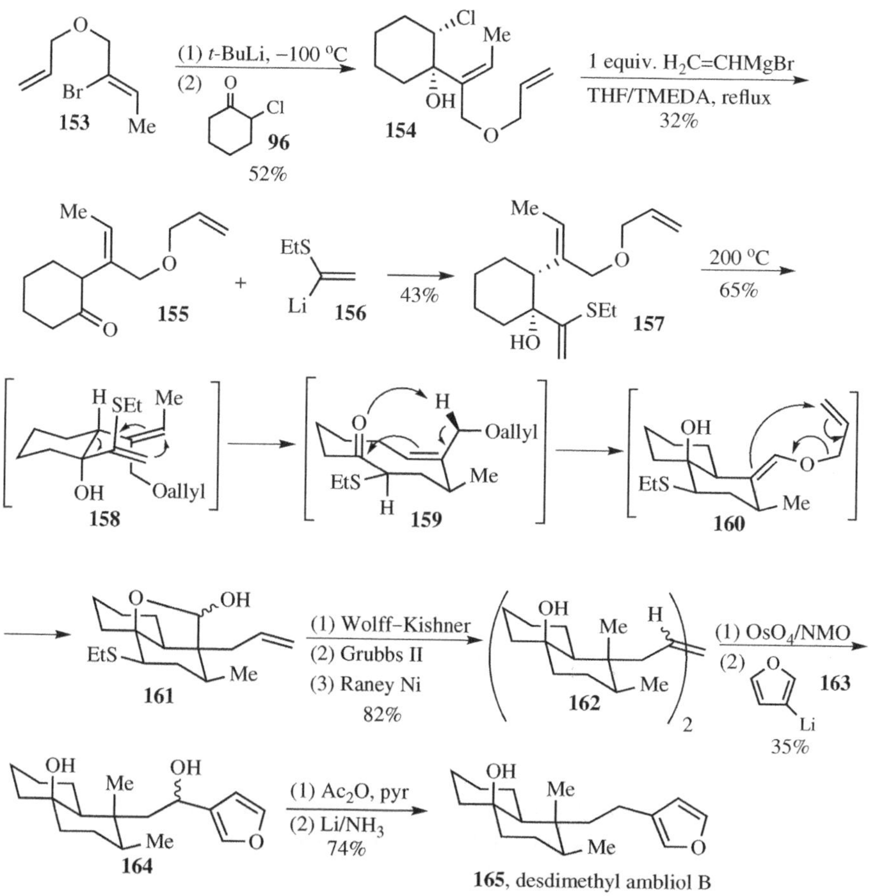

Scheme 7.38 Synthesis of desdimethyl ambliol B.

via ring expansion, a quasi-Favorskii rearrangement. This substrate underwent (4+3)-cycloaddition with cyclopentadiene in the presence of TEA in a mixture of trifluoroethanol and ether to afford the adduct **171** with good diastereoselectivity (10.4:1). Reduction of **171** with LAH produced the corresponding alcohol **172**. When this compound was treated with potassium hydride, a quasi-Favorskii rearrangement ensued, leading to **174** in 76% yield over two steps. This aldehyde was oxidized to the corresponding acid using Pinnick conditions, and thermodynamically controlled iodolactonization led to **175**. Reductive deiodination and allylic oxidation afforded the enone **176**. Hydrogenation and Wittig methylenation then gave alkene **177** in 75% yield. Although reduction of the exocyclic double bond was expected to be selective, with reagent approach occurring from the convex face of the molecule, this step turned out to be challenging. Ultimately, **178** was produced as the major diastereomer in the process (dr = 6.6:1). The lactone of **178** was sequentially reduced

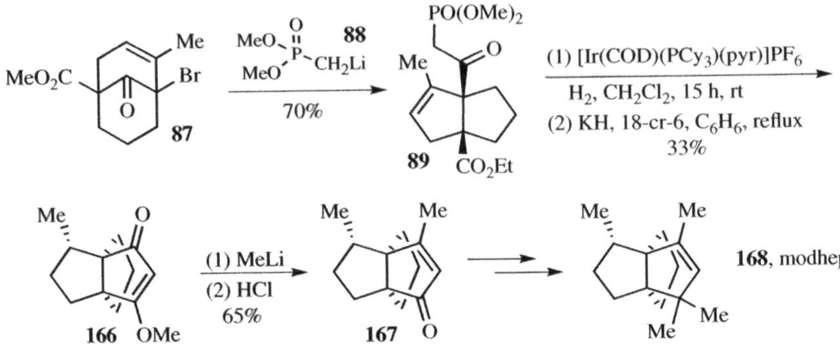

Scheme 7.39 Formal synthesis of modhephene.

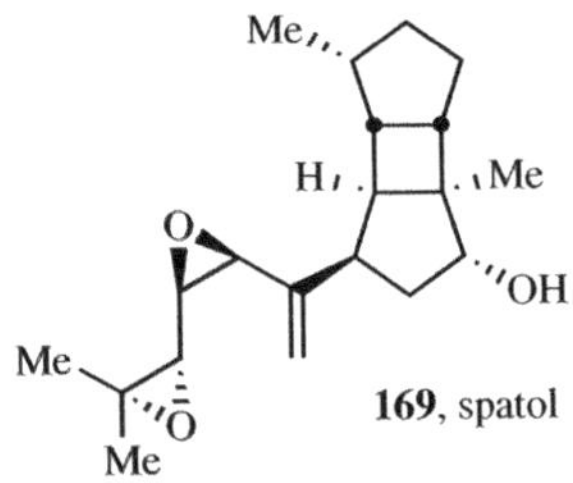

FIGURE 7.9 Structure of the antitumor agent spatol.

using DIBAL and a Wolff–Kishner reaction, and the resulting product was oxidized to produce the polycyclic ketone **179**. This compound had been transformed into spatol by Salomon *et al.*, thus completing our formal synthesis.[73]

7.5.2.7 Sterpurene (Harmata) Sterpurene (**188**, Scheme 7.41)[74] is one of a relatively small class of fungal sesquiterpenes whose unique 5-6-4 carbocyclic structure made it an interesting target for synthesis in the context of the (4+3)-cycloaddition/quasi-Favorskii methodology. The quaternary carbon embedded within the cyclobutane ring served, as it did for our approach to spatol, as a retron for the quasi-Favorskii rearrangement, with the assumption that the methyl group would be obtained through reduction of an aldehyde. Such a retrosynthetic construct then led naturally to a (4+3)-cycloaddition as a key carbon–carbon bond forming event in the synthesis.

Our synthesis began with the (4+3)-cycloaddition between diene **180** and 2,5-dibromocyclopentanone (**65**, Scheme 7.41).[75] This reaction was unique in that we used only a stoichiometric amount of diene and employed benzene as a co-solvent with trifluoroethanol. The adduct **181** was obtained in very good yield, especially given that we had used a stoichiometric amount of diene. The quasi-Favorskii sequence, initiated by LAH reduction of the ketone, included an additional reduction of the resulting aldehyde **182** to give alcohol **183**. Dehydroxylation was accomplished

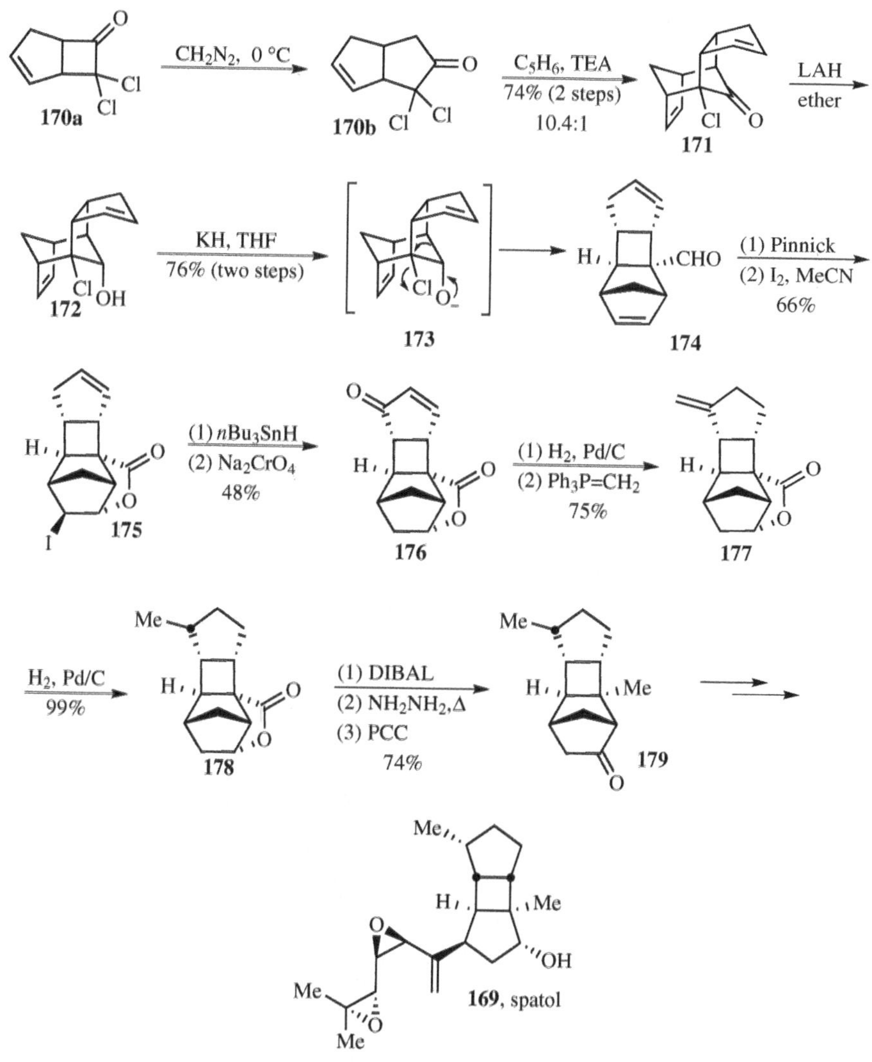

Scheme 7.40 Formal synthesis of spatol via a (4+3)-cycloaddition/quasi-Favorskii rearrangement sequence.

by conversion to the corresponding iodide followed by halogen–metal exchange and quenching with water, a process that afforded **184** in excellent yield. Allylic oxidation of **184** was extremely problematic and only small amounts of the desired ketone could be obtained in spite of intensive efforts. This ketone was converted to tosylhydrazone **185**. Reduction with allylic inversion gave the transposed alkene **186**. Hydroboration and oxidation with PCC led to the ketone **187**. The latter was treated with methyllithium and subsequently dehydrated to afford sterpurene (**188**).

7.5.2.8 Tricycloclavulone Core (Harmata) The natural product tricycloclavulone (Figure 7.10) was a clear candidate for a quasi-Favorskii strategy, given the fact that it possessed a cyclobutane ring with a quaternary center, just like spatol and sterpurene. We decided to approach the problem by first synthesizing the tricyclic core of the system using our methodology.[76] Attempts to find a route to prepare the natural product met up with a detour, however, and we decided to follow this alternative path in lieu of the total synthesis.[77]

We began with the (4+3)-cycloadduct **67**. Reaction of this ketone with organolithium **189** followed by warming gave **190** in excellent yield via addition and subsequent quasi-Favorskii rearrangement (Scheme 7.42). Ring-opening/ring-closing metathesis afforded the tricyclic core of the natural product (**191**), with an additional vinyl appendage that was removed by selective ozonolysis and decarbonylation using Wilkinson's catalyst. We demonstrated that the tertiary acetate of tricycloclavulone could be installed by radical bromination followed by silver-assisted acetolysis, resulting in the formation of **194**.

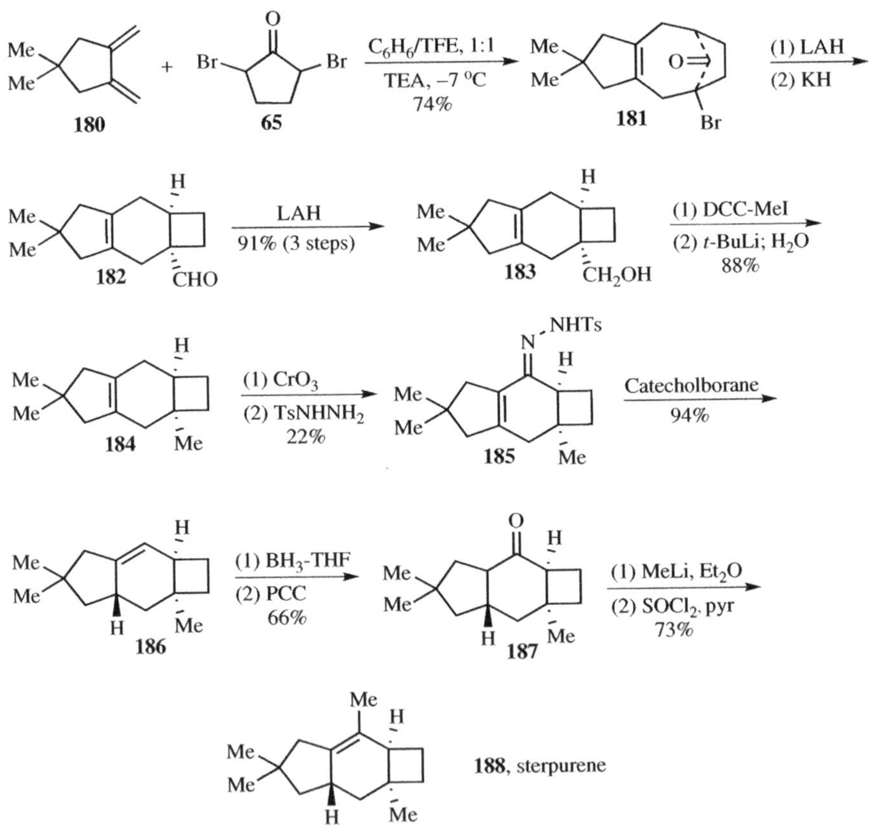

Scheme 7.41 Total synthesis of sterpurene via a (4+3)-cycloaddition/quasi-Favorskii rearrangement sequence.

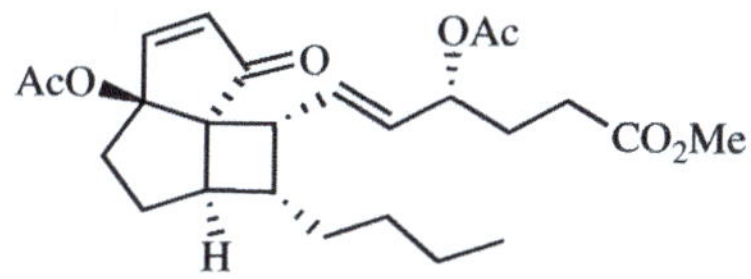

FIGURE 7.10 Structure of tricycloclavulone.

7.5.3 Unnatural Products

It is in the area of unnatural product synthesis, particularly polycarbocyclic systems, that the quasi-Favorskii rearrangement has seen its greatest application. A number of frameworks have been prepared,[78] and a selected few are discussed in the following text.

7.5.3.1 Cubane (Eaton) Perhaps the best-known application of the quasi-Favorskii rearrangement is also one of its earliest, leading to the synthesis of cubane, the organic chemist's realization of a Platonic solid. Eaton and Cole reported the first synthesis of this compound through an elegant combination of cyclopentadienone dimerization, photocyclization, and quasi-Favorskii rearrangements.[79]

The synthesis began with the 2-bromocyclopentadienone dimer **195** (Scheme 7.43). Protection of the carbonyl groups, selective hydrolysis, and photocyclization afforded **196** in excellent overall yield. A quasi-Favorskii rearrangement ensued, followed by conversion of the resultant carboxylic acid to the corresponding *t*-butyl perester. Heating this *t*-butyl perester in the hydrogen atom donor cumene resulted in a net reductive decarboxylation. Liberation of the protected ketone and another quasi-Favorskii rearrangement gave the carboxylic acid **198**. Reductive

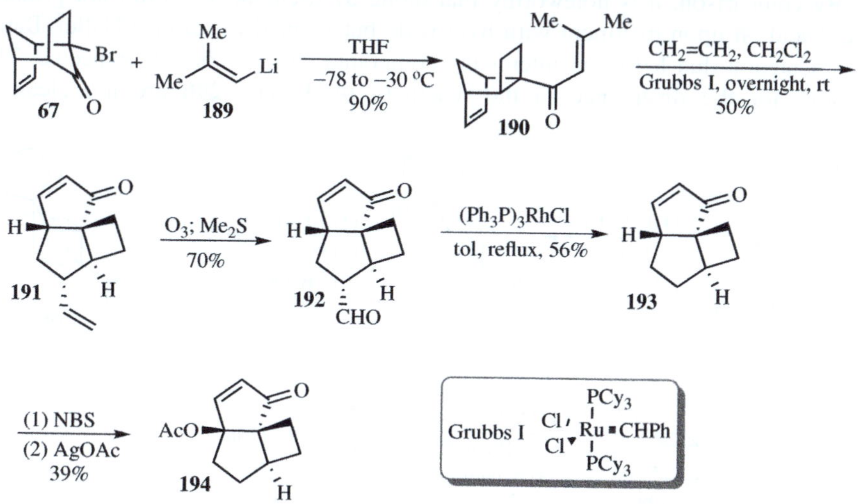

Scheme 7.42 Synthesis of the tricarbocyclic core of tricyclocalvulone.

decarboxylation as before afforded the target cubane (**199**). It is worth noting that the tertiary CH groups of cubane served as a partial retron for the two quasi-Favorskii rearrangements in the synthesis.

Eaton and Cole published a report prior to the cubane paper that ultimately constituted a formal total synthesis of this hydrocarbon.[80] The polycyclic dibromodiketone **200** was subjected to 50% aqueous KOH at reflux, affording the dicarboxylic acid **201**, the first cubane structure ever prepared. It was shown much later that reductive decarboxylation of a derivative of this diacid leads directly to cubane.[81]

The double quasi-Favorskii of **200** is interesting (Scheme 7.44). It appears that the two rearrangements are quite disconnected from one another. While 1% NaOH at room temperature suffices to produce **202** in essentially quantitative yield, the conversion of **202–201** requires 40% base, 110 °C, and produces less than a 50% yield of the final product (Philip E. Eaton, personal communication). Strain is the likely cause of the problem here, as there are limits to the amount of strain a quasi-Favorskii rearrangement can successfully establish before side reactions compete, lowering yields.

It should be noted that other substituted cubanes can be synthesized from cyclopentadienone dimers that bear substituents using the same quasi-Favorskii approach pioneered by Eaton.[82]

7.5.3.2 Cubane (Pettit) The approach to cubane by Pettit and coworkers also made use of two quasi-Favorskii rearrangements (Scheme 7.45).[83] The reaction between cyclobutadiene and 2,5-dibromoquinone afforded exclusively the endo cycloadduct **205** in 80% yield. Subsequent photolysis afforded the [2+2]-photocycloadduct **206**. When this compound was treated with aqueous KOH at 100 °C, an 80% yield of the diacid **38** was obtained. This diacid could be converted to cubane by conversion to the di-*t*-butyl perester followed by thermolysis, as per the method developed by Eaton and Cole.

By comparison, it is noteworthy that dione **37** (Scheme 7.12) did not produce the diacid **38** upon treatment with hydroxide but seemed to suffer a Haller–Bauer fragmentation that led to an interesting compound, but not the one expected. The reasons for the divergence in the reactivity of **37** and **206** are not clear. A

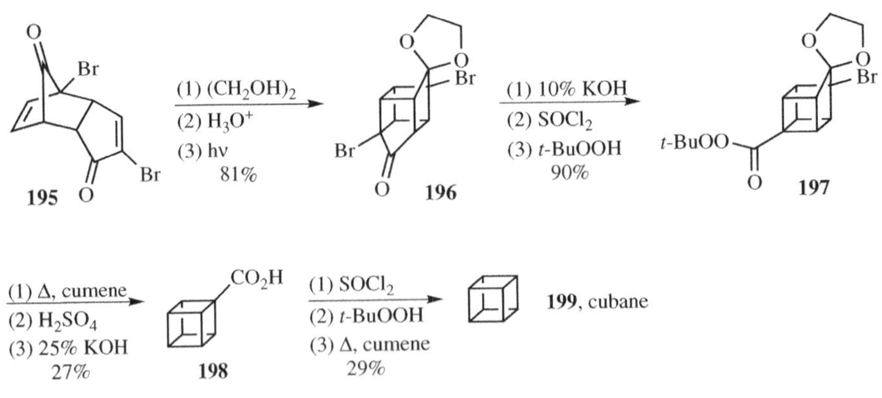

Scheme 7.43 Synthesis of cubane.

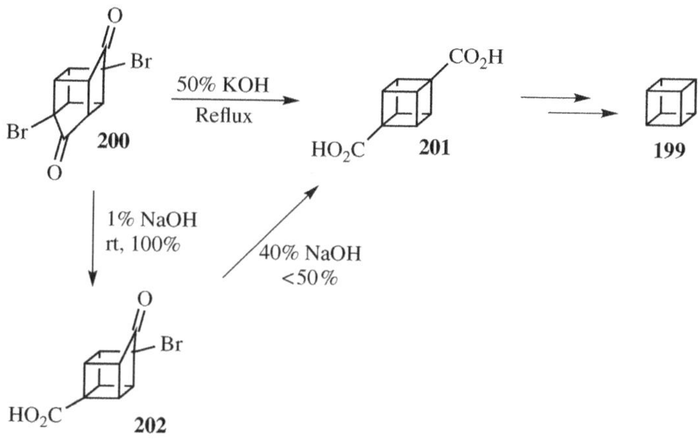

Scheme 7.44 Alternate synthesis of cubane.

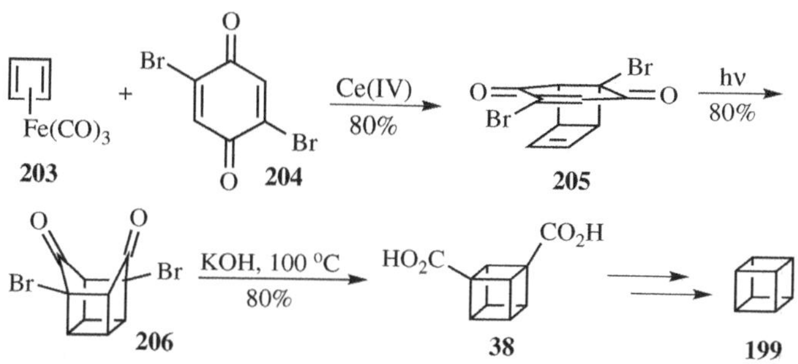

Scheme 7.45 Petit synthesis of cubane.

computational investigation might shed light on the subtle structural differences responsible for the different outcomes.

7.5.3.3 ***Almost a Prismane (Forman)*** This section describes a beautiful idea that fell victim to the strictures of Mother Nature, yet remains intriguing for that very reason.

Forman and coworkers attempted an entry into the [3]-prismane family using a quasi-Favorskii rearrangement.[84] Toward this end, the quadricyclanone **207** was treated with hydroxide under various conditions (Scheme 7.46). When decomposition did not occur, benzoic acid (**211**) was obtained in very high yields. Two pathways were proposed for the conversion to **211** – formation of the target prismane (**209**) followed by rearrangement under the reaction conditions or a Haller–Bauer rearrangement via **210**. The examination of transition structures for both pathways as well as attempts to accelerate the reaction using organometallics in lieu of hydroxide as the nucleophile might offer opportunities for making [3]-prismanes successfully. Even if this turns out to be impossible, this reaction serves to define boundaries for the quasi-Favorskii rearrangement.

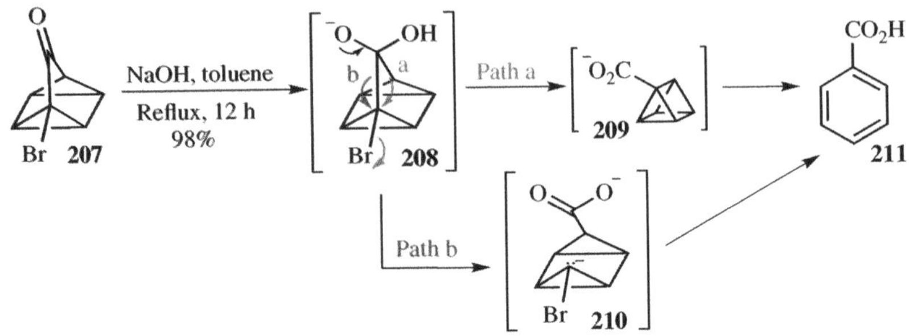

Scheme 7.46 A quasi-Favorskii approach to the [3]-prismane system.

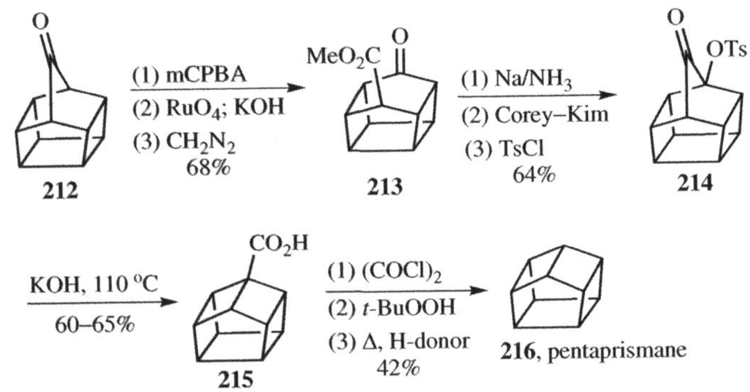

Scheme 7.47 Synthesis of pentaprismane.

7.5.3.4 Pentaprismane (Eaton) Pentaprismane represents the last of the prismanes that have been prepared, whether by a quasi-Favorskii rearrangement or other means. Eaton *et al.* reported the synthesis in 1981.[85]

A starting point for the synthesis was ketone **212**, readily available through a sequence that involved one thermal and two photochemical cycloadditions. The object of the first set of transformations shown in Scheme 7.47 was to prepare a substrate suitable for the quasi-Favorskii rearrangement. To that end, a Baeyer–Villiger oxidation was accomplished with *m*CPBA. This was followed by a CH oxidation with RuO_4, saponification, and ester formation using diazomethane to yield **213**. A pinacol-coupling reaction, alcohol oxidation using the Corey–Kim protocol, and tosylate formation afforded **214**, the substrate for the quasi-Favorskii rearrangement. These days one cannot help but wonder if **214** or its immediate precursor might be available in one step from **212** by direct CH activation.

In any case, the tosylate served well as the leaving group in the quasi-Favorskii reaction and the pentaprismane skeleton was established as the carboxylic acid **215**. Reductive decarboxylation took place in essentially the same manner as in the original cubane synthesis, though the H-atom donor in this case was 2,4,6-triisopropylnitrobenzene, giving pentaprismane (**216**) in 42% yield from **215**.

7.5.3.5 Secohexaprismane (Mehta) Mehta and Padma's synthesis of secohexaprismane was intimately associated with efforts to prepare hexaprismane itself.[86] Though conceptually elegant, many attempts to make hexaprismane failed, due to problems primarily associated with photocycloadditions that look as though they should work, but do not.

The path to secohexaprismane began with a Diels–Alder reaction between 1,5-cyclooctadiene (**217**) and **218**. This was followed by a bromination/dehydrobromination sequence to produce triene **219**. A singlet oxygen cycloaddition, reduction of the resulting peroxide, and acetylation gave the diester **220**. After photocyclization, the acetate functional groups were converted to the corresponding mesylates, leading to **221**. Treatment of **221** with iodide gave a monoelimination product, an alkene, whose hydroboration resulted in the formation of an organoborane that underwent Grob fragmentation upon reaction with hydroxide, affording the diene **222**. Another photocyclization, followed by acetal hydrolysis, gave **223** as the substrate for quasi-Favorskii rearrangement. When this compound was heated with sodium hydroxide in toluene, the major product was actually that derived from the Haller–Bauer reaction, but the desired quasi-Favorskii reaction did occur in relatively low yield. The product acid was esterified to give **224**. Hydrolysis of this ester, a Hunsdiecker decarboxylative bromination, and an exhaustive reductive halogenation using a dissolving metal reduction afforded secohexaprismane (**225**) (Scheme 7.48).

This synthesis is over 20 years old. No new prismane has appeared since the 1980s. It is not clear whether the will and resources exist for the pursuit of these types of

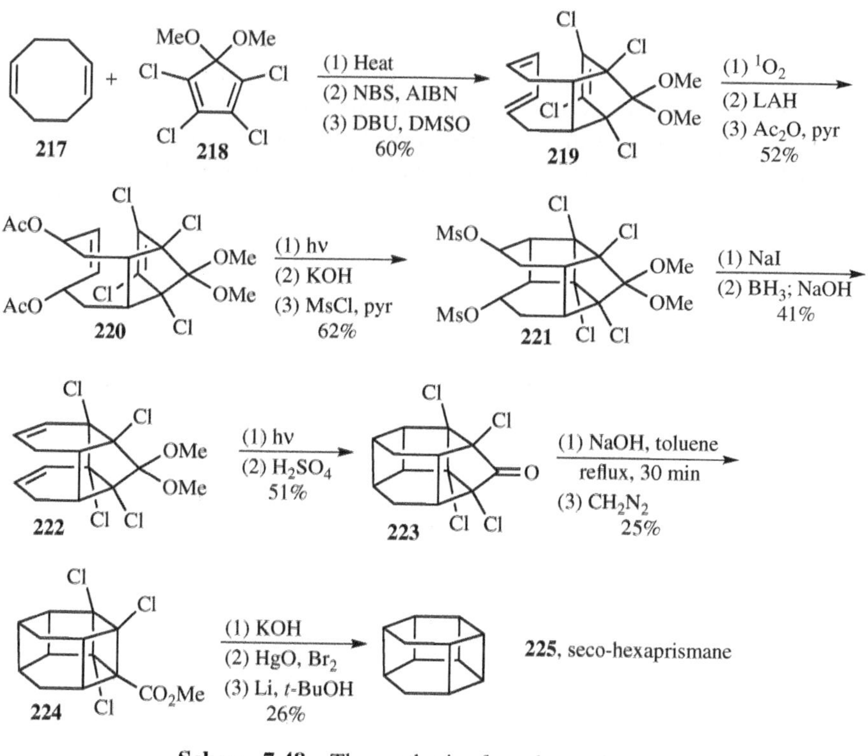

Scheme 7.48 The synthesis of secohexaprismane.

molecules, but they do represent exciting opportunities to examine concepts of bonding and strain energy, and they present a unique set of challenges that motivate the development of synthetic methodology.

7.5.3.6 Hexacyclo[6.4.2.0^{2,7}.0^{3,11}.0^{6,10}.0^{9,12}]tetradecane (Takeshita) This chapter's final example of caged hydrocarbon synthesis is one that further emphasizes the importance of cycloaddition reactions in creating substrates for the quasi-Favorskii rearrangement. This synthesis also showed, as many polycyclic hydrocarbon syntheses have, the limits that exist in intramolecular photochemical [2+2]-cycloaddition processes.

Takeshita *et al.* reported that the thermal cycloaddition between 2-chlorotropone and ethylene proceeded to give **227** in fair yield (Scheme 7.49).[87] Another cycloaddition with 1,3-cyclohexadiene afforded **228** in 85% yield. This compound was unreactive when exposed to ultraviolet light. However, its reaction with KOH in refluxing methanol for 12 h afforded the quasi-Favorskii rearrangement product **229** in 88% yield. Peroxyesterification of the carboxylic acid with *t*-BuOOH via the acid chloride and subsequent photolysis led to **230**. Thermolysis of the latter in 1,4-diisopropylbenzene gave the title hydrocarbon **231** in 50% yield.

7.6 CONCLUSIONS AND PROSPECTS

The quasi-Favorskii rearrangement is a reliable way to make rings, strained rings in particular. However, its scope is potentially quite broad, as suggested by its use to prepare α-arylketones, its application to macroexpansion methodology and associated natural product targets, and its incorporation into the (4+3)-cycloaddition/quasi-Favorskii reaction sequence. Not only are these areas still not adequately explored, it is easy to imagine many other ways to construct systems that are set up for the quasi-Favorskii reaction. For example, consider the two compounds shown in Figure 7.11. To the best of our knowledge, these classes of compounds have not been subjected to quasi-Favorskii conditions, yet they and very likely many

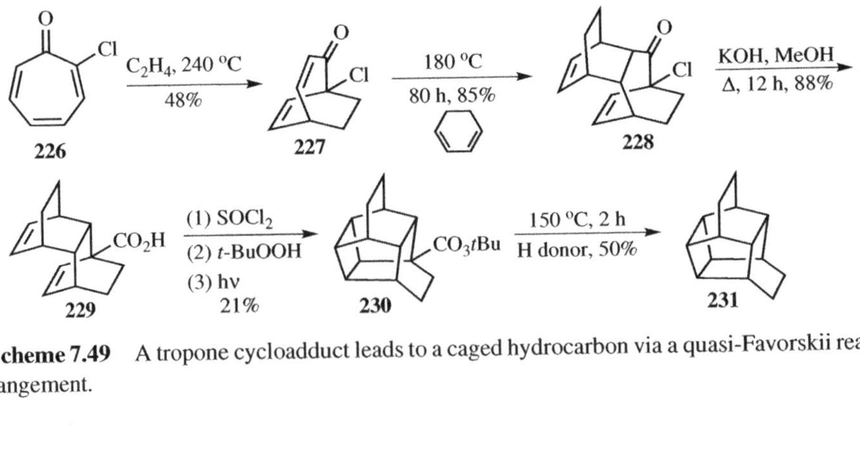

Scheme 7.49 A tropone cycloadduct leads to a caged hydrocarbon via a quasi-Favorskii rearrangement.

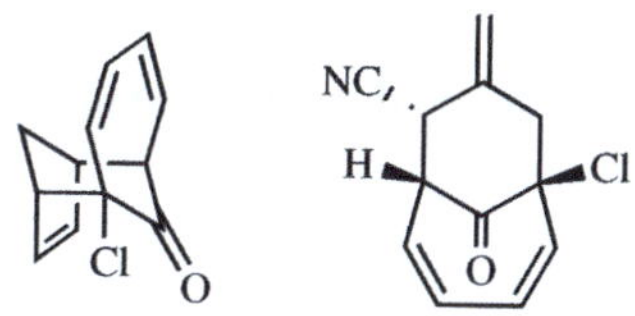

FIGURE 7.11 Possible substrates for the quasi-Favorskii rearrangement.

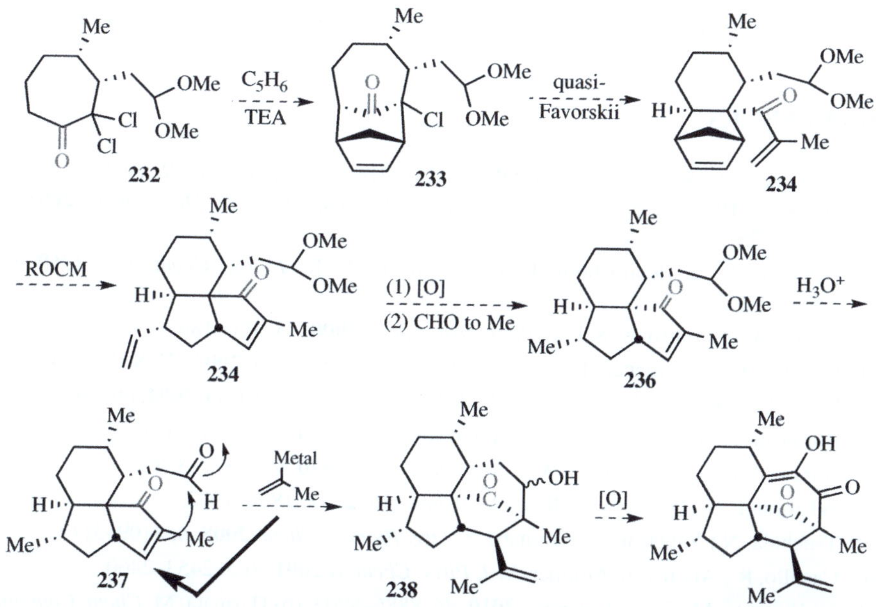

Scheme 7.50 Possible (4+3)-cycloaddition/quasi-Favorskii approach to elisapterosin B.

congeners are easily available.[88] They are ideal candidates for such exploration. Whether these ideal candidates for quasi-Favorskii exploration would satisfy the utilitarian desires of those who need applied research to justify intellectual inquiry is another story, but one that can be written in due time.

Similarly, synthetic application of the quasi-Favorskii process, especially with substrates arising from cycloaddition reactions, merits more attention. As a very simple example, one could consider a target like the antitumor agent elisapterosin B (**239**).[89] This compound bears a classic quasi-Favorskii retron – one ring fused to another with a quaternary carbon at an angular position. An entirely speculative synthesis is shown in Scheme 7.50, where the carbonyl group in the starting ketone becomes one of the ketones in the product by virtue of a (4+3)-cycloaddition/quasi-Favorskii sequence. Whether this synthetic scheme could be realized is less important than an emphasis on the idea that the retron for the quasi-Favorskii rearrangement can often be obvious, as in this case, or subtly hidden; but it offers opportunities for strategic disconnections and synthetic tactics that could bear fruit in the context of a complete synthetic effort.

ACKNOWLEDGMENTS

Thanks to Dr. Carissa S. Hampton for assistance in preparing this manuscript. We thank the NSF for its support of our research program, including efforts in the area of the (4+3)-cycloaddition/quasi-Favorskii rearrangement, over a number of years.

REFERENCES

1. (a) Kende, A. S. *Org. React.* **1960**, *11*, 261–316. (b) Baretta, A.; Waegell, B. *React. Intermed.* **1982**, *2*, 527–585. (c) Guijarro, D.; Yus, M. *Curr. Org. Chem.* **2005**, *9*, 1713–1735.
2. Bordwell, F. G.; Scamehorn, R. G.; Springer, W. R. *J. Am. Chem. Soc.* **1969**, *91*, 2087–2093.
3. Cramer, C. J.; Barrows, S. E. *J. Phys. Org. Chem.* **2000**, *13*, 176–186.
4. Hamblin, G. D.; Jimenez, R. P.; Sorensen, T. S. *J. Org. Chem.* **2007**, *72*, 8033–8045.
5. Faza, O. N.; López, C. S.; Álvarez, R.; de Lera, Á. R. *J. Org. Chem.* **2004**, *69*, 9002–9010.
6. Bordwell, F. G.; Carlson, M. W. *J. Am. Chem. Soc.* **1970**, *92*, 3370–3377.
7. Bordwell, F. G.; Carlson, M. W. *J. Am. Chem. Soc.* **1970**, *92*, 3377–3385.
8. Wharton, P. S.; Fritzberg, A. R. *J. Org. Chem.* **1972**, *37*, 1899–1902.
9. Tsuchida, N.; Yamazaki, S.; Yamabe, S. *Org. Biomol. Chem.* **2008**, *6*, 3109–3117.
10. Castillo, R.; Andrés, J.; Moliner, V. *J. Phys. Chem. B* **2001**, *105*, 2453–2460.
11. (a) Harmata, M. *Chem. Commun.* **2010**, *46*, 8886–8903. (b) Harmata, M. *Chem. Commun.* **2010**, *46*, 8904–8922.
12. Tchoubar, B.; Sackur, O. *Compt. Rend.* **1939**, *208*, 1020–1022.
13. (a) McPhee, W. D.; Klingsberg, E. *J. Am. Chem. Soc.* **1944**, *66*, 1132–1136. (b) Loftfield, R. B. *J. Am. Chem. Soc.* **1950**, *72*, 632–633. (c) Loftfield, R. B. *J. Am. Chem. Soc.* **1951**, *73*, 4707–4714.
14. Cope, A. C.; Synerholm, M. E. *J. Am. Chem. Soc.* **1950**, *72*, 5228–5232.
15. (a) Bredt, J. *Liebigs Ann. Chem.* **1924**, *437*, 1–13. (b) Shea, K. J. *Tetrahedron* **1980**, *36*, 1683–1715.
16. Warnhoff, E. W.; Wong, C. M.; Tai, W.-T. *J. Am. Chem. Soc.* **1968**, *90*, 514–515.
17. Hayes, C. J.; Simpkins, N. S.; Kirk, D. T.; Mitchell, L.; Baudoux, J.; Blake, A. J.; Wilson, C. *J. Am. Chem. Soc.* **2009**, *131*, 8196–8210.
18. (a) Wang, B.; Tu, Y. Q. *Acc. Chem. Res.* **2011**, *44*, 1207–1222. (b) Overman, L. E. *Tetrahedron* **2009**, *65*, 6432–6446. (c) Kirsch, S. F. *Nachr. Chem.* **2008**, *56*, 1228–1231. (d) Snape, T. J. *Chem. Soc. Rev.* *2007*, 36, 1823–1842. (e) Overman, L. E.; Pennington, L. D. *J. Org. Chem.* **2003**, *68*, 7143–7157.
19. (a) Burke, A. J.; Marques, C. S. *Mini-Rev. Org. Chem.* **2007**, *4*, 310–316. (b) Paquette, L. A.; Hofferberth, J. E. *Org. React.* **2003**, *62*, 477–567.

20. Song, Z.-L.; Fan, C.-A.; Tu, Y.-Q. *Chem. Rev.* **2011**, *111*, 7523–7556.
21. For examples of the use of a nitrogen nucleophile, see: (a) Komarov, I. V.; Kornilov, M. Y.; Gorichko, M. V. *Tetrahedron Lett.* **1999**, *40*, 3935–3936. (b) Stevens, C. L.; Pillai, P. M.; Taylor, K. G. *J. Org. Chem.* **1974**, *39*, 3158–3161.
22. Baudry, D.; Begue, J. P.; Charpentier-Morize, M. *Tetrahedron Lett.* **1970**, 2147–2150.
23. (a) Fetizon, M.; Golfier, M.; Louis, J. M. *Tetrahedron Lett.* **1973**, 1931–1933. (b) Fetizon, M.; Golfier, M.; Montaufier, M. T.; Rens, J. *Tetrahedron* **1975**, *31*, 987–991.
24. Smissman, E. E.; Hite, G. *J. Am. Chem. Soc.* **1960**, *82*, 3375–3381.
25. Hasegawa, T.; Nigo, T.; Kuwatani, Y.; Ueda, I. *Bull. Chem. Soc. Jpn.* **1993**, *66*, 2676–2681.
26. Snider, B. B.; Walner, M. *Tetrahedron* **1989**, *45*, 3171–3182.
27. Ross, A. G.; Townsend, S. D.; Danishefsky, S. J. *J. Org. Chem.* **2013**, *78*, 204–210.
28. (a) Conia, J. M.; Salaun, J. R. *Acc. Chem. Res.* **1972**, *5*, 33–40. (b) Moliner, V.; Castillo, R.; Safont, V. S.; Oliva, M.; Bohn, S.; Tunon, I.; Andres, J. *J. Am. Chem. Soc.* **1997**, *119*, 1941–1947.
29. For a broad review of cyclobutyl rearrangements, see: Leemans, E.; D'Hooghe, M.; De, K. N. *Chem. Rev.* **2011**, *111*, 3268–3333.
30. Harmata, M.; Shao, L. *Synthesis* **1999**, 1534–1540.
31. Harmata, M.; Wacharasindhu, S. *Chem. Commun.* **2003**, 2492–2493.
32. Harmata, M.; Wacharasindhu, S. *J. Org. Chem.* **2005**, *70*, 725–728.
33. Harmata, M.; Shao, L.; Kurti, L.; Abeywardane, A. *Tetrahedron Lett.* **1999**, *40*, 1075–1078.
34. Harmata, M.; Bohnert, G.; Kürti, L.; Barnes, C. L. *Tetrahedron Lett.* **2002**, *43*, 2347–2349.
35. Kraus, G. A.; Shi, J. *J. Org. Chem.* **1990**, *55*, 5423–5424.
36. Harmata, M.; Wacharasindhu, S. *Synthesis* **2007**, 2365–2369.
37. Moriarty, R. M.; Enache, L. A.; Zhao, L.; Gilardi, R.; Mattson, M. V.; Prakash, O. *J. Med. Chem.* **1998**, *41*, 468–477.
38. Tiffeneau, M. *C. R. Hebd. Séances Acad. Sci.* **1903**, *137*, 989–991.
39. (a) Tiffeneau, M. *Ann. Chim. Phys.* **1908**, *10*, 322–378. (b) Tiffeneau, M.; Tchoubar, M. B. *Compt. Rend.* **1934**, *198*, 941–943.
40. Tiffeneau, M. *Bull. Soc. Chim. Fr.* **1945**, *12*, 621–627.
41. Bouveault, L.; Chereau, F. *Compt. Rend.* **1906**, *142*, 1086–1087.
42. Bartlett, P. D.; Rosenwald, R. H. *J. Am. Chem. Soc.* **1934**, *56*, 1990–1994.
43. Sisti, A. J.; Vitale, A. C. *J. Org. Chem.* **1972**, *37*, 4090–4094.
44. (a) Newman, M. S.; Farbman, M. D. *J. Am. Chem. Soc.* **1944**, *66*, 1550–1552. (b) Newman, M. S.; Booth, Jr., W. T., *J. Org. Chem.* **1947**, *12*, 737–739. (c) Mueller, G. P.; May, R. *J. Am. Chem. Soc.* **1949**, *71*, 3313–3317.
45. Smith, M. B.; March, J. March's Advanced Organic Chemistry, 6th ed., Wiley: New York, **2007**, pp. 455–459, Chapter 10.
46. Bachman, W. E.; Fujimoto, G. I.; Wick, L. B. *J. Am. Chem. Soc.* **1950**, *72*, 1995–2000.
47. Huang, R. L. *J. Org. Chem.* **1954**, *19*, 1363–1369.
48. Hussey, A. S.; Herr, R. R. *J. Org. Chem.* **1959**, *24*, 843–845.
49. (a) Pei, T.; Chen, C.-Y.; Dormer, P. G.; Davies, I. W. *Angew. Chem. Int. Ed.* **2008**, *47*, 4231–4233. (b) Pei, T.; Tellers, D. M.; Streckfuss, E. C.; Chen, C.-Y.; Davies, I. W. *Tetrahedron* **2009**, *65*, 3285–3291.

50. Butsugan, Y.; Kadosaka, I.; Araki, S. *Chem. Lett.* **1979**, 527–528.
51. Araki, S.; Butsugan, Y. *Chem. Lett.* **1980**, 185–186.
52. (a) Paquette, L. A.; Shi, Y. J. *J. Org. Chem.* **1989**, *54*, 5205–5207. (b) Paquette, L. A.; Shi, Y. J. *J. Am. Chem. Soc.* **1990**, *112*, 8478–8489. (c) Grisé, C. M.; Barriault, L. *Org. Lett.* **2006**, *8*, 5905–5908.
53. Riehl, J. J.; Smolikiewicz, A.; Thil, L. *Tetrahedron Lett.* **1974**, 1451–1454.
54. Nagasawa, T.; Taya, K.; Kitamura, M.; Suzuki, K. *J. Am. Chem. Soc.* **1996**, *118*, 8949–8950.
55. Wender, P. A.; Sieburth, S. M.; Petraitis, J. J.; Singh, S. K. *Tetrahedron* **1981**, *37*, 3967–3975.
56. (a) Sisti, A. J.; Rusch, G. M. *J. Org. Chem.* **1974**, *39*, 1182–1186. (b) Sisti, A. J.; Meyers, M. *J. Org. Chem.* **1973**, *38*, 4431. (c) Sisti, A. J. *J. Org. Chem.* **1970**, *35*, 2670–2673. (d) Sisti, A. J.; Vitale, A. C. *Tetrahedron Lett.* **1969**, 2269–2270. (e) Sisti, A. J. *J. Org. Chem.* **1968**, *33*, 453–454. (f) Sisti, A. J. *J. Org. Chem.* **1968**, *33*, 3953–3955. (g) Sisti, A. J. *Tetrahedron Lett.* **1967**, 5327–5329.
57. Li, L.; Cai, P.; Guo, Q.; Xue, S. *J. Org. Chem.* **2008**, *73*, 3516–3522.
58. Wang, Z.; Li, M.; Zhang, W.; Jia, J.; Wang, F.; Xue, S. *Tetrahedron Lett.* **2011**, *52*, 5968–5971.
59. Examples related to this chemistry but probably not involving alkoxide formation include: (a) Agrawal, M. K.; Ghosh, P. K. *J. Org. Chem.* **2009**, *74*, 7947–7950. (b) Tandon, V. K.; Awasthi, A. K.; Singh, K. A.; Maurya, H. K.; Gautam, S. K. *Heterocycles* **2010**, *80*, 593–600.
60. Magnusson, G.; Thoren, S. *J. Org. Chem.* **1973**, *38*, 1380–1384.
61. Bergman, R.; Magnusson, G. *J. Org. Chem.* **1986**, *51*, 212–217.
62. Bruckner, R.; Harmata, M. Organic Mechanisms; Springer: Berlin, **2012**, pp. 146–147.
63. Ponten, F.; Magnusson, G. *J. Org. Chem.* **1997**, *62*, 7972–7977.
64. Wender, P. A.; Holt, D. A.; Sieburth, S. M. *J. Am. Chem. Soc.* **1983**, *105*, 3348–3350.
65. Wender, P. A.; Holt, D. A. *J. Am. Chem. Soc.* **1985**, *107*, 7771–7772.
66. Goel, G.; Makkar, H. P. S.; Francis, G.; Becker, K. *Int. J. Toxicol.* **2007**, *26*, 279–288.
67. Beans, E. J.; Fournogerakis, D.; Gauntlett, C.; Heumann, L. V.; Kramer, R.; Marsden, M. D.; Murray, D.; Chun, T.-W.; Zack, J. A.; Wender, P. A. *Proc. Natl. Acad. Sci. U. S. A.* **2013**, *110*, 11698–11703.
68. Paquette, L. A.; Sauer, D. R.; Edmondson, S. D.; Friedrich, D. *Tetrahedron* **1994**, *50*, 4071–4086.
69. Barriault, L.; Denissova, I.; Goulet, N. *Synthesis* **2012**, *44*, 1833–1840.
70. Kraus, G. A.; Shi, J. *J. Org. Chem.* **1991**, *56*, 4147–4151.
71. Jasperse, C. P.; Curran, D. P. *J. Am. Chem. Soc.* **1990**, *112*, 5601–5609.
72. Harmata, M.; Rashatasakhon, P. *Org. Lett.* **2001**, *3*, 2533–2535.
73. Salomon, R. G.; Basu, B.; Roy, S.; Sachinvala, N. D. *J. Am. Chem. Soc.* **1991**, *113*, 3096–3106.
74. Ayer, W. A.; Saeedi-Ghomi, M. H. *Can. J. Chem.* **1981**, *59*, 2536–2538.
75. Harmata, M.; Bohnert, G. J. *Org. Lett.* **2003**, *5*, 59–61.
76. Harmata, M.; Wacharasindhu, S. *Org. Lett.* **2005**, *7*, 2563–2565.
77. Harmata, M.; Huang, C.; Rooshenas, P.; Schreiner, P. R. *Angew. Chem. Int. Ed.* **2008**, *47*, 8696–8699.

78. Chenier, P. J. *J. Chem. Ed.* **1978**, *55*, 286–291.
79. Eaton, P. E.; Cole, T. W. *J. Am. Chem. Soc.* **1964**, *86*, 3157–3158.
80. Eaton, P. E.; Cole, T. W. *J. Am. Chem. Soc.* **1964**, *86*, 962–964.
81. (a) Eaton, P. E.; Nordari, N.; Tsanaktsidis, J.; Upadhyaya, S. P. *Synthesis* **1995**, 501–502. (b) Bliese, M.; Tsanaktsidis, J. *Aust. J. Chem.* **1997**, *50*, 189–192.
82. (a) Bliese, M.; Cristiano, D.; Tsanaktsidis, J. *Aust. J. Chem.* **1997**, *50*, 1043–1045. (b) Bliese, M.; Fanning, L.-A.; Gable, R. W.; Lowe, D. A.; Tsanaktsidis, J. *Aust. J. Chem.* **1998**, *51*, 593–597.
83. Barborak, J. C.; Watts, L.; Pettit, R. *J. Am. Chem. Soc.* **1966**, *88*, 1328–1329.
84. Boyer, L. E.; Brazzillo, J.; Forman, M. A.; Zanoni, B. *J. Org. Chem.* **1996**, *61*, 7611–7613.
85. (a) Eaton, P. E.; Or, Y. S.; Branca, S. J. *J. Am. Chem. Soc.* **1981**, *103*, 2134–2136. (b) Eaton, P. E.; Or, Y. S.; Branca, S. J.; Ravi Shankar, B. K. *Tetrahedron* **1986**, *42*, 1621–1631.
86. Mehta, G.; Padma, S. *Tetrahedron* **1991**, *47*, 7783–7806.
87. Takeshita, H.; Kawakami, H.; Ikeda, Y.; Mori, A. *J. Org. Chem.* **1994**, *59*, 6490–6492.
88. (a) Ito, S.; Sakan, K.; Fujise, Y. *Tetrahedron Lett.* **1969**, 775–778. (b) Trost, B. M.; McDougall, P. J.; Hartmann, O.; Wathen, P. T. *J. Am. Chem. Soc.* **2008**, *130*, 14960–14961.
89. Rodriguez, A. D.; Ramirez, C.; Rodriguez, I. I.; Barnes, C. L. *J. Org. Chem.* **2000**, *65*, 1390–1398.

8

THE RAMBERG–BÄCKLUND REACTION

JOANNE E. HARVEY

School of Chemical and Physical Sciences, Victoria University of Wellington, Wellington, New Zealand

8.1 INTRODUCTION

The Ramberg–Bäcklund reaction (RBR), in its most general form, generates an alkene, **2**, from a sulfone, **1** (Scheme 8.1). The process involves halogenation at the α-center of **1**, abstraction of a proton from the nonhalogenated α-center of the resulting halosulfone **3**, and cyclization to give an episulfone (thiirane 1,1-dioxide, **4**), followed by extrusion of sulfur dioxide.

The original report of the RBR was in 1940 by Ramberg and his student Bäcklund based on their observations that α-halosulfones, such as α-bromoethyl ethyl sulfone (**5**), are converted to the corresponding alkenes, for example, 2-butene (**6**), upon treatment with KOH (Scheme 8.2).[1] The (*Z*)-isomers were preferentially formed.

There have been a number of extensions and variations to the original method, most significantly the Meyers *et al.* extension,[2] which greatly expanded the synthetic utility of the RBR by including an *in situ* halogenation, thus allowing the more general application of sulfones as substrates. Meyers' conditions (Scheme 8.3a) involve use of carbon tetrachloride as the chlorinating agent, together with powdered KOH in *tert*-butanol to provide the base, either at room temperature for active (e.g., benzylic or allylic) sulfones or at 50–60 °C.[2] Chan *et al.*'s modification dating from 1994 (Scheme 8.3b) uses alumina-supported KOH (with *tert*-butanol) as the base, and dibromodifluoromethane as the halogenating reagent.[3] Due to the low boiling

Molecular Rearrangements in Organic Synthesis, First Edition. Edited by Christian M. Rojas.

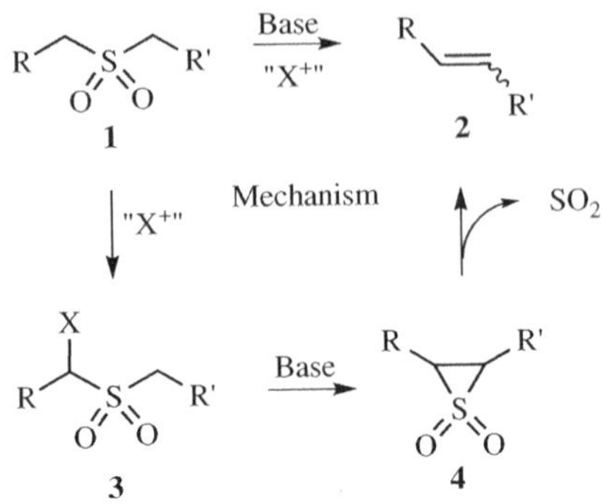

Scheme 8.1 Ramberg–Bäcklund reaction mechanism.

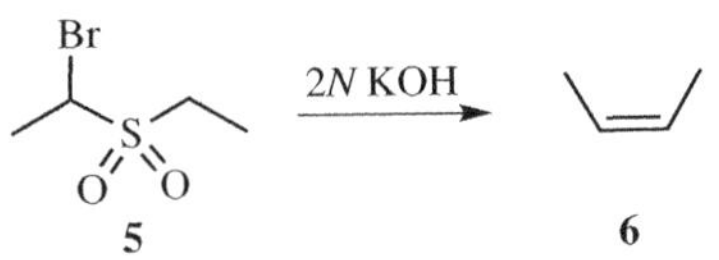

Scheme 8.2 First reported RBR.

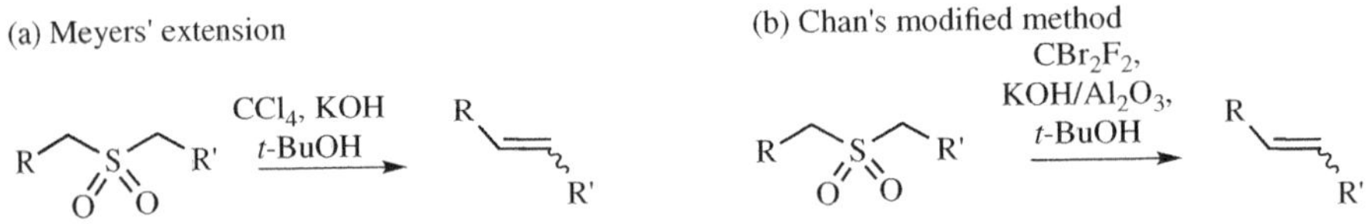

Scheme 8.3 Meyers' and Chan's modifications of the RBR.

point of dibromodifluoromethane (22–23 °C), the reaction needs to be performed at or below room temperature. This may not be suitable for unactivated sulfones, in which case a higher boiling alternative, such as 1,2-dibromotetrafluoroethane (47 °C), is employed.[4]

Meyers' extension and Chan's modification represent one-pot conversions of sulfones into alkenes. These are almost invariably used in modern synthetic applications due to the improved convenience and synthetic efficiency compared to the classic (two-step) RBR. The Chan method is recommended as the first choice when contemplating the use of the RBR in synthesis.[5] This is due to the highly dispersed nature of the solid-supported base, which provides more reliable reactivity.

In practice, there have been many other modifications of the RBR conditions, mostly differing in the choice of base used. The most popular base for the RBR is potassium *tert*-butoxide, either added directly or generated *in situ* from *tert*-BuOH and KOH (powdered or supported on alumina). Nonetheless, numerous other bases have been employed for the RBR, including lithium *tert*-butoxide, 1,8-diazabicyclo[5.4.0]undec-7-ene (DBU), lithium bis(trimethylsilyl)amide

(LiHMDS), and butyllithium (*n*-BuLi).[5–10] The stronger bases tend to favor formation of (*E*)-alkenes (see Section 8.4).

There are concerns about the environmental impact of the halogenating reagents used in the one-pot RBR because they are heavily halogenated ozone depletors. Indeed, CCl_4, CBr_2F_2, and $(BrF_2C)_2$ are being phased out or no longer available in some countries. Very recently, Söderman and Schwan have introduced 1,2-dibromotetrachloroethane as a nonozone-depleting, less expensive brominating agent, which is a solid at ambient conditions and, therefore, easier and safer to work with.[11]

The RBR has been thoroughly reviewed,[5–10] and readers are directed to these excellent expositions for comprehensive coverage of the literature until 2009. The remainder of this chapter provides an overview of the RBR, with pertinent recent examples, for practitioners of organic chemical synthesis. The focus is on strategies and considerations relevant to the application of the RBR in synthesis. By virtue of their greater efficiency and utility, most of the discussion will be on one-pot variants of the RBR, wherein the substrate is a nonhalogenated sulfone.

8.2 METHODS TO SYNTHESIZE SULFONES AS RBR PRECURSORS

Several preparative methods for sulfones exist, any of which can, in principle, be used for generating RBR precursors. Nonetheless, by far the most popular and straightforward route, particularly for complex target synthesis, is to introduce the sulfur atom into a carbon skeleton from a thiol salt such as sodium or potassium thioacetate, or by Mitsunobu reaction with thioacetic acid. A stepwise deacetylation, to release the monosubstituted thiol, and reaction with another carbon electrophile follows. The process typically involves substitution of an alkyl halide **7** by the thioacetate, methanolysis of the thioester in **8**, substitution of another alkyl halide **10** by the released thiol **9** to give a thioether **11**, and then oxidation to the sulfone **12** (Scheme 8.4). Typical oxidants for the latter step include *meta*-chloroperoxybenzoic acid (*m*-CPBA) or Oxone ($KHSO_5$ as a mixture with $KHSO_4$ and K_2SO_4).[8] Sulfur oxidation usually occurs faster than epoxidation of any alkenes present. Magnesium monoperoxyphthalate hexahydrate has also been used to oxidize thioethers to sulfones.[12]

A one-pot variation of the thioether generation, applicable to the formation of cyclic sulfones, involves reaction of a dihalide (or halide equivalent) with a divalent sulfide, such as sodium sulfide. This methodology was used in the synthesis of conduritols, whereby sodium sulfide reacted with diepoxy stereoisomers **13** in a one-step thioether formation (Scheme 8.5).[13] The resulting thiapane isomers **14** were separated (HPLC) and then individually methylated and oxidized with *m*-CPBA to a set of isomeric sulfones, **15**.

A similar double substitution in a single pot was employed for thioether formation in the RBR-based synthesis of eremantholide A (see Section 8.6.4).[14] In this case, the sulfide generated from hexamethyldisilathiane [bis(trimethylsilyl)sulfide] and sodium methoxide reacted with the bromoiodide **16** to form the 10-membered thioether **17** (Scheme 8.6). This was oxidized with Oxone to provide the sulfone **18**.

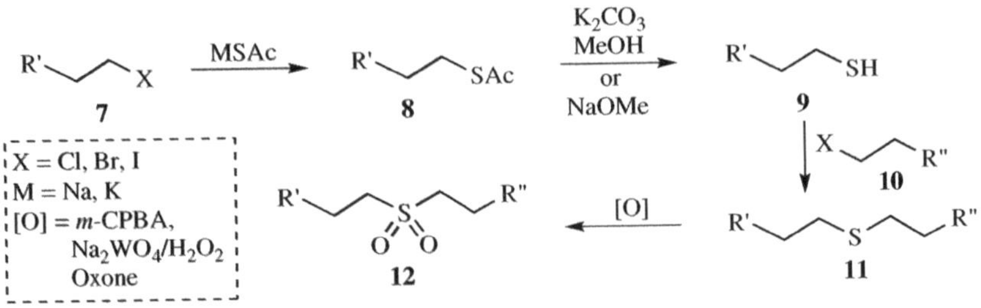

Scheme 8.4 Sulfone synthesis by stepwise substitution of two alkyl halides with thioacetate.

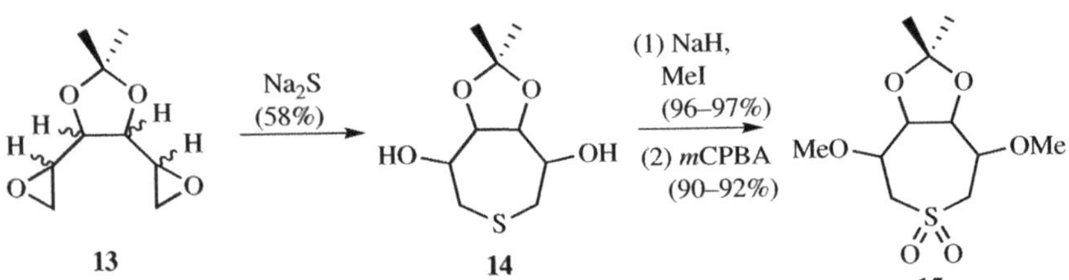

Scheme 8.5 Sulfone synthesis by one-pot substitution of a bis-epoxide with sulfide.

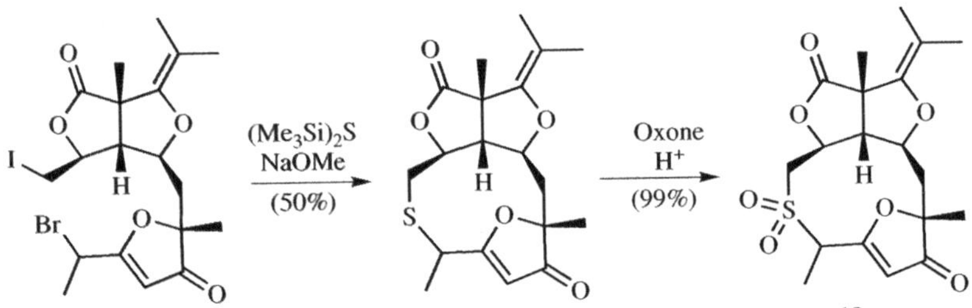

Scheme 8.6 Sulfone synthesis in preparation of eremantholide A.

An alternative method for incorporating a sulfone moiety into a carbon framework for the RBR involved an elegant four-component coupling between sulfur dioxide, silyl enol ethers **19** and **20**, and iodide **21** (Scheme 8.7).[15] The mechanism presumably comprises addition of 1-siloxy-1,3-diene **19** to sulfur dioxide, reaction of the resulting aldehyde with silyl enol ether **20**, and then attack of the resulting sulfinate on the alkyl iodide **21**. The product **22**, obtained as four diastereomers, was converted by the RBR to an intermediate that comprised a formal total synthesis of apoptolidin (see Section 8.6).

The installation of an unsaturated sulfone **25** can be achieved through Horner–Wadsworth–Emmons reaction between an aldehyde **23** and a sulfonylphosphonate **24** (Scheme 8.8).[16]

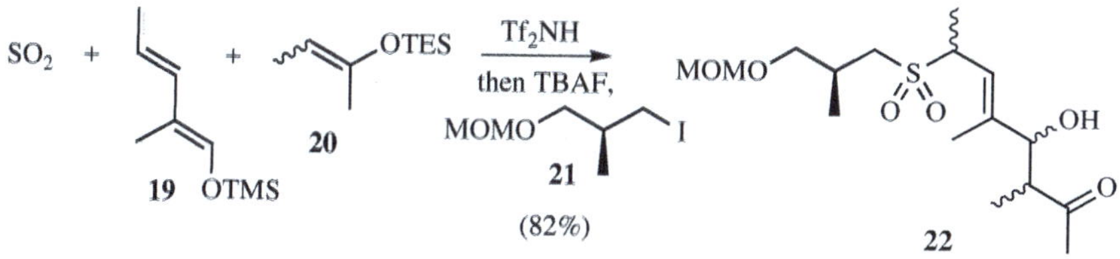

Scheme 8.7 Four-component coupling for sulfone preparation in formal synthesis of apoptolidin.

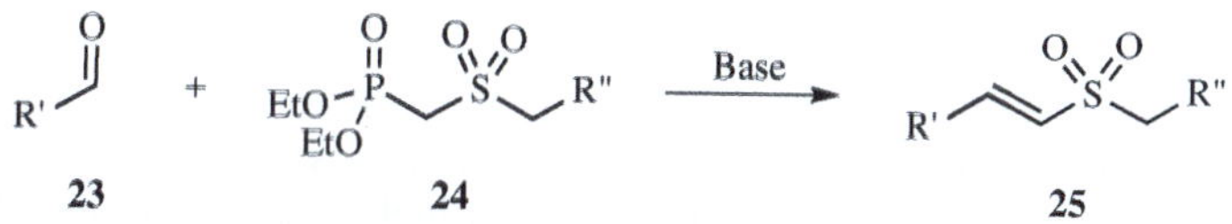

Scheme 8.8 Sulfone synthesis by Horner–Wadsworth–Emmons strategy.

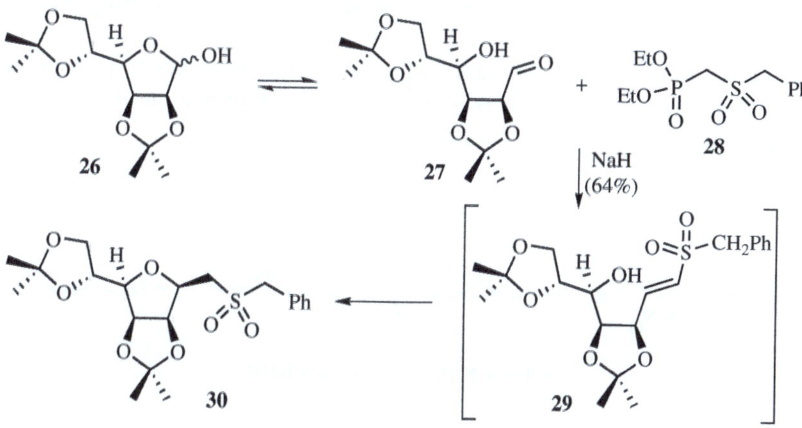

Scheme 8.9 Sulfone synthesis by Horner–Wadworth–Emmons reaction and cyclisation.

In the presence of an internal nucleophile, such as an alcohol, the unsaturated sulfone can spontaneously cyclize to afford a saturated sulfone linked to a heterocycle, which is an RBR precursor. This has been particularly valuable in preparation of highly functionalized *C*-glycosides via sulfones such as **30** (Scheme 8.9).[17] In this case, the hemiacetal **26** reacts, via its aldehyde form **27**, with the sulfonylphosphonate **28**. This proceeds through the intermediate unsaturated sulfone **29** to produce the exocyclic sulfone **30**.

8.3 VARIATIONS OF THE RBR

A great number of variations of the RBR have been developed and will be briefly described here.

The vinylogous RBR is a variation whereby either a vinyl(α-bromoalkyl)sulfone or a bromoallyl(alkyl)sulfone is treated with base, delivering, via a vinyl-substituted episulfone intermediate, a 1,3-diene (Scheme 8.10).[5,18]

The decarboxylative RBR involves saponification-induced decarboxylation of a β-sulfonyl ester, leading to α-sulfonyl anion formation. With *in situ* halogenation of the other α-center, the episulfone is formed, which produces the alkene (Scheme 8.11). This protocol works best when the doubly α-center is disubstituted, otherwise competing halogenation at that position results in a haloalkene product.[5,19]

The Michael-induced RBR exploits the electrophilic nature of vinyl-substituted sulfones.[5] Upon nucleophilic attack, the alkene double bond migrates to form an episulfone intermediate by displacement of an α-halide (Scheme 8.12). The product alkene, resulting from subsequent SO_2 extrusion, is transposed by one position from the original alkene. The α-halide may instead be positioned on the vinyl group (i.e., an α-bromovinylsulfone), in which case proton transfer occurs before episulfone formation. α-Bromovinylsulfones tend to be more reactive toward nucleophiles and so may benefit the overall reaction.[20] Employing dienes as the nucleophile acceptors is also successful and leads to conjugated (migrated) diene products.[21]

The epoxy-RBR utilizes epoxides formed from vinylsulfones to provide allylic alcohols (Scheme 8.13).[22] The leaving group ejected in formation of the episulfone is the epoxide oxygen rather than a halide.

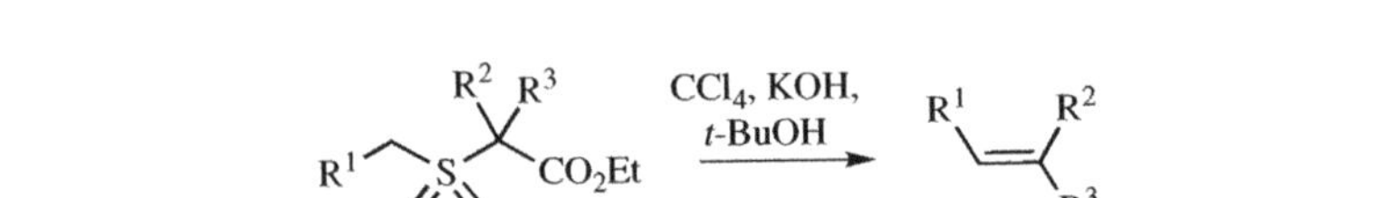

Scheme 8.10 Vinylogous RBR.

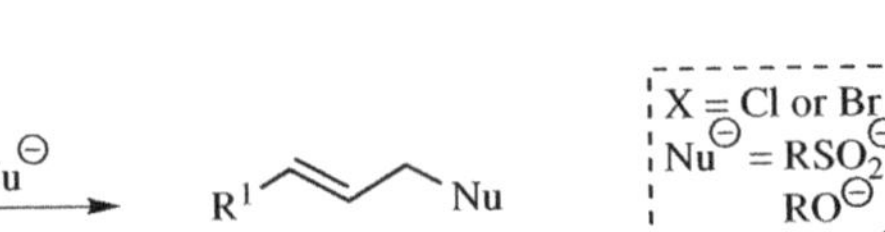

Scheme 8.11 Decarboxylative RBR.

Scheme 8.12 Michael-induced RBR.

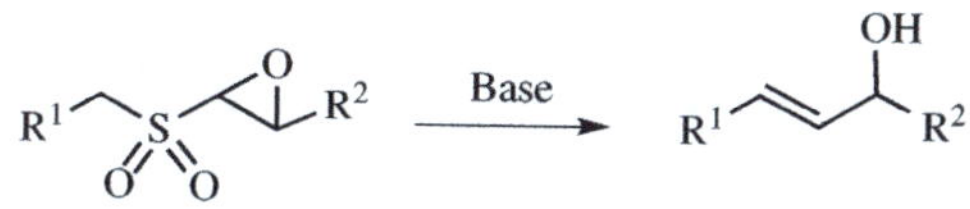

Scheme 8.13 Epoxy-RBR.

8.4 MECHANISTIC EVALUATION OF THE RBR

Significant advances have been made to the understanding of the RBR mechanism by Bordwell, Paquette, Taylor and others.[23–26] Aspects of the mechanism have been studied theoretically and the results agree with experimental outcomes.[27] The release of halide ion was found to be first order in both hydroxide ion and sulfone, lending credence to the proposed mechanism.[23] Deuteration experiments have shown that deprotonation is reversible and that all possible α-anions form (Scheme 8.14).[28,29]

The formation of the episulfone is thought to occur from a W-like arrangement of halide, sulfone and anion lobe that serves to optimize orbital overlap in displacement of the chloride (Scheme 8.15). The anion is formed by deprotonation, from a trajectory between the sulfone oxygens, of a sterically minimized starting material conformation. After rotation to the W-arrangement, this results in cis-substituents in the episulfone.[25] This means that (*Z*)-alkenes are usually the kinetic products. However, use of stronger bases (e.g., KO^{t}Bu) or substrates with additional anion-stabilizing groups (e.g., R′ = phenyl) leads to predominance of (*E*)-alkenes, which is thought to be due to epimerization of the episulfone.[5] The intermediacy of episulfones has been confirmed through their isolation, characterization, and subsequent conversion under RBR conditions into the alkene.[26]

The mechanism of sulfur dioxide extrusion has yet to be confirmed,[5] although it is known to be accelerated by base.[30] Decomposition of *cis*-episulfones has been calculated to be faster than for the corresponding *trans*-episulfones.[31]

While the RBR does not encompass a true rearrangement of the carbon skeleton, it has been likened, mechanistically, to the Favorskii rearrangement, which does.[32] Indeed, both involve a rate-limiting intramolecular ring-forming step that follows an acid–base equilibrium. The Favorskii intermediate is a cyclopropanone that is formed by α-deprotonation of an α-haloketone, analogous to the episulfone invoked in the RBR mechanism.

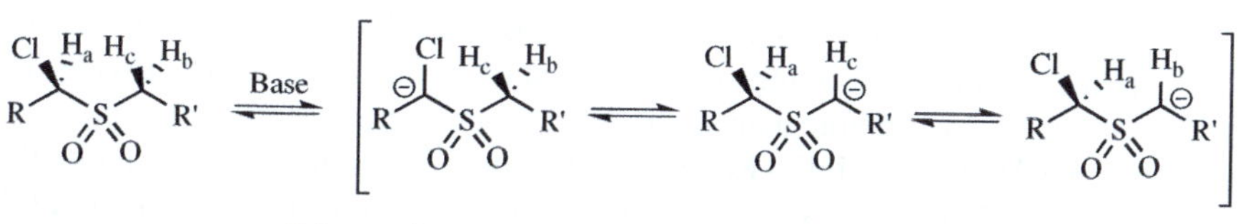

Scheme 8.14 RBR mechanism: deprotonation.

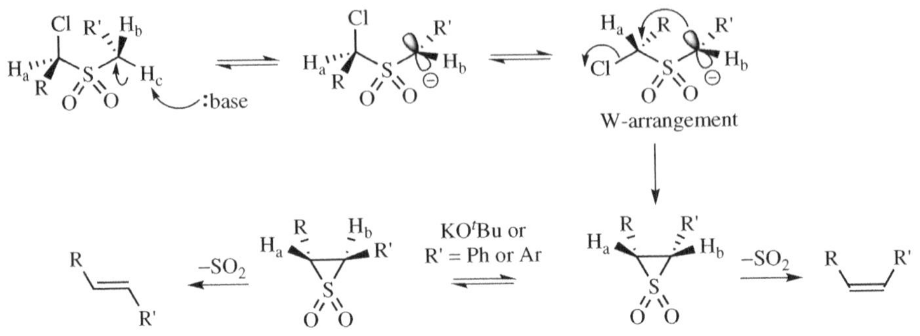

Scheme 8.15 RBR mechanism: episulfone formation.

8.5 STRATEGIC CONSIDERATIONS RELEVANT TO THE USE OF THE RBR IN SYNTHESIS

The RBR has become an important and versatile method for preparation of carbon–carbon double bonds within a variety of structural motifs, including a number of highly strained cyclic systems.[5–10] Taking a broad view of the overall reaction sequence that encompasses preparation of a sulfone precursor and the RBR itself, the alkene functionality in the product **31** is constructed from its two individual components **32** and **33** (Scheme 8.16). Thus, the retrosynthetic disconnection is through the alkene double bond. From this point of view, the RBR sequence occupies a similar position in organic synthesis to the Wittig-type reactions, including Julia, Julia–Kocienski, Horner–Wadsworth–Emmons and the (*Z*)-selective Still–Gennari and Ando variants, in that an alkene is assembled from two individual carbon-based fragments. As with these other olefination methods, the position of an alkene prepared by the RBR is unambiguous, a necessary feature in modern, target-oriented synthetic chemistry.

The RBR sequence is most often initiated by substitution of an organohalide **33** with a thiol **32**, affording a thioether **34** that contains the full carbon skeleton (see Scheme 8.16). Typically, this intermediate is isolated, as with the Julia reaction but unlike the Wittig-type protocols. The next step is oxidation to a sulfone **35**, followed by the key RBR. This makes the overall process a three-step sequence, which has disadvantages with regard to step economy.[33] Nonetheless, the multistep sequence has greater synthetic flexibility, which is beneficial in some scenarios (e.g., as demonstrated in the syntheses of integrastatin and aigialomycin D, Sections 8.6.2 and 8.7.4, respectively).

From an atom economy[34] viewpoint, the RBR involves only loss of hydrogen halide (in the episulfone-forming step) followed by extrusion of sulfur dioxide. This makes it more atom economic than several of the common alternative strategies, such as the popular Wittig and Julia–Kocienski reactions,[35] although the chosen method for forming the RBR precursor also needs to be considered if fully evaluating the synthetic efficiency.

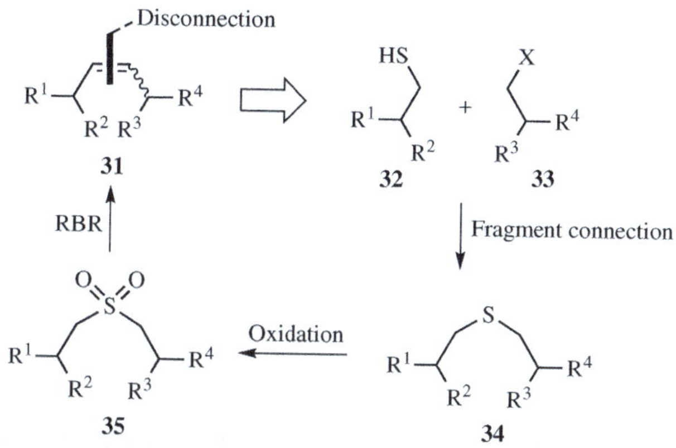

Scheme 8.16 The RBR as a retrosynthetic disconnection.

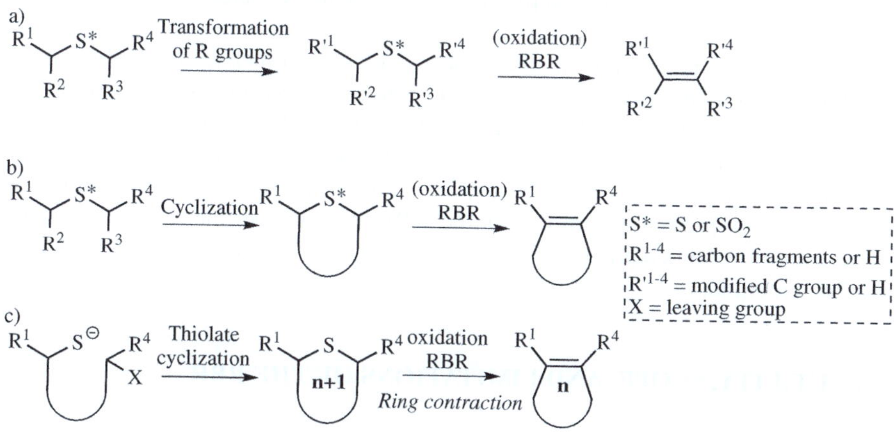

Scheme 8.17 Strategic considerations of the RBR as a multistep alkene formation.

The synthetic flexibility of a multistep alkene formation can be understood by noting that the thioether and sulfone intermediates represent masked alkenes. Further synthetic steps can be performed on the full carbon skeleton prior to conversion into the alkene, allowing for the use of chemistry that would be incompatible with the presence of an alkene. By connecting the two fragments through a thioether or sulfone linkage, one can perform reactions that are common to more than one substituent, thus decreasing the number of linear steps in the synthetic route compared to performing them separately before coupling (Scheme 8.17a).

Alternatively, for the synthesis of cyclic products, the thioether or sulfone can act as a tether, enabling an alternative reaction to serve as a cyclization step (Scheme 8.17b). If this intramolecular reaction is not orthogonal to alkene reactivity, such separation of the RBR from its predecessor steps is essential. This strategic advantage of the RBR has been underutilized in synthesis.

Additionally, the RBR has important advantages as a method to synthesize cyclic compounds.[10,36–41] In particular, the ring-forming step is often achieved through substitution involving a thiolate nucleophile. These nucleophiles are highly reactive by virtue of their polarizability, electron density, and flexibility in angle of attack, making them well suited to difficult cyclizations.[42] Additionally, using this methodology, the formation of a ring containing *n* atoms actually proceeds by ring contraction of a larger, $n + 1$, cycle (Scheme 8.17c). Therefore, for strained systems, the ring strain that must be overcome in the cyclization step is decreased relative to direct ring closure of the smaller ring. In terms of the ring-contraction step, sulfur dioxide extrusion is entropically favored, which offsets the development of strain in the smaller ring.[43]

In terms of synthetic strategy, the RBR can potentially be used for synthesis of any di- or trisubstituted alkene, although the geometry of the product cannot be firmly predicted in all cases. The *E:Z* ratio is determined by factors such as ring size (in a cyclic substrate), the strength of the base, and the presence of anion-stabilizing groups. It should also be noted that, because alkenes may be converted to alkanes by hydrogenation or other reductive means, an alkyl chain is a latent retron for an RBR strategy.

Despite some uncertainty in the expected olefin geometry,[5] the utility of the RBR is amply demonstrated by its application to the synthesis of many diverse targets.[5,10] These include dendrimers,[44] cyclophanes,[41] modified carbohydrates such as *C*-glycosides and exoglycals,[17,45,46] alkaloids,[40,47] terpene derivatives,[48] enediynes,[49] amino acids,[50] stilbenes,[51,52] and macrocycles.[40,53] The remainder of this chapter presents examples of the RBR that are pertinent to the strategic considerations outlined previously and illustrate its scope as a versatile and powerful synthetic method.

8.6 UTILITY, SCOPE, AND LIMITATIONS OF THE RBR

The RBR has been successfully employed in the synthesis of a range of acyclic di- and trisubstituted alkenes, and many cyclic alkenes, including some that display considerable ring strain. As mentioned earlier, one of the advantages of the RBR is the fact that the cyclization step proceeds through a larger ring than is present in the final target, and therefore, strained systems may be more facile to obtain by this procedure compared to other methods. This section will endeavor to provide an indication of the scope and limitations of the RBR using exemplary syntheses of complex targets to illustrate the points. Priority will be given to more recent work, as earlier reports have been well reviewed.[5–10]

8.6.1 (Sterically Crowded) Acyclic Alkenes

In preparing the second generation nona(phenylenevinylidene) dendrimer **37** by multiple RBR, Chow *et al.* illustrated the versatility and efficiency of this chemistry (Scheme 8.18).[44] The substrate nona(sulfide) **36** was prepared in a convergent manner by sequential substitution of benzylic bromide functional groups by benzylic

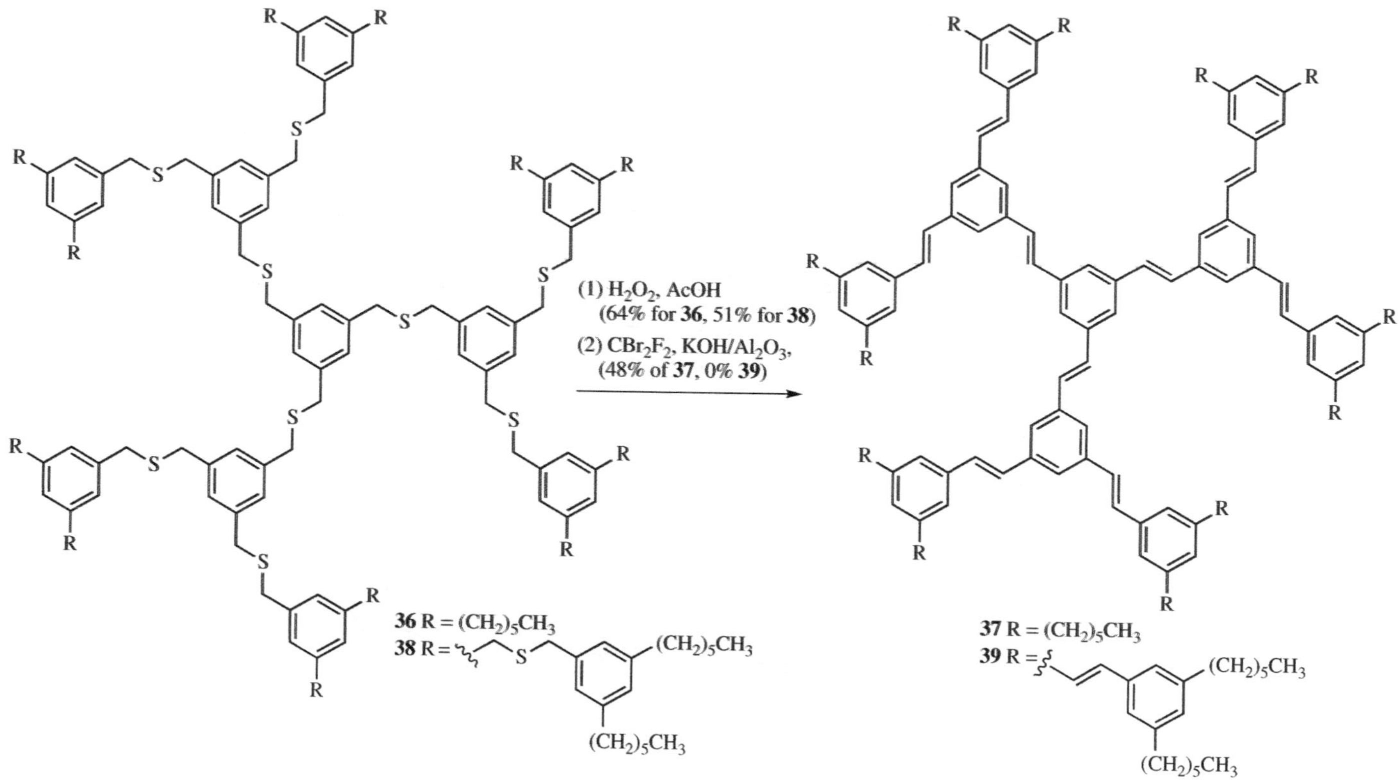

Scheme 8.18 Dendimer synthesis by RBR.

thiols. Oxidation of the nona(sulfide) **36** proceeded smoothly with hydrogen peroxide and acetic acid in dichloromethane at reflux. The subsequent RBR under Chan's conditions provided the all-(*E*)-nona(phenylenevinylene) **37** in a 48% isolated yield. A myriad of geometrical isomers was also formed in 24% yield. The combined yield of nonaenes indicates an average of better than 96% in each individual RBR process! These isomeric products could be hydrogenated to afford the same nona(phenyleneethylene) dendrimer.

In contrast to this success, the attempted RBR of the third-generation heneicos(sulfone), formed from the heneicos(sulfide) **38** in reasonable yield, did not proceed to completion but instead gave mixtures of partially reacted dendrimers. It was proposed that the extreme steric hindrance encountered in the interior of the dendrimer would prevent reaction with the solid-supported potassium hydroxide, particularly if the outermost sulfones reacted first, thus further crowding the periphery of the dendrimer.

Taken together, these two dendritic examples demonstrate the scope and limits of the RBR. While being very powerful for multiple transformations in a single operation, there are limits to the ability of the RBR to operate in cases of extreme steric encumbrance, at least with the solid-supported base developed by Chan.

Sulfones that are nonbenzylic and have no other activating groups can react sluggishly in the RBR, meaning undesired side reactions may occur or even dominate.[12] This can be a significant problem if the sulfone contains β-alkoxy and β-siloxy groups, because elimination can occur. For example, while the benzyl sulfone **40** reacted smoothly to provide the styrene **41**, the methyl sulfone **42** and isopropyl sulfone **43** only epimerized, presumably through deprotonation of the secondary α-center and reversible elimination to afford the thermodynamically favored isomers **44** and **45** (Scheme 8.19).[54]

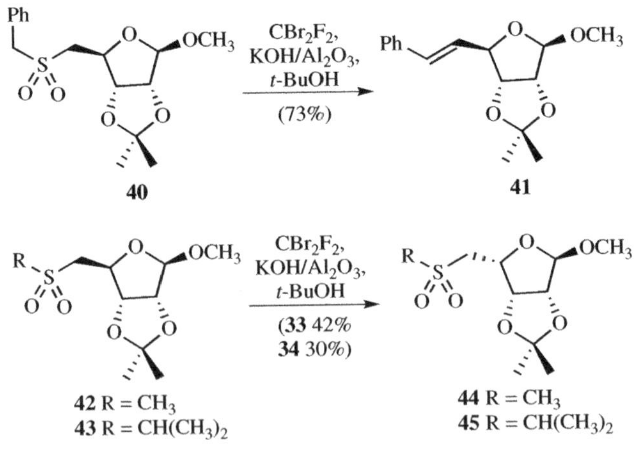

Scheme 8.19 Comparative reactivity of benzylic and non-benzylic sulfones.

8.6.2 Acyclic Trisubstituted Alkenes

The utility of the RBR in the synthesis of crowded trisubstituted alkenes is demonstrated by the synthesis of ampelopsin D (Scheme 8.20).[55] The thioether **48** was generated by acid-promoted cyclization of alcohol **46** and attack by thiol **47** on the benzylic cation. This compound had a trans-relationship between the aryl branches on the indane system and was produced as a 1:1 mixture of epimers at the thioether linkage. In preparation for the RBR, the thioether **48** was oxidized to sulfone **49**, which was treated with base and carbon tetrachloride to produce the RBR product **50** in 52% yield, along with 15% of its (*Z*)-isomer. Meyers' conditions were found to be optimal for stereoselectivity in the RBR, with other methods producing poorer selectivity and unreliable results. The natural product ampelopsin D (**51**) was obtained after global deprotection with boron tribromide.

The synthesis of the integrastatin core by Taylor's group also demonstrates the utility of the RBR in generating hindered alkenes.[56] It was envisaged that the trisubstituted olefin **57** could afford the bridged core through intramolecular acetal formation (Scheme 8.21). However, alkene **57** was inaccessible by Wittig, Julia, Grignard, or lithiation methodologies,[56] so the RBR was explored. Indeed, the sulfone **54**, formed by substitution of benzylic bromide **52** with secondary thiol **53** and sulfur oxidation, underwent a high-yielding RBR to provide a 1:1 mixture of isomeric alkenes **55**.

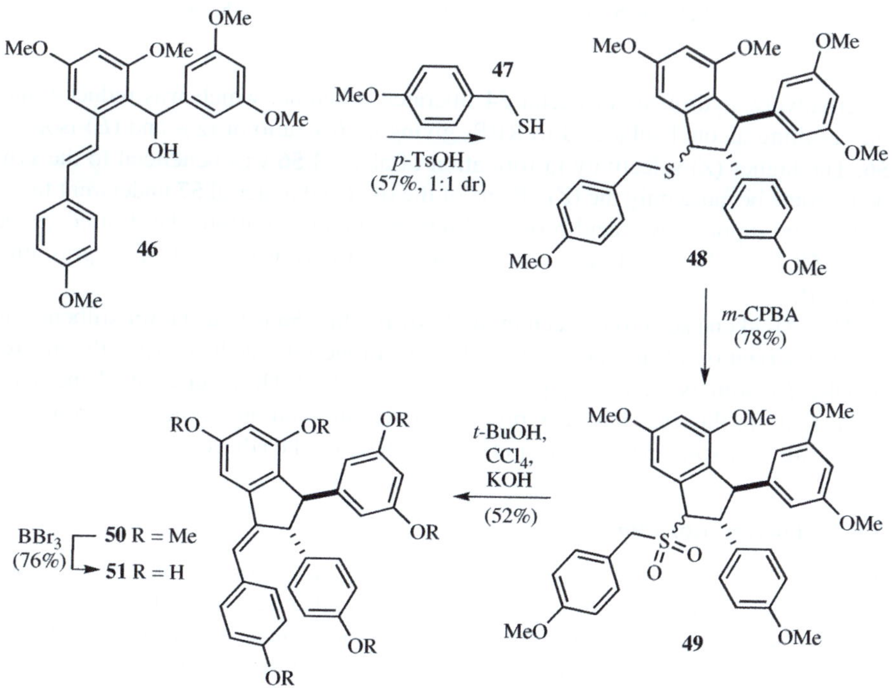

Scheme 8.20 Synthesis of ampelopsin D by RBR.

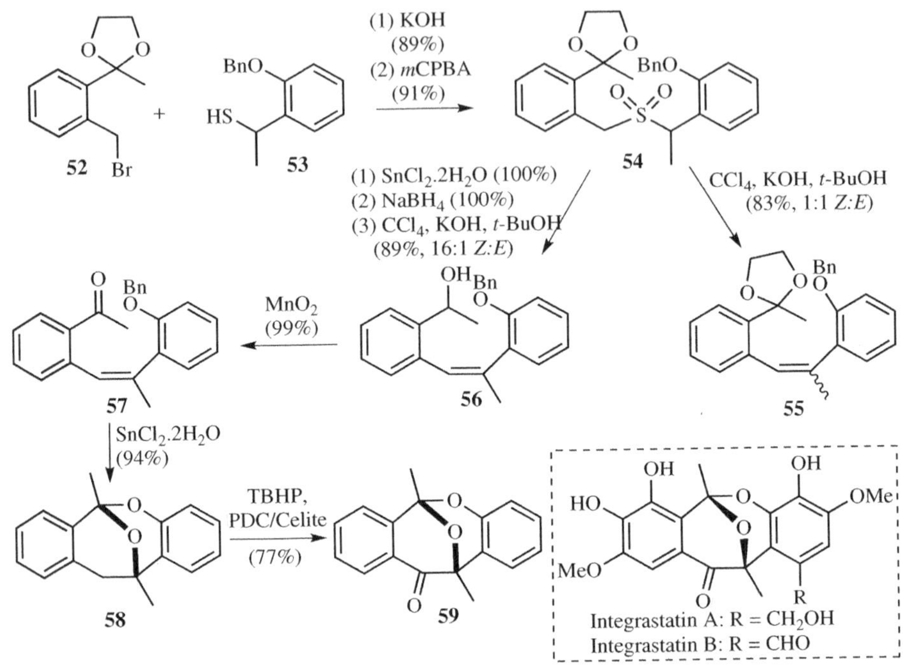

Scheme 8.21 Synthesis of integrastatin core by RBR.

Alternatively, hydrolysis of acetal **54** liberated a ketone, which was reduced, and the resulting alcohol subjected to RBR, giving a 16:1 ratio of (*Z*)- and (*E*)-isomers **56**. The higher (*Z*)-selectivity in formation of alcohol **56** was beneficial to the synthetic route because only the (*Z*)-alkene of the oxidized material **57** underwent facile cyclization, upon Lewis acid-promoted debenzylation, to afford the desired cyclic acetal **58**. Benzylic oxidation gave rise to the full tetracyclic core, **59**, of integrastatins A and B.

The (*Z*)-alkene selectivity seen in the RBR product **56** is unusual for stilbenes, in that isomerization of the relatively acidic episulfone intermediate typically ensures that the (*E*)-isomers exclusively prevail (see Section 8.4). The reversal of alkene stereoselectivity in this case has been proposed to result from intramolecular promotion of the sulfur dioxide extrusion from the episulfone by the adjacent alkoxide.[57]

8.6.3 Polyunsaturation

The RBR has been used extensively for the preparation of polyunsaturated compounds. The challenging synthesis of carotenoids has been achieved through a combination of Ramberg–Bäcklund and Julia-type methodologies.[58,59] The benefit of these two sulfone-based transformations is in the stability of the intermediates that function as masked alkenes. The convergent synthesis began with the anion of allylic sulfone **60** displacing allylic dichloride **61** to give the chain-extended sulfone **62**

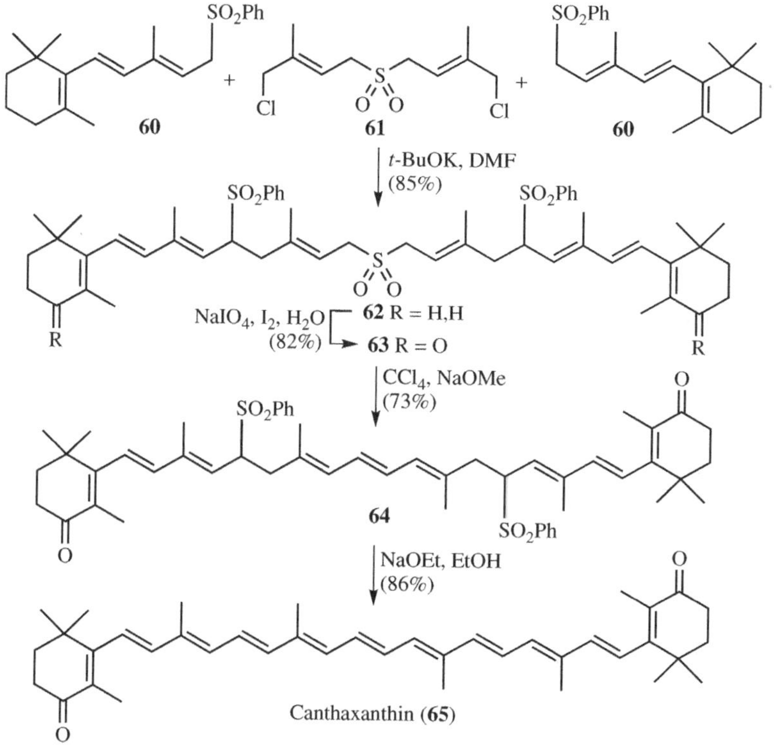

Scheme 8.22 Synthesis of canthaxanthin by RBR.

(Scheme 8.22). Allylic oxidation of the cyclohexene rings provided diketone **63** and was best performed at this stage because of the instability of subsequent compounds to the oxidative conditions. An RBR then revealed the central alkene in the conjugated triene **64**. The remaining two alkenes were unmasked through dehydrosulfonation, affording the colored antioxidant canthaxanthin, undecaene **65**.

A key fragment of the highly unsaturated natural product apoptolidin (**68**) invoked the RBR of sulfone **66** under Chan's conditions (Scheme 8.23).[15] The RBR precursor **66** was prepared from sulfone diastereoisomers **22** (see Section 8.2, Scheme 8.7) by acid-promoted dehydration, ketone–ester conversion, and alkynylation. It is notable that the sulfone group is stable to these diverse manipulations. The RBR was high yielding and stereoselective, affording the conjugated (*E,E,E*)-triene **67** in a 12:1 ratio with an (*E,E,Z*)-isomer. The preparation of **67** represents a formal synthesis of apoptolidin (**68**), as Nicolaou *et al.*'s total synthesis encompassed this intermediate.[60]

Several acyclic enediynes have been prepared in high yield by the one-pot RBR using Chan's conditions, albeit with poor stereoselectivity.[49,61] For example, the unsymmetrical doubly propargylic sulfone **69** afforded the enediyne **70** in high yield but with almost no *E*:*Z* selectivity (Scheme 8.24).

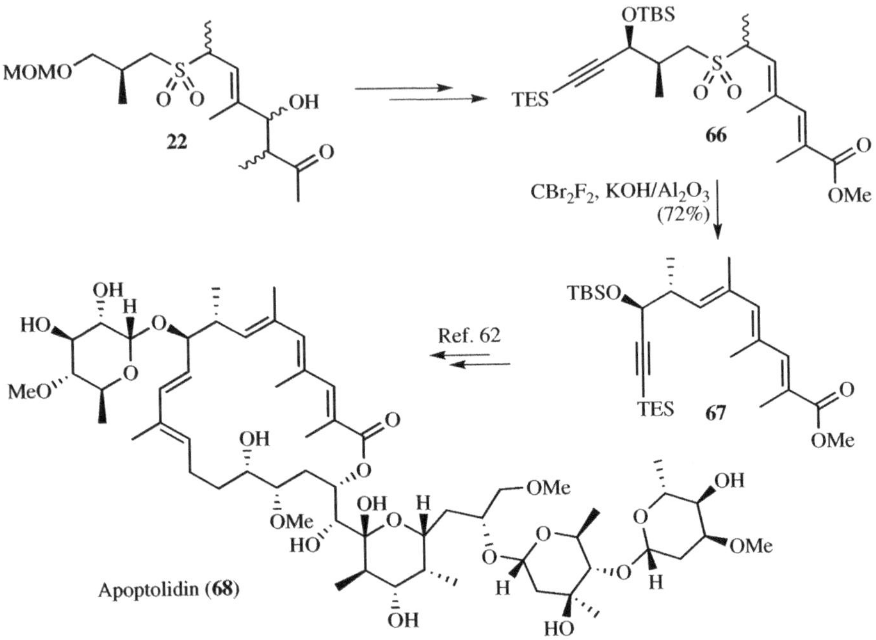

Scheme 8.23 Formal synthesis of apoptolidin by RBR.

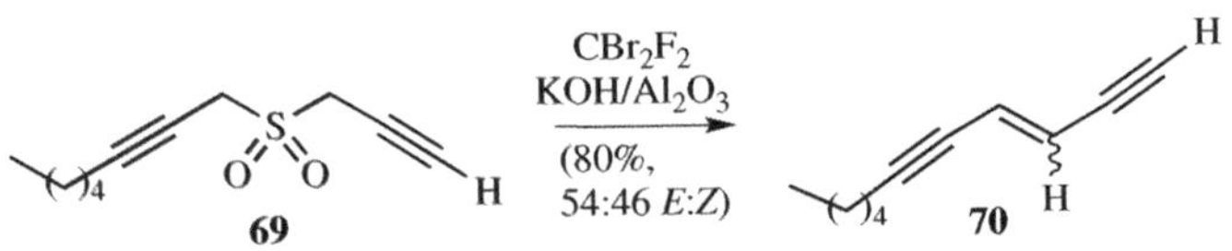

Scheme 8.24 Synthesis of an acyclic enediyne by RBR.

8.6.4 Strained Cyclic Alkenes

To illustrate the capacity of the RBR to generate strained ring systems, the case of Dewar benzene deserves mention. Notoriously difficult to prepare, dimethyl Dewar benzene **72** has been reported to form by the double RBR of chlorinated disulfone isomers **71** (Scheme 8.25).[62] It was detected by GCMS analysis of the worked-up reaction and was accompanied by the aromatic **73** (*p*-xylene), presumably the stable end product.

Similarly, a series of cyclic diyne chlorosulfones **74–77** were treated with base to provide enediynes **78–81** with consecutive ring sizes (Scheme 8.26).[36,63] These examples provide an indication of the boundaries of efficacy for the RBR. Thus, poor yields were obtained for cyclic enediynes **78** and **79** with tether lengths of 3 or 4 ($n = 1$ or 2), while the larger ring sizes provide modest-to-good results of cyclic enediynes **80** and **81**. More recently, Cao and coworkers have shown that the one-pot RBR using

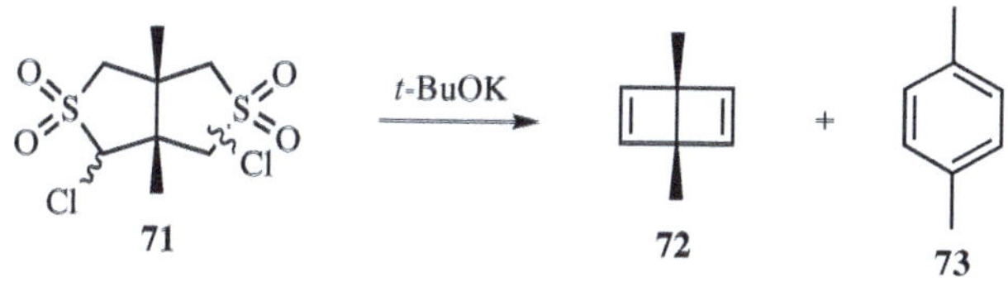

Scheme 8.25 Synthesis of dimethyl Dewar benzene by RBR.

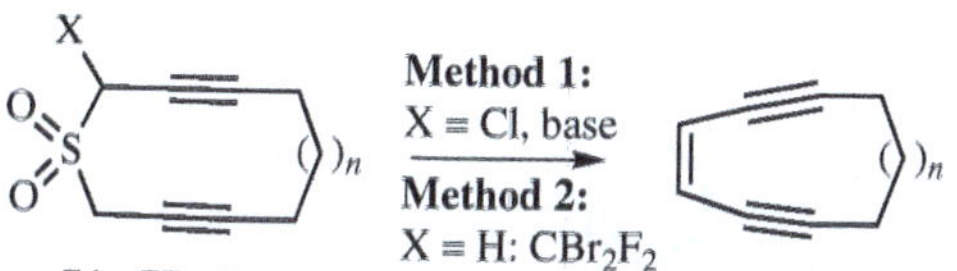

74a-77a X = Cl
74b-77b X = H
78-81

	Product	Method 1 Base	Method 1 Yield	Method 2 Yield
$n = 1$	**78**	MeLi	0%	-
$n = 2$	**79**	MeLi	12%	50%
$n = 3$	**80**	KO^{t}Bu	32%	65%
$n = 4$	**81**	KOH	62%	70%

Scheme 8.26 Synthesis of cyclic enediynes by RBR.

Chan's method provides better yields of the cyclic enediynes.[49] By this means, **79** was prepared in 50% yield, **80** in 65% yield, and **81** in 70% yield from the corresponding sulfones.

Cyclophanes are cyclic alkanes containing at least one bridged aromatic group. The small cyclophanes bear considerable strain in their ring, particularly *para*cyclophanes, sometimes to such an extent that the aromatic ring is bent out of plane.[43] Their rigid structures, taken together with rotational restriction leading to potential atropisomerism, makes these compounds particularly challenging from a synthetic viewpoint.[64] As a result of the ring strain in certain members of this class, the RBR could present benefits to their synthesis, in that it operates by ring contraction of a larger, less-constrained structure. Nonetheless, even this potent methodology has been found to fail or provide low yields in formation of *para*cyclophanes with short tethers. For example, the chlorine-substituted hexahydroparacyclophane **83** was obtained in low yields from an attempted RBR after 24 days! It presumably forms by over-chlorination of the sulfone **82** (or the monosulfone resulting from a single RBR) prior to the sluggish ring contraction (Scheme 8.27).[65]

In contrast, orthocyclophanes[66] and some of the larger-ring paracyclophanes[41] have been successfully prepared by the RBR (Scheme 8.28). Thus, the orthocyclophane **85** was prepared in reasonable yield from the disulfone **84** by the one-pot RBR under Meyers' conditions.[67] In addition, the [12][12]paracyclophane **87** was generated from the disulfone **86** in a reasonable yield. The RBR gave a mixture of alkene isomers **87**, which converged, on hydrogenation, to the saturated cyclophane **88**. These materials were optically active, having been resolved by crystallization of the menthyl xanthate diastereomers prior to incorporation of sulfur, although the absolute configuration was not known.[38]

The RBR met its match in the eight-membered ring of vinigrol (**95**) (Scheme 8.29).[68] Despite considerable efforts and adaptations of the structural

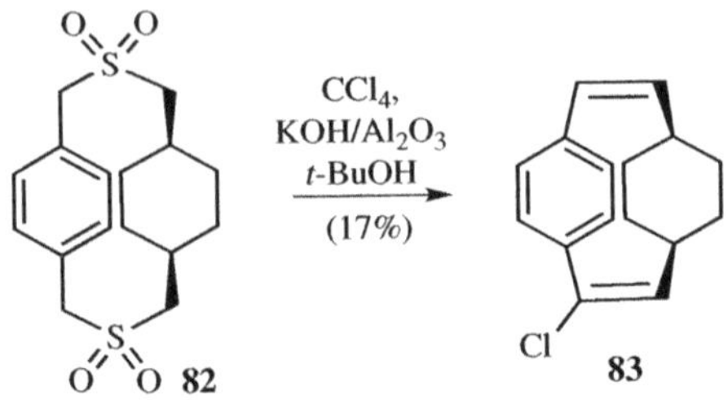

Scheme 8.27 Attempted synthesis of a [2.2]paracyclophane by RBR.

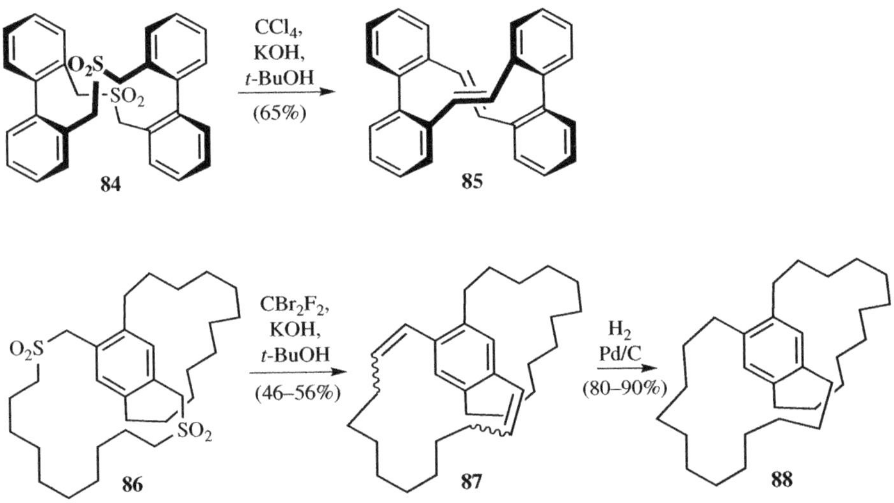

Scheme 8.28 Synthesis of diverse cyclophanes by RBR.

framework, the cis-bridged cyclooctene was not generated. To begin with, the dibromides **89–91** did not form the respective cyclic thioethers **92–94**. By increasing the conformational flexibility through unclipping one of the six-membered rings of the *cis*-decalin system, the double substitution of dibromide **96** did give a cyclic nine-membered ring. Surprisingly, despite the absence of oxidizer in the reaction of dibromide **96** with sodium sulfide, the product obtained was the sulfone **97**! Yields were unreliable, however, and ranged from 39% to 72%. Unfortunately, this sulfone was unreactive toward the ring-contracting RBR under Meyers' conditions, and none of the bicyclic skeleton **98** was isolated even after prolonged reaction times. These results showcase the difficulty in formation of this ring system. It should be noted that ring-closing metathesis (RCM), S_N2 displacement, pinacol coupling, and lactam sulfoxide ring contraction were likewise unsuccessful in generating the eight-membered ring of vinigrol.[69,70] However, the anionic oxy-Cope rearrangement was applied successfully in the vinigrol system,[71] and total syntheses have subsequently been reported that utilize Grob and Wharton fragmentations to prepare the eight-membered ring.[72]

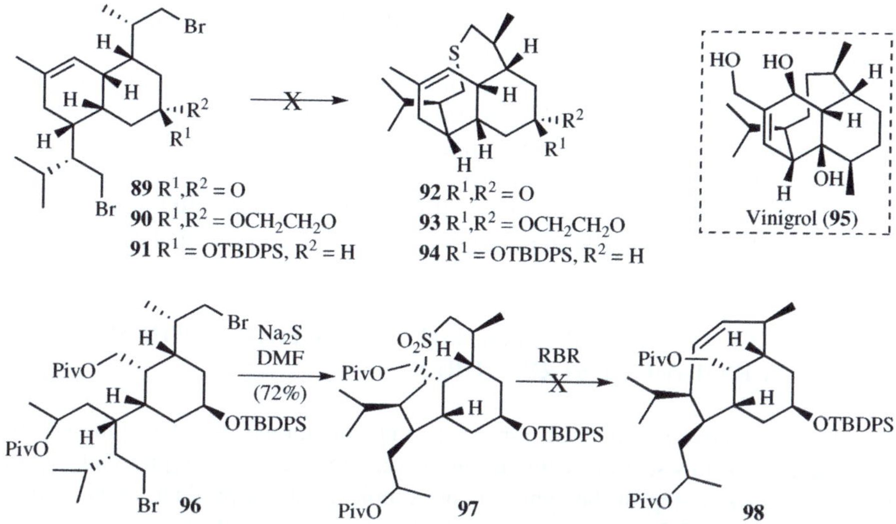

Scheme 8.29 Attempted synthesis of vinigrol by RBR.

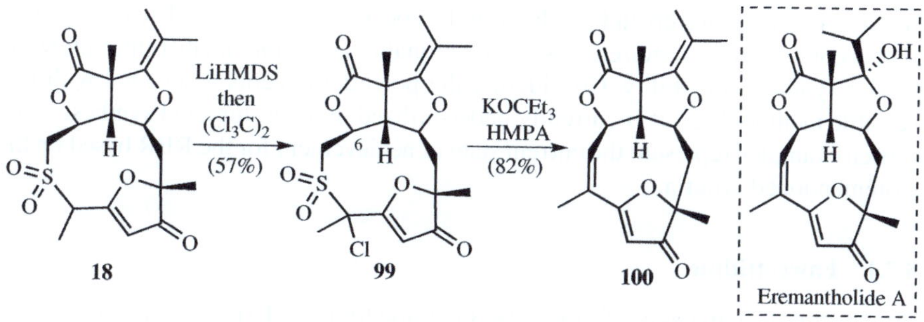

Scheme 8.30 Synthesis of eremantholide A by RBR.

The synthesis of eremantholide A by the RBR stands out as a landmark achievement due to the strained and highly functionalized nine-membered ring that is generated (Scheme 8.30).[14] The sulfone **18** was prepared as described earlier (see Section 8.2, Scheme 8.6) through double substitution of an α-bromo-ω-iodide by hexamethyldisilathiane. A classic (two-step) RBR was performed, by first chlorinating the α-center of sulfone **18** and then treating the α-chlorosulfone **99** with base. The chlorination was achieved by deprotonation of the γ-position of enone **18** with strong base (either LiHMDS or NaH), followed by addition of hexachloroethane to provide **99** as a single diastereomer (stereochemistry not determined). The RBR required the extremely hindered base potassium *tert*-heptoxide; otherwise, the product was contaminated with dechlorinated starting material. Under these conditions, the RBR afforded dehydroeremantholide A (**100**) in 5 min at 70 °C in a polar solvent system consisting of HMPA (10 equiv.) and dimethoxyethane. Despite the steric hindrance presumably encountered in displacement of the chloride from the tertiary α-center and the strain in the resulting ring, the reaction proceeded in high yield. Indeed, molecular modeling (MM2) had indicated that the ring contraction should be geometrically favored and that competing elimination of the β-oxygen at C6 should be unlikely due to the conformation. This highlights the utility of theoretical methods in predicting the facility of an RBR. The RBR product **100** was converted to eremantholide A in a single step by hydration of the exocyclic enol ether.

8.7 RECENT APPLICATIONS OF THE RBR IN THE SYNTHESIS OF COMPLEX TARGET STRUCTURES

The literature covering a wide range of synthetic targets has been thoroughly reviewed,[5–10] so this section will concentrate on a few recent examples that rely on the RBR as a key step in the synthesis of complex, natural product targets. These applications of the RBR all have noteworthy features in terms of strategy or synthetic challenge that will be highlighted.

The RBR occupies a privileged position in total synthesis because (1) it enables the formation of strained cyclic systems through ring closure to a larger cycle before the key ring-contraction step and (2) the overall process can be achieved sequentially or split according to the requirements for functionality and orthogonality, thus allowing greater synthetic flexibility. Considering the potential benefits associated with this versatile methodology, it is currently under-utilized in target-directed synthesis. The chosen examples represent the current state of achievement for the RBR based on the aforementioned criteria.

8.7.1 Fawcettidine

The complex, tetracyclic framework of fawcettidine (**101**), a member of the *Lycopodium* alkaloid family, provides a considerable challenge for synthesis, particularly in light of the additional ring strain caused by unsaturation at one of the ring junctions. The seminal total synthesis of (+)-fawcettidine by Kozak and Dake employed an elegant late-stage RBR to construct the seven-membered ring.[73] It should be noted that no alkene is present in the target compound at the position of the RBR disconnection, so the retron is latent and the unsaturation is introduced retrosynthetically (Figure 8.1).

The synthesis began from the trialkylated cyclohexanone **102**, which was prepared from the chiral pool starting material (*R*)-(+)-pulegone in six steps. The sulfur required for the RBR was installed as a thiocarbamate fairly early in the synthesis, during fusion of the second ring. This was achieved by reaction of the ketoester **102** with the ammonium salt **103** to form enamide **104** (Scheme 8.31). The tricycle **105** was generated by platinum-catalyzed enyne annulation. At this point, allylic oxidation to enone **106** was necessary to enable the fourth ring to be formed. Ring formation was achieved through the Michael addition of the thiol prepared by saponification

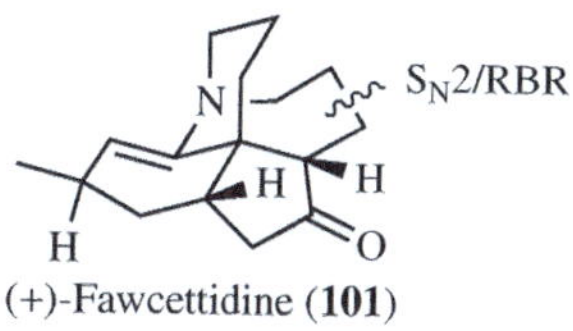

FIGURE 8.1 Fawcettidine

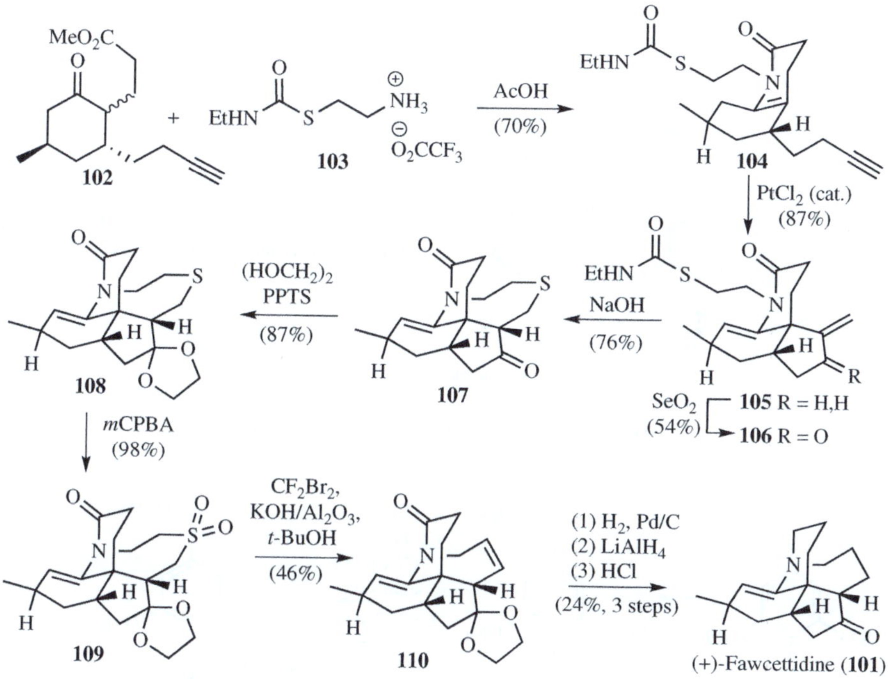

Scheme 8.31 RBR-based synthesis of fawcettidine.

of the thiocarbamate in **106**, thus providing the 8-membered bridged thioether **107**. After protection of the ketone as an acetal **108**, oxidation to the sulfone afforded the RBR precursor **109**. The ring contraction encompassed by the RBR proceeded in modest yield using Chan's conditions to provide the seven-membered ring of **110**, whose structure was confirmed by X-ray crystallography. The synthetic endgame also involved selective hydrogenation of the disubstituted alkene generated in the RBR, reduction of the enamide to an enamine and acetal deprotection to afford the natural product **101**.

8.7.2 Cylindrocyclophanes A and F

Given their intriguing chemical structures, the discovery of naturally occurring cyclophanes was an exciting development.[74] Cylindrocyclophanes are dimeric natural products that have succumbed to synthesis via the RBR (among other methods such as olefin metathesis and Wittig-type chemistry). Nicolaou has reported the synthesis of cylindrocyclophanes A and F (**111** and **112**) through a head-to-tail cyclodimerization involving sequential inter- and intramolecular S_N2 reactions of a thiol-containing mesylate, followed by an RBR (Figure 8.2).[75]

The synthesis commenced with 3,5-dimethoxy-4-bromobenzoic acid (**113**), which was converted to conjugated enone **114** through a sequence of redox reactions and carbonyl additions with lithiated species (Scheme 8.32). The stereogenic benzylic center was formed asymmetrically by, first, CBS reduction of the ketone, followed

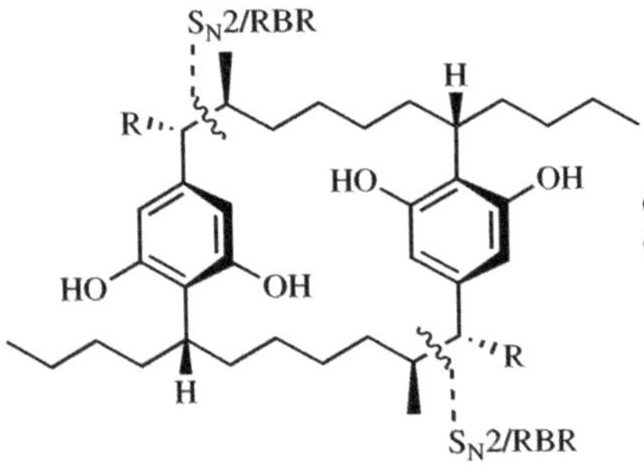

FIGURE 8.2 Cylindrocyclophanes A and F

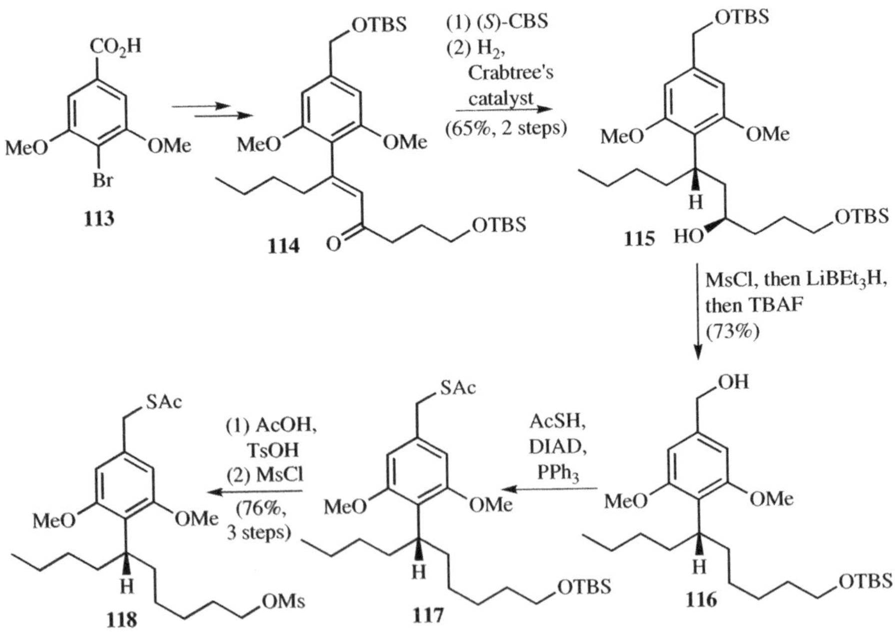

Scheme 8.32 Fragment synthesis for cylindrocyclophanes A and F.

by hydroxyl-directed hydrogenation using Crabtree's catalyst. Deoxygenation of the resultant alcohol **115** was achieved through mesylation and reduction with superhydride ($LiBEt_3H$). After selective desilylation to afford benzyl alcohol **116**, installation of the sulfur was achieved by Mitsunobu reaction with thioacetic acid. Subsequent desilylation of the thioacetate **117** and mesylation of the resulting primary alcohol provided the electrophilic partner **118** for the cyclodimerization.

The macrocylic [8.8]paracyclophane **119** was prepared in 64% yield by treatment of the thioacetate **118** with sodium methoxide, leading to dimerization through sequential double nucleophilic displacement of the mesylate by the resulting thiolate, first in an intermolecular sense and then intramolecularly (Scheme 8.33). Oxidation was performed with hydrogen peroxide and catalytic ammonium molybdate, providing the sulfone **120** in high yield (80%). This was subjected to Chan's conditions and underwent RBR to afford the diene **121** as a 12:1 mixture with its (*E*,*Z*)-isomer. The crude mixture was isomerized by heating with catalytic Pd(II) to produce the (*E*,*E*)-diene **121** in good yield (70%). Sharpless asymmetric dihydroxylation of **121** with AD-mix-ß was followed by Barton deoxygenation of the benzylic hydroxyls by way of the thionocarbonate, affording homobenzylic diol **122**. For the synthesis of cylindrocyclophane F, substitution of the remaining hydroxyl groups was carried out by mesylation and displacement with trimethylaluminum. Deprotection of the methyl groups with boron tribromide provided the natural product **112**. The diol **122** could alternatively be diverted to diene **123**, an intermediate in Hoye *et al.*'s synthesis of cylindrocyclophane A,[76] by oxidation, enolate triflation, and Kumada-type cross-coupling with methyl magnesium bromide. This represents, therefore, a formal total synthesis of cylindrocyclophane A (**111**).

8.7.3 Hirsutellones A–C

The highly strained and functionally decorated cyclophane ring of the hirsutellones provides a worthy challenge for any synthetic methodology. Nicolaou and coworkers have elegantly demonstrated the prowess of the RBR in the total syntheses of hirsutellones A–C by taking advantage of the ring-contracting nature of the RBR (Figure 8.3).[77,78] The hirsutellones are fungal secondary metabolites that display exciting antimicrobial activity against *Mycobacterium tuberculosis*, the causative pathogen of tuberculosis.

The polyfunctional cyclophane rings of the hirsutellones are 12 (in A) and 13 membered (in B and C). Their preparation by direct ring closure has proven very difficult, as attested by unsuccessful attempts[79] and the difficulty in performing RCM even on a less-constrained model.[80] However, a recent total synthesis of hirsutellone B successfully utilized the Ullmann coupling for ring formation, a step which proceeded in 42% yield.[81]

The fused tricyclic system of hirsutellones A–C, embodied by **125**, was generated by a concise and stereoselective intramolecular epoxide opening/Diels–Alder cascade reaction of epoxide **124** (Scheme 8.34), which was obtained from (+)-citronellal.[77] Conversion of the core tricycle **125** into the diol **126** was achieved through a series of redox and substitution steps. The thioacetate group was then introduced by

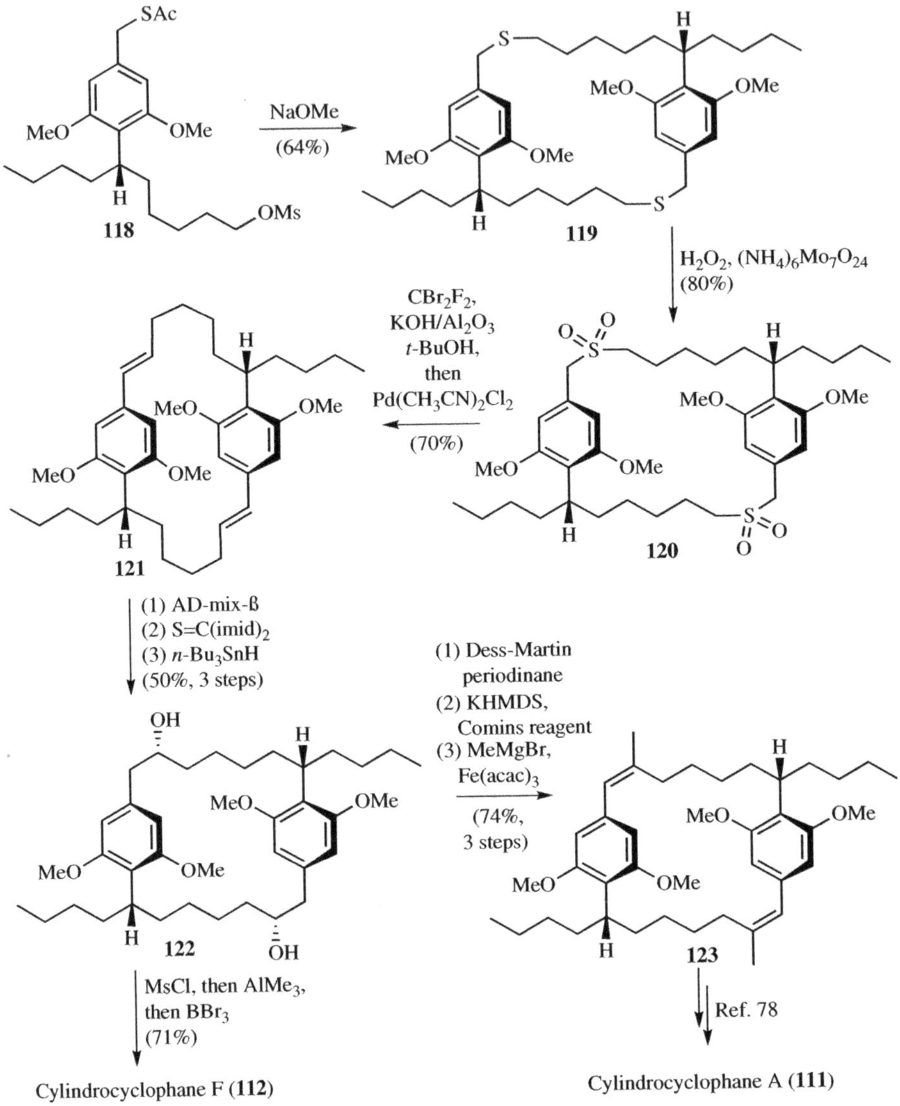

Scheme 8.33 RBR-based synthesis of cylindrocyclophanes A and F.

Lewis acid-promoted displacement of the benzylic alcohol in **126** by thioacetic acid, probably via an intermediate such as **128**. Concurrent substitution of the other hydroxyl group by iodide presumably occurred through zinc-promoted formation of an intermediate such as **129**. This cascade process led to the cyclization precursor **127** in a good yield. Deacetylation and ring closure to the 14-membered macrocycle was promoted by sodium methoxide in methanol/THF and proceeded in high yield,

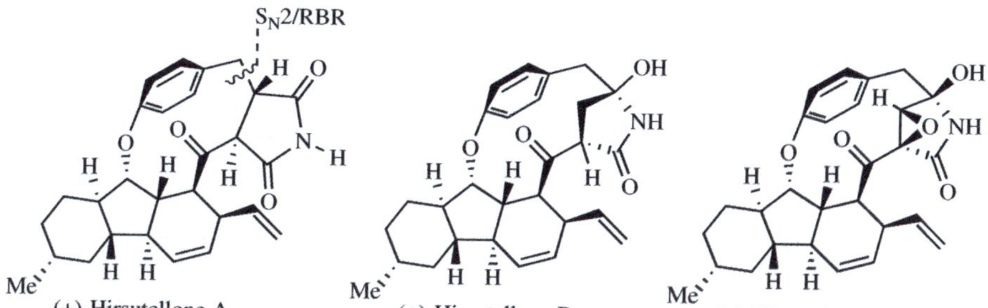

FIGURE 8.3 Hirsutellones A–C

followed by sulfur oxidation with hydrogen peroxide and sodium tungstate to afford sulfone **130**.

The stage was now set for the key RBR, which was carried out directly after the preparatory steps leading to sulfone **130**. Employment of the Chan protocol for the ring-contractive RBR produced the desired 13-membered paracyclophane ether **131** in an impressive 95% yield (Scheme 8.35), with suitable functional handles for conversion to the natural products. Hirsutellones A–C were generated from the common intermediate **132**, obtained from the RBR product **131** through carboxylation of the ketone α-center and Sharpless dihydroxylation of the styrenyl moiety.[77] The ensuing chemistry (shown for hirsutellone B) involved Barton–McCombie deoxygenation of the benzylic center and lactam ring formation.[77] For hirsutellone A, an additional rearrangement to the succinimide was conducted.[78]

This application of the RBR to the synthesis of hirsutellones A–C serves to delineate its impressive scope as a strategic connection in the total synthesis of highly complex, strained cyclic natural products. The high-yielding ring-formation and ring-contraction reactions deserve special note (Schemes 8.34 and 8.35).

8.7.4 Aigialomycin D

The synthetic sequence involving formation of a thioether, oxidation to the corresponding sulfone and the RBR need not be performed consecutively, although it almost invariably is when applied to the synthesis of complex targets. An example in which divorcing the sulfone preparation from the RBR was actually beneficial to the synthetic strategy is seen in the total synthesis of the resorcylic acid macrolactone (RAL) aigialomycin D by Harvey and coworkers.[82]

Aigialomycin D is a metabolite of the mangrove fungus *Aigialus parvus* and was shown to be a kinase inhibitor with modest antimalarial activity. The convergent synthesis showcased here assembled three fragments into the full carbon skeleton (Scheme 8.36). The ester-containing benzylic bromide **134** was prepared from methyl orsellinate, itself derived from methyl acetoacetate. Thioacetate **135** was formed from D-ribose as will be described, while homoallylic alcohol **136** was commercially available.

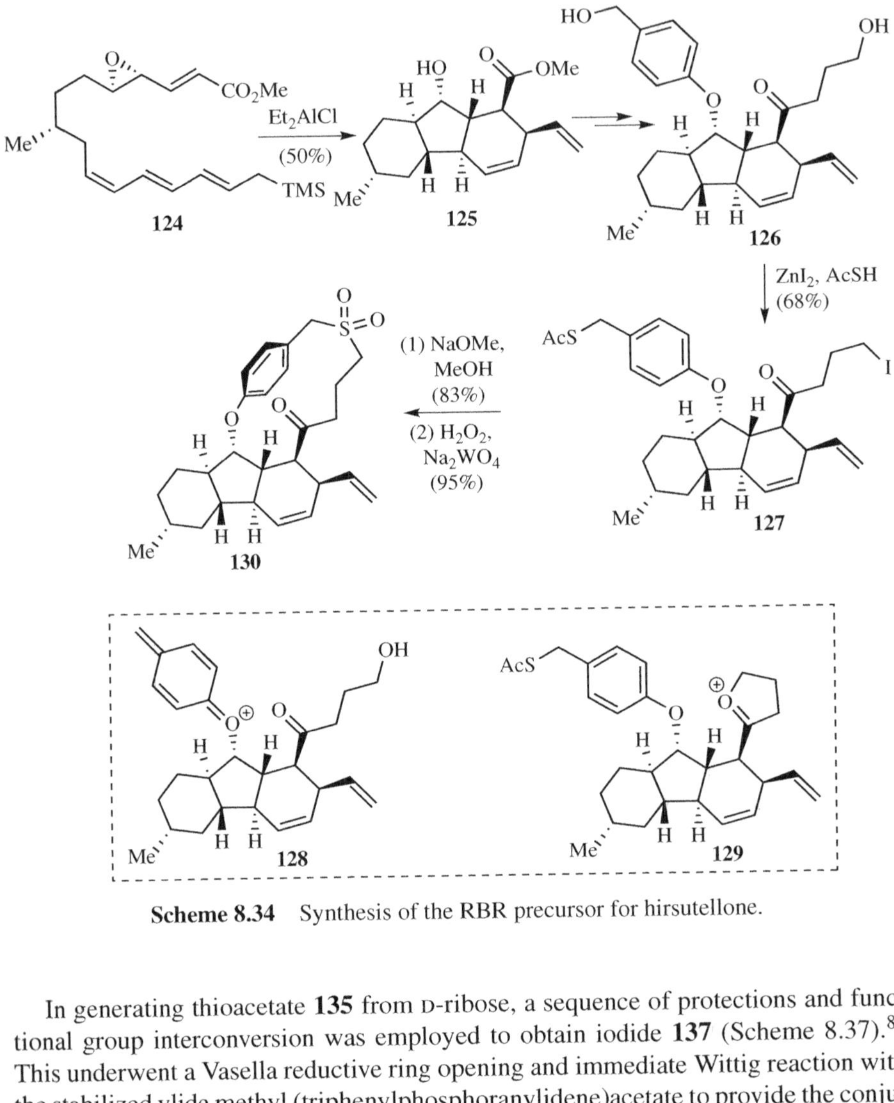

Scheme 8.34 Synthesis of the RBR precursor for hirsutellone.

In generating thioacetate **135** from D-ribose, a sequence of protections and functional group interconversion was employed to obtain iodide **137** (Scheme 8.37).[83] This underwent a Vasella reductive ring opening and immediate Wittig reaction with the stabilized ylide methyl (triphenylphosphoranylidene)acetate to provide the conjugated enoate **138** as a 5:1 mixture of (*Z*)- and (*E*)-alkenes. Reduction of the conjugated olefin double bond in the presence of the terminal alkene proved challenging. After considerable experimentation, employment of sodium borohydride in the presence of cuprous chloride, including cyclohexene as a sacrificial alkene, was found to provide the best results, with ester **139** isolated in high yields. Lithium aluminum hydride reduction of this ester, mesylation, and substitution with potassium thioacetate provided the sulfur-containing fragment **135** reliably and in high yield.

Displacement of the benzylic bromide in **134** with the thiol derived from thioacetate **135** was achieved using potassium carbonate in methanol (Scheme 8.38).[82] In the process, the phenolic protecting groups were cleaved, and they were replaced

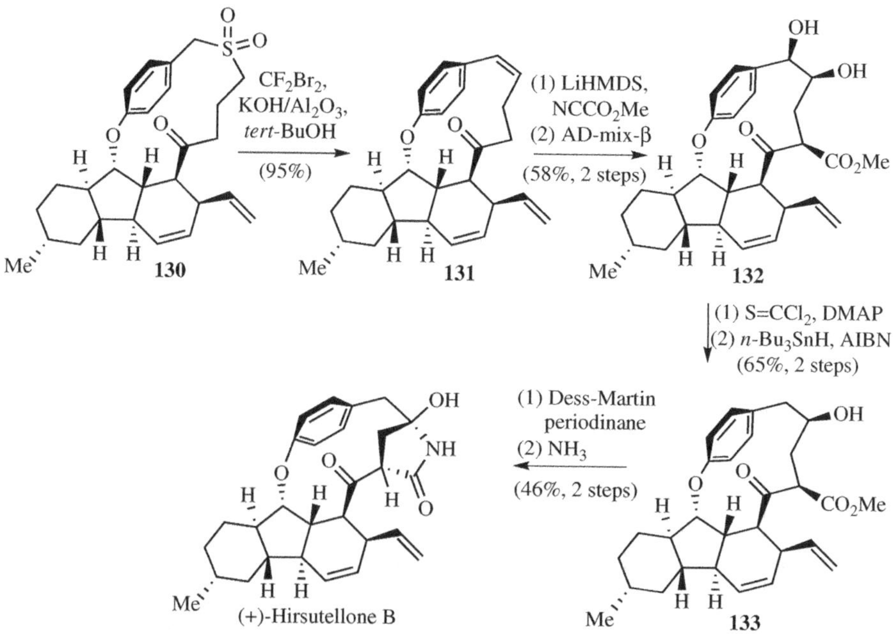

Scheme 8.35 RBR-based synthesis of hirsutellone B.

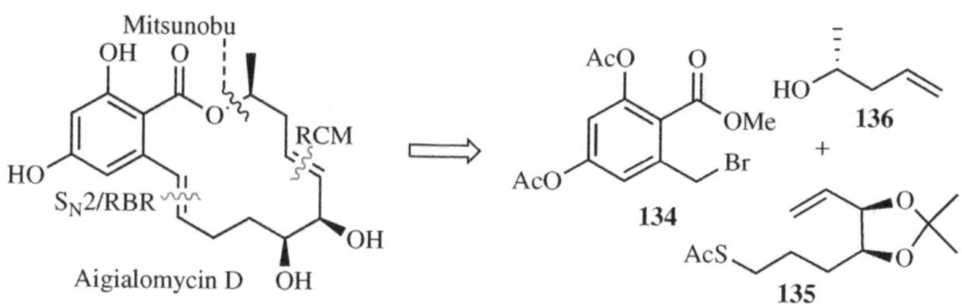

Scheme 8.36 Aigialomycin D and retrosynthetic analysis.

with MOM acetals that would be more compatible with the key RBR. Saponification provided the acid **140**, which was coupled with homoallylic alcohol **136**, using the Mitsunobu protocol, to give thioether **141**.

With the full carbon skeleton present, the synthetic endgame could be considered. There were several options for the order of the sulfone formation, RBR and RCM steps. However, it was determined that RCM should not be attempted before thioether oxidation due to the likelihood that the Ru catalyst would be poisoned by the low-valent sulfur. Thioethers are known to inactivate the Grubbs second-generation catalyst, while the corresponding sulfoxides and sulfones do not.[84] Furthermore, reports from other groups indicated that undertaking the RBR first, thus requiring RCM to be performed in the presence of a preexisting internal

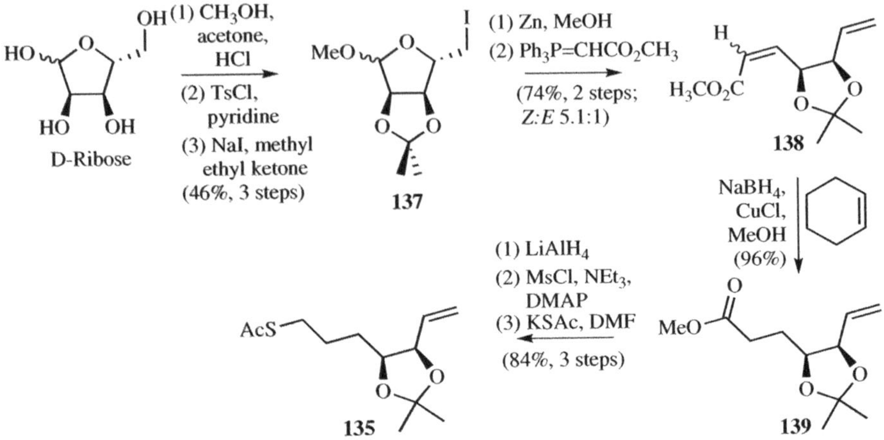

Scheme 8.37 Fragment synthesis for aigialomycin D.

alkene **145**, would lead to a competing metathesis process that would produce a cyclohexene **147** and a truncated aigialomycin D framework **146** (Scheme 8.39).[85] Therefore, the ideal strategy would invoke oxidation to the sulfone and then an RCM before the ring-contracting RBR.[86]

Indeed, the endgame sequence proceeded smoothly and in high yield using *m*-CPBA to oxidize the thioether **141** to sulfone **142** (see Scheme 8.38). The Grubbs second-generation catalyst was employed for RCM to give **143**, and Meyers' conditions induced the ring-contractive RBR. Only the (*E*)-alkene was observed to form in both the RCM and RBR steps. This sequence provided protected aigialomycin D, which was subjected to global deprotection to provide the natural product **144** in high yield.[82]

It should be emphasized that the sulfone **142**, on which the RCM is performed, represents a masked alkene. The sulfone is not susceptible to the unwanted (truncating) RCM seen in Scheme 8.39, while the corresponding alkene **145**, derived from prior RBR, would be. Similar considerations should have utility in other scenarios where an alkene needs to be introduced in masked form to avoid unwanted reactivity. For example, truncated rings have been obtained in attempts to synthesize dictyostatin analogues by RCM.[87] The multiply unsaturated precursors are diverted into a competing cyclohexene formation (related to that shown for the aigialomycin D precursor **145** in Scheme 8.39), which causes an undesired relay-type RCM to a smaller macrocycle. This is potentially an ideal setting for employing a sulfone as a masked alkene during the macrocyclic RCM and subsequent RBR.

8.8 CONCLUDING REMARKS

This chapter has highlighted the scope of the RBR and strategic considerations relevant to its application in the synthesis of natural products and other complex

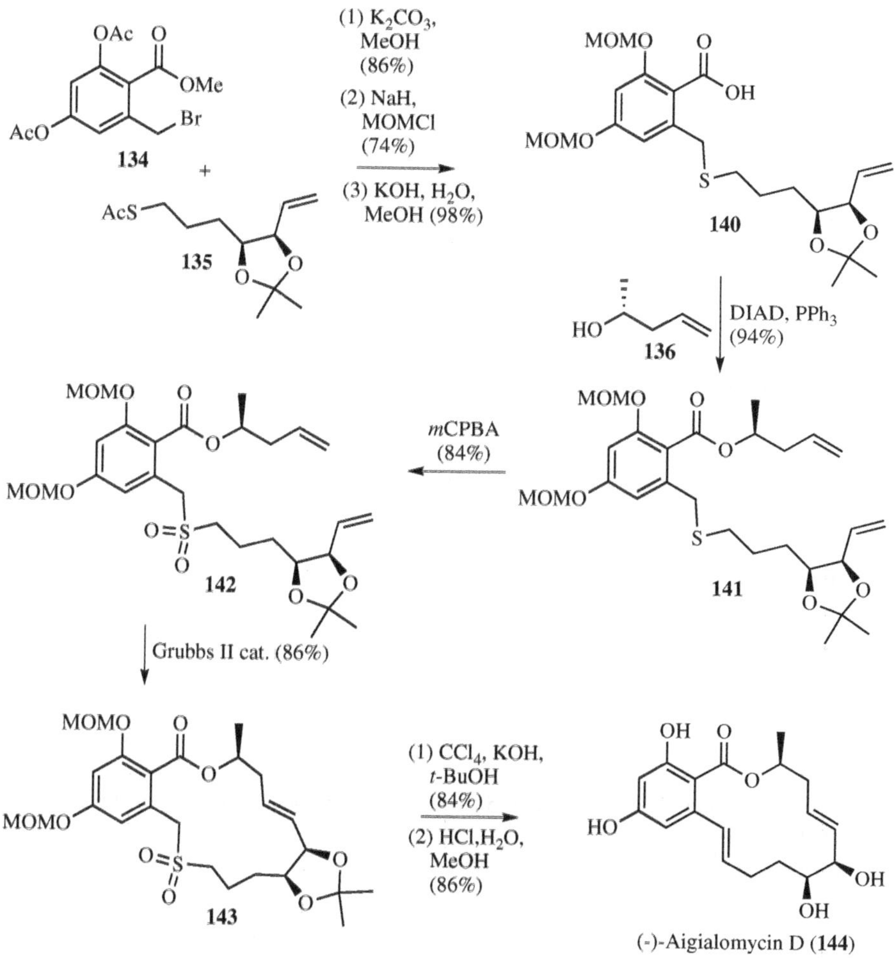

Scheme 8.38 RBR-based synthesis of aigialomycin D.

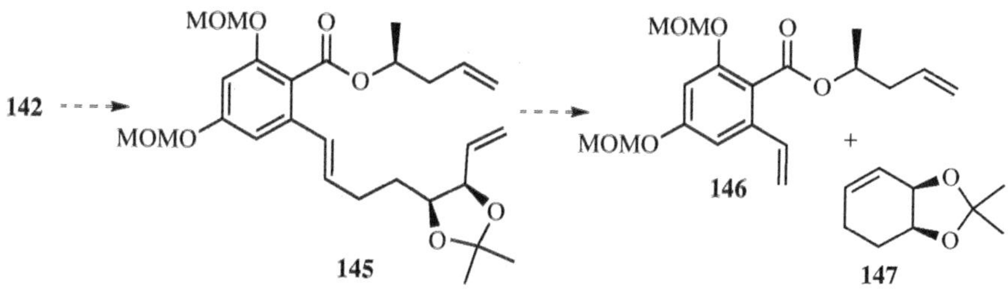

Scheme 8.39 Unwanted truncation previously observed with aigialomycin D synthesis.

targets. The versatility and potency of the RBR, particularly for alkene formation in strained, polyunsaturated, and macrocyclic settings, have been demonstrated through examples. The fact that the RBR of cyclic sulfones effects a ring contraction can facilitate the formation of small and strained medium-sized rings by allowing the typically challenging cyclization step to be performed on the larger system. Furthermore, segregating sulfone formation from the actual RBR step enables transformations to be performed on sulfone-containing intermediates that would be incompatible with the product alkene. Therefore, the sulfone represents a masked alkene. Future applications of the RBR to synthetic endeavors are likely to make use of the beneficial features of this reliable and versatile methodology.

ACKNOWLEDGMENTS

I would like to extend my thanks to Dr Mark Bartlett for the help with locating some references and to Professor Richard J. K. Taylor for editing the manuscript.

REFERENCES

1. Ramberg, L.; Bäcklund, B. *Ark. Kemi. Mineral. Geol.* **1940**, *27*, 1–50 (*Chem. Abstr.* **1940**, *34*, 4725).
2. Meyers, C. Y.; Malte, A. M.; Matthews, W. S. *J. Am. Chem. Soc.* **1969**, *91*, 7510–7512.
3. Chan, T.-L.; Fong, S.; Li, Y.; Man, T.-O.; Poon, C.-D. *J. Chem. Soc., Chem. Commun.* **1994**, 1771–1772.
4. Yang, G.; Franck, R. W.; Byun, H.-S.; Bittman, R.; Samadder, P.; Arthur, G. *Org. Lett.* **1999**, *1*, 2149–2151.
5. Taylor, R. J. K.; Casy, G. *Org. React.* **2003**, *62*, 357–475, and references cited therein.
6. Paquette, L. A. *Org. React.* **1977**, *25*, 1–71.
7. (a) Clough, J. M. In *Comprehensive Organic Synthesis*; Trost, B. M.; Fleming, I., Eds.; Pergamon Press: Oxford, **1991**; Vol. 3, pp. 861–886. (b) Braverman, S.; Cherkinsky, M.; Raj, P. *Sulfur Rep.* **1999**, *22*, 49–84.
8. Taylor, R. J. K. *Chem. Commun.* **1999**, 217–227.
9. Feng, J.-P.; Wang, X.-L.; Cao, X.-P. *Chin. J. Org. Chem.* **2006**, *26*, 158–167.
10. Harvey, J. E.; Bartlett, M. J. *Chemistry in New Zealand*, **2010**, *74*, 63–69.
11. Söderman, S. C.; Schwan, A. L. *J. Org. Chem.* **2012**, *77*, 10978–10984.
12. Pal, T. K.; Pathak, T. *Synlett* **2005**, 2263–2266.
13. (a) Cerè, V.; Minzoni, M.; Pollicino, S.; Ricci, A.; Gasparrini, F.; Ciogli, A.; D'Acquarica, I. *J. Comb. Chem.* **2006**, *8*, 74–78. (b) Cerè, V.; Peri, F.; Pollicino, S.; Ricci, A.; Devlin, F. J.; Stephens, P. J.; Gasparrini, F.; Rompietti, R.; Villani, C. *J. Org. Chem.* **2005**, *70*, 664–669.
14. Boeckman, Jr., R. K.; Yoon, S. K.; Heckendorn, D. K. *J. Am. Chem. Soc.* **1991**, *113*, 9682–9684.
15. Bouchez, L. C.; Vogel, P. *Chem. Eur. J.* **2005**, *11*, 4609–4620.

16. McAllister, G. D.; Paterson, D. E.; Taylor, R. J. K. *Angew. Chem. Int. Ed.* **2003**, *42*, 1387–1391.
17. Taylor, R. J. K.; McAllister, G. D.; Franck, R. W. *Carbohydr. Res.* **2006**, *341*, 1298–1311.
18. Block, E.; Aslam, M.; Eswarakrishnan, V.; Gebreyes, K.; Hutchinson, J.; Iyer, R.; Laffitte, J.-A.; Wall, A. *J. Am. Chem. Soc.* **1986**, *108*, 4568–4580.
19. Wladislaw, B.; Marzorati, L.; Torres Russo, V. F.; Zaim, M. H.; Di Vitta, C. *Tetrahedron Lett.* **1995**, *36*, 8367–8370.
20. Evans, P.; Taylor, R. J. K. *Synlett* **1997**, 1043–1044.
21. Burger, J. J.; Chen, T. B. R. A.; De Waard, E. R.; Huisman, H. O. *Tetrahedron* **1981**, *37*, 417–424.
22. Evans, P.; Johnson, P.; Taylor, R. J. K. *Eur. J. Org. Chem. 2006*, 1740–1754.
23. Bordwell, F. G.; Cooper, G. D. *J. Am. Chem. Soc.* **1951**, *73*, 5187–5190.
24. Bordwell, F. G.; Doomes, E. *J. Org. Chem.* **1974**, *39*, 2526–2531.
25. Paquette, L. A. *Acc. Chem. Res.* **1968**, *1*, 209–216.
26. Sutherland, A. G.; Taylor, R. J. K. *Tetrahedron Lett.* **1989**, *30*, 3267–3270.
27. Zhang, H.; Zhang, L.-X.; Yang, F.; Liu, F.-L. *J. Mol. Struct.* **2010**, *950*, 41–45.
28. Paquette, L. A. *J. Am. Chem. Soc.* **1964**, *86*, 4085–4089.
29. Neureiter, N. P.; Bordwell, F. G. *J. Am. Chem. Soc.* **1963**, *85*, 1209–1210.
30. Bordwell, F. G.; Williams, Jr., J. M.; Hoyt, Jr., E. B.; Jarvis, B. B. *J. Am. Chem. Soc.* **1968**, *90*, 429–435.
31. Suárez, D.; Sordo, J. A.; Sordo, T. L. *J. Phys. Chem.* **1996**, *100*, 13462–13465.
32. Ginsburg, H.; Le Ny, G.; Tournois, N.; Tchoubar, B.; Janot, M.-M. *Comptes Rendus Acad. Sc. Paris, Série C* **1969**, *269*, 339–341.
33. Wender, P. A.; Croatt, M. P.; Witulski, B. *Tetrahedron* **2006**, *62*, 7505–7511.
34. Trost, B. M. *Science* **1991**, *254*, 1471–1477.
35. Aïssa, C. *Eur. J. Org. Chem.* **2009**, 1831–1844 and references cited therein.
36. Nicolaou, K. C.; Zuccarello, G.; Ogawa, Y.; Schweiger, E. J.; Kumazawa, T. *J. Am. Chem. Soc.* **1988**, *110*, 4866–4868.
37. Mlinarić-Majerski, K.; Pavlović, D.; Marinić, Z. *Tetrahedron Lett.* **1996**, *37*, 4829–4832.
38. Chan, T.-L.; Hung, C.-W.; Man, T.-O.; Leung, M.-K. *J. Chem. Soc., Chem. Commun.* **1994**, 1971–1972.
39. Doomes, E.; McKnight, A. A. *J. Heterocycl. Chem.* **1995**, *32*, 1467–1471.
40. MaGee, D. I.; Beck, E. J. *Can. J. Chem.* **2000**, *78*, 1060–1066.
41. Wei, C.; Mo, K.-F.; Chan, T.-L. *J. Org. Chem.* **2003**, *68*, 2948–2951.
42. Edwards, J. O.; Pearson, R. G. *J. Am. Chem. Soc.* **1962**, *84*, 16–24.
43. Gulder, T.; Baran, P. S. *Nat. Prod. Rep.* **2012**, *29*, 899–934.
44. Chow, H.-F.; Ng, M.-K.; Leung, C.-W.; Wang, G.-X. *J. Am. Chem. Soc.* **2004**, *126*, 12907–12915.
45. (a) Pasetto, P.; Franck, R. W. *J. Org. Chem.* **2003**, *68*, 8042–8060. (b) Belica, P. S.; Franck, R. W. *Tetrahedron Lett.* **1998**, *39*, 8225–8228. (c) Jeanmart, S.; Taylor, R. J. K. *Tetrahedron Lett.* **2005**, *46*, 9043–9048. (d) Griffin, F. K.; Paterson, D. E.; Murphy, P. V.; Taylor, R. J. K. *Eur. J. Org. Chem.* **2002**, 1305–1322.
46. Zhu, X.; Jin, Y.; Wickham, J. *J. Org. Chem.* **2007**, *72*, 2670–2673.

47. Li, Y.; Zhang, Y.; Wang, X. L.; Huang, Z.; Cao, X. P. *Chin. Chem. Lett.* **2004**, *15*, 631–634.
48. Oh, S.; Jeong, I. H.; Shin, W.-S.; Lee, S. *Bioorg. Med. Chem. Lett.* **2004**, *14*, 3683–3686.
49. Cao, X.; Yang, Y.; Wang, X. *J. Chem. Soc., Perkin Trans. 1* **2002**, 2485–2489.
50. Hodgson, D. M.; Hachisu, S.; Andrews, M. D. *Synlett.* **2005**, 1267–1270.
51. Robinson, J. E.; Taylor, R. J. K. *Chem. Commun.* **2007**, 1617–1619.
52. McAllister, G. D.; Robinson, J. E.; Taylor, R. J. K. *Tetrahedron* **2007**, *63*, 12123–12130.
53. MaGee, D. I.; Beck, E. J. *J. Org. Chem.* **2000**, *65*, 8367–8371.
54. Pal, T. K.; Pathak, T. *Carbohydr. Res.* **2008**, *343*, 2826–2829.
55. Snyder, S. A.; Breazzano, S. P.; Ross, A. G.; Lin, Y.; Zografos, A. L. *J. Am. Chem. Soc.* **2009**, *131*, 1753–1765.
56. Foot, J. S.; Giblin, G. M. P.; Taylor, R. J. K. *Org. Lett.* **2003**, *5*, 4441–4444.
57. Foot, J. S.; Giblin, G. M. P.; Whitwood, A. C.; Taylor, R. J. K. *Org. Biomol. Chem.* **2005**, *3*, 756–763.
58. Choi, S.; Koo, S. *J. Org. Chem.* **2005**, *70*, 3328–3331.
59. Jeong, Y. C.; Ji, M.; Lee, J. S.; Yang, J.-D.; Jin, J.; Baik, W.; Koo, S. *Tetrahedron* **2004**, *60*, 10181–10185.
60. (a) Nicolaou, K. C.; Fylaktakidou, K. C.; Monenschein, H.; Li, Y.; Weyershausen, B.; Mitchell, H. J.; Wei, H.; Guntupalli, P.; Hepworth, D.; Sugita, K. *J. Am. Chem. Soc.* **2003**, *125*, 15433–15442. (b) Nicolaou, K. C.; Li, Y.; Sugita, K.; Monenschein, H.; Guntupalli, P.; Mitchell, H. J.; Fylaktakidou, K. C.; Vourloumis, D.; Giannakakou, P.; O'Brate, A. *J. Am. Chem. Soc.* **2003**, *125*, 15443–15454.
61. Cao, X.-P.; Chan, T.-L.; Chow, H.-F. *Tetrahedron Lett.* **1996**, *37*, 1049–1052.
62. Weinges, K.; Schwarz, G.-U.; Weber, M.; Schilling, G. *Chem. Ber.* **1977**, *110*, 2961–2968.
63. Nicolaou, K. C.; Zuccarello, G.; Riemer, C.; Estevez, V. A.; Dai, W.-M. *J. Am. Chem. Soc.* **1992**, *114*, 7360–7371.
64. Bodwell, G. J.; Nandaluru, P. R. *Isr. J. Chem.* **2012**, *52*, 105–138.
65. Tang, F.-M.; Lin, S.-T. *J. Org. Chem.* **1997**, *62*, 2727–2731.
66. Bestmann, H. J.; Schaper, W.; Holzmann, H. G.; Zimmermann, R. *Chem. Ber.* **1991**, *124*, 2773–2779.
67. Bestmann, H. J.; Schaper, W. *Tetrahedron Lett.* **1975**, *16*, 3511–3512.
68. Paquette, L. A.; Liu, Z.; Efremov, I. *J. Org. Chem.* **2005**, *70*, 514–518.
69. Paquette, L. A.; Efremov, I.; Liu, Z. *J. Org. Chem.* **2005**, *70*, 505–509.
70. Paquette, L. A.; Guevel, R.; Sakamoto, S.; Kim, I. H.; Crawford, J.; *J. Org. Chem.* **2003**, *68*, 6096–6107.
71. Gentric, L.; Hanna, I.; Ricard, L. *Org. Lett.* **2003**, *5*, 1139–1142.
72. (a) Maimone, T. J.; Shi, J.; Ashida, S.; Baran, P. S. *J. Am. Chem. Soc.* **2009**, *131*, 17066–17067. (b) Yang, Q.; Njardarson, J. T.; Draghici, C.; Li, F. *Angew. Chem. Int. Ed.* **2013**, *52*, 8648–8651.
73. Kozak, J. A.; Dake, G. R. *Angew. Chem. Int. Ed.* **2008**, *47*, 4221–4223.
74. (a) Moore, B. S.; Chen, J.-L.; Patterson, G. M. L.; Moore, R. E.; Brinen, L. S.; Kato, Y.; Clardy, J. *J. Am. Chem. Soc.* **1990**, *112*, 4061–4063. (b) Moore, B. S.; Chen, J.-L.; Patterson, G. M. L.; Moore, R. E. *Tetrahedron* **1992**, *48*, 3001–3006.
75. Nicolaou, K. C.; Sun, Y.-P.; Korman, H.; Sarlah, D. *Angew. Chem. Int. Ed.* **2010**, *49*, 5875–5878.

76. Hoye, T. R.; Humpal, P. E.; Moon, B. *J. Am. Chem. Soc.* **2000**, *122*, 4982–4983.
77. Nicolaou, K. C.; Sarlah, D.; Wu, T. R.; Zhan, W. *Angew. Chem. Int. Ed.* **2009**, *48*, 6970–6974.
78. Nicolaou, K. C.; Sun, Y.-P.; Sarlah, D.; Zhan, W.; Wu, T. R. *Org. Lett.* **2011**, *13*, 5708–5710.
79. Jiang, X.; Shi, L.; Liu H.; Khan, A. H.; Chen, J. S. *Org. Biomol. Chem.* **2012**, *10*, 8383–8392.
80. Huang, M.; Song, L.; Liu, B. *Org. Lett.* **2010**, *12*, 2504–2507.
81. Uchiro, H.; Kato, R.; Arai, Y.; Hasegawa, M.; Kobayakawa, Y. *Org. Lett.* **2011**, *13*, 6268–6271.
82. Baird, L. J.; Timmer, M. S. M.; Teesdale-Spittle, P. H.; Harvey, J. E. *J. Org. Chem.* **2009**, *74*, 2271–2277.
83. (a) Wender, P. A.; Bi, F. C.; Buschmann, N.; Gosselin, F.; Kan, C.; Kee, J.-M.; Ohmura, H. *Org. Lett.* **2006**, *8*, 5373–5376. (b) Paquette, L. A.; Bailey, S. *J. Org. Chem.* **1995**, *60*, 7849–7856.
84. Bates, D. K.; Li, X.; Jog, P. V. *J. Org. Chem.* **2004**, *69*, 2750–2754.
85. (a) Barluenga, S.; Dakas, P.-Y.; Ferandin, Y.; Meijer, L.; Winssinger, N. *Angew. Chem. Int. Ed.* **2006**, *45*, 3951–3954. (b) Bajwa, N.; Jennings, M. P. *Tetrahedron Lett.* **2008**, *49*, 390–393.
86. Yao, Q. *Org. Lett.* **2002**, *4*, 427–430.
87. Zhu, W.; Jimenez, M.; Jung, W.-H.; Camarco, D. P.; Balachandran, R.; Vogt, A.; Day, B. W.; Curran, D. P. *J. Am. Chem. Soc.* **2010**, *132*, 9175–9187.

9

APPLICATIONS OF DI-π-METHANE AND RELATED REARRANGEMENT REACTIONS IN CHEMICAL SYNTHESIS

MARTIN G. BANWELL AND DAVID J.-Y. D. BON

Research School of Chemistry, Institute of Advanced Studies, The Australian National University, Canberra, Australia

9.1 INTRODUCTION: THE BASIC PROCESS AND ITS VARIANTS

The di-π-methane (DPM) rearrangement reaction was first observed in 1966[1] and has since become one of the most studied photochemical transformations thanks, in large measure, to the sustained and extensive efforts of Zimmerman and his coworkers.[2] Indeed, these particularly reliable, quite versatile, and mechanistically rather well-understood reactions are sometimes described as Zimmerman rearrangements.[3] The basic process is shown in Scheme 9.1 and, as the di-π-methane name suggests, the required substrate is one in which two π-moieties are bonded to a common and normally sp^3-hybridized or "methane" carbon. In an overall sense, then, a 1,4-diene (**1**) is converted into an isomeric vinylcyclopropane (**2**). The initial step in what is a probably a four-stage process (but see Section 9.2) is the formation of a biradical, **3**. This rearranges to the corresponding cyclopropyl-containing 1,4-biradical **4** which, in turn, fragments in the indicated manner to afford the 1,3-isomer **5**. Cyclization of this last species then delivers the observed product **2**. The reaction is formally a $[\pi^2 + \sigma^2]$ rearrangement[4] of a β,γ-unsaturated alkene (1,4-diene) involving a 1,2-vinyl shift and σ-bond formation between the original α- and γ-carbons.[5]

Molecular Rearrangements in Organic Synthesis, First Edition. Edited by Christian M. Rojas.

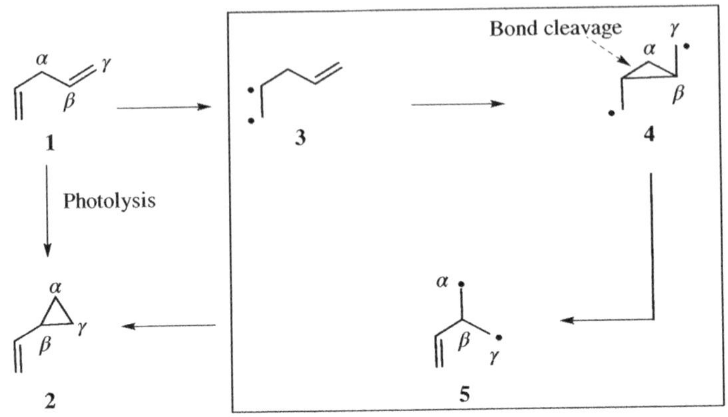

Scheme 9.1 The basic elements of the di-π-methane (DPM) rearrangement reaction.

The first example of the DPM rearrangement reaction was reported by Zimmerman and Grunewald[1] who observed that when a solution of barrelene (**6**) in *i*-pentane containing acetone was irradiated with a 450 W medium pressure mercury lamp, the fluxional molecule semibullvalene (**7**) (Scheme 9.2) was obtained. Only two of the three alkenyl groups participate in this process, but it is interesting to note that so-called tri-π-methane rearrangements involving tri-alkenyl-substituted methane-type substrates have since been reported.[6]

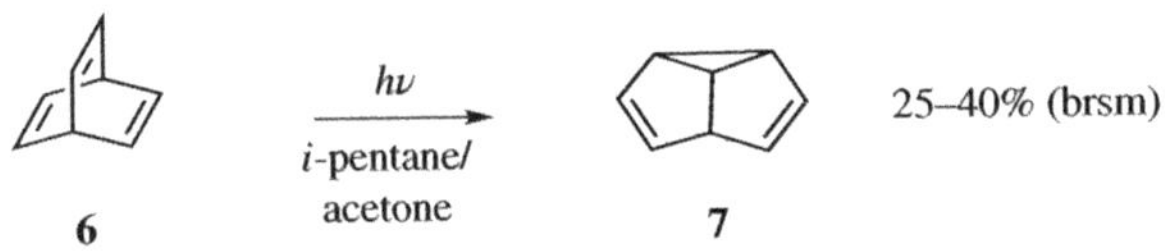

6 7

Scheme 9.2 The first reported example of a di-π-methane (DPM) rearrangement reaction.

Extensions of this type of conversion to the synthesis of benzo- and dibenzosemibullvalene, **8** and **9**, respectively, have been reported[7,8] with the reaction leading to the latter necessarily involving an aryl–alkenyl variant of the DPM rearrangement during the course of which the intermediate (and nonaromatic) biradical **10** is formed.[9]

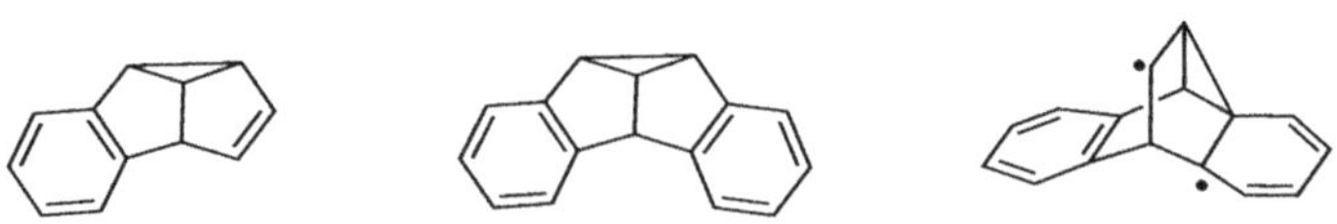

8 **9** **10**

Lest the reader think, given the examples presented thus far, that the reaction is confined to cyclic substrates, acyclic systems also participate rather effectively in the DPM process and thus highlighting its generality. Particularly notable open-chain substrates used in the early stages of the study of the DPM rearrangement were the hexa-substituted 1,4-dienes **11** (the Mariano diene)[10] and **12** (the Pratt diene).[11] These afforded, upon direct irradiation, cyclopropanes **13** and **14**, respectively, in near quantitative yield. Notably, the regioisomer, **15**, of the latter photoproduct is not observed, and a simple mechanistic explanation for this synthetically useful regioselectivity has been advanced (see Section 9.3).

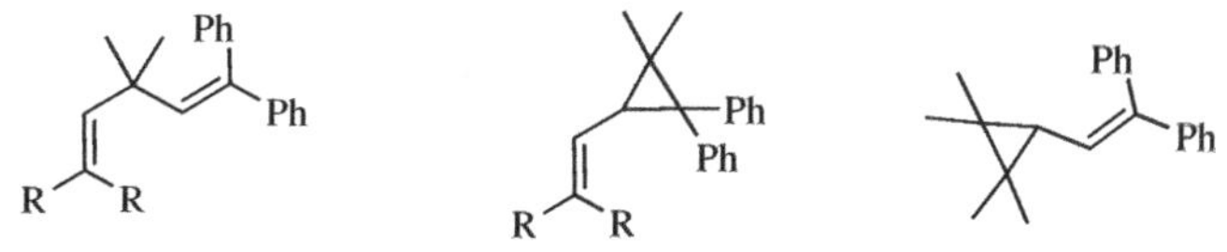

11 R = Ph
12 R = Me

13 R = Ph
14 R = Me

15

The oxa-di-π-methane (ODPM) rearrangement reaction is a particularly important variant of the parent process in which the 1,4-diene is replaced by a β,γ-unsaturated aldehyde or ketone **16** (Scheme 9.3). A cyclopropyl ketone or aldehyde (**17**) is the product of the reaction which involves, in overall terms, a 1,2-acyl shift as well as σ-bond formation between the original α- and γ-carbons. The reaction pathway may involve initial photoactivation of the carbonyl group with the resulting species **18** cyclizing to afford the cyclopropyl-containing 1,4-diradical **19** (but see Section 9.2). Fragmentation of **19** generates the isomeric 1,3-diradical **20** that is finally converted into the observed product through recombination. The first example of such a process was observed[12] by Lutz and coworkers in 1966, but it was only some 3–4 years later that these types of conversions were recognized as the oxa-variants of the parent DPM rearrangement reaction.[13] Since then, the ODPM rearrangement reaction has been recognized as a powerful synthetic protocol[14] and one that can be applied to both cyclic and acyclic substrates. It is being deployed with increasing frequency in complex chemical synthesis, especially because the reaction can provide access to compounds that are essentially unobtainable by thermal means.[15]

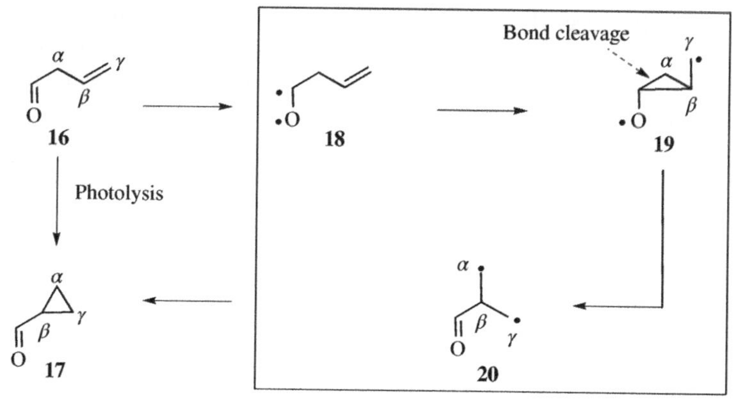

Scheme 9.3 The basic elements of the oxa-di-π-methane (ODPM) rearrangement reaction.

A prototypical example of the ODPM rearrangement reaction is the conversion of the bicyclo[2.2.2]oct-5-en-2-one **21** into the isomeric and cyclopropa-fused diquinane **22** (Scheme 9.4).[16] This conversion proceeds in 81% yield, and many variants of it have been used to great effect in the synthesis of polyquinane-containing natural products,[17] examples of which are presented in Section 9.4.

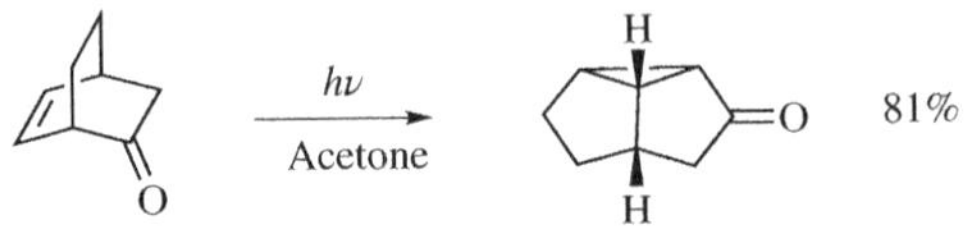

21 **22**

Scheme 9.4 A prototypical example of the ODPM rearrangement reaction.

Aza-di-π-methane (ADPM)[1] rearrangement reactions of both 1-aza- and 2-aza-1,4-dienes are known but not so common at the present time. Nevertheless, and largely due to the work of Armesto *et al.*,[18] the potential utility of such processes is becoming clearer. In the case of ADPM rearrangement reactions involving 1-aza-1,4-dienes, regioselective conversions are observed with, for example, the oxime **23** being converted into cyclopropane **24** in 74% yield after irradiation under triplet-sensitized conditions for just 30 min (Scheme 9.5).

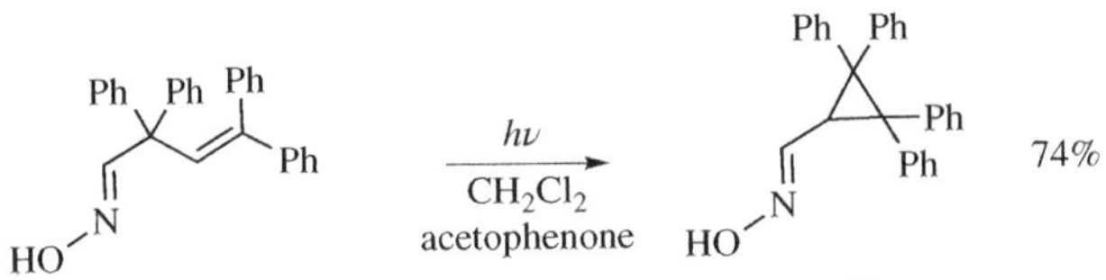

23 **24**

Scheme 9.5 An example of an aza-di-π-methane (ADPM) rearrangement reaction.

The ADPM rearrangement of 2-aza-1,4-dienes is a potentially more interesting transformation because, depending on the regioselectivity of the process, either *N*-alkenylaziridines or *N*-cyclopropylimines could be formed. The outcome of such processes is dictated, to some extent, by the mode of activation with the yield of the aziridines increasing under conditions of single-electron transfer (SET) and the cyclopropyl imines being favored under more conventional triplet-sensitized irradiation conditions. For example, reaction of imine **25** (Scheme 9.6) using acetophenone as the triplet sensitizer only proceeds to low conversions and affords iminocyclopropane **26** as the major product and with just traces of the aziridine **27** being observed. By comparison, when an acetonitrile solution of the same substrate containing 9,10-dicyanoanthracene (DCA) and biphenyl is irradiated a SET-mediated process takes place affording the aziridine **27** (11% at ca. 59% conversion) and 1-amino-1,2,2,3-tetraphenylcyclopropane (19% at ca. 59% conversion)

[1]The acronym ADPM has occasionally been used in referring to aryl di-π-methane rearrangements.

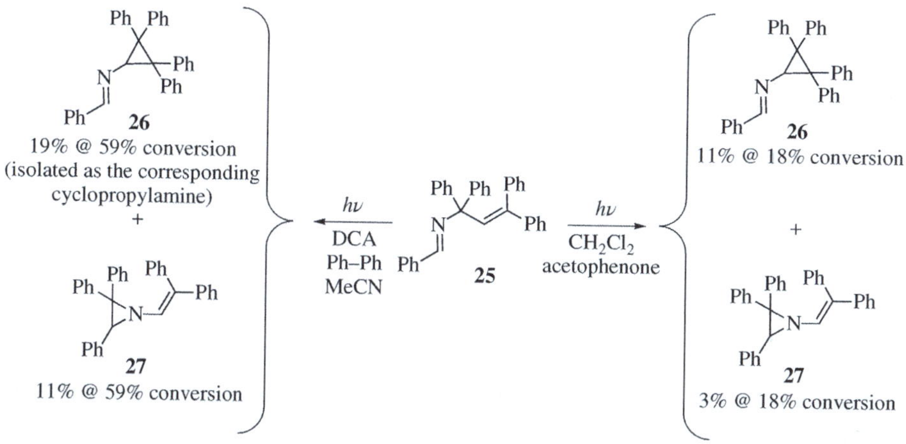

Scheme 9.6 Variations in the mode of the ADPM rearrangement reaction as a function of the method of activation.

as the notable products of reaction. The latter product is presumed to arise via a novel phenyl group migration within a radical-cation intermediate and hydrolysis of the imine **26** so formed on contact with silica gel.[18c]

A thermally promoted metalla-DPM rearrangement reaction has been reported,[19] but the utility and scope of this variant of the title reaction have yet to be fully investigated.

9.2 MECHANISTIC FEATURES AND COMPETING REACTIONS

The DPM rearrangement reaction can proceed via both singlet and triplet pathways, and which one is operational (or more effective) is very much substrate dependent.[2,4] Thus, the reaction usually takes place through a singlet pathway (S_1) in the excited state when acyclic 1,4-dienes are involved. The corresponding triplet process is often (but not always – see this section) ineffective because of competing cis–trans photoisomerization of the double bond(s) under the relevant conditions. In polycyclic systems, however, the reaction almost invariably proceeds via a triplet state. The involvement of the biradical corresponding to intermediate **4** in Scheme 9.1 has been established in certain instances through the synthesis and photochemical decomposition of diazenes that upon extrusion of nitrogen would lead to such a species. So, for example, independent and sensitized photolysis of either 1,4-diene **28** or diazene **29** leads (Scheme 9.7), presumably via the triplet biradical **30** in each instance, to the same mixture of the regioisomeric and bisannulated diquinanes **31** and **32**.[20] At 0 °C, the ratio of compounds **31** to **32** was ca. 12:1, while at 50 °C, this was 6:1, suggesting that intermediate **30** is a thermally equilibrated species.

While such studies demonstrate the involvement of biradicals of the general form **4** (Scheme 9.1) in the DPM rearrangement reaction of certain polycyclic systems, these may not be obligatory intermediates. A recent theoretical study[4] suggests the direct

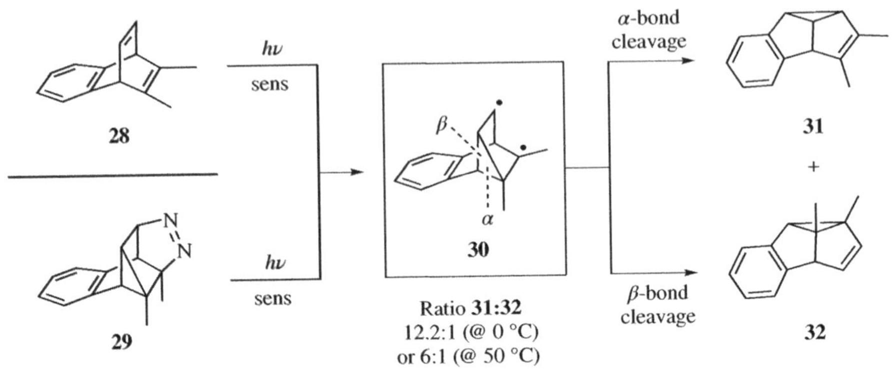

Scheme 9.7 Evidence for the intermediacy of cyclopropyl-containing 1,4-diradicals in the DPM rearrangement reaction.

formation of biradical **5** (Scheme 9.1) from precursor **3** (bypassing **4**) could occur in some instances and that competing one- and two-step mechanisms are possible on the triplet surface.

A striking illustration of the divergent behavior of a 1,4-diene as a function of the mode of photoactivation is shown in Scheme 9.8. Thus, irradiation of a benzene solution of the 1,4-diene **33** containing acetophenone as the triplet sensitizer gave cyclopropane **39** in ca. 96% yield. On the other hand, direct irradiation of a solution of the same substrate in *t*-butanol afforded the regioisomeric product **40** in ca. 95% yield.[21] Presumably, in the opening stages of either conversion, the initially formed diradical **34/35** rearranges to the isomeric and cyclopropane-containing congener **36**. Depending upon whether this last species is formed in the triplet or single state, it either fragments via homolysis of bond a (a triplet state event) to give diradical **37** (and thence product **39**) or via analogous cleavage of bond b (a singlet state event) to give diradical **38** (and thence product **40**).

The ODPM rearrangement reaction is almost invariably a triplet-mediated process.[14] Nevertheless, the reaction can proceed under conditions of direct irradiation because intersystem crossing is possible and which thus allows for the necessary singlet-to-triplet conversion.[14f] The crucial role of the sensitizer in the more conventional triplet-mediated variant of the reaction has been emphasized.[22,23] Most particularly, the triplet energies of the sensitizers used to effect the reaction must be close to those of the triplet states of the alkene moiety associated with the *β*,*γ*-unsaturated carbonyl compound serving as the substrate. So, for example, the triplet sensitizer thioxanthone (**41** – $E_T = 63$ kcal/mol) is rather effective in promoting the ODPM rearrangement of compound **42** ($E_T = 62$ kcal/mol) to cyclopropyl ketone **43** (50% at 80% conversion), while the more commonly used sensitizer acetophenone ($E_T = 74$ kcal/mol) is completely ineffective.

Three mechanistic scenarios beyond that shown in Scheme 9.3 have been invoked to explain the outcomes of various ODPM rearrangement reactions.[14f] The so-called "radicaloid" process involves homolysis of the carbonyl carbon to *α*-carbon bond within the triplet state and the simultaneous generation of an acyl and an allylic radical

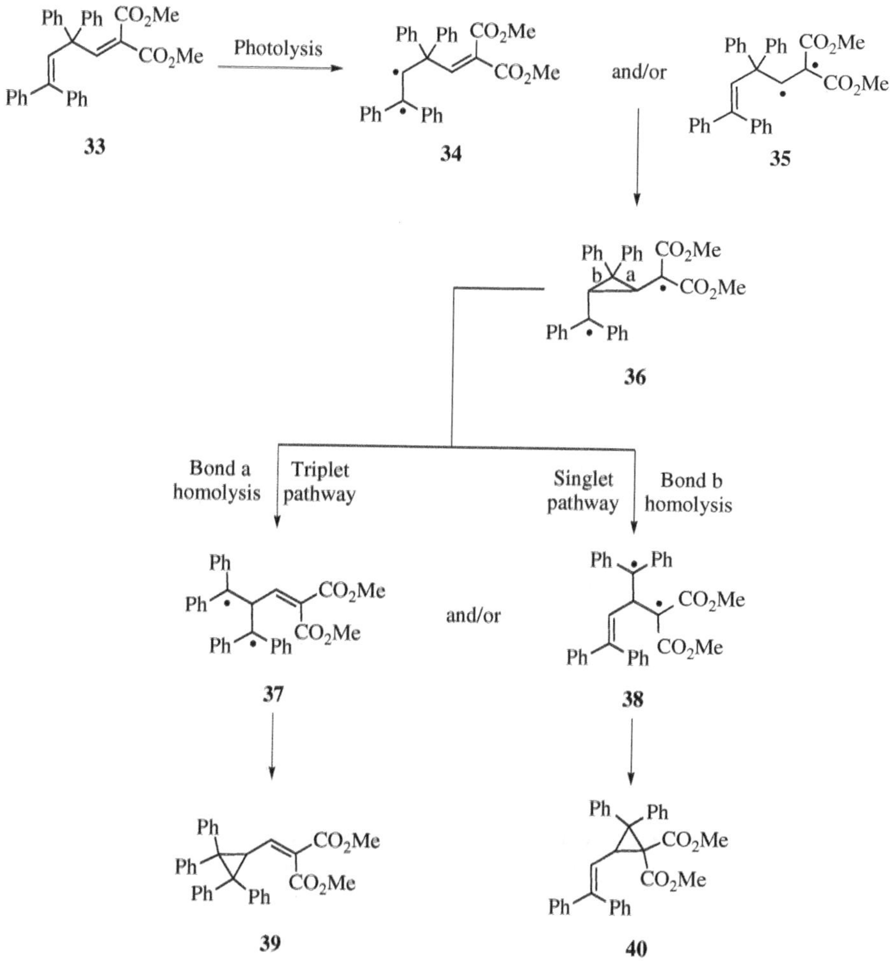

Scheme 9.8 Variations in the mode of the DPM rearrangement reaction as a function of the method of photoactivation.

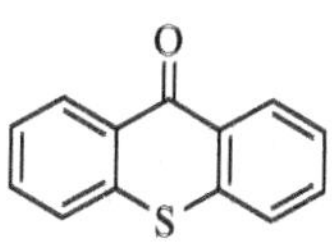

41

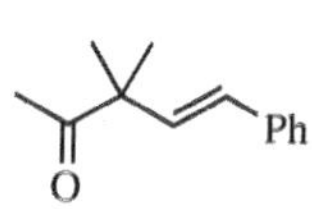

42

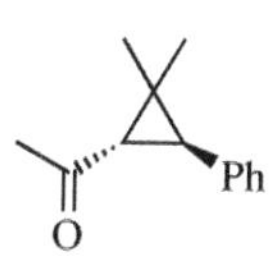

43

with the latter rearranging to the corresponding cyclopropyl radical. This last radical then recombines with its acyl counterpart to generate the observed/final product. A variation on this would involve the acyl radical adding to the cogenerated allylic radical with the ensuing 1,3-diradical then cyclizing to give the final product. A second scenario is a concerted process in which the triplet excited state of the substrate is converted directly into the product, while the third possibility is a mixed radicaloid and concerted process.[14f]

The ADPM rearrangement reactions are mechanistically more complicated than their DPM and ODPM counterparts because of the potential for the involvement of SET processes and the participation of radical cations. Given the thus far limited synthetic scope of the ADPM processes, no mechanistic discussion beyond that presented in Section 9.2 will be advanced here.

There are a number of processes that can compete with the di-π-rearrangement and related reactions. As indicated earlier, in the case of the parent DPM rearrangement, unproductive isomerization of the olefins within the substrate can occur when triplet-mediated conditions are applied to acyclic substrates. Electrocyclic processes can also compete.[7a,24] For example, while acetophenone-sensitized irradiation of benzobarrelene (**44**) affords benzosemibullvalene (**8**), direct irradiation of the same substrate yields benzocyclooctatetraene (**45**) via a [2+2] photocycloaddition/cycloreversion sequence. Such electrocyclic processes tend to proceed preferentially from the singlet excited state and take place exceptionally rapidly. In broad terms, cyclic substrates are more likely to suffer from competing electrocyclic reactions by comparison with their acyclic counterparts, one reason being the entropic advantage conferred on such processes by more conformationally rigid frameworks.

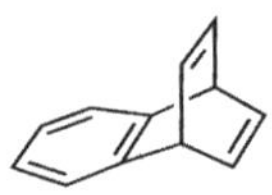

44

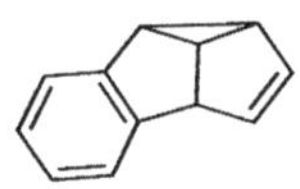

8

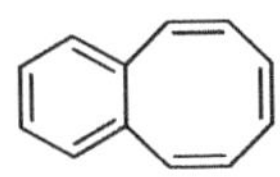

45

There are also a significant number of processes that can take place in parallel (competition) with the ODPM rearrangement reaction. Most notable among these are acyl migration, ketene formation, decarbonylation, and [2+2] photocycloaddition reactions.[14e] Alkene cis-/trans-isomerization can also occur. So, for example, during the course of their synthesis of the sesquiterpenoid natural product (−)-silphiperfol-6-en-5-one, Demuth and Hinsken observed[25] that triplet-sensitized irradiation of the cyclopentannulated bicyclo[2.2.2]octenone **46** (Scheme 9.9) afforded not only the desired ODPM rearrangement product **47** but the cyclobutanone **48** as well. The latter is presumably the product of a 1,3-acyl migration process.

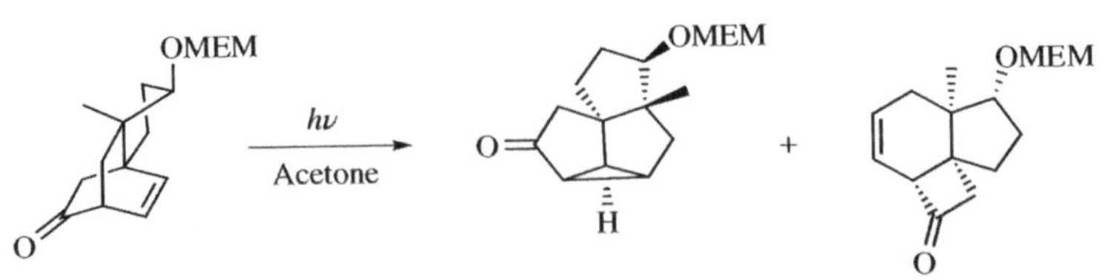

46 **47** (70%) **48** (4%)

Scheme 9.9 The pivotal ODPM rearrangement reaction associated with Demuth's synthesis of the angular triquinane-containing natural product (–)-silphiperfol-6-en-5-one.

In another case, irradiation of a solution of the potential ODPM substrate trimethylbicyclo[2.2.2]octadienone **49** in acetone containing acetophenone only resulted in a cycloreversion process leading to toluene (90%) and dimethylketene. The ketene could be trapped by added cinnamyl alcohol to give ester **50** (90%).[26]

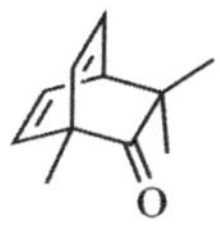

49

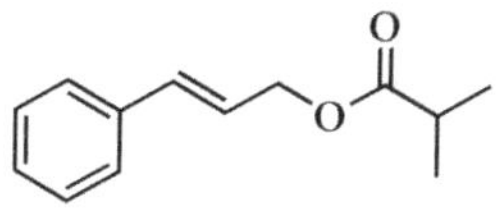

50

Armesto *et al.* have shown that Norrish type I reactions (wherein the carbonyl carbon to adjacent carbon bond of an aldehyde or ketone undergoes homolytic cleavage) compete with ODPM rearrangement reactions when β,γ-unsaturated methyl ketones bearing electron-withdrawing groups at the γ-position are subjected to direct irradiation at 254 nm.[27] The same group has reported that β,γ-unsaturated aldehydes undergo competing ODPM and decarbonylation reactions when 1,4-dimethoxynaphthalene is used as an electron donor/sensitizer.[28]

Reactions observed to "compete" with ADPM reactions include cleavage of the imine residue associated with the 2-aza-1,4-diene substrates,[18c] although such processes are almost certainly a result of the application of the chromatographic protocols used to separate the true reaction products from the recovered starting material. When the oximes and hydrazones of various β,γ-unsaturated carbonyl compounds (viz., 1-aza-1,4-dienes) are subjected to triplet-sensitized irradiation conditions, certain 5-exo-trig cyclization reactions can occur depending upon the substitution patterns involved, although not usually to the exclusion of the ADPM reaction.[18d] As an illustration, while substrate **51** affords the "expected" cyclopropyl oxime **52** (19%), congener **53** gives a mixture of cyclopropane **54** (10% of a 3:2 mixture of (*Z*)- and (*E*)-isomers) and the dihydroisoxazole **55** (8%) when subjected to analogous treatment.

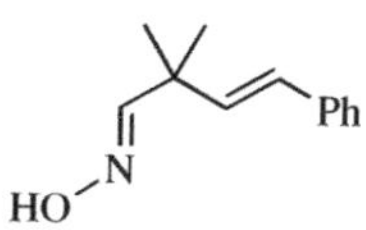

51

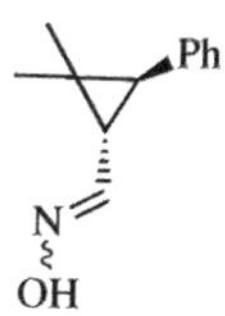

52

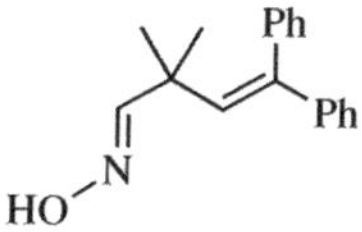

53

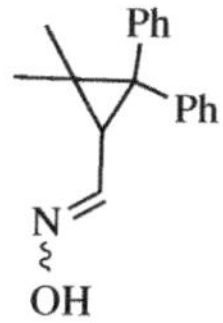

54

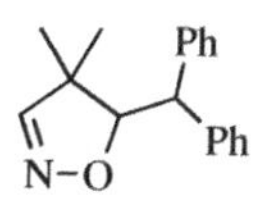

55

Perhaps the most intriguing "competing" reactions are those involving substrates wherein, in principle at least, two or more distinct DPM, ODPM, and/or ADPM rearrangement reactions can take place. For example, it has been shown that triplet-sensitized irradiation of triene **56** affords a mixture of photoproducts **57** and **58** arising from participation of the endocyclic double bonds of the substrate in the DPM rearrangement. No product arising from involvement of the exocyclic double in an analogous reaction was observed.[29]

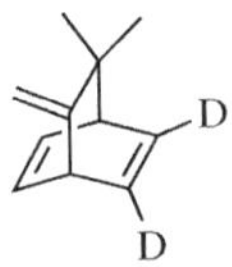

56

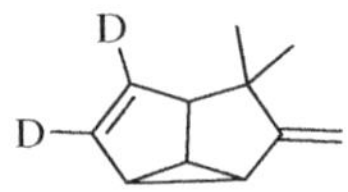

57

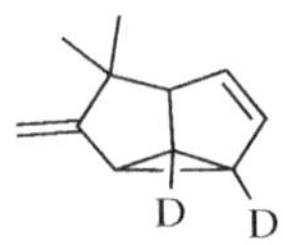

58

In a similar vein, irradiation of a benzene solution of dienone **59** in the presence of the triplet sensitizer 4-benzoylbiphenyl leads to the formation of a single DPM rearrangement product, namely compound **60** (52%), and two ODPM rearrangement products, the epimeric cyclopropanes **61** and **62** (ca. 17% of each).[30]

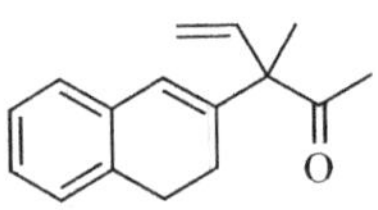

59

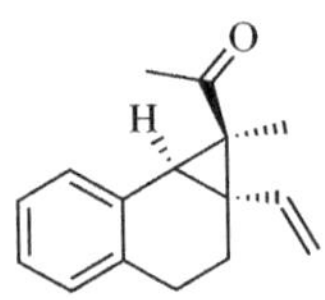

60

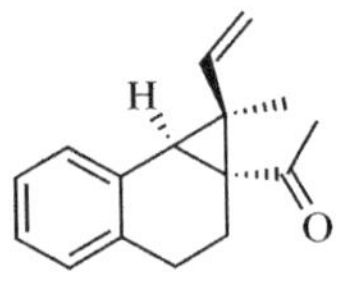

61

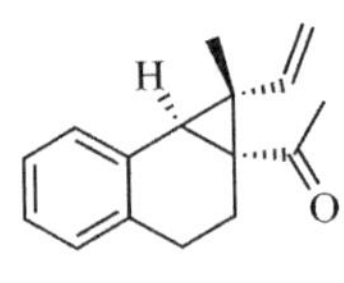

62

A detailed study on the photochemical behavior of 1,3,3-trimethylbicyclo[2.2.2] octa-5,7-dien-2-ones has revealed that under triplet-sensitized conditions, and despite the potential for competition from ODPM-based processes, only DPM-derived photoproducts are observed. In fact, it has been suggested that the rates and regioselectivities of the DPM reactions involved are enhanced by the presence of the carbonyl moiety embedded within the framework.[26] In a related vein, a study of the photochemistry of a series of 1-aza-2-vinyl-1,4-dienes has established that DPM- and ADPM-based rearrangement reactions can compete with one another, the precise outcome being dictated by the nature of the substituents attached to the substrate (by virtue of their influence on the relative stabilities of the 1,4-bridged biradical intermediates, e.g., **4**, that would be involved in the competing pathways).[18b]

Finally, any discussion of processes that compete with the DPM and related reactions must acknowledge that, under certain conditions, the DPM rearrangement is reversible. For example, it has been shown that irradiation of disilane **63** at 10 K in an argon matrix gave cyclopropane **64**, while photolysis of the latter in solution at room temperature regenerates its precursor.[31] The conversion **64** → **63** can also be effected thermally.

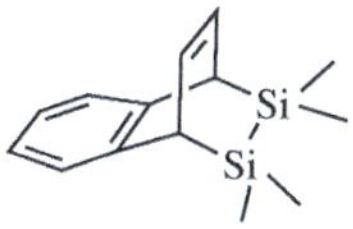

63

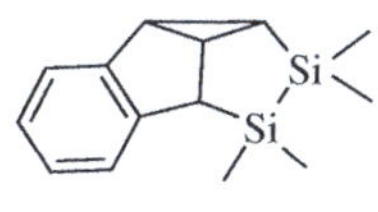

64

The preceding paragraphs have described a number of processes that compete with the DPM, ODPM, and ADPM rearrangement reactions. Nevertheless, it should be recognized that at least the DPM and ODPM rearrangements are particularly effective and can serve as highly reliable transformations for deployment in even quite extended synthetic sequences (as is shown in Section 9.4).

9.3 STRUCTURAL REQUIREMENTS OF SUBSTRATES AND MATTERS OF REGIO- AND STEREO-CHEMISTRY

9.3.1 Basic Structural Requirements

As the name implies, and as noted earlier, the DPM rearrangement reaction requires a substrate incorporating two olefinic residues (π-systems) linked to a common sp^3-hybridized carbon or, in other words, a 1,4-diene. Benzenoid units can replace one or both of these olefinic moieties, and the two units of unsaturation (be they olefinic or benzenoid in nature) can be linked in an open-chain (acyclic) arrangement or embedded within a cyclic or even a polycyclic framework. Naturally, any such frameworks must allow for appropriate orbital overlap of the reacting centers associated with the DPM rearrangement and the incorporation of a cyclopropane ring within the product. So, just as organic chemists view 1,3-dienes (at least in the *s*-cis conformation) as potential 4π-addends for Diels–Alder reactions and 1,5-dienes as substrates for [3,3]-sigmatropic processes, 1,4-dienes should be viewed as potential participants in DPM rearrangement reactions (Figure 9.1).

Given the need for photoactivation, an often encountered additional requirement for substrates participating in DPM reactions is the presence of a "strong" chromophore as provided, for example, by the presence of a phenyl group on at least one of the double bonds. In acyclic 1,4-dienes, the central sp^3-hybridized carbon normally needs to be tetrasubstituted; otherwise, competing 1,2-hydrogen shifts can occur. This requirement does not apply, however, to those cyclic substrates where isomerization would lead to an anti-Bredt olefin (as would be the case for the conversion of barrelene to semibullvalene, **6** → **7**, shown in Scheme 9.2).

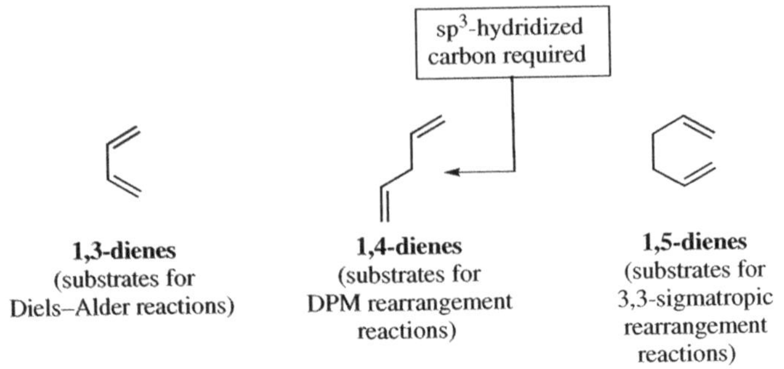

FIGURE 9.1 The signature reactions of 1,3-, 1,4-, and 1,5-dienes

Just as 1,4-dienes are the "signature" substrates for the DPM rearrangement reaction, β,γ-unsaturated aldehydes or ketones are the starting materials for the corresponding ODPM-based processes. Acyclic and cyclic variants can engage in the ODPM reaction (although, obviously, aldehyde-containing substrates are precluded in the cyclic case). A limitation is that α,α-disubstituted systems must be used where isomerization to the α,β-unsaturated congener would otherwise be possible. Significantly, there is normally no need to incorporate any additional chromophoric residues onto the β,γ-unsaturated aldehyde or ketone framework in order to effect an ODPM rearrangement (see, e.g., conversion **21** → **22** shown in Scheme 9.4), rendering it a generally more synthetically versatile process when compared with the equivalent DPM-based transformation.

The ADPM rearrangement reaction has two variants, involving either 1-aza-1,4-dienes or 2-aza-1,4-dienes as substrates. In the first, oximes or hydrazones derived from the corresponding β,γ-unsaturated aldehydes or ketones can be used as substrates. On the other hand, Schiff bases obtained from the relevant allylic amines are most commonly used in the second ADPM variant.[18] Both acyclic and cyclic substrates have been shown to engage in the ADPM rearrangement, and when open-chain 1-aza-1,4-dienes are involved, the process can be a very general one. Less is known about the substrate requirements for the successful engagement of 2-aza-1,4-dienes in ADPM reactions although the involvement of both acyclic and semicyclic systems (Scheme 9.10)[18c] have been reported. Almost invariably, the

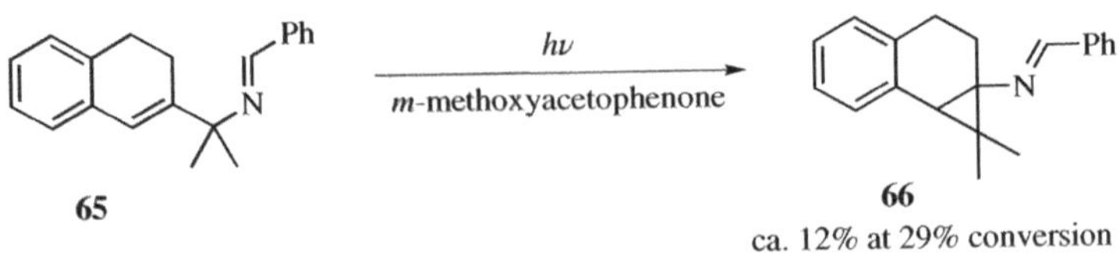

Scheme 9.10 The ADPM rearrangement reaction of a semi-cyclic 2-aza-1,4-diene.

substrates are "decorated" with one or more chromophoric residues (often phenyl groups).

9.3.2 Matters of Regioselectivity

As suggested earlier, the DPM rearrangement reaction can display high levels of regioselectivity with outcomes being dictated by both the nature of the substrate and the mode of photoactivation (Scheme 9.8). Some useful generalizations can be made. In broad terms, when unsymmetrical 1,4-dienes are involved as substrates, then the mode of opening of the pivotal 1,4-biradical (corresponding to structure **4** in Scheme 9.1) is dictated by which of the two alternative product 1,3-diradicals is more stable. To put matters another way, in such instances, the preferred product is normally the one that has the less delocalizing group(s) on the residual double bond. The outcome of the DPM rearrangement of Pratt diene **12** (Scheme 9.11) serves to highlight this guideline.[11] Thus, the 1,4-diradical **67** obtained from photoactivation of **12** undergoes a-bond cleavage to give 1,3-diradical **68** rather than the analogous b-bond cleavage process to give isomer **69** and so delivering the DPM rearrangement product **14** rather than congener **15**.

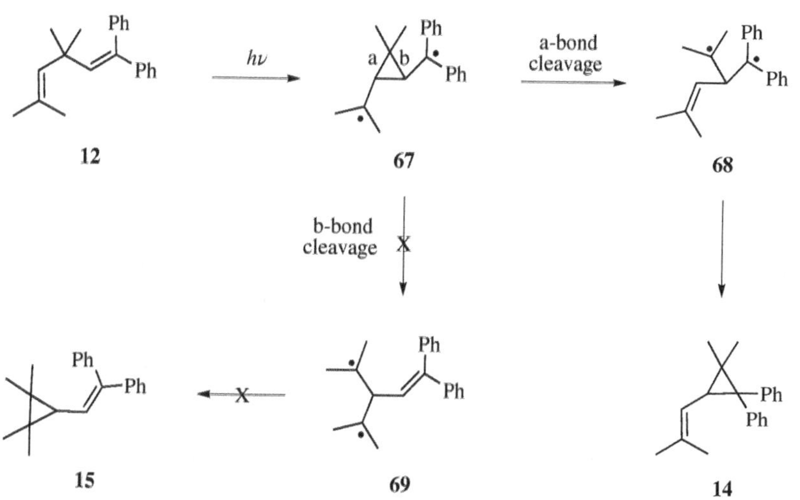

Scheme 9.11 The regioselective DPM rearrangement reaction of an unsymmetrically substituted 1,4-diene.

In those instances where there are electron-donating and/or electron-accepting groups associated with the π-systems of the substrate, there is a preference for the donor groups to be attached to the residual double bond. For example, direct irradiation of the cyano-bearing substrate **70** delivers predominantly the cyanocyclopropane **71** (36% of a mixture of cis- and trans-isomers), while analogous treatment of enol ether **72** affords compound **73** (36% of a mixture of (*E*)- and (*Z*)-isomers) as the major product.[32]

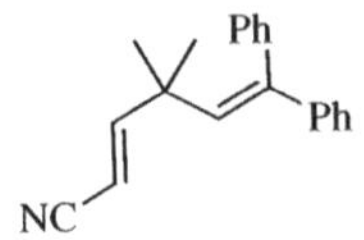

70

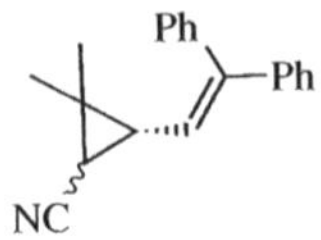

71

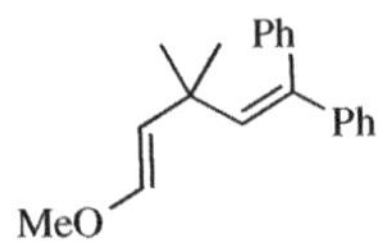

72

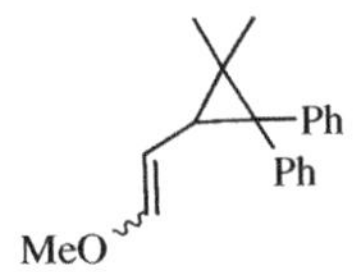

73

A further issue of regiocontrol emerges when three distinct π-systems are attached to a common sp^3-hybridized carbon. Such a situation is encountered not only in substrate **56** but also in cases like the cyano-substituted benzobarrelene **74** which upon direct irradiation rearranges, via intermediate **75**, to give semibullvalene **76** (8%) as the only observable product of a DPM rearrangement reaction.[33] Compound **75** is the more stable of the two possible 1,3-diradicals since the aromatic character of the benzene ring is preserved in this instance.

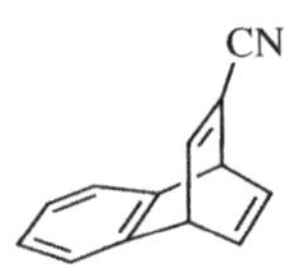

74

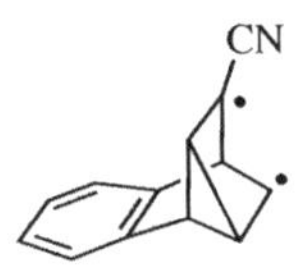

75

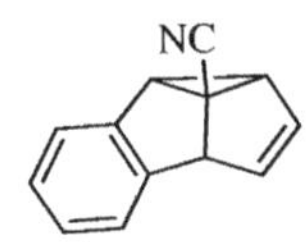

76

Similarly, when benzene solutions of the benzannulated norbornadienes **77** and **78** are subject to irradiation in the presence of acetophenone, the predominant products are the tetracyclic compounds **79** (ca. 52%) and **80** (ca. 90%), respectively.[34] Such outcomes, when considered in conjunction with those detailed in the preceding few paragraphs, lead to the general conclusion that "electron donors avoid positioning themselves in conjugation with the carbinyl centers of the cyclopropyldicarbinyl diradicals" (cf. **4**), while "electron-withdrawing groups lead to stabilization when so situated."[2c]

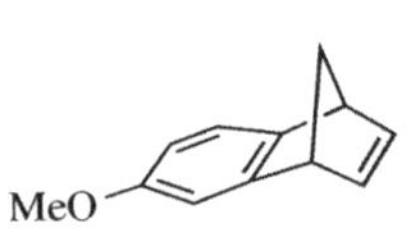

77

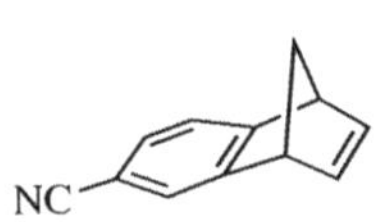

78

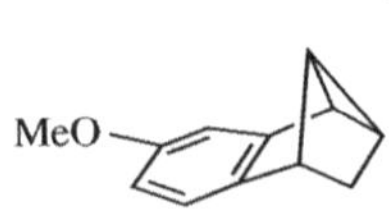

79

80

In contrast to the situation just described with respect to the parent DPM rearrangement reactions, the issues of regioselectivity that might apply in the case of the oxa-variant have been studied less extensively. Nevertheless, some interesting observations have been made. For example, the triplet-sensitized photolysis of the unsymmetrical bicyclo[2.2.2]octenedione **81** (1:2 mixture of epimers at C5) leads, in 70% yield, to the isomeric cyclopropannulated diquinane **82**, a precursor to the natural product (−)-coriolin, with only minor amounts (12%) of the undesired regioisomer **83** being obtained.[35]

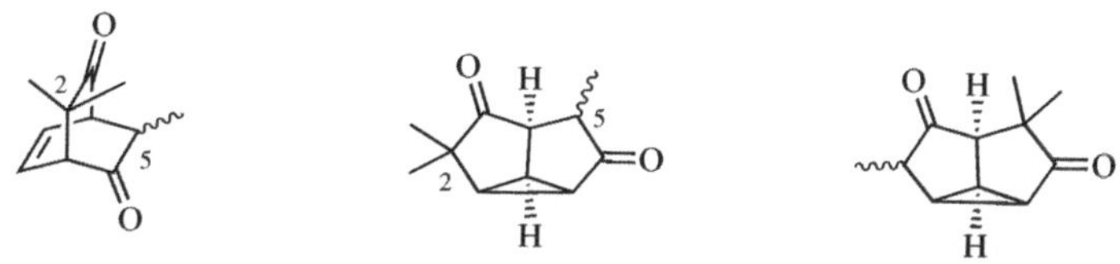

81 **82** **83**

Competing ODPM and vinylogous ODPM rearrangement reactions have been observed. For instance, when compound **84** (Scheme 9.12) is subjected to acetone-sensitized irradiation, a mixture of photoproducts **85**, **86**, and **87** is obtained, the first of these arising from a conventional ODPM rearrangement and the second and third from the vinylogous form of this process.[36]

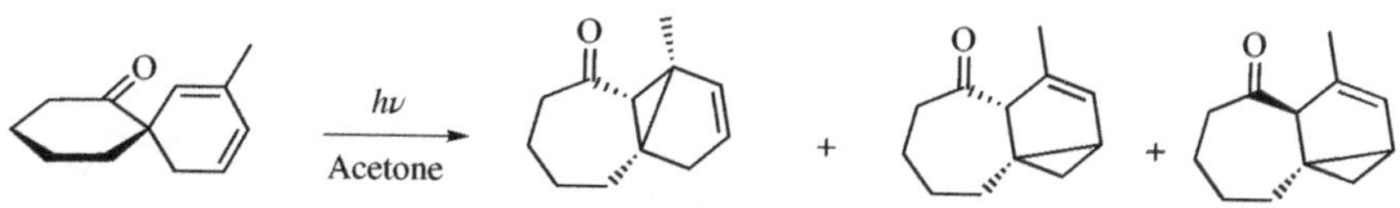

84 **85** (59%) **86** (16%) **87** (4%)

Scheme 9.12 Competing ODPM and vinylogous ODPM rearrangement reactions.

Matters of regioselectivity that might apply in ADPM rearrangement reactions seem to remain essentially unexplored, a reflection of the less mature nature of research in this area.

9.3.3 Matters of Stereoselectivity

The stereoselectivities of DPM rearrangement reactions involving acyclic substrates have been examined and clear outcomes observed. Thus, direct photolysis of the (*E*)-1,4-diene **88** affords the trans-configured cyclopropane **89** as the major product, while equivalent treatment of the (*Z*)-isomer **90** delivers the cis-configured cyclopropane **91**.[37] It has been suggested that the involvement of an excited singlet Möbius orbital array accounts for the observed selectivities.

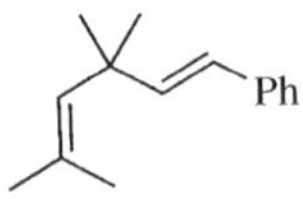

88

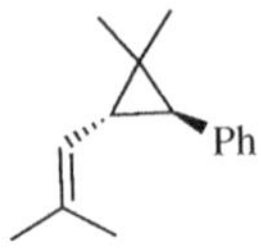

89

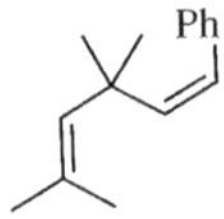

90

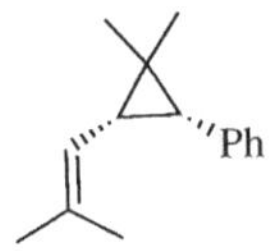

91

Meanwhile, and for the same reasons, inversion of configuration at the methane carbon of acyclic substrates has been observed during the course of the DPM rearrangement reaction.[2e]

Some interesting and potentially synthetically useful diastereoselectivities have been observed in the ODPM rearrangement of 2,4-cyclohexadienones incorporating chiral auxiliaries. For example, direct irradiation of compound **92** in the presence of NaY zeolite gave a mixture of photoproducts **93** and **94** in a ca. 4:1 ratio (predominant diastereoisomer undefined).[38] The sensitized rearrangement of aldehyde **95** also appears to be diastereoselective in that only the endo-configured photoproduct **96** is obtained, albeit in just 25% yield because of the intervention of other isomerization processes.[39] In related systems, mixtures of exo- and endo-products are typically observed.[40]

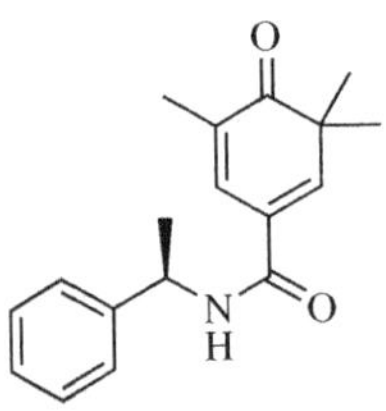

92

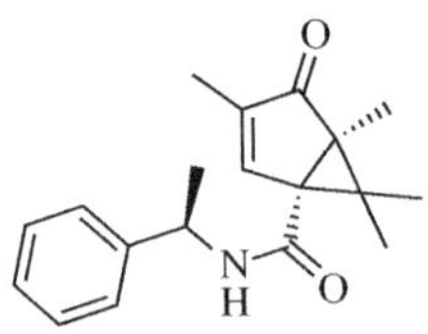

93

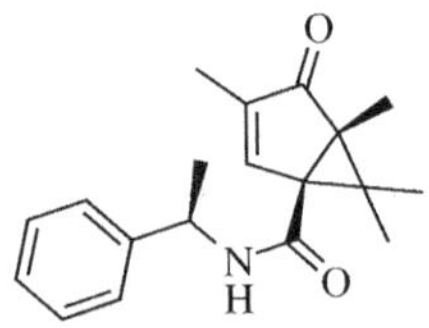

94

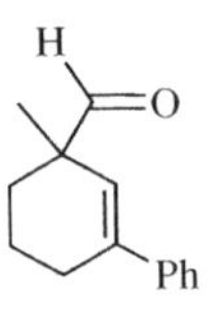

95

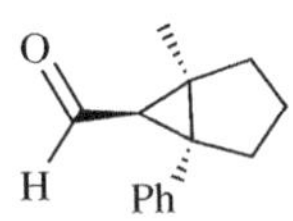

96

In cyclic systems, the constraints of the frameworks bearing the reacting π-systems tend to limit the stereochemical outcomes of the reaction in a synthetically beneficial manner. The preservation of enantiomeric purity associated with the ODPM rearrangement of the (+)- and (−)-forms of compound **21**[16] into the corresponding enantiomers of cyclopropa-fused diquinane **22** (Scheme 9.4) is particularly significant because of the extensive application of this basic process in the synthesis of a range of polyquinane-containing natural products (see Section 9.4).

Photoinduced epimerization reactions are sometimes encountered during the course of the ODPM rearrangement. Thus, it has been observed,[41] as part of a study directed toward the total synthesis of (+)-hirsutic acid and (−)-complicatic acid, that triplet-sensitized irradiation of compound **97** affords the anticipated primary photolysis product **98**, but this is accompanied by quantities of epimer **99**. Sustained irradiation leads to increasing quantities of the thermodynamically favored isomer **99**, presumably via either photoenolization of precursor **98** or through photoinduced C—C bond homolysis within the same compound followed by diradical recombination (to generate **99**).

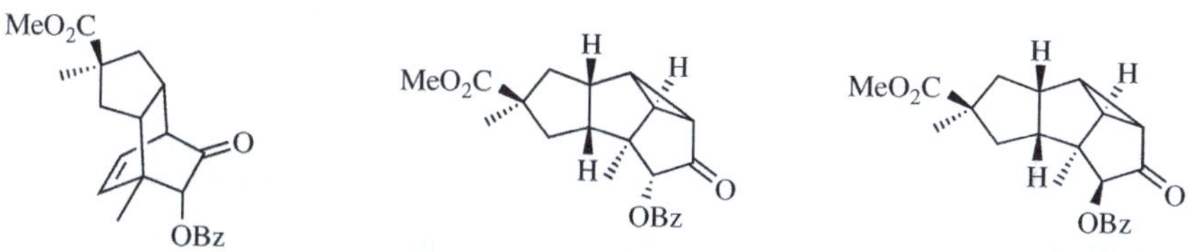

97 **98** **99**

In the rather less developed field of ADPM rearrangement chemistry, there has been limited opportunity to define the stereochemical features of such processes. One salient observation was made during the course of the synthesis of cyclopropanecarboxylic acids present in pyrethroids.[42] In particular, it was noted that acetophenone-sensitized irradiation of the 1-aza-1,4,6-triene **100** for 30 min afforded a 67% yield of a 3:1 mixture of the cis- and trans-configured cyclopropanes **101** (with the cis-isomer predominating). The origins of this selectivity have not been established. Interestingly, no products arising from the operation of a vinylogous ADPM pathway from substrate **100** seem to have been observed.

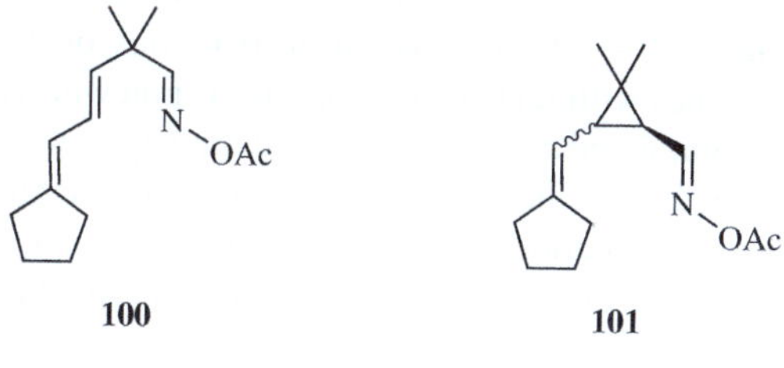

100 **101**

9.4 SYNTHETIC ROUTES TO SUBSTRATES AND APPLICATIONS IN SYNTHESIS

A range of methods is available for the synthesis of the acyclic 1,4-dienes required as substrates for the DPM rearrangement reaction,[43] and some of these are summarized in Figure 9.2. The olefination of β,γ-unsaturated aldehydes and ketones represents one of the key pathways for producing such substrates, with the carbonyl-containing

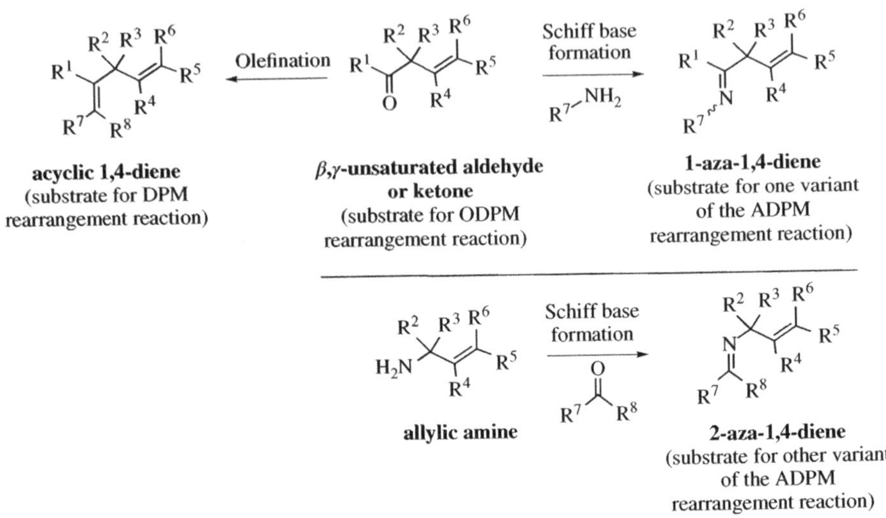

FIGURE 9.2 Overview of certain synthetic routes to substrates for the DPM, ODPM, and ADPM rearrangement reactions

precursors themselves being available via a large variety of routes.[44] Of course, the β,γ-unsaturated aldehydes and ketones are substrates for the ODPM rearrangement reaction, while their Schiff base-type condensation with relevant nitrogen-centered nucleophiles can provide the 1-aza-1,4-dienes required for one variant of the ADPM process. The analogous condensation of allylic amines with ketones and aldehydes provides a pathway to the isomeric 2-aza-1,4-dienes, the substrates for the second variant of the ADPM reaction.

Methods for the synthesis of cyclic substrates employed in the DPM and related rearrangements are many and varied. Nevertheless, given the dominant position of the bicyclo[2.2.2]octanes and, to a lesser extent, the bicyclo[2.2.1]heptanes as substrates, their preparation via Diels–Alder cycloaddition reactions of cyclohexa-1,3-dienes and cyclopenta-1,3-dienes with relevant dienophiles is noteworthy. In particular, barrelenes and their benzannulated analogues are generated by such means. For example, compound **74** is formed[33] by the reaction of benzyne (generated by diazotization of anthranilic acid) with bromobenzene, followed by the treatment of the resulting alkenyl bromide **102** with cuprous cyanide in refluxing DMF (Scheme 9.13).

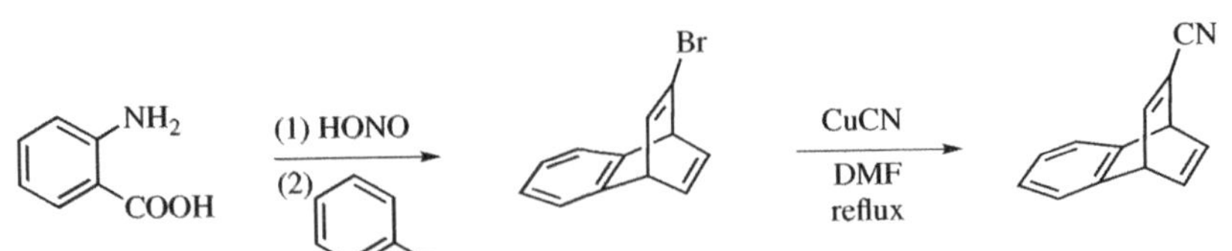

102 **74**

Scheme 9.13 A synthetic route to benzobarrelenes **102** and **74**.

As another illustration of this approach, the Diels–Alder reaction between cyclohexa-1,3-diene (**103**) and acrylonitrile (**104**; Scheme 9.14) affords the expected adduct that upon treatment with PCl_5 in pyridine delivers a diastereoisomeric mixture of the α-chloronitriles **105** and **106**.[45] This mixture can be hydrolyzed with aqueous KOH in water/DMSO to give compound **21**, an archetypal substrate for ODPM rearrangement reactions that affords isomer **22** on triplet-sensitized irradiation (Scheme 9.4). Ketones such as **21** can also be olefinated, for example, under Knoevenagel conditions with malononitrile, to give compounds such **107** (69%) that are themselves substrates for DPM rearrangement reactions. In fact, compound **107** affords cyclopropane **108** (63%) on direct irradiation in pentane.[46] Interestingly, the latter compound can be converted back into precursor **107** on heating or exposure to certain catalysts. Since the conversion of **108** → **107** is exothermic by 28.8 kcal/mol, as determined by DSC techniques, the interconversion of valence isomers **107** and **108** could be considered a model for a light energy conversion and storage system.

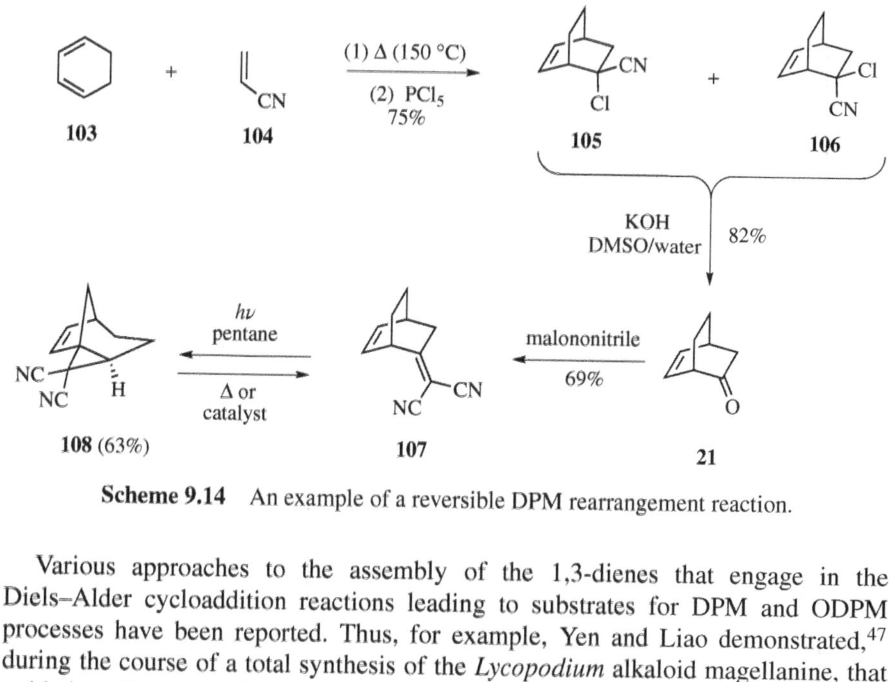

Scheme 9.14 An example of a reversible DPM rearrangement reaction.

Various approaches to the assembly of the 1,3-dienes that engage in the Diels–Alder cycloaddition reactions leading to substrates for DPM and ODPM processes have been reported. Thus, for example, Yen and Liao demonstrated,[47] during the course of a total synthesis of the *Lycopodium* alkaloid magellanine, that oxidation of acetovanillone (**109**) (Scheme 9.15) with diacetoxyiodobenzene (DAIB) in the presence of methanol afforded the *o*-benzoquinone monoketal **110**. The latter compound engaged in an *in situ* Diels–Alder reaction with added cyclopentadiene (**111**) and the resulting adduct **112** proved to be an excellent substrate for the ODPM rearrangement reaction. Thus, photolysis of **112** as a solution in acetone afforded the pivotal tetracyclic diketone **113** in 92% yield.

In an important variant of this oxidative dearomatization approach to the generation of cyclohexa-1,3-dienes, Singh *et al.*[48] subjected the α-hydroxymethylated

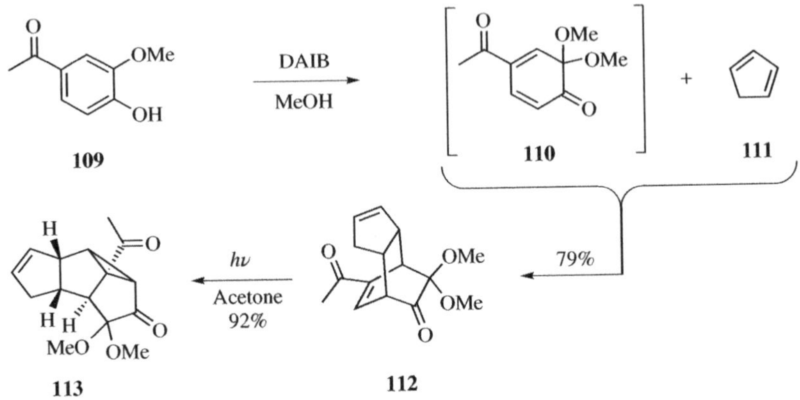

Scheme 9.15 Synthesis and ODPM rearrangement reaction of a cyclopentannulated bicyclo[2.2.2]octenone leading to a cyclopropannulated linear triquinane.

phenol **114** (Scheme 9.16) to $NaIO_4$-mediated Alder–Becker-type oxidation and then trapped the resulting spiroepoxycyclohexa-2,4-dione **115** with cyclopentadiene (**111**) to give adduct **116** in 50% yield. Various conventional manipulations then led to congener **117** that engaged in a triplet-sensitized ODPM rearrangement reaction and so providing the bis-cyclopropannulated triquinane **118** in 76% yield.

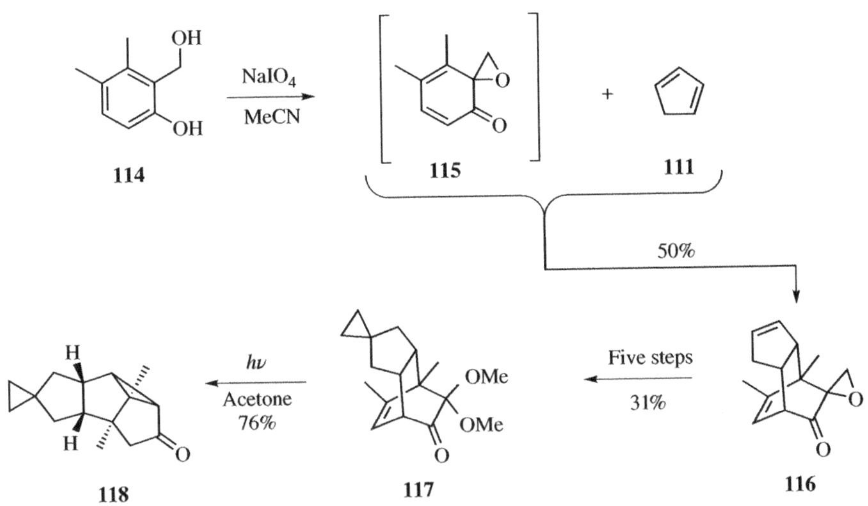

Scheme 9.16 Use of an Alder–Becker reaction for the synthesis of a substrate that participates in an ODPM rearrangement reaction.

Intramolecular Diels–Alder (IMDA) cycloaddition reactions of cyclohexa-1, 3-dienes have also provided a useful means of constructing substrates for the ODPM rearrangement reaction. Thus, the enantiomerically pure *cis*-1,2-dihydrocatechol **119** (which is obtained through the whole-cell biotransformation of iodobenzene) is converted (Scheme 9.17) over two steps,[49] into the triene **120** that upon heating affords, inter alia, the cyclopentannulated bicyclo[2.2.2]octenone **121** (39%). In five steps and 47% overall yield, **121** can be transformed into the β,γ-unsaturated ketone **122**. Finally, irradiation of **122** (as a solution in acetone) effected a ODPM rearrangement and so affording the angular triquinane **123** in 54% yield.

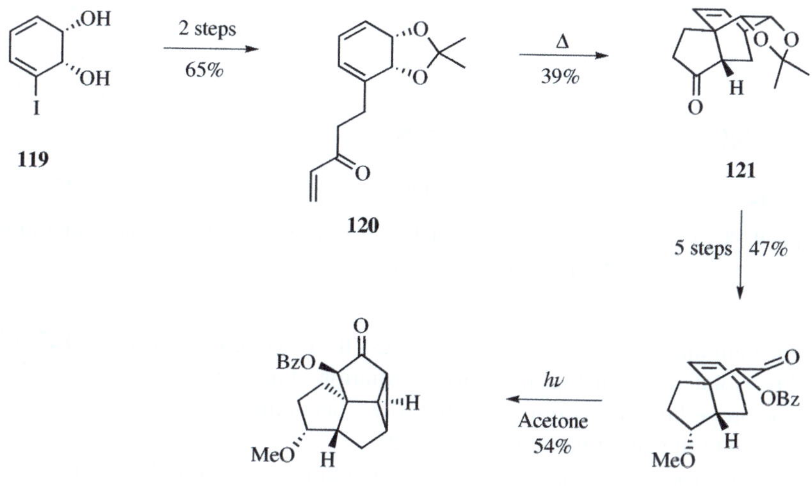

Scheme 9.17 Use of an intramolecular Diels–Alder reaction for the synthesis of a substrate that participates in an ODPM rearrangement reaction.

In reaction sequences closely related to those shown in Schemes 9.15 and 9.16, dienes such as **119** and the corresponding acetonides also participate in facially selective intermolecular Diels–Alder reactions with dienophiles such as 2-cyclopenten-1-one, and the resulting adducts have been elaborated, using straightforward techniques, to cyclopentannulated bicyclo[2.2.2]octenones that participate in ODPM rearrangement reactions. The resulting linear triquinanes have been exploited in total syntheses of a range of hirsutene-type sesquiterpenoid natural products including (+)-hirsutene, (−)-complicatic acid, (−)-phellodonic acid, (−)-connatusin A, (+)-connatusin B, and (−)-hypnophilin.[17b]

An elegant synthesis of the sesquiterpenoid natural product (±)-modhephene[50] started (Scheme 9.18) with the conversion, via a Grignard methylation/dehydration sequence, of the readily available bicyclic enone **124** into the cyclopentannulated cyclohexa-1,3-diene **125** (86% yield of an admixture with double-bond isomers). Diene **125** engaged in a Diels–Alder cycloaddition reaction with α-chloroacrylonitrile (a ketene equivalent) to afford a mixture of adduct **126** (obtained as a mixture of diastereoisomers) and the expected regioisomer (43% combined yield). Hydrolysis of this mixture of regioisomers then afforded the corresponding mixture of

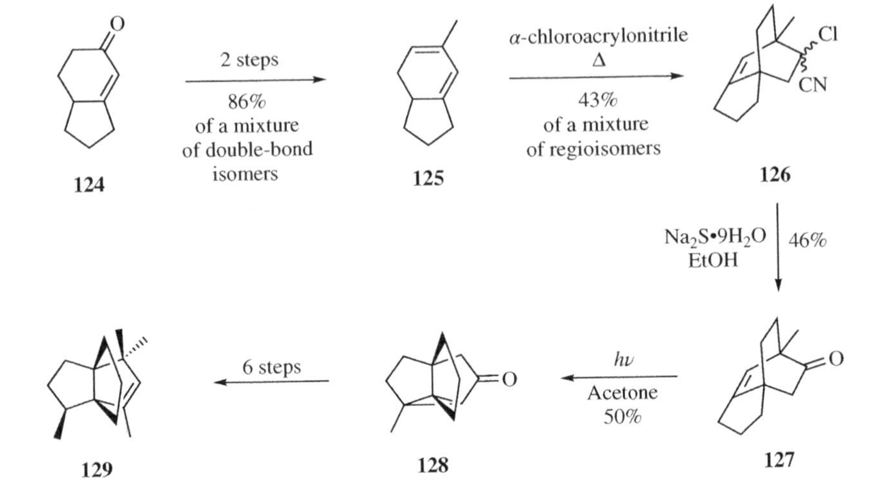

Scheme 9.18 Use of an intermolecular Diels–Alder reaction for the synthesis of a substrate that participates in an ODPM rearrangement reaction.

cyclopentannulated bicyclo[2.2.2]octenones from which compound **127** (46%) could be isolated chromatographically. Irradiation of an acetone solution of **127** then gave the propellane-like triquinane **128** (50%) that could be converted, over a further six steps, into the target natural product **129**.

Novel heterocyclic compounds have been obtained using ODPM rearrangements. En route to one substrate for such a rearrangement, potassium *tert*-butoxide-promoted Dieckmann cyclization (Scheme 9.19) of the readily available piperidine diester **130** afforded, after treatment of the initially formed product with HCl then, separately, with trimethyl orthoformate (and after transesterification to reduce volatility), 1-azabicyclo[2.2.2]octane carboxylic acid ester **131**[51] as a mixture of epimers. Compound **131** was readily dehydrogenated, using selenium-based chemistry, to afford the 1-azabicyclo[2.2.2]octenone **132**.[52] This last compound served as an excellent substrate for the ODPM rearrangement reaction. So when an acetone solution of **132** "spiked" with acetophenone was subjected to irradiation at 350 nm, the tricyclic compound **133** was obtained. This molecule is a potentially useful precursor to a range of pyrrolizidine alkaloids.

Cross-conjugated cyclohexadienones participate in the DPM rearrangement reaction, and such substrates can be prepared by a variety of techniques. The prototypical compound 4,4-diphenylcyclohexa-2,5-dien-1-one (**134**), which delivers cyclopropane **135** on direct photolysis, was generated using Robinson annulation techniques from methyl vinyl ketone and diphenylacetaldehyde with the resulting 4,4-diphenylcyclohexenone being oxidized to the final dienone using selenium dioxide.[53] Quinone monoketals, a special form of cyclohexadienone, are also often effective substrates for the DPM rearrangement reaction and can usually be obtained by a variety of techniques including by ketalization of the analogous quinone, Swenton oxidation/hydrolysis, or diol exchange from the dimethyl ketals.[54]

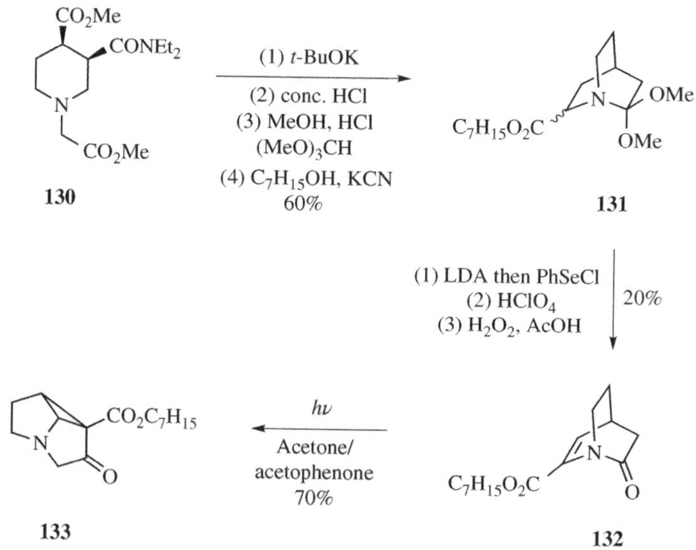

Scheme 9.19 Synthesis of a 1-azabicyclo[2.2.2]octenone that participates in an ODPM rearrangement reaction.

Similarly, 4*H*-thiopyran-1,1-dioxides such as **136**, which can be obtained by oxidation of the corresponding cyclic sulfides, undergo triplet-sensitized rearrangement to give the corresponding and epimerically related cyclopropanes **137** and **138**.[55]

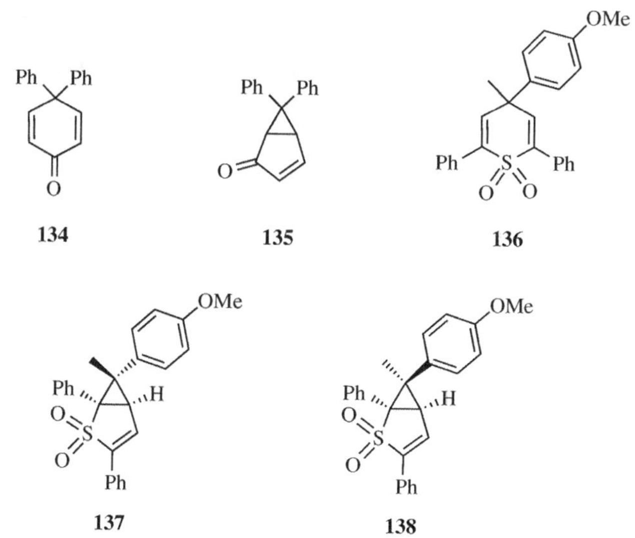

As described earlier, quinone monoketals (e.g., **139**) are not only good substrates for the DPM reaction, but they also participate in acid-catalyzed [5+2]cycloaddition reactions with styrene derivatives (e.g., *β*-methylstyrene) to give bicyclo[3.2.1]octenediones (e.g., **140**)[56] that can themselves participate in ODPM processes to give multifunctional, donor–acceptor cyclopropanes (e.g., **141**).[57] The last type of compound can participate in further reactions generating several additional and novel scaffolds that may be useful in drug discovery settings. Meanwhile, iminium ethers that are readily derived from [5+2] cycloadducts of type **140** participate in ADPM rearrangements.

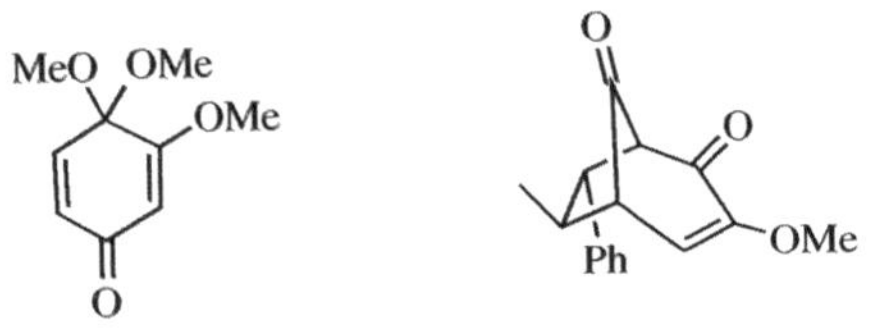

139 **140** **141**

The synthetic utility of DPM-type rearrangements now extends well beyond their application in the synthesis of natural products and biologically active systems. So, for example, dibenzobarrelenes annulated with a pyrrolinium unit (e.g., **142**) and including a benzophenone counterion **143** undergo a DPM rearrangement reaction in the solid state, and the product dibenzosemibullvalene **144** serves as a phase-transfer catalyst in alkylation reactions.[58]

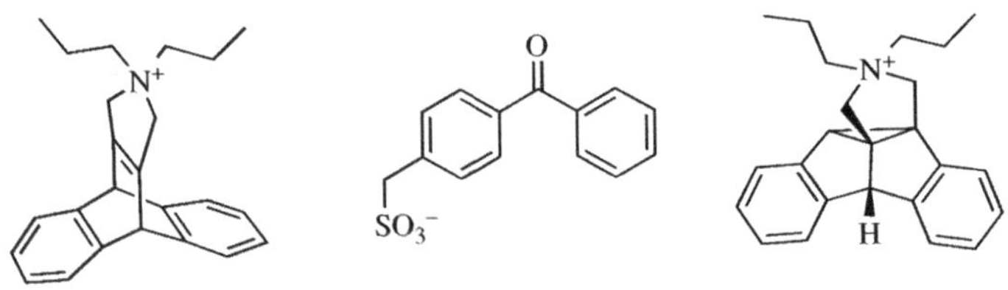

142 **143** *rac*-**144**

In related processes, the presence of chiral acid residues has been shown to exert some influence, albeit modest so far, in the formation of enantiomerically enriched dibenzosemibullvalenes.[59] In the materials science arena, it has recently been shown that a dibenzobarrelene-based polymer incorporating adjacent ester groups and 1,4-related alkyne units (which are capped by benzophenone residues in a polymerization process) engages in a smooth and regioselective DPM rearrangement reaction to afford a dibenzosemibullvalene-based polymer admixed with its precursor. This polymer has properties suitable for the development of new liquid crystals.[60]

Finally, a methylene-linked and readily accessible cyclopentannulated bis-bicyclo[2.2.2]octenone has been shown to undergo a two-fold ODPM rearrangement reaction and thereby affording bis-linear triquinanes possessing a "bird-shaped" architecture.[61]

9.5 OUTLOOK

The DPM rearrangement reaction and its oxa- and aza-variants frequently provide a highly effective and reliable means for generating cyclopropane-containing products that would be difficult if not completely unrealistic to prepare in other ways. While aficionados appreciate the utility of these processes, the rather unusual nature of this chemistry and the need to employ photochemical protocols mean that DPM rearrangements have not truly entered the mainstream of methods routinely considered by researchers engaged in chemical synthesis. This situation persists despite the exquisitely detailed mechanistic understanding of these processes provided by Zimmerman and others, as well as these workers' articulation of the apparatus and operational conditions required to "harness" the reactions.[1c]

Admittedly, some substrate frameworks seem more suited to participation in DPM, ODPM, and ADPM rearrangement reactions than others, but new substrate systems continue to be revealed (e.g., consider conversion **140** → **141**). Clearly, the enthusiasm of the chemical community for these transformations will benefit greatly from the discovery of further truly innovative permutations, most particularly those providing access to new molecular scaffolds from readily available starting materials. The emergence of tandem variants as well as enantioselective ones will also help drive the field forward. In addition, the identification of new cyclopropane ring-cleavage and/or ring-expansion protocols that exploit the three-membered ring automatically generated during the course of the DPM-type reactions could prove "game-changing." Furthermore, due attention must be given to Armesto's relatively recent observations[22,23] that the choice of sensitizer is pivotal to the success of the triplet-mediated version of the DPM reaction.

Finally, and as suggested in the opening parts of Section 9.3, there is a more pedagogical aspect to the further successful development of the title reactions. Organic chemists almost automatically think about Diels–Alder reactions when 1,3-dienes are mentioned and Cope-type rearrangements when 1,5-dienes are discussed. We also need to train ourselves (and our students) to think about DPM rearrangement reactions when 1,4-dienes are mentioned. Further, just as conjugate addition reactions come to mind when α,β-unsaturated aldehydes and ketones are considered, β,γ-unsaturated aldehydes and ketones should automatically prompt consideration of the ODPM rearrangement reaction.

REFERENCES

1. (a) Zimmerman, H. E.; Grunewald, G. L. *J. Am. Chem. Soc.* **1966**, *88*, 183–184. (b) Zimmerman, H. E.; Binkley, R. W.; Givens, R. S.; Sherwin, M. A. *J. Am. Chem. Soc.* **1967**, *89*, 3932–3933.
2. For some useful reviews and summaries of the di-π-methane rearrangement reaction see: (a) Hixson, S. S.; Mariano, P. S.; Zimmerman, H. E. *Chem. Rev.* **1973**, *73*, 531–551. (b) Zimmerman, H. E. The di-π-methane (Zimmerman) rearrangement. In *Rearrangements in the Ground and Excited States*; de Mayo, P., Ed.; Academic Press: New York, **1980**; Vol. 3, pp. 131–164. (c) Zimmerman, H. E.; Armesto, D. *Chem. Rev.* **1996**, *96*, 3065–3112. (d) Wang, Z. Di-π-methane rearrangement. In *Comprehensive Organic Name Reactions and Reagents*; Wiley Online Library, **2010**, pp. 910–915 (DOI: 10.1002/9780470638859.conrr195). (e) Zimmerman, H. E. Di-π-methane rearrangement. In *CRC Handbook of Organic Photochemistry and Photobiology*, 3rd ed.; Griesbeck, A.; Oelgemöller, M.; Ghetti, F., Eds.; CRC Press: Boca Raton, FL, **2012**; pp. 511–525.

3. See, for example: (a) Janssen, R. A. J.; Hummelen, J. C.; Wudl, F. *J. Am. Chem. Soc.* **1995**, *117*, 544–545. (b) Ref. 2d.
4. Matute, R. A.; Houk, K. N. *Angew. Chem. Int. Ed.* **2012**, *51*, 13097–13100.
5. The α, β, γ terminology is used here to facilitate comparisons with the closely related oxa-di-π-methane rearrangement reaction as defined in the IUPAC Gold Book, see: Verhoeven, J. W. *Pure Appl. Chem.* **1996**, *68*, 2223–2286.
6. Zimmerman, H. E.; Novak, T. *J. Org. Chem.* **2003**, *68*, 5056–5066 and references cited therein.
7. (a) Zimmerman, H. E.; Givens, R. S.; Pagni, R. M. *J. Am. Chem. Soc.* **1968**, *90*, 6096–6108. (b) Altundas, R.; Dastan, A.; Ünaldi, N. S.; Güven, K.; Uzun, O.; Balci, M. *Eur. J. Org. Chem.* **2002**, 526–533 and references cited therein. For extensions to pyrazino-, quinoxalino-, and benzoquinoxalinobarrelenes, see: (c) Chen, A.-C.; Lin, S.-Y.; Villarante, N. R.; Chuang, G. J.; Wu, T.-C.; Liao, C.-C. *Tetrahedron* **2008**, *64*, 8907–8921. (d) Hsieh, H.-P.; Chen, A.-C.; Villarante, N. R.; Chuang, G. J.; Liao, C.-C. *RSC Adv.* **2013**, *3*, 1165–1178.
8. For useful points-of-entry into the literature on the photoisomerization of dibenzobarrelenes and related compounds see: (a) Ramaiah, D.; Sajimon, M. C.; Joseph, J.; George, M. V. *Chem. Soc. Rev.* **2005**, *34*, 48–57. (b) Ohkura, K.; Akizawa, H.; Kudo, M.; Ishihara, T.; Oshima, N.; Seki, K.-I. *Heterocycles* **2012**, *84*, 1057–1065.
9. Triptycene does not engage in the equivalent processes: (a) Turro, N. J.; Tobin, M.; Friedman, L.; Hamilton, J. B. *J. Am. Chem. Soc.* **1969**, *91*, 516. (b) Iwamura, H. *Chem. Lett.* **1974**, *3*, 1205–1208.
10. Zimmerman, H. E.; Mariano, P. S. *J. Am. Chem. Soc.* **1969**, *91*, 1718–1727.
11. Zimmerman, H. E.; Pratt, A. C. *J. Am. Chem. Soc.* **1970**, *92*, 6259–6267.
12. Tenney, L. P.; Boykin, Jr., D. W.; Lutz, R. E. *J. Am. Chem. Soc.* **1966**, *88*, 1835–1836.
13. (a) Givens, R. S.; Oettle, W. F. *J. Chem. Soc. D* **1969**, 1164–1165. (b) Dauben, W. G.; Kellogg, M. S.; Seeman, J. I.; Spitzer, W. A. *J. Am. Chem. Soc.* **1970**, *92*, 1786–1787.
14. For useful reviews and summaries of the oxa-di-π-methane rearrangement reaction see: (a) Ref. 2a. (b) Dauben, W. G.; Lodder, G.; Ipaktschi, J. *Top. Curr. Chem.* **1975**, *54*, 73–114. (c) Houk, K. N. *Chem. Rev.* **1976**, *76*, 1–74. (d) Liao, C.-C.; Peddinti, R. K. Photochemistry of heteroarene-fused barrelenes. In *CRC Handbook of Organic Photochemistry and Photobiology*, 2nd ed.; Horspool, W. M.; Lenci, F., Eds.; CRC Press: New York, **2004**, pp. 32/1–32/17. (e) Ref. 2c. (f) Singh, V. Photochemical rearrangements in β,γ-unsaturated enones: the oxa-di-π-methane rearrangement. In *CRC Handbook of Organic Photochemistry and Photobiology*, 2nd ed.; Horspool, W. M.; Lenci, F., Eds.; CRC Press: Boca Raton, FL, **2004**, pp. 78/1–78/34. (g) Rao, V. J.; Griesbeck, A. G. Oxa-di-π-methane rearrangements. In *Synthetic Organic Photochemistry*; Griesbeck, A. G., Mattay, J., Eds.; Marcel Dekker: New York, **2005**; pp. 189–210. (h) Rao, V. J.; Srinivas, K. Oxa-di-π-methane rearrangement of β,γ-unsaturated ketones. In *CRC Handbook of Organic Photochemistry and Photobiology*, 3rd ed.; Griesbeck, A.; Oelgemöller, M.; Ghetti, F., Eds.; CRC Press: Boca Raton, FL, **2012**, pp. 527–548.
15. (a) Demuth, M.; Schaffner, K. *Angew. Chem. Int. Ed. Engl.* **1982**, *21*, 820–836. (b) Demuth, M. *Pure Appl. Chem.* **1986**, *58*, 1233–1238. (c) Singh, V.; Thomas, B. *Tetrahedron* **1998**, *54*, 3647–3692. (d) Singh, V. *Acc. Chem. Res.* **1999**, *32*, 324–333.
16. Demuth, M.; Raghavan, P. R.; Carter, C.; Nakano, K.; Schaffner, K. *Helv. Chim. Acta* **1980**, *63*, 2434–2439.

17. (a) Mehta, G.; Srikrishna, A. *Chem. Rev.* **1997**, *97*, 671–720. (b) Bon, D. J.-Y. D.; Lee, B.; Banwell, M. G.; Cade, I. A. *Chim. Oggi.* **2012**, *30* (No. 5 Chiral Technologies Supplement), 22–27.

18. (a) Armesto, D.; Ramos, A.; Mayoral, E. P. *Tetrahedron Lett.* **1994**, *35*, 3785–3788. (b) Armesto, D.; Ortiz, M. J.; Ramos, A.; Horspool, W. M.; Mayoral, E. P. *J. Org. Chem.* **1994**, *59*, 8115–8124. (c) Armesto, D.; Caballero, O.; Ortiz, M. J.; Agarrabeitia, A. R.; Martin-Fontecha, M.; Torres, M. R. *J. Org. Chem.* **2003**, *68*, 6661–6671. (d) Armesto, D.; Ramos, A.; Ortiz, M. J.; Horspool, W. M.; Mancheño, M. J.; Caballero, O.; Mayoral, E. P. *J. Chem. Soc., Perkin Trans. 1* **1997**, 1535–1542.

19. Ünaldi, S.; Aumann, R.; Fröhlich, R. *Chem. Eur. J.* **2003**, *9*, 3300–3309.

20. Zimmerman, H. E.; Kutateladze, A. G.; Maekawa, Y.; Mangette, J. E. *J. Am. Chem. Soc.* **1994**, *116*, 9795–9796.

21. Zimmerman, H. E.; Factor, R. E. *Tetrahedron* **1981**, *37*(Suppl. 1), 125–141.

22. Armesto, D.; Ortiz, M. J.; Agarrabeitia, A. R.; Martín-Fontecha, M.; El-Boulifi, N.; Duran-Sampedro, G.; Enma, D. *Org. Lett.* **2009**, *11*, 4148–4151.

23. Armesto, D.; Ortiz, M. J.; Agarrabeitia, A. R.; El-Boulifi, N. *Angew. Chem. Int. Ed.* **2005**, *44*, 7739–7741.

24. Zimmerman, H. E.; Givens, R. S.; Pagni, R. M. *J. Am. Chem. Soc.* **1968**, *90*, 4191–4193.

25. Demuth, M.; Hinsken, W. *Helv. Chim. Acta* **1988**, *71*, 569–576.

26. Chang, S.-Y.; Huang, S.-L.; Villarante, N. R.; Liao, C.-C. *Eur. J. Org. Chem.* **2006**, 4648–4657.

27. Armesto, D.; Ortiz, M. J.; Agarrabeitia, A. R.; Martín-Fontecha, M. *Org. Lett.* **2005**, *7*, 2687–2690.

28. Armesto, D.; Ortiz, M. J.; Agarrabeitia, A. R.; Martín-Fontecha, M. *Org. Lett.* **2004**, *6*, 2261–2264.

29. Luibrand, R. T.; Fujinari, E. M. *J. Org. Chem.* **1980**, *45*, 958–960.

30. van der Veen, R. H.; Kruk, C.; Cerfontain, H. *J. Am. Chem. Soc.* **1983**, *105*, 4367–4374.

31. (a) Rich, J. D.; Drahnak, T. J.; West, R.; Michl, J. *J. Organomet. Chem.* **1981**, *212*, C1–C3. For other examples of reversible DPM rearrangement reactions see: (b) Demuth, M.; Burger, U.; Mueller, H. W.; Schaffner, K. *J. Am. Chem. Soc.* **1979**, *101*, 6763–6765. (c) Lamberts, J. J. M.; Laarhoven, W. H. *J. Org. Chem.* **1983**, *48*, 2202–2206.

32. Zimmerman, H. E.; Klun, R. T. *Tetrahedron* **1978**, *34*, 1775–1803.

33. Bender, C. O.; Brooks, D. W.; Cheng, W.; Dolman, D.; O'Shea, S. F.; Shugarman, S. S. *Can. J. Chem.* **1978**, *56*, 3027–3037.

34. Paquette, L. A.; Cottrell, D. M.; Snow, R. A. *J. Am. Chem. Soc.* **1977**, *99*, 3723–3733.

35. Demuth, M.; Ritterskamp, P.; Weigt, E.; Schaffner, K. *J. Am. Chem. Soc.* **1986**, *108*, 4149–4154.

36. (a) Zizuashvili, J.; Abramson, S.; Shmueli, U.; Fuchs, B. *J. Chem. Soc., Chem. Commun.* **1982**, 1375–1378. (b) Oren, J.; Vardi, M.; Viskin, R.; Abramson, S.; Fuchs, B. *Helv. Chim. Acta* **1993**, *76*, 1182–1193.

37. (a) Zimmerman, H. E.; Baeckstrom, P.; Johnson, T.; Kurtz, D. W. *J. Am. Chem. Soc.* **1974**, *96*, 1459–1465. Detailed stereochemical analyses of the aromatic DPM rearrangement reaction have been carried out: (b) Scholl, B.; Hansen, H.-J. *Helv. Chim. Acta* **1986**, *69*, 1936–1958.

38. Uppili, S.; Ramamurthy, V. *Org. Lett.* **2002**, *4*, 87–90.

39. Armesto, D.; Ortiz, M. J.; Romano, S.; Agarrabeitia, A. R.; Gallego, M. G.; Ramos, A. *J. Org. Chem.* **1996**, *61*, 1459–1466.
40. Reimann, B.; Sadler, D. E.; Schaffner, K. *J. Am. Chem. Soc.* **1986**, *108*, 5527–5530.
41. Banwell, M. G.; Austin, K. A. B.; Willis, A. C. *Tetrahedron* **2007**, *63*, 6388–6403.
42. (a) Armesto, D.; Gallego, M. G.; Horspool, W. M.; Agarrabeitia, A. R. *Tetrahedron* **1995**, *51*, 9223–9240. (b) Armesto, D.; Ortiz, M. J.; Agarrabeitia, A. R.; El-Boulifi, N. *Tetrahedron* **2010**, *66*, 8690–8697.
43. See, for example: Oishi, S.; Hatano, K.; Tsubouchi, A.; Takeda, T. *Chem. Commun.* **2011**, *47*, 11639–11640 and references cited therein.
44. See, for example: (a) Mander, L. N.; Turner, J. V. *J. Org. Chem.* **1973**, *38*, 2915–2916. (b) Trofimov, B. A.; Schmidt, E. Y.; Zorina, N. V.; Ivanova, E. V.; Ushakov, I. A. *J. Org. Chem.* **2012**, *77*, 6880–6886 and references cited therein.
45. Kozikowski, A. P.; Schmiesing, R. J. *J. Org. Chem.* **1983**, *48*, 1000–1007.
46. Kumagai, T.; Murakami, K.; Hotta, H.; Mukai, T. *Tetrahedron Lett.* **1982**, *23*, 4705–4708.
47. Yen, C.-F.; Liao, C.-C. *Angew. Chem. Int. Ed.* **2002**, *41*, 4090–4093.
48. Singh, V.; Chandra, G.; Mobin, S. M. *Synlett* **2008**, 2267–2270.
49. Austin, K. A. B.; Elsworth, J. D.; Banwell, M. G.; Willis, A. C. *Org. Biomol. Chem.* **2010**, *8*, 751–754.
50. Mehta, G.; Subrahmanyam, D. *J. Chem. Soc., Chem. Commun.* **1985**, 768–769.
51. Snow, R. J.; Street, L. J. *Tetrahedron Lett.* **1989**, *30*, 5795–5798.
52. McClure, C. K.; Kiessling, A. J.; Link, J. S. *Org. Lett.* **2003**, *5*, 3811–3813.
53. Zimmerman, H. E.; Schuster, D. I. *J. Am. Chem. Soc.* **1962**, *84*, 4527–4540.
54. Pirrung, M. C.; Nunn, D. S. *Tetrahedron* **1996**, *52*, 5707–5738.
55. (a) Rezanejadebardajee, G.; Jafarpour, F.; Pirelahi, H. *J. Heterocycl. Chem.* **2006**, *43*, 167–170. (b) Jafarpour, F.; Ramezani, F.; Pirelahi, H. *Heteroat. Chem.* **2008**, *19*, 557–561. The triplet-sensitized DPM rearrangements of 4-phenyl-substituted 4*H*-pyrans have been studied, including by theoretical methods: (c) Mori, Y.; Takano, K. *J. Photochem. Photobiol. A* **2011**, *219*, 278–284.
56. Goodell, J. R.; McMullen, J. P.; Zaborenko, N.; Maloney, J. R.; Ho, C.-X.; Jensen, K. F.; Porco, Jr., J. A.; Beeler, A. B. *J. Org. Chem.* **2009**, *74*, 6169–6180.
57. Goodell, J. R.; Poole, J. L.; Beeler, A. B.; Aubé, J.; Porco, Jr., J. A. *J. Org. Chem.* **2011**, *76*, 9792–9800.
58. Ihmels, H.; Luo, J. *Beilstein J. Org. Chem.* **2011**, *7*, 119–126.
59. Ihmels, H.; Luo, J. *Beilstein J. Org. Chem.* **2011**, *7*, 278–289.
60. Cox, J. R.; Simpson, J. H.; Swager, T. M. *J. Am. Chem. Soc.* **2013**, *135*, 640–643.
61. Singh, D.; Deota, P. T. *Tetrahedron Lett.* **2012**, *53*, 6527–6530.

PART III

1,3-TRANSPOSITIONS

10

PAYNE REARRANGEMENT

Takashi Yamazaki

Division of Applied Chemistry, Institute of Engineering, Tokyo University of Agriculture and Technology, Tokyo, Japan

10.1 BACKGROUND ON THE PAYNE REARRANGEMENT

10.1.1 Payne Rearrangement

In 1962, Payne reported the results of an investigation on a unique base-mediated transposition of the 2,3-epoxypropan-1-ol structure **1** to the corresponding 1,2-epoxypropan-3-ol counterpart **2**,[1,2] a transformation presaged by two other independent reports in 1935[3] and 1957.[4] Moreover, Payne showed that the two isomers, **1** and **2**, were in an equilibrating relationship (Scheme 10.1) where aqueous alkaline conditions are usually employed for smooth progress of this conversion (0.5 M NaOH in H_2O at room temperature for 1 h in the original conditions[1]). In the aprotic solvent THF, treatment of **3** with NaH generated an equimolar amount of hydrogen gas, indicating quantitative formation of the corresponding sodium alkoxide, but **3** was recovered in 70% yield along with the detection of only 2% of the rearranged **4** even after stirring the mixture for 1.5 h at 10 °C.[1] And there is another instance where no reaction was observed with the 2,3-epoxy-1-alkanol structure under similar aprotic reaction conditions.[5] This low reactivity in nonaqueous solutions was explained as a consequence of tight ion pair formation of the resultant sodium alkoxide.

Payne further investigated the scope and limitations of the epoxy alcohol rearrangement under the standard conditions depicted in Scheme 10.1. A major quantity of **3** was found to be readily transformed to the isomeric **4**, attaining an equilibrium ratio of **3**:**4** = 8:92. The position of equilibrium for this structural transposition was further probed by using representative isomeric substrates; thus, only a

Molecular Rearrangements in Organic Synthesis, First Edition. Edited by Christian M. Rojas.

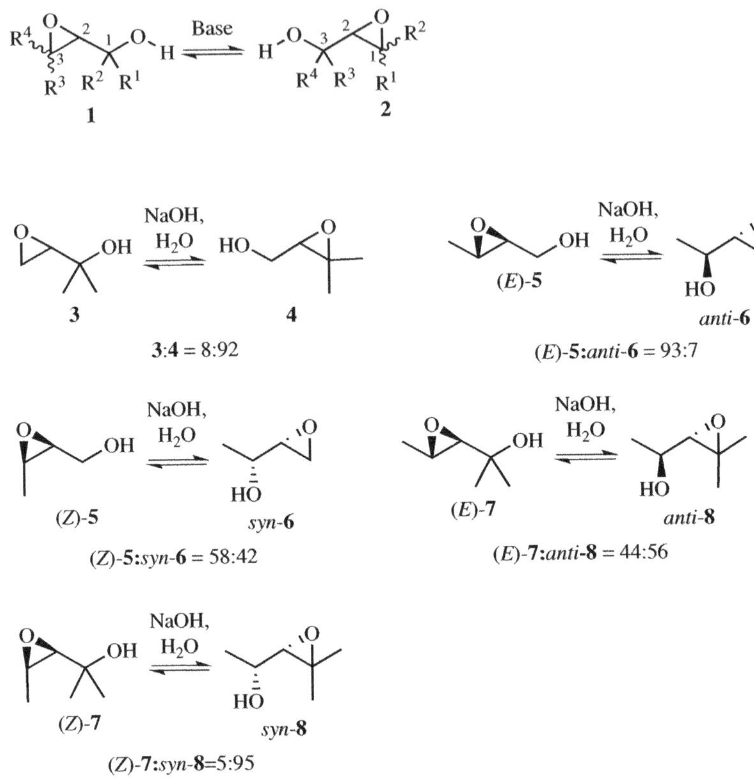

Scheme 10.1 Payne rearrangement of simple 2,3-epoxy alcohols.

small amount of *anti*-**6** was obtained from (*E*)-**5**, while the diastereomeric *syn*-**6** was formed from (Z)-**5** as a mixture of *syn*-**6**:(Z)-**5** = 42:58 after 1 h at ambient temperature. The same situation was noticed for (*E*)-**7**, which afforded *anti*-**8** in a ratio of (*E*)-7:*anti*-**8** = 44:56, and (Z)-**7**, which was preferentially converted to *syn*-**8**, leaving only 5% of (Z)-**7** in the mixture.

These results suggested that product selectivity is mainly controlled by the difference in stability of the resultant alkoxides under the conditions employed and by the steric environment of the epoxides. On the basis of the well-accepted pK_a values for the representative primary, secondary, and tertiary alcohols EtOH, *i*-PrOH, and *tert*-BuOH of ca. 16, 18, and 19, respectively, the stability of alkoxides is expected to be in the order of primary > secondary > tertiary. Moreover, because increasing the number of substituents on the three-membered-ring framework of epoxides is considered to increase thermodynamic stability,[1] the Payne rearrangement favors formation of the less substituted alcohols and more substituted epoxides. When both factors work in the same direction, as in the rearrangement from *anti*-**6** (*sec*-OH, monosubstituted epoxide) to (*E*)-**5** (*primary*-OH, disubstituted epoxide) or from (Z)-**7** (*tert*-OH, disubstituted epoxide) to *syn*-**8** (*sec*-OH, trisubstituted epoxide), a high preference of greater than 90% is usually observed. However, even with both elements working in concert, stereochemistry plays an important role. For example, employment of stereoisomers *syn*-**6** or (*E*)-**7** gave only approximately equal amounts of the rearranged isomers, (*Z*)-**5** or *anti*-**8**, respectively. This is due to the additional steric repulsion present between groups that are cis on an epoxide ring. Comparing the reactions of (*E*)-**5** and (*Z*)-**5**, the latter is higher in energy and correspondingly destabilized in its Payne equilibrium. Comparing the equilibria for (*E*)-**7** and (*Z*)-**7**, meanwhile, when (*E*)-**7** undergoes Payne rearrangement, a new *cis*-epoxide steric interaction results, while for (*Z*)-**7** there is a *cis*-epoxide interaction on both sides of the equilibrium.

It is noteworthy that the Payne rearrangement proceeds in a stereospecific manner, providing clean inversion of stereochemistry at the central 2-position as a result of an intramolecular S_N2-type mechanism. Thus, both configurations at this site are readily available from the same starting material by appropriate control of the rearrangement.

As with other rearrangements, the Payne rearrangement also has its variants. Similar transformations of substrates with nitrogen- or sulfur-based functionalities at the carbon atom adjacent to the epoxide are known as aza-[6] or thia-Payne[7,8] rearrangements, respectively, and there exist a number of noteworthy previously reported examples of each. Although study of these systems has been somewhat limited in recent years, they are briefly surveyed in the following sections.

10.1.2 Aza-Payne Rearrangement

Upon treatment with base, 2,3-epoxypropylamines **9** (Scheme 10.2) can be transformed to the isomeric aziridinemethanols **10**. A base stronger than NaOH is usually required for facile conversion, due to the difference in amine pK_a compared to the alcohol group involved in the regular Payne rearrangement. Because the pK_a values for secondary amines like Et_2NH are reported to be about 35 (roughly 20 units greater than the ones for primary alcohols), aza-Payne rearrangements without any special activating functions on nitrogen should proceed in the direction of producing compounds **10** from **9** under the strongly basic conditions typically employed: *n*-BuLi/*tert*-BuOK or *n*-BuLi/Me_3Al.[9] Lewis acidic conditions (TMSOTf,[10] $Ti(OPr^{-i})_4$,[11] or $BF_3{\cdot}OEt_2$[12]) have also been employed to promote the aza-Payne process.

The reverse mode from **10** to **9** is also operative when a good electron-withdrawing group is attached to the nitrogen atom, providing stabilization of the nitrogen anion. Ibuka *et al.* investigated the scope and limitations of this aza-Payne version by using substrates with a variety of substitution patterns.[13] They found that, in sharp contrast to the original epoxy alcohol system, smooth progress of the aza-Payne reaction was possible under not only protic but also aprotic conditions, sometimes furnishing different outcomes even starting from the same substrates (Scheme 10.3). The presence of a tosyl group as an anion-stabilizing protective moiety increased

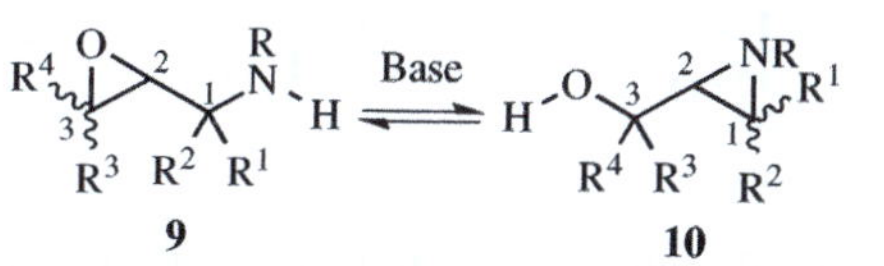

Scheme 10.2 Aza-Payne rearrangement of 2,3-epoxy amines **9**.

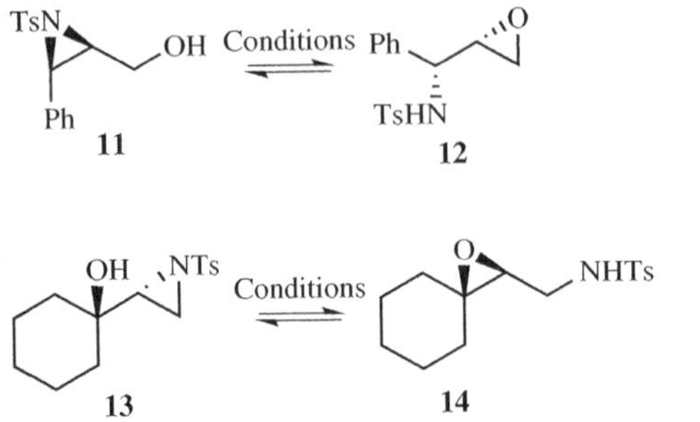

0.28 M NaOH, t-BuOH–H_2O
0 °C, 18 h **11**:**12**=61:39

KH (4 equiv.), THF–HMPA
0 °C, 1.5 h **12** only

0.36 M NaOH, t-BuOH–H_2O
0 °C, 18 h **13**:**14**=2:98

KH (4 equiv.), THF
rt, 18 h **14** only

Scheme 10.3 Aza-Payne rearrangement of representative 2,3-epoxy amines.

the acidity of the proton on the nitrogen atom in **12** to approximately $pK_a = 8$, which should thermodynamically favor the corresponding anion, compared to the conjugate base of the less acidic hydroxy group in **11** ($pK_a = 16$). Thus, conversion of **11** to the corresponding alkoxide should shift the equilibrium to the right because of this sizable pK_a gap. However, **11** was favored in an aqueous NaOH solution, giving, after 18 h stirring at 0 °C, a mixture of **11**:**12** = 61:39. This would be due to prompt quenching of the resultant less stable alkoxide by a water molecule, rendering the stability difference smaller. However, this was not the case in the aprotic THF–HMPA system, exclusively affording **12** due to the presence of the intermediates as anionic forms.

Application of the aza-Payne rearrangement was reported for the construction of highly substituted pyrrolidines starting from alkynylated aziridinemethanols as the substrates.[14] Thus, as described in Scheme 10.4, treatment of the stereospecifically prepared *syn*-**16** in the case of R^2, $R^3 \neq H$ with dimethylsulfoxonium methylide initiated the structural transposition to furnish the intermediary alkynyl epoxide Int-**2** which further cyclized to epoxy pyrrolidine **17** in a one-pot manner. Although the diastereomeric *anti*-**16** experienced the same rearrangement, but the following

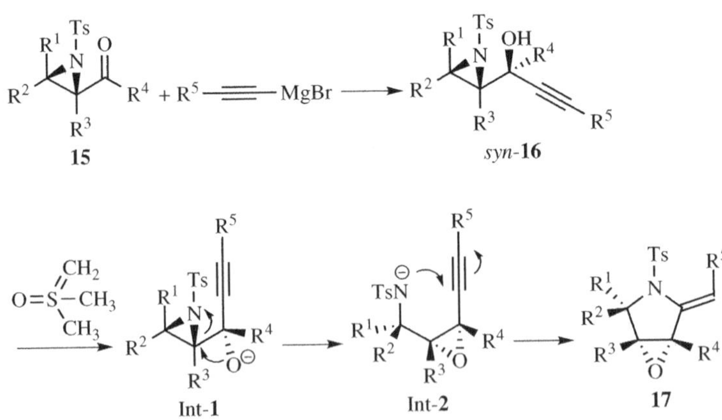

Scheme 10.4 Pyrrolidine formation via aza-Payne rearrangement.

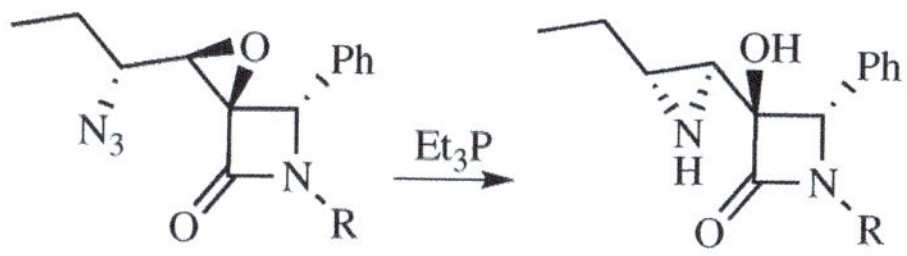

Scheme 10.5 Aza-Payne rearrangement of the β-lactam-based substrate.

cyclization was not possible due to trans relationship between the reacting two units, the N atom and alkynyl moiety.

A recent interesting example in this area is the rearrangement of epoxy azide **18** which, on exposure to triethylphosphine and after liberation of nitrogen gas, underwent the aza-Payne rearrangement (Scheme 10.5).[15] In this case, Et_3P was responsible for successful stabilization of the reactive nitrogen species by formation of an aza-ylide intermediate, enabling the direct synthesis of N—H aziridine **19**.

10.1.3 Thia-Payne Rearrangement

The central framework of the substrates in this version of the Payne rearrangement is analogous to the ones already described, with sulfur in place of the oxygen or nitrogen nucleophiles used in the *O*- or *N*-versions. However, no example is found using free thiols as starting materials. Instead, an *S*-acetyl 2,3-epoxypropanethiol derivative such as **20** is employed as the precursor of the thiolate which, in this case, was generated *in situ* by the deacetylation with NH_3. The resultant anion on the sulfur atom served as the nucleophile for the intramolecular S_N2-type ring opening to furnish **21** in 89% yield (Scheme 10.6a).[16] It is quite interesting to note that quantitative preparation of its epimer, *epi*-**21**, was also possible from the same compound **20** just by treatment with moisture-excluded silica gel. This stereochemical reversal was explained mechanistically as the result of a double inversion process involving acetyl oxygen participation for opening of the oxirane ring, activated by the Lewis acidic silica gel (Scheme 10.6b).

Other thia-Payne-type rearrangements involve activation of 2,3-epoxypropyl thioethers with trimethylaluminum as the Lewis acid (Scheme 10.6c). This activates the oxirane C—O bonds and leads to ready acceptance of intramolecular nucleophilic attack at the 2-position by the proximate sulfur atom.[17] Transposition of a methyl group from Me_3Al to the episulfonium intermediate Int-**3** furnished the product **23** with net retention of stereochemistry at C-2 due to double inversion at this site.

10.2 SYNTHETIC APPLICATIONS OF 2,3-EPOXY ALCOHOLS

10.2.1 Construction of Requisite Starting Materials for Payne Rearrangement

As pointed out previously, the 2,3-epoxypropan-1-ol structure like (*E*)-**24** in Scheme 10.7 is required for Payne rearrangement, and its convenient construction in optically active form can be approached along two routes proceeding from

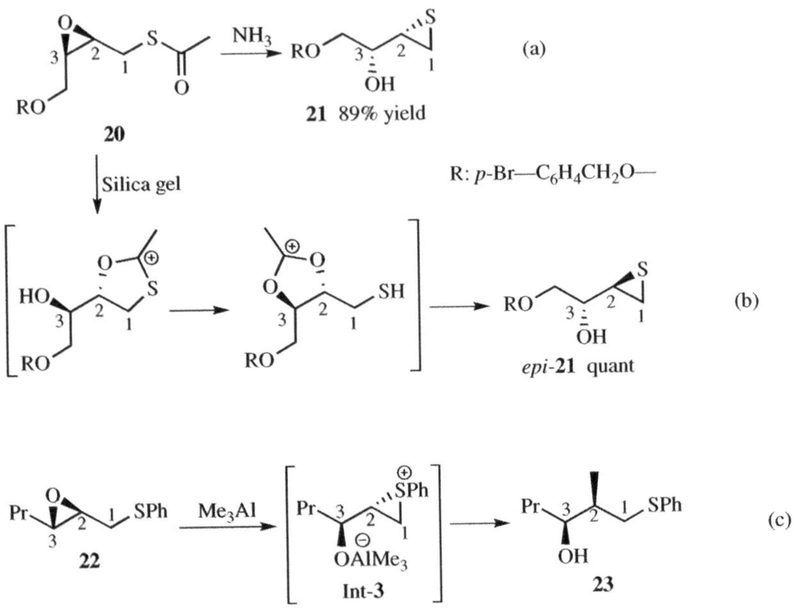

Scheme 10.6 Thia-Payne rearrangement of representative 2,3-epoxy sulfides.

α,β-unsaturated carbonyl compounds (*E*)-**27** as key starting materials. Stereoselective preparation of (*E*)-**27** is, in turn, crucial because double-bond stereoisomers should lead to the formation of (*E*)-**25** diastereomers if the same enantiofacial selection is operative for both isomers in the epoxidation step which is also true for (*E*)-**24**. Compounds of structure (*E*)-**27** are usually prepared by way of Wittig-type condensations of ketone- or ester-derived phosphorus ylides with ketones possessing the appropriate substituents (R^2 and R^3, Scheme 10.7). Another route to α,β-unsaturated carbonyl

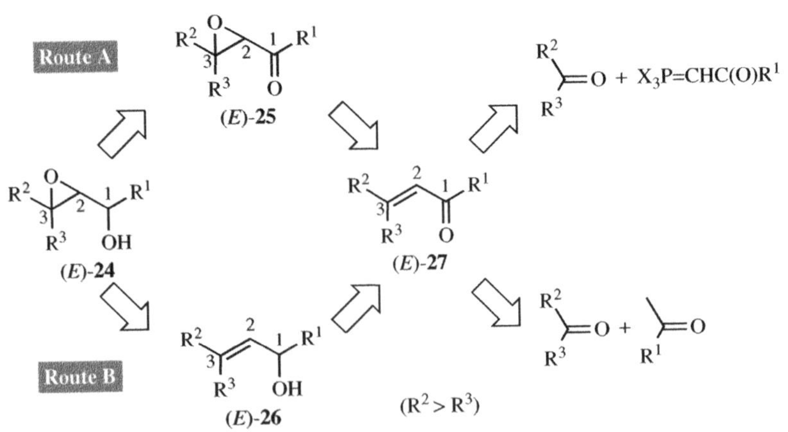

Scheme 10.7 Retrosynthetic schemes for construction of 2,3-epoxy alcohols.

compounds (*E*)-**27** would be the crossed aldol reaction between two carbonyl compounds, followed by (formal) dehydration. Although there are many alternative possibilities, such as Darzens reactions,[18] to directly synthesize (*E*)-**25**, these two pathways would be superior in terms of wide choice of substituents R^1, R^2, and R^3 and possible access to both (*E*)- and (*Z*)-isomers just by changing reaction conditions or reagents.[19]

Starting from (*E*)-**27** thus prepared, one route to access epoxy alcohols (*E*)-**24** is the diastereoselective reduction of the chiral epoxy ketones (*E*)-**25** after enantioselective epoxidation of (*E*)-**27** (Route A). The alternative method requires the opposite combination of the two procedures, namely, enantioselective reduction of (*E*)-**27** to (*E*)-**26**, followed by diastereoselective epoxidation (Route B). In the case of Route A, because attachment of the carbonyl group in (*E*)-**27** renders the C—C double-bond electron deficient, nucleophilic asymmetric epoxidation[20] should be the method of choice, and enantioselective versions are available for this purpose.[21] On the other hand, electrophilic reagents [22] are suitable for oxirane preparation in Route B.

One of the major advantages of Route A is the possibility of following two independent pathways to obtain either (*E*)-*syn*- or (*E*)-*anti*-**24** in a stereodivergent manner. This would be achieved by selection of reducing agents such as DIBAL for the *syn* isomers via a nonchelation control mechanism[23] or $Zn(BH_4)_2$ for the *anti* diastereomers by way of bidentate chelation[24].

On the other hand, racemic allylic alcohols (*E*)-**26** are possible candidates for use in Route B. After treatment of (*E*)-**27** with such usual reductants as $NaBH_4$ or $LiAlH_4$, the resultant (*E*)-**26** would be resolved by the well-known Sharpless epoxidation protocol[25] or via enzymatic methods before[26] or after[27] epoxidation.

Another modification of Route B requires enantioselective reduction of ketones (*E*)-**27**[28] or stereoselective carbon–carbon bond formation at C-1 of (*E*)-**27** ($R^1 = H$) with appropriate organometallic species in the presence of chiral additives,[29] both of which successfully supply the optically active (*E*)-**26**. The resulting chiral allylic alcohols (*E*)-**26** are subjected to hydrogen bond-directed epoxidation with *m*CPBA, leading to the diastereoselective formation of *syn*-epoxy alcohols.[30] In complementary fashion, *anti*-selective epoxidation is possible using the Sharpless protocol.[31]

10.2.2 Product Structures Accessible by Way of Payne Rearrangement

As described previously, 2,3-epoxypropan-1-ols **24** are typical substrates for Payne rearrangement, affording 1,2-epoxypropan-3-ols **30** as products (Scheme 10.8). Clearly, both these substrates as well as the rearrangement products possess high value for construction of 1,2- and 1,3-diols by regioselective epoxide ring opening with a wide variety of nucleophiles. Scheme 10.8 shows a concise map of compounds accessible from a single starting material (*E*)-**27**. Thus, after enantioselective nucleophilic epoxidation,[20] the resultant (*E*)-**25** is stereoselectively transformed to either (*E*)-*syn*- or (*E*)-*anti*-**24** by way of reduction under nonchelation[21] or chelation-controlled[22] conditions, respectively. These epoxy alcohols **24** are

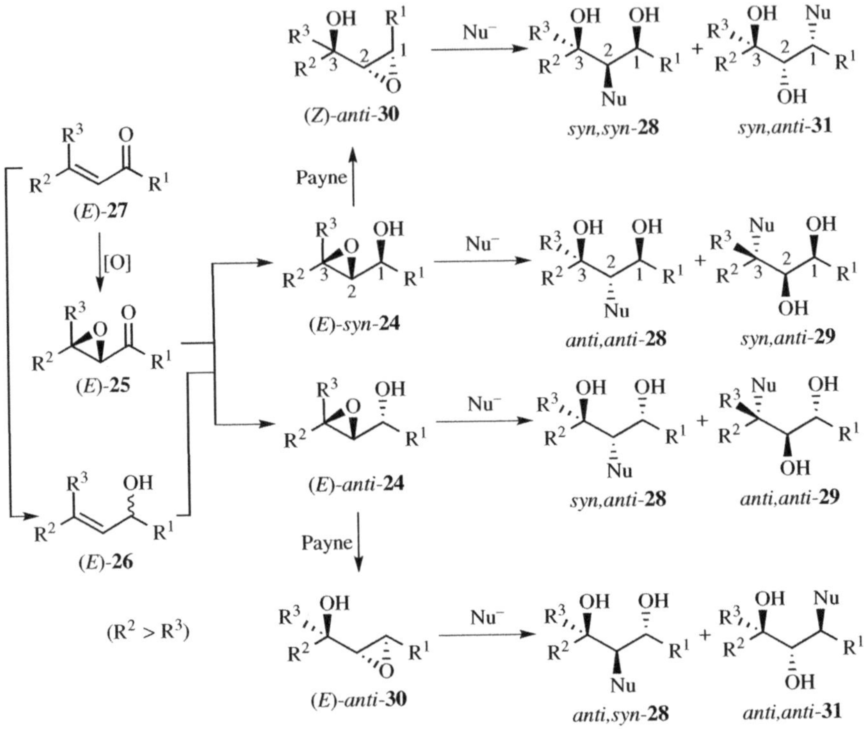

Scheme 10.8 Construction of 1,2- as well as 1,3-diols with a variety of stereochemical relationships.

alternatively synthesized by way of (*E*)-**26** after reduction of (*E*)-**27** or its reaction with organometallics ($R^1 = H$), followed by epoxidation.

At this point, nucleophilic epoxide opening, combined in some cases with the Payne rearrangement, opens routes to a rich array of stereo- and regioisomers. Supposing that regioselective introduction of nucleophiles at C-2 is possible, then (*E*)-*syn*-**24** could afford *anti,anti*-**28**, while (*Z*)-*anti*-**30**, the Payne rearrangement product from (*E*)-*syn*-**24**, leads to the formation of the diastereomeric *syn,syn*-**28**. Moreover, because *syn,anti*- and *anti,syn*-**28** can be constructed from (*E*)-*anti*-**24** in a similar manner, a single substrate like (*E*)-**25** allows us to obtain all possible stereoisomers of **28**. Additionally, half of the possible diastereomers for the isomeric 1,2-diols **29** and **31** would be accessible through site-selective C-3 or C-1 nucleophilic addition, respectively.

Regioselective oxirane ring opening was studied in detail by the Sharpless group[32] after disclosure of the asymmetric epoxidation procedure,[24] revealing that $Ti(i\text{-}OPr)_4$ nicely mediated the transformation, with nucleophilic attack uniformly occurring at the C-3 site. Results from epoxide C—O bond cleavage reactions using 2,3-epoxyhexan-1-ol with a wide range of nucleophiles including amines, alcohols, thiols, and carboxylic acids are shown in Table 10.1. The characteristic feature

TABLE 10.1 Reaction of Epoxy Alcohol 32 with a Variety of Nucleophiles

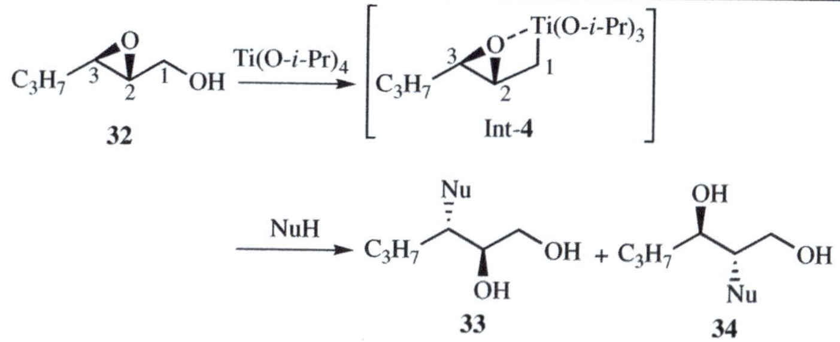

Entry	$Ti(i\text{-}OPr)_4$ (equiv.)	NuH	(equiv.)	Conditions	Isolated yield (%)	**33**:**34**
1	0	Et_2NH	Excess	Reflux, 18 h	4	3.7:1
2	1.5	Et_2NH	Excess	rt, 5 h	90	20:1
3	0	*i*-PrOH	Excess	Reflux, 18 h	0	–
4	1.5	*i*-PrOH	Excess	Reflux, 18 h	88	100:1
5	0	PhSH	5.0	Benzene, rt, 22 h	0	–
6	1.5	PhSH	1.6	Benzene, rt, 5 min	95	6.4:1
7	0	Me_3SiN_3	3.0	Benzene, reflux, 3 h	0	–
8	1.5	Me_3SiN_3	3.0	Benzene, reflux, 3 h	74	14:1
9[a]	0	KCN	2.0	DMSO, rt, 72 h	0	–
10[a]	2.2	KCN	2.0	DMSO, rt, 72 h	91	1.3:1
11	0	NH_4Cl	2.0	DMSO, rt, 18 h	0	–
12	1.5	NH_4Cl	2.0	DMSO, rt, 15 min	84	2.8:1
13	1.2	$PhCO_2H$	1.1	CH_2Cl_2, rt, 15 min	74	100:1

[a] 1.6 equiv. of tetra-*n*-butylammonium iodide (TBAI) was used.

highlighted by these experiments is the significance of the Lewis acid. In the absence of $Ti(i\text{-}OPr)_4$, a very low yield or none of the desired product was obtained, with recovery of unreacted **32** or its hydroxyl-silylated derivative. The role of this Lewis acid is considered to involve intramolecular chelation for effective activation of the epoxide, proceeding through the reactive intermediate Int-**4**. Use of $(MeO)_3B$ instead of $Ti(i\text{-}OPr)_4$ affected the regioselectivity of the reaction, giving a preference for **34**. For example, ratios of **33**:**34** = 16:84 (83% yield), 15:85 (97% yield), and 18:82 (97% yield) were obtained for NaCN, PhSH, and NaN_3 as nucleophiles, respectively.[33] This selectivity for **34** was experimentally as well as theoretically shown to arise by way of an intermediate with B···O chelation, and calculation of a model transition state with azide anion clarified that the ensuing S_N2-type nucleophilic attack at C-2 was favored by 3.5 kcal/mol over the opening at C-3.

10.2.2.1 Reaction of Epoxy Alcohols with Carbon Nucleophiles Scheme 10.9 describes two representative examples of reactions of epoxy alcohols with the cuprate

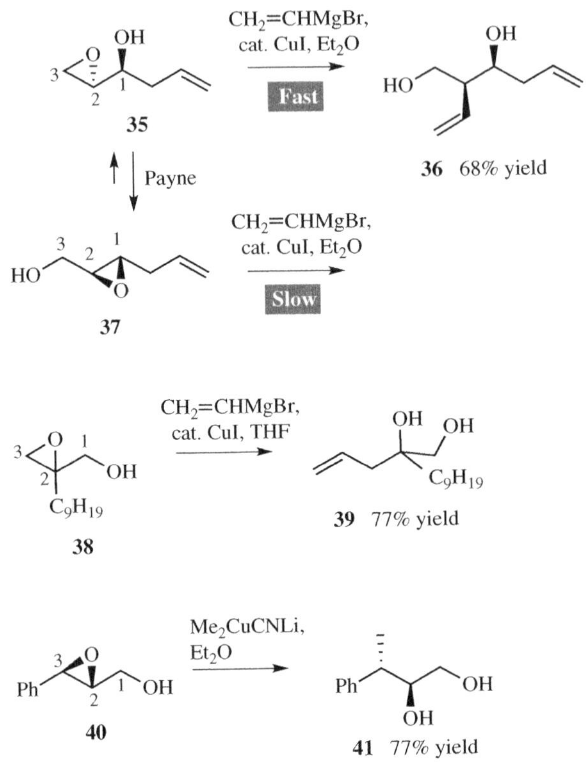

Scheme 10.9 Ring opening of 2,3-epoxy alcohols by organometallic species.

from vinylmagnesium bromide and a catalytic amount (10 mol% relative to the Grignard reagent) of CuI. It is quite interesting to note that, in spite of the very similar reaction conditions employed, the substrate **35** accepted the nucleophile at the 2-position,[34] while ring opening of **38** at the 3-position predominated to yield **39** as a single product. The latter result is dictated by the significant steric hindrance around the reaction site at C-2 when the cuprate attacks in an intramolecular manner after transformation to the corresponding ate complex at the free OH group in **38**.[35] The previous report[36] using cuprates as carbon nucleophiles demonstrated that the attack usually occurred at the sterically least hindered epoxy carbon atoms. So, in spite of the possible energetic preference of **37** to **35** as discussed in Scheme 10.1, the ring opening is considered to be the rate determining step and smooth reaction at the C-2 position of **35** yielded **36** as the sole product. When the substrate contains a phenyl group directly attached to the oxirane as in the case of **40**, reactions proceed exclusively at the benzylic position as a result of the stronger activation at this site.[37]

Another interesting example of ring opening via *C*-nucleophile addition involves the semipinacol rearrangement of the epoxy alcohol **42**.[38] In spite of a 70:30 diastereomeric mixture at C-1, this substrate **42** was smoothly converted into **43** as

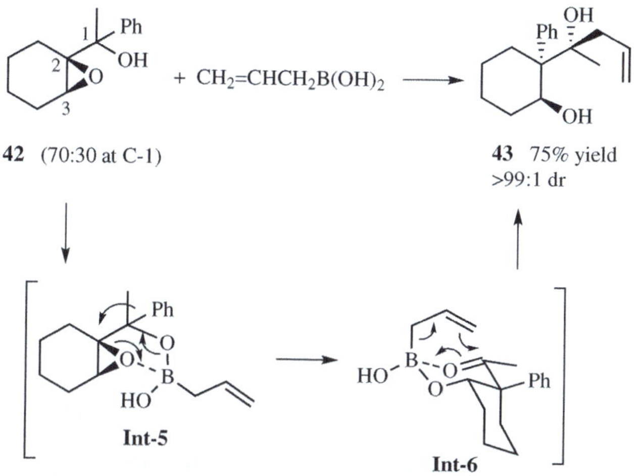

Scheme 10.10 Stereoselective formation of 1,3-diols by way of phenyl migration.

a single stereoisomer in high yield (Scheme 10.10). The mechanism of this process was proposed to be initiated by coordination of the boronic acid to the epoxy oxygen atom, followed by migration of the neighboring phenyl group to the backside of the activated tertiary carbon at the 2-position. This leads to the cleavage of the O—C-2 bond with clean inversion (Int-**5**). The resultant intermediary ketone would accept intramolecular carbon–carbon bond construction from the carbonyl *si* face from the axially located allyl group (Int-**6**), terminating the tandem semipinacol-alkylation sequence.

10.2.2.2 Reaction with Nitrogen Nucleophiles Ring opening by amine nucleophiles typically occurs at the 3-position of epoxy alcohols. For example, substrate **44** regioselectively afforded 3-aminopropane-1,2-diol **45** in quantitative yield, where the difference in steric congestion between the C-2 and C-3 sites determined the location of nucleophilic addition (Scheme 10.11).[39] The 1,2-*syn* stereochemistry of **44** should be another key for the excellent selectivity of **45** because its Payne rearrangement leads to formation of the energetically less favorable compound with two cis-disposed substituents at the oxirane ring. The same site selectivity was observed for **46**, which cyclized when the amino group was generated by deprotection of the Cbz moiety under hydrogenation conditions.[40] Addition of an appropriate Lewis acid is effective for the production of aminodiols, and a regioisomeric mixture of **49** and **50** was obtained from **48** in a ratio of 80:20, favoring the 1,2-diol.[41] However, a more sterically hindered substrate (the terminal allyl group in **48** was replaced with a *tert*-Bu moiety) exclusively furnished the corresponding 1,3-diol under similar conditions, probably because of steric blocking at the otherwise preferred reaction site.[42] In addition, other examples studied under Lewis acidic conditions indicate that epoxy ring opening is likely to proceed in a fashion to give 1,2-diols (attack at

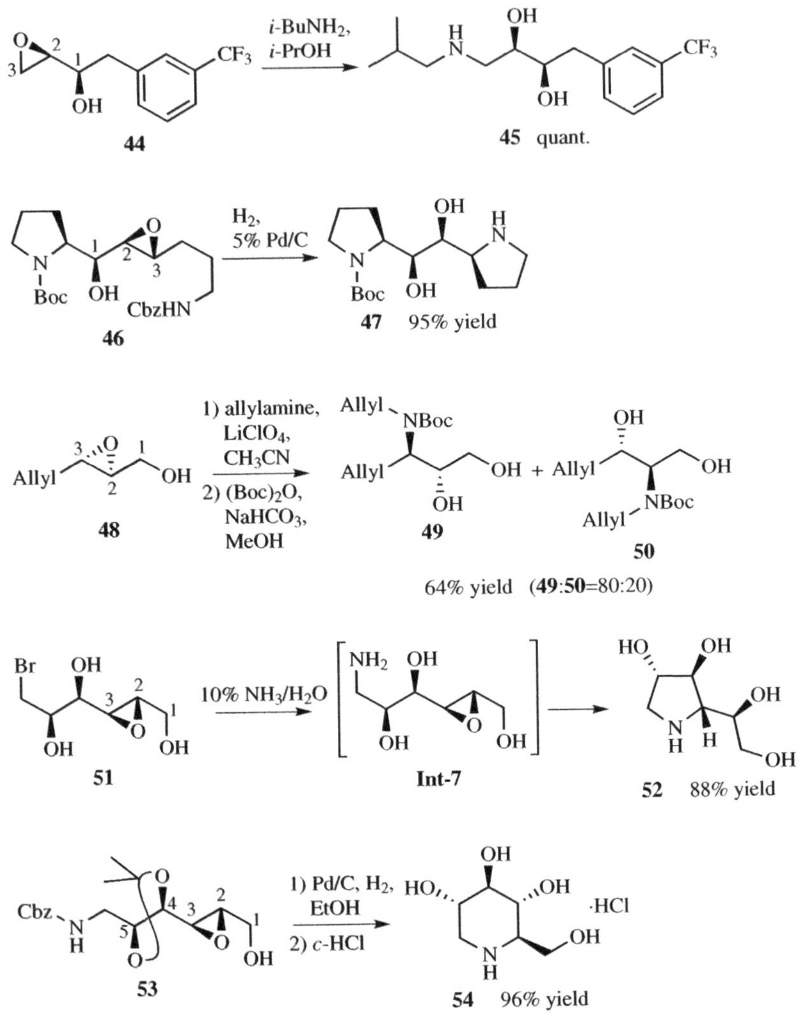

Scheme 10.11 Ring opening of 2,3-epoxy alcohols by amines.

C-3) as the major products when the Lewis acidic metal center is small, while attack at C-2 is preferred with larger Lewis acids.[43]

Another interesting set of examples involves intramolecular processes from closely related substrates furnishing different products. An aqueous ammonia solution initiated substitution of a bromine atom in **51** with an NH_2 group, leading directly to internal epoxide opening and furnishing the 5-membered pyrrolidine **52** in high yield as a consequence of the 5-exo-tet-type cyclization at the 3-position, rather than the possibly competing 6-endo-tet mode.[44,45] On the other hand, the intermediate after Cbz deprotection from **53**, structurally analogous to Int-**7**, resulted in the 6-membered piperidine **54** as the sole product by internal amine addition at the 2-position. This sharp difference might stem from the trans-disposed acetonide structure of **53**, enforcing cyclization from a specially restricted conformation, while this was not the case for the acyclic system **51** or its derivative Int-7. The same regiochemical preference was observed for the material in which the terminal CH_2OH unit in **53** was replaced with a $CH{=}CH_2$ group.[46]

10.2.2.3 Reaction with Oxygen Nucleophiles 2,3-Epoxy alcohols are also known to be regio- as well as stereospecifically opened at the 2-position in an S_N2 manner by a water molecule with the aid of a fluoride ion, enabling the formation of 1,2,3-triols **56** (Scheme 10.12).[47] It is interesting to note that the similarly basic tetra-*n*-butylammonium hydroxide was not as effective as TBAF and resulted in less than 20% yield of **56** under the same conditions. Water attenuates the basic nature of TBAF, suppressing any Payne rearrangement of **55** prior to the epoxy ring opening. Park and coworkers[48] employed epoxy alcohol **57** as a common intermediate which could be transformed into two isomeric triols **58** or **59** in diastereomerically specific fashion, simply by selection of reaction conditions using either KOH or H_2SO_4, respectively. Because the compound **58** is the enantiomer of the therapeutically valuable piperidine 1,6-dideoxynojirimycin,[49] choice of the appropriate stereoisomer of **57** should open a direct as well as concise routes to this bioactive material.

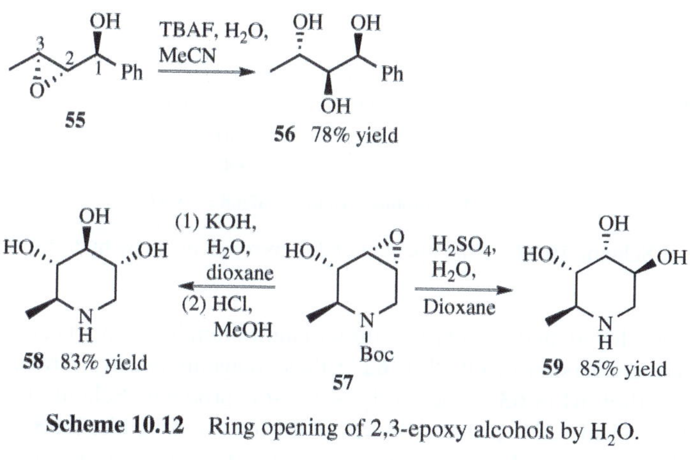

Scheme 10.12 Ring opening of 2,3-epoxy alcohols by H_2O.

Additionally, as shown in entry 13 of Table 10.1, free benzoic acid was proved to be an excellent nucleophile for the site-selective epoxy ring opening of **32**, leading almost exclusively to the corresponding C-3 benzoate ester in good yield. In the same publication,[32] installation of carboxylates was also realized by the action of 1.5 equiv. of the ammonium carboxylates in THF at ambient temperature for 15 min, producing the corresponding benzoate and acetate in 74% and 73% yields with 100:1 and 65:1 **33**:**34** selectivity, respectively (i.e., favoring C-3 over C-2 addition).[50]

10.2.2.4 Reaction with Hydride For ring opening of hydroxy epoxides by hydride, sodium bis(methoxyethoxy)aluminum hydride (Red-Al or Vitride) is the reagent most

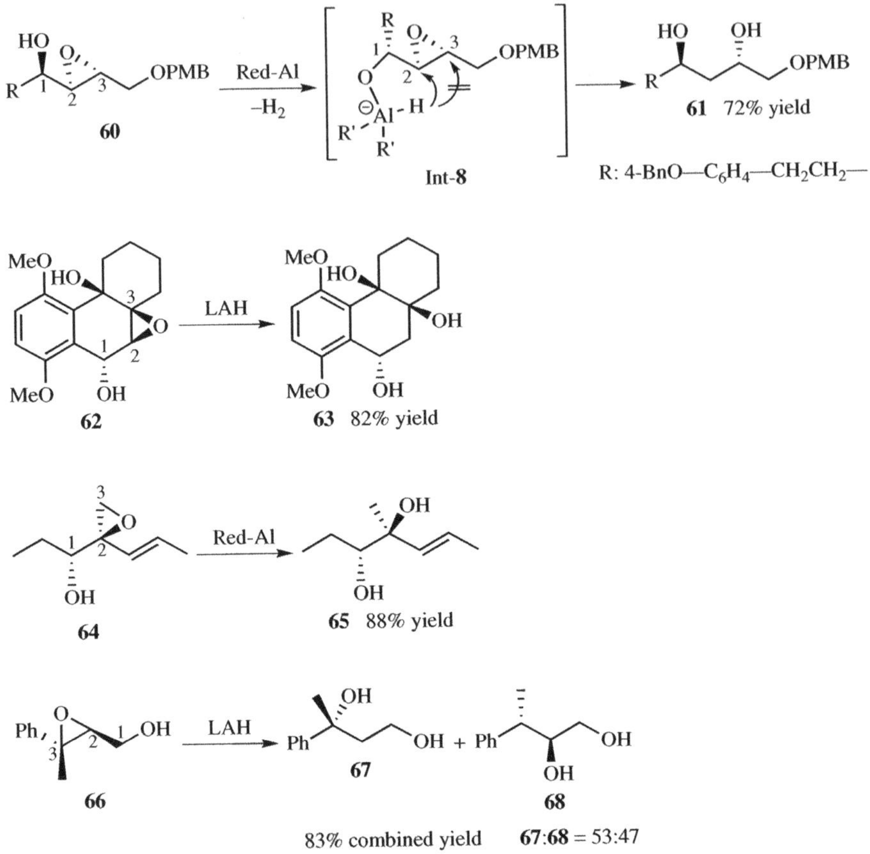

Scheme 10.13 Ring opening of 2,3-epoxy alcohols by hydride.

frequently employed, and its "parent" lithium aluminum hydride[51] is also useful. As shown in the conversion of **60**[31d] or **62**,[52] these reagents usually furnish the corresponding 1,3-diols **61** or **63**, respectively, as the sole products (Scheme 10.13). These reactions are considered to proceed by way of the intermediate like Int-**8**, and the final delivery of hydride would be preferentially occurred in an intramolecular manner to the closer C-2 reaction site rather than C-3. In the case of **62**, in spite of the presence of two hydroxy moieties, the secondary OH group at C-1 actually participated for the reductive epoxy ring opening to afford **63**. But there are some exceptions. For example, because the possible C-2 reaction site for hydride addition in **64** represents a quaternary carbon center with significant steric congestion, Red-Al circumvented this position to deliver hydride to the more accessible 3-position.[53,54] It is interesting to compare the two oxirane carbon atoms in compound **66**,[55] with one and no hydrogen atoms at the 2- and 3-positions, respectively. From the standpoint of steric factors, LAH should select the 2-position, but the 3-position is activated by a phenyl group. These conflicting factors led to a final product mixture of **67** and **68** almost in equal amounts. To an extent, the case for **64** was similar, having a steric preference for one end of the epoxide (C-3) but an electronic bias for the other epoxide carbon (C-2). However, the weaker level of electronic activation by an allyl moiety and the large steric difference in C-2/C-3 substitution patterns led to the formation of **65** as a single product in that case.[56]

10.2.2.5 Reaction with Fluoride Levoglucosan-based epoxide **69** was dissolved in a mixed solvent system of ethylene glycol/diethylene glycol diethyl ether (1:1), and treatment of the resultant solution with a combination of KHF_2 and KF (6 equiv. each) at 200 °C for 1 h afforded a 4:1 ratio of fluorinated diols **70** and **71** in 55% combined yield.[57] Formation of **70** was easily understood as the result of direct nucleophilic ring opening by a fluoride ion at the 3-position of **69**. On the other hand, it is apparent that isomeric product **71** was not formed by the simple S_N2-type fluoride displacement at the 2-position of **69** because such a process should afford the C-2 epimer of **71**. Instead, fluoride ion acted as a base to abstract the hydroxy proton in **69**, promoting Payne rearrangement of **69** to **72** as the first stage of the overall reaction. Coordination to the epoxy oxygen atom by acidic KHF_2 might also facilitate this isomerization. Fluoride opening of epoxide **72** at C-2 would then provide **71**. In support of this pathway, **72** could be isolated from the reaction mixture in low yield (Scheme 10.14).

Typically, amine-based hydrogen fluoride salts have been employed for introduction of a fluorine atom accompanied by epoxy ring opening. It is possible to anticipate the regioselectivity of such transformations on the basis of the number of HF molecules involved as well as the basicity of the amine component of the salt: increased equivalents of HF or lowered amine basicity contribute to a relatively acidic environment which promotes fluorination at the more substituted epoxy carbon atom. On the contrary, the relatively strong nucleophilicity of a fluoride ion would favor the S_N2-type reaction mechanism at the less congested position.

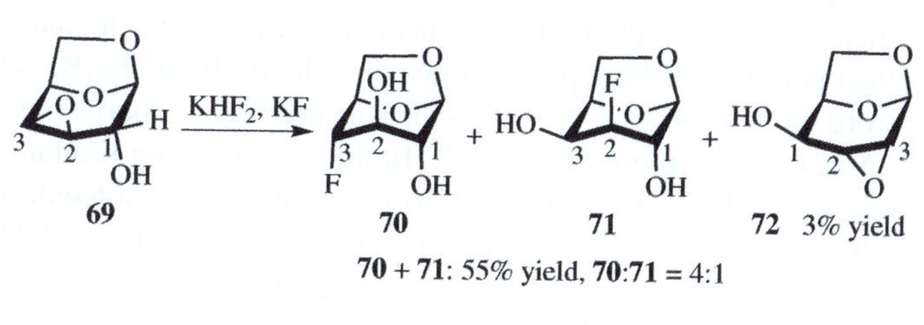

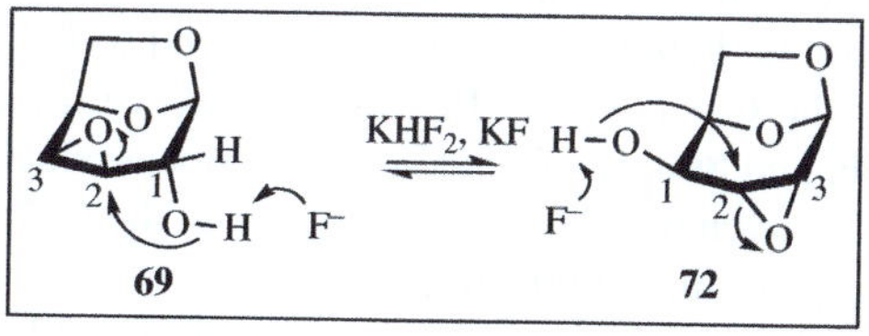

Scheme 10.14 Ring opening of 2,3-epoxy alcohols by fluoride (1).

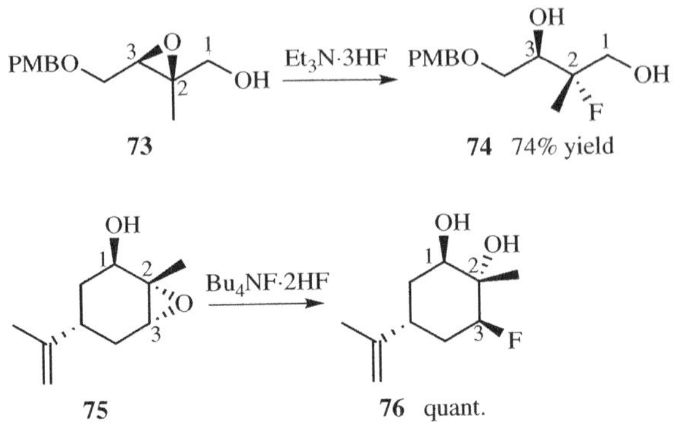

Scheme 10.15 Ring opening of 2,3-epoxy alcohols by fluoride (2).

In both cases, incorporation of fluorine usually occurs with clean inversion of stereochemistry. Scheme 10.15 shows two representative instances using unsymmetrically substituted epoxides as substrates. In one case, **73** was converted to 2-fluoro-1,3-diol **74** in 74% yield by the action of $Et_3N{\cdot}3HF$.[58] Activation by this reagent led to preferential cleavage of the O—C-2 rather than the O—C-1 due to partial S_N1 character in the reaction pathway, giving regioselective production of the more substituted fluorine-bearing center in **74**. On the other hand, the more nucleophilic fluoride source TBAF·2HF selected the less congested C-3 carbon atom in **75** as the site of the S_N2 reaction to produce **76** in quantitative yield.[59] Similar regioselectivity for fluoride addition to the less substituted epoxide carbon was also attained with TBAF·2HF in the presence of KHF_2.[60] In this instance, the role of KHF_2 was to convert TBAF·HF that resulted after fluorination back to the initial 2HF salt. Evidently, here, KHF_2 did not act as a Brønsted acid for epoxide activation, as shown by the purely S_N2 regiochemical outcome (displacement at the less substituted position).[61]

On the other hand, some reports have appeared on Lewis acid-mediated fluorinative epoxy ring opening. For example, as shown in Scheme 10.16, $Et_4NF{\cdot}4HF$ reacted with **77** to afford the 3-fluoro-1,2-diol **78** as the sole product, in sharp contrast to **75** in Scheme 10.15 in terms of regioselectivity.[62] In this instance, the product specificity observed could be predicted by assuming formation of an intramolecularly chelated intermediate by the Lewis acidic $(i\text{-PrO})_2TiF_2$ after reaction with the OH moiety in **77**, similar to the Sharpless case depicted in Table 10.1. Titanium coordination to the oxirane oxygen atom in **77** should cause partial cationic character at C-3 where positive charge should be better stabilized as compared to C-2, leading to predominant substitution at the C-3 position. Regioselectivity with the epoxy alcohol **79a** was found to be high, and **80a** was formed as the major product over its regioisomer **81a** in a ratio of 10:1. Given the similar environment at the C-2 and C-3 carbons, this selectivity discrepancy is quite interesting.

An analogous transformation was also possible with $Bu_4NF{\cdot}2HF$ as the fluorinating reagent in the presence of TiF_4 or HfF_4 as a Lewis acid. The use of TiF_4 afforded

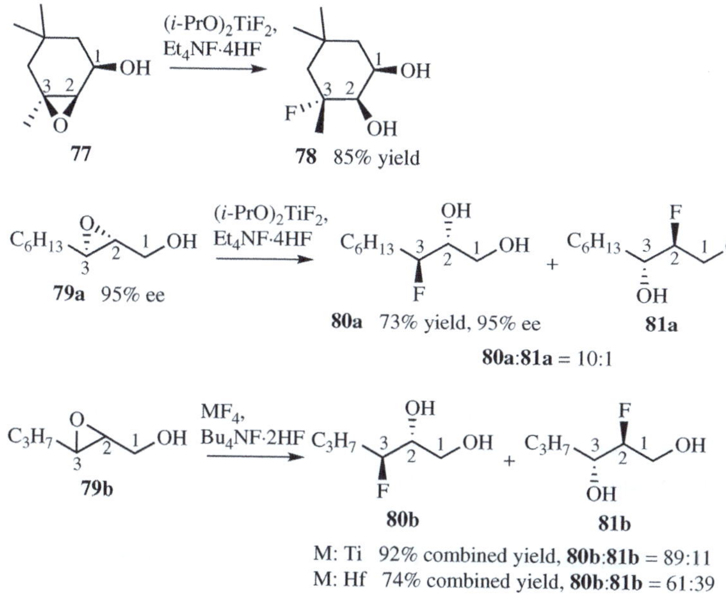

Scheme 10.16 Ring opening of 2,3-epoxy alcohols by fluoride (3).

a mixture of **80b** and **81b** in a ratio of 9:1, while the preference of **80b** decreased to 3:2 by employment of HfF_4 as the Lewis acid.[43] This difference was explained by the size of the central metal. Because the smaller Group 4 metal Ti would readily form a 5-membered dioxatitanacycle, fluoride might prefer the attack at the 3-position, while the larger Hf would have a tendency to accommodate a 6-membered ring upon epoxide opening, yielding **81b** in a relatively larger amount. In place of metal tetrafluorides, the corresponding triflates, $Hf(OTf)_4$ and $Sc(OTf)_3$, also worked in a similar fashion.[63]

Only one report was found in the literature on the fluorination of aziridinemethanols, as depicted in Scheme 10.17.[64] Olah's reagent[65] (pyridine·$(HF)_x$) was employed for fluorination, and the relative acidity of this reagent played a significant role in this transformation. Thus, successful introduction of a fluorine atom was realized for the N-protonated **82a** at the benzylic C-2 position. However, the aziridine ring in **82b**, lacking directly connected aromatic groups, did not open due to insufficient activation; instead, the hydroxy group was replaced by fluorine, furnishing **84**.

10.3 UTILIZATION OF THE PAYNE REARRANGEMENT FOR THE PREPARATION OF FLUORINE-CONTAINING COMPOUNDS

Because fluorine-containing compounds possess unique characteristics, their preparation has drawn significant attention from chemists, especially in the field of fine-chemical synthesis.[66] A variety of methods have been developed for the construction of these fluorinated molecules, and some of these routes are described

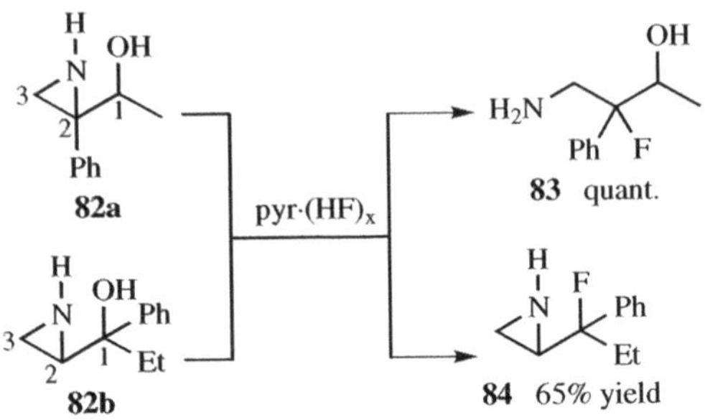

Scheme 10.17 Reaction of aziridine alcohols with HF·pyridine.

in recent review articles.[67] In this section, a very rare electronically promoted Payne rearrangement is discussed for unique substrates having a CF_3 group directly connected to the epoxy alcohol framework.

10.3.1 Payne Rearrangement for Trifluorinated Compounds

In our search of the literature, we find no approaches for direct incorporation of CF_3 or fluorine-containing alkyl moieties into the 2,3-epoxypropan-1-ol framework. Instead, the Seebach group disclosed in 1992 that epoxy alcohols **86**, prepared from the corresponding epoxy ester **85**, experienced Payne rearrangement in aqueous acetone or *tert*-BuOH, strongly favoring the rearranged epoxy alcohol isomer **87** at equilibrium (**86**:**87** = <1:100) (Scheme 10.18).[68] To the best of our knowledge, this is the first example of the Payne rearrangement using CF_3-containing substrates.

During our research directed toward the synthesis of a variety of trifluorinated sugars,[69] a unique formal ring expansion of furanose **88** to pyranose **89** was discovered when the former was treated with *tert*-BuOK in THF at −78 °C. After 3 h, the 5-membered ring **88** was successfully converted to the 6-membered **89** in quantitative yield (Scheme 10.19).[70] The alkoxide Int-**9** from **88** should be in equilibrium with the acyclic Int-**10**, allowing 1,4-*O*,*O*-migration of the TBS group by way of a pentavalent silyl anion. The silyl migration proceeds so as to form the energetically preferable anionic species Int-**11**, where the CF_3 group inductively stabilizes the negative charge by virtue of its electron-withdrawing nature.[71] Cyclization of Int-**11** provided its pyranose form, Int-**12**, which was quenched to give **89**.

This interesting migration process was also applicable to a 1,5-*O*,*O*-silyl transfer, using substrates **90**. Quite intriguingly, the TBS group in **90a** was

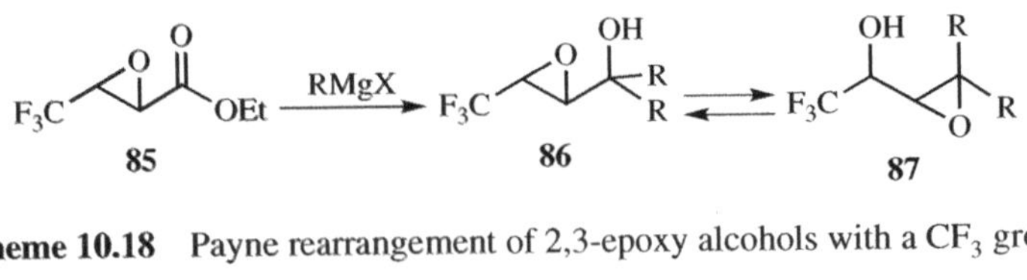

Scheme 10.18 Payne rearrangement of 2,3-epoxy alcohols with a CF_3 group.

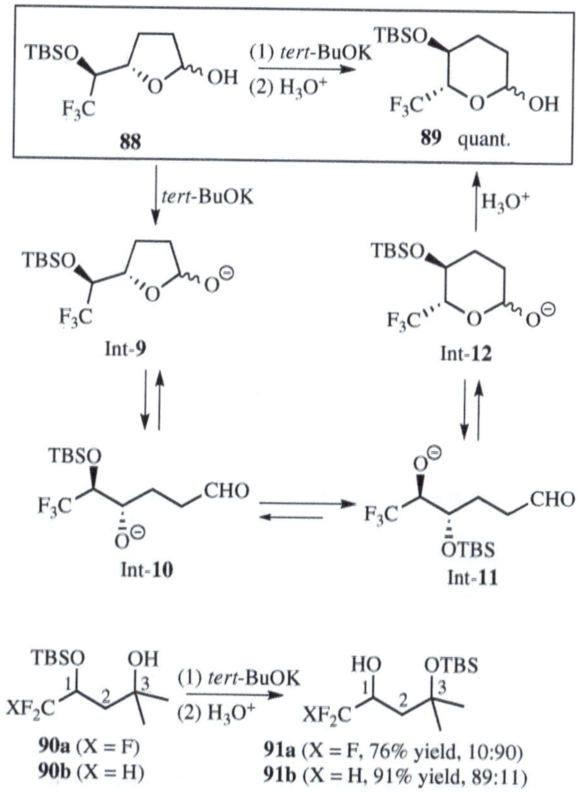

Scheme 10.19 Silyl migration of CF_3-containing diols.

displaced to the other hydroxy group at C-3 to give **91a** in 76% yield in a ratio of **90a**:**91a** = 10:90.[72] The 10% of remaining substrate **90a** would stem from the fact that in spite of the better stability of the alkoxide in the rearranged product **91a**, this compound is encumbered by the steric congestion at C-3. In other words, the strong electron-withdrawing characteristic of the CF_3 group effectively counterbalances the increased steric interactions in the migration product. Substitution of even a single fluorine atom in **90a** by hydrogen was enough to reverse the product selectivity, and a ratio of **90b**:**91b** = 89:11 was observed.

After exploring this silyl migration from experimental and theoretical points of view, our next aim was to apply the general reactivity principles involved to Payne rearrangements. We expected that isomerization of Int-**13**, prepared from the model compound (*E*)-*anti*-**92a**, to Int-**14** should proceed smoothly due to the stability difference between the anions, with Int-**14** preferred because of the alkoxide proximity to the CF_3 group (Scheme 10.20).[73] When (*E*)-*syn*-**92a** was employed as the substrate, the corresponding product after Payne rearrangement should be (*Z*)-*anti*-**93a** with methyl and 2,2,2-trifluoro-1-hydroxyethyl groups in a cis relationship, corresponding to an increase of steric interference. Prior to initiation of

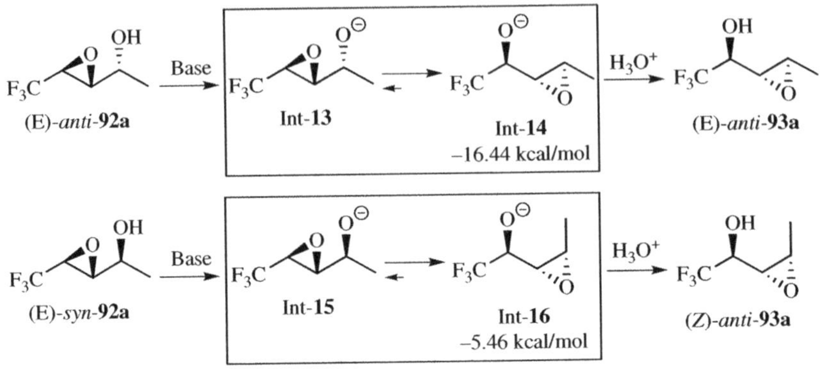

Scheme 10.20 Computational preference of 2,3-epoxy alcohol isomers with a CF_3 group.

our experimental investigation, we carried out theoretical computations[74] for Int-**13** to Int-**16** as their lithium salts. The calculations showed that Int-**14** was energetically more favorable than Int-**13** by 16.44 kcal/mol, which would be a direct reflection of the anion stability gap because both starting Int-**13** and rearranged Int-**14** include a trans-disubstituted epoxide and a secondary alkoxide. Meanwhile, the significant energetic preference for having the anionic site close to the CF_3 moiety would be the reason why Int-**16** with the less advantageous cis-disposed epoxide was still energetically preferable to Int-**15** by 5.46 kcal/mol. Additional evidence for the CF_3-anion-stabilizing effect was obtained by calculating the relative energies of the parent compounds **92a** and **93a**. Thus, (*E*)-*anti*-**93a** and (*E*)-*syn*-**92a** were found to be more stable than (*E*)-*anti*-**92a** and (*E*)-*syn*-**93a** by 0.45 and 1.77 kcal/mol, respectively, showing that the calculated differences in anion stability obtained previously did not stem from their inherent structural stability.

With computational results in hand that anticipated the success of the reactions shown in Scheme 10.20, two substrates (*E*)-**92** with either *anti* or *syn* configuration were investigated under Payne rearrangement conditions. The requisite substrates (*E*)-**92** (Scheme 10.21) were synthesized from the α,β-unsaturated ketones (*E*)-**94** by nucleophilic epoxidation mediated by the urea·H_2O_2 adduct in the presence of DBU,[75] followed by stereoselective reduction of the resultant (*E*)-**95** by *in situ*-prepared $Zn(BH_4)$[24,76] or DIBAL[23] for the construction of (*E*)-*anti*-**92** or (*E*)-*syn*-**92**, respectively. Payne rearrangement of these epoxy alcohols (*E*)-**92** was successfully carried out by the addition of NaH in the presence of 5 equiv. of HMPA in THF, giving complete reaction within 3 h and affording the corresponding isomers *anti*-**93** in a highly stereoselective manner. The requirement for HMPA as an additive was identified during investigation of the reaction condition. In the absence of this additive, (*Z*)-*anti*-**93b** was formed from (*E*)-*syn*-**92b** in only 49% yield, and an almost equimolar amount of the unexpected β-hydroxylated carboxylic acid **96** was isolated. (Yamazaki, T.; Ichige, T. unpublished results.) A possible β-hydroxyketone intermediate might be involved in the formation of **96** resulting from intramolecular chelation of a sodium cation by the epoxy alcohol oxygens and hydride migration

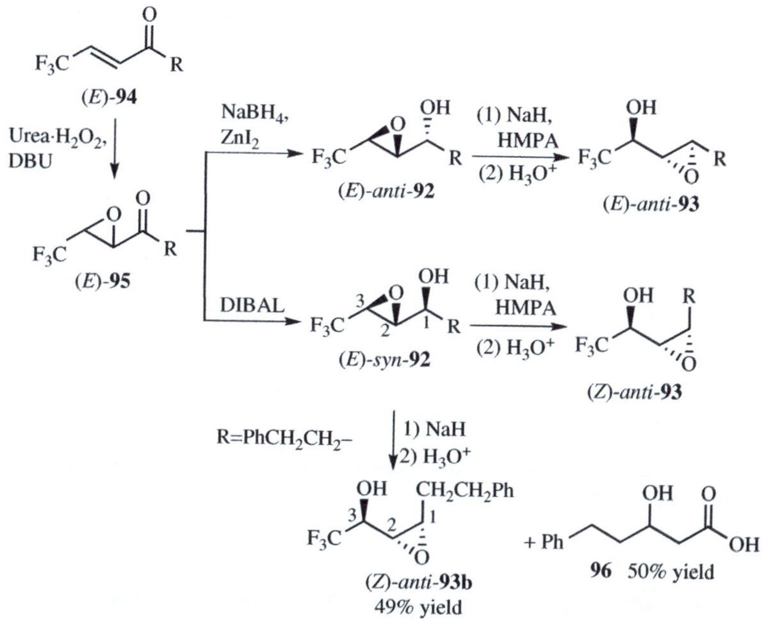

Scheme 10.21 Payne rearrangement of stereoisomeric 2,3-epoxy alcohols with a CF_3 group.

from C-3 to C-2. Hydrolysis of the resultant β-hydroxylated trifluoromethyl ketone with elimination of CF_3^- would then result in **96**.[77]

The results for stereoselective reductions of (*E*)-**95** to the both diastereomers of (*E*)-**92** and their subsequent Payne rearrangements are summarized in Table 10.2. Reduction of epoxy ketones (*E*)-**95b-e** was stereoselectively performed in a range of *anti*:*syn* = 86:14–95:5 by the $NaBH_4$–ZnI_2 system and 83–92% *syn* selectivity for the DIBAL conditions. The exception for the DIBAL case was as shown in entry 3, where (*E*)-**95d** gave a nearly equal amount of *syn* and *anti* isomers. The reason for the lack of selectivity in this case is not clear, and we could not find any instances of DIBAL reduction of nonfluorinated compounds possessing a similar structure to (*E*)-**95**.

Smooth and essentially stereospecific transformation was realized for the following Payne rearrangements of the (*E*)-**92** substrates. The limited (29%) conversion of (*E*)-*syn*-**92e** (Table 10.2, entry 4) reflects the presence of a tertiary carbon directly attached to the epoxide three-membered ring in the corresponding product (*Z*)-*anti*-**93e**, affording significant steric repulsion with the trifluoromethyl-containing epoxide substituent. In the case of the substrates (*E*)-*anti*-**92d** and (*E*)-*syn*-**92d**, contamination of the desired product with about 15% of unidentified materials was detected, but formation of these byproducts was nicely suppressed by the addition of 3 equiv. of EtOH to the reaction mixture.

This type of electronically promoted Payne rearrangement is very rare, and in spite of the similarity of a CF_3 group to carboxyl or ester functionalities in terms

TABLE 10.2 Stereodivergent Synthesis of Epoxy Alcohols (*E*)-92 and Their Payne Rearrangement (see Scheme 10.21)

Entry	R	Isolated yield (%)			
		(*E*)-*anti*-**92**[a,b]	(*E*)-*syn*-**92**[b,c]	(*E*)-*anti*-**93**[d]	(*Z*)-*anti*-**93**[d]
1	$PhCH_2CH_2$– (**b**)	97 (90:10)	91 (8:92)	91 (90:10)	96 (8:92)
2	C_6H_{13}– (**c**)	83 (89:11)	90 (12:88)	96 (89:11)	94 (9:91)
3	Ph– (**d**)	90 (86:14)	90 (46:54)	91 (90:10)[e]	88 (1:99)[e,f]
4	$PhCH_2C(CH_3)_2$– (**e**)	67 (95:5)	96 (17:83)	81 (98:2)	29 (17:83)[g]

[a]Reduction of (*E*)-**94** was carried out using a mixture of $NaBH_4$ and ZnI_2.
[b]The *anti*:*syn* ratios of the products are shown in parentheses.
[c]Reduction of (*E*)-**95** was carried out using DIBAL.
[d]Payne rearrangement was performed by the addition of NaH in THF in the presence of 5 equiv. of HMPA. The (*E*):(*Z*) ratios of the products are shown in parentheses.
[e]3 equiv. of EtOH was added.
[f]The starting material (*E*)-*syn*-**92d** was used as a substrate after recrystallization (*anti*:*syn* = 1:99).
[g]Yield determined by ^{19}F NMR.

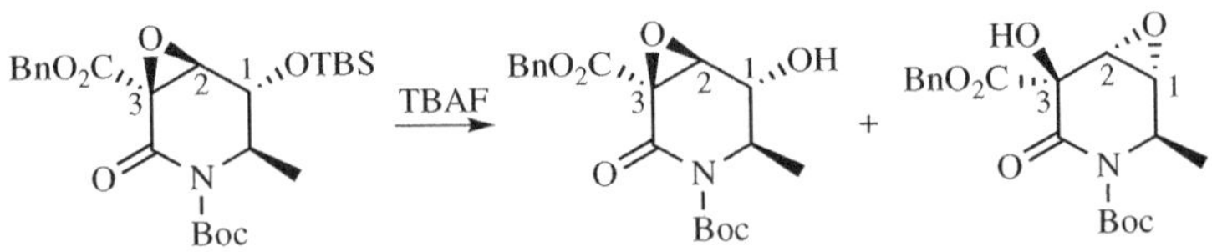

97 **98** 91% yield **99** 3% yield

Scheme 10.22 Payne rearrangement of substrates with an electron-withdrawing ester group.

of electron-withdrawing ability, we found only one related Payne precedent in the literature irrespective of the presence of fluorine atoms (Scheme 10.22).[78] Thus, when deprotection of a TBS group in **97** was carried out by the action of TBAF, a resultant anionic species would be ideally located on the back side of the epoxy group, rendering intramolecular attack easy. The two carbonyl-related functionalities at the C-3 position of **97** should provide an additional inductive advantage for driving the conversion forward to **99**. However, contrary to expectation, only 3% of **99** was obtained. This was likely due to the fact that the addition of TBAF only generated the pentavalent silyl anion from **97**, not the corresponding free alkoxide. Further attempt to promote the Payne rearrangement from **98** was not carried out in this study.

10.3.2 Ring Opening of CF_3-Containing Epoxy Alcohols by Nitrogen Nucleophiles

The Payne rearrangement substrates and products, CF_3-containing epoxy alcohols (*E*)-**92** and 2,3-*anti*-**93**, could be applied in synthetically useful oxirane ring-opening reactions with amine nucleophiles. Scheme 10.23 describes representative examples

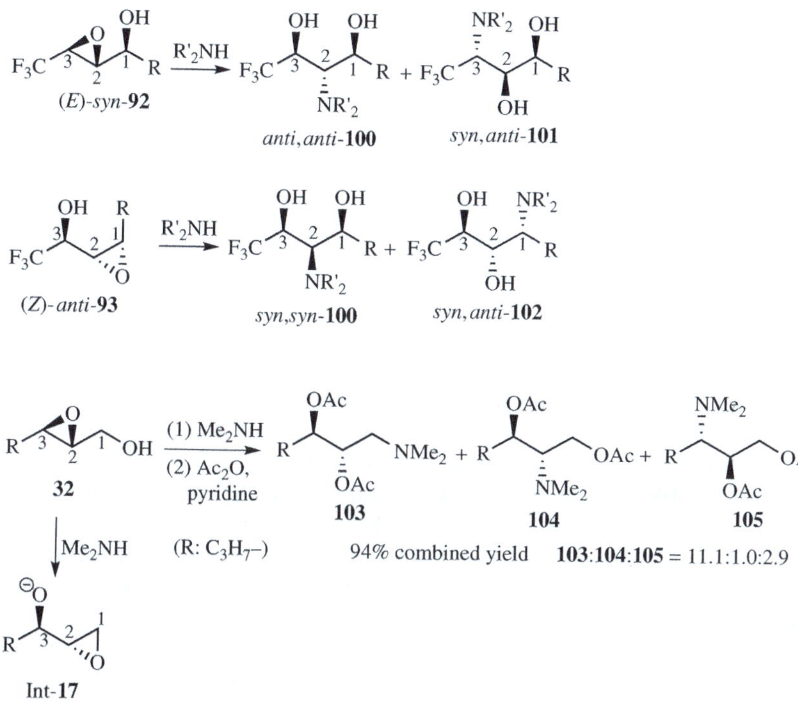

Scheme 10.23 Comparison of regioselective epoxy ring opening by amines.

where (*E*)-*syn*-**92** and (*Z*)-*anti*-**93** were independently treated with R'_2NH. A stronger O—C-3 bond compared to the O—C-2 bond in the former substrate was anticipated[66b,79] because the inductively electron-withdrawing characteristics of the CF_3 group should strengthen the proximal bond. Additionally, the CF_3 substituent, with nine lone pairs in total, should hamper the access of electron-rich nucleophiles to the C-3 position. On the basis of these two factors along with an S_N2 mode of ring opening, *anti,anti*-**100** would be obtained from (*E*)-*syn*-**92** as the major product. On the other hand, we anticipated lower selectivity from (*Z*)-*anti*-**93** because the electrostatic difference between C-1 and C-2 seemed to be small.

The reaction of **32** with an aqueous solution of dimethylamine was reported to produce a mixture of **103**, **104**, and **105** in a ratio of 11.1:1.0:2.9,[80] and this case served as a reference for our studies. Formation of **103** under these conditions resulted from C-1 attack of Me_2NH after base-mediated Payne rearrangement of **32**. In spite of the inherent instability of the rearranged intermediate Int-**17**, which contains the alkoxide of a *sec*-OH as well as a monosubstituted epoxide, a possible kinetic preference for reaction at the unhindered C-1 position of Int-**17** would rationalize why **103** was formed as the main product. The same situation was already discussed in Scheme 10.9. Compounds **104** and **105**, meanwhile, formed in a 1:3 ratio by direct ring opening of **32**. This analysis led to the inference that a part of (*E*)-*syn*-**92** would first be converted to the more favorable (*Z*)-*anti*-**93** in a basic media, so that the amine would attack either C-1 or C-2 of (*Z*)-*anti*-**93** as well as C-2 in (*E*)-*syn*-**92** giving a mixture of *syn,anti*-**102**, *syn,syn*-**100**, and *anti,anti*-**100**, respectively.

Initially, (*E*)-*anti*-**92d** and (*E*)-*anti*-**93d** were employed for the reaction with aqueous Me_2NH, and after stirring for 3 h at 70 °C, both substrates furnished the same product, *anti,anti*-**102da**, in an essentially stereospecific fashion and without detection of the possible product *anti,syn*-**100da** (Scheme 10.24a).[81] Considering that the nucleophilic epoxide-opening pathway follows an S_N2-type mechanism, determination of the stereochemical relationships within the product as 1,2-*anti* and 2,3-*anti* led to conclusion that, before ring opening by Me_2NH, (*E*)-*anti*-**92d** experiences Payne rearrangement to the more thermodynamically stable alkoxy anion corresponding to (*E*)-*anti*-**93d**. This is the intermediate that actually accepts the addition of the amine at the activated benzylic C-1 position to yield *anti,anti*-**102da**.

Subsequently, we have investigated a wide range of reaction conditions with $BnNH_2$ as the nucleophile and found that, as expected, C-2 attack was observed, giving *syn,anti*-**100db** selectively from (*E*)-*anti*-**92d** (Scheme 10.24b) when an aprotic solvent was employed (heating at 100 °C for 14 h in DMF was found to be ideal for this transformation). On the other hand, the course of the reaction was completely altered in an aqueous environment, and *anti,anti*-**102db** was obtained after 6 h at reflux, reflecting a pathway via Payne rearrangement prior to the reaction with $BnNH_2$. It is interesting to note that addition of an equal amount of H_2O to DMF afforded *anti,anti*-**102db** selectively and that, in spite of its possible hydrogen

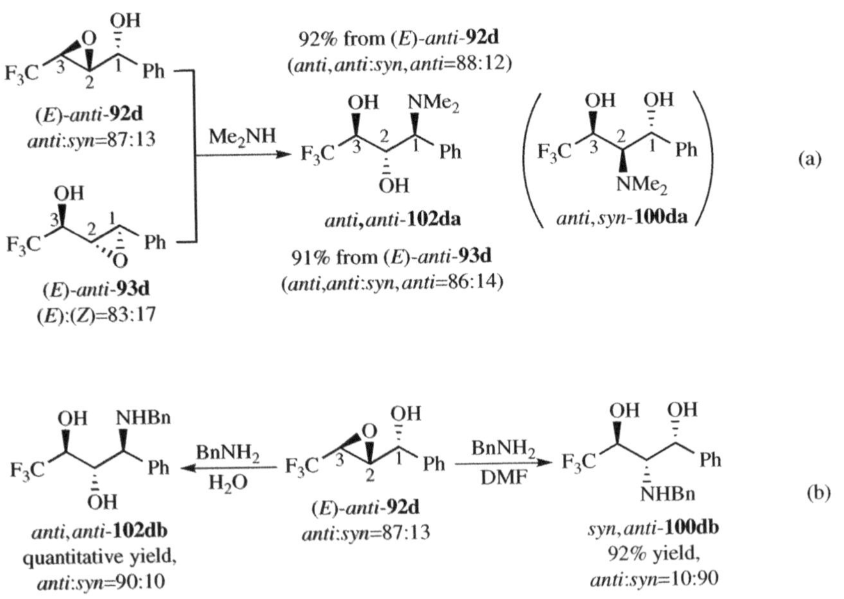

Scheme 10.24 Regioselective ring opening of CF_3-containing 2,3-epoxy alcohols by amines.

TABLE 10.3 Epoxy Ring Opening of (*E*)-92 and 2,3-*anti*-93 by $BnNH_2$

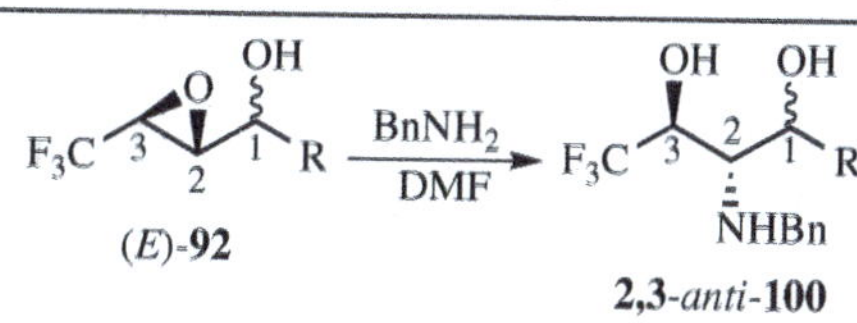

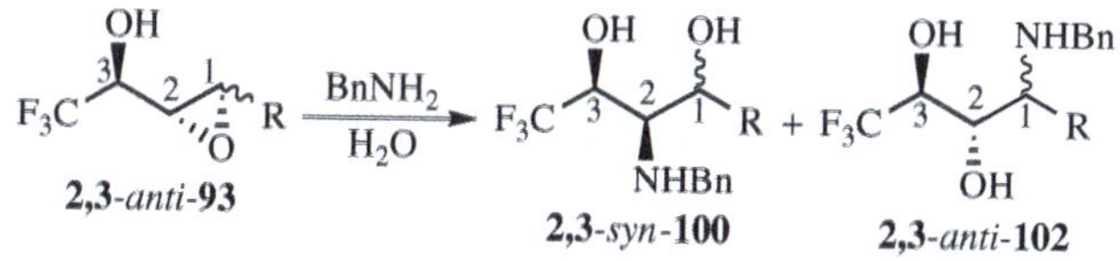

Entry	Substrate	R	*Anti*:*syn* or (*E*):(*Z*)[a]	Isolated yield of **100** [**102**] (%)	*Anti*:*syn* of **100** [**102**][b]
1	(*E*)-*anti*-**92b**	$PhCH_2CH_2$–	89:11	82	90:10
2	(*E*)-*anti*-**92c**	C_6H_{13}–	91:9	80	89:11
3	(*E*)-*anti*-**92d**	Ph–	87:13	92	90:10
4	(*E*)-*syn*-**92b**	$PhCH_2CH_2$–	9:91	95	9:91
5	(*E*)-*syn*-**92c**	C_6H_{13}–	12:88	49	8:92
6	(*E*)-*anti*-**93b**	$PhCH_2CH_2$–	89:11	[77]	[90:10]
7	(*E*)-*anti*-**93c**	C_6H_{13}–	83:17	6[63]	[86:14]
8	(*E*)-*anti*-**93d**	Ph–	90:10	[>99]	[90:10]
9	(*Z*)-*anti*-**93b**	$PhCH_2CH_2$–	11:89	[71]	[6:94]
10	(*Z*)-*anti*-**93c**	C_6H_{13}–	14:86	16[69]	[8:92]

[a] *Anti*:*syn* ratios between C-1 and C-2 of compounds (*E*)-**92** and (*E*):(*Z*) ratios at C-1 of compounds *anti*-**93**.

[b] *Anti*:*syn* ratios between C-1 and C-2 of **100** [**102**].

bonding ability and boiling point of 97 °C, use of propanol as the solvent gave results totally different from H_2O, exclusively yielding *syn*,*anti*-**100db**.

Because our preliminary experiments (as shown in Scheme 10.24) demonstrated that a nonaqueous solvent system was most likely to inhibit *in situ* Payne rearrangement, we next studied reactions of $BnNH_2$ with a variety of (*E*)-**92** and 2,3-*anti*-**93**, and the results are summarized in Table 10.3. High to excellent chemical yields starting from (*E*)-**92** were recorded for the preparation of 2,3-*anti*-**100** in a DMF solvent, and regio- as well as stereospecific C-2—O bond cleavage was observed as the major pathway. On the other hand, 1-aminopropane-2,3-diols 2,3-*anti*-**102** were conveniently synthesized by heating 2,3-*anti*-**93** to reflux in H_2O for 6 h, although the products from (*E*)- and (*Z*)-*anti*-**93c** were contaminated by a small amount of 2,3-*anti*-**100c** (entries 7 and 10).

The two-solvent systems described previously opened new routes to different structural isomers, 2,3-*anti*-**100** and 2,3-*anti*-**102**, but unfortunately, preparation of 2,3-*syn*-**100**, for example, could not be attained. To solve this problem, we utilized intramolecular epoxy ring opening starting from 2,3-*anti*-**93** (Scheme 10.25),

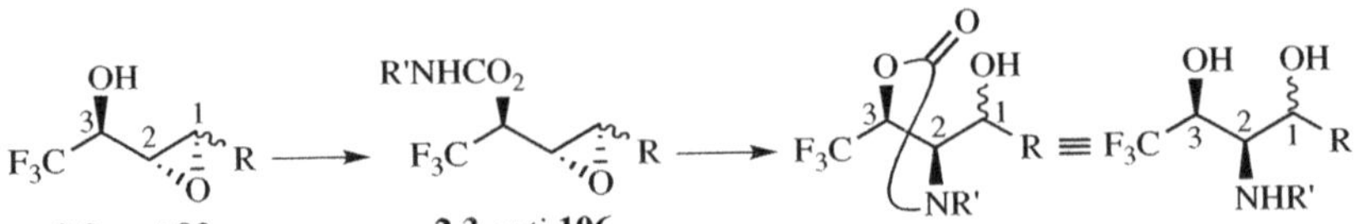

Scheme 10.25 Formation of CF_3-containing 2-amino-1,3-diols.

an approach based on previously reported results from our group.[82] Thus, after conversion of 2,3-*anti*-**93** to the corresponding carbamate 2,3-*anti*-**106**, treatment with an appropriate base would facilitate intramolecular attack to selectively furnish the oxazolidinones 2,3-*syn*-**107**[83] by way of a 5-exo-tet mode of displacement. 2,3-*Syn*-**107** thus synthesized are compounds structurally equivalent to 2,3-*syn*-**100**, which was not accessible by our previous method shown in Table 10.3. The expected product selectivity for 2,3-*syn*-**107** is based on the fact that the competing 6-endo-tet mode is known to be unfavorable.[45]

Implementing this concept, we have successfully found reaction conditions for two independent routes to access either 2,3-*syn*-**107** or 2,3-*syn*-**108** selectively, and our results are collected in Table 10.4. First of all, contrary to our expectation, addition of 2,3-*anti*-**106** to an aprotic THF solution of *tert*-BuOK at 0 °C (conditions A) led to formation of 2,3-*syn*-**108** rather than 2,3-*syn*-**107** in excellent chemical yields as well as selectivity starting from the epoxides with (*Z*)-geometry (entries 4–6). However, clean results were not obtained from (*E*)-epoxides (entries 1–3). On the other hand, switching to the protic solvent MeOH (conditions B) yielded the anticipated products 2,3-*syn*-**107** in high yields as well as excellent diastereoselectivity (entries 7–12).

The surprising formation of 2,3-*syn*-**108** could be explained mechanistically as depicted in Scheme 10.26. Thus, treatment of carbamates 2,3-*anti*-**106** with *tert*-BuOK should afford the anionic intermediate Int-**18a**, and its protonation by the solvent or upon quenching of the reaction would produce 2,3-*syn*-**107**, which was the case under conditions B. In contrast, in an aprotic solvent such as THF (conditions A), the anionic species should survive until quenching and would have a chance to attack the carbamate carbonyl carbon atom in an intramolecular fashion, enabling conversion to the electronically more stable Int-**19b**, possessing the negative charge on an oxygen closer to the CF_3 group.

Substrates (*E*)-*anti*-**106** led to the intermediate Int-**18a** with substituents at C-2 and C-3 in a trans relationship, and the corresponding Int-**19a** should have the sterically less favorable cis-disposed R and $CF_3CH(OH)$ groups at C-1 and C-2, respectively. In these cases, in spite of the preference for Int-**19a** from an electrostatic point of view, the dominant factor was apparently sterics so that the less crowded Int-**18a** was favored, resulting in *anti,syn*-**107** as the major products. However, starting from the corresponding (*Z*)-isomers, the C-1—C-2 trans geometry of substituents in Int-**19a** was sterically favorable, furnishing *syn,syn*-**108** as the sole products. Independently, we verified that isolated *syn,syn*-**107b** was cleanly and almost quantitatively transformed to *syn,syn*-**108b** under the reaction conditions, but this was not the

TABLE 10.4 Formation of Oxazolidinones 2,3-*syn*-107 and 2,3-*syn*-108 from 2,3-*anti*-106

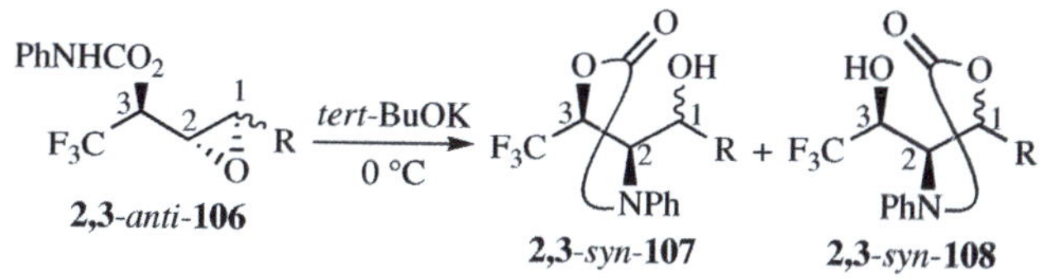

Entry	Substrate	R	(*E*):(*Z*)[a]	Isolated yield of **107** [**108**] (%)	*Anti*:*syn* of **107** [**108**][b,c]
Condition A[d]					
1	(*E*)-*anti*-**106b**	$PhCH_2CH_2$–	91:9	[31[c]]	[70:30]
2	(*E*)-*anti*-**106c**	C_6H_{13}–	88:12	[50[c]]	[73:27]
3	(*E*)-*anti*-**106d**[e]	Ph–	>99:1	Complex mixture	
4	(*Z*)-*anti*-**106b**[f]	$PhCH_2CH_2$–	1:>99	[95]	[1:>99]
5	(*Z*)-*anti*-**106c**[f]	C_6H_{13}–	1:>99	[94]	[1:>99]
6	(*Z*)-*anti*-**106d**[e]	Ph–	1:>99	[91]	[1:>99]
Condition B[d]					
7	(*E*)-*anti*-**106b**	$PhCH_2CH_2$–	91:9	87	95:5
8	(*E*)-*anti*-**106c**	C_6H_{13}–	88:12	83	92:8
9	(*E*)-*anti*-**106d**[e]	Ph–	>99:1	62	>99:1
10	(*Z*)-*anti*-**106b**[f]	$PhCH_2CH_2$–	1:>99	72	1:>99
11	(*Z*)-*anti*-**106c**[f]	C_6H_{13}–	1:>99	68	1:>99
12	(*Z*)-*anti*-**106d**[e]	Ph–	1:>99	59	1:>99

[a](*E*):(*Z*) ratios of compounds 2,3-*anti*-**106** are depicted.
[b]*Anti*:*syn* ratios between C-1 and C-2.
[c]Determined by ^{19}F NMR.
[d]A: in THF for 1 h. B: in MeOH for 3 h.
[e]An (*E*),(*Z*)-mixture of 2,3-*anti*-**93d** was chromatographically separable after conversion to the corresponding carbamate, 2,3-*anti*-**106d**.
[f]Recrystallization of these substrates furnished essentially pure (*Z*)-forms.

case for the diastereomeric *anti,syn*-**107d**, which was converted into the totally hydrolyzed 2-amino-1,3-diol in 70% yield instead of into *anti,syn*-**108d**.

10.4 CONCLUSION

In this chapter, the Payne rearrangement, which entails transposition of the 2,3-epoxypropan-1-ol structure **1** (Scheme 10.1) to the corresponding 1,2-epoxypropan-3-ol counterpart **2**, has been examined from various standpoints, including (1) structural and mechanistic characteristics of the Payne rearrangement,

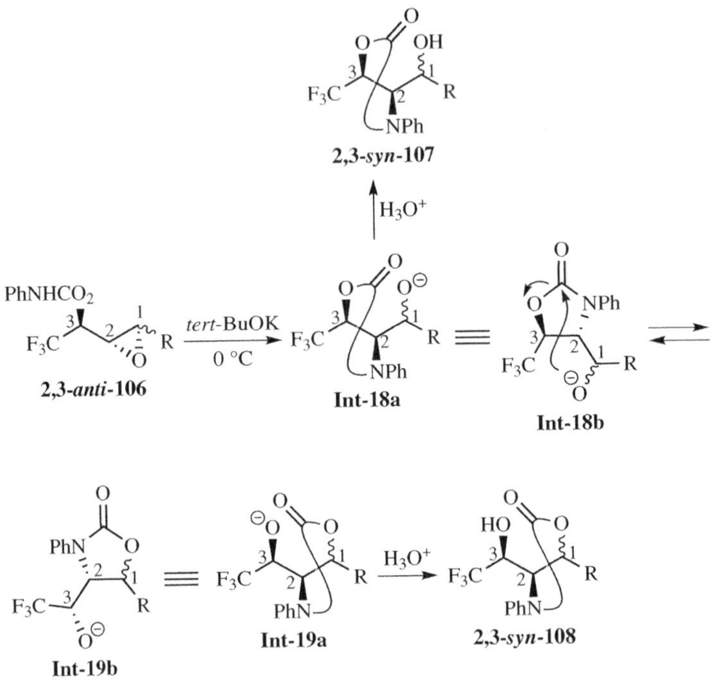

Scheme 10.26 Formation of CF_3-containing oxazolidinones and their migration.

(2) possible substrate structures required for this conversion, (3) routes used to access the rearrangement substrates, and (4) synthetic applications of substrates and products for this rearrangement. Moreover, on the basis of our group's previous work, utilization of the Payne rearrangement was also demonstrated for the construction of fluorine-containing compounds, and relevant results from other research groups in this area have also been included. It is our hope that this chapter offers insights to readers for approaching related synthetic problems and for discovering new and efficient routes for construction of useful organic compounds via the Payne rearrangement.

REFERENCES

1. Payne, G. B. *J. Org. Chem.* **1962**, *27*, 3819–3822.
2. For a recent review, see: Hanson, R. M. *Org. React.* **2002**, *60*, 1–156.
3. Kohler, E. P.; Bickel, C. L. *J. Am. Chem. Soc.* **1935**, *57*, 1099–1101.
4. Angyal, S. J.; Gilham, P. T. *J. Chem. Soc.* **1957**, 3691–3699.
5. Bajwa, J. S.; Sunay, U. B. *J. Labelled Compd. Radiopharm.* **1997**, *39*, 787–790.
6. Ibuka, T. *Chem. Soc. Rev.* **1998**, *27*, 145–154.

7. Brånalt, J.; Kvarnström, I.; Classon, B.; Samuelsson, B. *J. Org. Chem.* **1996**, *61*, 3604–3610.
8. Liu, C.-Q.; Hashimoto, Y.; Kudo, K.; Saigo, K. *Bull. Chem. Soc. Jpn.* **1996**, *69*, 2095–2105.
9. Najime, R.; Pilard, S.; Vaultier, M. *Tetrahedron Lett.* **1992**, *33*, 5351–5354.
10. Liu, Q.-Y.; Marchington, A. P.; Boden, N.; Rayner, C. M. *J. Chem. Soc., Perkin Trans. 1* **1997**, 511–526.
11. Urabe, H.; Aoyama, Y.; Sato, F. *Tetrahedron* **1992**, *48*, 5639–5646.
12. Latif, F.; Malik, A.; Völker, W. *Liebigs Ann. Chem.* **1987**, 717–720.
13. Ibuka, T.; Nakai, K.; Habashita, H.; Hotta, Y.; Otaka, A.; Tamamura, H.; Fujii, N.; Mimura, N.; Miwa, Y.; Taga, T.; Chounan, Y.; Yamamoto, Y. *J. Org. Chem.* **1995**, *60*, 2044–2058.
14. Schomaker, J. M.; Geiser, A. R.; Huang, R.; Borhan, B. *J. Am. Chem. Soc.* **2007**, *129*, 3794–3795.
15. Benfatti, F.; Cardillo, G.; Gentilucci, L.; Perciaccante, R.; Tolomelli, A.; Catapano, A. *J. Org. Chem.* **2006**, *71*, 9229–9232.
16. Brånalt, J.; Kvarnström, I.; Classon, B.; Samuelsson, B. *J. Org. Chem.* **1996**, *61*, 3604–3610.
17. Liu, C.-Q.; Hashimoto, Y.; Kudo, K.; Saigo, K. *Bull. Chem. Soc. Jpn.* **1996**, *69*, 2095–2105.
18. Rosen, T. In *Comprehensive Organic Synthesis*; Trost, B. M.; Fleming, I., Eds.; Pergamon: Oxford, UK, **1991**; Vol. 2, pp. 409–441.
19. For the Hörner–Wadsworth–Emmons reactions, see: (a) Ando, K. *J. Org. Chem.* **1999**, *64*, 8406–8408. (b) Ando, K.; Oishi, T.; Hirama, M.; Ohno, H.; Ibuka, T. *J. Org. Chem.* **2000**, *65*, 4745–4749.
20. (a) Lifchits, O.; Mahlau, M.; Reisinger, C. M.; Lee, A.; Farès, C.; Polyak, I.; Gopakumar, G.; Thiel, W.; List, B. *J. Am. Chem. Soc.* **2013**, *135*, 6677–6693. (b) Makó, A.; Rapi, Z.; Keglevich, G.; Szöllősy, A.; Drahos, L.; Hegedűs, L.; Bakó, P. *Tetrahedron: Asymmetry* **2010**, *21*, 919–925. (c) Yoo, M.-S.; Kim, D.-G.; Ha, M.-W.; Jew, S.-S.; Park, H.-G.; Jeong, B.-S. *Tetrahedron Lett.* **2010**, *51*, 5601–5603. (d) El Kadiri, M. Y.; Framery, E.; Andrioletti, B. *Tetrahedron Lett.* **2012**, *53*, 6335–6338.
21. (a) Kelly, D. R.; Meek, A.; Roberts, S. M., *Chem. Commun.* **2004**, 2021–2022. (b) Yoo, M.-S.; Kim, D.-G.; Ha, M.-W.; Jew, S.-S.; Park, H.-G.; Jeong, B.-S. *Tetrahedron Lett.* **2010**, *51*, 5601–5603. (c) Wang, B.; Miao, C.-X.; Wang, S.-F.; Xia, C.-G.; Sun, W. *Chem. – Eur. J.* **2012**, *18*, 6750–6753. (d) Lifchits, O.; Mahlau, M.; Reisinger, C. M.; Lee, A.; Farès, C.; Polyak, I.; Gopakumar, G.; Thiel, W.; List, B. *J. Am. Chem. Soc.* **2013**, *135*, 6677–6693.
22. (a) Li, Z.; Zhang, W.; Yamamoto, H. *Angew. Chem., Int. Ed.* **2008**, *47*, 7520–7522. (b) Barlan, A. U.; Zhang, W.; Yamamoto, H. *Tetrahedron* **2007**, *63*, 6075–6087.
23. For recent representative examples, see: (a) Hookins, D. R.; Burns, A. R.; Taylor, R. J. K. *Eur. J. Org. Chem.* **2011**, 451–454 (L-Selectride). (b) Mehta, G.; Roy, S.; Pan, S. C. *Tetrahedron Lett.* **2012**, *53*, 4093–4095 (DIBAL).
24. For recent representative examples, see: (a) Ref. 21d ($NaBH_4 + Zn(ClO_4)_2$). (b) Adachi, M.; Higuchi, K.; Thasana, N.; Yamada, H.; Nishikawa, T. *Org. Lett.* **2012**, *14*, 114–117 ($LiAlH(OBu^{-t})_3$. (c) Buchanan, G. S.; Cole, K. P.; Li, G.; Tang, Y.; You, L.-F.; Hsung, R. P. *Tetrahedron* **2011**, *67*, 10105–10118 ($NaBH_4 + CeCl_3$).
25. Katsuki, T.; Sharpless, K. B. *J. Am. Chem. Soc.* **1980**, *102*, 5974–5976.

26. (a) Underwood, B. S.; Tanuwidjaja, J.; Ng, S.-S.; Jamison, T. F. *Tetrahedron* **2013**, *69*, 5205–5220. (b) El-Behairy, M. F.; Sundby, E. *Tetrahedron: Asymmetry* **2013**, *24*, 285–289. (c) Yadav, J. S.; Sengupta, S. *Eur. J. Org. Chem.* **2013**, 376–388.
27. (a) Yu, Y.; Plettner, E. *Bioorg. Med. Chem.* **2013**, *21*, 1811–1822. (b) Wrona, I. E.; Lowe, J. T.; Turbyville, T. J.; Johnson, T. R.; Beignet, J.; Beutler, J. A.; Panek, J. S. *J. Org. Chem.* **2009**, *74*, 1897–1916. (c) Cortijos, A. M.; Snape, T. J. *Tetrahedron: Asymmetry* **2008**, *19*, 1761–1763. (d) Faigl, F.; Thurner, A.; Farkas, F.; Battancs, M.; Poppe, L. *Chirality* **2007**, *19*, 197–202.
28. (a) Chen, M. Z.; McLaughlin, M.; Takahashi, M.; Tarselli, M. A.; Yang, D.; Umemura, S.; Micalizio, G. C. *J. Org. Chem.* **2010**, *75*, 8048–8059.(b) Yang, D.-X.; Micalizio, G. C. *J. Am. Chem. Soc.* **2012**, *134*, 15237–15240. (c) He, P.; Liu, X.-H.; Zheng, H.-F.; Li, W.; Lin, L.-L.; Feng, X.-M. *Org. Lett.* **2012**, *14*, 5134–5137.
29. Hatano, M.; Mizuno, T.; Ishihara, K. *Chem. Commun.* **2010**, *46*, 5443–5445.
30. (a) Fernández-Mateos, A.; Teijón, P. H.; González, R. R. *Tetrahedron* **2013**, *69*, 1611–1616. (b) Nicolas, L.; Anderl, T.; Sasse, F.; Steinmetz, H.; Jansen, R.; Höfle, G.; Laschat, S.; Taylor, R. E. *Angew. Chem., Int. Ed.* **2011**, *50*, 938–941.
31. (a) Nuhant, P.; Kister, J.; Lira, R.; Sorg, A.; Roush, W. R. *Tetrahedron* **2011**, *67*, 6497–6512. (b) Krishna, P. R.; Ramana, D. V. *J. Org. Chem.* **2012**, *77*, 674–679. (c) Schmidt, B.; Kunz, O.; Petersen, M. H. *J. Org. Chem.* **2012**, *77*, 10897–10906. (d) Reddy, P. R.; Sudhakar, C.; Kumar, J. N.; Das, B. *Helv. Chim. Acta* **2013**, *96*, 289–295.
32. Caron, M.; Sharpless, K. B. *J. Org. Chem.* **1985**, *50*, 1557–1560.
33. Sasaki, M.; Tanino, K.; Hirai, A.; Miyashita, M. *Org. Lett.* **2003**, *5*, 1789–1791.
34. (a) Tsuna, K.; Noguchi, N.; Nakada, M. *Angew. Chem., Int. Ed.* **2011**, *50*, 9452–9455. (b) Takamura, H.; Iwamoto, K.; Nakao, E.; Kadota, I. *Org. Lett.* **2013**, *15*, 1108–1111.
35. (a) Mizutani, H.; Watanabe, M.; Honda, T. *Tetrahedron* **2002**, *58*, 8929–8936. (b) Singh, S.; Guiry, P. J. *Tetrahedron* **2010**, *66*, 5701–5706.
36. (a) Page, P. C. B.; Rayner, C. M.; Sutherland, I. O. *J. Chem. Soc., Chem. Commun.* **1988**, 356–358. (b) Page, P. C. B.; Rayner, C. M.; Sutherland, I. O. *J. Chem. Soc., Perkin Trans. 1* **1990**, 1375–1382.
37. Pastó, M.; Moyano, A.; Pericàs, M. A.; Riera, A. *J. Org. Chem.* **1997**, *62*, 8425–8431.
38. Hu, X.-D.; Fan, C.-A.; Zhang, F.-M.; Tu, Y.-Q. *Angew. Chem., Int. Ed.* **2004**, *43*, 1702–1705.
39. (a) Chiummiento, L.; Funicello, M.; Lupattelli, P.; Tramutola, F. *Org. Lett.* **2012**, *14*, 3928–3931. (b) Ohnliya, H.; Yoshida, M.; Sawamura, M. *Synlett* **2010**, 2136–2140.
40. Benedetti, F.; Berti, F.; Budal, S.; Campaner, P.; Dinon, F.; Tossi, A.; Argirova, R.; Genova, P.; Atanassov, V.; Hinkov, A. *J. Med. Chem.* **2012**, *55*, 3900–3910.
41. Reddipalli, G.; Venkataiah, M.; Fadnavis, N. W. *Tetrahedron: Asymmetry* **2011**, *22*, 1778–1783.
42. Jimeno, C.; Pastó, M.; Riera, A.; Pericàs, M. A. *J. Org. Chem.* **2003**, *68*, 3130–3138.
43. For a brief discussion about relationship between a size of a Lewis acid and selectivity, see: Mikami, K.; Ohba, S.; Ohmura, H. *J. Organomet. Chem.* **2002**, *662*, 77–82.
44. Lindström, U. M.; Ding, R.; Hidestål, O. *Chem. Commun.* **2005**, 1773–1774.
45. (a) Baldwin, J. E. *J. Chem. Soc., Chem. Commun.* **1976**, 734–736. (b) Baldwin, J. E.; Cutting, J.; Dupont, W.; Kruse, L.; Silberman, L.; Thomas, R. C. *J. Chem. Soc., Chem. Commun.* **1976**, 736–738.

46. Lindström, U. M.; Somfai, P. *Tetrahedron Lett.* **1998**, *39*, 7173–7176.
47. Mukerjee, P.; Abid, M.; Schroeder, F. C. *Org. Lett.* **2010**, *12*, 3986–3989.
48. Rengasamy, R.; Curtis-Long, M. J.; Seo, W.-D.; Jeong, S. H.; Jeong, I.-Y.; Park, K.-H. *J. Org. Chem.* **2008**, *73*, 2898–2901.
49. Nishimura, Y. *Curr. Top. Med. Chem.* **2003**, *3*, 575–591.
50. Velvadapu, V.; Paul, T.; Wagh, B.; Klepacki, D.; Guvench, O.; MacKerell, Jr., A.; Andrade, R. B. *ACS Med. Chem. Lett.* **2011**, *2*, 68–72.
51. (a) Cuccarese, M. F.; Wang, H.-Y. L.; O'Doherty, G. A. *Eur. J. Org. Chem.* **2013**, 3067–3075. (b) Fujiwara, K.; Suzuki, Y.; Koseki, N.; Murata, S.; Murai, A.; Kawai, H.; Suzuki, T. *Tetrahedron Lett.* **2011**, *52*, 5589–5592.
52. Vila-Gisbert, S.; Urbano, A.; Carreño, M. C. *Chem. Commun.* **2013**, *49*, 3561–3563.
53. Kim, H.-C.; Kang, S.-H. *Angew. Chem., Int. Ed.* **2011**, *50*, 1827–1829.
54. (a) For similar instances, see: (b) Dehn, R.; Katsuyama, Y.; Weber, A.; Gerth, K.; Jansen, R.; Steinmetz, H.; Höfle, G.; Müller, R.; Kirschning, A. *Angew. Chem., Int. Ed.* **2011**, *50*, 3882–3887. (c) Reddy, G. V.; Kumar, R. S. C.; Siva, B.; Babu, K. S.; Rao, J. M. *Synlett* **2012**, 2677–2681. (d) Li, J.-Y.; Fu, N.-K.; Zhang, L.; Zhou, P.-X.; Luo, S.-Z.; Cheng, J.-P. *Eur. J. Org. Chem.* **2011**, 6840–6849.
55. Rahaim, Jr., R. J.; Maleczka, Jr., R. E. *Org. Lett.* **2011**, *13*, 584–587.
56. If the both epoxy carbon atoms are trisubstituted and the 3 position is activated by an allyl group, the hydride regiospecifically attacks at the 3 position. See: Takamura, H.; Wada, H.; Lu, N.; Ohno, O.; Suenaga, K.; Kadota, I. *J. Org. Chem.* **2013**, *78*, 2443–2454.
57. Bernet, B.; Vasella, A. *Helv. Chim. Acta* **2007**, *90*, 1874–1888.
58. Ghilagaber, S.; Hunter, W. N.; Marquez, R. *Org. Biomol. Chem.* **2007**, *5*, 97–102.
59. Barbier, P.; Mohr, P.; Muller, M.; Masciadri, R. *J. Org. Chem.* **1998**, *63*, 6984–6989.
60. Chandra, G.; Majik, M. S.; Lee, J.-Y.; Jeong, L.-S. *Org. Lett.* **2012**, *14*, 2134–2137.
61. Landini, D.; Albanese, D.; Penso, M. *Tetrahedron* **1992**, *48*, 4163–4170.
62. Hara, S.; Hoshio, T.; Kameoka, M.; Sawaguchi, M.; Fukuhara, T.; Yoneda, N. *Tetrahedron* **1999**, *55*, 4947–4954.
63. Itoh, Y.; Jang, S.; Ohba, S.; Mikami, K. *Chem. Lett.* **2004**, 776–777.
64. Alvernhe, G.; Lacombe, S.; Laurent, A.; Rousset, C. *J. Chem. Res., Synop.* **1983**, 246–247.
65. Olah, G. A.; Welch, J. T.; Vankar, Y. D.; Nojima, M.; Kerekes, I.; Olah, J. A. *J. Org. Chem.* **1979**, *44*, 3872–3881.
66. (a) Hiyama, T. *Organofluorine Compounds, Chemistry and Applications*; Springer: Berlin, **2000**. (b) Smart, B. E. *J. Fluorine Chem.* **2001**, *109*, 3–11. (c) Kirsch, P. *Modern Fluoroorganic Chemistry: Synthesis, Reactivity, Applications*; Wiley-VCH: Weinheim, **2004**. (d) Uneyama, K. *Organofluorine Chemistry*; Blackwell: Oxford, UK, **2006**. (e) Bégué, J.-P.; Bonnet-Delpon, D. *Bioorganic and Medicinal Chemistry of Fluorine*; John Wiley & Sons, Inc.: New Jersey, **2008**. (f) Yamazaki, T.; Taguchi, T.; Ojima, I. In *Fluorine in Medicinal Chemistry and Chemical Biology*; Ojima, I., Ed.; John Wiley & Sons, Inc.: West Sussex, UK, **2009**, pp. 3–46.
67. (a) Valero, G.; Companyó, X.; Rios, R. *Chem. – Eur. J.* **2011**, *17*, 2018–2037. (b) Roy, S.; Gregg, B. T.; Gribble, G. W.; Le, V.-D.; Roy, S. *Tetrahedron* **2011**, *67*, 2161–2195. (c) Tomashenko, O. A.; Grushin, V. V. *Chem. Rev.* **2011**, *111*, 4475–4521. (d) Hollingworth, C.; Gouverneur, V. *Chem. Commun.* **2012**, *48*, 2929–2942. (e) Zhao, Y.-J.; Pan, Y.-H.; Sim, S.-B. D.;Tan, C.-H. *Org. Biomol. Chem.* **2012**, *10*, 479–485.

68. Von dem Bussche-Hünnefeld, C.; Seebach, D. *Chem. Ber.* **1992**, *125*, 1273–1281.

69. (a) Differding, E.; Frick, W.; Lang, R. W.; Martin, P.; Schmit, C.; Veenstra, S.; Greuter, H. *Bull. Soc. Chim. Belg.* **1990**, *99*, 647–671. (b) Bansal, R. C.; Dean, B.; Hakomori, S.; Toyokuni, T. *J. Chem. Soc., Chem. Commun.* **1991**, 796–798. (c) Hanzawa, Y.; Uda, J.; Kobayashi, Y.; Ishido, Y.; Taguchi, T.; Shiro, M. *Chem. Pharm. Bull.* **1991**, *39*, 2459–2461.

70. Yamazaki, T.; Mizutani, K.; Kitazume, T. *J. Org. Chem.* **1993**, *58*, 4346–4359.

71. The electron-withdrawing nature of a CF_3 group is sometimes compared with the one of an ester functional group. For example, the pK_a values of CHF_2CF_3 and CHF_2CO_2Me were reported to be 28.2 and 25, respectively. See Ref. 66b. Moreover, pK_a values of CF_3CO_2H and HO_2CCO_2H were 0.5 and 1.271. Dean, J. A. *Lange's Handbook of Chemistry*, 14th ed., McGraw-Hill: New York, **1992**; Section 8.

72. Yamazaki, T.; Oniki, T.; Kitazume, T. *Tetrahedron* **1996**, *52*, 11753–11762.

73. Yamazaki, T.; Ichige, T.; Kitazume, T. *Org. Lett.* **2004**, *6*, 4073–4076.

74. Computation was carried out by Gaussian 03W, version 6.0 using B3LYP/6-31+G* level of theory. Frisch, M. J.; Trucks, G. W.; Schlegel, H. B.; Scuseria, G. E.; Robb, M. A.; Cheeseman, J. R.; Montgomery, J. A.; Vreven, T.; Kudin, K. N.; Burant, J. C.; Millam, J. M.; Iyengar, S. S.; Tomasi, J.; Barone, V.; Mennucci, B.; Cossi, M.; Scalmani, G.; Rega, N.; Petersson, G. A.; Nakatsuji, H.; Hada, M.; Ehara, M.; Toyota, K.; Fukuda, R.; Hasegawa, J.; Ishida, M.; Nakajima, T.; Honda, Y.; Kitao, O.; Nakai, H.; Klene, M.; Li, X.; Knox, J. E.; Hratchian, H. P.; Cross, J. B.; Adamo, C.; Jaramillo, J.; Gomperts, R.; Stratmann, R. E.; Yazyev, O.; Austin, A. J.; Cammi, R.; Pomelli, C.; Ochterski, J. W.; Ayala, P. Y.; Morokuma, K.; Voth, G. A.; Salvador, P.; Dannenberg, J. J.; Zakrzewski, V. G.; Dapprich, S.; Daniels, A. D.; Strain, M. C.; Farkas, O.; Malick, D. K.; Rabuck, A. D.; Raghavachari, K.; Foresman, J. B.; Ortiz, J. V.; Cui, Q.; Baboul, A. G.; Clifford, S.; Cioslowski, J.; Stefanov, B. B.; Liu, G.; Liashenko, A.; Piskorz, P.; Komaromi, I.; Martin, R. L.; Fox, D. J.; Keith, T.; Al-Laham, M. A.; Peng, C. Y.; Nanayakkara, A.; Challacombe, M.; Gill, P. M. W.; Johnson, B.; Chen, W.; Wong, M. W.; Gonzalez, C.; Pople, J. A. Gaussian, Inc.: Pittsburgh, PA, **2003**.

75. Allen, J. V.; Bergeron, S.; Griffiths, M. J.; Mukherjee, S.; Roberts, S. M.; Williamson, N. M.; Wu, L. E. *J. Chem. Soc., Perkin Trans. 1* **1998**, 3171–3180.

76. Fustero, S.; Pina, B.; Torre, M. G.; Navarro, A.; Arellano, C. R.; Simon, A. *Org. Lett.* **1999**, *1*, 977–980.

77. Delgado, A.; Clardy, J. *Tetrahedron Lett.* **1992**, *33*, 2789–2790.

78. Fu, R.; Ye, J.-L.; Dai, X.-J.; Ruan, Y.-P.; Huang, P.-Q. *J. Org. Chem.* **2010**, *75*, 4230–4243.

79. Shinohara, N.; Haga, J.; Yamazaki, T.; Kitazume, T.; Nakamura, S. *J. Org. Chem.* **1995**, *60*, 4363–4374.

80. Behrens, C. H.; Ko, S.-Y.; Sharpless, K. B.; Walker, F. J. *J. Org. Chem.* **1985**, *50*, 5687–5696.

81. Obinata, R.; Kawasaki-Takasuka, T.; Yamazaki, T. *Org. Lett.* **2010**, *12*, 4316–4319.

82. Yamazaki, T.; Iwatsubo, H.; Kitazume, T. *Tetrahedron: Asymmetry* **1994**, *5*, 1823–1830.

83. (a) Schubert, J.; Schwesinger, R.; Knothe, L.; Prinzbach, H. *Liebigs Ann. Chem.* **1986**, 2009–2052. (b) Kuehlmeyer, R.; Seitz, B.; Weller, T.; Fritz, H.; Schwesinger, R.; Prinzbach, H. *Chem. Ber.* **1989**, *122*, 1729–1743.

11

VINYLCYCLOPROPANE–CYCLOPENTENE REARRANGEMENT

Ashish Thakur and Janis Louie
Department of Chemistry, University of Utah, Salt Lake City, UT, USA

11.1 INTRODUCTION

The synthesis of useful organic motifs via molecular rearrangement of strained molecules continues to be an area of profound interest to synthetic chemists. The strain energy of a variety of three-membered carbocycles has been utilized to develop new synthetic transformations.[1] One of the important reactions that has been explored within this class of molecules is the vinylcyclopropane (VCP)–cyclopentene (CP) rearrangement,[2] which involves the formation of a cyclopentene ring from the ring expansion of the VCP via C—C bond cleavage (Scheme 11.1).

Cyclopentene motifs are present in a number of biologically active molecules and natural products including terpenes and prostaglandins. As such, this rearrangement can provide rapid access to these cores in a step-economical process.[2a,f] Additionally, the use of substituted VCPs allows for the regio- and diastereoselective synthesis of cyclopentene building blocks that can be subjected to postreaction modification to access variety of complex molecular architectures.[2a,f,3] Finally, the introduction of heteroatoms in VCP moiety provides the opportunity to synthesize heterocyclic rings that are present in a variety of alkaloids, terpenoids, and other natural products.[1j,2f,4]

Neureiter first discovered this rearrangement reaction in 1959 by subjecting 1,1-dichloro-2-vinylcyclopropane to pyrolysis (>400 °C), affording 4,4-dichlorocyclopentene along with unidentified chlorinated hydrocarbons (Scheme 11.2, Eq. 1).[5a] The all-carbon VCP–CP rearrangement was discovered independently by Vogel and by Overberger and Borchert in 1960 (Scheme 11.2, Eq. 2),[5b–d] although

Molecular Rearrangements in Organic Synthesis, First Edition. Edited by Christian M. Rojas.

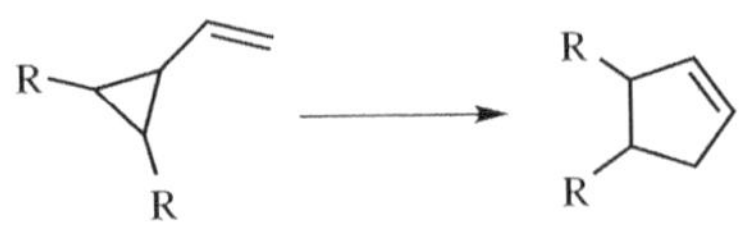

Scheme 11.1 VCP–CP rearrangement.

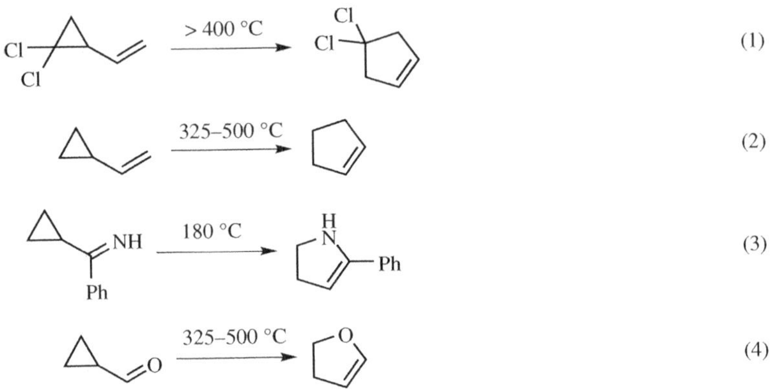

Scheme 11.2 Thermal rearrangement of (1) 1,1-dichloro-2-VCP, (2) unsubstituted VCP, (3) cyclopropylimine, and (4) cyclopropylcarbaldehyde.

two heteroatom variants of this rearrangement, a cyclopropylimine–pyrroline rearrangement[5e] and cyclopropylcarbaldehyde-2,3-dihydrofuran rearrangement,[5f] had been known in the literature for a few decades (Scheme 11.2, Eqs. 3 and 4).

Since the discovery of the carbon VCP–CP rearrangement, tremendous progress has been made in this field using thermal, acid-mediated/catalyzed, photochemical, and metal-catalyzed reaction conditions. This chapter will focus on the developments made in VCP–CP rearrangements, their heteroatom-substituted analogues, and the application of these rearrangements in syntheses.

11.2 THERMAL VCP–CP REARRANGEMENT

Despite the early discovery of a thermal VCP–CP rearrangement in 1959, the harsh reaction conditions required for the reaction significantly hampered its potential as a useful synthetic method. Corey and Walinsky reported in 1972 that the introduction of a dithiane group on the cyclopropyl ring of the VCP enabled its rearrangement to cyclopentene product at a significantly lower temperature (200 °C) and in high yields (Scheme 11.3).[6]

In 1973, Richey and Simpson discovered that the presence of methoxy and phenyl substituents on the cyclopropyl ring of the VCP lowered the activation energy for this process by 5–11 kcal/mol, thereby facilitating the rearrangement at 220 °C.[7] Trost and Bogdanowicz observed a similar effect, lowering the activation energy by the

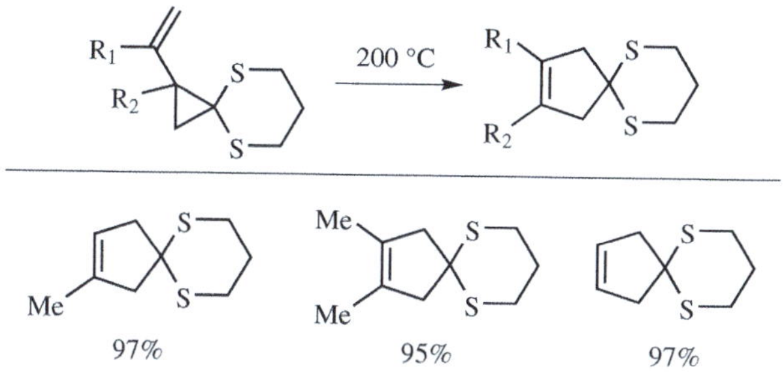

Scheme 11.3 Thermal rearrangement of 1,3-dithiane-substituted VCPs.

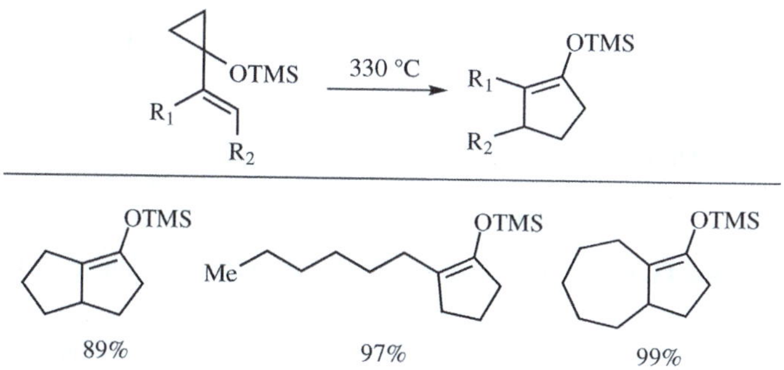

Scheme 11.4 Thermal rearrangement of siloxy-substituted VCPs.

use of siloxy-substituted VCPs to afford synthetically useful cyclopentene products in excellent yields (Scheme 11.4).[8a]

The Trost group successfully extended this concept to include the rearrangement of thioether-substituted VCPs, which are more readily synthesized, affording a variety of cyclopentenes with a broader substrate scope (Scheme 11.5).[8b]

Interestingly, the thermal rearrangement of a thioether–VCP containing a *tert*-butyl cyclohexene **1** proceeded in a stereoselective fashion to form the annelated cyclopentene ring **2** in high yield with an equatorial stereochemistry at the migration terminus (Scheme 11.6).[8b] The stereochemical outcome of this rearrangement could be rationalized by the fact that the formation of an equatorial bond at the migration terminus would allow the six-membered ring to be in a more favorable chair conformation, whereas axial bond formation would force the six-membered ring to adopt an unfavorable twist boat conformation.

Although the previously mentioned heteroatom-substituted VCPs (see Schemes 11.3–11.6) underwent conversion to cyclopentenes at somewhat lower temperatures than the original chloro-substituted and all-carbon systems, lowering reaction temperatures even further would make the rearrangement synthetically more useful. The Danheiser group demonstrated a significant rate enhancement of the VCP–CP rearrangement by the use of lithium salts of vinylcyclopropanols. Cyclopentenol formation proceeded effectively at room temperature in low to excellent yields (Scheme 11.7).[9]

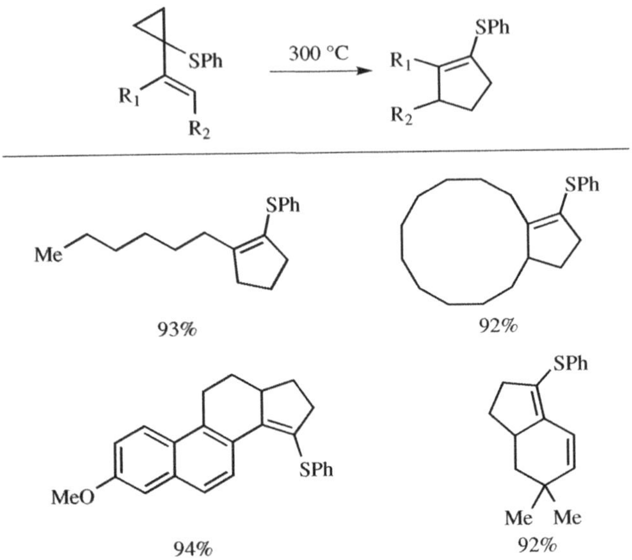

Scheme 11.5 Thermal rearrangement of thioether-substituted VCPs.

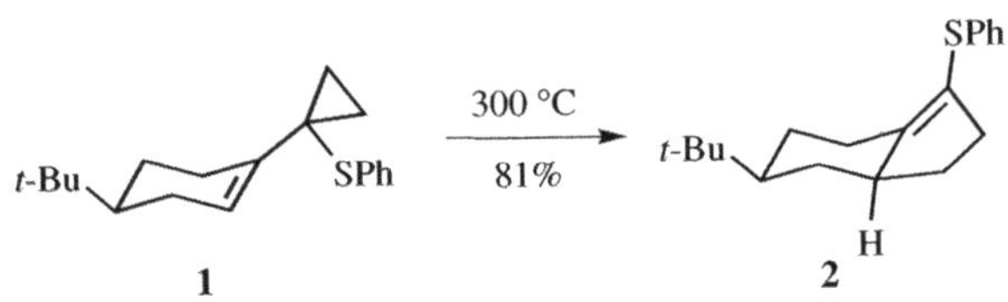

Scheme 11.6 Stereoselective thermal rearrangement of thioether-VCP **1**.

Larsen *et al.* reported an interesting way of synthesizing cyclopentenes from dihydrothiopyrans **I** (Scheme 11.8), that involved rearrangement of highly activated VCPs **III**, as one of the possible route to the product. The hetero Diels–Alder cycloadducts **I** were deprotonated at α-position to sulfur-forming carbanion intermediates **II** that can undergo either 1,2-addition or 1,4-addition to the thioaldehyde.[10a,b] 1,2-Addition followed by methylation with CH_3I would directly result in the cyclopentene product **IV**, whereas 1,4-addition would lead to an intermediate VCP **III**, that could undergo rearrangement, followed by methylation to form cyclopentene **IV**. Remarkably, the rearrangement occurred at −78 °C, affording a variety of cyclopentenes in good to excellent yields with high diastereoselectivity. Additionally, cyclopentene-containing contiguous quaternary centers **5** were synthesized using this methodology.

The Hudlicky group also took advantage of anion-accelerated VCP–CP rearrangements by treating (siloxyvinyl)cyclopropanes to the *tetra*-butylammonium fluoride (TBAF) or trimethylsilyl iodide in the presence of hexamethyldisilazane (TMSI/HMDS). These conditions promoted rearrangement to form [5–5] and [5–6] annulated cyclopentenes at low temperatures with high selectivities (Scheme 11.9).[11]

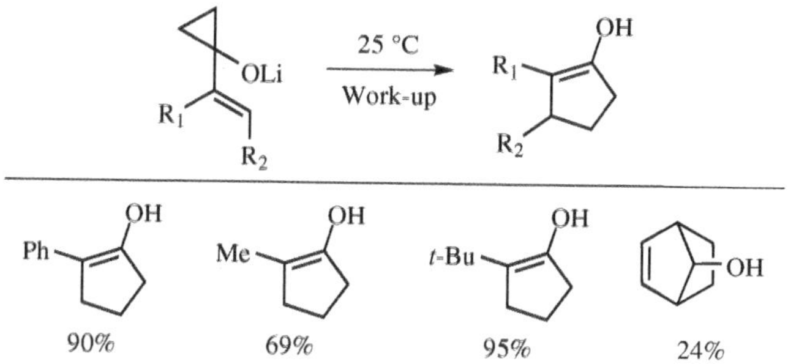

Scheme 11.7 Thermal rearrangement of lithium salts of vinylcyclopropanols.

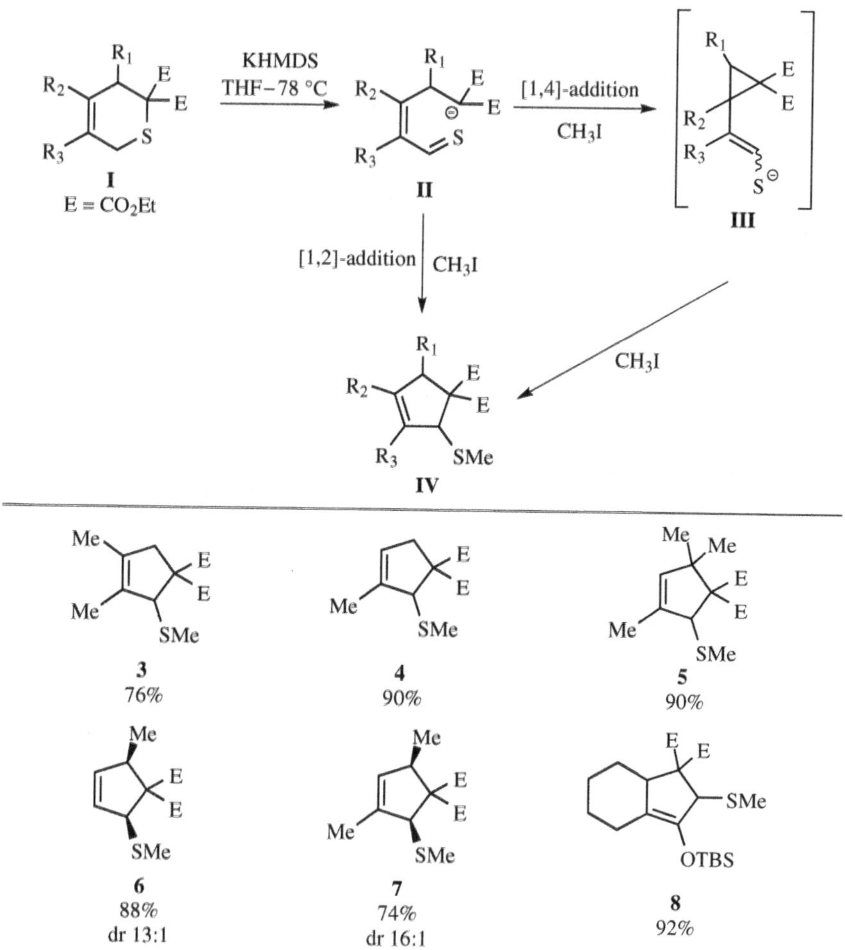

Scheme 11.8 *In situ* generation of activated VCPs from dihydrothiopyrans and their thermal rearrangement.

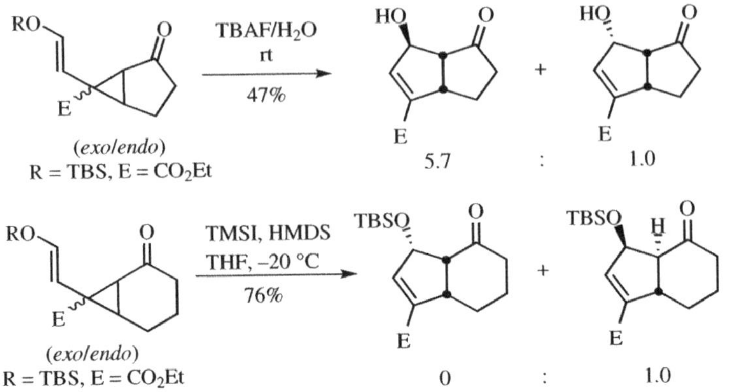

Scheme 11.9 TBAF or TMSI-promoted rearrangement of (siloxyvinyl)cyclopropanes.

11.3 ACID-MEDIATED VCP–CP REARRANGEMENT

VCPs bearing acceptor groups on the cyclopropyl ring can be activated by acid, resulting in cleavage of the cyclopropyl ring to form zwitterionic species that ultimately collapse to form cyclopentenes. In 1986, Suzukamo's group reported that 2-styrylcyclopropanecarbonyl chlorides undergo BBr_3-catalyzed isomerization to form substituted cyclopentenes at room temperature in good yields. The reaction, however, had a limited substrate scope, and only VCPs having electron-rich phenyl groups, for example, **9**, rearranged to give substituted cyclopentenes **10** (Scheme 11.10).[12]

Davies and coworkers developed a Et_2AlCl-promoted stereoselective rearrangement of donor–acceptor VCPs **VI**, which were synthesized by the $Rh_2(Oct)_4$-catalyzed cyclopropanation of ethyl vinyl ethers **IV**, using vinyldiazoesters **V** (Scheme 11.11). This process led to highly substituted cyclopentenes in good to excellent yields.[13a]

This methodology has a broad substrate scope, and both monocyclic and bicyclic VCPs were successfully isomerized to the corresponding cyclopentenes with excellent diastereoselectivity. Interestingly, Davies's group discovered that these VCP rearrangements are highly dependent on the nature of the ester group on the cyclopropyl ring. While the methylester group remained intact in the Lewis acid-mediated ring

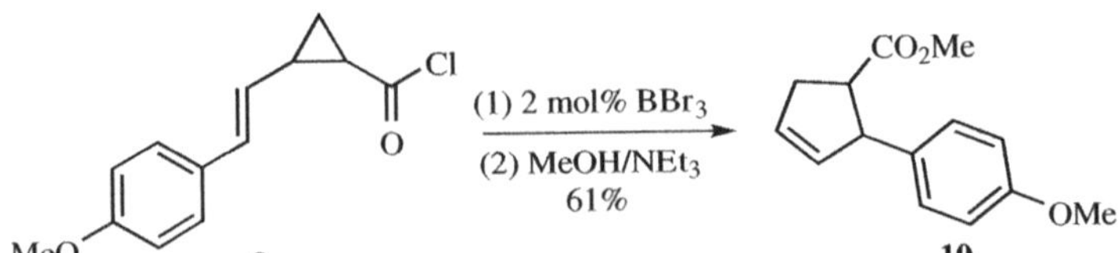

9 **10**

Scheme 11.10 BBr_3-catalyzed rearrangement of 2-styrylcyclopropanecarbonyl chlorides.

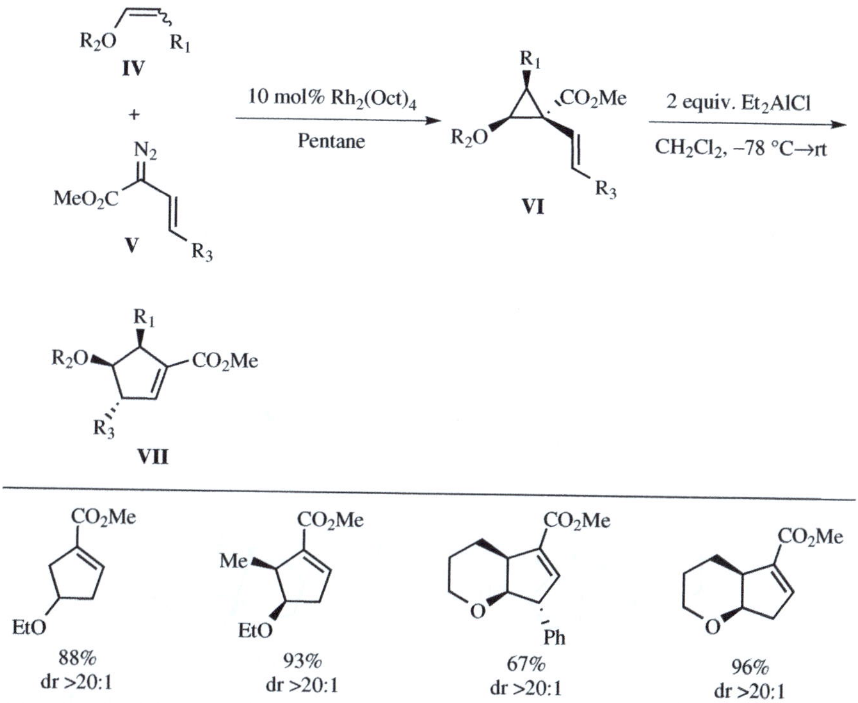

Scheme 11.11 Rh-catalyzed stereoselective synthesis of VCPs and their Et_2AlCl-promoted rearrangement.

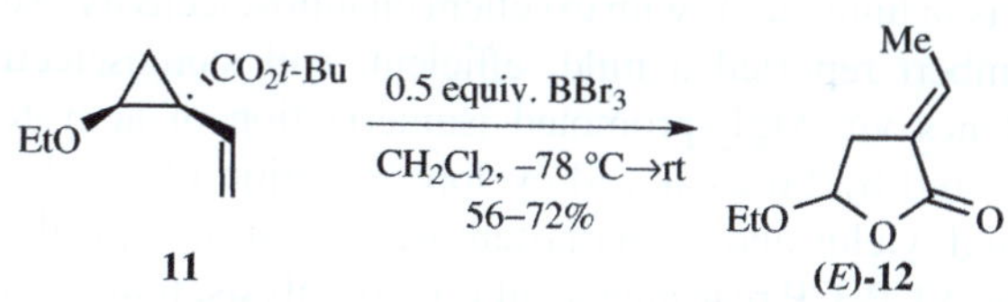

Scheme 11.12 BBr_3-mediated rearrangement of *tert*-butylester-substituted VCP.

opening of methylester-substituted VCPs **VI** (Scheme 11.11), the rearrangement of corresponding *tert*-butylester containing VCP **11** afforded butyrolactone (*E*)-**12** in good yield (Scheme 11.12).[13b]

This divergent outcome in the case of *tert*-butyl containing VCP **11** was rationalized by the formation of an intermediate vinylcyclopropyl carboxylic acid **VIII** due to the labile nature of *tert*-butyl group under Lewis acidic conditions. This carboxylic acid would then undergo a proton transfer-induced ring opening to form a zwitterionic intermediate **IX** that would collapse to yield the corresponding butyrolactone **X** (Scheme 11.13).[13b]

Davies's group also extended this methodology toward the asymmetric syntheses of substituted cyclopentenes. Although chiral monocyclic VCPs isomerized to

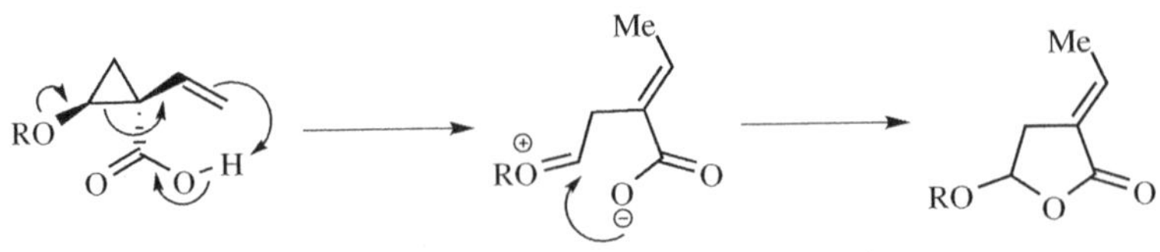

Scheme 11.13 Mechanistic hypothesis for the rearrangement of *tert*-butylester-substituted VCP to form butyrolactone.

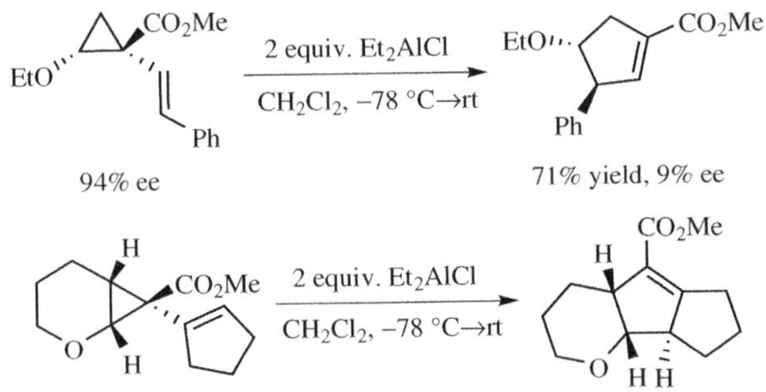

Scheme 11.14 Et_2AlCl-promoted rearrangement of chiral monocyclic and bicyclic VCPs.

racemic cyclopentenes under the reaction conditions, the use of fused chiral VCPs led to the products in high yields with excellent enantioselectivity (Scheme 11.14).[13c]

In 2009, Lambert reported a mild, efficient, and stereoselective synthesis of fused cyclopentenes via MgI_2-promoted isomerization of activated VCPs.[14] The VCPs were prepared by Pd-catalyzed cyclopropanation of 1,3-dienyl β-ketoesters in moderate to high yields and good diastereoselectivities. Typical standard reaction conditions for the VCP–CP rearrangement (i.e., pyrolysis, transition metal catalysis, standard Lewis acid catalysis) were found to be ineffective for these substrates. However, the use of 1.5 equiv. of MgI_2 led to complete conversion, providing bicyclic cyclopentenes in high yields (Scheme 11.15).

11.4 MECHANISMS

11.4.1 Mechanism of the Thermal Vinylcyclopentene Rearrangement

Thermal VCP–cyclopentene rearrangements are generally believed to operate by two different mechanistic pathways.[2a–c,e] The first of these involves the homolytic fission of the cyclopropyl ring to generate a biradical species, which then recombines to form the cyclopentene ring. The second mechanistic alternative involves a concerted, orbital symmetry-controlled process to form the cyclopentene ring (Scheme 11.16).

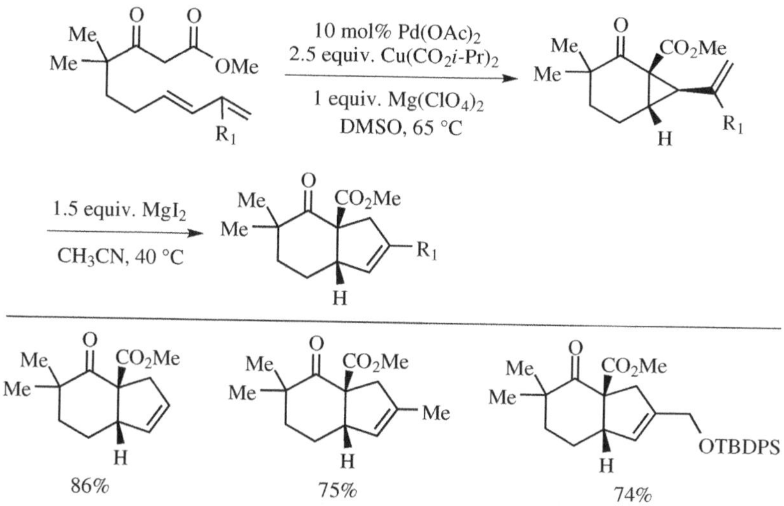

Scheme 11.15 Pd-catalyzed cyclopropanation of 1,3-dienyl β-ketoesters to form VCPs and their subsequent MgI_2-promoted rearrangement.

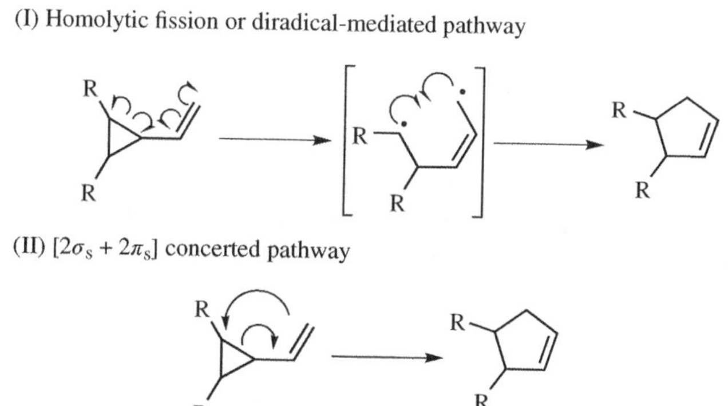

Scheme 11.16 Mechanistic pathways for thermal VCP–cyclopentene rearrangement.

The activation energy, E_a, of the thermal parent VCP–CP rearrangement is approximately 50 kcal/mol and is ~13 kcal/mol less than the activation energy required for the *cis–trans* isomerization of the 1,2-d_2-cyclopropanes.[2a–c,e] Interestingly, the realization that the resonance energy of the allyl radical (about 13 kcal/mol) might contribute to lowering the E_a for this rearrangement led to the proposal of the diradical pathway.[2a–c,e] Furthermore, substituents on the cyclopropyl ring appear to lower the activation barrier for this rearrangement, which provides support for the diradical mechanism. Specifically, the E_a values determined for the thermal rearrangement of various *trans*-C2-substituted VCPs suggest that the reaction

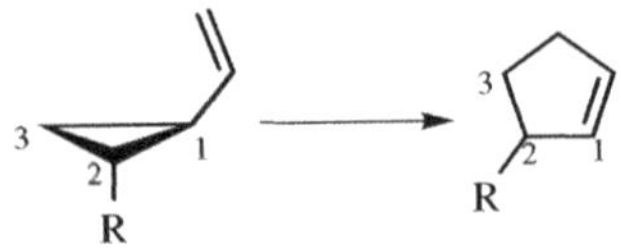

R	E_a (kcal/mol)
Me	48.7
SMe	41.5
OMe	38.7
OEt	38.4
NMe_2	31.2

Scheme 11.17 Activation energy (E_a) values for *trans*-C2-substituted VCPs.

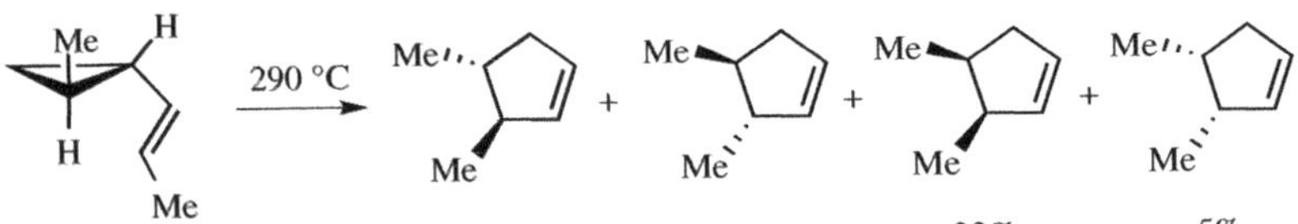

65%	8%	22%	5%
13	**14**	**15**	**16**

Scheme 11.18 Formation of four stereoisomeric products from the thermal rearrangement of *trans*-methyl-substituted VCP.

might follow a diradical pathway along which heteroatom substituents stabilize the radicals, leading to a lower activation energy (Scheme 11.17).[15]

Despite these data, the high level of stereoselectivity usually observed in the rearrangement of enantiopure VCPs suggests that the concerted, symmetry-controlled pathway might be operational as well.[2a,c,e] For example, Baldwin's group reported that the rearrangement of *trans*-methyl-substituted VCP led to four different stereoisomeric cyclopentene products. The two *trans* products, **13** and **14**, are symmetry-allowed and their formation is indicative of the concerted symmetry-allowed pathway being operational in this reaction. However, the formation of symmetry-forbidden *cis* products **15** and **16** could be the result by the stepwise diradical mechanism (Scheme 11.18).[16]

Since this report in 1976, numerous studies of stereoselective ring opening reactions of variously substituted and enantioenriched VCPs have been reported.[2c,e] The formation of different stereoisomeric cyclopentene products suggests that the mechanism of the rearrangement might involve both the concerted and diradical pathways to different extents, depending upon the nature of substituents and their precise orientation on the corresponding VCP.[2c,e]

Interestingly, in the case of the anion-accelerated vinylcyclopentane–CP rearrangement,[9–11,17a] the involvement of ylide or 1,3-zwitterionic intermediates may facilitate the process at low temperatures and in a stereoselective manner. For example, Takeda proposed that the rearrangement of vinylcyclopropylacetates **17**

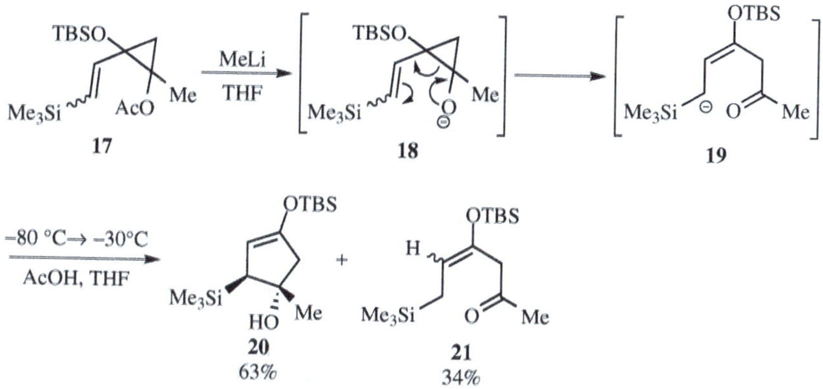

Scheme 11.19 Mechanism of MeLi-promoted rearrangement of vinylcyclopropylacetates.

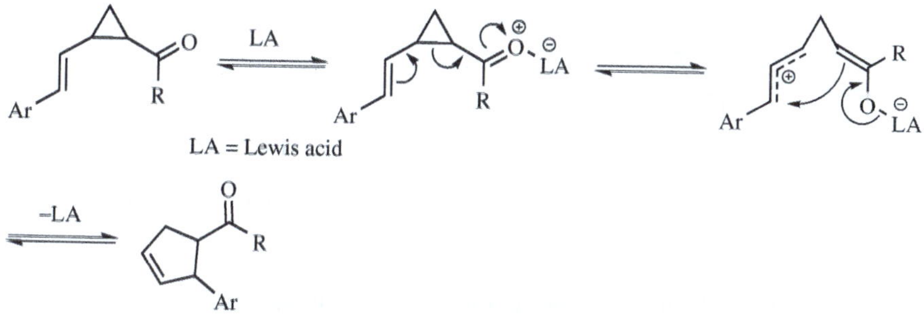

Scheme 11.20 Mechanism of Lewis-acid mediated/catalyzed rearrangement of activated VCPs.

using MeLi might involve the formation of a silyl-stabilized carbanion **19** which either cyclizes to form the cyclopentene product or is protonated to form the acyclic ketone product (Scheme 11.19).[17b]

11.4.2 Mechanism of the Acid-Mediated VCP–CP Rearrangement

The general mechanism of acid-promoted/catalyzed isomerization of activated VCPs to cyclopentenes is shown in Scheme 11.20. The initial activation of the acceptor group on the cyclopropyl ring by acid leads to cleavage of the cyclopropyl ring to form a zwitterionic intermediate, which eventually collapses to form the cyclopentene ring.[12,18a]

Another mechanistic possibility that has been proposed with MgI_2- and LiI-promoted annulations of *β*-ketoester cyclopropanes **22** involves the bis-coordination of the *β*-ketoester group to magnesium followed by homoconjugate addition of iodide ion to the VCP, forming an allyl iodide intermediate **23**.[14,18b] The (*E*)-isomer of this intermediate can only revert back to the VCP via an S_N2' pathway, whereas

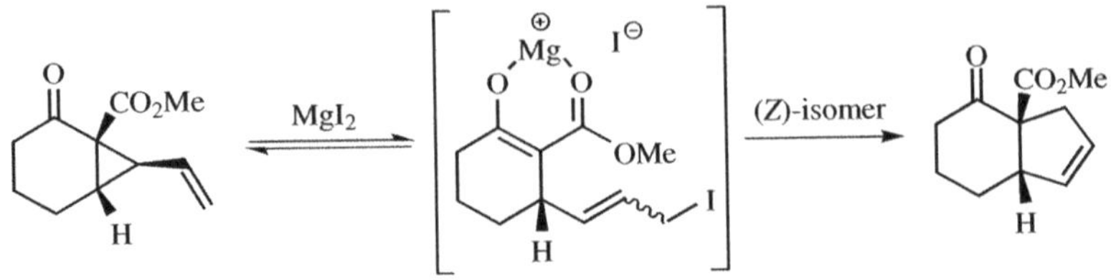

22 23 24

Scheme 11.21 Mechanism of MgI_2-promoted rearrangement of fused VCPs.

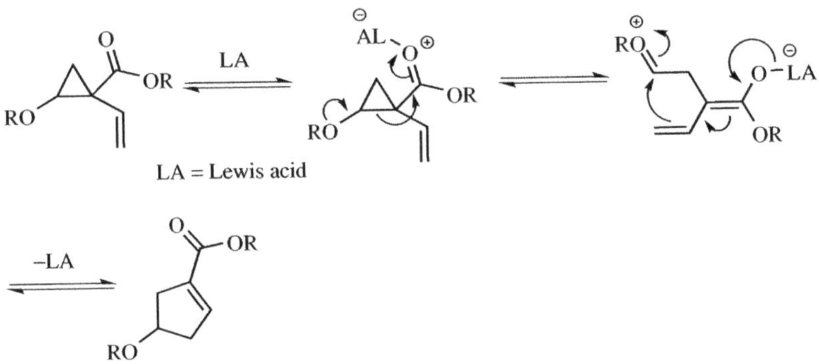

Scheme 11.22 Mechanism for the Lewis-acid promoted rearrangement of donor-acceptor VCPs.

the (*Z*)-isomer would undergo intramolecular S_N2 alkylation to afford the fused cyclopentene product (Scheme 11.21).

For donor–acceptor VCPs, the acid-mediated mechanism involves initial activation of the acceptor group, followed by donor group-assisted cyclopropyl ring opening to form a zwitterionic intermediate, which recloses to give the cyclopentene ring. (Scheme 11.22).[1d,13a–c,18c]

11.5 HETEROATOM-CONTAINING ANALOGUES OF THE VCP–CP REARRANGEMENT

Heterocyclic variants of VCP–CP rearrangement have also been explored to synthesize five-membered heterocyclic rings, although to a lesser extent than their carbocyclic counterparts. The ring expansion of vinylaziridines to pyrroline products was reported independently by Atkinson and Rees in 1967 and by Lwowski and coworkers in 1968,[19a,b] whereas the corresponding thermal rearrangement of vinyloxiranes to dihydrofurans was disclosed by Paladini and Chuche in 1971 (Scheme 11.23).[19c]

Some of the most extensive efforts in this area have come from Hudlicky's laboratory, including an elegant strategy to synthesize pyrrolizidine alkaloids by the pyrolysis of vinylaziridines, which were prepared via the decomposition of conjugated azidodienes as shown in Scheme 11.24.[20a,b]

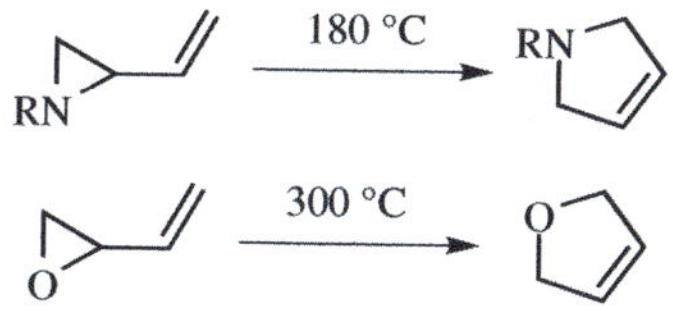

Scheme 11.23 Thermal rearrangement of vinyl aziridines and vinyl oxiranes.

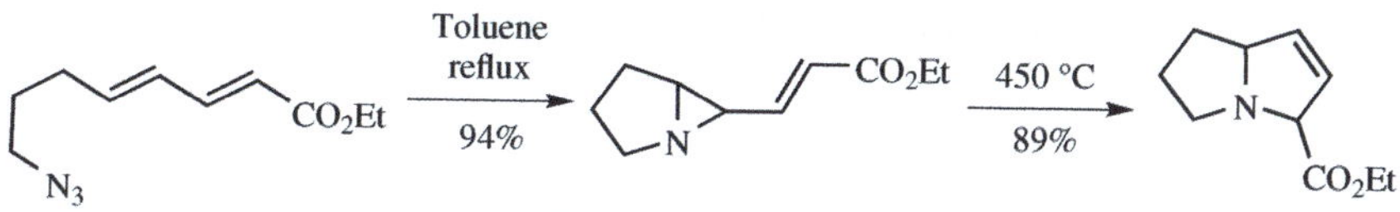

Scheme 11.24 Synthesis of fused vinyl aziridine and its thermal rearrangement.

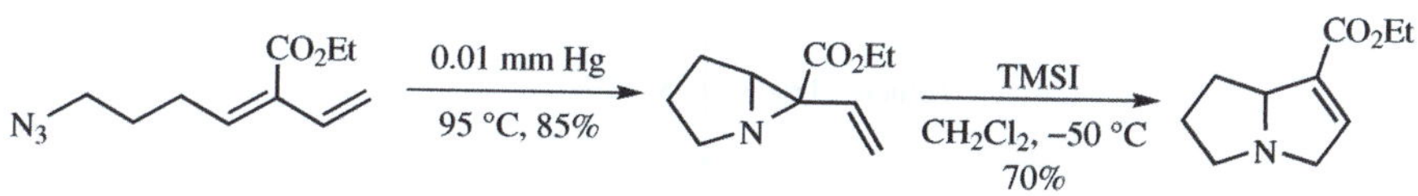

Scheme 11.25 TMSI-promoted rearrangement of vinyl aziridines.

However, as illustrated in Scheme 11.24, this method was limited to the use of linearly conjugated azidodienes and required harsh pyrolytic conditions for the rearrangement. To extend upon this existing strategy for cross-conjugated azidodienes, Hudlicky *et al.* developed a TMSI-mediated ring expansion of vinylaziridines, which afforded the annulated pyrrolines in high yield under mild conditions (Scheme 11.25).[20c]

The analogous vinyloxirane–dihydrofuran rearrangement has also been developed.[21a,b] Hudlicky *et al.* reported the thermal decomposition of vinyloxiranes, which were prepared by low-temperature addition of the lithium dienolate of ethyl-2-bromocrotonate to aldehydes. Subsequent pyrolytic rearrangement afforded the corresponding 2,3-dihydrofurans in variable yields (Scheme 11.26).[21c,d]

Steel and coworkers, meanwhile, disclosed the ring expansion of chiral auxiliary-bound vinyloxirane **25** to afford the substituted 2,5-dihydrofurans **26–28** in modest diastereoselectivity. Interestingly, the diastereoselectivity obtained in the reaction was found to be independent of the *cis–trans* ratio of the vinyloxirane and depends only on the chiral auxiliary used (Scheme 11.27).[21e]

11.5.1 Mechanism

The heteroatom variants of the VCP–CP rearrangement generally operate either by a zwitterionic or a diradical-type mechanism. The zwitterionic mechanism is more common and involves the heteroatom-assisted C—C cleavage of the three-membered

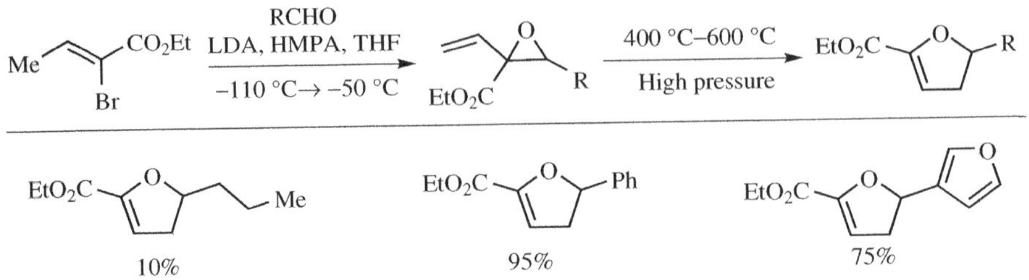

Scheme 11.26 Synthesis of vinyl oxiranes and their thermal rearrangement.

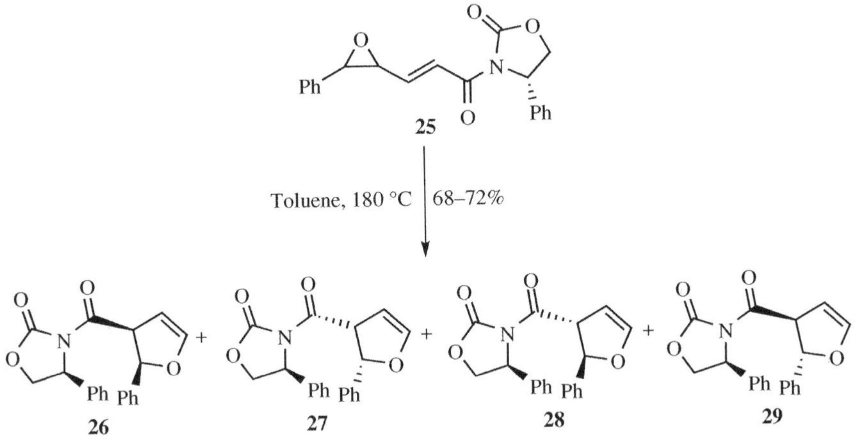

Scheme 11.27 Thermal rearrangement of chiral auxiliary-bound vinyloxirane.

ring to form an ylide, which undergoes electrocyclic ring closure to form the product (Scheme 11.28, Eq. 1).[20a,b,21a,b,22a,b] Alternatively, the three-membered ring can undergo homolytic fission to form a diradical species that will recombine to form the five-membered product (Scheme 11.28, Eq. 2).[19a,20a–22c]

In contrast, the mechanism of LiI or TMSI-mediated hetero-VCP–CP rearrangement involves the nucleophilic ring opening followed by ring closure to form the product (Scheme 11.29).[20a,c,22d]

11.6 APPLICATIONS IN SYNTHESIS

The VCP–CP rearrangement has been utilized extensively to synthesize a variety of natural products and biologically active molecules. Trost *et al.* applied the siloxy-accelerated VCP–CP rearrangement developed in their laboratory for the total synthesis of the antibiotic aphidicolin.[23a] The tricyclic core of the molecule was

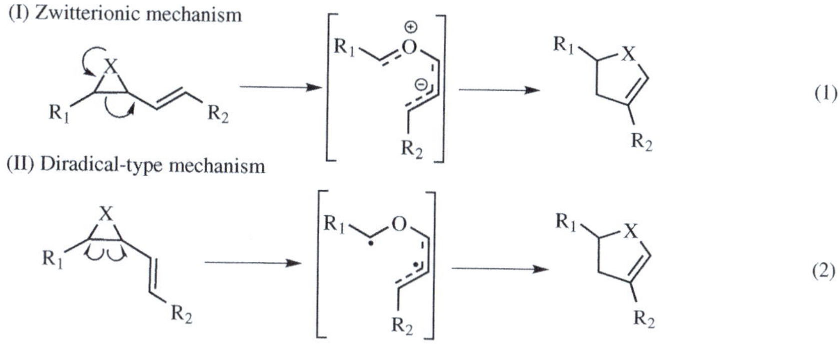

Scheme 11.28 The two mechanistic pathways for the rearrangement of heteroatom-variants of VCPs.

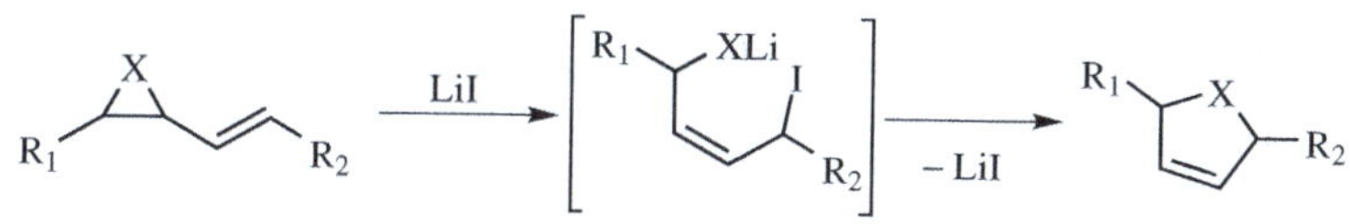

Scheme 11.29 Mechanism of LiI or TMSI-mediated rearrangement of heteroatom-variants of VCPs.

synthesized using the late-stage flash vacuum pyrolysis (FVP) of the corresponding siloxy-substituted VCP **30** (Scheme 11.30). The resulting annulated cyclopentene product **31** was formed in excellent yield, however, as a 2:1 mixture of epimers at C-8 with undesired epimer as the major product. In order to obtain the desired epimer as major isomer, Trost group subjected the epimeric mixture **31** to Saegusa–Ito oxidation, affording enone **32**. Birch reduction of enone **32** followed by silylation afforded the desired enol silane **33**, as a single diastereomer and was rapidly converted in a few steps to the complex natural product (Scheme 11.30).

Hudlicky's group applied the VCP–CP rearrangement to the synthesis of a variety of complex natural products. For example, their total synthesis of the terpenoid (±)-hirsutene utilized the thermal rearrangement of activated VCP **34** to diastereoselectively form the advanced tricyclic core **35** of the molecule (Scheme 11.31).[23b]

Hudlicky and coworkers also reported an elegant synthesis of the iridoid sesquiterpene (−)-specionin, by utilizing the low-temperature anion-accelerated VCP–CP rearrangement developed in their lab. The precursor siloxyvinylcyclopropane **38** was synthesized as a mixture of *exo*/*endo* isomers by the cyclopropanation of substituted cyclopentenone **36** with the lithium dienolate **37** derived from 4-(dimethyl-*tert*-butylsiloxy)-2-bromocrotonate (Scheme 11.32). Rearrangement of the diastereomeric VCP substrates was achieved by the use of TMSI/HMDS at −78 °C to afford the tricyclic ketone in high yield as a mixture of diastereomers, which was then converted to the natural product.[23]

Corey and Kigoshi disclosed the total synthesis of the major antheridiogen from *Anemia phyllitidis* (+)-antheridic acid by applying the Lewis acid-mediated

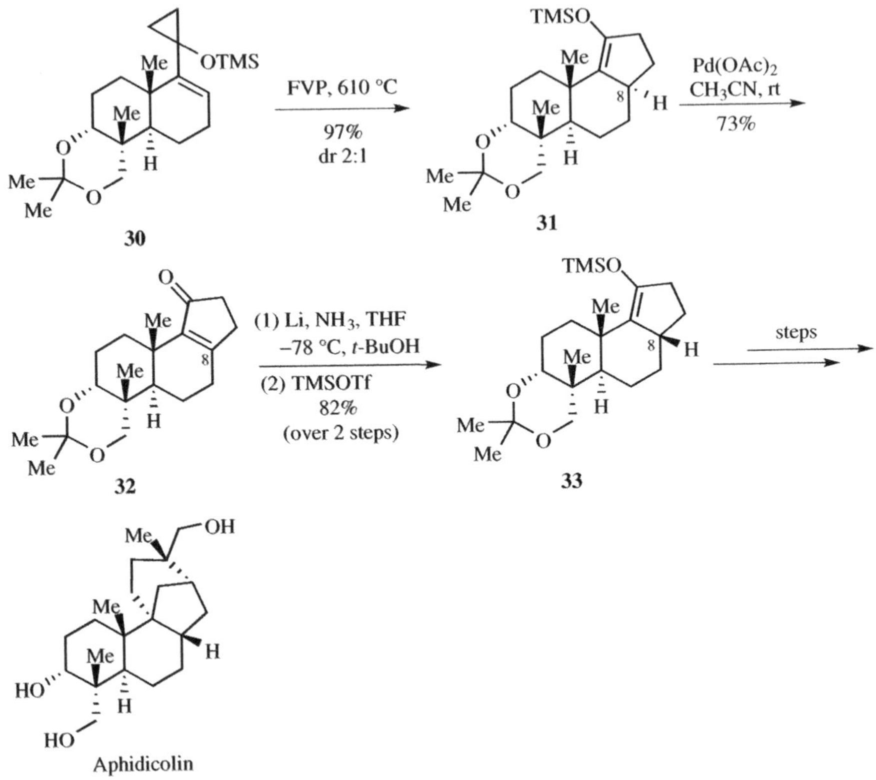

Scheme 11.30 Trost's synthesis of aphidicolin involving the thermal rearrangement of VCP 30 as the key step.

VCP–CP rearrangement. The VCP substrate **41**, synthesized using a Cu(II)-catalyzed intramolecular cyclopropanation of diazoacetic ester **40** was subjected to Et_2AlCl-promoted rearrangement, affording the highly functionalized cyclopentene **41**. The thermal reaction conditions, however, were found to be ineffective for this key transformation (Scheme 11.33).[23d]

Somfai's group demonstrated the use of a LiI-promoted vinylaziridine-3-pyrroline rearrangement in the formal synthesis of the antibiotic (−)-anisomycin.[23e] The commercially available 2-(4-methoxyphenyl)acetaldehyde **43** was subjected to Brown allylation followed by aminolysis of the resulting chlorohydrin to afford the enantioenriched aminoalcohol **44** (Scheme 11.34). The enantiopure *cis*-vinylaziridine **45** was then prepared by O- and N-tosylation of the chiral aminoalcohol followed by KOH-promoted ring closure. Microwave-assisted rearrangement of vinylaziridine **45** using LiI as an additive afforded enantioenriched 3-pyrroline **46** in excellent yield, and this was converted to the natural product in several steps (Scheme 11.34).[23e]

In 2008, Majetich *et al.* reported an elegant synthesis of diterpenoid (−)-salviasperanol by utilizing a vinyloxirane–dihydrofuran rearrangement as the key step.[23f] The vinyloxirane precursor was prepared by subjecting the tricyclic

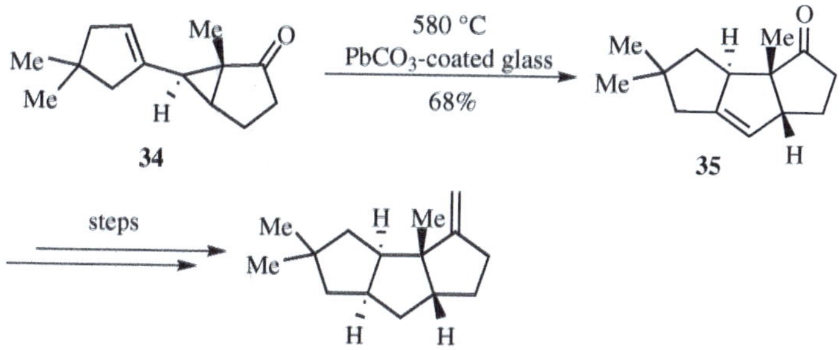

(±)-Hirsutene

Scheme 11.31 Hudlicky's total synthesis of (±)-hirsutene utilizing the thermal rearrangement of fused VCP **34**.

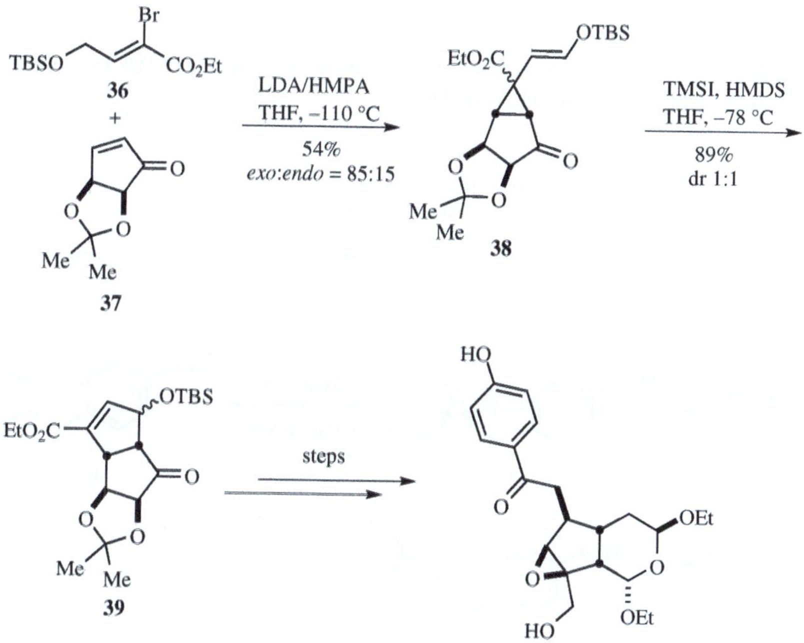

(−)-Specionin

Scheme 11.32 Hudlicky's synthesis of sesquiterpene (−)-specionin involving the TMSI-promoted rearrangement of fused siloxy VCP **38**.

enone **47** to asymmetric CBS reduction, followed by the hydroxyl-directed *m*-CPBA epoxidation (Scheme 11.35).

The acid-sensitive hydroxyl group was protected as an *O*-thiocarbamate, and treatment of the resulting vinyloxirane **49** with catalytic amounts of trifluoroacetic acid (TFA) afforded the substituted-dihydrofuran derivative **50** in high yield. Free-radical deoxygenation followed by the cleavage of the methyl ethers completed the synthesis.

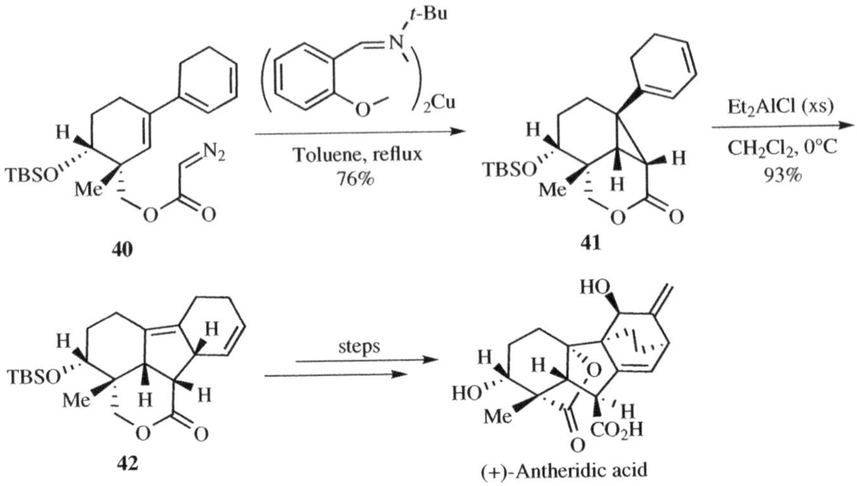

Scheme 11.33 Corey's synthesis of (+)-antheridic acid involving the Et_2AlCl-promoted rearrangement of highly functionalized VCP **41**.

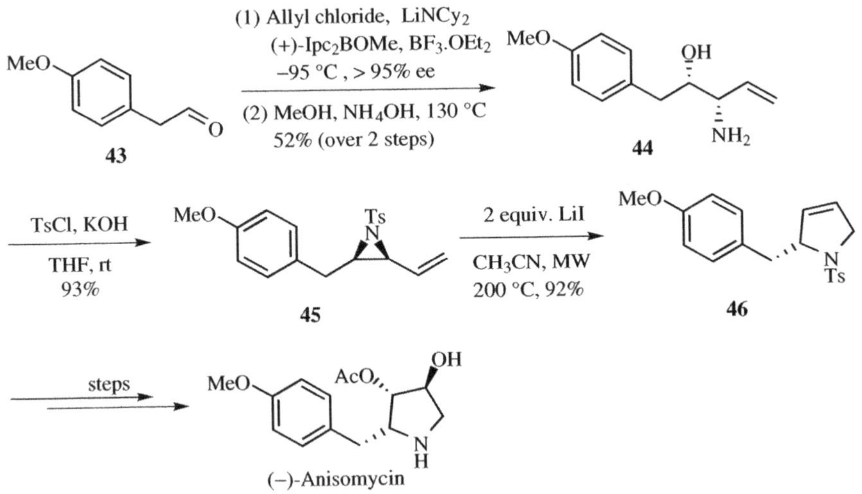

Scheme 11.34 Somfai's synthesis of (−)-anisomycin utilizing a LiI-mediated vinylaziridine–pyrroline rearrangement.

11.7 PHOTOCHEMICAL VCP–CP REARRANGEMENT

Two years after the discovery of the parent thermal VCP–CP rearrangement in 1960, Frey observed the first photochemical version of this process.[24] Since then, appreciable efforts have been made to investigate the photochemical behavior of VCPs and to apply these processes in synthesis.[25] The absorption maxima for the $\pi \rightarrow \pi^*$

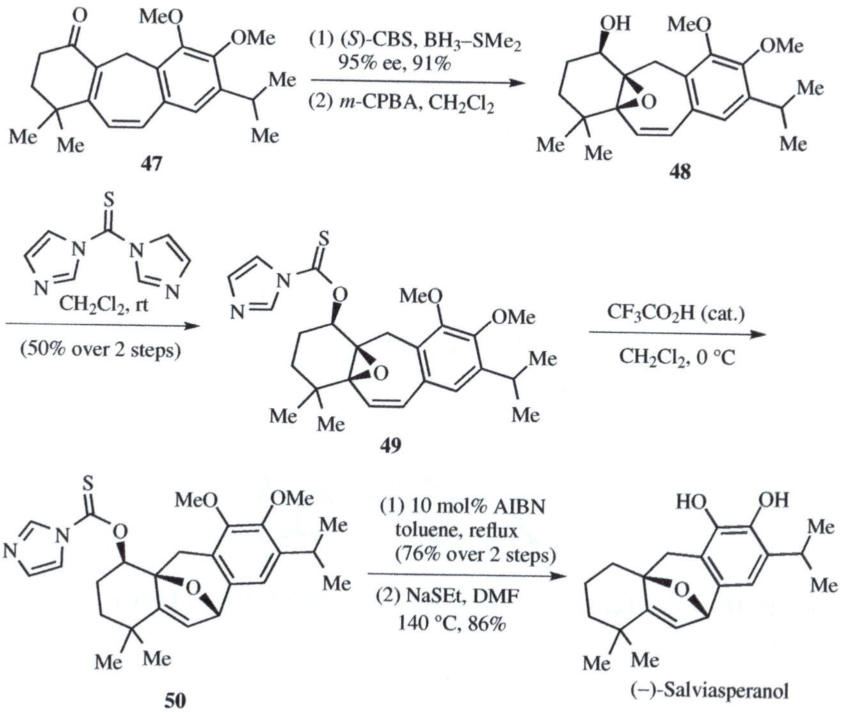

Scheme 11.35 Majetich's synthesis of (−)-salviasperanol involving a TFA-catalyzed vinyloxirane–dihydrofuran rearrangement.

olefin band in the parent VCP has been 192 nm, which clearly indicates the requirement of additional chromophores for its excitement in the more accessible UV/Vis region.[2a,b,26] The photochemical VCP–CP rearrangement can be achieved by both direct and sensitized methods. Direct irradiation of VCPs having additional chromophores results in relatively low-energy $n \rightarrow \pi^*$ excitations and, therefore, is more selective for cyclopentene formation. In the case of sensitized irradiations, the energy of the generated triplet species ($E_T \sim 80$ kcal/mol) is quite high compared to the activation energy for VCP–CP rearrangement ($E_{act} \sim 50$ kcal/mol); therefore, these processes are usually less selective and result in a mixture of products.[2a,b]

In 1970, Cooke reported an example of isomerization of isopropenylcyclopropane to methylcyclopentene upon direct irradiation in hexane (Scheme 11.36).[27a] The rearrangement presumably occurs from the singlet excited state of the isopropenylcyclopropane, as photosensitized conditions using acetone, benzene, or naphthalene as triplet sensitizers were either found to be ineffective or produced only a trace amount of methylcyclopentene under reaction conditions (Scheme 11.36).

Paquette *et al.* demonstrated the use of both direct and sensitized reaction conditions to synthesize [5,5]-ring-fused cyclopentenes in good yields by the rearrangement of bicyclic VCPs. The high yield obtained using acetone as a photosensitizer

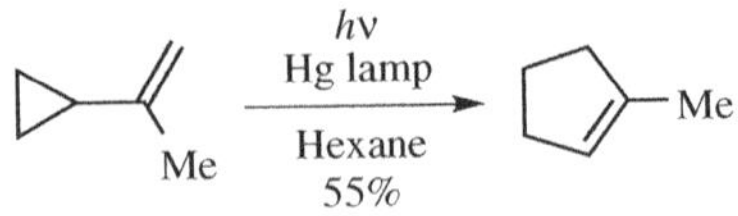

Scheme 11.36 Photochemical rearrangement of isopropenylcyclopropane to methylcyclopentene.

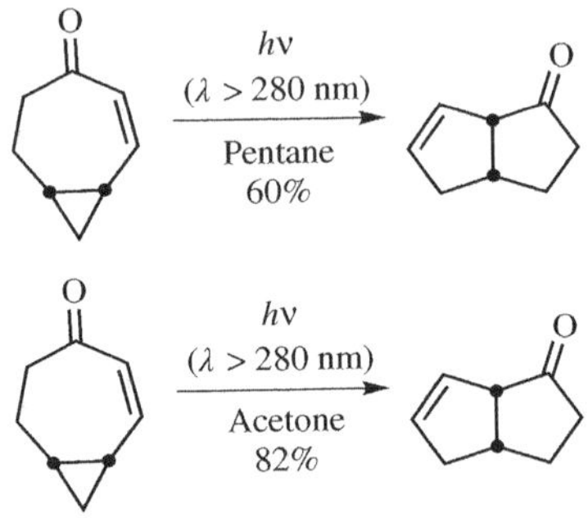

Scheme 11.37 Photochemical rearrangement of bicyclic VCP under both direct and sensitized conditions.

suggests that the rearrangement proceeds efficiently via triplet excited state of the corresponding VCP (Scheme 11.37).[27b]

Farneth's group used infrared multiphoton irradiation of VCP with a CO_2–TEA (transversely excited atmospheric) laser to form cyclopentene in low yield (~32%). Unfortunately, the remainder of the VCP isomerized to a mixture of cyclic and acyclic dienes.[27c,d]

Sonawane's group, meanwhile, disclosed the photosensitized irradiation of carene-derived VCPs to [4,5]-ring-fused cyclopentenes using toluene as a photosensitizer. A variety of VCPs were applied to afford enantiopure cyclopentenes in high yields and moderate diastereoselectivities. Direct irradiation, however, did not result in cyclopentene formation (Scheme 11.38).[28a]

Armesto *et al.* studied the photochemical rearrangement of 1-substituted-3-(2,2-diphenylvinyl)-2,2-dimethylcyclopropanes under both triplet-sensitized irradiation as well as direct irradiation conditions. Unfortunately, the majority of the VCP starting materials were recovered as mixtures of *cis–trans* stereoisomers, and only low yields of cyclopentene products were obtained (Scheme 11.39, Eq. 1).[28b]

Interestingly, direct irradiation of ester- or carboxylic acid-bearing VCP afforded five-membered lactones in low yields with unreacted VCP (Scheme 11.39, Eq. 2). Formation of the cyclic lactone was presumably due to resonance delocalization of the biradical intermediate into the carbonyl group, revealing an O-centered radical species **53**, which recombines to afford dihydrofuran **54**. This dihydrofuran undergoes addition of water, followed by elimination of alcohol, to yield the cyclic lactone product **56** (Scheme 11.40).[28b]

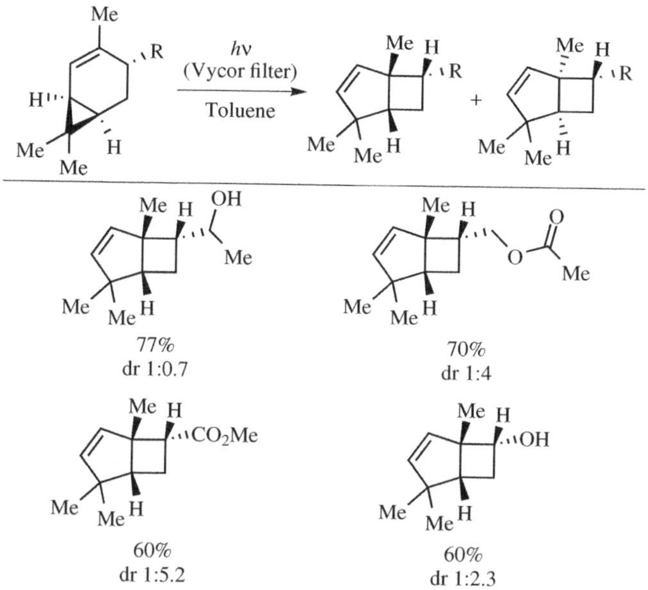

Scheme 11.38 Photochemical rearrangement of fused VCPs under photosensitized condtions.

The photochemical route has also been applied to a heterocyclic variant of the VCP–CP rearrangement. Campos's group reported an efficient method to synthesize pyrrolines by the direct irradiation of *N*-cyclopropylimines in acetonitrile (Scheme 11.41). A variety of imines containing both alkyl and aryl groups underwent smooth ring expansion to yield pyrrolines **57–62** in high yields. The reaction is also regioselective as 2-phenyl-substituted imine undergoes selective C_1—C_2 bond cleavage to afford pyrroline **61** as a single regioisomer (Scheme 11.41).[2]

11.7.1 Mechanism

The mechanism of the majority of photo VCP–CP rearrangements involves the formation of diradical intermediates, although the involvement of zwitterionic species has also been proposed for highly activated VCPs.[2a,b] These diradical or zwitterionic intermediates photochemical rearrangements can take place from either singlet or triplet excited states. As an example, the mechanism of photochemical cyclopropylimine–pyrroline rearrangement has been investigated both experimentally and computationally.[29,30] Direct irradiation of *N*-cyclopropylimine promotes it to a singlet excited state S_2, corresponding to the $\pi \rightarrow \pi^*$ electronic transition. This excited state molecule relaxes by undergoing C—C σ-bond cleavage of the cyclopropyl ring in the S_2 state and decays to S_1 excited state. Rotation of iminic moiety **63** stabilizes S_1 state, lowering the energy barrier between S_1 and S_0 and ultimately forms the photoproduct in ground state S_0 (Scheme 11.42). Additionally,

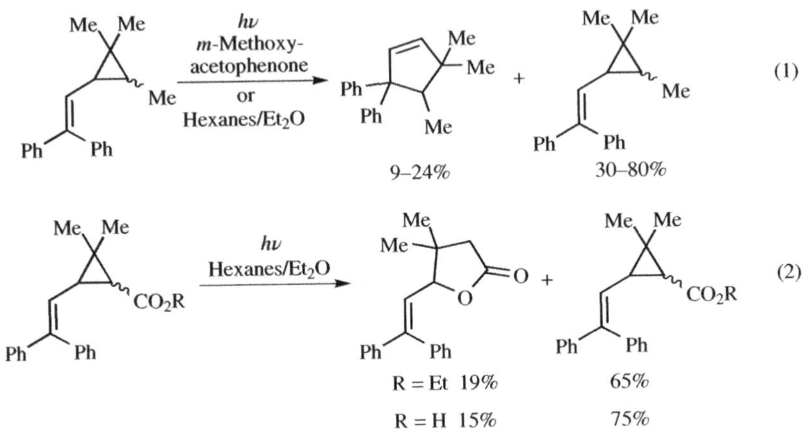

Scheme 11.39 Photochemical rearrangement of (diphenylvinyl)cyclopropanes under both direct and sensitized conditions.

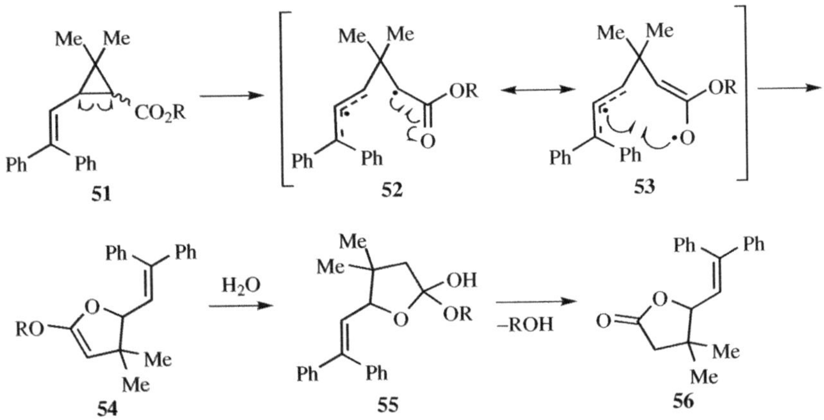

Scheme 11.40 Mechanistic rationale for cyclic lactone formation from the photochemical rearrangement of ester- or carboxylic acid-bearing VCP.

use of triplet state quenchers such as molecular oxygen and *Z*-piperylene had no influence on the reaction, suggesting that only singlet transition states are involved in this photochemical process.

11.7.2 Applications in Synthesis

Due to the relatively high energy content of the intermediates involved in photochemical VCP–CP reactions, these processes are often less selective as compared to their thermal counterparts and, therefore, have not been widely utilized for natural product synthesis.[2a,30] Nevertheless, a few representative examples come from

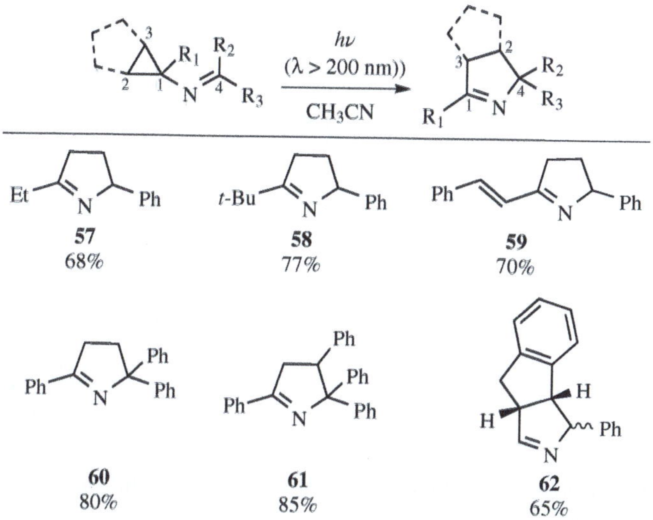

Scheme 11.41 Photochemical rearrangement of *N*-cyclopropylimines under direct sensitized conditions.

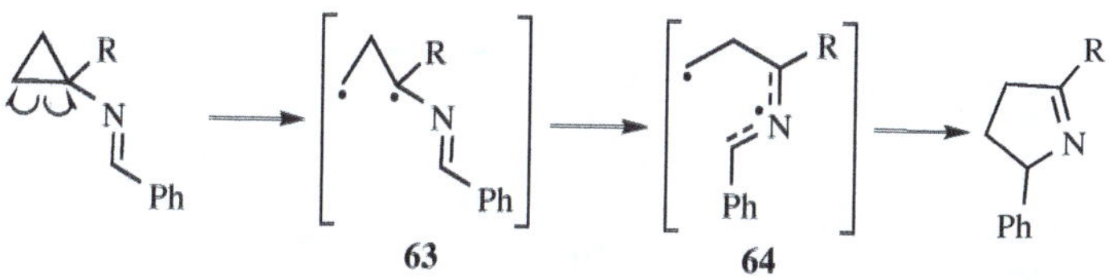

Scheme 11.42 Diradical mechanism for *N*-cyclopropylimine-pyrroline rearrangement.

Sonawane's group, which reported the synthesis of the marine natural product, $\Delta^{9(12)}$-(−)-capnellene as well as its unnatural enantiomer $\Delta^{9(12)}$-(+)-capnellene, by utilizing the photochemical VCP–CP rearrangement.[31a,b] The enantiopure VCP precursor **66** was synthesized by subjecting (+)-Δ^3-carene **65** to epoxidation, followed by base-promoted ring opening. The photosensitized irradiation of the VCP **66** led to a mixture of two diastereomeric cyclopentene products (Scheme 11.43). The major product **67** was transformed to furnish the natural product, while the minor diastereomer **68** led to the capnellene enantiomer.

Sonawane *et al.* further utilized the photoinduced VCP–CP rearrangement in a formal synthesis of the sex pheromone of the male cotton boll weevil, (±)-grandisol.[31a,c] The toluene-sensitized irradiation of enantiopure (+)-Δ^2-carene **69** afforded the annulated cyclopentene (±)-**72** in good yield, but with a complete loss of optical activity. Racemization en route to substituted cyclopentene was presumably due to the involvement of a diradical pathway, which was supported by the concomitant loss of optical rotation in the unreacted (+)-Δ^2-carene. Nevertheless, the cyclopentene intermediate (±)-**72** obtained in the rearrangement could

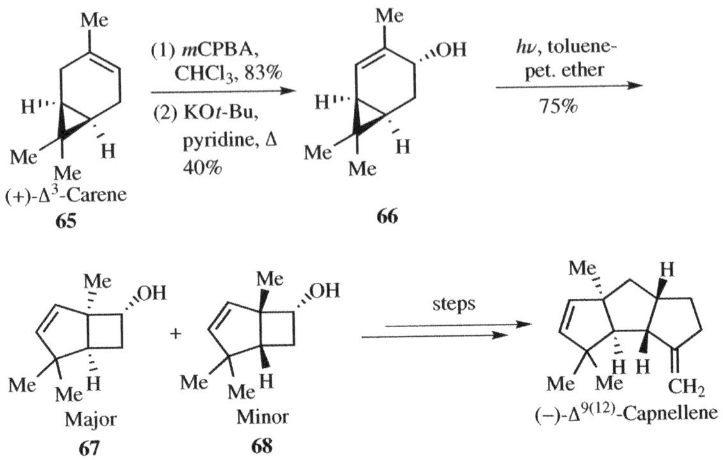

Scheme 11.43 Sonowane's synthesis of $\Delta^{9(12)}$-(−)-capnellene utilizing photochemical rearrangement of VCP **66**.

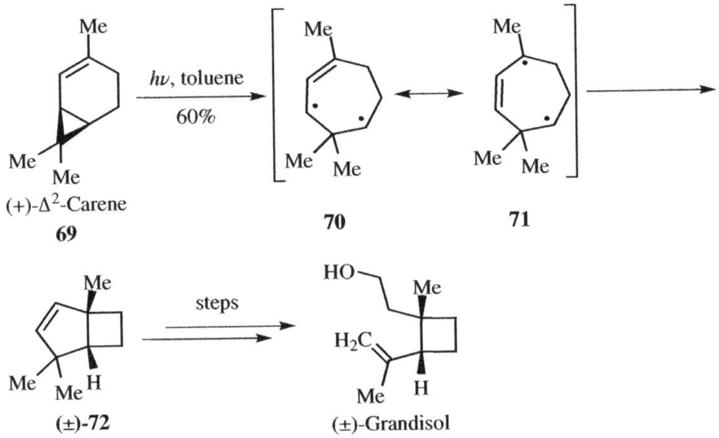

Scheme 11.44 Sonowane's synthesis of (±)-grandisol involving the photochemical rearrangement of VCP **69**.

be converted via a previously reported sequence of reactions to furnish the natural product (Scheme 11.44).

11.8 METAL-CATALYZED VCP–CP REARRANGEMENT

One of the most significant advances in VCP–CP rearrangement chemistry has been the discovery of metal catalysts that allow these reactions to occur under particularly mild conditions.[1j,2a,b,d,f] Although a number of metal-mediated rearrangements of

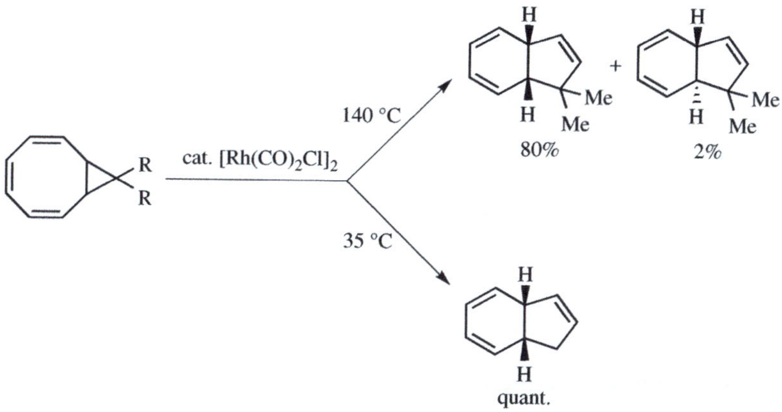

Scheme 11.45 Rh(I)-catalyzed rearrangement of 9,9-bicyclononatrienes.

VCPs (Rh, Pd, Ni, Cu, Cr, Mo, Fe) have been developed over the past few decades, only a subset of these metals (Rh, Pd, Ni) can promote these processes in a catalytic manner.[2d] Since the transition metal-*promoted* rearrangements have been thoroughly reviewed,[2a,b,d] this section will cover only the transition metal-*catalyzed* rearrangements of VCPs and their heteroatom-substituted analogues.

11.8.1 Rh-Catalyzed VCP–Cyclopentene Rearrangement

Only a few reports describing the Rh-catalyzed VCP–CP rearrangement are currently in the literature.[2d] Grigg *et al.* developed a Rh(I)-catalyzed rearrangement of 9,9-bicyclononatriene VCPs to form the *cis*-dihydroindene product predominantly. However, the reaction was limited to the use of activated substrates, that is, dienyl- or trienylcyclopropanes, and the products were often contaminated with low amounts of unidentified impurities (Scheme 11.45).[32a]

Saigo's group disclosed the Rh(I)-catalyzed rearrangement of allenylcyclopropanes to form methylenecyclopentene products. The rearrangement proceeded smoothly, affording methylenecyclopentenes **74-a** and **74-b** in high yields and in a regioselective manner depending on the nature of the substituents (Scheme 11.46).[32b]

Unfortunately, the majority of known Rh(I) catalysts are highly active for the isomerization of VCPs to dienes rather than forming cyclopentenes.[32c,d]

11.8.2 Mechanism

A variety of stoichiometric complexation studies on Rh(I)-mediated VCP–CP rearrangements as well as divinylcyclopropane rearrangements have been reported.[32c,33] In addition, some proposed reactive intermediates have been isolated and characterized.[32a,33] Based on these studies, the proposed mechanism involves the formation of an η^2-alkene complex **75** (Scheme 11.47) of Rh

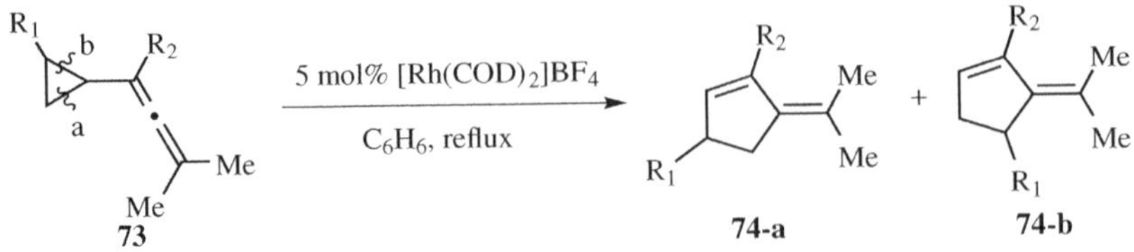

Entry	R_1	R_2	Isolated yield	**74a:74b**
18	*n*-Pr	CO_2Bn	9>	99:1
2	Ph	CO_2Me	95	5:95
3	Ph	H	78	3:97

Scheme 11.46 Rh(I)-catalyzed rearrangement of allenylcyclopropanes.

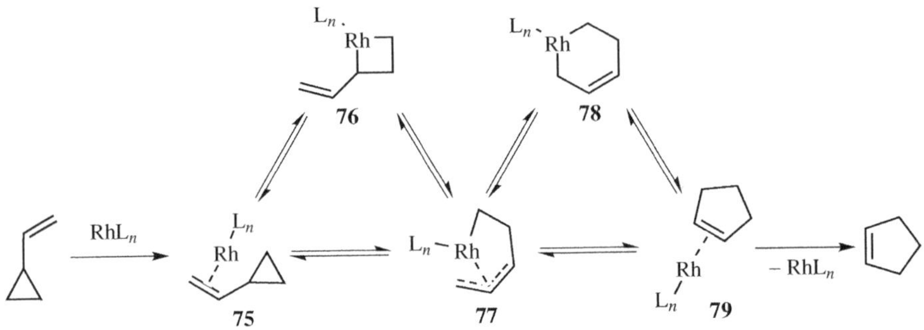

Scheme 11.47 Mechanism of Rh-catalyzed VCP-CP rearrangement.

followed by oxidative addition to form a η^1-alkyl/η^3-allyl-Rh complex **77** either directly or through the formation of an intermediate vinylmetallabutane **76**. The η^1-alkyl/η^3-allyl-Rh complex **77** then undergoes rearrangement and reductive elimination, either directly or via a metallacyclohexene **78**, to form a π-complex of cyclopentene **79**, which regenerates the Rh catalyst upon release of the cyclopentene product.

11.8.3 Pd-Catalyzed VCP–Cyclopentene Rearrangement

Oshima and coworkers reported the use of catalytic amounts of $Pd(PPh_3)_4$ to effect the rearrangement of dienylcyclopropanes to vinylcyclopentene products in high yields.[34a] The reaction was amenable to the use of both cyclic and acyclic dienylcyclopropanes. However, the success of this reaction required the presence of dienyl unit as well as two electron-withdrawing groups on the cyclopropane ring (Scheme 11.48).

Hiroi *et al.* developed a Pd-catalyzed asymmetric rearrangement of chiral dienylcyclopropanes that were synthesized from the Horner–Wadsworth–Emmons reaction of the cyclopropylacrylaldehydes and phosphonates bearing a chiral sulfinyl group. The reaction afforded vinylcyclopentenes in good yield and high

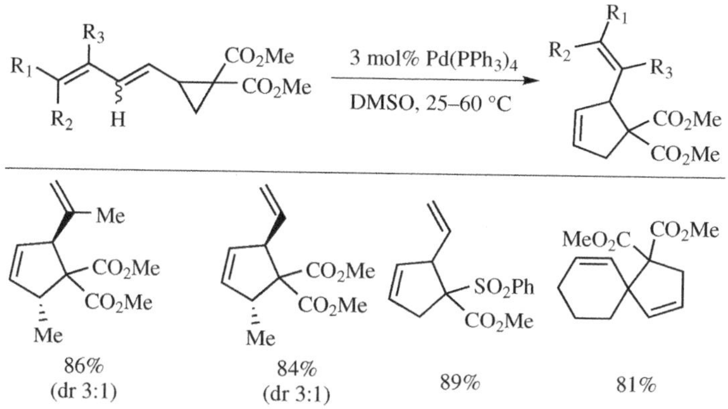

Scheme 11.48 Pd(0)/PPh$_3$-catalyzed rearrangement of dienylcyclopropanes.

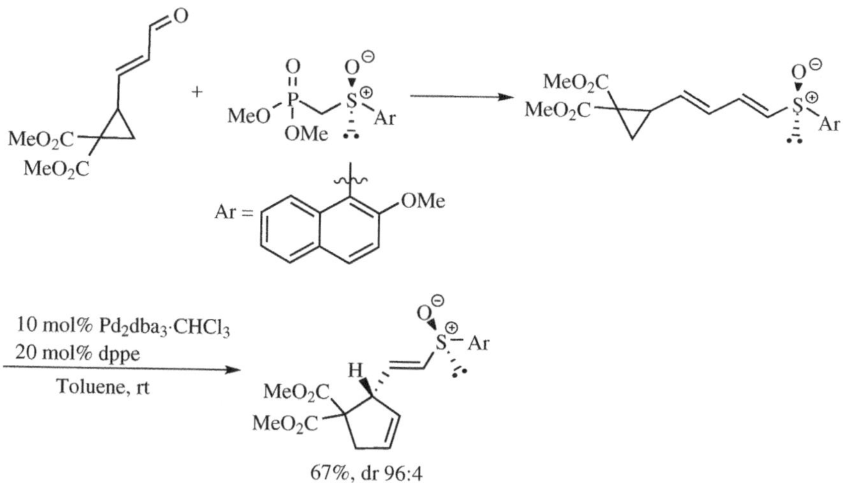

Scheme 11.49 Pd(0)/dppe-catalyzed diastereoselective rearrangement of dienylcyclopropanes bearing chiral sulfinyl group.

diastereoselectivities.[34b] Interestingly, the HPLC analysis of the recovered starting VCPs before the completion of the reaction showed that the asymmetric carbon center on VCP remained racemic during the course of the reaction. This observation suggests that asymmetric induction in product does not result from the kinetic resolution but rather from the direct participation of the chiral sulfinyl group in the formation π-allyl palladium complexes. Unfortunately, this methodology has a limited substrate scope and high catalyst loadings and activated substrates are necessary for the success of this Pd-catalyzed reaction (Scheme 11.49).

Recently, in 2013, Shanmugam's group reported the efficient synthesis of 3-spirocyclopentene-2-oxindoles by Pd-catalyzed rearrangement of

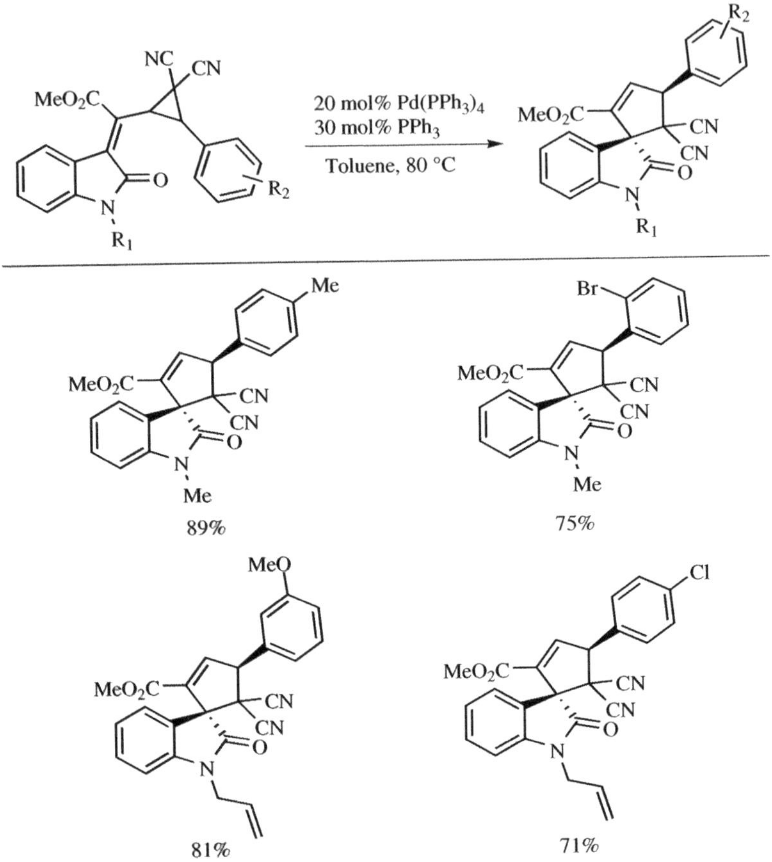

Scheme 11.50 Pd(0)/PPh_3-catalyzed rearrangement of 3-vinylcyclopropane-2-oxindoles.

3-vinylcyclopropane-2-oxindole derivatives (Scheme 11.50).[34c] A variety of VCPs with different protecting groups on the indole and different substituents on the phenyl ring underwent rearrangement to give the spirocyclopenteneoxindoles as single diastereomers. The reaction, however, is still limited to activated VCPs and requires a high loading of Pd catalyst.

11.8.4 Mechanism

The proposed mechanism for the Pd(0)-catalyzed VCP–CP rearrangement involves the nucleophilic attack by Pd on vinyl or dienyl unit (Scheme 11.51), resulting in the cleavage of the cyclopropyl ring and forming a zwitterionic intermediate **80** having a π-pentadienyl palladium along with a stabilized anion. This intermediate undergoes allylic substitution to generate the vinylcyclopentene product and the Pd catalyst.[2d,34]

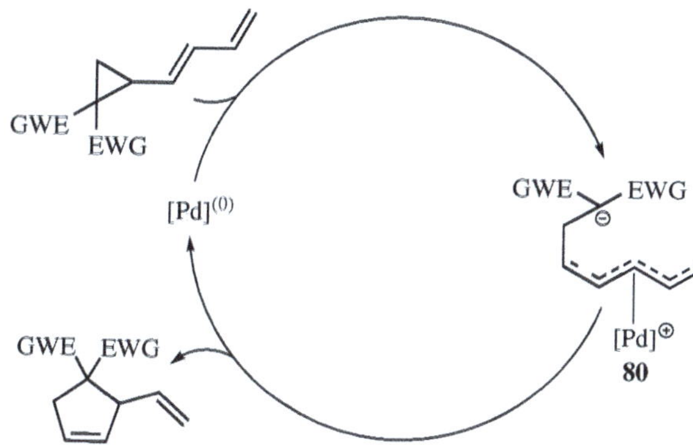

Scheme 11.51 Mechanism of Pd(0)-catalyzed rearrangement of dienylcyclopropanes.

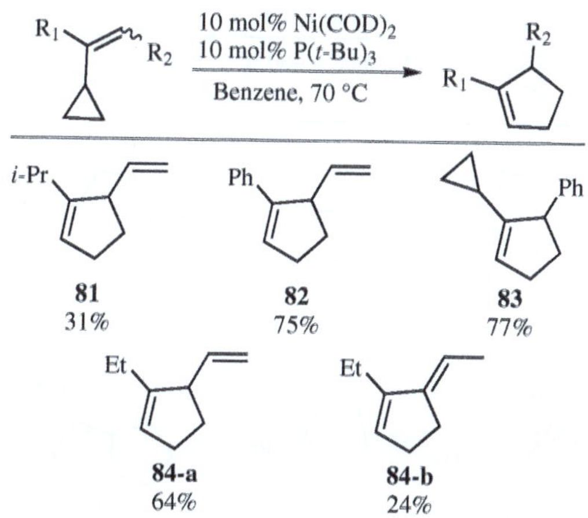

Scheme 11.52 Ni(0)/P(t-Bu)$_3$-catalyzed rearrangement of activated VCPs.

11.8.5 Ni-Catalyzed VCP–Cyclopentene Rearrangement

A variety of Ni catalysts have been reported in the literature to effect the VCP–CP rearrangement. Murakami's group demonstrated the use of a Ni(0)/P(t-Bu)$_3$ catalyst[35a] which was effective only with activated VCPs bearing dienyl or styrenyl substituents. The isomerization of the vinylcyclopentene products to methylenecyclopentenes was also observed in some cases under the reaction conditions (**84a** and **84b**, Scheme 11.52).

In 1994, Ryu's group reported the Ni–phosphine catalyzed rearrangement of 1-siloxy-vinylcyclopropanes to the corresponding siloxycyclopentenes.[35b] Although this methodology provides an attractive alternative to the thermal rearrangement of siloxy-VCPs, there is a limited substrate scope and relatively high temperatures and catalyst loadings are required (Scheme 11.53).

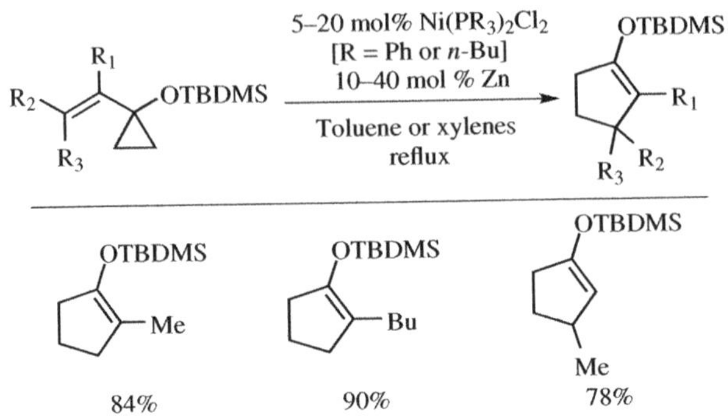

Scheme 11.53 Ni(0)/PR$_3$-catalyzed rearrangement of 1-siloxy VCPs.

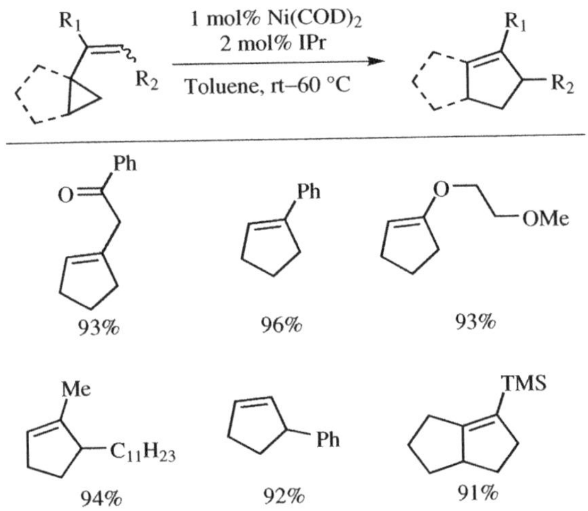

Scheme 11.54 Ni(0)/IPr-catalyzed rearrangement of activated and unactivated VCPs.

In 2004, our research group disclosed that combinations of Ni(COD)$_2$ with highly σ-donating *N*-heterocyclic carbene (NHC) ligands effectively catalyze the rearrangement of VCPs to cyclopentenes under mild conditions and with low catalyst loadings.[35c] This methodology has a broad substrate scope, as a variety of VCPs bearing electron-withdrawing groups, heteroatoms, phenyl groups, and even simply unactivated alkyl groups underwent smooth rearrangement to afford cyclopentene products in excellent yields (Scheme 11.54). The success of this methodology is believed to arise from the highly electron-rich nature of Ni/NHC catalyst, which would facilitate nucleophilic attack of nickel on the olefin unit of VCP, and therefore, cleave the cyclopropyl ring.

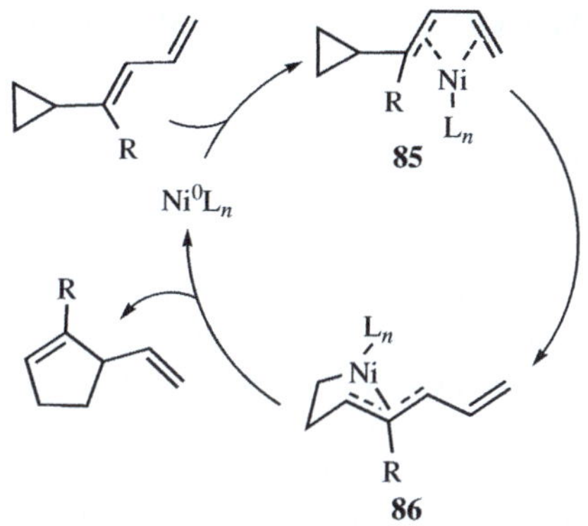

Scheme 11.55 Mechanism for Ni(0)-catalyzed rearrangement of dienylcyclopropanes.

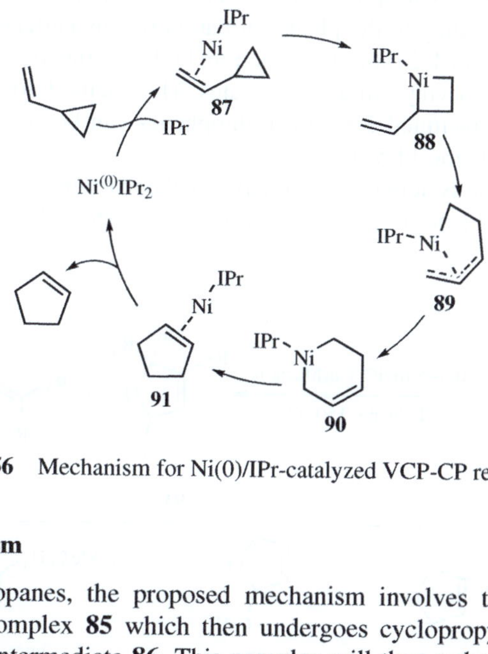

Scheme 11.56 Mechanism for Ni(0)/IPr-catalyzed VCP-CP rearrangement.

11.8.6 Mechanism

For dienylcyclopropanes, the proposed mechanism involves the formation of a butadiene–Ni(0) complex **85** which then undergoes cyclopropyl ring opening to form a σ, π-allyl intermediate **86**. This complex will then reductively eliminate to form the cyclopentene product and regenerate the catalyst (Scheme 11.55).[2d,35a]

Our recent kinetic and computational studies for the Ni/IPr-catalyzed VCP–CP rearrangement[36] suggest that the VCP undergoes precomplexation with nickel to generate η^2-alkene complex **87**, which upon oxidative addition yields a vinyl nickellacyclobutane **88**. This complex rearranges to nickellacyclohexene **90** via an η^1-alkyl, η^3-allyl-nickel intermediate **89**. The nickellacyclohexene complex then undergoes reductive elimination to form the cyclopentene product through a π-cyclopentene complex of nickel **91** (Scheme 11.56).

11.9 HETEROATOM VARIANTS OF THE METAL-CATALYZED VCP–CP REARRANGEMENT

In recent years, efficient transition metal-catalyzed processes have been developed to synthesize five-membered heterocycles via heteroatom-containing analogues of the VCP–CP rearrangement. In 2006, Njardarson's group reported an efficient $Cu(hfacac)_2$ [hfacac = hexafluoroacetylacetonate] catalyst that could selectively isomerize vinyl oxiranes **92** to the corresponding 2,5-dihydrofuran products **93** while avoiding the hydride-shift pathway to form β,γ-unsaturated carbonyl compounds **94** (Scheme 11.57).[37a] Vinyl oxiranes with a variety of substitution patterns underwent ring expansion to dihydrofuran products in a highly selective manner. The reaction is also amenable to the use of fused vinyloxiranes, providing oxabicyclic products in high yields and selectivity.

Njardarson and coworkers successfully extended this methodology to rearrange vinylthiiranes to 2,5-dihydrothiophenes.[37b] The electrophilic $Cu(hfacac)_2$ catalyst is not only efficient for dihydrothiophene formation but also effective for suppressing the desulfurization and 1,2-dithiine formation, highly competitive pathways that are usually observed with vinylthiirane substrates. This methodology has a broad substrate scope, and substituted 2,5-dihydrothiophenes can be synthesized in good to excellent yields (Scheme 11.58).

The same catalytic system is also effective for the rearrangement of vinylaziridines to 3-pyrrolines (Scheme 11.59).[37c] A variety of monocyclic as well as bicyclic

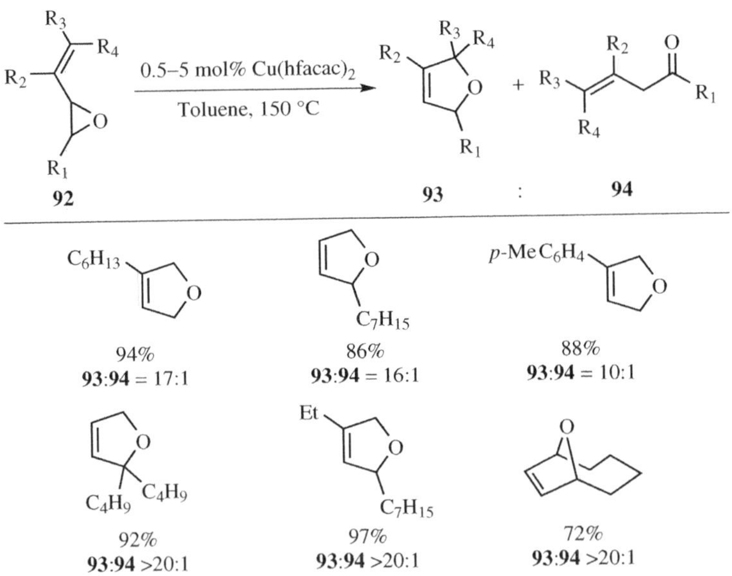

Scheme 11.57 $Cu(hfacac)_2$-catalyzed vinyloxirane-dihydrofuran rearrangement.

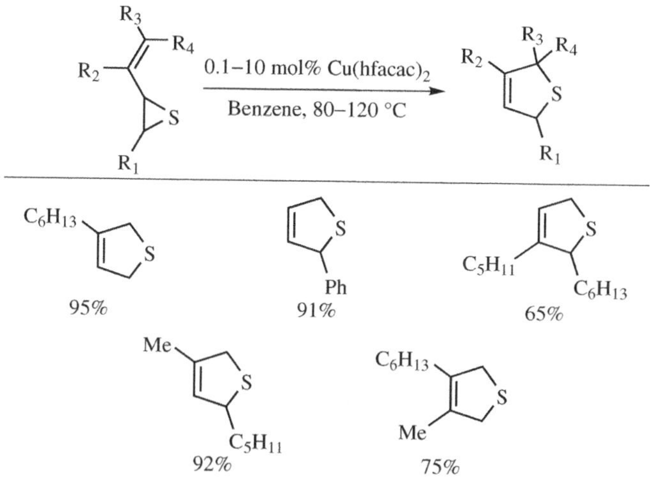

Scheme 11.58 Cu(hfacac)$_2$-catalyzed vinylthiirane-dihydrothiophene rearrangement.

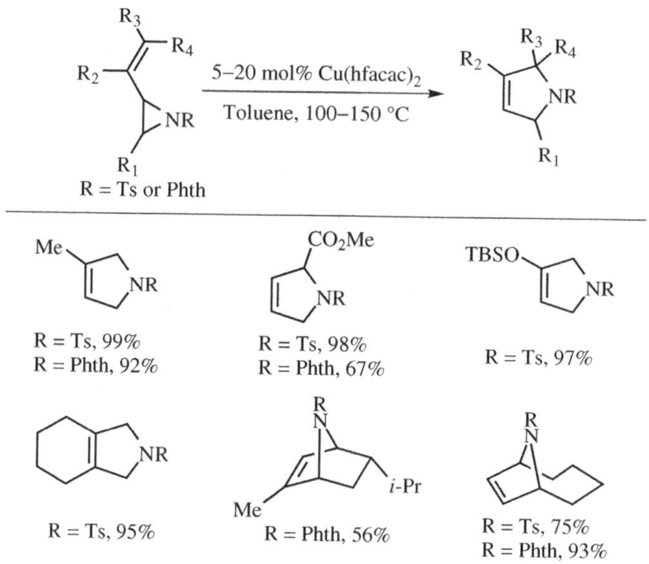

Scheme 11.59 Cu(hfacac)$_2$-catalyzed vinylaziridine-pyrroline rearrangement.

tosyl- or phthalimido-protected aziridines were efficiently isomerized to substituted pyrrolines in moderate to excellent yields.

Gratifyingly, chiral vinylaziridines underwent stereospecific rearrangement to afford the chiral 3-pyrroline products in high yields and diastereoselectivities (Scheme 11.60).[37d]

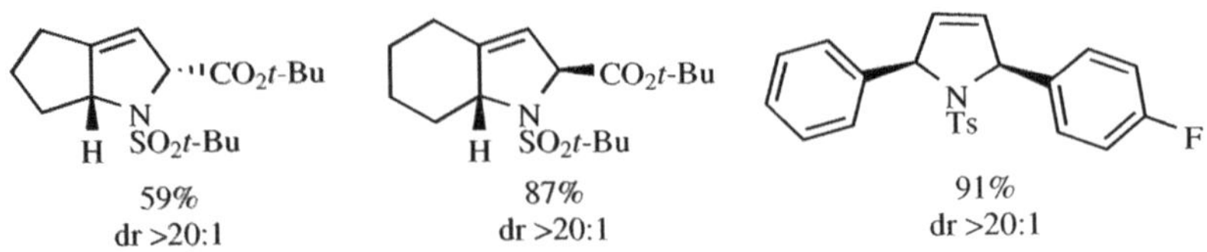

Scheme 11.60 Cu(hfacac)$_2$-catalyzed stereospecific rearrangement of chiral vinylaziridines.

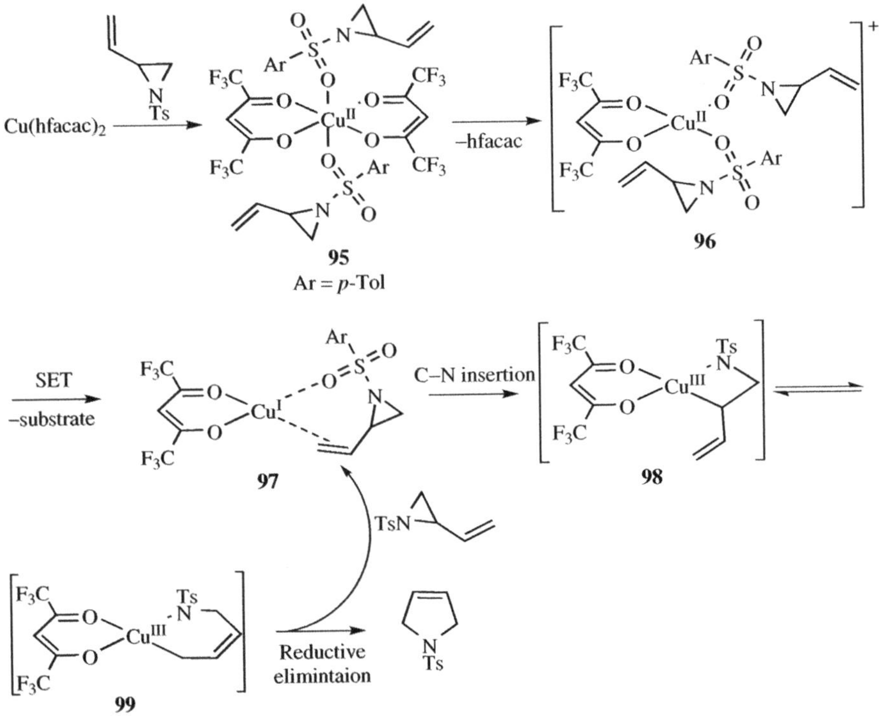

Scheme 11.61 Mechanism of Cu(hfacac)$_2$-catalyzed rearrangement of heteroatom-variant of VCP-CP rearrangement.

11.9.1 Mechanism

The proposed mechanism based on experimental and computational studies involves the bis-coordination of the substrate (vinyloxirane, vinylthiirane, or vinylaziridine) to the Cu(hfacac)$_2$ catalyst resulting in the formation of *trans*-octahedral-Cu(II)-complex **95** in Scheme 11.61 (In the case of vinyloxirane and vinylthiirane, O and S, respectively, will coordinate to Cu).[4,38] This complex then loses a hfacac ligand to form a cationic Cu(II)-complex **96**, which undergoes single-electron transfer (SET) reduction and loss of substrate molecule to generate the catalytic-active Cu(I) complex **97**. The possible reductant in this mechanism

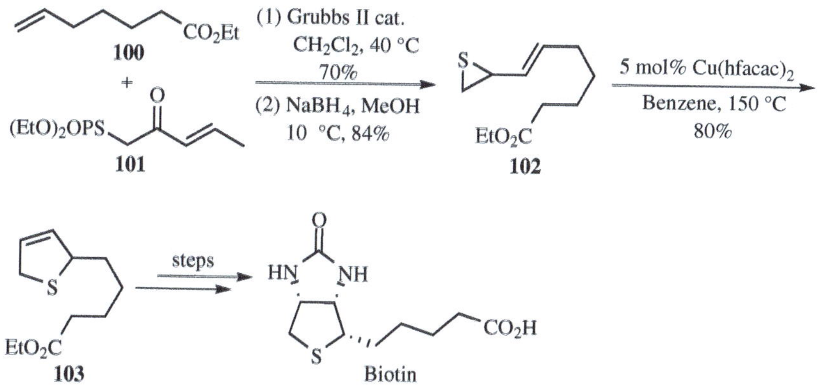

Scheme 11.62 Njardarson's synthesis of biotin utilizing a Cu-catalyzed vinylthiirane-dihydrothiophene rearrangement as the key step.

is speculative as the electron could come from Cu, hfacac ligand, or the olefin of the substrate. The Cu(I) complex undergoes C—N insertion followed by allylic transposition to form Cu(III) complex **99**. Reductive elimination gives the pyrroline product and regenerates Cu(I) complex **97** upon binding of another substrate molecule.

11.9.2 Applications in Synthesis

Njardarson and coworkers utilized the Cu-catalyzed ring expansion of thiiranes developed in their lab as the key step in syntheses of the cores of Biotin and Plavix (Schemes 11.62 and 11.63).[37b] The vinylthiirane precursor **102** for Biotin's core was prepared by cross metathesis of ethyl 6-heptenoate **100** with enone thiophosphate **101**, followed by selective reduction using $NaBH_4$ and subsequent cyclization (Scheme 11.61). Rearrangement of **102** using 5 mol% Cu(hfacac)$_2$ in benzene afforded dihydrothiophene **103** in high yield and can be converted to Biotin in a few steps (Scheme 11.62).[37b]

For Plavix core, aldehyde **104** was converted to an epoxide **105** using chloroiodomethane and *n*-BuLi, followed by its treatment with potassium thiocyanate to give vinyl thiirane **106**. Ring expansion of **106** was achieved by heating with 5 mol% Cu(hfacac)$_2$ in benzene, to afford dihydrothiophene core **107**, which can be transformed to Plavix by a previously reported procedure (Scheme 11.63).[37b]

The Cu(hfacac)$_2$-catalyzed stereoselective rearrangement of vinyl oxiranes to 2,5-dihydrofurans was successfully applied by Njardarson's group to the total synthesis of anticancer agent (+)-goniothalesdiol. The enantiopure vinyloxirane precursor **110** was synthesized using Jorgenson's organocatalytic asymmetric epoxidation of cinnamaldehyde **108**, followed by vinyl Grignard addition and Johnson–Claisen rearrangement. The stereoselective ring expansion of vinyloxirane **110** was successfully achieved using relatively low catalyst loadings [1 mol %

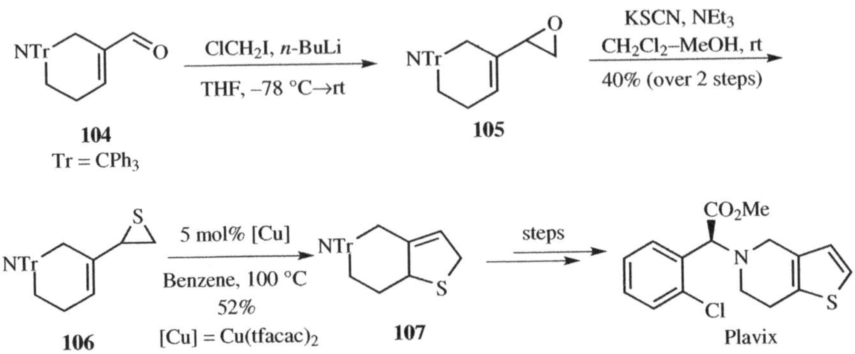

Scheme 11.63 Njardarson's synthesis of dihydrothiophene core **107** of Plavix using a Cu-catalyzed vinylthiirane-dihydrothiophene rearrangement.

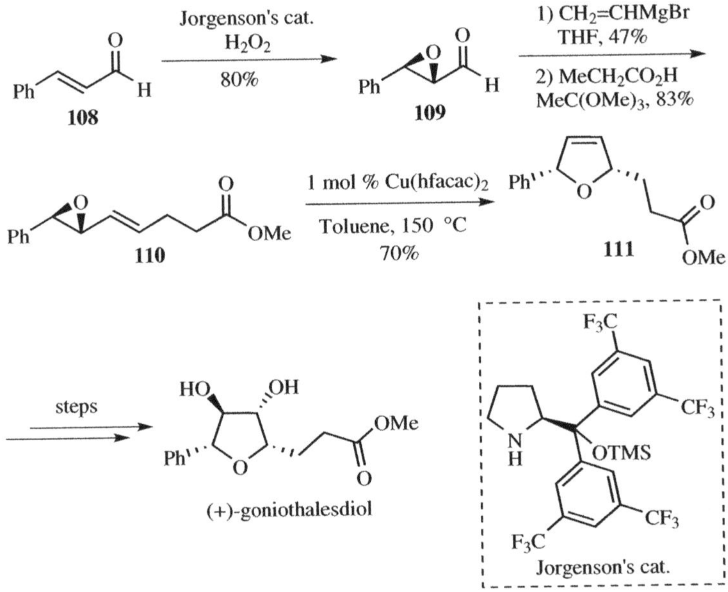

Scheme 11.64 Njardarson's synthesis of (+)-goniothalesdiol utilizing a stereoselective Cu-catalyzed vinyloxirane-dihydrofuran rearrangement.

$Cu(hfacac)_2$] to afford *cis*-2,5-dihydrofuran **111** in good yield, followed by conversion in a few steps to (+)-goniothalesdiol (Scheme 11.64).[39a]

Recently, in 2013, the Njardarson group utilized the stereoselective vinyloxirane–dihydrofuran rearrangement to synthesize labdane diterpenoids.[39b] The decalin core structure **112** obtained in enantiopure form from inexpensive juniper berries was subjected to Shi's epoxidation to form the vinyloxirane **113** in a stereoselective fashion.

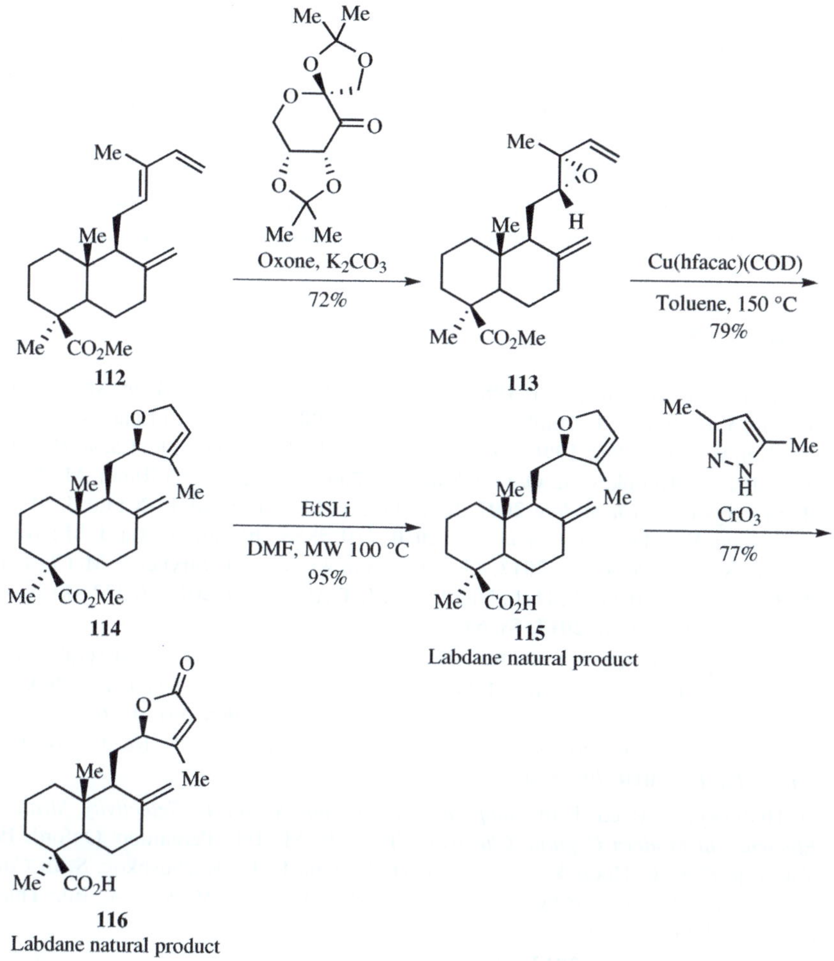

Scheme 11.65 Njardarson's synthesis of labdane diterpenoids involving a stereoselective Cu-catalyzed vinyloxirane-dihydrofuran rearrangement.

The vinyloxirane was then treated with Cu(hfacac)(COD) catalyst to afford dihydrofuran **114**, which was rapidly converted to labdane natural products, **115** and **116** (Scheme 11.65).

11.10 SUMMARY AND OUTLOOK

This chapter describes advances made in methods for the VCP–CP rearrangement, including heteroatom analogues, and the applications of these rearrangements in organic synthesis. The thermal VCP–CP rearrangement has been explored and utilized extensively in natural product synthesis for over 40 years. Recent progress in this field has been made by introduction of new transition metal catalysts that can promote these rearrangements selectively and under relatively mild conditions. While impressive methodologies have been developed with the use of transition metal catalysts, their application in complex-molecule synthesis is still underdeveloped compared to the thermal VCP–CP rearrangement. Future advances will involve the discovery of new approaches, which have broader substrate scope, high functional group compatibility, and, most importantly, should be application oriented.

REFERENCES

1. (a) Paquette, L. A. *Chem. Rev.* **1986**, *86*, 733. (b) Wong, H. N. C.; Hon, M. Y.; Tse, C. W.; Yip, Y. C.; Tanko, J.; Hudlicky, T. *Chem. Rev.* **1989**, *89*, 165. (c) Lautens, M.; Klute, W.; Tam, W. *Chem. Rev.* **1996**, *96*, 49. (d) Reissig, H.-U.; Zimmer, R. *Chem. Rev.* **2003**, *103*, 1151. (e) Kulinkovich, O. G. *Chem. Rev.* **2003**, *103*, 2597. (f) Rubin, M.; Rubina, M.; Gevorgyan, V. *Chem. Rev.* **2007**, *107*, 3117. (g) Shi, M.; Shao, L.-X.; Lu, J.-M.; Wei, Y.; Mizuno, K.; Maeda, H. *Chem. Rev.* **2010**, *110*, 5883. (h) Shi, M.; Lu, J. M.; Wei, Y.; Shao, L. X. *Acc. Chem. Res.* **2012**, *45*, 641. (i) Ylijoki, K. E. O.; Stryker, J. M. *Chem. Rev.* **2013**, *113*, 2244. (j) Mack, D. J.; Njardarson, J. T. *ACS Catal.* **2013**, *3*, 272. (k) Jiao, L.; Yu, Z.-X. *J. Org. Chem.* **2013**, *78*, 6842.
2. (a) Hudlicky, T.; Kutchan, T. M.; Naqvi, S. M. *Org. React.* **1985**, *33*, 247. (b) Goldschmidt, Z.; Crammer, B. *Chem. Soc. Rev.* **1988**, *17*, 229. (c) Baldwin, J. E. *Chem. Rev.* **2003**, *103*, 1197. (d) Wang, S. C.; Tantillo, D. J. *J. Organomet. Chem.* **2006**, *691*, 4386. (e) Baldwin, J. E.; Leber, P. A. *Org. Biomol. Chem.* **2008**, *6*, 36. (f) Hudlicky, T.; Reed, J. W. *Angew. Chem., Int. Ed.* **2010**, *49*, 4864.
3. (a) Hudlicky, T.; Reed, J. In *Comprehensive Organic Synthesis: Selectivity, Strategy & Efficiency in Modern Organic Chemistry*; Trost, B. M., Ed.; Pergamon: Oxford, **1991**; Vol. 5, p. 899. (b) Hudlicky, T.; Becker, D. A.; Fan, R. L.; Kozhushkov, S. In *Carbocyclic Three-Membered Ring Compounds (Houben–Weyl)*; de Meijere, A., Ed.; Thieme: Stuttgart, **1996**; Vol. E 17c, p. 2538.
4. Njardarson, J. T. *Synlett* **2013**, *24*, 787.
5. (a) Neureiter, N. P. *J. Org. Chem.* **1959**, *24*, 2044. (b) Vogel, E. *Angew. Chem.* **1960**, *4*, 72. (c) Overberger, C. G.; Borchert, A. E. *J. Am. Chem. Soc.* **1960**, *82*, 1007. (d) Overberger, C. G.; Borchert, A. E. *J. Am. Chem. Soc.* **1960**, *82*, 4896. (e) Cloke, J. B.; Borchert, A. E. *J. Am. Chem. Soc.* **1929**, *51*, 1174. (f) Wilson, C. L.; Borchert, A. E. *J. Am. Chem. Soc.* **1947**, *69*, 3002.
6. Corey, E. J.; Walinsky, S. W. *J. Am. Chem. Soc.* **1972**, *94*, 8932.
7. Simpson, J. M.; Richey, H. G. *Tetrahedron Lett.* **1973**, *14*, 2545.
8. (a) Trost, B. M.; Bogdanowicz, M. J. *J. Am. Chem. Soc.* **1973**, *95*, 5311. (b) Trost, B. M.; Keeley, D. E. *J. Am. Chem. Soc.* **1976**, *98*, 248.
9. Danheiser, R. L.; Martinez-Davila, C.; Morin, J. M. *J. Org. Chem.* **1980**, *45*, 1340.
10. (a) Larsen, S. D. *J. Am. Chem. Soc.* **1988**, *110*, 5932. (b) Larsen, S. D.; Fisher, P. V.; Libby, B. E.; Jensen, R. M.; Mizsak, S. A.; Watt, W. *J. Org. Chem.* **1996**, *61*, 4725.
11. Hudlicky, T.; Heard, N. E.; Fleming, A. *J. Org. Chem.* **1990**, *55*, 2570.
12. Sakito, Y.; Suzukamo, G. *Chem. Lett.* **1986**, 621.

13. (a) Davies, H. M. L.; Hu, B. *J. Org. Chem.* **1992**, *57*, 3186. (b) Davies, H. M. L.; Hu, B. *J. Org. Chem.* **1992**, *57*, 4309. (c) Davies, H. M. L.; Kong, N.; Churchill, M. R. *J. Org. Chem.* **1998**, *63*, 6586.

14. Coscia, R. W.; Lambert, T. H. *J. Am. Chem. Soc.* **2009**, *131*, 2496.

15. McGaffin, G.; Grimm, B.; Heinecke, U.; Michaelsen, H.; deMeijere, A.; Walsh, R. *Eur. J. Org. Chem.* **2001**, 3559.

16. Andrews, G. D.; Baldwin, J. E. *J. Am. Chem. Soc.* **1976**, *98*, 6705.

17. (a) Hudlicky, T.; Price, J. D. *Chem. Rev.* **1989**, *89*, 1467. (b) Takeda, K.; Sakurama, K.; Yoshi, E. *Tetrahedron Lett.* **1997**, *38*, 3257.

18. (a) Satyanarayana, J.; Rao, M. V. B.; Ila, H.; Junjappa, H. *Tetrahedron Lett.* **1996**, *37*, 3565. (b) Hashimoto, S.-I.; Shinoda, T.; Ikegami, S. *Tetrahedron Lett.* **1986**, *27*, 2885. (c) Harvey, D. F.; Brown, M. F. *Tetrahedron Lett.* **1991**, *32*, 2871.

19. (a) Atkinson, R. S.; Rees, C. W. *Chem. Commun.* **1967**, 1232. (b) Mishra, A.; Rice, S. N.; Lwowski, W. *J. Org. Chem.* **1968**, *33*, 481. (c) Paladini, J. C.; Chuche, J. *Tetrahedron Lett.* **1971**, *12*, 4383.

20. (a) Hudlicky, T.; Frazier, J. O.; Seoane, G.; Tiedje, M.; Seoane, A.; Kwart, L. D.; Beal, C. *J. Am. Chem. Soc.* **1986**, *108*, 3755. (b) Hudlicky, T.; Seoane, G.; Lovelace, T. C. *J. Org. Chem.* **1988**, *53*, 2094. (c) Hudlicky, T.; Gurudas, S.-Z.; Seoane, G. *Synth. Commun.* **1987**, *17*, 1155.

21. (a) Eberbach, W.; Trostmann, U. *Chem. Ber.* **1981**, *114*, 2979. (b) Eberbach, W.; Trostmann, U. *Chem. Ber.* **1985**, *118*, 4035. (c) Hudlicky, T.; Fleming, A.; Lovelace, T.C. *Tetrahedron* **1989**, *45*, 3021. (d) Hudlicky, T.; Fleming, A.; Lovelace, T.C. *Synth. Commun.* **1990**, *20*, 1721. (e) Batsanov, A. S.; Byerley, A. L. J.; Howard, J. A. K.; Steel, P. G. *Synlett* **1996**, 401.

22. (a) Crawford, R. J.; Lutener, S. B.; Cockcroft, R. D. *Can. J. Chem.* **1976**, *54*, 3364. (b) Eberbach, W.; Burchardt, B. *Chem. Ber.* **1978**, *111*, 3665. (c) Muxfeldt, H.; Schneider, R. S.; Mooberry, J. B. *J. Am. Chem. Soc.* **1966**, *88*, 3670. (d) Braslavsky, S.; Heicklen, J. *Chem. Rev.* **1977**, *77*, 473.

23. (a) Trost, B. M.; Nishimura, Y.; Yamamoto, K.; McElvain, S. S. *J. Am. Chem. Soc.* **1979**, *101*, 1328. (b) Hudlicky, T.; Kutchan, T. M.; Wilson, S. R.; Mao, D. T. *J. Am. Chem. Soc.* **1980**, *102*, 6351. (c) Hudlicky, T.; Natchus, M. G. *J. Org. Chem.* **1992**, *57*, 4740. (d) Corey, E. J.; Kigoshi, H. *Tetrahedron Lett.* **1991**, *32*, 5025. (e) Hirner, S.; Somfai, P. *Synlett* **2005**, 3099. (f) Majetich, G.; Zou, G.; Grove, J. *Org. Lett.* **2008**, *10*, 85.

24. Frey, H. M. *Trans. Faraday Soc.* **1962**, *58*, 516.

25. Sonawane, H. R.; Bellur, N. S.; Kulkarni, D. G.; Ahuja, J. R. *Synlett* **1993**, 875.

26. (a) Heathcock, C. H.; Poulter, S. R. *J. Am. Chem. Soc.* **1968**, *90*, 3766. (b) Hixon, S. S. *Org. Photochem.* **1976**, *4*, 218.

27. (a) Cooke, R. S. *Chem. Commun.* **1970**, 454. (b) Paquette, L. A.; Meehan, G. V.; Henzel, R. P.; Eizember, R. F. *J. Org. Chem.* **1973**, *38*, 3250. (c) Farneth, W. E.; Thomsen, M. W.; Berg, M. A. *J. Am. Chem. Soc.* **1979**, *101*, 6468. (d) Farneth, W. E.; Thomsen, M. W.; Schultz, N. L.; Davies, M. A. *J. Am. Chem. Soc.* **1981**, *103*, 401.

28. (a) Sonawane, H. R.; Nanjundiah, B. S.; Kumar, M. V. *Tetrahedron Lett.* **1985**, *26*, 1097. (b) Armesto, D.; Ortiz, M. J.; Agarrbeitia, A. R. *J. Org. Chem.* **1999**, *64*, 1056.

29. (a) Campos, P. J.; Soldevilla, A.; Sampedro, D.; Rodríguez, M. A. *Org. Lett.* **2001**, *3*, 4087. (b) Soldevilla, A.; Sampedro, D.; Campos, P. J.; Rodríguez, M. A. *J. Org. Chem.* **2005**, *70*, 697.

30. Sampedro, D.; Soldevilla, A.; Rodríguez, M. A.; Campos, P. J.; Olivucci, M. *J. Am. Chem. Soc.* **2005**, *127*, 441.

31. (a) Sonawane, H. R.; Naik, V. G.; Bellur, N. S.; Shah, V. G.; Purohit, P. C.; Kumar, M. U.; Kulkarni, D. G.; Ahuja, J. R. *Tetrahedron* **1991**, *47*, 8259. (b) Sonawane, H. R.; Bellur, N. S.; Shah, V. G.; Kulkarni, D. G.; Ahuja, J. R. *Tetrahedron Lett.* **1991**, *32*, 1107. (c) Sonawane, H. R.; Nanjundiah, B. S.; Kumar, M. U. *Tetrahedron Lett.* **1984**, *25*, 2245.

32. (a) Grigg, R.; Hayes, R.; Sweeney, A. *Chem. Commun.* **1971**, 1248. (b) Hayashi, M.; Ohmatsu, T.; Meng, Y.-P.; Saigo, K. *Angew. Chem., Int. Ed.* **1998**, *37*, 837. (c) Salomon, R. G.; Salomon, M. F.; Kachinski, J. L. C. *J. Am. Chem. Soc.* **1977**, *99*, 1043. (d) Paquette, L. A.; Detty, M. R. *Tetrahedron Lett.* **1978**, *19*, 713.

33. (a) Alcock, N. W.; Brown, J. M.; Conneely, J. A.; Williamsom, D. H. *J. Chem. Soc., Chem. Commun.* **1975**, 792. (b) Aris, V.; Brown, J. M.; Conneely, J. A.; Golding, B. T.; Williamson, D. H. *J. Chem. Soc., Perkin Trans. 2* **1975**, 4. (c) Alcock, N. W.; Brown, J. M.; Conneely, J. A.; Williamsom, D. H. *J. Chem. Soc., Perkin Trans. 2* **1979**, 962.

34. (a) Morizawa, Y.; Oshima, K.; Nozaki, H. *Tetrahedron Lett.* **1982**, *23*, 2871. (b) Hiroi, K.; Yoshida, Y.; Kaneko, Y. *Tetrahedron Lett.* **1999**, *40*, 3431. (c) Lingam, K. A. P.; Shanmugam, P. *Tetrahedron* **2013**, *54*, 4202.

35. (a) Murakami, M.; Nishida, S. *Chem. Lett.* **1979**, 927. (b) Ryu, I.; Ikura, K.; Tamura, Y.; Maenaka, J.; Ogawa, A.; Sonoda, N. *Synlett* **1994**, 941. (c) Zuo, G.; Louie, J. *Angew. Chem., Int. Ed.* **2004**, *43*, 2227.

36. Wang, S. C.; Troast, D. M.; Conda-Sheridan, M.; Zuo, G.; LaGarde, D.; Louie, J.; Tantillo, D. J. *J. Org. Chem.* **2009**, *74*, 7822.

37. (a) Batory, L. A.; McInnis, C. E.; Njardarson, J. T. *J. Am. Chem. Soc.* **2006**, *128*, 16054. (b) Rogers, E.; Araki, H.; Batory, L. A.; McInnis, C. E.; Njardarson, J. T. *J. Am. Chem. Soc.* **2007**, *129*, 2768. (c) Brichacek, M.; Lee, D.; Njardarson, J. T. *Org. Lett.* **2008**, *10*, 5023. (d) Brichacek, M.; Villalobos, M. N.; Plichta, A.; Njardarson, J. T. *Org. Lett.* **2011**, *13*, 1110.

38. (a) Mack, D. J.; Njardarson, J. T. *Chem. Sci.* **2012**, *3*, 3321. (b) Mustard, T. J. L.; Mack, D. J.; Njardarson, J. T.; Cheong, P. H.-Y. *J. Am. Chem. Soc.* **2013**, *135*, 1471.

39. (a) Brichacek, M.; Batory, L. A.; Njardarson, J. T. *Angew. Chem., Int. Ed.* **2010**, *49*, 1684. (b) Mack, D. J.; Njardarson, J. T. *Angew. Chem., Int. Ed.* **2013**, *52*, 1543.

12

FERRIER CARBOCYCLIZATION REACTION

Noritaka Chida

Department of Applied Chemistry, Keio University, Kohoku-ku, Yokohama, Japan

12.1 INTRODUCTION

Ferrier has reported two important rearrangement reactions concerning the transformation of carbohydrates (Scheme 12.1). One is an acid-catalyzed rearrangement of glycal **1** in which various nucleophiles are introduced at C-1 position in **1** with migration of double bond to give **2**.[1] Although the allylic rearrangement of glycal with water was first reported by Emil Fischer, Ferrier generalized the reaction and showed the utility of the rearrangement in the 1960s. This reaction has been called "Ferrier rearrangement," "Ferrier I rearrangement," or "Type I Ferrier rearrangement." In this chapter, this rearrangement of glycals is called "Ferrier rearrangement." The other reaction is a transformation of 6-deoxy-hex-5-enopyranoside **3** into cyclohexanone derivative **4** by the action of a Hg(II) salt, first reported by Ferrier in 1979.[2] This reaction has been called "Ferrier II rearrangement," "Ferrier II reaction," "Type II Ferrier rearrangement," or "Ferrier carbocyclization." Since the reaction provides a highly substituted carbocycle, it will be called the "Ferrier carbocyclization reaction" in this chapter. While a brief introduction to the synthetic utility of the Ferrier rearrangement and a beautiful example in natural product synthesis are shown in the following, this chapter focuses on the Ferrier carbocyclization reaction.

The Ferrier rearrangement is a useful reaction for conversion of glycals to many types of glycosides. Hydrogen-, fluorine-, oxygen-, nitrogen-, sulfur-, and carbon-linked products were obtained with relatively high stereoselectivity.[3] One of the important applications of the rearrangement is the formation of *C*-glycosides,

Molecular Rearrangements in Organic Synthesis, First Edition. Edited by Christian M. Rojas.

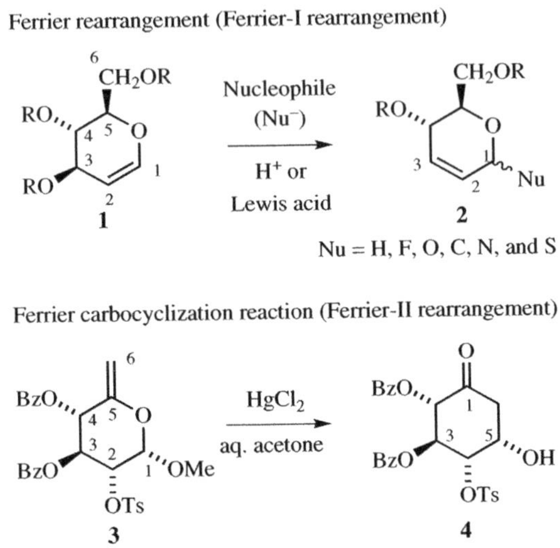

Scheme 12.1 Two types of Ferrier rearrangement reactions.

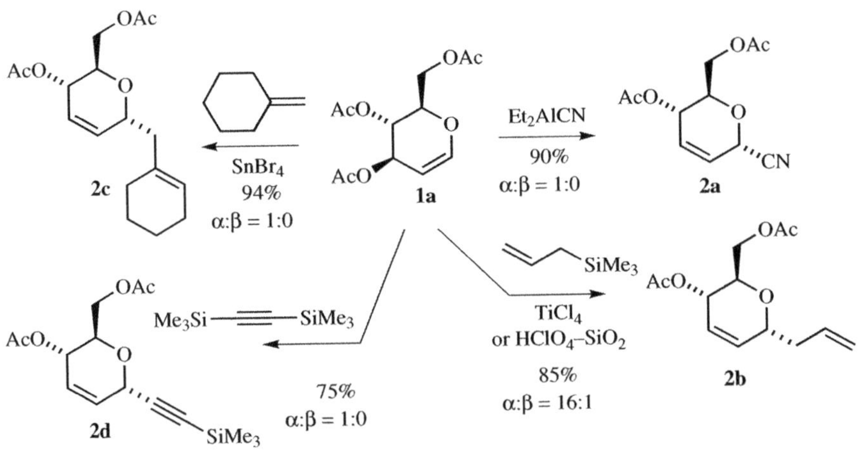

Scheme 12.2 *C*-Glycoside formation by the Ferrier rearrangement.

where various nucleophilic carbon sources could be introduced at C-1 position to give 2,3-unsaturated *α*-*C*-glycosides in good yield. Some representative examples of *C*-glycosides obtained by the rearrangement of tri-*O*-acetyl-D-glucal **1a** are shown in Scheme 12.2.

In 2003, Williams and Heidebrecht reported the total synthesis of marine diterpene, 4,5-deoxyneodolabelline **10**, where subunits cyclic acetal **5** and allylsilane **6** were coupled by the Ferrier rearrangement (Scheme 12.3).[4] The product **7**, obtained as a single isomer, was transformed into ketoaldehyde **8**. Intramolecular McMurry

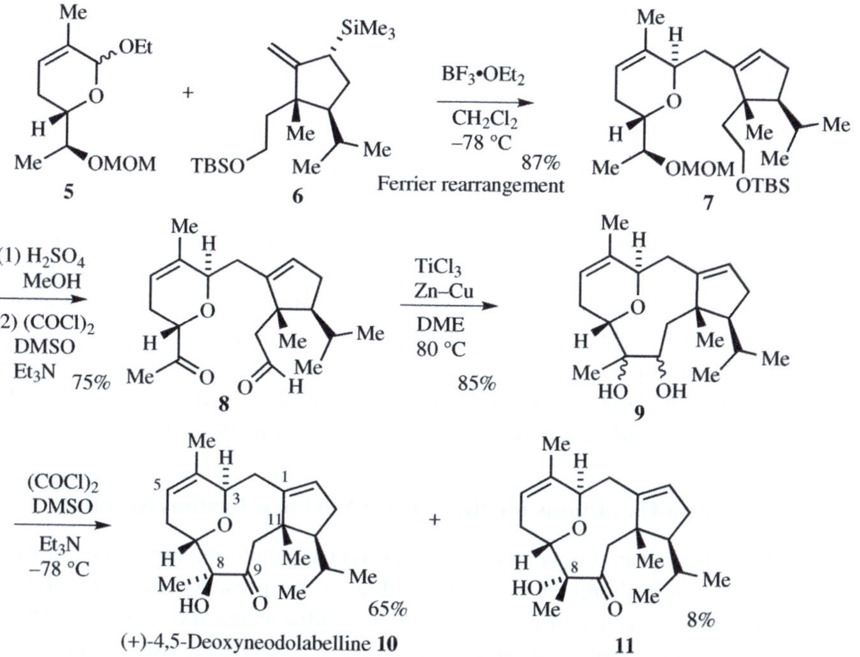

Scheme 12.3 Application of the Ferrier rearrangement to the synthesis of 4,5-deoxyneodolabelline.

coupling of **8** with $TiCl_3$ and Zn–Cu successfully provided ring-closed product **9**, oxidation of which afforded 4,5-deoxyneodolabelline **10** and its C-8 epimer **11**. Although Williams and coworkers synthesized glycal-type substrate **5** from a noncarbohydrate precursor, their elegant achievement revealed the potential of the Ferrier rearrangement in the synthesis of structurally complex natural products.

The Ferrier carbocyclization reaction, meanwhile, is another important carbohydrate-based transformation in organic synthesis. The reaction is highly effective for the conversion of readily available aldohexoses into enantiomerically pure cyclohexanones.[5,6] Chiral and highly functionalized cyclohexanones obtained by the Ferrier carbocyclization are potentially useful chiral building blocks, so the reaction has been frequently employed in the synthesis of structurally complex natural products containing a cyclohexane unit. In this chapter, we will focus on the Ferrier carbocyclization reaction and its applications to natural product synthesis.

12.2 GENERAL DISCUSSION AND MECHANISTIC FEATURES

In 1979, Ferrier reported the conversion of 6-deoxy-hex-5-enopyranoside **3** into cyclohexanone **4** by the action of a stoichiometric amount of $HgCl_2$ in aqueous acetone (Scheme 12.4).[2,5,6] This rearrangement attracted much attention from the

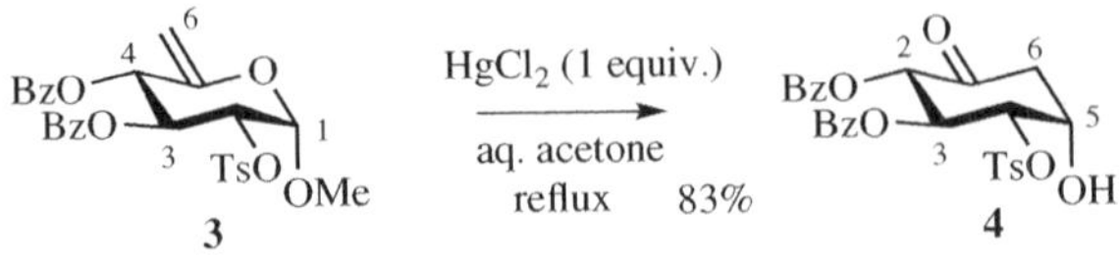

Scheme 12.4 Ferrier carbocyclization reaction.

synthetic community for the following reasons: (1) the substrates, 5-enopyranosides, are readily available from aldohexoses; (2) the reaction proceeds in aqueous media in high yields; (3) high diastereoselectivities concerning the newly formed hydroxy group at C-5 are observed; and (4) densely functionalized cyclohexanones are obtained in enantiomerically pure form.[5] This reaction, now called the Ferrier carbocyclization reaction, greatly expands the possibility of carbohydrates as chiral building blocks, providing a new concept in "chiral pool synthesis."[7]

12.2.1 Reaction Conditions for the Ferrier Carbocyclization Reaction

Although the Ferrier carbocyclization is an effective reaction, sometimes, difficulty is encountered in isolating pure products due to the presence of excess Hg salts or in suppressing side reactions caused by the acidic reaction conditions. After the first report of Ferrier, many studies concerning improved reaction conditions have appeared (Scheme 12.5). Lukacs and coworkers reported that the rearrangement of **12** took place with a catalytic amount (2 mol%) of $HgSO_4$ in aqueous 1,4-dioxane containing sulfuric acid at 60–80 °C to give rearranged product **13** as a single isomer.[8] In 1991, Chida *et al.* reported that when $Hg(OCOCF_3)_2$ was employed as a catalyst (5–10 mol%), the rearrangement of **14** smoothly proceeded at room temperature in a neutral solvent to afford products **15a** and **15b** in excellent overall yield.[9] In 1988, Adam disclosed that the carbocyclization (**16** → **17**) works well with $PdCl_2$ or $Pd(OAc)_2$ (20 mol%) in aqueous 1,4-dioxane containing sulfuric acid at 60 °C.[10] Later, in 1996, Ikegami and coworkers reported that the rearrangement of **18** proceeds with 5 mol% of $PdCl_2$ in the absence of sulfuric acid to provide **19a** and **19b**.[11]

At present, the procedure using a catalytic amount of $Hg(OCOCF_3)_2$ or $PdCl_2$ in a neutral solvent system (aq. acetone or aq. 1,4-dioxane) provides the mildest conditions, allowing the presence of acid-labile functionalities such as esters, methoxymethyl ethers, *p*-methoxybenzyl ethers, and silyl ethers in the substrates. In the reaction of **20**, however, a *tert*-butyldimethylsilyl ether (OTBS) was partially deprotected during the reaction. Use of acetate buffer (0.1 M, pH 4.8) as a cosolvent suppressed the cleavage of the OTBS to give the rearranged product **21** in good yield, although an increased amount (10–30 mol%) of $Hg(OCOCF_3)_2$ was required (Scheme 12.6).[12,13] 5-Enopyranoside derivative **22**, which has no electron-withdrawing substituent, was found to be unstable to acid. Addition of $Hg(OCOCF_3)_2$ to the solution of **22** did not give any cyclized product, but inverse addition (slow addition of a solution of **22** to the solution of Hg salt) gave **23** in 86% yield.[14]

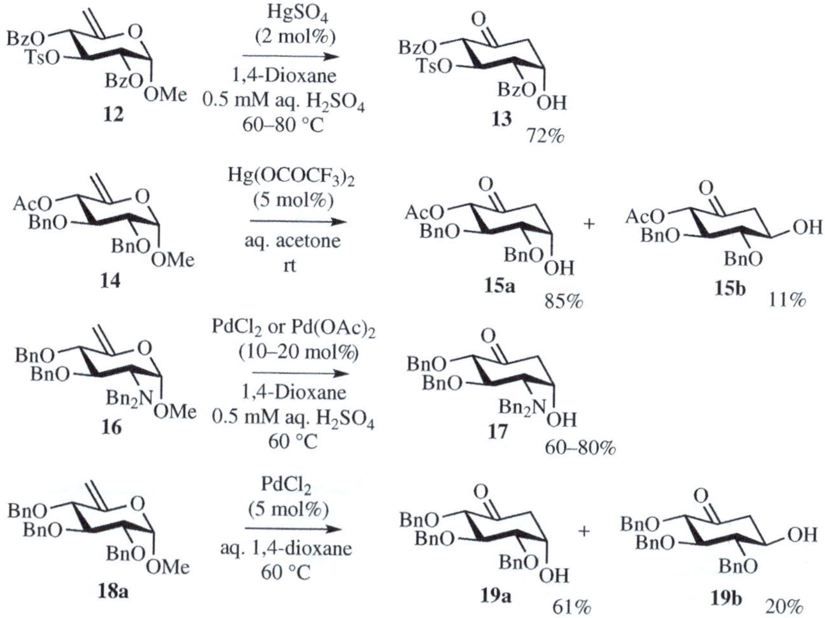

Scheme 12.5 Catalytic Ferrier carbocyclization reaction.

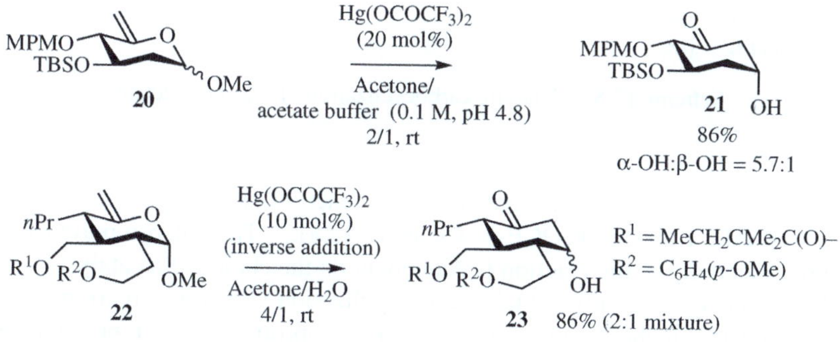

Scheme 12.6 Improved reaction conditions for the carbocyclization.

Two synthetically important variants of the Ferrier carbocyclization reaction have been reported. One is a rearrangement of enol acetate **24** (Scheme 12.7). Reaction of **24** with a stoichiometric amount of Hg salt afforded an organomercurial intermediate **25**, which was then treated with NaCl to induce the cyclization affording inosose derivatives **26a** and **26b** with good stereoselectivity.[15,16] As biologically important *myo*-inositol derivatives, such as D-*myo*-inositol phosphates, are optically active, the enol-acetate version of the Ferrier carbocyclization would be effective for the preparation of enantiomerically pure inositol derivatives.

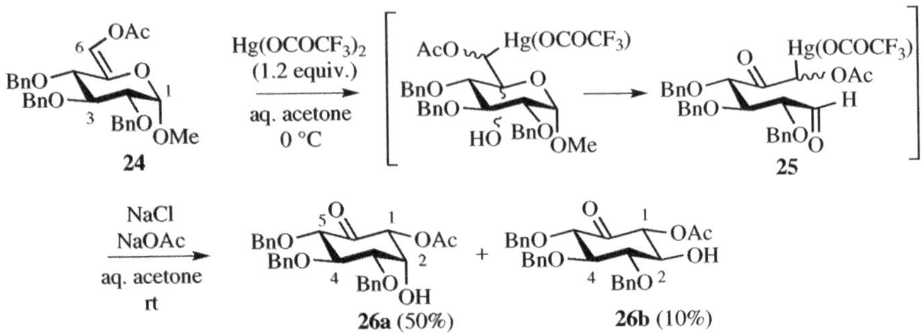

Scheme 12.7 Ferrier carbocyclization of an enol-acetate.

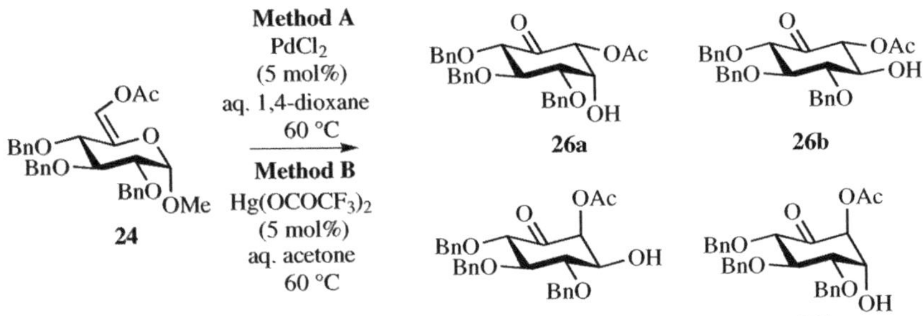

	26a	26b	26c	26d
Method A	40%	8%	19%	14%
Method B	16%	0%	4%	0%

Scheme 12.8 Catalytic carbocyclization of the enol-acetate.

Later, Ikegami *et al.* found that the same reaction proceeded using a catalytic amount of $PdCl_2$ or $Hg(OCOCF_3)_2$ (Scheme 12.8).[17] They also reported that the diastereoselectivity of the reaction is dependent on the choice of a catalyst.

The second synthetically valuable carbocyclization variant is a rearrangement of 5-enopyranosides in the presence of Lewis acids in aprotic solvents reported by Sinaÿ and coworkers (Scheme 12.9).[18] The "normal" Ferrier carbocyclization afforded the cyclohexanone derivatives in which the anomeric substituents (aglycon) were lost and the stereochemical information at the anomeric center was also partially lost (**18a** → **19a** + **19b**),[11] whereas the Lewis acid conditions gave products with retention of the anomeric substituent as well as its stereochemistry. For example, reaction of **18a** with excess $(i\text{-Bu})_3Al$ afforded cyclohexanol derivative **28** where $(i\text{-Bu})_3Al$ had reduced the ketone carbonyl formed in the rearrangement. When $Ti(Oi\text{-Pr})Cl_3$ was employed, the corresponding cyclohexanone **27** was obtained. The same reaction of *β*-pyranoside **18b** afforded **29** as the major isomer.

The important feature of the Lewis acid-catalyzed rearrangement is that these reactions are applicable to *C*-glycosides (Scheme 12.10, **30** → **31**) when the substrates

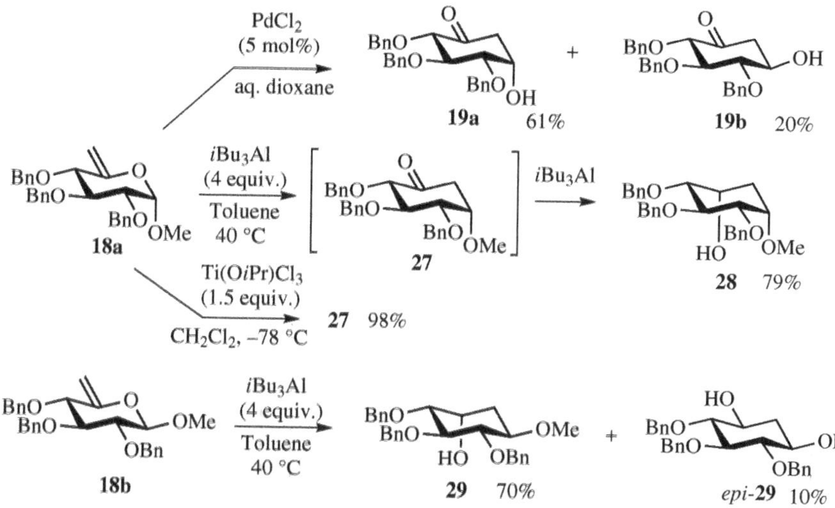

Scheme 12.9 Lewis acid-mediated Ferrier-type carbocyclization.

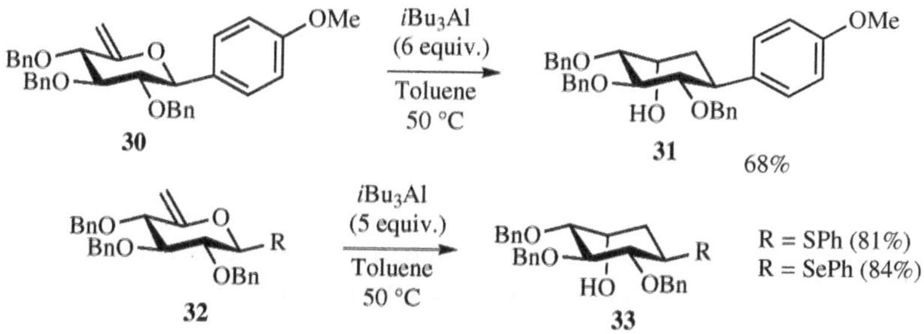

Scheme 12.10 Carbocyclization of *C*-, *S*-, and *Se*-glycosides.

possess electron-donating aglycons. *S*- and *Se*-Glycosides also afforded the corresponding carbocycles (**32** → **33**).[19]

If the Ferrier carbocyclization reaction could be applied to 4-enofuranoside substrates, the corresponding cyclopentanones, which are useful chiral building blocks in natural product synthesis, would be obtained. However, it has been reported that the attempted Ferrier carbocyclization of furanoside **34** with a stoichiometric amount of $Hg(OAc)_2$ gave no cyclopentanone product **35**, but gave organomercurial compound **36** instead (Scheme 12.11).[20] On the other hand, formation of cyclopentane derivative **38** has been reported when γ-lactone **37** was treated with $LiAlH(OtBu)_3$ in THF.[21] It was observed that the aldol-type cyclization had occurred during the acidic work-up process. β-Elimination of the hydroxy group in **38** gave cyclopentenone **39** in 74% yield from **37**. Likewise, reaction of **37** with PhMgBr followed by β-elimination afforded cyclopentenone **41** in 40% yield.[21]

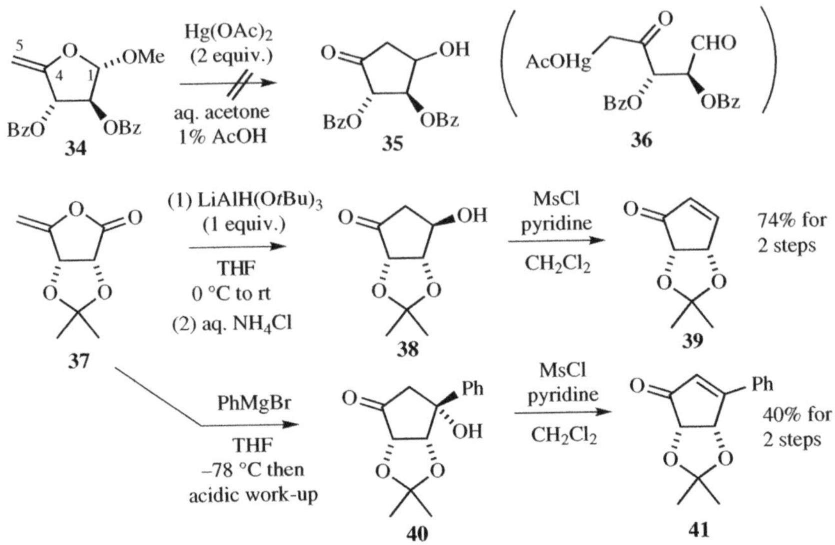

Scheme 12.11 Ferrier-type carbocyclization of 4-enofuranosides.

12.2.2 Mechanism and Stereoselectivity of the Ferrier Carbocyclization Reaction

The Ferrier carbocyclization reaction is a reliable transformation of 5-enopyranosides into carbocycles. As shown in Scheme 12.12 (the carbon numberings in substrates and products have been changed; the numbering of substrates is based on the nomenclature of carbohydrates, and the numbering of products is that of cyclohexanones), irrespective of the kinds of substituents and patterns of stereochemistry in the substrates, the rearranged products (**15** and **42–46**) were obtained in moderate to good yields with relatively high diastereoselectivity.[22] An exception is a reaction of a 4-deoxy-5-enopyranoside derivative, which resulted in a low yield of product **47**.

There is a strong correlation between the orientation of hydroxy groups at C-5 and the substituents at C-3 in the major products. The newly formed hydroxy group at C-5 and the substituent at C-3 are transrelated to each other: substrates possessing C-3 substituents directed upward gave rearranged products having hydroxy group directed downward as the major isomer.

The proposed mechanism of the Ferrier carbocyclization reaction is outlined in Scheme 12.13.[3,5,23] First, oxymercuration of the *exo*-olefin in **48** affords mercurial–hemiacetal **49**, whose aglycon moiety (–OMe) eliminates to give mercurial–aldehyde derivative **50**. This mercurial intermediate **50** was isolable when a stoichiometric amount of Hg salt was employed at low temperature.[23] Intramolecular aldol-type cyclization of **50** provides product **51**.

For an explanation of the stereoselectivity, the chair-like transition structures (**52A** and **52B**) in which the mercury is chelated by the enolate oxygen and the aldehydic oxygen have been proposed by Machado *et al.*[24] and Ferrier and Middleton[5]

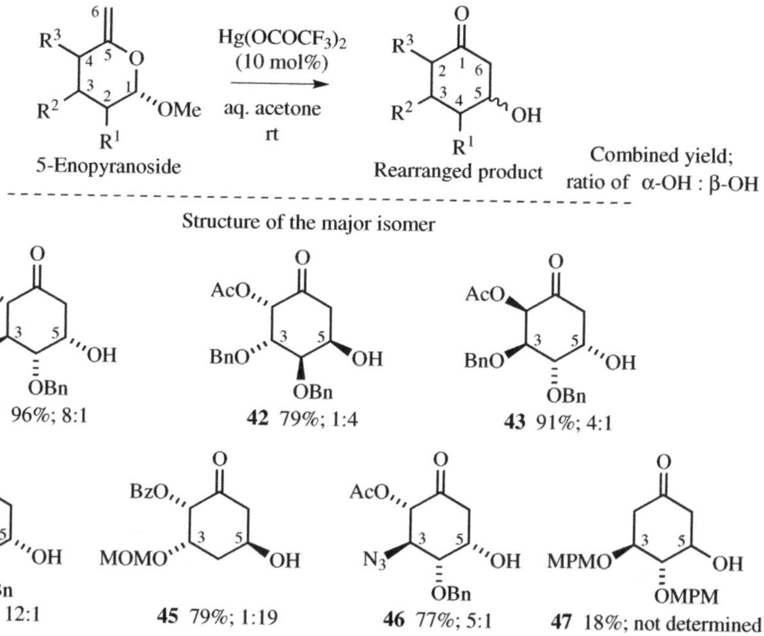

Scheme 12.12 Ferrier carbocyclization of various substrates.

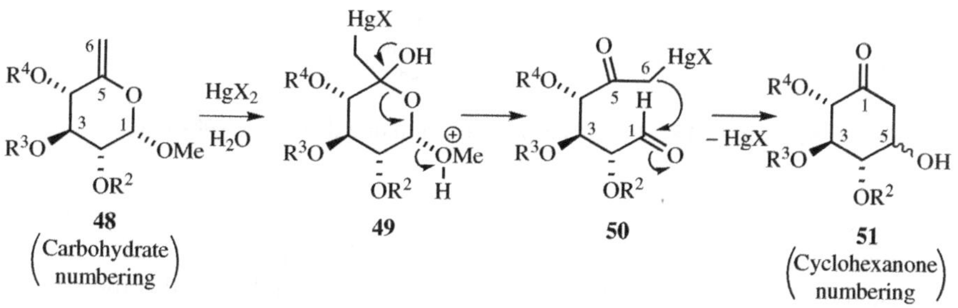

Scheme 12.13 Proposed mechanism of the Ferrier carbocyclization reaction.

(Scheme 12.14). This transition model would account for the observed stereoselectivity concerning the newly formed hydroxy group at C-5. In the transformation of **14** → **15a**, while the transition structure **52A** giving **15a** has no severe steric repulsion, that of **52B** giving the 5-epimer of **15a** suffers 1,3-diaxial interactions. In the reaction of **53** → **45b**, the favored transition structure would be **54B**, giving 3,5-trans-isomer **45b** as the major product.

Careful experiments using deuterated substrate **55** by Kakinuma and Eguchi revealed that (1) the Ferrier carbocyclization of **55** involved no retro-aldol process and was thus an irreversible reaction and (2) the overall cyclization reaction was nonstereoselective with respect to the C-6 methylene site (Scheme 12.15).[25] Based

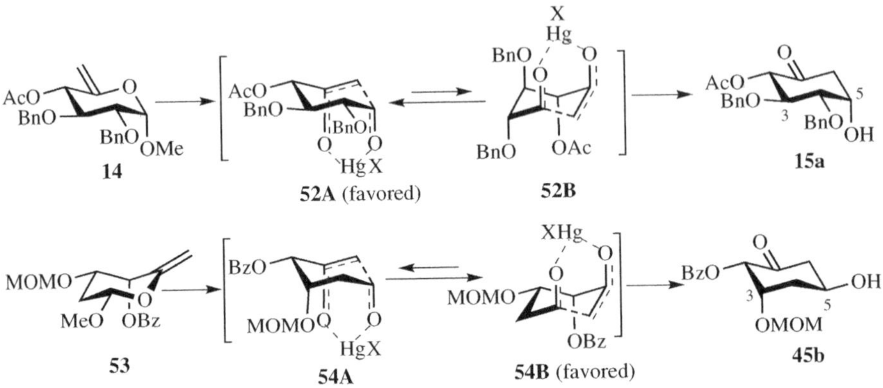

Scheme 12.14 Possible transition structures of the cyclization.

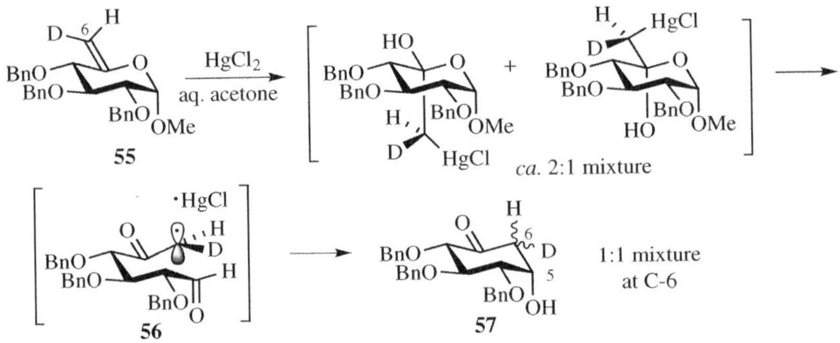

Scheme 12.15 Ferrier carbocyclization of a deuterated substrate.

on the results of complete stereochemical scrambling at C-6 in cyclized product **57**, Kakinuma and Eguchi proposed the presence of radical intermediate **56**. They also suggested that the cyclization might be a radical process, although this model could not explain the stereoselectivity at C-5.

The true mechanism of the cyclization process as well as the origin of the diastereoselectivity in the Ferrier carbocyclization reaction has not been fully clarified at present. More detailed investigations will be required for a more complete understanding of the mechanism of the Ferrier carbocyclization reaction.

The mechanism of the Lewis acid-mediated rearrangement in aprotic solvent is somewhat different from that under the aqueous conditions, and this reaction is best classified as a carbohydrate version of the Petasis–Ferrier rearrangement.[26] The reaction is initiated by coordination of the aluminum with the endocyclic oxygen to give **58** (Scheme 12.16).[27] The activation of the endocyclic oxygen induces ring opening of the pyranoside to generate a zwitterionic aluminum enolate intermediate **59**, which may undergo direct cyclization via a twist-chair transition structure to give cyclohexanone **27′**. The chair-like transition structure **60**, generated by the bond rotation

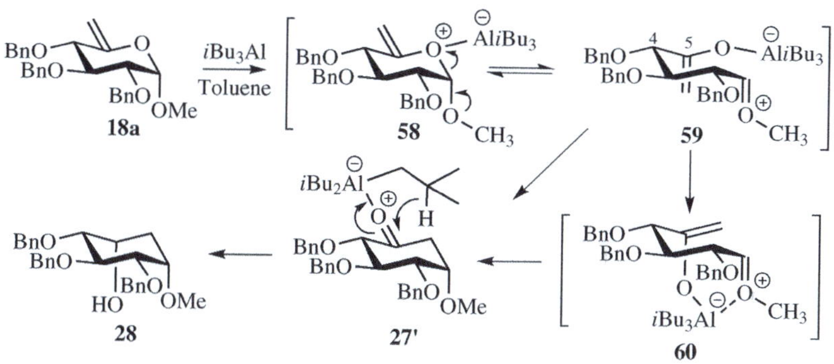

Scheme 12.16 Proposed mechanism of the Lewis acid-mediated cyclization.

between C-4 and C-5 in **59**, would also be plausible. Intramolecular hydride delivery in **27′** from the less hindered *β*-side provides alcohol **28**.

12.3 SYNTHETIC STRATEGIES BASED ON THE FERRIER CARBOCYCLIZATION REACTION

Chiral cyclohexanones obtained by the Ferrier carbocyclization reaction are useful precursors for the synthesis of cyclitols and aminocyclitols, some of which are found in clinically important aminoglycoside antibiotics.[28] Additionally, highly substituted cyclohexenones, prepared by the Ferrier carbocyclization followed by *β*-elimination, can undergo various further transformations, also making these compounds potential chiral building blocks for the preparation of structurally complex compounds having cyclohexane unit(s). This section provides an overview of the reported synthetic strategies toward various types of natural products based on utilization of the Ferrier carbocyclization reaction.

12.3.1 Inositol and Aminocyclitol Synthesis

Strategic planning for the synthesis of the aminocyclitol moiety of antibiotic hygromycin A **61** based on the Ferrier carbocyclization reaction is shown in Scheme 12.17[29] (carbons that come from carbohydrates are shown in bold lines). A key structural feature of **61** is the presence of a methylene ketal located at the C-4/C-5 position, which makes this molecule optically active. For the chiral synthesis of **61**, cyclohexene **62**, possessing an amino group at C-3, would be a promising precursor. In turn, synthesis of **62** was planned to proceed from **15**. The Ferrier carbocyclization of 5-enopyranoside **14** would generate cyclohexanone **15**. Substrate **14** would be derived by *β*-elimination of primary bromide **63**. As compound **63** has a D-gluco configuration, D-glucose would be the choice of the starting material.

The synthetic plan for an antitumor alkaloid, lycoricidine **64**, based on the Ferrier carbocyclization is shown in Scheme 12.18.[30] The tricyclic phenanthridone skeleton was envisioned to be constructed by C—C bond formation between C-10a and

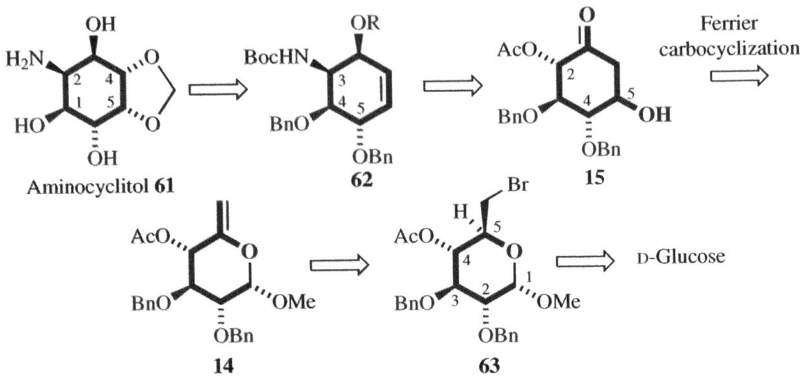

Scheme 12.17 Synthetic plan for an aminocyclitol in hygromycin A.

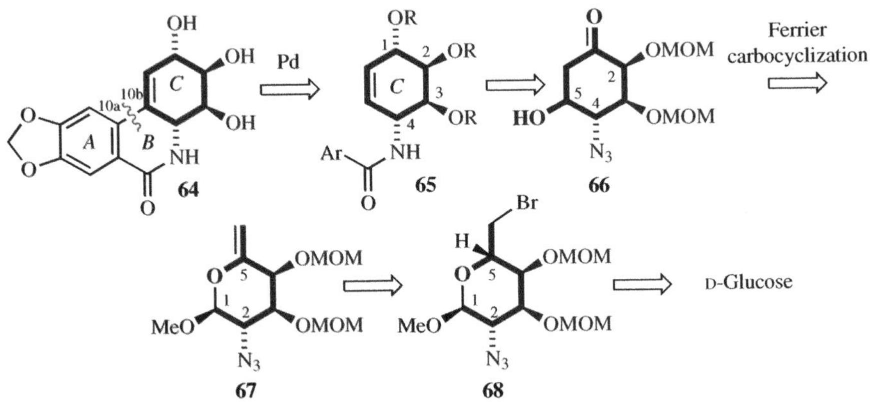

Scheme 12.18 Synthetic plan for lycoricidine.

C-10b by a transition metal-catalyzed sp^2–sp^2 carbon coupling reaction. The C-ring **65** possessing an amido function at C-4 was planned to be derived from 4-azido cyclohexenone **66**, which would be prepared from 5-enopyranoside **67** by the Ferrier carbocyclization. The substrate **67**, possessing the D-altro configuration, would be derived from D-glucose by well-known interconversion reactions of carbohydrates. Thus, the primary bromide **68** would be a suitable starting material.

12.3.2 Combination of Ferrier Carbocyclization and Three-Component Coupling Reaction

A cyclohexenone easily obtained by the Ferrier carbocyclization followed by β-elimination readily serves as a Michael acceptor. For the synthesis of actinoboline **69**, a natural product possessing antibacterial activity, a three-component coupling

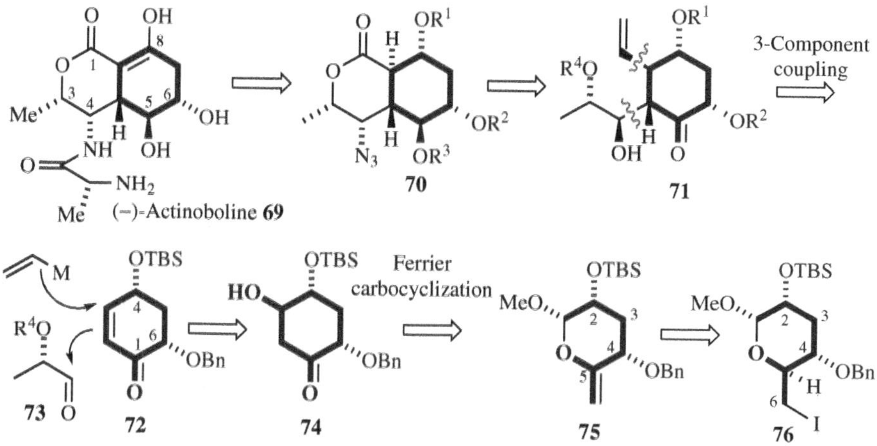

Scheme 12.19 Synthetic plan for actinoboline.

strategy (1,4-addition of organometallics to a cyclohexenone, followed by trapping of the resulting enolate with an electrophile in one pot)[31] was employed to assemble the target molecule (Scheme 12.19).[13] For the preparation of actinoboline precursor **71**, a three-component coupling reaction of cyclohexenone **72** with a vinyl copper reagent and aldehyde **73** was envisioned. The substituents on **72** would control the facial selectivity of the initial 1,4-addition, and the stereoselectivity in the subsequent aldol process would be governed by chelation or nonchelation control. Cyclohexenone **72** was planned to be synthesized from **74**, which would be derived from 5-enopyranoside **75** by the Ferrier carbocyclization. 3,6-Dideoxy-6-iodo-D-*erythro*-hexopyranoside **76** is an appropriate starting material.

12.3.3 Combination of Ferrier Carbocyclization and Sigmatropic Rearrangement

[3,3]-Sigmatropic rearrangements of chiral allylic alcohol derivatives are known to proceed with chirality transfer via the six-membered chair-like transition structure.[32,33] As cyclic allylic alcohols are easily synthesized in enantiomerically pure form by the Ferrier carbocyclization reaction, a sigmatropic rearrangement such as the Claisen rearrangement could then generate a new C—C bond with chirality transfer. This transformation effectively transfers the C—O chirality of sugars into the C—C chirality in the target molecule.

Retrosynthetic analysis of galanthamine **77**, an alkaloid used as drug for Alzheimer's disease, based on a combined strategy of Ferrier carbocyclization and Claisen rearrangement is shown in Scheme 12.20.[34] (+)-Galanthamine **77** was planned to be derived from **78**. For the stereoselective construction of the quaternary carbon chiral center in **78**, Claisen rearrangement of cyclohexenol **79** possessing an aryl substituent emerged as a synthetically powerful transformation. Cyclohexenol **79**, in turn, was envisioned to be prepared by 1,2-addition of aryl metal species to

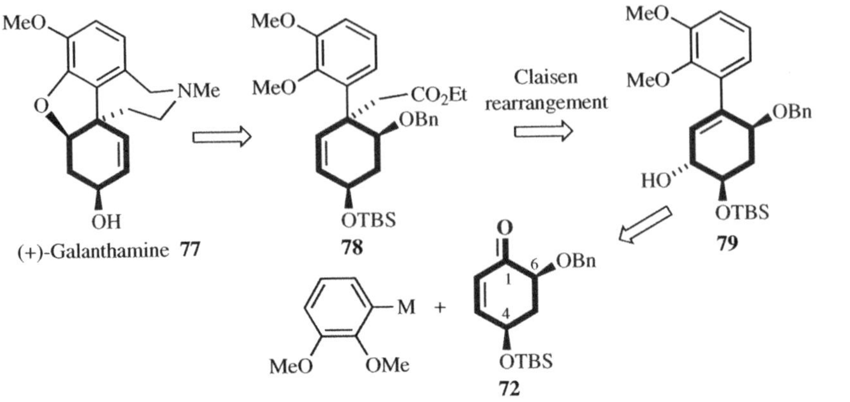

Scheme 12.20 Synthetic plan for galanthamine.

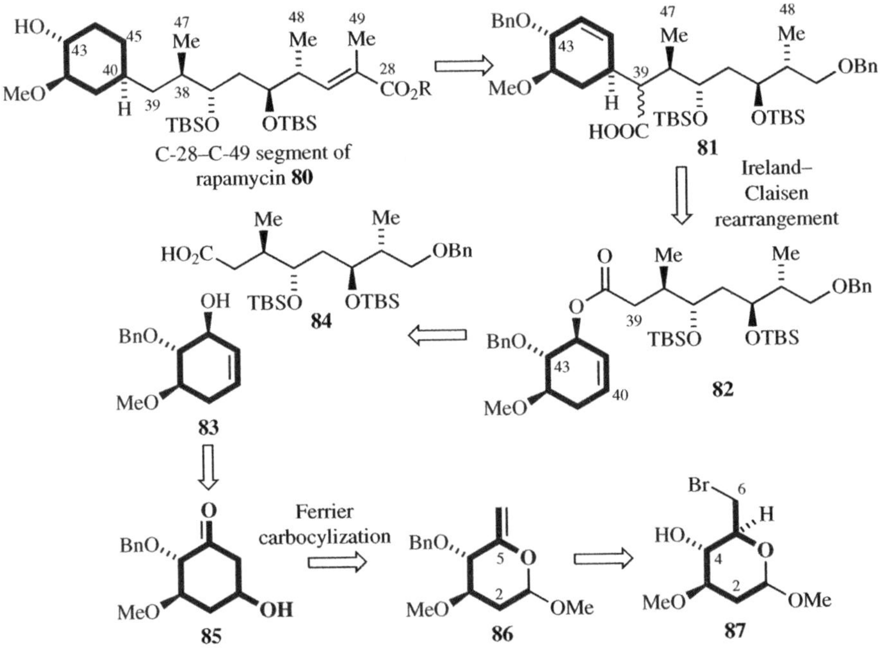

Scheme 12.21 Synthetic plan for C-28–C-49 segment of rapamycin.

cyclohexenone **72**, which is the same compound that was planned as the precursor in the actinoboline synthesis (Section 12.3.2).

In Scheme 12.21, strategic planning for synthesis of the C-28–C-49 part of the immunosuppressive natural product rapamycin is depicted.[35] The target compound **80** possessing a C-28–C-49 side chain on a cyclohexane ring was planned to be prepared by the decarboxylation of **81**. For the stereoselective synthesis of **81**,

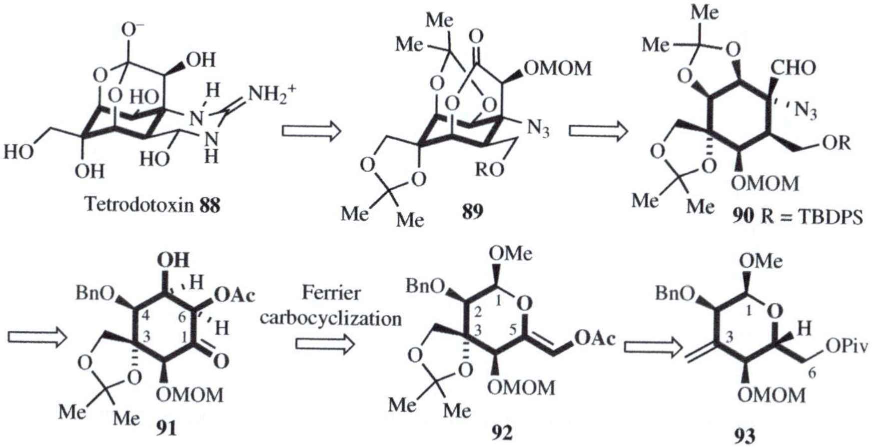

Scheme 12.22 Synthetic plan for tetrodotoxin.

Ireland–Claisen rearrangement of ester **82** with chirality transfer was envisioned. Ester **82** would be derived by the condensation of **83** with carboxylic acid **84**. Cyclohexenol **83** was planned to be derived from cyclohexanone **85** by β-elimination of the hydroxy group followed by reduction of the ketone carbonyl. The Ferrier carbocyclization of 5-enopyranoside **86** would deliver cyclohexanol **85**. As compound **86** has a 2-deoxy-D-arabino configuration, it could be derived from 2,6-dideoxy-6-bromo-D-glucopyranoside derivative **87**.

12.3.4 Enol-Acetate Version of the Ferrier Carbocyclization Reaction

The Ferrier carbocyclization reaction of an enol-acetate substrate gives an α,β-dihydroxy-cyclohexanol derivative (see Schemes 12.7 and 12.8). This transformation would be effective for the chiral synthesis of inositol derivatives. A retrosynthetic plan for the marine natural product tetrodotoxin **88** based on the enol-acetate version of Ferrier carbocyclization is shown in Scheme 12.22.[36] Tetrodotoxin **88** was planned to be synthesized from lactone **89**, the precursor of which would be highly functionalized cyclohexane **90**. Cyclohexane **90** was envisioned to arise from cyclohexanone **91**. For the preparation of **91**, Ferrier carbocyclization of enol acetate **92** would be a suitable transformation. D-Glucose derivative **93** possessing an *exo*-methylene at C-3 would serve as a promising precursor of **92**.

12.4 METHODOLOGIES FOR ASSEMBLING THE FERRIER CARBOCYCLIZATION REACTION SUBSTRATES

The substrates for the Ferrier carbocyclization reaction are 6-deoxy-hex-5-enopyrano side derivatives. The chemistry of the interconversions of carbohydrates has a long history; hence, a number of reliable methods for the preparation of 5-enopyranosides

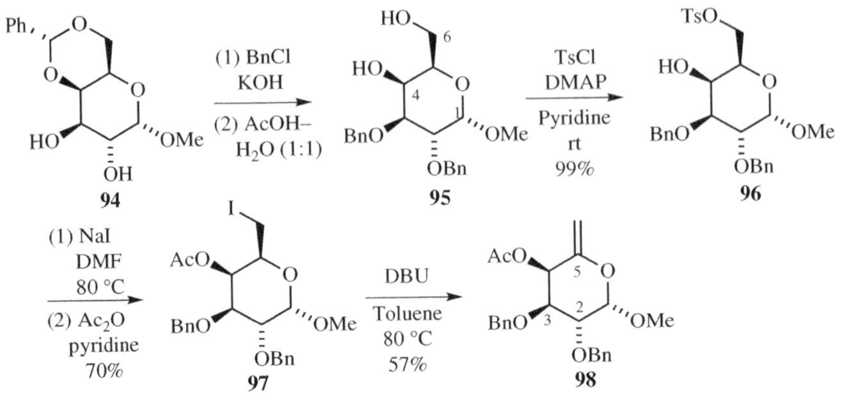

Scheme 12.23 Preparation of substrates for the Ferrier carbocyclization (1).

have been established.[37] A typical procedure for the preparation of 5-enopyranosides is *β*-elimination of 6-deoxy-6-halo-aldohexopyranosides by the action of base. This section will provide an overview of some representative syntheses of 5-enopyranosides.

Conversion of a D-galactose derivative into the corresponding enopyranoside is depicted in Scheme 12.23.[9] The primary hydroxy group in **95**, prepared from methyl 4,6-*O*-benzylidene-*α*-D-galactopyranoside **94** by O-benzylation followed by acid hydrolysis of the benzylidene acetal, was selectively tosylated to give **96**. Displacement of the OTs group with iodide followed by O-acetylation of the secondary alcohol provided **97**. Treatment of **97** with DBU afforded methyl 4-*O*-acetyl-2,3-di-*O*-benzyl-6-deoxy-*β*-L-*arabino*-hex-5-enopyranoside **98**.

In a streamlined approach, the primary hydroxy group could be converted into the iodide in a one-step reaction in the presence of a secondary hydroxyl. Protection of the secondary alcohols in methyl 4,6-*O*-benzylidene-*α*-D-glucopyranoside **99** gave **100**, and acid hydrolysis afforded **101** (Scheme 12.24).[38] Treatment of diol **101** with I_2 in the presence of Ph_3P provided primary iodide **102**, which was converted into 5-enopyranoside **103** by the action of base, followed by O-acetylation.

The 4,6-*O*-benzylidene acetal group, which is frequently used as a protecting group in aldohexoses, can be directly transformed into 4-*O*-benzoyl-6-bromo derivatives by a radical-mediated cleavage reaction, and 4-*O*-benzyl-6-hydroxy derivatives by a reductive cleavage reaction, respectively. For example, reaction of 2-deoxy-glucopyranoside derivative **104** with *N*-bromosuccinimide (NBS) smoothly afforded 6-bromo-4-*O*-benzoyl derivative **105** (Scheme 12.25).[35,39] Methanolysis of **105** afforded a secondary alcohol, whose treatment with NaH and benzyl chloride induced the elimination of HBr as well as O-benzylation to provide 4-*O*-benzyl-5-enopyranoside **86**. Compound **86** was used in the Danishefsky's rapamycin synthesis (Section 12.5.5).

Reaction of 3-deoxy-glucopyranoside **106**, prepared from **99** by selective O-tosylation, followed by $LiAlH_4$ reduction, with DIBAL-H gave 4-*O*-benzyl-6-hydroxy derivative **107** (Scheme 12.26).[13] Selective introduction of iodide at the

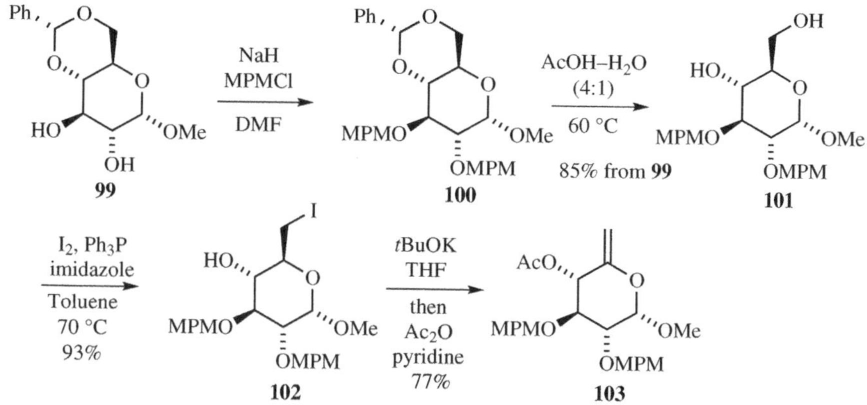

Scheme 12.24 Preparation of substrates for the Ferrier carbocyclization (2).

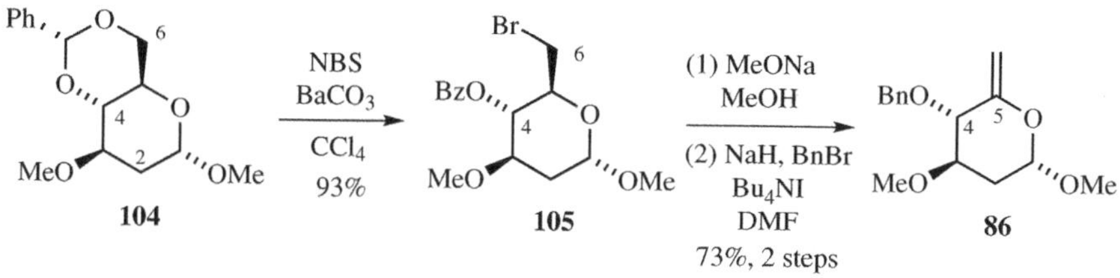

Scheme 12.25 Preparation of substrates for the Ferrier carbocyclization (3).

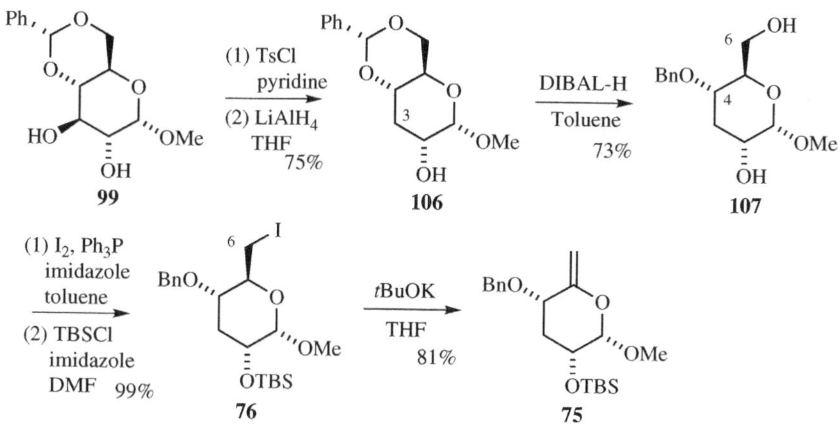

Scheme 12.26 Preparation of substrates for the Ferrier carbocyclization (4).

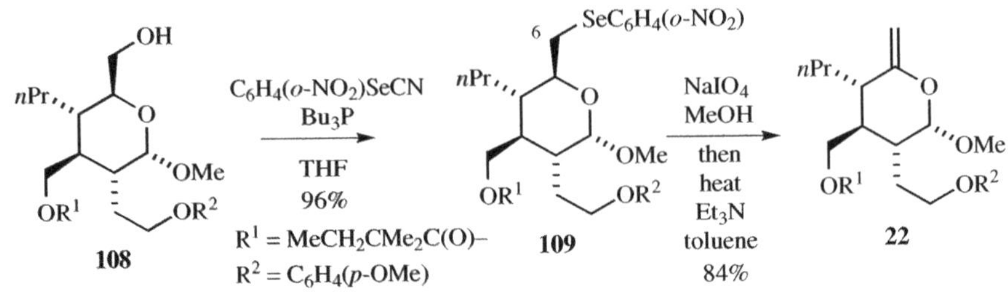

Scheme 12.27 Preparation of substrates for the Ferrier carbocyclization (5).

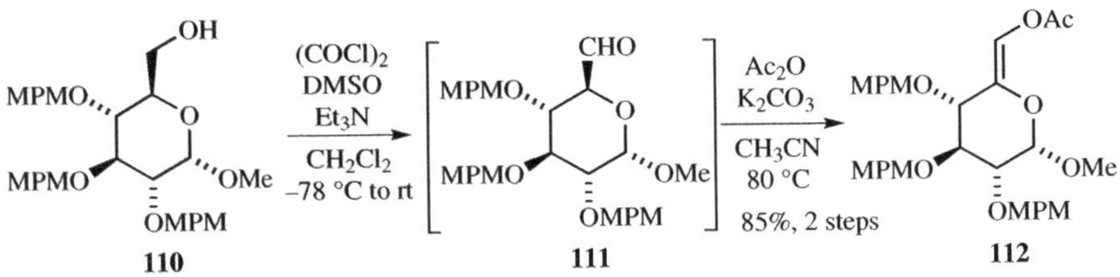

Scheme 12.28 Synthesis of an enol-acetate.

primary carbon in **107** and subsequent protection of the secondary alcohol gave **76**, treatment of which with *t*-BuOK provided 5-enopyranoside **75**. Compound **75** was employed in the synthesis of actinoboline and galanthamine (Sections 12.5.3 and 12.5.4).

Preparation of a 5-enopyranoside by the elimination of a selenoxide has been reported.[14,25] The primary hydroxy group in **108** was converted into 6-seleno derivative **109** (Scheme 12.27).[14] Oxidation of **109** and subsequent thermolysis afforded *exo*-methylene **22**.

A substrate for the enol-acetate version of the Ferrier carbocyclization was prepared by the treatment of aldehyde **111** with acetic anhydride and K_2CO_3 (Scheme 12.28).[16] Only the (*Z*)-enol acetate **112** was obtained under these conditions. The aldehyde **111** was synthesized from primary alcohol **110** by Swern oxidation and quickly used in the next reaction without purification.

12.5 APPLICATIONS OF THE FERRIER CARBOCYCLIZATION REACTION IN NATURAL PRODUCT SYNTHESIS

12.5.1 Preparation of an Aminocyclitol: Synthesis of Hygromycin A

Hygromycin A **120** (Scheme 12.30) is an antibiotic isolated from *Streptomyces* in 1953. In 1991, Ogawa and Chida reported the first total synthesis of this antibiotic employing the Ferrier carbocyclization as the key reaction.[29] Based on the retrosynthetic analysis discussed in Section 12.3.1 (Scheme 12.17), the

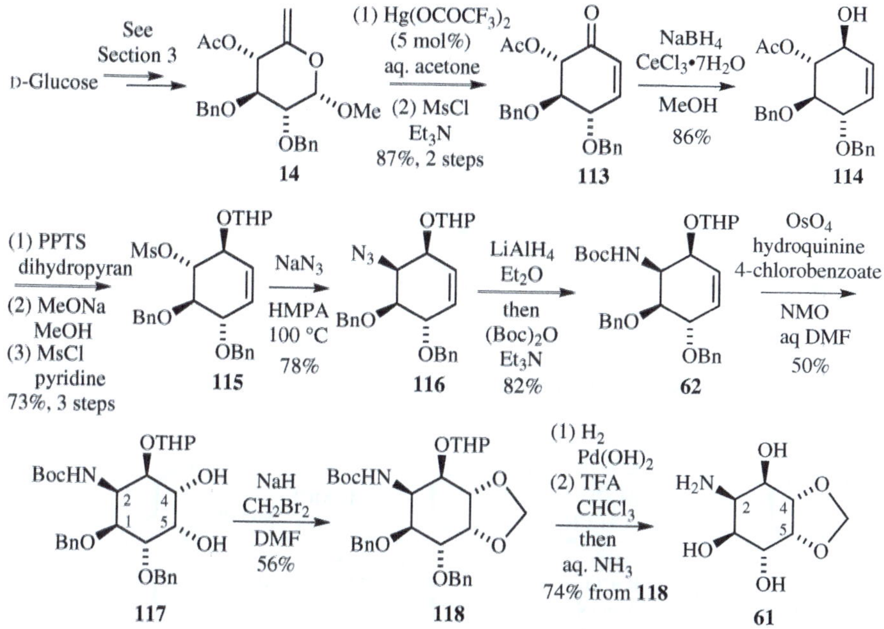

Scheme 12.29 Synthesis of an aminocyclitol in hygromycin A.

aminocyclitol moiety **61** was synthesized from D-glucose (Scheme 12.29). The Ferrier carbocyclization reaction of **14** followed by *β*-elimination of the newly formed hydroxy group afforded **113**. 1,2-Reduction of the ketone carbonyl in **113** under Luche conditions gave **114** stereoselectively. Conventional protecting group manipulation afforded mesylate **115**, and S_N2 reaction with NaN_3 successfully introduced a nitrogen function to give **116**. After conversion of the azido function in **116** into a *tert*-butyl carbamate, OsO_4 oxidation of the resulting **62** gave diol **117** in 50% yield along with its C-4/C-5 isomer (32%). Treatment of **117** with NaH and CH_2Br_2 provided methylene ketal **118**, and global deprotection afforded 1L-2-amino-2-deoxy-4,5-*O*-methylene-*neo*-inositol **61**.

The sugar moiety of hygromycin A **119** was synthesized from D-glucose. Condensation of **119** with **61** by the Shioiri procedure followed by O-acetylation provided a condensate in 75% yield (Scheme 12.30). Removal of the protecting groups completed the total synthesis of hygromycin A **120**.

12.5.2 Construction of a Polycyclic Alkaloid: Synthesis of Lycoricidine

The phenanthridone alkaloid, lycoricidine **64** (Scheme 12.32), is known to possess cytotoxic activity. Based on the retrosynthetic analysis as discussed in Section 12.3.2 (Scheme 12.18), the Chida group reported the total synthesis of lycoricidine starting from D-glucose in which the Ferrier carbocyclization reaction was employed

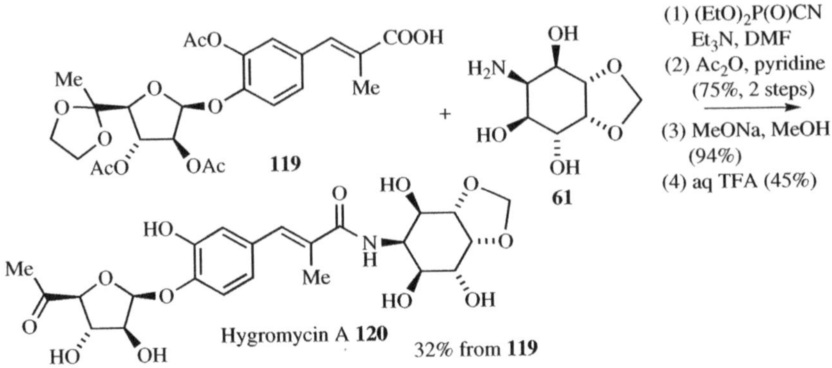

Scheme 12.30 Total synthesis of hygromycin A.

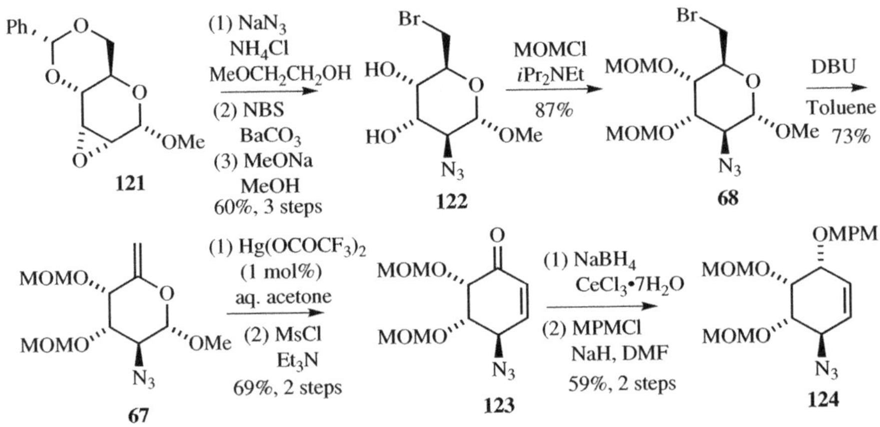

Scheme 12.31 Preparation of a cyclohexene unit in lycoricidine.

as the key reaction.[30] For the preparation of the cyclohexene unit in lycoricidine, D-glucose was converted into 2-azido-D-altropyranoside derivative **122** via epoxide **121** (Scheme 12.31). Protection of the diol moiety in **122** afforded **68**, which was transformed into 5-enopyranoside **67** by the action of DBU. Ferrier carbocyclization of **67** with 1 mol% of $Hg(OCOCF_3)_2$, followed by β-elimination, provided cyclohexenone **123**. Stereoselective Luche reduction of **123** and subsequent protection of the hydroxy function afforded **124**.

Reduction of the azido function in **124** gave an amine, which was condensed with 6-bromopiperonylic acid under Shioiri conditions (Scheme 12.32). Protection of the amide nitrogen with a *p*-methoxybenzyl group afforded **125**. The crucial ring closure reaction was successful when **125** was treated with $Pd(OAc)_2$ and 1,2-bis(diphenylphosphino)ethane (DPPE) in the presence of TlOAc to provide **126**. It was found that the protection of the amide nitrogen was essential for this Heck-type cyclization. The *O*-MPM protecting group in **126** was

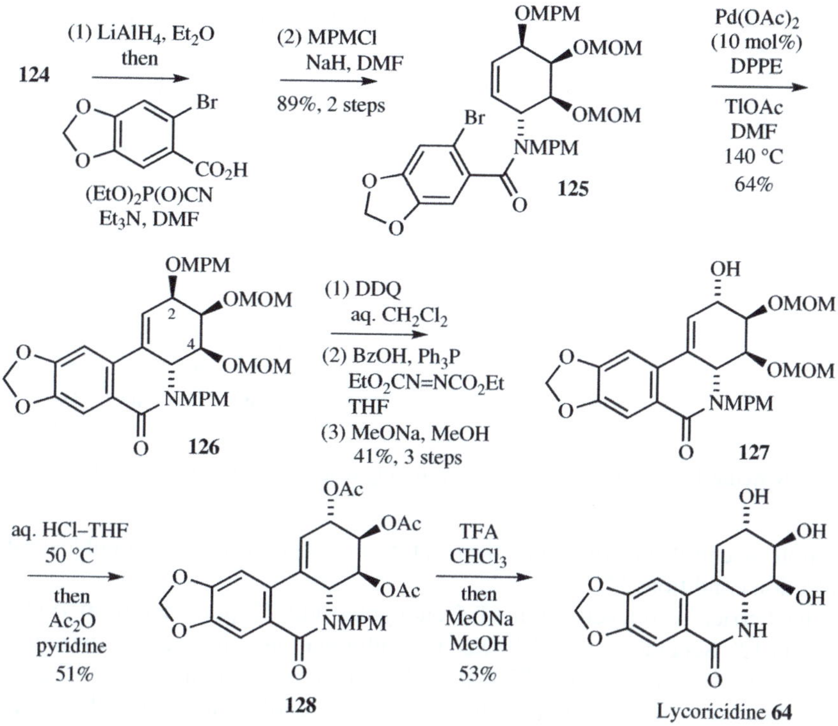

Scheme 12.32 Total synthesis of lycoricidine.

removed to give an allylic alcohol, which was converted into inverted alcohol **127** by the Mitsunobu reaction with benzoic acid, followed by de-O-benzylation. The *O*-MOM groups in **127** were cleaved by acid hydrolysis, and subsequent O-acetylation gave triacetate **128**. Finally, deprotection of the *N*-MPM group in **128** by the action of CF_3CO_2H, followed by basic methanolysis, furnished lycoricidine **64**.

12.5.3 Three-Component Coupling Reaction: Synthesis of (−)- and (+)-Actinoboline

The three-component coupling reaction of an α,β-unsaturated carbonyl compound with a nucleophile and an electrophile is a useful transformation in organic synthesis as two C—C bonds can be formed stereoselectively in a one-pot reaction.[31] In 2006, the Chida group disclosed the synthesis of (−)-actinoboline **69**[13] (the antipode of the natural product) based on the three-component coupling reaction of chiral cyclohexenone **72**, derived from D-glucose by the Ferrier carbocyclization, with a vinyl metal species and chiral aldehyde **73** (Section 12.3.2, Scheme 12.19). The aldol process in the coupling reaction of an enolate generated from **72** with a chiral aldehyde was expected to have the potential for high stereoselectivity, as both partners in the

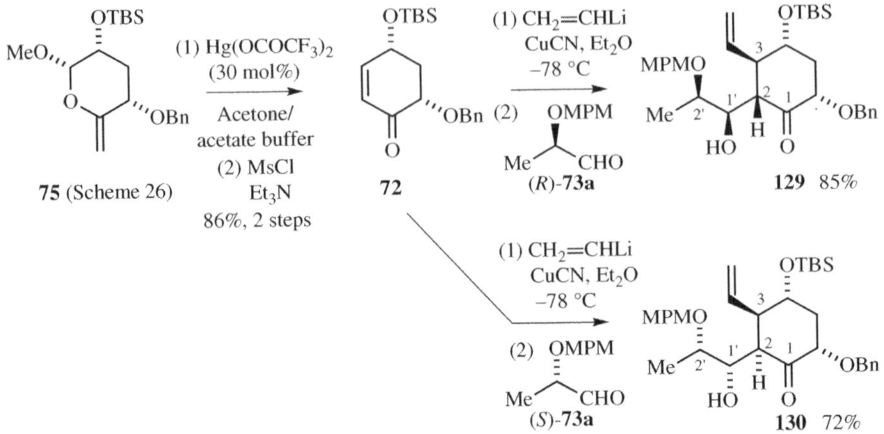

Scheme 12.33 Three-component coupling reaction with *O*-MPM aldehydes.

reaction are chiral, and, therefore, the reaction would proceed under "double diastereoselection conditions."[40]

The Ferrier carbocyclization of 5-enopyranoside **75** prepared from D-glucose (Section 12., Scheme 12.26), followed by β-elimination, afforded cyclohexenone **72** (Scheme 12.33). Treatment of **72** with higher-order vinyl cuprate at −78 °C gave the 1,4-adduct stereoselectively, and the intermediate enolate was trapped with chiral aldehyde (*R*)-**73a** to give **129** having the desired trans-stereochemistry at C-2 and C-3 in 85% yield. On the other hand, in the reaction with enantiomeric aldehyde (*S*)-**73a**, the aldol process was found to proceed much slowly than the reaction with (*R*)-**73a**, and diastereomer **130**, having the cis-stereochemistry at C-2 and C-3, was formed in 72% yield.

The predominant formation of 1′,2′-*syn* isomers (**129** and **130**) in the coupling reactions suggested that chelation control was an important factor in the aldol process.[40] Reaction of the enolate with chelated (*R*)-**73a** would proceed in a "matched pair" manner where no severe steric hindrance is present (Route A in Scheme 12.34) to give 2,3-trans-product **129** smoothly, whereas combination of (*S*)-**73a** and the enolate would be "mismatched." The steric repulsion between chelated (*S*)-**73a** and the enolate rendered the aldol process much sluggish (Routes C and D) but gave 2,3-cis-adduct **130** stereoselectively via the less crowded pathway (Route D).

Compound **129** has correct stereochemistry at C-2 and C-3 for the synthesis of actinoboline, and the required transformations were (1) introduction of a nitrogen function at C-1′ via S_N2 reaction and (2) formation of the δ-lactone with inversion of the configuration at C-2′. If a three-component coupling reaction with an aldehyde possessing an (*S*)-hydroxy group proceeded via nonchelation (Felkin–Anh) conditions, the product with proper stereochemistry at C-2′ was anticipated as the major isomer, which would then not require inversion at that center. Toward this end, three-component coupling of **72**, vinyl cuprate, and aldehyde (*S*)-**73b**,

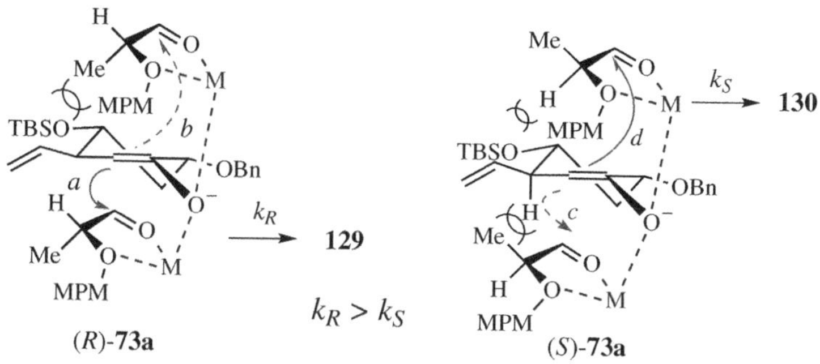

Scheme 12.34 Transition structure models of the three-component coupling.

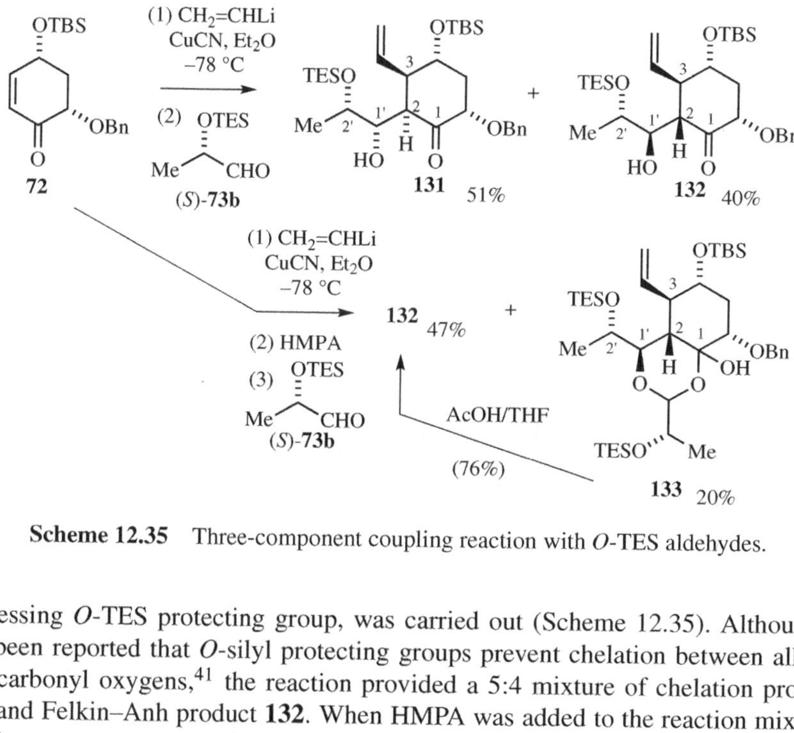

Scheme 12.35 Three-component coupling reaction with *O*-TES aldehydes.

possessing *O*-TES protecting group, was carried out (Scheme 12.35). Although it has been reported that *O*-silyl protecting groups prevent chelation between alkoxy and carbonyl oxygens,[41] the reaction provided a 5:4 mixture of chelation product **131** and Felkin–Anh product **132**. When HMPA was added to the reaction mixture, the three-component reaction gave only Felkin–Anh products **132** and **133** in 47% and 20% yields, respectively. The ketal product **133** was converted into **132** by the treatment with acetic acid.

Reduction of the ketone carbonyl in **132** with $LiBH_4$ gave **134** (Scheme 12.36). Treatment of **134** with BuLi, followed by TsCl generated 1′-OTs derivative **135**, and the remaining hydroxy group was masked as a MOM ether to afford **135**. Ozonolysis of **136**, prepared by removal of the *O*-TES protecting group in **135**, gave a lactol, which was further oxidized to provide δ-lactone **137**. Azidolysis of **137** with NaN_3 cleanly afforded **70**. The *O*-TBS group was removed to give an alcohol, whose Swern

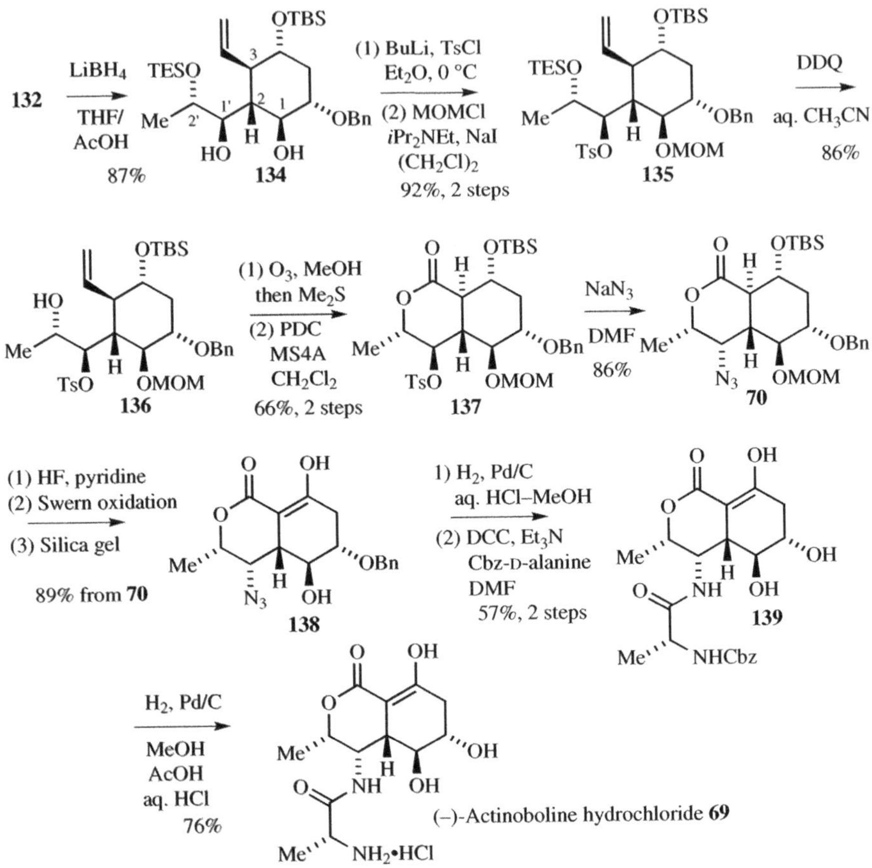

Scheme 12.36 Total synthesis of actinoboline.

oxidation gave *β*-ketoester **138**. Interestingly, the MOM group was unexpectedly removed during the purification process with silica gel chromatography. Treatment of **138** with H_2 in the presence of Pd/C and aqueous HCl reduced the azide function as well as removed the *O*-benzyl group to provide an amine hydrochloride, which was condensed with Cbz-D-alanine to give protected actinoboline **139**. Finally, removal of the Cbz group furnished (-)-actinoboline hydrochloride **69**.

The enantiomer of **72** (*ent*-**72**) was synthesized from the same starting material **106** (Scheme 12.37).[13] Benzylation of a hydroxy group in **106** followed by acetal hydrolysis afforded **140**, which was converted into 5-enopyranoside **141** by the conventional method. The Ferrier carbocyclization of **141** generated **142** as a diastereomeric mixture in 83% yield. Protection of the hydroxy group in **142** as a THP ether and subsequent reduction of the ketone carbonyl gave **143**. O-Mesylation of **143** followed by acidic work-up afforded **144**. Swern oxidation of **144** was accompanied by the *β*-elimination of the OMs group to furnish *ent*-**72** in 93% yield. Cyclohexenone *ent*-**72** could be used for the synthesis of natural enantiomer of actinoboline.

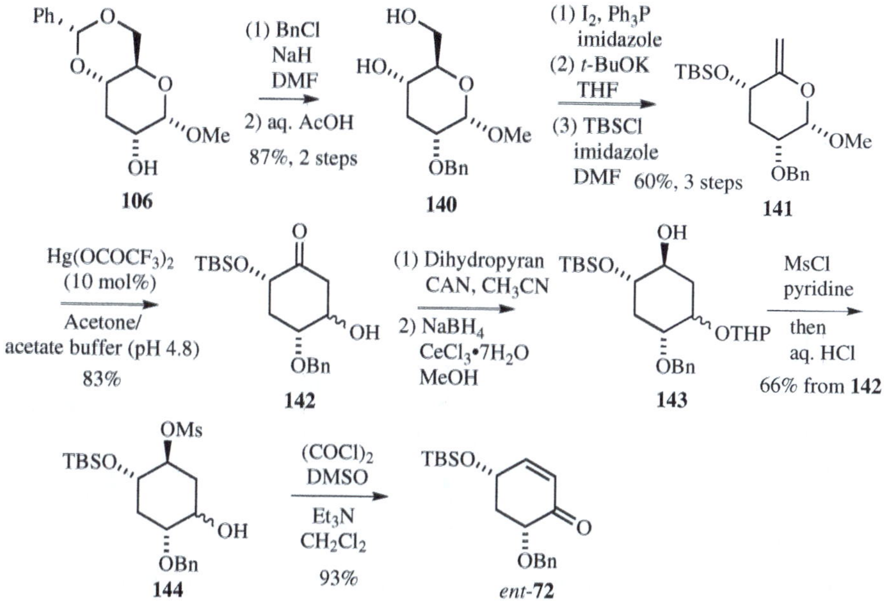

Scheme 12.37 Synthesis of an enantiomeric cyclohexenone from D-glucose.

12.5.4 Sigmatropic Rearrangement with Chirality Transfer (1): Synthesis of Galanthamine

[3,3]-Sigmatropic rearrangement such as the Claisen rearrangement of chiral allylic alcohols can generate a new C—C bond with high levels of chirality transfer. As chiral cyclohexenols can be easily prepared from carbohydrates by the Ferrier carbocyclization, a methodology employing a combination of the Ferrier carbocyclization with sigmatropic rearrangement would be effective for the synthesis of natural products possessing highly functionalized cyclohexane units.

In 2007, the Chida group reported the total synthesis of galanthamine **77**,[34] an *Amaryllidaceae* alkaloid showing potent acetylcholinesterase inhibitory activity, based on the Ferrier/Claisen methodology (see Section 12.3.3, Scheme 12.20). Reaction of chiral cyclohexenone **72**, which was synthesized by the Ferrier carbocyclization (see Section 12.5.3, Scheme 12.33) and used as the starting material in actinoboline synthesis,[13] with 2,3-dimethoxyphenylmagnesium bromide gave 1,2-adduct **145** in 85% yield as a 4:1 mixture of diastereomers (Scheme 12.38). PCC oxidation of **145** afforded cyclohexenone **146**, and the carbonyl group was stereoselectively reduced under Luche conditions to give cyclohexenol **79**. Johnson–Claisen rearrangement of **79** with triethyl orthoacetate in the presence of 2-nitrophenol[42] successfully constructed a benzylic quaternary carbon with complete chirality transfer to provide **78** in 80% yield as a single isomer. Overall, the Ferrier/Claisen procedure generated the sterically congested quaternary carbon of key intermediate **78** in a stereoselective manner with good chemical yield.

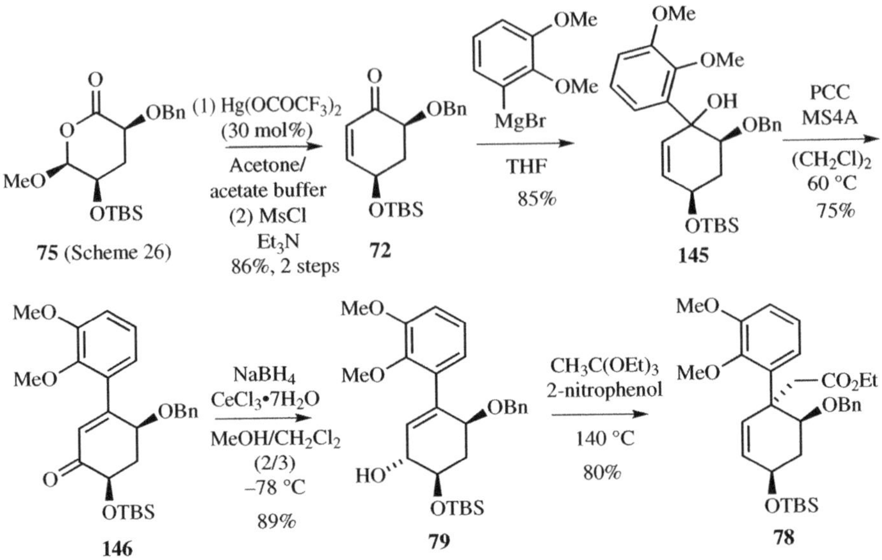

Scheme 12.38 Construction of a quaternary carbon by the Claisen rearrangement.

Treatment of **78** with NBS induced formation of the dibenzofuran skeleton to give **147** (Scheme 12.39). Hydrogenolysis of **147** afforded **148**, whose hydroxy group was eliminated to give cyclohexene **149**. The ester group in **149** was converted into an amide to afford **150**. After deprotection of the *O*-TBS group in **150**, the resulting amide was subjected to Pictet–Spengler conditions to provide tetracyclic lactam **151**. Reduction of **151** with $LiAlH_4$ afforded (+)-galanthamine **77**. Meanwhile, starting from *ent*-**72** (Scheme 12.37) resulted in the total synthesis of (−)-galanthamine (the natural enantiomer).

12.5.5 Sigmatropic Rearrangement with Chirality Transfer (2): Synthesis of Rapamycin

In 1991, the Danishefsky group disclosed the synthesis of the C-28–C-49 subunit of rapamycin utilizing the combination of the Ferrier carbocyclization reaction and an Ireland–Claisen rearrangement (see Section 12.3.3, Scheme 12.21).[35] The Ferrier carbocyclization of 5-enopyranoside **86**, prepared from 2-deoxy-D-glucose derivative (Section 12., Scheme 12.25), followed by elimination of the *β*-hydroxy group gave cyclohexenone **152** (Scheme 12.40). Luche reduction of **152** afforded cyclohexenol **83** stereoselectively. Condensation of **83** with carboxylic acid **84**, prepared from (*R*)-3-(benzyloxy)-2-methylpropanal, provided ester **82** in 75% yield.

Generation of a ketene acetal was accomplished by treatment of **82** with LDA in a THP/HMPA mixture at −78 °C and quenching of the resultant enolate with TBSCl (Scheme 12.41). Thermolysis of the ketene acetal induced Ireland–Claisen rearrangement, and subsequent hydrolysis of the resulting TBS ester afforded

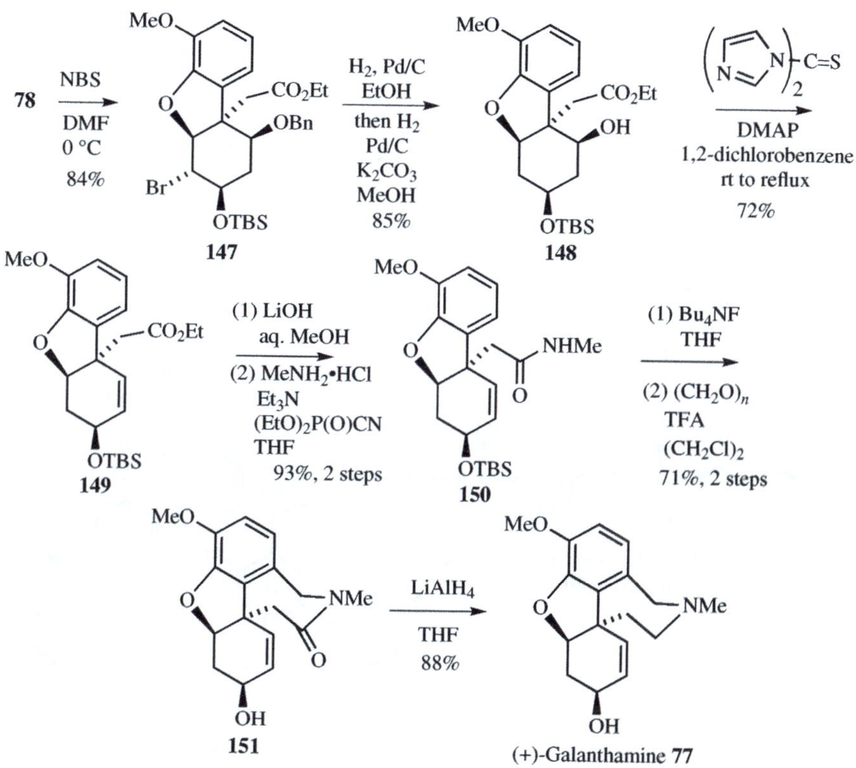

Scheme 12.39 Total synthesis of galanthamine.

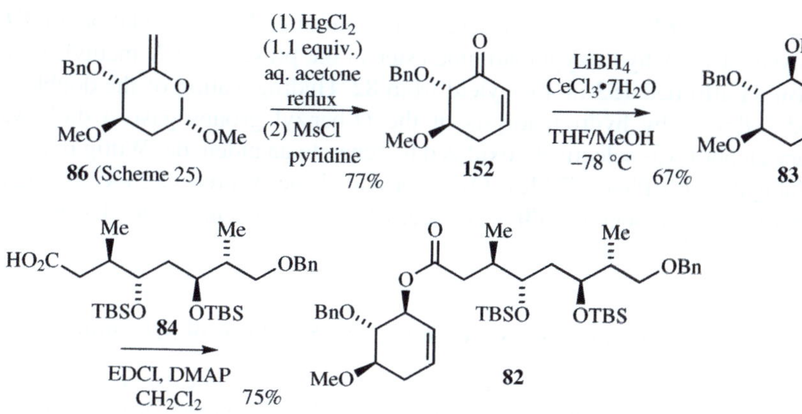

Scheme 12.40 Preparation of a cyclohexene unit of rapamycin.

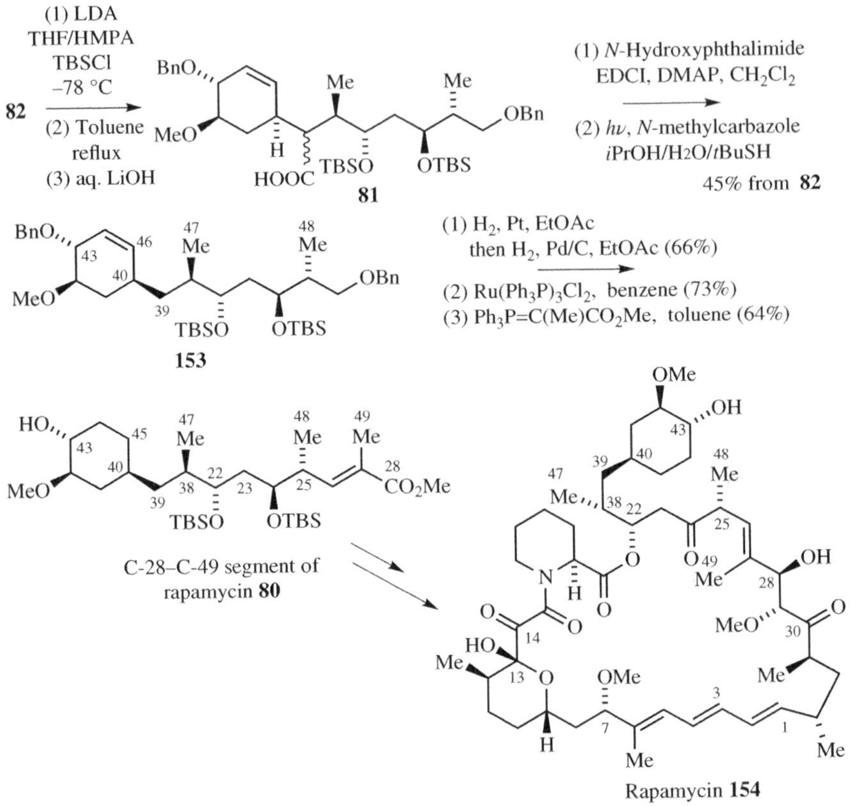

Scheme 12.41 Synthesis of C-28–C-49 segment of rapamycin.

carboxylic acid **81** as a 3:1 diastereomeric mixture. Decarboxylation of **81** by photolysis of its *N*-hydroxyphthalimide ester in the presence of *N*-methylcarbazole and *t*BuSH afforded **153** in 45% yield from **82**. Hydrogenation of the double bond in **153** followed by hydrogenolysis of the *O*-benzyl groups gave a diol, whose primary alcohol was selectively oxidized to generate an aldehyde. Wittig reaction of the aldehyde with $Ph_3P{=}C(Me)CO_2Me$ provided the desired C-28–C-49 segment of rapamycin. The segment **80** was successfully utilized in the total synthesis of rapamycin **154**.[35b]

12.5.6 Cascade Sigmatropic Rearrangement: Synthesis of Morphine

Morphine (**172**) is a well-known alkaloid and has been clinically used as an analgesic. In 2008 and 2013, the Chida group reported the synthesis of morphine based on methodology employing the combination of Ferrier carbocyclization and Claisen rearrangement.[12a] The key feature of the synthesis is the cascade Claisen rearrangement of an allylic vicinal diol, creating the vicinal tertiary and quaternary carbons

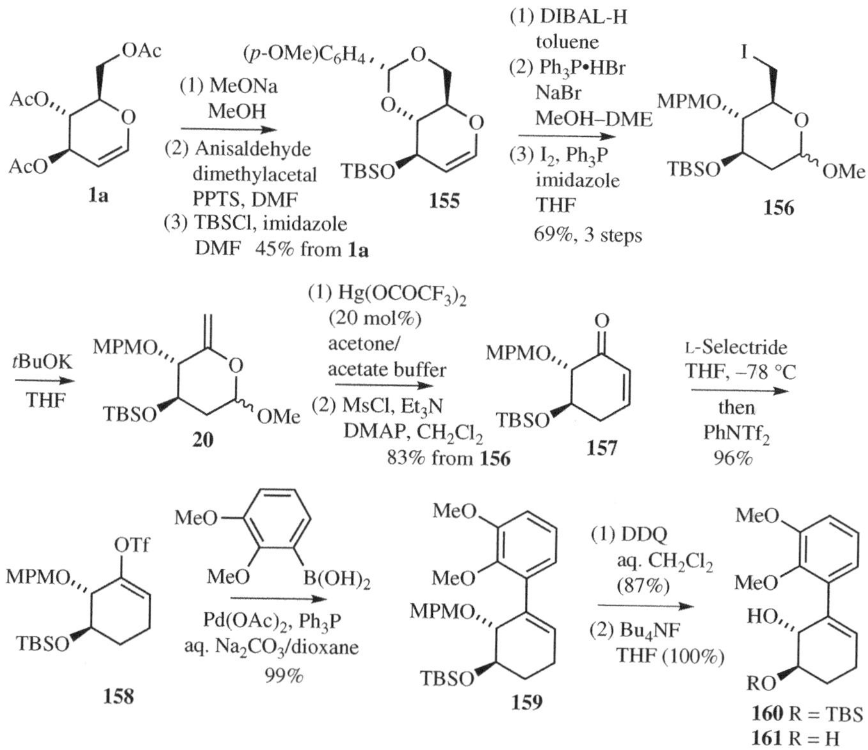

Scheme 12.42 Preparation of a cyclohexene unit of morphine.

stereoselectively in a one-pot reaction. Overall, the contiguous C—O chiralities in D-glucal were effectively transferred to C—C chiralities in morphine.

To implement this strategy, tri-*O*-acetyl-D-glucal **1a** was converted into fully protected derivative **155** by three reactions (Scheme 12.42). Reductive opening of the benzylidene acetal and subsequent treatment with $Ph_3P{\cdot}HBr$ and MeOH provided a methyl glycoside, which was transformed into primary iodide **156**. Elimination of HI provided **20**. Ferrier carbocyclization of **20** with $Hg(OCOCF_3)_2$ gave a cyclohexanone (see also Scheme 12.6), which was converted into cyclohexenone **157**. 1,4-Reduction of **157** and trapping of the intermediate enolate with $PhNTf_2$ gave enol triflate **158**. Suzuki–Miyaura coupling of **158** with 2,3-dimethoxyphenylboronic acid afforded **159**. Deprotection of the *O*-MPM group gave allylic alcohol **160**, and further deprotection of the *O*-TBS group provided diol **161**.

Claisen rearrangement of allylic alcohol **160** in the presence of triethyl orthoacetate and propionic acid gave **162** in 87% yield (Scheme 12.43). After deprotection of the *O*-TBS group in **162**, the second Claisen rearrangement of the resulting **163** with 2-nitrophenol as the acid catalyst successfully constructed the benzylic quaternary carbon to provide **164** in 57% yield. On the other hand, when a solution of diol **161**

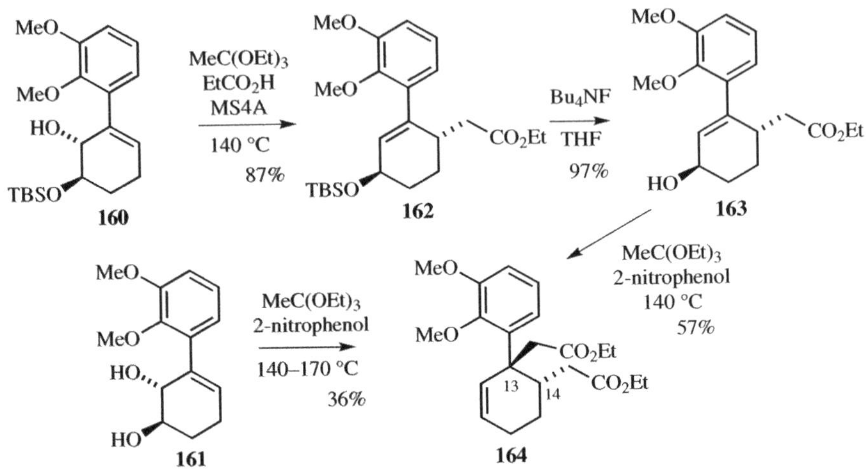

Scheme 12.43 The cascade Claisen rearrangement.

in triethyl orthoacetate in the presence of 2-nitrophenol was heated at 140 °C, the cascade Claisen rearrangement took place to afford the doubly rearranged product **164** in 36% yield in a one-pot process. By the cascade sigmatropic rearrangements, the vicinal tertiary and quaternary carbon centers in morphine (C-14 and C-13) were stereoselectively constructed via sequential chirality transfer.

Treatment of **164** with *m*-CPBA induced dibenzofuran formation to provide **165**, and protection of the resulting hydroxy group in **165** afforded **166** (Scheme 12.44). The ester functions in **166** were reduced with DIBAL-H to give dialdehyde **167**. When **167** was reacted with montmorillonite K-10 clay, Friedel–Crafts-type cyclization and elimination of the resulting benzylic alcohol took place to give **168** in 81% yield after resilylation of the C-6 hydroxy group. Reductive amination of **168** with methylamine gave **169**, whose amino group was tosylated and TBS group deprotected to give tosylamide **170**. Intramolecular reductive hydroamination by Parker's method[43] afforded dihydroisocodeine **171**. As **171** had been previously converted into morphine **172** in nine steps,[43] the formal synthesis of morphine starting from D-glucal has been completed.

In 2013, the Chida group reported its second-generation synthesis of morphine, in which the requisite C-7/C-8 double bond was introduced at an early stage (Scheme 12.45).[12b] Oxidation of the hydroxy group in **165** gave a ketone, which was converted into cyclohexenone **173** by the action of PhSeCl and AgOTf in the presence of $BF_3{\cdot}OEt_2$, followed by oxidation with Davis reagent (2-tosyl-3-phenyloxaziridine). Stereoselective 1,2-reduction of **173** and subsequent protection of the hydroxy function gave **174**. The diester functionality in **174** was transformed into dialdehyde **175** by a reduction–oxidation sequence. Friedel–Crafts-type cyclization and deprotection of the *O*-TBS group then successfully provided tetracyclic compound **176**. Reductive amination of **176** followed by N-tosylation afforded tosylamide **177**, which is the known synthetic intermediate and could be converted to morphine **172** in two steps.[44]

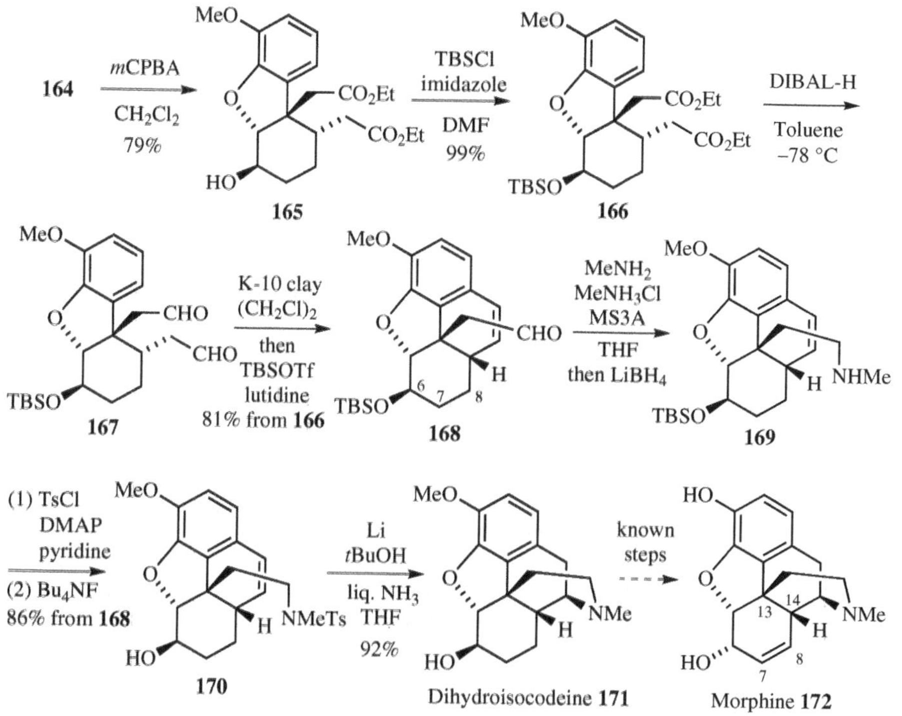

Scheme 12.44 Synthesis of morphine.

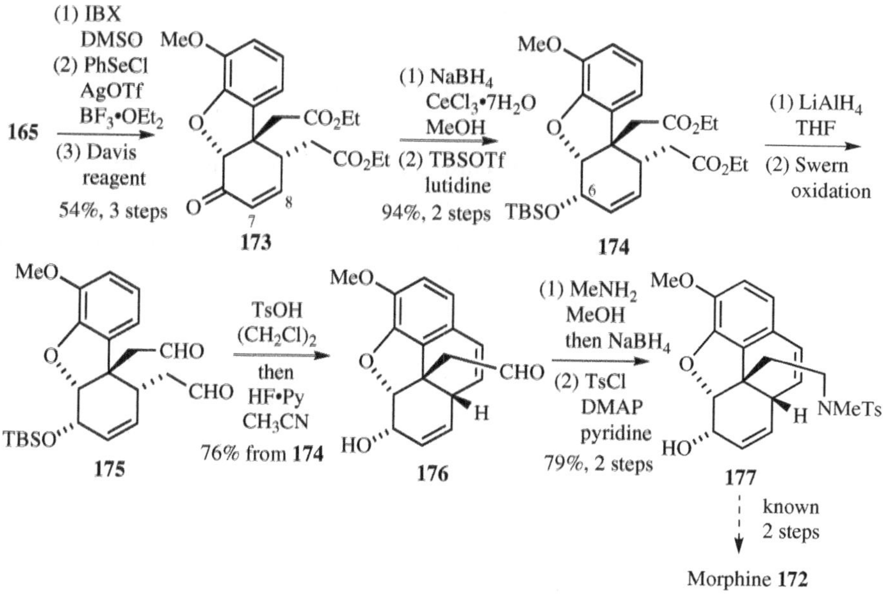

Scheme 12.45 Second-generation synthesis of morphine.

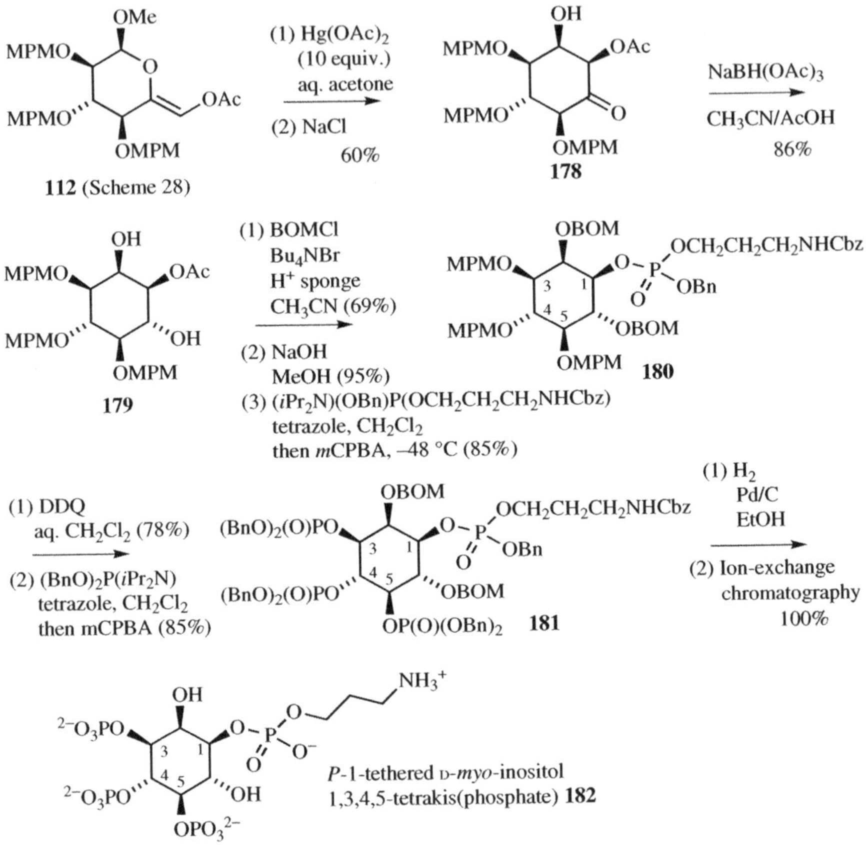

Scheme 12.46 Synthesis of inositol tetrakis(phosphate).

12.5.7 Rearrangement of Enol Acetates (1): Synthesis of an Inositol Tetrakis(phosphate)

D-*myo*-Inositol 1,4,5-tris(phosphate) (IP_3) is an intracellular second messenger that mediates the release of calcium from nonmitochondrial stores. Other inositol polyphosphates have also been implicated in the regulation of calcium levels. In 1991, Estevez and Prestwich reported the synthesis of optically pure D-*myo*-inositol 1,3,4,5-tetrakis(phosphate) (IP_4) derivative **182**, which was utilized in the isolation and purification of the receptor protein of IP_4.[16a] Estevez and Prestwich employed the enol-acetate version of the Ferrier carbocyclization reaction for the preparation of this optically pure *myo*-inositol derivative (Scheme 12.46). (*Z*)-Enol acetate **112** was synthesized from D-glucose (see Section 12., Scheme 12.28). The Ferrier carbocyclization of **112** with excess $Hg(OAc)_2$ followed by NaCl treatment gave cyclohexanone **178** as the major isomer in 60% yield. Reduction of **178** with $NaBH(OAc)_3$ stereoselectively afforded *myo*-inositol derivative **179**. After protection of the diol moiety in **179**, de-O-acetylation and subsequent condensation with

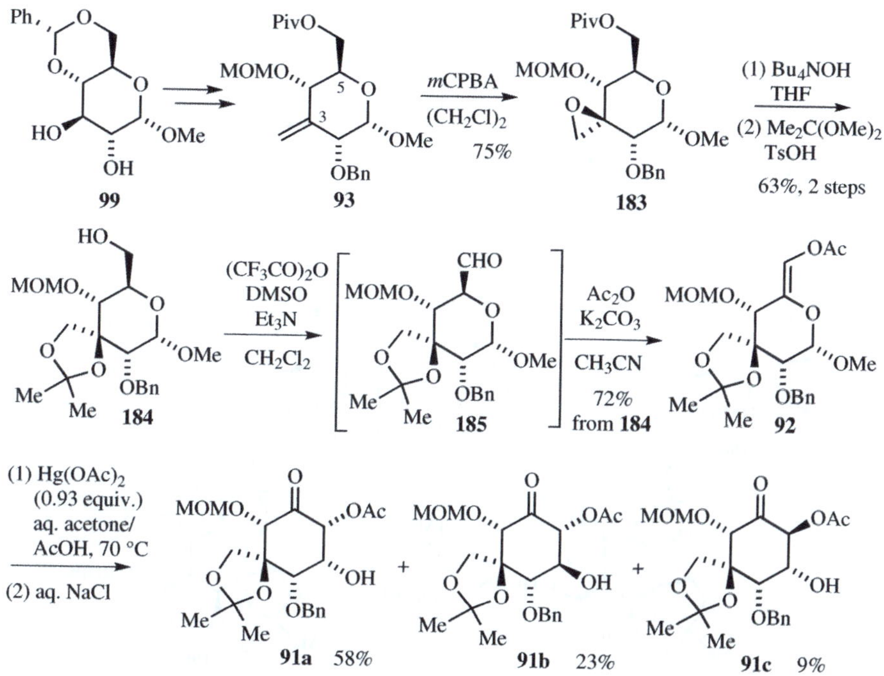

Scheme 12.47 Preparation of a cyclohexane unit of tetrodotoxin (1).

the appropriate phosphoramidite followed by *m*-CPBA oxidation provided fully protected aminopropyl-tethered inositol **180**. Deprotection of *O*-MPM group in **180** gave a triol. Phosphitylation of the triol followed by oxidation gave protected IP_4 derivative **181**. Hydrogenolysis of **181** induced the global deprotection to give P-1 aminopropyl-tethered D-*myo*-inositol 1,3,4,5-tetrakis(phosphate) **182**.

12.5.8 Rearrangement of Enol Acetates (2): Synthesis of Tetrodotoxin

In 2010, the Sato group disclosed the synthesis of tetrodotoxin **86** starting from D-glucose, in which the Ferrier carbocyclization of enol acetate **92** was employed for the construction of the cyclohexane core of tetrodotoxin.[36] 4,6-*O*-Benzylidene derivative **99** was converted into pyranoside **93** possessing an *exo*-methylene group at C-3 (Scheme 12.47). *m*-CPBA oxidation of **93** stereoselectively provided epoxide **183**. Alkaline hydrolysis of the epoxide in **183** followed by acetonide formation gave primary alcohol **184**. Oxidation of **184** afforded aldehyde **185**, which was converted into (*Z*)-enol acetate **92** by the action of acetic anhydride and potassium carbonate. When enol acetate **92** reacted with $Hg(OAc)_2$, followed by NaCl treatment, the Ferrier carbocyclization reaction successfully took place to afford a mixture of cyclized products **91a**, **91b**, and **91c** in 58%, 23%, and 9% yields, respectively. Use

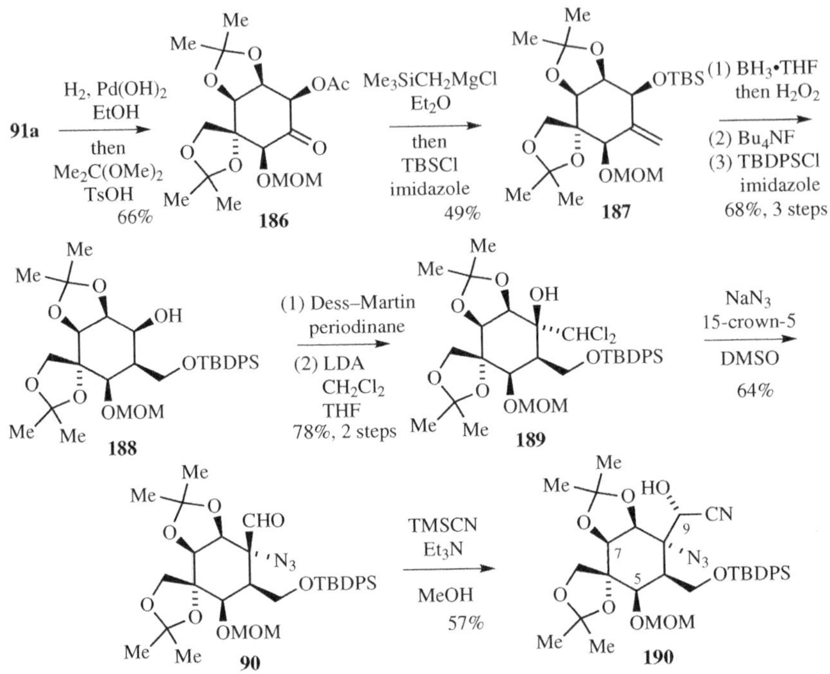

Scheme 12.48 Preparation of a cyclohexane unit of tetrodotoxin (2).

of $PdCl_2$ was ineffective for the reaction of **92** and resulted in the formation of a complex mixture.

Hydrogenolysis of cyclohexanone **91a**, possessing the correct stereochemistry for tetrodotoxin, followed by acetonide formation, provided diacetonide **186** (Scheme 12.48). Peterson olefination of ketone **186** generated an *exo*-methylene, and an exchange of the protecting group of the product (OAc → OTBS) afforded **187**. Hydroboration of the methylene moiety in **187** stereoselectively afforded a hydroxymethyl derivative, which was converted into **188**. Oxidation of **188** with Dess–Martin periodinane gave a ketone, which reacted with lithiated dichloromethane to give **189** stereoselectively. Treatment of **189** with NaN_3 in DMSO induced a Darzens-type reaction to give an epoxide intermediate, whose oxirane ring was opened with azide ion to provide **90** with complete stereo- and regioselectivity. The reaction of aldehyde **90** with TMSCN in methanol gave an equilibrium mixture of cyanohydrin **190** and its 9-epimer in 56% and 17% yields, respectively.

After protection of the hydroxy group in **190** as a MOM ether, the resulting nitrile was reduced with DIBAL-H to afford an aldehyde, which was then treated with Jones reagent (Scheme 12.49). Under these acidic conditions, the MOM protecting group at C-5 was first cleaved to give a lactol, which underwent oxidation to provide lactone **89**. The azide group in **89** was reduced to give an amine, and

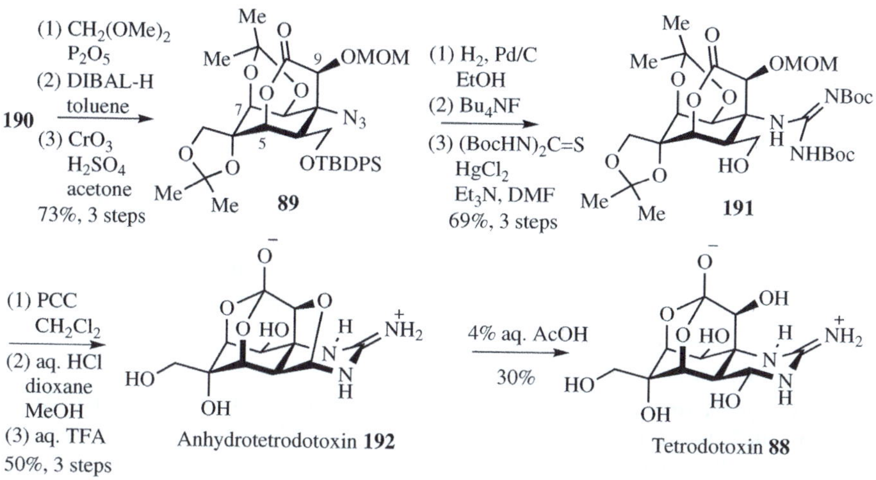

Scheme 12.49 Total synthesis of tetrodotoxin.

the silyl protecting group was removed to afford a hydroxyamine. Guanidinylation of the amine gave **191**. Oxidation of a primary alcohol in **191**, followed by acid hydrolysis and subsequent treatment with aq. TFA, afforded anhydrotetrodotoxin **192** in 50% yield. Treatment of **192** with aq. AcOH provided a mixture of **192** and tetrodotoxin **88** (ca. 1:3). Purification of the mixture afforded tetrodotoxin **88** in 30% yield.

12.6 CONCLUSION

Since R. J. Ferrier's discovery in 1979 of an interesting rearrangement of 5-enopyranosides, the Ferrier carbocyclization reaction has become one of the most synthetically powerful transformations of carbohydrates to carbocycles. At the early stage of its development, this reaction was applied to prepare cyclitol and aminocyclitol derivatives. Later, more complex molecules such as polycyclic and macrocyclic natural products possessing cyclohexane units with multiple stereocenters were synthesized utilizing the Ferrier cyclization. Another of Ferrier's findings, the Ferrier rearrangement of a glycal system, has also played an important role in the transformation of carbohydrates. Carbohydrates, the most abundant class of compounds in the biological world, are relatively inexpensive and readily available molecules possessing multiple well-defined stereogenic centers, making them useful and important organic raw materials as renewable and nonfossil carbon resources. The Ferrier carbocyclization as well as the Ferrier rearrangement will continue to serve as efficient reactions in carbohydrate chemistry, expanding the possibilities for the synthesis of important molecules by the "chiral pool approach from carbohydrates."[7]

REFERENCES

1. (a) Ferrier, R. J.; Overend, W. G.; Ryan, A. E. *J. Chem. Soc. C* **1962**, 3667–3670. (b) Ferrier, R. J. *J. Chem. Soc. C* **1964**, 5443–5449.
2. Ferrier, R. J. *J. Chem. Soc., Perkin Trans. 1* **1979**, 1455–1458.
3. (a) Ferrier, R. J.; Zubkov, O. A. In *Organic Reactions*; Overman, L. E., Ed.; John Wiley & Sons, Inc.: New York, **2003**; Vol. *62*, pp. 569–736. (b) Ferrier, R. J. *Top. Curr. Chem.* **2001**, *215*, 153–175.
4. Williams, D. R.; Heidebrecht, Jr., R. W. *J. Am. Chem. Soc.* **2003**, *125*, 1843–1850.
5. (a) Ferrier, R. J.; Middleton, S. *Chem. Rev.* **1993**, *93*, 2779–2831. (b) Ferrier, R. J. *Top. Curr. Chem.* **2001**, *215*, 277–291.
6. Collins, P.; Ferrier R. *Monosaccharides*; John Wiley & Sons: Chichester, **1995**.
7. (a) *Comprehensive Chirality*; Carreira, E. M.; Yamamoto, H., Eds.; Elsevier: Amsterdam, **2012**; Vol. 2. (b) Hanessian, S. *Total Synthesis of Natural Products: The 'Chiron' Approach*; Pergamon Press: Oxford, **1983**.
8. Machado, A. S.; Olesker, A.; Lukacs, G. *Carbohydr. Res.* **1985**, *135*, 231–239.
9. Chida, N.; Ohtuka, M.; Ogura, K.; Ogawa, S. *Bull. Chem. Soc. Jpn.* **1991**, *64*, 2118–2121.
10. Adam, S. *Tetrahedron Lett.* **1988**, *29*, 6589–6592.
11. Iimori, T.; Takahashi, H.; Ikegami, S. *Tetrahedron Lett.* **1996**, *37*, 649–652.
12. (a) Tanimoto, H.; Saito, R.; Chida, N. *Tetrahedron Lett.* **2008**, *49*, 358–362. (b) Ichiki, M.; Tanimoto, H.; Miwa, S.; Saito, R.; Sato, T.; Chida, N. *Chem. – Eur. J.* **2013**, *19*, 264–269.
13. (a) Imuta, S.; Ochiai, S.; Kuribayashi, M.; Chida, N. *Tetrahedron Lett.* **2003**, *44*, 5047–5051. (b) Imuta, S.; Tanimoto, H.; Momose, M. K.; Chida, N. *Tetrahedron* **2006**, *62*, 6926–6944.
14. Ermolenko, M. S.; Olesker, A.; Lukacs, G. *Tetrahedron Lett.* **1994**, *35*, 711–714.
15. Bender, S. L.; Budhu, R. J. *J. Am. Chem. Soc.* **1991**, *113*, 9883–9885.
16. (a) Estevez, V. A.; Prestwich, G. D. *J. Am. Chem. Soc.* **1991**, *113*, 9885–9887. (b) Prestwich, G. D. *Acc. Chem. Res.* **1996**, *29*, 503–513. (c) Chen, J.; Feng, L.; Prestwich, G. D. *J. Org. Chem.* **1998**, *63*, 6511–6522.
17. (a) Takahashi, H.; Kittaka, H.; Ikegami, S. *Tetrahedron Lett.* **1998**, *39*, 9703–9706. (b) Takahashi, H.; Kittaka, H.; Ikegami, S. *J. Org. Chem.* **2001**, *66*, 2705–2716.
18. (a) Das, S. K.; Mallet, J.-M.; Sinaÿ, P. *Angew. Chem., Int. Ed.* **1997**, *36*, 493–496. (b) Sollogoub, M.; Sinaÿ, P. *Molecules* **2005**, *10*, 843–858.
19. (a) Sollogoub, M.; Mallet, J.-M.; Sinaÿ, P. *Angew. Chem., Int. Ed.* **2000**, *39*, 362–364. (b) Jia, C.; Pearce, A. J.; Blériot, Y.; Zhang, Y.; Zhang, L.-H.; Sollogoub, M.; Sinaÿ, P. *Tetrahedron: Asymmetry* **2004**, *15*, 699–703.
20. Ferrier, R. J.; Haines, S. R. *J. Chem. Soc., Perkin Trans. 1* **1984**, 1689–1692.
21. Bélanger, P.; Prasit, P. *Tetrahedron Lett.* **1988**, *29*, 5521–5524.
22. Chida, N.; Ogawa, S. *J. Synth. Org. Chem., Jpn.* **1995**, *53*, 858–868.
23. (a) Blattner, R.; Ferrier, R. J.; Hains, S. R. *J. Chem. Soc., Perkin Trans. 1* **1985**, 2413–2416. (b) Ferrier, R. J.; Prasit, P. *Carbohydr. Res.* **1980**, *82*, 263–272.
24. Machado, A. S.; Dubreuil, D.; Cleophax, J.; Gero, S. D.; Thomas, N. F. *Carbohydr. Res.* **1992**, *233*, C5–C8.
25. Yamauchi, N.; Terachi, T.; Eguchi, T.; Kakinuma, K. *Tetrahedron* **1994**, *50*, 4125–4136.

26. (a) Petasis, N. A.; Lu, S.-P. *J. Am. Chem. Soc.* **1995**, *117*, 6394–6395. (b) Petasis, N. A.; Lu, S.-P. *Tetrahedron Lett.* **1996**, *37*, 141–144. (c) Smith, III, A. B.; Fox, R. J.; Razler, T. M. *Acc. Chem. Res.* **2008**, *41*, 675–687.
27. Dalko, P. I.; Sinaÿ, P. *Angew. Chem., Int. Ed.* **1999**, *38*, 773–777.
28. (a) Sakairi, N.; Hayashida, M.; Amano, A.; Kuzuhara, H. *J. Chem. Soc., Perkin Trans. 1* **1990**, 1301–1313. (b) Sato, K.; Sakuma, S.; Murmatsu, S.; Bokura, M. *Chem. Lett.* **1991**, 1473–1474. (c) Barton, D. H. R.; Géro, S. D.; Cléophax, J.; Machado, A. S.; Quiclet-Sire, B. *J. Chem. Soc., Chem. Commun.* **1988**, *1184–1186*. (d) Köhn, A.; Schmidt, R. R. *Liebigs Ann. Chem.* **1987**, 1045–1054.
29. Chida, N.; Ohtsuka, M.; Nakazawa, K.; Ogawa, S. *J. Org. Chem.* **1991**, *56*, 2976–2983.
30. (a) Chida, N.; Ohtsuka, M.; Ogawa, S. *J. Org. Chem.* **1993**, *58*, 4441–4447. (b) Chida, N.; Ohtsuka, M.; Ogawa, S. *Tetrahedron Lett.* **1991**, *32*, 4525–4528.
31. (a) Stork, G.; Isobe, M. *J. Am. Chem. Soc.* **1975**, *97*, 6260–6261. (b) Suzuki, M.; Kawagishi, T.; Yanagisawa, A.; Suzuki, T.; Okamura, N.; Noyori, R. *Bull. Chem. Soc. Jpn.* **1988**, *61*, 1299–1312. (c) Taylor, R. J. K. *Synthesis* **1985**, 364–392.
32. (a) Martin Castro, A. M. *Chem. Rev.* **2004**, *104*, 2939–3002. (b) Majumdar, K. C.; Alam, S.; Chattopadhyay, B. *Tetrahedron* **2008**, *64*, 597–643.
33. Overman, L.E.; Carpenter, N. E. In *Organic Reactions*; Overman, L. E., Ed.; John Wiley & Sons, Inc.: New York, **2005**; Vol. 66, pp. 1–107.
34. (a) Tanimoto, H.; Kato, T.; Chida, N. *Tetrahedron Lett.* **2007**, *48*, 6267–6270. (b) Kato, T.; Tanimoto, H.; Yamada, H.; Chida, N. *Heterocycles* **2010**, *82*, 563–579.
35. (a) Fisher, M. J.; Myers, C. D.; Joglar, J.; Chen, S.-H.; Danishefsky, S. J. *J. Org. Chem.* **1991**, *56*, 5826–5834. (b) Hayward, C. M.; Yohannes, D.; Danishefsky, S. J. *J. Am. Chem. Soc.* **1993**, *115*, 9345–9346.
36. Akai, S.; Seki, H.; Sugita, M.; Kogure, T.; Nishizawa, N.; Suzuki, K.; Nakamura, Y.; Kajihara, Y.; Yoshimura, J.; Sato, K. *Bull. Chem. Soc. Jpn.* **2010**, *83*, 279–287.
37. (a) *The Carbohydrates*; Pigman, W.; Horton, D., Eds.; Academic Press: New York, **1972**; Vol. IA, 1980; Vol. IB, 1970; Vol. IIA, and 1970; Vol. IIB. (b) *Preparative Carbohydrate Chemistry*; Hanessian, S., Ed.; Marcel Dekker, Inc.: New York, **1997**.
38. Amano, S.; Ogawa, N.; Ohtsuka, M.; Chida, N. *Tetrahedron* **1999**, *55*, 2205–2224.
39. (a) Hanessian, S.; Plessas, N. R. *J. Org. Chem.* **1969**, *34*, 1035–1044. (b) Hanessian, S.; Plessas, N. R. *J. Org. Chem.* **1969**, *34*, 1045–1053.
40. Heathcock, C. H. In *Comprehensive Organic Synthesis*; Trost, B. M.; Fleming, I., Eds.; Pergamon: Oxford, **1991**; Vol. 2, pp. 181–238.
41. (a) Reetz, M. T.; Hüllmann, M. *J. Chem. Soc., Chem. Commun.* **1986**, 1600–1602. (b) Stanton, G. R.; Johnson, C. N.; Walsh, P. J. *J. Am. Chem. Soc.* **2010**, *132*, 4399–4408.
42. (a) Bohno, M.; Sugie, K.; Imase, H.; Yusof, Y. B.; Oishi, T.; Chida, N. *Tetrahedron* **2007**, *63*, 6977–6989. (b) Fukazawa, T.; Shimoji, Y.; Hashimoto, T. *Tetrahedron: Asymmetry* **1996**, *7*, 1649–1658.
43. Parker, K. A.; Fokas, D. *J. Org. Chem.* **2006**, *71*, 449–455.
44. Varin, M.; Barré, E.; Iorga, B.; Guillou, C. *Chem. – Eur. J.* **2008**, *14*, 6606–6608.

PART IV

[3,3]- AND [2,3]-SIGMATROPIC REARRANGEMENTS

13

THE CLAISEN REARRANGEMENT

Donald Craig

Department of Chemistry, Imperial College London, South Kensington Campus, London, UK

13.1 INTRODUCTION

Since it was first reported in 1912,[1] the Claisen rearrangement has become a mainstay of organic synthesis and a classic among methods for the assembly of small and large molecules. Central to the remarkable longevity of this strategy-level pericyclic transformation is that it transforms C—heteroatom bonds into C—C bonds regiospecifically and almost invariably in a highly stereoselective manner. In addition, substrates for the Claisen rearrangement are usually straightforward to synthesize from commonplace precursors, and in many instances, the substrates may be generated *in situ*, with rearrangement taking place spontaneously upon their formation. Thus, the Claisen rearrangement offers unrivaled opportunities for the creation of complexity and the generation of high added-value molecular structures in a rapid and reliable fashion.[2]

13.1.1 Fundamentals of the Claisen Rearrangement Reaction

In its simplest manifestation, the Claisen rearrangement reaction involves the [3,3]-sigmatropic rearrangement of an allyl vinyl ether **1**, giving a γ,δ-unsaturated aldehyde or ketone product **2**. The reaction takes place with the breaking of a carbon–oxygen σ-bond and two carbon–carbon π-bonds and the formation of a transposed carbon–carbon π-bond together with a new carbon–carbon σ-bond and a carbon–oxygen π-bond which is part of the carbonyl group (Scheme 13.1). Thus, depending on the degree and position of substitution within the substrate,

Molecular Rearrangements in Organic Synthesis, First Edition. Edited by Christian M. Rojas.

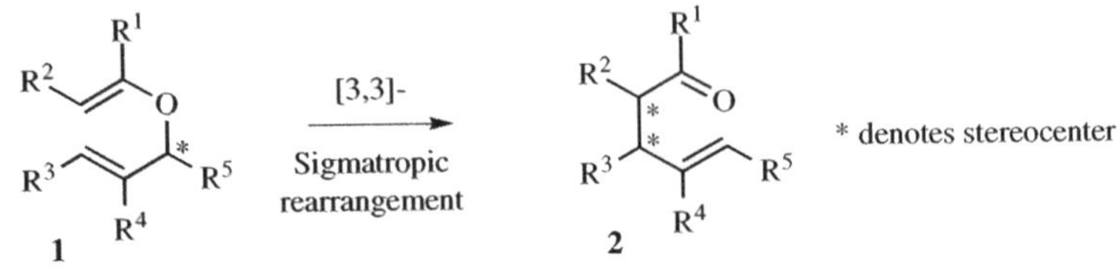

Scheme 13.1 The Claisen rearrangement.

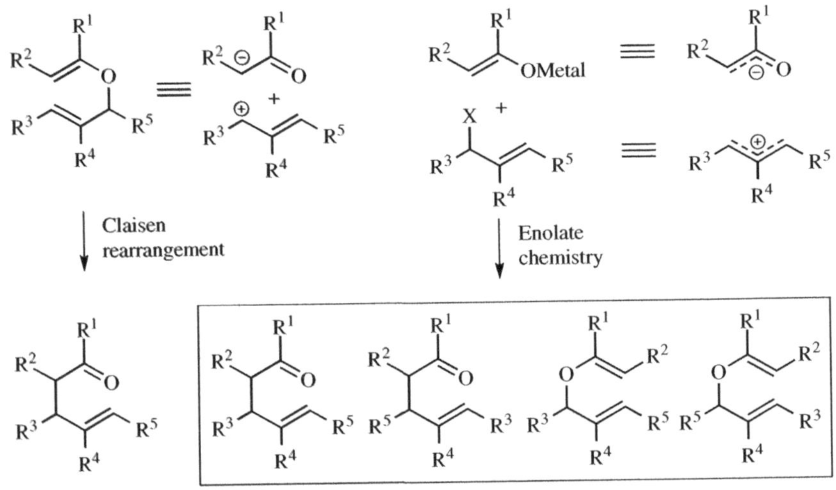

Scheme 13.2 Strategic value of the Claisen rearrangement.

the rearrangement enables the allylic transposition of a C—O bond to a C—C bond, with a corresponding increase in stereochemical complexity. The major driving force of the rearrangement is the formation of the carbonyl functional group, whose thermodynamic stability is such that the intrinsically reversible allyl vinyl ether→γ,δ-unsaturated carbonyl transformation is almost invariably a one-way process for the purposes of synthesis.[3]

From a synthetic strategy standpoint, Claisen rearrangement reactions give rise to the products of C—C bond formation by formal allylation at the α-position of a carbonyl compound. Such products may also be accessed by using enolate chemistry in conjunction with an allylating agent. While the latter transformations are usually relatively efficient because of the high mutual S_N2 reactivity of the enolate anion with the allylic electrophile, both nucleophile and electrophile are ambident in nature. In contrast, use of the Claisen rearrangement reaction allows regiospecific allylation because of the reaction's pericyclic nature: C—C rather than C—O bond formation is bound to take place, and this occurs at only one of the allylic termini (Scheme 13.2). In addition, the intramolecular nature and correspondingly less unfavorable activation entropy of the Claisen rearrangement enable its application to the synthesis of highly substituted, sterically congested systems.

13.1.2 Functional Group Versatility: Claisen Rearrangement Reaction Variants

Although the very first Claisen rearrangement substrate was synthesized by O-allylation of phenol,[1] this approach is not general because of the issues of enolate allylation regiochemistry alluded to previously. Instead, substrates are typically synthesized using strategies in which the allylic moiety is part of a nucleophilic reactant, and using this approach, the Claisen rearrangement has been extended to include a range of substrates which give rise to γ,δ-unsaturated carbonyl-containing products possessing a variety of functional groups. The most significant substrate types are (1) allyl vinyl ethers **3**, (2) *O,O*-dialkylketene acetals **4**, (3) ketene aminals **5**, and (4) *O*-alkyl-*O*-silyl ketene acetals (commonly, these are referred to simply as silyl ketene acetals) **6**, as well as reactive zwitterionic intermediates such as **7** arising from the combination of allylic amines and thioethers with ketenes generated *in situ*.

Common to the majority of these processes is the use of an allylic alcohol for substrate generation, whether in a separate step or *in situ*; meanwhile, the source of the vinylic portion of the substrate varies according to the Claisen rearrangement variant. In the classical Claisen rearrangement, catalyzed condensation of an allylic alcohol with a vinyl ether gives the allyl vinyl ether **3**, which typically rearranges *in situ*.[4] The Johnson–Claisen[5] and Eschenmoser–Claisen[6] variants are closely related mechanistically, involving the condensation of allylic alcohols with more highly oxidized precursors to give substrates **4** and **5**, respectively, which rearrange to deliver carboxylic acid derivatives rather than aldehydes or ketones. In the Ireland–Claisen rearrangement,[7] allylic esters synthesized using standard methods are converted at low temperatures into silyl ketene acetals **6**, which undergo [3,3]-sigmatropic rearrangement upon warming. In the Belluš–Claisen variant,[8,9] allylic amines and thioethers combine with ketenes generated *in situ* to give zwitterionic species **7** with $X = NR_2$, S, which undergo spontaneous rearrangement to give amides and thiol esters, respectively. Propargylic alcohols may be used in place of allylic alcohols in many of these Claisen rearrangement variants, giving functionalized allenes **8** (Scheme 13.3).[10] Further, detailed discussion of methods, reactants, and reagents for substrate assembly may be found in Section 13.4.

13.1.3 Stereochemistry

The Claisen rearrangement is a concerted, three-component pericyclic transformation comprising two π-components and one σ-component. As such, it may be described either as a ${}_{\pi}2_s + {}_{\sigma}2_s + {}_{\pi}2_s$ or as a ${}_{\pi}2_a + {}_{\sigma}2_a + {}_{\pi}2_s$ process, and since in both analyses the total number of $(4q+2)$ suprafacial and $4r$ components is an odd number (three and one, respectively), the transformation is thermally allowed.[11] However,

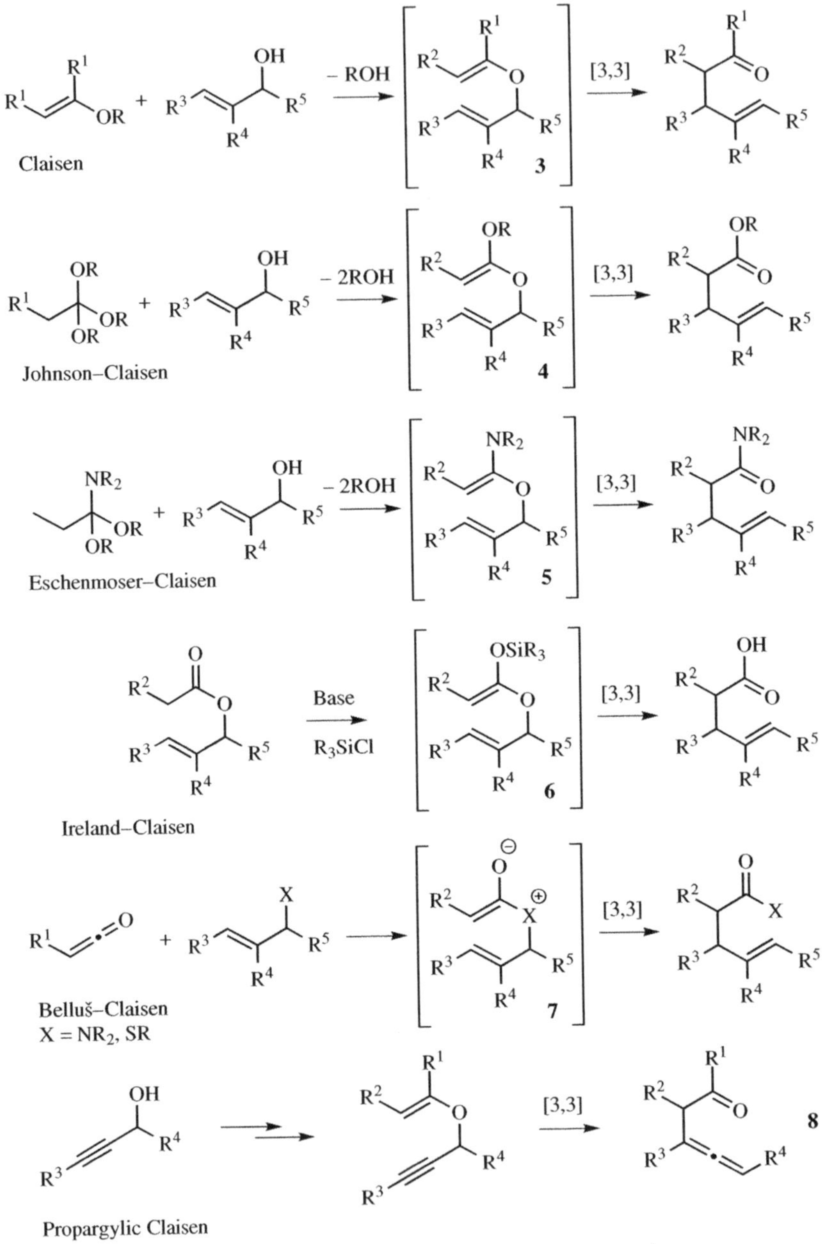

Scheme 13.3 Claisen rearrangement variants.

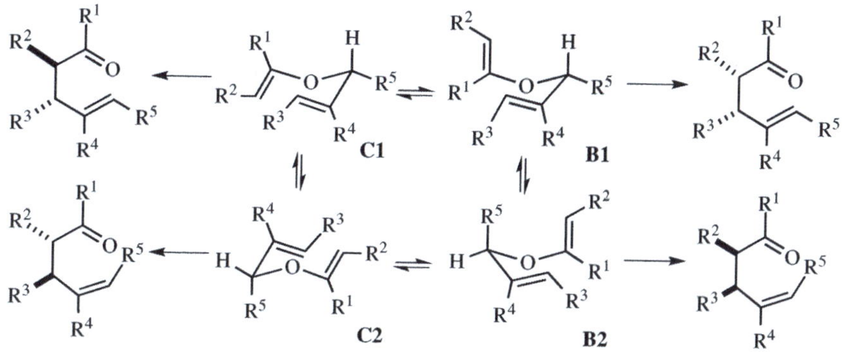

Scheme 13.4 Competing reacting conformations in the Claisen rearrangement.

this orbital symmetry conservation-based analysis is of less predictive value than it is for cycloaddition or electrocyclic processes because of the greater inherent flexibility of the Claisen rearrangement substrate, which means that more than one diastereoisomeric outcome is possible even though the reactions are always stereospecific. This is because competing reactive geometries lead to different topicities of mutual approach of the reacting π-systems, with alternative geometries arising through the interconversion of chair- and boat-like conformations and by the preferred disposition of substituents on the pericyclic array. The analysis depicted in Scheme 13.4 indicates that Claisen rearrangement of any acyclic substrate may proceed via two competing chair-like reactive conformations, **C1** and **C2**, and two competing boat-like conformations, **B1** and **B2**. These competing conformers give rise to products varying in a complementary way with respect to α,β relative stereochemistry and alkene geometry. In the chair-like conformation **C1**, pseudoequatorially disposed R^5 leads to the formation of an *E*-configured γ,δ-alkene with the α,β relative stereochemistry shown, while in **C2** a pseudoaxial R^5 group gives the corresponding *Z*-isomer having the same relative α,β stereochemistry but in the opposite enantiomeric sense. Similar arguments apply for the competing boat-like conformers **B1** and **B2**, with α,β relative stereochemistry now in the opposite diastereoisomeric sense (Scheme 13.4).

Conformationally more constrained substrates possessing cyclic allylic moieties undergo Claisen rearrangements in which allylic alkene topicity is determined by the configuration of the ring carbon atom bearing the vinyloxy motif, with variable topicity associated with the vinylic olefin (Scheme 13.5). A more in-depth discussion of the factors affecting the nature and degree of stereoselectivity which may be expected in these reactions is provided in Section 13.3.

13.2 STRATEGIC PLANNING FOR THE CLAISEN REARRANGEMENT REACTION

The Claisen rearrangement enables the synthesis of multifunctional molecules through two-carbon chain extension starting from an allylic alcohol or a derivative

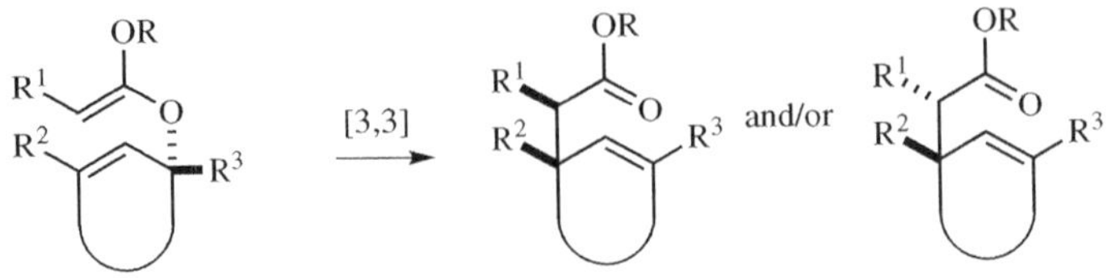

Scheme 13.5 Effect of allylic configuration on product stereochemistry.

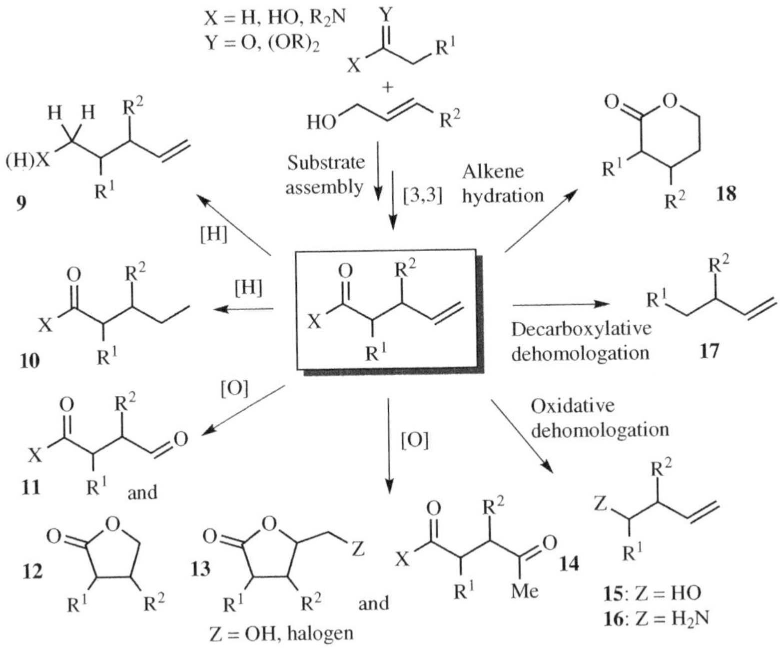

Scheme 13.6 Synthetic transformations of Claisen rearrangement products.

thereof. The range of Claisen variants described earlier (cf. Scheme 13.3) allows access to products having a variety of functional motifs, all of which contain modifiable heteroatom-containing groups. This combination of functional group versatility and regiospecificity offers multiple opportunities for selective synthesis.

13.2.1 Identification of Structural Elements

In principle, any target structure possessing a heteroatom linearly connected to a chain possessing five contiguous aliphatic carbon atoms may be accessed using the Claisen rearrangement. Post-rearrangement, additional standard functional group transformations include oxidation level adjustment, homologation and dehomologation, alkene hydration, and cyclization processes. Scheme 13.6 depicts the synthetically most useful options, starting from a simple rearrangement product congener possessing a terminal alkene, which is derived ultimately from a primary allylic alcohol. Thus, unsaturated amines and alcohols **9** are formed by straightforward carbonyl group oxidation level adjustment, and alkene hydrogenation gives the saturated analogues **10**. Oxidative cleavage of the alkene provides valuable 1,4-dicarbonyl compounds **11** or γ-lactones **12**, depending upon work-up conditions. Oxidation without C—C bond cleavage yields more highly functionalized γ-lactones **13**, by electrophile-assisted cyclization,[12] or homologous 1,4-dicarbonyl compounds **14**, by Wacker oxidation[13] of the terminal alkene. Depending on the nature of X in the rearrangement product, varying dehomologation reactions are possible. Where X = Me, chemoselective Baeyer–Villiger oxidation[14] gives secondary alcohols **15** following acetate hydrolysis. When X = OH, Curtius rearrangement[15–17] of the derived acyl azide gives amines **16** upon hydrolytic cleavage of the carbamic acid/ester products, and analogous aminative dehomologation may be achieved by Schmidt reaction[18] of ketone-containing rearrangement products or by Beckmann rearrangement[19] of their oxime derivatives. Notably, these dehomologation transformations are stereospecific, such that the stereochemistry of the Claisen rearrangement products is retained. Decarboxylation of the Claisen rearrangement products is facile when X=Me OH and R^1 is an electron-withdrawing group, giving dehomologated products **17** without concomitant oxidation. Hydration of the terminal alkene in the rearrangement products using a hydroboration–oxidation sequence gives the products of overall hydration, which may readily be converted into the δ-lactones **18** by intramolecular esterification. These generic transformations are collected in Scheme 13.6.

13.3 MECHANISTIC FEATURES OF THE CLAISEN REARRANGEMENT REACTION

As stated in Section 13.1, the Claisen rearrangement is a concerted, pericyclic reaction. In this [3,3]-sigmatropic rearrangement process, an allylic or propargylic C—O σ-bond is exchanged for an allylic or allenic C—C σ-bond with concomitant transposition of the alkene π-bond or one of the two π-bonds present in the alkyne. One of the major advantages of this C—C bond-forming process is its intramolecularity, which imparts regio- and stereospecificity together with diastereoselectivity, as well as a less negative activation entropy.

13.3.1 Diastereoselectivity: Acyclic Substrates

For most acyclic Claisen rearrangement substrates, reactions may be viewed as taking place via a chair-like reactive conformation, with diastereoselectivity dictated by vinylic and allylic double-bond geometry. For example, isomeric ester-containing substrates cleanly rearranged to give stereocomplementary, diastereoisomeric products (Scheme 13.7).[20]

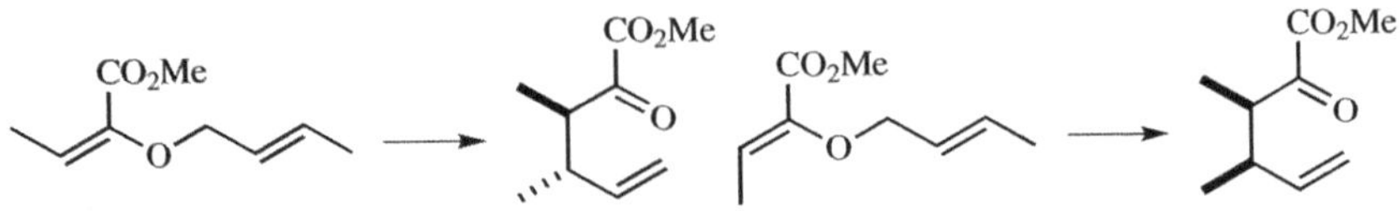

Scheme 13.7 Effect of vinylic geometry on product stereochemistry.

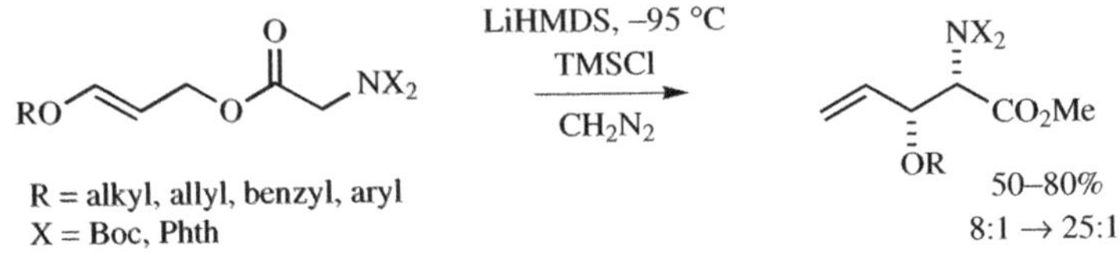

Scheme 13.8 Synthesis of aldol-type products by the Ireland–Claisen rearrangement.

When the allylic motif in the rearrangement substrate contains an enol ether and the vinylic alkene is substituted on the terminal position, the products of rearrangement correspond to those of selective aldol reactions. Glycine-derived substrates possessing enol ethers underwent highly selective Claisen rearrangements to give α-amino-β-alkoxy and α-amino-β-aryloxy products; this transformation is of particular value since it allows access to aryloxy analogues which are less straightforward to assemble using aldol chemistry (Scheme 13.8).[21]

For enantiomerically pure substrates possessing a stereocenter within the pericyclic array, absolute stereochemistry is governed by the preferred equatorial disposition of the sterically more demanding group on the stereocenter. This also leads to pronounced selectivity for *E*-alkene geometry in the product. Thus, if allylic and vinylic geometry can be controlled independently, all possible diastereoisomers of the γ,δ-unsaturated carbonyl/carboxylic product may be accessed (Scheme 13.9). Inspection of the putative transition-state structures in Scheme 13.9 shows that asymmetric induction arising from the R^5-bearing stereocenter should in principle be maximized when the R^3 and R^4 groups on the allylic double bond are mutually *anti*, because such alkene geometry maximizes the allylic 1,3-interactions of R^3 with R^5 in the unfavorable alternative chair-like conformations leading to products possessing isomeric alkenes.

Diastereoselectivity has also been observed for acyclic substrates where the stereocenter is positioned outside the pericyclic array, though levels of stereocontrol are variable (Scheme 13.10).[22]

The Ireland–Claisen rearrangement of allylic azides gave carboxylic acids which were subjected to decarboxylation under basic conditions to give azide-substituted homoallylic sulfones with varying diastereoselectivity.[23] This work also represented an unusual example of the use of interconverting substrates in the Claisen rearrangement, where only one of the equilibrating allylic azide isomers[24] has the necessary allylic double-bond position for rearrangement to take place (Scheme 13.11).

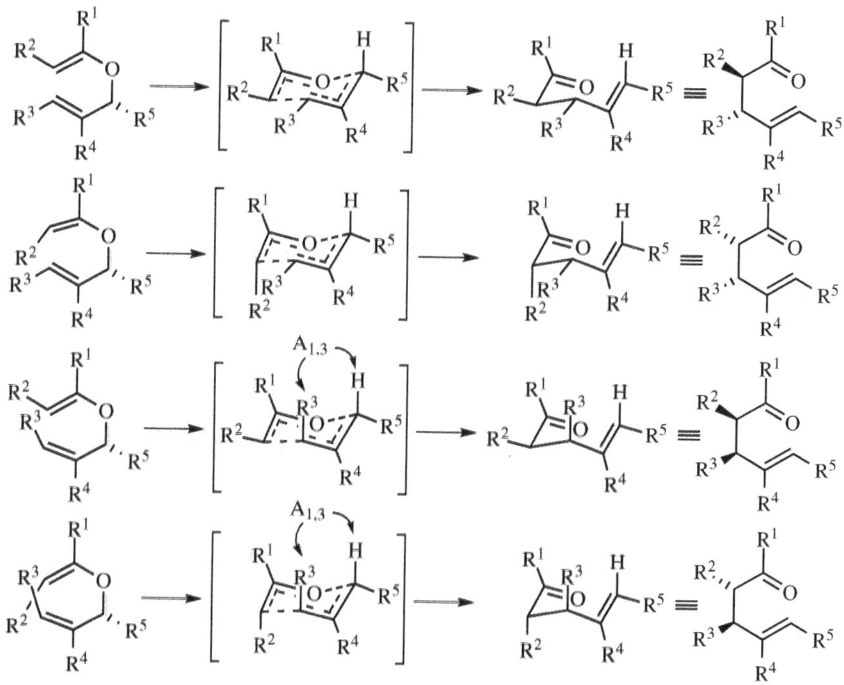

Scheme 13.9 Effect of a stereocenter on the allylic moiety.

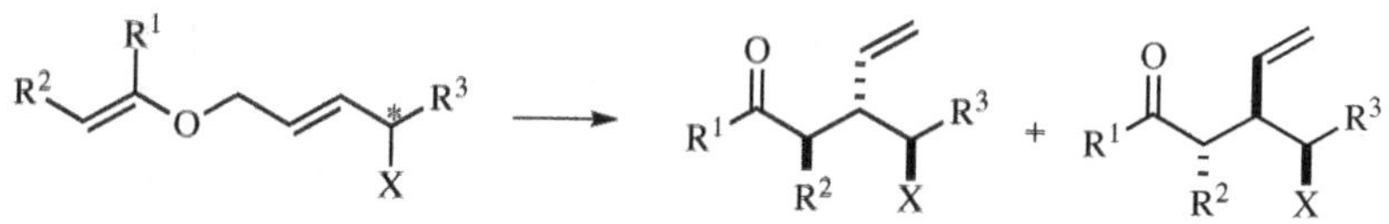

Scheme 13.10 Effect of a stereocenter outside the pericyclic array.

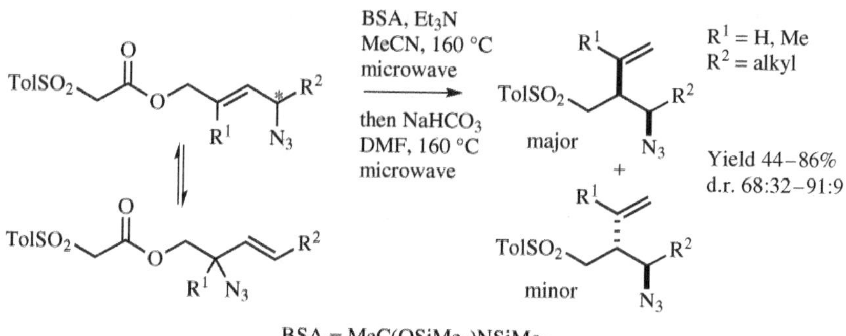

BSA = $MeC(OSiMe_3)NSiMe_3$

Scheme 13.11 Claisen rearrangement of equilibrating allylic azides.

13.3.2 Diastereoselectivity: Cyclic Substrates

Claisen rearrangement reactions of cyclic, chiral substrates show diastereoselectivity depending on the location of the stereocenter, which may be within or external to the pericyclic array. For the former class of substrates, the allylic stereocenter is part of a ring, and this effectively dictates the topicity of delivery of the vinylic moiety to the allylic terminus. Such reactions show a tendency to proceed via a boat-like rather than a chair-like transition state. Scheme 13.12 depicts the Johnson–Claisen rearrangement[5] reaction of a chiral cyclohexenol with an ortholactone, which gave a product with relative stereochemistry consistent with a boat-like reactive conformation of the *in situ*-generated ketene acetal.[25] This may be explained in terms of a minimization of steric interactions between the nonparticipating atoms in the vinyl- and allyl-containing moieties.

Schemes 13.13 and 13.14 show examples of Ireland–Claisen rearrangements of substrates possessing cyclic allylic motifs. In Scheme 13.13, a glycal-derived substrate is used; this transformation has become established as a useful method for the generation of *C*-glycosides from readily available carbohydrate derivatives.[26] Again, the boat-like arrangement is adopted so as to minimize destabilizing interactions between the silyloxy group and the dihydropyran ring. In Scheme 13.14, formation of

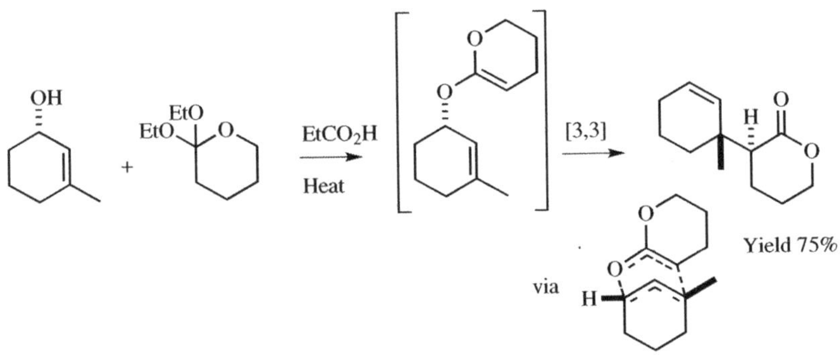

Scheme 13.12 Johnson–Claisen rearrangement of an ortholactone-derived substrate.

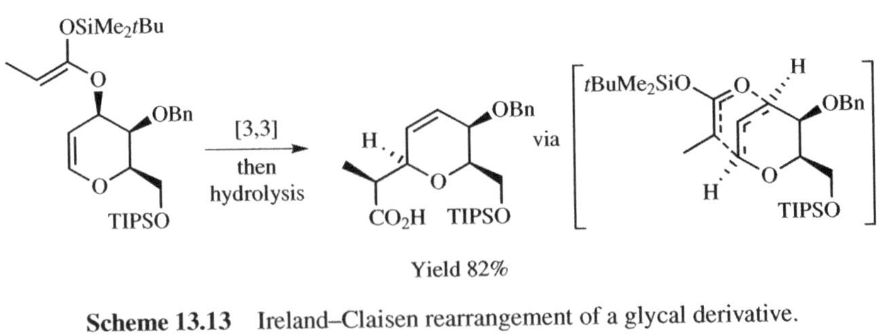

Scheme 13.13 Ireland–Claisen rearrangement of a glycal derivative.

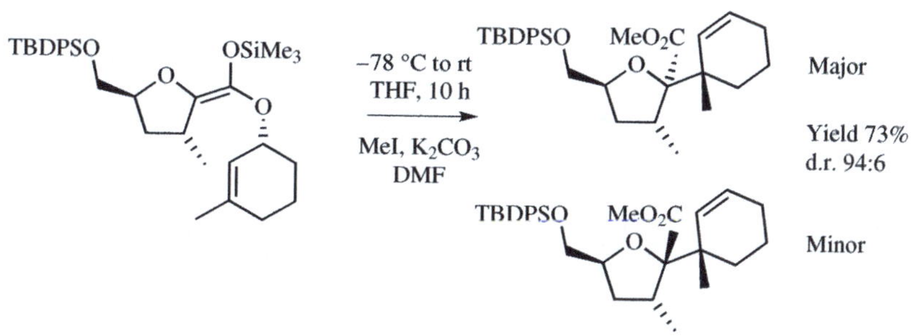

Scheme 13.14 Generation of products possessing contiguous quaternary centers.

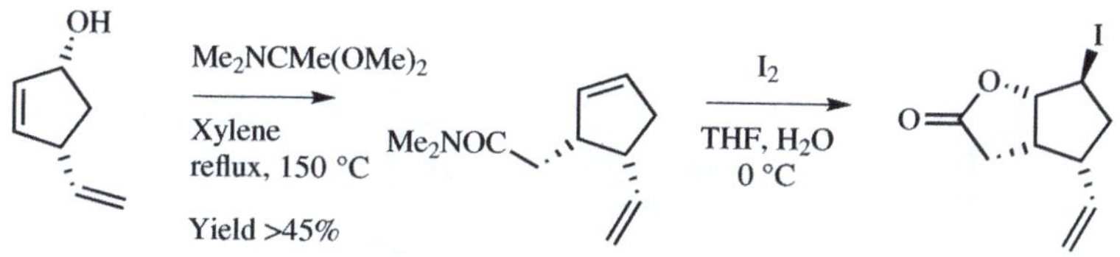

Scheme 13.15 Eschenmoser–Claisen rearrangement of a cyclopentenol derivative.

the silyl ketene acetal under standard low-temperature conditions (see Section 13.4) and warming to ambient temperature gave in good yield and high selectivity the acids, which were separated and characterized as the corresponding methyl esters. Formation of the major isomer was explained in terms of the adoption of the more favorable boat-like transition state, although in this case preferential formation of the major isomer may also be interpreted by a consideration of the directing effect of the methyl substituent α- to the vinylic alkene.[27] All of these transformations additionally demonstrate the usefulness of the Claisen rearrangement for the assembly of structures possessing contiguous sterically congested centers.

Cyclopentenol derivatives show similar behavior. Combination of an ethenyl-substituted substrate with dimethylacetamide dimethyl acetal under thermal conditions gave the product of Eschenmoser–Claisen rearrangement, with delivery of the C—C bond taking place in a *syn* fashion with respect to the ethenyl group (Scheme 13.15). The product amide was converted via iodolactonization into the bicyclic lactone, demonstrating the utility of the rearrangement product in subsequent transformations.[28]

For cyclic substrates having stereocenters external to the pericyclic array, a number of scenarios may be envisaged, in which one or other of the alkene double bonds is partly or wholly embedded in the ring. In general, whether it is the allylic or vinylic component, approach of the acyclic alkene takes place in such a way as to minimize steric interactions with substituents on the ring, resulting in bond formation in an *anti* sense (Scheme 13.16).

In Scheme 13.17, decarboxylative Claisen rearrangement (dCr) reaction of myrtenyl tosylacetate **19** is shown. The sole product obtained for this tandem

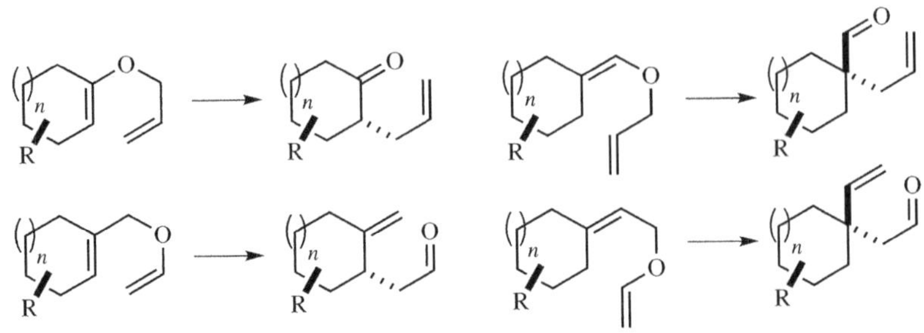

Scheme 13.16 Stereoselectivity arising from cyclic stereochemistry.

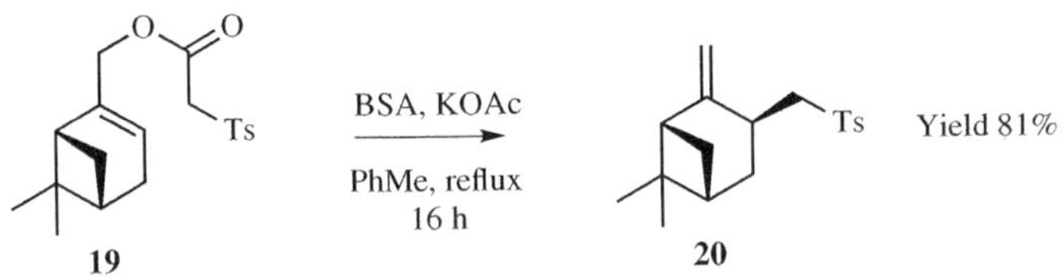

Scheme 13.17 Decarboxylative Claisen rearrangement of a myrtenol derivative.

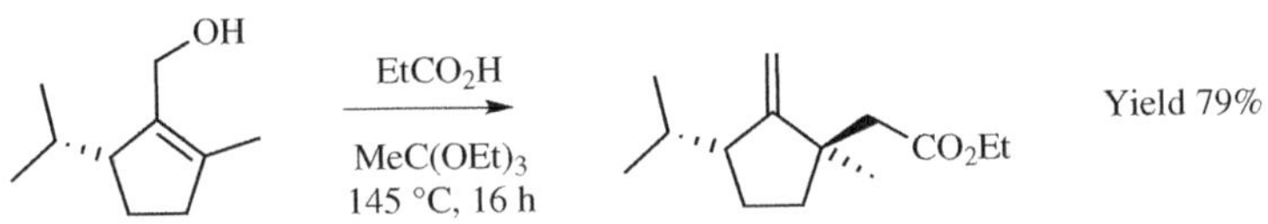

Scheme 13.18 Stereoselective Johnson–Claisen rearrangement of a cyclopentenol derivative.

transformation was that of delivery of the ketene acetal to the allylic double bond in a sense *anti* to the more sterically demanding dimethyl-substituted methano bridge. Subsequent decarboxylation *in situ* gave the homoallylic sulfone **20**. This reaction gave similar yields when carried out under microwave conditions.[29]

Scheme 13.18 shows a reaction in which a substituted cyclopentenemethanol derivative is converted stereoselectively into a quaternary center-containing product, again demonstrating the power of the transformation for the generation of sterically congested products. Heating of the substrate under classical Johnson–Claisen conditions gave the methylene cyclopentane in high yield as a single diastereoisomer.[30]

13.3.3 Reactivity: Effect of Substrate Substitution

The Claisen rearrangement reaction has been deployed on numerous substrate types possessing a wide range of structural motifs and degrees of substitution. As stated previously, the intramolecularity of the [3,3]-sigmatropic rearrangement confers

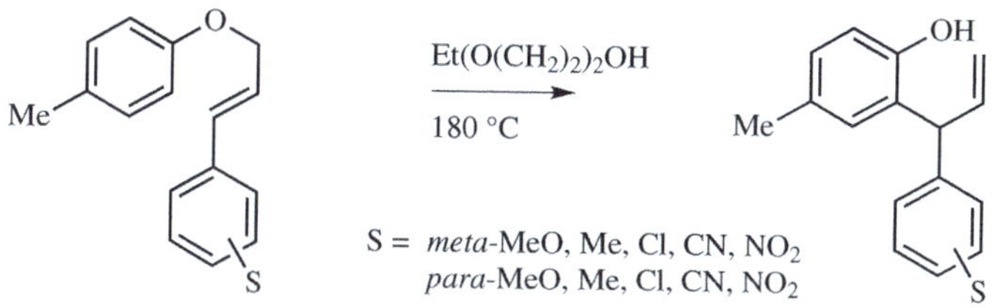

Scheme 13.19 Phenyl ether–allyl phenol rearrangement.

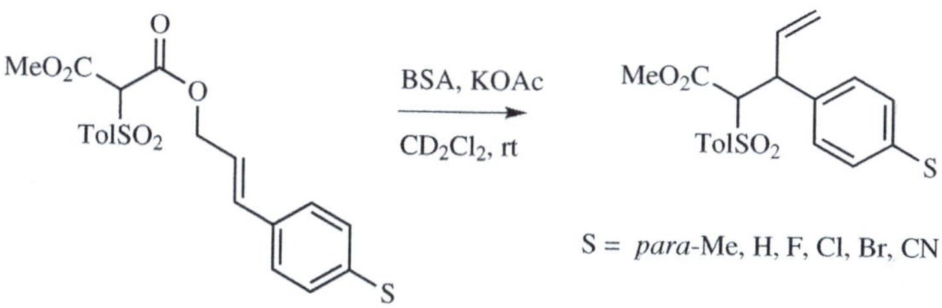

Scheme 13.20 Decarboxylative Claisen rearrangement of allyl 2-tosylmalonates.

high intrinsic substrate reactivity, and this is manifested in the efficiency with which sterically congested products may be accessed, including those possessing contiguous quaternary centers. Despite the proven utility and versatility of the Claisen rearrangement, scant quantitative data has been published concerning the relationship of reaction rate and substrate substitution. In early work, Fife and coworkers uncovered a linear relationship between the Hammett $\sigma+$ and first-order rate constant for the ortho-Claisen rearrangement of a series of meta- and para-substituted *p*-tolyl cinnamyl ethers (Scheme 13.19).[31]

Subsequently, Craig and Slavov carried out a systematic study of the dCr reactions of a series of mixed cinnamyl methyl tosylmalonic esters. In these reactions, initial *O*-silylation triggers [3,3]-sigmatropic rearrangement, after which decarboxylation occurs spontaneously to give α-tosyl-γ,δ-unsaturated esters. Consumption of the silyl ketene acetal substrates formed *in situ* was monitored by ^{1}H NMR spectroscopy, and first-order rate constants were derived from analysis of their exponential decay.[32] In this study, the rate enhancement observed for substrates possessing electron-rich Ar supported the conclusion that breakage of the allylic C—O bond is significantly more advanced than formation of the new C—C bond. This induces positive charge character on the distal allylic carbon atom, which is stabilized by more electron-donating substituents attached at that position (Scheme 13.20). Acceleration of the Ireland–Claisen rearrangement by the incorporation of electron-donating substituents at the distal position of the allylic moiety has been described and discussed elsewhere.[33]

Steric hindrance and stereoelectronics also exert significant effects on the rate of Claisen rearrangement reactions, though again quantitative rate data is not readily

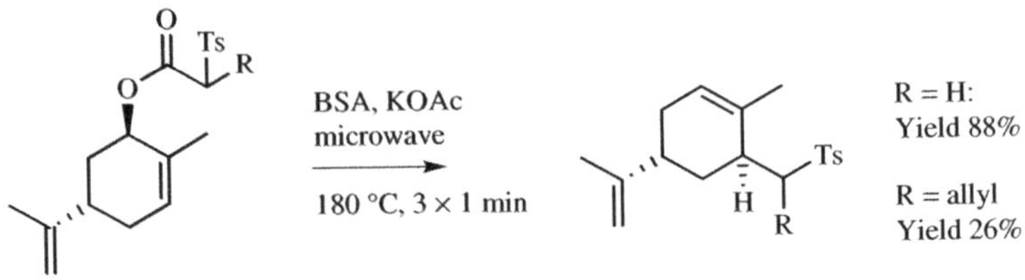

Scheme 13.21 Stereoselective decarboxylative Claisen rearrangement.

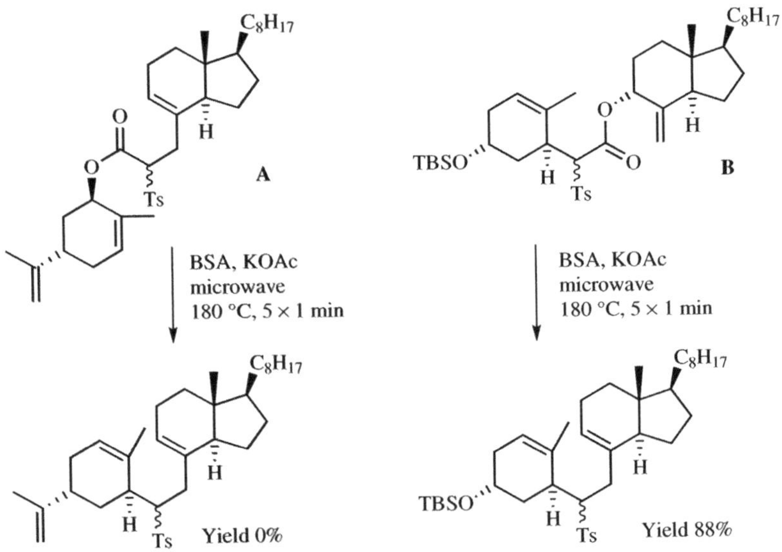

Scheme 13.22 Contrasting reactivity in decarboxylative Claisen rearrangements.

available. Scheme 13.21 shows comparative reactions of two homologous substrates in the dCr reaction. Simple substitution of a hydrogen atom with an allyl group in the *in situ*-generated silyl ketene acetal caused a significant loss of reactivity, as evidenced by a sharp drop in the yield of homoallylic sulfone product obtained under the same reaction conditions.[34]

Scheme 13.22 depicts related reactions of two steroid-derived substrates which converge on similar dienylsulfone products upon [3,3]-sigmatropic rearrangement and decarboxylation *in situ*.[32] Whereas substrate **A** was unreactive even under relatively forcing thermal microwave conditions, substrate **B** reacted smoothly and efficiently. This again reflects the rate-diminishing effect of increasing ketene acetal substitution and also shows that reactivity may be maximized by consideration of stereoelectronic effects when designing rearrangement substrates. While the products of rearrangement of substrates **A** and **B** are very similar in terms of connectivity and stereochemistry, substrate **B** possesses an axial allylic C—O bond. This maximizes orbital overlap with the allylic π-system and is such that there is substantial

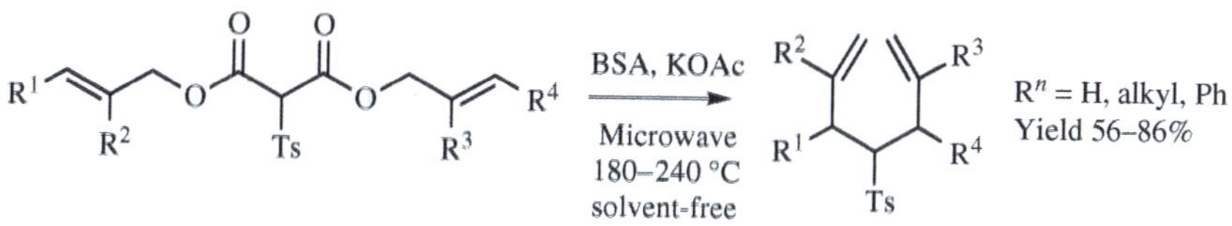

Scheme 13.23 Double decarboxylative Claisen rearrangements of bis(allylic) 2-tosylmalonates.

relief of 1,3-diaxial interactions upon allylic transposition. Related, double dCr reactions of bis(allyl) tosylmalonic esters showed decreasing reactivity with increasing substitution (Scheme 13.23).[35]

13.3.4 Reactivity: Choice of Claisen Variant

The majority of Claisen rearrangement reactions take place on substrates in which the vinylic ether moiety is generated *in situ*. As depicted in Scheme 13.3, Claisen rearrangement variants have been developed which give rise to aldehyde/ketone,[36] carboxylic acid,[37] carboxylic ester,[38] and carboxylic amide[39] derivatives, as well as thiol esters. In addition, carboxylic acid products bearing electron-withdrawing groups may be subjected to decarboxylation either *in situ* or in a separate step, extending the utility of the rearrangement to encompass noncarbonyl-containing products. As well as the rearrangement reactions taking place on substrates formed *in situ*, many Claisen rearrangements are also characterized by the use of thermal conditions, which facilitate the condensation of the allylic alcohol with the reagent containing the latent vinylic group. Methods for the formation and *in situ* rearrangement of these substrates are discussed in greater detail in Section 13.4.

The vinyl ether and Ireland–Claisen rearrangements are distinct from the other variants in that the [3,3]-sigmatropic rearrangement reactants or their precursors already possess the requisite σ-frameworks, which are formed in separate steps by etherification and esterification, respectively, using one of the well-established methods available. Ireland–Claisen rearrangements are notable also for high substrate reactivity, and this combination of simple substrate assembly and high reactivity has made this the method of choice for the assembly of multifunctional structures and in target-oriented synthesis.[37] Applications to target-oriented synthesis of the Claisen rearrangement reaction and its variants are discussed in further detail in Section 13.5.

13.4 METHODOLOGIES FOR SYNTHESIS OF CLAISEN REARRANGEMENT SUBSTRATES

13.4.1 Substrates Possessing Vinylic Ethers

The classical method for allyl vinyl ether synthesis is by the thermal and/or catalyzed combination of allylic alcohols with simple vinyl ethers such as ethoxyethene. This leads to reversible *trans*-enol etherification with concomitant generation of ethanol;

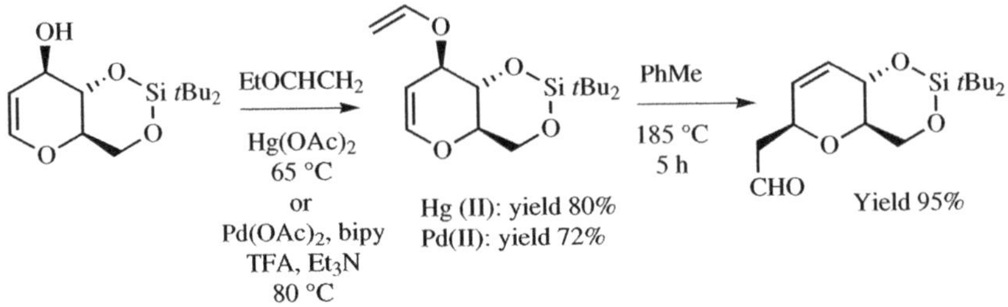

Scheme 13.24 Claisen rearrangements of vinyl ethers.

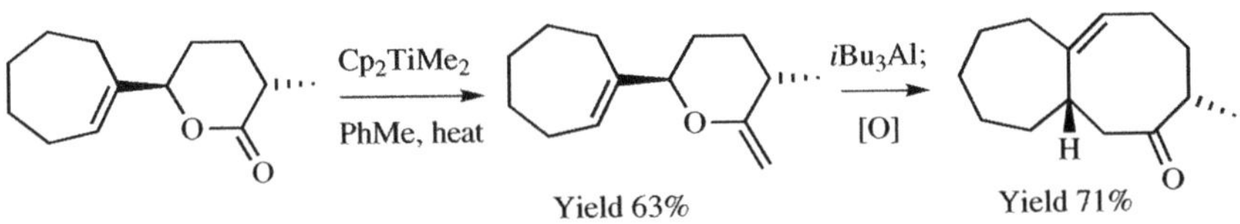

Scheme 13.25 Claisen rearrangement mediated by a trialkylaluminum.

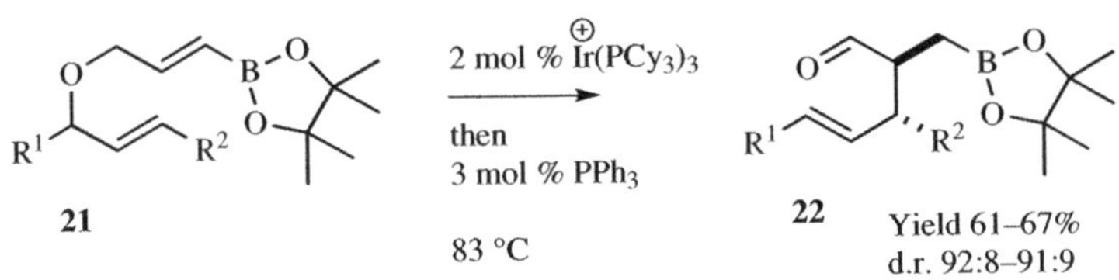

Scheme 13.26 Tandem alkene isomerisation–Claisen rearrangement.

reactions may be driven toward the allyl vinyl ether product by the use of an excess of ethoxyethene, often as the solvent. Sugar derivatives such as glycals have been used in this context. Reaction of the cyclically protected glycal shown in Scheme 13.24 with ethoxyethene under Hg(II) or Pd(II) catalysis gave the corresponding vinyl ether, which underwent stereospecific [3,3]-sigmatropic rearrangement upon thermolysis in toluene. Interestingly, both Johnson–Claisen and Eschenmoser–Claisen reactions of the hydroxyl-containing glycal under standard conditions were low yielding or gave none of the desired product.[40]

Olefination reactions using the Tebbe and Petasis reagents have been employed to generate vinylic ethers for Claisen rearrangement reactions. Scheme 13.25 shows the formation of a Claisen substrate by reaction of a δ-lactone with the Petasis reagent, which was subjected to iBu$_3$Al-mediated Claisen rearrangement with *in situ* reduction—also mediated by iBu$_3$Al[41]—followed by reoxidation to yield the fused-ring product.[42]

Isomerization reactions of bis(allylic) ethers have been used as a method for the generation of Claisen rearrangement substrates. Scheme 13.26 depicts the iridium-catalyzed isomerization of vinyl boronates **21**, which gives rise to Claisen rearrangement substrates which undergo [3,3]-sigmatropic rearrangement directly. The vinyl boronate substrates are themselves the products of chemoselective hydroboration of allyl propargyl ethers, which are readily accessed from allylic alcohols.[43] Chemoselectivity in the allyl vinyl ether isomerization is governed by the greater propensity of iridium to insert into the sterically more accessible C—H bond on the CH_2 rather than the CHR^1 group in substrates **21**. The boronate moiety in the rearrangement products **22** served as a latent hydroxyl group and as a means of forming further C—C bonds post-rearrangement through the use of Pd-catalyzed coupling reactions.

Vinylic ether-containing Claisen rearrangement substrates may be generated using *syn*-elimination reactions of sulfoxides, selenoxides, and selenones. 2-(Arylsulfinyl)ethyl ethers are particularly useful substrates in these reactions because of their ready availability by nucleophilic addition of allylic alcohols to commercially available phenylsulfinylethene. Scheme 13.27 shows a typical synthetic context for this chemistry, involving the stereospecific introduction of quaternary centers from easily accessed allylic alcohol precursors.[44]

Vinylic ether-containing Claisen rearrangement substrates may be generated in some cases by O-allylation reactions of ketones if there is an intrinsic bias in the reactivity of the derived enolate. Scheme 13.28 depicts the O-crotylation of a β-ketoester

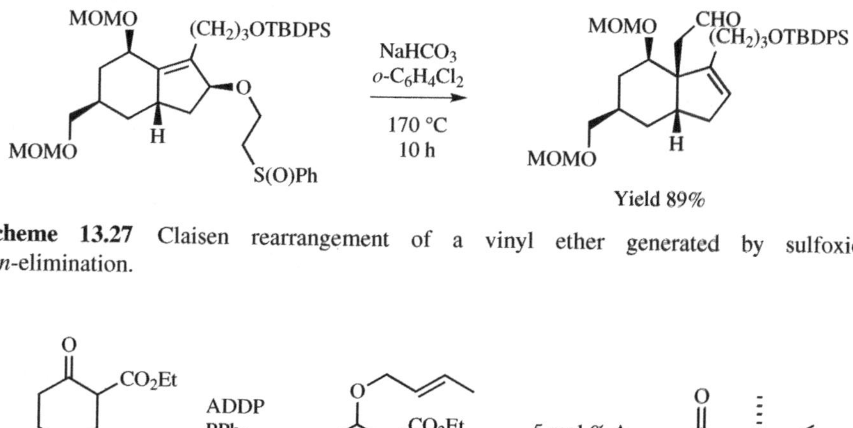

Scheme 13.27 Claisen rearrangement of a vinyl ether generated by sulfoxide *syn*-elimination.

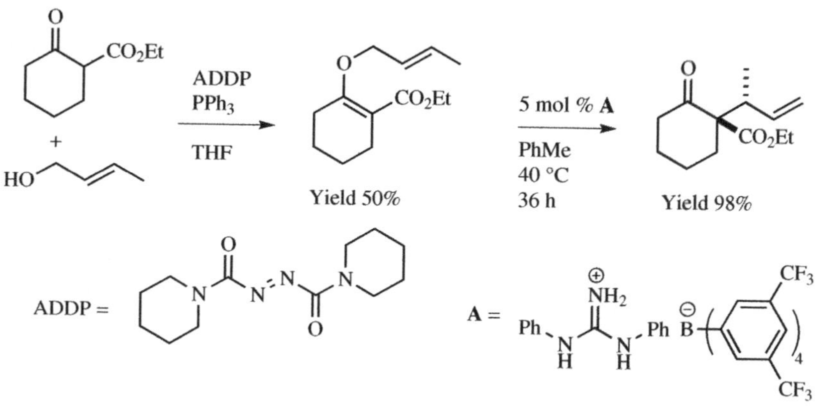

Scheme 13.28 Claisen rearrangement catalysed by a guanidinium salt.

under Mitsunobu conditions to give a Claisen rearrangement substrate, which undergoes guanidinium-catalyzed rearrangement in high yield.[45] These reactions were subsequently carried out enantioselectively using chiral analogues of the catalyst **A**.[46]

13.4.2 Substrates Possessing Ketene Acetals and Ketene Aminals

Two classical variants of the Claisen rearrangement involve *in situ* generation of the substrate. In the Johnson–Claisen rearrangement (often termed the Johnson–Claisen orthoester rearrangement),[36] ketene acetals are generated by the acid-catalyzed condensation of the allylic alcohol precursor with an orthoester (or ortholactone, as shown in Scheme 13.12). Typically, simple orthoester reactants such as triethyl orthoacetate are employed as the solvent in these processes, in conjunction with high-boiling-point acid catalysts such as propanoic acid. The reversible formation of the ketene acetal substrate for these transformations is driven by the large excess of orthoester and by the removal through distillation of the ethanol by-product. Scheme 13.29 shows a typical procedure, which was used to synthesize substrates in a recent study of intramolecular alkene amidoarylation reactions.[47]

The Eschenmoser–Claisen rearrangement is a closely related transformation which delivers carboxylic amide rather than ester products, via the intermediacy of the analogous ketene aminals.[39] Scheme 13.30 depicts the reaction of a cyclic, unsaturated *anti*-1,2-diol which was subjected to sequential Johnson–Claisen and Eschenmoser–Claisen rearrangements to give an intermediate in a synthesis approach to the opioid agonist morphine.[48]

As stated in Section 13.4, the mild conditions and resulting functional group tolerance of the Ireland–Claisen rearrangement are such that this is often the variant

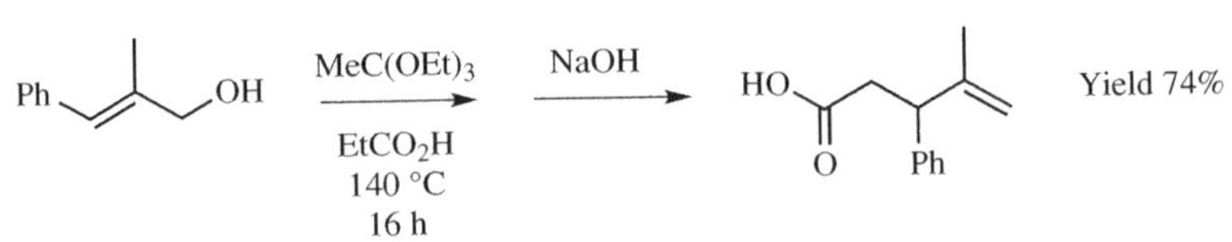

Scheme 13.29 Synthesis of a γ,δ-unsaturated acid by Johnson–Claisen rearrangement and hydrolysis.

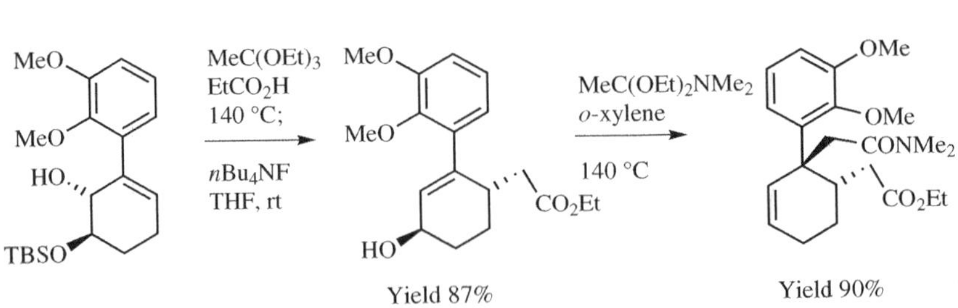

Scheme 13.30 Sequential Johnson–Claisen and Eschenmoser–Claisen rearrangements.

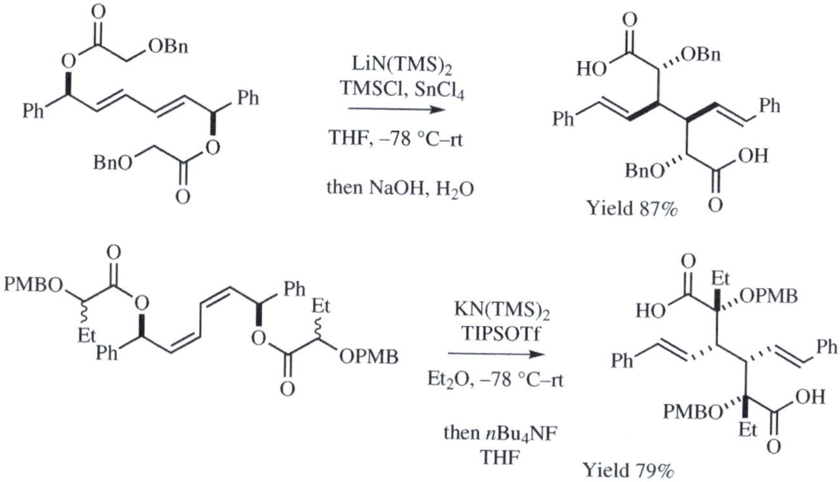

Scheme 13.31 Double Ireland–Claisen rearrangement reactions.

of choice for complex-molecule synthesis, where chemo- and stereoselectivity are vital. Shown in Scheme 13.31 is an example of the deployment of the Ireland–Claisen rearrangement within a two-directional chain extension strategy, whereby the two ends of a symmetrical substrate are elaborated simultaneously.[49] Formation of both products may be understood in terms of *Z*-configured ketene acetals rearranging via chair-like transition states, with the switch from *E*- to *Z*-allylic alkene geometry accounting for the differing topicity and the differing relative stereochemistry at the allylic positions of the two products shown. These impressive transformations underscore the power of the Ireland–Claisen rearrangement for the assembly of sterically congested structures with high levels of predictable stereoselectivity.

13.5 APPLICATIONS OF THE CLAISEN REARRANGEMENT REACTION IN TARGET-ORIENTED SYNTHESIS

The power of the Claisen rearrangement reaction in terms of ease of substrate assembly, functional group versatility, substrate tolerance, reactivity, and reliable stereoselectivity based on readily understood models and principles is such that over several decades it has emerged as one of the most important strategy-level transformations available to organic synthesis chemists. The following case studies highlight some of its key features.

13.5.1 Cananodine

Cananodine **23**[50] (Scheme 13.32) is a guaipyridine sesquiterpene alkaloid isolated from the perfumery tree *Cananga odorata*, commonly known as ylang ylang. Cananodine possesses a trisubstituted pyridine which is annulated to a seven-membered

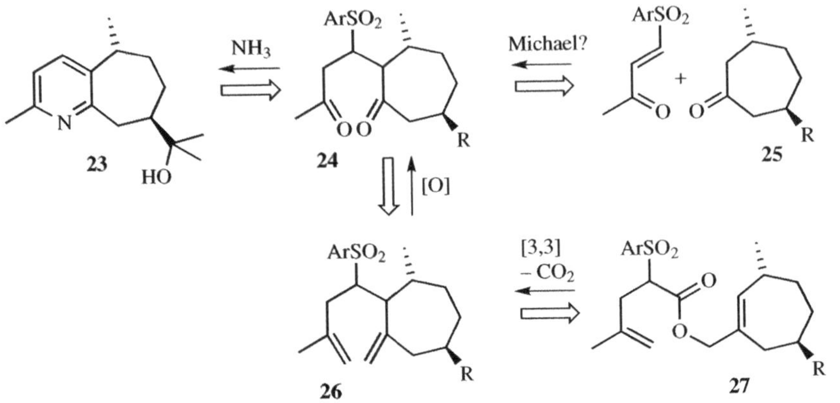

Scheme 13.32 Retrosynthetic analysis of cananodine.

ring and shows sub-micromolar activity against Hep G2 human hepatocarcinoma cell lines. During studies on pyridine synthesis via oxidative cleavage of 4-(arylsulfonyl)-1,6-heptadienes and ammonolysis of the resulting 1,5-dicarbonyl compounds,[35] Craig and coworkers recognized that diketone **24** could serve as a late-stage intermediate in the assembly of the natural product. While in principle **24** could be accessed by Michael addition of an enolate derived from cycloheptanone **25** to a 3-substituted 3-buten-2-one or a synthetic surrogate, the regiochemistry of alkylation of **25** would be problematic, because of the similar α- and α'-positions. In addressing this issue, it was recognized that a methylenecycloheptane such as **26** is a latent cycloheptanone and that control of the positioning of the required four-carbon side chain could be achieved using [3,3]-sigmatropic rearrangement chemistry provided that the ester **27** could be synthesized (Scheme 13.32). Substrate **27** would be subjected to dCr,[29] and the latent carbonyl groups revealed by oxidative cleavage of the methylene groups.

Scheme 13.33 shows the realization of this idea. dCr of **27** (Ar = Tol, R = CO_2Me) in the presence of BSA and substoichiometric KOAc under microwave conditions gave 1,6-diene **28**, which upon ozonolysis with triphenylphosphine work-up and treatment with ethanolic ammonia gave pyridine **29**. Finally, treatment with excess MeMgBr gave cananodine **23** (Scheme 13.33).[51]

From a strategic perspective, the synthesis of cananodine **23** described earlier highlights the utility and versatility of the Claisen rearrangement in that it may be tailored to deliver noncarbonyl-containing products by judicious incorporation of functionality in the substrates. Given that **28** may in principle be made by double dCr of a bis(allylic)tosylmalonic ester,[35] a wide range of annulated pyridines should be available using this approach (Scheme 13.34).

13.5.2 Basiliolides

The basiliolides are members of family of natural products isolated from plants belonging to the genus *Thapsia*. They are believed to inhibit calcium ATPases located

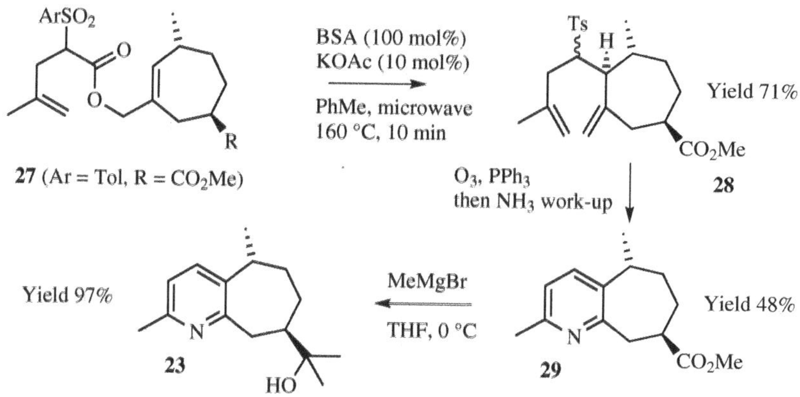

Scheme 13.33 Decarboxylative Claisen rearrangement in the synthesis of cananodine.

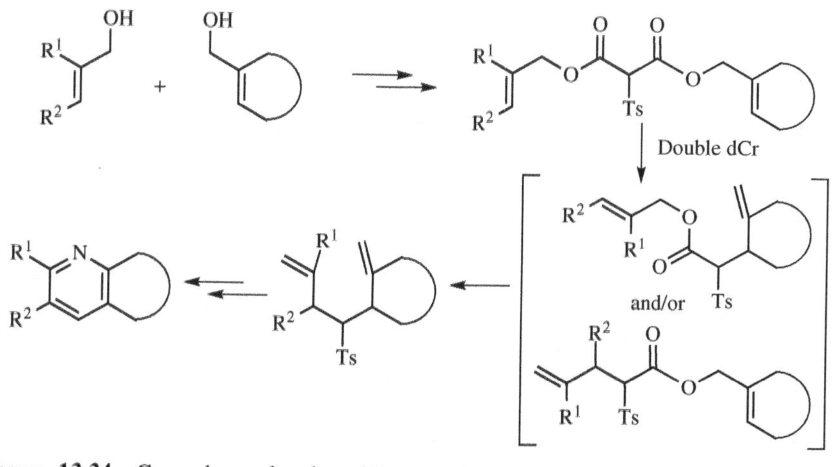

Scheme 13.34 General annulated pyridine synthetic strategy using the decarboxylative Claisen rearrangement.

within the sarco-/endoplasmic reticulum (SERCA-ATPases), which in turn induces rapid mobilization of intracellular Ca^{2+} stores. Studies carried out independently in the laboratories of Johannson[52] and Stoltz[53] demonstrated the effectiveness of the Ireland–Claisen rearrangement in bringing two mutually reactive functional groups into sufficiently close proximity to undergo a second pericyclic reaction following the [3,3]-sigmatropic rearrangement step. Thus, treatment of pyrone **30** (Scheme 13.35) with silylating agent and base followed by heating resulted in silyl ketene acetal formation, Claisen rearrangement, and intramolecular Diels–Alder reaction to give a diastereoisomeric mixture of the tricyclic products **31**, in which the configuration of the stereocenter α- to the acid controlled facial selectivity in the intramolecular cycloaddition. The epimers **31** were subsequently coupled with

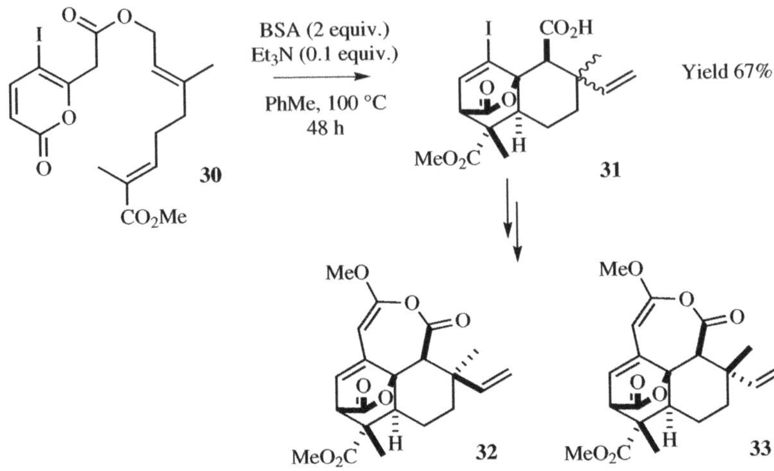

Scheme 13.35 Synthesis of basiliolides using a Claisen rearrangement–intramolecular Diels–Alder strategy.

a difunctionalized alkyne to provide the natural product basiliolide B **32** and its C8 epimer **33**.

13.5.3 Ene–Diynes in Bergman Cyclization Reactions

The Claisen rearrangement has also been used to generate substrates for Bergman cyclization (Scheme 13.36). Ketene acetals **35**, made from enolization–silylation of **34**, underwent Ireland–Claisen rearrangement and *in situ* Bergman cyclization of intermediates **36** in the presence of 1,4-cyclohexadiene as the source of hydrogen, yielding the epimeric tetrahydronaphthalenes **37** in good yield for this cascade process. The Claisen rearrangement step served to bring the ene–diyne termini into sufficiently close mutual proximity for cyclization to occur via a presumed 1,4-diyl species.[54]

13.5.4 Azadirachtin

Azadirachtin **38** (Scheme 13.37) is a tetranortriterpenoid first isolated in 1968[55] from the neem tree, *Azadirachta indica* A. Juss., a fast-growing species indigenous to subtropical regions.[56] While azadirachtin has pronounced antifeedant and growth-inhibiting properties toward over 200 insect species, it is nontoxic to higher organisms and mammalian species. From a synthesis perspective, it presents numerous challenges, perhaps most notably the formation of the exceptionally congested C8—C14 bond, which links two quaternary centers in an extremely encumbered steric environment. Ley and coworkers successfully overcame this significant synthetic hurdle through use of the propargylic Claisen rearrangement.

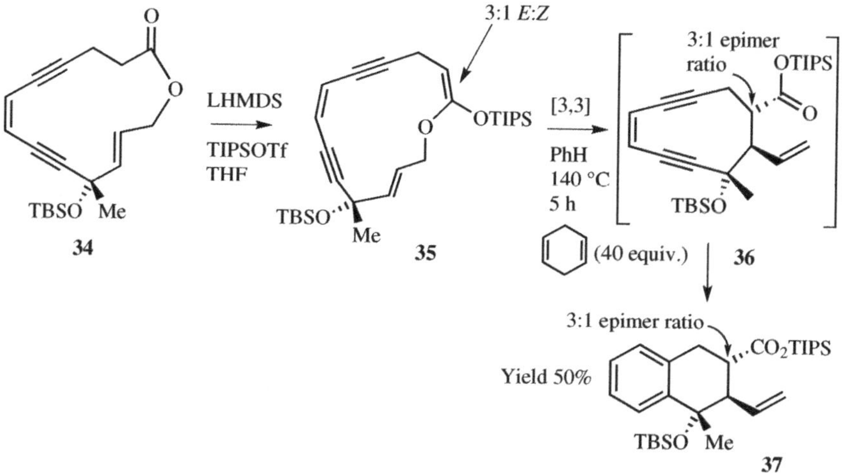

Scheme 13.36 Tandem Bergman cyclisation–Claisen rearrangement.

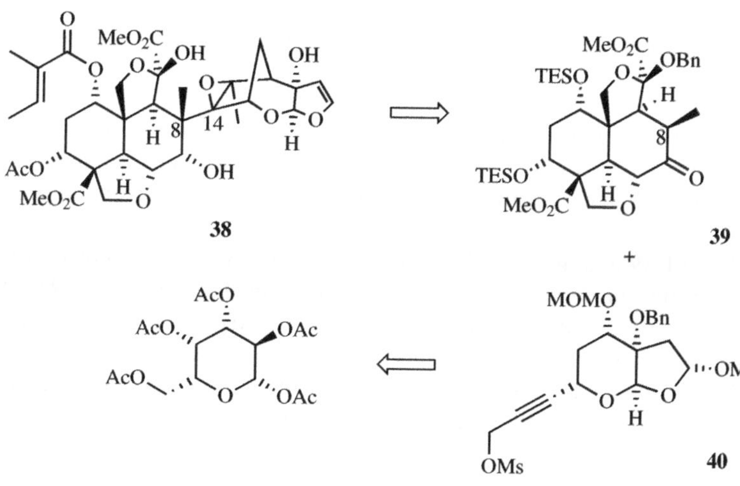

Scheme 13.37 Azadirachtin synthetic strategy.

Implicit in the Ley group strategy was the recognition that the extreme steric congestion at the C8–C14 linkage meant that an intramolecular transformation involving the least sterically hindered substrate possible would be essential for success in this step. Also key to their strategy was the incorporation of the requisite functionality for the late-stage elaboration of the epoxide and acetal functionality, and therefore, a propargylic Claisen rearrangement was selected for the crucial C8—C14 bond-forming step, with ketone **39** and propargylic

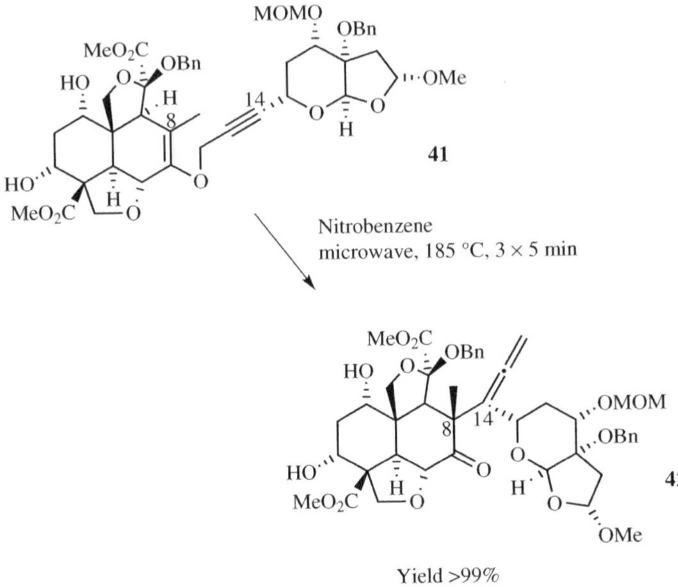

Scheme 13.38 Formation of the azadirachtin C8–C14 bond using the Claisen rearrangement.

mesylate **40** as the coupling partners (Scheme 13.37). Ketone **39** could be made via a sequence involving as key steps an intramolecular Diels–Alder reaction and intramolecular Michael addition, while **40** was synthesized from a D-galactose derivative.

Coupling of **39** and **40** using NaH–15-crown-5 followed by removal of the silyl protecting groups gave the propargylic Claisen rearrangement substrate **41**. This was subjected to microwave irradiation to give the rearranged product **42** in virtually quantitative yield (Scheme 13.38).[57] This success of this key transformation stems from its perfect design, in that C8 is bonded to C14 with the latter carbon atom as sterically unencumbered as possible while still bearing the functionality necessary for elaboration of the remaining carbocycle and epoxide.

13.6 CONCLUSIONS

During the century which has elapsed since its discovery, the Claisen rearrangement has evolved into one of the most powerful organic synthesis transformations known. The research reported through many thousands of publications has demonstrated the power of the reaction, which combines great efficiency, versatility, selectivity, functional group tolerance with ease of substrate synthesis, and high levels of predictability. As such, the Claisen rearrangement has a bright future: further advances may be anticipated in innovative methods for substrate synthesis, incorporation of the rearrangement within cascade processes, catalyzed variants including enantioselective reactions, multicomponent reactions, and total synthesis applications. All of these new discoveries will enhance still further the attractiveness and usefulness of this classic reaction.

REFERENCES

1. Claisen, L. *Chem. Ber.* **1912**, *45*, 3157–3166.
2. (a) For a seminal review, see: Castro, A. M. M. *Chem. Rev.* **2004**, *104*, 2939–3002. See also: (b) *The Claisen Rearrangement, Methods and Applications*; Hiersemann, M.; Nubbemeyer, U., Eds.; Wiley-VCH Verlag GmbH: Weinheim, **2007**. (c) Majumdar, K. C.; Nandi, R. K. *Tetrahedron* **2013**, *69*, 6921–6957. (d) Rehbein, J.; Hiersemann, M. *Synthesis* **2013**, *45*, 1121–1159.
3. For a recent review of the retro-Claisen rearrangement, see: Jukic, M.; Sterk, D.; Casar, Z. *Curr. Org. Synth.* **2012**, *9*, 488–512.
4. Burgstahler, A. W.; Nordin, I. C. *J. Am. Chem. Soc.* **1961**, *83*, 198–206.
5. Johnson, W. S.; Werthemann, L.; Bartlett, W. R.; Brockson, T. J.; Li, T.; Faulkner, D. J.; Petersen, M. R. *J. Am. Chem. Soc.* **1970**, *92*, 741–743.
6. (a) Wick, A. E.; Felix, D.; Steen, K.; Eschenmoser, A. *Helv. Chim. Acta* **1964**, *47*, 2425–2429. (b) Wick, A. E.; Felix, D.; Gschwend-Steen, K.; Eschenmoser, A. *Helv. Chim. Acta* **1969**, *52*, 1030–1042.
7. (a) Ireland, R. E.; Mueller, R. H. *J. Am. Chem. Soc.* **1972**, *94*, 5897–5898. (b) Ireland, R. E.; Mueller, R. H.; Willard, A. K. *J. Am. Chem. Soc.* **1976**, *98*, 2868–2877.
8. (a) Malherbe, R.; Belluš, D. *Helv. Chim. Acta* **1978**, *61*, 3096–3099; (b) Malherbe, R.; Rist, G.; Belluš, D. *J. Org. Chem.* **1983**, *48*, 860–869; (c) Nubbemeyer, U.; Oehrlein, R.; Gonda, J.; Ernst, B.; Belluš, D. *Angew. Chem., Int. Ed.* **1991**, *30*, 1465–1467; (d) Ernst, B.; Gonda, J.; Jeschke, R.; Nubbemeyer, U.; Oehrlein, R.; Belluš, D. *Helv. Chim. Acta* **1997**, *80*, 876–891.
9. For a review of the Belluš–Claisen rearrangement, see: Gonda, J. *Angew. Chem., Int. Ed.* **2004**, *43*, 3516–3524.
10. For a review of the propargylic Claisen rearrangement, see: Tejedor, D.; Mendez-Abt, G.; Cotos, L.; Garcia-Tellado, F. *Chem. Soc. Rev.* **2013**, *42*, 458–471. See also: Hashmi, A. S. K. In *Science of Synthesis*; Thieme: Stuttgart, **2007**; Vol. 44, pp. 287–299.
11. Alder, R. W.; Baker, R.; Brown, J. M. *Mechanism in Organic Chemistry*; John Wiley & Sons: London, **1971**, pp. 248–249.
12. For a review of iodolactonization reactions of non-conjugated unsaturated carboxylic acids, see: Laya, M. S.; Banerjee, A. K.; Cabrera, E. V. *Curr. Org. Chem.* **2009**, *13*, 720–730.
13. For a review of the mechanism of the Wacker oxidation, and leading references, see: Keith, J. A.; Henry, P. M. *Angew. Chem., Int. Ed.* **2009**, *48*, 9038–9049.
14. For a review of the hydrogen peroxide-mediated Baeyer–Villiger oxidation, and leading references, see: Uyanik, M.; Ishihara, K. *ACS Catal.* **2013**, *3*, 513–520.
15. Hart, D. J.; Yang, T. K. *J. Org. Chem.* **1985**, *50*, 235–242.
16. Lysenko, I. L.; Lee, H. G.; Cha, J. K. *Org. Lett.* **2009**, *11*, 3132–3134; see Footnote 9 in this paper.

17. For a review of the Curtius rearrangement, see: Rojas, C. M. In *Name Reactions for Homologations*; Li, J. J., Ed.; John Wiley & Sons, Inc.: Hoboken, NJ, **2009**; Pt. 2, pp. 136–163.
18. For a review of the Schmidt reaction, see: Wrobleski, A.; Coombs, T. C.; Huh, C. W.; Li, S.-W.; Aubé, J. *Org. React.* **2012**, *78*, 1–320.
19. For a review of the Beckmann rearrangement, and leading references, see: Kaur, N.; Sharma, P.; Kishore, D. *J. Chem. Pharm. Res.* **2012**, *4*, 1938–1946.
20. (a) Rehbein, J.; Leick, S.; Hiersemann, M. *J. Org. Chem.* **2009**, *74*, 1531–1540. (b) Troendlin, J.; Rehbein, J.; Hiersemann, M.; Trapp, O. *J. Am. Chem. Soc.* **2011**, *133*, 16444–16450.
21. Tellam, J. P.; Carbery, D. R. *J. Org. Chem.* **2010**, *75*, 7809–7821.
22. (a) Cha, J. K.; Lewis, S. C. *Tetrahedron Lett.* **1984**, *25*, 5263–5266. (b) Suzuki, T.; Sato, E.; Kamada, S.; Tada, H.; Unno, K.; Kametani, T. *J. Chem. Soc., Perkin Trans. 1* **1986**, 387–391. (c) Takano, S.; Seijo, K.; Hatakeyama, S. *Tetrahedron Lett.* **1985**, *26*, 865–868. (d) Mulzer, J.; Shanyoor, M. *Tetrahedron Lett.* **1993**, *34*, 6545–6548. (e) Hauske, J. R.; Julin, S. M. *Tetrahedron Lett.* **1993**, *34*, 4909–4912. (f) Fleming, I.; Betson, M. S. *Org. Biomol. Chem.* **2003**, *1*, 4005–4016. (g) Ellwood, A. R.; Mortimer, A. J. P.; Tocher, D. A.; Porter, M. J., *Synlett* **2008**, 2199–2203. (h) Craig, D.; Harvey, J. W.; O'Brien, A. G.; White, A. J. P. *Chem. Commun.* **2010**, *46*, 6932–6934.
23. Craig, D.; Harvey, J. W.; O'Brien, A. G.; White, A. J. P. *Org. Biomol. Chem.* **2011**, *9*, 7057–7061.
24. Padwa, A.; Sá, M. M. *Tetrahedron Lett.* **1997**, *38*, 5087–5090. For an example of rearrangement via an ionic transition state, see: Closs, G. L.; Harrison, A. M. *J. Org. Chem.*, **1972**, *37*, 1051–1052. See also: (b) Trost, B. M.; Pulley, S. R. *Tetrahedron Lett.* **1995**, *36*, 8737–8740; (c) Klepper, F.; Jahn, E.-M.; Hickmann, V.; Carell, T. *Angew. Chem., Int. Ed.* **2007**, *46*, 2325–2327; (d) Guo, H.; O'Doherty, G. A. *Org. Lett.* **2006**, *8*, 1609–1612; (e) Lauzon, S.; Tremblay, F.; Gagnon, D.; Godbout, C.; Chabot, C.; Mercier-Shanks, C.; Perreault, S.; DeSève, H.; Spino, C. *J. Org. Chem.* **2008**, *73*, 6239–6250.
25. Bouchard, H.; Renard, P. Y.; Lallemand, J. Y. *Tetrahedron Lett.* **1991**, *32*, 5953–5956.
26. Chen, C.-L.; Namba, K.; Kishi, Y. *Org. Lett.* **2009**, *11*, 409–412.
27. Akahori, Y.; Yamakoshi, H.; Sawayama, Y.; Hashimoto, S.; Nakamura, S. *J. Org. Chem.* **2014**, *79*, 720–735.
28. Nonaka, H.; Ogawa, N.; Maeda, N.; Wang, Y.-G.; Kobayashi, Y. *Org. Biomol. Chem.* **2010**, *8*, 5212–5223.
29. Bourgeois, D.; Craig, D.; Grellepois, F.; Mountford, D. M.; Stewart, A. J. W. *Tetrahedron* **2006**, *62*, 483–495.
30. Carman, L.; Kwart, L. D.; Hudlicky, T. *Synth. Commun.* **1986**, *16*, 169–182.
31. White, W. N.; Fife, W. K. *J. Am. Chem. Soc.* **1961**, *83*, 3846–3853.
32. Craig, D.; Slavov, N. K., *Chem. Commun.* **2008**, 6054–6056.
33. For experimental and theoretical studies of this effect, see: (a) Curran, D. P.; *Tetrahedron Lett.* **1982**, *23*, 4309–4312. (b) Curran D. P.; Suh, Y. *Tetrahedron Lett.* **1984**, *25*, 4179–4182. (c) Curran D. P.; Suh, Y. *J. Am. Chem. Soc.* **1984**, *106*, 5002–5004. (d) Curran D. P.; Suh, Y. *Carbohydr. Res.* **1987**, *171*, 161–191. (e) Coates, R. M.; Rogers, B. D.; Hobbs, S. J.; Curran, D. P.; Peck, D. R. *J. Am. Chem. Soc.* **1987**, *109*, 1160–1170. (f)

Curran, D. P.; Kuo, L. H. *Tetrahedron Lett.* **1995**, *36*, 6647–6650. (g) Yoo, H. Y.; Houk, K. N. *J. Am. Chem. Soc.* **1997**, *119*, 2877–2884. (h) Aviyente, V.; Houk, K. N. *J. Phys. Chem. A* **2001**, *105*, 383–391. (i) Aviyente, V.; Yoo, H. Y.; Houk, K. N. *J. Org. Chem.* **1997**, *62*, 6121–6128. (j) Denmark, S. E.; Dappen, M. S. *J. Org. Chem.* **1984**, *49*, 798–806, and references therein. (k) McMichael, K. D.; Korver, G. L. *J. Am. Chem. Soc.* **1979**, *101*, 2746–2747. (l) Gajewski, J. J.; Conrad, N. D. *J. Am. Chem. Soc.* **1979**, *101*, 2747–2748. (m) Gajewski, J. J.; Emrani, J. *J. Am. Chem. Soc.* **1984**, *106*, 5733–5734. (n) Gajewski, J. J. *Acc. Chem. Res.* **1997**, *30*, 219–225. (o) Meyer, M. P.; DelMonte, A. J.; Singleton, D. A. *J. Am. Chem. Soc.* **1999**, *121*, 10865–10874.

34. Partridge, A. S., PhD thesis, Imperial College London, **2012**.
35. Craig, D.; Paina, F.; Smith, S. C. *Chem. Commun.* **2008**, 3408–3410.
36. Ichikawa, H.; Maruoka, K. In *The Claisen Rearrangement, Methods and Applications*; Hiersemann, M.; Nubbemeyer, U., Eds.; Wiley-VCH Verlag GmbH: Weinheim, **2007**, pp. 45–85.
37. McFarland, C. M.; McIntosh, M. C. In *The Claisen Rearrangement, Methods and Applications*; Hiersemann, M.; Nubbemeyer, U., Eds.; Wiley-VCH Verlag GmbH: Weinheim, **2007**, pp. 117–210.
38. Langlois, Y. In *The Claisen Rearrangement, Methods and Applications*; Hiersemann, M.; Nubbemeyer, U., Eds.; Wiley-VCH Verlag GmbH: Weinheim, **2007**, pp. 301–366.
39. Gradl, S.; Trauner, D. In *The Claisen Rearrangement, Methods and Applications*; Hiersemann, M.; Nubbemeyer, U., Eds.; Wiley-VCH Verlag GmbH: Weinheim, **2007**, pp. 367–396.
40. Pazos, G.; Pérez, M.; Gándara, Z.; Gómez, G.; Fall, Y. *Tetrahedron* **2012**, *68*, 8994–9003.
41. Paquette, L. A.; Friedrich, D.; Rogers, R. D. *J. Org. Chem.* **1991**, *56*, 3841–3849.
42. Zhang, Y.-D.; Ren, W.-W.; Lan, L.; Xiao, Q.; Wang, K.; Xu, J.; Chen, J.-H.; Yang, Z. *Org. Lett.* **2008**, *10*, 665–668.
43. Stevens, B. D.; Bungard, C. J.; Nelson, S. G. *J. Org. Chem.* **2006**, *71*, 6397–6402.
44. Nakayama, A.; Kogure, N.; Kitajima, M.; Takayama, H. *Angew. Chem., Int. Ed.* **2011**, *50*, 8025–8028. See also: Palanichamy, K.; Subrahmanyam, A. V.; Kaliappan, K. P. *Org. Biomol. Chem.* **2011**, *9*, 7877–7886.
45. Uyeda, C.; Jacobsen, E. N. *J. Am. Chem. Soc.* **2008**, *130*, 9228–9229.
46. Uyeda, C.; Rötheli, A. R.; Jacobsen, E. N. *Angew. Chem., Int. Ed.* **2010**, *49*, 9753–9756.
47. Yip, K.-T.; Yang, D. *Org. Lett.* **2011**, *13*, 2134–2137.
48. Ichiki, M.; Tanimoto, H.; Miwa, S.; Saito, R.; Sato, T.; Chida, C. *Chem. Eur. J.* **2013**, *19*, 264–269.
49. Feldman, K. S.; Selfridge, B. R. *J. Org. Chem.* **2013**, *78*, 4499–4511.
50. Hsieh, T.-J.; Chang, F.-R.; Chia, Y.-C.; Chen, C.-Y.; Chiu, H.-F. Wu, Y.-C. *J. Nat. Prod.* **2001**, *64*, 616–619.
51. Craig, D.; Henry, G. D. *Eur. J. Org. Chem.* **2006**, 3558–3561.
52. Larsson, R.; Sterner, O.; Johansson, M. *Org. Lett.* **2009**, *11*, 657–660.
53. (a) Nelson, H. M.; Stoltz, B. M. *Tetrahedron Lett.* **2009**, *50*, 1699–1701. (b) Nelson, H. M.; Murakami, K.; Virgil, S. C.; Stoltz, B. M. *Angew. Chem., Int. Ed.* **2011**, *50*, 3688–3691.
54. Magriotis, P. A.; Kim, K. D. *J. Am. Chem. Soc.* **1993**, *115*, 2972–2973.
55. Butterworth, J. H.; Morgan, E. D. *Chem. Commun.* **1968**, 23–24.

56. Mordue, A. J.; Blackwell, A. *J. Insect Physiol.* **1993**, *39*, 903–1000.

57. Ley, S. V.; Abad-Somovilla, A.; Anderson, J. C.; Ayats, C.; Bänteli, R.; Beckmann, E.; Boyer, A.; Brasca, M. G.; Brice, A.; Broughton, H. B.; Burke, B. J.; Cleator, E.; Craig, D.; Denholm, A. J.; Denton, R. M.; Durand-Reville, T.; Gobbi, L. B.; Gçbel, M.; Gray, B. L.; Grossman, R. B.; Gutteridge, C. E.; Hahn, N.; Harding, S. L.; Jennens, D. C.; Jennens, L.; Lovell, P. J.; Lovell, H. J.; de la Puente, M. L.; Kolb, H. C.; Koot, W.-J.; Maslen, S. L.; McCusker, C. F.; Mattes, A.; Pape, A. R.; Pinto, A.; Santafianos, D.; Scott, J. S.; Smith, S. C.; Somers, A. Q.; Spilling, C. D.; Stelzer, F.; Toogood, P. L.; Turner, R. M.; Veitch, G. E.; Wood, A.; Zumbrunn, C. *Chem. Eur. J.* **2008**, *14*, 10683–10704.

14

[3,3]-SIGMATROPIC REARRANGEMENTS WITH HETEROATOM–HETEROATOM BONDS

LAURA L. ANDERSON
Department of Chemistry, University of Illinois at Chicago, Chicago, IL, USA

14.1 INTRODUCTION

[3,3]-Sigmatropic rearrangements that include heteroatom–heteroatom bonds in the 6-atom fragment are a unique class of transformations that are often used in cascade processes for the preparation of heterocycles. These transformations are usually thermodynamically favorable since they result in the cleavage of a weak N—N or N—O bond and the formation of a new C—C bond. The Fischer-indole reaction is a well-known method for heterocycle synthesis that uses a [3,3]-rearrangement of a hydrazine fragment as the key C—C bond-forming step (Scheme 14.1a and b).[1] A classic example of the use of the Fischer-indole reaction in Woodward's synthesis of strychnine is illustrated in Scheme 14.1c.[2] A recent example of an interrupted Fischer-indole reaction which also utilizes an analogous [3,3]-rearrangement followed by intramolecular nucleophilic trapping to form an indoline is described in Scheme 14.1d.[3] Due to the availability of extensive reviews on the synthetic applications of the Fischer-indole reaction,[1] this transformation will not be covered in this chapter; however, a discussion of the Fischer-indole synthesis will be used as an introduction to the reaction characteristics that will be considered when assessing related methods that exploit similar heteroatom–heteroatom [3,3]-sigmatropic rearrangements.

There are several key aspects of the Fischer-indole reaction to consider when evaluating potential synthetic applications: (1) the ease of preparing the

Molecular Rearrangements in Organic Synthesis, First Edition. Edited by Christian M. Rojas.

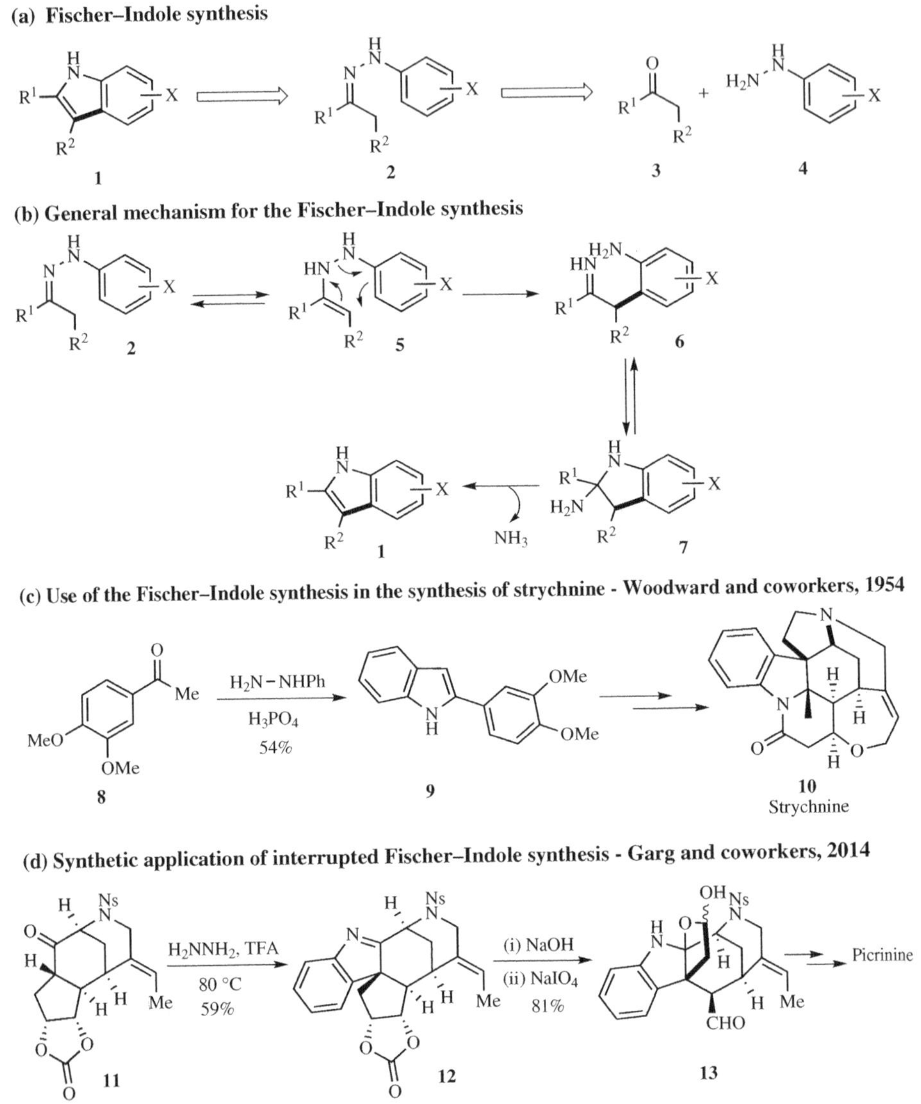

Scheme 14.1 Examples of Fischer-indole and interrupted Fischer-indole synthesis.

N-arylhydrazone starting materials; (2) the energetic cost and regioselectivity associated with tautomerization of the hydrazone to form the rearrangement precursor; (3) the regioselectivity of the C—C bond-forming event that occurs during the [3,3]-rearrangement with respect to both the enamine and the arene functionalities; and (4) the reactivity of the initial rearrangement product that leads to heterocycle formation. The simplicity of condensation reactions that form hydrazones from aldehydes or ketones and *N*-arylhydrazines facilitates the use of the Fischer-indole reaction in synthetic processes. More challenging condensations with electron-rich

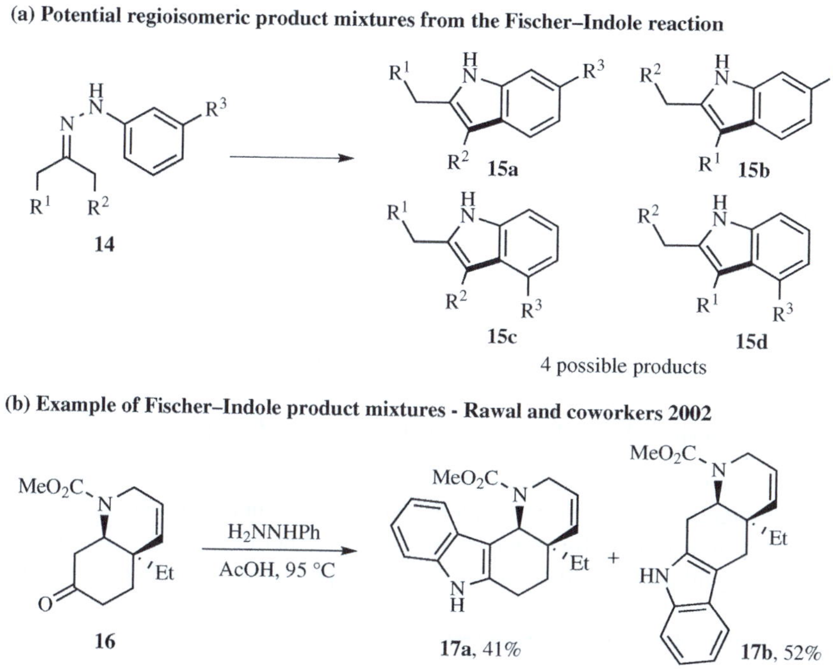

Scheme 14.2 Regioisomeric product mixtures of Fischer-indole reactions.

carbonyl compounds can be promoted with Brønsted or Lewis acids. The tautomerization of hydrazones to their enamine form is energetically uphill but can also be promoted by acid additives.[4] The regioselectivity of the tautomerization event is important for the avoidance of product mixtures, which is why most applications of the Fischer-indole reaction use hydrazones with only one enolizable position (Scheme 14.2).[5] The substitution pattern of the *N*-aryl functionality also needs to be considered in reaction design since unsymmetrical examples can also lead to product mixtures during the C—C bond-forming event.[6] The [3,3]-rearrangement of the cascade process is the rate-determining step for hydrazines such as **2** that give N—H indoles such as **1** (Scheme 14.1b).[7] The energy required to reach the transition state for these transformations is sufficient to promote the subsequent cyclization and elimination to form the corresponding heterocycle. Only in situations where elimination is prevented by specific substitution patterns or when the imine is protected with an electron-withdrawing group can the subsequent cyclization be prevented or intercepted.[8] All four of these reaction attributes need to be evaluated when considering whether or not to use a Fischer-indole synthesis in a synthetic scheme. Due to similar mechanistic characteristics, analogous considerations are also important for the applications of other heteroatom–heteroatom [3,3]-sigmatropic rearrangements.

This chapter is organized into three major sections: (1) the [3,3]-rearrangements of N—O bond fragments; (2) the [3,3]-rearrangements of N—N bond fragments; and (3) the [3,3]-rearrangements of N—N bond fragments that result in the elimination of

N_2. The descriptions of the transformations in each section will include (1) options for starting material preparation; (2) a discussion of the regio- and stereochemical issues involved in forming new C—C and C—O bonds during the [3,3]-rearrangement processes; (3) an explanation of the potential cascade reactivity of the initial rearrangement products; as well as (4) applications in synthesis. This presentation is intended to give the reader a better overall picture of where these powerful rearrangements can be applied and to provide an overview of how to incorporate these transformations in retrosynthetic analyses. While older literature will be mentioned where appropriate, the chapter will focus on new transformations, advances, and applications.

14.2 [3,3]-SIGMATROPIC REARRANGEMENTS OF N—O BONDS

14.2.1 Synthesis of Benzofurans from *O*-Aryloxime Ethers

The preparation of benzofurans from *O*-aryloxime ethers occurs through a mechanism analogous to the Fischer-indole synthesis (Scheme 14.3a and b). This transformation was initially reported in 1966 by Sheradsky and was applied to the formal synthesis of aflatoxin B2 by Rapoport and Castellino in 1986.[9,10]

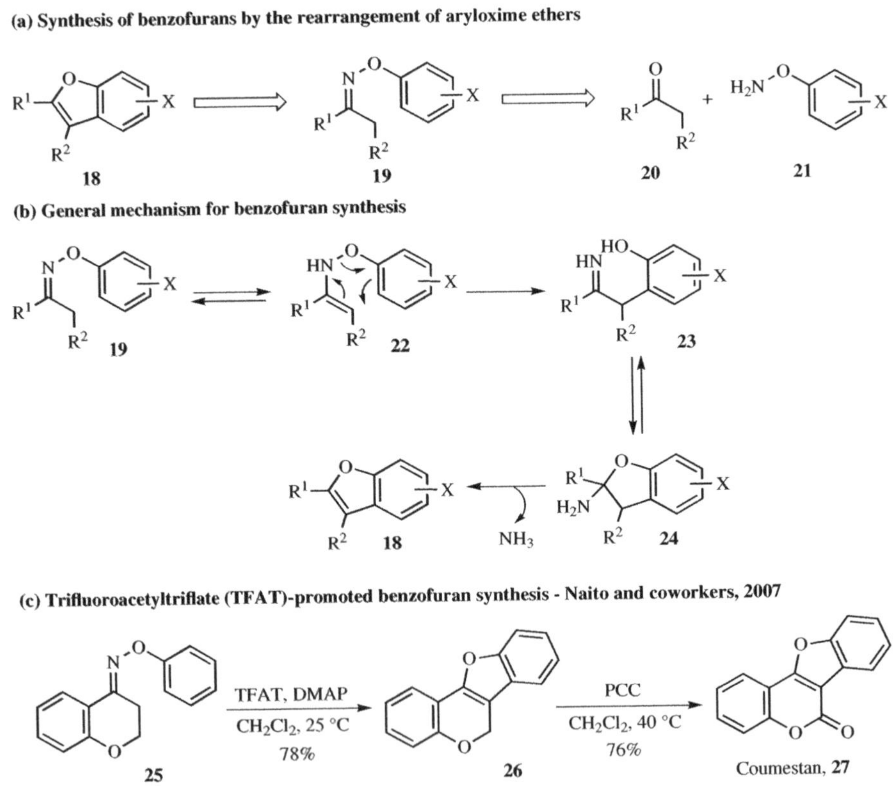

Scheme 14.3 Synthesis of benzofurans by [3,3]-rearrangements of aryl oxime ethers.

While this [3,3]-rearrangement and condensation process has similar limitations to the Fischer-indole synthesis with respect to the regioselectivity of C—C bond formation, it is equally effective for aryloxime ethers with only one enolizable position and symmetrically substituted *O*-aryl groups.[11] Traditional benzofuran syntheses using rearrangements of aryloxime ethers require the use of harsh, acidic, and high-temperature reaction conditions.[11] Consistent with the similarity of the benzofuran synthesis to the Fischer-indole reaction, the initial condensation, [3,3]-rearrangement, and dehydration steps can all be promoted in the presence of an acidic reagent or catalyst. Recent improvements to increase the scope and efficiency of these transformations have included the use of trifluoroacetyltriflate and methanesulfonic acid to promote the [3,3]-rearrangement under milder conditions and lower temperatures with greater functional group compatibility (Scheme 14.3c).[12,13] Further notable advances to the generality of these transformations involve the development of new methods for the preparation of aryloxyamines.

Aryloxime ethers can be prepared by the condensation of aryloxyamines with a ketone or an aldehyde; however, traditional methods for the preparation of aryloxyamines are mostly limited to nucleophilic aromatic substitution processes with *N*-hydroxyphthalimide followed by deprotection through hydrazination.[14] The scope of available aryloxyamines was significantly expanded by Sharpless and Kelly who determined conditions for the preparation of *N*-aryloxyphthalimides through a Chan–Lam–Evans coupling of *N*-hydroxyphthalimide and arylboronic acids (Scheme 14.4a).[15] A method for the direct arylation of oximes through an Ullmann coupling has also been reported by Wailes and coworkers and is effective

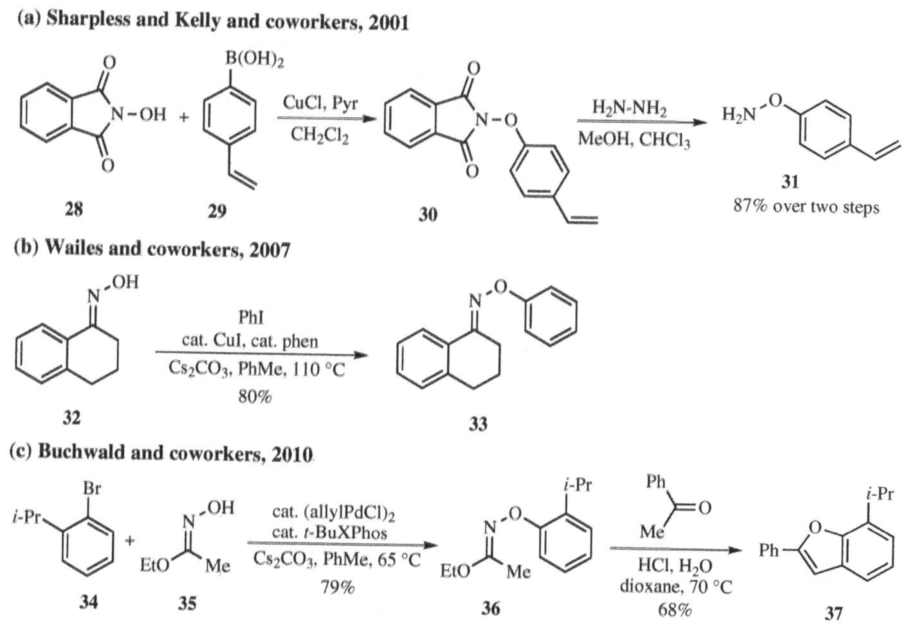

Scheme 14.4 New routes to aryloxyamines and aryloxime ethers that expand the scope of benzofuran synthesis.

for a range of acetophenone- and cyclohexanone-based oximes as well as meta- and para-substituted aryl halides (Scheme 14.4b).[16] Alternatively, the Buchwald group has shown that a Pd-catalyzed arylation method is effective for the preparation of oximes such as **36** which can also function as aryloxyamine transfer reagents (Scheme 14.4c).[17] Although this method provides a less direct route than the Ullmann coupling, the palladium-catalyzed transformation allows for the preparation of benzofurans such as **37**. These improvements in aryloxyamine synthesis have significantly increased the general applicability of the [3,3]-rearrangements of aryloxime ethers as tools for the preparation of benzofurans and will facilitate their use in new synthetic strategies toward the preparation of complex molecules containing these important fragments.

14.2.2 Rearrangements of *O*-Vinyl Oximes to Pyrroles

[3,3]-Rearrangements of vinyloxime ethers have been used for the synthesis of pyrroles. The first example of this transformation was reported by Sheradsky in 1970 but has been extensively explored by Trofimov and coworkers.[18,19] The Trofimov reaction is a cascade process that involves the *in situ* generation of a vinyloxime ether by the addition of an oxime to acetylene under superbasic conditions, followed by a [3,3]-rearrangement and Paal–Knorr cyclization and elimination to form the corresponding pyrrole (Scheme 14.5a and b). Due to the reaction conditions, these transformations often form the corresponding *N*-vinylated heterocycles or mixtures of the N—H and *N*-vinyl products. Consistent with its similarity to the Fischer-indole synthesis, the Trofimov reaction also provides mixtures of products for oximes with two enolizable positions. Two unique limitations of the Trofimov reaction are the harsh reaction conditions as well as the lack of regioselectivity and decreased efficiency observed for substituted alkynes (Scheme 14.5c).[20] Camp and coworkers have recently published two reports where the addition of an amine catalyst facilitates the addition of oximes to activated alkynes under much milder conditions and with high regioselectivity for propiolates (Scheme 14.5d). The same group has also determined appropriate microwave conditions for the rearrangement of these vinyloxime ethers to pyrroles as well as a gold-catalyzed procedure that promotes heterocycle formation at 60 °C.[21,22] These contributions provide access to highly substituted pyrroles with reactive functional groups that are not accessible under the traditional Trofimov reaction conditions.

As an alternative entry into [3,3]-rearrangements of vinyloxime ethers that avoids the challenges and limitations associated with the addition of oximes to alkynes, Anderson and coworkers have recently shown that *O*-propenyl oximes can be prepared by an iridium-catalyzed isomerization of allyloxime ethers and are stable to isolation (Scheme 14.6a).[23] Examination of the reactivity of these compounds showed that *O*-propenyl oximes **54** undergo [3,3]-rearrangements to give 4-methyl pyrroles **52** if tautomerization is facilitated by a cyano group at the α-position of the oxime and 5-methyl pyrroles **53** if tautomerization is disfavored with respect to a competing [1,3]-rearrangement (Scheme 14.6b). Identification of these mechanistic trends inspired the development of alternative rearrangement conditions in the presence of a

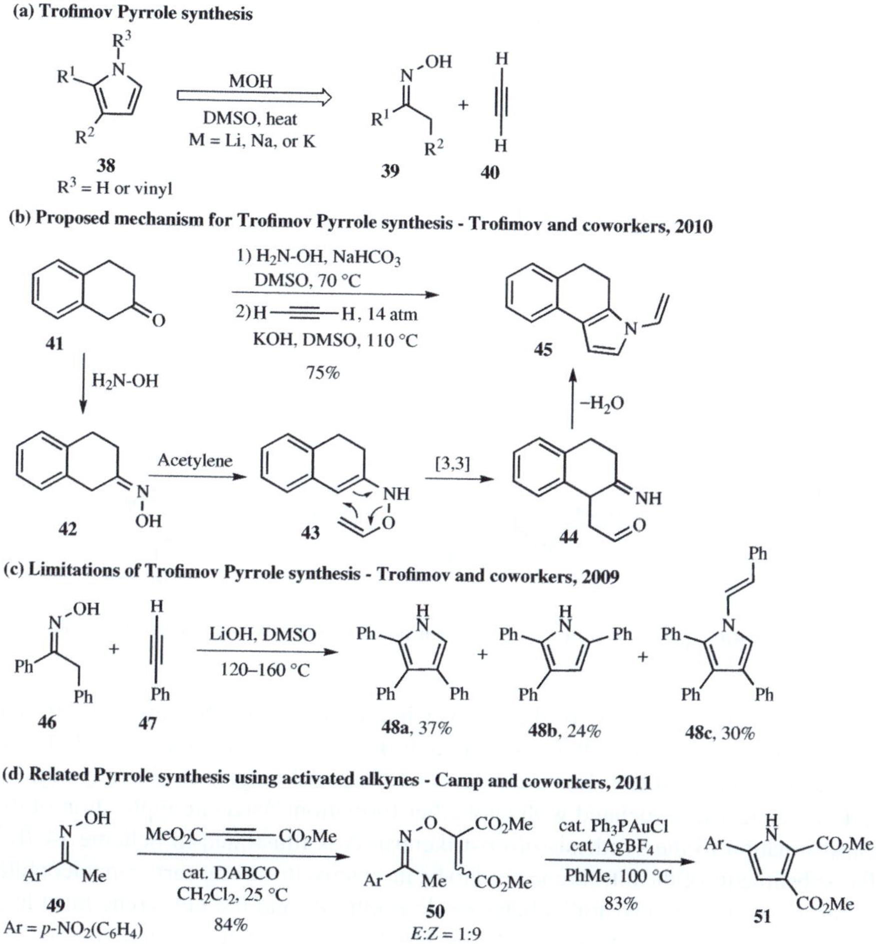

Scheme 14.5 The Trofimov reaction and recent advances in substrate scope.

basic additive that could reverse the substrate selectivity for the [1,3]-rearrangement and provide the corresponding 4-methyl pyrrole for both β-ketoester- and α-arylated ketone-derived oximes. These transformations allow for the selective conversion of ketones to two distinct pyrrole regioisomers by a simple isomerization reaction to access the necessary rearrangement precursors and by controlling the tautomerization equilibrium of these reactive intermediates.

14.2.3 Rearrangements of *O*-Vinyl-*N*-Arylhydroxylamines to Indoles, 2-Amino-2′-Hydroxy-1,1′Biaryls, and α-Arylated Carbonyl Compounds

The Bartoli-indole synthesis utilizes the [3,3]-rearrangement of an *O*-vinyl-*N*-arylhydroxylamine intermediate to form indoles through a similar rearrangement

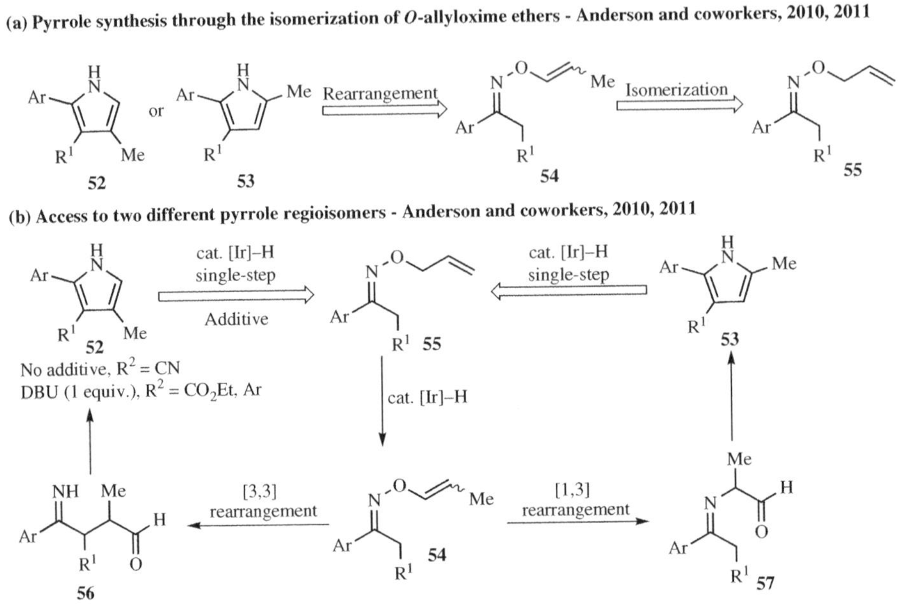

Scheme 14.6 pK_a-controlled pathways for the regioselective synthesis of pyrroles.

and condensation process as the Fischer-indole synthesis (Scheme 14.7a). The rearrangement precursor **59** is generated from the addition of a vinyl Grignard reagent to a nitroarene.[24] The use of a vinyl Grignard reagent avoids any regioselectivity concerns associated with enol ether formation. A recent application of this transformation to the synthesis of *cis*-trikentrin A is illustrated in Scheme 14.7b.[25] The substituents of the nitroarene need to be insensitive to strong carbon nucleophiles to participate effectively in this heterocycle synthesis, and the nitroarene must have a substituent at the 2-position. Dobbs *et al.* were able to expand the method beyond just the preparation of 7-substituted indoles by developing conditions to remove a 7-bromo-substituent after heterocycle formation (Scheme 14.7c).[26] Recent modifications of the Bartoli-indole synthesis include an expansion of the method to include the preparation of 1,1′-biaryl compounds as well as alternative methods of accessing the *O*-vinyl-*N*-arylhydroxylamine rearrangement precursors to be able to target the isolation and synthetic utility of intermediates along the cascade pathway.

Kürti and coworkers have reported that when *ortho*-halonitroarenes such as **71** are treated with aryl Grignard reagents **70**, the resulting *N*,*O*-diarylhydroxylamine addition products such as **72** undergo a similar [3,3]-rearrangement to the C—C bond-forming step of the Bartoli-indole synthesis to give 2-amino-2′-hydroxy-1,1′biaryls such as **73**.[27] The advantage of this new transformation is that it provides access to 2-amino-2′-hydroxy-1,1′biaryls under simple, transition metal-free conditions. Alternative methods for forming these compounds usually involve multistep procedures.[28] The corresponding [5,5]-rearrangement product **74** was also observed with the same reaction mixture but can be suppressed by raising the reaction

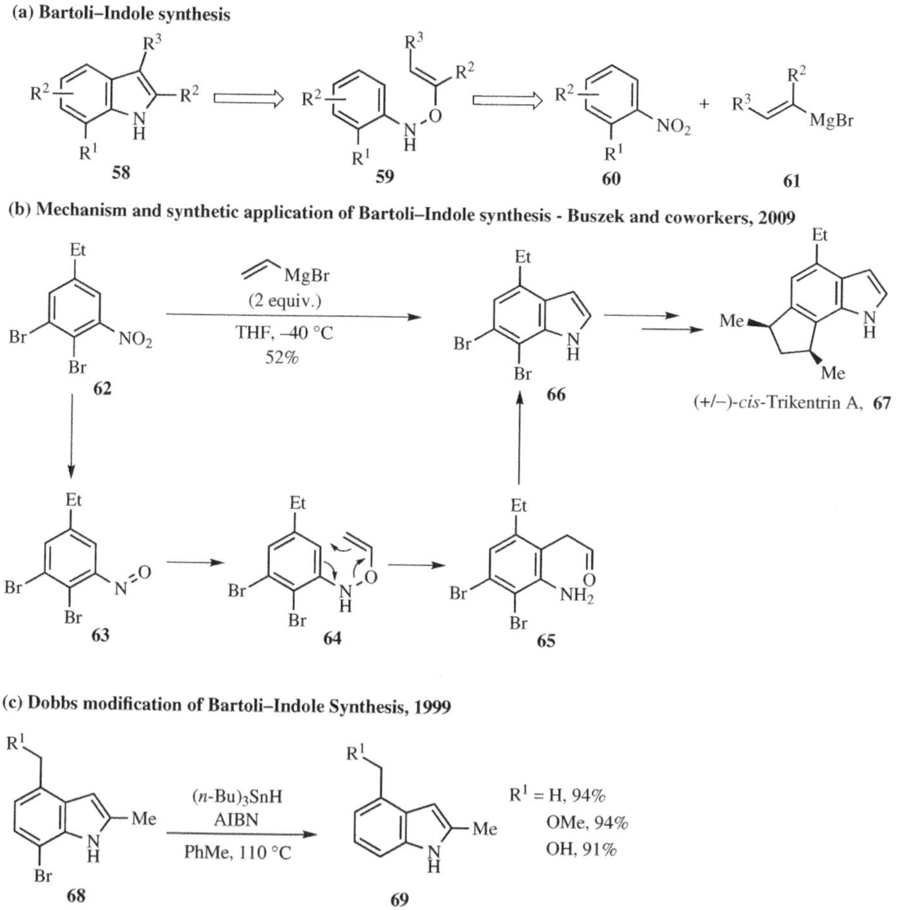

Scheme 14.7 Mechanism and scope of Bartoli-indole synthesis.

temperature from −25 to 0 °C. In analogy to the Bartoli-indole synthesis, all substrates require an ortho-halosubstituent to exhibit the desired reactivity; however, the versatility of this functional group for chemical library synthesis was demonstrated for alkyne annulation, heterocycle synthesis, diazonium formation and displacement, as well as cross coupling. Although the yields of this 2-amino-2′-hydroxy-1,1′biaryl synthesis are moderate, the new method provides an exceptionally efficient route to these compounds from simple nitroarene starting materials. (Scheme 14.8).

Rearrangements related to the Bartoli-indole synthesis have also been designed for the preparation of ortho-alkylated anilides and oxindoles from *O*-acetyl-*N*-arylhydroxylamines. The starting materials for these transformations are readily prepared by the protection of *N*-arylhydroxamic acids or *N*-arylhydroxylamines with substituted acetyl chlorides (Scheme 14.9a). The key C—C bond-forming [3,3]-sigmatropic rearrangement is initiated by deprotonation of the substituted

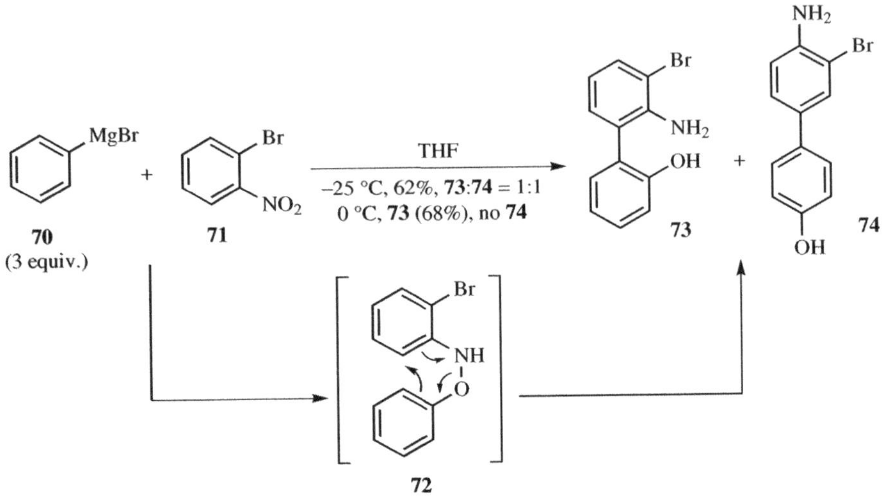

Scheme 14.8 Modified Bartoli-indole synthesis for the preparation of 2-amino-2'-hydroxy-1,1-Biaryls.[27]

acetyl group to form a ketene acetal such as **76**. Subsequent intramolecular cyclization of the initially formed α-anilido carboxylic acid can provide the corresponding oxindole **75**. The use of symmetrical substitution patterns at the 3- and 5-positions of the arene is preferred to avoid regioisomeric product mixtures. Early examples of this transformation were reported by the research groups of both Prabhakar and Endo (Scheme 14.9b).[29,30] Recently, more challenging examples have been optimized for the preparation of spirocyclic oxindoles such as **84**, and similar intermediates have been accessed by the addition of chiral nitrones to ketenes for the synthesis of enantioenriched oxindoles such as **87** (Scheme 14.9c and d).[31,32]

A Chan–Lam–Evans coupling has also been designed as a new method to access *O*-vinyl-*N*-arylhydroxylamines under mild conditions that avoid the use of strong bases as described previously for the Bartoli-indole synthesis and related preparations of oxindoles (see Section 14.2.3). Anderson and coworkers have shown that treatment of *N*-arylhydroxamic acids with vinylboronic acids leads directly to the synthesis of α-(*o*-anilido)ketones such as **90** (Scheme 14.10).[33] These transformations likely proceed through the formation of *O*-vinyl-*N*-arylhydroxamates such as **89** followed by a rapid [3,3]-rearrangement. The use of vinylboronic acid reagents selectively affords single enol ether isomers due to the conservation of stereochemistry in the cross-coupling process[34]; however, 3-substituted aryl groups do lead to regioisomeric mixtures of products. Surprisingly, the conditions required to promote this C—O bond coupling and rearrangement process are mild enough that further cyclization to form the corresponding indole is not observed, and these interrupted Fischer-indole intermediates can be isolated and functionalized. This modular method for the preparation of α-(*o*-anilido)ketones provides opportunities for the consideration of *N*-arylhydroxylamines and vinyl boronic acids as precursors to α-arylated ketones that are challenging to access by transition metal-catalyzed α-arylation procedures.[35]

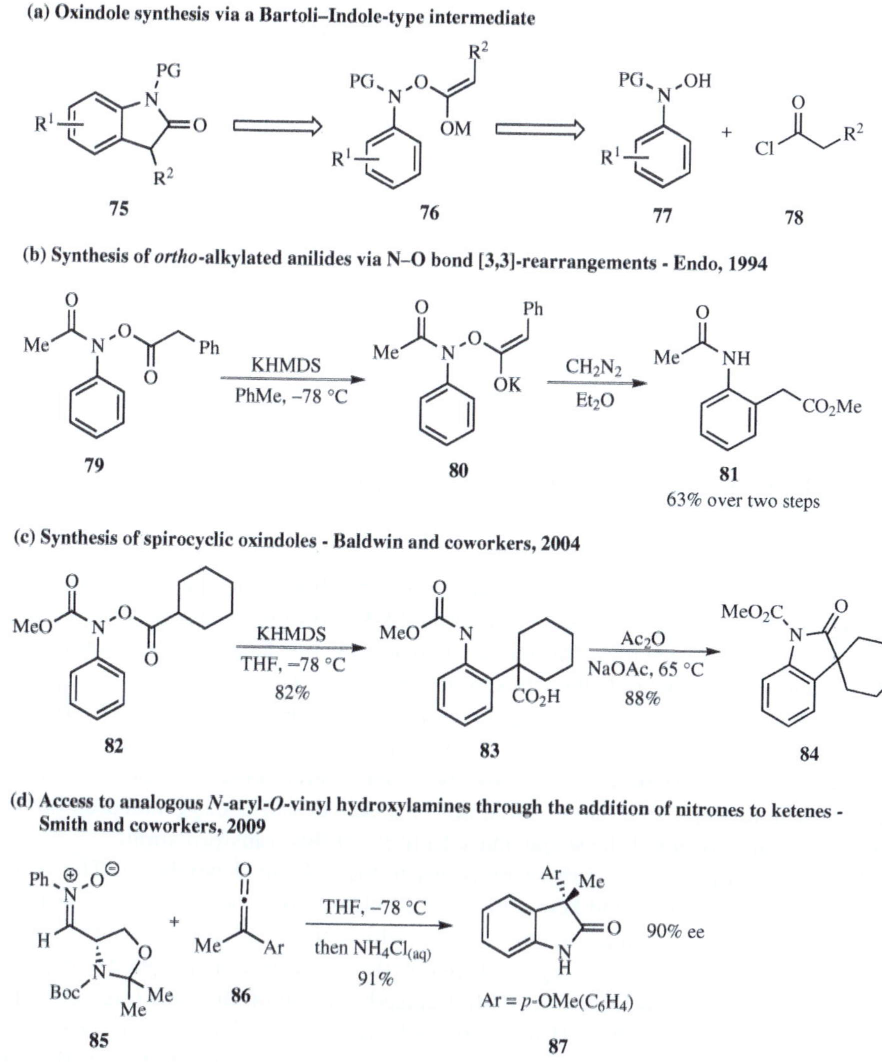

Scheme 14.9 Oxindole synthesis via the [3,3]-rearrangement of *O*-acetyl-*N*-arylhydroxymines.

14.2.4 α-Oxygenation Using [3,3]-Rearrangements of *O*-Protected Oximes and *O*-Vinyl Hydroxamates

A common characteristic of the methods described earlier is the use of N—O bond rearrangements for the formation of new C—C bonds in heterocycle-forming cascades. While C—C bond formation has been the primary synthetic application for [3,3]-sigmatropic rearrangements of N—O bond fragments, several elegant methods for C—O bond construction have also been designed using similar transformations.

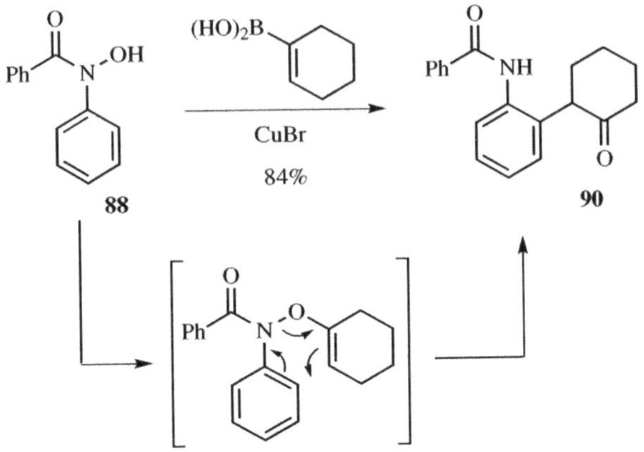

Scheme 14.10 Rearrangements of *O*-vinyl-*N*-arylhydroxylamines for the preparation of interrupted Fischer-indole intermediates.[33]

Tomkinson and coworkers have developed methods for the preparation of α-oxygenated ketones and aldehydes using [3,3]-rearrangements of *O*-protected *N*-hydroxyenamines such as **93** generated from the condensation of *O*-protected hydroxylamines **91**, **96**, and **100** with ketones and aldehydes (Scheme 14.11a–c).[36–40] This work was inspired by an analogous stepwise transformation reported by House and coworkers in 1959 and a similar stepwise transformation used in the synthesis of fumagillol.[41,42] Tomkinson and coworkers have streamlined the process to an efficient single-step conversion of ketones or aldehydes to α-oxygenated carbonyl compounds and explored the scope and selectivity of this transformation.

One interesting aspect of these reactions is the solvent dependence. The rearrangements of aldehyde- and ketone-derived condensation products of acetylated or benzoylated hydroxylamine reagents are most efficient when run in DMSO at 50 °C (Scheme 14.11a),[37] while sulfonyl-protected hydroxylamines such as **96** decompose in polar solvents such as DMSO or DMF but lead to clean product mixtures in THF, EtOAc, or CH_2Cl_2 with MsOH at 25 °C (Scheme 14.11b).[38] Carbonate-protected hydroxylamines undergo equally efficient rearrangements in DMSO or THF, but while transformations run in THF allow for the isolation of the initial α-oxygenated product, reactions run in DMSO lead to the conversion of these products to the corresponding oxazolidinones (Scheme 14.11c).[39]

Another interesting trend is the regioselectivity of oxygenation: the rearrangements of acetyloxy-, benzoyloxy-, or carbonate-protected hydroxylamine condensation products strongly favor oxygenation at secondary carbon centers over primary alkyl groups (Scheme 14.11a and c).[37,39] In contrast, the rearrangements of the condensation products of sulfonyl hydroxylamines prefer oxygen transfer to primary carbon substituents over secondary (Scheme 14.11b).[38]

These transformations offer a simple alternative to enolate α-oxygenation procedures, enabling the formation of α-oxygenated ketones using simple reagents such

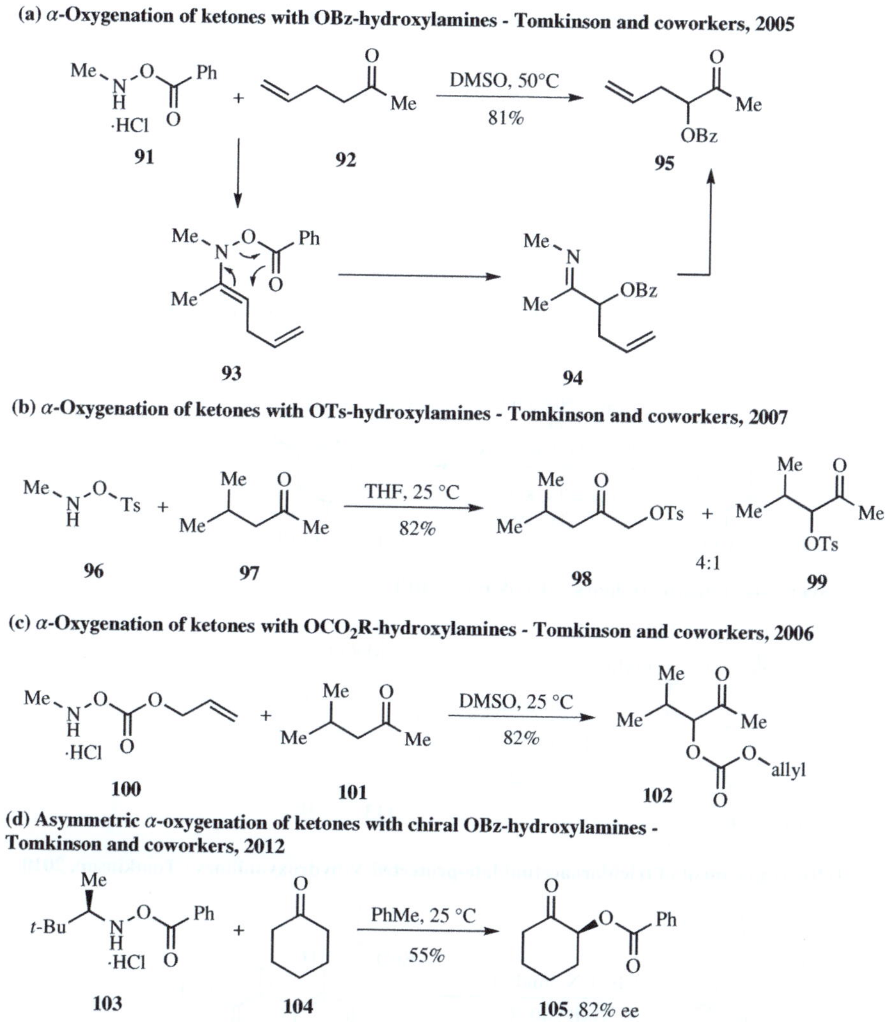

Scheme 14.11 Preparation of α-oxygenated ketones and aldehydes by the condensation and rearrangement of *N*-methyl-*O*-protected hydroxylamines and carbonyl compounds.

as **91**, **96**, and **100** instead of electrophilic oxygen sources such as benzoylperoxide or nitrosobenzene.[43] More recently, the Tomkinson group has rendered these transformations asymmetric with the introduction of a chiral substituent at nitrogen (Scheme 14.11d).[40] Substituted cyclohexanone derivatives also show high diastereoselectivity for conformationally well-defined substrates such as 4-*t*-butylcyclohexanone.[40]

In addition to the α-oxygenation of ketones, the Tomkinson group has also explored the preparation of α-aminophenols through similar [3,3]-rearrangements of *N*-arylhydroxylamines (Scheme 14.12a). Rearrangements of acetylated and benzoylated *N*-hydroxycarbamates smoothly proceed at elevated temperatures to give the

(a) General scheme for aryl halide amino oxygenation - Tomkinson, 2008

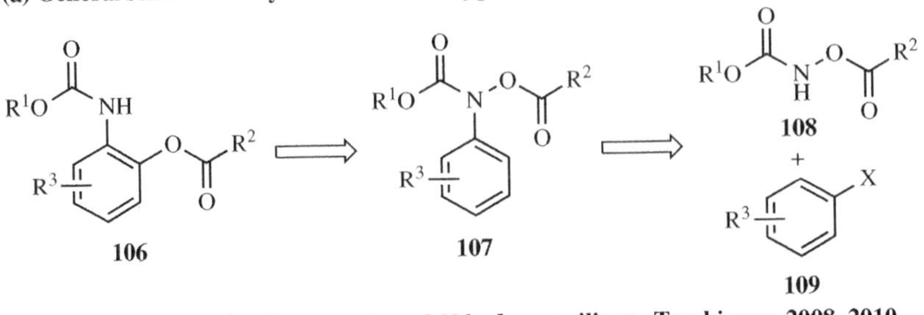

(b) Rearrangement of sulfonyl-protected *N*-hydroxyanilines - Tomkinson, 2008, 2010

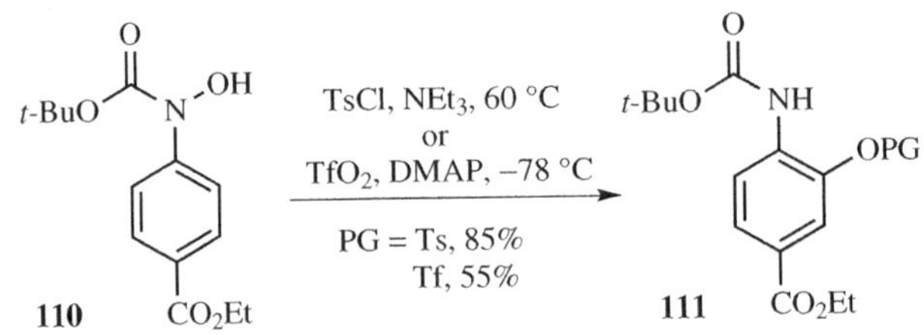

(c) Synthesis of benzoxazolones - Tomkinson, 2010

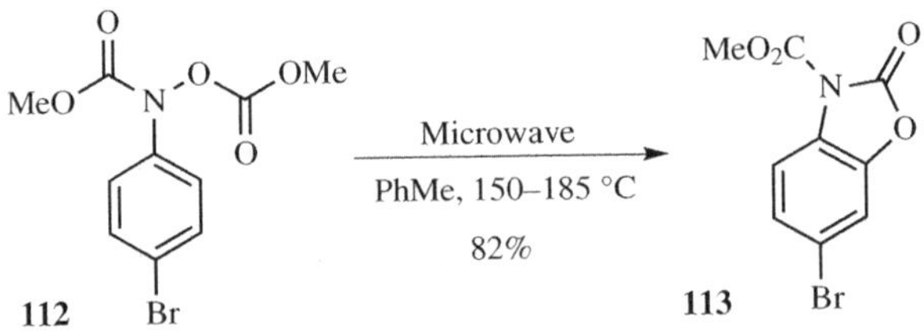

(d) Rearrangement of trichloroacetimidate-protected *N*-hydroxyanilines - Tomkinson, 2010

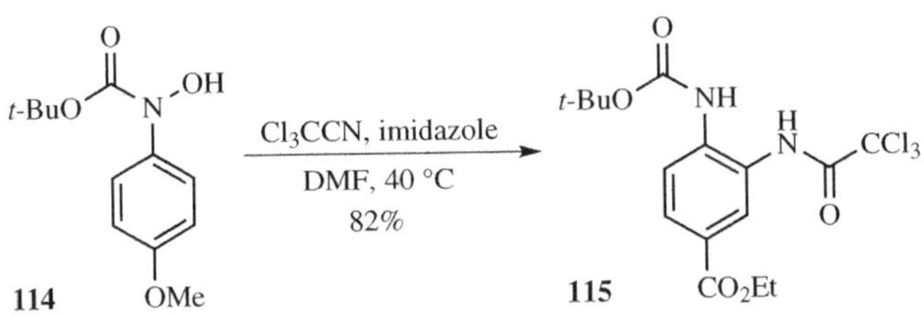

(e) Metal-catalyzed *N*-arylation of *N*-alkoxycarbamates - Tomkinson, 2008, 2009

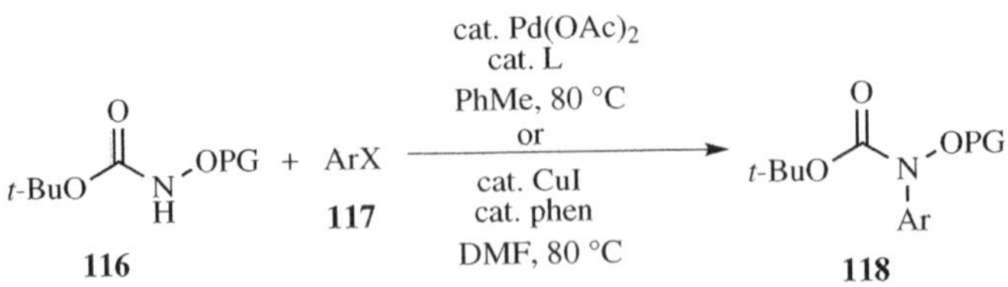

PG = benzyl, methyl, *t*-butyl, allyl, THP, SiR_3
Ar = 4-nitrophenyl, 4-methoxylphenyl, 3-halo, 3-ester

Scheme 14.12 Preparation and rearrangement of *O*-protected-*N*-arylhydroxylamines.

desired products.[44] These transformations are more facile for sulfonyl-substituted hydroxylamines which can be generated *in situ* and rearrange spontaneously as shown in Scheme 14.12b.[45] The rearrangements of carbonate-protected hydroxylamines such as **112** have also been used to prepare the corresponding benzoxazolones (Scheme 14.12c), and an analogous amination procedure has been developed through the *in situ* generation of imidoyl-substituted hydroxylamines from trichloroacetonitrile (Scheme 14.12d).[46,47]

To facilitate the use of these methods, the Tomkinson group has also developed three different ways to synthesize the *N*-arylhydroxylamine precursors. The first two methods employ transition metal-mediated C—N bond coupling reactions which tolerate carbamate-protected N-functionalities and alkoxy or silyloxy O-protecting groups (Scheme 14.12e).[48,49] Additionally, Tomkinson and coworkers have shown that traditional nitroarene reduction is equally effective for the preparation of similar compounds.[50]

The Anderson group has taken a different approach to the construction of C—O bonds through the [3,3]-rearrangements of N—O bonds. Using a Chan–Lam–Evans reaction, they have shown that *N*-enoxyphthalimides such as **121** and **125** can be prepared through the C—O bond coupling of *N*-hydroxyphthalimide and vinylboronic acids (Scheme 14.13a).[51] When heated, these compounds undergo a [3,3]-rearrangement to form α-oxygenated ketones such as **122** and **126**. These imidate rearrangement products can then be hydrolyzed to give the corresponding α-hydroxy- or α-benzoyloxyketones. This transformation is distinct from the Tomkinson method for α-oxygenation in that the rearrangement precursor is prepared by a C—O bond coupling of *N*-hydroxyphthalimide and a vinylboronic acid instead of a condensation between a ketone or aldehyde and a hydroxylamine. The cross-coupling method places the vinyl functionality on the O-atom of the hydroxylamine instead of the N-atom, and the stereochemistry of the vinylboronic acid is transferred to the *N*-enoxyphthalimide. Preparation of the rearrangement precursor by this method avoids the challenges of selective deprotonation approaches to alkyl-substituted vinyl ethers such as **125**. Retrosynthetically, this transformation provides a new route to α-oxygenated ketones that avoids the need for the oxidation of an enolate with an electrophilic oxygen source and provides an opportunity for the direct conversion of alkynes to α-oxygenated ketones without the need for the intermediacy of a carbonyl compound (Scheme 14.13b). This unique transformation can also be considered as a dioxygenation of an alkenyl boronic acid.

14.3 [3,3]-SIGMATROPIC REARRANGEMENTS OF N—N BONDS

14.3.1 Piloty–Robinson Pyrrole Synthesis

The Piloty–Robinson synthesis is an appealingly simple method for the preparation of symmetrical pyrroles via [3,3]-rearrangement of a divinylhydrazine that is accessed through the double condensation of hydrazine with two equivalents of a ketone or aldehyde (Scheme 14.14a).[52] To avoid product mixtures, only one ketone or aldehyde reagent can be used for the condensation, and ketones must have only one

(a) Preparation and [3,3] rearrangement of *N*-enoxyphthalimides

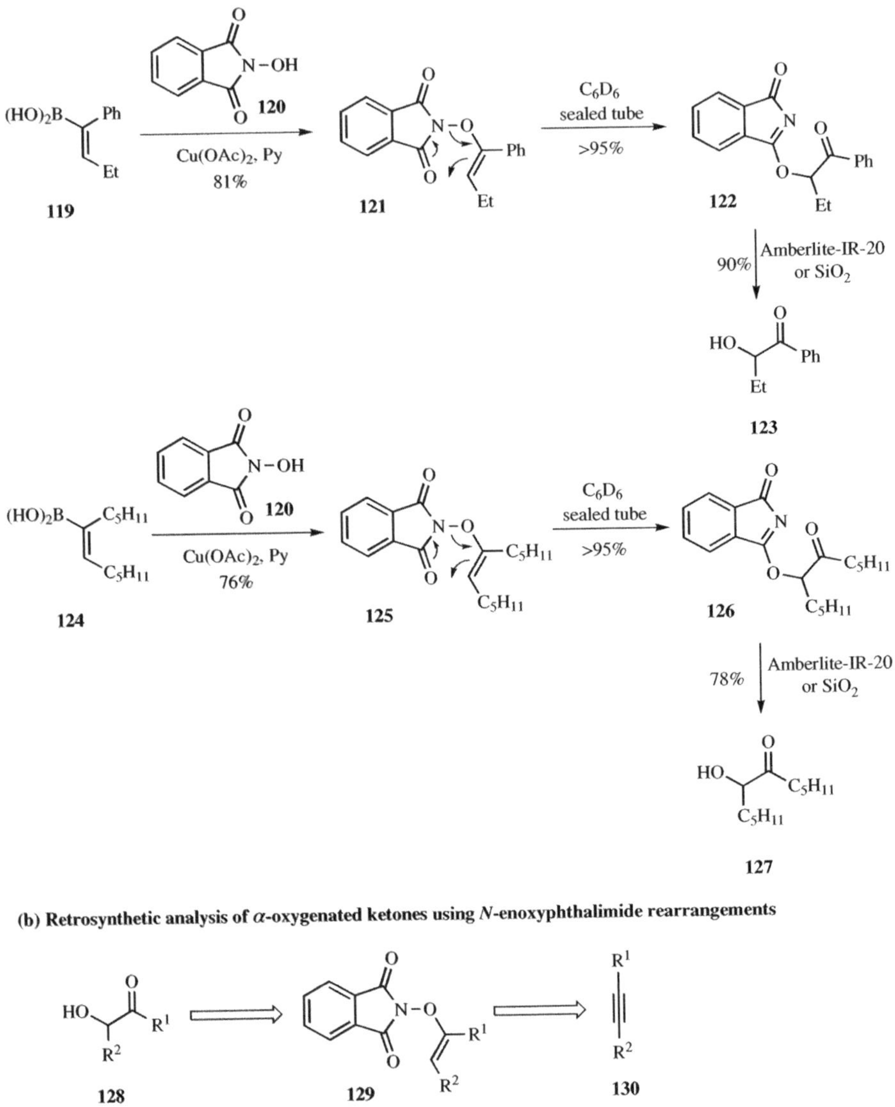

(b) Retrosynthetic analysis of α-oxygenated ketones using *N*-enoxyphthalimide rearrangements

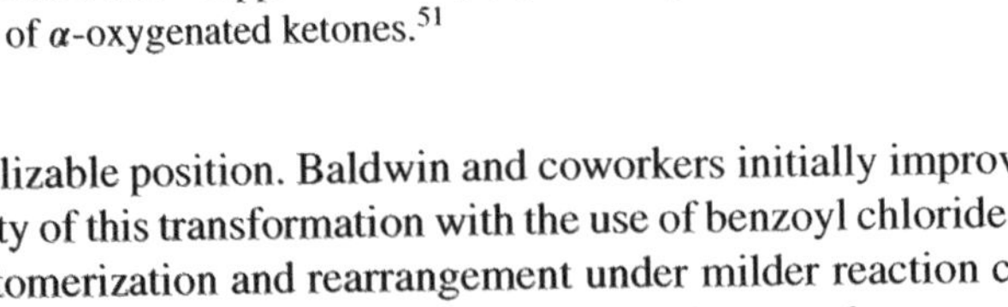

Scheme 14.13 Application of [3,3]-rearrangements of *N*-enoxyphthalimides to the preparation of α-oxygenated ketones.[51]

enolizable position. Baldwin and coworkers initially improved the synthetic applicability of this transformation with the use of benzoyl chloride to promote the necessary tautomerization and rearrangement under milder reaction conditions than originally reported.[53] However, practical reaction times and rearrangement efficiencies have

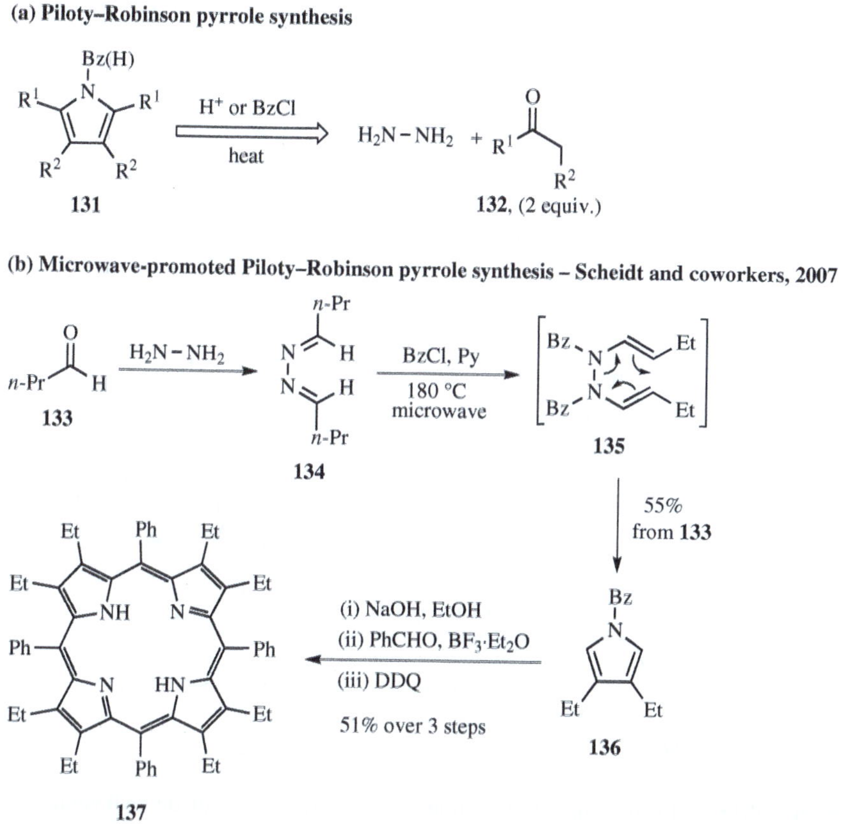

Scheme 14.14 Microwave-promoted Piloty–Robinson pyrrole synthesis and application to the synthesis of porphyins.[54]

only recently been achieved by Scheidt and coworkers who employed microwave heating for rapid access to 3,4-disubstituted pyrroles from aldehydes, hydrazine, and benzoyl chloride (Scheme 14.14b).[54] These transformations occur in 30 min–1 h in moderate yield. This rapid synthesis was shown to be particularly useful for the preparation of octaethylporphyrins. Scheidt and coworkers further demonstrated that *N*-benzoylpyrroles such as **136** could be deprotected, condensed with either formaldehyde or benzaldehyde, and oxidized to give porphyrins such as **137** in good yield.[54] This synthetic application showcased the advantages of using this method for the preparation of symmetrically substituted pyrroles.

Alternative access to Piloty–Robinson rearrangement precursors can also be achieved by transition metal-mediated C—N bond coupling (Scheme 14.15a). Modular synthesis of divinylhydrazines was demonstrated for bis-Boc-hydrazine and vinyl iodide reagents by Buchwald and coworkers under copper-catalyzed amination conditions.[55] The preparation of divinylhydrazines by an Ullmann coupling

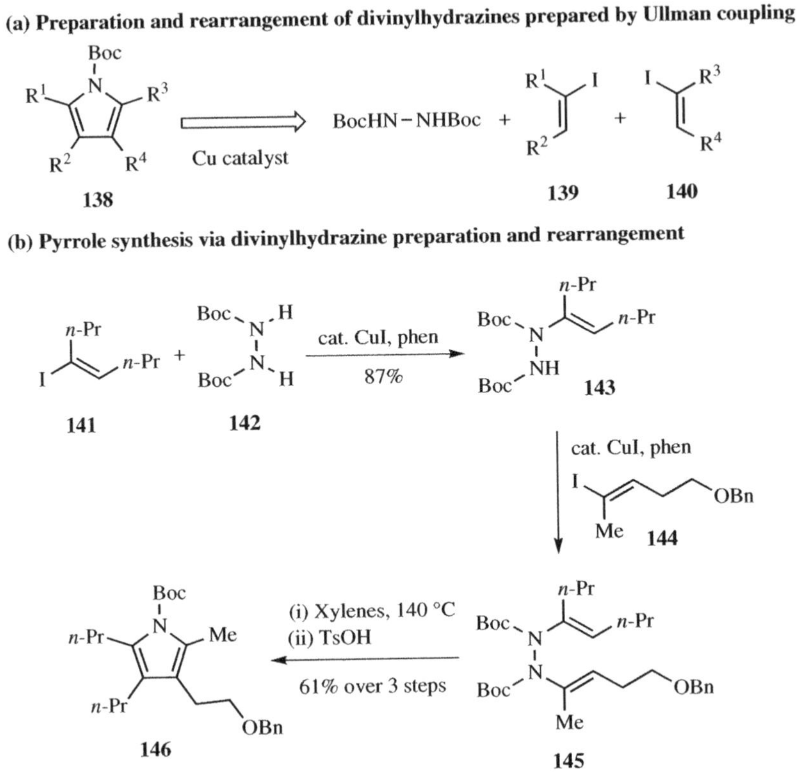

Scheme 14.15 Ullman coupling conditions for the preparation of divinylhydrazines and application to tetrasubstituted pyrrole synthesis.[55]

approach provides two distinct advantages over the condensation methods described for the traditional Piloty–Robinson pyrrole syntheses discussed previously: (1) unsymmetrical divinylhydrazines can be prepared by sequential coupling reactions with two different vinyl iodide reagents, and (2) regioselective tautomerization is not required to avoid product mixtures. These two characteristics of the copper-catalyzed construction of divinylhydrazine reagents have provided a significant advance in the scope of pyrroles accessible by Piloty–Robinson-type syntheses by allowing for the preparation of unsymmetrical, tetrasubstituted pyrroles as well as pyrroles with alkyl substituents at the 2- and 5-positions in good yield and with high selectivity (Scheme 14.15b).

14.3.2 The Use of Hydrazones for Enolate Coupling

[3,3]-Rearrangements of diacylhydrazides have been used for the preparation of 1,4-bisamides and succinimides through a formal enolate coupling processes (Scheme 14.16a). This type of transformation was originally reported by both Endo

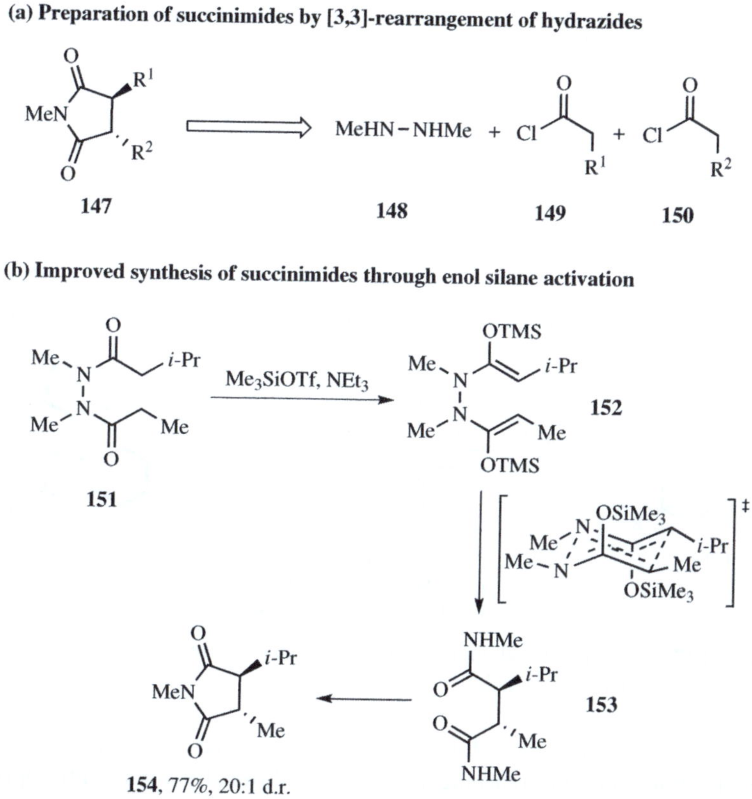

Scheme 14.16 Preparation of succinimides by the [3,3]-rearrangement of diacylhydrazides.[58]

and Magedov but had limited substrate scope and poor yields due to the use of a strong base to promote bisamide enolate formation from the diacylhydrazide.[56,57] Miller and Bayne have demonstrated that a similar transformation can be achieved under milder conditions by forming the corresponding bis(enolsilanes).[58] These TMSOTf-promoted reactions give succinimide products in good yield for a wider variety of substitution patterns with moderate to high diastereoselectivity (Scheme 14.16b). Initial formation of the bis(enolsilane) also suppresses any alternative rearrangement activity through the carbonyl functionality of the amides. This method provides an appealing alternative to oxidative enolate coupling methods for the synthesis of 1,4-bisamides and related derivatives.

14.3.3 Benzidine Rearrangement

Both acid-catalyzed and thermolytic preparations of 2,2′-diamino-1,1′-binaphthyl (BINAM) compounds from 1,2-dinaphthylhydrazines were shown to occur through concerted [3,3]-rearrangements by Shine, Banthrope, and Trisler in the 1960s.[59]

(a) New C–N bond coupling methods for the preparation of unsymmetrical diarylhydrazine rearrangement precursors – Cho and coworkers, 2011

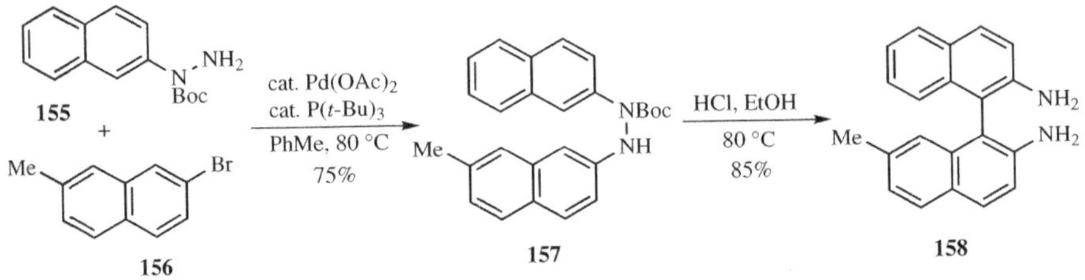

(b) Chiral Brønsted acid promoted rearrangements of diarylhydrazines – Kürti and coworkers, 2013.

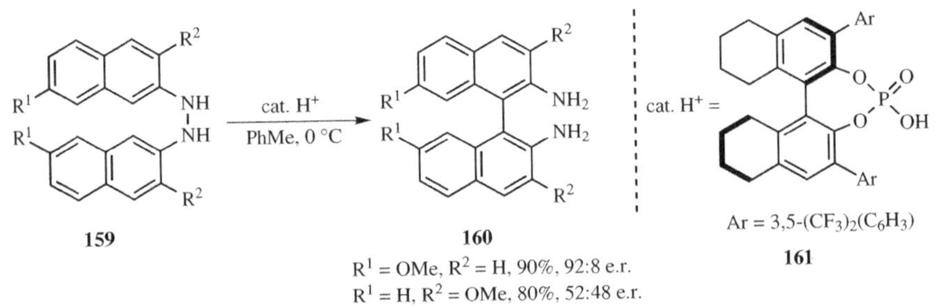

Scheme 14.17 Unsymmetrical and asymmetric benzidine rearrangements through new advances in cross coupling and acid-catalysis.

Due to a lack of methods for preparing substituted 1,2-dinaphthylhydrazine precursors, however, this rearrangement has traditionally found little application in organic synthesis despite strong demand for facile routes to chiral, nonracemic 2,2′-diamino-1,1′-binaphthyl ligands. Recently, however, the Cho and Kürti groups have each disclosed important advances for this rearrangement which will increase the applicability of 1,2-dinaphthylhydrazines as precursors for BINAM derivatives. Cho and coworkers have optimized palladium-catalyzed coupling conditions for the facile preparation of a range of 1,2-dinaphthylhydrazines with a variety of substitution patterns (Scheme 14.17a).[60,61] These compounds smoothly underwent mild acid-catalyzed [3,3]-rearrangements to give 2,2′-diamino-1,1′-binaphthyl compounds with a wider range of substitution patterns than tolerated by oxidative coupling methods. The method was also shown to be equally effective for the preparation of unsymmetrical dinaphthylhydrazines and 1-aryl-2-naphthylhydrazines. Kürti and coworkers further contributed to the advancement of the benzidine rearrangement method by identifying a chiral phosphoric acid catalyst for the asymmetric preparation of 2,2′-diamino-1,1′-binaphthyl compounds through the asymmetric rearrangement of 1,2-dinaphthylhydrazines (Scheme 14.17b).[62] Phosphoric acid **161** was shown to be the most effective catalyst for this transformation and provided the rearrangement products of the parent binaphthylhydrazine as well as several other electron-rich and electron–neutral binaphthylhydrazines in good yield and enantioselectivity, The enantioselectivity of this process is highly dependent on the structure of the chiral phosphoric acid catalyst as illustrated in Scheme 14.17b.

14.4 [3,3]-REARRANGEMENTS OF N—N BOND FRAGMENTS THAT ELIMINATE N_2

While the majority of the [3,3]-rearrangements discussed previously in this chapter involve the cleavage of an N—O or an N—N bond at the center of a 6-atom fragment, this section will focus on a unique [3,3]-rearrangement of *N*-allyl hydrazones developed by Thomson and coworkers that achieves an elegant C—C bond fragment coupling through cleavage of a C—N bond and provides access to homoallylic structures after release of N_2 (Scheme 14.18a). While the loss of N_2 is a common mode of reactivity for hydrazone reagents with appropriate leaving groups, the combination of this process with an intramolecular $C(sp^3)$—$C(sp^3)$ bond-forming event has provided an innovative and "traceless" method for the preparation of functionalized homoallylic motifs from a nonobvious retron. A report of this type of rearrangement was originally described by Stevens and coworkers in 1973, but the low yield and harsh conditions required for this transformation prevented further application in synthesis.[63]

Thomson and coworkers have recently developed two distinct ways to access the [3,3]-rearrangements of *N*-allyl hydrazones under either acidic or oxidizing conditions (Scheme 14.18b): the oxidant-promoted reaction allows for the addition of several different nucleophiles to the rearrangement product, while the Brønsted

(a) Traceless attachment of homoallylic fragments by [3,3]-rearrangements of allylic hydrazones

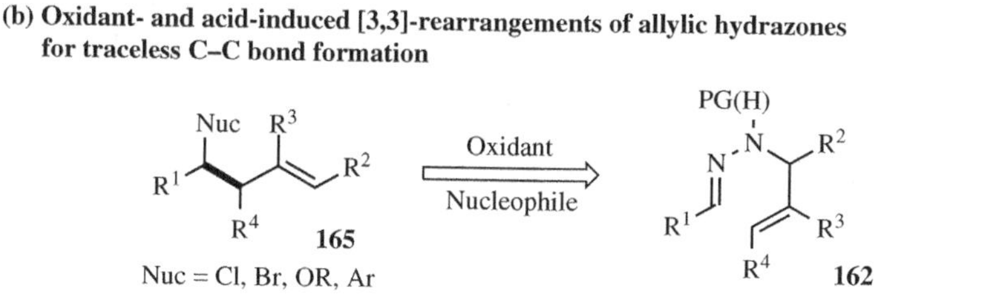

(b) Oxidant- and acid-induced [3,3]-rearrangements of allylic hydrazones for traceless C–C bond formation

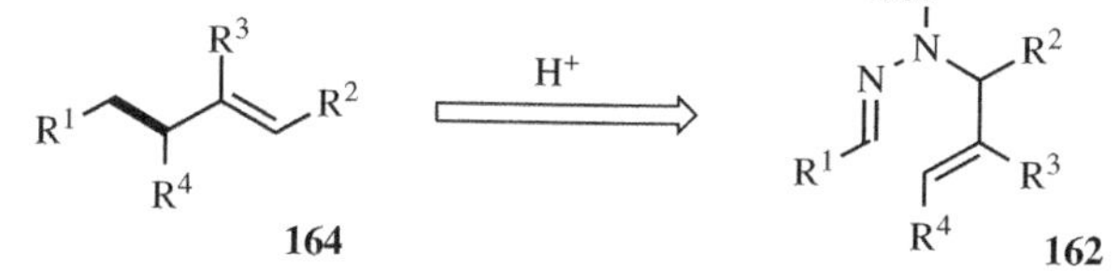

Scheme 14.18 [3,3]-Rearrangements of *N*-allyl hydrazones.[64–66]

acid-promoted process provides the parent C—C bond connection.[64–69] Both transformations give a high degree of stereocontrol, and procedures have been developed for the direct conversion of aldehydes to the desired products in good yields over 2–3 steps without intermediate purification. These new methods offer an alternative approach to the formation of the C(sp^3)—C(sp^3) bond between carbons 1 and 4, which would be challenging to achieve by other methods.[68]

14.4.1 Oxidant-Promoted *N*-Allylic Hydrazone Rearrangements

Examples of three different oxidant-promoted *N*-allylic hydrazone rearrangements developed by the Thomson group are illustrated in Scheme 14.19a. These transformations convert *N*-allylic hydrazones to homoallylic halides in the presence of either $CuCl_2$ or NBS and to homoallylic ethers in the presence of $PhI(OTf)_2$ and an alcohol.[64–66] The NBS halogenation process was cleverly combined with a subsequent elimination to form 1,4-dienes. Yields for the diene and homoallylic ether syntheses have been calculated from the corresponding aldehydes because intermediate purification is not needed. Proposed mechanistic pathways for these transformations are illustrated in Scheme 14.19b. Electrophilic activation of the hydrazone promotes a [3,3]-rearrangement that controls formation of the allylic stereocenter of the product through a chair-like transition state **175**. Displacement of N_2 with bromide followed by elimination gives the corresponding diene. In contrast, the nature of the weakly coordinating triflate ion allows for the interception of an alcohol nucleophile, which approaches the carbocation through a trajectory that minimizes strain with the allylic position. The diene synthesis shows high *E,E*-selectivity for R^1 = alkyl, R^2, R^3 = H or R^2 = alkyl, R^1, R^3 = H. The homoallylic ether and homoallylic chloride syntheses show high *E*-selectivity for R^1 = alkyl, R^2, R^3 = H or R^2 = alkyl, R^1, R^3 = H.

The Thomson group has also recently adapted the iodonium-promoted rearrangement of *N*-allylic hydrazones to the synthesis of several naphthyl-type lignan natural products that display promising antimalarial activity.[67] As shown in Scheme 14.20a, hydrazone rearrangement products that have been trapped with MeOH can undergo a Friedel–Crafts displacement with an electron-rich arene in the presence of an acid to give benzhydryl homoallylic fragments such as **181**. Importantly, the diastereoselectivity of the rearrangement can be controlled by the allylic stereocenter of the hydrazone, and methanol displacement with the arene proceeds with high diastereoselectivity due to minimization of $A_{1,2}$ strain. This method was used to prepare **182–186** and indicates significant generality for the synthesis of these types of compounds (Scheme 14.20b).

14.4.2 Acid-Promoted *N*-Allylic Hydrazone Rearrangements

A triflimide-promoted rearrangement of *N*-allylic hydrazones has been designed to provide new diastereoselective and enantioselective routes to unfunctionalized homoallylic fragments through an C(sp^3)—C(sp^3) fragment coupling similar to the oxidative rearrangement discussed earlier (Scheme 14.21a). Thomson and coworkers have shown that this acid-catalyzed transformation is general for a variety of aryl,

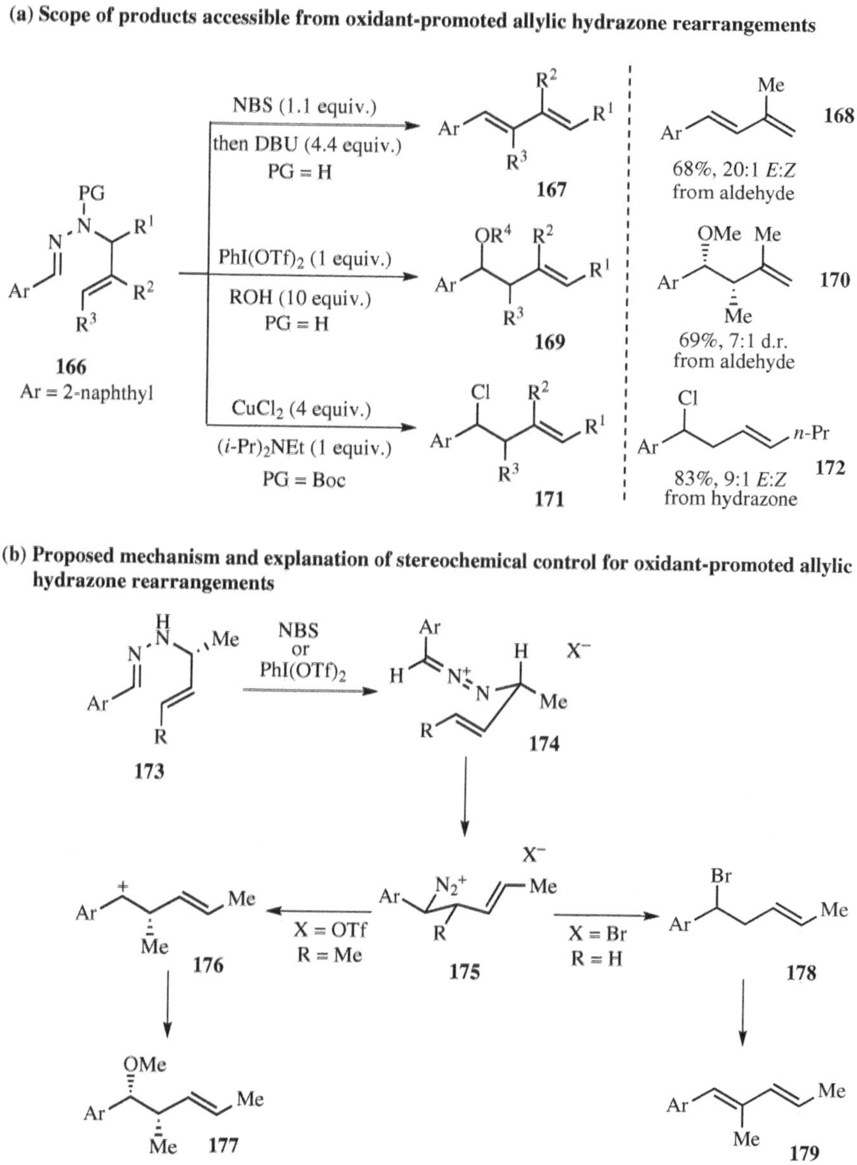

Scheme 14.19 Oxidative [3,3]-rearrangements of *N*-allyl hydrazones.[64–66]

heteroaryl, and alkyl aldehydes and provides a fragment coupling product that would be difficult to achieve with transition metal-mediated coupling processes.[68] The enantiomeric ratios of chiral substituents installed on either the hydrazine or the aldehyde were retained throughout the rearrangement process. To better understand the greater efficiency of Boc-protected hydrazones in these transformations in

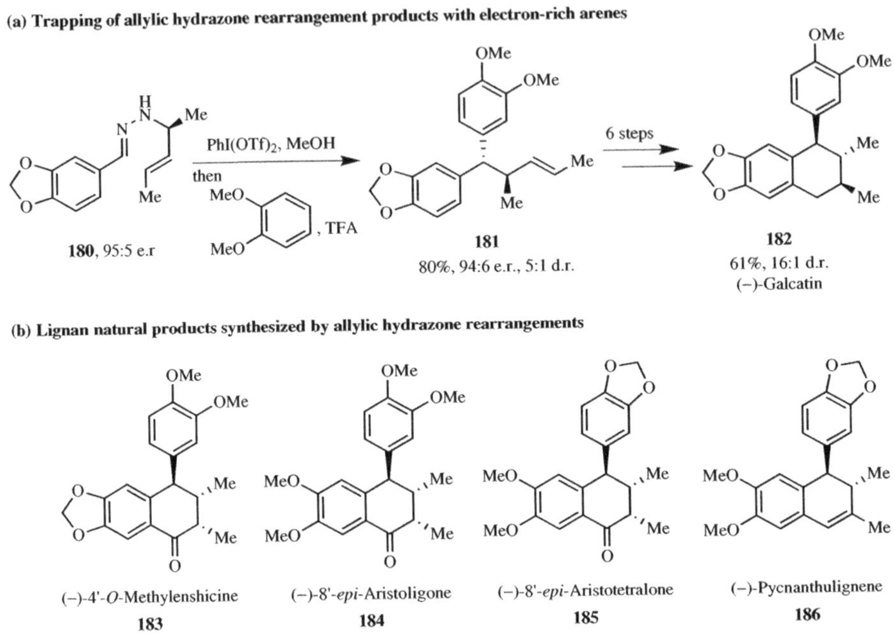

Scheme 14.20 Synthesis of Lignan natural products through [3,3]-rearrangements of *N*-allyl hydrazones.[67]

comparison to N—H hydrazones, as well as the lack of rearrangement activity for terminally substituted allylic groups, the Thomson and Tantillo groups collaborated in a series of computational and synthetic experiments.[69] These investigations showed that Boc protection avoids pathways that lead to decomposition products and that electron-withdrawing groups at the electrophilic carbon of the hydrazone and electron-donating groups at the allylic position lower the barrier for rearrangement. These trends were validated by showing that methyl glyoxylate-derived hydrazones underwent the [3,3]-rearrangement at much lower temperatures than alkyl- or aryl-substituted hydrazones and that a terminal substituent on the allylic group was tolerated for these accelerated rearrangement substrates (Scheme 14.21b).

14.5 SUMMARY

As highlighted in this chapter, [3,3]-rearrangements of N—N and N—O bond fragments participate in a variety of cascade processes to generate important heterocyclic structures. Recent advances in this area have focused primarily on the development of improved general methods for accessing the rearrangement precursors in order to expand the utility of the heterocycle syntheses. In addition to these practical improvements, exploration of these transformations has also led to the discovery of new applications such as the α-oxygenation of carbonyl compounds using [3,3]-rearrangements of oximes and hydroxylamines as well as traceless $C(sp^3)$—$C(sp^3)$ bond construction

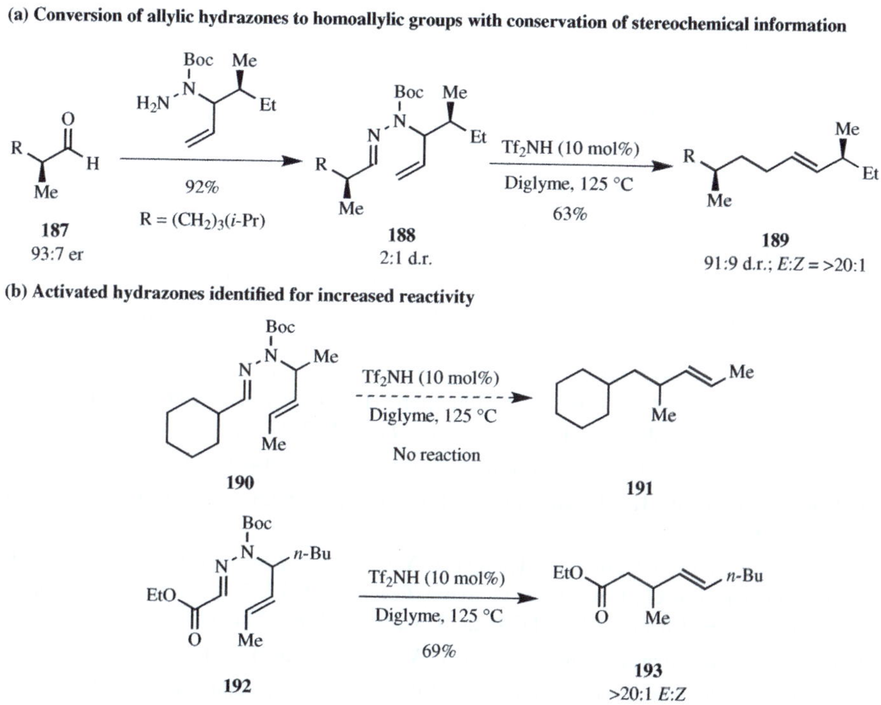

Scheme 14.21 Enantio- and diastereoselective synthesis of homoallylic fragments.[68,69]

through the rearrangements of allylic hydrazones. These developments suggest that while heterocyclic syntheses employing these methods will continue to be refined, new applications in asymmetric synthesis and catalysis are likely to be the next frontier for this expanding field of organic synthesis.

REFERENCES

1. (a) Downing, R. S.; Dunkeler, P. J. The Fischer indole synthesis. In *Fine Chemicals through Heterogeneous Catalysis*; Sheldon, R. A.; Van Bekkum, H. Eds.; Wiley-VCH Verlag GmbH: Weinheim, **2001**; 178–183. (b) Hughes, D. L. *Org. Prep. Proced. Int.* **1993**, *25*, 607–632. (c) Ambekar, S. Y. *Curr. Sci.* **1983**, *52*, 578–582. (d) Robinson, B. *The Fischer Indole Synthesis*; John Wiley & Sons, Inc.: Hoboken, **1983**.
2. Woodward, R. B.; Cava, M. P.; Ollis, W. D.; Hunger, A.; Daeniker, H. U.; Schenker, K. *J. Am. Chem. Soc.* **1954**, *76*, 4749–4751.
3. (a) Smith, J. M.; Moreno, J.; Boal, B. W.; Garg, N. K. *J. Am. Chem. Soc.* **2014**, *136*, 4504–4507. (b) Schammel, A. W.; Chiou, G.; Garg, N. K. *Org. Lett.* **2012**, *14*, 4556–4559. (c) Schammel, A. W.; Chiou, G.; Garg, N. K. *J. Org. Chem.* **2012**, *77*, 725–728. (d) Zu, L.; Boal, B. W.; Garg, N. K. *J. Am. Chem. Soc.* **2011**, *133*, 8877–8879.
4. Celebi-Ölcüm, N.; Boal, B. W.; Huters, A. D.; Garg, N. K.; Houk, K. N. *J. Am. Chem. Soc.* **2011**, *133*, 5752–5755.

5. Kozmin, S. A.; Iwama, T.; Huang, Y.; Rawal, V. H. *J. Am. Chem. Soc.* **2002**, *124*, 4628–4641.
6. (a) Schammel, A. W.; Boal, B. W.; Zu, L.; Mesganaw, T.; Garg, N. K. *Tetrahedron* **2010**, *66*, 4687–4695. (b) Phillips, R. R. *Org. React.* **1959**, *10*, 1143.
7. Hughes, D.; Zhao, D. *J. Org. Chem.* **1993**, *58*, 228–233.
8. (a) Douglas, A. W. *J. Am. Chem. Soc.* **1978**, *100*, 6463–6469; (b) Douglas, A. W. *J. Am. Chem. Soc.* **1979**, *101*, 5676–5678.
9. Sheradsky, T. *Tetrahedron Lett.* **1966**, *7*, 5225–5227.
10. Castellino, A. J.; Rapoport, H. *J. Org. Chem.* **1986**, *51*, 1006–1011.
11. (a) Guzzo, P. R.; Buckle, R. N.; Chou, M.; Dinn, S. R.; Flaugh, M. E.; Kiefer, Jr., A. D.; Ryter, K. T.; Sampognaro, A. J.; Tregay, S. W.; Xu, Y.-C. *J. Org. Chem.* **2003**, *68*, 770–778. (b) Liao, Y.; Kozikowski, A. P.; Guidotti, A.; Costa, E. *Bioorg. Med. Chem. Lett.* **1998**, *8*, 2099–2102. (c) Castellino, A. J.; Rapoport, H. *J. Org. Chem.* **1984**, *49*, 4399–4404. (d) Mooradian, A.; Dupont, P. E. *Tetrahedron Lett.* **1967**, *8*, 2867–2870.
12. Takeda, N.; Miyata, O.; Naito, T. *Eur. J. Org. Chem.* **2007**, 1491–1509.
13. Contiero, F.; Jones, K. M.; Matts, E. A.; Porzelle, A.; Tomkinson, N. C. O. *Synlett* **2009**, 3003–3006.
14. (a) Miyazawa, E.; Sakamoto, T.; Kikugawa, Y. *Org. Prep. Proced. Int.* **1997**, *29*, 594–600. (b) Cadogan, J. I. G.; Rowley, A. G. *Synth. Commun.* **1977**, *7*, 365–366. (c) Baldoli, C.; Del Bettero, P.; Licandro, E.; Maiorana, S. *Synthesis* **1988**, 344–345.
15. Petrassi, H. M.; Sharpless, K. B.; Kelly, J. W. *Org. Lett.* **2001**, *3*, 139–142.
16. De, P.; Pandurangan, N. K.; Maitra, U.; Wailes, S.; *Org. Lett.* **2007**, *9*, 2767–2770.
17. Maimone, T. J.; Buchwald, S. L. *J. Am. Chem. Soc.* **2010**, *132*, 9990–9991.
18. Sheradsky, T. *Tetrahedron Lett.* **1970**, *11*, 25–26.
19. (a) Mikhaleva, A. I.; Zaitsev, A. B.; Trofimov, B. A. *Russ. Chem. Rev.* **2002**, *71*, 563–591. (b) Trofimov, B. A. *Curr. Org. Chem.* **2002**, *6*, 1121–1162. (c) Trofimov, B. A.; Mikhaleva, A. I.; *Heterocycles* **1994**, *37*, 1193–1232. (d) Trofimov, B. A. *Adv. Heterocycl. Chem.* **1990**, *51*, 177–301. (e) Vasil'tsov, A. M.; Ivanov, A. V.; Mikhaleva, A. I.; Trofimov, B. A. *Tetrahedron Lett.* **2010**, *51*, 1690–1692.
20. (a) Petrova, O. V.; Sobenina, L. N.; Ushakov, I. A.; Mikhaleva, A. I.; Hyun, S. H.; Trofimov, B. A., *ARKIVOC* **2009**, 14–20. (b) Trofimov, B. A.; Tarasova, O. A.; Mikhaleva, A. I.; Kalinina, N. A.; Sinegovskya, L. M.; Henkelmann, J., *Synthesis* **2000**, 1585–1590.
21. Ngwerume, S.; Camp, J. E. *Chem. Commun.* **2011**, *47*, 1857–1859.
22. Ngwerume, S.; Camp, J. E. *J. Org. Chem.* **2010**, *75*, 6271–6274.
23. (a) Wang, H.-Y.; Mueller, D. S.; Sachwani, R. M.; Londino, H. N.; Anderson, L. L. *Org. Lett.* **2010**, *12*, 2290–2293. (b) Wang, H.-Y.; Mueller, D. S.; Sachwani, R. M.; Kapadia, R.; Londino, H. N.; Anderson, L. L. *J. Org. Chem.* **2011**, *76*, 3203–3221.
24. (a) Dalpozzo, R.; Bartoli, G. *Curr. Org. Chem.* **2005**, *9*, 163–178. (b) Bartoli, G.; Palmieri, G.; Bosco, M.; Dalpozzo, R. *Tetrahedron Lett.* **1989**, *30*, 2129–2132. (c) Bartoli, G.; Bosco, M.; Dalpozzo, R.; Palmieri, G.; Marcantoni, E. *J. Chem. Soc., Perkin Trans. 1* **1991**, 2757–2761.
25. Buszek, K. R.; Brown, N.; Luo, D. *Org. Lett.* **2009**, *11*, 201–204.
26. Dobbs, A. P.; Voyle, M.; Whitall, N., *Synlett* **1999**, 1594–1596.
27. Gao, H.; Ess, D. H.; Yousuffuddin, M.; Kürti, L. *J. Am. Chem. Soc.* **2013**, *135*, 7086–7089.

28. (a) Lin, S.; Danishefsky, S. J. *Angew. Chem., Int. Ed.* **2002**, *41*, 512–515. (b) Albrecht, B. K.; Williams, R. M. *Proc. Natl. Acad. Sci. U. S. A.* **2004**, *101*, 11949–11954. (c) Ashburn, B. O.; Rathbone, L. K.; Camp, E. H.; Carter, R. G. *Tetrahedron* **2008**, *64*, 856–865. (d) Masters, K. S.; Brase, S. *Angew. Chem., Int. Ed.* **2013**, *52*, 866–869.
29. Endo, Y.; Uchida, T.; Hizatate, S.; Shudo, K. *Synthesis* **1994**, *1096–1105*.
30. Almeida, P. S.; Prabhakar, S.; Lobo, A. M.; Marcelo-Curlo, M. J. *Tetrahedron Lett.* **1991**, *32*, 2671–2674.
31. Mao, Z.; Baldwin, S. W. *Org. Lett.* **2004**, *6*, 2425–2428.
32. Duguet, N.; Slawin, A. M. Z.; Smith, A. D. *Org. Lett.* **2009**, *11*, 3858–3861.
33. Wang, H.-Y.; Anderson, L. L. *Org. Lett.* **2013**, *15*, 3362–3365.
34. (a) Ridgway, B. H.; Woerpel, K. A. *J. Org. Chem.* **1998**, *63*, 458–460. (b) Matos, K.; Soderquist, J. A. *J. Org. Chem.* **1998**, *63*, 461–470.
35. (a) Johansson, C. C. C.; Colacot, T. J. *Angew. Chem., Int. Ed.* **2010**, *49*, 676–707. (b) Mazet, C. *Synlett* **2012**, *23*, 1999–2004. (c) Kawatsura, M.; Hartwig, J. F. *J. Am. Chem. Soc.* **1999**, *121*, 1473–1478. (d) Fox, J. M.; Huang, X.; Chieffi, A.; Buchwald, S. L. *J. Am. Chem. Soc.* **2000**, *122*, 1360–1370. (e) Xie, X.; Chen, Y.; Ma, D. *J. Am. Chem. Soc.* **2006**, *128*, 16050–16051.
36. Beshara, C. S.; Hall, A.; Jenkins, R. L.; Jones, T. C.; Parry, R. T.; Thomas, S. P.; Tomkinson, N. C. O. *Chem. Commun.* **2005**, 1478–1480.
37. Beshara, C. S.; Hall, A.; Jenkins, R. L.; Jones, K. L.; Jones, T. C.; Killeen, N. M.; Taylor, P. H.; Thomas, S. P.; Tomkinson, N. C. O. *Org. Lett.* **2005**, *7*, 5729–5732.
38. John, O. R. S.; Killeen, N. M.; Knowles, D. A.; Yau, S. C.; Bagley, M. C.; Tomkinson, N. C. O. *Org. Lett.* **2007**, *9*, 4009–4012.
39. Hall, A.; Jones, K. L.; Jones, T. C.; Killeen, N. M.; Pörzig, R.; Taylor, P. H.; Yau, S. C.; Tomkinson, N. C. O. *Synlett* **2006**, 3435–3438.
40. Smithen, D. A.; Mathews, C. J.; Tomkinson, N. C. O. *Org. Biomol. Chem.* **2012**, *10*, 3756–3762.
41. House, H. O.; Richey, F. A. *J. Org. Chem.* **1969**, *34*, 1430–1439.
42. (a) Vosburg, D. A.; Weiler, S.; Sorensen, E. *Chirality* **2003**, *15*, 156–166. (b) Cummins, C. H.; Coates, R. M. *J. Org. Chem.* **1983**, *48*, 2070–2076.
43. (a) Smith, A. M. R.; Hii, K. K. *Chem. Rev.* **2011**, *111*, 1637–1656. (b) Janey, J. M. *Angew. Chem., Int. Ed.* **2005**, *44*, 4292–4300. (c) Brown, S. P.; Brochu, M. P.; Sinz, C. J.; MacMillan, D. W. C. *J. Am. Chem. Soc.* **2003**, *125*, 10808–10809. (d) Lifchits, O.; Demoulin, N.; List, B. *Angew. Chem., Int. Ed.* **2011**, *50*, 9680–9683.
44. Porzelle, A.; Woodrow, M. D.; Tomkinson, N. C. O. *Eur. J. Org. Chem.* **2008**, 5135–5143.
45. Porzelle, A.; Cooper, A. W. J.; Woodrow, M. D.; Tomkinson, N. C. O. *Synlett* **2010**, 2471–2473.
46. Porzelle, A.; Woodrow, M. D.; Tomkinson, N. C. O. *Org. Lett.* **2010**, *12*, 812–815.
47. Porzelle, A.; Woodrow, M. D.; Tomkinson, N. C. O. *Org. Lett.* **2010**, *12*, 1492–1495.
48. Jones, K. L.; Porzelle, A.; Hall, A.; Woodrow, M. D.; Tomkinson, N. C. O. *Org. Lett.* **2008**, *10*, 797–800.
49. Porzelle, A.; Woodrow, M. D.; Tomkinson, N. C. O. *Org. Lett.* **2009**, *11*, 233–236.
50. Porzelle, A.; Woodrow, M. D.; Tomkinson, N. C. O. *Synlett* **2009**, 798–802.
51. Patil, A. S.; Mo, D.-L.; Wang, H.-Y.; Mueller, D. S.; Anderson, L. L. *Angew. Chem., Int. Ed.* **2012**, *51*, 7799–7803.

52. (a) Piloty, O.; *Chem. Ber.* **1910**, *43*, 489–498. (b) Robinson, R.; Robinson, G. M. *J. Chem. Soc.* **1918**, *43*, 639–644. (c) Posvic, H.; Dombro, R.; Ito, H.; Telinski, T. *J. Org. Chem.* **1974**, *39*, 2575–2580.
53. Baldwin, J. E.; Bottaro, J. C. *J. Chem. Soc., Chem. Commun.* **1982**, 624–625.
54. Milgram, B. C.; Eskildsen, K.; Richter, S. M.; Scheidt, W. R.; Scheidt, K. A. *J. Org. Chem.* **2007**, *72*, 3941–3944.
55. Rivero, M. R.; Buchwald, S. L. *Org. Lett.* **2007**, *9*, 973–976.
56. Endo, Y.; Shudo, K. *Tetrahedron Lett.* **1991**, *32*, 4517–4520.
57. Magedov, I. V.; Smushkevich, Y. I. *J. Chem. Soc., Chem. Commun.* **1990**, 1686–1687.
58. Miller, S. J.; Bayne, C. D. *J. Org. Chem.* **1997**, *62*, 5680–5681.
59. (a) Banthorpe, D. V.; Hughes, E. M. *J. Chem. Soc.* **1964**, *2860–2864*. (b) Shine, H. J.; Trisler, J. C. *J. Am. Chem. Soc.* **1960**, *82*, 4054–4058. (c) Shine, H. J.; Gruszecka, E.; Subotkowski, W.; Brownawell, M.; San Filippo, Jr., J. *J. Am. Chem. Soc.* **1985**, *107*, 3218–3223. (d) Shine, H. J.; Kupczuk-Subotkowski, L.; Subotkowski, W. *J. Am. Chem. Soc.* **1985**, *107*, 6674–6678.
60. Suh, S.-E.; Park, I.-K.; Lim, B.-Y.; Cho, C.-G. *Eur. J. Org. Chem.* **2011**, 455–457.
61. Lim, B.-Y.; Choi, M.-K.; Cho, C.-G. *Tetrahedron Lett.* **2011**, *52*, 6015–6017.
62. Li, G.-Q.; Gao, H.; Keene, C.; Devonas, M.; Ess, D. H.; Kürti, L. *J. Am. Chem. Soc.* **2013**, *135*, 7414–7417.
63. Stevens, R. V.; McEntire, E. E.; Barnett, W. E.; Wenkert, E. *J. Chem. Soc., Chem. Commun.* **1973**, 662–663.
64. Mundal, D. A.; Lee, J. J.; Thomson, R. J. *J. Am. Chem. Soc.* **2008**, *130*, 1148–1149.
65. Mundal, D. A.; Lutz, K. E.; Thomson, R. J. *Org. Lett.* **2009**, *11*, 465–468.
66. Lutz, K. E.; Thomson, R. J. *Angew. Chem., Int. Ed.* **2011**, *50*, 4437–4440.
67. Reddel, J. C. T.; Lutz, K. E.; Diagne, A. B.; Thomson, R. J. *Angew. Chem., Int. Ed.* **2014**, *53*, 1395–1398.
68. Mundal, D. A.; Avetta, Jr., C. T.; Thomson, R. J. *Nat. Chem.* **2010**, *2*, 294–297.
69. Gutierrez, O.; Strick, B. F.; Thomson, R. J.; Tantillo, D. J. *Chem. Sci.* **2013**, *4*, 3997–4003.

15

[2,3]-REARRANGEMENTS OF AMMONIUM ZWITTERIONS

HONGLI BAO

Key Laboratory of Coal to Ethylene Glycol and Its Related Technology, Fujian Institute of Research on the Structure of Matter, Chinese Academy of Sciences, Fuzhou, Fujian, China

UTTAM K. TAMBAR

Department of Biochemistry, The University of Texas Southwestern Medical Center at Dallas, Dallas, TX, USA

15.1 INTRODUCTION

Sigmatropic rearrangements are powerful reactions for the synthesis of complex natural products and other target molecules. While [3,3]-rearrangements have received the most attention,[1,2] [2,3]-rearrangements have also been well studied because of the synthetic utility of the products and the unique mechanistic features of the transformations.[3] In particular, [2,3]-rearrangements of ammonium zwitterions **1** are useful for the synthesis of nitrogen-containing chiral molecules, such as *O*-allylhydroxylamines **2a** and unnatural amino acid derivatives **2b** (Scheme 15.1). These rearrangements have served as key steps in the synthesis of complex natural products, which will be the subject of this review. Zwitterions **1** can also undergo [1,2]-rearrangements to generate constitutional isomers of **2a** and **2b**, and these processes have been reviewed elsewhere.[4] We focus on the sigmatropic [2,3]-rearrangements of reactive ammonium zwitterions, which proceed through five-membered pericyclic transition states.

Molecular Rearrangements in Organic Synthesis, First Edition. Edited by Christian M. Rojas.

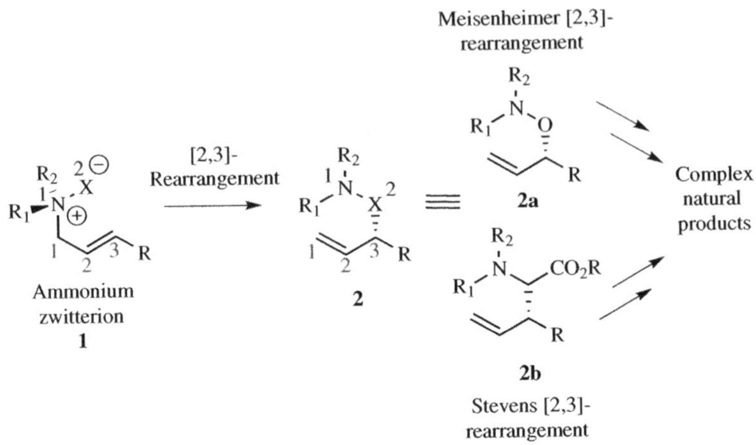

Scheme 15.1 Sigmatropic [2,3]-rearrangements of reactive ammonium zwitterions.

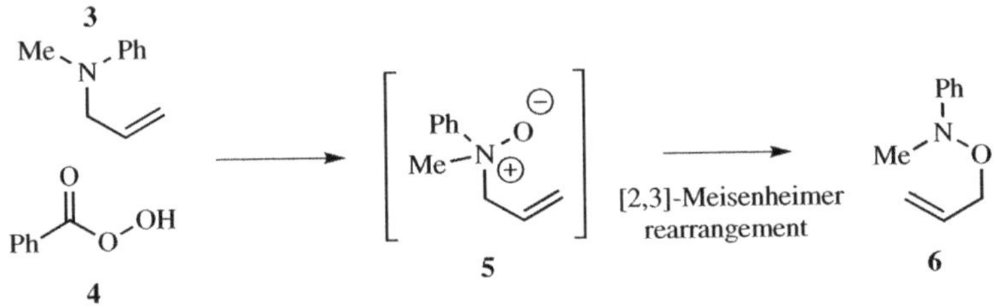

Scheme 15.2 Meisenheimer's original report of an amine N-oxide rearrangement.

15.2 [2,3]-MEISENHEIMER REARRANGEMENT OF AMINE N-OXIDES

15.2.1 General Discussion of [2,3]-Meisenheimer Rearrangement

In 1919, Meisenheimer reported a detailed account of an unusual transformation of methyl-allyl-aniline **3** in the presence of benzoic peracid **4**, which yielded the N,N,O-trisubstituted hydroxylamine product **6** through amine *N*-oxide **5** (Scheme 15.2).[5] Although the mechanism of the process was not fully appreciated at the time, this discovery was one of the earliest examples of a [2,3]-sigmatropic rearrangement, and it served as the foundation for the discovery and development of many [2,3]-rearrangements of reactive zwitterionic substrates.

Today, it is believed that [2,3]-sigmatropic rearrangements of ammonium zwitterions such as **7** proceed through a five-membered pericyclic transition state **8** of a doubly suprafacial migration (Scheme 15.3).[3c] The transition state is of the Hückel type, and since six electrons participate, the reaction is expected to be thermally allowed in accordance with the Woodward–Hoffmann orbital symmetry rules.[6] Amine *N*-oxides can also undergo [1,2]-rearrangements that proceed by a radical-pair

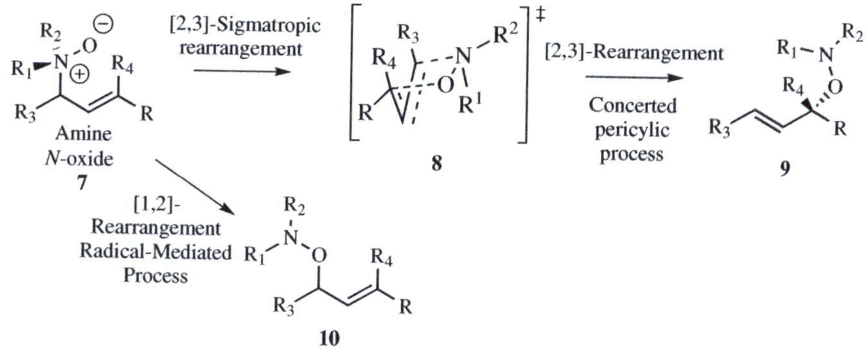

Scheme 15.3 [2,3]- and [1,2]-Meisenheimer rearrangements of amine *N*-oxides.

mechanism (**7** → **10**). Although this secondary pathway usually possesses a higher energy of activation, the likelihood of this nonpericyclic pathway depends on the specific structure of the amine *N*-oxide substrate and the reaction conditions. For example, in cases where the substrate contains an allyl group on nitrogen, the concerted [2,3]-rearrangement competes favorably with the radical-mediated [1,2]-rearrangement. In fact, the [2,3]-rearrangement is usually kinetically favored.[7] In contrast, the rearrangement of non-allylic compounds proceeds through a radical dissociation–recombination event, as demonstrated in the [1,2]-rearrangement of benzylamine oxides.[8]

15.2.2 Synthesis Strategies Based on [2,3]-Meisenheimer Rearrangement

The [2,3]-rearrangement of amine *N*-oxides has become a powerful strategy for the total synthesis of natural products and other complex molecules. From a retrosynthetic perspective, there are several retrons associated with this process (Scheme 15.4). For example, the identification of an *O*-allylhydroxylamine **9** in the target molecule can guide the utilization of the Meisenheimer rearrangement in a total synthesis. In addition, the synthetic versatility of the olefin functional group in *O*-allylhydroxylamine **9** provides an opportunity to synthesize other *O*-alkylhydroxylamines (**11a–c**) from amine *N*-oxide **7**. Chiral alcohol structures **12** and **13a–c** represent more latent retrons for the [2,3]-rearrangement of amine *N*-oxides, because several methods exist for the mild and chemoselective cleavage of N—O bonds in the presence of other functional groups. As a result, the Meisenheimer rearrangement is a truly versatile method for the synthesis of nitrogen- and/or oxygen-containing chiral building blocks.

15.2.3 Methodologies for Assembling Amine *N*-Oxide Substrates

N-allyl amine *N*-oxides **1a** are usually synthesized by the electrophilic oxidation of tertiary allylic amines **14** (Scheme 15.5). Metal-oxo complexes can be generated

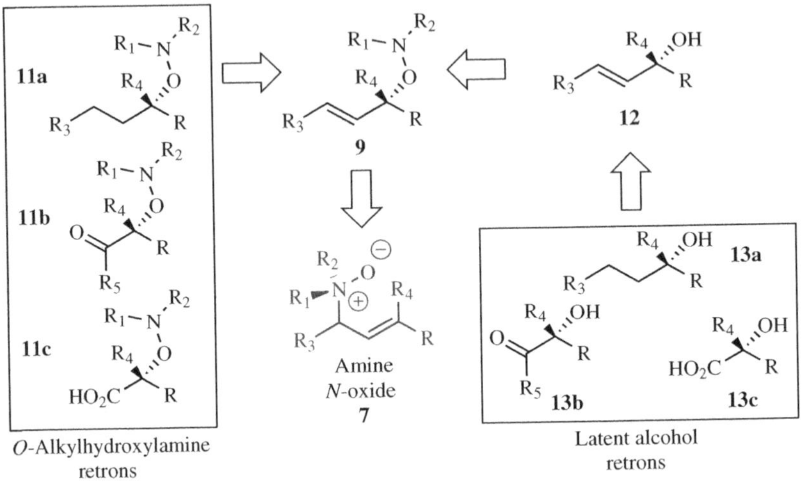

Scheme 15.4 Retrosynthetic analysis with the [2,3]-Meisenheimer rearrangement.

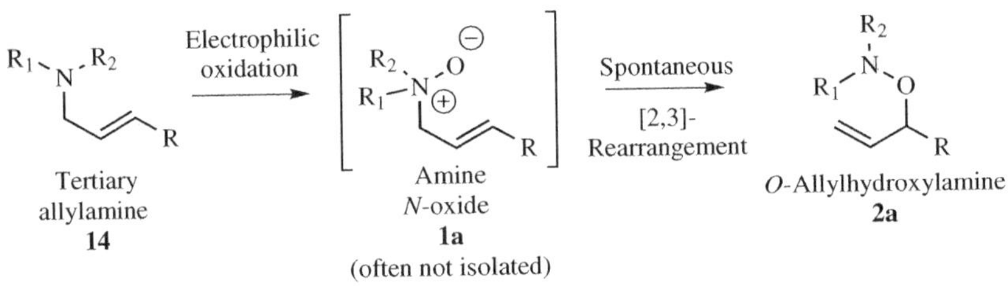

Electrophilic oxidation:
- ML_n (e.g., $Mo(CO)_6$, $VO(acac)_2$, V_2O_5, $W(CO)_6$), H_2O_2 or *t*-BuOOH, THF or *t*-BuOH
- *m*-CPBA, CH_2Cl_2 or $CHCl_3$, 0 °C

Scheme 15.5 Synthesis of amine *N*-oxides from tertiary amines.

in the presence of high-valent metals and peroxides. These electrophilic reagents can oxidize the nitrogen atom without oxidizing the pendant olefin. More commonly, nonmetal oxidants such as peracids are utilized in the selective oxidation of tertiary allylic amines into amine *N*-oxides. Usually, the resulting reactive zwitterions undergo spontaneous [2,3]-rearrangements at ambient temperature to yield *O*-allylhydroxylamines **2a**. However, in some cases, the amine *N*-oxides are stable at ambient temperature and are recalcitrant to rearrangement. In these cases, elevated temperatures can facilitate the [2,3]-rearrangement to *O*-allylhydroxylamines.

15.2.4 Applications of [2,3]-Meisenheimer Rearrangements in Synthesis

The presence of an N—O bond in a target molecule is often a retron for the [2,3]-rearrangement of amine *N*-oxides. For example, the medicinal agent **18**, an

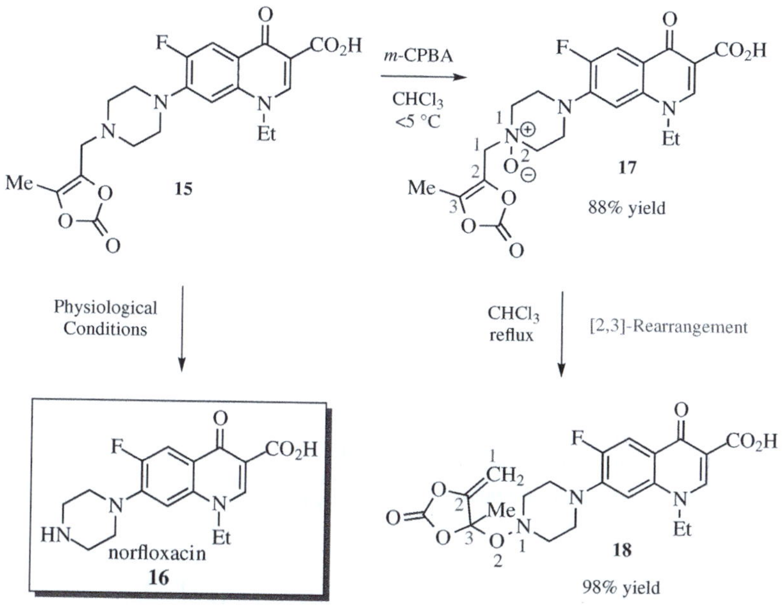

Scheme 15.6 Synthesis of an analog of norfloxacin.

analogue of norfloxacin **16**, was assembled through a selective [2,3]-rearrangement of amine *N*-oxide **17** (Scheme 15.6).[9] Norfloxacin is a functionalized quinolone with antibacterial activity. The *m*-CPBA-mediated oxidation of its N-masked derivative **15**, which can be transformed into norfloxacin under physiological conditions, afforded amine *N*-oxide **17**, which was converted to hydroxylamine **18**. *In vivo*, the activity of **18** was higher than that of norfloxacin for both Gram-positive and Gram-negative bacteria. Bioavailability studies in mice showed that **18** liberated a higher concentration of norfloxacin in plasma than norfloxacin itself when administered orally. From these data, **18** obtained by the chemical oxidation of **15** functioned as an effective prodrug of norfloxacin. The mechanism of the formation of **18** from amine *N*-oxide **17** was interpreted as a concerted [2,3]-sigmatropic rearrangement, based on the lack of any observed effect on the reaction upon addition of radical scavengers such as 1-butanethiol.

The eudistomins are a class of polycyclic alkaloids that contain a unique 1,3,7-oxathiazepine ring system as well as an *N,N,O*-trisubstituted hydroxylamine substructure (Scheme 15.7). These natural products were isolated by Rinehart *et al.* from the colonial tunicate *Eudistoma olivaceum* in 1984, and they exhibited strong antiviral activity against the Herpes simplex virus (HSV-1), some *in vivo* antitumor activity, and calmodulin antagonist activity.[10]

Although the eudistomins have not been synthesized through [2,3]-rearrangements of amine *N*-oxides, the related compounds **20–27** were synthesized by Kurihara *et al.* from 2-vinylazetopyridoindoles **30** and **31** via the [2,3]-Meisenheimer rearrangement of the corresponding *N*-oxides (Scheme 15.8).[11] More specifically, tertiary allylic

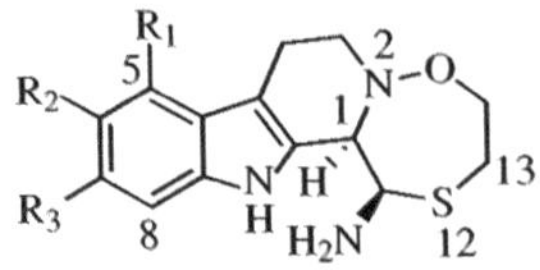

Eudistomins (**19**)

L: R_1=R_3=H, R_2=Br
K: R_1=R_2=H, R_3=Br
C: R_1=H, R_2=OH, R_3=Br
E: R_1=Br, R_2=OH, R_3=H

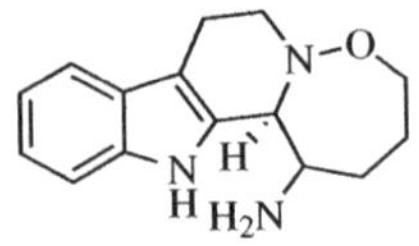

20: R=H, β-NH_2
(12-carbaeudistomin)
21: R=H, α-NH_2

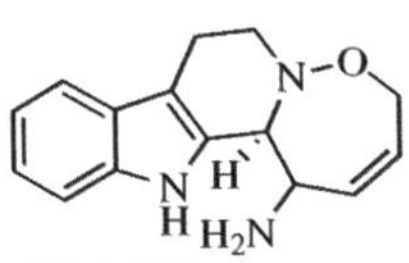

22: β-NH_2
(11,12-didehydro-
12-carbaeudistomin)
23: α-NH_2

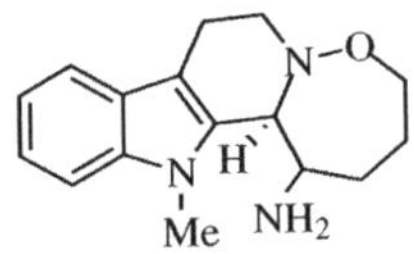

24: β-NH_2
(9-methyl-12-carbaeudistomin)
25: α-NH_2

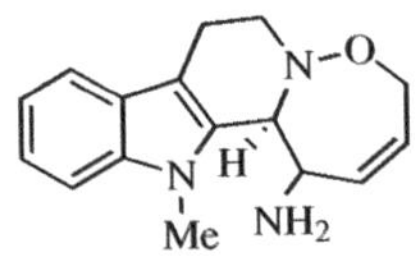

26: β-NH_2
(9-methyl-11,12-didehydrom-
12-carbaeudistomin)
27: α-NH_2

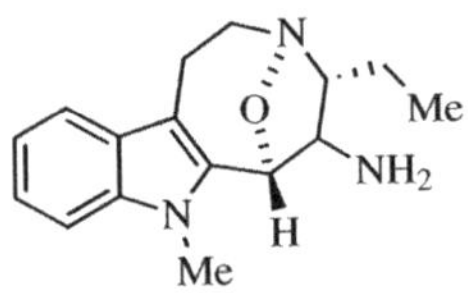

28: β-NH_2
29: α-NH_2

Scheme 15.7 Structures of the eudistomin natural products.

amines **30** and **31** were transformed into the amine *N*-oxides **32** upon treatment with *m*-CPBA. These ammonium zwitterions underwent selective [2,3]-rearrangements to furnish *O*-allylhydroxylamines **33**. These functionalized products were then transformed into the various analogues of the eudistomins (**20–27**).

[2,3]-Rearrangements have also been employed in the synthesis of analogues of tropanone alkaloids such as physoperuvine (Scheme 15.9). For example, Bremner *et al.* transformed the natural product scopolamine **37** in three steps into allylic amine **38**, which was selectively oxidized to diastereomeric amine *N*-oxides **39a** and **39b** in the presence of hydrogen peroxide.[12] *N*-oxide **39b** was then refluxed in butyronitrile to furnish *O*-allylhydroxylamine **40**. This intermediate was hydrogenated in two stages to the saturated analogue **42**, which was transformed into the 3-hydroxy analogue **43** of physoperuvine (**44**).

While most synthetic applications of [2,3]-Meisenheimer rearrangements result in the formation of N—O-containing target molecules, chiral alcohols can also be latent retrons for these rearrangement processes because of the ease of cleaving N—O bonds. For example, the natural product sulcatol **50** was synthesized through a

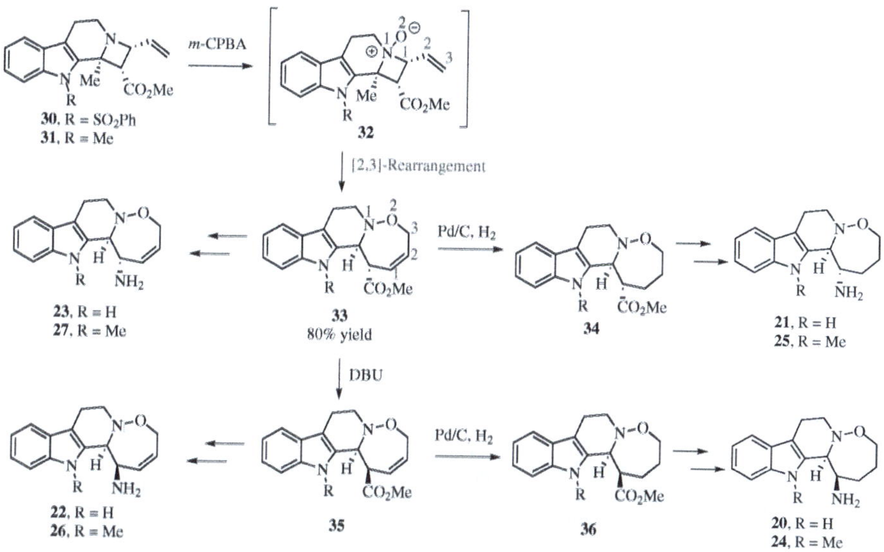

Scheme 15.8 Synthesis of analogs of the eudistomin natural products.

stereoselective [2,3]-Meisenheimer rearrangement of amine *N*-oxide **47** (Scheme 15.10). Sulcatol is a male-produced aggregation pheromone of the ambrosia beetle, a pest of economic significance in Western North America. Interestingly, many studies have shown that beetle species respond differently to sulcatol, depending on the compound's enantiomeric composition. As a result, there has been much interest in synthesizing sulcatol in enantiomerically pure form for more detailed biological studies and as a potential agent for trapping beetles in pest control programs. Davies and Smyth achieved an asymmetric synthesis of (*R*)-sulcatol via a route involving stereoselective conjugate addition, Grignard addition, and stereoselective [2,3]-Meisenheimer rearrangement.[13] The stereochemical information from the two stereocenters in amine *N*-oxide **47** was transferred to the carbinol stereocenter in *O*-allylhydroxylamine **48** through a diastereoselective [2,3]-rearrangement. Subsequent dehydration, olefin reduction, and N—O bond cleavage revealed sulcatol **50**.

Linalool **54** was similarly synthesized through a [2,3]-Meisenheimer rearrangement by taking advantage of the facile cleavage of N—O bonds in the *O*-allylhydroxylamine rearrangement products (Scheme 15.11).[14] Aminoalcohol **51** was converted to amine *N*-oxide **52**, which was transformed to *O*-allylhydroxylamine **53** under thermal conditions. This intermediate was subjected to reductive conditions, which selectively cleaved the N—O bond to unveil linalool **54**.

The [2,3]-rearrangement of amine *N*-oxides was utilized in an efficient synthesis of 2,6-dimethyl-1,5-heptadien-3-ol acetate **61**, a pheromone of the insect *Pseudococcus comstocki* (Scheme 15.12).[15] Dimethylpyridine **55** was converted in two steps into silylated piperidine **56**, which was oxidized with *m*-CPBA to generate cyclic amine *N*-oxide **57**. Sila-Cope elimination furnished *O*-silylhydroxylamine **58**, which was methylated and desilylated in the presence of MeI and CsF to yield acyclic amine

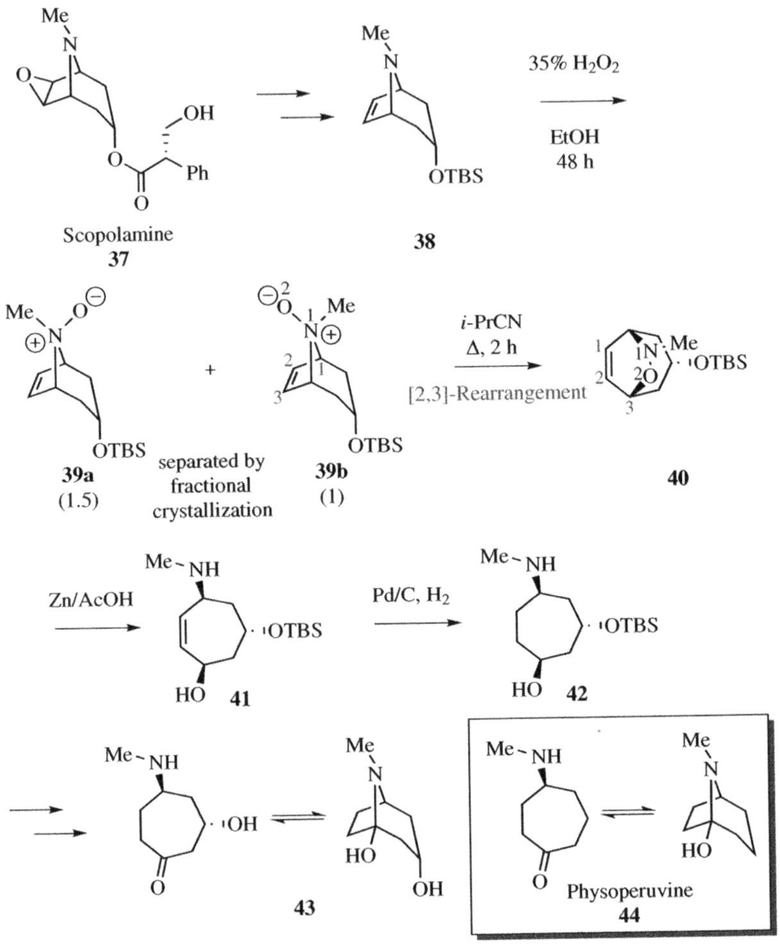

Scheme 15.9 Synthesis of physoperuvine.

N-oxide **59**. Heating this ammonium zwitterion facilitated the [2,3]-Meisenheimer rearrangement to *O*-allylhydroxylamine **60**, which was transformed to the desired pheromone **61** in three steps.

The [2,3]-Meisenheimer rearrangement has also been utilized to interconvert biosynthetically related natural products *β*-santalene to *β*-santalol (Scheme 15.13).[16] Initially, *β*-santalene **62** was subjected to a selective allylic chlorination followed by displacement with aqueous dimethylamine to generate allylic amine **63**. This intermediate was then oxidized in the presence of peracetic acid to yield amine *N*-oxide **64**, which underwent a [2,3]-rearrangement to *O*-allylhydroxylamine **65**. Reductive cleavage of the N—O bond unveiled the (*E*)-isomer of the natural product *β*-santalol **66**.

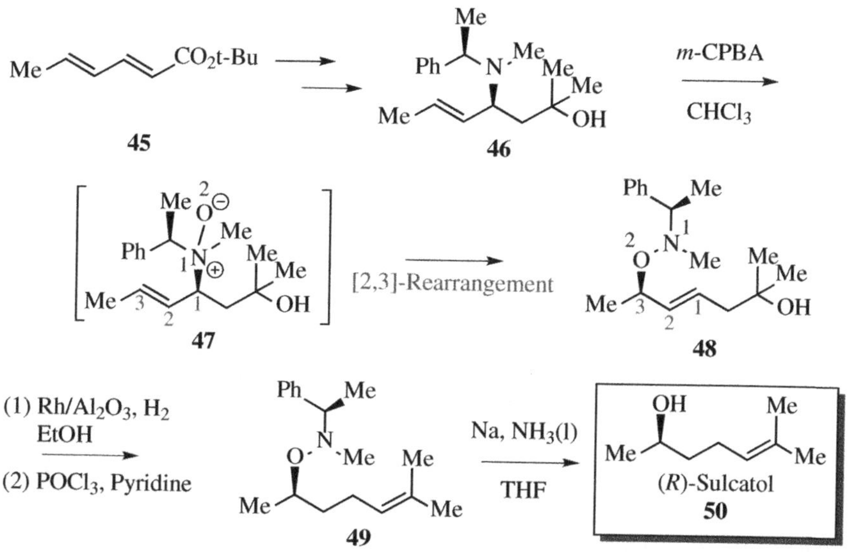

Scheme 15.10 Synthesis of sulcatol.

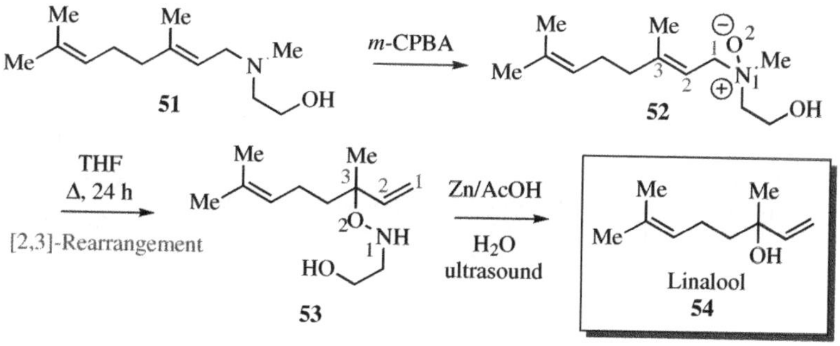

Scheme 15.11 Synthesis of linalool.

15.2.5 Mechanistic Features Crucial for Synthetic Utility: Stereoselectivity

For many synthetic applications, the [2,3]-Meisenheimer rearrangement is utilized in complex settings where there is at least one stereocenter already present. However, this rearrangement could be more generally applied as an enantioselective process earlier in a synthetic sequence for the assembly of enantioenriched alcohol building blocks. This would require the generation of the carbinol stereocenter selectively by the rearrangement process. Therefore, much of the recent work on the [2,3]-rearrangement of amine *N*-oxides has focused on developing stereoselective methods for the generation of enantioenriched products.[17]

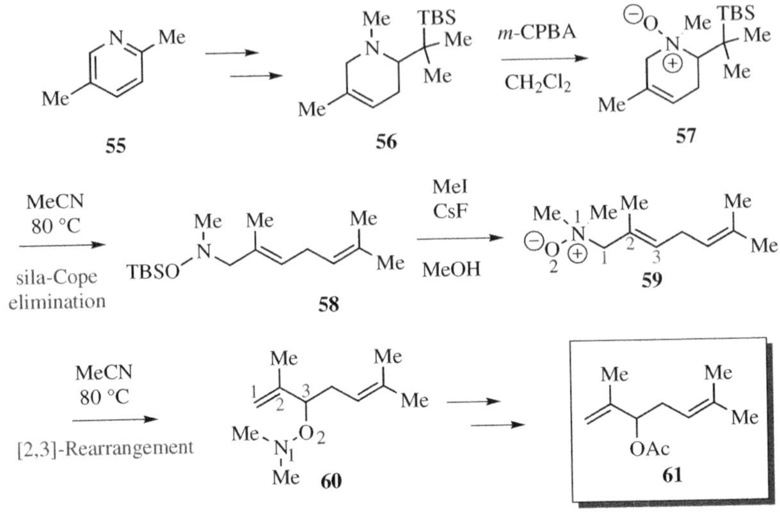

Scheme 15.12 Synthesis of the pheromone 2,6-dimethyl-1,5-heptadiene-3-ol acetate.

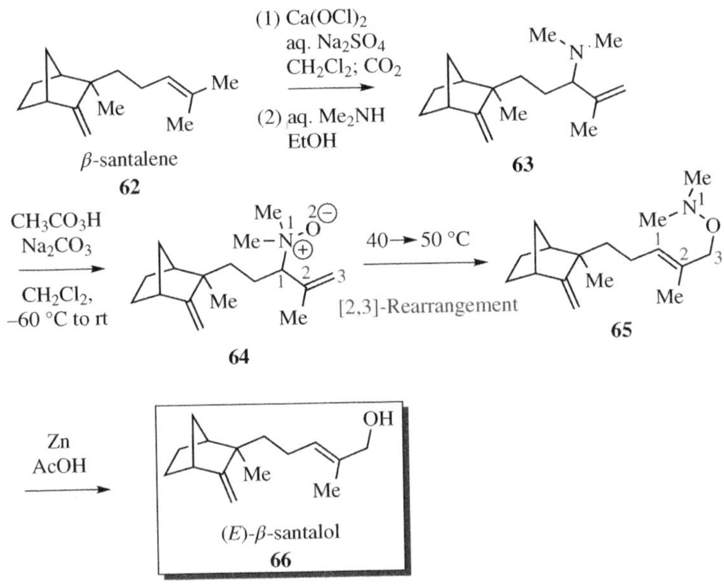

Scheme 15.13 Synthesis of *β*-santalol.

15.2.5.1 Diastereoselective [2,3]-Meisenheimer Rearrangements Early examples of stereoselective [2,3]-Meisenheimer rearrangements were based on diastereoselective reactions of chiral amine *N*-oxides such as **68** (Scheme 15.14). Stereochemical information was transferred from the carbon stereocenter in

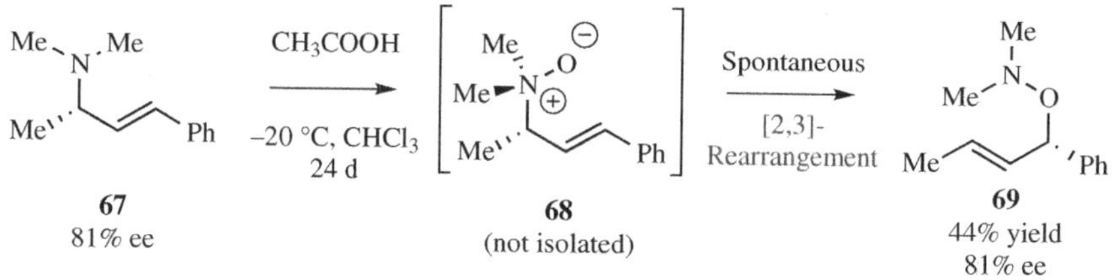

Scheme 15.14 Early example of a diastereoselective [2,3]-Meisenheimer rearrangement of enantioenriched amine *N*-oxides with carbon stereocenters.

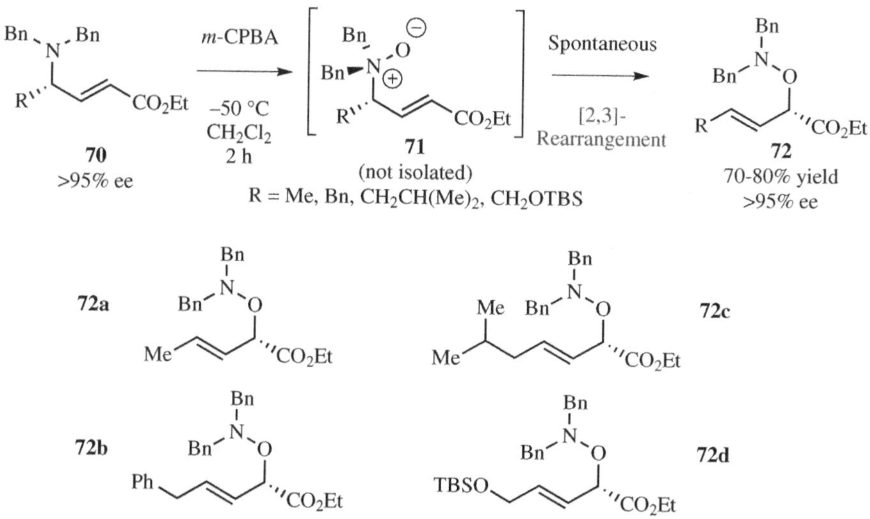

Scheme 15.15 Reetz's diastereoselective [2,3]-Meisenheimer rearrangement of enantioenriched amine *N*-oxides with carbon stereocenters.

enantioenriched ammonium zwitterion **68** to the single carbon stereocenter in the rearrangement product **69**. While the selective [2,3]-sigmatropic rearrangement proceeded at −20 °C, the radical path prevailed at higher temperatures to yield exclusively the [1,2]-rearrangement product in considerably lower enantiomeric excess.[18]

Similarly, Reetz and Lauterbach studied the diastereoselective [2,3]-rearrangement of chiral amine *N*-oxides **71**, which were generated from enantioenriched chiral allylic amines **70** (Scheme 15.15).[19] These sigmatropic rearrangements generated *O*-allylhydroxylamine products **72a–d** with complete 1,3-transfer of chirality.

In an effort to develop a more efficient method to synthesize enantioenriched allylic amine substrates for [2,3]-rearrangements, Davies and Smyth reported an asymmetric conjugate addition with chiral lithium amide **74** (Scheme 15.16).[20] The resulting aminoester **75** was reduced to alcohol **76**, which was treated with *m*-CPBA to furnish chiral amine *N*-oxide **77**. This ammonium zwitterion underwent

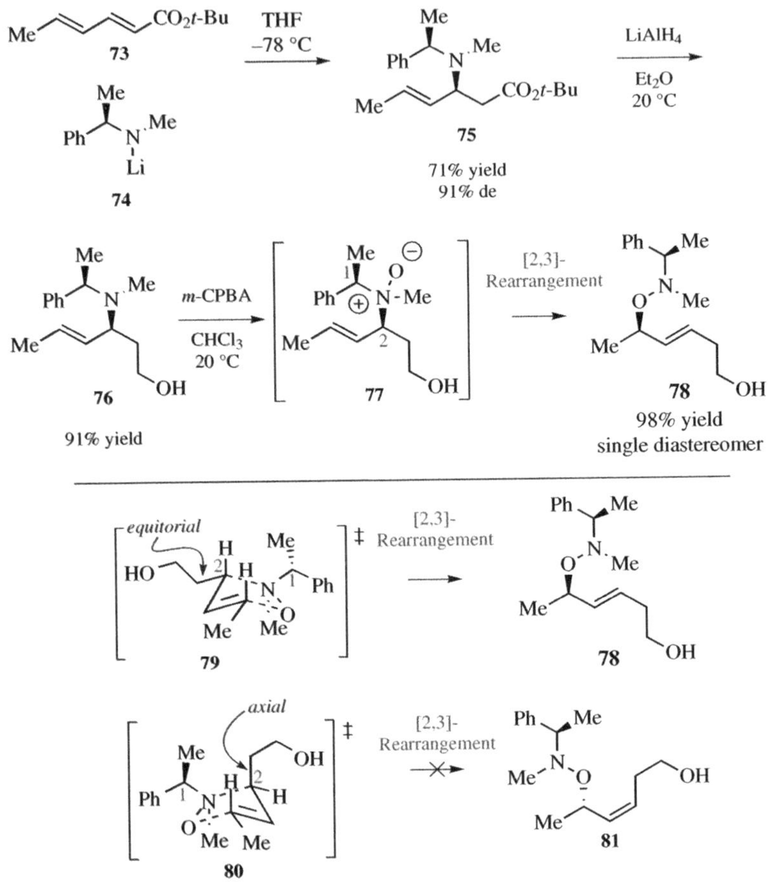

Scheme 15.16 Davie's diastereoselective [2,3]-Meisenheimer rearrangement of enantioenriched amine *N*-oxides with chiral benzyl amine.

a diastereoselective [2,3]-Meisenheimer rearrangement at room temperature to form *O*-allylhydroxylamine **78** as a single diastereomer. Although the absolute stereochemistry of C1 in amine *N*-oxide **77** may have affected the stereochemical outcome of the [2,3]-rearrangement, it is more likely that the absolute stereochemistry at C2 governed the sense of stereoinduction in the formation of *O*-allylhydroxylamine **78**. The preference for forming *O*-allylhydroxylamine **78** instead of *O*-allylhydroxylamine **81** may be due to the greater stability of transition state **79** over transition state **80**. Interestingly, a new stereocenter at nitrogen was formed during oxidation of allylic amine **76** to amine *N*-oxide **77**, but the stereoselectivity in forming this nitrogen stereocenter was not discussed.

While these three diastereoselective processes were seminal for the elucidation of models of stereoinduction in the [2,3]-Meisenheimer rearrangement, they necessitated the enantioselective synthesis of chiral allylic amines such as **67**, **70**, and **76**.

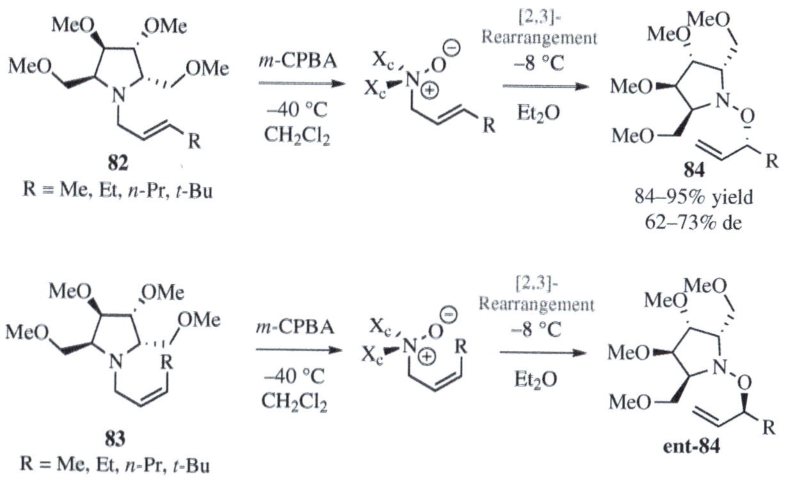

Scheme 15.17 Ender's [2,3]-Meisenheimer rearrangement with C2-symmetric chiral auxiliaries.

15.2.5.2 **Enantioselective [2,3]-Meisenheimer Rearrangements with *C_2-Symmetric Chiral Auxiliaries*** The use of C_2-symmetric chiral auxiliaries attached to the nitrogen atom of amine *N*-oxides can obviate the need for synthesizing substrates bearing an allylic amine stereocenter for stereoselective [2,3]-Meisenheimer rearrangements. For example, Enders and Kempen developed a C_2-symmetric pyrrolidine auxiliary **82** that guided the selective formation of the carbon stereocenter in *O*-allylhydroxylamine **84** (Scheme 15.17).[21] Several C_2-symmetric amines furnished the rearrangement product in low diastereomeric excess (30–35% de), and some pyrrolidine auxiliaries showed nearly no asymmetric induction. However, the best results (62–73% de) were obtained with the C_2-symmetric amine **82**. Oxidation and rearrangement of the corresponding (*Z*)-allylic amine **83** led to the inverse absolute configuration of the new carbon stereocenter in the rearrangement product (*ent*-**84**), which was expected from a concerted sigmatropic rearrangement through a well-ordered transition state.

This early example by Enders demonstrated the efficient transfer of stereochemical information of the carbon stereocenters in the auxiliary to the carbon stereocenter in the hydroxylamine product. The model for stereoinduction in this [2,3]-Meisenheimer rearrangement can be explained through the competing diastereoselective pericyclic transition states **85** and **86** (Scheme 15.18). This model also accounts for the observed formation of *ent*-**84** from the (*Z*)-allylic amine **83** via transition state **87**.

15.2.5.3 Enantioselective [2,3]-Meisenheimer Rearrangements via Nitrogen Stereocenters An alternative approach for developing a stereoselective Meisenheimer rearrangement is based on the transfer of chirality from the nitrogen atom of the ammonium zwitterion to the carbon stereocenter of the *O*-allylhydroxylamine

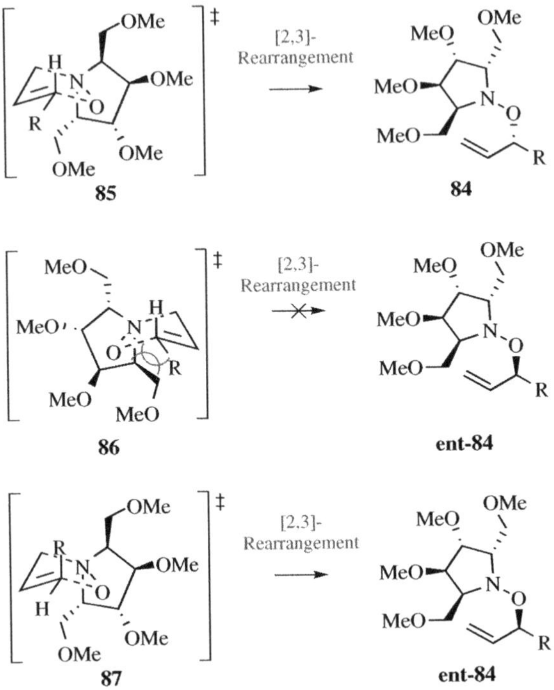

Scheme 15.18 Stereochemical model for Ender's [2,3]-Meisenheimer rearrangement with C2-symmetric chiral auxiliaries.

product (Scheme 15.19). The quaternized nitrogen stereocenter in an amine *N*-oxide is configurationally stable, which allows the possibility of transferring chirality from nitrogen to carbon through a stereospecific sigmatropic rearrangement. Inouye and coworkers examined this approach by classically resolving chiral amine *N*-oxide **88** to an enantiomeric excess of 16%.[7] This slightly enantioenriched ammonium zwitterion underwent a stereospecific [2,3]-rearrangement followed by a reductive N—O bond cleavage to generate allylic alcohol **90** in 13.6% ee. Although this was a landmark result for the development of an enantioselective Meisenheimer rearrangement, the amine *N*-oxide could not be classically resolved to higher and more synthetically useful levels of enantiopurity.

To generate nitrogen stereocenters with greater stereoenrichment, several groups examined non-C_2-symmetric chiral auxiliaries. In Coldham group's early examination of this approach, the expectation was that the stereochemical information of the carbon stereocenters in the auxiliary would lead to a diastereoselective oxidation of allylic amine substrates such as **91** in the presence of an achiral oxidant such as *m*-CPBA (Scheme 15.20).[14] The resulting amine *N*-oxide **92** would possess a nitrogen stereocenter, which was hopefully generated stereoselectively in the oxidation step, and this intermediate would then undergo a stereospecific [2,3]-rearrangement to yield diastereomerically enriched *O*-allylhydroxylamine **93**. The amine *N*-oxides

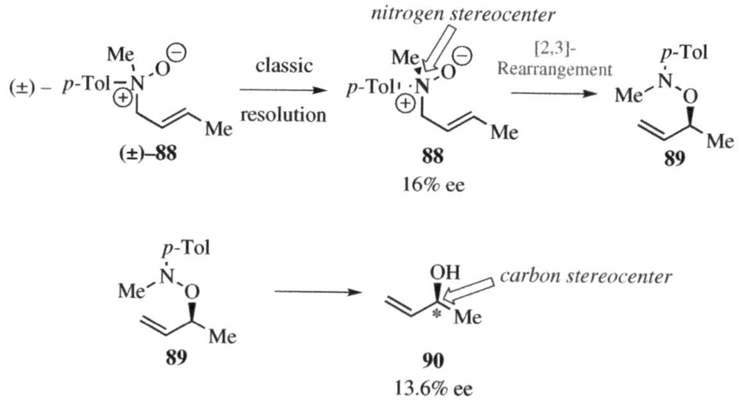

Scheme 15.19 Configurationally stable quaternized nitrogen stereocenter in amine *N*-oxides.

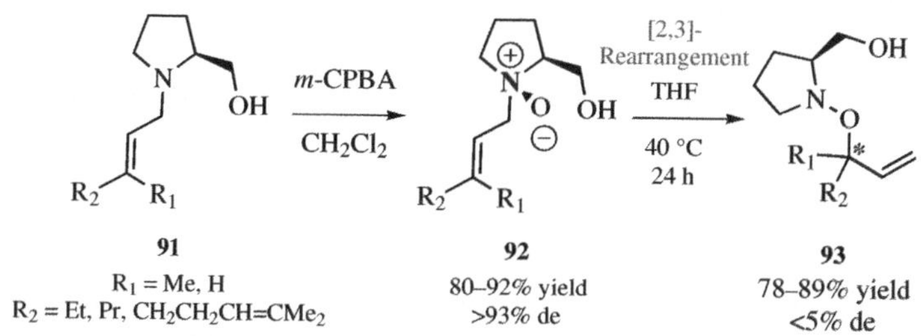

Scheme 15.20 [2,3]-Meisenheimer rearrangement with a proline-based chiral auxiliary.

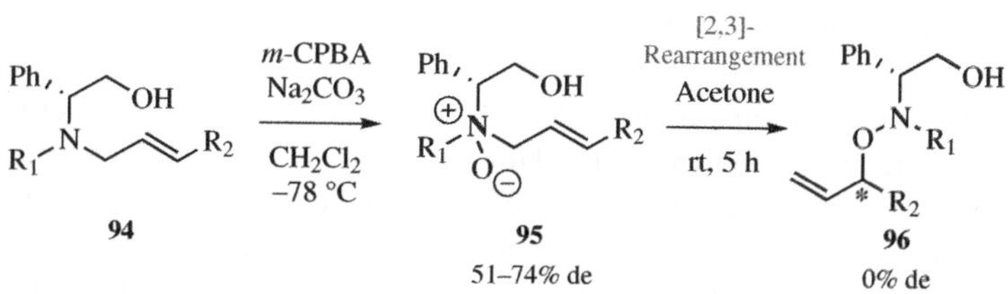

Scheme 15.21 [2,3]-Meisenheimer rearrangement with an aminoalcohol-based chiral auxiliary.

92 were generated with high diastereoselectively, as confirmed by NOE NMR studies and X-ray crystallography. Unfortunately, the rearrangement products **93** were isolated in very low diastereoselectivity (<5% de).

To compare the transfer of chirality in a [2,3]-rearrangement from carbon to carbon versus nitrogen to carbon, Bonin and coworkers devised a clever series of experiments (Scheme 15.21).[22] Allylic amines **94** were prepared from N-protected

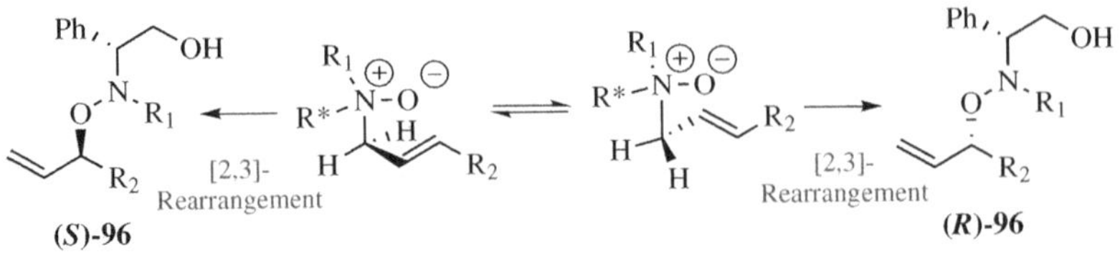

Scheme 15.22 Explanation for low stereoselectivity in [2,3]-Meisenheimer rearrangement with an aminoalcohol-based chiral auxiliary.

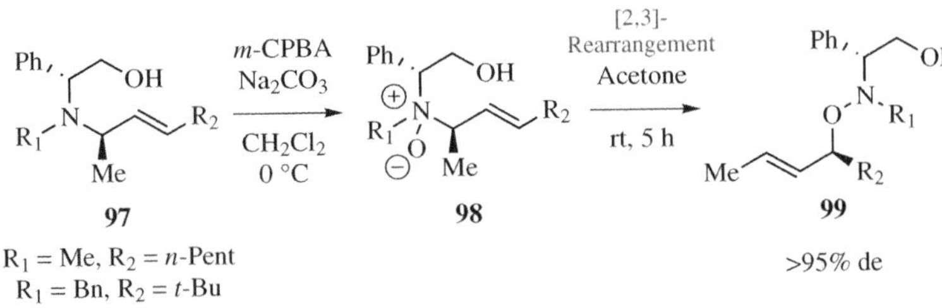

Scheme 15.23 Matched case for double diastereoselective [2,3]-Meisenheimer rearrangement.

phenylglycinol. Oxidation with *m*-CPBA yielded stable amine *N*-oxides **96**, which were isolated in 51–74% de (confirmed by ^{1}H NMR analysis). Ammonium zwitterions **95** proved to be highly stable in dichloromethane or chloroform, presumably because of an intramolecular hydrogen bond of the oxygen of the amine *N*-oxide and the pendant hydroxyl group. Increasing the polarity of the solvent led to an increase in rate of the [2,3]-rearrangement. In acetone, *O*-allylhydroxylamines **96** were obtained in quantitative yield after less than 5 h at room temperature. However, in each case, no diastereoselectivity was observed after the rearrangement. Lowering the reaction temperature (−20 °C) led to a large decrease of chemical yield, without any selectivity improvement. This loss of selectivity was attributed to the reversibility of the Meisenheimer rearrangement, which could lead to epimerization of the hydroxylamine product via the N-inverted diastereomer of **95**.

Another explanation for the low diastereoselectivity of the [2,3]-rearrangement of **95** is the presence of two reactive conformations of the amine *N*-oxide, which lead to different diastereomers of the product (Scheme 15.22). These two conformations are in equilibrium by a simple bond rotation. Even if the nitrogen stereocenter in amine *N*-oxide is generated stereoselectively, if the two conformations of the amine *N*-oxide are equally populated, the rearrangement product will be obtained in low diastereoselectivity, assuming the rates of conversion to product from each of the conformers are comparable.

Bonin designed related amine *N*-oxides **98** that would not be able to adopt multiple energetically similar conformations (Scheme 15.23). The presence of the allylic stereocenter α to the nitrogen led to a high yielding [2,3]-rearrangement with complete

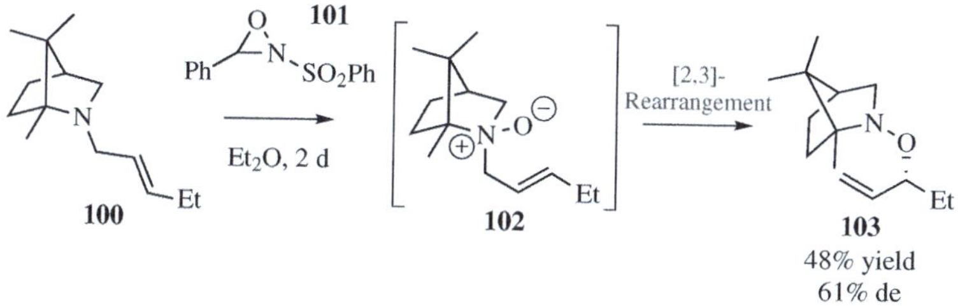

Scheme 15.24 [2,3]-Meisenheimer rearrangement with a camphor-based chiral auxiliary.

chirality transfer. The introduction of the α-substituent also resulted in a dramatic increase of rearrangement rate. Although *N*-oxides could not be isolated for such derivatives, it is highly probable that the formation of amine *N*-oxides **98** was not fully stereoselective at nitrogen. Bonin and coworkers therefore concluded that in this case, the configuration at nitrogen in amine *N*-oxide **98** is not important for the stereochemical outcome of the [2,3]-rearrangement. Rather, the excellent transfer of chirality occurs from the allylic carbon stereocenter in **98** to the allylic carbon stereocenter in **99**.

Although the configuration at the nitrogen stereocenter of many ammonium ylides does not influence the stereoselectivity of [2,3]-Meisenheimer rearrangements, several groups hypothesized that cyclic, rigid chiral auxiliaries would lead to nitrogen stereocenters with more well-defined stereochemical environments that may facilitate a diastereoselective rearrangement. For example, Coldham and coworkers utilized camphor-based auxiliaries **100** with oxaziridine **101** to generate amine *N*-oxides **102** with high diastereoselectivity. These underwent [2,3]-rearrangements with reasonably efficient transfer of chirality from nitrogen to carbon in forming rearrangement products **103** (Scheme 15.24).[23]

The effectiveness of non-C_2-symmetric, cyclic, rigid chiral auxiliaries was confirmed by Guarna *et al.*, using a system derived from tartaric acid and amino acids (Scheme 15.25).[24] While the formation of both *N*-oxides **99** and **102** was highly diastereoselective, the asymmetric induction in the rearrangement of **99** was generally low (e.g., *O*-allylhydroxylamine **100**). However, the interaction between a benzyl group at the 4-endo position of the chiral auxiliary and a substituent on the allylic moiety in amine *N*-oxide **102** allowed for a more efficient transfer of chirality in the [2,3]-sigmatropic process. *O*-allylhydroxylamine **103** was isolated in 65% de.

15.2.5.4 Catalytic Enantioselective [2,3]-Meisenheimer Rearrangements Our group was interested in developing a catalytic enantioselective [2,3]-Meisenheimer rearrangement, wherein a chiral catalyst (as opposed to a chiral auxiliary) would govern the formation of enantioenriched chiral hydroxylamine products. One approach would be the development of a catalytic enantioselective oxidation of allylic amine **14** to generate the nitrogen stereocenter in amine *N*-oxide **1a** in a stereoselective manner (Scheme 15.26, Eq. 1). This chiral nonracemic ammonium zwitterion could then undergo a stereospecific [2,3]-rearrangement to generate

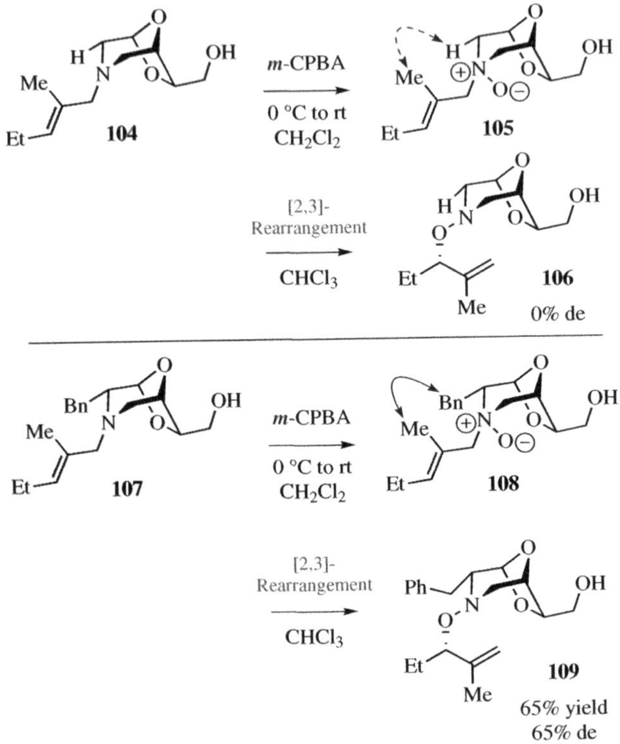

Scheme 15.25 [2,3]-Meisenheimer rearrangement with Guarna's bicylic chiral auxiliary.

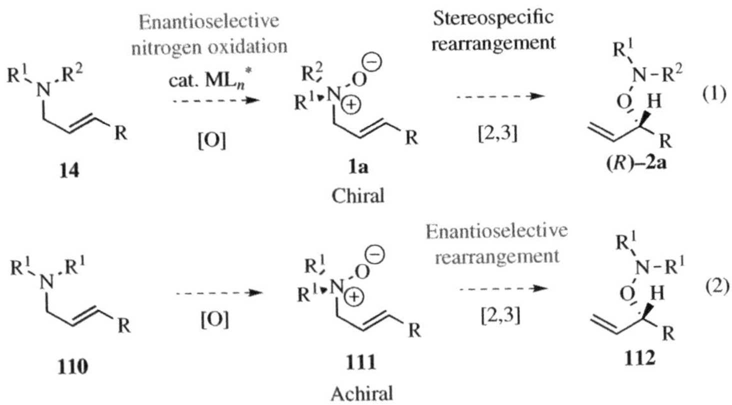

Scheme 15.26 General strategies for a catalytic enantioselective [2,3]-Meisenheimer rearrangement.

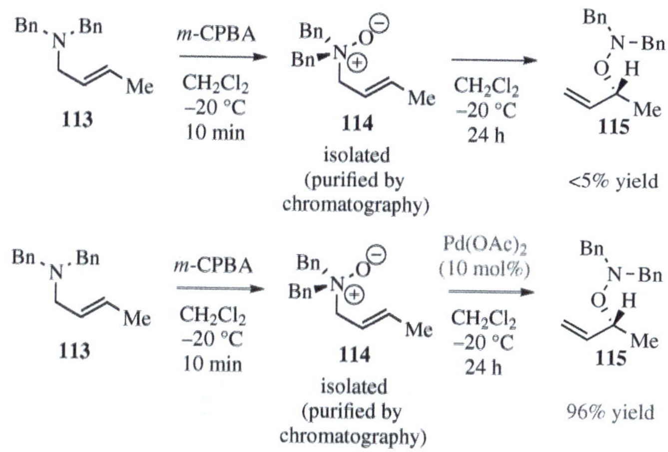

Scheme 15.27 Proof of catalysis for the [2,3]-rearrangement of amine *N*-oxides.

enantioenriched hydroxylamine (*R*)-**2a**. Unfortunately, all attempts by our group and other groups to develop an enantioselective oxidation of tertiary allylic amines for the Meisenheimer rearrangement have been unsuccessful.[14]

An alternative strategy for the development of a catalytic enantioselective Meisenheimer rearrangement is based on the conversion of achiral amine *N*-oxide **111** into chiral *O*-allylhydroxylamine **112** via an enantioselective [2,3]-rearrangement (Scheme 15.26, Eq. 2). A chiral catalyst would govern the conversion of achiral **111** into enantioenriched **112**.

We began our studies of this alternative strategy by synthesizing and isolating stable achiral allylic amine *N*-oxide **114** from oxidation of dibenzyl allylic amine **113** with *m*-CPBA (Scheme 15.27). Amine *N*-oxide **114** was stable to column chromatography. We explored the feasibility of using metal complexes to catalyze the [2,3]-rearrangement of this ammonium zwitterion. While the amine *N*-oxide was relatively stable to rearrangement at low temperatures (−20 °C) in the absence of catalyst and in the presence of several low-valent metal catalysts, we discovered that $Pd(OAc)_2$ catalyzed the [2,3]-rearrangement to furnish *O*-allylhydroxylamines **115** in high yield.

Based on the ability of $Pd(OAc)_2$ to accelerate the [2,3]-rearrangement of amine *N*-oxide **114**, we explored chiral palladium(II) salts to catalyze the enantioselective rearrangement. When we treated amine *N*-oxides **116** with $Pd(OAc)_2$ and chiral phosphoramidite **118**, chiral *O*-allylhydroxylamines **117** were isolated in 80–85% ee (Scheme 15.28). Subsequent optimization revealed the beneficial effect of methanol and *meta*-chlorobenzoic acid (*m*-CBA) as additives, allowing the isolation of chiral nonracemic *O*-allylhydroxylamines **117** with greater than 90% enantioenrichment. This palladium-catalyzed enantioselective [2,3]-rearrangement can tolerate a wide variety of functional groups in the amine *N*-oxide substrate.[25] For example, we can

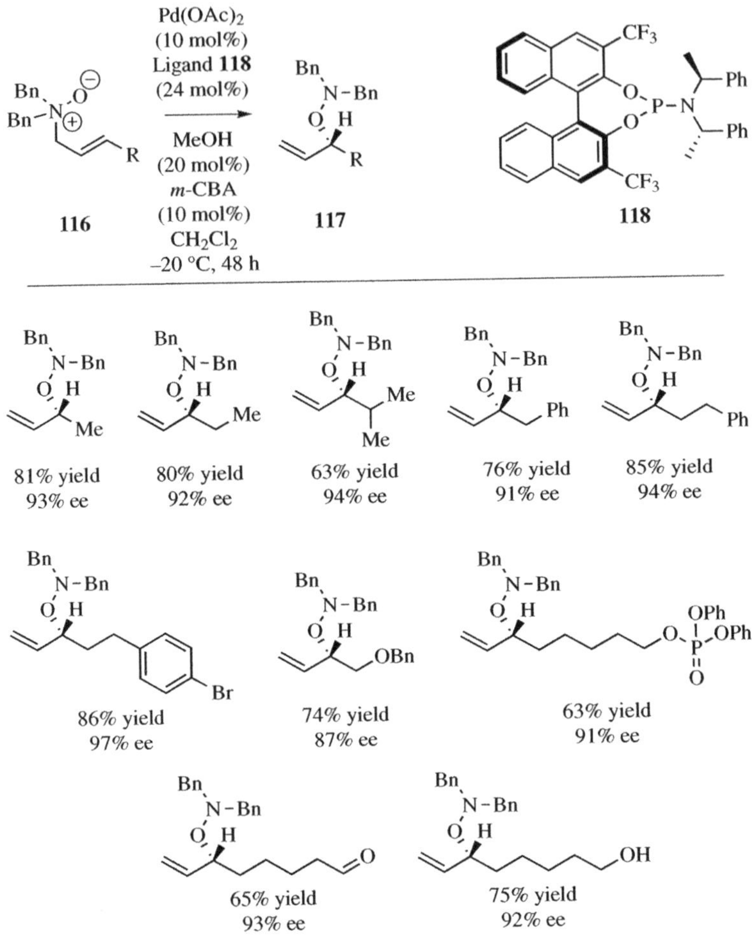

Scheme 15.28 Palladium-catalyzed enantioselective [2,3]-rearrangement of amine *N*-oxides.

synthesize chiral allylic hydroxylamine products with reactive functional groups such as alcohols and aldehydes, which are incompatible with many other methods for the synthesis of chiral alcohol derivatives.

As a preliminary mechanistic proposal, we hypothesize that the palladium(II)-phosphoramidite catalyst acts as a chiral π-acid to activate the amine *N*-oxide substrate (Scheme 15.29), similar to the mechanism proposed for the Overman rearrangement of allylic trichloroacetimidates.[26] While it is not clear whether the reactive species is oxide-bound complex **119a** or olefin-bound complex **119b**, we propose heterocycle **120** as an intermediate in this cyclization-induced mechanism. Grob-type fragmentation eventually reveals *O*-allylhydroxylamine **117** and the palladium(II)-phosphoramidite catalyst, which can reenter the catalytic cycle.

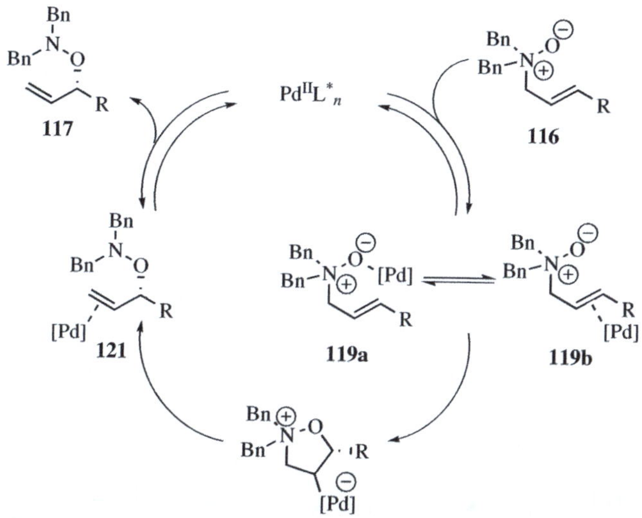

Scheme 15.29 Catalytic cycle of palladium-catalyzed [2,3]-rearrangement of amine *N*-oxides.

15.3 [2,3]-STEVENS REARRANGEMENT OF AMMONIUM YLIDES

15.3.1 General Discussion of [2,3]-Stevens Rearrangement

Ammonium ylides undergo sigmatropic rearrangements that proceed through well-ordered pericyclic transition states. Although there are several variants of this reaction, the [2,3]-Stevens rearrangement of allyl ammonium enolates **122** in particular has become a synthetically powerful transformation for accessing unnatural α,β-substituted α-amino acid derivatives **124**, and it will be the subject of this section of the chapter (Scheme 15.30).[4,27] In analogy to the rearrangement of amine *N*-oxides, ammonium ylides can also undergo [1,2]-rearrangements to **125** that proceed by a radical-pair mechanism. This higher-energy secondary pathway is unlikely if the substrate is an allylic ammonium ylide. The [1,2]-rearrangement is most commonly observed with non-allylic compounds, such as benzyl ammonium ylides.[27a,28]

15.3.2 Synthesis Strategies Based on [2,3]-Stevens Rearrangement

The [2,3]-rearrangement of ammonium ylides has become a general strategy for the total synthesis of natural products and other target molecules. Retrosynthetically, α-allyl acid derivatives **124** and **127** are common retrons for this sigmatropic process (Scheme 15.31). Given the ubiquity of noncanonical amino acid substructures in naturally occurring small molecules, the Stevens rearrangement has emerged as a synthetically powerful transformation. Moreover, the synthetic versatility of carboxyl

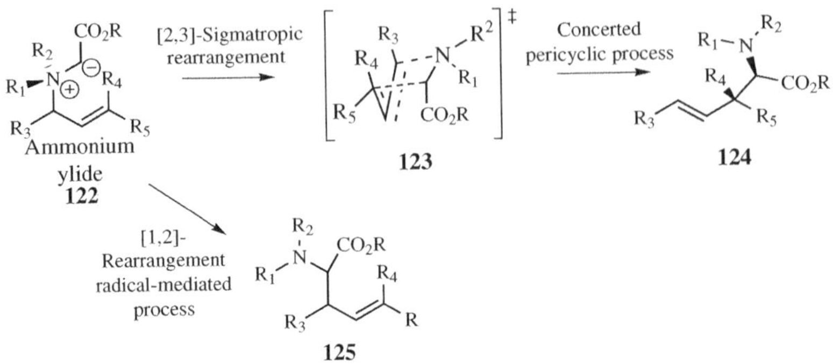

Scheme 15.30 [2,3]- and [1,2]-Stevens rearrangements of ammonium ylides.

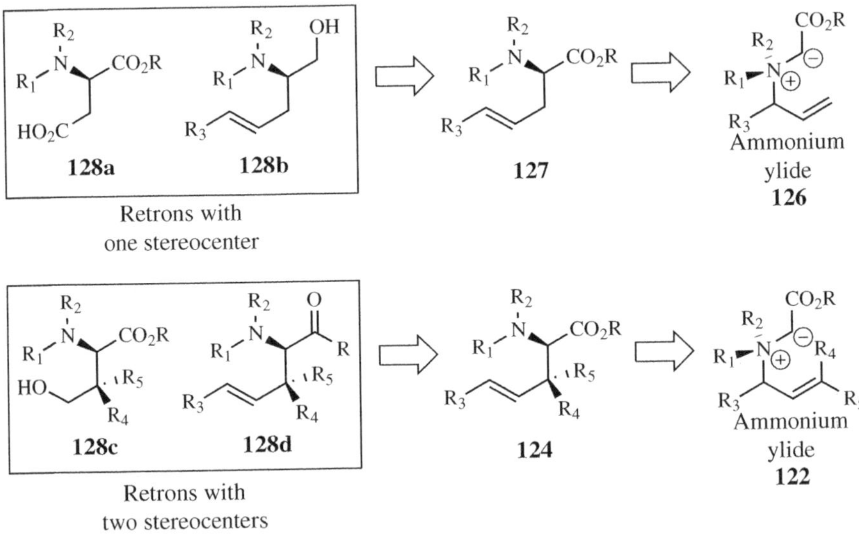

Scheme 15.31 Retrosynthetic analysis with the [2,3]-Stevens rearrangement.

groups and olefins represents an opportunity to utilize the Stevens rearrangement to access structures such as **128a–d**. While terminally unsubstituted allyl ammonium enolates such as **126** rearrange to generate products with one stereocenter (**127**, **128a**, **128b**), terminally substituted ammonium enolates such as **122** ($R_4, R_5 \neq H$) can generate products with two contiguous stereocenters (**124**, **128c**, **128d**), which provides a unique opportunity to access unnatural α-amino acid derivatives with multiple stereocenters.

15.3.3 Methodologies for Assembling the Ammonium Ylide Substrates

Traditionally, allyl ammonium enolates are synthesized by treating tertiary allyl aminoesters **129** with reactive electrophiles **130** under harsh conditions

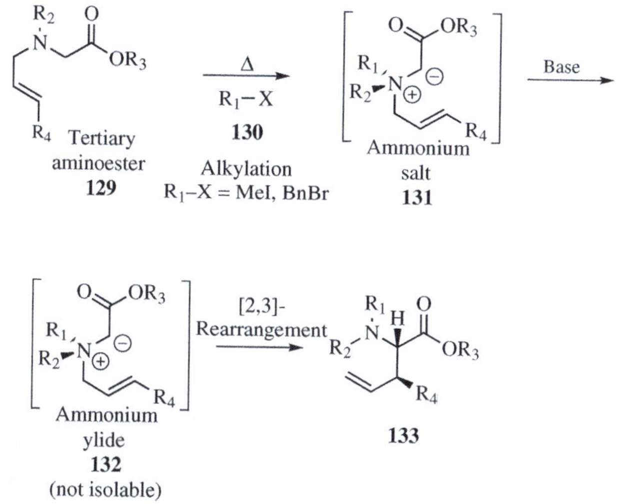

Scheme 15.32 Synthesis of ammonium ylides by deprotonation of ammonium salts.

(Scheme 15.32).[29] The resulting ammonium salts **131** are deprotonated by a base to generate ammonium ylides **132**, which cannot be observed or isolated. These zwitterions undergo spontaneous [2,3]-rearrangement to furnish aminoesters **133**. Although this method has been utilized in several structurally complex examples for the synthesis of target molecules, the potentially harsh conditions for alkylation often limit the functional groups that can be incorporated into the final aminoester products **133**. Moreover, the ammonium salts are not easily isolated and purified, which further limits the synthetic utility of this process in complex molecular settings.

The development of a mild and general metal-catalyzed tandem ammonium ylide generation/[2,3]-rearrangement could obviate the need to synthesize and isolate ammonium salts as reactive intermediates and would provide an opportunity for enantioselective catalysis. In this context, metal carbenoid-mediated couplings between tertiary allylic amines **134** and diazoesters **135** represent a powerful strategy for directly generating ammonium ylides **132** (Scheme 15.33).[30–32] However, the challenge of synthesizing diazoesters with diverse functionalities reduces the potential scope of this approach.

To address the need for a more general catalytic method of synthesizing allylic ammonium ylides for [2,3]-rearrangements that is amenable to enantioselective catalysis, our group recently developed a palladium-catalyzed tandem allylic amination/[2,3]-Stevens rearrangement of tertiary amines that proceeds through a palladium-π-allyl intermediate (Scheme 15.34).[33] Metal-catalyzed allylic aminations between primary or secondary amines and allylic electrophiles have become one of the most versatile methods for synthesizing nitrogen-containing compounds.[34] As an extension to these protocols, we coupled tertiary aminoesters **136** with allylic carbonates **137** in the presence of a palladium catalyst and Cs_2CO_3 to

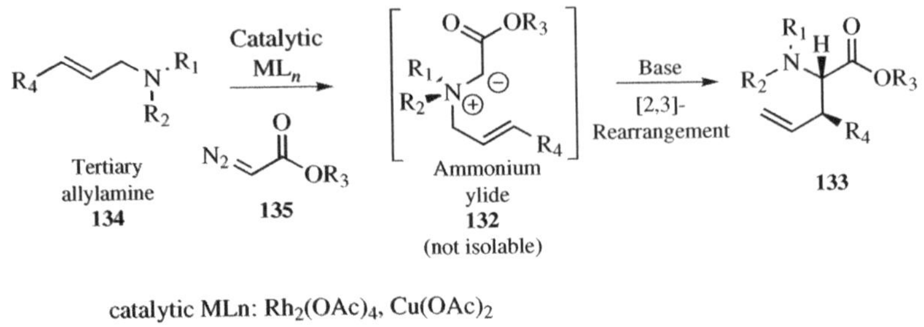

Scheme 15.33 Synthesis of ammonium ylides via metal carbenoids.

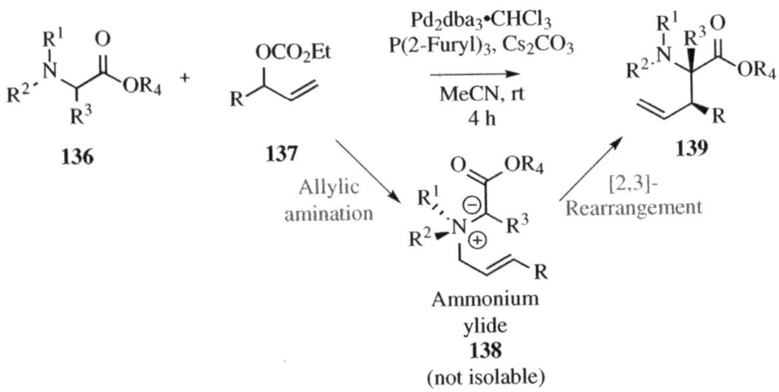

Scheme 15.34 Synthesis of ammonium ylides by allylic amination with tertiary amines.

generate ammonium ylides **138**. These intermediates then underwent spontaneous diastereoselective [2,3]-rearrangements to α-substituted aminoesters **139** with two contiguous stereocenters. This discovery is the first example of using tertiary amines as intermolecular nucleophiles in metal-catalyzed allylic substitution chemistry (some examples of intramolecular allylic amination with tertiary amines were known prior to our work[35]). In addition, this palladium-catalyzed Stevens rearrangement exhibits unprecedented substrate scope for ammonium ylide functionality and high diastereoselectivity for products with two stereocenters.

Based on early mechanistic experiments, we propose that aminoester **139** and palladium(II)-π-allyl complex **140** establish an unfavorable equilibrium with palladium(0) and ammonium salt **141** (Scheme 15.35). As soon as this unstable ammonium intermediate is formed, it undergoes a rapid deprotonation to generate ammonium ylide **142**, which is transformed into the observed [2,3]-rearrangement product **143** through an exo transition state. An unfavorable equilibrium for the palladium-catalyzed ammonium salt formation, in conjunction with the facile conversion of ammonium salts into the [2,3]-rearrangement products, could explain the difficulty in observing any ammonium intermediates. This mechanistic proposal also accounts for why catalytic intermolecular allylic amination with tertiary amines has never been reported before.

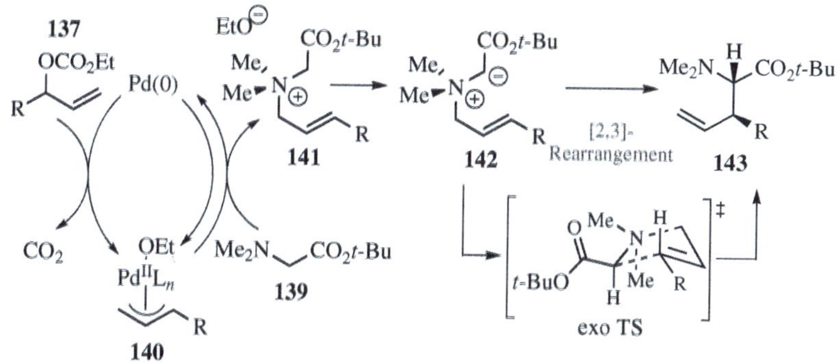

Scheme 15.35 Mechanism of the palladium-catalyzed tandem allylic amination/[2,3]-Stevens rearrangement of tertiary amines.

15.3.4 Applications of [2,3]-Stevens Rearrangements in Synthesis

The [2,3]-Stevens rearrangement has become a generally useful reaction in the synthesis of natural products because the products of this transformation contain three versatile functional groups: an amine, a carbonyl-containing group (ester or ketone), and an olefin (see Scheme 15.31).

In the 1990s, Honda, Inoue, and coworkers utilized the [2,3]-rearrangement of ammonium ylides to synthesize a series of polyisoprenoid natural products. For example, *β*-sinensal **150** was assembled from the hydrocarbon natural product *β*-myrcene **144** in a short sequence of steps (Scheme 15.36).[36] Initially, *β*-myrcene was converted into allylic amine **145**, which was alkylated to yield ammonium salt **146**. Subsequent deprotonation formed ammonium ylide **147**, which underwent a [2,3]-rearrangement. The resulting aminoester **149** was generated as a 3:1 mixture of *E*:*Z* olefins, which can be explained by the preferable equatorial orientation of the large diene-containing substituent in transition state **148**. Aminoester **149** was then transformed into *β*-sinensal **150** by reducing the ester to an aldehyde and reductively removing the dimethylamine group.

A similar strategy was employed for the synthesis of 13-*cis*-retinol **158** (Scheme 15.37).[37] Silyl-protected isoprenol **151** was converted through multiple steps into ammonium ylide **154**, which spontaneously rearranged to *γ*-aminoester **156** in 71% yield as predominantly the (*Z*)-olefin, which differed from the selective formation of the (*E*)-olefin in rearrangement product **149** (Scheme 15.36). Although the change in reaction conditions may explain the switch in olefin stereochemistry, more likely, the selective formation of the (*Z*)-olefin in **156** is due to the preferable equatorial orientation of the methyl group over the silyl ether-containing substituent in transition state **155**. Reductive removal of the dimethylamine group, deprotection of the silyl ether, and conversion of the ester into the cyclic diene side chain eventually furnished the natural product 13-*cis*-retinol **158**.

Analogously, the polyisoprenoid plaunotol **164** was synthesized from allyl dimethylamine **159** (Scheme 15.38).[38] Alkylation with ethyl bromoacetate and deprotonation with KO*t*-Bu yielded ammonium ylide **161**, which rearranged to

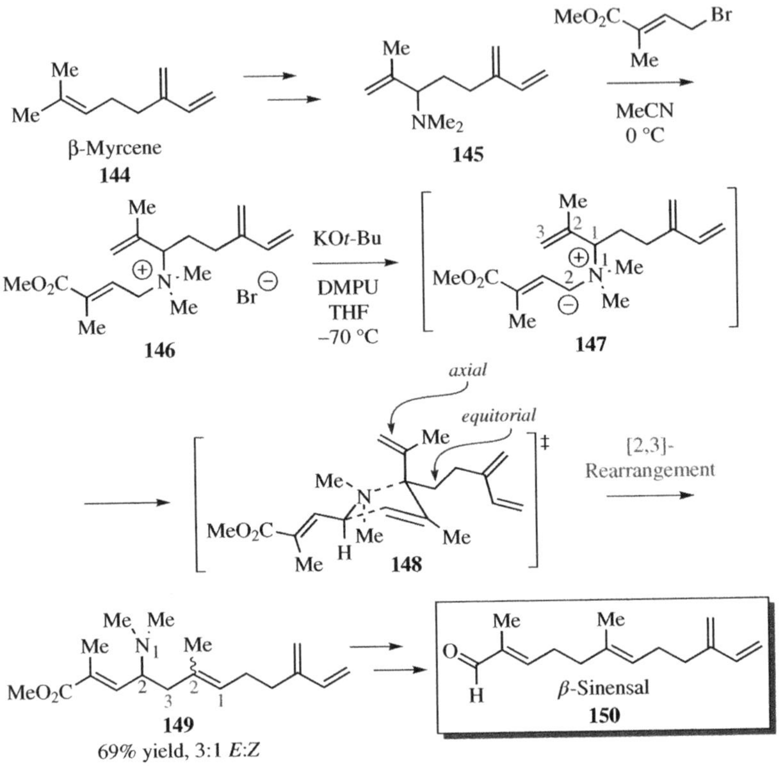

Scheme 15.36 Synthesis of *β*-sinensal.

α-aminoester **163**. The exclusive formation of a single olefin isomer in **163** presumable occurs via transition state **162**. Reductive removal of the dimethylamine group and benzyl ethers, along with conversion of the ester into the hydrocarbon side chain, yielded the natural product plaunotol **163**.

[2,3]-Stevens rearrangements have been employed in the synthesis of structurally more complex natural products, including polycyclic alkaloids. For example, in 2003, Li and Wang utilized an ammonium ylide [2,3]-rearrangement as the key step in their synthesis of cephalotaxine **172** (Scheme 15.39).[39] Primary amine **165** was transformed into aminoester **166**, which was treated with allyl bromide to furnish ammonium salt **167**. Treatment with strong base resulted in the formation of ammonium ylide **168**, which underwent a spontaneous [2,3]-rearrangement to yield *α*-allyl aminoester **169**. The olefin in this intermediate was oxidized to a ketone, and the two esters were reductively cyclized, resulting in aminoketone **170**. This advanced intermediate was then converted to cephalotaxine **172** in a short sequence of steps.

Several years later, Liu and coworkers reported another synthesis of cephalotaxine **172** that relied on a distinct [2,3]-Stevens rearrangement (Scheme 15.40).[40] Proline derivative **173** was transformed into ammonium ylide **174** in the presence of allyl

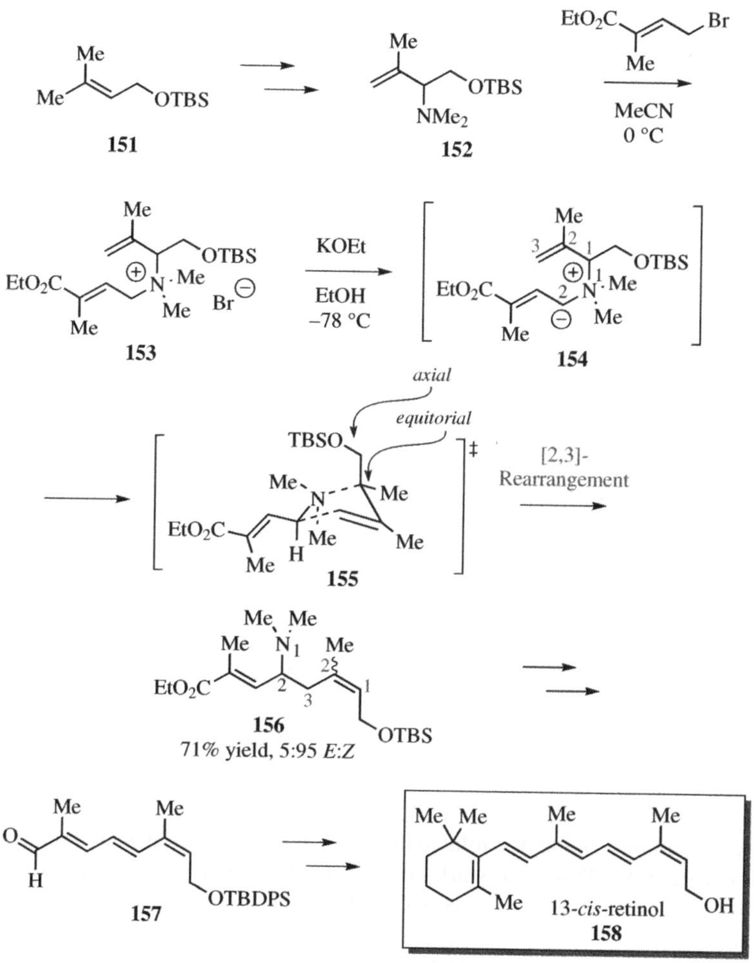

Scheme 15.37 Synthesis of 13-*cis*-retinol.

bromide and K_2CO_3. This zwitterion rearranged to α-allyl aminoester **175**. Hydration of the olefin and reduction of the ester furnished diol **176**, which was converted to aminoketone **177** via oxidation and aldol condensation. The assembly of this spirocyclic intermediate represented a formal synthesis of cephalotaxine **172**.[41]

15.3.5 Mechanistic Features Crucial for Synthetic Utility: Stereoselectivity

15.3.5.1 Diastereoselective [2,3]-Stevens Rearrangements in Synthesis As described in the previous section, the use of [2,3]-Stevens rearrangements in the total synthesis of natural products has been mainly limited to the generation of one carbon stereocenter. [2,3]-Stevens rearrangements that generate two contiguous

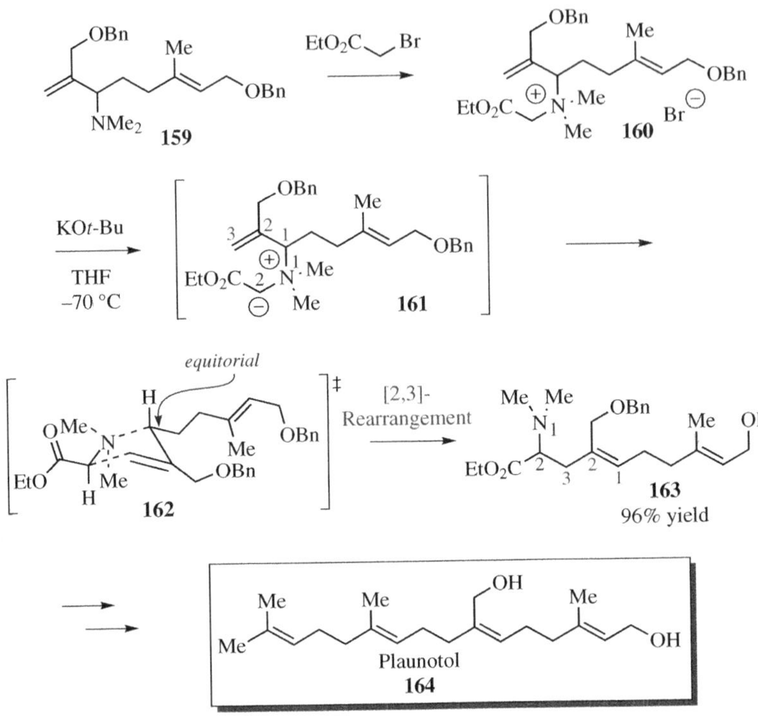

Scheme 15.38 Synthesis of plaunotol.

stereocenters in the course of a total synthesis often lead to modest and unpredictable diastereoselectivity. This is in stark contrast to the general application of diastereoselective [3,3]-rearrangements to assemble contiguous stereocenters in complex molecular settings.

The synthesis of platynecine **184** is one seminal example of a diastereoselective [2,3]-Stevens rearrangement for the generation of two contiguous stereocenters (Scheme 15.41).[31d] Monoprotected diol **178** was converted into diazoaminoketone **179**. With ruthenium catalysis, this diazo intermediate furnished cyclic ammonium ylide **180**, which rearranged to aminoketones **181a** and **181b**. The desired diastereomer **181a** was isolated as the major product. Reduction of the ketone and hydration of the olefin yielded alcohol **183**, which was transformed into platynecine **184**. Although no explanation was provided for the observed preference for **181a** in the [2,3]-rearrangement, **181b** may have been disfavored because of the unfavorable interaction between the benzyl ether substituent and the *N*-benzyl group in transition state **185b**.

Our group has recently reported another example of a diastereoselective [2,3]-Stevens rearrangement for the generation of two contiguous stereocenters in the context of a total synthesis of amathaspiramide F **195** (Scheme 15.42) (A. Soheili, U.K. Tambar, unpublished). We utilized our palladium-catalyzed tandem allylic

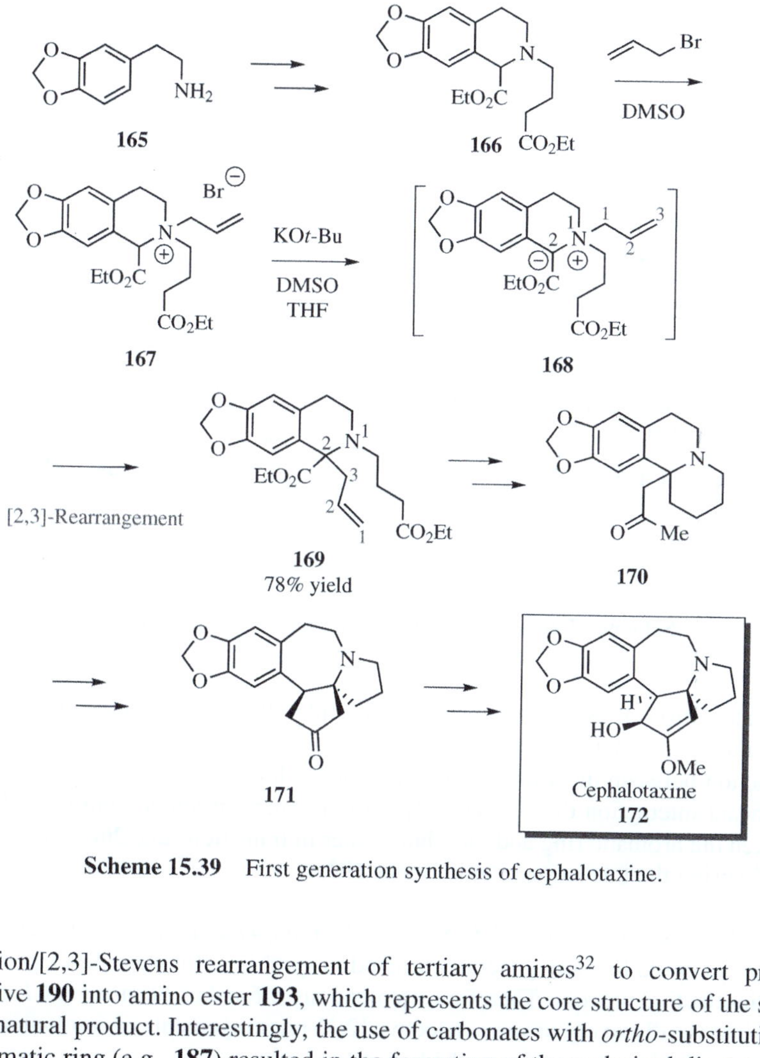

Scheme 15.39 First generation synthesis of cephalotaxine.

amination/[2,3]-Stevens rearrangement of tertiary amines[32] to convert proline derivative **190** into amino ester **193**, which represents the core structure of the spirocyclic natural product. Interestingly, the use of carbonates with *ortho*-substitution in the aromatic ring (e.g., **187**) resulted in the formation of the undesired diastereomer as the major product (**189**), whereas carbonates that lacked *ortho*-substitution in the aromatic ring (e.g., **191**) furnished the desired diastereomer as the major product (**193**). Aminoester **193** was converted to aminophenol **194**, which represented a formal synthesis of amathaspiramide F **195**.[42]

Although the origin of the switch in diastereoselectivity for the [2,3]-rearrangement is still not completely understood, it is believed that the unique torsional strain effect of the ortho-substituted aromatic ring in ammonium ylide **188** may be pivotal (Scheme 15.43). Ammonium ylide **188** favored formation of product **189** (3:1 dr) because the unfavorable interaction between the nonplanar ortho-substituted aromatic ring and the proline ring in transition state **196** was more destabilizing than the steric interaction between the aromatic ring and the *t*-butyl ester in transition state

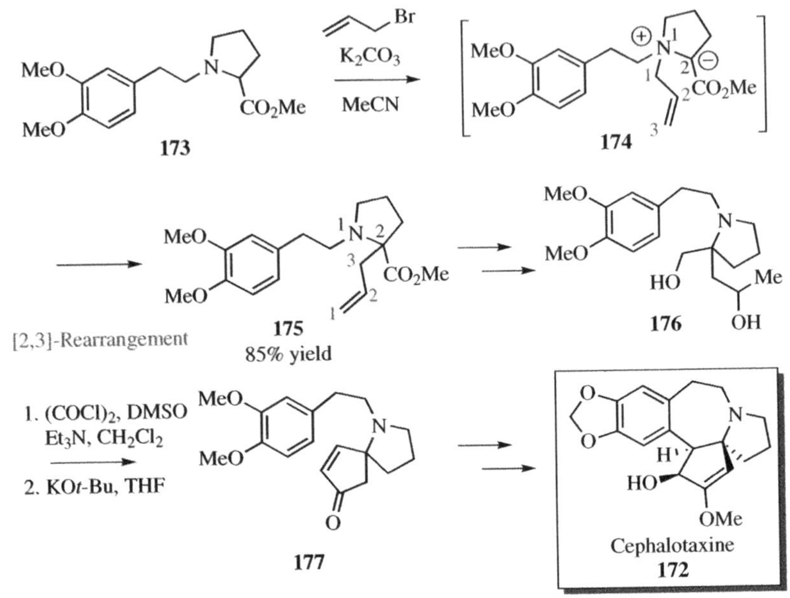

Scheme 15.40 Second generation synthesis of cephalotaxine.

197 (interactions *a* vs *b*). With ammonium ylide **192**, the metasubstituted aromatic ring in transition state **199** could adopt a planar conformation, mitigating the steric interaction between the aromatic ring and the proline ring (interaction *c*). The most significant interaction during the [2,3]-rearrangement of ammonium ylide **192** was between the aromatic ring and the *t*-butyl ester in transition state **200** (interaction *d*), thus favoring the formation of aminoester **193**.

15.3.5.2 Enantioselective [2,3]-Stevens Rearrangement The [2,3]-Stevens rearrangement could be more generally utilized in synthesis if it were developed into an enantioselective process. Therefore, tremendous effort has been invested in recent years into the discovery of stereoselective [2,3]-rearrangements of ammonium ylides for the generation of enantioenriched products.

Chiral auxiliary approaches have been examined for the production of enantioenriched products through [2,3]-rearrangements.[43] For example, Sweeney and coworkers treated *N,N*-dimethylammonium salts **186** with sodium hydride to generate ammonium ylides bearing Oppolzer's camphorsultam, which underwent diastereoselective [2,3]-Stevens rearrangements to enantioenriched products **187** (Scheme 15.44, Eq. 1).[29d] Oppolzer's camphorsultam was also compatible with our group's palladium-catalyzed tandem allylic amination/[2,3]-Stevens rearrangement of tertiary amines (Scheme 15.44, Eq. 2).[32]

Catalytic enantioselective [2,3]-Stevens rearrangements are rare, partially because of difficulties associated with synthesizing and isolating ammonium ylides. Ammonium ylides undergo spontaneous [2,3]-rearrangement as soon as they are formed

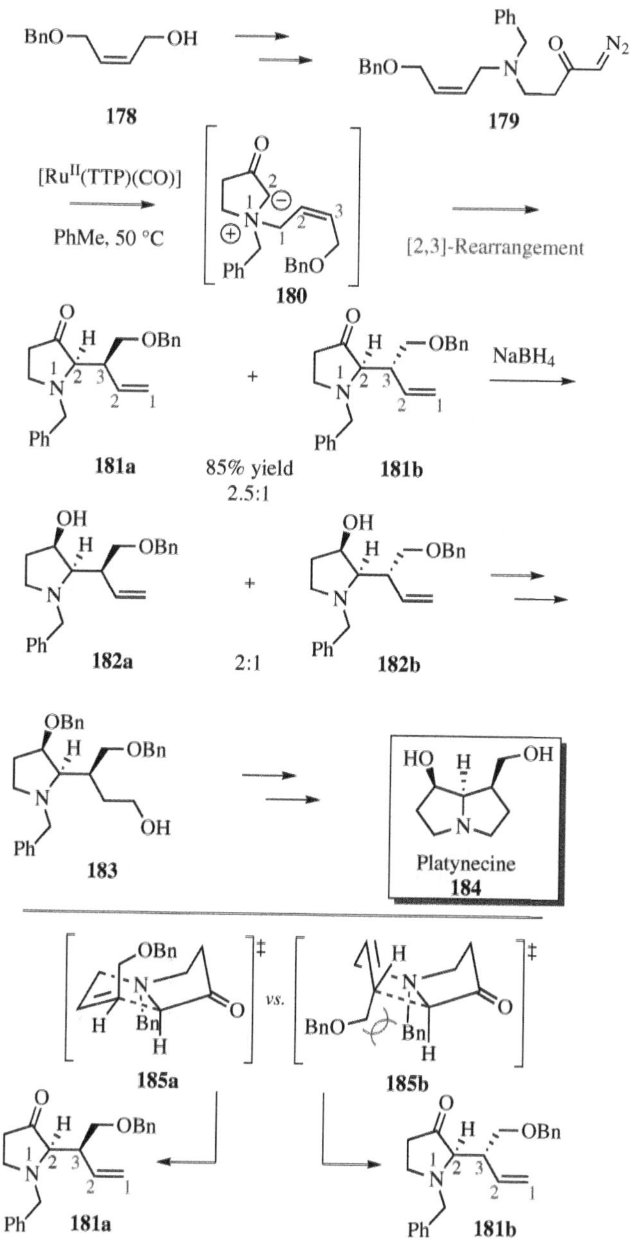

Scheme 15.41 Diastereoselective [2,3]-Stevens rearrangement for the synthesis of platynecine.

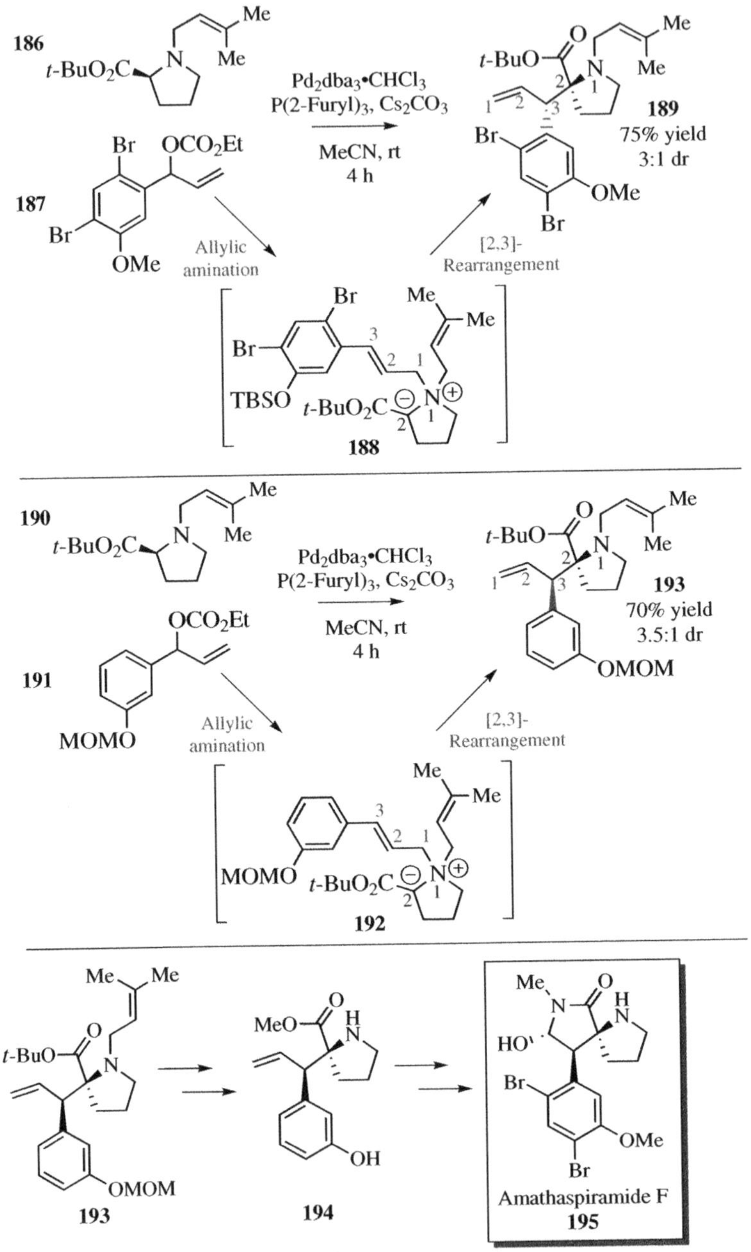

Scheme 15.42 Diastereoselective [2,3]-Stevens rearrangement for the synthesis of amathaspiramide F.

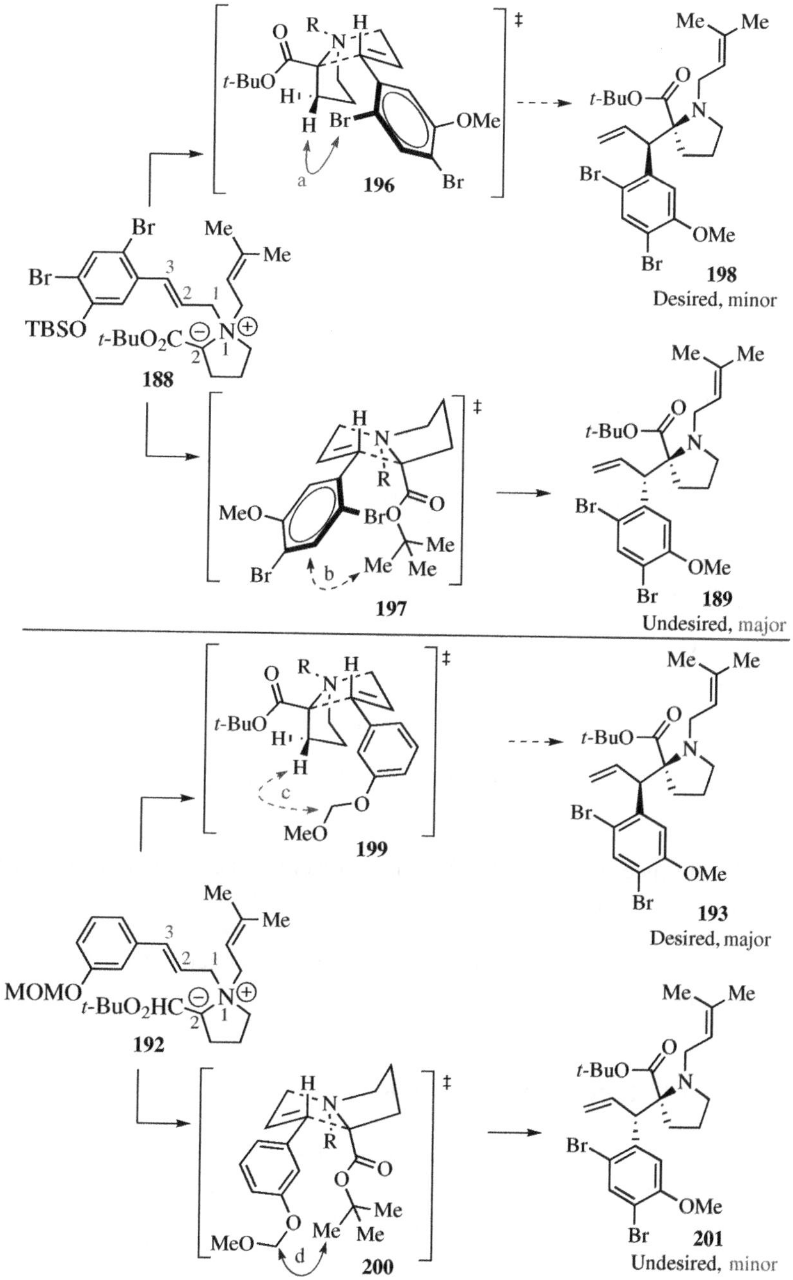

Scheme 15.43 Switch in diastereoselectivity for palladium catalyzed allylic amination/[2,3]-rearrangement.

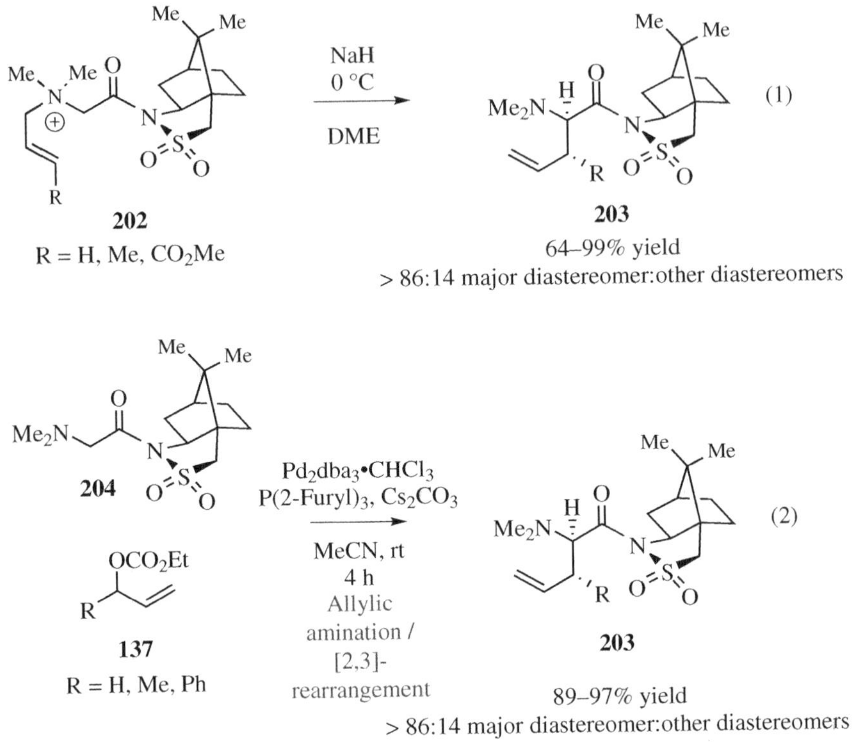

Scheme 15.44 [2,3]-Stevens rearrangement with Oppolzer's chiral auxiliary.

(even in the absence of catalyst), which has prevented the development a catalytic enantioselective Stevens rearrangement. Conceptually, new strategies are required to realize this goal.[44] For example, Smith and coworkers recently reported an elegant benzotetramisole catalyzed enantioselective [2,3]-rearrangement of allylic quaternary ammonium salts (Scheme 15.45).

15.4 CONCLUSION AND OUTLOOK

Since Meisenheimer's landmark discovery in 1919 of an unusual rearrangement of allylic amine *N*-oxides, the [2,3]-rearrangements of ammonium zwitterions have grown into a synthetically powerful class of transformations for the synthesis of complex molecules. Although a detailed picture is emerging for the mechanism and synthetic potential of these processes, a greater understanding of the factors that control the stereoselectivity of [2,3]-rearrangements is required to ultimately utilize these reactions in a more predictable manner for natural product synthesis. Furthermore, these future studies will shed light on fundamental issues of reactivity

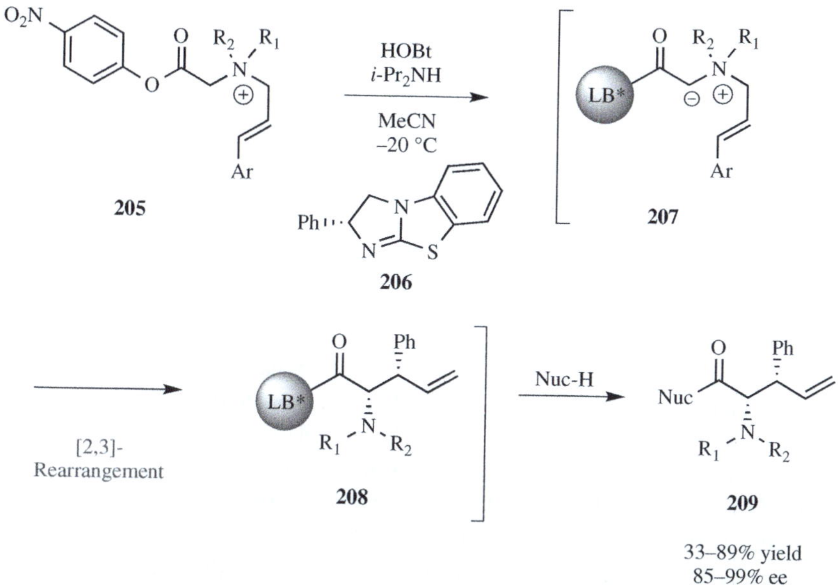

Scheme 15.45 Smith's catalytic enantioselective [2,3]-Stevens rearrangement.

in the context of sigmatropic rearrangements, which will have an overall impact on the synthetic utility of this general class of reactions.

REFERENCES

1. For general reviews of [3,3]-rearrangements, see: (a) Ziegler, F. E. *Chem. Rev.* **1988**, *88*, 1423–1452. (b) Martín Castro, A. M. *Chem. Rev.* **2004**, *104*, 2939–3002. (c)*The Claisen Rearrangement*: Hiersemann, M.; Nubbemeyer, U., Eds.; Wiley-VCH Verlag GmbH: Weinheim, **2007**. (d) Nubbemeyer, U. *Synthesis* **2003**, 961–1008. (e) Hiersemann, M.; Abraham, L. *Eur. J. Org. Chem.* **2002**, *2002*, 1461–1471. (f) Paquette, L. A. *Tetrahedron* **1997**, *53*, 13971–14020. (g) Davies, H. M. L. *Tetrahedron* **1993**, *49*, 5203–5223. (h) Wilson, S. R. *Organic Reactions*; Wiley: New York, **1993**; Vol. 43, pp. 93–250. (i) Blechert, S. *Synthesis* **1989**, 71–82. (j) Lutz, R. P. *Chem. Rev.* **1984**, *84*, 205–247. (k) Rhoads, S. J.; Raulins, N. R. *Organic Reactions*; Wiley: New York, **1975**; Vol. 22, pp. 1–252. (l) Enders, D.; Knopp, M.; Schiffers, R. *Tetrahedron: Asymmetry* **1996**, *7*, 1847–1882. (m) Wipf, P. In *Comprehensive Organic Synthesis*; Trost, B. M.; Fleming, I., Eds.; Pergamon Press: Oxford, **1991**; Vol. 5, pp. 827–873, Chapter 7.2. (n) Majumdar, K. C.; Bhattacharyya, T.; Chattopadhyay, B.; Sinha, B. *Synthesis* **2009**, 2117–2142.
2. For some examples of enantioselective [3,3]-rearrangements, see: (a) Corey, E. J.; Lee, D. H. *J. Am. Chem. Soc.* **1991**, *113*, 4026–4028. (b) Maruoka, K.; Saito, S.; Yamamoto, H. *J. Am. Chem. Soc.* **1995**, *117*, 1165–1166. (c) Yoon, T. P.; MacMillan, D. W. C. *J. Am. Chem. Soc.* **2001**, *123*, 2911–2912. (d) Abraham, L.; Czerwonka, R.; Hiersemann, M. *Angew. Chem., Int. Ed.* **2001**, *40*, 4700–4703. (e) Anderson, C. E.; Overman, L. E. *J. Am.

Chem. Soc. **2003**, *125*, 12412–12413. (f) Burger, E. C.; Tunge, J. A. *Org. Lett.* **2004**, *6*, 4113–4115. (g) Uyeda, C.; Jacobsen, E. N. *J. Am. Chem. Soc.* **2008**, *130*, 9228–9229. (h) Linton, E. C.; Kozlowski, M. C. *J. Am. Chem. Soc.* **2008**, *130*, 16162–16163. (i) Kaeobamrung, J.; Mahatthananchai, J.; Zheng, P.; Bode, J. W. *J. Am. Chem. Soc.* **2010**, *132*, 8810–8812.

3. For general reviews of [2,3]-rearrangements, see: (a) Brückner, R. In *Comprehensive Organic Synthesis*; Trost, B. M.; Fleming, I., Eds.; Pergamon Press: New York, **1991**; Vol. 6, pp. 873–908. (b) Markó, I. E. In *Comprehensive Organic Synthesis*; Trost, B. M.; Fleming, I., Eds.; Pergamon Press: New York, **1991**; Vol. 3, pp. 913–974. (c) Hoffman, R. W. *Angew. Chem. Int. Ed.* **1979**, *18*, 563–572. (d) Pine, S. H. In *Organic Reactions*; Wiley: New York, **1970**; Vol. 18, pp. 403–464. (e) Sweeney, J. B. *Chem. Soc. Rev.* **2009**, *38*, 1027–1038.
4. (a) Vanecko, J. A.; Wan, H.; West, F. G. *Tetrahedron* **2006**, *62*, 1043–1062. (b) Hodgson, D. M.; Pierard, F. Y. T. M.; Stupple, P. A. *Chem. Soc. Rev.* **2001**, *30*, 50–61. (c) Li, A.-H.; Dai, L.-X.; Aggarwal, V. K. *Chem. Rev.* **1997**, *97*, 2341–2372.
5. Meisenheimer, J. *Ber. Dtsch. Chem. Ges.* **1919**, *52*, 1667–1677.
6. Woodward, R. B.; Hoffmann, R. *Angew. Chem. Int. Ed.* **1969**, *8*, 781–853.
7. Moriwaki, M.; Yamamoto, Y.; Oda, J.; Inouye, Y. *J. Org. Chem.* **1976**, *41*, 300–303.
8. Lepley, A. R.; Cook, P. M.; Willard, G. F. *J. Am. Chem. Soc.* **1970**, *92*, 1101–1102.
9. Kondo, H.; Sakamoto, F.; Uno, T.; Kawahata, Y.; Tsukamoto, G. *J. Med. Chem.* **1989**, *32*, 671–674.
10. (a) Rinehart, K. L.; Kobayashi, J.; Harbour, G. C.; Hughes, R. G.; Mizsak, S. A.; Scahill, T. A. *J. Am. Chem. Soc.* **1984**, *106*, 1524–1526. (b) Rinehart, K. L.; Kobayashi, J.; Harbour, G. C.; Gilmore, J.; Mascal, M.; Holt, T. G.; Shield, L. S.; Lafargue, F. *J. Am. Chem. Soc.* **1987**, *109*, 3378–3387.
11. (a) Kurihara, T.; Sakamoto, Y.; Takai, M.; Tsukamoto, K.; Sakai, T.; Harusawa, S.; Yoneda, R. *Chem. Pharm. Bull.* **1994**, *42*, 31–38. (b) Kurihara, T.; Sakamoto, Y.; Matsumoto, H.; Kawabata, N.; Harusawa, S.; Yoneda, R. *Chem. Pharm. Bull.* **1994**, *42*, 475–480.
12. Bremner, J. B.; Smith, R. J.; Tarrant, G. J. *Tetrahedron Lett.* **1996**, *37*, 97–100.
13. Davies, S. G.; Smyth, G. D. *Tetrahedron: Asymmetry* **1996**, *7*, 1005–1006.
14. Buston, J. E. H.; Coldham, I.; Mulholland, K. R. *Tetrahedron: Asymmetry* **1998**, *9*, 1995–2009.
15. Fall, Y.; Vanbac, N.; Langlois, Y. *Tetrahedron Lett.* **1986**, *27*, 3611–3614.
16. Unnikrishnan, P. A. *Org. Prep. Proced. Int.* **1994**, *26*, 488–491.
17. Bao, H.; Qi, X.; Tambar, U. K. *Synlett* **2011**, *2011*, 1789–1792.
18. Yamamoto, Y.; Oda, J.; Inouye, Y. *J. Org. Chem.* **1976**, *41*, 303–306.
19. Reetz, M. T.; Lauterbach, E. H. *Tetrahedron Lett.* **1991**, *32*, 4481–4482.
20. Davies, S. G.; Smyth, G. D. *Tetrahedron: Asymmetry* **1996**, *7*, 1001–1004.
21. Enders, D.; Kempen, H. *Synlett* **1994**, 969–971.
22. Blanchet, J.; Bonin, M.; Micouin, L.; Husson, H. P. *Tetrahedron Lett.* **2000**, *41*, 8279–8283.
23. (a) Buston, J. E. H.; Coldham, I.; Mulholland, K. R. *Synlett* **1997**, 322–324. (b) Buston, J. E. H.; Coldham, I.; Mulholland, K. R. *J. Chem. Soc., Perkin Trans. 1* **1999**, 2327–2334.
24. Guarna, A.; Occhiato, E. G.; Pizzetti, M.; Scarpi, D.; Sisi, S.; van Sterkenburg, M. *Tetrahedron: Asymmetry* **2000**, *11*, 4227–4238.

25. Bao, H.; Qi, X.; Tambar, U. K. *J. Am. Chem. Soc.* **2011**, *133*, 1206–1208.

26. (a) Donde, Y.; Overman, L. E. *J. Am. Chem. Soc.* **1999**, *121*, 2933–2934. (b) Overman, L. E.; Carpenter, N. E. *Organic Reactions*; Wiley: New Jersey, **2005**; Vol. 66, pp. 1–107. (c) Overman, L. E. *Tetrahedron* **2009**, *65*, 6432–6446. (d) Watson, M. P.; Overman, L. E.; Bergman, R. G. *J. Am. Chem. Soc.* **2007**, *129*, 5031–5044.

27. (a) Stevens, T. S.; Creigton, E. M.; Gordon, A. B.; MacNicol, M. *J. Chem. Soc.* **1928**, 3193–3197. (b) Honda, K.; Inoue, S.; Sato, K. *J. Am. Chem. Soc.* **1990**, *112*, 1999–2001. (c) Arboré, A. P. A.; Cane-Honeysett, D. J.; Coldham, I.; Middleton, M. L. *Synlett* **2000**, *2*, 236–238.

28. For a recent discussion of the competition between [1,2]-rearrangements and [2,3]-rearrangements of allylic ammonium ylides, see: Biswas, B.; Collins, S. C.; Singleton, D. A. *J. Am. Chem. Soc.* **2014**, *136*, 3740–3743.

29. (a) Coldham, I.; Middleton, M. L.; Taylor, P. L. *J. Chem. Soc., Perkin Trans. 1* **1997**, 2951–2952. (b) Coldham, I.; Middleton, M. L.; Taylor, P. L. *J. Chem. Soc., Perkin Trans. 1* **1998**, 2817–2821. (c) Honda, K.; Tabuchi, M.; Kurokawa, H.; Asami, M.; Inoue, S. *J. Chem. Soc., Perkin Trans. 1* **2002**, 1387–1396. (d) Workman, J. A.; Garrido, N. P.; Sancüon, J.; Roberts, E.; Wessel, H. P.; Sweeney, J. B. *J. Am. Chem. Soc.* **2005**, *127*, 1066–1067. (e) Gawley, R. E.; Moon, K. *Org. Lett.* **2007**, *9*, 3093–3096.

30. (a) Clark, J. S.; Hodgson, P. B. *J. Chem. Soc., Chem. Commun.* **1994**, 2701–2702. (b) Clark, J. S.; Hodgson, P. B.; Goldsmith, M. D.; Street, L. J. *J. Chem. Soc., Perkin Trans. 1* **2001**, 3312–3324. (c) Clark, J. S.; Hodgson, P. B.; Goldsmith, M. D.; Blake, A. J.; Cooke, P. A.; Street, L. J. *J. Chem. Soc., Perkin Trans. 1* **2001**, 3325–3337. (d) Clark, J. S.; Middleton, M. D. *Org. Lett.* **2002**, *4*, 765–768. (e) Clark, J. S.; Middleton, M. D. *Tetrahedron Lett.* **2003**, *44*, 7031–7034. (f) Padwa, A.; Beall, L. S.; Eidell, C. K.; Worsencroft, K. J. *J. Org. Chem.* **2001**, *66*, 2414–2421. (g) Padwa, A.; Snyder, J. P.; Curtis, E. A.; Sheehan, S. M.; Worsencroft, K. J.; Kappe, C. O. *J. Am. Chem. Soc.* **2000**, *122*, 8155–8167. (h) Roberts, E.; Sancüon, J. P.; Sweeney, J. B. *Org. Lett.* **2005**, *7*, 2075–2078. (i) West, F. G.; Naidu, B. N. *J. Am. Chem. Soc.* **1993**, *115*, 1177–1178. (j) Glaeske, K. W.; Naidu, B. N.; West, F. G. *Tetrahedron: Asymmetry* **2003**, *14*, 917–920.

31. (a) Roberts, E.; Sancüon, J. P.; Sweeney, J. B.; Workman, J. A. *Org. Lett.* **2003**, *5*, 4775–4777. (b) Heath, P.; Roberts, E.; Sweeney, J. B.; Wessel, H. P.; Workman, J. A. *J. Org. Chem.* **2003**, *68*, 4083–4086.

32. (a) Doyle, M. P.; Tamblyn, W. H.; Bagheri, V. *J. Org. Chem.* **1981**, *46*, 5094–5102. (b) Aviv, I.; Gross, Z. *Chem. Eur. J.* **2008**, *14*, 3995–4005. (c) Honda, K.; Shibuya, H.; Yasui, H.; Hoshino, Y.; Inoue, S. *Bull. Chem. Soc. Jpn.* **2008**, *81*, 142–147. (d) Zhou, C.-Y.; Yu, W.-Y.; Chan, P. W. H.; Che, C.-M. *J. Org. Chem.* **2004**, *69*, 7072–7082. (e) Del Zotto, A.; Baratta, W.; Miani, F. Verardo, G.; Rigo, P. *Eur. J. Org. Chem.* **2000**, 3731–3735.

33. Soheili, A.; Tambar, U. K. *J. Am. Chem. Soc.* **2011**, *133*, 12956–12959.

34. (a) Johannsen, M.; Jørgensen, K. A. *Chem. Rev.* **1998**, *98*, 1689–1708. (b) Hartwig, J. F.; Stanley, L. M. *Acc. Chem. Res.* **2010**, *43*, 1461–1475. (c) Trost, B. M.; Zhang, T.; Sieber, J. D. *Chem. Sci.* **2010**, *1*, 427–440.

35. (a) Grellier, M.; Pfeffer, M.; van Koten, G. *Tetrahedron Lett.* **1994**, *35*, 2877–2880. (b) van der Schaaf, P. A.; Sutter, J.-P.; Grellier, M.; van Mier, G. P. M.; Spek, A. L.; van Koten, G.; Pfeffer, M. *J. Am. Chem. Soc.* **1994**, *116*, 5134–5144.

36. Honda, K.; Tabuchi, M.; Inoue, S. *Chem. Lett.* **1996**, 385–386.

37. Honda, K.; Yoshii, I.; Inoue, S. *Chem. Lett.* **1996**, 671–672.

38. Honda, K.; Igarashi, D.; Asami, M.; Inoue, S. *Synlett* **1998**, 685–686.

39. Li, W.-D. Z.; Wang, Y.-Q. *Org. Lett.* **2003**, *5*, 2931–2934.

40. Sun, M.; Lu, H.; Wang, Y.; Yang, H.; Liu, H. *J. Org. Chem.* **2009**, *74*, 2213–2216.

41. Li, W.-D. Z.; Wang, X.-W. *Org. Lett.* **2007**, *9*, 1211–1214.

42. Sakaguchi, K.; Ayabe, M.; Watanabe, Y.; Okada, T.; Kawamura, K.; Shiada, T.; Ohfune, Y. *Org. Lett.* **2008**, *10*, 5449–5452.

43. (a) Stoichiometric amounts of chiral Lewis acids can also accelerate aza-Wittig-type rearrangements of aminoesters and aminoamides via the generation of allylic ammonium ylides and subsequent [2,3]-rearrangement: Blid, J.; Panknin, O.; Somfai, P. *J. Am. Chem. Soc.* **2005**, *127*, 9352–9353. (b) Blid, J.; Panknin, O.; Tuzina, P.; Somfai, P. *J. Org. Chem.* **2007**, *72*, 1294–1300.

44. For a recent report of benzotetramisole catalyzed enantioselective [2,3]-rearrangements of allylic quaternary ammonium salts, see: West, T. H.; Daniels, D. S. B.; Slawin, A. M. Z.; Smith, A. D. *J. Am. Chem. Soc.* **2014**, *136*, 4476–4479.

16

OXONIUM YLIDE REARRANGEMENTS IN SYNTHESIS

GRAHAM K. MURPHY

Department of Chemistry, University of Waterloo, Waterloo, ON, Canada

F. G. WEST

Department of Chemistry, University of Alberta, E3-43 Gunning-Lemieux Chemistry Centre, Edmonton, AB, Canada

16.1 INTRODUCTION

Oxonium ylides are high-energy intermediates able to undergo two valuable types of molecular rearrangements: [2,3]-sigmatropic rearrangements and [1,2]-shifts (Stevens rearrangement). Typically generated from carbenoid precursors, the vast majority of examples of oxonium ylides in synthesis involve *cyclic* ylides, formed from intramolecular reaction of metallocarbenes with pendent ether oxygens. This first step forms a new C—O bond and a cyclic ether that will typically be retained in the eventual product (Scheme 16.1). Subsequent rearrangement transfers a group from a trivalent oxonium oxygen to the neighboring carbon, in the process forming a second strategic C—C bond and (usually) a new stereogenic center. This chapter will discuss the scope and limitations of oxonium ylide rearrangements and their unique applicability in the synthesis of various heterocyclic and carbocyclic targets.

16.1.1 Oxonium Ylides: Generation and Rearrangement Reactions

The commonly encountered ylides of nitrogen, sulfur, or phosphorus can be generated by deprotonation of the corresponding onium salts. In contrast, since dealkylation of

Molecular Rearrangements in Organic Synthesis, First Edition. Edited by Christian M. Rojas.

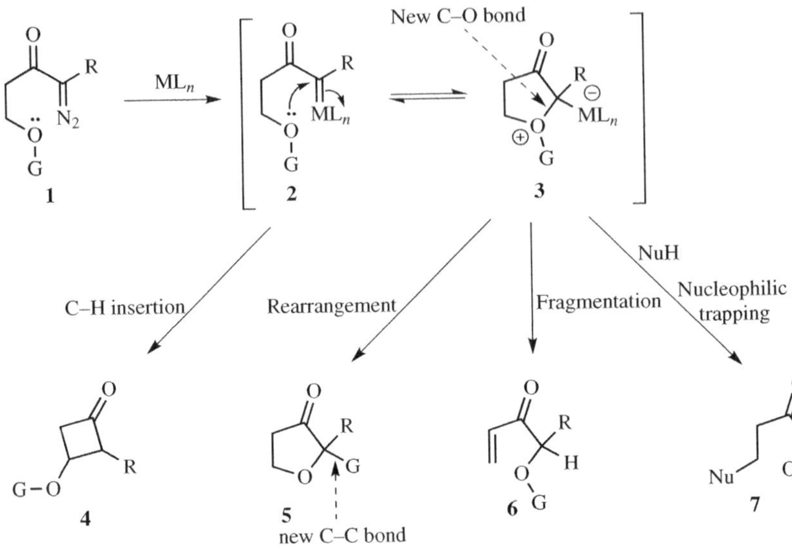

Scheme 16.1 Generation and potential reaction pathways of oxonium ylides.

oxonium salts is facile, oxonium ylides are typically formed by intercepting a catalytically generated metallocarbene **2** with a Lewis basic lone pair of electrons from an ethereal oxygen (Scheme 16.1). Most often, this is an intramolecular capture to maximize the desired ylide formation over alternative reactions of the metallocarbene. Once formed, the oxonium ylide **3** can undergo rearrangements, fragmentations, or nucleophilic displacement, furnishing products such as **5–7**. If a viable decomposition pathway is not available, the oxonium ylide can revert to the metallocarbene, which can decompose by alternate pathways, such as C—H insertion to give **4**. Thus, the initial synthetic challenge is the design of a suitable precursor which will undergo ylide formation in preference to other options and whose subsequent rearrangement is energetically feasible.

From the ylide, one of two possible rearrangement pathways can be utilized to generate heterocyclic products of the type exemplified by **5** in Scheme 16.1. The first is the [2,3]-sigmatropic rearrangement, a symmetry-allowed process that typically occurs with a relatively low-energy barrier.[1] When a substituent on the oxonium oxygen atom is allylic or propargylic (Scheme 16.2), a five-center, six-electron rearrangement can occur, delivering the former oxygen substituent to the ylide carbon, with allylic/propargylic inversion. Substitution at the terminus of the allyl group (R^3) results in the formation of two contiguous stereocenters, while substitution at the initial allylic position (R^2) affords a substituted olefin in the product. In many cases, the substrate substitution pattern can have a substantial effect on the outcome of the rearrangement (see Schemes 16.15 and 16.16). For simplicity, the free ylides are depicted; however, it is possible (and in some cases, likely) that the transition metal catalyst remains associated with the ylide at the point of [2,3]-rearrangement.

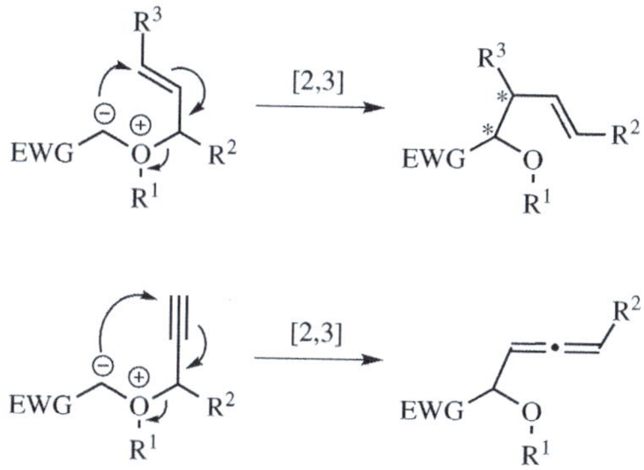

Scheme 16.2 [2,3]-Sigmatropic shifts of oxonium ylides.

On the other hand, certain oxonium substituents can support an alternative rearrangement pathway, in which the group migrates directly from oxygen to carbon without inversion, known as a [1,2]-shift or Stevens rearrangement. Whereas the [2,3]-rearrangement is a concerted, pericyclic process, the [1,2]-shift appears to proceed by a stepwise mechanism. Most typically, this is thought to occur through an initial homolytic bond dissociation to afford a radical pair, followed by recombination to form the new C—C bond (Scheme 16.3).[2] In light of this mechanistic pathway, the [1,2]-shift occurs only when a suitable radical-stabilizing group (Ar, C=O, heteroatom) is bonded to the migrating carbon. Since alkenes also provide significant levels of radical stabilization, a competing [1,2]-shift by allylic substituents can be seen in addition or in preference to the [2,3]-shift in some cases. Notably, as with the [2,3]-rearrangement, two contiguous stereocenters can be formed when the migrating carbon bears more than one substituent, though preservation of the original stereocenter (migration with retention) requires that radical recombination be faster than reorientation of relative positioning of the radical pair.

16.1.2 Strategic Planning

Effective use of oxonium ylide rearrangements as strategic steps in synthesis requires the recognition that the fundamental cyclic ether structure formed by sequential intramolecular addition of ether oxygen to metallocarbene followed by [2,3]- or [1,2]-shift may be buried within a larger, complex skeleton, and that the newly formed C—C bond can be an important attachment point for two key fragments, or part of a larger ring. Careful design of a suitable precursor can lead to large increases in molecular complexity. Specific cases are found in the "Applications" section, but general examples are shown in the following text (Scheme 16.4).

16.1.3 Selectivity Aspects

The first systematic investigations into the rearrangements of oxonium ylides appeared in 1965 in an article by Nozaki *et al.*[3] Though there were a small number

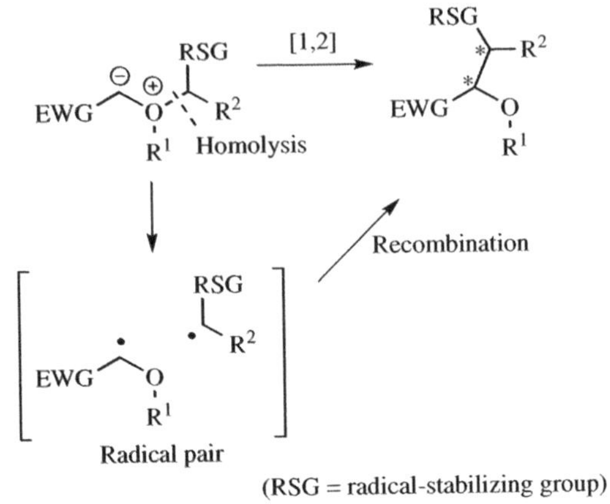

Scheme 16.3 [1,2]-Stevens rearrangement of oxonium ylides.

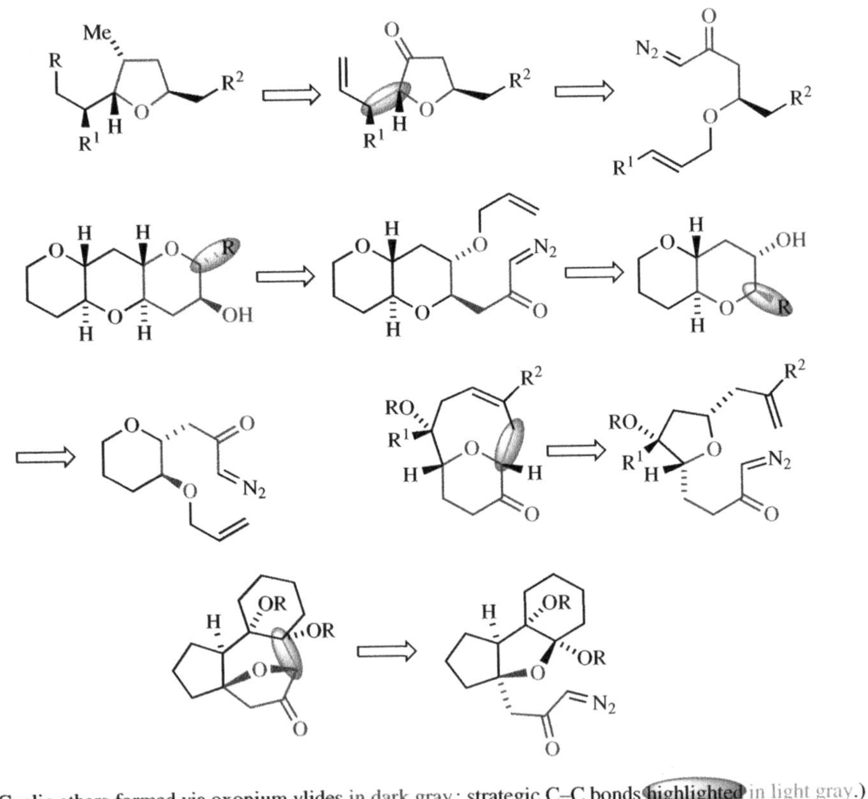

Scheme 16.4 Cyclic ether scaffolds within complex targets via oxonium ylides.

of intervening reports,[4] it wasn't until 1986 that the field took off in earnest when the groups of Pirrung and Werner[5] and Johnson and Roskamp[6] reported a variety of rearrangements of intramolecularly derived oxonium ylides in back-to-back communications. Their work collectively reported a variety of [1,2]-shifts and [2,3]-sigmatropic rearrangements of cyclic oxonium ylides generated by intramolecular reaction of metallocarbenes with ethers. Many examples in these seminal reports formed the basis for subsequent methodological studies on the rearrangements of oxonium ylides, as they highlighted key concepts that are essential to understanding and employing these transformations in advanced synthesis.

16.1.3.1 Chemoselectivity For example, metallocarbenes tend to favor reaction via five-membered transition states on kinetic grounds. Many examples using oxonium ylides in synthesis rely on this kinetic preference to achieve good chemoselectivity (Scheme 16.5). Pirrung and Werner demonstrated that allyl ether **8** will form a five-membered cyclic oxonium ylide upon exposure to a transition metal catalyst, which undergoes ready [2,3]-rearrangement to give **9**.[5] They showed the reaction of the homologous allyl ether **10** to be far less chemoselective, as the six-membered ylide formation and rearrangement pathway to give **11** must compete with a five-membered C—H insertion reaction pathway to give **12**. The effect of cyclic ylide ring size was shown by West and coworkers[7] to be even more pronounced for the case of a [1,2]-shift. While the formation of **14** via a five-membered transition state is quite

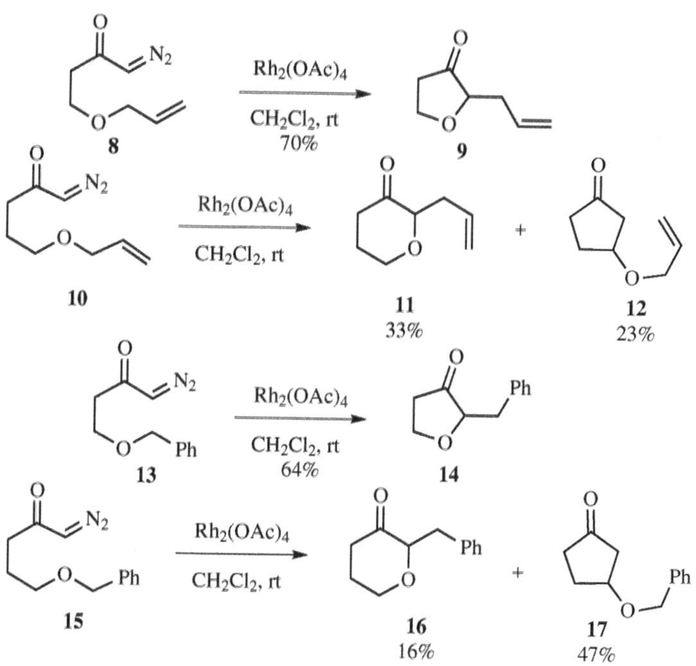

Scheme 16.5 Ring size effects on ylide formation versus C—H insertion.

chemoselective, the reaction of the homologous substrate **15** is not. A possible explanation is the increased activation barrier associated with the [1,2]-shift (as compared with the corresponding [2,3]-shift) due to an initial bond homolysis step, allowing reversion of the ylide to acyclic carbene to compete and predominant reaction via the five-membered C—H insertion pathway to give **17** as the major product.

The Clark group discovered a solution to the chemoselectivity problem associated with larger cyclic oxonium ylides (Scheme 16.6).[8] They showed that fluorinated copper acetylacetonate (acac) catalysts (e.g., tfacac = trifluoroacetylacetonate; hfacac = hexafluoroacetylacetonate) favor ylide formation over C—H insertion, and they were able to generate six-, seven-, and eight-membered ethers (**19a**, **19b**, and **19c**, respectively) with good chemoselectivity. Clark and coworkers attributed this chemoselectivity to an increase in electrophilicity of copper carbenoids containing electron-deficient fluorinated acac ligands, which should make attack by the Lewis basic ethereal oxygen (to generate an ylide) more favorable. Consistent with this proposal, increasing fluorine content on the ligands led to higher ratios of ylide-derived products to C—H insertion product in all cases.

An additional factor influencing the partition between ylide rearrangement and C—H insertion is the degree to which an initially formed cyclic ylide can revert to the metallocarbene precursor. If a low-energy decomposition pathway does not exist for the oxonium ylide, equilibration with the metallocarbene becomes more likely. Though ylide reversibility can account for the product distribution in the aforementioned cases (cf. Scheme 16.5, **11** vs **12**; **16** vs **17**), other examples better illustrate the effect (Scheme 16.7). When diazoketone **21** was treated with catalytic $Rh_2(OAc)_4$, the initially formed rhodium carbenoid was expected to undergo kinetically favored five-membered oxonium ylide formation. Though the substrate was designed to undergo a [1,2]-shift of a stabilized benzyl radical to give **22**, no rearrangement occurred.[9] The West group suggested that the radical resulting from homolytic bond cleavage does not experience any stabilization in the homolysis transition state, as this scissile bond is orthogonal to the π-system of the phenyl ring.

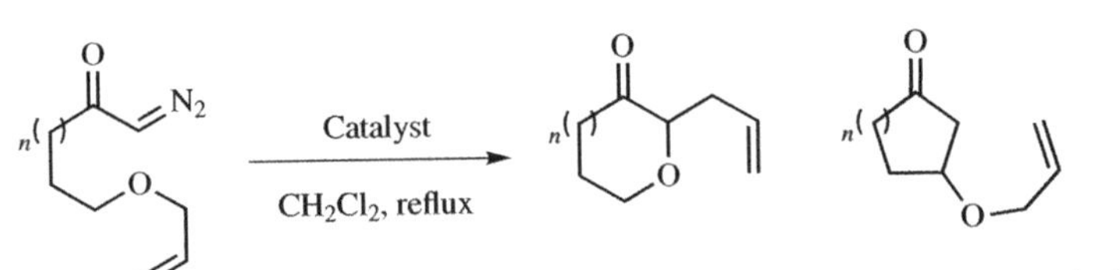

18a–c **19a–c**([2,3]-shift) **20a–c** (C–H insertion)

Substrate	Catalyst	Product (yield %)	Product (yield %)
18a ($n = 1$)	$Cu(acac)_2$	**19a** (61)	**20a** (12)
"	$Cu(tfacac)_2$	**19a** (78)	**20a** (0)
"	$Cu(hfacac)_2$	**19a** (83)	**20a** (0)
18b ($n = 2$)	$Cu(acac)_2$	**19b** (18)	**20b** (61)
"	$Cu(tfacac)_2$	**19b** (38)	**20b** (41)
"	$Cu(hfacac)_2$	**19b** (76)	**20b** (3)
18c ($n = 3$)	$Cu(hfacac)_2$	**19c** (40)	**20c** (0)

Scheme 16.6 Enhanced chemoselectivity for ylide formation using fluorinated ligands.

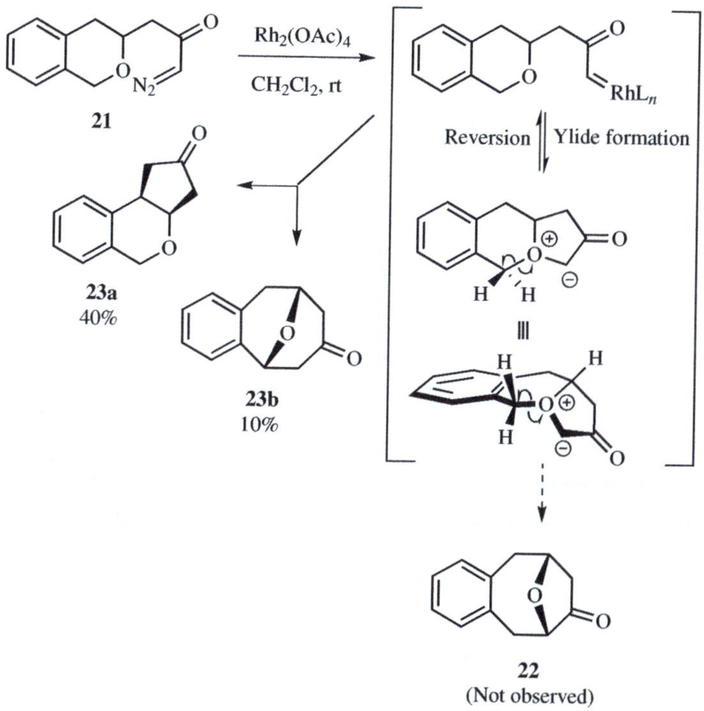

Scheme 16.7 Reversible ylide formation and preferential C—H insertion.

The ylide can then revert to the metallocarbene, which decomposes through two C—H insertion pathways to afford **23a** and **23b**.

16.1.3.2 Stereoselectivity In addition to chemoselectivity, stereoselectivity in oxonium ylide rearrangements is critical for their effective application to chemical synthesis. When a stereogenic center already exists on a substrate undergoing intramolecular oxonium ylide formation/rearrangement, diastereocontrolled formation of the new stereogenic center becomes possible. Often, this selectivity arises from conformational preferences in the transition state. For the case of five-membered ylides, Yakura *et al.* has shown excellent levels of diastereocontrol (Scheme 16.8).[10] The [2,3]-rearrangements of substituted allyl ethers **24a–c** illustrate this effect, with the rearrangements occurring to place the migrated allyl group at C-2 trans to the directing alkyl substituent at C-5 with good to excellent diastereoselectivity and in good yield.

Clark has shown that rearrangements of the homologous series of substituted allyl ethers (**27a,b**) follow the same trend (Scheme 16.9).[11] Catalysis with Cu(hfacac)$_2$ (to favor six-membered ylide formation over C—H insertion) afforded products **28** and **29** in good yield, again favoring the trans-isomers **28**. The six-membered oxonium ylides are believed to organize themselves in chair-like transition states, with

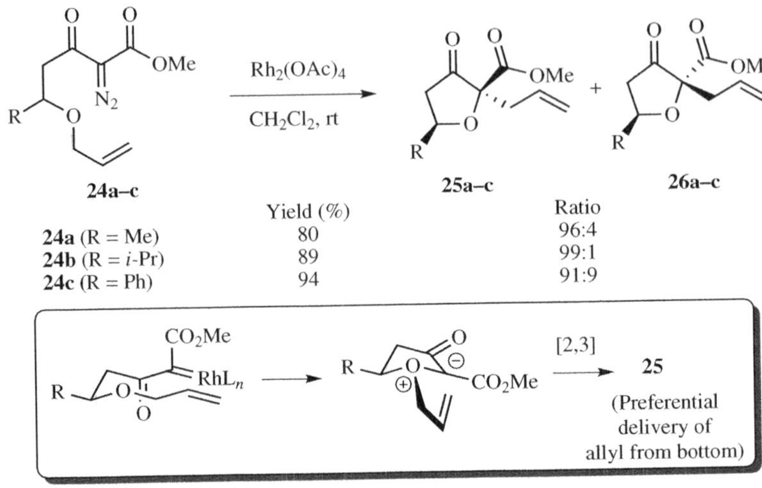

Scheme 16.8 Diastereocontrol in [2,3]-rearrangement of five-membered oxonium ylides.

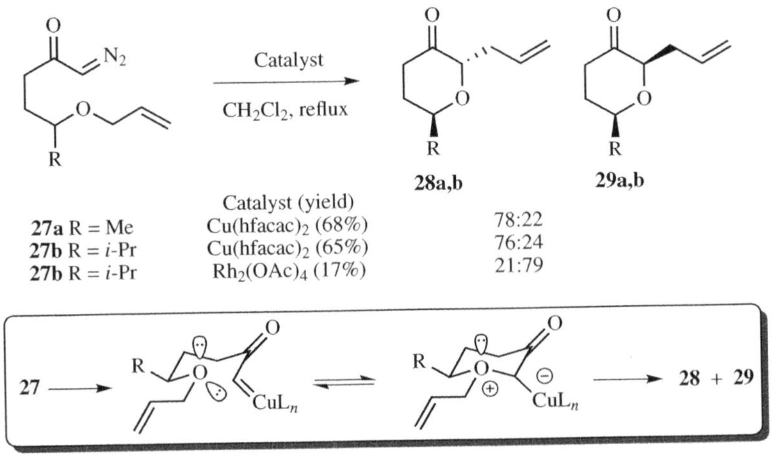

Scheme 16.9 Diastereocontrol in [2,3]-rearrangement of six-membered oxonium ylides.

substituents in the lowest energy conformation. The decreased levels of diastereocontrol relative to the case of five-membered oxonium ylides (cf. Scheme 16.8) suggest a possibly slower rearrangement, enabling oxonium inversion. Moreover, when $Rh_2(OAc)_4$ was used as the catalyst, the combined yield of ylide products **28** and **29** was greatly reduced due to competing C—H insertion, and the diastereoselectivity was reversed to favor cis-isomer **29**.

16.1.3.3 Role of Catalyst As a final point, the aforementioned results (catalyst-dependent selectivity for **28** or **29**) highlight the important role that choice of catalyst can play in the product distribution. Another example is seen for diazoketone **21**, for which the rearrangement under rhodium catalysis fails, providing only C—H insertion products **23a** and **23b** in moderate yield (Schemes 16.7 and 16.10), a result that was attributed to reversible ylide formation. West has shown that under copper catalysis, reaction via the oxonium ylide becomes predominant, providing the desired [1,2]-shift product **22** in near-quantitative yield (Scheme 16.10).[12] Such catalyst-dependent changes in selectivity strongly suggest continued association of the metal complex with the ylide during the rearrangement with either copper or rhodium catalysts or perhaps both. (Scheme 16.11)

Ligand variation on the metal center can also modulate product distribution. West and coworkers evaluated diazoketone **30**, which is capable of both five- and six-membered ylide formation.[13] Under copper catalysis, where both five- and six-membered ylides are readily formed, they found that $Cu(tfacac)_2$ favored the six-membered ylide formation/[2,3]-rearrangement pathway to afford **31**.

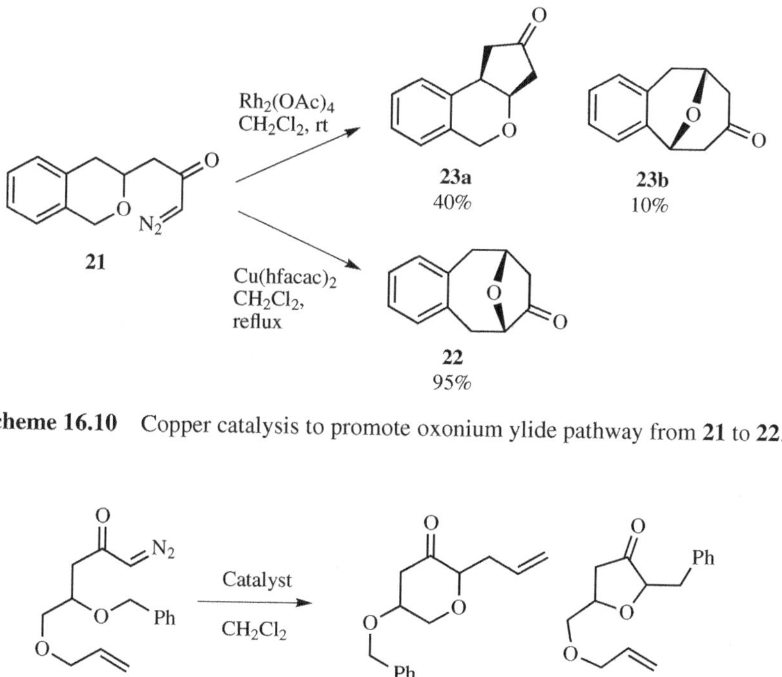

Scheme 16.10 Copper catalysis to promote oxonium ylide pathway from **21** to **22**.

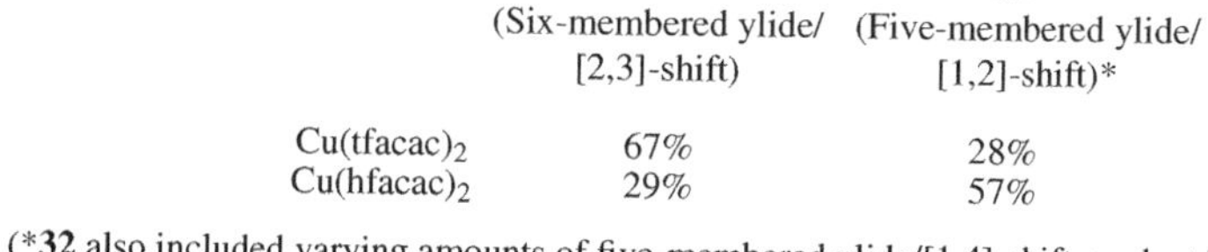

	31	32
$Cu(tfacac)_2$	67%	28%
$Cu(hfacac)_2$	29%	57%

(***32** also included varying amounts of five-membered ylide/[1,4]-shift product.)

Scheme 16.11 Effect of copper(II) ligand on ylide selectivity.

Conversely, $Cu(hfacac)_2$ favored the five-membered ylide formation/[1,2]-shift pathway, providing **32**. Though the differing reactivity preference for $Cu(tfacac)_2$ is not understood, the result is consistent and can be applied synthetically where appropriate.

16.1.4 Substrate Preparation

As mentioned previously, oxonium ylides are usually formed by intramolecular reaction of metallocarbenes with ether oxygens. For ammonium, sulfonium, and phosphonium ylides, an alternative stepwise approach involving initial formation of an onium salt, followed by base-mediated deprotonation, is feasible. However, in the case of oxonium ylides, this is not practical for two reasons. First, it is relatively difficult to generate trivalent oxonium salts by alkylation of ethers, due to the electronegativity of the oxygen atom. Second, dealkylation of oxonium salts – even very hindered ones – is facile with a variety of nucleophiles[14] and can be a major side process in many cases. Thus, direct ylide formation via carbene or metallocarbene precursors is used almost exclusively for oxonium ylides.

The metallocarbene intermediates are most often formed from thermal, photolytic, or metal-catalyzed decomposition of diazocarbonyl compounds,[15] with concomitant loss of dinitrogen. Under transition metal catalysis, the initially formed species is a metallocarbene rather than a free carbene, and this is usually desirable due to the moderated reactivity (and, hence, fewer undesired side reactions) of the metal-complexed carbene. The two most common methods for introduction of the diazo group are acylation of diazoalkanes with suitably activated carboxylic acid derivatives and diazo transfer reactions[16] in the case of more acidic active methylene substrates (Scheme 16.12).

Metallocarbene precursors other than diazocarbonyls have been examined sporadically for the generation of oxonium ylides, though not in the context of complex target synthesis. Among these, iodonium ylides[17] and oxidative activation of alkynes in the presence of gold catalysts[18] have shown significant promise.

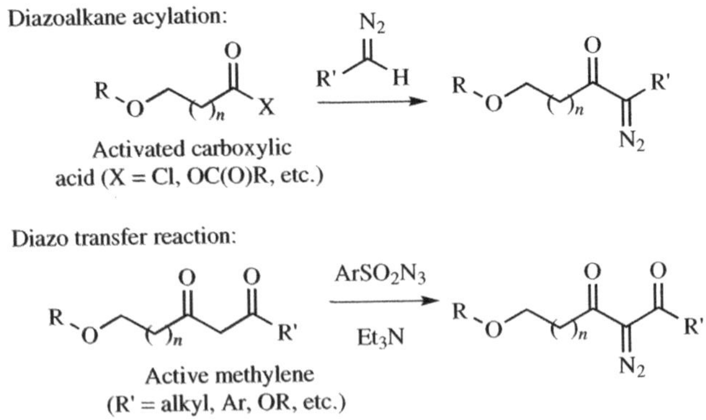

Scheme 16.12 Common methods for synthesis of diazocarbonyl compounds.

16.2 APPLICATIONS IN SYNTHESIS: OXONIUM YLIDE [2,3]-SIGMATROPIC REARRANGEMENTS

Synthetic applications of oxonium ylide rearrangements will be presented in two sections, the first concerning [2,3]-sigmatropic rearrangements and the second covering [1,2]-shifts. To begin, a number examples which assemble furan-3-one and pyran-3-one rings via [2,3]-migration of an oxonium ylide allyl substituent will be discussed.

16.2.1 Tetrahydrofuran-Containing Targets

16.2.1.1 Norhalichondrin B Phillips and coworkers have applied a [2,3]-rearrangement strategy to construct a 10-carbon tetrahydrofuran-containing fragment as part of their synthesis of norhalichondrin B (Scheme 16.13).[19] An asymmetric Noyori reduction on *β*-ketoester **33** was followed by a Pd-catalyzed allylation reaction with ethyl allyl carbonate. Saponification and diazoacylation gave diazoketone **36**, which was treated with 20 mol% Cu(hfacac)$_2$ in THF at reflux to give oxonium ylide **37**, which rearranged to afford trans-2,5-disubstituted furan-3-one **38** in 91% yield as a single diastereomer. Wittig olefination and hydroboration/oxidation gave **39**, which completed the C14–C23 fragment. The [2,3]-rearrangement occurred in accord with the predictive model, with the allyl group rearranging on the opposite face relative to the existing alkyl group (cf. **37**), giving the trans relationship between groups on the resulting furan-3-one.

16.2.1.2 Amphidinolides T1–T5 Using a similar strategy, Clark assembled the C1–C12 fragment of amphidinolides T1–T5.[20] A common structural feature within the amphidinolide T family of marine natural products is the polysubstituted tetrahydrofuran system, which the Clark group targeted by employing a [2,3]-rearrangement of a five-membered cyclic oxonium ylide. The synthesis began with *β*-ketoester **40**, which was subjected to a Noyori reduction and O-allylation reaction (Scheme 16.14) to afford **42**. Saponification and diazoacylation (via the mixed carbonic anhydride) gave diazoketone **43**. Treating **43** with Cu(hfacac)$_2$ in THF at reflux gave the desired trans-fused furan-3-one **44** in 90% yield. Removal of the PMB group and esterification (with **45**) gave diene **46**, which was converted to macrolactone **47** by Z-selective ring closing metathesis. The ketone was olefinated and then reduced with Wilkinson's catalyst to give the *β*-Me substituent in 4.5:1 ratio. Lactone cleavage gave **48**, which was dehydrated and saponified to give **49**, the C1–C12 fragment of amphidinolides T1–T5.

16.2.1.3 Gambieric Acid A Increased complexity arises in the [2,3]-process when a substituent is incorporated as part of the rearranging allyl group. As shown previously, in cases involving five-membered ylides, the product with a 2,5-trans relationship between furan-3-one substituents will predominate; however, an additional exocyclic stereocenter can arise when the allyl terminus is substituted. Clark's group set out to establish a predictive model for the relative configuration of the products of these rearrangements (Scheme 16.15) in conjunction with studies toward the synthesis of the marine ladder toxin gambieric acid A.[21] For a terminally substituted substrate **50**, two transition structures were envisioned, with the migrating allyl group either endo or exo to the ring of the cyclic ylide. These transition states would lead to diastereomers **51** and **52**, respectively, epimeric at the exocyclic stereocenter adjacent to the tetrahydrofuranone ring.

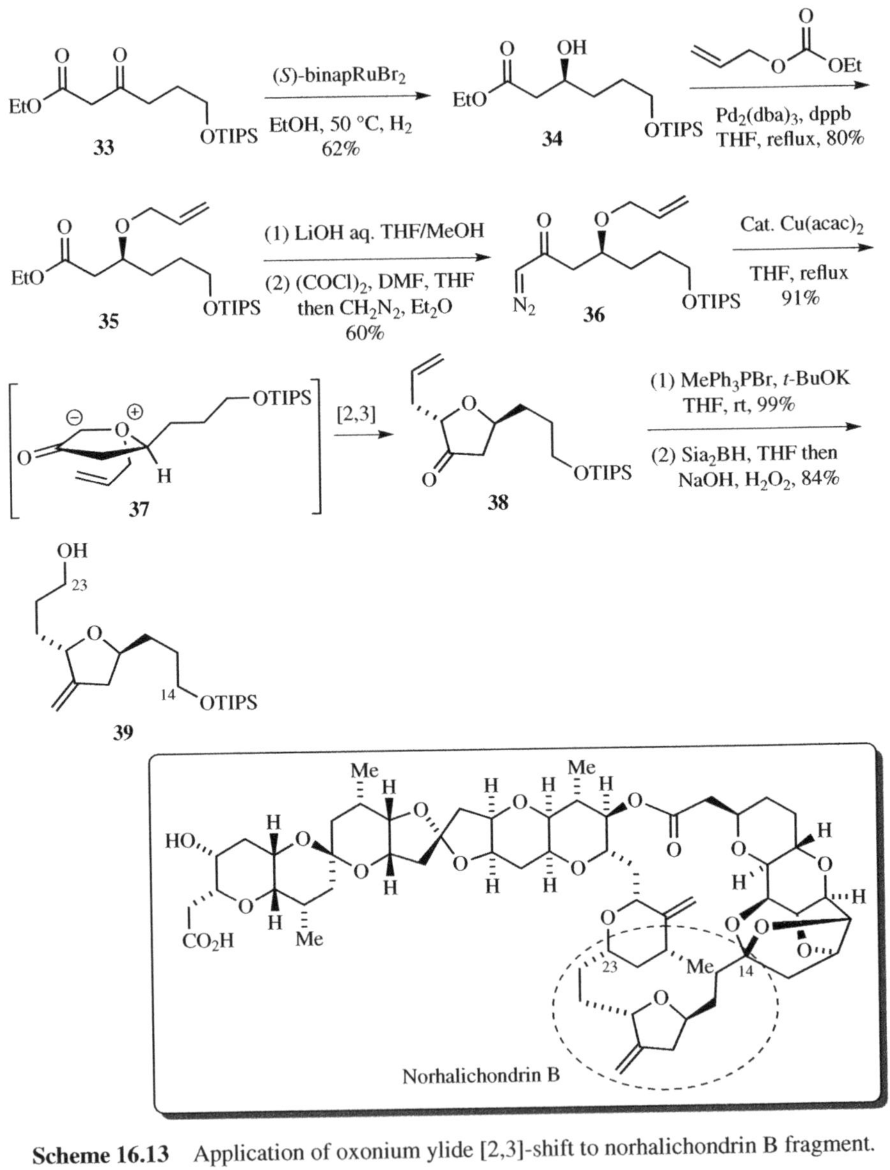

Scheme 16.13 Application of oxonium ylide [2,3]-shift to norhalichondrin B fragment.

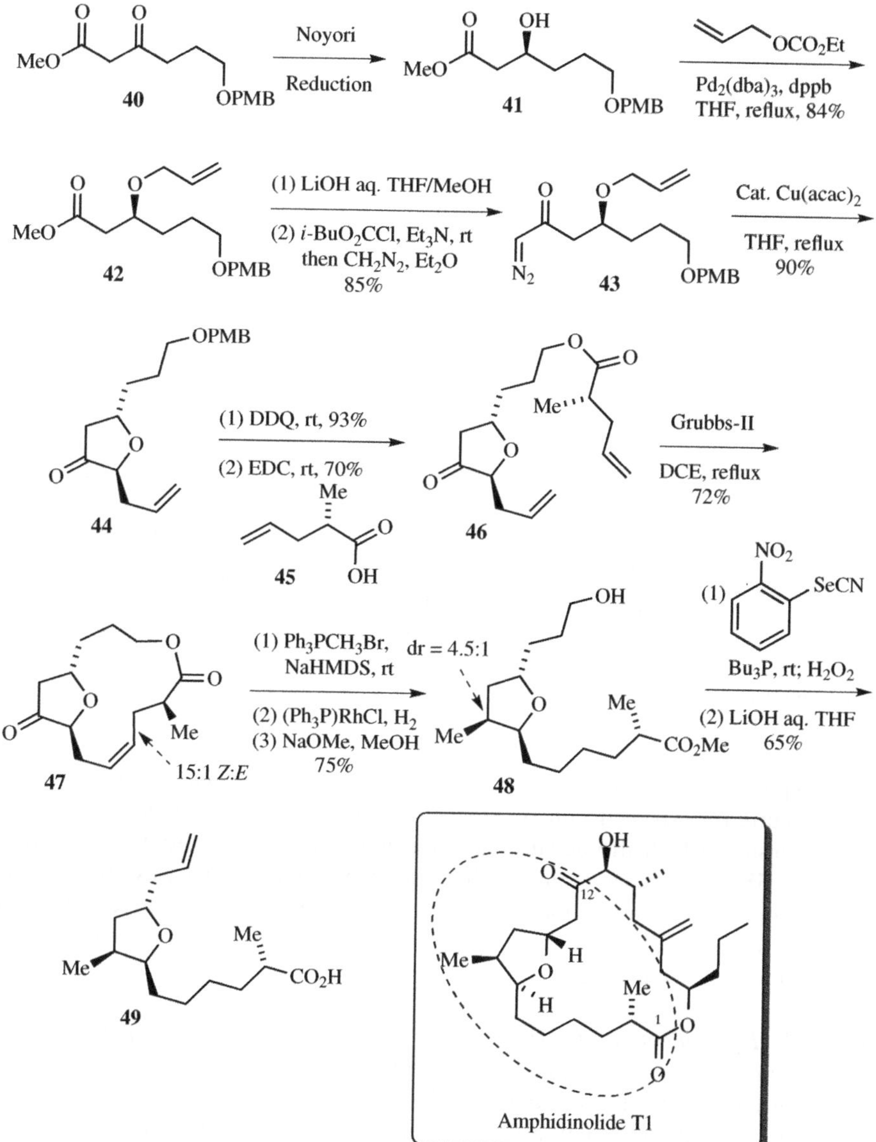

Scheme 16.14 Application to the substituted tetrahydrofuran of amphidinolides T1–T5.

Clark group's synthesis of the isolated tetrahydrofuran ring of gambieric acid began with **53**, derived from (*S*)-malic acid, which was selectively reduced and protected as the silyl ether to give **54** (Scheme 16.16). A short sequence afforded diazoketone **55a**, with a (*Z*)-crotyl ether migrating group. Treatment with 2 mol% $Cu(acac)_2$ in THF at reflux gave exclusively the trans furan-3-one product **56a** in 75% yield and

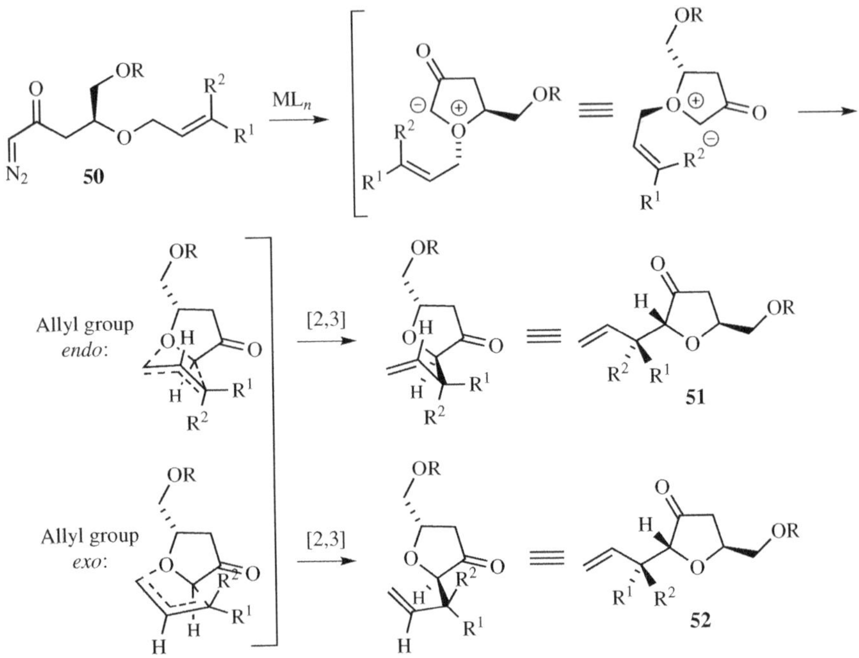

Scheme 16.15 Models for acyclic stereocontrol with substituted allyl migrating groups.

>19:1 epimer ratio at the methyl-substituted exocyclic stereocenter. The rearrangement occurred as expected with excellent diastereoselectivity, apparently via the endo transition state, which afforded the undesired configuration. Therefore, the (*E*)-crotyl substrate **55b** was prepared, and in this case, the major product was **56b** with the correct configuration at the methylated center. A short sequence of steps afforded tetrahydrofuranone **58**, corresponding to the tetrahydrofuran ring of gambieric acid A with four stereocenters set.

16.2.1.4 Hyperolactone C Hodgson *et al.* have reported multiple syntheses of the spirocyclic natural product hyperolactone C, based on variations of a [2,3]-rearrangement of an oxonium ylide (Scheme 16.17).[22] This compound possesses a spirolactone with a vicinal quaternary center, making it a structurally interesting synthetic challenge. Hodgson group's initial investigation started with known diazoketoester **59**, which was converted to the corresponding silyl enol ether and subjected to a Mukaiyama aldol reaction with the allyl acetal of benzaldehyde. The resulting allyl ether **60** rearranged under rhodium catalysis to afford the furan-3-one **61** in 99% yield and >10:1 dr, favoring the trans-isomer. Given the structural similarity between **61** and hyperolactone C, the use of a more functionalized "allyl" moiety would permit an expedient synthesis of the natural product. By subjecting **60** to a cross metathesis reaction with methacrolein (100 equiv.) using the Grubbs second-generation catalyst to afford **62**, followed immediately

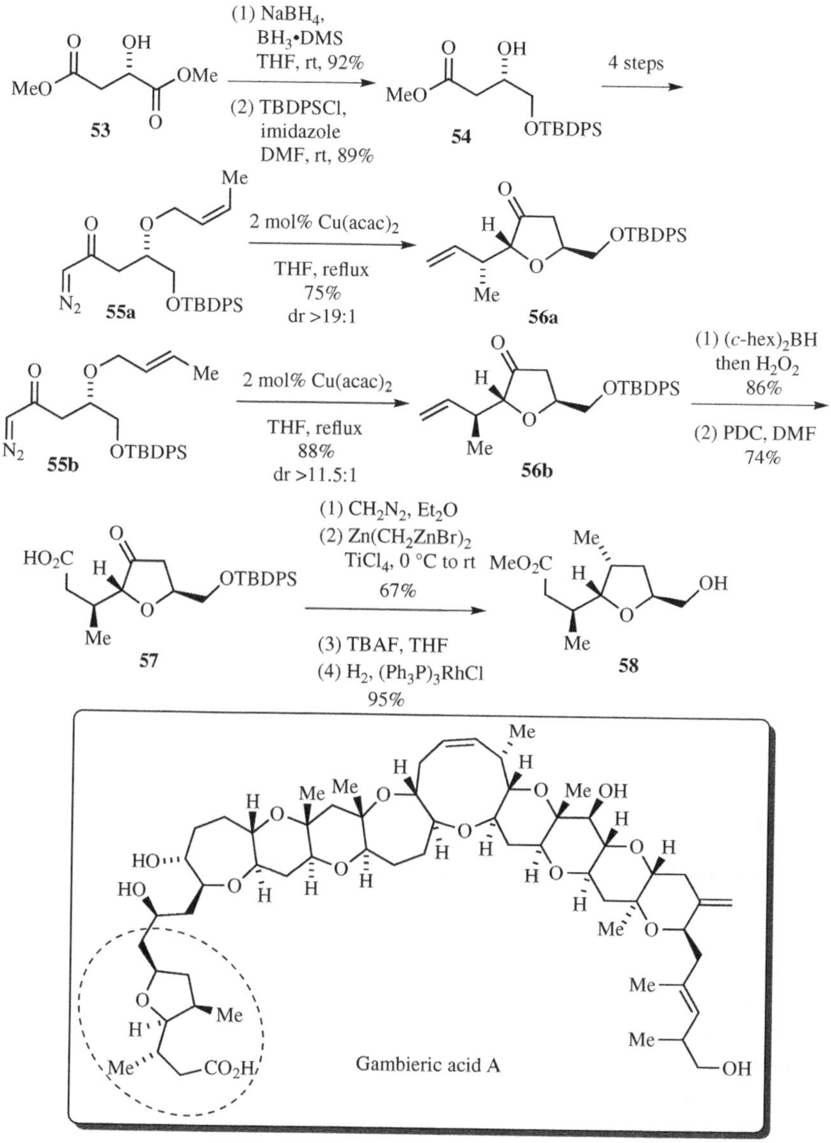

Scheme 16.16 Clark's route to the gambieric acid tetrahydrofuran ring.

by a rhodium-mediated [2,3]-rearrangement, the desired tetrahydrofuranone **63** was prepared. As a result of isolation challenges, the Hodgson approach used an *in situ* cyanoborohydride reduction, which enabled recovery of hemiketal **64** in 26% yield from **60**. The synthesis of hyperolactone C was completed by effecting a lactonization at elevated temperature and installing the unsaturation by treatment with DDQ. Consistent with Clark's observation, the [2,3]-rearrangement leading to

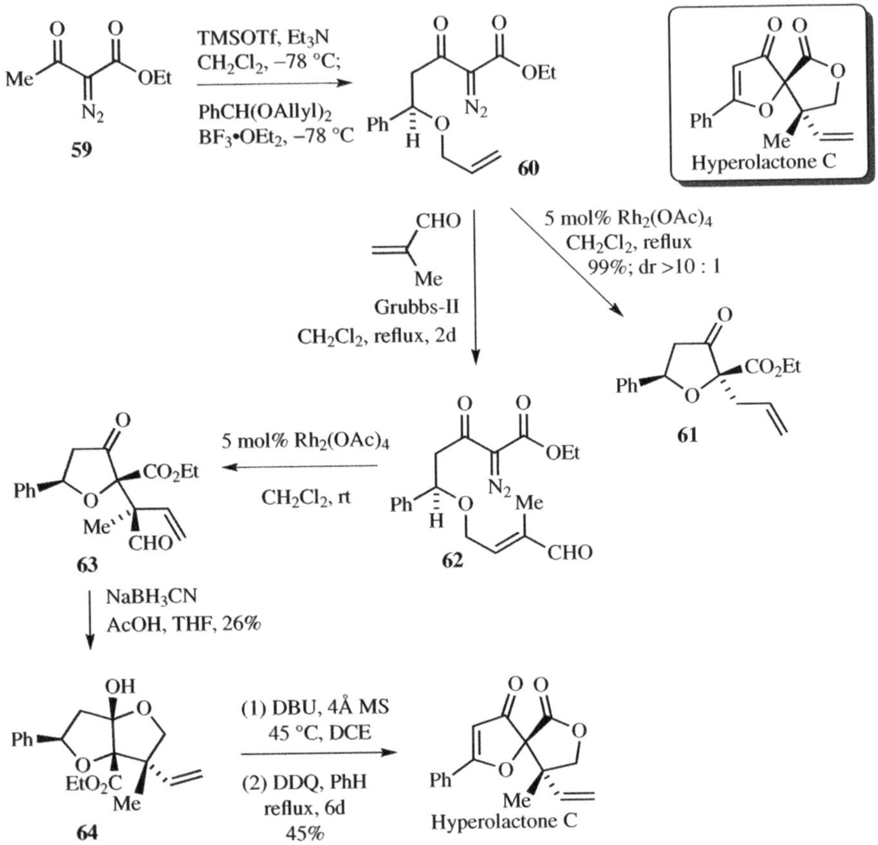

Scheme 16.17 Initial hyperolactone C study by Hodgson and coworkers.

63 occurred with preference for the endo transition state. Unfortunately, the overall yield of this multistep transformation was poor, which the authors attributed to a low-yielding cross metathesis step.

A higher yielding, asymmetric total synthesis was subsequently disclosed by the same group (Scheme 16.18).[23] Beginning with enantiopure styrene oxide, opening by cyanide anion and O-allylation furnished **65** in 91% yield. A Reformatsky addition of ethyl bromoacetate to the nitrile followed by hydrolysis gave β-ketoester **66** in 81% yield. Diazotransfer with *p*-ABSA (4-acetamidobenzenesulfonyl azide) then afforded (*R*)-**60**, which was converted to (*R*)-**62** by a two-step ozonolysis/Wittig olefination process, avoiding the low-yielding cross metathesis step. Subjecting (*R*)-**62** to $Rh_2(OAc)_4$ provided an intractable mixture of [1,2]- and [2,3]-rearrangement products, so the aldehyde was first reduced and protected as TMS ether **67**. The rhodium-catalyzed rearrangement occurred in this case to give a mixture of four diastereomers (**68**) that lactonized upon silyl deprotection to give

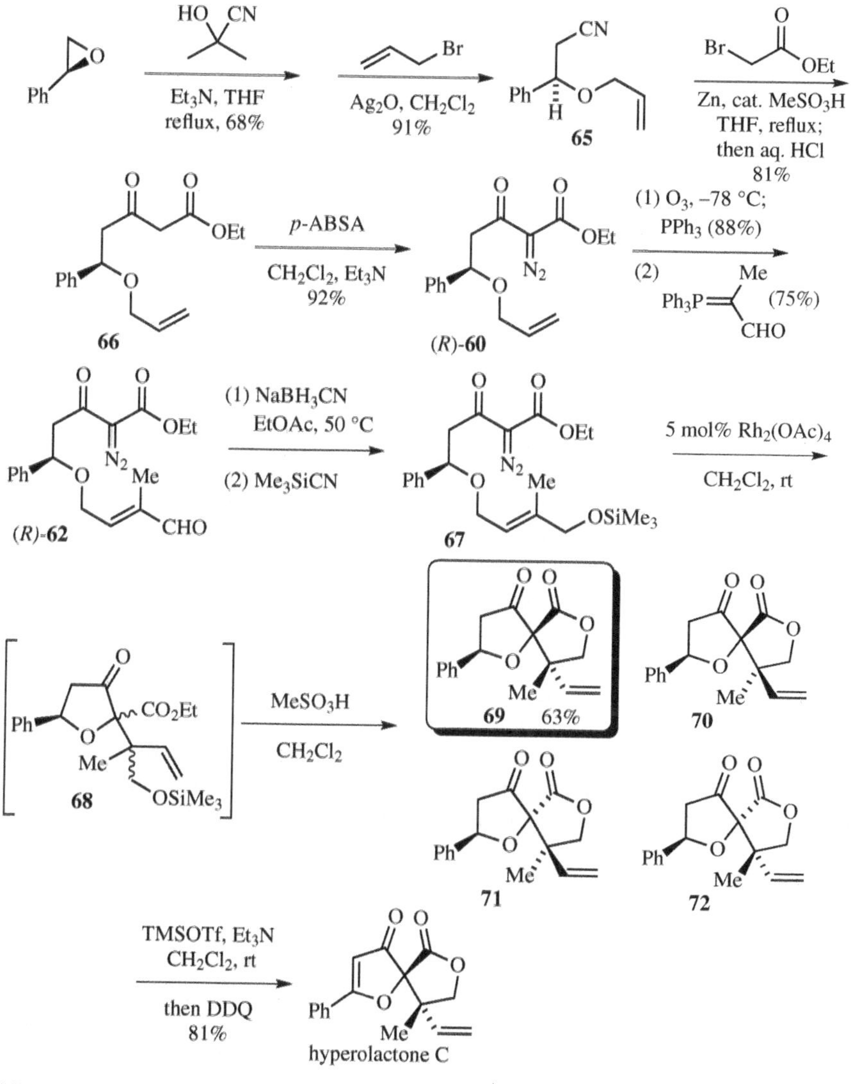

Scheme 16.18 Enantioselective hyperolactone C synthesis by Hodgson and coworkers.

69–72. The [2,3]-shift occurred predominantly as expected, with allyl delivery anti- to the phenyl group, and via the expected endo transition state. Through these combined preferences, the desired product **69** was recovered in 63% isolated yield, and completion of the asymmetric total synthesis was accomplished by installation of the unsaturation. Hodgson and coworkers recently published a third synthesis of hyperolactone C by a conceptually similar route but with late introduction of the phenyl substituent to avoid forming four stereoisomers in the key [2,3]-shift step.[24]

16.2.2 Benzofuranone-Containing Targets

16.2.2.1 Griseofulvin Like furan-3-ones, benzofuran-3-ones are readily constructed via oxonium ylide intermediates. This was first demonstrated by Pirrung and Werner when they showed how both diazoketone **73a** and α-diazo-β-ketoester **73b** would rearrange to give benzofuran-3-one products **74a** and **74b** upon exposure to $Rh_2(OAc)_4$ (Scheme 16.19).[5] The longer reaction time for **73b**, which could be reduced by heating the reaction, was attributed to the presence of two stabilizing carbonyl groups in the α-diazo-β-ketoester. The Pirrung group set out to apply this transformation as the key step in the synthesis of the tricyclic natural product (+)-griseofulvin.[25] This densely functionalized compound served as an excellent proving ground for the use of substrate control in a highly diastereoselective rearrangement of an oxonium ylide. The authors used an existing stereogenic center to control the relative stereochemistry during the rearrangement and in doing so found that the experimentally observed preference for endo transition states can be outweighed by steric effects.

The Pirrung synthesis began with the chlorination and acetylation 3,5-dimethoxy phenol. Fries rearrangement of acetate **75** gave a phenol, which was subjected to a Mitsunobu coupling with allyl alcohol **76**. The methyl ketone of **77** was deprotonated and trapped with Mander's reagent to give a β-ketoester that was then subjected to a diazo transfer reaction with MsN_3, giving diazoester **78**. Exposure of **78** to 5 mol% $Rh_2(piv)_4$ in benzene at reflux gave the desired [2,3]-rearrangement product **79** in 62% yield. The alkene was cleaved by ozonolysis, and the resulting aldehyde was olefinated to give **80** in 70% yield after ester cleavage. A Curtius rearrangement was then used to degrade the unsaturated acid to methyl ketone **81** via the enamine. Dieckmann-type condensation and methylation of the resulting enol completed the synthesis of (+)-griseofulvin (see Scheme 16.20).

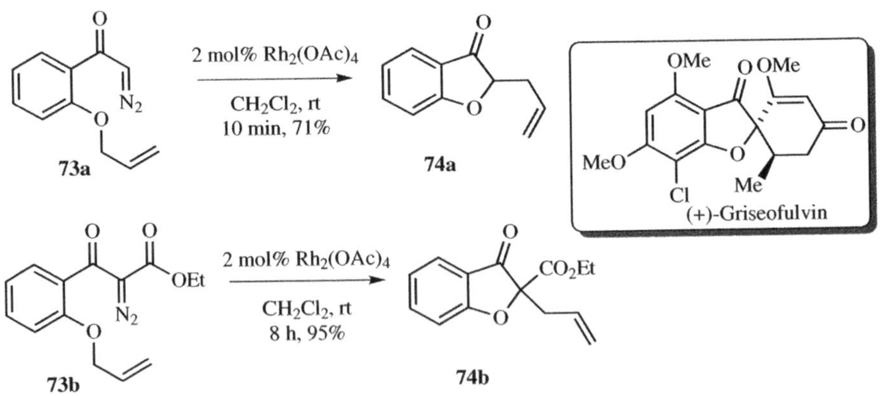

Scheme 16.19 Benzofuranones via oxonium ylide [2,3]-rearrangement.

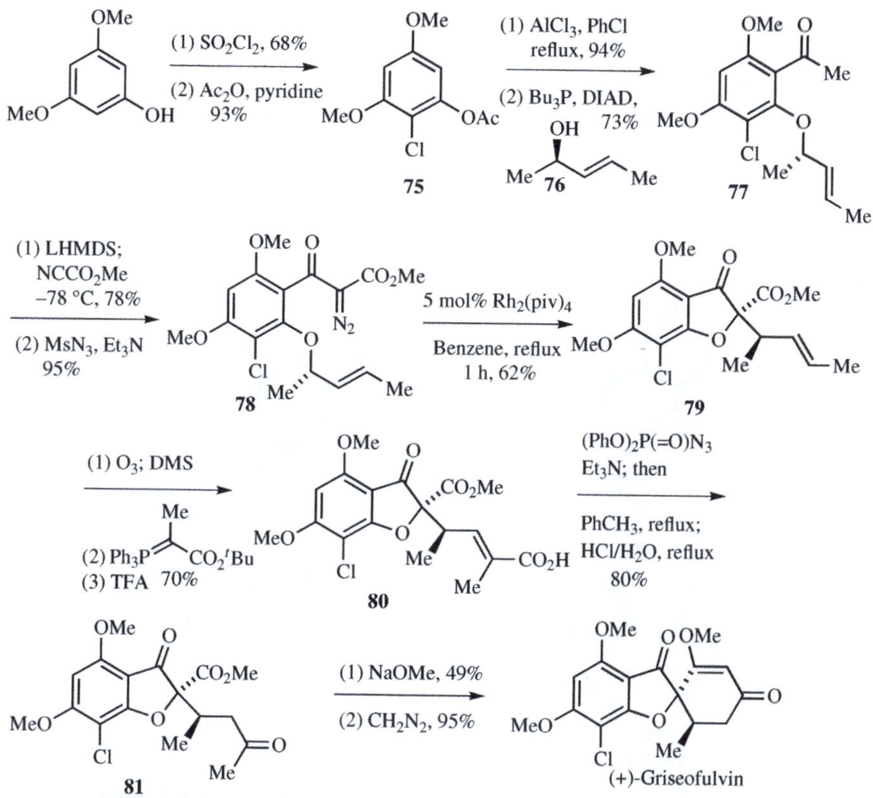

Scheme 16.20 Synthesis of (+)-griseofulvin by Pirrung and coworkers.

The [2,3]-rearrangement in the key step of this synthesis is a complex example for the predictive model outlined previously. The use of the disubstituted allyl tether was necessary to relay the stereogenicity through the rearrangement. However, the additional methyl branch adjacent to the oxonium oxygen caused a reversal in the typically observed preference for rearrangement via the endo conformation. The stereochemical outcome of the formation of oxonium ylide **82** (Scheme 16.21) is dictated by a preference for placement of the allyl substituent in a β orientation on the five-membered ylide, perhaps to minimize steric interactions with the neighboring methine carbon. The Newman projections for the exo and endo conformations looking from the α face of the ylide ring (**82a** and **82b**) illustrate the steric demand experienced by the substituents on the chiral allylic center. On inspection, endo transition state **82b** appears to be less favorable than exo transition state **82a**, since **82a** allows placement of the sterically least demanding substituent (H) over the ylide ring, whereas **82b** requires the much larger propenyl group to occupy this

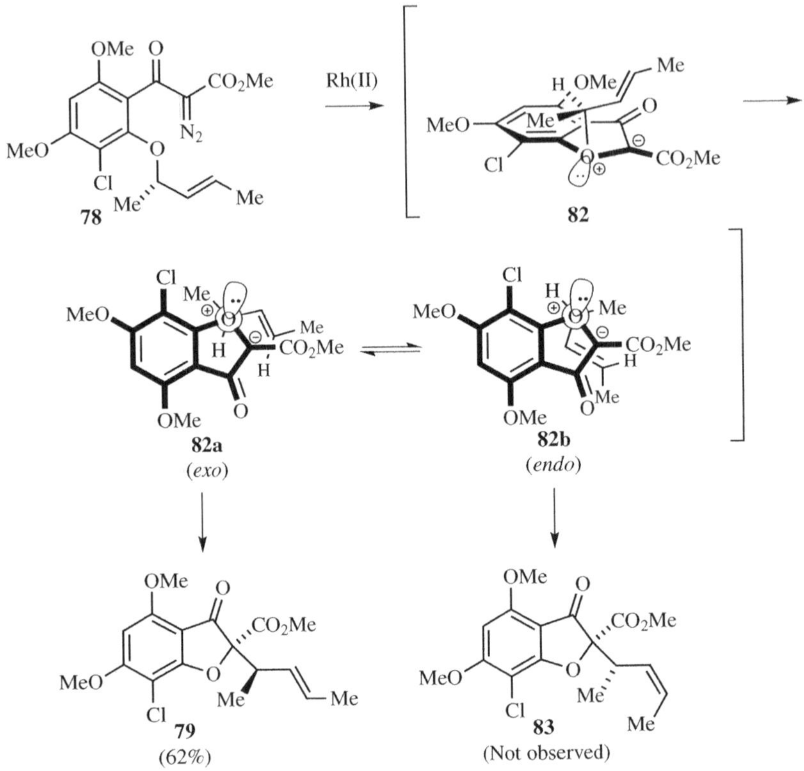

Scheme 16.21 Relay of stereochemical information in griseofulvin synthesis.

position. As a result, only product **79** formed, and the hypothetical product **83** of endo transition state **82b** was not observed.

16.2.3 Tetrahydropyran-Containing Targets

16.2.3.1 Decarestricine L As described earlier, the 3-pyranones are also easily accessed stereoselectively by the [2,3]-rearrangement of cyclic oxonium ylides derived from homologous diazoketone precursors (e.g., Scheme 16.9). Clark and Whitlock used a [2,3]-rearrangement of a six-membered oxonium ylide to complete a racemic synthesis of the pyran-containing natural product decarestricine L (Scheme 16.22).[26] The trans relationship between alkyl substituents flanking the ether oxygen of decarestricine L makes it an ideal target, given the intrinsic stereochemical preferences for ylide formation and rearrangement. The synthesis started with monoprotected diol **84**, which was subjected to allylation and silyl group removal to give **85**, followed by oxidation and diazoacylation, providing diazoketone **86**. Treatment with $Cu(hfacac)_2$ in dichloromethane at reflux gave, in 68% yield, a 77:23 mixture of *trans*-**87a** and *cis*-**87b** isomers. Reduction of the major isomer

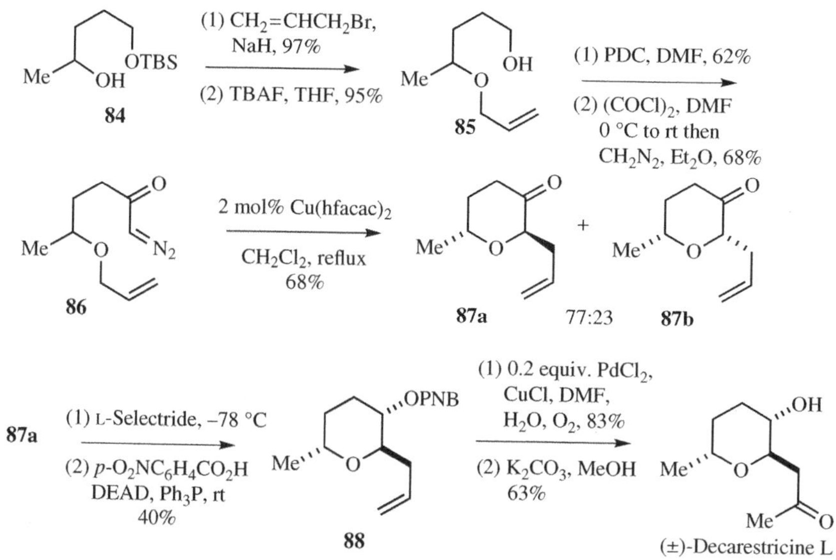

Scheme 16.22 Racemic synthesis of decarestricine L by Clark and Whitlock.

87a and Mitsunobu inversion gave **88**, which underwent a Wacker oxidation to give decarestricine L after ester cleavage.

An enantioselective synthesis of the same target was disclosed by the Clark group in 2006 (Scheme 16.23).[27] In this route, the enantiopure alcohol **89** was allylated

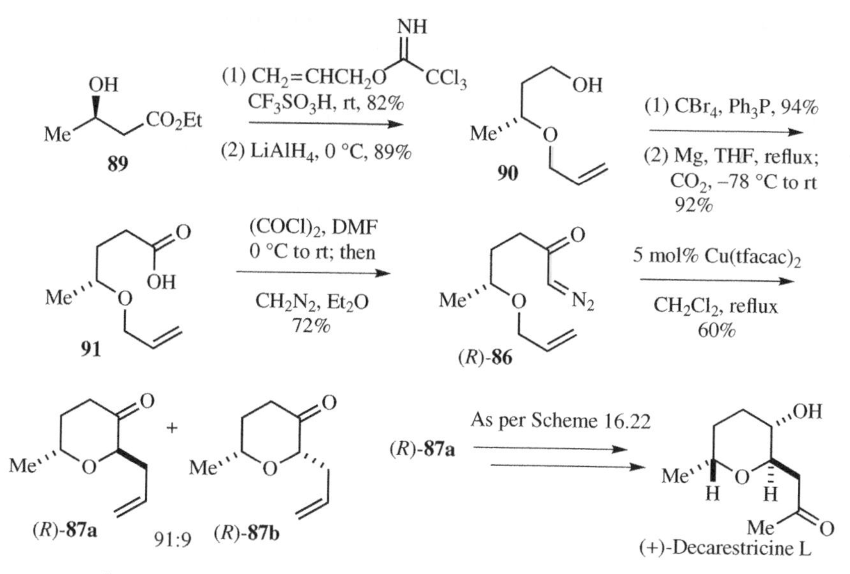

Scheme 16.23 Enantioselective synthesis of (+)-decarestricine L.

and the ethyl ester reduced to give alcohol **90**. Conversion to the bromide and subsequently to the Grignard reagent enabled a one-carbon homologation by quenching with CO_2, giving **91**. Diazoacylation gave (*R*)-**86**, which rearranged upon exposure to $Cu(tfacac)_2$ to give the desired product (*R*)-**87** in 60% yield as a 91:9 mixture of trans- and cis-isomers. The synthesis was then completed in analogy to the earlier racemic route. The change in catalyst was based on the observation by Clark[27] and others[28] that $Cu(tfacac)_2$ affords the highest diastereoselectivity for the trans product in rearrangements of six-membered oxonium ylides.

16.2.3.2 ***Laulimalide*** Yakura *et al.* applied a similar strategy to synthesize the C3–C12 fragment of the marine natural product (−)-laulimalide,[29] again taking advantage of the facile formation of the thermodynamically less stable trans-2,6-disubstituted pyran fragment via a rearrangement of an oxonium ylide. The synthesis began with protection of alcohol **92** as the PMB ether, followed by ester reduction to provide primary alcohol **93** (Scheme 16.24). A one-carbon homologation was followed by cyanide reduction with Dibal-H to give the aldehyde **94**. An asymmetric allylboration at low temperature gave **95** in excellent yield and diastereoselectivity. This was followed by a hydroboration/oxidation sequence, and protection of the resulting alcohol gave silyl ether **96**. A high-yielding allylation of the secondary alcohol gave **97**, which was deprotected and oxidized to afford carboxylic acid **98**. Diazoacylation via a mixed carbonic anhydride gave diazoketone **99** in 85% yield. Upon exposure to $Cu(tfacac)_2$, the desired rearrangement occurred to give **100** in 82% yield, as a 97:3 mixture favoring the desired trans-isomer. A series of functional group manipulations gave aldehyde **101**, which was converted into pyran-3-one **102**. Trapping the derived kinetic enolate with Comins' reagent gave an enol triflate that could be reduced to provide **103**, the C_3–C_{12} fragment of (−)-laulimalide. It is notable that significantly higher diastereoselectivity was obtained in the rearrangement of **99** as compared with (*R*)-**86** (cf. Scheme 16.22). The enhanced stereoselectivity is presumably the result of the greater steric demand of the branching group adjacent to the ether oxygen in **99** and a consequently greater enforcement of the preferred chair conformation in the transition state for cyclic oxonium ylide formation.

16.2.3.3 ***Polycyclic Ethers*** The polyether structural motif, such as the trans-fused polypyran unit often found in the skeletons of the marine ladder toxins (e.g., see gambieric acid A, Scheme 16.16), is an appropriate choice of target for application of oxonium ylide [2,3]-rearrangement methodology. Such a strategy theoretically provides access to the alternating oxygenation pattern, and enables an iterative approach to construction of successive pyran units. The pyranone resulting from the [2,3]-rearrangement should possess substituents at C2 and C6 with a trans relationship; however, the requisite cis-isomer should be available by epimerization facilitated by the neighboring carbonyl unit. In a feasibility study of this approach to polypyrans, Marmsäter and West prepared hydroxypyran **104** from DHP (Scheme 16.25).[28] The ketal was reduced (**105**), the allyl group converted to a carboxylic acid, and the alcohol allylated to give **106**. Diazoacylation (via the acid chloride) gave **107**, which smoothly underwent the [2,3]-rearrangement to

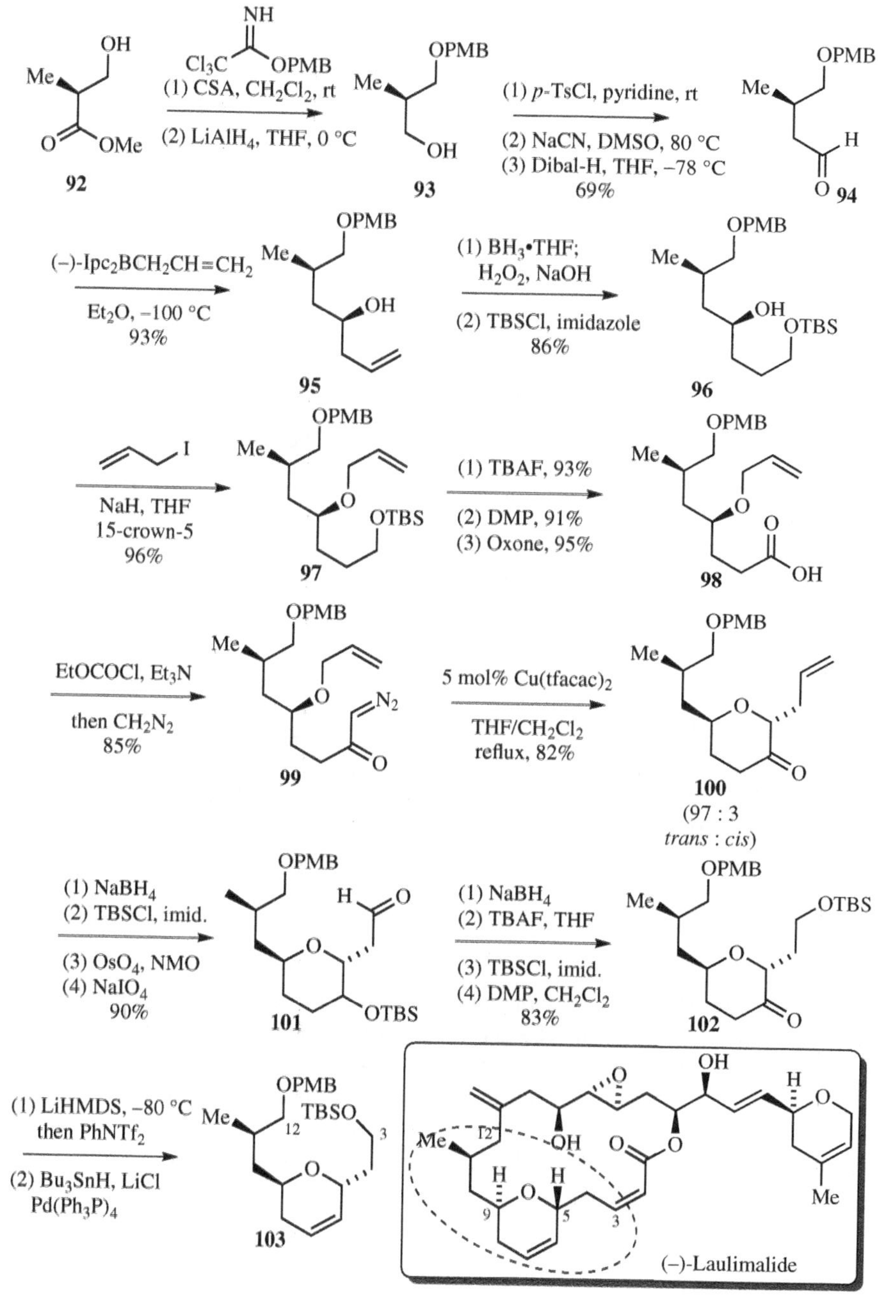

Scheme 16.24 Synthesis of (−)-laulimalide C3–C12 fragment by Yakura and coworkers.

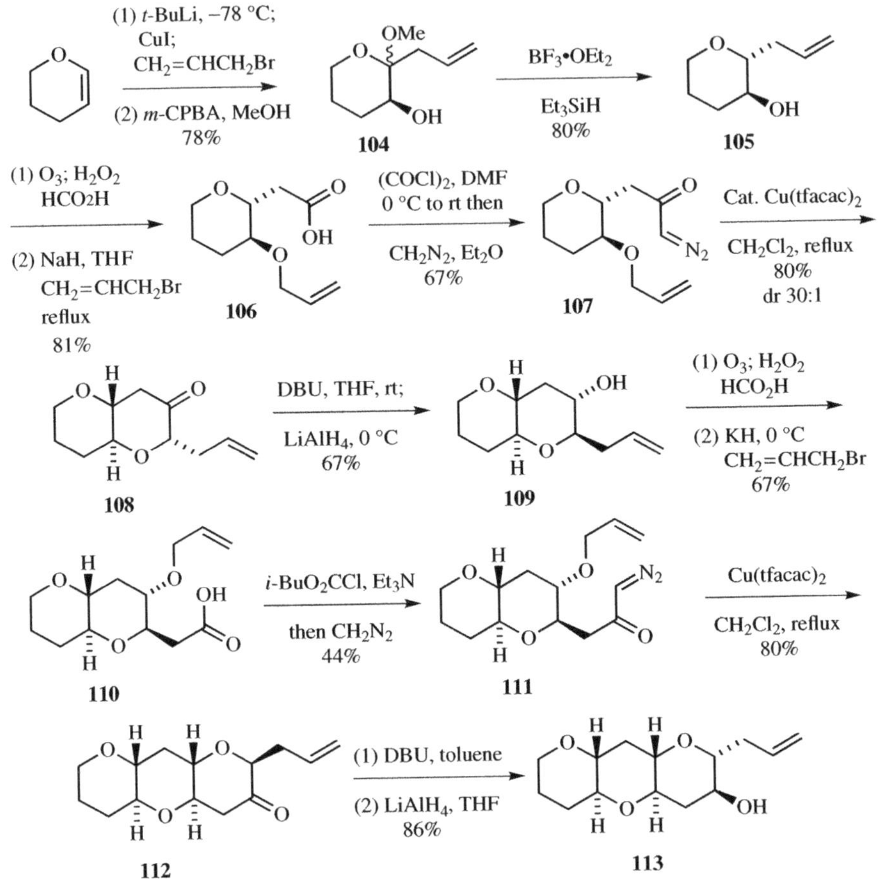

Scheme 16.25 Iterative approach to polypyrans via oxonium ylide [2,3]-shifts.

afford bis-pyran **108** in 80% yield and 30:1 dr. Epimerization to the cis-isomer and reduction gave bis-pyran **109** with a trans relationship between hydroxyl and allyl substituents. Resubjecting **109** to another iteration of the synthetic sequence gave diazoketone **111**, which rearranged upon treatment with Cu(tfacac)$_2$ to give **112** in 80% yield as a single diastereomer. Epimerization to the cis-isomer and reduction gave tris-pyran **113** with the correct trans ring junctions associated with the polypyran units found in polyether natural products. The high trans-selectivity in the rearrangement may derive from a preorganized conformation of the uncyclized metallocarbene in analogy to the previously described examples. In the cases shown in Scheme 16.25, the added rigidity of the fused bicyclic skeletons may be responsible for even greater stereoselectivity.

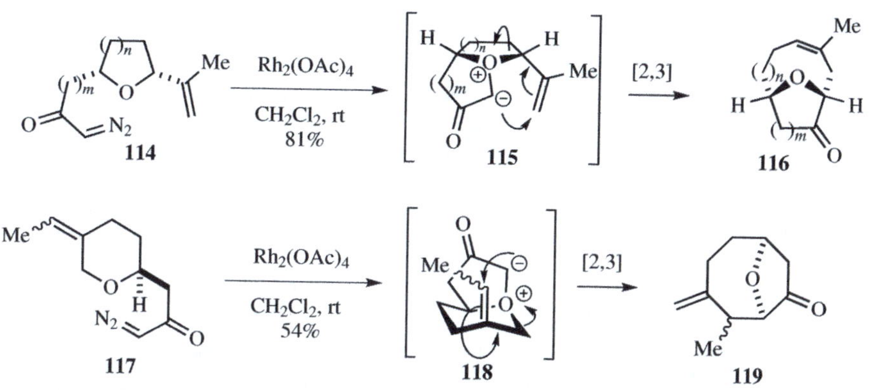

Scheme 16.26 Bridged bicyclic structures via [2,3]-shift of fused bicyclic oxonium ylides.

16.2.4 Reaction via Bicyclic Oxonium Ylides

There are numerous examples of metallocarbene capture by lone pairs of *cyclic* ether oxygens, generating fused bicyclic oxonium ylides in which the oxonium oxygen resides at a bridgehead position. If the migrating allyl group is part of the initial cyclic ether, rearrangement will lead to a bridged bicyclic product. An early example of this was reported by Pirrung where vinyl-substituted tetrahydrofuran **114** (m, $n = 1$) was treated with $Rh_2(OAc)_4$ to provide the bicyclic ylide **115** (Scheme 16.26).[5] The ensuing [2,3]-rearrangement occurred to provide the ether-bridged medium-ring structure **116** in good yield, as a single isomer. By altering the size of the cyclic ether ($n = 1, 2, 3, \ldots$) or the length of tether to the diazo group ($m = 1, 2, 3, \ldots$), numerous oxabicyclo[$n.m.$1] scaffolds can in theory be targeted. An additional example was shown by the West group, in which one of the alkenyl carbons of the allyl group was incorporated into the starting pyran ring.[9] When diazoketone **117** was exposed to $Rh_2(OAc)_4$, the initially formed oxonium ylide (**118**) underwent the [2,3]-rearrangement with allylic transposition, giving O-bridged cyclooctanone **119** in 54% yield. The bicyclic molecular skeletons provided by these methodologies correspond to a large number of natural products, making this oxonium ylide approach synthetically valuable. In cases where steric congestion or conformational bias precludes the allyl terminus from coming into bonding proximity with the ylidic carbon, the high-energy ylide will decompose by alternate pathways, such as the vinyl-stabilized [1,2]-shift.

16.2.4.1 Neoliacinic Acid Clark and coworkers have used bicyclic oxonium ylide methodology in studies that culminated in a completion of the total synthesis of the plant-derived sesquiterpene natural product neoliacinic acid. The natural product consists of an oxabicyclo[5.3.1]undecane scaffold with a bridging lactone

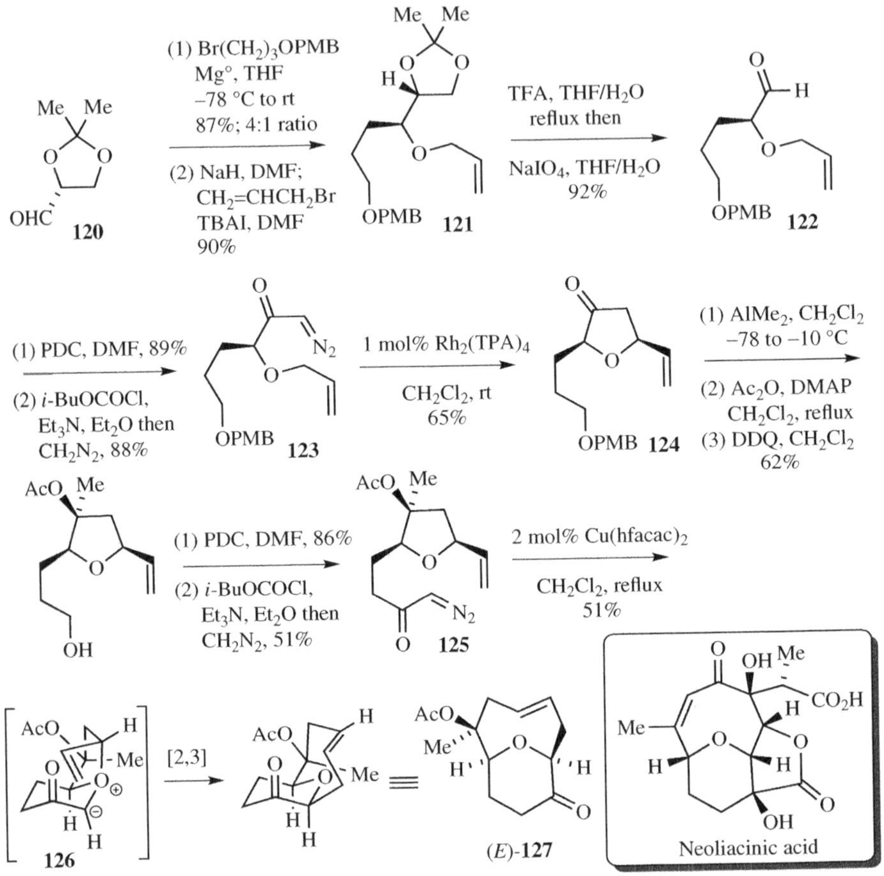

Scheme 16.27 Model study toward neoliacinic acid by Clark and coworkers.

and α-branched carboxylic acid side chain (Scheme 16.27). Accessing this scaffold via oxonium ylide rearrangement would require a tetrahydrofuran whose oxygen is flanked by a vinyl group and a four-carbon diazoketone side chain. Clark's model study began with stereoselective addition of a Grignard reagent to protected glyceraldehyde **120**, followed by O-allylation to give **121** (Scheme 16.27).[30] Removal of the acetonide and cleavage of the resulting diol gave **122**. The aldehyde was oxidized and the resulting acid converted to diazoketone **123**. The diazoketone was treated with $Rh_2(TPA)_4$ to synthesize the required furan-3-one **124** as a single diastereomer via a five-membered C—H insertion reaction. A straightforward conversion of the tethered alcohol of **124** to a diazoketone gave cyclization precursor **125**. In the key step, treatment with $Cu(hfacac)_2$ gave the bicyclic oxonium ylide **126** that rearranged to give (*E*)-**127** in 51% yield.

A more elaborate model study was subsequently disclosed in which the α-branched carboxylic acid side chain of neoliacinic acid was incorporated.[31] Given this substituent's placement in the natural product, its installation at the internal allyl carbon would provide the correct structure upon rearrangement. The rearrangement precursor was prepared by a chelation-controlled cuprate addition to protected glyceraldehyde **128**, giving **129** and setting the correct bridgehead stereochemistry for neoliacinic acid (Scheme 16.28). Allylation of the alcohol with functionalized allyl bromide **130** gave acetonide **131**, which was subjected to a series of steps including stereoselective tetrahydrofuranone cyclization by C—H insertion (**133–134**), various protecting group manipulations and redox modification, and finally a diazoacylation to provide diazoketone **136**. Treatment of **136** with Cu(hfacac)$_2$ gave a 61% yield of **138** (3:2, *E*/*Z*), as well as 7% of **137**, the vinyl-stabilized [1,2]-shift product. The mixture of *E*/*Z* products **138** was equilibrated to the *cis*-alkene and then epoxidized on the α-face to give **139**, the neoliacinic acid core. Clark subsequently demonstrated the conversion of **139–140**, which possesses all of the required rings and stereogenic centers of neoliacinic acid.[32]

16.2.4.2 Litophynins E and I Litophynins E and I are highly oxygenated marine natural products that belong to the cladiellin family and that share an oxabicyclo[6.2.1]undecane skeleton (Scheme 16.29). As part of their work on the oxabicyclo[5.3.1]undecane scaffold of neoliacinic acid, the Clark group discovered a selenium-mediated rearrangement of (*E*)-**127** (Scheme 16.27) to the oxabicyclo[6.2.1]undecane skeleton that corresponds to the core of the litophynins E and I. As the required rearrangement precursor **147** is the enantiomer of **125**, the Clark synthesis began with the (*R*)-glutamic acid derivative **141** (Scheme 16.29).[33] Transesterification and O-allylation gave **142**, which was saponified and cyclized to anhydride **143**. Treatment with diazomethane resulted in a regioselective opening of the anhydride to provide the diazoketone **144** as a single regioisomer. A C—H insertion was effected under rhodium catalysis, followed by a methyl addition/acylation sequence to afford **145**. Selective hydrolysis of the methyl ester was followed by conversion of the carboxylic acid to diazoketone **147**. Upon treatment with Cu(hfacac)$_2$, the substrate underwent a tandem oxonium ylide formation/[2,3]-rearrangement to afford (*E*)-**148** in 50% yield. The unstable *E*-alkene reacted with electrophilic PhSe(O_2CCF_3) to undergo a modest-yielding rearrangement that via an intermediary bridged tricyclic oxonium ion that is opened via anchimeric assistance from the neighboring acetate group. The resulting O-bridged carbocycle **149a** possessed the required absolute and relative configuration at four oxygenated stereogenic centers, though it exists primarily as the transannular hemiketal **149b**.

16.2.4.3 Labiatin and Australin The natural products labiatin A and australin A (Scheme 16.30) are marine diterpene natural products that also belong to the cladiellin family. Although these compounds are structurally related to neoliacinic acid, possessing the oxabicyclo[5.3.1]undecane core, the fused, substituted cyclohexane adds an additional level of complexity when planning a synthetic approach involving an oxonium ylide rearrangement. Analysis of the [2,3]-rearrangement precursor and product reveals that the cyclohexane unit must be part of the rearranging allyl moiety.

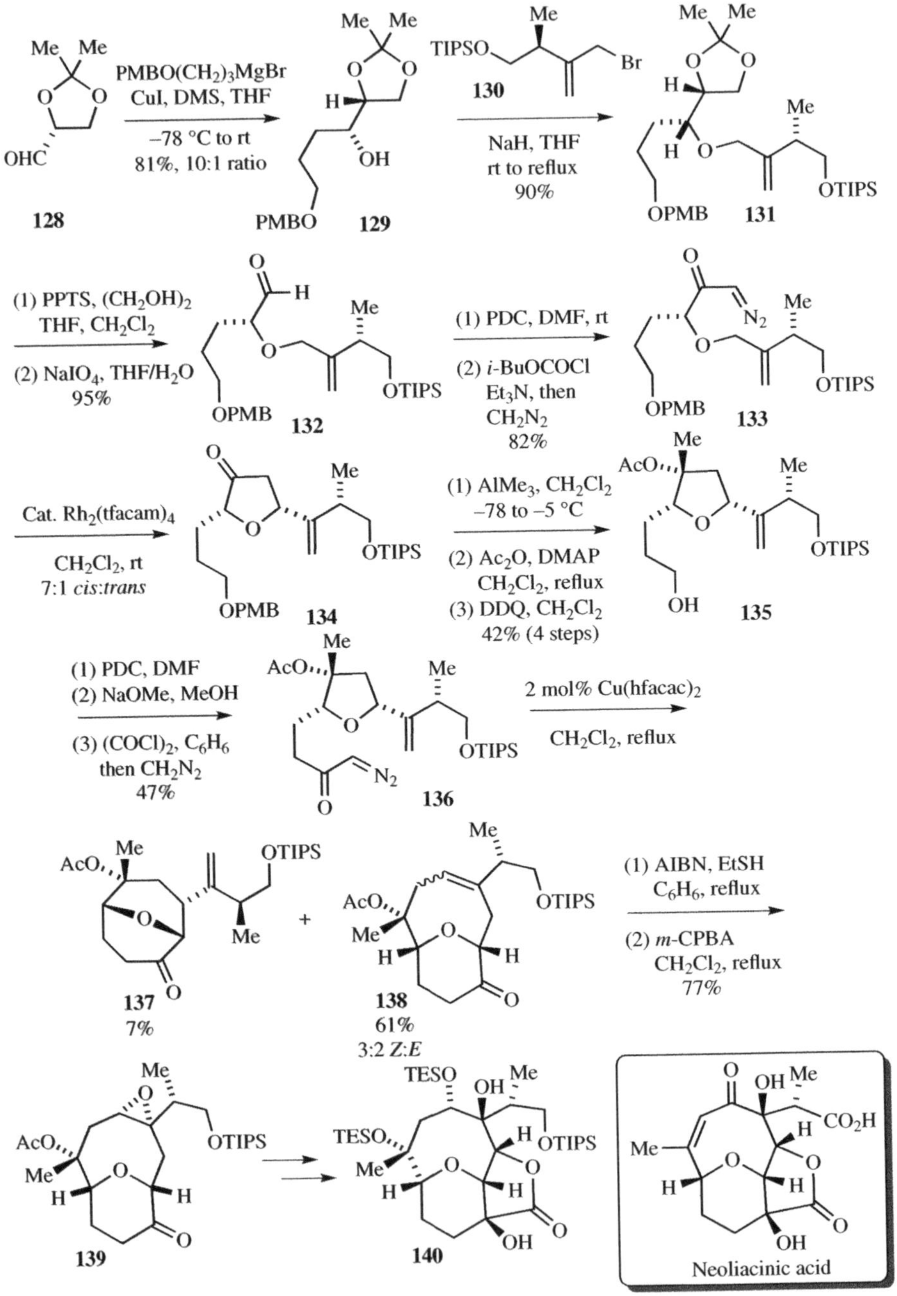

Scheme 16.28 Second-generation approach to neoliacinic acid by Clark and coworkers.

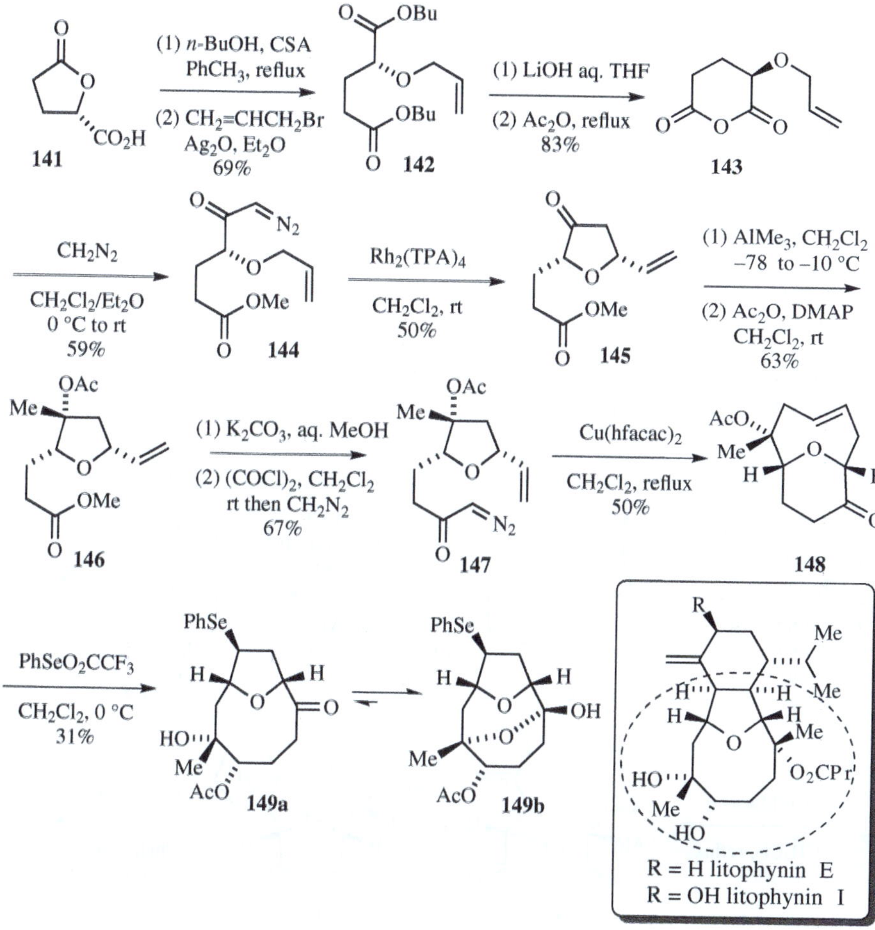

Scheme 16.29 Route to oxabicyclo[6.2.1]undecane litophynin skeleton by Clark and Wong.

This additional substitution is likely to increase steric interactions in the rearrangement transition state, which could possibly lead to rearrangement by alternative, less sterically demanding pathways.

Nevertheless, the Clark group conducted a model study which showed the feasibility of the oxonium ylide rearrangement approach.[34] O-Alkylation of alcohol **150** with (bromomethyl)cyclohexene gave **151**, after which the acetonide was removed and the resulting diol cleaved to give aldehyde **152** (Scheme 16.30). A Pinnick oxidation gave a carboxylic acid that was diazoacylated via the mixed carbonic anhydride to give diazoketone **153**. Exposure to rhodium catalysis gave the desired C—H insertion product **154** in 60% yield, as a 7:1 mixture favoring the desired cis-isomer. Conversion to diazoketone **156** was carried out via a multistep route analogous to earlier examples. Treatment of **156** with $Cu(hfacac)_2$ gave a 2:1 mixture of the desired product **158**

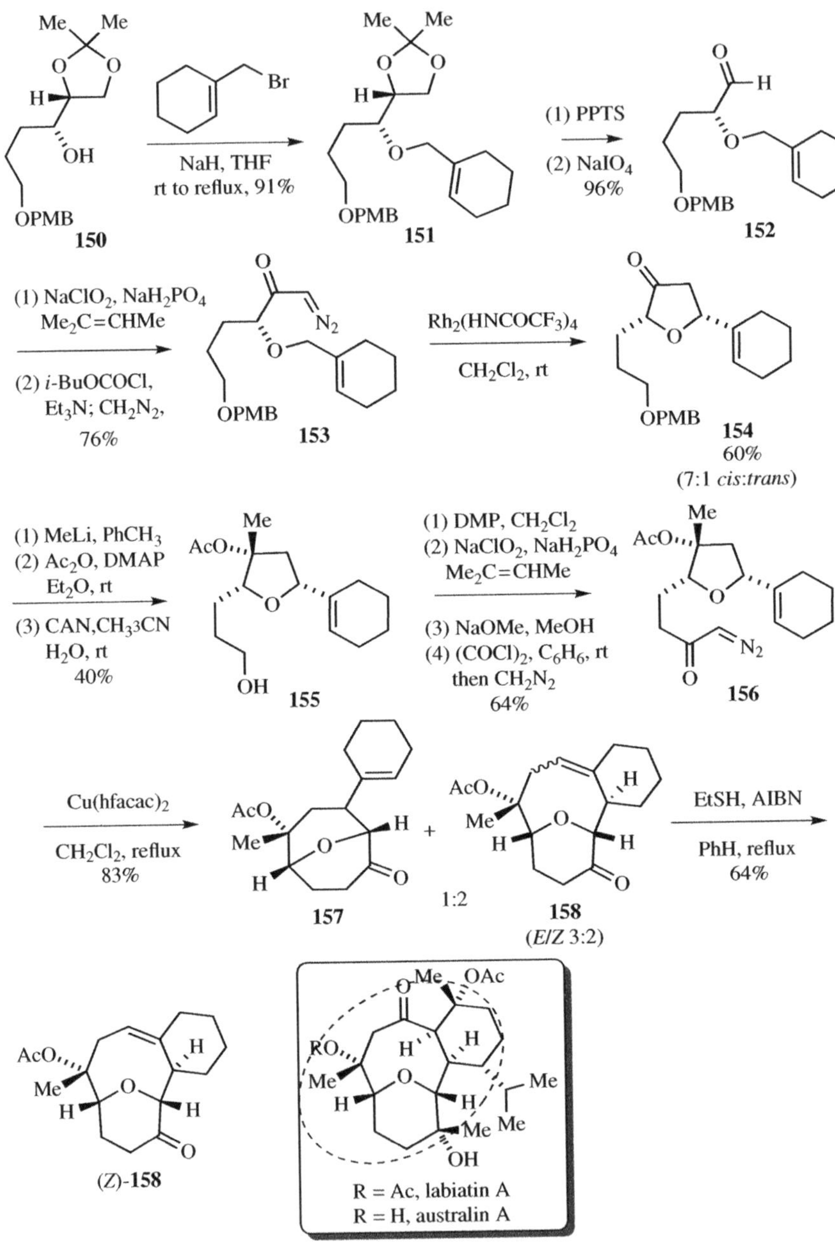

Scheme 16.30 Synthesis of the labiatin A tricyclic core by the Clark group.

(3:2 *E/Z*) and side product **157**, derived from a vinyl-stabilized [1,2]-shift. The mixture of rearrangement products **158** was isomerized to the (*Z*)-isomer in 64% yield.

Using an analogous route, the diazoketone **160** bearing a fully functionalized cyclohexenyl moiety was prepared via acetonide **150** and cyclohexenyl bromide **158** (Scheme 16.31).[35] Treatment with $Cu(hfacac)_2$ gave the desired rearrangement product **161** in 76% yield as a single isomer. Addition of MeLi to this intermediate effected both deacylation and methylation, providing advanced intermediate **162** *en route* to labiatin A and australin A. It is especially notable that this more functionalized example was not subject to the undesired [1,2]-shift side reaction seen in the simpler model, nor was the alkene formed as a mixture of geometrical isomers.

16.2.4.4 Vigulariol The aforementioned syntheses demonstrate the ease with which a [2,3]-rearrangement of an oxonium ylide can deliver the oxabicyclo[5.3.1]undecane ring system, as well as a method for further transformation to the oxabicyclo[6.2.1]undecane skeleton. Owing to the versatility of structural types that

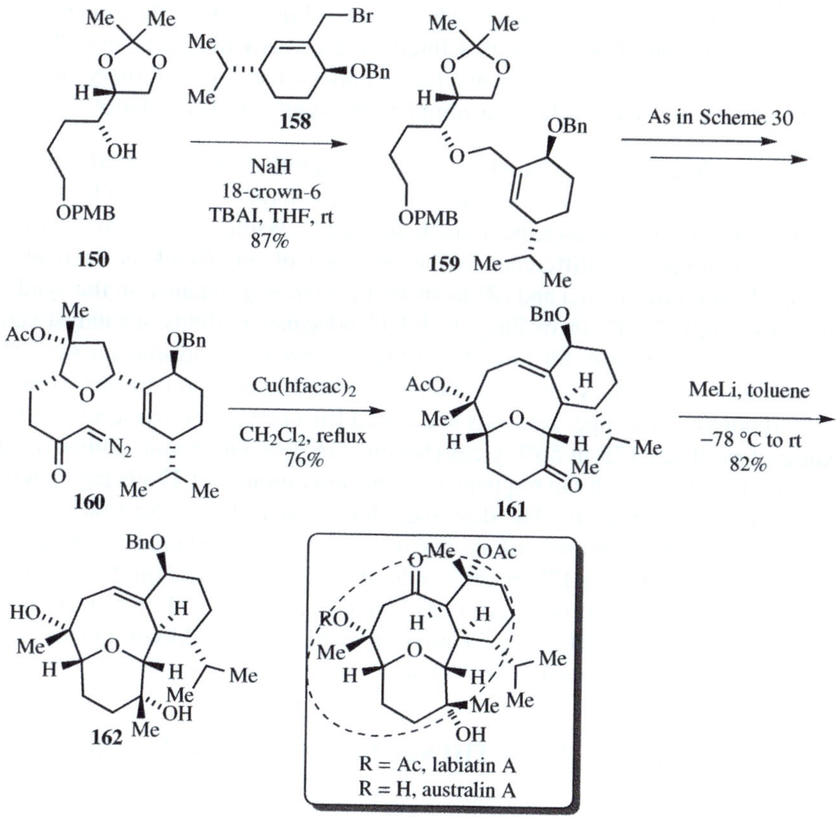

Scheme 16.31 Synthesis of a more elaborated labiatin A/australin A tricyclic core by the Clark group.

can be employed in this chemistry, Clark also investigated the [2,3]-rearrangement of a tetrahydropyran-derived oxonium ylide to give the oxabicyclo[6.2.1]undecane scaffold directly. By altering the tether length and ether ring size, Clark *et al.* developed a new application of the [2,3]-rearrangement methodology in his total synthesis of vigulariol, an ether-bridged, polyoxygenated marine natural product.[36] The synthesis began with the addition of the Grignard reagent from bromide **163** into methacrolein, giving **164** (Scheme 16.32). Conjugate addition of the resulting alcohol to ethyl propiolate, followed by deprotection and Swern oxidation gave aldehyde **165**. The required tetrahydropyran precursor was prepared by a SmI_2-induced cyclization,[37] giving the trisubstituted heterocycle **166** in 76% yield as a single isomer. O-Protection and diazoacylation gave the rearrangement precursor **167** in 60% yield. Upon exposure to $Cu(hfacac)_2$, the initially formed oxonium ylide **168** underwent the desired rearrangement in 96% yield to give **169** as a 5:1 (*Z*/*E*) mixture. The (*Z*)-isomer was converted to the enol triflate and coupled with vinyl stannane **170** to give diene **171**. Heating in toluene with methyl vinyl ketone gave, after epimerization of the exocyclic keto group, enol ether **172** in 58% yield. Ketone olefination, enol ether hydrolysis, and chemoselective reduction gave **173**. The *exo*-methylene was installed by Wittig olefination, which was followed by a three-step methyl incorporation, giving **174**. When this intermediate was exposed to *m*-CPBA, the initial epoxidation product underwent a transannular attack by the tertiary alcohol to close the remaining ether bridge and complete the synthesis of vigulariol.

16.2.4.5 Cladiellins The cladiellins comprise another family of ether-bridged marine natural products whose total syntheses have been completed using [2,3]-oxonium ylide rearrangement methodology.[38] While structurally similar to vigulariol, an important difference is the presence of an (*E*)-alkene in cladiellin skeleton. As mixtures of (*E*) and (*Z*) isomers are routinely obtained in the synthesis of the oxabicyclo[6.2.1]- or oxabicyclo[5.3.1]undecane scaffolds, a catalyst screen was conducted to optimize for formation of the necessary (*E*)-alkene during the key rearrangement step. Clark and coworkers subjected cyclization precursor **167** to various rhodium and copper catalysts and found $Rh_2(TPA)_4$ to provide a 6.3:1 (*E*/*Z*) mixture of products **175** in 56% yield (Scheme 16.33). The major isomer (*E*)-**175** was carried forward, with the requisite cyclohexenyl moiety of **176** being appended using a strategy similar to that described for vigulariol (see Section 16.2.4.4). Conversion of **176** to an enol triflate was followed by methylation under Kumada conditions to give alcohol **177** in 68% yield after silyl deprotection. Oxidation and stereoselective methyl addition gave (−)-cladiella-6,11-dien-3-ol, its acylation gave (−)-3-acetoxycladiella-6,11-diene in modest yield, and its dihydroxylation gave (−)-cladiell-11-ene-3,6,7-triol in 66% yield.

16.3 APPLICATIONS IN SYNTHESIS: OXONIUM YLIDE [1,2]-STEVENS REARRANGEMENTS

The [1,2]-shift, or Stevens rearrangement, is also a versatile process for the synthesis of a variety of cyclic ethers or oxygen-bridged carbocycles. An essential requirement

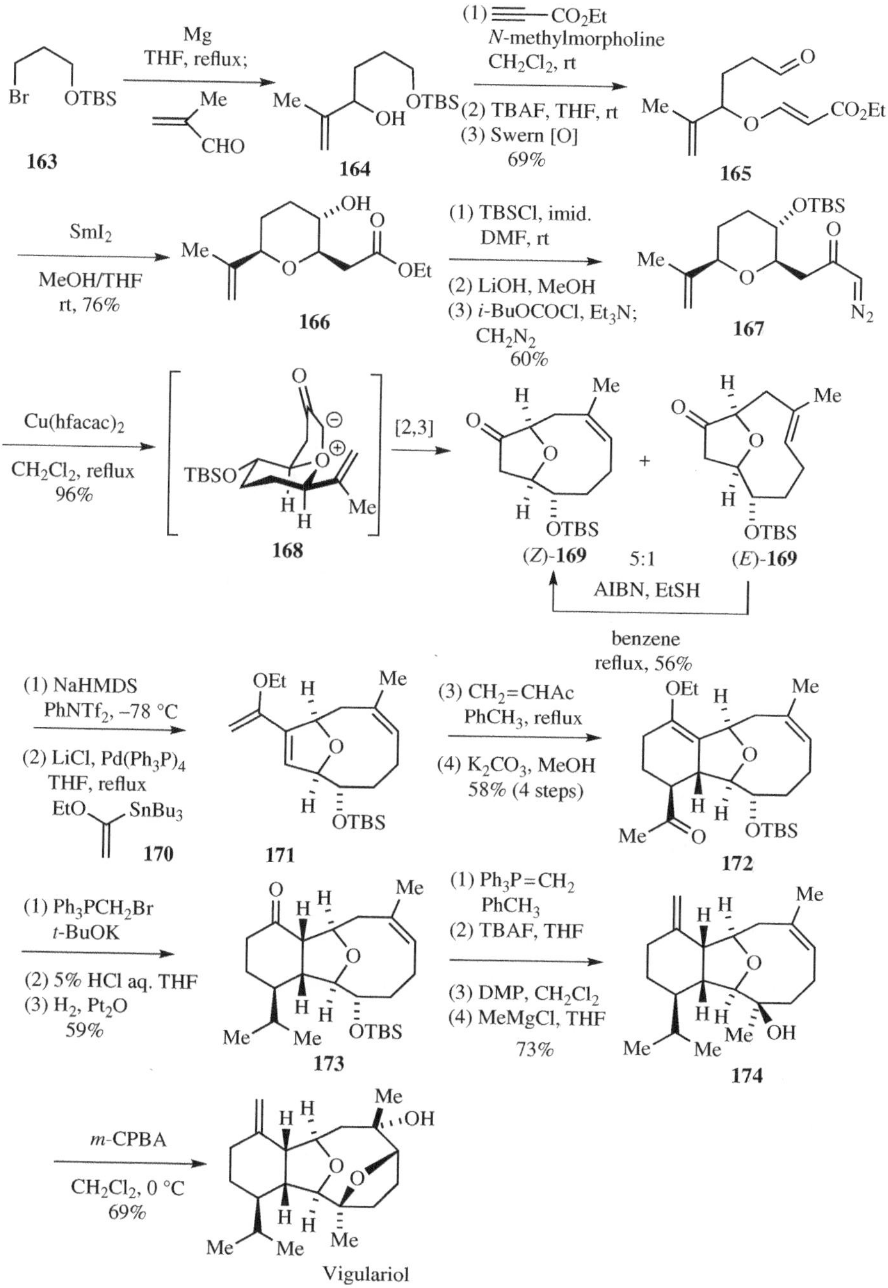

Scheme 16.32 Total synthesis of vigulariol by the Clark group.

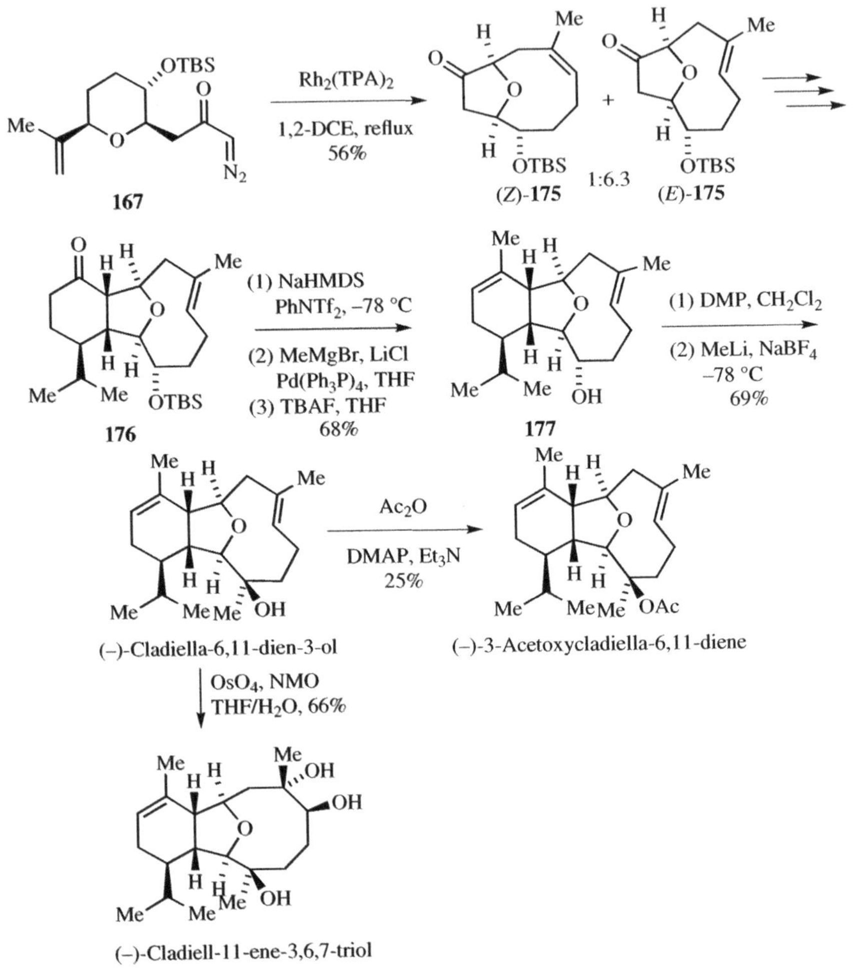

Scheme 16.33 Total synthesis of cladiellins by the Clark group utilizing an (*E*)-selective [2,3]-shift.

for a successful [1,2]-shift is that the migrating carbon be substituted with a suitable stabilizing group. As the [1,2]-shift proceeds via a likely homolytic dissociation/radical recombination pathway, groups that are effective radical stabilizers can facilitate the overall process and serve to direct which bond is cleaved. Numerous methodological studies have been undertaken to explore the scope of this transformation, and the breadth of possible stabilizing groups, starting with the seminal report by Johnson and Roskamp.[6] Investigations into the rearrangements of oxonium ylides formed from simple acyclic benzyl ethers (**178**) or α-phenyl substituted ethers (**179, 180**) showed that the phenyl group is an effective stabilizing group, readily promoting the [1,2]-shift (Scheme 16.34).[7] Consistent with expectation, changing catalyst can result in differing product distributions: under rhodium catalysis, **178**

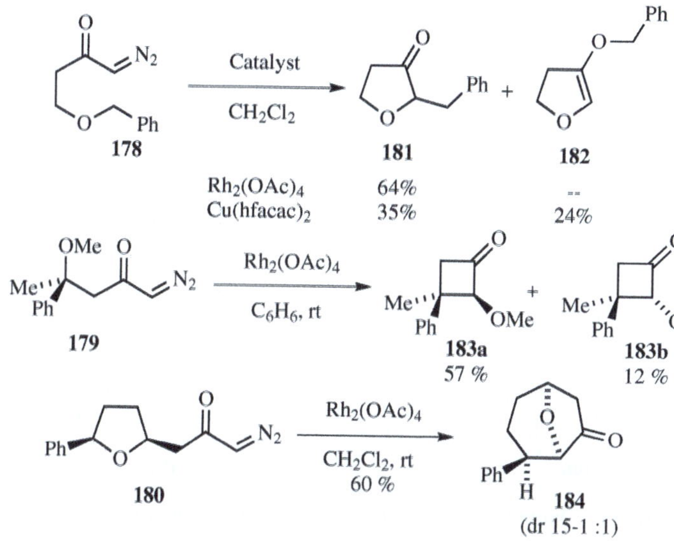

Scheme 16.34 Effective [1,2]-shift of oxonium ylides bearing benzylic migrating groups.

gives the [1,2]-shift product **181** in 64% yield, whereas under copper catalysis, competition from the [1,4]-shift (giving **182** in addition to **181**) is significant.[12]

Allylic ethers will typically undergo a [2,3]-rearrangement unless the allyl terminus cannot enter into bonding proximity with the ylide carbon, in which case the ylide will undergo a vinyl-stabilized [1,2]-shift (occasionally seen as a side reaction in some of the synthetic applications discussed in the previous section). A straightforward example involves **185**, which was treated under rhodium catalysis, gave a mixture of [2,3]-rearrangement product **186** and [1,2]-shift product **187** as the major component (Scheme 16.35).[6] Though tertiary alkyl radicals are stabilized, the alkyl substitution of **188** was insufficient to stabilize the putative radical intermediate of the [1,2]-shift, and the ylide instead underwent decomposition by fragmentation to give **189**. The West group has shown *O*-α-acyl substitution to be a modestly effective promoter in the [1,2]-shift, with the *O*-phenacyl-substituted oxonium ylide derived from **190** undergoing the [1,2]-shift in 17% yield.[7]

Later investigations centered on the formation and rearrangement of oxonium ylides derived from cyclic acetals. Tester and West showed the oxonium ylide derived from benzylidene acetal **192** to smoothly undergo the [1,2]-shift in excellent yield and diastereoselectivity (Scheme 16.36).[39] Phenyl substitution on the migrating carbon was shown to be unnecessary when α-heteroatom substitution was present, as acetonide centers also migrated. The oxonium ylides derived from *cyclic* acetals **194a–d** also underwent the [1,2]-shift when the α-carbon was oxygen- or sulfur-substituted, giving the 1,2-shift products **195a–d**.[40] This development was pivotal to the synthesis of oxygen-bridged medium rings and served as the basis for West group's investigations into preparation of medium ring-containing natural products with bridgehead oxygenation.

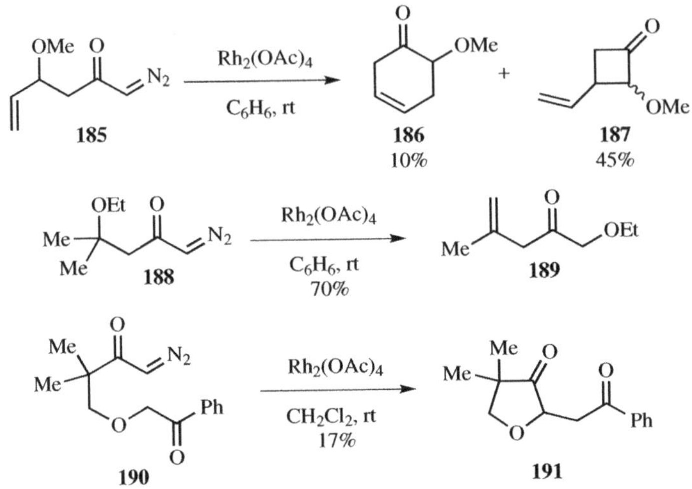

Scheme 16.35 Oxonium ylide [1,2]-shifts and possible alternative reactivities.

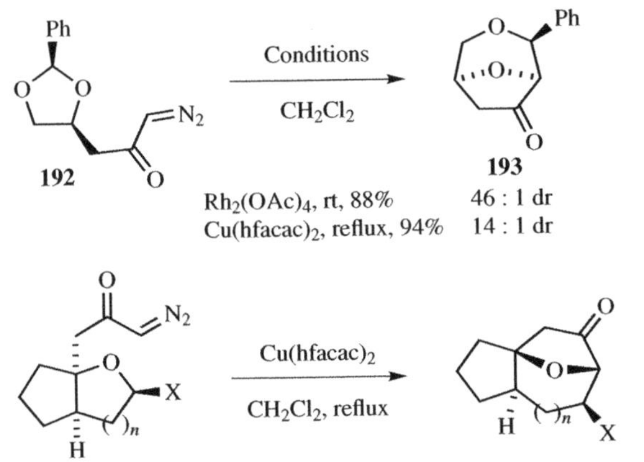

194a $n = 1$, X = OMe
194b $n = 1$, X = STol
194c $n = 2$, X = OMe
194d $n = 2$, X = STol

195a 85%, 2.4 : 1 dr
195b 70%, 4.8 : 1 dr
195c 82%, 20 : 1 dr
195d 80%, 11 : 1 dr

Scheme 16.36 [1,2]-Shifts of acetal-derived oxonium ylides.

16.3.1 Dactylol

As part of their work on acetal-derived oxonium ylides, West and coworkers have explored the total synthesis of the terpenoid natural product (+)-dactylol (Scheme 16.37).[41] The oxygen-bridged, fused bicyclic system of key intermediate

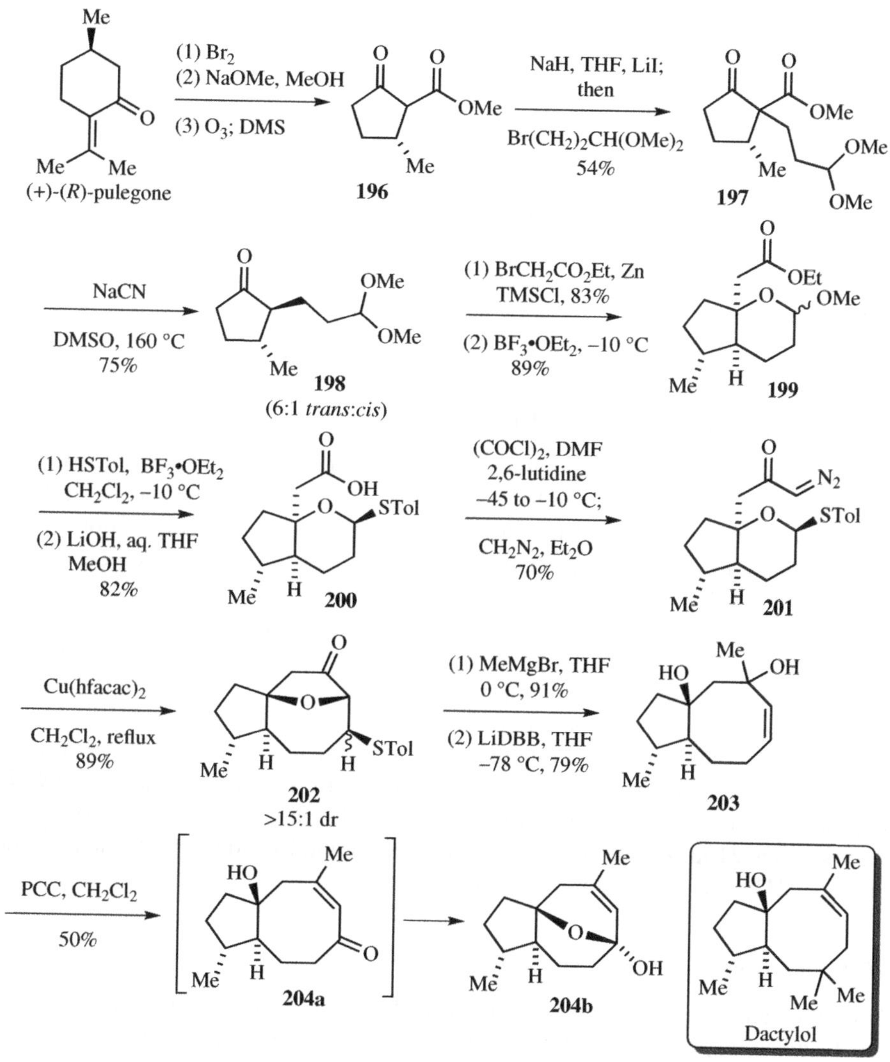

Scheme 16.37 Dactylol approach using oxonium ylide [1,2]-shift.

202 could be prepared by sulfur-stabilized [1,2]-shift of a mixed thioacetal-derived oxonium ylide, analogous to the synthesis of **195d**, after which a reductive desulfurization would cleave the bridging ether. The synthesis began with the chiral pool reagent (+)-(*R*)-pulegone, which was converted to the *α*-methyl-substituted cyclopentanone **196** in three steps, including a Favorskii ring contraction (Scheme 16.37). The *β*-ketoester was C-alkylated in 54% yield and then decarboxylated to give a 6:1 (trans–cis) mixture of acetals **198**. A stereoselective

Reformatsky addition was followed by a Lewis acid-mediated acetal closure to give **199**. Transacetalization with thiocresol and saponification afforded carboxylic acid **200**, which was diazoacylated to give diazoketone **201**, the key rearrangement substrate. Exposure to $Cu(hfacac)_2$ gave the O-bridged bicycle **202** in 89% yield with >15:1 dr. Treatment with MeMgBr installed the second methyl group, after which reductive ether cleavage with LiDBB gave diol **203**. The tertiary allylic alcohol underwent an oxidative rearrangement with PCC to give enone **204a**, providing a functionality handle by which the remaining *gem*-dimethyl substituents could be installed. The oxidation product was isolated as transannular hemiketal **204b**, leaving installation of the *gem*-dimethyl unit and removal of the carbonyl as the remaining steps for completion of (+)-dactylol.

16.3.2 Tigliane/Daphnane Skeleton

West and coworkers have also shown the [1,2]-shift of an acetal-derived oxonium ylide to be applicable to the synthesis of the tricyclic core of daphnane and tigliane natural products, exemplified by phorbol (Scheme 16.38).[42] Incorporation of the cyclohexyl moiety in the rearrangement precursor **210** enabled an efficient construction of a polyoxygenated, fused medium-ring scaffold. West group's synthesis began by a combination of cyclohexene oxide and cyclopentanone to give alcohols **205** in 75% yield with 5:1 dr (Scheme 16.38). The alcohols were acetylated and separated, and the major isomer **206** was treated with the enolate of ethyl acetate to give diol **207** as a single isomer upon acetate hydrolysis. Oxidation and dehydration of the resulting hemiketal gave dihydrofuran **208**, which was cleanly dihydroxylated from the convex face and protected to give anomeric benzylidene acetals **209a,b**. Diazoacylation via the mixed carbonic anhydride gave a separable mixture of diazoketones **210a,b**, each of which rearranged in good yield to afford the tricyclic products **211a,b** as single diastereomers.

16.3.3 Zaragozic Acid Core

Formation and rearrangement of oxonium ylides derived from cyclic ketals in which the diazoketone precursor is attached via the anomeric carbon can be used to furnish interesting bridged bicyclic systems. Zercher and coworkers carried out numerous investigations into the rearrangements of ketal-derived oxonium ylides and found ketal-bridged carbons to also be suitable migrating groups in [1,2]-shifts if they are substituted with sufficiently stabilizing groups.[43] Zercher used this strategy as an entry into the carbocyclic core of zaragozic acid, beginning with ketalization of methyl acetoacetate with C_2-symmetric diol **212** to give ester **213** (Scheme 16.39).[44] Saponification and acylation gave a β-ketoester that was subjected to diazotransfer, giving **214**. Treating the diazoketone with $Cu(hfacac)_2$ resulted in oxonium ylide **215** that decomposed by both [1,2]- and [2,3]-rearrangement pathways. However, under $Rh_2(OAc)_4$ catalysis the oxonium ylide reacted chemoselectively by the vinyl-stabilized [1,2]-shift pathway to provide the bridged, bicyclic core of zaragozic acid (**216**) in 64% yield.

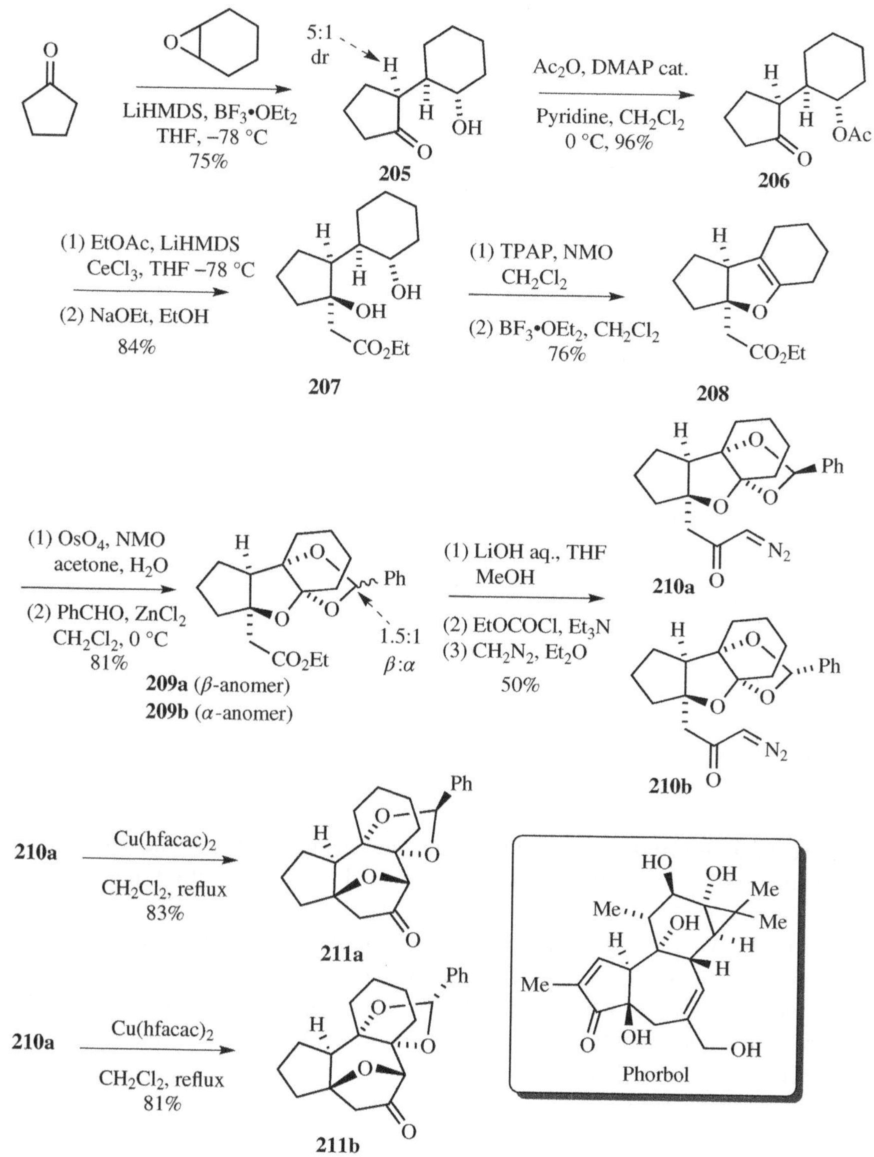

Scheme 16.38 Approach to the tricyclic daphnane/tigliane skeleton via [1,2]-shift.

16.4 CONCLUDING REMARKS

Oxonium ylides are transient, high-energy intermediates that can undergo useful rearrangements (i.e., [2,3]-sigmatropic rearrangement or [1,2]-Stevens rearrangement). These processes result in the formation of a new carbon–carbon bond, often with

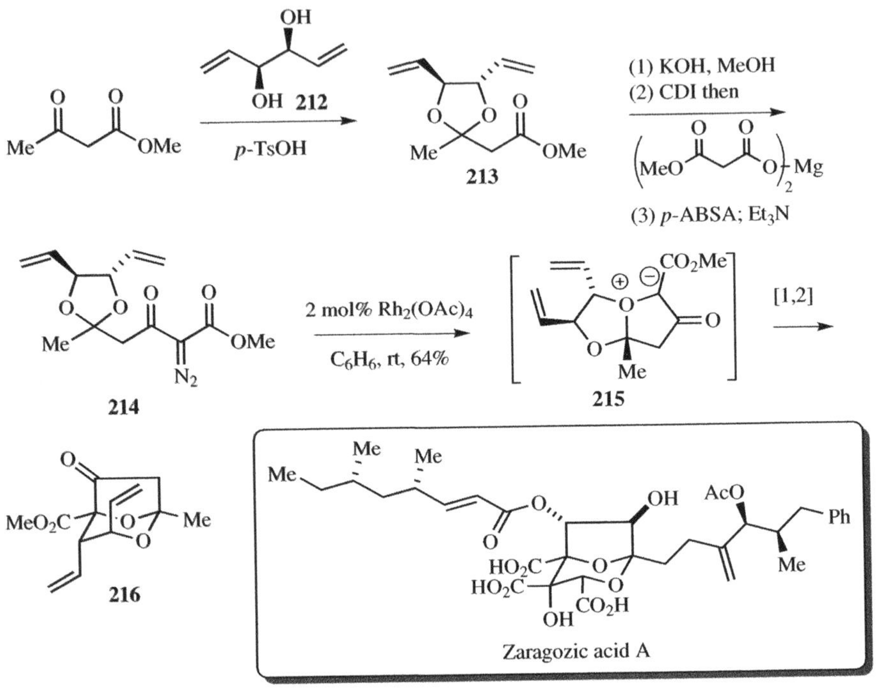

Scheme 16.39 Construction of zaragozic acid core by Brogan and Zercher via [1,2]-shift of ketal-derived oxonium ylide.

effective diastereocontrol as a result of other stereogenic centers in the oxonium ylide intermediate. Most typically, the ylides are formed from the addition of an ether oxygen to an electrophilic, catalytically generated metallocarbene, and this process is usually intramolecular, resulting in a cyclic oxonium ylide. Rearrangements of these oxonium ylides have been applied to the synthesis of a variety of complex heterocyclic and carbocyclic targets, typically by very direct routes owing to the major bond reorganization that takes place from the metallocarbene precursors to the final rearrangement products. While there has been considerable work in the synthetic applications of oxonium ylide intermediates, they remain ripe with potential in the construction of oxygenated natural products.

REFERENCES

1. For recent reviews, see: (a) Sweeney, J. B. *Chem. Soc. Rev.* **2009**, *38*, 1027–1038. (b) Weibel, J.-M.; Blanc, A.; Pale, P. In Silver in Organic Chemistry, Harmata, M., Ed.; Wiley: Hoboken, NJ, **2009**, 83–116.
2. (a) Ollis, W. D.; Rey, M.; Sutherland, I. O. *J. Chem. Soc., Perkin Trans. 1* **1983**, 1009–1027. (b) Review: Vanecko, J. A.; Wan, H.; West, F. G. *Tetrahedron* **2006**, *62*, 1043–1062.

3. (a) Nozaki, H.; Takaya, H.; Noyori, R. *Tetrahedron Lett.* **1965**, 2563–2567. (b) Nozaki, H.; Takaya, H.; Noyori, R., *Tetrahedron* **1966**, *22*, 3393–3401.
4. (a) Ando, W.; Yamato, H.; Kohoda, H.; Migita, T.; Kondo, S.; Ichibori, K.; Imai, I.; Nakaido, S.; Nakayama, K. *J. Am. Chem. Soc.* **1972**, *94*, 3870. (b) Friedrich, K.; Jansen, U.; Kirmse, W. *Tetrahedron Lett.* **1985**, *26*, 193–196. (c) Kirmse, W.; Vanchiem, P.; Schurig, V. *Tetrahedron Lett.* **1985**, *26*, 197–200.
5. Pirrung, M. C.; Werner, J. A. *J. Am. Chem. Soc.* **1986**, *108*, 6060–6062.
6. Roskamp, E. J.; Johnson, C. R. *J. Am. Chem. Soc.* **1986**, *108*, 6062–6063.
7. Eberlein, T. H.; West, F. G.; Tester, R. W. *J. Org. Chem.* **1992**, *57*, 3479–3482.
8. Clark, J. S.; Krowiak, S. A.; Street, L. J. *Tetrahedron Lett.* **1993**, *34*, 4385–4388.
9. West, F. G.; Eberlein, T. H.; Tester, R. W. *J. Chem. Soc., Perkin Trans. 1* **1993**, 2857–2859.
10. Yakura, T.; Matsui, K.; Matsuzaka, K.; Yamashita, M. *Heterocycles* **2009**, *79*, 353–358.
11. Clark, J. S.; Whitlock, G.; Jiang, S.; Onyia, N. *Chem. Commun.* **2003**, 2578–2579.
12. West, F. G.; Naidu, B. N.; Tester, R. W. *J. Org. Chem.* **1994**, *59*, 6892–6894.
13. Marmsäter, F. P.; Vanecko, J. A.; West, F. G. *Org. Lett.* **2004**, *6*, 1657–1660.
14. (a) Olah, G. A.; Doggweiler, H.; Felberg, J. D. *J. Org. Chem.* **1984**, *49*, 2112–2116. (b) Pichlmair, S. *Synlett* **2004**, 195–196. (c) Mascal, M.; Hafezi, N.; Toney, M. D. *J. Am. Chem. Soc.* **2010**, *132*, 10662–10664.
15. Doyle, M. P.; McKervey, M. A.; Ye, T. *Modern Catalytic Methods for Organic Synthesis with Diazo Compounds*; John Wiley & Sons, Inc.: New York, **1998**.
16. Regitz, M. *Synthesis* **1972**, 351–373.
17. Murphy, G. K.; West, F. G. *Org. Lett.* **2006**, *8*, 4359–4361.
18. (a) Xu, M.; Ten, T.-T.; Li, C.-Y. *Org. Lett.* **2012**, *14*, 4902–4905. (b) Fu, J.; Shang, H.; Wang, Z.; Chang, L.; Shao, W.; Yang, Z.; Tang, Y. *Angew. Chem., Int. Ed.* **2013**, *52*, 4198–4202.
19. Jackson, K. L.; Henderson, J. A.; Motoyoshi, H.; Phillips, A. J. *Angew. Chem., Int. Ed.* **2009**, *48*, 2346–2350.
20. Clark, J. S.; Labre, F.; Thomas, L. H. *Org. Biomol. Chem.* **2011**, *9*, 4823–4830.
21. Clark, J. S.; Fessard, T. C.; Wilson, C. *Org. Lett.* **2004**, *6*, 1773–1776.
22. Hodgson, D. M.; Angrish, D.; Erickson, S. P.; Kloesges, J.; Lee, C. H. *Org. Lett.* **2008**, *10*, 5553–5556.
23. Hodgson, D. M.; Man, S. *Chem. – Eur. J.* **2011**, *17*, 9731–9737.
24. Hodgson, D. M.; Moreno-Clavijo, E.; Day, S. E.; Man, S. *Org. Biomol. Chem.* **2013**, *11*, 5362–5369.
25. Pirrung, M. C.; Brown, W. L.; Rege, S.; Laughton, P. *J. Am. Chem. Soc.* **1991**, *113*, 8561–8562.
26. Clark, J. S.; Whitlock, G. A. *Tetrahedron Lett.* **1994**, *35*, 6381–6382.
27. Clark, J. S.; Fessard, T. C.; Whitlock, G. A. *Tetrahedron* **2006**, *62*, 73–78.
28. Marmsäter, F. P.; West, F. G. *J. Am. Chem. Soc.* **2001**, *123*, 5144–5145.
29. Yakura, T.; Muramatsu, W.; Uenishi, J. *Chem. Pharm. Bull.* **2005**, *53*, 989–994.
30. Clark, J. S.; Dossetter, A. G.; Whittingham, W. G. *Tetrahedron Lett.* **1996**, *37*, 5605–5608.
31. Clark, J. S.; Dossetter, A. G.; Blake, A. J.; Li, W. S.; Whittingham, W. G. *Chem. Commun.* **1999**, 749–750.

32. Clark, J. S.; Baxter, C. A.; Dossetter, A. G.; Poigny, S.; Castro, J. L.; Whittingham, W. G. *J. Org. Chem.* **2008**, *73*, 1040–1055.
33. Clark, J. S.; Wong, Y. S. *Chem. Commun.* **2000**, 1079–1080.
34. Clark, J. S.; Baxter, C. A.; Castro, J. L. *Synthesis* **2005**, 3398–3404.
35. Clark, J. S.; Vignard, D.; Parkin, A. *Org. Lett.* **2011**, *13*, 3980–3983.
36. Clark, J. S.; Hayes, S. T.; Wilson, C.; Gobbi, L. *Angew. Chem., Int. Ed.* **2007**, *46*, 437–440.
37. Matsuo, G.; Kadohama, H.; Nakata, T. *Chem. Lett.* **2002**, 148–149.
38. Clark, J. S.; Berger, R.; Hayes, S. T.; Thomas, L. H.; Morrison, A. J.; Gobbi, L. *Angew. Chem., Int. Ed.* **2010**, *49*, 9867–9870.
39. Tester, R. W.; West, F. G. *Tetrahedron Lett.* **1998**, *39*, 4631–4634.
40. (a) Marmsäter, F. P.; Murphy, G. K.; West, F. G. *J. Am. Chem. Soc.* **2003**, *125*, 14724–14725. (b) Murphy, G. K.; West, F. G. *Org. Lett.* **2005**, *7*, 1801–1804.
41. (a) Murphy, G. K.; Johnston, J. R.; West, F. G. unpublished results. (b) See also: Murphy, G. K. *Oxonium Ylides In the Synthesis of Fused, Medium-Sized Rings*, Ph.D. Dissertation, University of Alberta, **2007**.
42. Stewart, C.; McDonald, R.; West, F. G. *Org. Lett.* **2011**, *13*, 720–723.
43. (a) Brogan, J. B.; Bauer, C. B.; Rogers, R. D.; Zercher, C. K. *Tetrahedron Lett.* **1996**, *37*, 5053–5056. (b) Brogan, J. B.; Zercher, C. K.; Bauer, C. B.; Rogers, R. D. *J. Org. Chem.* **1997**, *62*, 3902–3909.
44. Brogan, J. B.; Zercher, C. K. *Tetrahedron Lett.* **1998**, *39*, 1691–1694.

17

THE [2,3]-WITTIG REARRANGEMENT

MINORU ISOBE AND CHATCHAWAN PLOYSUK
Department of Chemistry, National Tsing Hua University, Hsinchu, Taiwan, ROC

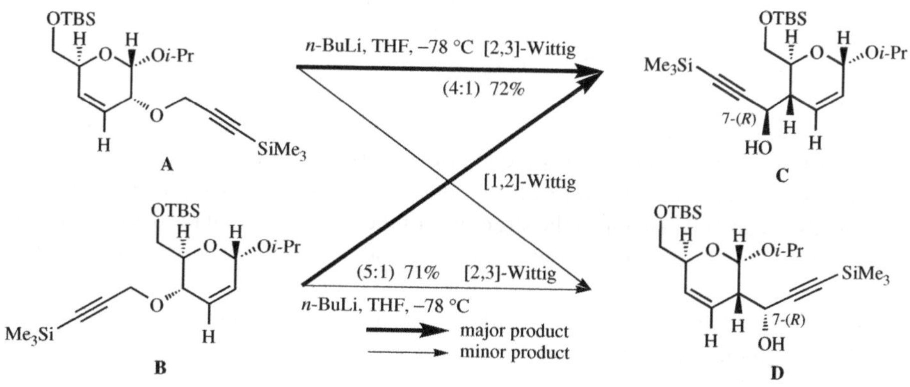

17.1 INTRODUCTION

The [2,3]-Wittig rearrangement is not only a historically significant reaction but is still widely used as an important C—C bond-forming process in current organic synthesis. In fact, many synthetic papers have reported on the application of Wittig rearrangements for constructing complex molecules or natural products. Among these, the current chapter focuses on recently published examples of the [2,3]-Wittig rearrangement in synthesis. The [2,3]-Wittig rearrangement has been reviewed by Nakai and Mikami[1] and Marshall.[2] Occasionally the same anionic

Molecular Rearrangements in Organic Synthesis, First Edition. Edited by Christian M. Rojas.

species involved in the [2,3]-Wittig rearrangement can undergo [1,2]-Wittig rearrangement to produce similar homoallylic alcohols, but that mode of reactivity is not covered extensively in this paper.

One of the typical examples of a [2,3]-sigmatropic reaction is the [2,3]-Wittig rearrangement (Figure 17.1), where X is oxygen and Y is the corresponding carbanion, which is often propargylic. This reaction is related to [3,3]-sigmatropic reactions for stereoselective C—C bond formation process such as Claisen-type rearrangements. It is also related to the aza-Wittig reaction (Figure 17.1a, with X = N) and the Overman rearrangement (Figure 17.1b, with Y = N) in the sense of C—N bond analogy. In many cases, the [3,3]-sigmatropy proceeds under thermal conditions (approximately 100 °C), and the reaction equilibrates toward the less sterically congested system.[3] Consequently, the reaction sometimes does not go to completion and gives lower yields under equilibrating conditions. On the other hand, the [2,3]-Wittig reaction involves a carbanion species, which pushes the reaction forward at temperatures as low as −78 °C to afford products with the negative charge better accommodated on an alkoxide group.

[2,3]-Wittig rearrangements are well known to proceed from (1) allyl propargyl ethers **1** (Figure 17.2, Eq. 1), (2) allyl trialkyltin-methyl ethers **4** (Wittig–Still rearrangement),[4] (3) allyloxy alkoxycarbonylmethyl ethers **5**, (4) diallyl ethers **6**, (5) allyl benzyl ethers **7**, or even dipropargyl ethers and via tandem or sequential [2,3]- and [3,3]- rearrangements. The stereochemical outcome in Eq. 1 represents many reported cases providing *anti* **2** or *syn* **3** isomer from the (*E*)- or (*Z*)-olefin, respectively (Figure 17.2). There are, however, exceptions so that the outcome of some reactions is difficult to predict. For example, when R_3 is H or an acetylenic terminal carbanion (dianion), the corresponding diastereomeric results are quite different.[5] Additional examples with stereodivergent results are discussed in the following text. In general, one needs to carefully establish the structures of the rearrangement products. Comparison of the stereochemical outcome in acyclic and cyclic systems provides valuable information on the course of this rearrangement. The first two examples include the [2,3]-Wittig rearrangement as the key step in the authors' own studies toward the total synthesis of natural products.

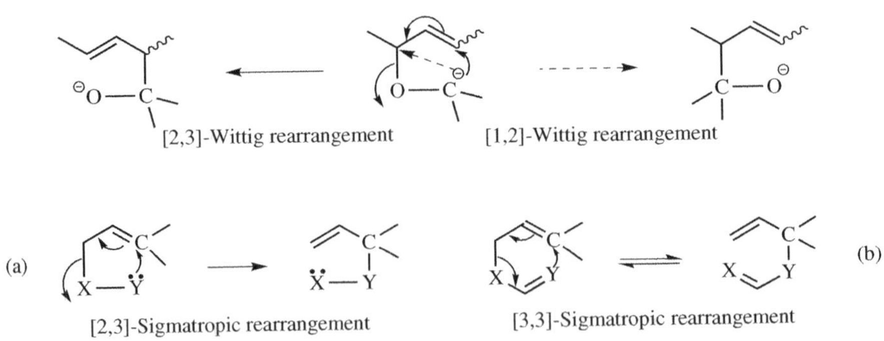

FIGURE 17.1 Wittig rearrangement and [2,3]- and [3,3]-sigmatropic rearrangement

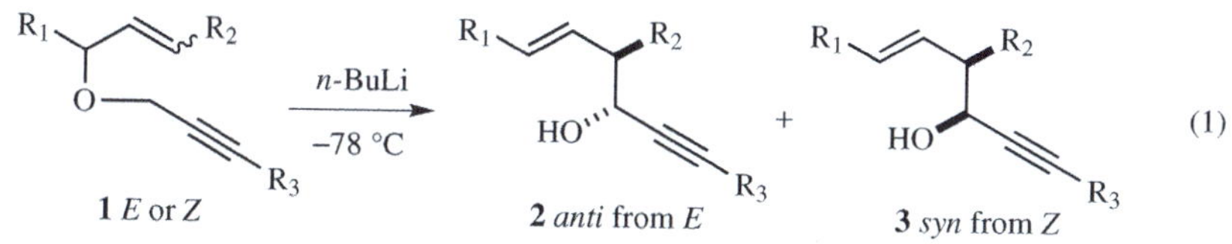

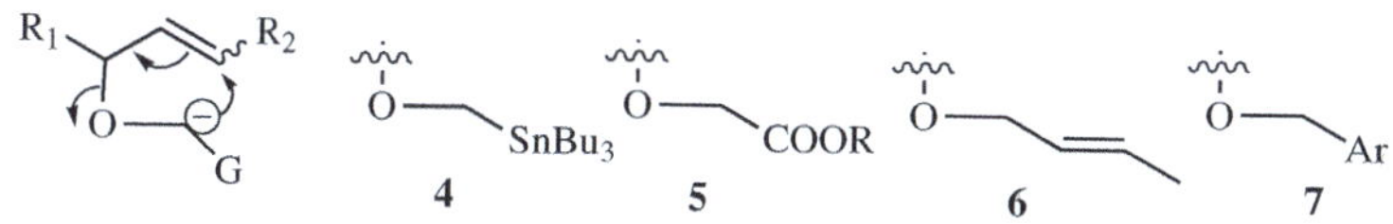

FIGURE 17.2 Various [2,3]-Wittig rearrangements

17.2 [2,3]-WITTIG REARRANGEMENT OF ALLYL PROPARGYL ETHERS

Our research group became particularly interested in the construction of the tricyclo[5.2.1.0^{1,6}]decene ring system of solanoeclepin A **8**, and the right segment of the natural product was synthesized in the form of **14**, containing the tricyclic substructure (Scheme 17.1). The key step was the [2,3]-Wittig rearrangement of **10** via the *axial* C3-oxygen.[6] This reaction provided **11** with a trans-substituted perhydroindene ring and the rearranged chain opposite to the ring-junction methyl group. Cyclization to form the cyclobutane via **12a** to **13** took place between the Nicholas cation as electrophile, which was stabilized by the acetylene–dicobalthexacarbonyl complex, and a Hosomi–Sakurai-type allyltrimethylsilane nucleophile. Notably, gold-catalyzed cyclization from **11** afforded a different tricyclic compound **18** through an sp^2-cation intermediate **11a**. The configuration of C13-(*R*) **11** was proved by X-ray diffraction analysis of PNB derivative **18** (Figure 17.3).

The [2,3]-Wittig rearrangement from **10** to **11** with *n*-BuLi (via one-pot silylation) resulted in high yield and high selectivity to give exclusively the 13-(*R*)-configuration. This may be due to proper orientation of the 3-*axial*-oxypropargyl relative to what is required for the transition state. Three model compounds **15a–c** were compared (Scheme 17.1) to determine the outcome with compounds having different-size substituents at the C8 position; thus, C8-β-H (**15a**), C8-β-alkynyltrimethylsilyl (**15b**), and C8-sp^2 (**15c**).

Two substrates, **15a** and **15c**, gave the [2,3]-Wittig rearrangement products **16a** and **16c** in high yields and as single diastereoisomers. But **15b** afforded **16b** in low yield (9%) giving largely elimination to the 3,5-diene **17b** (75%). This significant change due to remote steric effects led us to consider a plausible steric course for the [2,3]-Wittig rearrangement that would explain why the C13-(*R*) configuration of **11** arose from this reaction. The transition state of this [2,3]-Wittig rearrangement would include a 3-*quasi-axial* alkoxy group on the twist chair conformation as shown for **10a**, **15c**, **15a**, and **15b** (Scheme 17.2). The allyl propargyl ether **10** has two prochiral

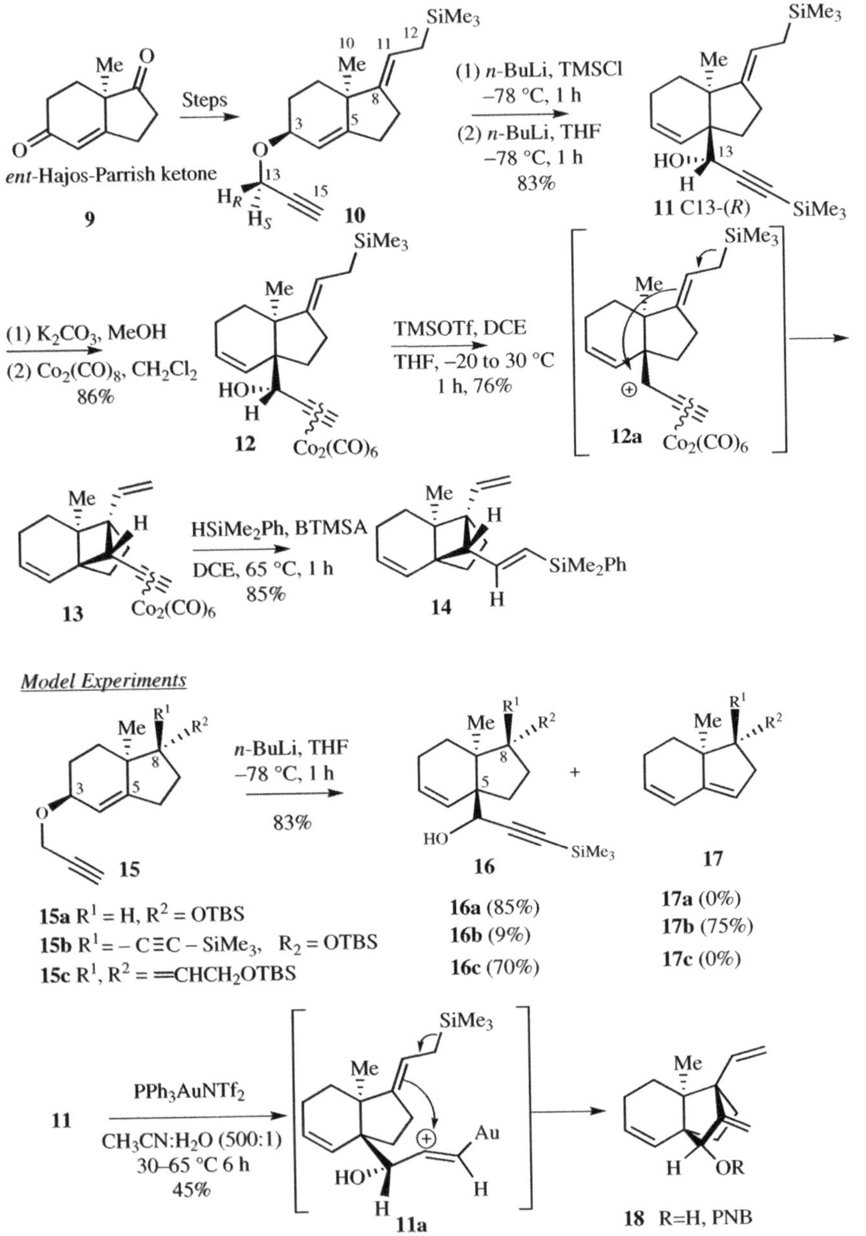

Scheme 17.1 Solanoeclepin A right segment synthesis and [2,3]-Wittig rearrangement.

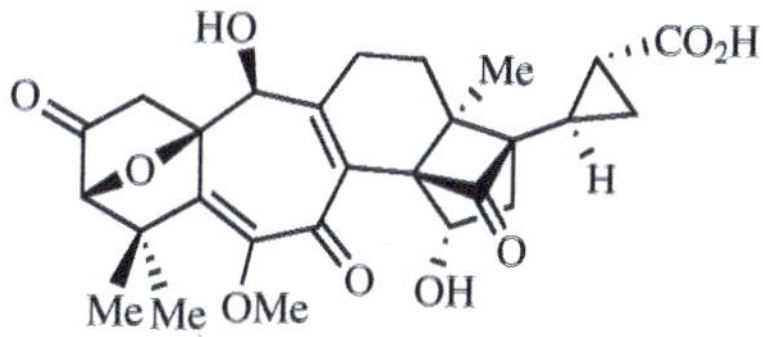

FIGURE 17.3 Solanoeclepin A **8**

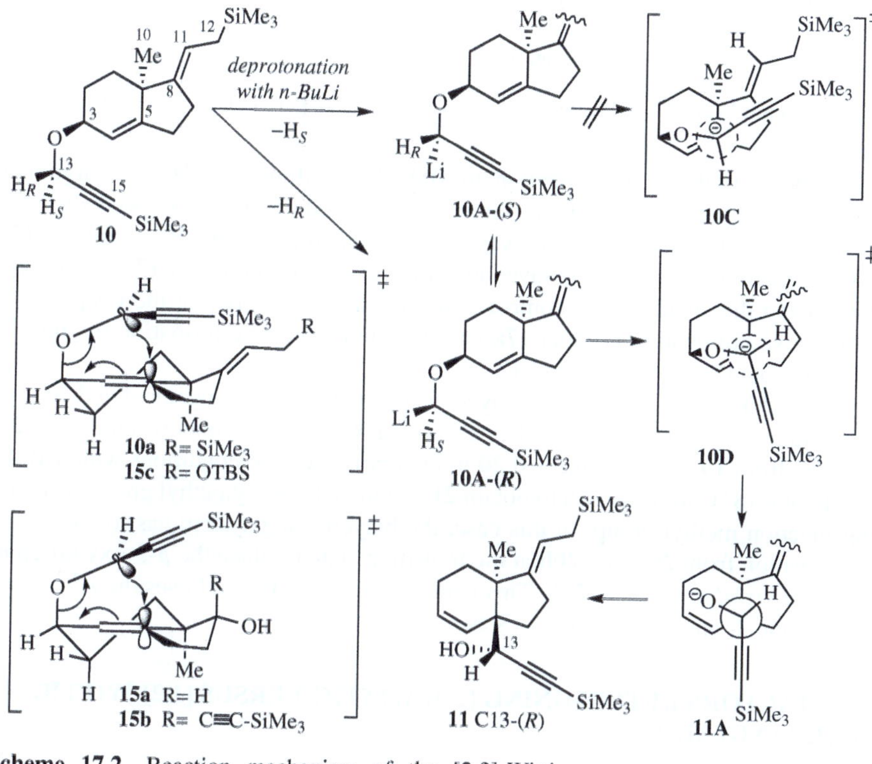

Scheme 17.2 Reaction mechanism of the [2,3]-Wittig rearrangement on a *3β-axial* propargyloxy-hexahydroindene ring system. The Newman-like projections in the dashed circles indicate transition state bond formation.

protons *pro-S* and *pro-R* at the propargylic position. Deprotonation of H_S or H_R from **10** by treatment with *n*-BuLi at −78 °C gives 13-(*S*)-lithium carbanion **10A**-(*S*) or 13-(*R*)-isomer **10A**-(*R*), respectively.

As each anion went to the transition state **10C** and **10D**, respectively, the steric interaction of **10C** would be greater than for **10D**, so that conversion of **10D** to **11A** would be faster. Experimentally, only product **11** with configuration 13-(*R*) was isolated. If the steric interaction of **10C** is too large, the anion **10A**-(*S*) should

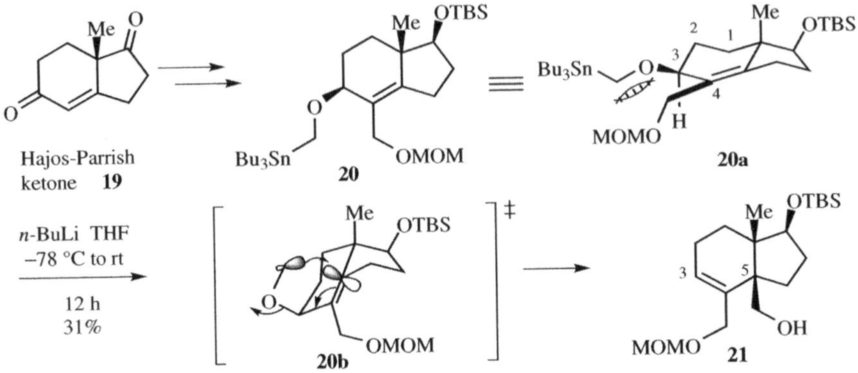

Scheme 17.3 [2,3]-Wittig rearrangement on Hajos–Parrish ketone derivative.

have time to undergo isomerization to **10A**-(*R*), leading to **10D**. This argument is supported by the fact that the three propargyl ethers **15a–c** showed different results; thus, the rearrangement was highly hindered by the R^1-group of **15b** (C8-*β*-trimethylsilylethynyl) to give largely the elimination product **17**. By contrast, **15a**, having the smaller H as compared to the ethynyl R group at the C8 position, did not give elimination product **17a** but only stereoselective formation of **16a**, the 13(*R*) product (Schemes 17.1 and 17.2).

Danishefsky and coworkers employed the [2,3]-Wittig rearrangement in their synthesis studies toward the xenibellol core through tributylstannylmethyl allyl ether **20**, prepared from **19**.[7] They subjected **20** to the rearrangement conditions with *n*-BuLi in THF at −78 °C to rt for 12 h to obtain **21**, having a hydroxymethyl group cis to the ring-junction methyl group. In this case, the bicyclic ring system has to change its conformation from **20a** into **20b** at the transition state to place the *β*-3-oxy group in the *pseudoaxial* orientation **20b**. This conformational change relieves the $A^{1,2}$ strain between the 3- and 4-substituents (Scheme 17.3).

17.3 FACTORS DETERMINING [2,3]-WITTIG VERSUS [1,2]-WITTIG REARRANGEMENT

During the course of our second-generation tetrodotoxin (TTX) synthesis from D-glucose, a Claisen rearrangement of a 2-*α*-propenyl ether was carried out at high temperature.[8] In addition, this synthesis required many steps for introducing the hydroxyl groups. To address these issues, we changed to a [2,3]-Wittig rearrangement strategy instead of the Claisen rearrangement route. Scheme 17.4 shows the 2-*α*-allyl propargyl ether **22**, also having the 3-alkynyl group. Its 4-*α*-isomer **23** was also prepared and examined since compounds **22** and **23** should lead to pseudoenantiomeric isomers for TTX's cyclohexane ring.[9] Treatment of **22** and **23** with *n*-BuLi at −78 °C in THF afforded [2,3]-Wittig rearrangement products **24** and **25**, respectively, in less than 30 min. The product **24** was obtained as a mixture of 9-(*R*) and 9-(*S*) diastereoisomers (2:1), while **25** was a single 9-(*R*) product. The products are *pseudosymmetric* and show such similar NMR spectra that their structures had to be carefully analyzed. All the absolute configurations of stereocenters located on the C_9 chain were determined by the Mosher–Kusumi method.[10] In the reaction of Scheme 17.4, no [1,2]-Wittig rearrangement was observed at all.

Compounds **26** and **27**, lacking the 3-substituent, were prepared in order to study the stereochemical outcome of these reactions (Scheme 17.5). Each of **26** and **27** was

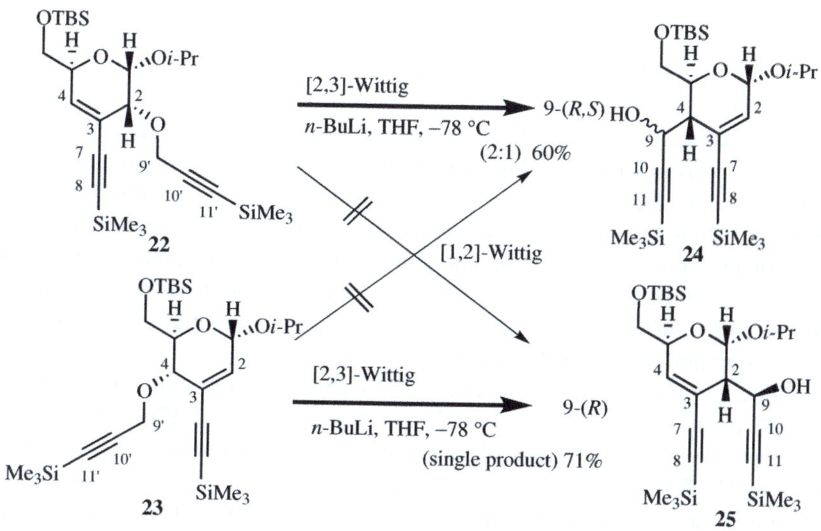

Scheme 17.4 [2,3]-Wittig rearrangement on a 3-alkynyl dihydropyran ring.

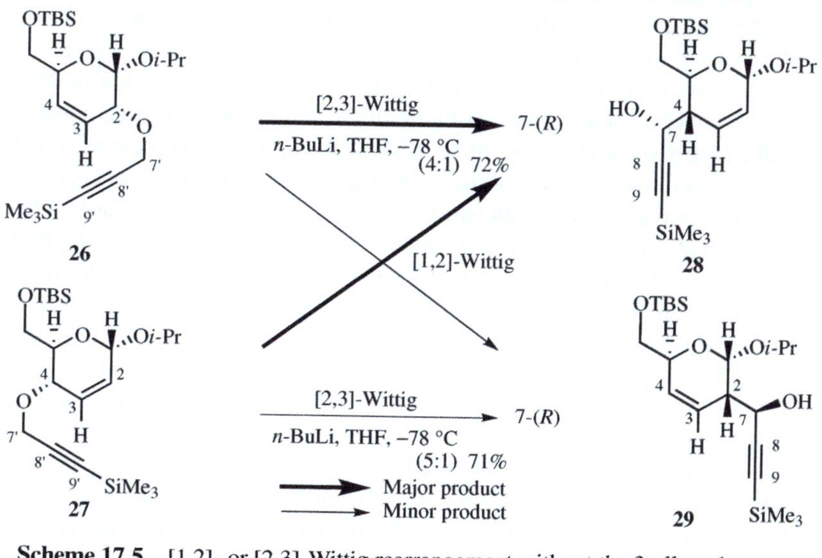

Scheme 17.5 [1,2]- or [2,3]-Wittig rearrangement without the 3-alkynyl group.

subjected to the same conditions with *n*-BuLi at −78 °C in THF to yield two products with similar *major/minor* ratios of 4:1 and 5:1, respectively. The reaction pathways were strikingly different; thus, **26** gave largely the [2,3]-Wittig rearrangement C7-(*R*)-product **28**, while **27** afforded the same C7-(*R*)-product **28** via [1,2]-Wittig rearrangement as the major product. Conversely, the minor products from **26** and **27** were both C7-(*R*)-product **29** via [1,2]-Wittig and [2,3]-Wittig rearrangement, respectively. Interestingly, both **28** and **29** had the C7-(*R*)-configuration as shown in Scheme 17.5.

In the case with the 3-alkynyl group (**22** and **23**, Scheme 17.4), both of the major products were derived via [2,3]-Wittig rearrangement, and majority of the stereogenic center was C9-(*R*) in both cases. On the contrary, C3-H compounds behaved differently; thus, **26** showed largely [2,3]-Wittig rearrangement but **27** underwent largely [1,2]-Wittig rearrangement, and both of them gave the C7-(*R*) configuration. We analyzed these results according to the following reaction steps: (1) deprotonation at the propargylic site to form the lithiated anion (*R*/*S*) and/or (2) isomerization between the diastereomeric anions (in the case where the ensuing rearrangements were expected to be slow) and then (3) [1,2]-Wittig rearrangement without inversion of the dihydropyran ring conformation or (4) [2,3]-Wittig rearrangement after inversion of the conformation and finally (5) O-protonation (workup).

Scheme 17.6 shows the reaction process of **23** and **27** in more detail using compound **30** as the summary structure (**23** has R = alkynyl and **27** has R = H). First, *pro-R* deprotonation would take place through chelate **30a** to form C7-(*R*) lithiated carbanion **31**. When the C3-R group is alkynyl, $A^{1,2}$ strain with the 4-α-equatorial group would favor conformational inversion to **33**, facilitating the [2,3]-Wittig rearrangement. In this case, the transition state **33a** leads to the product (*R*)-**34**, which is formed rapidly without opportunity for isomerization of the anion to **33b**. (Compare the cases **10C** and **10D** in Scheme 17.2).

With **22** and **26** having the 2-α-*equatorial* propargyloxy ethers, deprotonation took place differently as judged from the formation of a C9-(*R*/*S*)-mixture (2:1) of **24**. This was in contrast to the case of C7-(*R*) selective deprotonation in **26**. However, the [2,3]-Wittig rearrangement mode of **22** was highly selective compared to that of **26**, which showed a ratio of [2,3]:[1,2]-Wittig rearrangement of 4:1. Comparing this ratio to the 1:5 ratio in the case of **27** suggested that the energy required for conformational inversion in **26** was much lower compared to the case of **27**. This may be the result of a smaller gauche effect between the 1,2-substituents of **26** due to having one *axial* group and one *equatorial* group both before and after the conformational inversion. There is the loss of the anomeric effect in one conformer even if this effect is small (0.9 kcal/mol). The situation may be related to the case of a system having a 1-alkynyl-2-alkoxy dihydropyran (figure not shown), as reported that the equilibrium of the dicobalt complex analogs depends on the size of the alcohol protective groups.[11]

As summarized in the energy diagram in Figure 17.4, the *pro-R* deprotonation from **30** produces the (*R*)-*carbanion* **30a**. When R = H, this anion facilitates the [1,2]-Wittig rearrangement to afford **32**. Meanwhile the anion **30a** (R = alkynyl) go to the [2,3]-Wittig rearrangement to give **34** probably due to a lower energy barrier

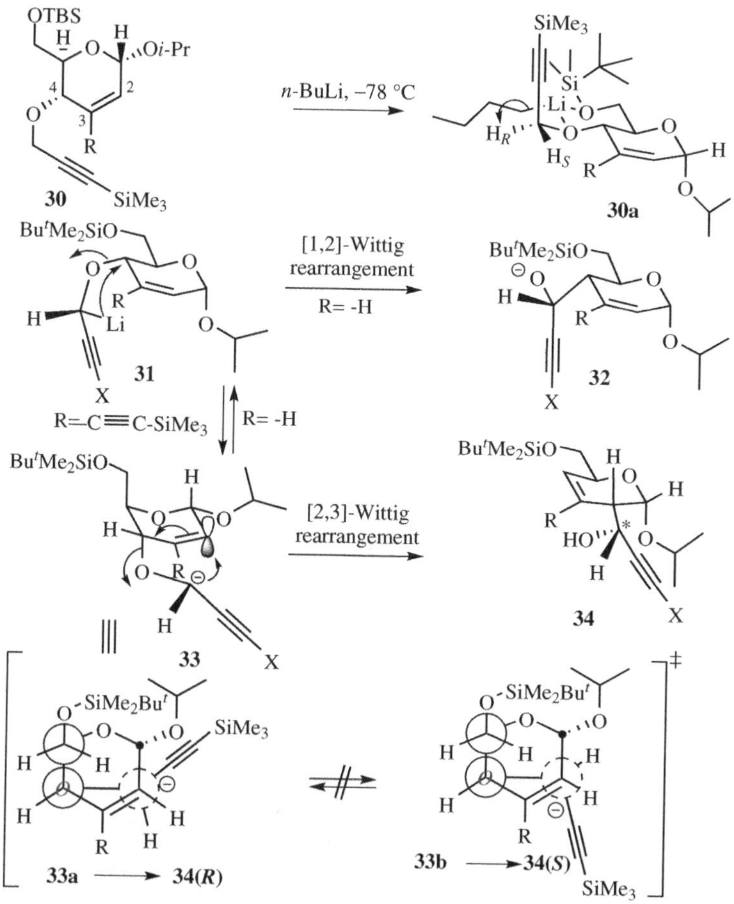

Scheme 17.6 Plausible mechanism of the Wittig rearrangements from 4-α-*O*-propargyl ether.

33 required for conformational inversion contributed by $A^{1,2}$-strain (or gauche effect). Notably, in the case that either the (*R*)- or (*S*)-*propargylic anion* from **10** needs to go over a higher energy transition state due to steric constraints as shown in Scheme 17.2, a relatively faster equilibrium took place between **10a**-(*S*) and **10A**-(*R*) to give a single product **11**-(13-*R*). This analysis is also informed by additional details from recently published DFT calculations.[9]

17.4 ACYCLIC [2,3]-WITTIG REARRANGEMENT OF PROPARGYL-ALLYL ETHERS

The phomactins are a series of diterpenes with unique 12-membered cyclic structures such as **35** and **36**, which have activity as antagonistic platelet-activating factors.

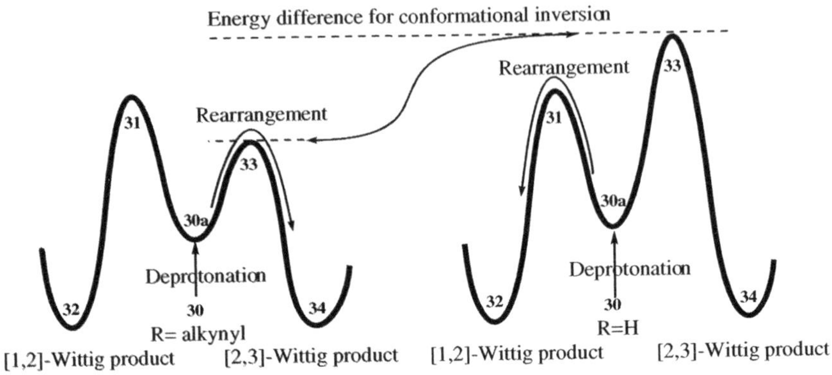

FIGURE 17.4 Plausible energy diagram from **30** to [1,2]- and [2,3]-Wittig rearrangement due to different energy barrier for the conformational inversion (see DFT calculations in Ref. 9)

Thomas and coworkers have focused their attention on the total synthesis of these compounds using the [2,3]-Wittig rearrangement as the key step.[12] The Thomas group prepared the cyclohexene derivative **37** containing an allyl propargyl ether, which was deprotonated by *n*-BuLi. The products **38** were obtained in good yield but as a mixture of diastereoisomers (Scheme 17.7).

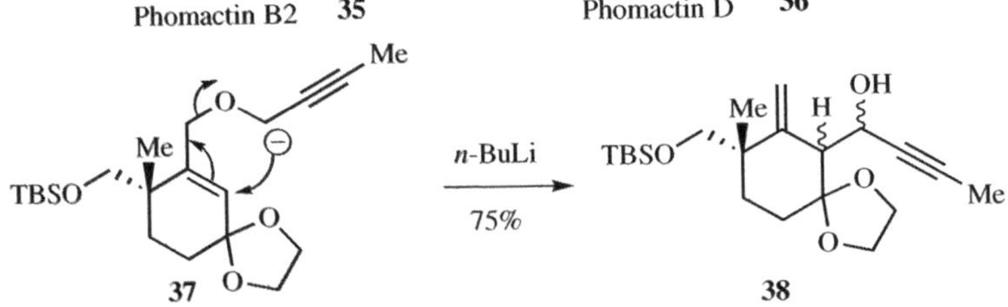

Scheme 17.7 Phomactin synthesis and the key [2,3]-Wittig rearrangement.

Attempted modification of the rearrangement substrate by alkylating the substituents on the cyclohexane ring as in vinyl ether **39**, ethylene ketal **41**, and silylether **43** afforded products nonselectively. The diene **39** gave allenes **40** in 50% yield due to dominant deprotonation at the allylic position (see **39a**). On the other hand, ethylene ketal **41** afforded the *exo*-methylene compounds **42** by deprotonation of the propargylic proton, which allowed the [2,3]-Wittig rearrangement (**41a**) to give

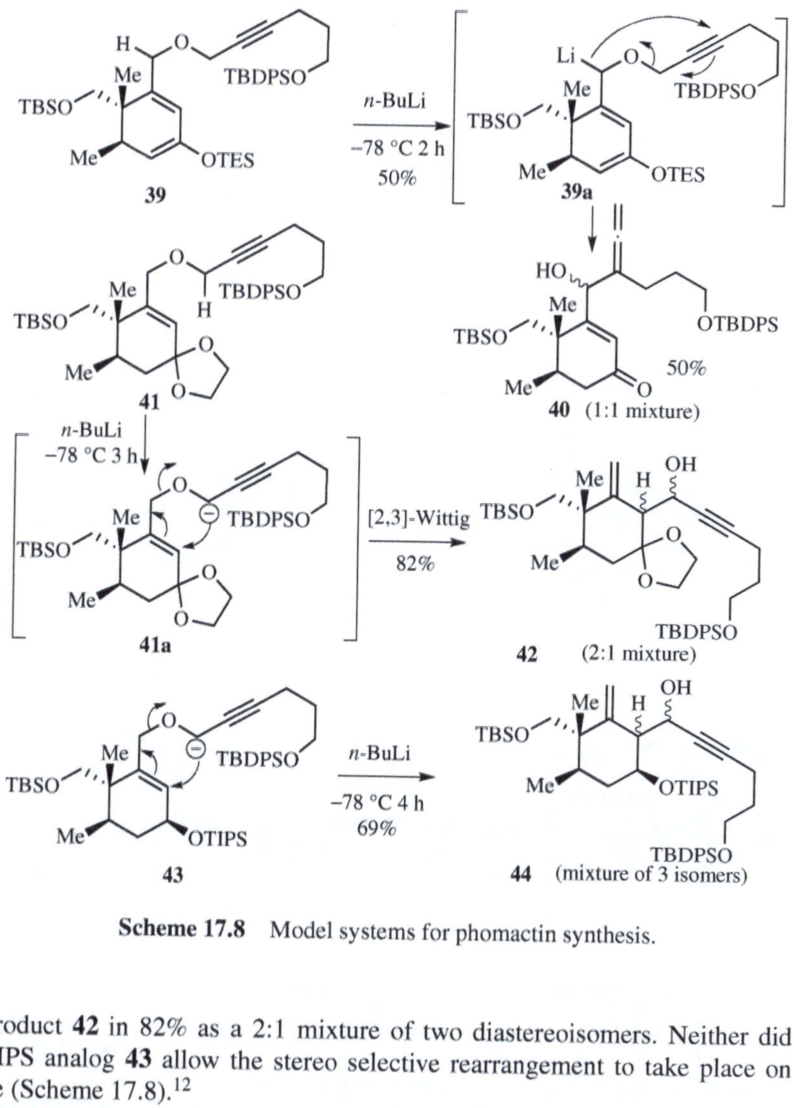

Scheme 17.8 Model systems for phomactin synthesis.

the product **42** in 82% as a 2:1 mixture of two diastereoisomers. Neither did the β-OTIPS analog **43** allow the stereo selective rearrangement to take place on the α-side (Scheme 17.8).[12]

To set the stage for macrocyclization through a C—C bond-forming step, the primary *O*-TBS was converted to the phenyl sulfone group of **45**, and then an allylic bromide was introduced to give **46** (Scheme 17.9). It was indeed possible to obtain the macrocyclic compound **47** through the sulfonyl carbanion, but further [2,3]-Wittig rearrangement did not proceed upon treatment with *n*-BuLi. However, the sulfonyl carbanion of **48** could undergo a second deprotonation followed by α-selective rearrangement. In the event, the deprotonation of **48** took place with 2 equiv. of *n*-BuLi through a chelation-assisted mechanism from the orientation shown as **48a**. This gave rearrangement predominantly from α-face, resulting mainly in **49** rather than **50** from reaction on the β-face.

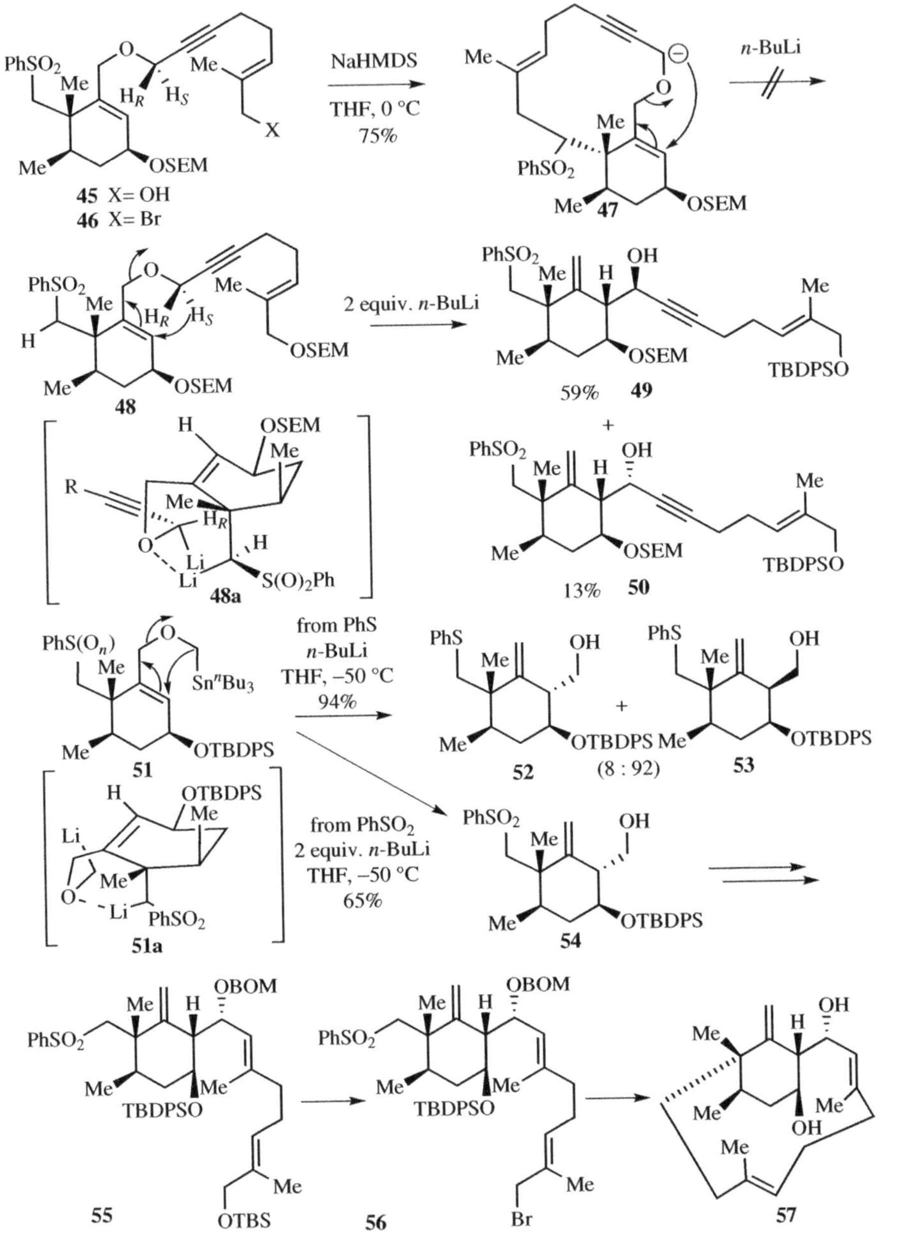

Scheme 17.9 Sulfonyl group-assisted deprotonation for [2,3]-Wittig rearrangement directed toward phomactin synthesis.

The [2,3]-Wittig–Still rearrangement of the phenyl sulfide **51** ($n = 0$) afforded a mixture of **52** and **53** from α and β approach in 8:92 ratio (Scheme 17.9). From the corresponding sulfone **51** ($n = 2$), use of 2 equiv. *n*-BuLi selectively yielded **54** in 65% yield with the α stereochemistry, resulting from lithium-ion chelation with the sulfonyl methyl and oxymethyl lithium (**51a**). The synthesis was continued to attach the side chain, and alkylative macrocyclization gave diol **57** (Scheme 17.9).[12]

During their synthesis of phoslactomycin B (**58**, Figure 17.5), Cossy and coworkers employed an acyclic [2,3]-Wittig rearrangement of a propargyl-allyl ether.[13] The selective *pro-R* deprotonation of **60** generated propargylic (*R*)-lithium carbanion and proceeded through transition state **60a** due to $A^{1,3}$ strain of the (*Z*)-olefin. Formation of product **61** was highly selective for the (*R*) configuration of the secondary alcohol, with syn-diastereoselectivity and (*E*)-olefin geometry. A high level of chirality transfer from starting material **60** was also observed. Wittig rearrangement product **61** was further converted to phoslactomycin B precursor **63** (Scheme 17.10).

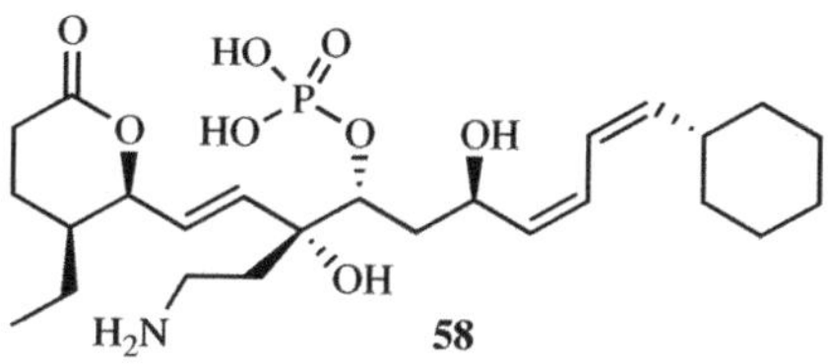

58

FIGURE 17.5 Phoslactomycin B

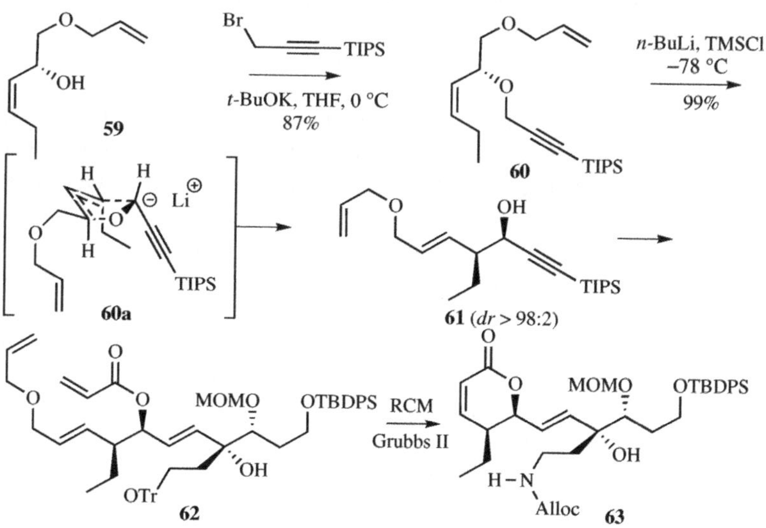

Scheme 17.10 Phoslactomycin B synthesis through [2,3]-Wittig rearrangement.

17.5 [2,3]-WITTIG–STILL REARRANGEMENT

Clive and coworkers employed the [2,3]-Wittig rearrangement during their synthetic studies toward CP-225,917 and related natural products.[14] The critical allyl-tinmethyl ether **69** was prepared from the unsaturated ester **66**. The mixture of methyl acetals (10:1) was inseparable, so **69** was subjected to the Wittig–Still rearrangement conditions, and the products were converted to triethylsilyl (TES) ethers for separation. The major isomer **70** was obtained as the rearranged product in 77% yield, and minor products were allylic alcohol **71** and methyl ether **72**. Attempted [3,3]-sigmatropic rearrangement (Ireland–Claisen) with the more functionalized species **73** was not successful due to steric impediments (Figure 17.6; Scheme 17.11).[14]

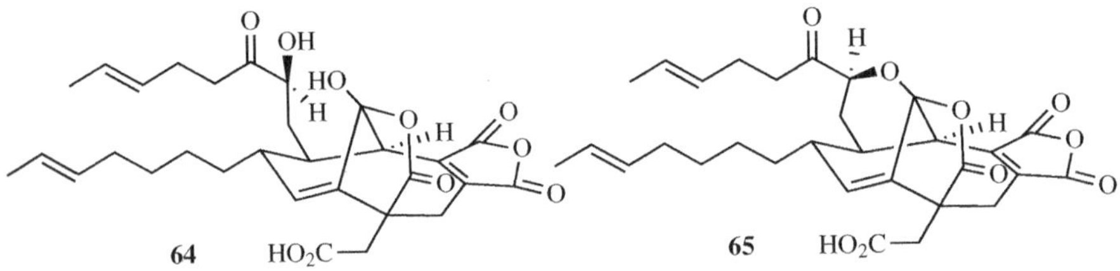

FIGURE 17.6 Synthetic target natural products CP-225,917 (**64**) and CP-263,114 (**65**)

In the total synthesis of anisatin, Fukuyama and coworkers employed the Wittig–Still rearrangement of **76** to establish the all-carbon quaternary stereocenter.[15] The face selectivity resulted from choice of chelation **76** but not from steric hindrance **77**. The tricyclic ring construction was initiated via intramolecular Diels–Alder cycloaddition to the orthoquinone equivalent from phenol **78** to give **79**. Horner–Wittig reaction of the ketone was followed by reduction to allylic alcohol **80**, which was converted to the tributyltinmethyl ether for the [2,3]-Wittig–Still rearrangement (Scheme 17.12). The oxymethyl lithium reacted from the less sterically congested *β*-face. Sterics controlled the outcome, since chelation through lithium is possible on both faces. Selective ozonolysis and further functional group manipulation led Fukuyama and coworkers to the successful total synthesis of (−)-anisatin (Figure 17.7).

The total synthesis of potential immunosuppressive agent (−)-subglutinol A **85** was achieved by employing a [2,3]-Wittig–Still rearrangement as the key step (Scheme 17.13). The strategy was to form the key intermediate **88** with the hydroxymethyl group in an *α*-axial position, which was designed to arise through the stereoelectronically preferred pathway as shown in **87a**. The allylic alcohol **86** was converted to the O-methyltributyltin ether **87**, followed by generation of the *O*-methylene anion. As planned, **87** was converted largely to **88** through **87a**, but some amount of **89** was obtained together with **90**.[16,17]

Meanwhile, Nishikawa and coworkers employed the [2,3]-Wittig–Still rearrangement for a partial synthesis of sespendole (Scheme 17.14).[18] Treatment of the tributyltinmethyl ether **91** with *n*-BuLi led to migration on the *β*-face (**91a**) to give

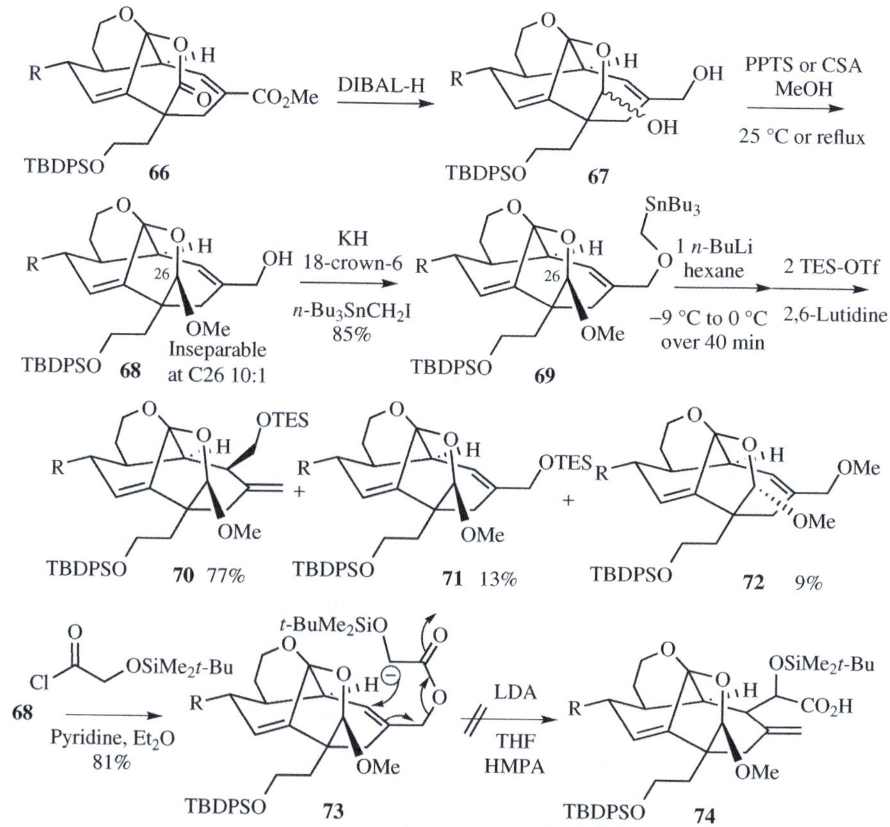

Scheme 17.11 [2,3]-Wittig–Still rearrangement en route to the CP framework.

the equatorial hydroxymethyl derivative **92**. It is interesting that this is the opposite stereochemical course as compared to **87a** (Scheme 17.13).

The phytotoxin, pyricuol (**93**) was isolated from the rice blast disease fungus. Kiyota and coworkers synthesized both of the enantiomers of **93** from optically active (*R*)- and (*S*)-*sec*-allylic alcohol **94** (Scheme 17.15). Each enantiomer was originally prepared from (*R*)- or (*S*)-lactic acid and converted via cross coupling with **95** to afford (*R*)- and (*S*)-**96**, respectively.[19] The chirality transfer process from **98** to **99** during [2,3]-Wittig–Still rearrangement was achieved through transition state **98a** to give each of the pyricuol enantiomers, (*S*)- and (*R*)-**99**. $A^{1,3}$ strain played an important role given the (*Z*)-olefin geometry in **98a** to furnish (*R*)-**99**. Each enantiomer of **99** was further converted to **93**, and only (*R*)-**93** showed the same biological activity as the natural product.

Smith employed a [2,3]-Wittig–Still rearrangement to install the trisubstituted (*Z*)-20-double bond in his group's total synthesis of (−)-hennoxazole A (**100**), an antiviral marine natural product.[20] The transition state conformation **101a** is similar to the ground state due to $A^{1,2}$ strain (Scheme 17.16).

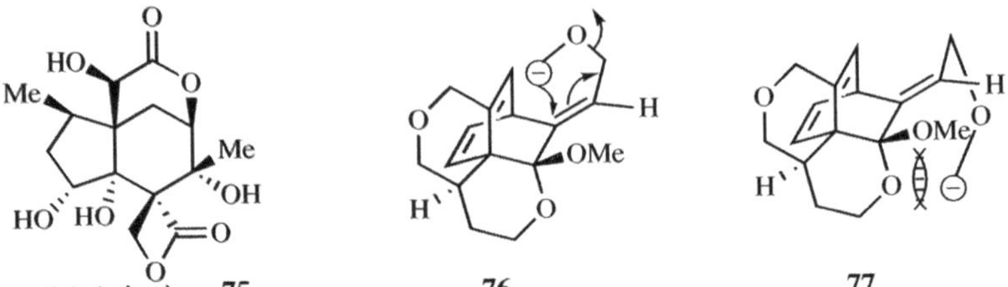

FIGURE 17.7 Anisatin and β-face bond-forming strategy via Wittig–Still rearrangement

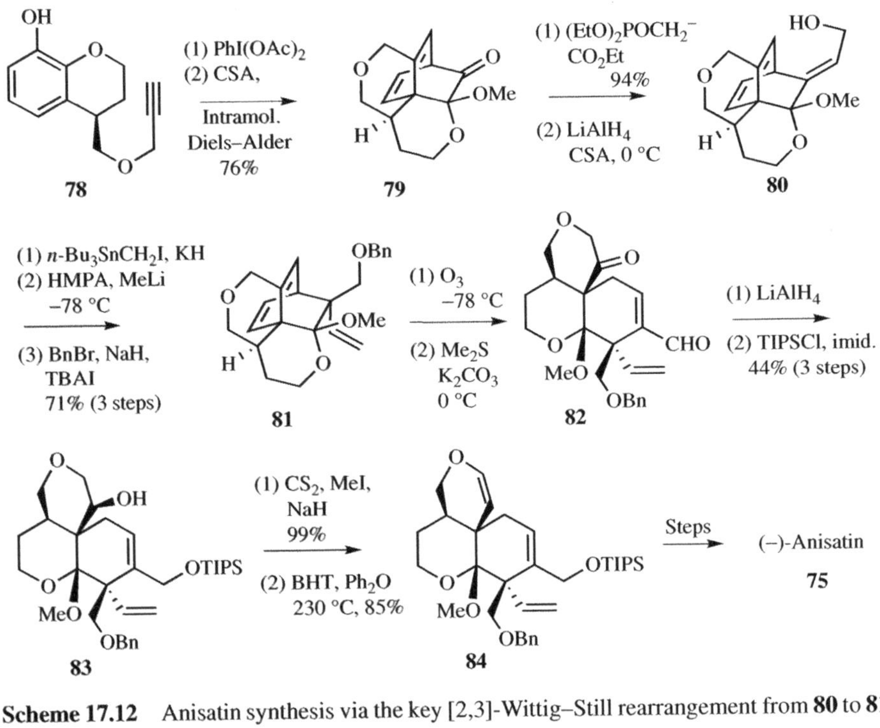

Scheme 17.12 Anisatin synthesis via the key [2,3]-Wittig–Still rearrangement from **80** to **81**.

17.6 ASYMMETRIC [2,3]-WITTIG REARRANGEMENT

Enantioselective [2,3]-Wittig rearrangement was applied to the total synthesis of eupomatilones by Maezaki and coworkers[21] Formation of the two stereogenic centers on the δ-lactone ring of eupomatilone 2 (**104**) was first studied with a model benzyl-allyl ether **105** (Figure 17.8). Selective deprotonation of the *pro-R* benzyl hydrogen was facilitated with *t*-BuLi in the presence of chiral ligand (*S*,*S*)-Box-*t*-Bu **106** at −78 °C (Table 17.1). The syn-product **107** was obtained selectively in good yield and high ee. The effect of the chiral ligand **106** on the asymmetric [2,3]-Wittig rearrangement is summarized in Table 17.1.

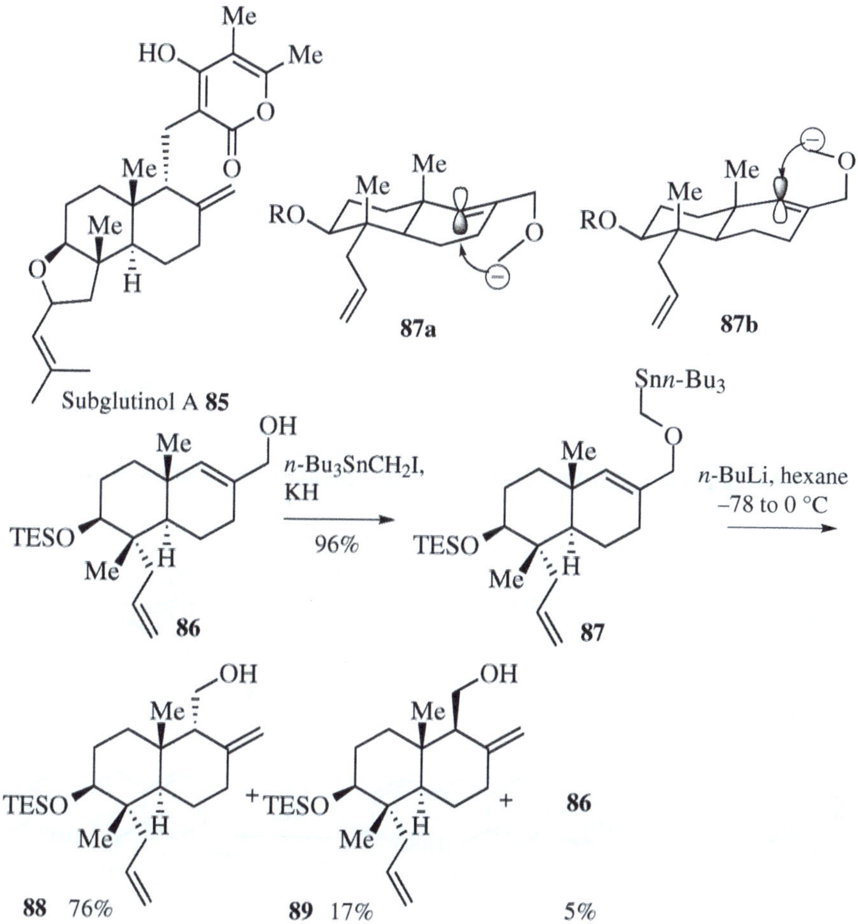

Scheme 17.13 Synthesis of subglutinol A via [2,3]-Wittig–Still rearrangement.

Allyl benzyl ether **105** was subjected to the asymmetric [2,3]-Wittig rearrangement conditions with 1 equiv. of the chiral ligand **106** and 5 equiv. of base at −78 °C for 2 h. As shown in Table 17.1, the ligand structure (**106a–e**) significantly affected the enantiomeric excess. Optimization of the base, solvent, and cosolvent for this reaction resulted in the production of **107** in 98% yield and 89% ee (entry 7).[22] The absolute configuration of **107** was determined by completing the total synthesis and comparison with the natural product, (+)-eupomatilone 2 (Figure 17.8).

17.7 AZA-[2,3]-WITTIG REARRANGEMENT

An example of aza-[2,3]-Wittig rearrangement through a chelation-controlled mechanism used the benzylic or allylic ether of an *N*-Boc-amino alcohol as in **108**

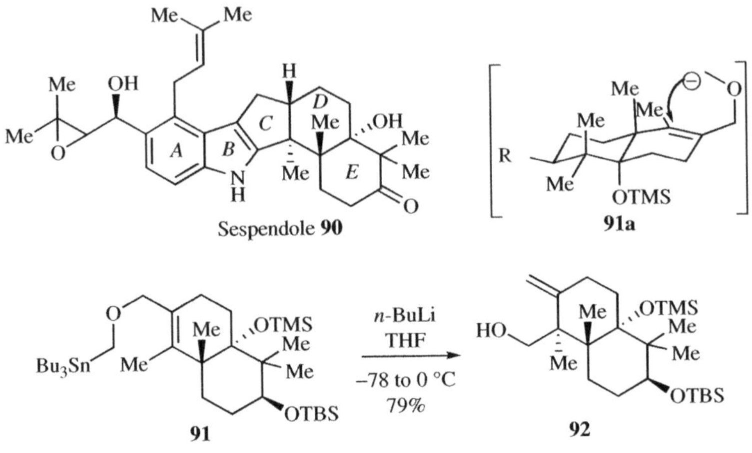

Scheme 17.14 [2,3]-Wittig–Still rearrangement for synthesis of sespendole.

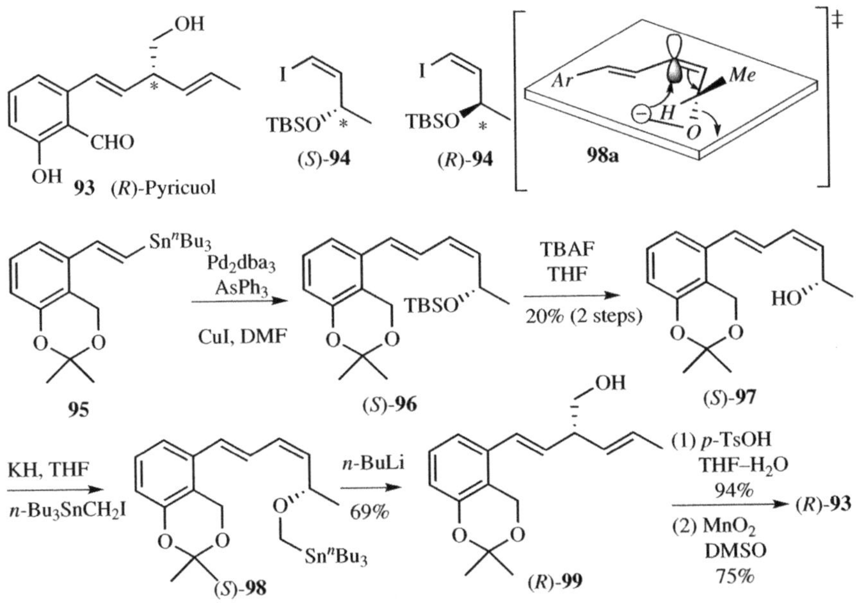

Scheme 17.15 Synthesis of pyricuol via [2,3]-Wittig–Still rearrangement.

(Scheme 17.17). The rearrangement proceeded through the chelate shown in **108a**, which allowed chirality transfer to give 1,5-amino alcohols **109**. These compounds were used for the synthesis of various cyclic amino alcohol compounds such as azasugar-*C*-glycoside **110** and indolizidine **111**.[23]

Amide α-carbanions such as **112** can take part in [2,3]-Wittig rearrangements through the preferred endo approach **112a**, which provided syn-product **113**, while

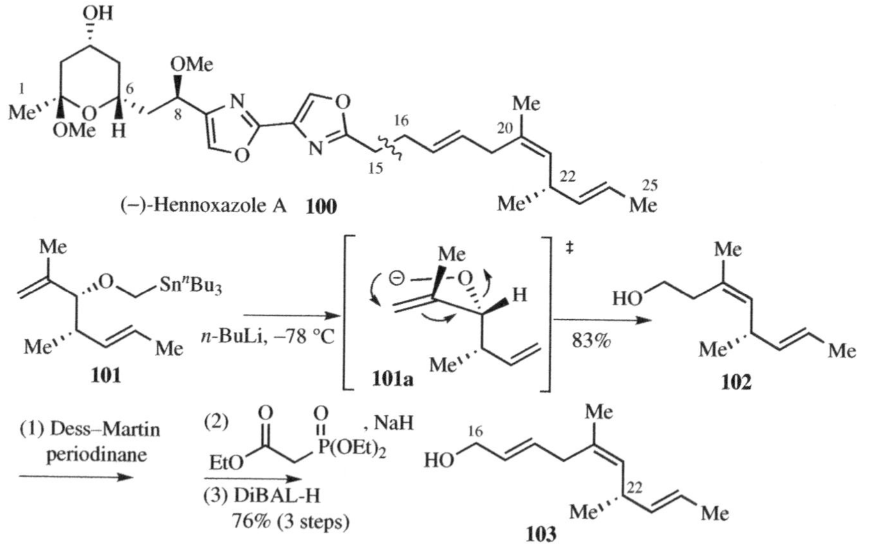

Scheme 17.16 Synthesis of the hennoxazole A C16–C25 segment via [2,3]-Wittig–Still rearrangement.

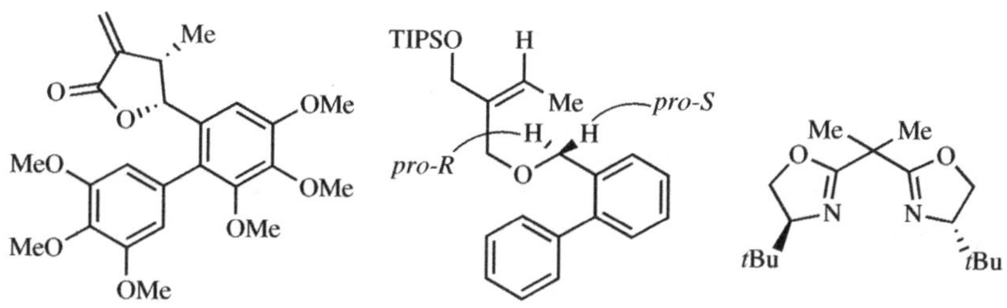

FIGURE 17.8 Asymmetric deprotonation using a chiral ligand for eupomatilone 2 synthesis

the less favorable exo arrangement **112b** provided anti-product **114**. Cossy and coworkers reported various rearrangements of 3-(*N*-tosylamino)allylic alcohol derivatives such as **115** and **118** as well.[24] Combination of the [2,3]-Wittig rearrangement and RCM with Grubbs's catalyst afforded 7-membered unsaturated lactam **117**. The chiral amide **118** afforded the 1,2-amino alcohol **119** with good selectivity (Figure 17.9; Scheme 17.18).

Li *et al.* made judicious application of the [2,3]-Wittig rearrangement in their synthesis of (+)-eldanolide **122** through the strategy shown as **123**.[25] The chiral auxiliary (1*R*,2*S*,4*S*)-2-hydroxy-7,7-dimethylbicyclo[2,2,1]heptan-1-carboxylic acid diisopropylamide was selected for this asymmetric synthesis, which differentiated the sp^2 faces at the transition state of the [2,3]-Wittig rearrangement. When **125** was treated with LDA, the rearrangement proceeded selectively via conformation **125b**

TABLE 17.1 Selective Deprotonation of *pro-R/S* using Chiral Bis-Oxazoline Ligands 106a–e

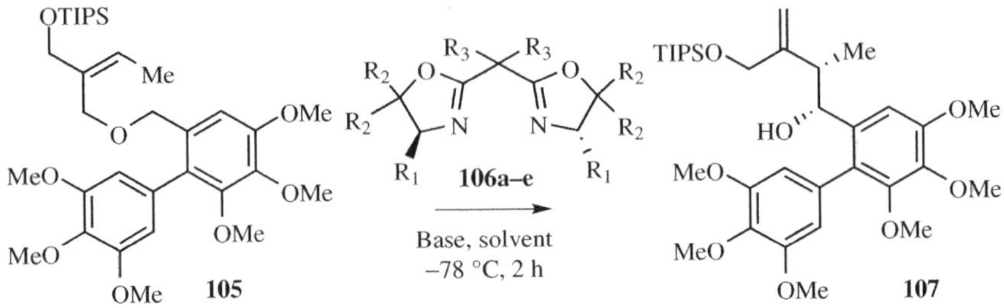

Entry	Ligand	R1	R2	R3	Conditions	Yield (%)	ee (%)
1	**106a**	*t*-Bu	H	Me	*t*-BuLi, hexane	92	68
2	**106b**	*i*-Pr	H	Me	*t*-BuLi, hexane	72	38
3	**106c**	*t*-Bu	Me	Me	*t*-BuLi, hexane	0	–
4	**106d**	*i*-Pr	Me	Me	*t*-BuLi, hexane	38	22
5	**106e**	*t*-Bu	H	Et	*t*-BuLi, hexane	43	12
6	**106a**	*t*-Bu	H	Me	*n*-BuLi, hexane	77	77
7	**106a**	*t*-Bu	H	Me	*n*-BuLi, hexane/ether (4:1)	98	89
8	**106a**	*t*-Bu	H	Me	*n*-BuLi, hexane/THF(4:1)	7	5

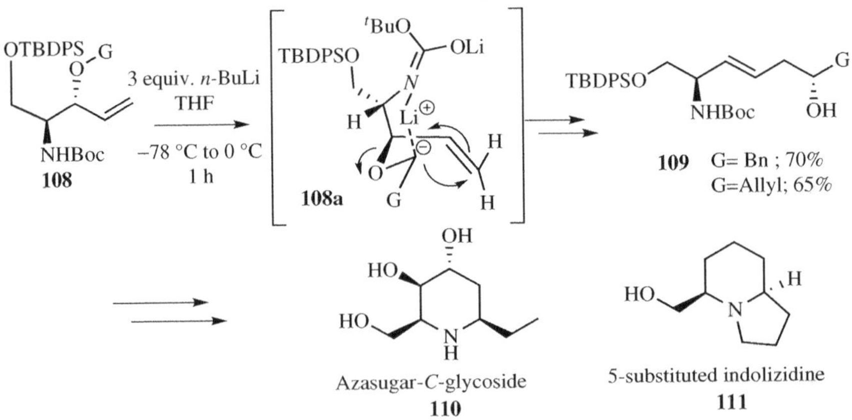

Scheme 17.17 [2,3]-Wittig rearrangement to 1,5-amino alcohols en route to *N*-heterocycles.

of the carbanion intermediate. This *reface* approach led to the *R* product **125c**, which lactonized to **126** with removal of the chiral auxiliary. Vinylogous carbamate **126** was further converted to eldanolide **122** (Scheme 17.19).

Here is an example of a [1,5]-anion relay leading to a C6-chain compound through [2,3]-Wittig rearrangement (Scheme 17.20). The allyl vinyl ether **128**

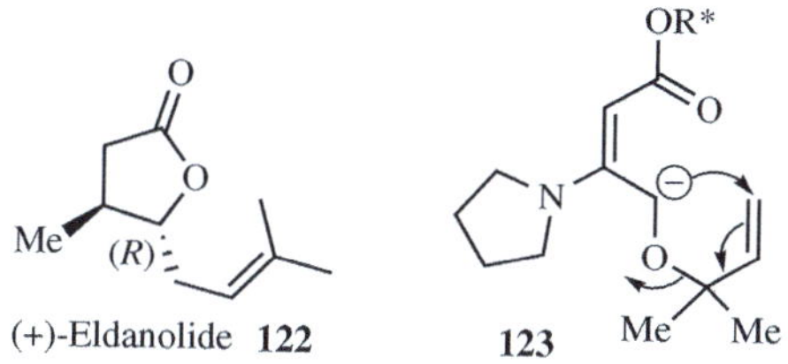

FIGURE 17.9 Eldanolide and possible rearrangement

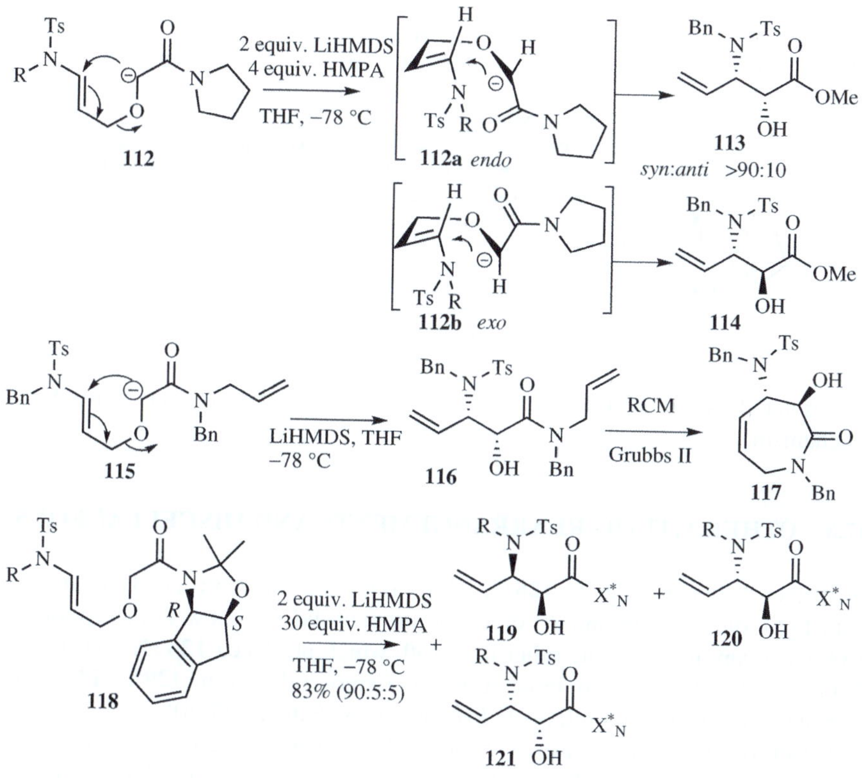

Scheme 17.18 Diastereoselective synthesis of amino alcohols via [2,3]-Wittig rearrangement, including tetrahydroazepinones.

(X = H), having a bis-TES-methyl group at the end, was deprotonated to the lithiated anion **130**, a process for which two courses are possible. One is direct deprotonation from **128** (X = H). The other path is indirect deprotonation, first at the allylic site next to the oxygen atom to form **129**, followed by intramolecular proton transfer to **130** via [1,5]-anion relay. As shown by resonance form **131**, this anion can undergo [2,3]-Wittig rearrangement to give **132**, and the product **133** was isolated in 80% yield after workup. When **128** (X = D) was treated in the same way, **133** (X = D)

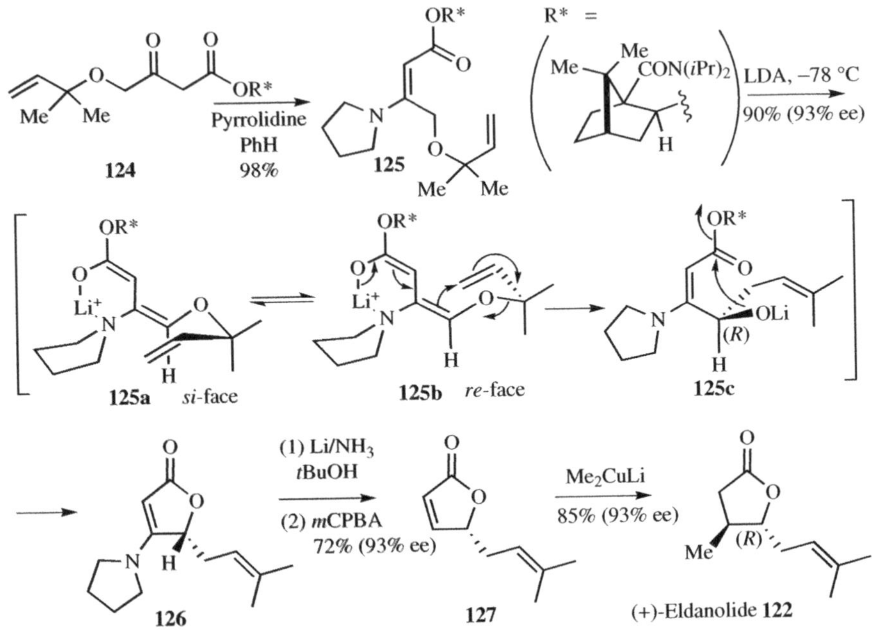

Scheme 17.19 Synthesis of (+)-eldanolide.

was isolated as *E/Z* = 1:1 mixture,[26] consistent with the indirect proton-transfer mechanism.

17.8 OTHER WITTIG REARRANGEMENTS AND MISCELLANEOUS

In another set of experiments, **128** (X = H) was treated with *t*-BuLi and then quenched with D_2O after 15 min and 30 min. The product did not show D-incorporation in **133** nor in any intermediate other than **130**, which produced **128** (X = D) upon the D_2O quench. These results indicate that the anion relay from **129** to **130** and the rearrangement to **132** are relatively fast processes (Scheme 17.20).

Effects of solvent and additives on the [2,3]-Wittig rearrangement were reported by Takeda and coworkers.[27] They studied the steric course of the chiral allyloxy-benzyl ether (*S*)-**137**, upon treatment with *t*-BuLi in various solvents. The rearranged product **138** was obtained in fair yields and with varying degrees of racemization: depending on the solvent (*R*/*S*) = 92:8 (hexane), 86:14 (toluene/1,4-dioxane), and ~50:50 (THF) in 40–70% yield. The differences in chirality transfer can be attributed to the configurational stability of the chiral carbanions **137a**, which depends on the solvent, additive, and temperature. Optimum conditions with *n*-BuLi gave **138** (*R*/*S*) = 84:16 in 87% yield (Scheme 17.21).

Anderson and Davies reported aza-Wittig rearrangement[28] of variously substituted allylamines **139** to give products **140** and/or **141**. The stereoselectivity depended on the R group due to steric interactions at the transition state **139a**,

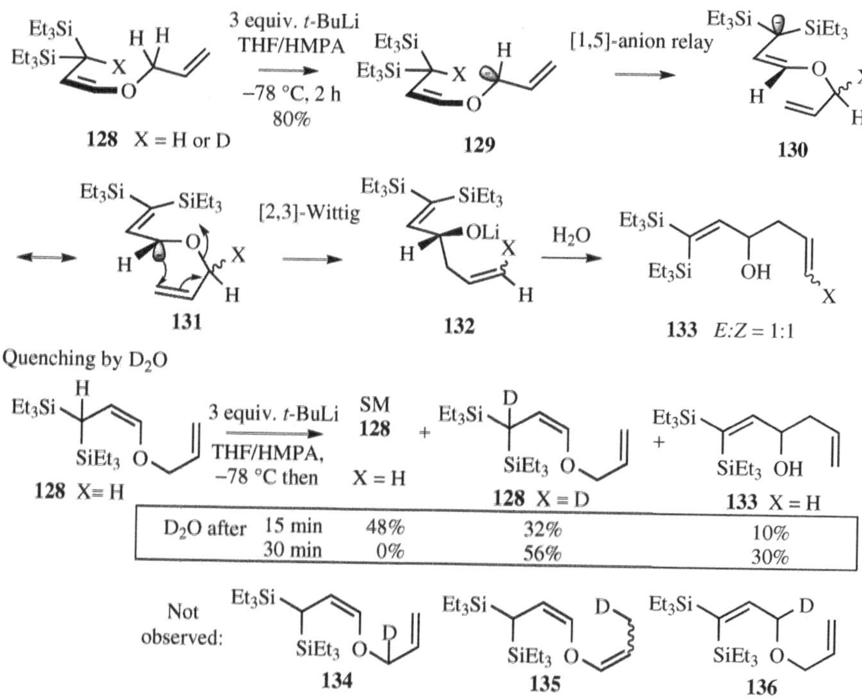

Scheme 17.20 [1,5]-Anion relay leading to the [2,3]-Wittig rearrangement.

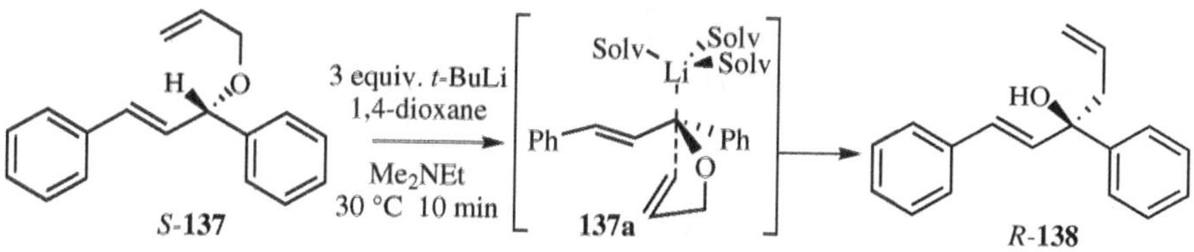

Scheme 17.21 Solvent effect on the steric course of [2,3]-Wittig rearrangement.

yielding diastereoisomers in good to fair yields with different ratios (R = Me, 77%, 7:11; R = *i*-Pr, 72%, 20:1; R = Bn, 75%, 1:1). Prolonging the reaction times led one of the diastereoisomers **140'** to cyclize into pyrrolidine **142** (Scheme 17.22).

Wolfe and Everett reported an unusual enolate Wittig rearrangement with alkylative cyclization by treatment of methyl *O*-(alkynylmethyl)glycolates **143** with dialkylboron triflates and Hünig's base to afford highly substituted 3-hydroxy-2-furanones **144**.[29] This transformation appeared to proceed via an unusual mechanism involving initial [2,3]-Wittig rearrangement of a boron ester enolate, followed by an alkylative cyclization reaction leading to incorporation of an alkyl group from the boron reagent into the product. Wolfe and coworkers proposed two possible mechanisms: (1) a radical cage mechanism for conversion of **147** to

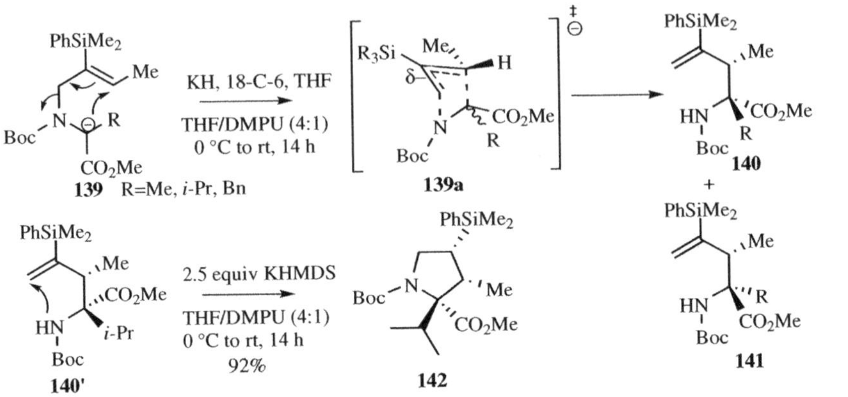

Scheme 17.22 Aza-[2,3]-Wittig rearrangement.

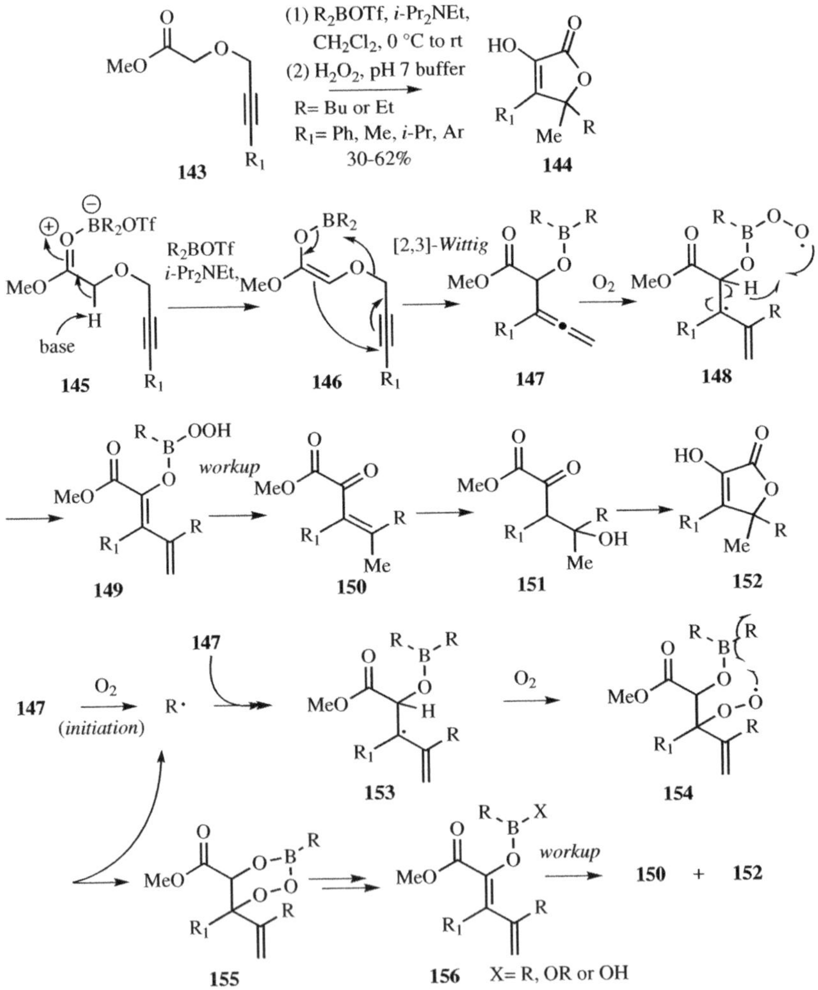

Scheme 17.23 Furanone via Wittig rearrangement.

149 or (2) a radical chain mechanism from **147** to **156**. To resolve the question of cage versus chain mechanisms, they carried out a crossover experiment (not shown in Scheme 17.23) in which the rearrangement of **143** was carried out in the presence of 3.2 equiv of Et_3B. Upon workup, protonation of the enolate would generate **149**, which can then undergo conjugate addition of water or hydroxide to provide alcohol **151**. Intramolecular acylation of the alcohol then yielded the product **152** (=**144**). This reaction provided a ca. 1:1 mixture of the product **152** R = Bu and Et, supporting the radical chain scenario.

Thomas and Hoegenauer achieved aliphatic 1,8-stereocontrol in the synthesis of racemic patulolide C **157** and epipatulolide C **158** via [2,3]-Wittig rearrangement of bisallylethers such as **160**.[30] The stereochemistry of the racemic product was confirmed by completion of the synthesis and comparison with natural epipatulolide C. Thomas reported several other stereoselective examples of the rearrangement (Scheme 17.24).

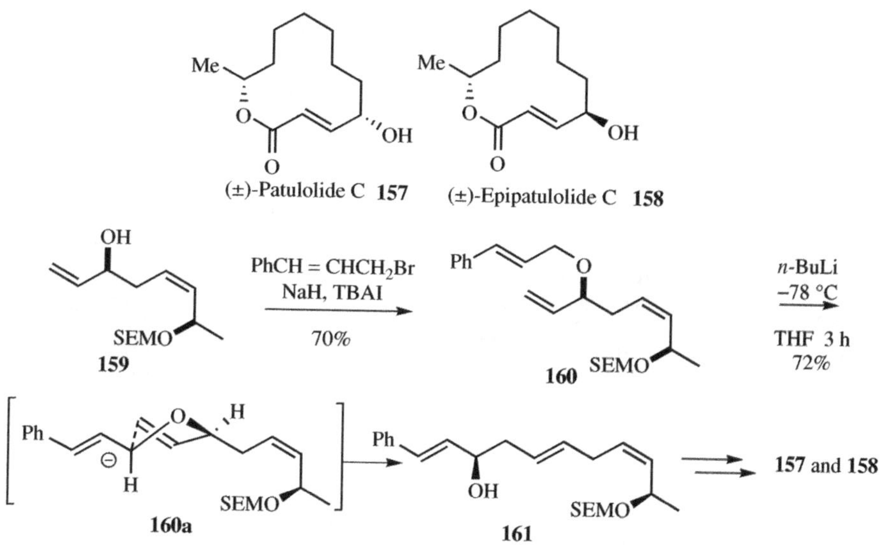

Scheme 17.24 Patulolide C via Wittig rearrangement.

Hanessian *et al.* recently reported the generation of functionalized quaternary carbon centers relying on Wittig and Wittig–Still allylic ether anionic transpositions.[31] Formation of the lithium anion from **162** by treatment with *n*-BuLi, *s*-BuLi, or *t*-BuLi in THF at −78 °C for 4 h led only to the ether cleavage product **165** in 35–38% yield. Maintaining this mixture at −78 °C for 4 h and then at 0 °C for 1 h afforded **163** (22%), **164** (13%), and **165** (40%), suggesting **165** was obtained by ether cleavage decomposition. Quenching the reaction mixture with D_2O led to **163** and **164** with no D-incorporation. The diene **166** afforded [2,3]-product **167** and [1,2]-product **168** only. The *axial-O*-tinmethyl compound **169** gave the quaternary [2,3]-product **170** in 17% yield but [1,2]-product **171** in 73% yield.

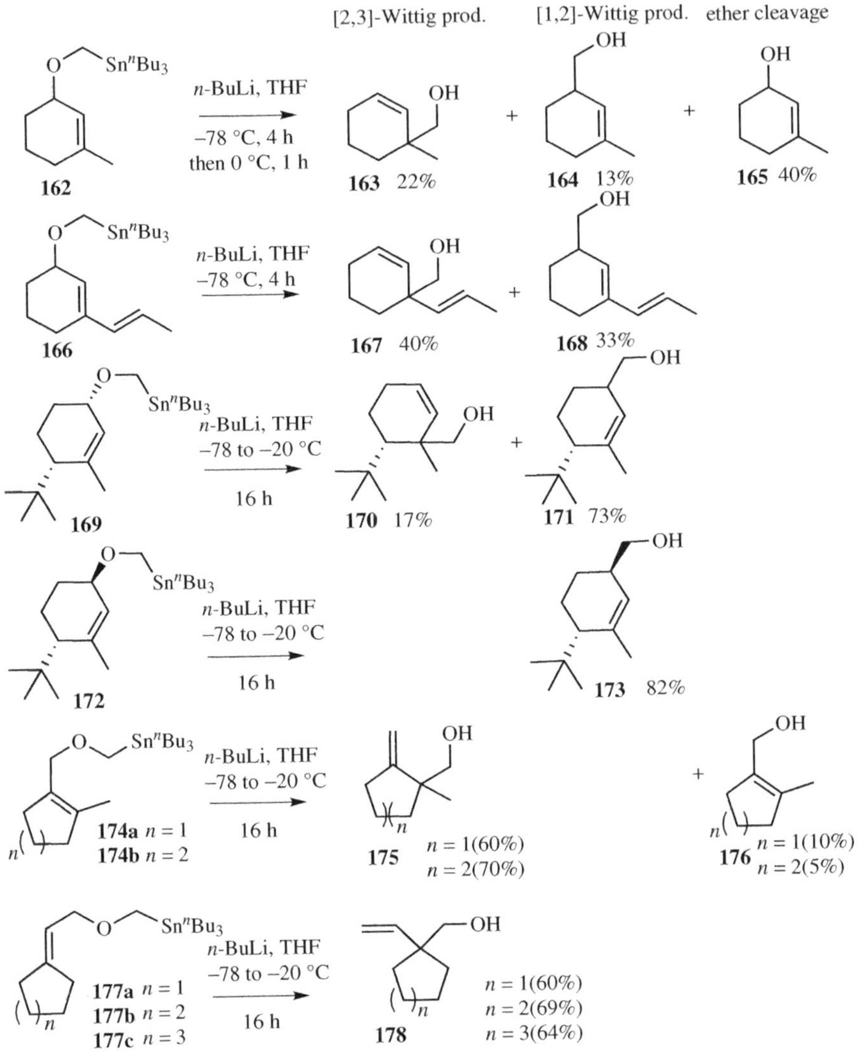

Scheme 17.25 Wittig–Still allylic ether anionic transpositions.

The *equatorial* isomer **172**, meanwhile, gave no [2,3]-product. To further probe the influence of geometric constraints, Hanessian subjected the stannyl ethers **174a** and **174b** to rearrangement. The [2,3]-products **175** were obtained in 60 and 70% yields, accompanied by ether cleavage by-products **176**. The exocyclic stannylmethyl allylic ethers **173abc** gave the [2,3]-products **178** in 60%, 69%, and 64% yields, respectively (Scheme 17.25).

The possible mechanism proposed by Hanessian is in Scheme 17.26 accounting for the [2,3]-process and [1,2]-process. The oxymethyl lithium **162a** located in the *equatorial orientation* has to undergo a conformation change of the cyclohexene ring

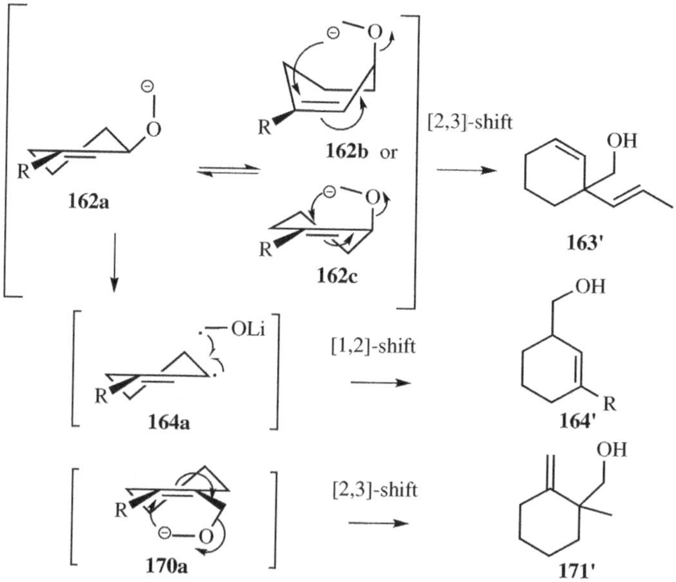

Scheme 17.26 Possible mechanisms accounting for [2,3]- versus [1,2]-Wittig rearrangement.

to either a boat form (**162b**) or to the alternative half-chair conformation **162c** for the [2,3]-Wittig rearrangement. As discussed earlier, when this conformational inversion requires high energy, the anion might dissociate to the radical pair **164a**, leading to go to the [1,2]-product **164**.

17.9 CONCLUSION

The Wittig rearrangement continues to be of great importance in synthetic organic chemistry as shown by the examples reviewed in this chapter as well as in other recent applications of the [2,3]-Wittig rearrangement in synthesis.[32] Of particular note is the utility of this rearrangement reaction toward synthesis of complex molecules, including natural products.

REFERENCES

1. (a) Nakai, T.; Mikami, K. *Chem. Rev.* **1986**, *86*, 885–902. (b) Nakai, T.; Mikami, K. *Org. React.* **1994**, *46*, 105–209.
2. Marshall, J. A. The Wittig Rearrangement. In Comprehensive Organic Synthesis, **1991**; Vol. 3, pp. 975–1014.
3. In the case of Overman rearrangements, the C–N bond-forming process is difficult if the attempted bond making is toward a congested *tert*-position. Using an $A^{1,3}$-strain effect,

however, could force the alpha substituent(s) into an axial orientation before the rearrangement, so that the rearrangement did in fact proceed via the diaxial intermediate, followed by an immediate inversion to the diequatorial orientation to drive the equilibrium, allowing successful formation of the *tert-N*-amide. For more details, see: (a) Isobe, M.; Fukuda, Y.; Nishikawa, T.; Chabert, P.; Kawai, T.; Goto, T. *Tetrahedron Lett.* **1990**, *31*, 3327–3330. (b) Overman, L. E. *J. Am. Chem. Soc.* **1976**, *98*, 2901–2910.

4. Still, W. C.; Mitra, A. *J. Am. Chem. Soc.* **1978**, *100*, 1927–1928.
5. Mikami, K.; Kawamoto, K.; Nakai, T. *Tetrahedron Lett.* **1985**, *26*, 5790–5802.
6. Tsao, K.-W.; Cheng, C.-Y.; Isobe, M. *Org. Lett.* **2012**, *14*, 5274–5277.
7. Kim, W.-H.; Angeles, A. R.; Lee, J.-H.; Danishefsky, S. J. *Tetrahedron Lett.* **2009**, *50*, 6440–6441.
8. Ohyabu, N.; Nishikawa, T.; Isobe, M. *J. Am. Chem. Soc.* **2003**, *125*, 8798–8805.
9. Isobe, M.; Chang, W.-C.; Tsou, P.-K.; Ploysuk, C.; Yu, C.-H. *J. Org. Chem.* **2015**, *80*, 6222–6237.
10. Ohtani, I.; Kusumi, T.; Kashman, T.; Kakisawa, H. *J. Am. Chem. Soc.* **1991**, *113*, 4092–4096.
11. $A^{1,2}$-Strain was reported for the equilibrium of 1-alkynyldicobalthexacarbonyl compounds having 2-protected hydroxyl groups; the larger protecting groups gave a greater amount of the equilibrated isomer. See: (a) Tanaka, S.; Isobe, M. *Tetrahedron* **1994**, *50*, 5633–5644. (b) Chang, W.-C.; Isobe, M. *Tetrahedron* **2014**, *70*, 8324–8333.
12. (a) Shapland, P. D. P.; Thomas, E. J. *Tetrahedron* **2009**, *65*, 4201–4211. (b) McGowan, G.; Thomas E. J. *Org. Biomol. Chem.* **2009**, *7*, 2576–2590. (c) Blackburn, T. J.; Helliwell, M.; Kilner, M. J.; Lee, A. T. L.; Thomas, E. J. *Tetrahedron Lett.* **2009**, *50*, 3550–3554.
13. (a) Druais, V.; Hall, M. J.; Corsis, C.; Wendeborn, S. V.; Meyer, C.; Cossy, J. *Tetrahedron* **2010**, *66*, 6358–6375. (b) Druais, V.; Hall, M. J.; Corsi, C.; Wendeborn, S. V.; Meyer, C.; Cossy, J. *Org. Lett.* **2009**, *11*, 935–938.
14. Malihi, F.; Clive, D. L. J.; Chang, C.-C.; Minaruzzaman *J. Org. Chem.* **2013**, *78*, 996–1013.
15. Ogura, A.; Yamada, K.; Yokoshima, S.; Fukuyama, T. *Org. Lett.* **2012**, *14*, 1632–1635.
16. Kikuchi, T.; Mineta, M.; Ohtaka, J.; Matsumoto, K.; Katoh, T. *Eur. J. Org. Chem.* **2011**, 5020–5030.
17. (a) Oguchi, T.; Watanabe, K.; Ohkubo, K.; Abe, H.; Katoh, T. *Chem. Eur. J.* **2009**, *15*, 2826–2845. (b) Watanabe, K.; Iwasaki, K.; Abe, T.; Inoue, M.; Ohkubo, K.; Suzuki, T.; Katoh, T. *Org. Lett.* **2005**, *7*, 3745–3748.
18. Sugino, K.; Nakazaki, A.; Isobe, M.; Nishikawa, T. *Synlett* **2011**, 647–650.
19. Tanaka, K.; Nakamura, Y.; Sasaki, A.; Ueda, R.; Suzuki, Y.; Kuwahara, S.; Kiyota, H. *Tetrahedron* **2009**, *65*, 6115–6122.
20. Smith, T. E.; Kuo, W. H.; Balskus, E. P.; Bock, V. D.; Roizen, J. L.; Theberge, A. B.; Carroll, K. A.; Kurihara, T.; Wessler, J. D. *J. Org. Chem.* **2008**, *73*, 142–150.
21. Hirokawa, Y.; Kitamura, M.; Kato, C.; Kurata, Y.; Maezaki, N. *Tetrahedron Lett.* **2011**, *52*, 581–584.
22. Hirokawa, Y.; Kitamura, M.; Mizubayashi, M.; Nakatsuka, R.; Kobori, Y.; Kato, C.; Kurata, Y.; Maezaki, N. *Eur. J. Org. Chem.* **2013**, *4*, 721–727.
23. Chandrasekhar, B.; Prasada Rao, J.; Venkateswara Rao, B.; Naresh, P. *Tetrahedron Lett.* **2011**, *52*, 5921–5925.

24. (a) Barbazanges, M.; Meyer, C.; Cossy, J. *Org. Lett.* **2007**, *9*, 3245–3248. (b) Barbazanges, M., Christophe, M.; Cossy, J. *Tetrahedron Lett.* **2008**, *49*, 2902–2906. (c) Barbazanges, M.; Meyer, C.; Cossy, J.; Turner, P. *Chem. Eur. J.* **2011**, *17*, 4480–4495.

25. Li, Y.-J.; Ho, G.-M.; Chen, P.-Z. *Tetrahedron: Asymmetry* **2009**, *20*, 1854–1863.

26. Sun, X.-W.; Lei, J.; Sun, C.-Z.; Song, Z.-L.; Yan, L.-J. *Org. Lett.* **2012**, *14*, 1094–1097.

27. (a) Ikemoto, H.; Sasaki, M.; Takeda, K. *Eur. J. Org. Chem.* **2010**, 6643–6650. (b) Sasaki, M.; Ikemoto, H.; Kawahata, M.; Yamaguchi, K.; Takeda, K. *Chem. Eur. J.* **2009**, *15*, 4663–4666.

28. Anderson, J. C.; Davies, E. A. *Tetrahedron* **2010**, *66*, 6300–6308.

29. Everett, R. K.; Wolfe, J. P. *Org. Lett.* **2013**, *15*, 2926–2929.

30. Hoegenauer, E. K.; Thomas, E. J. *Org. Biomol. Chem.* **2012**, *10*, 6995–7014.

31. Hanessian, S.; Dorich, S.; Chattopadhyay, A. K.; Büschleb, M. *J. Org. Chem.* **2013**, *78*, 8915–8921.

32. (a) Wieczorek, A.; Hammerschmidt, F. *J. Org. Chem.* **2012**, *77*, 10021–10034. (b) Miles, J. A.; Mitchell, L.; Percy, J. M.; Singh, K.; Uneyama, E. *J. Org. Chem.* **2007**, *72*, 1575–1587. (c) Kapeller, D. C.; Brecker, L.; Hammerschmidt, F. *Chem. Eur. J.* **2007**, *13*, 9582–9588. (d) Goto, A.; Fujiwara, K.; Kawai, A.; Kawai, H.; Suzuki, T. *Org. Lett.* **2007**, *9*, 5373–5376. (e) Tayama, E.; Sugai, S.; Hara, M. *Tetrahedron Lett.* **2006**, *47*, 7533–7735. (f) Sato, Y.; Fujisawa, H.; Mukaiyama, T. *Chem. Lett.* **2006**, *35*, 124–125. (g) Parker, K. A.; Cao, H. *Org. Lett.* **2006**, *8*, 3541–3544. (h) Modica, E.; Compostella, F.; Colombo, D.; Franchini, L.; Cavallari, M.; Mori, L.; De Libero, G.; Panza, L.; Ronchetti, F. *Org. Lett.* **2006**, *8*, 3255–3558. (i) McNally, A.; Evans, B.; Gaunt, M. J. *Angew. Chem., Int. Ed.* **2006**, *45*, 2116–2119.

18

THE MISLOW–EVANS REARRANGEMENT

Christian M. Rojas
Department of Chemistry, Barnard College, New York, NY, USA

18.1 INTRODUCTION

By the mid-to-late 1960s, it had become clear that something interesting occurred when allylic alcohols were converted to their sulfenate esters. A decade earlier, Sosnovsky had reported the apparently prosaic preparation of the (putative) allyl alcohol trichloromethyl sulfenyl derivative **2a** ($R = CCl_3$, Scheme 18.1, and see also Scheme 18.2).[1] However, by 1966, studies from Kurt Mislow's laboratory at Princeton University had revealed that the product from this type of sulfenate ester synthesis was generally not the sulfenate **2** itself but a rearranged allylic sulfoxide **3**.[2] Moreover, through an elegant and careful series of mechanistic experiments, including kinetic measurements and isotope-labeling studies, Mislow provided convincing evidence for a concerted [2,3]-sigmatropic rearrangement pathway for the sulfenate-to-sulfoxide conversion. These investigations also probed equilibration between sulfenate and sulfoxide forms via the [2,3] path.[3]

At about the same time, Samuel Braverman's group at Bar-Ilan University reported that Sosnovsky's product was actually the trichloromethyl allyl sulfoxide **3a** and pointed out the mechanistic possibility of sigmatropic rearrangement in its formation (Scheme 18.2). Braverman and Stabinsky also investigated the conversion of cinnamyl sulfenate **4** to sulfoxide **5**, noting that the more vigorous conditions required compared to the case of **2a** were consistent with a radical fragmentation route for **4** → **5**,without allylic rearrangement.[4] Unlike Mislow's studies from the year before,

Molecular Rearrangements in Organic Synthesis, First Edition. Edited by Christian M. Rojas.

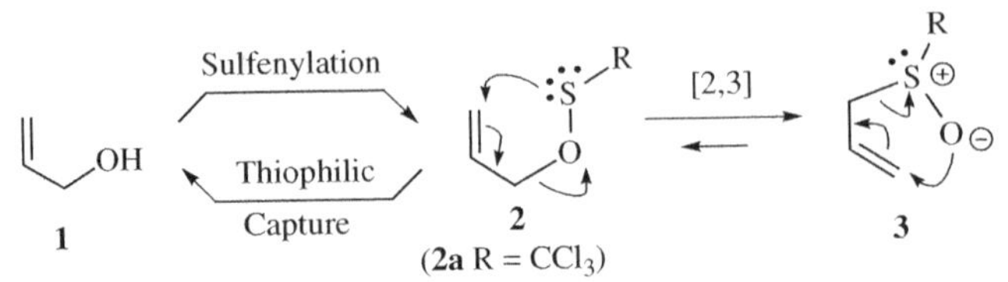

Scheme 18.1 Allylic alcohol/allyl sulfenate ester/allyl sulfoxide interconversions.

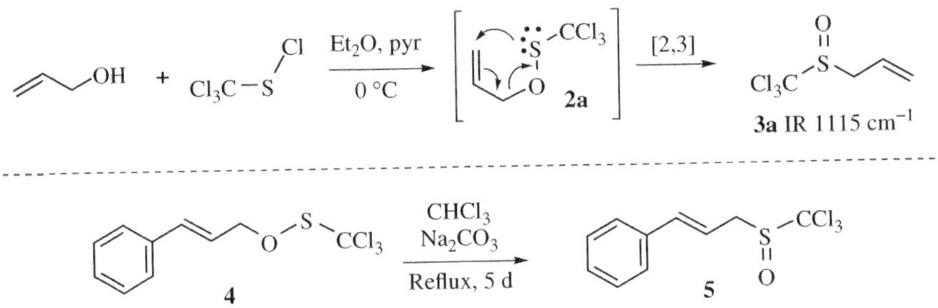

Scheme 18.2 Sulfenate-to-sulfoxide rearrangements of differing facility.[4]

Braverman's 1967 communication did not include mechanistic studies confirming the [2,3]-sigmatropic nature of the allyl sulfenate ester-to-allylic sulfoxide conversion.

The synthetic potential of this rearrangement, particularly under conditions that enabled trapping the typically less thermodynamically favorable sulfenate ester (**2** → **1**; Scheme 18.1), was brought to the fore through the efforts of David Evans's group at Harvard University in the early 1970s.[5] As a result of their pioneering studies on the nature of the allyl sulfenate–allyl sulfoxide interconversion and exploration of the rearrangement's utility in organic synthesis, the reaction is termed the Mislow–Evans rearrangement or, less frequently, the Mislow–Braverman–Evans rearrangement. A number of review articles that include mechanistic aspects and applications of the Mislow–Evans rearrangement have been published,[6] and the rearrangement is included in a spate of recent volumes on name reactions in organic chemistry.[7]

Creative new applications and variants of the Mislow–Evans rearrangement continue to highlight its utility in organic synthesis and to extend the range of structures available through this [2,3]-transposition reaction. This chapter aims to provide sufficient background to give the reader context for a presentation of applications of the Mislow–Evans rearrangement and closely related variants. Examples highlight control in formation of new stereogenic elements, including stereocenters at tetrahedral carbon as well as alkene *E*/*Z* selectivity. Interestingly, the chirality element at the stereogenic sulfoxide S-center can be an important feature in these [2,3]-rearrangements. The Mislow–Evans rearrangement provides clever routes to a number of useful synthons, including vinyl anion surrogates, and enables the synthetic equivalent of challenging transformations such as aldehyde α-hydroxylation. The rearrangement has also been incorporated into sequential reaction schemes (tandem-, domino-, or cascade-reaction processes) and applied in the preparation of complex natural products and other medicinally active compounds.

This chapter is organized by theme rather than chronologically, and while the most recent applications certainly highlight the continuing significance of the Mislow–Evans rearrangement in synthesis endeavors, classic examples are also included to paint a full picture of the reaction's significance and as a source of inspiration for contemporary practitioners of our art.

PART 1. MECHANISTIC ASPECTS AND THE [2,3] NATURE OF THE REARRANGEMENT

18.2 CONFIGURATIONAL LABILITY OF ALLYLIC SULFOXIDES

18.2.1 Racemization of *S*-Chiral Sulfoxides

The facile racemization of optically active sulfoxides (enantiomerically enriched in configuration at sulfur) provided early evidence for the concerted, [2,3]-sigmatropic nature of the allylic sulfoxide-to-sulfenate rearrangement (Scheme 18.3).[3a] Mislow found that racemization occurred readily at 50–70 °C, conditions considerably milder than the 130–150 °C required for the radical–cleavage pathway associated with the corresponding benzyl sulfoxide and well below the temperature needed for pyramidal inversion at the sulfoxide sulfur center (190–220 °C).

By measuring the temperature dependence of rates of racemization for optically enriched allyl sulfoxide **6**, Mislow and coworkers determined activation parameters of $\Delta H^{\ddagger} = 23$ kcal/mol and $\Delta S^{\ddagger} = -5$ eu in benzene. By comparison, this enthalpy of activation was about 20 kcal/mol lower than for the *p*-tolyl benzyl sulfoxide analogue. These results were consistent with a relatively low-barrier [2,3]-rearrangement for the allyl sulfoxide as shown in Scheme 18.3, rather than the radical fragmentation route involved in the benzyl sulfoxide case. Racemization would occur through the effectively achiral sulfenate ester intermediate **7** (discounting conformational memory of the asymmetry originally inherent in the chiral sulfoxide group). Additionally, Mislow found that (+)-allylmethyl sulfoxide also racemized readily, showing that the aryl group on sulfur was not required for the sigmatropic rearrangement. Mislow concluded his landmark 1968 full account, writing "It would appear that the mechanism of racemization proposed in this paper is applicable in general to compounds containing an allylic group attached to sulfoxide sulfur."[3a]

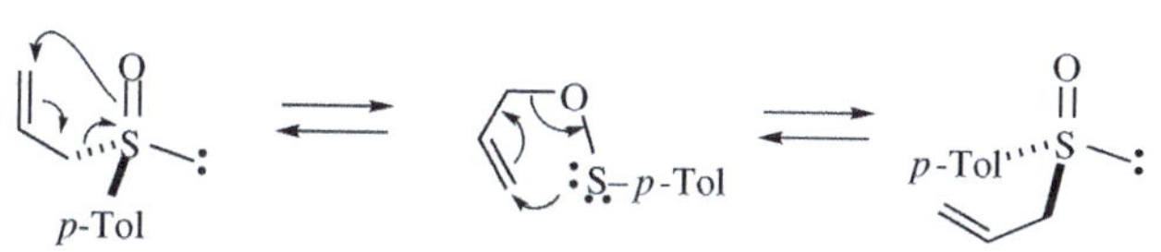

6 (*R*) configuration　　**7**　　**6**-ent (*S*) configuration

Scheme 18.3 Racemization of an optically enriched allyl sulfoxide.[3a]

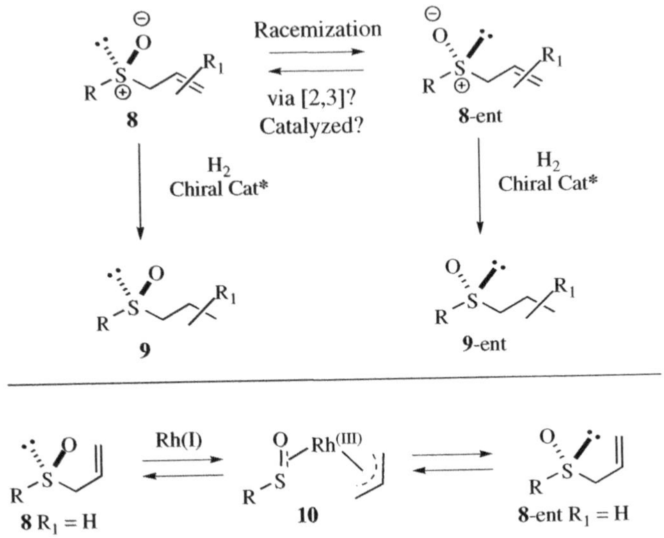

Scheme 18.4 Dynamic kinetic resolution via allylic sulfoxide racemization–alkene hydrogenation and a non-[2,3] Rh(I)-catalyzed mechanism for sulfoxide racemization.[8]

18.2.2 Opportunities for Dynamic Kinetic Resolution

To jump forward to a contemporary example, the ready racemization of allylic sulfoxides via reversible [2,3]-sigmatropic rearrangement could enable a dynamic kinetic resolution if one of the sulfoxide enantiomers were to be intercepted selectively. Dong's group examined this possibility using a chiral Rh(I) hydrogenation catalyst that distinguished between the enantiomeric allylic sulfoxides **8** (Scheme 18.4).[8] Interestingly, these studies revealed that the hydrogenation catalyst also accelerated

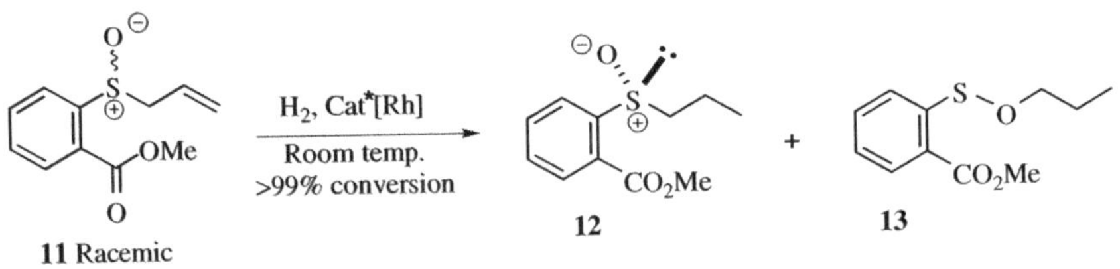

		12	:	**13**
Nonpolar	19:1 $PhMe/CH_2Cl_2$ 1 mol% [Rh] H_2 (1 atm)	2.4 (90% ee)	:	1
Polar	MeOH 2 mol% [Rh] H_2 (0.1 atm)	92 (73%) 88% ee	:	8

Scheme 18.5 Under non-polar conditions at least part of sulfoxide interconversion is due to Mislow–Evans rearrangement.[8]

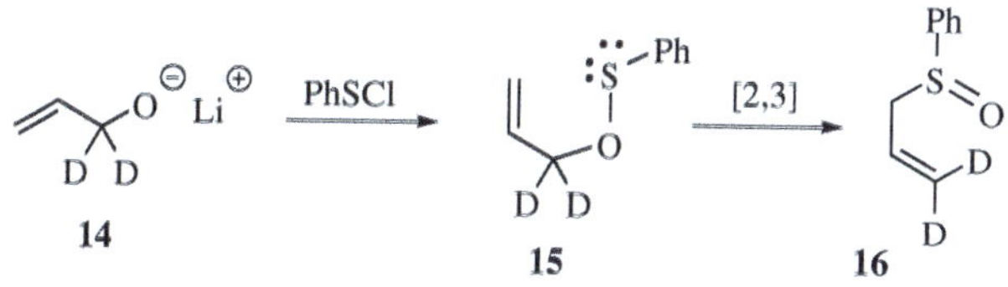

Scheme 18.6 Rearrangement of deuterium-labeled sulfenate ester.[3a]

the sulfoxide racemization in a solvent-dependent manner. For example, 2 mol% of a [Rh(*S*,*S*)-Ph-BPE](COD)BF_4 complex increased the racemization rate for phenyl allyl sulfoxide by 33 times in methanol but only by 10% in a mixture of toluene and methylene chloride. Moreover, deuterium-labeling studies, combined with computational investigations done in collaboration with Houk's group, indicated that racemization in methanol occurred via a Rh-catalyzed, π-allyl-mediated mechanism (see **10**; Scheme 18.4), which had a substantially lower overall barrier (16.6 kcal/mol) compared to a [2,3]-sigmatropic rearrangement pathway (23.5 kcal/mol). However, under nonpolar conditions, the sulfoxide racemization occurred at least in part via the Mislow–Evans rearrangement. With a mechanistic working model, the reaction conditions were adjusted to obtain optically active *n*-propyl aryl sulfoxide **12** in high enantiomeric excess starting from racemic **11**, while minimizing formation of the corresponding alkene-reduced sulfenate ester **13** (Scheme 18.5).

18.3 DEUTERIUM LABELING TO TRACK [2,3] PATHWAY

As in Dong's present-day investigations, early deuterium-labeling studies allowed Mislow's group to verify the allylic transposition nature of the sulfenate-to-sulfoxide rearrangement.[3a] For example, in one of several isotope-labeling experiments, reaction of α,α-dideuterated allylic alkoxide **14** with phenylsulfenyl chloride led to allylic sulfoxide **16**, dideuterated at the terminal vinyl positions (Scheme 18.6). This outcome reflected a [2,3]-rearrangement mechanism rather than a radical fragmentation–recombination route.

18.4 TRANSITION STATE FEATURES

18.4.1 Transition State Structure

Stereochemical analysis is a powerful probe for building predictive models of transition state geometry in sigmatropic rearrangement reactions. In a Mislow–Evans rearrangement such as that shown in Scheme 18.7, product stereochemistry can depend on the stereoarrangement at the sulfur of the allylic sulfoxide as well as at the allylic C-stereocenter. In studies by Evans, high selectivity for the *E*-alkene product (*E*)-**18** was observed starting from a diastereomeric mixture of sulfoxides.[9] Jorgensen later investigated this experimental result computationally, finding that for one of the

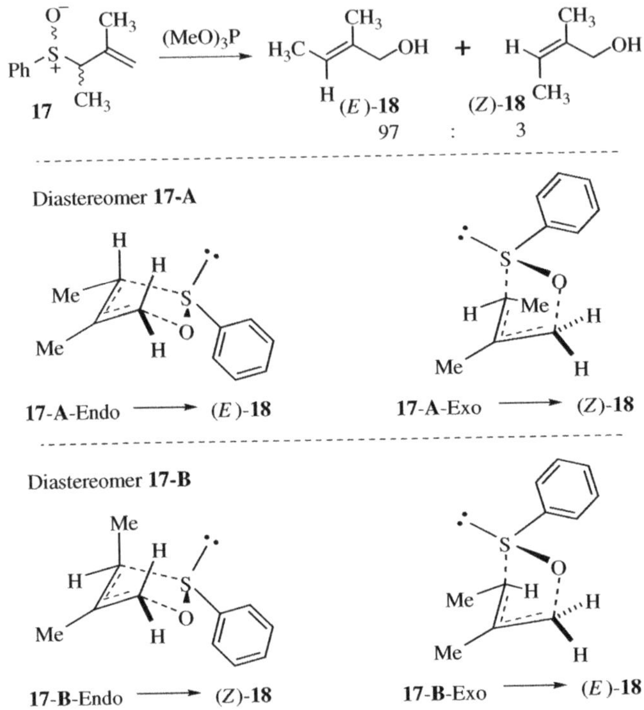

Scheme 18.7 Origins of product stereochemical features in Mislow–Evans rearrangements.[10,12]

starting diastereomers, the endo transition state (phenyl group oriented toward the allyl fragment) was energetically favorable, while the low-energy reaction pathway for the other diastereomer proceeded via the exo arrangement (phenyl group oriented away from the allyl fragment).[10] In both of the preferred pathways, the allylic methyl group adopts a pseudo-equatorial position in the 5-membered ring transition structure (i.e., **17-A**-endo and **17-B**-exo).[11] In this way, the diastereomeric sulfoxide starting materials converged on the *E* product in the presence of the thiophile trimethylphosphite.

A closely related stereochemical feature of the allylic sulfoxide-to-sulfenate rearrangement involves sulfur-to-carbon (S → C) chirality transfer. Trapping of the sulfenate with a thiophile has the potential to relay stereochemical information from the sulfoxide chiral center to the carbinol center of the allylic alcohol product. This stereochemical transfer originates from a preference for either the exo or the endo transition state, with the latter typically predominating.[12] Thus, as shown in Scheme 18.8, for a fixed configuration at the sulfoxide S-center and the *E*-alkene geometry, the exo and endo transition states provide enantiomeric products upon migration of the S—O moiety to the prochiral terminus of allylic sulfoxide **19**.

Finally, calculations show that the transition state for the Mislow–Evans rearrangement is sulfenate-like. This was consistent with the fact that the reaction rate

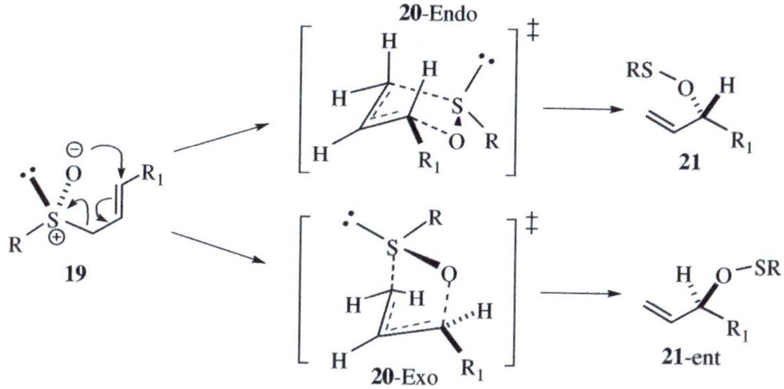

Scheme 18.8 Transfer of S → C chirality determined by endo versus exo transition states.

in the sulfoxide-to-sulfenate direction slowed in solvents of increasing polarity, presumably because a polar solvent stabilizes the sulfoxide starting material more than the sulfenate-like transition state.[3b] Indeed, Jorgensen's calculations indicated that the sulfoxide forms a stronger hydrogen bond to water (~10 kcal/mol) than does the sulfenate (~6 kcal/mol). Nevertheless, in a calculation that included methanol solvation, the Mislow–Evans transition state maintained a sulfoxide-like hydrogen bond to methanol.[12]

18.4.2 Antibody Catalysis

Insights into the nature of the Mislow–Evans rearrangement transition state helped enable the production of catalytic antibodies for this [2,3]-sigmatropic rearrangement (Scheme 18.9).[13] These antibody catalysts could distinguish allyl sulfoxide enantiomers, enabling antibody-catalyzed S → C chirality transfer in some cases. Additionally, diastereomeric substrates also reacted at significantly different rates in the presence of the antibodies. The cyclic five-membered transition structure for the Mislow–Evans rearrangement was mimicked by the cis- and trans-pyrrolidines **22** (cis shown), motifs that the Hilvert group had previously used as haptens in developing antibody catalysts for selenoxide elimination reactions.

For the Mislow–Evans rearrangement study, the haptens were coupled to BSA and thyroglobulin for eliciting immune response, leading to isolation of the respective antibodies. Three antibodies were found that accelerated the Mislow–Evans rearrangement of substrates such as **23** and **25** by factors of 100–1000. Enantiomer specificity in the use of racemic substrate **23** was observed using one of the antibodies, trans-28F8, while another of the antibodies could process either sulfoxide enantiomer. The same antibody, trans-28F8, also distinguished diastereomeric substrates, as only one pair of enantiomers of **25** underwent the catalyzed Mislow–Evans rearrangement, producing the *E*-allylic alcohol product **26**. Additionally, one of the enantiomers of the reactive diastereomer of **25** rearranged 10 times faster than the other enantiomer at low substrate concentration. Control experiments showed that BSA alone also

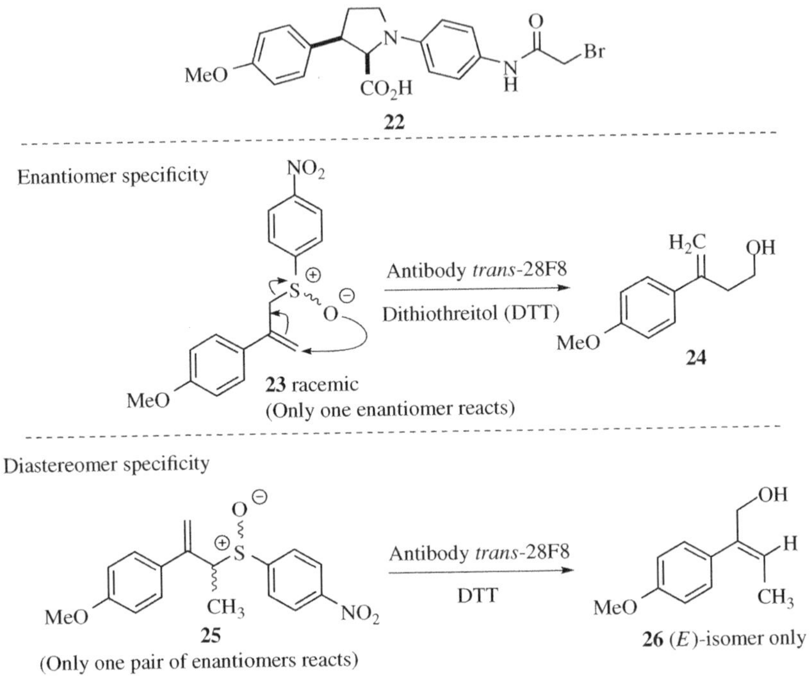

Scheme 18.9 Antibody-catalyzed Mislow–Evans rearrangement.[13]

accelerated the Mislow–Evans rearrangement but without enantiomer or diastereomer specificity. These results suggested that antibody catalysis originated in part from a nonspecific "medium effect," perhaps reflecting the polarity properties of the binding pocket, but also by favoring the reactive conformation required for reaching the sulfenate-like [2,3]-rearrangement transition state.

18.5 EQUILIBRIUM BETWEEN SULFOXIDE AND SULFENATE

18.5.1 Position of Equilibrium

Kinetically, the Mislow–Evans rearrangement is readily reversible under relatively mild conditions, but, at equilibrium, many allyl sulfenates are effectively completely converted to the rearranged allyl sulfoxides. For example, the allyl trichloromethyl sulfenate **2a** investigated in Mislow's[2,3a] and Braverman's[4] early studies equilibrates to a large preponderance of the allylic sulfoxide **3a** (Scheme 18.10). Likewise, Mislow found that allyl *p*-tolyl sulfenate, crotyl *p*-tolyl sulfenate, and methallyl *p*-tolyl sulfenate all gave >95% of the corresponding sulfoxides (structures not shown). An important factor is the bond stabilization of the sulfoxide group (approx. 90 kcal/mol), which favors the sulfoxide side of the equilibrium. (This may be contrasted with, for instance, the [2,3]-Meisenheimer rearrangement in which an allylic amine *N*-oxide rearranges to the thermodynamically preferred *N*-allyloxyamine.)

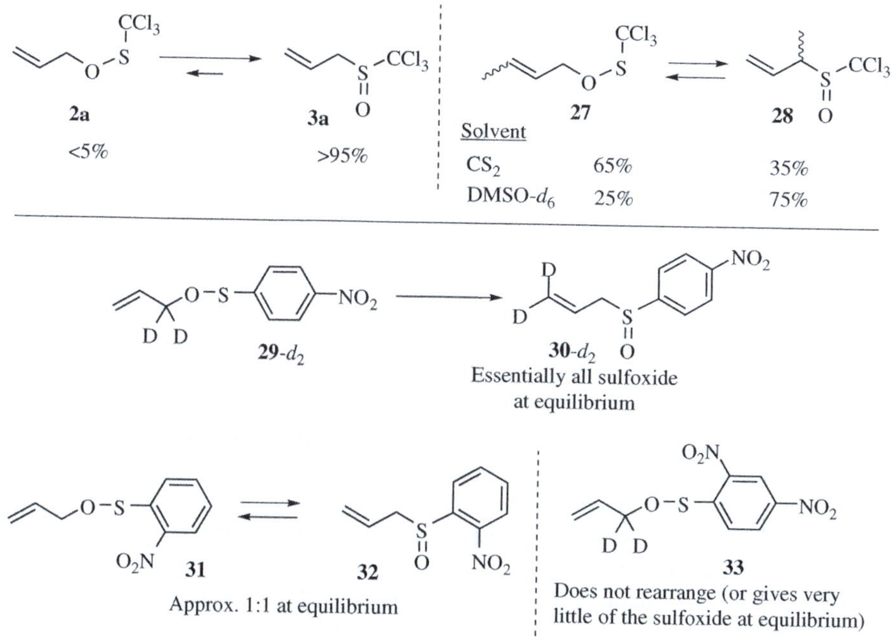

Scheme 18.10 Position of equilibrium is solvent and structure dependent.[3,4]

However, perturbations in structure can have a substantial effect on the equilibrium. For example, both crotyl trichloromethyl sulfenate **27** and methallyl trichloromethyl sulfoxide **28** are present in appreciable quantities at equilibrium. Additionally, solvents of increasing polarity favor the sulfoxide. More subtle electronic and steric effects may also be at play. Intriguingly, the *p*-nitro derivative **29** rearranges completely to sulfoxide **30**, while the ortho isomer **31** provides a balanced distribution of the sulfenate and sulfoxide. But with both *o*- and *p*-nitro groups, sulfenate **33** either does not rearrange at all or gives an undetectable amount of the sulfoxide at equilibrium.[3a]

Mislow and coworkers probed the origins of the equilibrium distribution by measuring activation parameters for the sulfenate ester-to-sulfoxide conversion.[3b] For example, using allyl *p*-nitrophenyl sulfenate **29** (Scheme 18.11) and monitoring conversion to the corresponding allyl sulfoxide by UV–visible spectroscopy, rate measurements over a range of temperatures gave $\Delta H^{\ddagger} = 17.6$ kcal/mol and $\Delta S^{\ddagger} = -9.6$ eu. Additionally, for a different substrate **34**, the activation parameters for the conversion to sulfoxide were determined ($\Delta H^{\ddagger} = 18.8$ kcal/mol and $\Delta S^{\ddagger} = -4.8$ eu) and combined with the activation parameters for the reverse reaction (sulfoxide to sulfenate), determined from racemization rate studies. From the difference in the barriers between the forward and reverse reactions, $\Delta G°$ was determined as +2.9 kcal/mol at 25 °C for the sulfoxide-to-sulfenate direction. In other words, the sulfoxide **35** was heavily favored (>99%) at equilibrium.

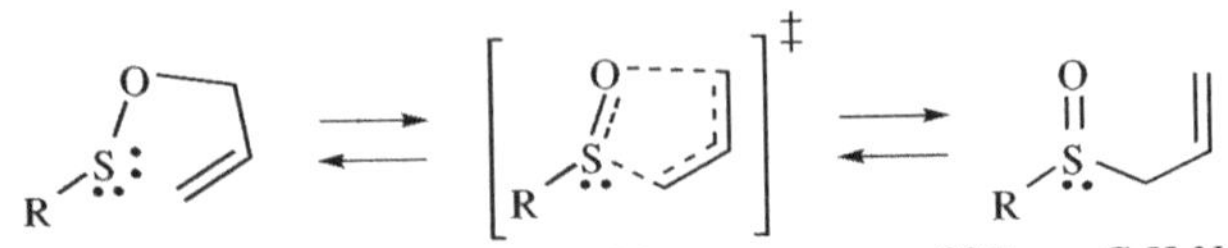

29 R = *p*-$C_6H_4NO_2$
34 R = *p*-$C_6H_4CF_3$

36

30 R = *p*-$C_6H_4NO_2$
35 R = *p*-$C_6H_4CF_3$

Scheme 18.11 Probing the [2,3]-sigmatropic rearrangement transition state.[3b]

As suggested by the behavior of sulfenates **31** and **33** (Scheme 18.10), electron-withdrawing substituents on the S-aryl group can shift the equilibrium toward the sulfenate ester. This is apparently an inductive effect (cf. **29**-d_2) due to destabilization of the sulfoxide ground state. A corresponding sulfenate-favoring effect was observed for the pentafluorophenyl system (not shown). Mislow also noted that the rate of rearrangement also increases with the electron-deficient S-aryl groups. Since the S=O dipole diminishes in going to the transition state for the [2,3]-sigmatropic rearrangement, the rate enhancement is apparently due to raising of the sulfoxide ground-state energy, while the transition state is relatively insensitive to electronic perturbation.[3b]

More recently, calculations on a variety of [2,3]-rearrangement transition states, including that for the Mislow–Evans rearrangement, highlighted the electron-delocalized aromatic character of the transition states (e.g., **36**; Scheme 18.11), befitting concerted processes. The magnetic character of the transition states was also computed and found to be consistent with their having an aromatic nature.[14]

18.5.2 Conversion of Allylic Alcohols to Allylic Sulfoxides

Because allylic alcohols are readily transformed into their sulfenate esters, conversion to the corresponding allylic sulfoxide is possible and usually occurs spontaneously under particularly mild conditions (**37** → **38** → **39**; Scheme 18.12, top). These interconversions, succinctly promulgated in Evans's 1971 study,[5a] form one basis for the panoply of synthetically valuable, creative, and efficient applications discussed in this chapter.

18.5.3 Thiophilic Capture of the Sulfenate Ester

Given the typical thermodynamic preference for allylic sulfoxides upon equilibration with sulfenate esters via [2,3]-rearrangement, synthetic utility in the sulfoxide-to-sulfenate direction requires trapping of the sulfenate by including a thiophilic nucleophile in the reaction mixture; thiolates, secondary amines, phosphines, and phosphites are particularly useful in this regard (Scheme 18.12, bottom). This thiophilic capture strategy was suggested by early work from Abbott and Stirling[15] and others,[16] and its synthetic potential was elevated by Evans's group, beginning in 1971.[5a] Thiophilic interception of the sulfenate leads to an overall transformation in which an allylic sulfoxide is parlayed into a 1,3-transposed allylic alcohol (e.g., **40** → **41** → **42**).

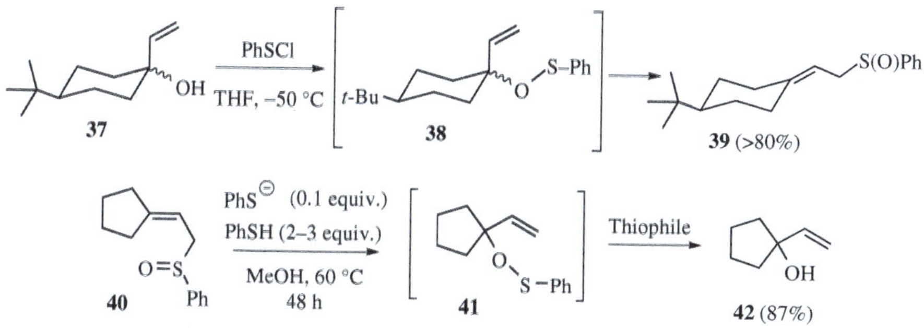

Scheme 18.12 Allylic alcohol → sulfoxide and allylic sulfoxide → alcohol transformations.[5a]

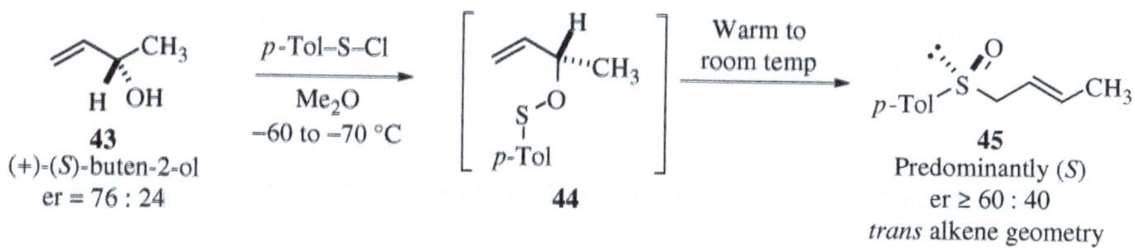

Scheme 18.13 Carbon → sulfur chirality transfer via [2,3] rearrangement.[3a]

18.6 CHIRALITY TRANSFER

18.6.1 Carbon to Sulfur

With a preexisting carbinol stereocenter, transfer of chirality to the resulting allylic sulfoxide is possible. Early studies by Mislow indicated that the sulfenate ester derived from optically enriched (+)-(*S*)-buten-2-ol (**43**) rearranges at low temperature with appreciable stereochemical fidelity to the (*S*)-configured sulfoxide product **45** (Scheme 18.13). Here, one of the diastereotopic lone pairs of the sulfenate **44** engages selectively in the [2,3]-rearrangement. An additional stereochemical feature of the rearrangement is the predominant formation of the trans-alkene product. Highlighting the stereochemical lability of allyl sulfoxides, the enantiomerically enriched sulfoxide **45** racemized with a half-life of 51 minutes at 29 °C and equilibrated to a 77:23 ratio of trans–cis olefin isomers.[3a]

18.6.2 Sulfur to Carbon

Starting with single enantiomers of vinyl sulfoxides such as **46** (Scheme 18.14), Miura and coworkers carried out an isomerization–rearrangement sequence that transferred the chirality at sulfur to the resulting allylic alcohol **47** with high fidelity.[17] With DBU as the base, *in situ* isomerization of the vinyl sulfoxide to the allylic sulfoxide enabled Mislow–Evans rearrangement. Triphenylphosphine presumably acted as the thiophilic nucleophile, intercepting the allyl sulfenate ester intermediate,

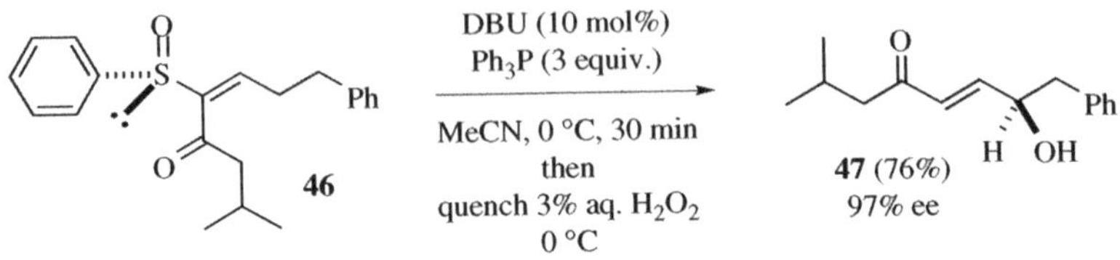

Scheme 18.14 Use of single enantiomer of vinyl sulfoxide with isomerization and S→C chirality transfer.[17]

providing the corresponding chiral γ-hydroxy-α-enones with excellent enantiomeric enrichment and exclusive trans-alkene geometry. An aqueous oxidative workup improved yields, perhaps by oxidizing remaining sulfenate ester to better free the allylic alcohol product or by oxidizing sulfur-containing by-products that can interfere with product isolation. Twelve examples using the α-sulfinylenone substrates proceeded in 56–97% yield and with 58–99% *ee*.

PART 2. SYNTHETIC CONSIDERATIONS AND APPLICATIONS

18.7 ALKENE STEREOSELECTIVITY

Importantly for synthetic planning using the Mislow–Evans rearrangement within acyclic frameworks, in products containing a stereogenic alkene, the allylic alcohol group reliably ends up trans to the more sterically demanding substituent at the other end of the double bond. This is particularly valuable in a sequence originally explored by Evans, whereby an allylic sulfoxide (e.g., **48**, Scheme 18.15) is α-alkylated, followed by [2,3]-rearrangement in the presence of a thiophile.[5b] For disubstituted-alkene allylic alcohol products such as **49**, the trans-alkene geometry strongly predominates (typically >98% trans). Stereoselective preparation of trisubstituted alkenes is also possible via this approach, starting either from β-vinyl-substituted allylic sulfoxides (e.g., **50** and via **51**) or by using a doubly α-alkylated substrate (e.g., **54**). Stereocontrol was excellent for *E* products (*E*)-**52**, while for rearrangement of sulfoxides of type **54**, better *E* selectivity resulted with a larger alkyl group at the α-position of the allylic sulfoxide. Low yields for certain products **52** and **55** were due to competing γ-alkylation of the allylic anions. An additional limitation was that the allylic anions, stabilized by the sulfoxide group, necessitated reactive allylic or primary alkylating agents.

Enhancing the utility of the alkylation–rearrangement sequence in synthesis, the Evans group addressed the problems of α-versus-γ alkylation as well as low anion reactivity by employing heterocyclic sulfides as the alkylation substrates (Scheme 18.16). For instance, allylic imidazolyl sulfide **56** could be alkylated efficiently, reaction at the α-position being favored by a chelated but reactive allyl lithium intermediate. Oxidation of **57** to the allylic sulfoxide and treatment with a secondary amine thiophile provided allylic alcohol **58** in high yield and

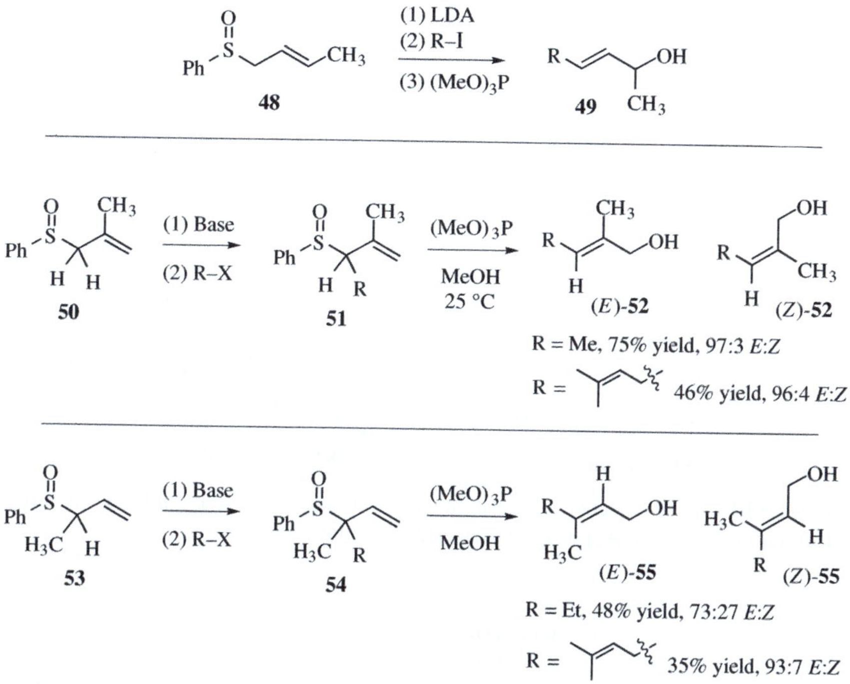

Scheme 18.15 In acyclic cases, allylic alcohol products are predominantly *E*-configured.[5b]

with excellent stereoselectivity at the trisubstituted alkene. Allylic oxidation with manganese dioxide completed a synthesis of the sesquiterpene nuciferal (**59**).[5b]

Control of alkene geometry was also an important factor in Otera's development of the [2,3]-rearrangement of γ-alkoxy allylic sulfoxides to give α,β-unsaturated aldehyde products.[18] This method was applicable in complex-molecule settings, including for prostaglandin synthesis (Scheme 18.17). For example, Michael-type reaction of the allylic anion derived from **60** to enone **61** gave carbon–carbon bond formation α to the thioether, and after transmetallation to a tin enolate, alkylation provided the highly functionalized allylic sulfide **62** for subsequent Mislow–Evans rearrangement. Oxidation to the sulfoxide with periodate under aqueous conditions led directly to α,β-unsaturated aldehyde **63**, exclusively as the trans-alkene isomer. Here, the sulfenyl product from the [2,3]-rearrangement of the allylic sulfoxide was part of a mixed acetal, which was hydrolyzed directly under the reaction conditions, so addition of a thiophile was not required to drive the Mislow–Evans rearrangement to completion.

Additionally, as elucidated through other investigations by the Otera group, the steric environment of the sulfoxide moiety had an impact on the *E* selectivity of these rearrangements (Scheme 18.18).[19] For the general case **64** to **65**, Otera found that high *E* selectivity occurred when R^1 was significantly larger than R^2. Formation of disubstituted α,β-unsaturated aldehydes such as **67** was therefore highly trans-selective,

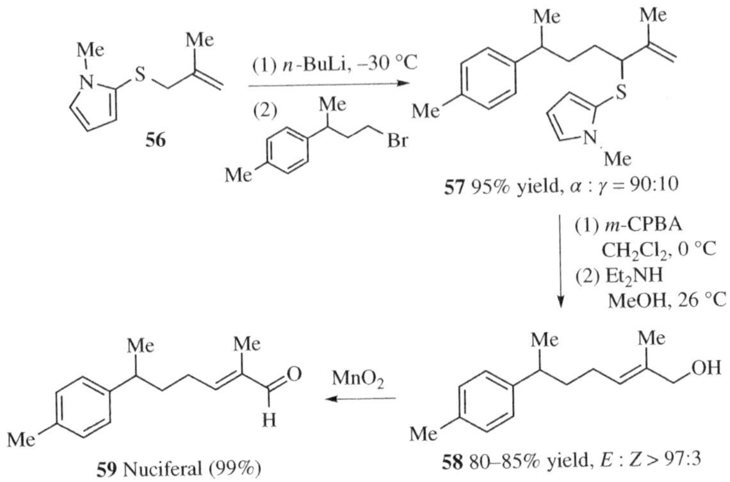

Scheme 18.16 Regioslective alkylation of heterocyclic allyl sulfides streamlines Mislow–Evans approach. Application to nuciferal synthesis.[5b]

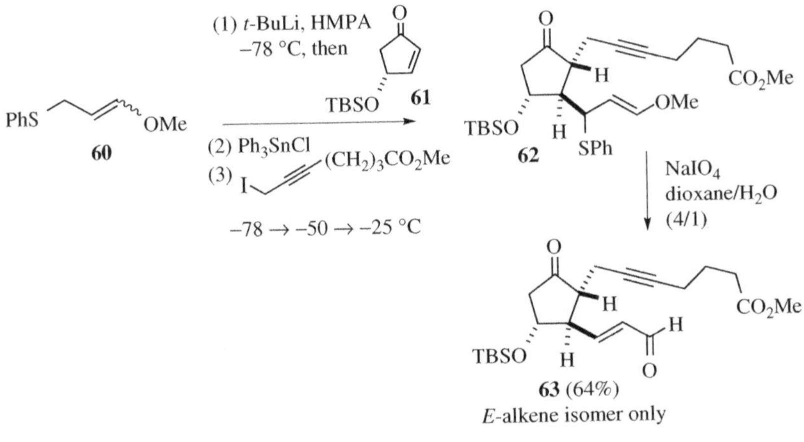

Scheme 18.17 Rearrangement of γ-alkoxy allylic sulfoxides to α,β-unsaturated aldehydes as applied to prostaglandin synthesis.[18b]

while in comparing **68** and **70**, the higher degree of branching β to the sulfur provided enhanced E selectivity in the formation of trisubstituted alkenes **69** and **71**, respectively. This high trisubstituted alkene stereoselectivity also occurred with transformations of non-vinyl ether substrates such as **72** → **73**. The conclusion was that β branching at the allylic sulfoxide group increased E selectivity in the Mislow–Evans rearrangement. Subsequently, the Otera group also discovered the longer-range effect of a hydroxyl γ to the sulfinyl group for engendering E selectivity in the rearrangement (not shown).[20]

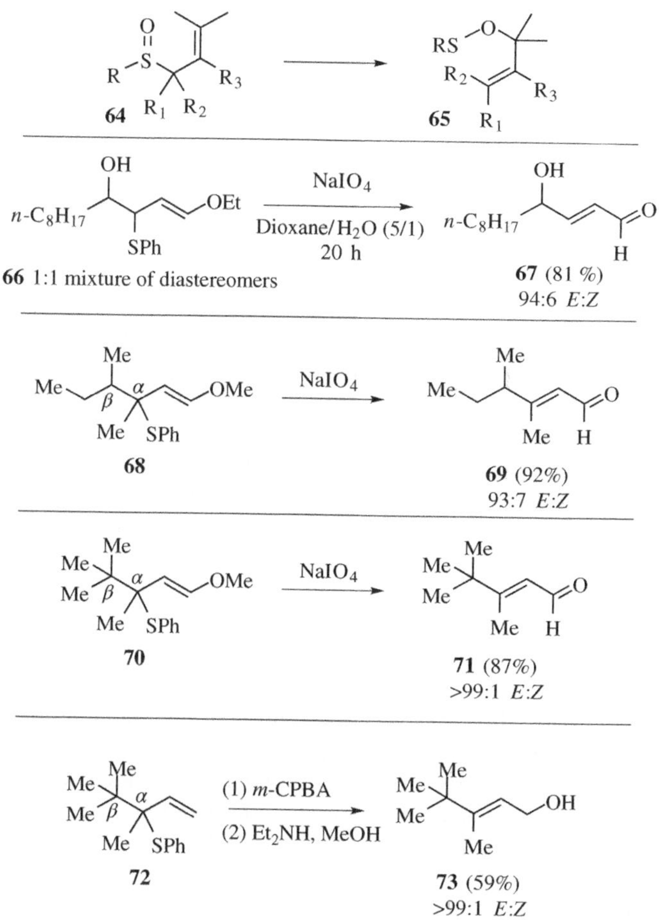

Scheme 18.18 *β*-Branching of allylic sulfoxide substituent helps impart high *E* selectivity.[19]

18.8 DIASTEREOFACE SELECTIVITY IN THE REARRANGEMENT

18.8.1 Diastereoselective Applications

In general, diastereoface selectivity in the allylic sulfoxide-to-sulfenate Mislow–Evans rearrangement is possible in cases such as **74** (Scheme 18.19), having the sulfoxide attached to a chiral center, or with substrates of type **75**, where the alkene faces are rendered diastereotopic by a structural element distal from the allyl sulfoxide. These diastereoselective [2,3]-sigmatropic rearrangements arise within acyclic as well as cyclic contexts. There are also cases in which both of these motifs **74** and **75** are present within one Mislow–Evans substrate. Additionally, the diastereo relationship between the sulfoxide sulfur center and other stereocenters in the molecule can influence the outcome of the reaction (discussed further in Section 18.8.3).

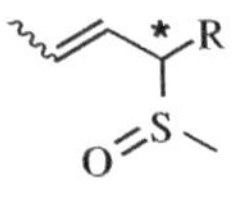

74

For examples, see:
Scheme 18.20 (Danishefsky)
Scheme 18.27 (Myers)
Scheme 18.31 (Smith)

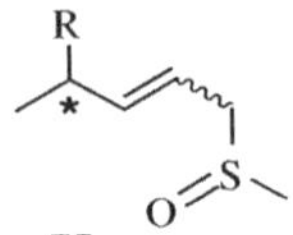

75

For examples, see:
Scheme 18.29 (Ohfune, Shinada)
Scheme 18.33 (Majetich)
Scheme 18.48 (Chida)

Effect of stereochemistry at sulfur?

Scheme 18.19 Structural scenarios for alkene diastereoface selectivity in allylic sulfoxide → sulfenate rearrangement.

The stereofacial reliability of the Mislow–Evans rearrangement within cyclic systems enabled Danishefsky's development of a synthetic route to glycals bearing the relatively unusual allal (**76**) and gulal (**77**) configurations (Scheme 18.20), both of which have pseudo-axial C-3—OH groups.[21] These investigations were prompted by the need to prepare the thiomethyl-bearing sugar **78**, a subunit of the trisaccharide from the DNA-cleaving agent esperamicin. Ferrier rearrangement of tri-*O*-acetyl-D-glucal (**79**) gave predominantly the *α*-thiopseudoglycal **80**. The stereochemical identity of this intermediate was crucial to the subsequent [2,3]-rearrangement, as oxidation to sulfoxide **81**, suprafacial migration, and interception of the sulfenate with piperidine led to **82**, wherein the 4*O*-acetyl group had transferred to the 3*O* position. Danishefsky considered an alternative mechanism in which the 4*O*-acetyl adds in intramolecular S_N2' fashion to the allylic sulfoxide in **81**. However, without addition of the thiophile, **82** did not form, supporting the [2,3]-sigmatropic rearrangement pathway. Moreover, starting from the corresponding galactal precursor **83**, the gulal **85** resulted, without any acetyl transfer. This overall sequence, including the transposition of stereochemistry from C-1 in sulfoxides **81** and **84** to C-3 in **82** and **85**, offers a succinct approach to site-selective inversion of the C-3 stereocenter in readily available glucal and galactal precursors, providing access to allal and gulal systems.

In applying this approach to the esperamicin sugar synthesis, modification of the Ferrier rearrangement product **84** provided 1,4-bis-thio pseudoglycal **86** (Scheme 18.21) for oxidation and rearrangement. Double oxidation to the bis-allylic sulfoxide **87** led to regioselective Mislow–Evans rearrangement involving the C-1 sulfoxide. Preferential reaction of this sulfoxide was due to its the axial disposition, better positioning it for rearrangement, and a weakening of the C-1—S bond from the anomeric effect, leading to alcohol **88**. Reduction of the remaining sulfoxide to the thiomethyl group gave **89**, which could be advanced to methyl glycoside **78**. The sugar unit **78** prepared via Danishefsky's route, starting from a D-galactal precursor, was identical to material from degradation of esperamicin, conclusively confirming the absolute configuration of that portion of the natural product.[21]

An approach avoiding oxidation-state adjustments at the C-4 thioether also utilized a Mislow–Evans rearrangement. Selective oxidation of the C-1 thiophenyl ether in bis-allylic sulfide **90** (Scheme 18.22) occurred with the electron-withdrawing 2,4-dinitrophenyl group at the 4*S* position. Sigmatropic rearrangement and sulfenate trapping with a secondary amine provided **91** as a saccharide building block in

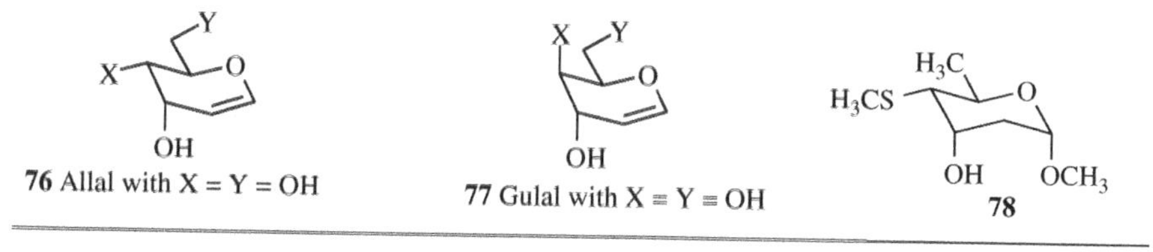

76 Allal with X = Y = OH

77 Gulal with X = Y = OH

78

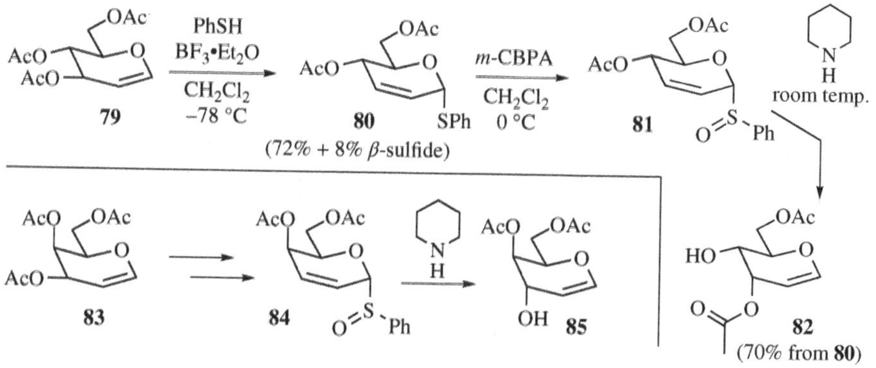

Scheme 18.20 Stereochemical transposition at C-3 of glycal substrates via Mislow–Evans rearrangement.[21]

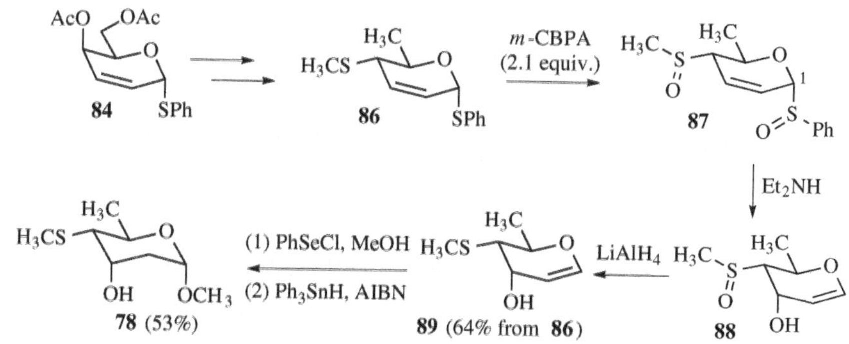

Scheme 18.21 Regioselective [2,3] rearrangement of bis-allylic sulfoxide 87 in synthesis of esperamicin sugar.[21]

Danishefsky's successful total synthesis efforts toward the calicheamicin–esperamicin class of antitumor enediynes.[22]

18.8.2 Chemoselective Oxidation of Allylic Sulfides to Sulfoxides

As highlighted in Scheme 18.22, when starting with allylic sulfides, application of the Mislow–Evans rearrangement in synthesis requires chemoselective oxidation to the sulfoxide. Selective oxidation is necessary not only when additional thioether functionality is present as in **90** but also because of potential oxidation of the olefin of

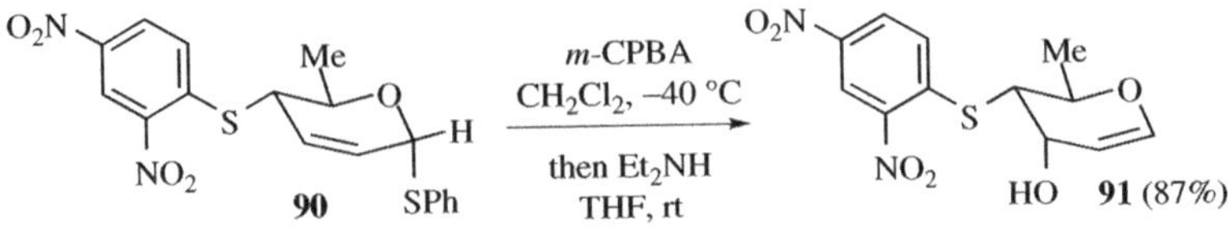

Scheme 18.22 Selective oxidation of bisallylic sulfide in synthesis of the calicheamicin saccharide.[22]

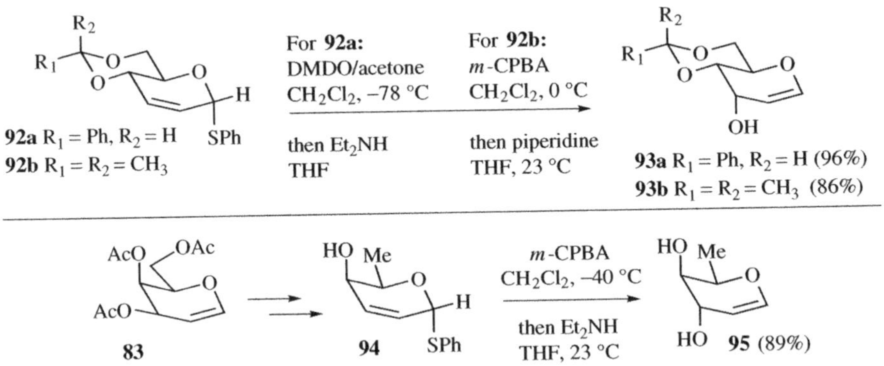

Scheme 18.23 Oxidant for pseudothioglycal oxidation varies with protecting groups (Danishefsky, 1991; Rojas, 2001) and application of Mislow–Evans rearrangement with a galactal-derived 4-hydroxy-6-deoxy derivative.[25]

the allylic thioether and any other oxidation-prone functionality within the molecule. In this regard, dimethyldioxirane (DMDO) is a valuable alternative to peroxyacids (typically *m*-CPBA) for the sulfide-to-sulfoxide conversion. Despite being more experimentally demanding (it must be freshly prepared by oxidation of acetone and isolated by codistillation), DMDO can be used at low temperature and enables oxidation–Mislow–Evans routes when *m*-CPBA provides unsatisfactory results. For example, benzylidene-protected **92a** (Scheme 18.23) required DMDO for the efficient production of allal derivative **93a**, an intermediate in Danishefsky's total synthesis of the chitinase inhibitor allosamidin.[23] However, our own group determined that the corresponding acetonide **92b** provided a high yield of **93b** using the more traditional peroxyacid oxidant.[24] Evidently, the activated benzylic site of the protecting group in **92a** is susceptible to oxidation under the *m*-CPBA conditions. The free secondary C-4—OH group of **94**, meanwhile, was compatible with peroxyacid oxidation, followed by Mislow–Evans rearrangement and trapping with diethylamine providing **95** in Gin's total synthesis of (+)-pyrenolide D.[25]

18.8.3 Influence of Sulfoxide Stereochemistry at Sulfur

Sulfide oxidation generally produces a mixture of diastereomers at sulfur within frameworks such as **92** and **94** (Scheme 18.23), bearing other stereocenters. In

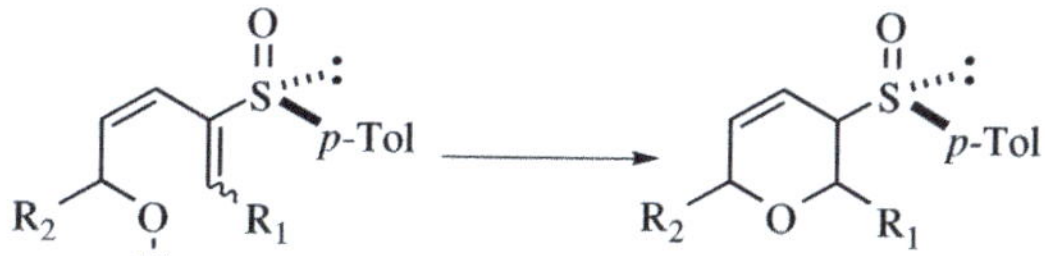

Scheme 18.24 Preparation of diastereomeric dihydropyran sulfoxides for rearrangement studies.[26]

most cases, either sulfoxide is competent to undergo the [2,3]-rearrangement. In our group's studies with oxidation of thiopseudoglycals of type **92**, the diastereomeric sulfoxides were sometimes distinguishable by thin-layer chromatographic analysis, both converting to the 1,3-transposed allylic alcohol products upon treatment with the thiophile (usually piperidine or diethylamine).[24] Examples where the Mislow–Evans reactivity of the diastereomeric sulfoxides is different enough to have a synthetically relevant impact are relatively rare and are highlighted in the following work of Fernández de la Pradilla (Schemes 18.25 and 18.26) and Myers (Scheme 18.27).

Fernández de la Pradilla *et al.* studied Mislow–Evans rearrangements of allylic sulfoxides **97**, within dihydropyran frameworks established by cyclization of hydroxysulfinyl dienes **96** (Scheme 18.24).[26] Stereocontrol in the cyclization to **97** depended on the reaction conditions and the alkene geometry of the starting dienes **96**, enabling preparation of different diastereomers for further study. Importantly, the allylic sulfoxides **97** were prepared as single epimers at the sulfur stereocenter and were configurationally stable to isolation and purification at ambient temperature. The effect of the configuration at sulfur in Mislow–Evans reactions can be difficult to study because of the possibility for epimerization of the S-center in allylic sulfoxides via reversible [2,3]-rearrangement. However, in this study, the barrier to rearrangement was high enough to identify the impact of the S-stereochemistry.

Comparison of the [2,3]-sigmatropic rearrangements of **97a** and **97a-dst**, having the same relative stereochemistry of the ring substituents but differing in the relationship of those two stereoelements to the sulfoxide center, indicated that the configuration at sulfur relative to the rest of the molecule had a marked impact on the efficiency of the allylic sulfoxide to allylic alcohol transposition (Scheme 18.25). Fernández de la Pradilla proposed that the better rearrangement substrate maintained a minimum of nonbonded interactions while properly orienting the sulfoxide group in a pseudo-axial position for interaction with the distal end of the alkene π system (cf. Danishefsky's results in Scheme 18.21).

Importantly, while a favored ground-state conformation may have the sulfoxide group in a pseudo-equatorial position, analysis of the reactive conformation with the pseudo-axial sulfoxide is required to rationalize the observed reactivity differences. So, although ^{1}H NMR measurements revealed that **97a** favored the conformation with the pseudo-equatorial sulfoxide group (**97a-eq**), in the other half-chair arrangement (**97a-ax**), the geometry for [2,3]-rearrangement is well accommodated while orienting the *S*-*p*-tolyl group away from the ring. Meanwhile, despite having a

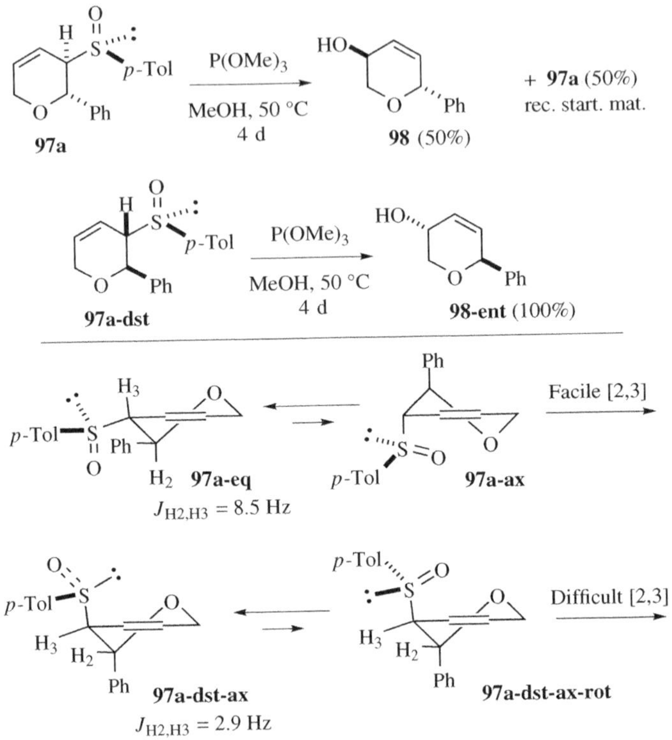

Scheme 18.25 [2,3]-Rearrangement efficacy depends on stereo relationship between sulfoxide and other stereocenters.[26]

preferred ground-state arrangement with the pseudo-axial sulfoxide, **97a-dst** requires an unfavorable C—S rotamer (**97a-dst-ax-rot**) with placement of the *S*-*p*-tolyl moiety toward the ring for the rearrangement to occur.

Similarly, **97b** and **97b-dst**, enantiomeric but for the fixed sulfoxide stereocenter, gave strikingly contrasting results (Scheme 18.26). The [2,3]-sigmatropic rearrangement of **97b**, favorably accommodated via conformation **97b-ax** (pseudo-axial sulfoxide and the less sterically demanding lone pair pointed toward the ring), occurred under the conditions shown in Scheme 18.26, giving allylic alcohol **99** in reasonable yield. On the other hand, **97b-dst** would rearrange via **97b-dst-ax** (pseudo-axial sulfoxide, but requiring the *p*-tolyl group toward the ring), an arrangement apparently not attainable under these same conditions.

Myers reported a case in which the sulfoxide stereochemistry needed to be matched with respect to other stereocenters for effective Mislow–Evans rearrangement (Scheme 18.27).[27] In developing a concise and modular synthetic route to tetracycline antibiotic analogues, Myers and coworkers found that only one of the two sulfoxide diastereomers **103** reacted smoothly in the presence of trimethylphosphite to give the desired allylic alcohol **104**. While the configuration at sulfur of **103** was

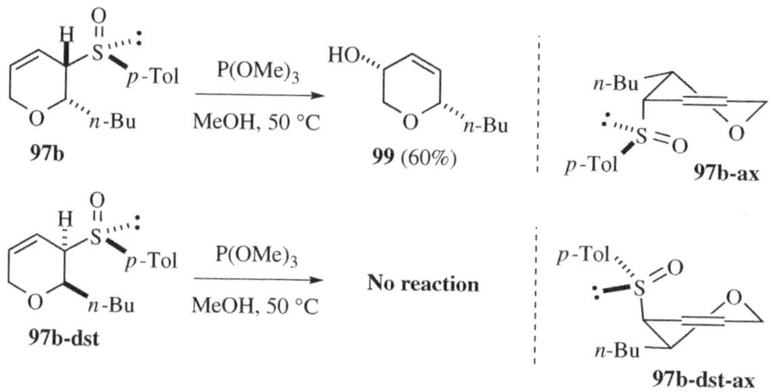

Scheme 18.26 Another example highlights importance of conformational features for Mislow–Evans rearrangement.[26]

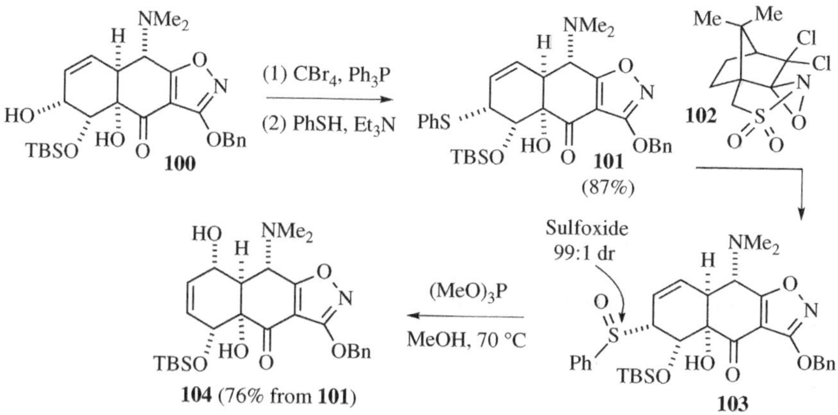

Scheme 18.27 Only one sulfoxide diastereomer undergoes Mislow–Evans rearrangement in modular approach to tetracycline analogue synthesis.[27]

not determined, oxidation of allylic sulfide **101** with Davis's chiral oxaziridine **102**[28] provided a very high excess of the diastereomer necessary for [2,3]-sigmatropic rearrangement. While in many cases within chiral frameworks, including in cyclic systems, either allyl sulfoxide diastereomer can lead to successful rearrangement (see the Danishefsky, Rojas, and Gin examples of Schemes 18.20–18.23 and also further examples later), there are situations in which only one of the diastereomers is an optimal or even competent [2,3] substrate. In these cases, as Myers's work demonstrates, use of a chiral oxidant for diastereoselective sulfide-to-sulfoxide conversion can rescue an otherwise untenable synthetic approach.

Geometrical constraints, however, can be insurmountable in attempts to apply the Mislow–Evans rearrangement for 1,3-allylic transposition. For example, during investigations toward a versatile synthesis of cortistatin A and congeners, the Baran

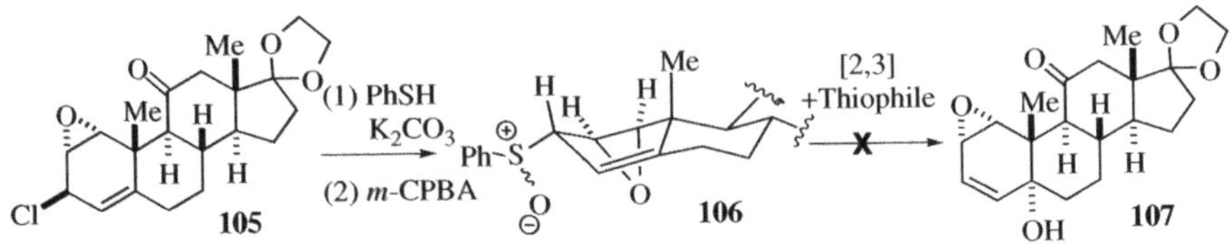

Scheme 18.28 Geometrical constraints preclude Mislow–Evans rearrangement.[29]

group planned to install an angular C-5 hydroxyl via [2,3]-sigmatropic rearrangement of allyl sulfoxide **106** (Scheme 18.28).[29] While this Mislow–Evans precursor was readily prepared by thiolate displacement of chloride **105** followed by peroxyacid S-oxidation, subsequent rearrangement did not take place. The authors attributed this lack of reactivity to the [2,3]-unfavorable pseudo-equatorial position of the sulfoxide group within a conformationally rigid scaffold. Interestingly, ^{1}H NMR data for **106** indicated that it was isolated mainly as one diastereomer (>5:1 dr);[29] perhaps this was the rearrangement-mismatched sulfoxide, which, in combination with a preferred pseudo-equatorial orientation, doomed the hoped-for Mislow–Evans reaction.

18.8.4 Additional Diastereoselective Examples

Anchoring the allyl sulfoxide at an exocyclic carbon, with the alkene situated within a stereochemically defined ring, provides the opportunity for diastereoselective installation of a hydroxyl group on the ring, along with an exocyclic olefin that serves as a handle for further elaboration. Ohfune and Shinada applied this Mislow–Evans-based strategy effectively in establishing the stereochemically dense core of two trideoxytetrodotoxin derivatives (Scheme 18.29).[30] Site-selective conversion of the primary alcohol group of diol **108** to the corresponding phenylthioether followed by oxidation at sulfur gave **109** as an equimolar mixture of sulfoxide diastereomers. Sigmatropic rearrangement in the presence of the thiophile triethylphosphite provided efficient conversion to **110** with high α-face stereoselectivity, attributed to β-face blocking by the axial carbomethoxy group. The free hydroxyl group of **109** did not interfere with the Mislow–Evans reaction. En route to the deoxytetrodotoxin analogues, the exocyclic olefin of **110** underwent subsequent stereoselective hydrogenation to a β-methyl group (not shown).

Shinada *et al.* used a similar Mislow–Evans approach in their synthesis of the gabosines, DNA-binding carbasugars containing extensively oxidized cyclohexane cores.[31] Sulfide **111** (Scheme 18.30), which was available in enantiomerically pure form starting from (−)-quinic acid, gave a mixture of sulfoxide diastereomers **112** upon peroxyacid S-oxidation. When heated in ethanol with triethylphosphite, both diastereomers of **112** underwent smooth conversion to a single stereoisomer of allylic alcohol **113**, bearing an exocyclic olefin. The diastereoselective sigmatropic rearrangement occurred on the face of cyclohexene **112** opposite the *tert*-butyldimethylsilyl ether.

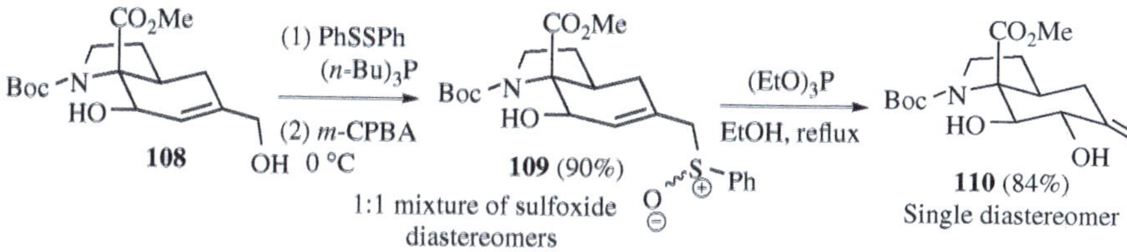

Scheme 18.29 High face slectivity in Mislow–Evans rearrangement en route to trideoxytetrodotoxin derivatives.[30]

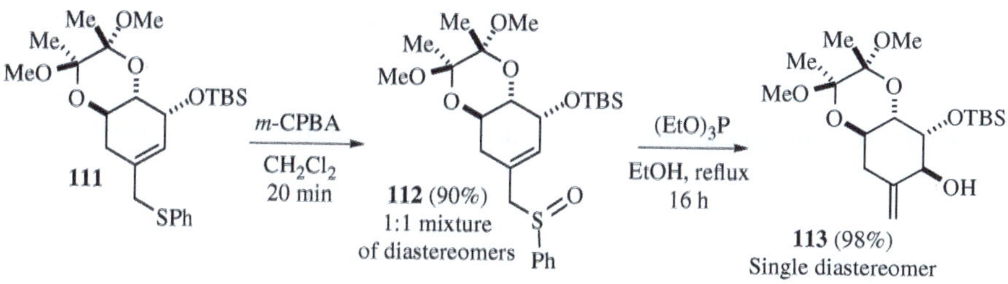

Scheme 18.30 Diastereoselective Mislow–Evans rearrangement in the synthesis of gabosine carbasugars.[31]

18.9 EPIMERIZATIONS VIA MISLOW–EVANS REARRANGEMENT SEQUENCES

The stereochemically labile nature of the position α to an allylic sulfoxide group can be used in combination with a pair of [2,3]-sigmatropic rearrangements to converge an epimeric mixture of allylic alcohols to a single diastereomer. In Smith's studies on the total synthesis of diterpenes in the jatrophone family, a 3:1 $\alpha:\beta$ mixture of tertiary allylic alcohols **114** resulted from addition of a dithiane anion to a cyclopentenone, followed by unmasking of the ethyl ketone carbonyl.[32] This epimeric mixture **114** was converted to the pure β-alcohol stereoisomer **114-β** by the Mislow–Evans-mediated route shown in Scheme 18.31. Formation of sulfenate ester **115** gave the [2,3]-rearranged allylic sulfoxide **116** as a mixture of diastereomers which equilibrated to the thermodynamically favored trans sulfoxide **116-β** upon treatment with base. Here, epimerization α to the sulfoxide was facilitated by the vinylogous relationship to the ketone. Finally, Mislow–Evans rearrangement in the sulfoxide-to-sulfenate direction, driven by trapping of the sulfenate with triethylphosphite, returned the original deconjugated tertiary α-hydroxy-β,γ-unsaturated ketone **114-β**, now as a single stereoisomer.

Evans had demonstrated a similar sequence of events in the interconversion of allylic alcohol diastereomers **37**-trans and **37**-cis (Scheme 18.32) by rearrangement of the derived sulfenates to allyl sulfoxide **39** (see top of Scheme 18.12 for this transformation). In turn, sulfoxide **39** could be permuted back to the tertiary carbinol, generally favoring diastereomer **37**-trans with the equatorial hydroxyl group.

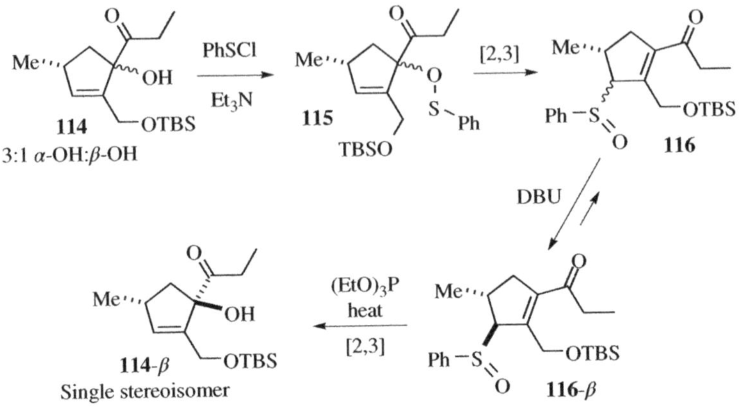

Scheme 18.31 Sulfenate to sulfoxide and back again to convert a mixture of epimers into a single diastereomer.[42]

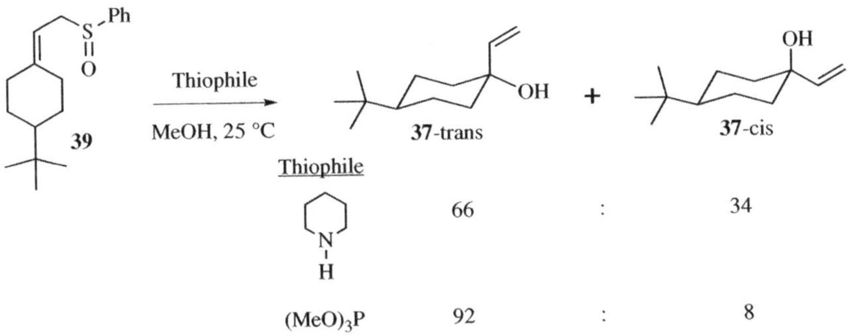

Scheme 18.32 Relative rates of sulfoxide–sulfenate interconversion and sulfenate trapping can affect stereochemical outcome.[5b,33]

Interestingly, the ratio of alcohol diastereomers depended on the nature of the thiophile; trimethylphosphite, for example, gave a high proportion of **37**-trans. Apparently, the distribution of product stereoisomers reflects the interplay among the rates of rearrangement of **39** to each of the two diastereomeric sulfenate esters and the rates of thiophilic trapping of those sulfenates to give either **37**-trans or **37**-cis.[5b,33]

Majetich and coworkers utilized Mislow–Evans rearrangements to invert the C-2 allylic alcohol configuration in the final stage of their synthesis of the dolastane marine diterpene (±)-deoxyisoamijiol **117**.[34] The sulfenate ester derived from the starting *α*-hydroxy epimer **117**-epi rearranged to allylic sulfoxide **118**, canceling the C-2 stereocenter. Diastereoselective Mislow–Evans rearrangement of **118** in the presence of trimethylphosphite reestablished the C-2 hydroxy group but with the *β*-configuration required for the (±)-deoxyisoamijiol architecture **117** (Scheme 18.33).

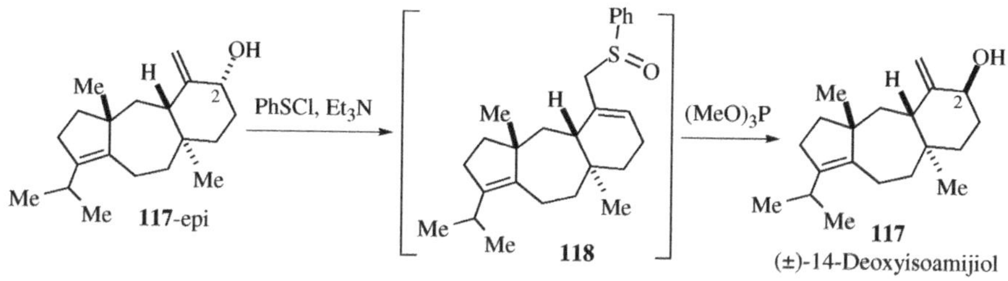

Scheme 18.33 Inversion of allylic alcohol stereochemistry via two [2,3] rearrangements.[34]

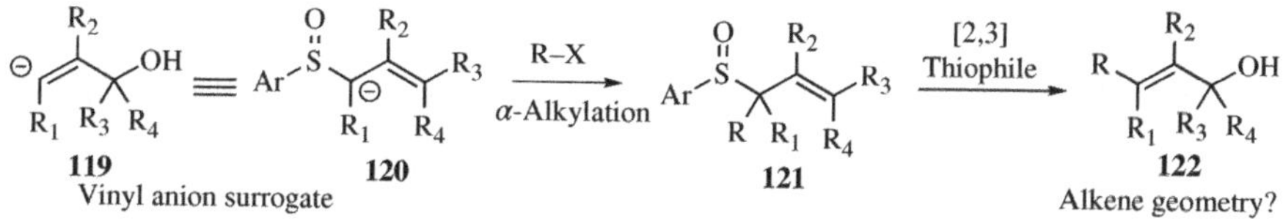

Scheme 18.34 Sulfoxide alkylation-rearrangement as *β*-hydroxy vinyl anion equivalent.[5a]

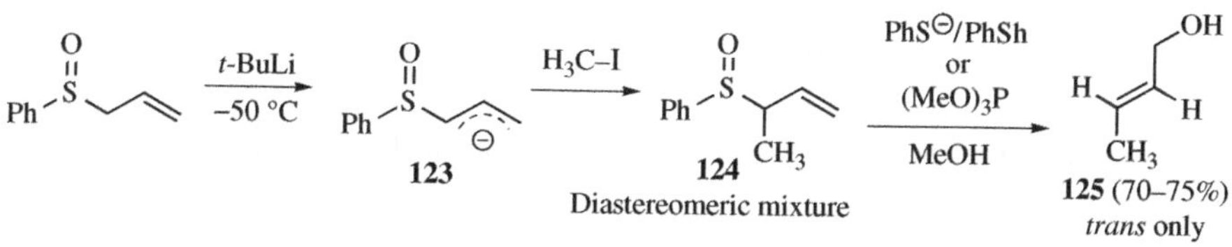

Scheme 18.35 Regiocontrolled *α*-alkylation and trans-selective rearrangement.[5a]

18.10 VINYL ANION SYNTHONS ACCESSIBLE VIA MISLOW–EVANS REARRANGEMENT

18.10.1 Vinyl Anion Equivalents

As Evans pointed out, from a synthesis strategy point of view, regioselective alkylation of a sulfoxide-stabilized allylic anion (**120** → **121**; Scheme 18.34), followed by [2,3]-rearrangement to the corresponding sulfenate ester and trapping with a thiophile (**121** → **122**), provides a synthetic equivalent for alkylation of vinyl anion **119**. As in the examples discussed previously in Section 18.7, an important consideration is stereoselectivity in formation of the alkene in the allylic alcohol product **122**. In practice, Evans showed that both *α*-selective alkylation of anions such as **123** (Scheme 18.35) and trans-selective rearrangement to allylic alcohols (e.g., **125**) occurred readily using thiophiles such as thiolates or phosphites. Both diastereomers of the alkylated sulfoxide **124** converged to the same final trans product **125** (cf. a similar outcome in Scheme 18.7).[5a]

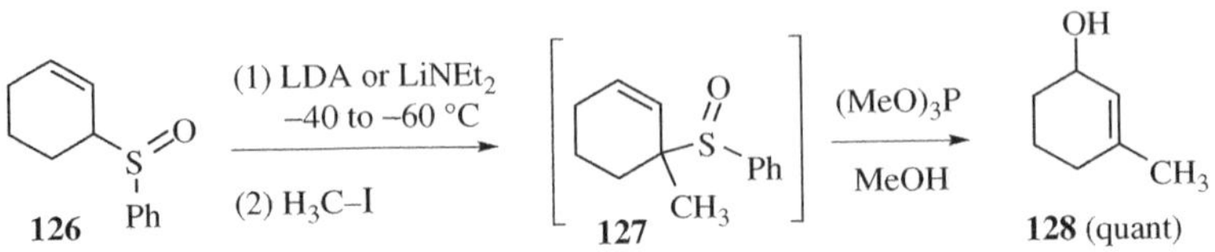

Scheme 18.36 Alkylation to giveα,α-disubstituted allyl sulfoxide, which rearranges rapidly.[5b,35]

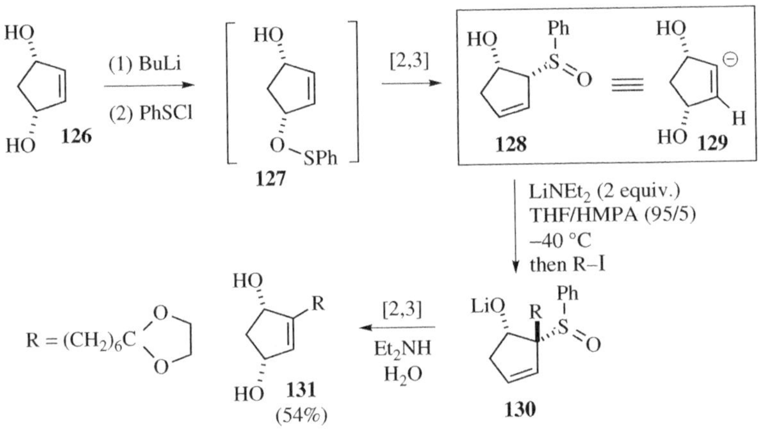

Scheme 18.37 Vinyl anion surrogate for prostaglandin synthesis.[5b]

Alkylation of an allylic sulfoxide followed by Mislow–Evans rearrangement was particularly facile within a cyclic system. Deprotonation and methylation of cyclohexenyl substrate **126** led to the α,α-disubstituted allyl sulfoxide **127**, which rearranged readily under mild conditions, perhaps aided by relief of steric crowding. Trimethylphosphite was employed as the thiophile, providing the allylic alcohol **128**, the overall equivalent of methylation of a 3-hydroxycyclohexenyl vinyl anion.[5b,35] (Scheme 18.36).

Similarly, a Mislow–Evans rearrangement strategy provided a valuable approach to prostaglandin synthesis, employing β-hydroxy allyl sulfoxides such as **128** as synthetic equivalents for dihydroxycyclopentenyl vinyl anions (e.g., **129**) as shown in Scheme 18.37. In turn, the vinyl anion synthon **128** is available from 1,4-bis-allylic alcohol **126** via [2,3]-rearrangement of the derived sulfenate monoester **127**. Regioselective alkylation of the derived dianion with a protected ω-iodoheptanal and protic quench in the presence of diethylamine gave the substituted cyclopentene-*cis*-diol **131** upon [2,3]-rearrangement. The cis relationship between the two hydroxyl groups in the meso starting material **126** was maintained in the final product **131** through (1) diastereoselective alkylation of β-hydroxysulfoxide **128**, perhaps due to chelation in the anionic intermediate, and (2) suprafacial migration of the allyl sulfoxide **130** to the sulfenate (not shown), which was intercepted by diethylamine as the thiophile. While

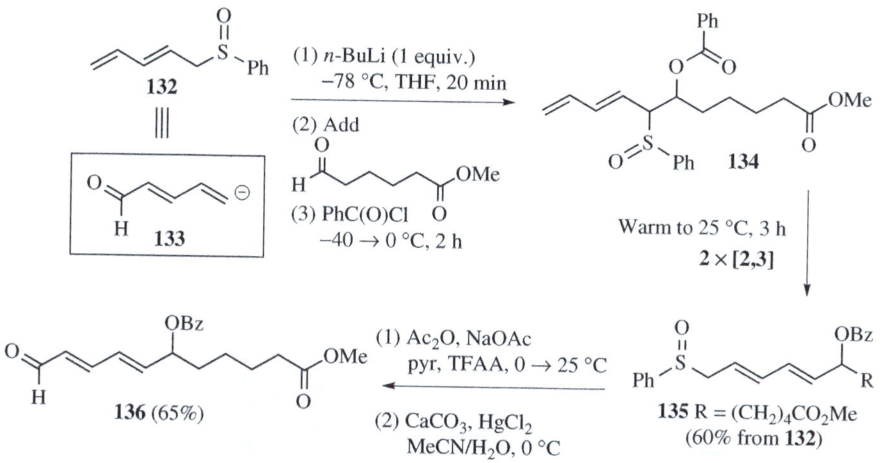

Scheme 18.38 1,5-Sulfinyl transposition via sequential Mislow–Evans rearrangement provides an α,β-γ,δ-dienyl aldehyde δ-anion equivalent.[36]

the initial sulfenate-to-sulfoxide rearrangement (**127** → **128**) was also suprafacial, the diastereomeric identity of the intermediate **128** was not essential, as the stereocenter α to the sulfoxide was likely reset in the alkylation event (**128** → **130**, assuming the sulfoxide-stabilized allyl anion does not retain its configuration).

18.10.2 Dienal Vinyl Anion Synthon in a Double Mislow–Evans Process

Corey and Hoover creatively recognized the synthetic equivalence between pentadienyl sulfoxide **132** and dienal vinyl anion **133**, established via two sequential Mislow–Evans rearrangements (Scheme 18.38).[36] This strategy was implemented in a total synthesis of 5-desoxyleukotriene D as part of a program to elucidate leukotriene binding modes. Addition of the anion from **132** to the aldehyde group of methyl 5-formylpentanoate, followed by *O*-benzoylation, provided β-acyloxysulfoxide **134**. Upon warming to room temperature, **134** rearranged via sequential, double [2,3]-sigmatropic rearrangement to the isomeric sulfoxide **135**. The preference for the internal diene of **135** versus the terminal diene moiety in **134** was sufficient to engender a >20:1 preference for the rearranged product **135**. Overall, this represents a 1,5-sulfinyl rearrangement, the synthetic utility of which is further enhanced by subsequent Pummerer rearrangement to a dienyl acetoxysulfide that, upon hydrolysis, yields the final dienal product **136**, corresponding to the addition of archetypal **133** to an aldehyde electrophile.

Schreiber's inspired application of **132** (Scheme 18.39) as the equivalent of the 1-formyl butadienyl anion **133** (see Scheme 18.38) enabled a total synthesis of (±)-asteltoxin, an oxidative phosphorylation inhibitor isolated from cultures of toxic maize.[37] Fittingly, the trans-pentadienyl sulfoxide **132** arose via Mislow–Evans rearrangement, stemming from the sulfenate ester of divinylcarbinol **137**. Metallation

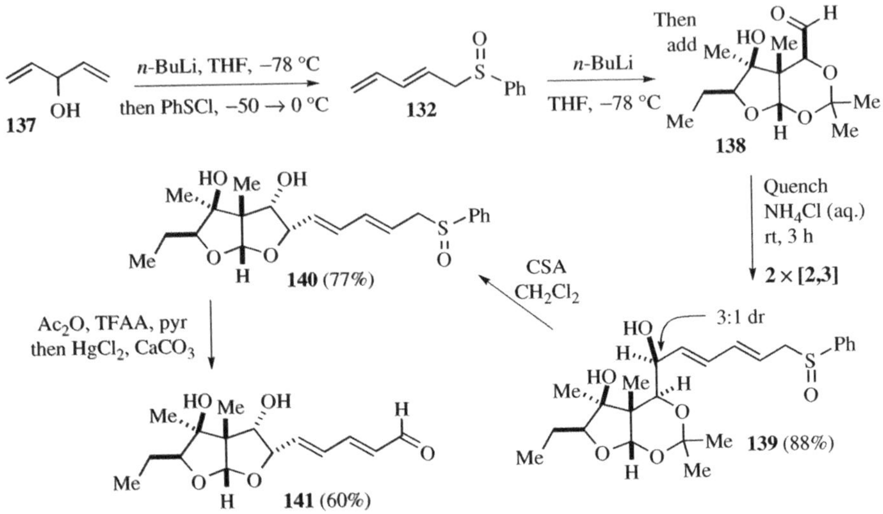

Scheme 18.39 Use of the 1-formyl butadienyl anion equivalent in the total synthesis of (±)-asteltoxin.[37]

of **132** and addition to functionally complex aldehyde **138**, which included a free tertiary alcohol, led directly to the double-[2,3]-rearrangement product **139** upon protic quench and warming of the reaction mixture. The major epimer of **139** could be obtained in pure form and was suitable for advancement to the natural product upon acetal exchange and the Pummerer–hydrolysis sequence for transforming the sulfoxide group of **140** to the aldehyde of **141**, a key intermediate en route to asteltoxin.

Ketones can also serve as electrophiles in this double-Mislow–Evans strategy, and the choice of metal counterion affects the regioselectivity of the initial organometallic addition, as Florio and coworkers demonstrated.[38] Regioselective addition of metallated pentadienyl sulfoxide **132** to cyclopentanone, for example, provided α-addition adduct **142**, which underwent spontaneous 1,5-transposition of the sulfinyl group to **143** (Scheme 18.40, top). The metal counterion affected the yield of **143**, with transmetallation from lithium to copper (using CuI) improving the efficiency, while the zinc species (generated using $ZnBr_2$) gave only a low yield. Reaction of the copper-complexed phenylsulfinyl pentadienyl anion from **132** with 2,6-dichlorobenzaldehyde gave analogous results, producing the secondary alcohol **144** upon the two sequential [2,3]-rearrangements (Scheme 18.40, bottom).

Recently, Markó combined the pentadienyl sulfoxide anion addition–sequential Mislow–Evans approach with a Julia-type olefination in the synthesis of a trienyl subunit of the marine toxin polycavernoside A (Scheme 18.41).[39] The requisite sulfoxide **132** was prepared using Corey's method[36]: thiolate displacement on pentadienyl bromide **145**, followed by oxidation at sulfur. Low-temperature metallation, addition to isobutyraldehyde, and protic workup enabled the double-[2,3]-mediated 1,5-sulfinyl migration (**146** → **147**). Chemoselective oxidation of dienyl sulfoxide **147** provided

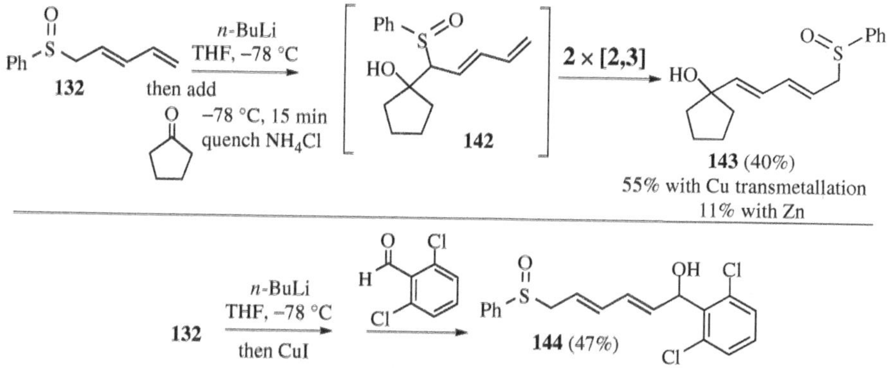

Scheme 18.40 Regioselective additions of metallated sulfinyl pentadienyl anions to ketones or aldehydes and double Mislow–Evans rearrangement.[38]

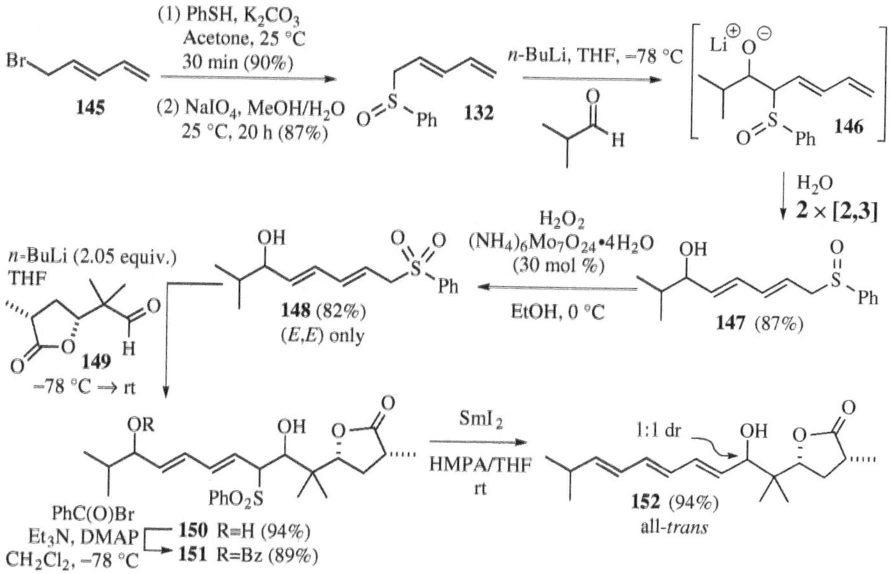

Scheme 18.41 Application of 1,5-sulfinyl rearrangement followed by Julia-type coupling in synthesis of polycavernoside A subunit.[39]

(*E*,*E*)-pentadienyl hydroxysulfone **148** for the Julia olefination sequence. Addition of the dianion from **148** to aldehyde **149** gave 1,7-diol **150**. Benzoylation of the hydroxyl distal from the sulfonyl group occurred efficiently, apparently due to deactivation of the *β*-OH group through hydrogen bonding to the sulfone. With the hydroxyl groups differentiated in **151**, 1,6-reductive elimination provided the all-trans heptatrienol unit of olefination product **152**. A shortcoming of this approach was that **152** was a 1:1 mixture of epimers at the alcohol stereocenter, originally set in the addition of the sulfone anion to aldehyde **149**. Nevertheless, this example demonstrates the utility of the terminal pentadienyl sulfoxide **147**, generated by two sequential Mislow–Evans rearrangements, in a subsequent Julia-type olefination. The overall sequence provides a concise and stereoselective triene synthesis from readily available precursors.

18.11 SEQUENTIAL PROCESSES INCORPORATING THE MISLOW–EVANS REARRANGEMENT

Further demonstrating the synthetic value and versatility of the Mislow–Evans rearrangement, it has been incorporated into reaction series that can be termed consecutive, sequential, tandem, or domino in nature.[40] The Mislow–Evans rearrangement has been combined with reactions including cycloaddition, other sigmatropic rearrangements, condensation, and isomerization, providing sequences of organic transformations capable of rapidly accessing complex molecular structures. This section highlights some of these reaction series, including classic examples as well as contemporary variants.

18.11.1 Diels–Alder/Mislow–Evans

During his group's early development of the synthetic potential of the allylic sulfoxide-to-sulfenate sigmatropic rearrangement, Evans recognized that a Diels–Alder/[2,3]-rearrangement combination could provide the synthetic equivalent of a corresponding but perhaps unfavorable cycloaddition approach (Scheme 18.42).[41] For example, an electronically matched inverse-electron-demand Diels–Alder reaction between dienyl sulfoxide **153** and electron-rich dienophile **154** (EDG = electron-donating group), followed by [2,3]-sigmatropic rearrangement of the resulting allyl sulfoxide **155** and thiophilic capture, would deliver 3,6-disubstituted cyclohexene **156**, a structure that by direct application

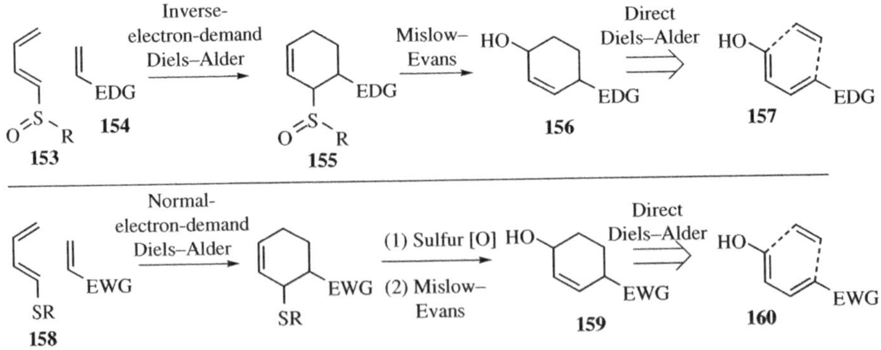

Scheme 18.42 Diels–Alder/Mislow–Evans sequences offer synthetic equivalents of potentially dubious Diels–Alder reactions.[41]

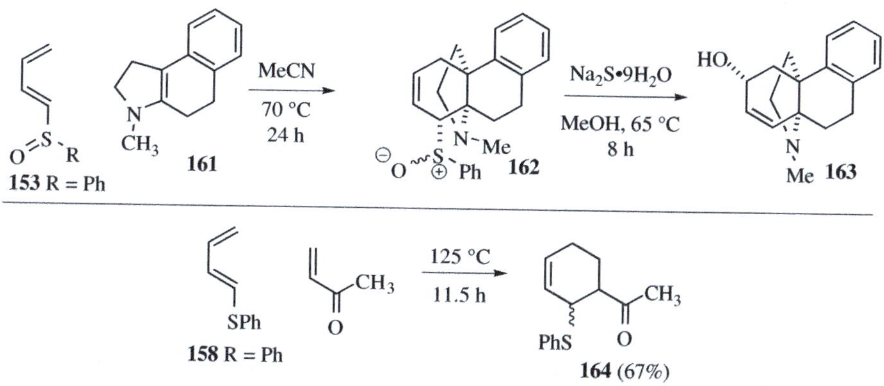

Scheme 18.43 Diels–Alder/Mislow–Evans approach in synthesis of hasubanan alkaloid framework and an electronically complementary approach to a rearrangement substrate.[41]

of the Diels–Alder transform would require the diene–dienophile pairing shown in **157**. Additionally, because a thioether can serve as a latent sulfoxide, a normal-electron-demand Diels–Alder between a now electron-rich diene **158** and an electron-poor dienophile, followed by oxidation and [2,3]-rearrangement, would access cyclohexene **159**. Here, a direct Diels–Alder route (see **160**) would require an electronically conflicted diene, bearing an electron-releasing group at one terminus and an electron-withdrawing substituent at the other.

Evans provided an impressive demonstration of the earlier dienyl sulfoxide cycloaddition approach in the synthesis of a hasubanan alkaloid derivative (Scheme 18.43, top).[41] Reaction of **153** with tetrahydrobenzindole **161** as the dienophile under relatively mild conditions provided allyl sulfoxide **162** (mixture of sulfoxide diastereomers) as the direct cycloaddition product. Under the Diels–Alder reaction conditions, a certain amount of the sulfoxide underwent rearrangement and sulfenate trapping to give the allylic alcohol **163**, a process driven to completion upon treatment of the crude reaction mixture with sodium sulfide in hot methanol. The single stereoisomer of product **163** corresponded to an endo-selective Diels–Alder followed by suprafacial [2,3]-sigmatropic rearrangement. An alternative direct Diels–Alder approach to the cyclohexene moiety of **163** would be both electronically and topologically unlikely. In the same study, Evans also demonstrated the viability of sulfide-bearing diene **158** in cycloaddition with methyl vinyl ketone, giving allyl thioether **164**, poised for the oxidation–rearrangement sequence.

While the promise of the Diels–Alder/Mislow–Evans strategy has yet to be fully realized, particularly in complex-molecule synthesis, there are a number of additional cases to stimulate the chemical mind.[6f,42] For example, Posner employed a regio- and stereoselective high-pressure Diels–Alder reaction between pyrone sulfoxide **165** and phenyl vinyl ether, followed by methanolysis of the resulting bicyclic lactone and spontaneous Mislow–Evans rearrangement (Scheme 18.44, top) in an elegant synthesis of (±)-chorismic acid.[43] Notably, this route provided streamlined access to the stereochemically defined, tetrasubstituted cyclohexene framework **166**, not readily

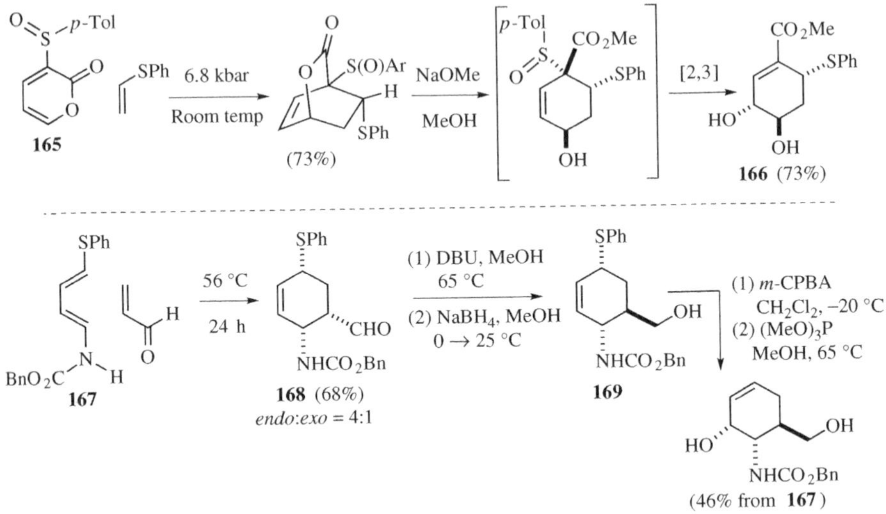

Scheme 18.44 Sequential [4+2]/[2,3] transformations in cyclohexenol synthesis (top[43]; bottom[44]).

accessible via direct Diels–Alder cycloaddition. The 1,4-acylamino/thiophenyl substitution pattern of diene **167** (Scheme 18.44, bottom) provides an electron-rich diene counterpoint, as explored by Overman and coworkers.[44] Endo-selective Diels–Alder cycloaddition with acrolein provided **168** in 68% yield after crystallization of the major product from the reaction mixture. Epimerization and reduction led to hydroxymethyl derivative **169**, which, upon oxidation at sulfur and Mislow–Evans rearrangement, enabled stereocontrolled access to a final diol product. Interestingly, sulfoxides from the unepimerized diastereomer of **169** (α hydroxymethyl group) did not undergo [2,3]-rearrangement, apparently because the required pseudo-axial position of the sulfoxide group would lead to a destabilizing 1,3-diaxial steric clash.

Carreño *et al.* have explored enantiomerically pure diene sulfoxides in Diels–Alder/Mislow–Evans rearrangements.[45] As shown in Scheme 18.45, a slow, endo-selective Diels–Alder reaction took place between diene **170** and *N*-methylmaleimide (used in excess since it also functioned as the thiophile), followed by [2,3]-sigmatropic rearrangement of the resulting allylic sulfoxide to cyclohexenol **171**, which was isolated in optically active form. Unfortunately, the yield of **171** was low because it was prone to dehydration, with the resulting cyclohexadiene being trapped by the maleimide to give the meso by-product **172**. Attempts to accelerate the Diels–Alder reaction by addition of a Lewis acid ($SnCl_4$) did give faster reaction but led to exclusive formation of the bicyclo[2.2.2]octene **172**.

Five-membered-ring allylic alcohol derivatives are also accessible by combining Diels–Alder cycloaddition with the Mislow–Evans rearrangement. In Weinreb's total synthesis of the antitumor alkaloid agelastatin A, a hetero-Diels–Alder reaction between cyclopentadiene and *N*-sulfinyl methyl carbamate provided bicyclic intermediate **173** (Scheme 18.46). Addition of phenyl Grignard cleaved the S—N

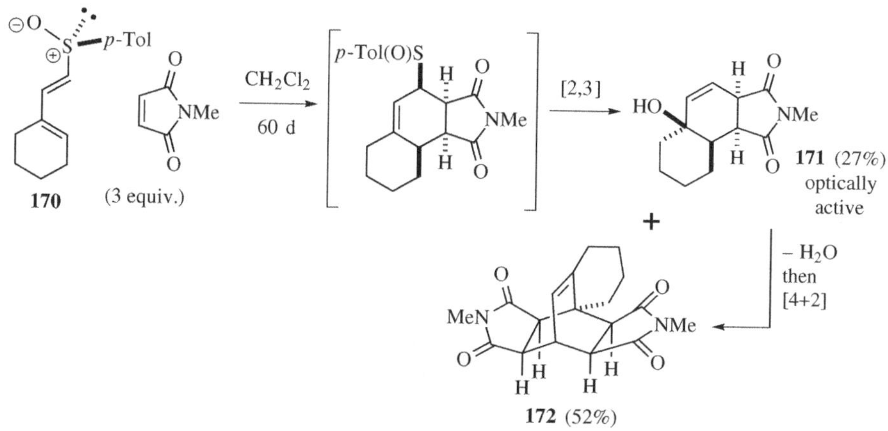

Scheme 18.45 Optically active cyclohexenol from enantiomerically pure 1-sulfinyl-1,3-butadienyl starting material.[45]

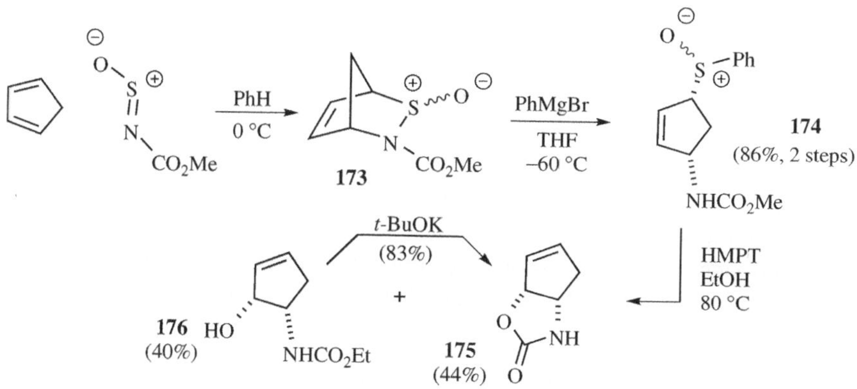

Scheme 18.46 Hetero-Diels–Alder/Mislow–Evans sequence in total synthesis of agelastatin A.[46]

bond and enabled [2,3]-rearrangement of the resulting allylic sulfoxide **174**. The Mislow–Evans step led to a mixture of oxazolidinone **175** from cyclization of the corresponding allylic alcohol as well as uncyclized ethyl carbamate **176**. The latter was readily converted to the oxazolidinone upon treatment with base.[46]

18.11.2 Attempted 1,3-Dipolar Cycloaddition/Mislow–Evans

Pearson advanced an inspired synthetic strategy in studies toward preparation of the homoerythrina alkaloids, whereby intramolecular [3+2] cycloaddition of an azomethine ylide with the terminal alkene of a sulfoxide- (or sulfide-)substituted diene would provide the tricyclic 6-5-7 core **178**, poised for Mislow–Evans rearrangement to install the C-2 hydroxyl group of homoerythratine (**177**) with the correct

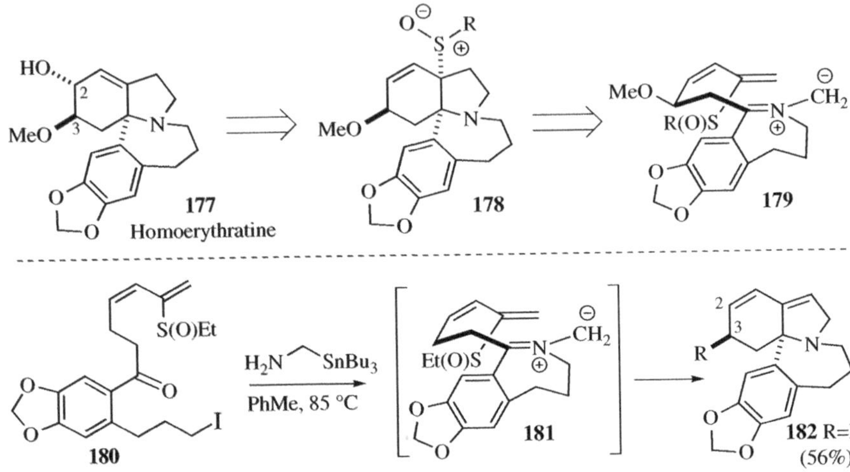

Scheme 18.47 Attempted 1,3-dipolar cycloaddition/Mislow–Evans rearrangement toward homoerythrina alkaloids.[47]

configuration (Scheme 18.47).[47] In the event, the azomethine ylide **181** could be generated from (tributylstannyl)methylamine and iodoketone **180**. Dipolar cycloaddition was successful, but the resulting allylic sulfoxide underwent elimination rather than rearrangement, providing conjugated diene **182**, a structure closely related to 3-*epi*-schelhammeridine (**182** with R = OMe). Although the 1,3-dipolar cycloaddition/Mislow–Evans sequence was not attained in this system, the use of dienyl sulfoxides (or the corresponding sulfides) in such cycloadditions may enable subsequent [2,3]-rearrangement in cases where elimination is less competitive.

18.11.3 Double Overman Rearrangement/Mislow–Evans

A remarkable cascade sequence of two sequential Overman rearrangements set up a subsequent Mislow–Evans rearrangement in Chida's total synthesis of the marine sponge-derived alkaloid (−)-agelastatin A (**189**; Scheme 18.48).[48] Diol **183**, derived in enantiomerically pure form from D-tartaric acid, was converted to the bis-trichloroacetimidate **184**, which underwent double Overman rearrangement, with the first [3,3]-sigmatropic event setting up the second. Besides establishing the syn stereochemistry at the vicinal nitrogen-bearing centers of **185**, the Overman rearrangements transposed alkene functionality into position with respect to the phenylthioether, enabling the ensuing Mislow–Evans step. Oxidation gave sulfoxides **186** as a 1:1 mixture of sulfoxide diastereomers. Rearrangement in the presence of trimethylphosphite provided an epimeric mixture of the corresponding allylic alcohols **187**. Interestingly, in this case, the two diastereomers of sulfoxide **186** could be separated by HPLC and were configurationally stable enough to each be characterized and individually subjected to the Mislow–Evans rearrangement conditions. Both sulfoxide diastereomers gave the same nonstereoselective outcome:

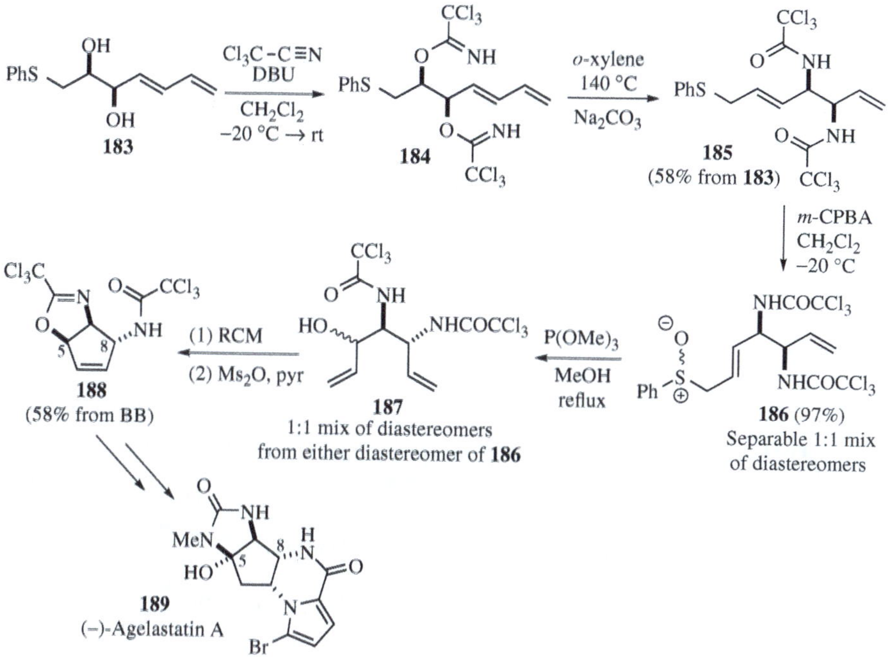

Scheme 18.48 Sequential double-Overman rearrangement/Mislow–Evans sequence in agelastatin total synthesis.[48]

a 1:1 mixture of epimers **187**. Nevertheless, both alcohol diastereomers could be advanced via ring-closing metathesis, followed by stereoconvergent cyclization to oxazoline **188**, a key intermediate for elaboration toward (−)-agelastatin A.

18.11.4 Claisen Rearrangement/Mislow–Evans

While the Chida example in Scheme 18.48 required a sulfur oxidation step between the [3,3]- and [2,3]-sigmatropic rearrangements, a more truly domino-type[49] sequence of rearrangements occurred in Zakarian's preparation of the spiro-fused imine portion of the pinnatoxin framework (Scheme 18.49).[50] Here, (*E*)-vinyl sulfoxide **191** (prepared by addition of the anion of methyl-*p*-tolyl sulfoxide to the aldehyde derived by oxidation of primary alcohol **190** and subsequent elimination) underwent an efficient Claisen rearrangement/Mislow–Evans cascade process when heated in the presence of triethylphosphite and *sym*-collidine. This high-yielding reaction sequence provided stereocontrolled synthesis of the cyclohexene **193**, bearing an all-carbon quaternary stereocenter at one allylic site and a tertiary alcohol stereocenter at the other allylic position. By successively using both (*S*)- and (*R*)-methyl-*p*-tolyl sulfoxides to generate either sulfur epimer of vinyl sulfoxide **191**, Zakarian and coworkers determined that the diastereomeric sulfoxides each gave **193** in comparable yield. Mechanistically, Claisen rearrangement of **191** evidently

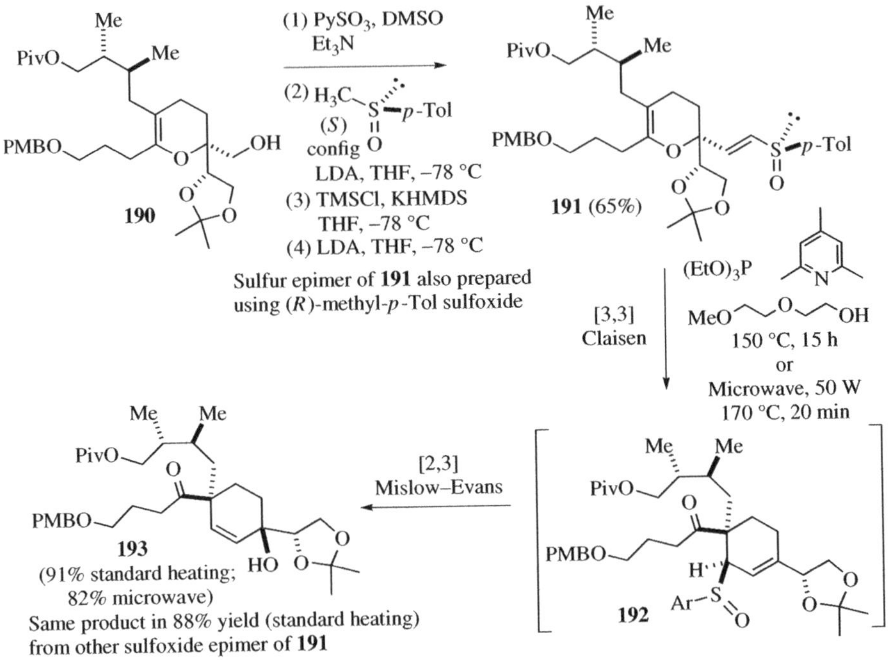

Scheme 18.49 Sequential Claisen/Mislow–Evans rearrangement in synthetic studies toward pinnatoxin.[50,51]

proceeded to give allyl sulfoxide **192**, followed by the Mislow–Evans rearrangement and *in situ* sulfenate trapping, arriving at the final allylic alcohol product **193**. Microwave heating could be used to drastically shorten the reaction time for the tandem rearrangement sequence while still maintaining a high yield of **193**.[51]

Zakarian's group encountered contrasting reactivity, however, when attempting to carry out a similar Claisen/Mislow–Evans cascade toward the preparation of (−)-joubertinamine and (−)-mesembrine, psychoactive alkaloids from plants of the sceletium family (Scheme 18.50).[52] In this case, only starting material was recovered when either sulfoxide diastereomer of vinyl sulfoxide **194** was heated under conditions that had given smooth rearrangement of **191** in the pinnatoxin approach (compare Schemes 49 and 50), and at higher temperatures, vinyl sulfoxide **194** decomposed. These results indicated that the initial Claisen rearrangement was not taking place.

In this case, the ability to use a sulfide in place of a sulfoxide, with the oxidation-state adjustment and subsequent Mislow–Evans to occur later, provided a solution to the recalcitrant Claisen rearrangement. This strategy (Scheme 18.51) was particularly appealing in this case, because terminal thioethers are known to be strongly activating for [3,3]-sigmatropic anionic oxy-Cope rearrangements.[53]

Vinyl sulfide **195** (Scheme 18.51) was prepared as a mixture of alkene isomers favoring the *Z* geometry, which, based on transition state considerations, would give an undesired product diastereomer. Fortunately, because (*Z*)-**195** underwent Claisen

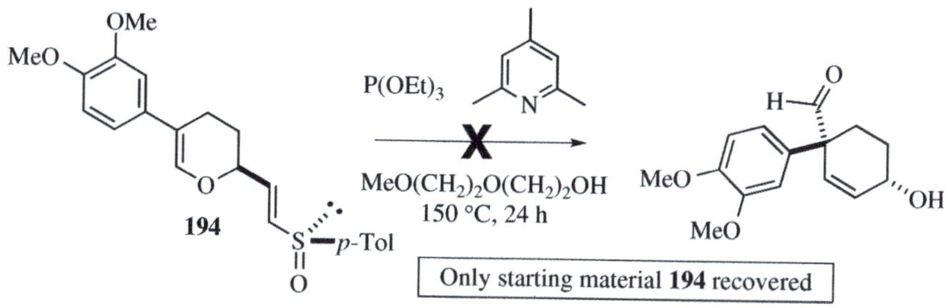

Scheme 18.50 Vinyl sulfoxide fails as Claisen rearrangement substrate in joubertinamine and mesembrine synthesis.[52]

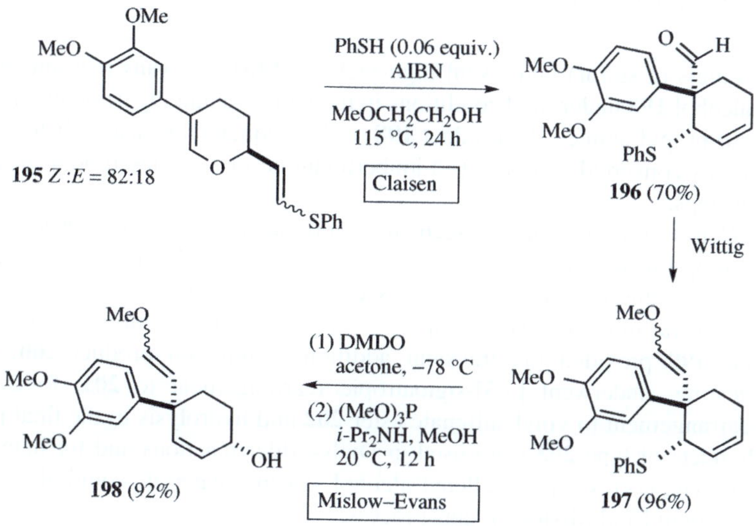

Scheme 18.51 Claisen rearrangement of the *E*-vinyl sulfide followed by Mislow–Evans provides the desired outcome.[52]

rearrangement much more slowly than (*E*)-**195**, radical-mediated equilibration of the alkene isomers under the rearrangement conditions enabled selective conversion of the *E* isomer to Claisen rearrangement product **196**. As with the pinnatoxin approach, a quaternary carbon center was established in the Claisen rearrangement step. In this way, starting from a mixture substantially enriched in (*Z*)-**195**, a useful yield of **196** resulted via [3,3]-rearrangement of (*E*)-**195**. Indeed, when the reaction was stopped prior to full conversion to **196**, the vinyl sulfide, now enriched in (*E*)-**195**, was reisolated.

Olefination of the aldehyde of **196** installed an additional carbon necessary for advancement to the natural products, giving methyl vinyl ether **197**. Treatment with dimethyldioxirane at low temperature accomplished oxidation at sulfur in the presence of the electron-rich vinyl ether, and the resulting sulfoxides (mixture of

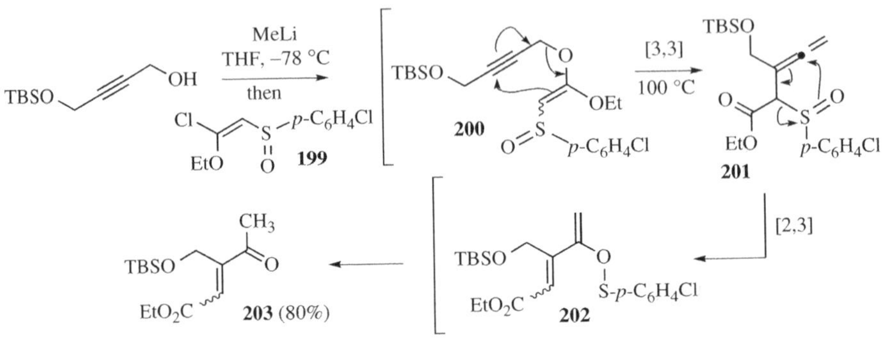

Scheme 18.52 Tandem Claisen-type [3,3]/Mislow–Evans rearrangement of sulfoxide-substituted propargyl ketene acetals.[54]

diastereomers at sulfur) underwent high-yielding Mislow–Evans rearrangement to allylic alcohol **198** under mild conditions in methanol containing trimethylphosphite and di-*iso*-propyl amine. The product **198** of this sequential Claisen/Mislow–Evans process was converted to either (−)-joubertinamine or (−)-mesembrine in a short series of steps.

The Claisen-type [3,3]-rearrangement of propargyl vinyl ethers bearing a sulfoxide group on the vinyl portion led directly to Mislow–Evans rearrangement in a synthesis of unsaturated 1,4-ketoesters developed by Posner (Scheme 18.52).[54] Addition of the lithium salts of propargyl alcohols to the β-chloro-β-ethoxy vinyl sulfoxide **199** provided the transient addition–elimination product **200**, which, upon heating, underwent [3,3]-sigmatropic rearrangement to **201**, followed by [2,3]-rearrangement to vinyl sulfenate ester **202** and hydrolysis to the final product **203**. Products of type **203** were used in Diels–Alder reactions and for heterocycle synthesis. The corresponding allenyl alcohol starting materials could also be used for this tandem [3,3]/Mislow–Evans process.

18.11.5 Aldol/Ireland–Claisen/Mislow–Evans

Heathcock developed an aldol/Ireland–Claisen/Mislow–Evans sequence to construct 1,2,5-stereoarrays containing a 3,4-*trans*-alkene and applied this reaction series to the synthesis of the polyene antibiotic myxalamide A.[55] The sulfur group that would eventually participate in the Mislow–Evans step originated from the β-*tert*-butyl thioether of trisubstituted α,β-unsaturated aldehyde **204** (Scheme 18.53). Evans *syn*-aldol reaction using the boron enolate of *N*-propionyl oxazolidinone **205**, followed by reduction, provided diol **206**, which was silylated at the primary hydroxyl and then esterified in advance of the Ireland–Claisen rearrangement. Formation of the *E*-silyl ketene acetal from ester **207** and warming to room temperature led to smooth and stereospecific [3,3]-sigmatropic rearrangement. The methyl ester **208** was isolated after hydrolysis of the intermediate silyl ester and treatment of the resulting acid with diazomethane. The well-ordered Ireland–Claisen transition structure parlayed three elements of stereochemistry in **207**, vinyl sulfide geometry,

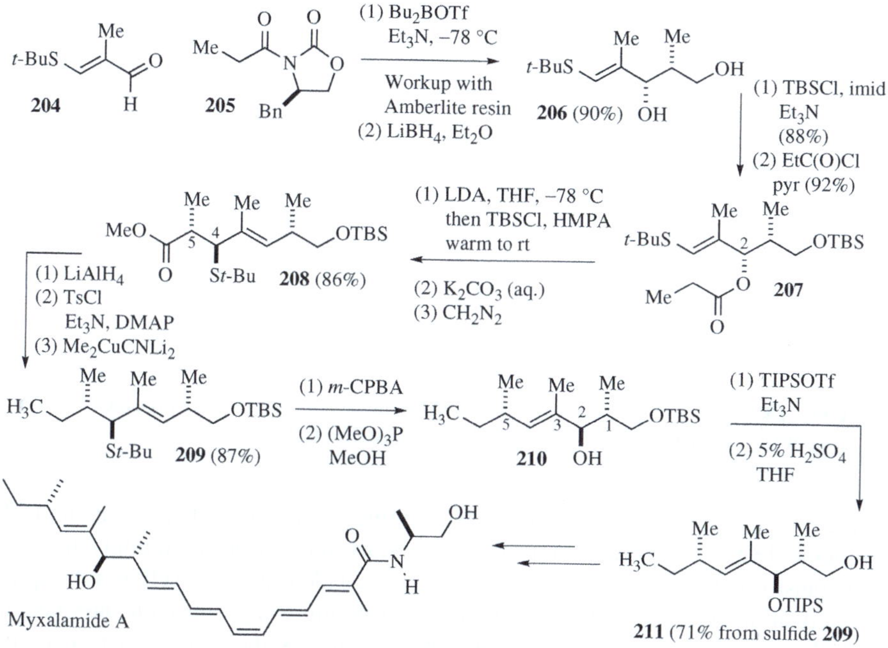

Scheme 18.53 Aldol/Ireland–Claisen/Mislow–Evans sequence in the total synthesis of myxalamide A.[55a]

silyl ketene acetal geometry, and C-2 configuration, into establishing the 4,5-anti relationship and the *E*-alkene geometry in rearrangement product **208** (atom numbering selected to highlight the 1,2,5 relative stereochemistry in **210**; see further discussion in the following).

The methyl ester of **208** was reduced and the resulting primary alcohol activated and homologated to yield **209**. Oxidation of the sulfide gave a mixture of sulfoxide diastereomers, and Evans–Mislow rearrangement in the presence of trimethylphosphite yielded the 1,2-anti–2,5-anti product **210**. Mechanistically, the stereochemical outcome is analogous to that investigated computationally by Jorgensen in a simpler system (see Scheme 18.7).[10] In such a scenario, the diastereomeric sulfoxides proceed stereoconvergently to the same stereoisomer of product but via different transition states (endo for one and exo for the other). Notably, the 1,2-relationship in **210** corresponds to what can be an otherwise synthetically elusive anti aldol motif. En route to the natural product, protecting group manipulations gave primary alcohol **211**, which was elaborated to myxalamide A.

18.11.6 Knoevenagel Condensation/Vinyl Sulfoxide Isomerization/Mislow–Evans

A particularly popular and useful cascade process that includes the Mislow–Evans rearrangement starts with the Knoevenagel condensation of an aldehyde (or

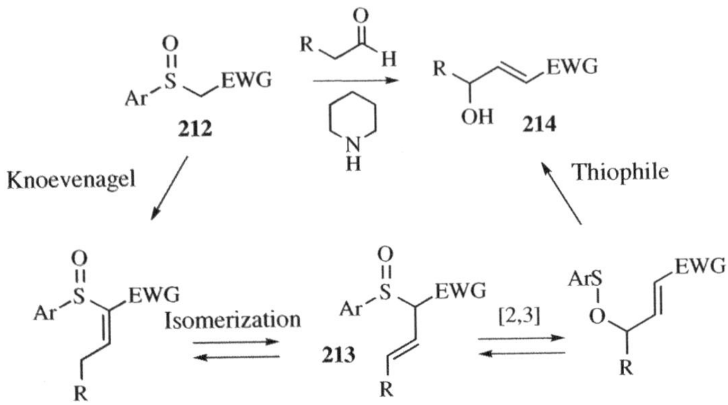

Scheme 18.54 Knoevenagel/isomerization/Mislow–Evans sequence provides synthetic equivalent for olefination of α-hydroxyaldehydes (EWG = CN).[56]

sometimes a ketone) with an α-sulfinyl-substituted active methylene compound **212** (Scheme 18.54). Piperidine and other secondary amines are commonly used for this transformation, as these bases facilitate the condensation as well as subsequent isomerization of the vinyl sulfoxide to the corresponding allyl sulfoxide intermediate **213**. *In situ* [2,3]-sigmatropic rearrangement converts the allyl sulfoxide to the sulfenate ester, which is captured by a thiophile (piperidine can play this role as well), producing final product **214**. Nokami reported this one-pot sequence in 1981, using 2-arylsulfinyl acetonitrile (EWG = CN),[56] and the use of sulfinylacetates (EWG = CO_2R)[57] or bis-sulfoxides (EWG = S(O)Ar)[58] is also possible. From a synthesis planning perspective, this transformation provides the equivalent of aldehyde (or ketone) α-hydroxylation, followed by olefination, giving γ-hydroxy-α, β-unsaturated products **214**.

Nokami extended this methodology to the use of β-ketosulfoxides as well.[59] For example, Knoevenagel condensation of β-ketosulfoxide **215** (Scheme 18.55), bearing a terminal alkene group, with hexanal provided a good yield of the γ-hydroxyl-α,β-unsaturated ketone product **216**, having the *E*-alkene geometry. Adding a substoichiometric amount of acetic acid improved reaction efficiency, apparently by promoting the dehydrative part of the Knoevenagel condensation step. The transformation tolerated additional ketone and ester functionality, acetal and silyl protecting groups, as well as a free hydroxyl group in one β-ketosulfoxide substrate. Finally, Nokami and coworkers used the Knoevenagel/isomerization/Mislow–Evans cascade to prepare the cytotoxic (*E*)-oxooctadecenoic acid **217**.

An advance that makes this reaction sequence easier to conduct, while reducing the need for organic solvents, involves the use of silica-supported tertiary amines. Hagiwara and coworkers used *N*,*N*-diethylaminopropyl-grafted silica gel (NDEAP) in water to carry out the Knoevenagel/isomerization/Mislow–Evans reaction sequence between α-arylsulfinyl acetonitriles and various aldehydes.[60] Workup with diethylamine improved yields by driving the conversion of the sulfenate ester intermediates to the final allylic alcohol products. With the *m*-chlorophenyl

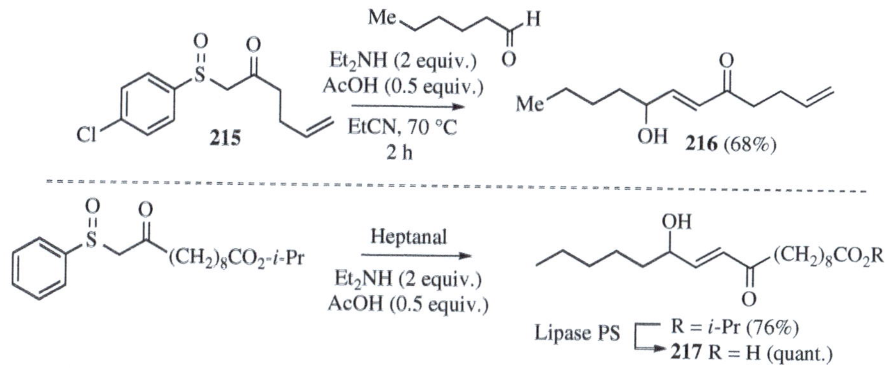

Scheme 18.55 γ-Hydroxy-α,β-unsaturated ketones via Knoevenagel/isomerization/Mislow–Evans sequence.[59b]

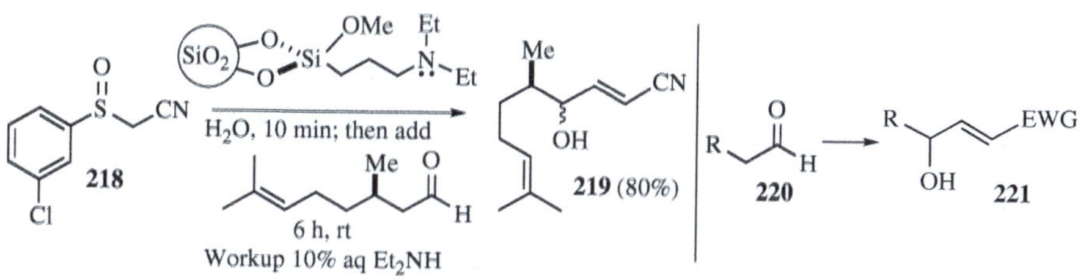

Scheme 18.56 Knoevenagel/Mislow–Evans sequence mediated by silica-supported amines under aqueous conditions; equivalent of aldehyde α-hydroxylation/olefination.[60]

sulfoxide **218** (Scheme 18.56), for example, condensation with citronellal, followed by isomerization, [2,3]-sigmatropic rearrangement, and trapping of the sulfenate, gave γ-hydroxy-α,β-unsaturated nitrile **219**, exclusively as the *trans*-alkene isomer but without diastereoselectivity at the hydroxyl-bearing center. Again, this transformation corresponds to olefination of the α-hydroxy derivative of the aldehyde component (**220** → **221**).

A diastereoselective example of the Knoevenagel/Mislow–Evans process included cyclization of the resulting alcohol via transesterification. In Trost's total synthesis of the antibacterial and antifungal cyathin-family diterpene (+)-allocyathin B_2, aldehyde **222** provided essentially a single diastereomer of lactone **223** upon treatment with α-phenylsulfinyl acetonitrile and piperidine (Scheme 18.57).[61] This key Mislow–Evans-based hydroxylation–olefination set the stage for rapid elaboration to the natural product via site-selective alkene hydrogenation, nitrile-to-aldehyde and lactone-to-lactol reduction, and intramolecular aldol condensation.

Lacôte and coworkers developed a fascinating variant of the Knoevenagel/Mislow–Evans domino reaction, starting by condensation of chiral, enantiopure bis-sulfoxide **224** (Scheme 18.58) with cyclic ketones.[62] In the presence of a thiophile (best results were obtained with *in situ* generated lithium phenylthiolate), subsequent

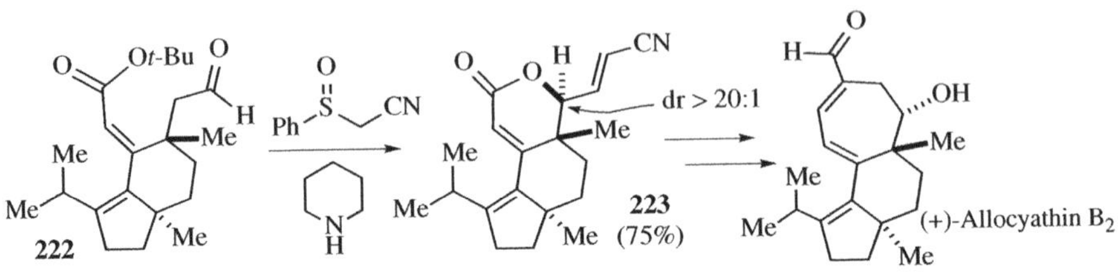

Scheme 18.57 Diastereoselective Knoevenagel/Mislow–Evans approach in (+)-allocyathin B_2 total synthesis.[61]

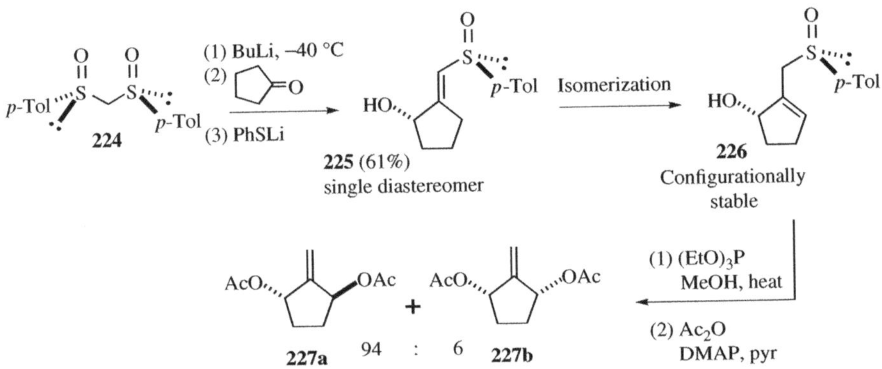

Scheme 18.58 Condensation of a chiral bis-sulfoxide with cyclopentanone followed by a double Mislow–Evans rearrangement and ketones.[62]

[2,3]-rearrangement and sulfenate trapping led diastereoselectively to cyclic allylic alcohols bearing an exocyclic vinyl sulfoxide. For example, the reaction of cyclopentanone provided a single diastereomer of **225**, the relative stereochemistry of which was determined by X-ray diffraction. With the five-membered-ring product **225** (though not with the corresponding cyclohexanone-derived system), some isomerization of the double bond into the ring occurred, producing allylic sulfoxide **226**. As discussed earlier in this chapter (cf. Scheme 18.3), the sulfur center of allylic sulfoxides often epimerizes readily via reversible [2,3] sigmatropy, but the sulfur stereocenter in the allylic sulfoxide **226** possessed unusual configurational stability, with a >1 day half-life for epimerization in THF solution at 30 °C and complete long-term stability when kept at freezer temperature. A second Mislow–Evans rearrangement, using **226** in the presence of triethylphosphite, followed by acetylation to simplify product purification, gave a preponderance of the optically active product **227a** over the meso diastereomer **227b**. Interestingly, the high stereoselectivity in [2,3]-rearrangement of **226** was attributed to a synergy between the configurations at the preexisting allylic alcohol and at the sulfoxide sulfur.

With α-chloro vinyl sulfoxides and an excess of *N*-lithio 2-piperidone base, an interesting variant of the isomerization/Mislow–Evans rearrangement occurs,

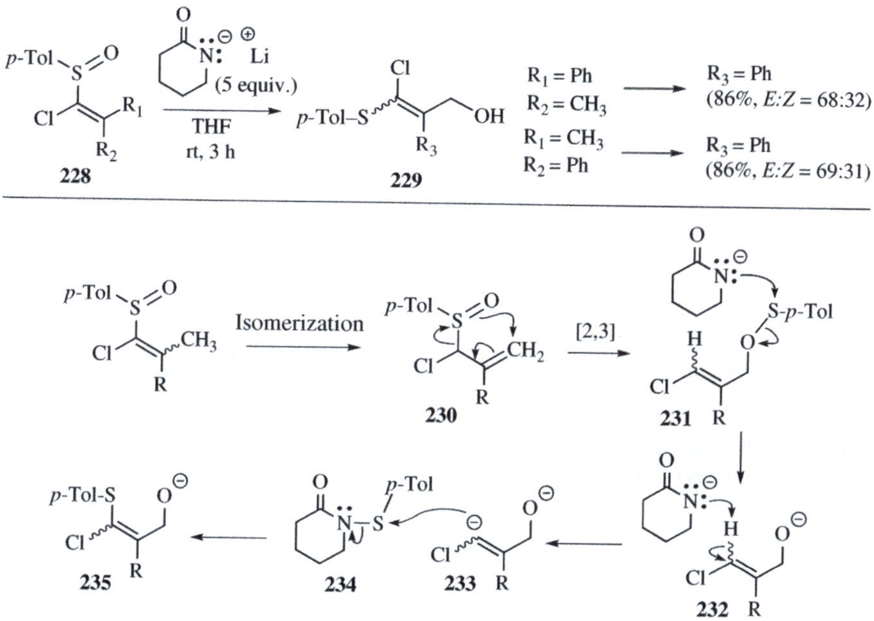

Scheme 18.59 Mislow–Evans process with return of sulfur to its original site.[63]

whereby a sulfide group derived from the original sulfoxide is returned to its original carbon site of attachment (**228** → **229**; Scheme 18.55).[63] Choice of base is crucial (e.g., use of LDA gave only decomposition), and at least one of the alkene substituents, R_1 or R_2, must be methyl. The double-bond geometry of the product 1-chloro-2-(hydroxymethyl)alkenyl sulfides **229** is independent of the alkene stereochemistry of the starting vinyl sulfoxide. Mechanistically, Satoh and coworkers proposed that initial isomerization of the vinyl sulfoxide to the allyl sulfoxide **230** enabled [2,3]-sigmatropic rearrangement. The 2-piperidone anion, used in excess, intercepted the sulfenate ester **231** and also deprotonated vinyl chloride **232**, generating vinyl anion **233**. The *N*-sulfenyl 2-piperidone (**234**) produced earlier served as the sulfenylating agent for the vinyl anion, giving the final vinyl sulfide product **235** (Scheme 18.59).

Zard has studied the isomerization/Mislow–Evans rearrangement of vinyl sulfoxides such as **237**, arising from enolate addition to alkynyl sulfoxides (Scheme 18.60).[64] Isomerization of **237** to the allylic sulfoxide **238** enabled the [2,3]-sigmatropic rearrangement to α-hydroxy-α-vinyl ketone **239**. In this case, diastereoselectivity was low in formation of the carbinol center within a steroid framework. Additions to allenyl sulfoxides provide a similar sequence, leading to 2-propenyl substitution at the tertiary alcohol center (not shown).

Vinyl sulfoxide isomerization followed by Mislow–Evans rearrangement was also central to a synthetic route toward the hydroazulene moiety of the antibiotic fungal metabolite guanacastepene A (Scheme 18.61).[65] In this case, a diastereomeric

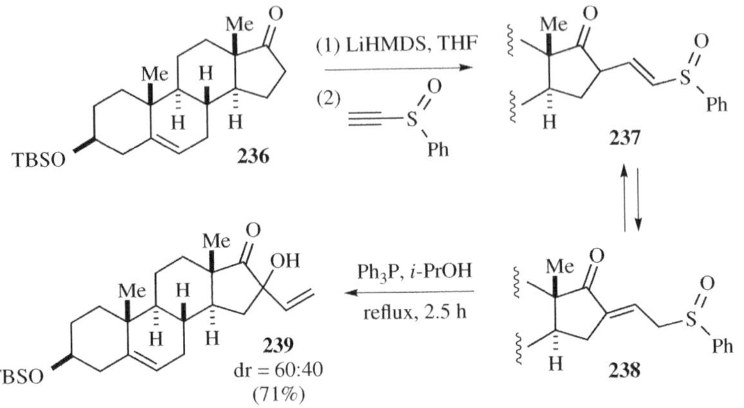

Scheme 18.60 Isomerization/Mislow–Evans rearrangement of vinyl sulfoxides generated by addition of enolates to alkynyl sulfoxides.[64]

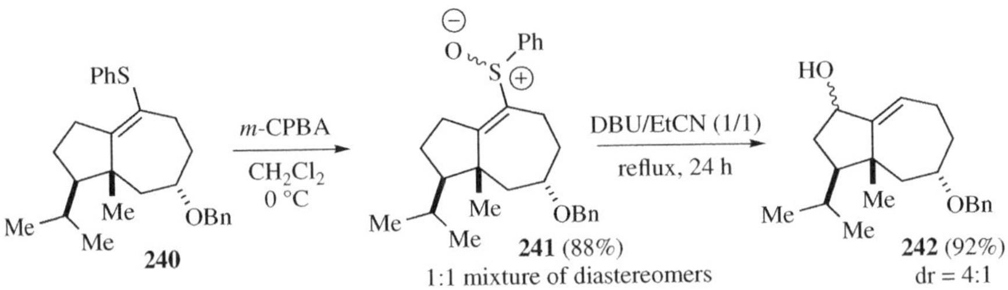

Scheme 18.61 Isomerization of a vinyl sulfoxide and rearrangement in an approach toward the guanacastepene A hydroazulene.[65]

mixture of vinyl sulfoxides **241** resulted upon oxidation of the starting vinyl sulfide **240**. Subsequent treatment with DBU led to the sequential isomerization/[2,3]-rearrangement process. Under these conditions, the intermediate sulfenate was converted to the allylic alcohol **242**, produced as a 4:1 mixture of epimers. Here, the modest selectivity in formation of the allylic stereocenter was of no synthetic consequence, as the alcohol was subsequently oxidized to the corresponding enone. Notably, the overall conversion from **240** to **242** represents a 1,3-vinyl-to-allyl heteroatom transposition.

18.11.7 Sulfenate-to-Sulfoxide Conversion and Olefination

Allylic sulfoxide functionality, established via Mislow–Evans rearrangement starting from allylic alcohol **243** (Scheme 18.62), served as a site for olefination. Oxidation of the [2,3]-rearrangement product **244** to the allylic sulfone **245**, followed by alkylation with iodotrimethylsilylmethane, gave *β*-silyl sulfone **246**. Fluoride-induced elimination according to Kocienski's protocol[66] provided the desired conjugated diene

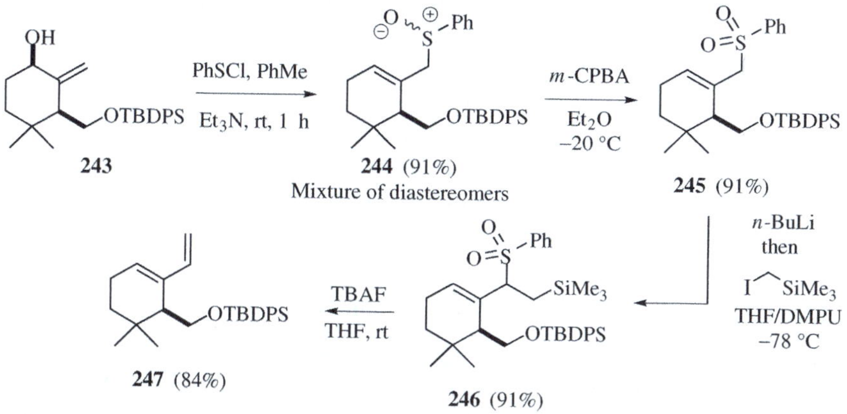

Scheme 18.62 Allylic alcohol converted via Misow–Evans to sulfoxide for olefination to a conjugated diene.[67]

product **247**, prepared by Vidari and coworkers during studies using α- and γ-ionone analogues to elucidate mechanisms of olfactory G protein-coupled receptor binding.[67]

18.11.8 Generation of Sulfenate Anions for Pd-Mediated Cross-Coupling

Madec and coworkers demonstrated the use of allylic sulfoxides as sulfenate anion precursors in a synthesis of diaryl sulfoxides (Scheme 18.63).[68] Mislow–Evans rearrangement provided some of the corresponding sulfenate, which was trapped by the Pd(0) catalyst to generate π-allyl Pd(II) complex **248**. Trapping of the allyl fragment with *tert*-butoxide as the nucleophile provided a source of sulfenate anion and regenerated Pd(0) for subsequent cross-coupling, giving diaryl sulfoxide products **249**. For example, using this Mislow–Evans-enabled methodology, sulfoxide **250** was prepared in 60% yield.

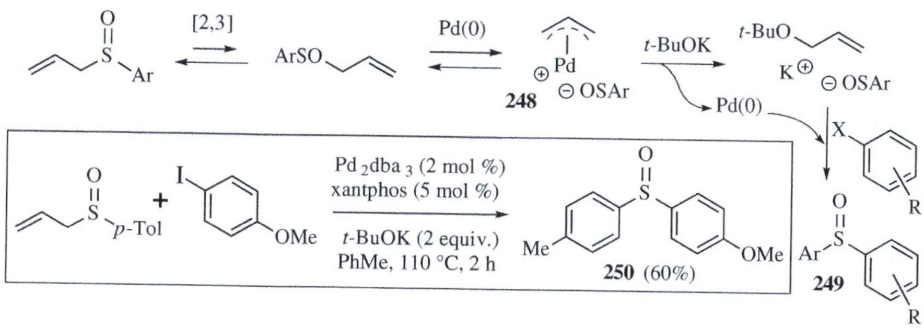

Scheme 18.63 Allyl sulfoxide as a sulfenate anion source in cross-coupling route to aromatic sulfoxides.[68]

18.12 HETEROATOM [2,3]-REARRANGEMENT VARIANTS

There are a number of [2,3]-sigmatropic rearrangements related to the Mislow–Evans rearrangement but differing in the heteroatom moiety. While these are not the main focus of the current chapter, they present unique opportunities for synthetic applications, and several recent examples are outlined in the following. Another closely related process, the [2,3]-Meisenheimer rearrangement of allylic *N*-oxides, is discussed separately in Chapter 15 on ammonium ylide rearrangements.

18.12.1 Allylic Selenoxide [2,3]-Rearrangements

The most widely used heteroatom variant of the Mislow–Evans rearrangement involves the use of allylic selenoxides in place of allyl sulfoxides. Early studies on allylic selenoxide rearrangements were carried out by Sharpless and Lauer[69] and Reich,[70] and several recent natural product syntheses have featured impressive applications of the method. As Reich pointed out, capture of the selenate esters resulting from the rearrangement can be easier than in the sulfur analogues because the selenates are easily hydrolyzed. Nevertheless, the addition of a selenophilic trap is often beneficial as highlighted in a number of the examples in this section.

En route to a total synthesis of the anticancer compound FR901464, Koide and coworkers carried out a diastereoselective allylic selenoxide rearrangement upon oxidation of either allyl selenide **251** (Scheme 18.64).[71] Optimization studies using the preformed allyl selenide **251** identified the *o*-nitrophenyl selenide as an effective aryl substituent and *N,N*-dimethylaminopyridine as the best selenophilic base additive in the formation of rearrangement product **252** (see top of Scheme 18.64). Reactions were slower and diastereoselectivity, a crucial parameter here, was lower using other bases or with less than 3 equiv of DMAP. Using the optimized conditions, a one-pot method for overall 1,3-allylic alcohol transposition was achieved. Starting from primary allylic alcohol **253**, formation of the allyl selenide and *in situ* oxidation–rearrangement–selenate ester trapping occurred to give the secondary allylic alcohol **252** in excellent yield and synthetically useful diastereoselectivity.

Increasing the synthetic versatility of this methodology, the allyl selenide precursor for selenoxide [2,3]-rearrangement can be carried through a number of synthetic steps, including ones involving oxidative conditions. This feature, along with diastereoselective, suprafacial migration in the sigmatropic rearrangement, was crucial in Kigoshi's total synthesis of (−)-13-oxyingenol, bearing the inside–outside architecture in its carbonyl-bridged and stereochemically dense tricyclic framework.[72] Introduction of the phenyl selenide group with inversion of configuration (**254** → **255**; Scheme 18.65) set the stage for transposition of the C-7 stereochemistry to a C-5 allylic alcohol. Moreover, the selenide was compatible with chemistry required for conversion of **255** to **256**, including oxidative removal of the methoxyphenylmethyl (MPM) protecting groups, Parikh–Doering oxidation at C-2, and OsO_4 dihydroxylation of a C-3—C-4 olefin. In the key [2,3]-sigmatropic rearrangement event, peroxyacid oxidation of the selenide and treatment of the allylic

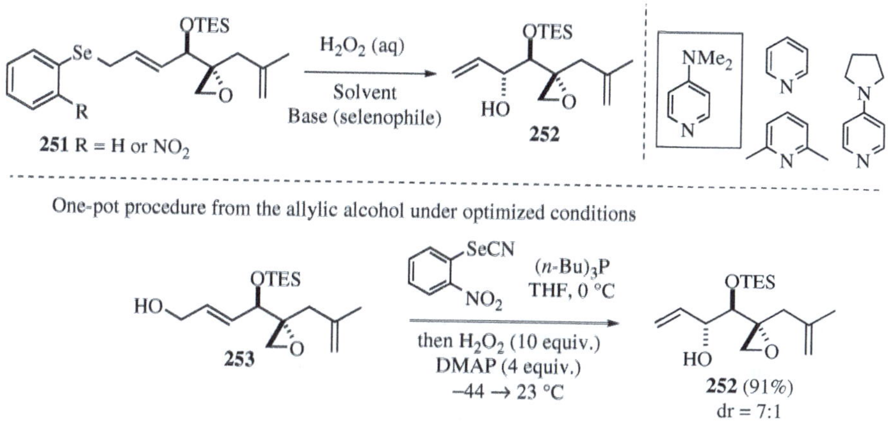

Scheme 18.64 Selenoxide [2,3]-rearrangement in the total synthesis of FR901464.[71]

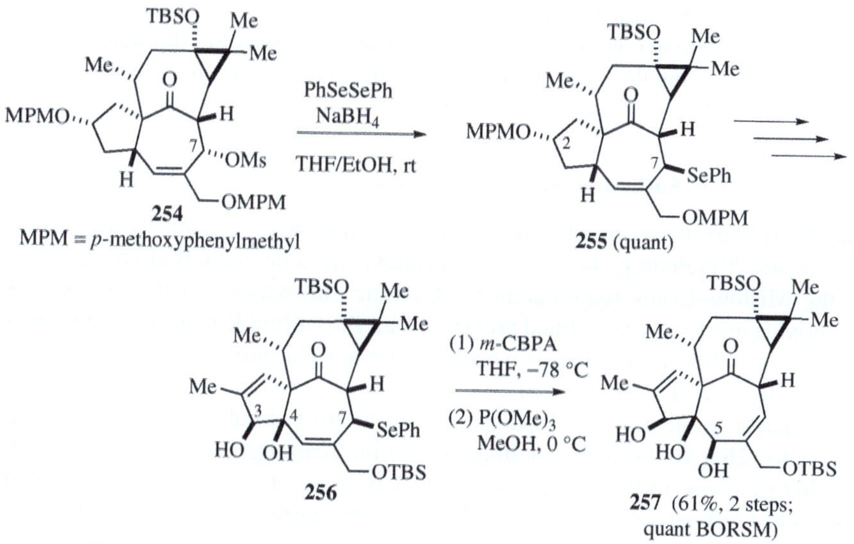

Scheme 18.65 Installation of allylic alcohol functionality in a highly complex setting in total synthesis of (−)-13-oxyingenol.[72]

selenoxides with trimethylphosphite provided triol **257** as a single diastereomer at C-5.

An allylic selenoxide rearrangement also proceeded with high diastereocontrol, but within an acyclic allyl system, in the course of Floreancig's total synthesis of the sponge-derived macrolactone natural product (+)-dactylolide.[73] Conversion of C-9 allylic alcohol **258** (Scheme 18.66) to the double-bond-migrated C-7 alcohol **259** was achieved via Mitsunobu-like incorporation of the selenide, oxidation to selenoxide **260**, and rearrangement with added pyridine. Product **259** was obtained as a single

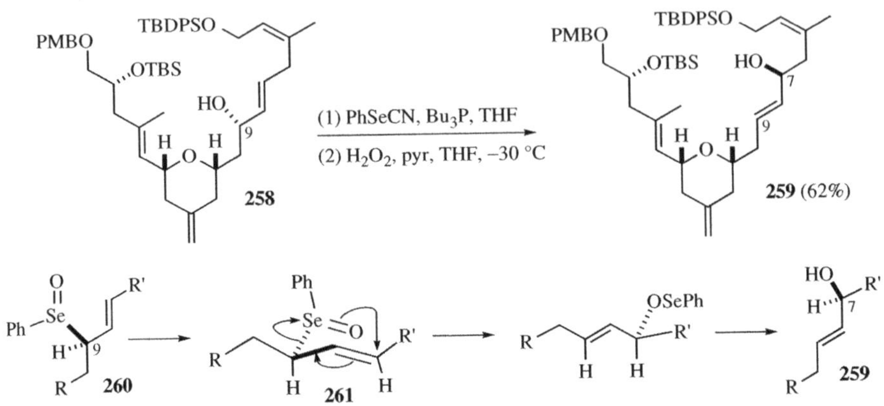

Scheme 18.66 Seleno Mislow–Evans applied for allylic transposition in total synthesis of (+)-dactylolide.[73]

epimer at C-7 and with the trans geometry at the C-8—C-9 alkene. This stereochemical outcome is consistent with rearrangement proceeding through a transition structure **261** having the large group at the allylic stereocenter in a pseudo-equatorial orientation.[74,75]

18.12.2 Allylic Sulfimide Rearrangements

The [2,3]-rearrangement of allylic sulfimides provides a route to allylic amine derivatives,[76] including alkaloids,[77] β-lactams,[78] and nonnatural amino acids.[79] As in the Mislow–Evans rearrangement of allylic sulfoxides and the corresponding selenoxide reactions exemplified previously, allylic sulfimide rearrangements present opportunities for chirality transfer and the preparation of enantioenriched products. In the amino acid area, Armstrong's group has used chiral proline derivative **262** as a catalyst in the enamine-mediated preparation of chiral, nonracemic α-sulfenyl aldehydes **263**, followed by olefination to give rearrangement precursors such as (*E*)-**264** (Scheme 18.67).[79] Transfer of an N-Boc nitrene fragment from oxaziridine **266** generated sulfimides of type **267**, which rearranged *in situ* to the allylic *N*-sulfenylcarbamate products **265**, with complete chirality transfer and *E*-specific alkene stereochemistry. Armstrong explained this outcome by invoking preferential transition assembly **267a** versus the $A^{1,3}$-disfavored alternative **267b** (i.e., the R group pseudo-equatorial in the five-membered-ring transition structure). Consistent with this transition state model, complementary allylic stereochemistry was obtained starting from the *Z*-alkene allylic sulfimide generated from sulfide (*Z*)-**264**.

Two additional features of the sulfimide rearrangement are noteworthy in the context of a chapter on the Mislow–Evans rearrangement. First, the sulfimides typically form as a mixture of diastereomers at sulfur, so at least in the examples shown earlier, the stereochemistry at sulfur does not appear to affect the outcome of the [2,3]-sigmatropic rearrangement. The same is often (but not always; see, for example, Fernández de la Pradilla's and Myers's studies in Schemes 18.25–18.27)

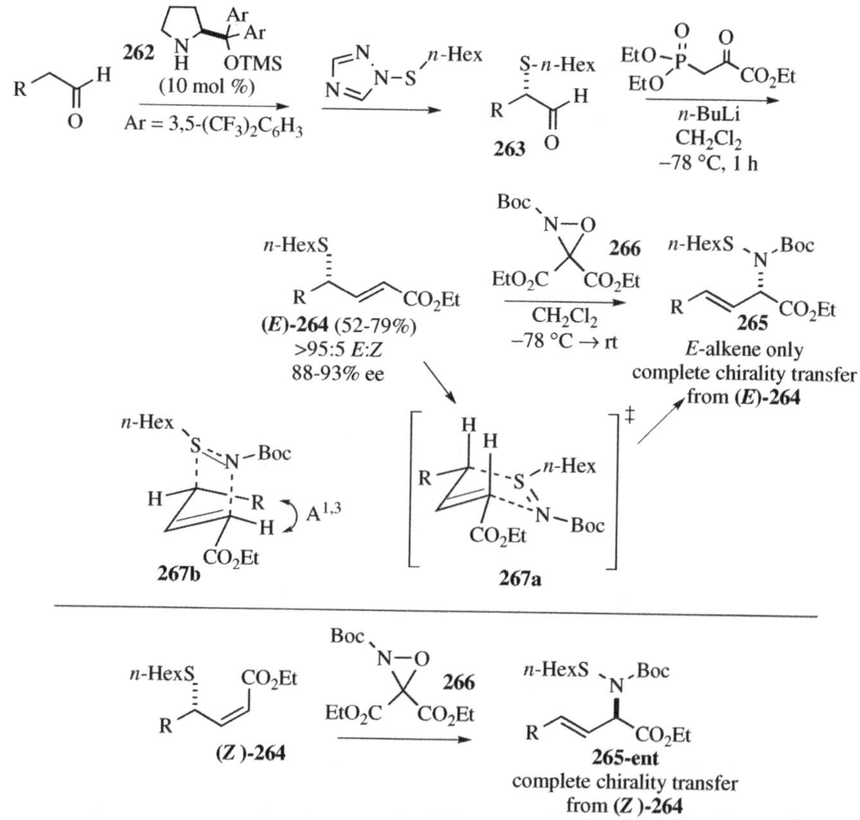

Scheme 18.67 Preparation and rearrangement of chiral allylic sulfimides.[79]

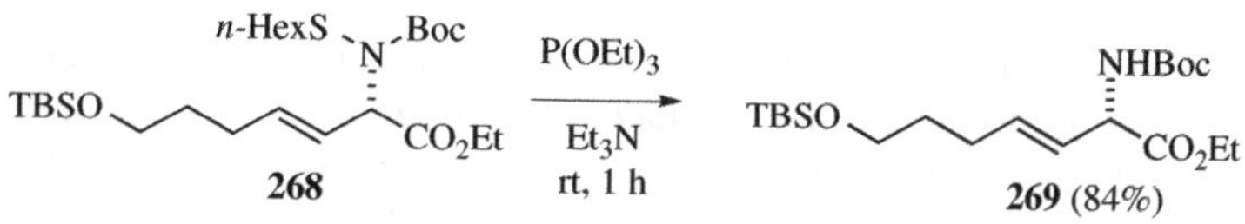

Scheme 18.68 Cleavage of the N—S bond occurs in a separate step.[79]

true of the Mislow–Evans rearrangement of allylic sulfoxides that are diastereomeric at the sulfur center and implies that the two sulfimide diastereomers proceed to the same major product but one via an endo and the other through an exo transition state arrangement (cf. Jorgensen's findings; Scheme 18.7). Second, in contrast to the usual thermodynamics of the allylic sulfoxide–allylic sulfenate interconversion, which favor the sulfoxide, for the sulfimide rearrangements shown in Scheme 18.67, the product **265** forms in high yield without *in situ* trapping by a thiophile, so cleavage of the N—S bond occurs in a subsequent and also high-yielding step (e.g., **268** → **269**; Scheme 18.68).

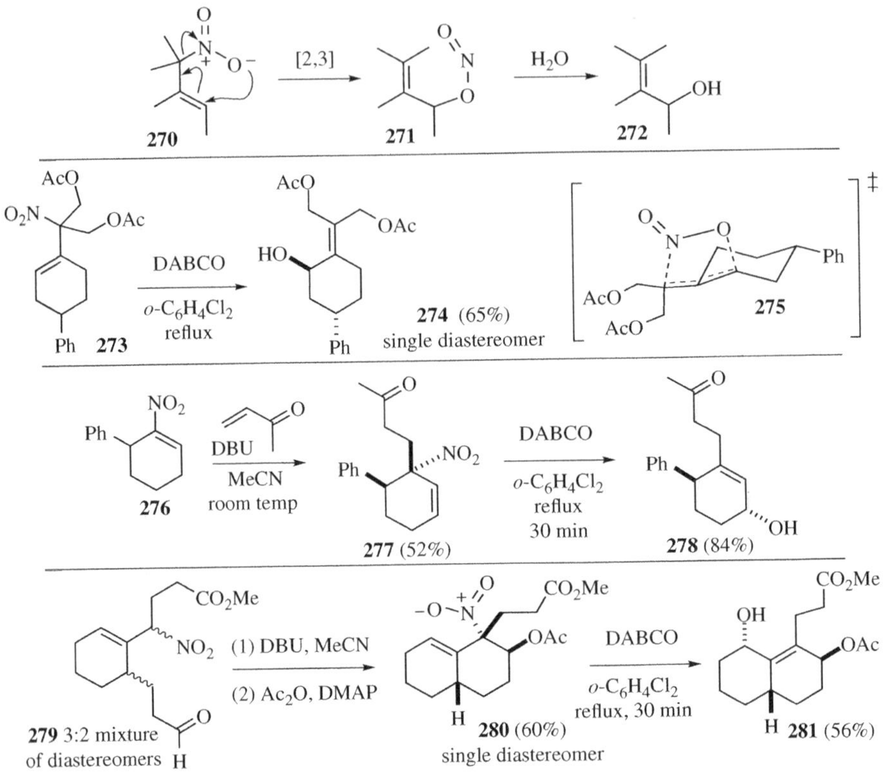

Scheme 18.69 Rearrangement of allylic nitro compounds via a [2,3] process.[80]

18.12.3 Rearrangement of Allylic Nitro Compounds

Allylic nitro derivatives can undergo Mislow–Evans-like [2,3]-sigmatropic rearrangement, giving allylic alcohol products upon hydrolysis of the resulting nitrite esters (**270** → **271** → **272**; Scheme 18.69). In exploring synthetic applications of this rearrangement, the Zard group found that inclusion of DABCO as the optimal base was crucial in order to suppress unwanted elimination side reactions.[80] The reaction was diastereoselective, as exemplified in the reaction of **273** to a single diastereomer of **274**, a result rationalized by invoking transition structure **275**, wherein there is pseudo-axial C—O bond formation within a chair-like arrangement and the phenyl group occupies an equatorial position.

The utility of this method also stems from the fact that the nitro group enables C—C bond formation prior to the rearrangement. Michael addition of the anion derived from **276** to methyl vinyl ketone led to allylic nitro compound **277**, which rearranged to allylic alcohol **278** in high yield and excellent diastereoselectivity. Meanwhile, a single diastereomer of the bicyclic framework **280** was available under thermodynamic control from nitro aldehyde **279** via reversible Henry reaction, and

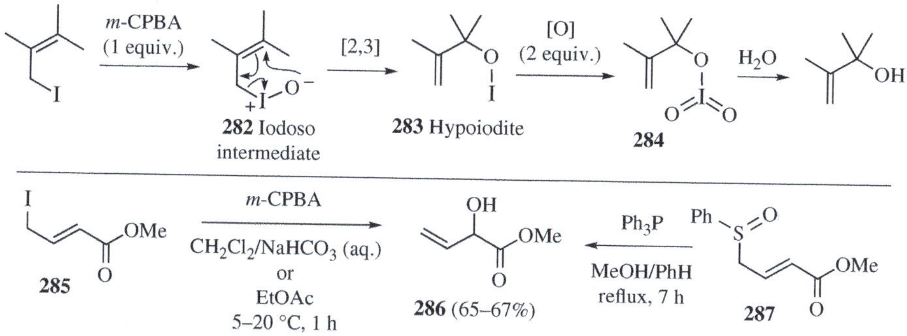

Scheme 18.70 Halogen oxide analogue of Mislow–Evans rearrangement.[81]

transposition of the allylic nitro stereocenter to allylic alcohol **281** resulted from the suprafacial nature of the ensuing [2,3]-rearrangement.

18.12.4 Allylic Halogen Oxide [2,3]-Rearrangements

An allylic halogen oxide [2,3]-sigmatropic rearrangement is another intriguing analogue of the Mislow–Evans process. In this scenario (Scheme 18.70), oxidation of an allylic iodide provides iodoso intermediate **282**, which rearranges to the corresponding allylic hypoiodite **283**. Yamamoto *et al.* found that the reaction worked best with excess peracid oxidant, as the initially formed hypoiodite was further oxidized to the iodate ester (**284**), an intermediate readily hydrolyzed to the final allylic alcohol product.[81] Examples included oxidative rearrangement of *trans*-γ-iodocrotonate ester **285**, giving allylic alcohol **286**, wherein the double bond was out of conjugation with the ester carbonyl. The same product could also be prepared in comparable yield from Mislow–Evans rearrangement of allylic sulfoxide **287**, but considerably more vigorous reaction conditions were required, highlighting the potential synthetic utility of the halogen oxide rearrangement.

Recently, this potential has been realized in Shoji's stereoselective preparation of methyl epi-anhydroquinate **290** using the allylic iodoso rearrangement strategy (Scheme 18.71).[82] The initial synthetic plan was to utilize either the sulfoxide (**289a**) or selenoxide (**289b**) as a [2,3]-rearrangement substrate (Scheme 18.71, top). However, attempted Mislow–Evans rearrangement of **289a** resulted in decomposition, while the corresponding selenoxide **289b** was capricious to prepare from epoxide precursor **288**. On the other hand, regio- and stereocontrolled epoxide opening of **288** using magnesium iodide resulted in allylic iodide **291**, which, upon oxidation, led to the desired product **290** with excellent diastereoselectivity. Shoji and coworkers found that in addition to the conditions pioneered by Yamamoto *et al.*,[81] addition of aqueous sodium potassium tartrate in the oxidation step was necessary to preserve high diastereoselectivity in formation of **290**. Apparently, **291** is prone to epimerization at the iodine-bearing stereocenter, a process that is suppressed by the added tartrate salts.[83]

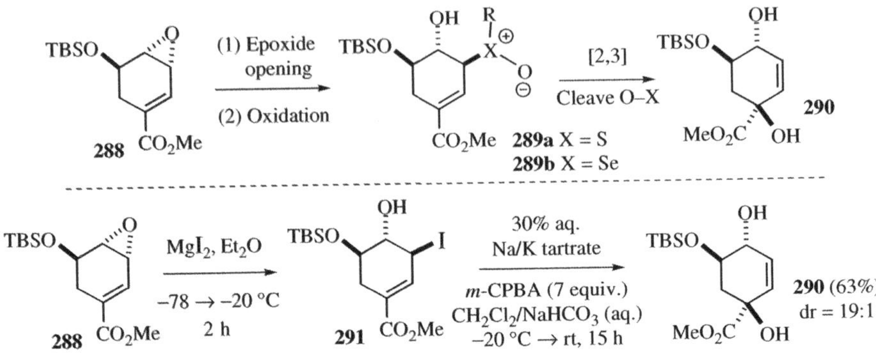

Scheme 18.71 Rearrangement of iodoso alkene in synthesis of quinic acid derivative.[82]

18.13 [2,3]-REARRANGEMENTS OF PROPARGYL AND ALLENYL SULFENATES AND SULFOXIDES

While they lie for the most part outside the scope of this chapter, [2,3]-sigmatropic rearrangements of sulfoxides and sulfenates within propargyl and allenyl systems present important opportunities for synthesis. Braverman has long been in the forefront of exploring this chemistry,[84] and numerous examples are highlighted in a 2007 review article.[6h] To give the reader at least a brief view of this panorama of reactions, several scenarios are shown in Scheme 18.72. In the first (Eq. 1), a propargylic sulfenate **292** rearranges to a sulfinyl-substituted allene **293**. These 1-allenyl sulfoxides are useful synthetically, for example, in further rearrangements and cycloadditions. Propargylic sulfoxides **294** (Eq. 2), meanwhile, provide the sulfenylated allen-1-ol **295** upon [2,3]-rearrangement en route to the final α,β-unsaturated carbonyl products of type **296**. Enones also result starting from β-allenyl sulfoxides **297**. In this

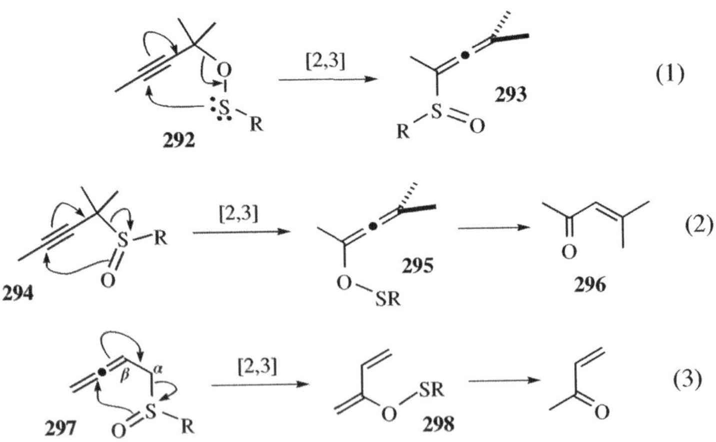

Scheme 18.72 Propargyl and allenyl sulfenate and sulfoxide [2,3] rearrangements.[6h]

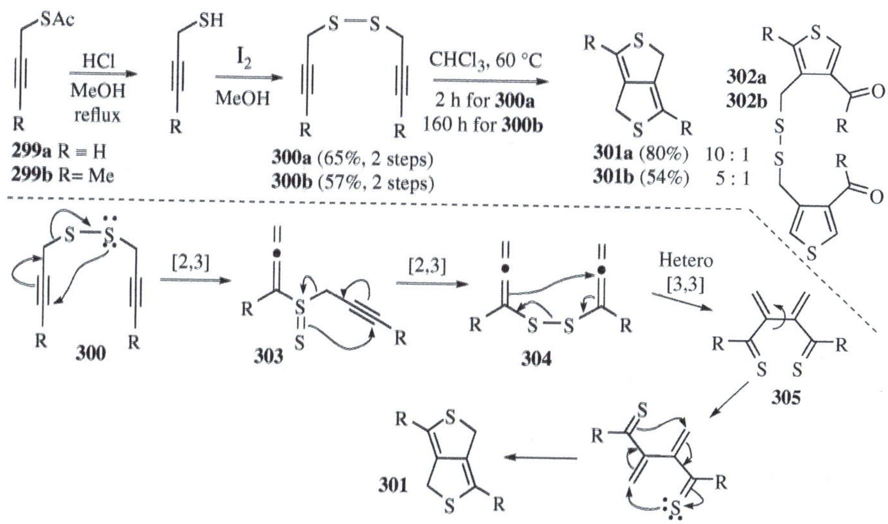

Scheme 18.73 Propargylic disulfide cascade rearrangement/cyclization process.[85]

case, the result of [2,3]-sigmatropic rearrangement is a 1,3-diene **298** with the sulfenate group at the 2-position, leading to the final enone product upon ketonization. This was the reaction used in tandem with an initial Claisen-type rearrangement in Posner's study presented earlier in this chapter (see Scheme 18.52).

Braverman recently reported a fascinating set of examples incorporating some of these themes but emanating from dipropargylic disulfides such as **300** (Scheme 18.73).[85] These starting materials were prepared from propargylic thioacetates **299** in two steps and can be viewed as sulfur analogues of propargylic sulfenates (cf. **292**; Scheme 18.72). Warming of **300** in chloroform provided thieno-thiophenes **301** along with some of oxidative dimerization by-products **302**, with the terminal alkynyl substrate **300a** reacting markedly faster than the methyl-substituted **300b** and giving less of the by-product (**301a**:**302a** = 10:1; **301b**:**302b** = 5:1). Mechanistically, Braverman and coworkers proposed that initial [2,3]-rearrangement of **300** provides thiosulfoxide **303**, which has functionality reminiscent of both sulfinyl allene **293** and propargylic sulfoxide **294** (cf. Eqs. 1 and 2; Scheme 18.72). A second [2,3] process then gives bis(allenyl) disulfide **304**. Remarkably, compared to the starting **300**, this represents an overall 1,3-transpostion of the disulfide linkage with concomitant migration of the π-bonds. Finally, a hetero-[3,3]-rearrangement, followed by double conjugate addition of the resulting α,β-unsaturated thioketone or thioaldehyde **305**, would explain the formation of the final product **301**. Braverman has also investigated related reactions of bis(propargyloxy) disulfides that provide dithiabicyclic architectures.[86] These examples beautifully illustrate the rich array of [2,3]-rearrangement-based transformations available to unsaturated sulfenate, sulfoxide, and disulfide starting materials, including tandem and cascade reactions.

18.14 CONCLUSION

This chapter has aimed to provide a broad survey of the Mislow–Evans rearrangement, starting from early work establishing the [2,3]-sigmatropic nature of the reaction and other key mechanistic features and ranging through classic synthetic applications and onto extensions at the forefront of current-day challenges in organic synthesis. Rather than providing a comprehensive review, the goal has been to communicate a sense of the breadth of this area and to stimulate thought on possible applications of the Mislow–Evans reaction and related rearrangements in future synthetic endeavors. Myriad opportunities remain for creative uses of these processes for the preparation of stereochemically rich and structurally complex molecular frameworks. Hopefully, this chapter will help bring these possibilities to the continuing attention of synthetic organic chemists.

REFERENCES

1. Sosnovsky, G. *J. Chem. Soc.* **1956**, 3139–3141.
2. (a) Rayner, D. R.; Miller, E. G.; Bickart, P.; Gordon, A. J.; Mislow, K. *J. Am. Chem. Soc.* **1966**, *88*, 3138–3139. (b) Miller, E. G.; Rayner, D. R.; Mislow, K. *J. Am. Chem. Soc.* **1966**, *88*, 3139–3140.
3. (a) Bickart, P.; Carson, F. W.; Jacobus, J.; Miller, E. G.; Mislow, K. *J. Am. Chem. Soc.* **1968**, *90*, 4869–4876. (b) Tang, R.; Mislow, K. *J. Am. Chem. Soc.* **1970**, *92*, 2100–2104.
4. Braverman, S.; Stabinsky, Y. *Chem. Commun.* **1967**, 270–271.
5. (a) Evans, D. A.; Andrews, G. C.; Sims, C. L. *J. Am. Chem. Soc.* **1971**, *93*, 4956–4957. (b) Evans, D. A.; Andrews, G. C. *Acc. Chem. Res.* **1974**, *7*, 147–155.
6. (a) Hoffmann, R. W. *Angew. Chem., Int. Ed. Engl.* **1979**, *18*, 563–572. (b) Braverman, S. In *The Chemistry of Sulphones and Sulphoxides*; Patai, S.; Rappoport, Z., Stirling, C., Eds.; John Wiley & Sons, Inc.: New York, **1988**, pp. 717–757. (c) Altenbach, H.-J. In *Comprehensive Organic Synthesis*; Trost, B. M., Fleming, I., Eds.; Pergamon: Oxford, **1991**; Vol. 6, pp. 829–871. (d) Brückner, R. In *Comprehensive Organic Synthesis*; Trost, B. M., Fleming, I., Eds.; Pergamon: Oxford, **1991**; Vol. 6, pp. 873–908. (e) Mikolajczk, M.; Drabowicz, J.; Kielbasinsky, P. *Chiral Sulfur Reagents*; CRC Press: Boca Raton, **1997**, pp. 78–87. (f) Prilezhaeva, E. N. *Russ. Chem. Rev.* **2001**, *70*, 897–920. (g) Reggelin, M. *Top. Curr. Chem.* **2007**, *275*, 1–65. (h) Braverman, S.; Cherkinsky, M. *Top. Curr. Chem.* **2007**, *275*, 67–101.
7. (a) Hassner, A.; Stumer, C. *Organic Syntheses Based on Name Reactions and Unnamed Reactions*; Elsevier: Amsterdam, **1994**; p. 266. (b) Li, J. J. *Name Reactions*; Springer: Berlin, **2002**; p 237. (c) Kürti, L.; Czakó, B. *Strategic Applications of Named Reactions in Organic Synthesis*; Elsevier: Burlington, MA, **2005**; pp. 292–293.
8. Dornan, P. K.; Kou, K. G. M.; Houk, K. N.; Dong, V. M. *J. Am. Chem. Soc.* **2014**, *136*, 291–298.
9. Evans, D. A.; Andrews, G. C.; Fujimoto, T. T.; Wells, D. *Tetrahedron Lett.* **1973**, *14*, 1389–1392.
10. Jones-Hertzog, D. K.; Jorgensen, W. L. *J. Org. Chem.* **1995**, *60*, 6682–6683.
11. (a) Rautenstrauch, V. *Chem. Commun.* **1970**, 526–527. (b) Goodwin, T. E.; Ratcliff, D. G.; Crowder, C. M.; Seitzinger, N. K. *J. Org. Chem.* **1982**, *47*, 815–820.

12. Jones-Hertzog, D. K.; Jorgensen, W. L. *J. Am. Chem. Soc.* **1995**, *117*, 9077–9078.
13. Zhou, Z. S.; Flohr, A.; Hilvert, D. *J. Org. Chem.* **1999**, *64*, 8334–8341.
14. Fokin, A. A.; Kushko, A. O.; Kirij, A. V.; Yurchenko, A. G.; Schleyer, P. v. R. *J. Org. Chem.* **2000**, *65*, 2984–2995.
15. (a) Abbott, D. J.; Stirling, C. J. M. *J. Chem. Soc. C* **1969**, 818–821. (b) Parker, A. J.; Kharasch, N. *Chem. Rev.* **1959**, *59*, 583–628.
16. Phosphites had been previously used to trap sulfenic acids: Cooper, R. D. G.; José, F. L. *J. Am. Chem. Soc.* **1970**, *92*, 2575–2576.
17. Miura, M.; Toriyama, M.; Kawakubo, T.; Yasukawa, K.; Takido, T.; Motohashi, S. *Org. Lett.* **2010**, *12*, 3882–3885.
18. (a) Sato, T.; Okazaki, H.; Otera, J.; Nozaki, H. *Tetrahedron Lett.* **1988**, *29*, 2979–2982. (b) Sato, T.; Shima, H.; Otera, J. *J. Org. Chem.* **1995**, *60*, 3936–3937.
19. Sato, T.; Otera, J.; Nozaki, H. *J. Org. Chem.* **1989**, *54*, 2779–2780.
20. Sato, T.; Otera, J. *Tetrahedron Lett.* **1994**, *35*, 6701–6704.
21. Wittman, M. D.; Halcomb, R. L.; Danishefsky, S. J. *J. Org. Chem.* **1990**, *55*, 1979–1981.
22. Halcomb, R. L.; Boyer, S. H.; Wittman, M. D.; Olson, S. H.; Denhart, D. J.; Liu, K. K. C.; Danishefsky, S. J. *J. Am. Chem. Soc.* **1995**, *117*, 5720–5749.
23. Griffith, D. A.; Danishefsky, S. J. *J. Am. Chem. Soc.* **1991**, *113*, 5863–5864.
24. (a) Kan, C.; Long, C. M.; Paul, M.; Ring, C. M.; Tully, S. E.; Rojas, C. M. *Org. Lett.* **2001**, *3*, 381–384. (b) Levites-Agababa, E.; Menhaji, E.; Perlson, L. N.; Rojas, C. M. *Org. Lett.* **2002**, *4*, 863–865.
25. Engstrom, K. M.; Mendoza, M. R.; Navarro-Villalobos, M.; Gin, D. Y. *Angew. Chem., Int. Ed.* **2001**, *40*, 1128–1130.
26. Fernández de la Pradilla, R.; Lwoff, N.; del Aguila, M. Á.; Tortosa, M.; Viso, A. *J. Org. Chem.* **2008**, *73*, 8929–8941.
27. Charest, M. G.; Lerner, C. D.; Brubaker, J. D.; Siegel, D. R.; Myers, A. G. *Science* **2005**, *308*, 395–398.
28. Davis, F. A.; Weismiller, M. C.; Murphy, C. K.; Reddy, R. T.; Chen, B.-C. *J. Org. Chem.* **1992**, *57*, 7274–7285.
29. Shi, J.; Manolikakes, G.; Yeh, C.-H.; Guerrero, C. A.; Shenvi, R. A.; Shigehisa, H.; Baran, P. S. *J. Am. Chem. Soc.* **2011**, *133*, 8014–8027 and Supporting Information.
30. Umezawa, T.; Hayashi, T.; Sakai, H.; Teramoto, H.; Yoshikawa, T.; Izumida, M.; Tamatani, Y.; Hirose, T.; Ohfune, Y.; Shinada, T. *Org. Lett.* **2006**, *8*, 4971–4974.
31. Shinada, T.; Fuji, T.; Ohtani, Y.; Yoshida, Y.; Ohfune, Y. *Synlett* **2002**, 1341–1343.
32. Smith, A. B., III; Adams, C. M. *Acc. Chem. Res.* **2004**, *37*, 365–377.
33. Evans, D. A.; Andrews, G. C. *J. Am. Chem. Soc.* **1972**, *94*, 3672–3674.
34. Majetich, G.; Song, J. S.; Ringold, C.; Nemeth, G. A. *Tetrahedron Lett.* **1990**, *31*, 2239–2242.
35. Evans, D. A.; Andrews, G. C.; Fujimoto, T. T.; Wells, D. *Tetrahedron Lett.* **1973**, *14*, 1385–1388.
36. Corey, E. J.; Hoover, D. J. *Tetrahedron Lett.* **1982**, *23*, 3463–3466.
37. Schreiber, S. L.; Satake, K. *J. Am. Chem. Soc.* **1984**, *106*, 4186–4188.
38. Epifani, E.; Florio, S.; Ingrosso, G.; Ronzini, L.; Sgarra, R.; Troisi, L. *Tetrahedron* **1991**, *47*, 7489–7500.

39. Jourdain, P.; Philippart, F.; Dumeunier, R.; Markó, I. E. *Tetrahedron Lett.* **2009**, *50*, 3366–3370.
40. For a discussion of the precise implications of each of these terms, including several examples of Mislow–Evans processes, see: Jones, A. C.; May, J. A.; Sarpong, R.; Stoltz, B. M. *Angew. Chem., Int. Ed.* **2014**, *53*, 2556–2591.
41. Evans, D. A.; Bryan, C. A.; Sims, C. L. *J. Am. Chem. Soc.* **1972**, *94*, 2891–2892.
42. A review that includes Diels–Alder/Mislow–Evans examples: Neuschütz, K.; Velker, J.; Neier, R. *Synthesis* **1998**, 227–255.
43. Posner, G. H.; Haces, A.; Harrison, W.; Kinter, C. M. *J. Org. Chem.* **1987**, *52*, 4836–4841.
44. Overman, L. E.; Petty, C. B.; Ban, T.; Huang, G. T. *J. Am. Chem. Soc.* **1983**, *105*, 6335–6337.
45. Carreño, M. C.; Cid, M. B.; Colobert, F.; García Ruano, J. L.; Solladié, G. *Tetrahedron: Asymmetry* **1994**, *5*, 1439–1442.
46. (a) Anderson, G. T.; Chase, C. E.; Koh, Y.-h.; Stien, D.; Weinreb, S. M. *J. Org. Chem.* **1998**, *63*, 7594–7595. (b) Stien, D.; Anderson, G. T.; Chase, C. E.; Koh, Y.-h.; Weinreb, S. M. *J. Am. Chem. Soc.* **1999**, *121*, 9574–9579.
47. Pearson, W. H.; Kropf, J. E.; Choy, A. L.; Lee, I. Y.; Kampf, J. W. *J. Org. Chem.* **2007**, *72*, 4135–4148.
48. Hama, N.; Matsuda, T.; Sato, T.; Chida, N. *Org. Lett.* **2009**, *11*, 2687–2690.
49. Tietze, L. F.; Beifuss, U. *Angew. Chem., Int. Ed. Engl.* **1993**, *32*, 131–163.
50. Pelc, M. J.; Zakarian, A. *Org. Lett.* **2005**, *7*, 1629–1631.
51. Pelc, M. J.; Zakarian, A. *Tetrahedron Lett.* **2006**, *47*, 7519–7523.
52. Ilardi, E. A.; Isaacman, M. J.; Qin, Y.-c.; Shelly, S. A.; Zakarian, A. *Tetrahedron* **2009**, *65*, 3261–3269.
53. (a) Paquette, L. A.; Reddy, Y. R.; Haeffner, F.; Houk, K. N. *J. Am. Chem. Soc.* **2000**, *122*, 740–741. (b) Paquette, L. A.; Reddy, Y. R.; Vayner, G.; Houk, K. N. *J. Am. Chem. Soc.* **2000**, *122*, 10788–10794
54. Posner, G. H.; Carry, J.-C.; Crouch, R. D.; Johnson, N. *J. Org. Chem.* **1991**, *56*, 6987–6993.
55. (a) Mapp, A. K.; Heathcock, C. H. *J. Org. Chem.* **1999**, *64*, 23–27. (b) For an additional application of this method in natural-product synthesis, see: Munchhof, M. J.; Heathcock, C. H. *J. Org. Chem.* **1994**, *59*, 7566–7567.
56. Nokami, J.; Mandai, T.; Imakura, Y.; Nishiuchi, K.; Kawada, M.; Wakabayashi, S. *Tetrahedron Lett.* **1981**, *22*, 4489–4490.
57. Tanikaga, R.; Nozaki, Y.; Tamura, T.; Kaji, A. *Synthesis* **1983**, 134–135.
58. Guerrero-de la Rosa, V.; Ordoñez, M.; Alcudia, F.; Liera, J. M. *Tetrahedron Lett.* **1995**, *36*, 4889–4892.
59. (a) Nokami, J.; Nishimura, A.; Sunami, M.; Wakabayashi, S. *Tetrahedron Lett.* **1987**, *28*, 649–650. (b) Nokami, J.; Kataoka, K.; Shiraishi, K.; Osafune, M.; Hussain, I.; Sumida, S.-i. *J. Org. Chem.* **2001**, *66*, 1228–1232.
60. (a) Hagiwara, H.; Isobe, K.; Numamae, A.; Hoshi, T.; Suzuki, T. *Synlett* **2006**, 1601–1603. (b) Minakata, S.; Komatsu, M. *Chem. Rev.* **2009**, *109*, 711–724.
61. (a) Trost, B. M.; Dong, L.; Schroeder, G. M. *J. Am. Chem. Soc.* **2005**, *127*, 2844–2845. (b) For a study on diastereoselectivity in the Knoevenagel/Mislow–Evans sequence, see: Trost, B. M.; Mallart, S. *Tetrahedron Lett.* **1993**, *34*, 8025–8028.

62. Brebion, F.; Nàjera, F.; Delouvrié, B.; Lacôte, E.; Fensterbank, L.; Malacria, M. *Synthesis* **2007**, 2273–2278.

63. Satoh, T.; Miyagawa, T. *Tetrahedron Lett.* **2006**, *47*, 1981–1983.

64. Debien, L.; Zard, S. Z. *Org. Lett.* **2013**, *15*, 6066–6069.

65. Magnus, P.; Ollivier, C. *Tetrahedron Lett.* **2002**, *43*, 9605–9609.

66. Kocienski, P. J. *Tetrahedron Lett.* **1979**, *20*, 2649–2650.

67. Luparia, M.; Legnani, L.; Porta, A.; Zanoni, G.; Toma, L.; Vidari, G. *J. Org. Chem.* **2009**, *74*, 7100–7110.

68. Bernoud, E.; Le Duc, G.; Bantreil, X.; Prestat, G.; Madec, D.; Poli, G. *Org. Lett.* **2010**, *12*, 320–323.

69. Sharpless, K. B.; Lauer, R. F. *J. Am. Chem. Soc.* **1972**, *94*, 7154–7155.

70. (a) Reich, H. J. *Acc. Chem. Res.* **1979**, *12*, 22–30. (b) Reich, H. J.; *J. Org. Chem.* **1975**, *40*, 2570–2572.

71. (a) Albert, B. J.; Sivaramakrishnan, A.; Naka, T.; Koide, K. *J. Am. Chem. Soc.* **2006**, *128*, 2792–2793. (b) Albert, B. J.; Sivaramakrishnan, A.; Naka, T.; Czaicki, N. L.; Koide, K. *J. Am. Chem. Soc.* **2007**, *129*, 2648–2659.

72. Ohyoshi, T.; Funakubo, S.; Miyazawa, Y.; Niida, K.; Hayakawa, I.; Kigoshi, H. *Angew. Chem., Int. Ed.* **2012**, *51*, 4972–4975.

73. Aubele, D. L.; Wan, S.; Floreancig, P. E. *Angew. Chem., Int. Ed.* **2005**, *44*, 3485–3488.

74. Reich, H. J.; Yelm, K. E. *J. Org. Chem.* **1991**, *56*, 5672–5679.

75. Hill, R. K. In *Asymmetric Synthesis*; Morrison, J. D., Ed.; Academic Press: Orlando, Fla., **1984**; Vol. 3, pp. 503–572 (see, in particular, pp. 554–558).

76. (a) See, for example: Bach, T.; Körber, C. *J. Org. Chem.* **2000**, *65*, 2358–2367. (b) The [2,3]-sigmatropic rearrangements of the corresponding selenium derivatives, selenilimines, have also been studied: Shea, R. G.; Fitzner, J. N.; Frankhauser, J. E.; Spaltenstein, A.; Carpino, P. A.; Peevey, R. M.; Pratt, D. V.; Tenge, B. J.; Hopkins, P. B. *J. Org. Chem.* **1986**, *51*, 5243–5252.

77. (a) Dolle, R. E.; Li, C.-S.; Shaw, A. N. *Tetrahedron Lett.* **1989**, *30*, 4723–4726. (b) Dolle, R. E.; Osifo, K. I.; Li, C.-S. Tetrahedron Lett. **1991**, *32*, 5029–5030.

78. Dolle, R. E.; Li, C.-S.; Novelli, R.; Kruse, L. I.; Eggleston, D. *J. Org. Chem.* **1992**, *57*, 128–132.

79. Armstrong, A.; Challinor, L.; Moir, J. H. *Angew. Chem., Int. Ed.* **2007**, *46*, 5369–5372.

80. Alameda-Angulo, C.; Quiclet-Sire, B.; Schmidt, E.; Zard, S. Z. *Org. Lett.* **2005**, *7*, 3489–3492.

81. Yamamoto, S.; Itani, H.; Tsuji, T.; Nagata, W. *J. Am. Chem. Soc.* **1983**, *105*, 2908–2909.

82. Inomata, S.; Ueda, M.; Sugai, T.; Shoji, M. *Chem. Lett.* **2013**, *42*, 1273–1275.

83. For additional examples of the allylic isodosyl rearrangement in synthesis, see: (a) Pearson, A. J.; Khan, M. N. I. *Tetrahedron Lett.* **1984**, *25*, 3507–3510. (b) Pearson, A. J.; Khan, M. N. I.; Clardy, J. C.; Cun-heng, H. *J. Am. Chem. Soc.* **1985**, *107*, 2748–2757.

84. Braverman, S.; Stabinsky, Y. *Isr. J. Chem.* **1967**, *5*, 125–126.

85. Braverman, S.; Cherkinsky, M.; Meridor, D.; Sprecher, M. *Tetrahedron* **2010**, *66*, 1925–1930.

86. Braverman, S.; Pechenick-Azizi, T.; Gottlieb, H. E.; Sprecher, M. *Synthesis* **2011**, 1741–1750.

PART V

IPSO REARRANGEMENTS

19

SMILES REARRANGEMENTS

LAURENT EL KAIM

Laboratoire de Synthèse Organique, CNRS, Ecole Polytechnique, ENSTA ParisTech – UMR 7652, Universtité Paris-Saclay, Palaiseau, France

LAURENCE GRIMAUD

Département de Chimie, Ecole Normale Supérieure-PSL Research University, Paris, France; CNRS, UMR 8640, PASTEUR, Sorbonne Universités, UPMC Univ Paris 06, Paris, France

19.1 INTRODUCTION

Smiles rearrangements are intramolecular nucleophilic aromatic substitutions (S_NAr) involving skeletal rearrangements of the type $ArX(CR_1R_2)_nYH \rightarrow ArY(C)_nXH$ with X and Y holding for heteroatoms (Scheme 19.1). Though the first indications of such migrations came at the end of the 19th century (Henriques and Hinsberg),[1] it was Smiles and Warren who later recognized the nature of the migration occurring on bis-naphthyl β-hydroxythioether **1** (Scheme 19.2)[2] and demonstrated that a spiro intermediate **4** could be trapped under oxidative conditions starting from sulfone **3**.[3]

Smiles extensively studied these rearrangements and showed that migrations were mainly observed between N, O, and S atoms. An important extension came in 1958 with the report by Truce *et al.* of a carbanionic Smiles-type rearrangement leading to new carbon–carbon bond formation under treatment of sulfone **5** with *n*-BuLi (Scheme 19.3).[4]

The ability to perform these reactions on nitrogenated heterocycles is a key to the synthetic utility of the Smiles rearrangements. Six-membered heterocycles such as pyridines or pyrimidines participate in Smiles rearrangements without requiring a further activating group. As a consequence, these reactions have been widely used

Molecular Rearrangements in Organic Synthesis, First Edition. Edited by Christian M. Rojas.

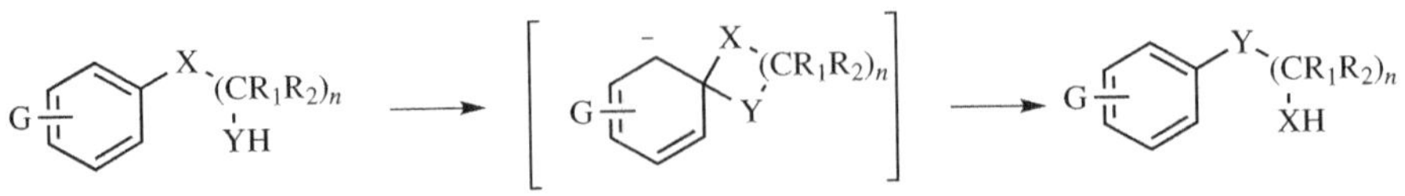

Meisenheimer complex

Scheme 19.1 Smiles rearrangements.

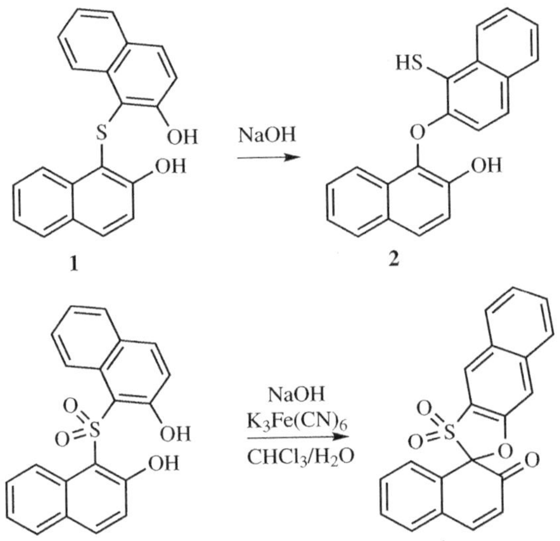

Scheme 19.2 Smiles rearrangements of bis-naphthyl β-hydroxythioether derivatives.

in medicinal chemistry, giving access to biologically relevant aminopyridine and aminopyrimidine derivatives.

Beyond the medicinal chemistry sphere, the most striking use of the Smiles rearrangements in synthesis came with the work of Julia and coworkers who transformed the rather tedious two-step olefination developed by his brother, Marc Julia, into an elegant one-step transformation under milder conditions.[5] Further improved by Kocienski and coworkers,[6] this olefination is now recognized as the Julia–Kocienski reaction. An illustrative example may be found in the conversion of the tetrazolylsulfone **7** into olefin **8** and tetrazole **9** (Scheme 19.4).

Smiles and Truce–Smiles rearrangements have been the object of several reviews.[7] The latest one on the Smiles rearrangement with an extensive coverage of the Julia–Kocienski process was by Wicha and coworkers in 2007. More focused reviews on the Julia–Kocienski olefination are also available.[8] In this chapter, we will focus on the Smiles rearrangements reported during the last 10 years without the goal of comprehensive coverage but highlighting recent integration of Smiles rearrangements in complex synthetic cascades. We include results from our

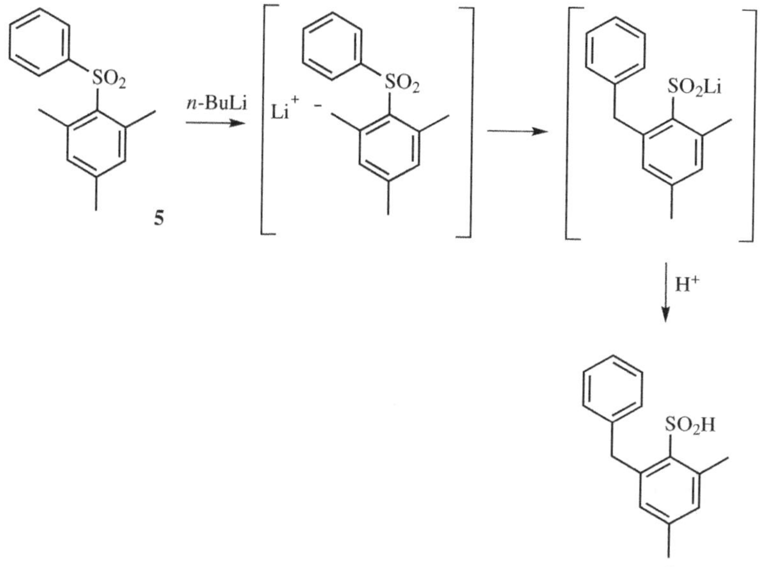

Scheme 19.3 Truce–Smiles rearrangements.

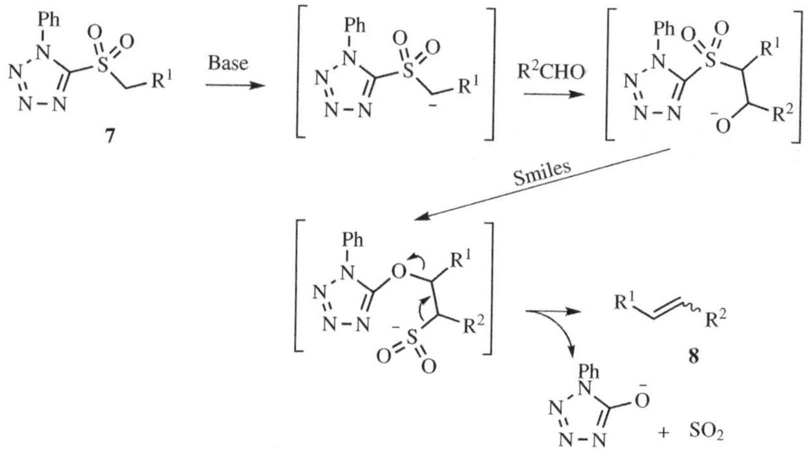

Scheme 19.4 Julia–Kocienski reaction of tetrazolylsulfone **7**.

research group and also emphasize relevant experimental details for implementation of Smiles-based strategies. Extensions of the Smiles rearrangement such as the Truce–Smiles rearrangement as well as the aryl migrations involving radical spirocyclization[9] will be outside the scope of this review. We begin with a discussion of the general principles controlling Smiles rearrangements to provide necessary background for readers.

19.2 SCOPE AND MECHANISTIC FEATURES

With hindsight over the last 80 years, most factors controlling the aryl shift in Smiles rearrangements are now well understood (a fast survey in the journals of the Royal Society of Chemistry and the American Chemical Society gave more than a thousand references on this rearrangement). Migrations are usually classified according to the leaving and entering atoms, an S—O shift, for example, standing for a transfer of the aryl group from a sulfur to an oxygen atom. When designing any synthetic plan involving a Smiles rearrangement, among the various elements that should be considered, the first one is certainly the nature of the aryl group and its substitution pattern. However, the ability of the aryl moiety to migrate between the two heteroatoms is not sufficient for synthetic efficiency as it only secures the existence of an equilibrium between the starting material and the desired product. Ultimately, what matters is the position of the equilibrium and the ability to displace it through changes in reaction conditions (e.g., pH, solvent, temperature). Thus, Smiles rearrangements must be analyzed in both thermodynamic and kinetic terms, taking into account all the actors in the reaction sequence. The following general trends are gleaned from the literature on Smiles processes.

19.2.1 Nature and Substitution of the Aryl Ring

As intramolecular aromatic nucleophilic substitutions, Smiles rearrangements involve spiro Meisenheimer complexes (Scheme 19.1) which are strongly stabilized by an electron-withdrawing substituent on the aromatic ring. It is not surprising that most of the early examples were observed with nitro, ester, or sulfone groups ortho or para to the leaving group. Substitution by halogen atoms (Cl, Br, F) as well as trifluoromethyl groups also activates the transfer, but efficient reactions often require more than one substituent on the aromatic ring. Gas-phase studies have shown that anionic Smiles rearrangement could occur without any activating group on the aryl ring, and this has been confirmed in solution under chelating conditions (cryptan or HMPA as solvent) which enhance the reactivity of the anion (Scheme 19.5).[10]

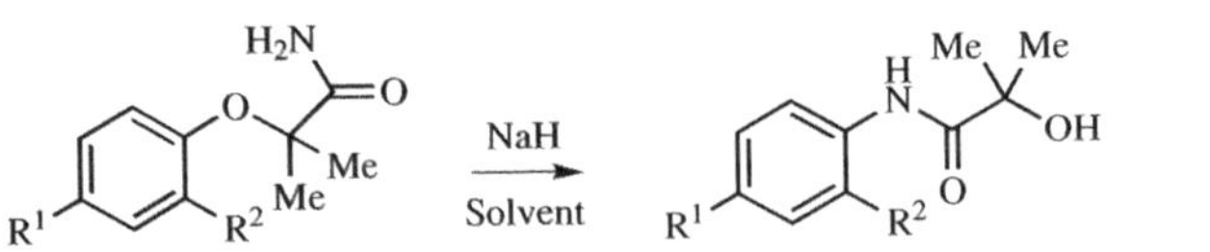

10 11

$R_1, R_2 = H$	DMF (100 °C/16 h): 80%
	HMPT (100 °C/1 h): 85%
$R_1 = Cl, R_2 = H$	DMF (100 °C/16 h): 80%
	HMPT (25 °C/16 h): 85%
$R_1 = H, R_2 = Me$	DMF (100 °C/16 h): 45%
	HMPT (100 °C/1 h): 60%

Scheme 19.5 Solvent effect in Smiles rearrangements.

Six-membered nitrogenated heterocycles such as pyridines and pyrimidines are inherently electron deficient, and a further activating group is not required for most transfers. The substitution patterns for these heterocycles follow the general rules observed for intermolecular S_NAr reactions on heterocycles.[7a] In the case of pyridines, even if substitution at the 2- and 4-positions is preferred, it is also possible to observe Smiles rearrangements at the less activated 3- and 5-positions. The situation is more complex for five-membered ring heterocycles. Being relatively electron rich, their analogous Meisenheimer complexes are of higher energy, and most Smiles rearrangements with theses heterocycles require more than one heteroatom in the cycle (e.g., tetrazoles, thiazole, oxadiazoles) allowing a decrease of the electron density together with lower aromaticity. In any case, the nature of the leaving and entering groups is expected to have a strong influence, and the electronic nature of the aromatic core should be adapted to the nucleophilicity of the heteroatom on which the aryl moiety is transferred.

19.2.2 Properties of the Nucleophile and Leaving Group

Smiles rearrangements involve mostly S, O, and N atoms. In the potential equilibrium established during the reaction, *N*-aryl derivatives are usually favored over their *O*- and *S*-aryl isomers. Though this behavior may be first explained by a simple analysis of the dissociation energies of various C_{aryl}-heteroatom bonds (Ph-OEt = 76 kcal/mol, Ph-NMe_2 = 100.6 kcal/mol, Ph-SMe = 85.4 kcal/mol, and Ph-SO_2Me = 82 kcal),[11,12] a more careful examination is always needed because of the strong influence of the experimental conditions, the importance of the chain tethered to the aryl, and the incompletely available thermodynamic data.

Most Smiles rearrangements are performed under basic conditions in protic polar solvents from room temperature to 120 °C. The base strength normally matches with the nature of the nucleophilic group involved. Amides are among the best nucleophiles for Smiles rearrangements, leading to the desired rearranged products from arylether and thioether starting materials as well as from the S-oxidized sulfinyls and sulfones. Treatment with potassium *tert*-butoxide, potassium hydroxide, or cesium carbonate in polar solvents (e.g., DMSO, DMF, CH_3CN) activates the amide into a reactive anionic species which efficiently forms the Meisenheimer complex. Amines may form the related spiro complex under neutral conditions, but the resulting ammonium salt normally reverts back to the starting amine, and additional bases have a strong activating effect. Table 19.1 underlines the general trends observed with nucleophiles and leaving groups in the Smiles rearrangement. If the sulfinate appears as one of the most efficient leaving groups, its effect should also be analyzed in conjunction with the electronic nature of the aryl core before and after the rearrangement. Indeed, the replacement of the electron-withdrawing sulfonyl with a π-donating O or N atom ensures the irreversibility of the Smiles process.

The importance of the experimental conditions in Smiles rearrangements may be further stressed by different studies showing a strong influence of the base/solvent couple on the chemoselectivity of synthetic cascades displaying a Smiles rearrangement. This is particularly true for Smiles reactions involving similar chemical

TABLE 19.1 General Trends in Smiles Rearrangements

YH	X
CONHR or NHCOR	SO_2, SO, O
NHR	SO_2, O, sometimes S, SO
SH	SO_2, O
OH	SO_2

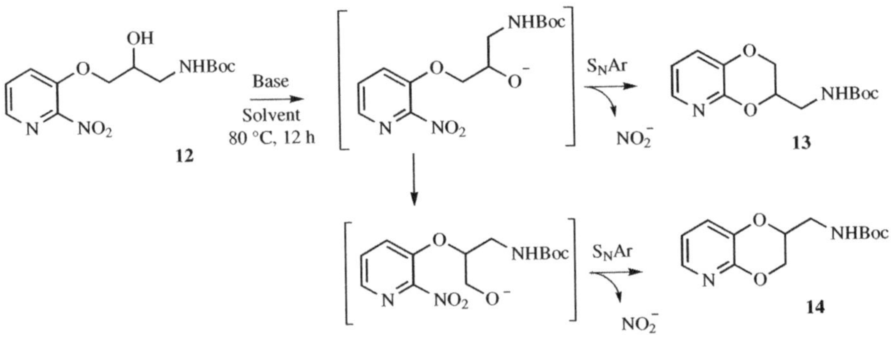

NaH/DME: (**13**/**14**)= 100/0 (92%)
t-BuOK/*t*-BuOH: (**13**/**14**)= 40/60 (94%)

Scheme 19.6 Influence of the base/solvent couple in Smiles rearrangements.

functionality in both leaving and entering groups as shown by the conversion of pyridine **12** into the mixture of dioxinopyridines **13** and **14** (Scheme 19.6).[13] Product **13** is the result of a direct intramolecular S_NAr on the nitro-tethered position, whereas a prior Smiles rearrangement is observed for **14**. The potential Smiles rearrangements involving the amide groups in **12**, **13**, and **14** are not productive due to the steric hindrance of the Boc group together with a faster 5-exo-trig cyclization of the alcoholate and final cyclization to more stable six-membered fused dioxines.

19.2.3 Chain Length of the Tether

Most examples of Smiles rearrangements involve five-membered ring spiro intermediates (with usually two carbon atoms between the heteroatoms). Though less common, migration of aryl moieties via six-membered spiro rings may still be observed under similar experimental conditions. When increasing the size of the spiro system, the reactions become more difficult to achieve and display properties closer to standard S_NAr. Smiles migrations involving small rings are known processes and are often categorized as different name reactions. These include the Chapman rearrangement of *O*-arylimidates[14] and the Schönberg or the related Newman–Kwart reactions of

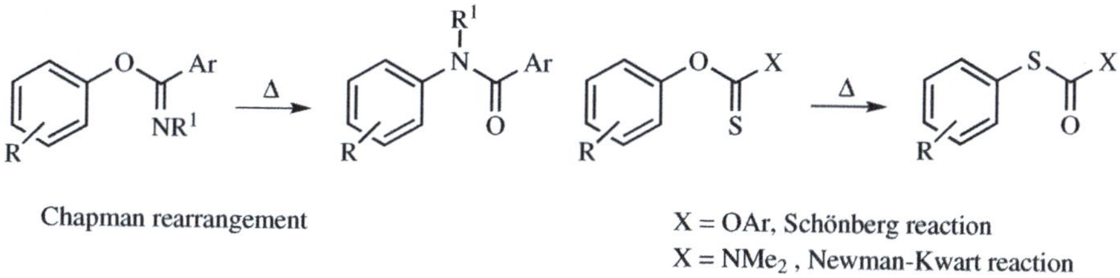

Scheme 19.7 Chapman, Schönberg, and Newman–Kwart reactions.

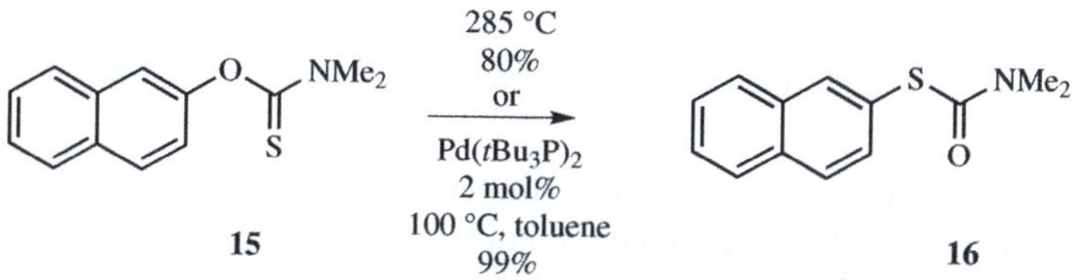

Scheme 19.8 Newman–Kwart reaction under palladium catalysis.

O-aryl thionocarbonates and thiocarbamates (Scheme 19.7).[15] These reactions have been used extensively for the conversion of phenols into anilines and thiophenols. They proceed via four-membered ring spiro intermediates and show, with respect to the aryl moieties, the same reactivity trends as Smiles rearrangements. However, the required experimental conditions are much harsher because of the strained spiro intermediate required for the process. Often performed under neat and neutral conditions, temperatures between 200 °C and 300 °C are often required for the 1,3-shift of the aryl ring, whereas traditional Smiles transfer occurs under 100 °C. A noteworthy improvement of the Newman–Kwart reactions is the disclosure of a catalytic effect of palladium, allowing the transformation of **15** into **16** to occur at 100 °C (Scheme 19.8).[16]

Smiles processes involving three-membered cyclic transition states are much less frequent under ionic conditions.[17] However, similar Truce–Smiles processes under radical conditions are well documented.[18] Known as neophilic rearrangements, they lead to efficient carbon–aryl bond formation with frequent extrusion of small molecules such as sulfur dioxide. These Smiles-type reactions involving cyclic transition states with a small ring size will not be evaluated further in this chapter.

19.3 APPLICATION OF SMILES REARRANGEMENTS

19.3.1 Synthesis of Fused Heterocycles through Cascades Involving Smiles Rearrangements

One of the most commonly encountered applications of Smiles rearrangements in synthesis is to prepare fused aromatic systems via three-step cascades displaying

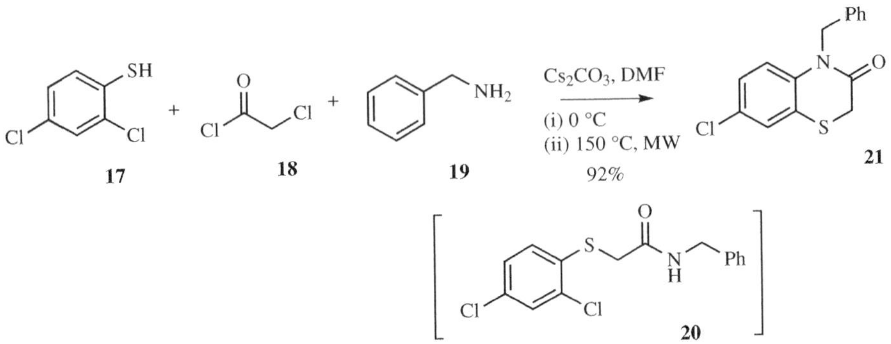

Scheme 19.9 Application of Smiles rearrangements to benzothiazine synthesis.

two nucleophilic substitutions with a Smiles rearrangement in between. These sequences have been adapted into more complex three-component couplings such as the benzothiazine synthesis shown in Scheme 19.9.[19] In this process, **19** reacts first with acyl chloride **18** leading to **20** after substitution with **17**. Heating the mixture under microwave conditions triggers the Smiles rearrangement and the final S_NAr toward **21**.

Whereas such cascades often involve an intermolecular nucleophilic substitution as the first step, more original preparations of precursors for Smiles rearrangements (mainly *O*-aryl derivatives) have been obtained through copper-catalyzed processes. Snieckus and coworkers have recently reported a copper-triggered formation of an intermediate diarylether **24**, which spontaneously evolves through a Smiles rearrangement followed by a cyclization to give oxazepinone **25** (Scheme 19.10).[20] A similar transformation of indole **26** into **28** has been described by Zhang and coworkers (Scheme 19.11).[21]

In terms of mechanism, the last step of these processes deserves particular attention. Traditionally represented as a simple aromatic nucleophilic substitution occurring after the Smiles rearrangement (Scheme 19.12), this final nucleophilic displacement involves a position on the aromatic core not usually activated for such substitution. Indeed, whereas in **29** the electron-withdrawing group E helps the Smiles rearrangement occur, its effect in the subsequent S_NAr leading to **30** is probably not strong enough to counterbalance the presence of an electron-donating N atom ortho to the leaving group.

Two recent DFT and MP2 studies on such cyclizations give a better understanding of the process. The first set of calculations involves a Smiles rearrangement of 3-hydroxypyridine derivative **31** (Scheme 19.13).[22] Whereas the theoretical study shows the preference for the Smiles rearrangement over direct S_NAr, it also demonstrates that the transition state **32** for the final substitution has a structure close to the spiro Smiles intermediate with a lengthening of both C—O spiro and C—X bonds (Scheme 19.13). Indeed, the elimination of the ortho leaving group may be considered as concerted with the ring opening of the spiro Smiles intermediates. The second DFT study on *ortho*-fluorothiophenol derivatives confirms this behavior.[23] It is

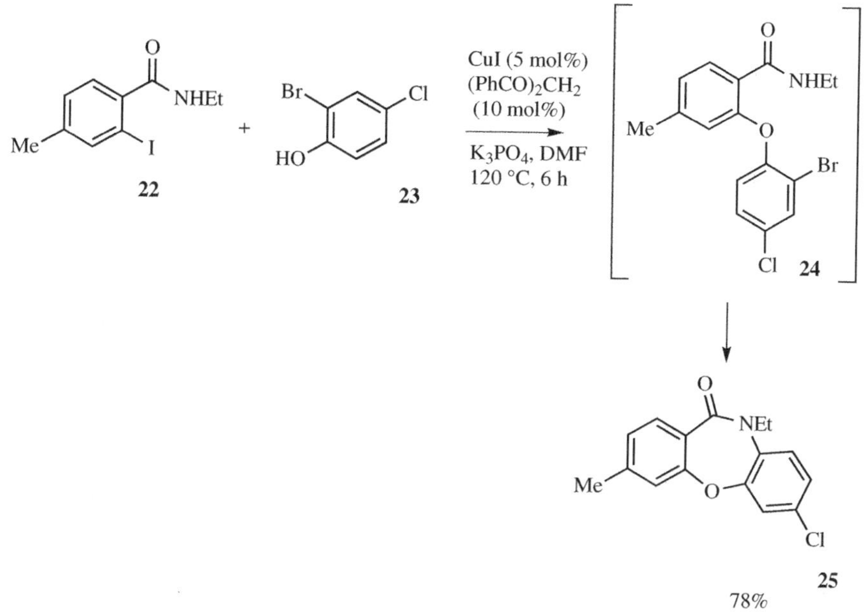

Scheme 19.10 Copper-triggered formation oxazepinone involving an intermediate Smiles rearrangement.

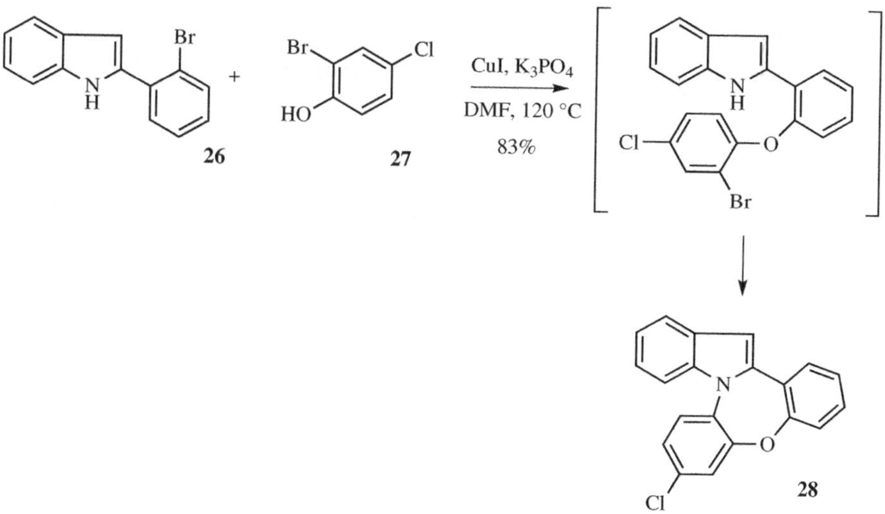

Scheme 19.11 Smiles rearrangements in Zhang's polycyclic indole synthesis.

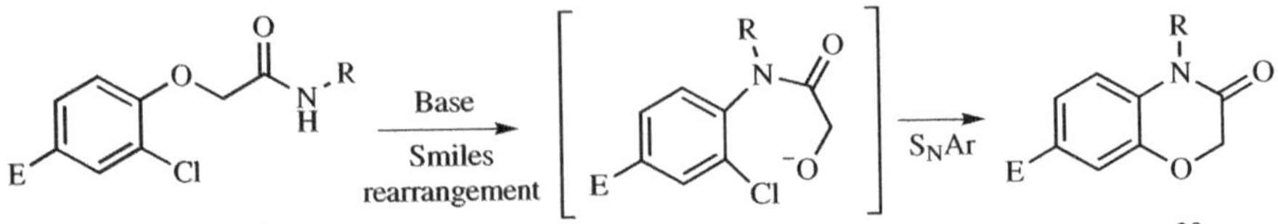

Scheme 19.12 Formal Smiles rearrangement/S_NAr cascade.

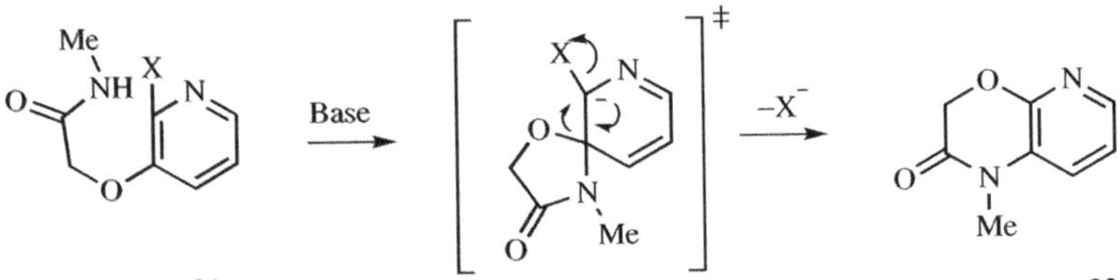

Scheme 19.13 Concerted ring opening/substitution in Smiles rearrangement of 3-hydroxypyridine derivative **31**.

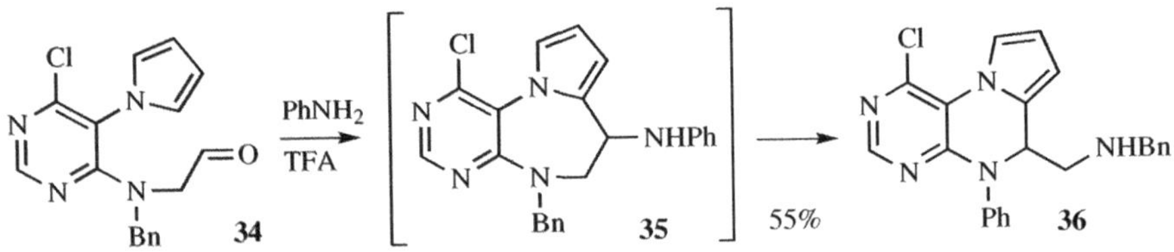

Scheme 19.14 Pictet–Spengler/Smiles cascade by Bai and coworkers.

possible that further theoretical studies may consider transient spirocarbene derivatives as intermediates or transition states for such processes.

Several elegant synthetic cascades have been disclosed with *in situ* formation of an amino derivative amenable to Smiles rearrangement. A good illustration involves Pictet–Spengler cyclizations of properly functionalized aldehydes. Upon addition of a primary amine, trapping of the resulting iminium forms a secondary amine, triggering a Smiles rearrangement. In the piperazinopyrimidine synthesis of Bai and coworkers (Scheme 19.14),[24] the observed Smiles transfer involves a substitution of one amine by another amine residue, traditionally a difficult reaction to control. In this case, the selectivity is partly managed by the conversion of a seven-membered ring intermediate **35** into a more stable six-membered final compound **36**. In addition, the potential equilibrium established during the Smiles rearrangement is probably controlled by the acidity of the reaction mixture, allowing protonation of the final more basic secondary dialkylamine. This is confirmed by the much lower yields observed when aliphatic amines are used instead of anilines. The same group has also disclosed similar Pictet–Spengler/Smiles cascades involving O—N Smiles shifts.[25]

19.3.2 Julia–Kocienski and Related Reactions

In the early 1970s, the Smiles rearrangement received renewed interest due to the development of the Julia olefination procedure.[8] Kende and Mendoza first introduced 1-methylimidazol-2-yl sulfones to avoid functionalization of the hydroxysulfone in the olefination process,[26] but Julia and coworkers introduced the benzothiazol-2-yl sulfones (BT-sulfones) to perform the reaction in a single step.[27] The olefination is generally performed in THF, sometimes with additives such as HMPT. The sulfone **37** is treated by a nonnucleophilic base, such as LDA or LiHMDS, at −78 °C, and then, the carbonyl compound is added to the sulfone anion at low temperature (Scheme 19.15). Control of the temperature is very important in this reaction to prevent intra- or intermolecular nucleophilic addition onto the carbon of the thiazolyl moiety. Indeed, due to the great electrophilicity of this carbon, self-condensation of **38** is difficult to avoid. The olefination mechanism involves addition of the sulfone anion **38** to the aldehyde, and the resulting hydroxysulfone **39** spontaneously undergoes a Smiles rearrangement, giving the desired alkene after concomitant SO_2 extrusion and hydroxybenzothiazole elimination (Scheme 19.15).

Various heterocyclic sulfones were investigated in this coupling, including pyridyl and pyrimidyl derivatives[27] as well as electron-poor aryl[28–31] sulfones. Due to lower electrophilicity of the carbon center linked to the sulfur atom, these are more stable than BT-sulfones. For instance, the pyridylsulfone carbanion does not self-condense at room temperature over several minutes. These compounds have been recently used in the preparation of glycosidic vinyl ethers **42** through the condensation of β-glycosidyl-substituted pyridylsulfones **41** with aldehydes (Scheme 19.16).[32]

Barbier conditions have been developed to improve the olefination yields, but they are not always compatible with highly functionalized carbonyl derivatives. Among the various heterocyclic sulfones studied in this reaction, the most important variant was introduced by Kocienski *et al.*, who used bulky tetrazolylsulfones

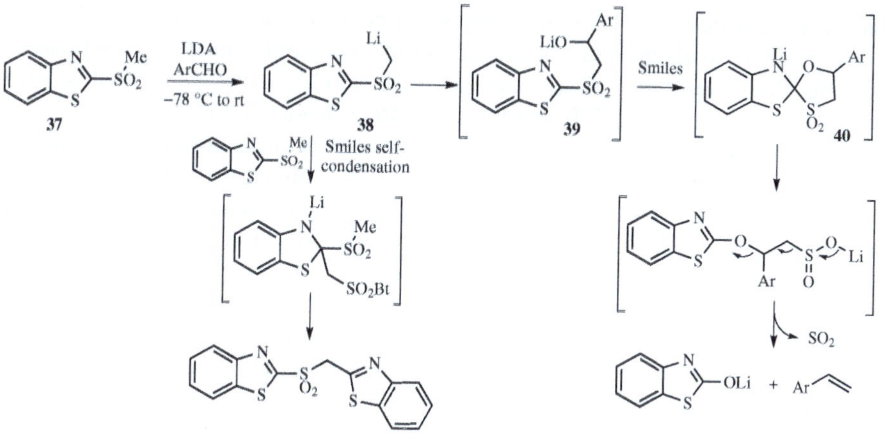

Scheme 19.15 Benzothiazo-2-yl sulfones in Julia–Kocienski reactions.

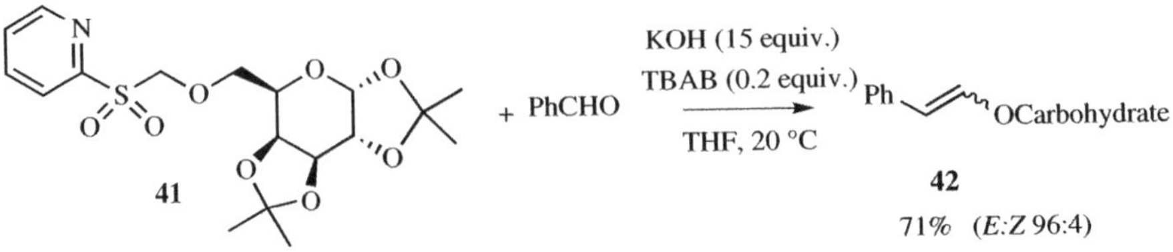

41

42

Scheme 19.16 Pyridylsulfones in Julia–Kocienski reactions.

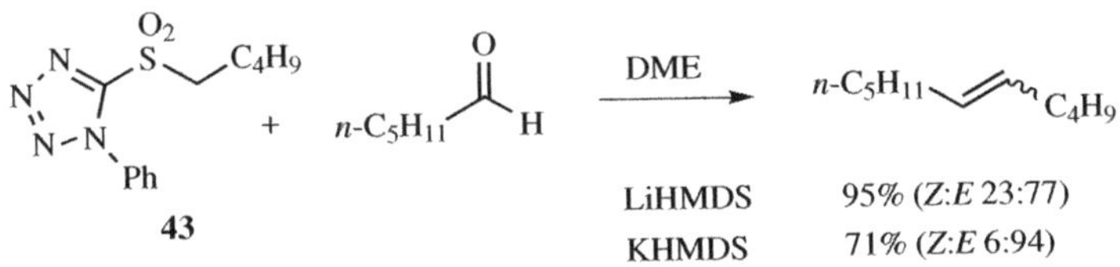

43

Scheme 19.17 Tetrazolylsulfones in Julia–Kocienski reactions.

with low propensity toward self-condensation: 1-phenyl-1*H*-tetrazol-5-yl sulfone **43** (PT-sulfone, Scheme 19.17)[6] and 1-*tert*-butyl-1*H*-tetrazol-5-yl sulfones (TBT-sulfone).[33] Their stability under basic conditions allows sulfone premetallation, thereby broadening the scope of compatible carbonyl derivatives for the condensation.

More recently, the use of electron-poor aryl sulfones has been successfully reinvestigated. For instance, 3,5-bis(trifluoromethyl)phenyl (BTFP)-sulfones **44** treated with a base such as KOH at room temperature or phosphazene (P_4-*t*Bu) at −78 °C have proved to react efficiently with a wide range of carbonyl compounds (Scheme 19.18).[29] Similarly, nitrophenylsulfones (NP-sulfones) **45a** react with aromatic aldehydes in the presence of sodium hydride in DMF at room temperature to give stilbene derivatives in high yields (Scheme 19.19).[31] Pentachlorophenylsulfones have been reported as well for the synthesis of benzylidenecyclopropenes.[30]

A number of alkene classes are accessible via Smiles-mediated olefinations. Among them, methylenation reactions give moderate to good yields using tetrazolylsulfones such as **45b** or arylsulfone **46** (Scheme 19.20).[29,34,35]

The synthesis of trisubstituted alkenes has scarcely been disclosed; some examples may be found in the work of Nájera and coworkers using electron-deficient aryl sulfones **47** combined with phosphazene (Scheme 19.21).[36] The scope is even more limited for tetrasubstituted olefins as the yields are quite low. Acylsilanes[37] afford trisubstituted vinylsilanes such as **48** mainly as the *E* isomers, but the stereoselectivity can be controlled by modulating the nature of the silyl group (Scheme 19.22). *E*-configured conjugated esters[38,39] and Weinreb amides[39] have been synthesized successfully as well by Nájera and coworkers starting from conveniently substituted arylsulfone **49** (Scheme 19.23).

The stereoselectivity of the olefination reaction strongly depends on both starting materials and experimental conditions. Indeed, the nature of the sulfone

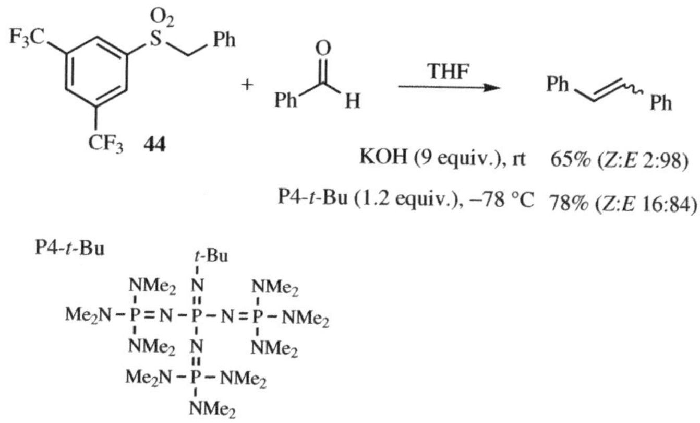

Scheme 19.18 3,5-Bis(trifluoromethyl)phenyl sulfones in Julia–Kocienski reactions.

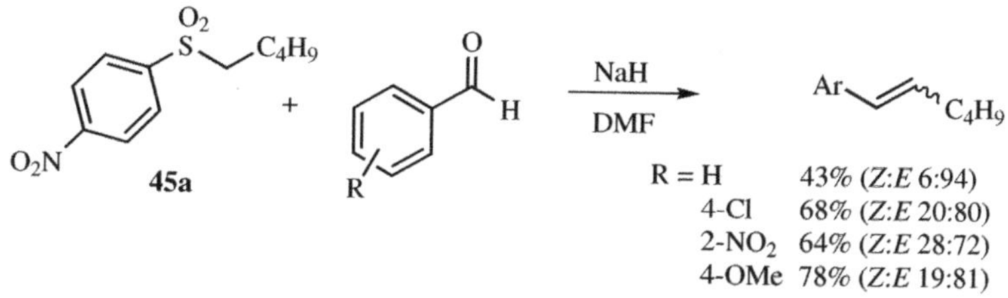

Scheme 19.19 Nitrophenylsulfones in Julia–Kocienski reactions.

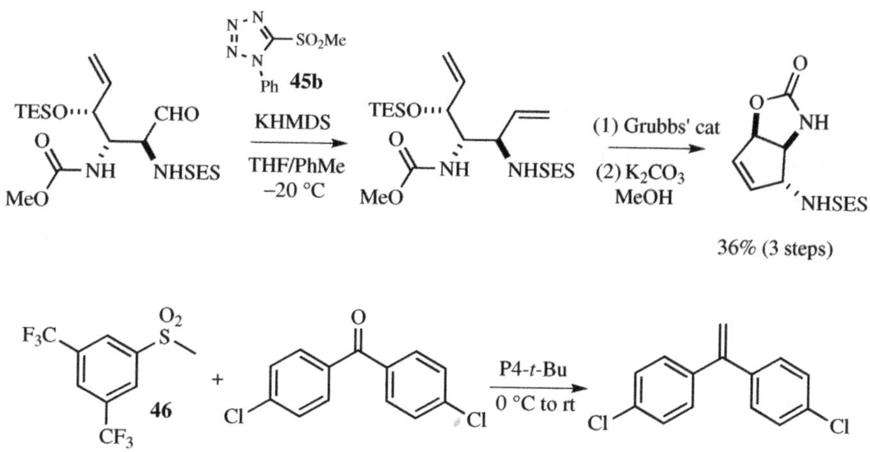

Scheme 19.20 Methylenation through Julia–Kocienski reactions.

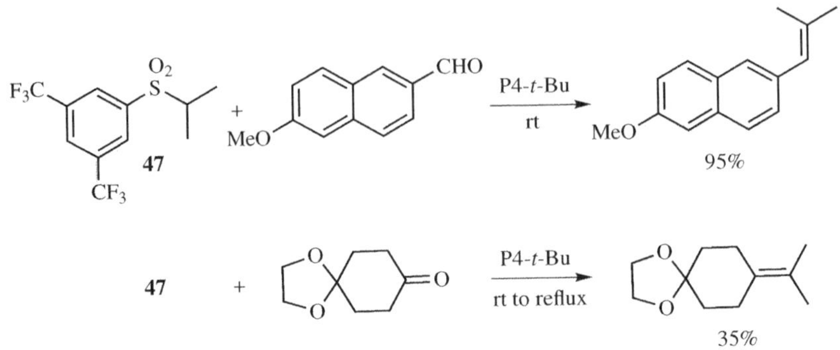

Scheme 19.21 Julia–Kocienski reactions leading to tri- and tetra-substituted alkenes.

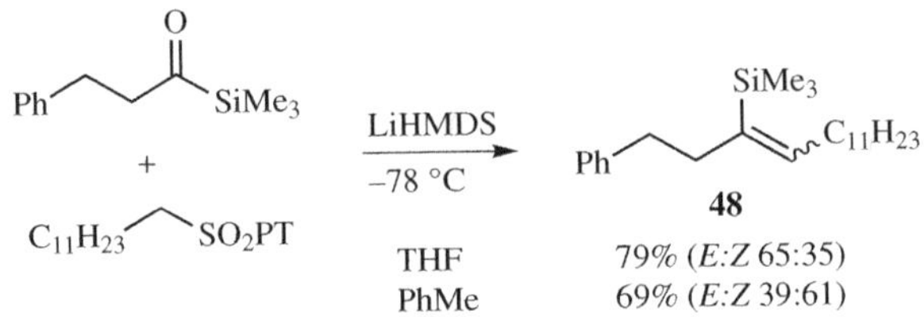

Scheme 19.22 Acylsilanes in Julia–Kocienski reactions.

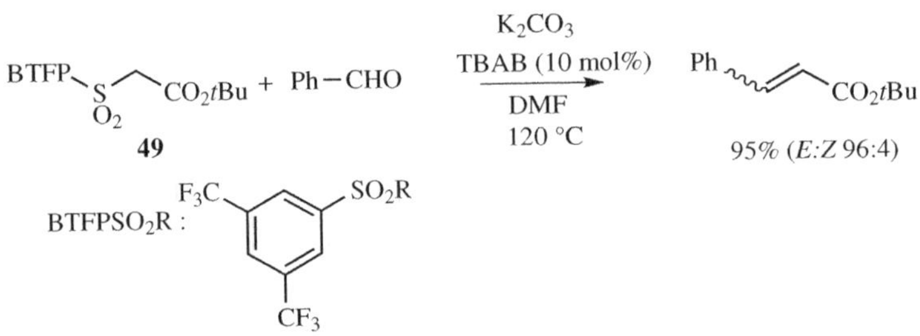

Scheme 19.23 Julia–Kocienski reactions leading to (*E*)-configured conjugated esters.

substituent – aliphatic, allylic, or benzylic – is quite important, as is the aromatic moiety of the sulfone. For instance, γ-allenylallyl BT-sulfones give a high yield of (*Z*)-allenyltrienes and polyenes with α,β-unsaturated aldehydes.[40] TBT-sulfones give best yields of 1,2-substituted alkenes compared with the PT-sulfone, but the *E* selectivity is generally lower. Finally, the counteranion and solvent polarity have a strong impact on the stereochemical outcome of these reactions. For aliphatic PT-sulfones **50** reacting with aliphatic aldehydes, the best *E* selectivities were achieved with KHMDS in DME[5] or by adding 18-crown-6 in THF (Scheme 19.24).[41–43]

The complexity of the stereoselectivity issue arises from the reversibility of the first step of the Julia–Kocienski mechanism as well as the stability of the resulting

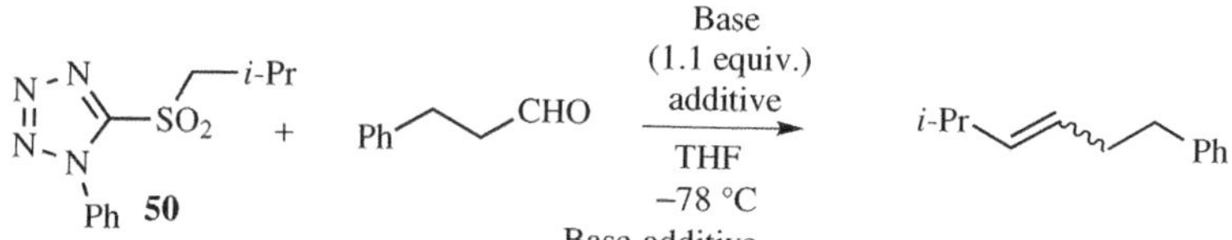

Base-additive	Yield (*E*:*Z*)
LiHMDS	90% (2.1:1)
NaHMDS, 18-cr-6 (2 equiv.)	78% (4:1)
KHMDS	88% (4.3:1)
KHMDS, 18-cr-6 (1.1 equiv.)	86% (15:1)
KHMDS, 18-cr-6 (2 equiv.)	84% (> 50:1)

Scheme 19.24 Effect of the base on the stereoselectivity of the Julia–Kocienski olefination.

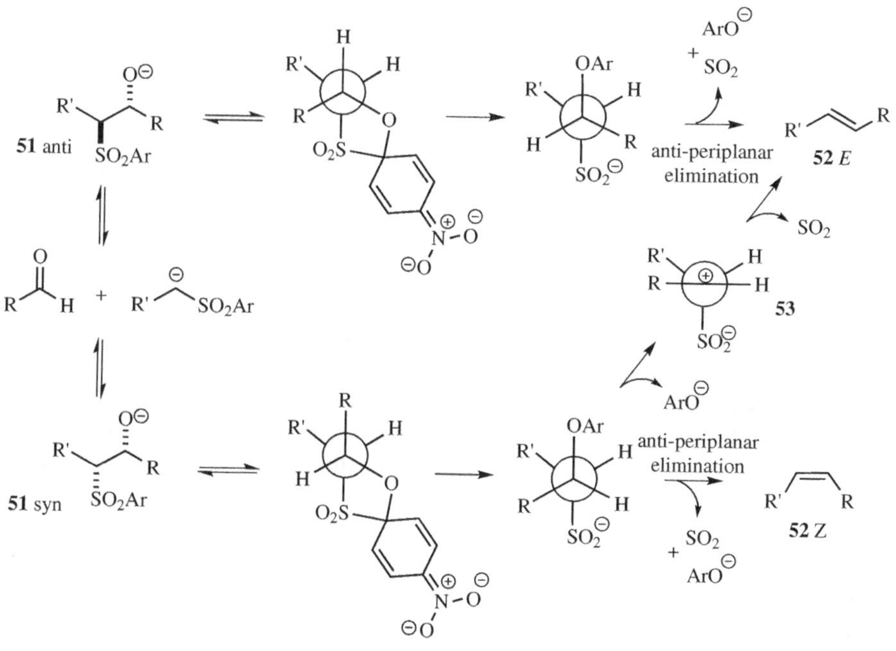

Scheme 19.25 General mechanistic pathway of the Julia–Kocienski olefination.

hydroxysulfones. As displayed in the general mechanistic pathway (Scheme 19.25), addition of the sulfone anion on the aldehyde gives two isomers **51** – syn or anti. If the addition is not reversible, the stereoselectivity of the overall olefination reaction mainly depends on this initial syn/anti ratio. If reversible, however, the relative rates of Smiles rearrangement for both isomers **51** strongly impact the selectivity of the whole process. Due to steric hindrance (Scheme 19.25), the isomer **51** anti is less prone to evolve toward the spiro intermediate, giving the (*Z*)-alkene **52*Z*** as the major product. Nevertheless, predominant formation of the (*E*)-isomer **52*E*** in some cases is generally explained by a competitive zwitterionic pathway, in which the loss of the

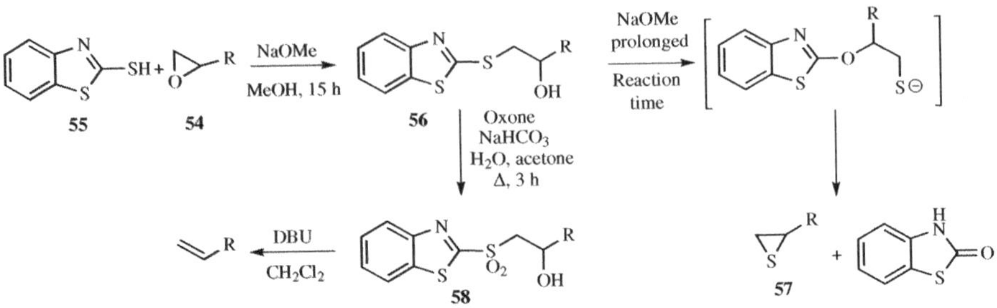

Scheme 19.26 Conversion of epoxides to alkenes related to Julia–Kocienski olefination.

alkoxyheterocycle occurs prior to sulfur dioxide elimination to form **53** as an intermediate. Such behavior has been particularly observed in the case of BTFP-[29] and NP-sulfones.[31] More recently, computational studies for the synthesis of conjugated esters or amides starting from BTFP-sulfones have indicated that the Smiles process is the selectivity-determining step and that SO_2 extrusion occurred before the loss of the phenolate.[39]

The Julia–Kocienski olefination has inspired McGeary and coworkers to develop the transformation of epoxides **54** into the corresponding alkenes (Scheme 19.26).[44] In a first step, the nucleophilic addition of the BT-thiol **55** to the epoxide **54** afforded the *β*-hydroxy BT-thioether **56**, which can further evolve under prolonged reaction time to give the episulfide **57** through direct Smiles rearrangement. However, controlled oxidation of the intermediate *β*-hydroxy BT-thioether **56** into **58** promoted the desired Smiles rearrangement with concomitant SO_2 extrusion to give mono-, di-, or trisubstituted alkenes (Scheme 19.26).

Another clever application of the Julia–Kocienski reaction was recently reported by Jorgensen and coworkers based on the combination of organocatalysis and the Smiles rearrangement for the functionalization of *α*,*β*-unsaturated carbonyl compounds.[45–47] Indeed, the organocatalytic conjugate addition of *β*-keto BT-sulfones **59** to cycloalkenone provides valuable diketosulfones **60** for further transformations. For instance, the subsequent reduction of the latter allows the formation of enantioenriched alkenyl alcohols **61** in good yields and enantioselectivities via the Smiles rearrangement of the resulting hydroxysulfones (Scheme 19.27, Path A).[45] Desulfurylation of the diketosulfones **60** occurs under basic treatment in protic media, giving access to enantioenriched 1,5-dicarbonyl compounds **62** (Scheme 19.27, Path B). Similarly, after protection of the other ketone group, basic treatment of the *β*-ketosulfone gives the corresponding enolate, which further evolves through a Smiles rearrangement to form, after deprotection, alkynyl cycloalkanones **63** (Scheme 19.27, Path C).[45] In the case of *α*,*β*-unsaturated aldehydes, the reactions can be performed in one pot when using PT-sulfones, affording a very efficient and stereoselective strategy for alkynylation and alkenylation of conjugated aldehydes.[47]

β-Ketosulfones **60** can also be transformed into enantioenriched substituted bicyclo[2.2.2]oct-5-en-2-ones **64** via a tandem aldol cyclization/Smiles rearrangement (Scheme 19.28).[46]

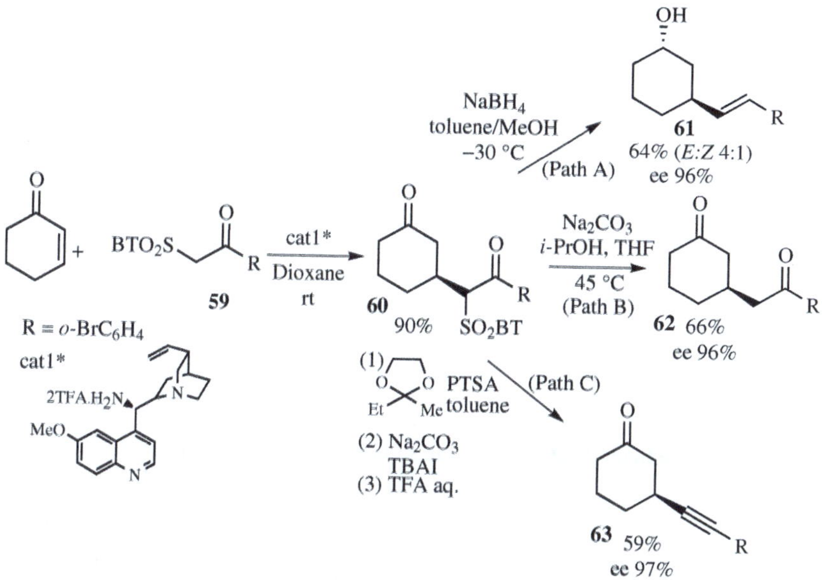

Scheme 19.27 Organocatalyzed Michael addition of β-keto BT-sulfones followed by Julia–Kocienski olefinations.

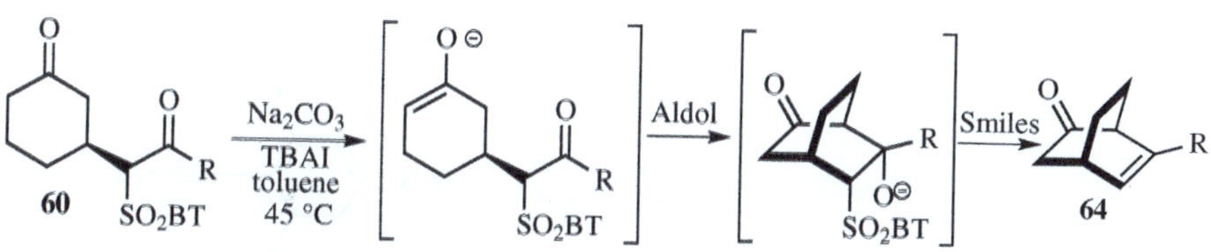

Scheme 19.28 Tandem aldol cyclization/Smiles rearrangement.

In a similar strategy, the organocatalyzed addition of β-keto BT-sulfones to *N*-Boc-protected imines, followed by reduction or desulfurization of the resulting adducts, affords optically active *trans*-allylic amines **65** or β-aminoketones (Scheme 19.29).[48] This reaction was extended to *N*-aryl imines and BT-sulfones bearing an ester, an amide, or a ketone at the β-position to give the corresponding Michael acceptor **66** in good to excellent yields.[49]

Considering these β-ketoheteroarylsulfones as highly valuable starting materials, Jørgensen also reported a metal-free functionalization of electron-deficient aromatic derivatives under basic conditions. An intermolecular nucleophilic aromatic substitution provides an intermediate β-aryl-β-ketoheteroarylsulfone which can be readily transformed into α-arylated ketones via desulfurylation or into arylated alkynes **67** via the Smiles rearrangement for elimination from the derived enolate **68**. The route to the alkyne product represents a formal metal-free Sonogashira coupling (Scheme 19.30).[50]

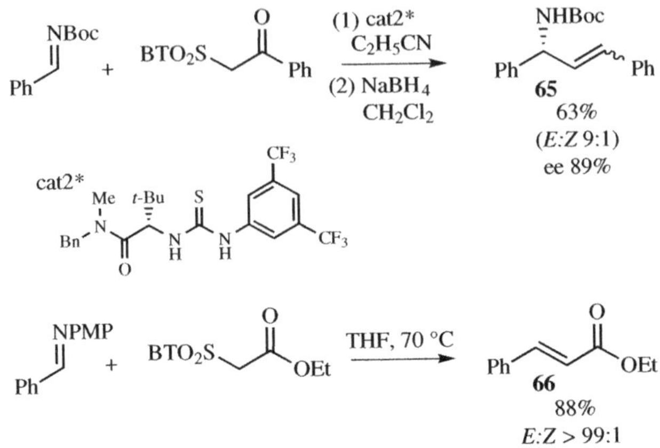

Scheme 19.29 Imines in Julia–Kocienski olefinations.

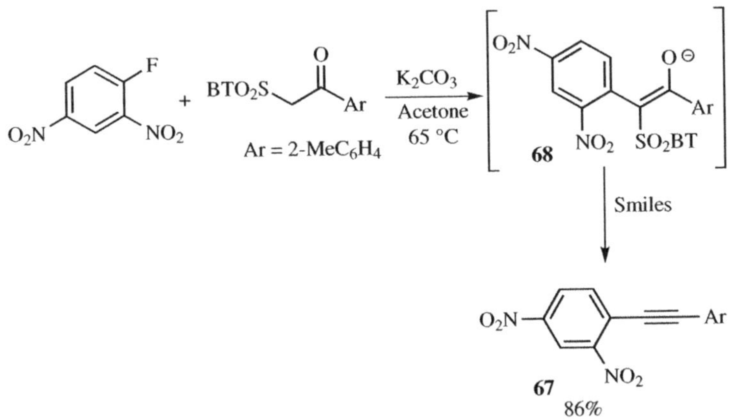

Scheme 19.30 Julia–Kocienski reactions leading to alkynes.

19.3.3 Ugi–Smiles Couplings

The Smiles rearrangement was recently involved in an Ugi-type reaction described by El Kaïm and Grimaud, by replacing the carboxylic acid classically used in Ugi couplings with a nitrophenol.[51–53] In this process, the coupling of an aldehyde, an amine, an isocyanide, and an electron-deficient phenol in a 1 M solution of methanol or toluene at 60 °C gives an aryl imidoyl intermediate **69**, which further evolves through a Smiles rearrangement to form the corresponding *N*-aryl carboxamide derivatives **70** (Scheme 19.31). Through the Smiles process, the aryl moiety is transferred from the oxygen atom to the nitrogen of the amine. Formation of the C—O double bond probably constitutes the driving force of the process.

The Smiles-mediated aryl transfer requires the presence of strong electron-withdrawing substituents such as nitro on the phenol, yet no reaction was

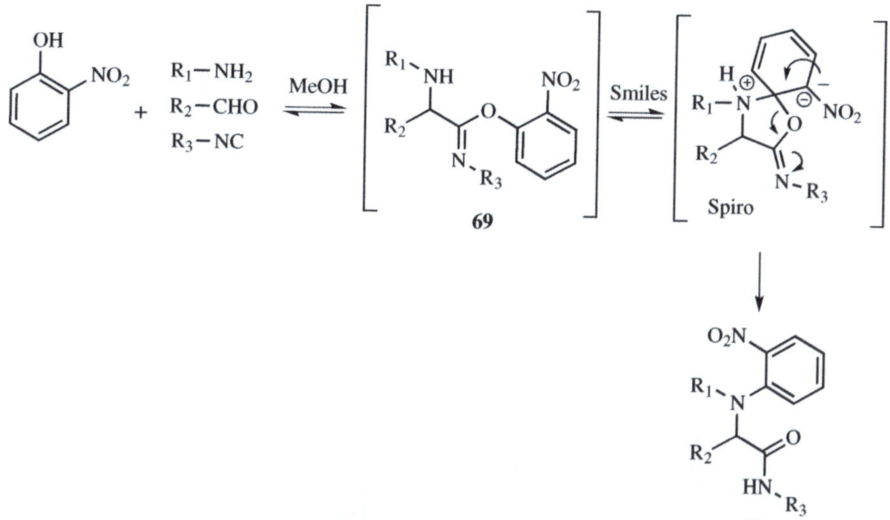

Scheme 19.31 Ugi–Smiles coupling of 2-nitrophenol.

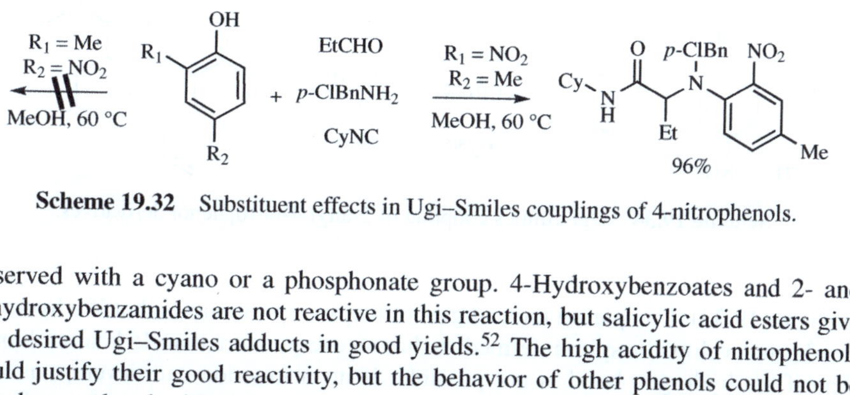

Scheme 19.32 Substituent effects in Ugi–Smiles couplings of 4-nitrophenols.

observed with a cyano or a phosphonate group. 4-Hydroxybenzoates and 2- and 4-hydroxybenzamides are not reactive in this reaction, but salicylic acid esters give the desired Ugi–Smiles adducts in good yields.[52] The high acidity of nitrophenols could justify their good reactivity, but the behavior of other phenols could not be simply correlated with their pK_a. For instance, 4-hydroxybenzoic acid methylester (pK_a 8.5) is more acidic than the corresponding 2-hydroxy derivative (pK_a 9.8). Neither could simple electronic effects be invoked to explain the difference in reactivity observed between the 2-methyl-4-nitrophenol and the 4-methyl-2-nitrophenol (Scheme 19.32).[54]

Inspired by the broad synthetic utility of phenols, various substituents have been introduced, and DFT calculations have been done to rationalize these experimental results.[55] It appears that the efficiency of the Smiles rearrangement can be correlated with the possibility of a hydrogen bond between the ortho substituent and the NH of the former amine. For instance, in the case of 2-nitrophenol (Scheme 19.33, relative energies determined by computational calculations carried out at the M06-2X/6-31+G(d,p) level of theory in methanol), two types of spiro intermediates can be drawn depending on the relative positions of the nitro group and the NH

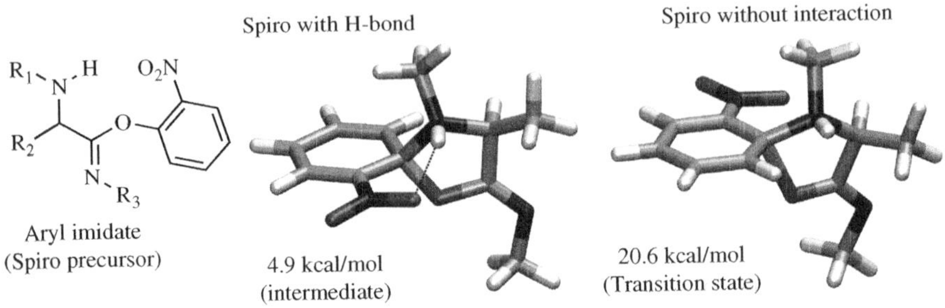

Scheme 19.33 Energies and structures of spiro intermediates in Ugi–Smiles coupling of 2-nitrophenol.

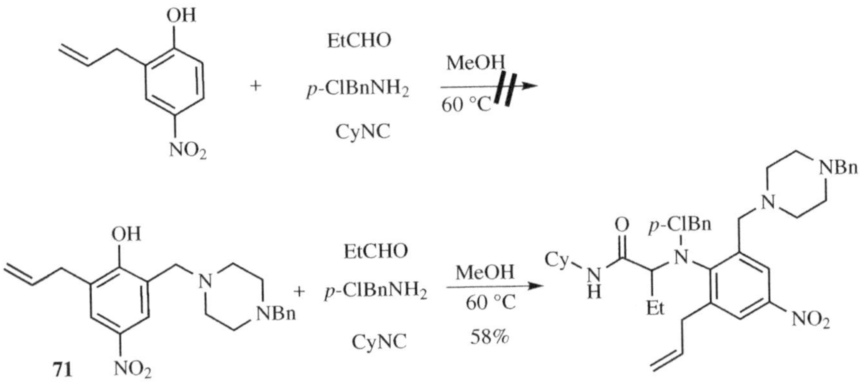

Scheme 19.34 Ugi–Smiles coupling of 2-allyl-4-nitrophenol derivatives.

(Scheme 19.33). Comparing the energies of the two spiros shows that if a hydrogen bond can develop, the spiro intermediate has an energy more than 15 kcal/mol lower than if the hydrogen bond is not possible. If no interaction can develop, as with *para*-nitrophenol, the spiro structure turned out to be a transition state instead of an intermediate. With ortho substitution, steric repulsion predominates, and the reaction fails when there is no hydrogen bonding.

The importance of the hydrogen bond is highlighted when comparing the behavior of 2-allyl-4-nitrophenol and its Mannich adduct **71**, which is much more hindered (Scheme 19.34). Indeed, a nitrogen atom of the Mannich adduct probably interacts with the NH of the amine to favor the spiro cyclization, allowing thus the Smiles rearrangement to proceed.

The scope of the Ugi–Smiles coupling was extended to the use heteroaromatic derivatives, allowing a rapid access to *N*-heteroarylcarboxamides.[56] As already mentioned, the presence of a nitrogen atom in the aromatic system decreases the electron density on the aromatic carbon atoms. Although 2-hydroxypyridine failed to react, the introduction of a nitro-substituent, or even a less electron-withdrawing group such

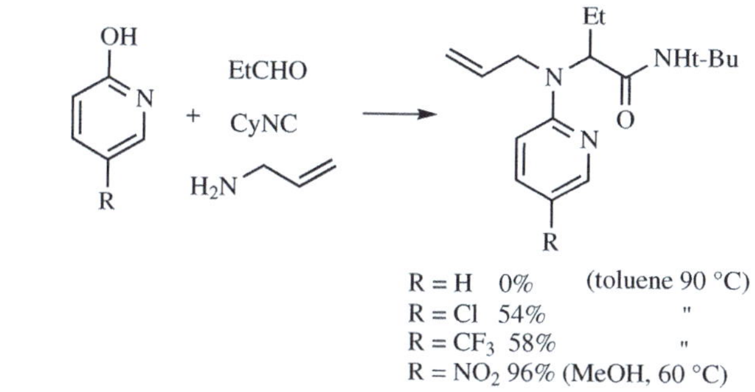

Scheme 19.35 Behavior of 2-hydroxypyridines in Ugi–Smiles couplings.

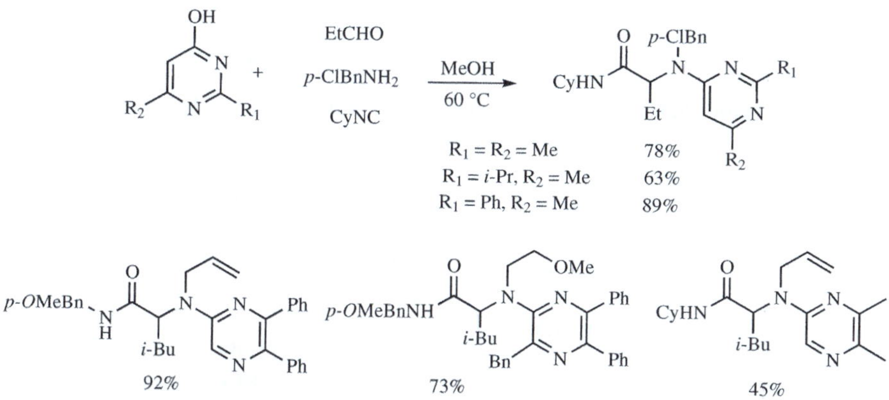

Scheme 19.36 Hydroxypyrimidines and hydroxypyrazines in Ugi–Smiles couplings.

as a chlorine atom or a trifluoromethyl group, induces the Smiles-driven coupling (Scheme 19.35).

As expected, the introduction of a second nitrogen atom facilitates the coupling, and an electron-withdrawing substituent is not required for hydroxypyrimidine and pyrazine substrates (Scheme 19.36).[56,57]

Due to their lower pK_a and the higher nucleophilicity of the resulting anion, thiols were expected to be more efficient in Ugi–Smiles couplings than their phenol analogues. However, *para*-nitrothiophenol (pK_a 5.1) fails to react under Ugi–Smiles conditions, probably due to the lower efficiency of the Smiles rearrangement in the case of thiols. The best results are obtained using the 2-nitro-4-trifluoromethylphenylmercaptan in toluene at 90 °C, which gave the expected thioamide **72** in 26% yield after 2 days (Scheme 19.37). Indeed, due to the lower energy of the C═S compared to the C═O bond, the thermodynamic stability of the attendant *N*-aryl thiocarboxamide **72** is probably comparable to that of the corresponding thioimidate **73**. In this case, the Smiles rearrangement is less prone to occur, and thioimidates **74** turned out to be stable enough to be isolated (Scheme 19.38).[56,58]

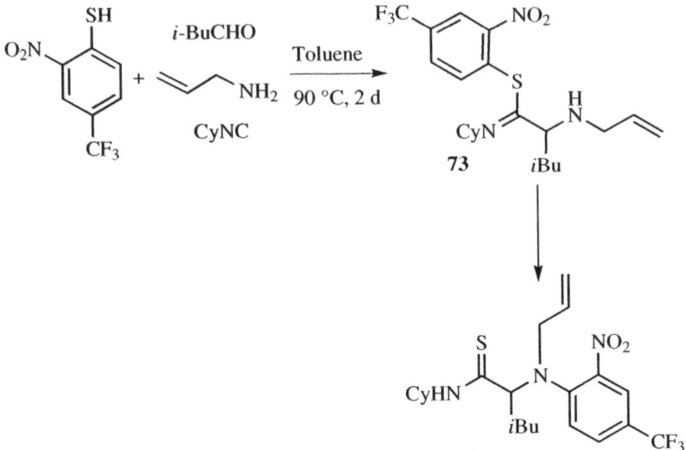

73

72

26%

Scheme 19.37 Ugi–Smiles couplings of 2-nitrothiophenol derivatives.

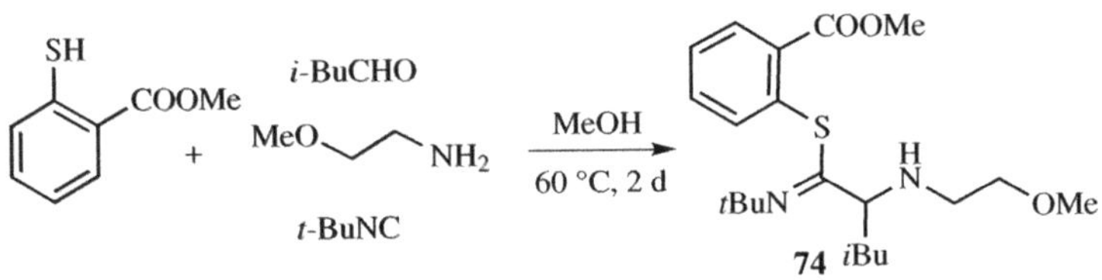

74

84%

Scheme 19.38 Thioimidate formation in attempted Ugi–Smiles coupling of 2-mercaptobenzoic acid derivative.

Nevertheless, heteroaromatic mercapto derivatives give much more interesting results in these couplings, allowing rapid access to highly functionalized thioamides. As observed for hydroxypyridine, substituted 2-mercaptopyridines react smoothly to give the corresponding *N*-pyridinothiocarboxamides **75** in moderate to good yields (Scheme 19.39). However, for 2- and 4-mercaptopyrimidines, solvent-free conditions were required for the coupling (Scheme 19.39).[56] In all the cases, their efficiency remains moderate compared to the hydroxy analogues, and mercapto pyrazines and quinoxalines yield rather low amounts of products.[57]

The most interesting results using mercapto derivatives emerge from the use of five-membered ring heteroaromatics, whose 2-hydroxy counterparts fail to react. Indeed, five-membered ring heterocycles such as pyrroles, pyrazoles, and oxazoles are electron rich because of the delocalization of the lone pair of one heteroatom, disfavoring the Smiles rearrangement. However, mercapto benzothiazoles and benzoxazoles (related N-alkylated or N—H benzimidazoles fail to react in these couplings) turned out to promote efficient couplings (Scheme 19.40).[59] Stabilization

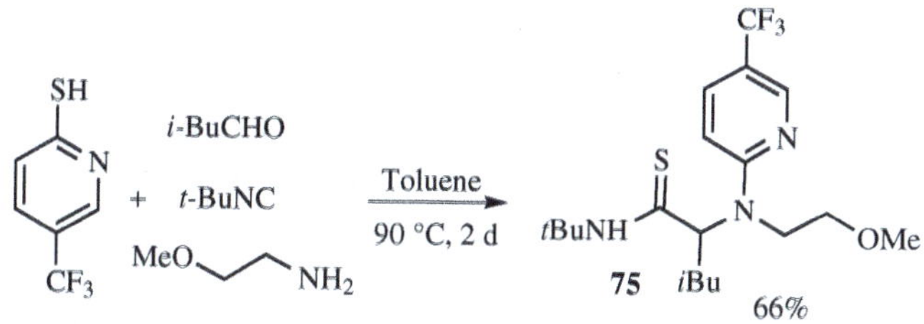

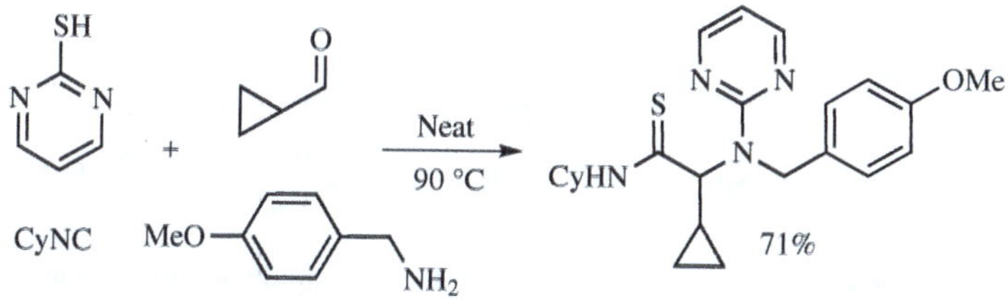

Scheme 19.39 2-Mercaptopyridines and pyrimidines in Ugi–Smiles couplings.

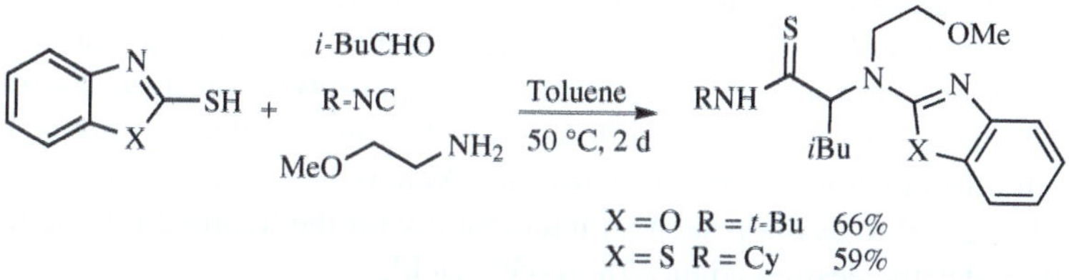

Scheme 19.40 Mercaptobenzothiazoles and benzoxazoles in Ugi–Smiles couplings.

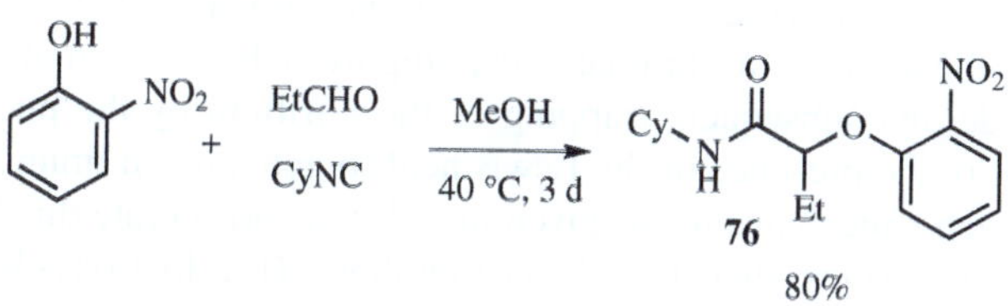

Scheme 19.41 Passerini–Smiles reaction of 2-nitrophenol.

of the negative charge that develops on the nitrogen atom of the five-membered ring should be aided by the fused aromatic system, favoring the Smiles rearrangement. No reaction is observed with monocyclic five-membered ring systems – it is worth noting that 3-mercapto-1,2,4-triazole gives the corresponding thioimidates in moderate yields, as observed with thiosalicylate.[60]

The related Passerini–Smiles reaction, an analogous three-component coupling performed in the absence of the amine partner, affords *O*-aryl carboxamides **76** via *O* to *O* aryl transfer (Scheme 19.41).[52,60]

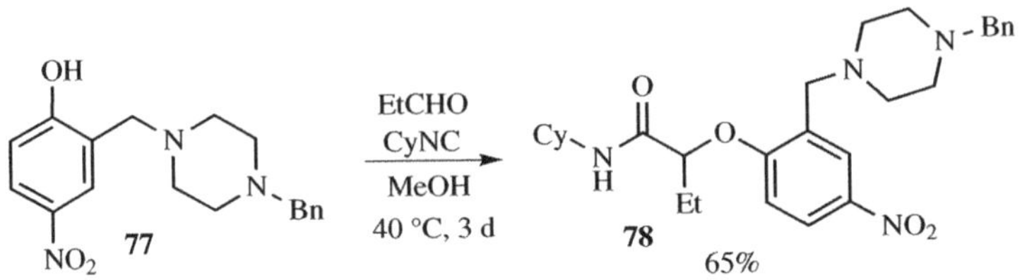

Scheme 19.42 Passerini–Smiles coupling of 4-nitrophenol derivative.

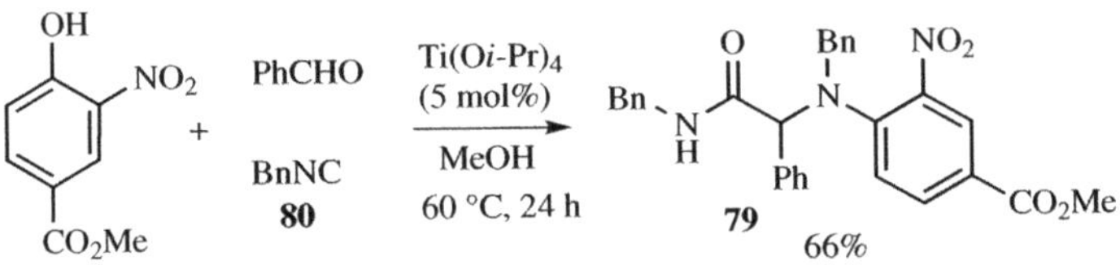

Scheme 19.43 Formal Ugi–Smiles with the isocyanide acting as an amine.

The scope of this reaction is narrower than the Ugi–Smiles coupling as usually observed for Passerini couplings due to the lower electrophilicity of the carbonyl compound compared with the corresponding iminium. Additionally, no reaction is observed for *para*-nitrophenol unless there is an appropriate substituent at the ortho position (Scheme 19.42). This reinforces the hypothesis that hydrogen bonding enables the Smiles transfer. For instance, the Mannich base **77**, with one nitrogen atom well located to develop such an interaction with the hydrogen of the hydroxyl group, provides the desired adduct **78** in 65% yield.

If a Lewis acid such as $Ti(Oi\text{-}Pr)_4$ is added to enhance the electrophilicity of the carbonyl compound, the product **79** is of Ugi type with two isocyanide molecules **80** incorporated, one playing the role of an amine (Scheme 19.43).[61] The authors suggested that the *β*-alkoxy arylimidate, resulting from the addition of the isocyanide on the aldehyde and subsequent trapping of the nitrilium by the phenolate, reacts with methanol in the presence of the Lewis acid to generate an imine. However, no evidence for such a mechanism was given in order to corroborate this hypothesis.

Hydroxy heterocycles, such as hydroxyquinolines **81**, 2-hydroxy-3-nitropyridine, or 4-hydroxypyrimidines **82**, promote Passerini–Smiles couplings in good to moderate yields (Scheme 19.44).[52] In these cases, hydrogen bonding probably develops to favor the whole process.

Disappointingly, the related thio Passerini–Smiles couplings do not proceed efficiently as the yields exceeded 50% only for the case shown in Scheme 19.45.[58] This lack of reactivity is probably linked to the weaker energy of C═S bond compared to the C═O one.

Westermann and coworkers have recently reported on tandem Ugi–Mumm/Ugi–Smiles reactions using various phenol-substituted carboxylic acids (Scheme 19.46). As the Ugi–Mumm is very rapid, it is possible to perform sequential couplings to reach products with seven points of diversity.[62]

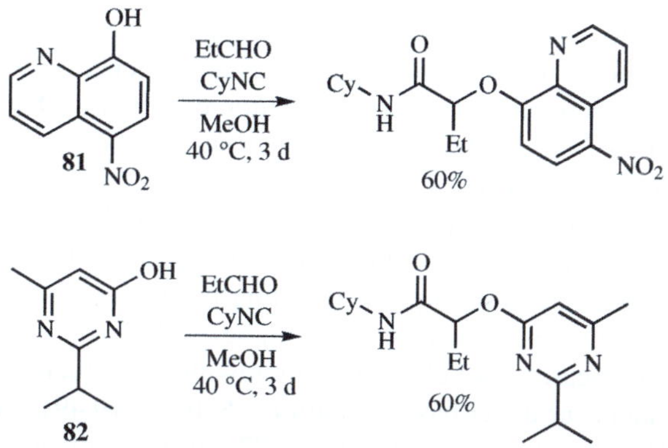

Scheme 19.44 Hydroxyheterocycles in Passerini–Smiles reactions.

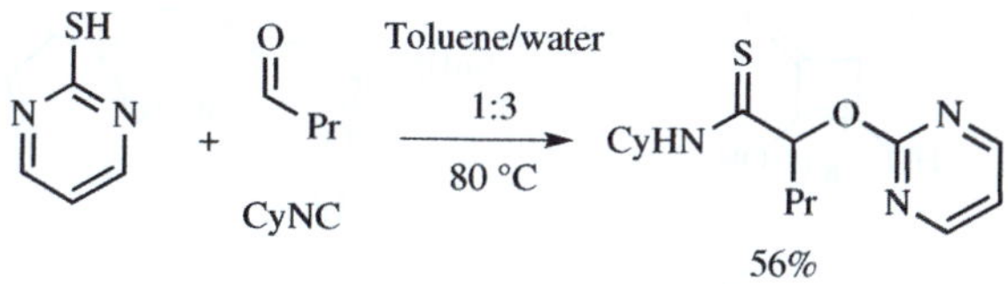

Scheme 19.45 Thio-Passerini–Smiles coupling.

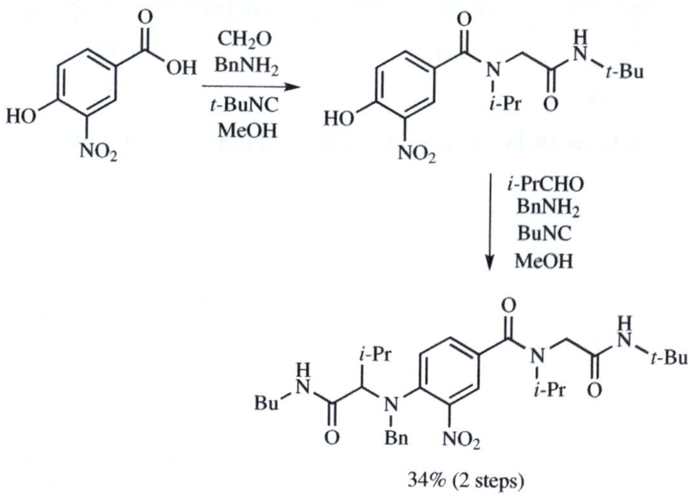

Scheme 19.46 Tandem Ugi–Mumm/Ugi–Smiles reactions with hydroxybenzoic derivatives.

Recently, different enol sources have been investigated to replace the starting phenol. For instance, Charton and coworkers reported a similar reaction using squaric acid **83** instead of an electron-deficient phenol (Scheme 19.47).[63] Similarly, Ramazani and coworkers described the application of tropolone **84** in Ugi–Smiles[64] or Passerini–Smiles[65] couplings to synthesize substituted cycloheptatrien-1-one derivatives **85** (Scheme 19.48).

As presented earlier, the scope of these Ugi-type couplings involving a Smiles rearrangement has been extensively studied in forming highly functionalized amides and thioamides. More interestingly, the resulting products constitute an extremely rich source of precursors in heterocyclic synthesis. Indeed, when using a more activated phenol **86** bearing two electron-withdrawing substituents, a second Smiles rearrangement can occur under basic conditions (Scheme 19.49).[66] Evidently, when treated with DBU, the 4-component adduct **87** further evolves via a spiro [3,6]

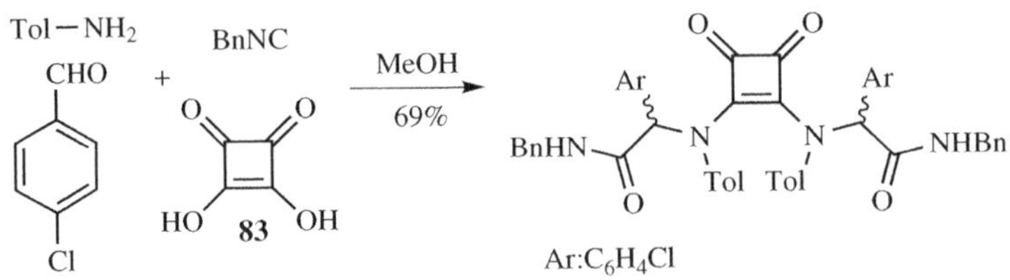

Scheme 19.47 Ugi–Smiles reaction of squaric acid.

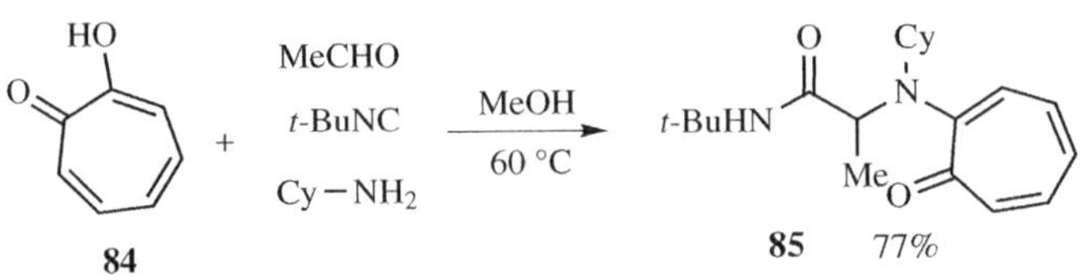

Scheme 19.48 Ugi–Smiles reaction of tropolone derivatives.

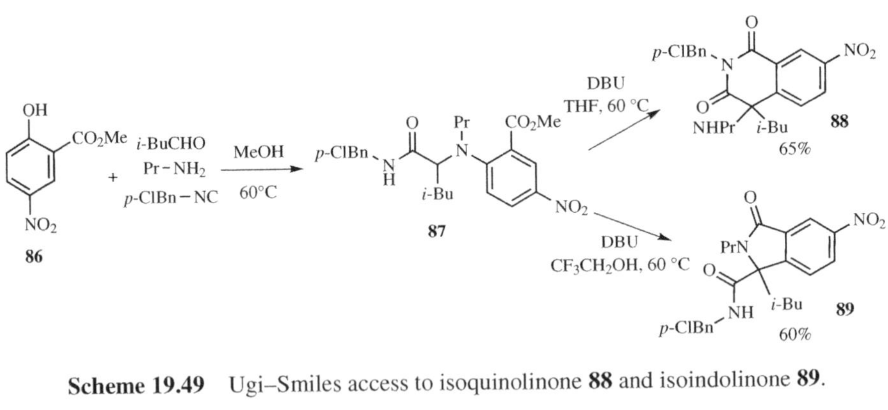

Scheme 19.49 Ugi–Smiles access to isoquinolinone **88** and isoindolinone **89**.

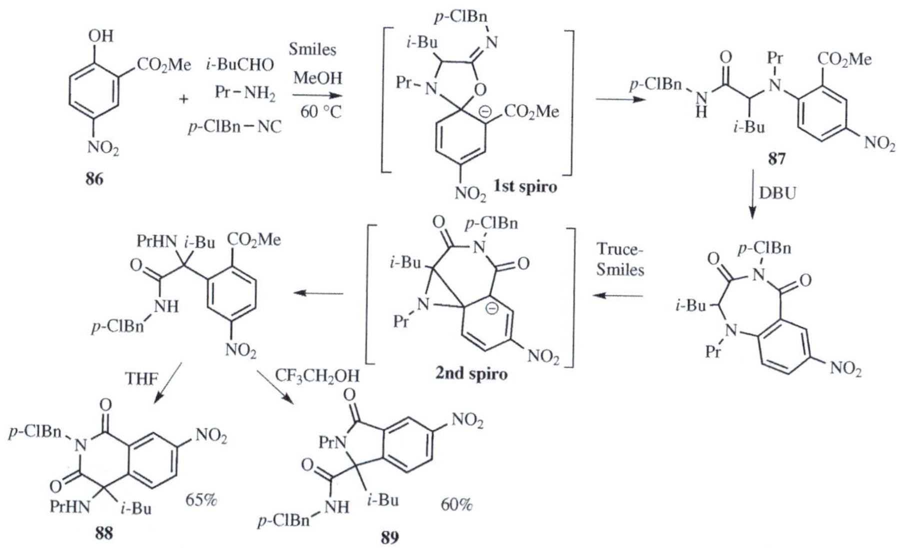

Scheme 19.50 Consecutive Ugi–Smiles/Truce–Smiles rearrangements in the formation of **88** and **89**.

intermediate to give the corresponding C-arylated product, affording, after cyclization of the amide moiety on the ester, the final isoquinolinone **88** (Scheme 19.50). Depending on the solvent used in these cascades, the amino residue can add onto the ester affording an isoindolinone **89**. The second aryl transfer in which a carbanion is involved as the nucleophile represents an interesting example of a Truce–Smiles rearrangement.[4]

A wide range of postcondensation transformations could be imagined to further transform Ugi–Smiles adducts. For instance, a one-pot two-step sequence has been reported for indole formation starting from 2-iodo-4-nitrophenol as the acidic partner in Ugi–Smiles reactions.[67] When the 4-component coupling is completed, a substoichiometric amount of trifluoroacetic acid is added to destroy the remaining isocyanide before addition of the palladium catalyst for a subsequent Heck–isomerization process (Scheme 19.51). Such a cascade allows the formation of various N-substituted indoles **90** in good yields, especially considering the number of bonds formed during the whole sequence.

Due to the high efficiency of these Ugi–Smiles/palladium coupling reactions, combined with the resulting rapid access to quite complex structures, such sequences have been used to develop various metal-catalyzed reactions. Indeed, this approach assembles in one-step functionalities that classical methods would require several steps to gather. For instance, furfurylamine **91** coupled with an iodinated hydroxy aromatic affords starting materials **92** for a new ring opening of furans catalyzed by palladium salts (Scheme 19.52).[68] This sequence provides easy and efficient access to indoles or pyrrolopyrimidines **93** substituted with a pendant α,β-unsaturated carbonyl moiety. Similarly, cyclopropropane ring opening has been studied using cyclopropylamine as

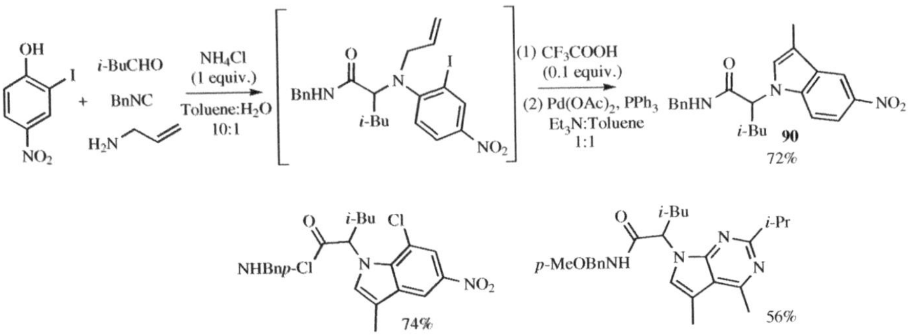

Scheme 19.51 Consecutive Ugi–Smiles/Heck reaction towards indole derivatives.

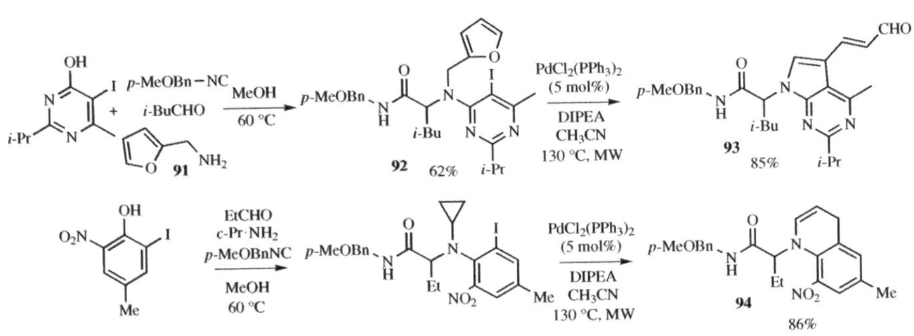

Scheme 19.52 Ugi–Smiles couplings followed by palladium-triggered ring cleavage of furans and cyclopropanes.

amine partner in Ugi–Smiles couplings, followed by palladium-catalyzed formation of dihydroquinolines **94** (Scheme 19.52).[69]

19.3.4 Miscellaneous Use of the Smiles Rearrangement in Synthesis

Among Smiles processes involving S-to-O transfer, the thermal rearrangement of sulfur ylides **95** has been recently reported under microwave irradiation (Scheme 19.53). The reaction proceeds through a Smiles rearrangement, leading to the (*Z*)-tetrasubstituted olefin **96*Z***, which undergoes a final isomerization to give the (*E*)-isomer **96*E*** in excellent to good yields.[70]

An unprecedented I-to-O aryl shift has been described for the synthesis of glycyrol, a benzofurancoumarin.[71] A hydroxycoumarin **97** reacts with diacetoxyiodobenzene to give the corresponding iodonium **98**, which undergoes a Smiles rearrangement in refluxing DMF to yield a 3-iodo-4-phenoxycoumarin **99** (Scheme 19.54). The latter cyclizes under palladium catalysis to build the benzofuran core.

Based on a related mechanistic concept, a transformation of acyl chlorides into nitriles was reported 15 years ago using 2,4-dinitrobenzenesulfonamide **100** under basic conditions (Scheme 19.55).[72] In this reaction, the intermediate sulfonimidamide

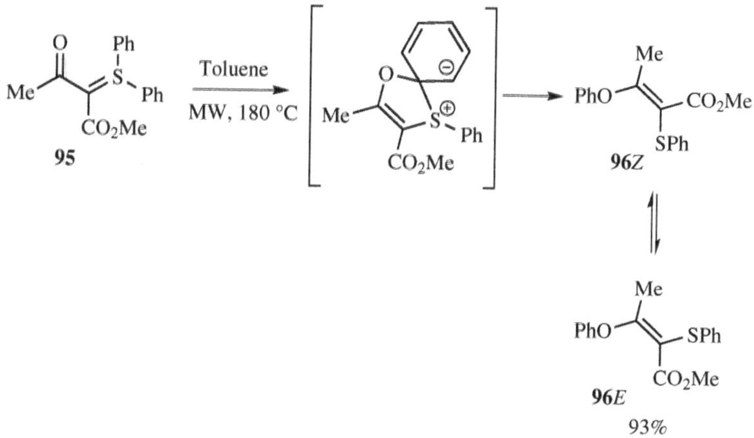

Scheme 19.53 Smiles rearrangement of sulfur ylides **95**.

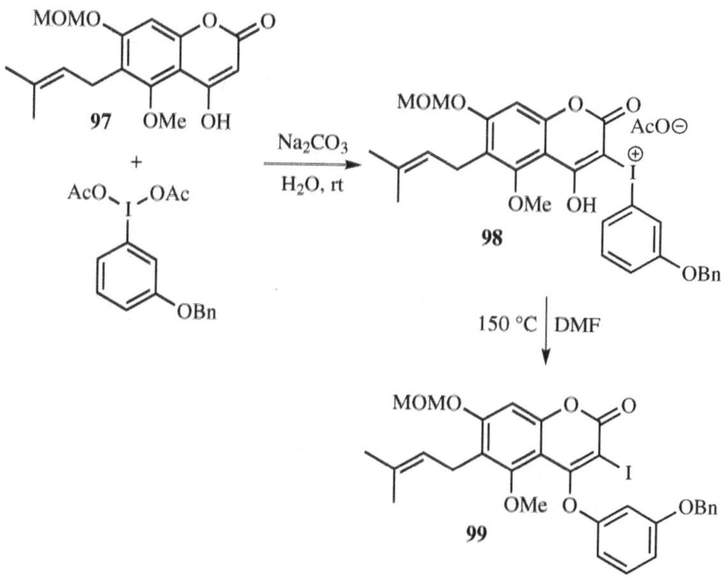

Scheme 19.54 Synthesis of glycyrol through I-to-O Smiles rearrangement.

anion **101** undergoes a Smiles rearrangement, forming the nitrile functional group after SO_2 extrusion and loss of the phenolate moiety.

19.4 CONCLUSION

First recognized in the early 1930s and extensively studied in the following decades, studies on the Smiles rearrangement still represent an exciting field. Besides the disclosure of new conditions triggering the reaction (e.g., photochemical, radical, and

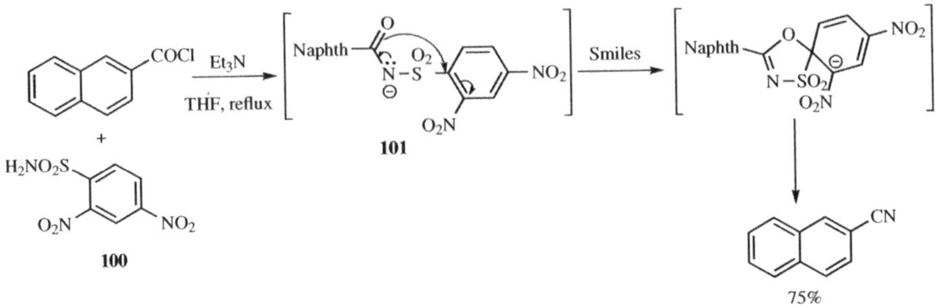

Scheme 19.55 Smiles rearrangement of sulfonimidamide anion **101**.

microwave irradiation), the most interesting developments on the Smiles rearrangement certainly involve its use in complex synthetic sequences. Cascades toward fused heterocyclic systems represent early and rather simple examples of this approach. Their use in more elegant and complex synthetic schemes really began with the Julia–Kocienski olefination. The Ugi–Smiles and Passerini–Smiles couplings show further how these rearrangements may enrich and expand some well-established reactions. Further important applications of the Smiles rearrangement certainly remain to be discovered.

REFERENCES

1. (a) Henriques, R. *Ber.* **1894**, *27*, 2993–3005. (b) Hinsberg, O. *J. Pr. Chem.* **1914**, *90*, 345. (c) Hinsberg, O. *J. Pr. Chem.* **1915**, *91*, 307. (d) Hinsberg, O. *J. Pr. Chem.* **1916**, *93*, 277.
2. Warren, L. A.; Smiles, S. *J. Chem. Soc.* **1930**, 935–963.
3. Warren, L. A.; Smiles, S. *J. Chem. Soc.* **1930**, 1327–1331.
4. Truce, W. E.; Ray, W. J.; Norman, O. L.; Eickemeyer, D. B. *J. Am. Chem. Soc.* **1958**, *80*, 3625–3629.
5. Baudin, J. B.; Hareau, G.; Julia, S. A.; Ruel, O. *Tetrahedron Lett.* **1991**, *32*, 1175–1178.
6. Blakemore, P. R.; Cole, W. J.; Kocienski, P. J.; Morley, A. *Synlett* **1998**, 26–28.
7. (a) Bunnet, J. F.; Zahler, R. E. *Chem. Rev.* **1951**, *49*, 273–308. (b) Truce, W. E.; Kreider, E. M.; Brand, W. W. *Org. React. (NY)* **1970**, *18*, 99–215. (c) Gerasimova, T. N.; Kolchina, E. F. *Russ. Chem. Rev.* **1994**, *66*, 69–74. (d) Gerasimova, T. N.; Kolchina, E. F. *Russ. Chem. Rev.* **1995**, *64*, 133–140. (e) Plesniak, K.; Zarecki, A.; Wicha, J. *Top. Curr. Chem.* **2007**, *275*, 163–250. (f) Snape, T. J. *Chem. Soc. Rev.* **2008**, *37*, 2452–2458.
8. (a) Aïssa, C. *Eur. J. Org. Chem.* **2009**, 1831–1844. (b) Zajc, B.; Kumar, R. *Synthesis* **2010**, *11*, 1822–1836.
9. (a) Studer, A.; Bossart, M. *Tetrahedron* **2001**, *57*, 9649–9667. (b) Bowman, W. R.; Fletcher, A. J.; Potts, G. B. S. *J. Chem. Soc., Perkin Trans. 1* **2002**, 2747–2762. (c) Quiclet-Sire, B.; Zard, S. Z. *Beilstein J. Org. Chem.* **2013**, *9*, 557–576.
10. Bayles, R.; Johnson, M.C.; Maisey, R.F.; Turner, R. W. *Synthesis* **1977**, 33–34.
11. McMillen, D.F.; Golden, D.M. *Annu. Rev. Phys. Chem.* **1982**,*33*, 493–532.
12. Pedley, J.B.; Naylor, R.D.; Kirby, S.P. *Thermochemical Data of Organic Compounds*, 2nd ed., Chapman and Hall: New York, **1986**.

13. Soukri, M.; Lazar, S.; Akssira, M.; Guillaumet, G. *Org. Lett.* **2000**, *2*, 1557–1560.
14. Schulenberg, J. W.; Archer, S. In *The Chapman Rearrangement in Organic Reactions*; Cope, A.C., Ed.; John Wiley & Sons, Inc., **1965**; Vol. 14, pp. 1–43.
15. Lloyd-Jones, G. C.; Moseley, J. D.; Renny, J. S. *Synthesis* **2008**, 661–689.
16. Harvey, J. N.; Jover, J.; Lloyd-Jones, G. C.; Moseley, J. D.; Murray, P.; Renny, J. S. *Angew. Chem. Int. Ed.* **2009**, *48*, 7612–7615.
17. For a possible Smiles rearrangement of N-aryl sulfonamide involving a three-membered spiro intermediate see: Muller, P.; Phuong, N.-T. M. *Helv. Chim. Acta* **1979**, *62*, 494–496.
18. (a) Urry, W. H.; Kharasch, M. S. *J. Am. Chem. Soc.* **1944**, *66*, 1438–1440. (b) Baroudi, A.; Alicea, J.; Flack, P.; Kirincich, J.; Alabugin, I. V. *J. Org. Chem.* **2011**, *76*, 1521–1537. (c) Gheorghe, A.; Quiclet-Sire, B.; Vila, X.; Zard, S. Z. *Tetrahedron* **2007**, *63*, 7187–7212. See also Ref. 9a.
19. Zuo, H.; Li, Z.-B.; Ren, F.-K.; Falck, J.R.; Lijuan, M.; Ahn, C.; Shin, D.-S. *Tetrahedron* **2008**, *64*, 9669–9674.
20. Kitching, M. O.; Hurst, T. E.; Snieckus, V. *Angew. Chem. Int. Ed.* **2012**, *51*(12), 2925–2929.
21. Sang, P.; Yu, M.; Tu, H.; Zou, J.; Zhang, Y. *Chem. Commun.* **2013**, *49*, 701–703.
22. Sun, H.; Li, J.; Zhang, D.; Ma, C.; Liu, C. *J. Phys. Org. Chem.* **2008**, *21*, 215–218.
23. Zhao, Y.; Wu, Y.; Jia, J.; Zhang, D.; Ma, C. *J. Org. Chem.* **2012**, *77*, 8501–8506.
24. Xiang, J.; Zheng, L.; Chen, F.; Bai, X. *Org. Lett.* **2007**, *9*, 765–767.
25. (a) Xiang, J.; Xie, H.; Wen, D.; Dang, Q.; Bai, X. *J. Org. Chem.* **2008**, *73*, 3281–3283. (b) Xiang, J.; Zheng, L.; Xie, H.; Hu, X.; Dang, Q.; Bai, X. *Tetrahedron* **2008**, *64*, 9101–9107. (c) Xiang, J.; Zhu, T.; Dang, Q.; Bai, X. J. *Org. Chem.* **2010**, *75*, 8147–8154.
26. Kende, A. S.; Mendoza, J. S. *Tetrahedron Lett.* **1990**, *31*, 7105–7108.
27. (a) Baudin, J. B.; Hareau, G.; Julia, S. A.; Ruel, O. *Bull. Soc. Chim. Fr.* **1993**, *130*, 336–357. (b) Baudin, J. B.; Hareau, G.; Julia, S. A.; Lorne, R.; Ruel, O. *Bull. Soc. Chim. Fr.* **1993**, *130*, 856–878.
28. Alonso, D. A.; Nájera, C.; Varea, M. *Tetrahedron Lett.* **2004**, *45*, 573–577.
29. Alonso, D. A.; Fuensanta, M.; Nájera, C.; Varea, M. *J. Org. Chem.* **2005**, *70*, 6404–6416.
30. Mąkosza, M.; Bujok, R. *Synlett* **2008**, 586–588.
31. Mirk, D.; Grassot, J.-M.; Zhu, J. *Synlett* **2006**, 1255–1259.
32. Sánchez, I. P.; Turos, E. *Tetrahedron: Asym.* **2009**, *20*, 1646–1660.
33. Kocienski, P. J.; Bell, A.; Blakemore, P. R. *Synlett* **2000**, 365–366.
34. Hale, K. J.; Domostoj, M. M.; Tocher, D. A.; Irving, E.; Scheinmann, F. *Org. Lett.* **2003**, *5*, 2927–2930.
35. Aissa, C. *J. Org. Chem.* **2006**, *71*, 360–363.
36. Alonso, D. A.; Fuensanta, M.; Nájera, C. *Eur. J. Org. Chem.* **2006**, 4747–4754.
37. Jankowski, P.; Plesniak, K.; Wicha, J. *Org. Lett.* **2003**, *5*, 2789–2792.
38. Bourdon, B.; Corbet, M.; Fontaine, P.; Goekjian, P. G.; Gueyrard, D. *Tetrahedron Lett.* **2008**, *49*, 747–749.
39. Alonso, D. A.; Fuensanta, M.; Gómez-Bengoa, E.; Nájera, C. *Eur. J. Org. Chem.* **2008**, 2915–2922.
40. Vaz, B.; Alvarez, R.; Souto, J. A.; De Lera, A. R. *Synlett* **2005**, 294–298.
41. Ishigami, K.; Watanabe, H.; Kitahara, T. *Tetrahedron* **2005**, *61*, 7546–7553.

42. Nakatani, Y.; Oshita, J.; Ishigami, K.; Watanabe, H.; Kitahara, T. *Tetrahedron* **2006**, *62*, 160–165.
43. Pospíšil, J. *Tetrahedron Lett.* **2011**, *52*, 2348–2352.
44. Wu, F.-L.; Ross, B. P.; McGeary, R. P. *Eur. J. Org. Chem. 2010*, 1989–1998.
45. Paixão, M. W.; Holub, N.; Vila, C.; Nielsen, M.; Jørgensen, K. A. *Angew. Chem. Int. Ed.* **2009**, *48*, 7338–7342.
46. Holub, N.; Jiang, H.; Paixão, M. W.; Tiberi, C.; Jørgensen, K. A. *Chem. Eur. J.* **2010**, *16*, 4337–4346.
47. Nielsen, M.; Jacobsen, C. B.; Paixão, M. W.; Holub, N.; Jørgensen, K. A. *J. Am. Chem. Soc.* **2009**, *131*, 10581–10586.
48. Jacobsen, C. B.; Lykke, L.; Monge, D.; Nielsen, M.; Ransborg, L. K.; Jørgensen, K. A. *Chem. Commun.* **2009**, *4*, 6554–6556.
49. Shao, Y.-D.; Wu, X.-S.; Tian, S.-K. *Eur. J. Org. Chem.* **2012**, 1590–1596.
50. Prüger, B.; Hofmeister, G. E.; Jacobsen, C. B.; Alberg, D. G.; Nielsen, M.; Jørgensen, K. A. *Chem. Eur. J.* **2010**, *16*, 3783–3790.
51. El Kaïm, L.; Grimaud, L.; Oble, J. *Angew. Chem. Int. Ed.* **2005**, *44*, 7961–7964.
52. El Kaïm, L.; Gizolme, M.; Grimaud, L.; Oble, J. *J. Org. Chem.* **2007**, *72*, 4169–4180.
53. El Kaïm, L.; Grimaud, L. *Mol. Div.* **2010**, *14*, 855–867.
54. Chéron, N.; Ramozzi, R.; El Kaïm, L.; Grimaud, L.; Fleurat-Lessard, P. *J. Org. Chem.* **2012**, *77*, 1361–1366.
55. Chéron, N.; El Kaïm, L.; Grimaud, L.; Fleurat-Lessard, P. *Chem. Eur. J.* **2011**, *17*, 14929–14934.
56. El Kaïm, L.; Gizolme, M.; Grimaud, L.; Oble, J. *Org. Lett.* **2006**, *8*, 4019–4021.
57. Barthelon, A.; Dos Santos, A.; El Kaïm, L.; Grimaud, L. *Tetrahedron Lett.* **2008**, *49*, 3208–3211.
58. Barthelon, A.; El Kaïm, L.; Gizolme, M.; Grimaud, L. *Eur. J. Org. Chem.* **2008**, *2008*, 5974–5987.
59. El Kaïm, L.; Gizolme, M.; Grimaud, L.; Oble, J. *Synlett* **2007**, 465–469.
60. El Kaïm, L.; Gizolme, M.; Grimaud, L. *Org. Lett.* **2006**, *8*, 5021–5023.
61. Dai, W.; Li, H. *Tetrahedron* **2007**, *63*, 12866–12876.
62. Brauch, S.; Gabriel, L.; Westermann, B. *Chem. Commun.* **2010**, *46*, 3387–3389.
63. Aknin, K.; Gauriot, M.; Totobenazara, J.; Deguine, N.; Deprez-Poulain, R.; Deprez, B.; Charton, J. *Tetrahedron Lett.* **2012**, *53*, 458–461.
64. Massoudi, A.; Amini, I.; Ramazani, A.; Nasrabadi, F. Z.; Ahmadi, Y. *Bull. Korean Chem. Soc.* **2012**, *33*, 39–42.
65. Massoudi, A.; Amini, I.; Ramazani, A. *Turk. J. Chem.* **2012**, *36*, 537–544.
66. El Kaïm, L.; Grimaud, L.; Le Goff, X. F.; Schiltz, A. *Org. Lett.* **2011**, *13*, 534–536.
67. El Kaïm, L.; Gizzi, M.; Grimaud, L. *Org. Lett.* **2008**, *10*, 3417–3419.
68. El Kaïm, L.; Grimaud, L.; Wagschal, S. *Chem. comm.* **2011**, *47*, 1887–1889.
69. Dos Santos, A.; El Kaïm, L.; Grimaud, L.; Ramozzi, R. *Synlett* **2012**, *23*, 438–442.
70. Huang, X.; Goddard, R.; Maulide, N. *Angew. Chem. Int. Ed.* **2010**, *49*, 8979–8983.
71. Jin, Y. L.; Kim, S.; Kim, Y. S.; Kim, S.-A.; Kim, H. S. *Tetrahedron Lett.* **2008**, *49*, 6835–6837.
72. Huber, V. J.; Bartsch, R. A. *Tetrahedron* **1998**, *54*, 9281–9288.

20

PUMMERER-TYPE REACTIONS AS POWERFUL TOOLS IN ORGANIC SYNTHESIS

DIEGO GAMBA-SÁNCHEZ AND FABIÁN GARZÓN-POSSE

Laboratory of Organic Synthesis, Bio and Organocatalysis, Chemistry Department, Universidad de los Andes, Bogotá, Colombia

20.1 INTRODUCTION

The Pummerer rearrangement, also called the Pummerer reaction, and its variants (Pummerer-type transformations) are extremely useful tools in organic synthesis. From a classical point of view, a Pummerer rearrangement is the transformation of a sulfoxide with at least one α-hydrogen into an α-substituted sulfide (via a thionium ion) as shown in Scheme 20.1. The substituent group at the α position can be inserted by either intra- or intermolecular addition, an important feature that increases the scope of the reaction. In other words, a new C—nucleophile bond is generated at the α center. Because these nucleophiles can be carbons or heteroatoms, an impressive number of structures are accessible using Pummerer chemistry. Additionally, the initially obtained sulfide may be easily hydrolyzed, increasing even further the structural variety of the Pummerer products.

More modern definitions give the Pummerer name to the addition of nucleophiles to intermediate thionium ions, which are generated by different procedures and not necessarily from sulfoxides (connective pathway in Scheme 20.1). Depending on how this thionium intermediate is generated, and on whether it undergoes nucleophile addition, several types of Pummerer reactions exist. This chapter will focus on recent (in the last 6 years) developments and applications of Pummerer chemistry

---

Molecular Rearrangements in Organic Synthesis, First Edition. Edited by Christian M. Rojas.

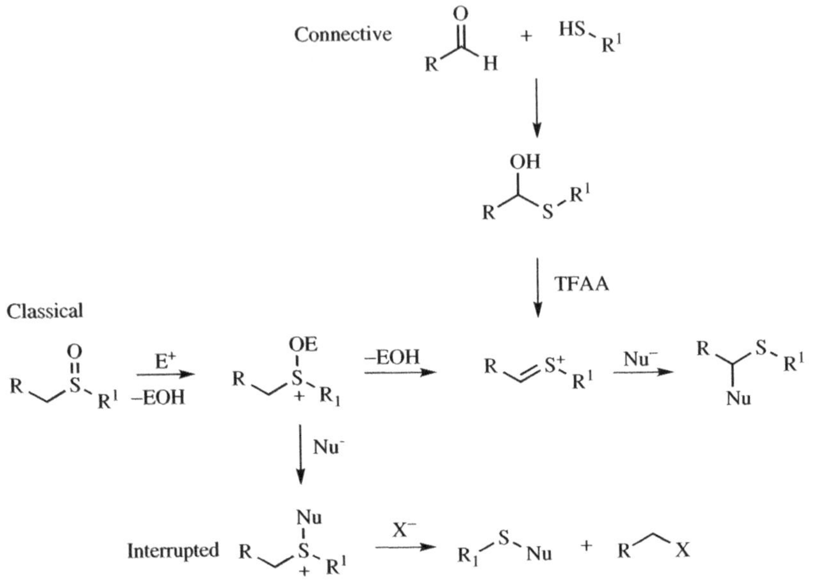

Scheme 20.1 Some Pummerer transformations.

in the synthesis of complex natural products and other structures that are difficult to prepare. Mechanistic studies that provide useful information to better plan a synthesis using Pummerer transformations are also covered.

The chapter is divided into six different sections: The first four describe the five types of Pummerer reactions, then a few examples of cascades or tandem processes with the participation of the Pummerer reaction are presented. Section 20.6 is used to highlight recent applications in selenium-Pummerer variants, and finally, a summary and outlook is provided. The examples presented herein showcase the enormous variety of structures that can be obtained by Pummerer chemistry, while highlighting that, even though this reaction has been known for more than one hundred years, there is still much work to be done.

20.2 CLASSICAL PUMMERER REACTION

20.2.1 Mechanistic Considerations

The "classical Pummerer reaction" is by far the most known and applied Pummerer variant. The thionium ion **3** is generated by activation of the sulfoxide **1** with an electrophilic reagent, and this ion reacts with a nucleophile to generate α-substituted sulfides **4** which can be hydrolyzed to afford thiols **5** and new carbon skeletons **6** (Scheme 20.2).

The mechanism of the Pummerer reaction has been studied by labeling and kinetic experiments.[1] The commonly accepted mechanistic route is shown in Scheme 20.3.

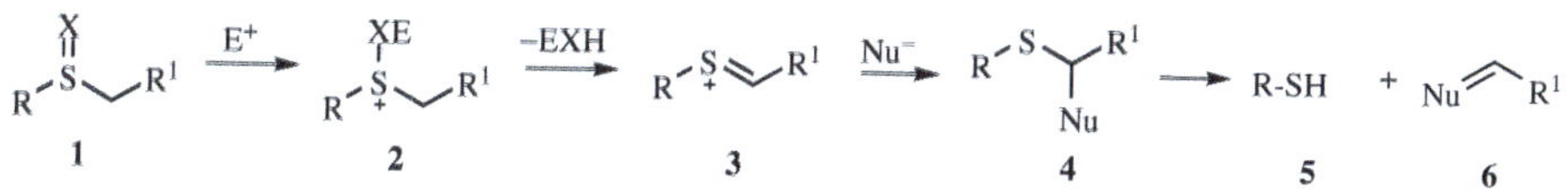

Scheme 20.2 Classical Pummerer reaction.

Since one of the classical activating agents for sulfoxides is acetic anhydride, we will use it to better explain this proposed mechanism.

The first step is acetylation of the sulfoxide oxygen and the release of an acetate ion (**7** → **8** Scheme 20.3). Recent computational studies[2] have shown that the initial acetylation step is rate determining and that the release of the acetate ion proceeds by stabilization of the acetate by coordination with a positively charged sulfur, which yields an achiral sulfurane **11**. It is generally accepted that the formation of this sulfurane is responsible for the poor enantioselectivity shown in the reactions of chiral sulfoxides. However, sulfurane formation can be avoided by some additives, for example, *N,N*-dimethylacetamide (DMAC), *N*-methyl-2-pyrrolidone (NMP), or the acetate trap DCC.[3]

The next step in the sulfurane-mediated mechanism is the elimination of acetic acid, providing a racemic mixture of ylides **12**. Alternatively, and avoiding the sulfurane, the chiral ylide **9** is formed by the action of acetate as a base on the sulfonium salt **8**. Different pathways are possible from **9/12** to **14**. Internal 1,2-acetoxy transfer may occur to afford the α-oxygenated sulfide **14** via cyclic transition state **13**; in the presence of an external or internal (different from acetate) nucleophile, α-Nu sulfides are obtained.[2] Another possible pathway involves the loss of acetate ion from **9** to form a close-fitting ion pair **16** between the thionium and the acetate; the thionium can then be trapped by internal or external nucleophiles with the possibility of enantioselectivity. However, the tight ion pair **16** can dissociate, generating an isolated thionium **15** which reacts with nucleophiles to afford a racemic mixture of the addition product.[1]

From a stereochemical point of view, it is important to control sulfurane formation, which is apparently responsible for the loss of stereochemistry, because it is the only achiral intermediate present in the proposed mechanism. As was already mentioned, the acetate ion can be trapped by DCC, and also by TMS^+, generated from a mixture of TMSOTf and DMAC. If acetate ion is removed, the sulfurane pathway in Scheme 20.3 is suppressed and good stereoselectivities can be obtained, because the original configuration at sulfur is maintained in intermediates **9** and **10** and even tight ion pair **16**.

20.2.2 Substrates

As was evident in the previous section, the main substrates for classical Pummerer reactions are sulfoxides. These are commonly obtained by the oxidation of sulfides. The oxidation can be performed with traditional oxidizing agents, such as $NaIO_4$, *m*-CPBA, and H_2O_2. Furthermore, the use of catalytic asymmetric methods for the selective oxidation of sulfides is well documented, and several synthetically applicable methods are described in the literature.[4]

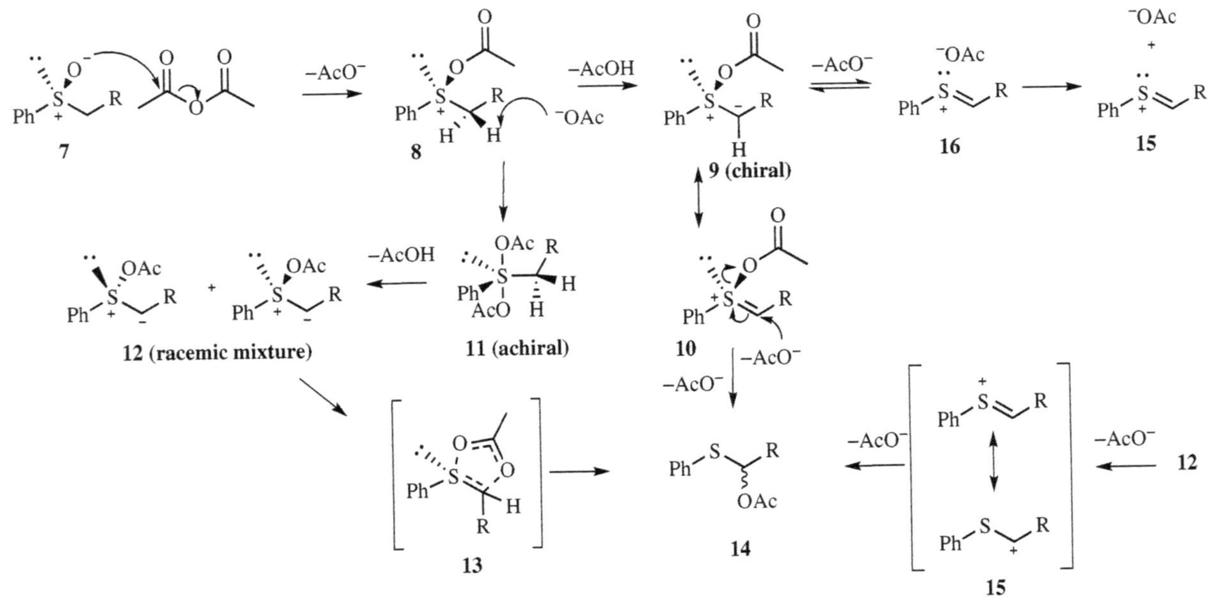

Scheme 20.3 Mechanism of classical Pummerer rearrangement.

Schemes 20.4 and 20.5 summarize some of the most useful methods for obtaining Pummerer substrates. As sulfides are generally the starting materials for preparing sulfoxides, Scheme 20.4 also highlights methods to generate those sulfide precursors.[4e]

Scheme 20.5 shows four of the most commonly used methods to produce enantioenriched chiral sulfoxides starting from sulfides. These can be used in addition to traditional oxidation methods mentioned previously.[4a–c]

Meanwhile, β-functionalized sulfoxides **20** can be easily obtained by Michael addition to α,β-unsaturated sulfoxides. The α,β-unsaturated sulfoxide starting materials **19** can be prepared by the oxidation of vinyl sulfides **21** and also by the Horner–Wadsworth–Emmons reaction,[5] using commercially available sulfoxy phosphonates **18** (Scheme 20.6a). Alternatively, cross metathesis on vinyl sulfoxides can be carried out using one equivalent of a Grubbs II catalyst in order to obtain complete conversion, although vinyl sulfides have proven to be unreactive.[6] Meanwhile, selective α addition of allyl and crotyl sulfides **22** to aldehydes **17** has been described and provides a useful way to access more functionalized vinyl sulfides **23** (Scheme 20.6b).[7]

20.2.3 Applications of Classical Pummerer Rearrangement

Scheme 20.7 shows different useful ways to use Pummerer chemistry. Although the Pummerer rearrangement has typically been used to synthesize α-substituted sulfides **25**, the substrate **24** can be viewed more generally as an $ACH_2S(O)B$ molecule, and either of the components A or B can be the principal reaction product, depending upon the synthesis strategy applied. Here, we present four different categories of examples, each of them focused on obtaining a different product from sulfoxides. The first category consists of the synthesis of α-oxygenated aldehydes starting with β-oxygenated sulfoxides, a process that corresponds formally to loss of the sulfur-containing moiety BSH (**24** → **26**, Scheme 20.7).

Recently, we described a useful sequence where Michael-acceptor sulfoxides **30** were obtained in two steps from homopropargylic alcohols **29** by radical addition of thiophenol and oxidation with sodium periodate.[8] The unsaturated sulfoxides were used in a highly stereoselective intramolecular oxa-Michael reaction. The sequence provided stereoselective functionalization of the sulfoxide moiety, and the products **31** proved to be useful in the synthesis of modified furanosides **32**. This represents a good example where sugars are prepared from acyclic precursors. The Michael addition was followed by a hydrolytic Pummerer reaction, yielding protected α-hydroxy aldehydes (Scheme 20.8) that upon acidic treatment afforded 3-substituted ribofuranoses.

Other recent reports on the synthesis of α-hydroxy aldehydes using Pummerer chemistry can be found in the synthesis of analogues of mannostatin A and in the synthesis of nelfinavir **38**, as described by Raghavan *et al.* (Scheme 20.9).[9]

The second set of examples involves the use of thionium ions as electrophiles in inter- and intramolecular processes to obtain α-substituted sulfides (see **24** → **25**, Scheme 20.7), which is the most common type of Pummerer reaction. Applications

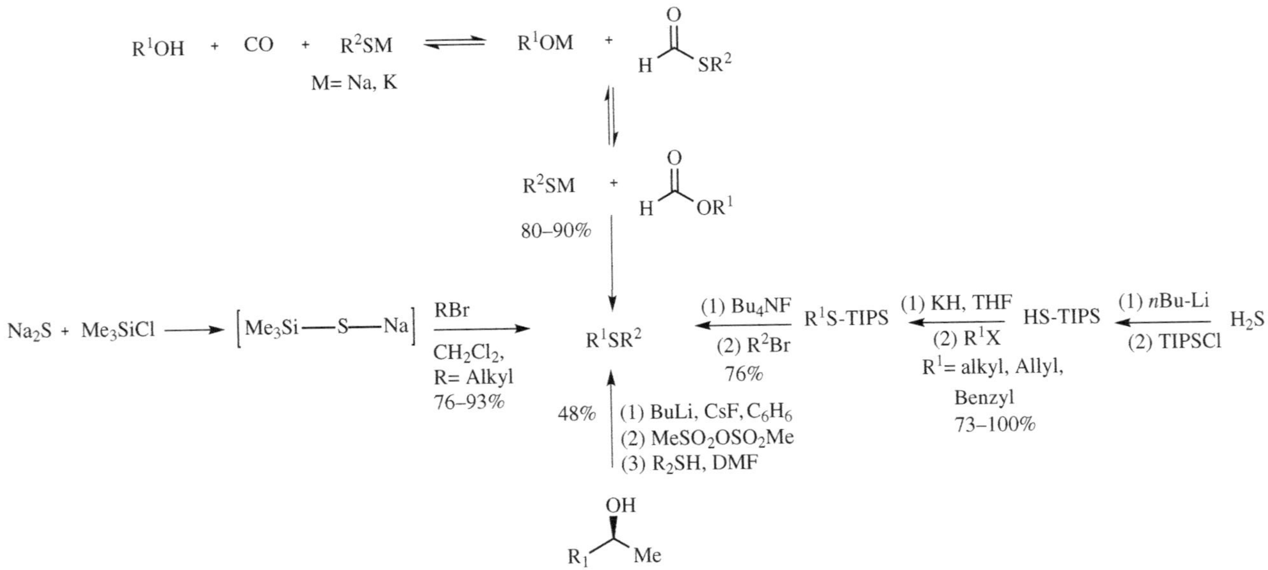

Scheme 20.4 Synthesis of sulfides.

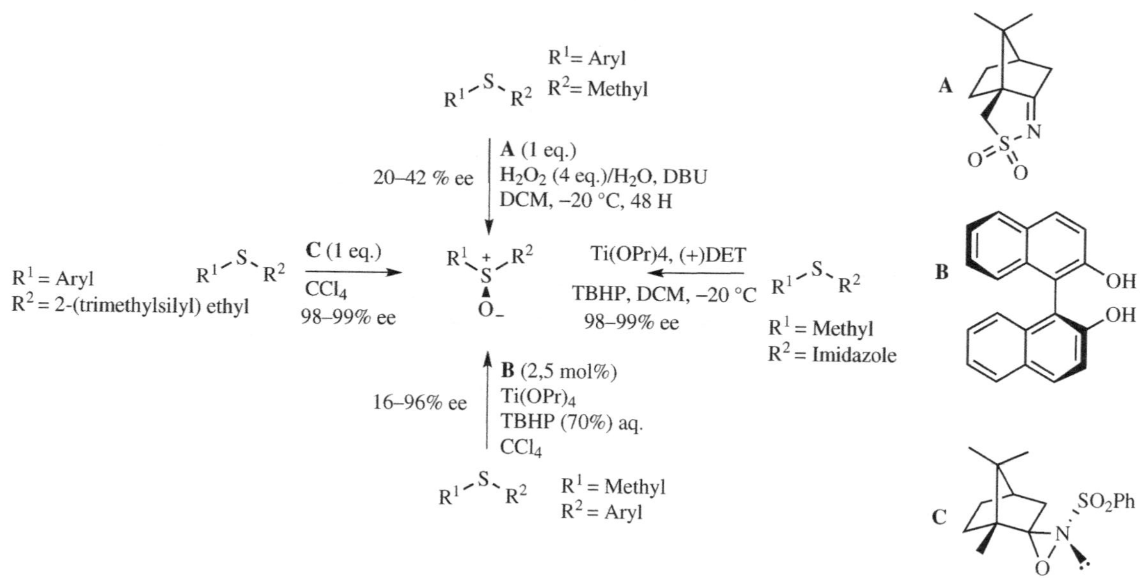

Scheme 20.5 Asymmetric synthesis of chiral sulfoxides.

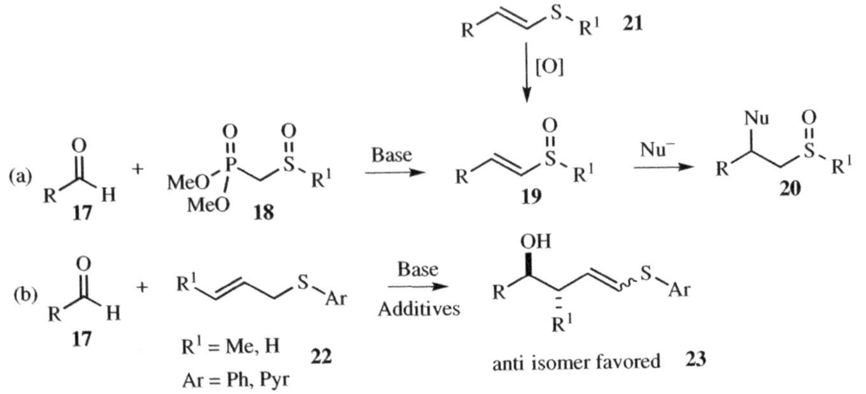

Scheme 20.6 Alternative synthesis of unsaturated sulfides and sulfoxides.

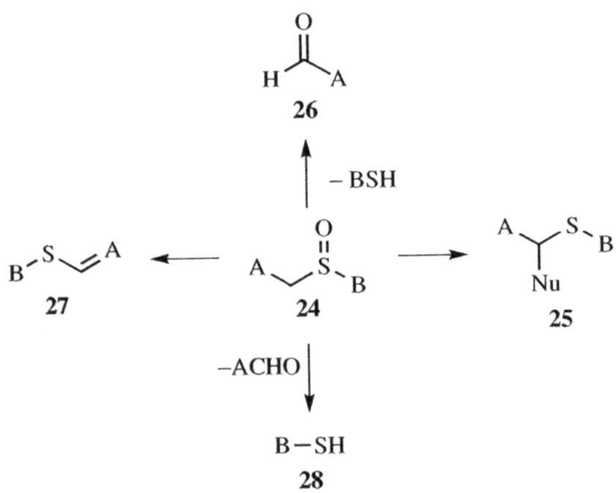

Scheme 20.7 Four different varieties of Pummerer products.

of this classical Pummerer rearrangement are exemplified in the synthesis of *trans*-solamin, the synthesis of indolizidine alkaloids, and the synthesis of the CDE ring of erinacine E. The first example (Scheme 20.10) uses Pummerer chemistry in the generation of a thionium ion, which reacts in an intermolecular tin-mediated ene reaction; the second one (Scheme 20.11) uses Pummerer chemistry to introduce a nitrogen-containing heterocycle by intramolecular addition to form the coniceine core; and the third example (Scheme 20.12) is an intramolecular silicon-induced Pummerer reaction with oxygenated nucleophiles applied to the synthesis of a precursor of erinacine. Details of these Pummerer-based strategies are discussed below.

Scheme 20.10 summarizes the application of the Pummerer rearrangement to the formal synthesis of *trans*-solamin. This is a good example to illustrate the introduction of an external nucleophile and to show the functional group tolerance

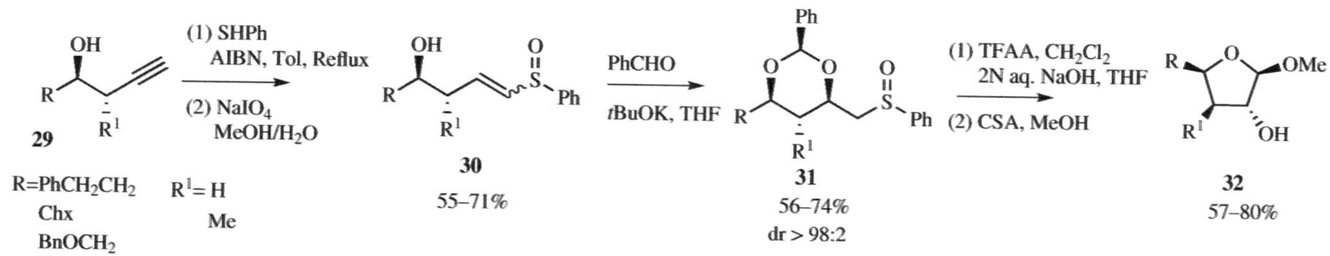

Scheme 20.8 Application of Pummerer rearrangement to the synthesis of modified furanosides.

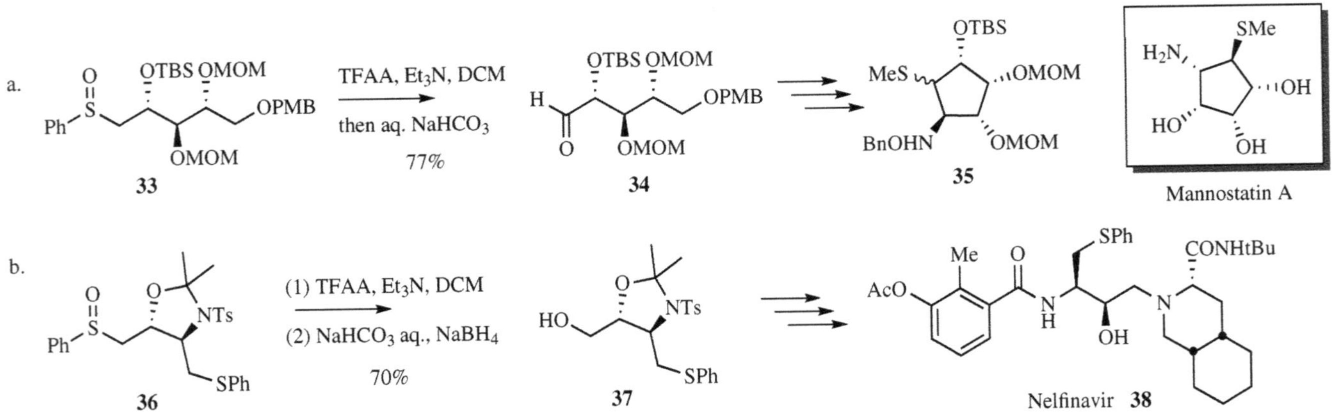

Scheme 20.9 Hydroxy aldehydes as intermediates in the synthesis of nelfinavir and analogues of mannostatin A.

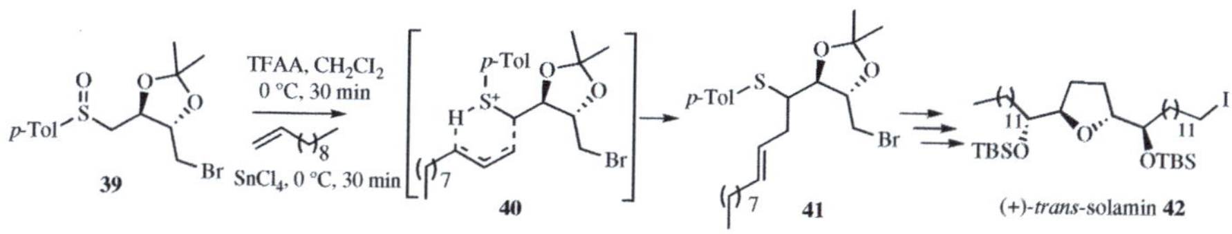

Scheme 20.10 Thionium ion generated by Pummerer chemistry as substrate in ene reaction.

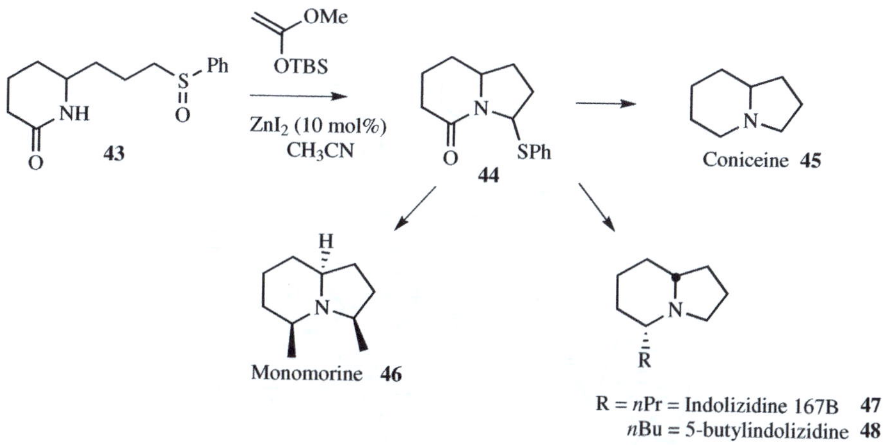

Scheme 20.11 Pummerer reaction in the synthesis of indolizine alkaloids.

of Pummerer chemistry and the impressive scope of thionium intermediates. It is known that thionium ions react with even extremely poor nucleophiles such as benzene[10] and the reaction with unactivated alkenes proceeds smoothly, generating the ene products. In 2008, Raghavan *et al.* published a formal convergent synthesis of (+)-*trans*-solamin.[11] In that paper, the Raghavan group used a Pummerer-type reaction to generate a thionium ion that reacted in the presence of anhydrous $SnCl_4$ and undecene to afford the homoallylic sulfide **41** (ene product) in an acceptable yield. Some stereoselectivity was observed in this addition; however, as the final product was obtained after the elimination of the sulfur moiety, the major isomer was not identified or isolated. The most important feature of this example is the reactivity shown by the thionium ion. An excellent electrophile, this ion reacts readily as an enophile and serves as a precursor of highly functionalized alkenes.

Additionally, Scheme 20.11 shows the versatility of Pummerer products. Under classical Pummerer conditions, heterocyclic compounds behave as good nucleophiles, including for the production of tertiary bicyclic lactams **44**. Kuhakarn *et al.*[12] used a silicon-induced Pummerer reaction (this activation method will be explained in more detail later in this section) to generate the thionium intermediate, which was intramolecularly trapped by the lactamic nitrogen. Even when a mixture of S-stereoisomers was used, the reaction proved to be diastereoselective, and the

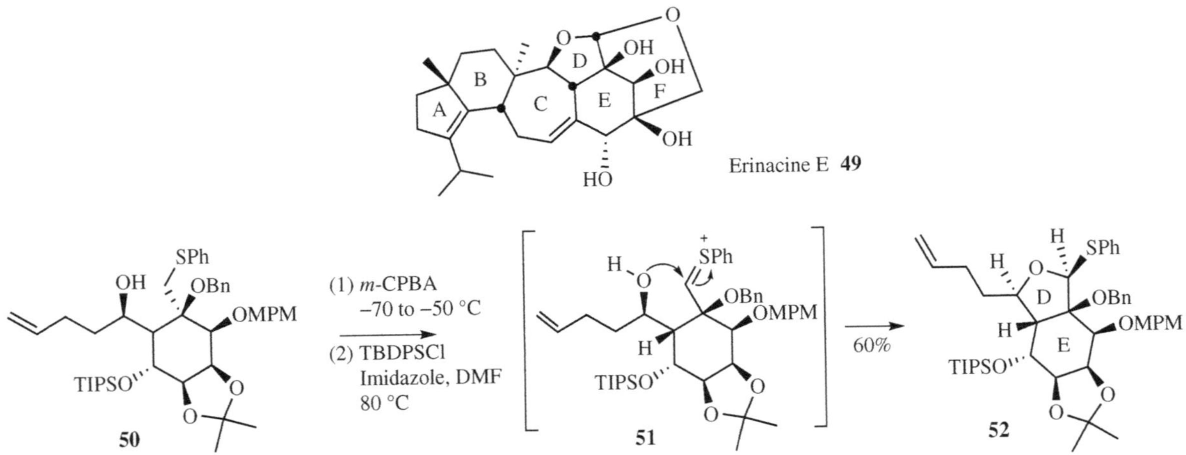

Scheme 20.12 Synthesis of CDE core of erinacine E.

lactam sulfide **44** was obtained as 5:1 mixture of trans : cis products. Subsequent transformations led to the synthesis of a variety of bicyclic alkaloids **45–48** starting from the same sulfide lactam intermediate **44**.

The last example of this series is Kobayashi *et al.*'s approach to the synthesis of erinacine E (Scheme 20.12).[13] Here, the thionium ion **51** was trapped by an internal hydroxyl nucleophile, forming the DE rings as a precursor of the CDE core system. This application shows two important aspects of Pummerer chemistry. First, while the activation of a sulfoxide is usually achieved by treatment with an anhydride, in this example, the sulfoxide activation competes with acylation of the OH group. In these cases, or when the substrate is very acid sensitive, milder activating agents (commonly silylation reagents) can be used to afford excellent results. The second important aspect in this example is stereochemical control. As discussed earlier, good stereocontrol in Pummerer reactions can be difficult to achieve without an additive that prevents loss of configuration at sulfur in the sulfurane intermediate. Nevertheless, highly substituted substrates with other pre-existing stereocenters do not need any additive for production of a configurationally stable thionium intermediate; the stereochemistry can be completely substrate controlled.

So far, the more common ways to use Pummerer chemistry have been described (i.e., **24** → **26** and **24** → **25**, Scheme 20.7). However, even if it usually is the A fragment (Scheme 20.7) of the sulfoxide which is the important component of the target, some researchers have used the classical Pummerer reaction to obtain thiols or complex vinyl sulfides (**24** → **28** or **24** → **27**, Scheme 20.7). Clearly, the convenience of using Pummerer chemistry in the synthesis of thiols is closely related to the structure of the desired product. Simple thiols can be obtained by Pummerer reactions, but in most cases, other shorter and cheaper methods are available. Consequently, the use of the Pummerer rearrangement to produce thiol products is most favorable in the synthesis of very complex thiols.

A good example of this application to thiol preparation can be seen in Scheme 20.13. Cryptophanes are synthetic supramolecular structures used principally in chemical encapsulation. Their synthesis usually starts with cyclotriguaiacylene (Figure 20.1). Sanseverino *et al.*[14] were interested in a thio analogue of the cryptophane, hypothesizing that it could be used as a building block in the self-assembly of more complex cryptophanes by disulfide bond formation. The required analogue **54** is a difficult-to-access trithiol. Sanseverino's group used a classical Pummerer rearrangement to produce the desired thiol in 80% yield. The starting sulfoxide **53** was easily accessible by oxidation of the corresponding sulfide, and the sulfide was obtained by condensation of benzylic alcohols as described by Collet and coworkers.[15] In order to avoid purification problems, it is best to use small A fragments in the **24** → **28** approach shown in Scheme 20.7. As pertains to the example of Scheme 20.13, the role of A is played by a hydrogen, with loss of formaldehyde in formation of the desired thiols.

The last example for an application of the classical Pummerer reaction corresponding to the **24** → **27** strategy of Scheme 20.7 is in the synthesis of pentenomycin analogues. Pohmakotr *et al.*[16] used base-mediated elimination of trifluoroacetate to

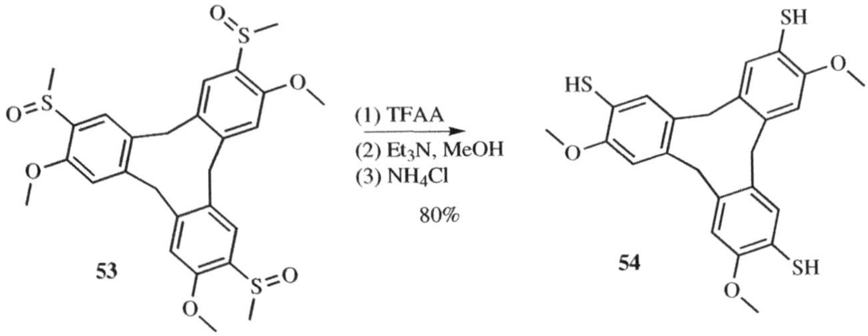

Scheme 20.13 Application of Pummerer rearrangement to the synthesis of complex thiols.

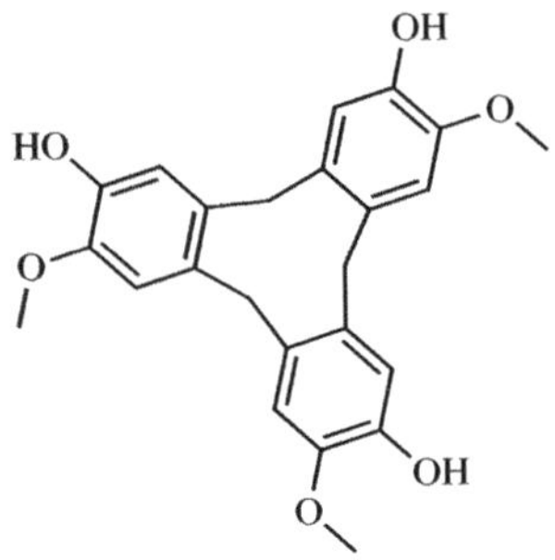

FIGURE 20.1 Cyclotriguaiacylene.

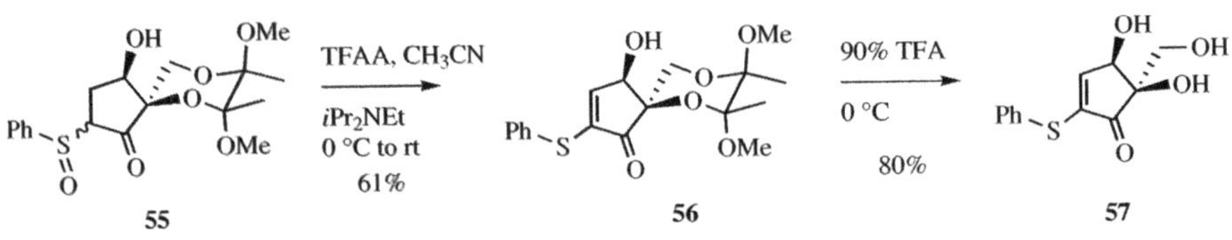

Scheme 20.14 Synthesis of vinyl sulfides by Pummerer rearrangement.

generate a cyclic vinyl sulfide **56**, which was described as an epipentenomycin analogue (Scheme 20.14). This transformation can be seen as a suitable methodology to reduce sulfoxides into sulfides by the Pummerer reaction with concomitant formation of a vinyl sulfide.

20.3 VINYLOGOUS PUMMERER REACTION

One of the most successful variants of the Pummerer rearrangement is the corresponding vinylogous reaction. In this modification, unsaturated sulfoxides **58** are activated

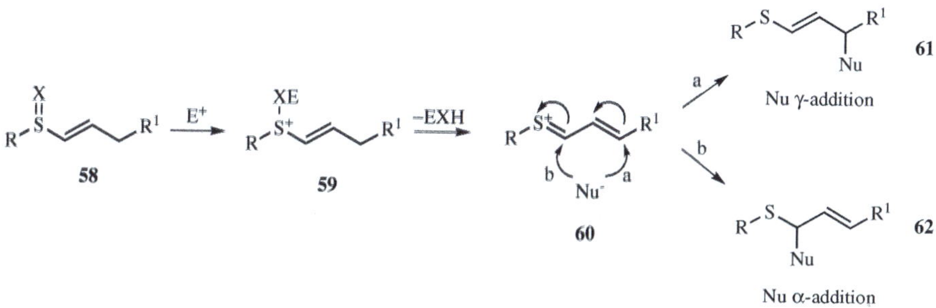

Scheme 20.15 Vinylogous Pummerer reaction.

by electrophiles, just as with saturated substrates. The resulting conjugated thionium ions **60** have two electrophilic positions, giving one of two products, that is, the α- or γ-isomer (Scheme 20.15).

Some of the most important applications of this reaction have been toward the synthesis of oridine-derived alkaloids. These routes have been extensively studied by Feldman's team[17] and received special attention in previous reviews.[18] In this chapter, we will focus on three more recent applications of the vinylogous Pummerer reaction and also on one recent report highlighting an unexpected and potentially useful result.

Lyconadin A (**66**) was isolated in 2001 by Kobayashi's group,[19] and its biological properties have been determined.[20] The particularities of the lyconadin structure, including its unprecedented nitrogen-bridged pentacyclic core, have made this alkaloid an attractive target for researchers. In 2011, Fukuyama and coworkers[21] published the third total synthesis of lyconadin A; they used two different approaches, one of which employed a vinylogous Pummerer reaction as the strategic step.

The key intermediate in this synthesis was the conjugated exocyclic enone **65**, which is not a common product of Pummerer chemistry. It provides, however, an excellent example that shows the versatility of oxygenated sulfides and how they can be transformed to obtain complex and not readily apparent structures.

As shown in Scheme 20.16, the α,β-unsaturated sulfoxide **63** was activated under classical Pummerer conditions, generating a conjugated thionium ion. This might undergo α or γ nucleophilic attack, but probably due to steric hindrance, the γ product **64** was isolated as the sole regioisomer in very good yield. The bifunctional

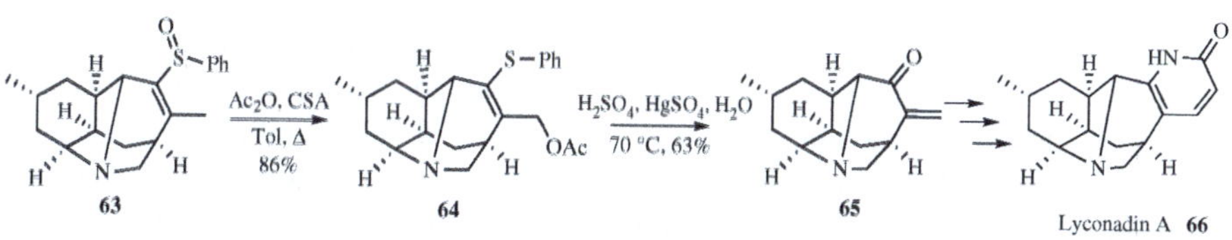

Scheme 20.16 Application of vinylogous Pummerer rearrangement in the total synthesis of lyconadin A.

sulfide **64** was exposed to acidic hydrolysis in the presence of mercury sulfate, yielding the expected α,β-unsaturated ketone **65**. This was used in subsequent steps to obtain lyconadin A (**66**).

It is necessary to emphasize that competition between α- and γ-isomers is usually observed, although this was not the case in Fukuyama's work. On the other hand, if the α-isomer is obtained as the major product, α,β-unsaturated ketones can be obtained as well by hydrolytic elimination of the sulfur-containing moiety (see Scheme 20.18).

As was already mentioned, we decided to highlight the Fukuyama lyconadin synthesis because ketones are unusual Pummerer products. Hopefully, the reader will perceive that Pummerer chemistry is rich, not only in terms of the primary products (functionalized sulfides), but also for the wide variety of functions that can be derived from it.

Continuing with the expediency of classical activation in vinylogous Pummerer chemistry and focusing on the different possibilities of applications for γ-esterified hydroxyl sulfides, it is logical to think of a mild hydrolysis to yield the free γ-hydroxy product. A recent example of this transformation was applied to the total synthesis of *ent*-hyperforin. Shibasaki and coworkers[22] described optimization studies on the synthesis of α-unsaturated-γ-hydroxy sulfides and their application in total synthesis of this natural product enantiomer **69**.

As shown in Scheme 20.17, the intermediate γ-trifluoroacetate was hydrolyzed *in situ* under mild conditions, affording the desired hydroxyl derivative **68**. Shibasaki also presented optimization studies where the competition between α- and γ-isomers was evident. The enone **73** was obtained by the hydrolysis of the α-product **71** (Scheme 20.18). These studies revealed the influence of the solvent and base. The combination of hindered bases with relatively polar solvents afforded best results for the desired γ-isomer. In contrast, unhindered bases and less polar solvents produced an almost equimolar mixture of **73** and **74**.

Clearly, Pummerer chemistry provides a useful way to access complex structures. To further illustrate this point in an even more complex setting, we present Feldman's approach to the synthesis of a palau'amine system. Palau'amine, as well as mono- and dibromo derivatives, has been the target of different synthetic studies.[23] As mentioned previously, Feldman's group has studied the synthesis of oridine alkaloids; their synthesis used a vinylogous Pummerer reaction, starting from a common intermediate (Scheme 20.19).

The application of the reaction to the more complex palau'amine structure assumed the use of an additional ring as in **80** (Scheme 20.20) in order to provide access to the pentacyclic system of palau'amine. As expected, Pummerer chemistry could be applied, and the authors described the successful extension of their previously developed methodology to the synthesis of these complex frameworks (Scheme 20.20). The activation of the sulfur moiety must be underlined. Usually, sulfoxides are thought to be the only advantageous Pummerer substrates; however, thionium ions are the real Pummerer starting material, and their generation can be carried out in different ways. Stang's reagent has proved to be practical in different

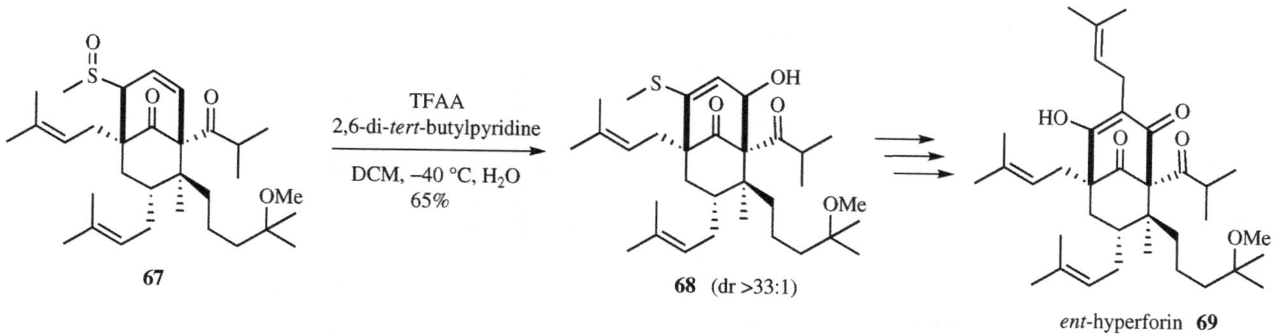

Scheme 20.17 Vinylogous Pummerer hydroxylation in the synthesis of *ent*-hyperforin.

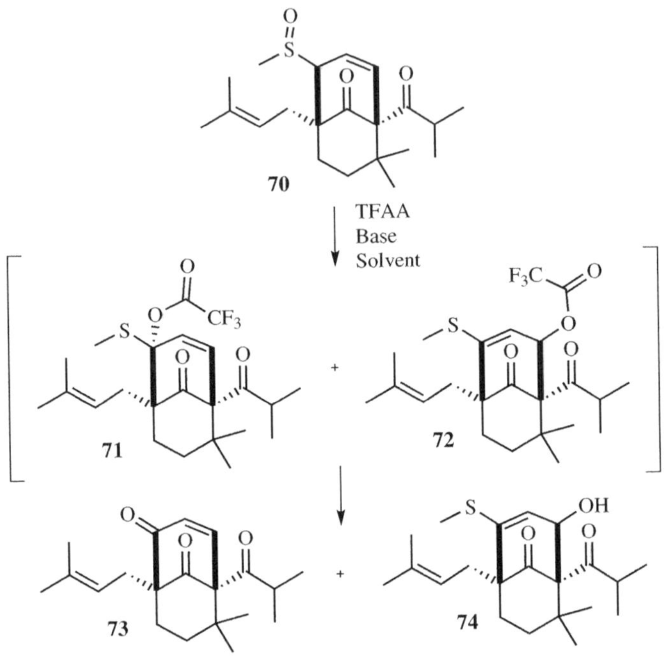

Solvent	Base	73:74
Toluene	Et_3N	1.1:1
Toluene	Pyridine	1:3
DCM	2,6 Lutidine	1:3.6
DCM	2,6 Di-Tert-butylpyridine	1:8.3

Scheme 20.18 Optimization studies on selectivity for vinylogous Pummerer reaction.

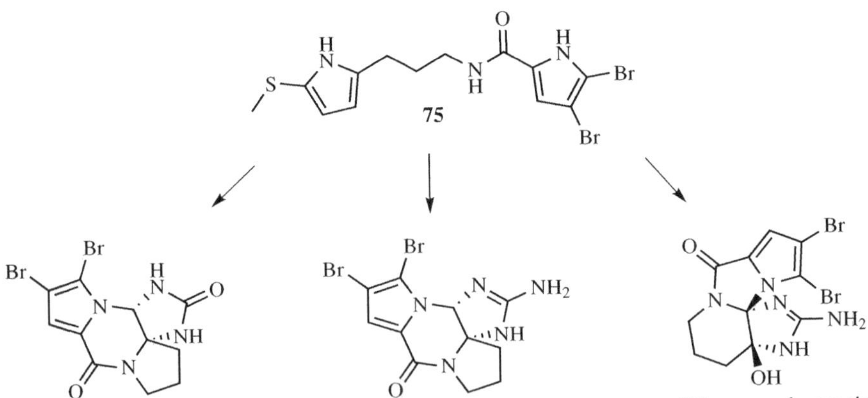

Dibromophakellin **76** Dibromophakellstatin **77** Dibromoagelaspongin **78**

Scheme 20.19 Oridine alkaloids from a common intermediate.

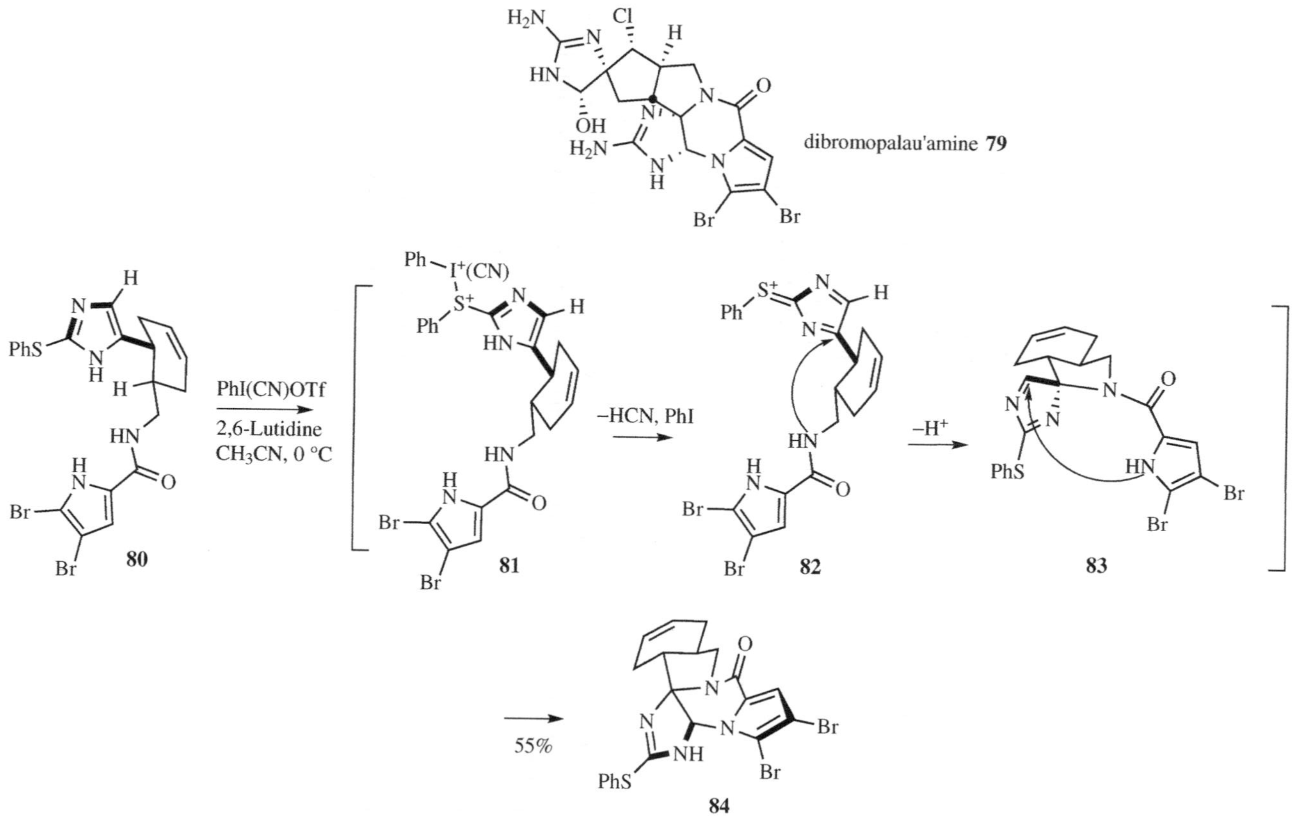

Scheme 20.20 Vinylogous Pummerer chemistry in the synthesis of palau'amine derivatives.

applications starting from sulfides or sulfoxides as shown for conversion of **80** to thionium ion **82**.[24] The authors speculated that, upon thionium formation, a double nucleophilic addition occurred. The first one is the vinylogous addition of the amide nitrogen to the conjugated thionium ion, and the next step is pyrrole addition to the second electrophilic site in the substituted imidazole ring. The stereochemistry of the only-formed isomer was confirmed by X-ray analysis.

To this point in our chapter, examples of γ addition to Pummerer intermediates have been reviewed; nevertheless, the usual query is whether conjugate addition can succeed with a more extended π system. Fortunately, an example that addresses this question was published in 2011.[25] Sulindac **85** is a nonsteroidal anti-inflammatory drug, and its treatment with oxalyl chloride under basic conditions gave the sulindac lactone derivative **88** (Scheme 20.21). The authors suggested a "longer-range" vinylogous Pummerer rearrangement caused by the addition of carboxylate **87** to the thionium ion generated by the activation of the sulfoxide with oxalyl chloride.

These examples show that the vinylogous Pummerer reaction is a potent tool in the synthesis of complex heterocyclic cores and unsaturated ketones. Moreover, α,β-unsaturated sulfoxides, when treated with an electrophilic activator, can undergo two different reactions. The first pathway (Scheme 20.22) involving elimination from activated intermediate **90** to give unsaturated thionium **91** was discussed earlier and is known as the vinylogous Pummerer reaction. The second path (**90** → **92** → **95**, Scheme 20.22) is believed to occur in competition and is called an additive Pummerer rearrangement. It involves the elimination of the activating agent promoted by a conjugate nucleophilic addition, better described as an S_N2 mechanism. In some cases, the same product can be obtained by either of these pathways, so competition does not represent an important issue. This was extensively studied by Feldman *et al.*[17c] in studies toward the synthesis of phakellin skeleton and is exemplified in Scheme 20.23.

Applications of the additive mechanism have often remained speculative because both vinylogous and additive pathways can provide the same product and because the preferred mechanism can depend on delicate changes in reaction conditions. Nonetheless, some examples exist in the literature (see next section) concerning the application of additive Pummerer reactions, and we present an innovative application of this Pummerer variant which is preceded by an interrupted Pummerer activation in the next section of this chapter.

20.4 INTERRUPTED AND ADDITIVE PUMMERER REACTIONS

As was introduced earlier, once the sulfoxide is activated, the resulting sulfonium salt **102** can undergo nucleophilic attack at the sulfur atom. This is known as an interrupted Pummerer reaction and is outlined in Scheme 20.24. The nucleophilic substitution on the sulfur atom yields a new sulfonium salt **103**. Depending on the structure of R, Nu, and R^1 in **103**, this can have three α electrophilic positions, which means that a nucleophilic substitution of sulfide will produce the loss of one of the three fragments. Usually, the first nucleophile is present in the same molecule, reacting in an intramolecular way. However, some examples using interrupted Pummerer

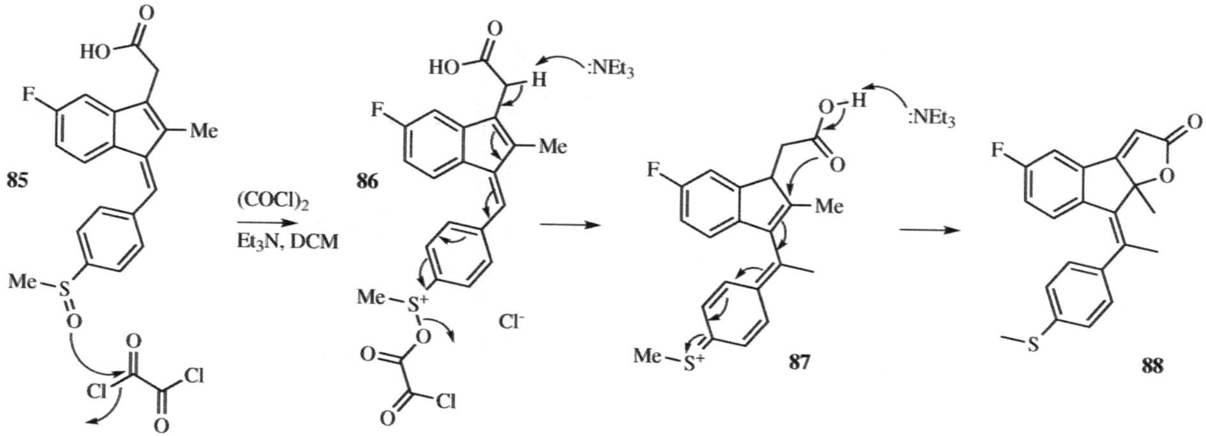

Scheme 20.21 A longer-range Pummerer reaction.

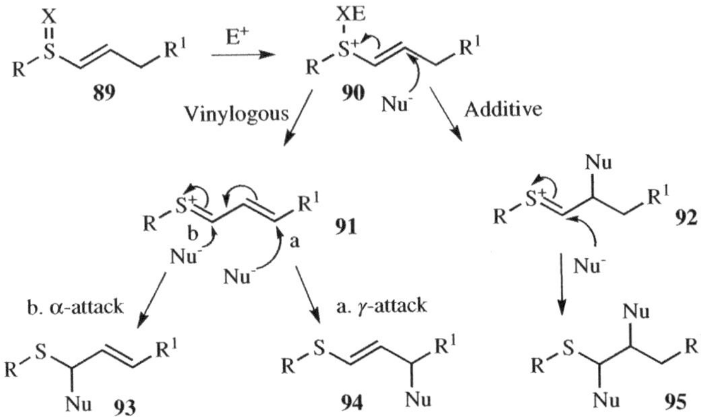

Scheme 20.22 The additive Pummerer reaction, competition with vinylogous addition.

reactions in intermolecular processes are described in the literature, and importantly, the reaction can be used to activate relatively unreactive positions in heterocyclic rings.

Scheme 20.25 describes the work published by Kawasaki's group[26] where they used a combination of dimethylsulfoxide and trifluoroacetic anhydride as source of trifluoroacetylated sulfonium ion **109** which reacted with **108** generating the new sulfonium salt **110** that underwent the loss of the sulfur-containing moiety promoted by nucleophilic attack. The nucleophile could be an alcohol, thiol, amine or organometallic species, or even another heterocyclic substrate. In cases where the nucleophile was a sulfoxide, the reaction led to an overall CH_2 oxidation (Scheme 20.25). Kawasaki's results suggest that this transformation, based on an interrupted Pummerer rearrangement, could be applied in the synthesis of biologically active tetrahydrocarbazoles and analogues (Figure 20.2).

Intramolecular substitution with enols or double bonds has been applied in the synthesis of fused heterocyclic compounds using the interrupted Pummerer strategy[27] (Scheme 20.26). These transformations follow the same principle described before, but the first nucleophile is added by an intramolecular process, and the second nucleophile is the acetate ion, generating methyl acetate as a reaction product. The substitution promoted by acetate can be an intra- or intermolecular process as was suggested before by Yuste, García Ruano, and coworkers.[28]

As was introduced earlier in this chapter (Scheme 20.22 and previous section), the additive pathway in Pummerer chemistry is in competition with the vinylogous path, and its application is still in some cases unpredictable. Fortunately, Yorimitsu, Oshima, and coworkers have published a sequence of papers[29] providing excellent applications of the additive Pummerer pathway. They used a combination of two Pummerer approaches, such that the activation of sulfoxides was done as usual by an electrophilic reagent, but they used the beginning of a so-called interrupted Pummerer reaction (*vide infra*) to generate the α,β-unsaturated sulfonium salt **121**. This intermediate did not follow the traditional

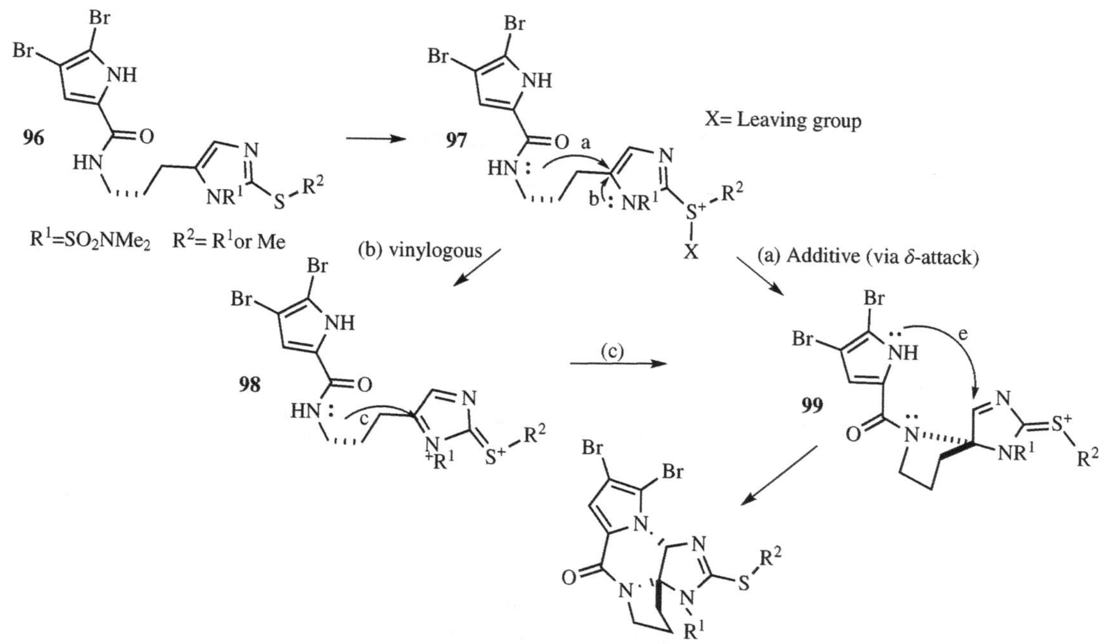

100 Phakellin skeleton

Scheme 20.23 The additive Pummerer reaction in the synthesis of phakellin skeleton.

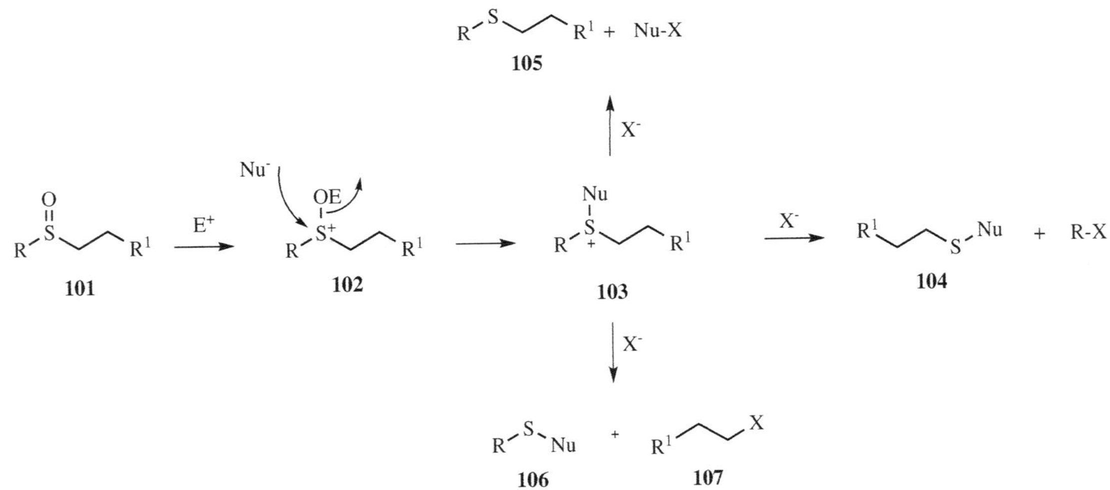

Scheme 20.24 The interrupted Pummerer reaction.

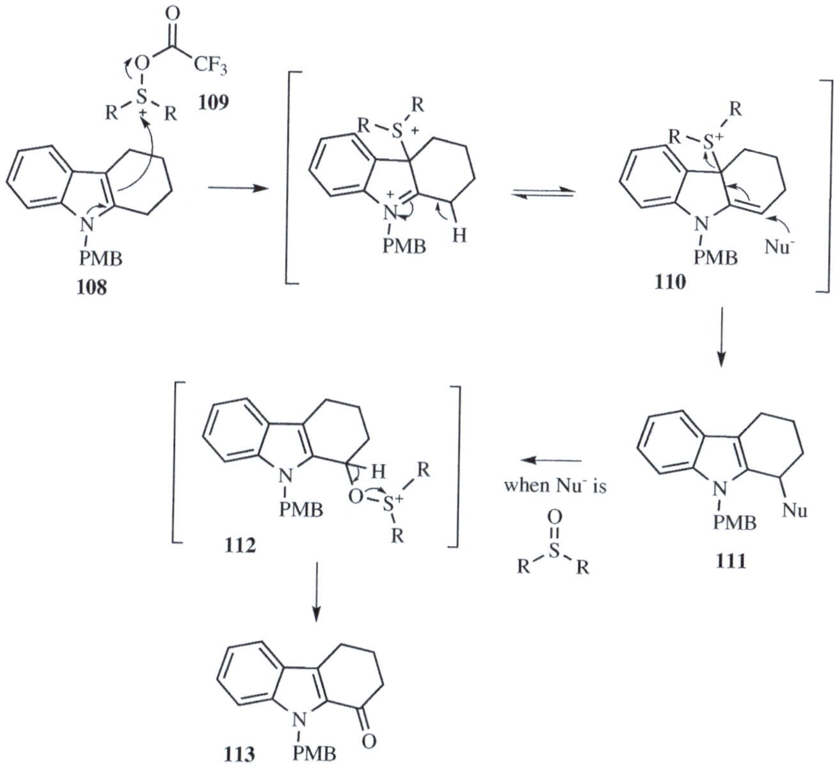

Scheme 20.25 Allylic oxidation promoted by interrupted Pummerer chemistry.

interrupted pathway, but rather underwent a [3,3]-sigmatropic rearrangement (Scheme 20.27). Overall, the transformation can be seen as an addition of a carbon nucleophile at the β-position of thionium **121** (cf. **90** → **92**; Scheme 20.22).

Yorimitsu and Oshima's procedure, with clever modification, has been applied to the synthesis of different structures, summarized in Scheme 20.28. Notably, thioacetals **137** were obtained by a copper-promoted additive Pummerer reaction. This is, in fact (to the best of our knowledge), the first example of a metal-catalyzed Pummerer reaction with direct activation of the sulfoxide and also the first direct and noncompetitive additive Pummerer rearrangement.[29c]

This combination of reactions provides tremendous scope for new transformations. Various authors, including Maulide and coworkers,[30] Procter and coworkers,[31] and Oshima and coworkers,[32] have described useful ways to apply this sequence with aromatic compounds, providing the possibility of obtaining ortho-substituted sulfides and benzofurans (Scheme 20.29).

Scheme 20.29 exemplifies four different approaches to carry out an interrupted Pummerer reaction [3,3]-sigmatropic rearrangement sequence. Oshima's work use phenol **138** as nucleophile generating the optimal intermediate **140** for the sigmatropic rearrangement; the formed product **141** can rearrange to afford

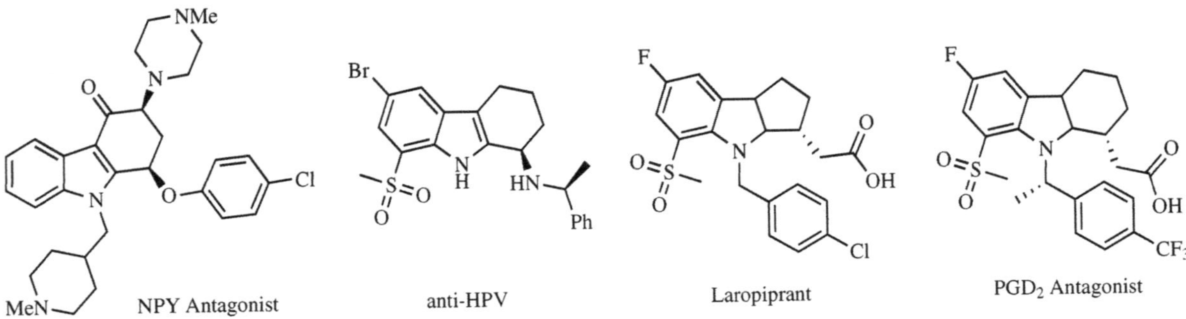

NPY Antagonist — anti-HPV — Laropiprant — PGD_2 Antagonist

FIGURE 20.2 Some bioactive tetrahydrocarbazoles and analogues.

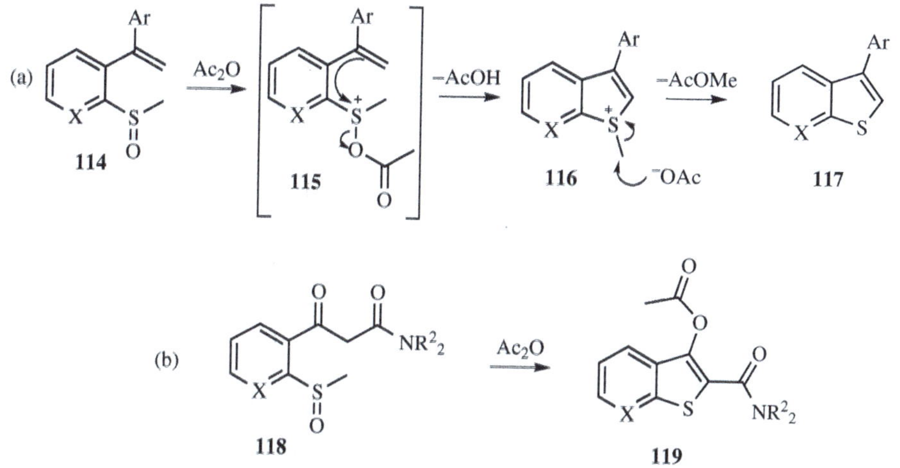

Scheme 20.26 Fused heterocyclic systems by interrupted Pummerer chemistry.

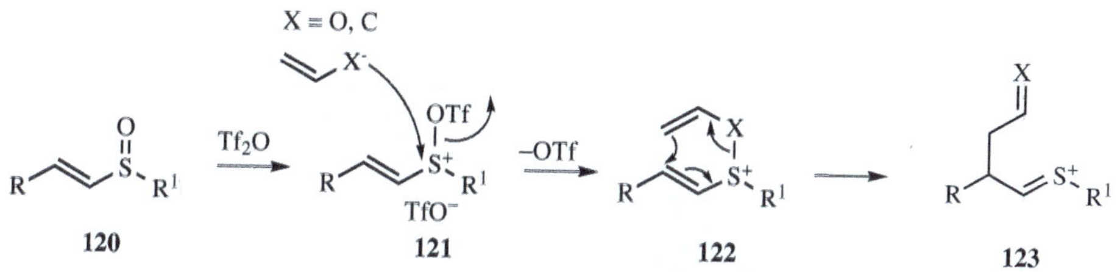

Scheme 20.27 Additive Pummerer reaction followed by [3,3]-sigmatropic rearrangement.

sulfur-substituted benzofurans **142**. Maulide used a similar intermediate but generated by a nucleophilic ketone **145** and the diphenylsulfonium salt; in this case, the final products were ortho-substituted phenyl sulfides **146**. Procter's approaches are based on the use of nucleophilic carbon atoms in silylated unsaturated structures like **148** or **154**; these strategies also afford substituted aryl sulfides.

20.5 CONNECTIVE PUMMERER REACTION

Connective Pummerer reactions are transformations where the thionium ion is generated by the addition of a thiol **158** to an aldehyde **157**, generating a hemithioacetal **159**, which undergoes elimination, yielding the corresponding thionium ion **160**. This ion can be trapped by a nucleophile as already described (Scheme 20.30). The nucleophile can be introduced by an inter- or intramolecular process, thereby increasing the applicability of this methodology.

As shown so far in this chapter, sulfoxides are the most commonly used Pummerer substrates. Sulfides are less commonly used than their oxygenated partners, but they are still useful (see Scheme 20.20).

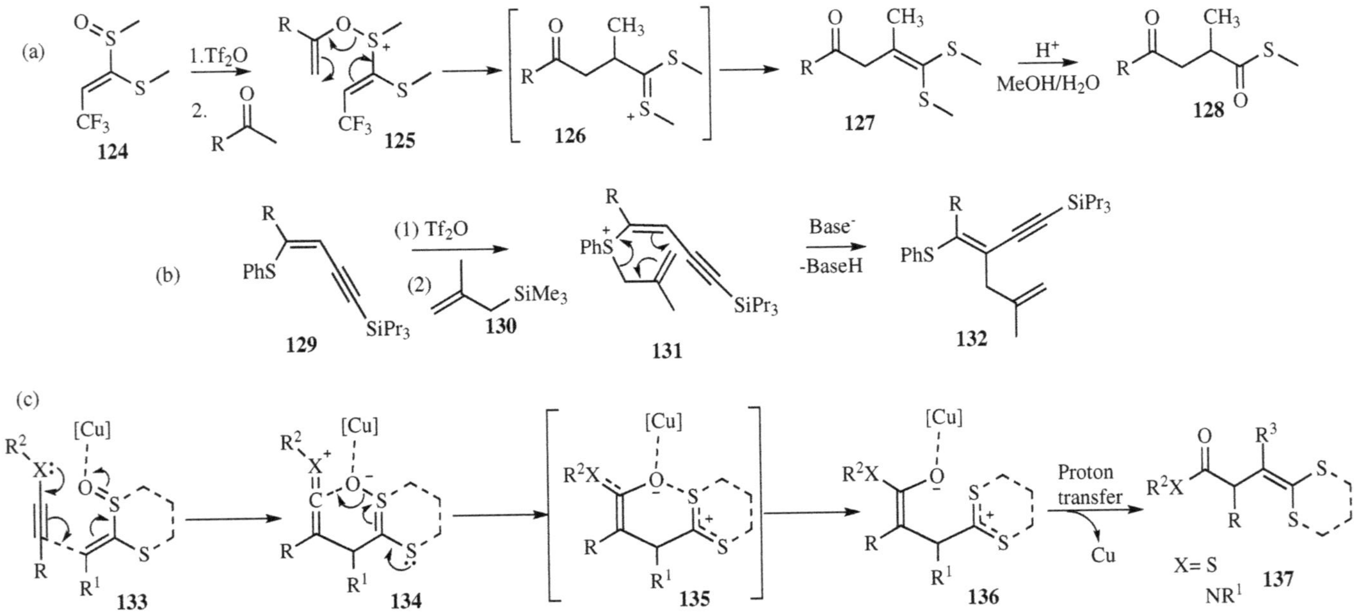

Scheme 20.28 Structures obtained by additive Pummerer reaction.

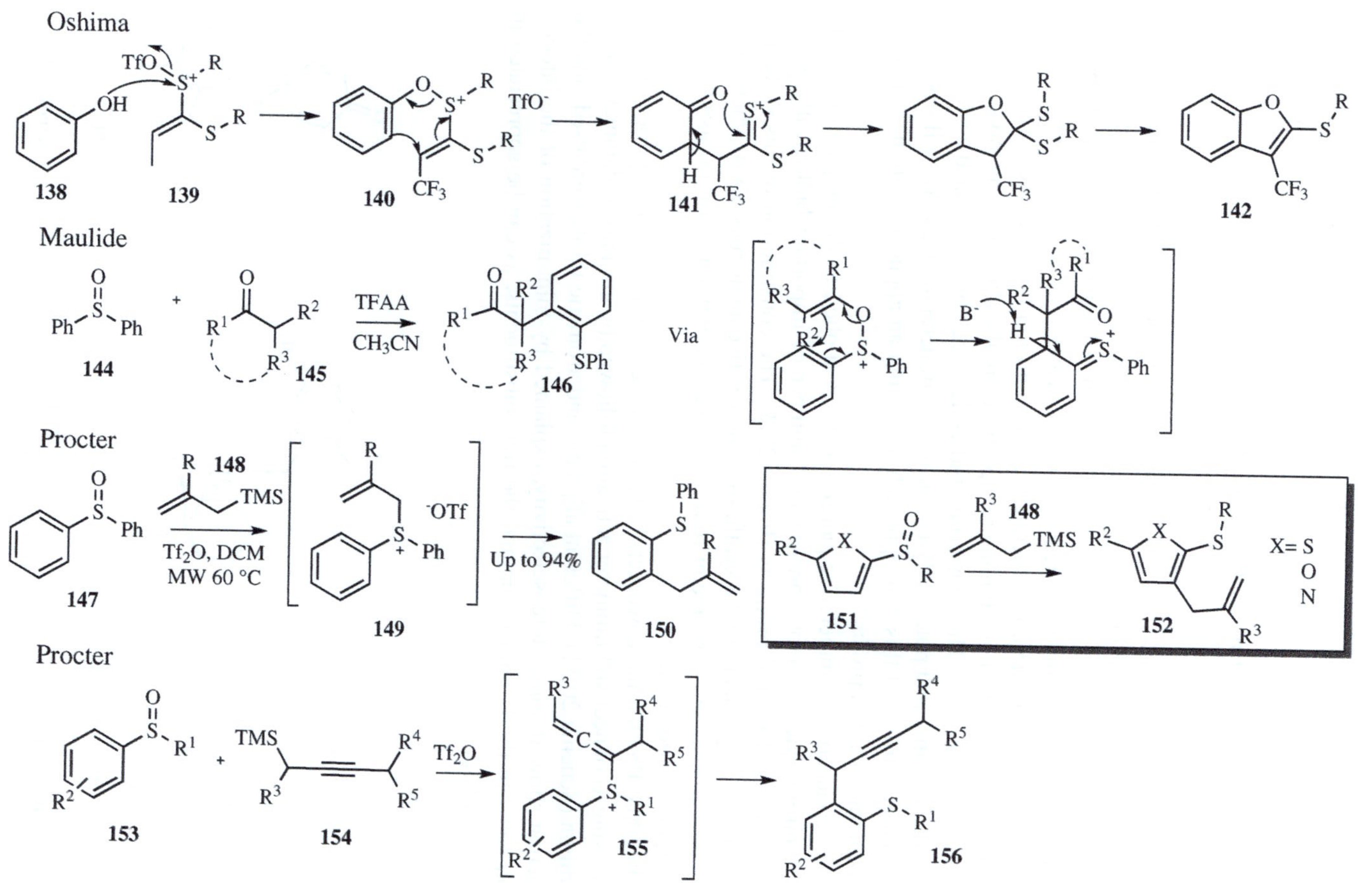

Scheme 20.29 Applications of interrupted Pummerer reaction – [3,3]-sigmatropic rearrangement sequence.

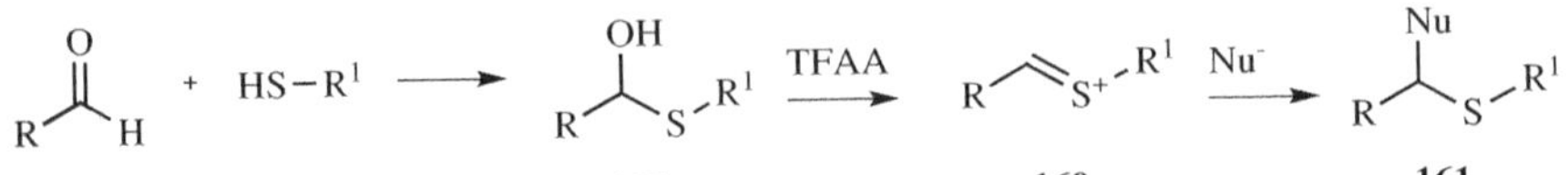

157 **158** **159** **160** **161**

Scheme 20.30 Connective Pummerer reaction.

We have also outlined how Pummerer reactions are classified, depending on the mode of nucleophile interaction with the thionium ion and on the way that the thionium is generated in the first place. In terms of nucleophile interaction, the so-called connective Pummerer reaction is in the category that involves the reaction of a nonclassical substrate to generate the thionium ion (see Scheme 20.30). What distinguishes the connective Pummerer approach are the alternative methods used to generate thionium ions. The pioneering studies described by Procter and coworkers[33] showed a versatile and potentially useful new Pummerer strategy. We will focus on the synthetic utility of this new approach, including an application in the synthesis of a complex natural product.

The connective Pummerer sequence has been applied to the preparation of the 2-azaspiro[4.5]decane motif present in a number of important biologically active compounds and natural products (Figure 20.3). The construction of this system is consequently significant in the synthesis of new active analogues. Even where other methods are available, Pummerer routes offer new opportunities for the assembly of structurally varied congeners.

In 2008, Procter and coworkers[34] described the synthesis of this spirocyclic system using a connective Pummerer activation followed by intramolecular nucleophilic addition (Scheme 20.31). Different aliphatic and aromatic thiols were used, and the reaction showed some stereoselectivity, explained by minimization of the interaction between the R thiol group and the ring substituents, as can be appreciated in Scheme 20.32.

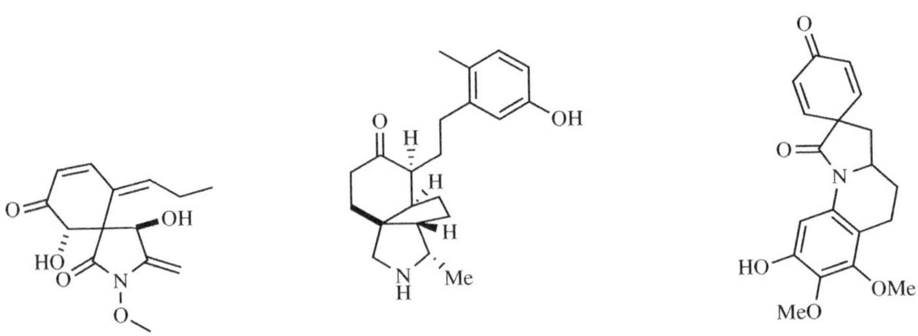

Spirostaphylotrichin A

Alkaloid from *didymeles madagascariensis*

Annosqualine

FIGURE 20.3 2-Azaspiro[4.5]decane in some natural products.

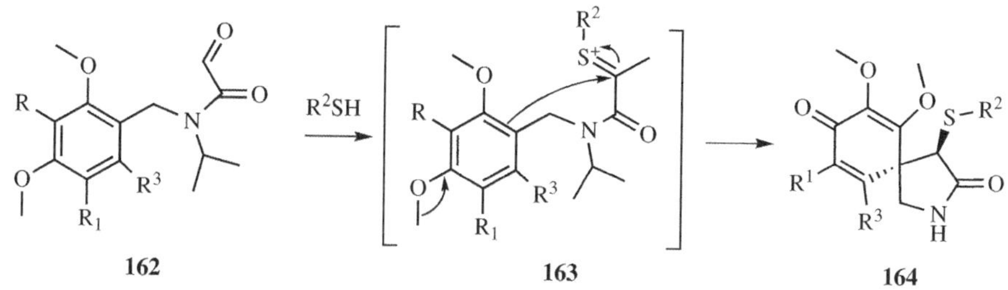

Scheme 20.31 Procter approach to 2-azaspiro[4.5]decane by interrupted Pummerer activation.

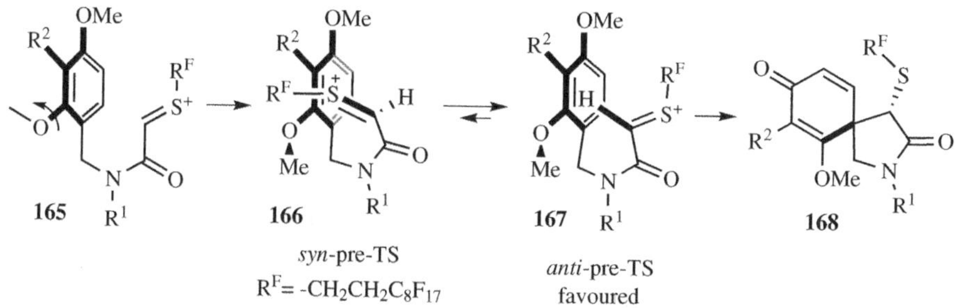

Scheme 20.32 Transition states and stereochemical explanation.

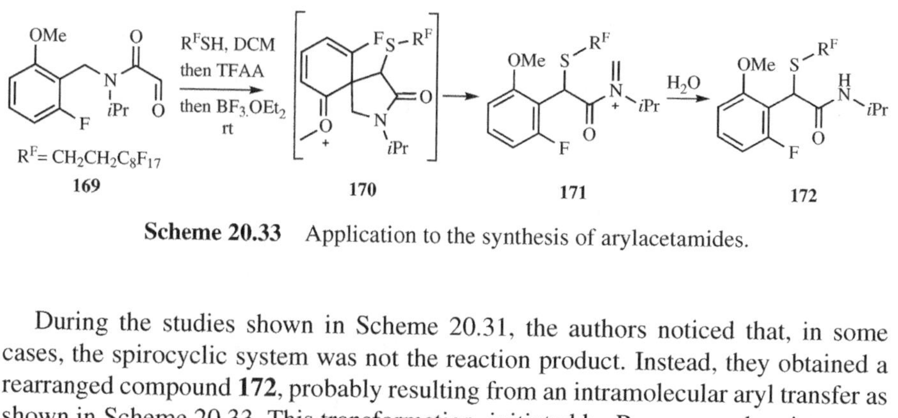

Scheme 20.33 Application to the synthesis of arylacetamides.

During the studies shown in Scheme 20.31, the authors noticed that, in some cases, the spirocyclic system was not the reaction product. Instead, they obtained a rearranged compound **172**, probably resulting from an intramolecular aryl transfer as shown in Scheme 20.33. This transformation, initiated by Pummerer chemistry provides a new synthesis of α-arylacetamides, which are present in a large number of active molecules.[35]

An extension of this methodology was published in 2009,[36] involving the use of glyoxamides with less substituted aromatic rings as the precursors of fused heterocyclic systems. The general reaction conditions are shown in Scheme 20.34, as well as the different kinds of complex fused rings obtained by this approach.

Procter and coworkers also applied their connective Pummerer methodology to the total synthesis of the fused-ring alkaloid cryptotackieine (neocryptolepine).

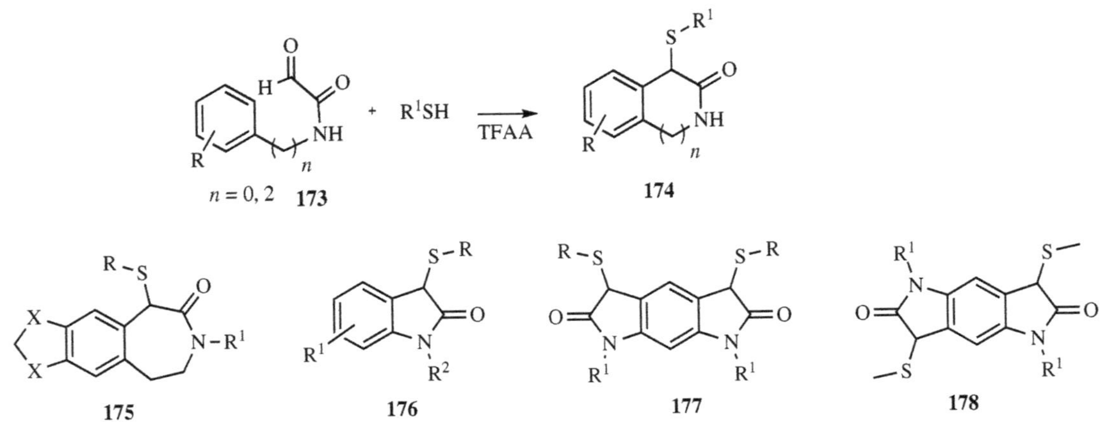

Scheme 20.34 Different structures obtained by the Procter connective approach.

To this end, they followed the synthetic route shown in Scheme 20.35, which includes a connective Pummerer cyclization from **179** followed by a pyridine-forming reaction sequence mediated by samarium, also developed by the Procter group.[37]

Notably, Procter's team also published the first synthetic approach to more complex structures using the connective Pummerer methodology described previously. Ecteinascidin 597 belongs to a family of structurally intricate molecules with an extensive range of antitumor and antimicrobial activities. It is licensed as Yondelis® for the treatment of sarcoma, and it is under study for other clinical applications.[38]

In Scheme 20.36, we present the synthesis of the ABH ring system of ecteinascidin 597 (**187**) developed by Procter and coworkers.[39] As expected, the extension of the connective Pummerer approach with glyoxamides and cysteine derivatives proved to be convenient and was applied successfully in the synthesis. After the preliminary cyclization step, a 1:1 mixture of stereoisomers **185** was isolated, showing that, in this case, the substituents present on the aromatic ring did not impart a steric bias (compare Scheme 20.32). Fortunately, that stereocenter could be subsequently equilibrated, and the desired product **187** was obtained upon lactonization.

20.6 PUMMERER REARRANGEMENT IN MULTIPLE-REACTION PROCESSES

Tandem, cascade, and other kinds of combined chemical transformations have been shown to be extremely useful in the synthesis of complex structures.[40] The Pummerer rearrangement can also be combined with different known reactions in order to produce structures more complex than the usual Pummerer products. Here, we describe two recent applications of Pummerer chemistry in combination strategies.

The first example is the use of chloranil in C—H oxidation of sulfides, leading to the *in situ* formation of the electrophilic thionium ion, which was trapped by 1,3-dicarbonyl compounds in a Knoevenagel reaction.[41] The products obtained were α-acyl-β-sulfurated carbonyl compounds **192**. This transformation and the authors' proposed activation mechanism are shown in Scheme 20.37. Elimination of the sulfur-containing moiety promoted by chloranil afforded the usual Knoevenagel products **189**.

The second example is a new synthetic approach to the synthesis of nakadomarin A **202** (Scheme 20.38). To this end, Winkler and coworkers developed a tandem Pummerer–Michael process.[42] The generation of the activated thionium ion was performed by the elimination of ethanethiol from thioacetal **198**. The intermediate **199** showed double reactivity, that is, a nucleophilic position on the furan to perform the Michael addition and the usual electrophilic thionium, which was in fact trapped by the enamine carbon generating tetracyclic product **200**. Winkler presented the synthesis of a tetracyclic model system and suggested that this reaction sequence can also be applied to the synthesis of the natural product.

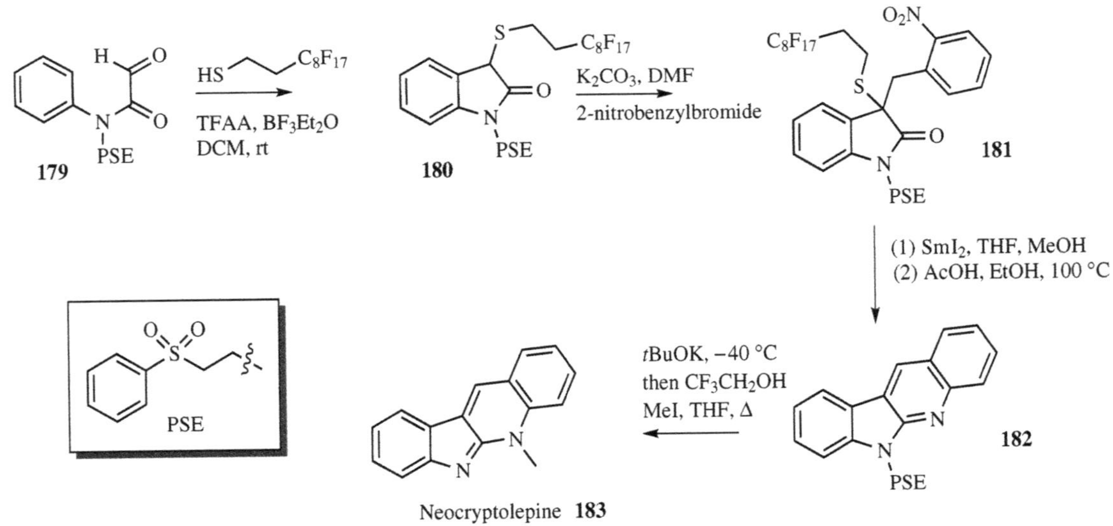

Scheme 20.35 Application to the total synthesis of neocryptolepine.

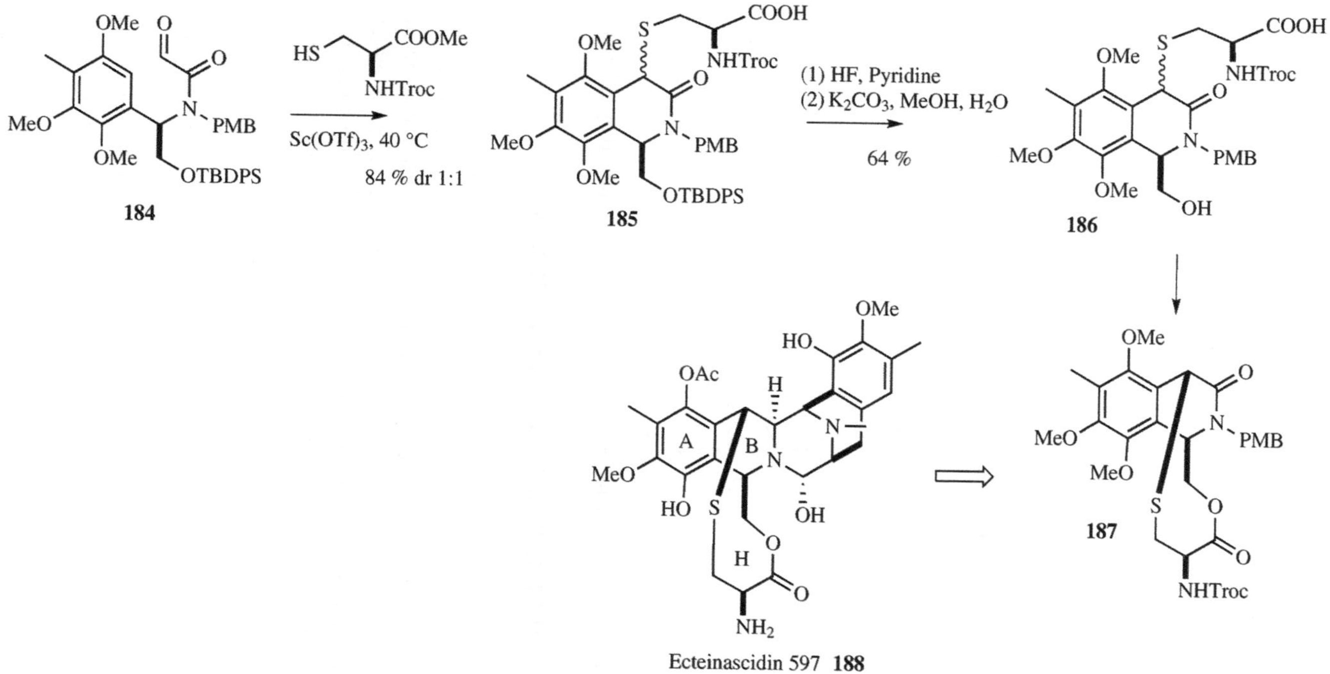

Scheme 20.36 Procter approach to the synthesis of ecteinascidin **597**.

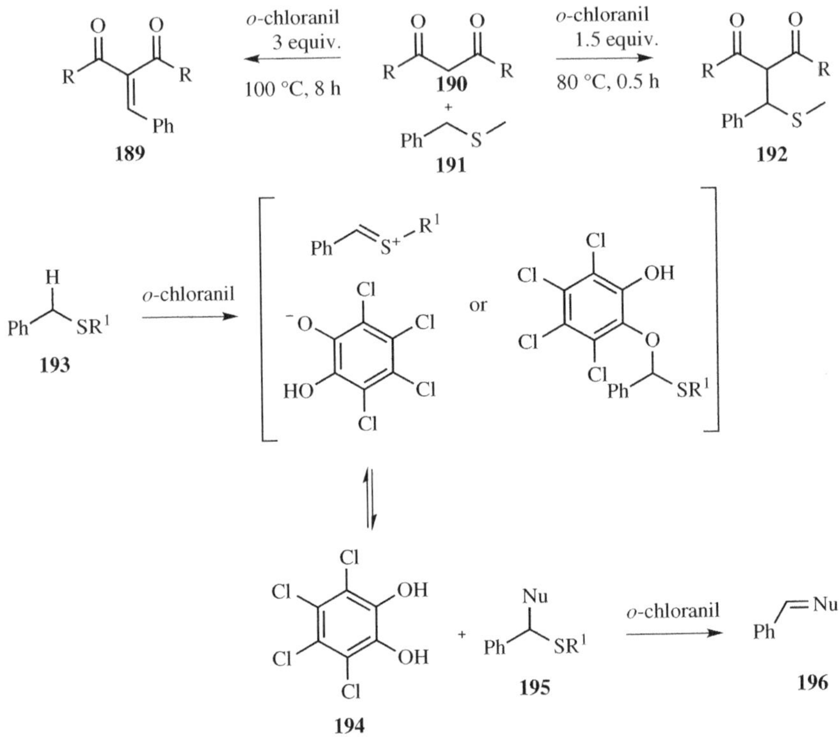

Scheme 20.37 Pummerer–Knoevenagel sequence promoted by chloranil.

20.7 OTHER PUMMERER REARRANGEMENTS

The literature on Pummerer chemistry is quite extensive, and multiple variants are described. However, looking deeply into these alternatives, it becomes apparent that most of them are based on the same principle, and the only variation is the nucleophile source. One example of this is the so-called fluoro-Pummerer reaction, which is in fact a fluoride addition to the normal Pummerer intermediate. The distinguishing feature is the generation of the thionium intermediate, using (diethylamino)sulfur trifluoride (DAST) in combination with metals (Scheme 20.39). DAST is known as a source of nucleophilic fluoride and has been widely used for this purpose. A review on the fluoro-Pummerer reaction was published by Haufe and coworkers in 2012.[43]

Given that selenium and sulfur have similarities in their chemical behavior, it is logical that another commonly used Pummerer variant involves the extension from sulfur to selenium. In this regard, the seleno-Pummerer reaction has been widely applied in selenonucleoside synthesis,[44] in the same manner that normal Pummerer chemistry is used in the synthesis of thionucleosides. One recent example that illustrates the use of seleno-Pummerer chemistry is the work of Jeong and coworkers.[45] Oxidation of selenium was accomplished with *m*-CPBA, and activation of the selenoxide with acetic anhydride proceeded well, generating the

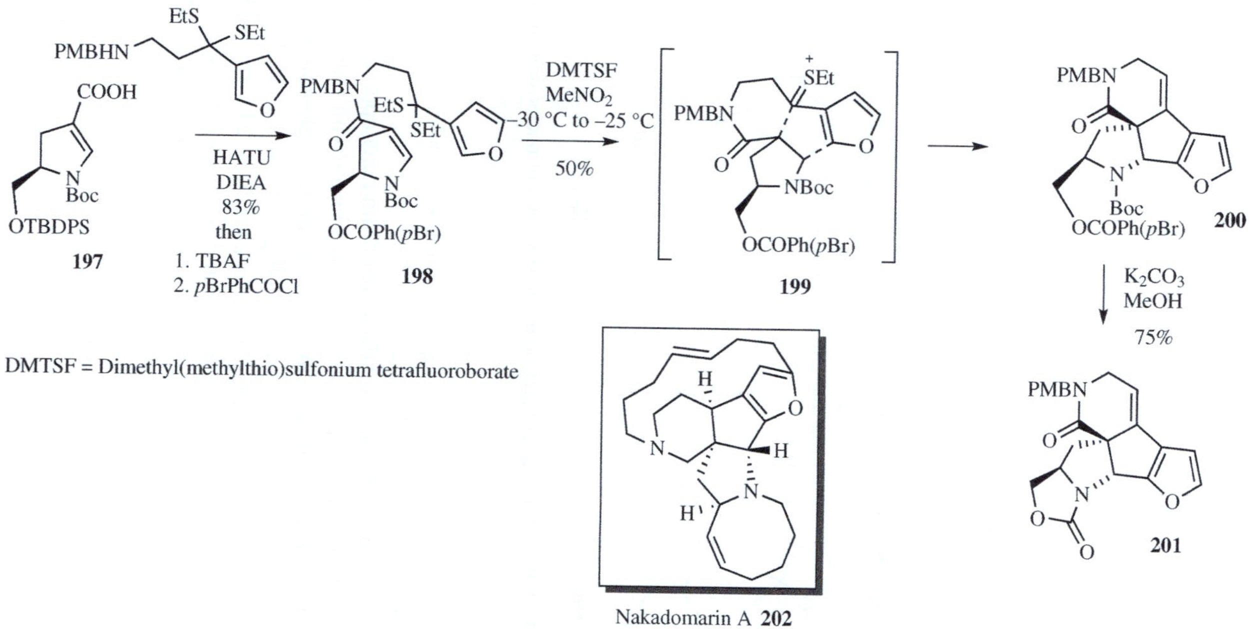

Scheme 20.38 Synthesis of a model system of nakadomarin A.

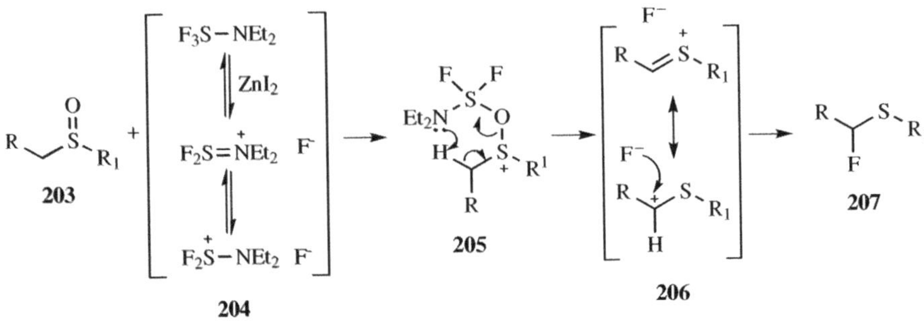

Scheme 20.39 Mechanism of fluoro-Pummerer reaction with DAST and ZnI_2.

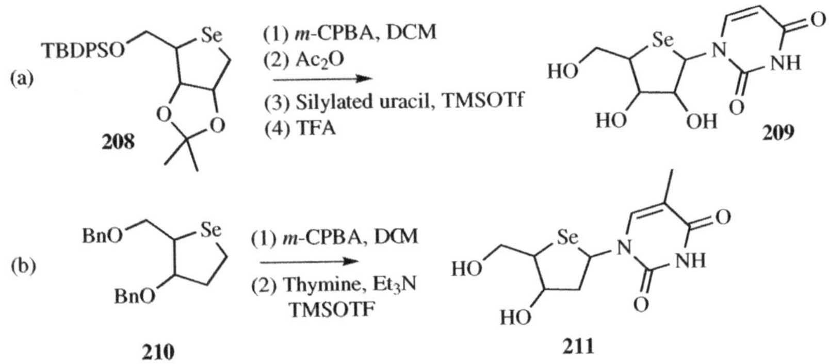

Scheme 20.40 Seleno-Pummerer reaction in the synthesis of nucleoside analogues.

electrophilic species which was then trapped by the nucleophile, in most cases a nitrogen-containing base (Scheme 20.40).

A recent and innovative application of the seleno-Pummerer approach is the use of *O,Se*-acetals in a radical-generating process that culminates with radical addition to double bonds. This method was used in the synthesis of prostaglandin analogues and the zedoarondiol cyclic core **221**.[46] This methodology is illustrated in Scheme 20.41. It is based on the possibility of generating oxygenated radicals **214** from *O,Se*-acetals **213**; the resulting radical reacts with an unsaturated compound **215** and, assisted by tributyltin hydride, affords the oxygenated product **216** and regenerates the radical source. These oxygenated radicals were successfully used in two- or three-component reactions, as shown by the formation of complex structures. Notably, the application of this reaction sequence with *O,Se*-acetal **217** (formed by seleno-Pummerer reaction) and cyclopentenone **218** as the unsaturated compound provided the highly functionalized alkene **220** which was used as a precursor of the zedoarondiol cyclic core **221** as indicated in Scheme 20.41.

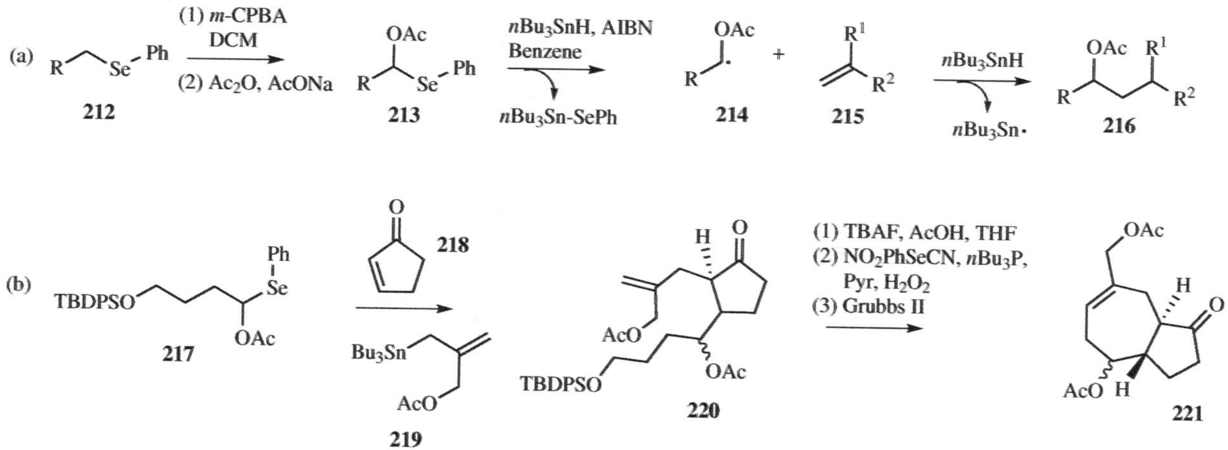

Scheme 20.41 Seleno-Pummerer in radical processes.

20.8 SUMMARY AND OUTLOOK

Even though long known, Pummerer rearrangements or Pummerer reactions have been shown to be extremely versatile transformations for organic synthesis. The most common way to use Pummerer chemistry is for the preparation of α-substituted sulfides. However, this is only the first step toward accessing an impressively wide array of structures.

In this chapter, we have presented the general principles and specific examples of Pummerer reactions, hoping to provide researchers with inspiration in planning their future syntheses. Applications include complex natural products and the synthesis of unusual structures. Moreover, new developments have opened further avenues for research. For example, the generation of thionium ions from thiols and aldehydes represents a new transformation where innovative work can be done. Additionally, even if in a large number of cases the Pummerer reaction is diastereoselective, the requirement for a general and readily applicable asymmetric version still exists. We expect that researchers around the world will continue developing new Pummerer-based transformations and will also apply this chemistry to the synthesis of complex natural products and other bioactive compounds.

REFERENCES

1. Feldman, K. S. *Tetrahedron* **2006**, *62*, 5003–5034 and references therein.
2. Patil, M.; Loerbroks, C.; Thiel, W. *Org. Lett.* **2013**, *15*, 1682–1685.
3. (a) Numata, T.; Itoh, O.; Oae, S. *Tetrahedron Lett.* **1979**, *20*, 1869–1870. (b) Kita, Y.; Shibata, N.; Kawano, N.; Fukui, S.; Fujimori, C. *Tetrahedron Lett.* **1994**, *35*, 3575–3576. (c) Shibata, N.; Matsugi, M.; Kawano, N.; Fukui, S.; Fujimori, C.; Gotanda, K.; Murata, K.; Kita, Y. *Tetrahedron: Asymmetry* **1997**, *8*, 303–310. (d) Kita, Y.; Shibata, N.; Yoshida, N. *Tetrahedron Lett.* **1993**, *34*, 4063–4066. (e) Kita, Y.; Shibata, N.; Kawano, N.; Tohjo, T.; Fujimori, C.; Ohishi, H. *J. Am. Chem. Soc.* **1994**, *116*, 5116–5121. (f) Nagao, Y.; Miyamoto, S.; Miyamoto, M.; Takeshige, H.; Hayashi, K.; Sano, S.; Shiro, M.; Yamaguchi, K.; Sei, Y. *J. Am. Chem. Soc.* **2006**, *128*, 9722–9729.
4. (a) Fernández, I.; Khiar, N. *Chem. Rev.* **2003**, *103*, 3651–3706. (b) Wojaczyńska, E.; Wojaczyński, J. *Chem. Rev.* **2010**, *110*, 4303–4356. (c) Carreno, M. C. *Chem. Rev.* **1995**, *95*, 1717–1760. (d) Rayner, C. M. *Contemp. Org. Synth.* **1995**, *2*, 409–440. (e) Koval, I. *Russ. Chem. Rev.* **1994**, *63*, 323–344 and references therein.
5. (a) van Steenis, J. H.; van Es, J. J. G. S.; van der Gen, A. *Eur. J. Org. Chem.* **2000**, 2787–2793. (b) Maezaki, N.; Izumi, M.; Yuyama, S.; Sawamoto, H.; Iwata, C.; Tanaka, T. *Tetrahedron* **2000**, *56*, 7927–7945.
6. Michrowska, A.; Bieniek, M.; Kim, M.; Klajn, R.; Grela, K. *Tetrahedron* **2003**, *59*, 4525–4531.
7. Gamba-Sanchez, D.; Oriez, R.; Prunet, J. *Tetrahedron Lett.* **2009**, *50*, 883–885.
8. Gamba-Sanchez, D.; Prunet, J. *J. Org. Chem.* **2010**, *75*, 3129–3132.
9. (a) Raghavan, S.; Kumar, C. N. *Indian J. Chem., Sec B.* **2011**, *50B*, 821–828. (b) Raghavan, S.; Rathore, K.; Sridhar, B. *Indian J. Chem., Sec B.* **2011**, *50B*, 559–579.
10. Amici, R. R.; Di Vitta, C.; Marzorati, L. *Synthesis* **2013**, *45*, 798–802.

11. Raghavan, S.; Ganapathy, S.; Tony, K. A. *Tetrahedron Lett.* **2008**, *49*, 1601–1604.

12. Kuhakarn, C.; Seehasombat, P.; Jaipetch, T.; Pohmakotr, M.; Reutrakul, V. *Tetrahedron* **2008**, *64*, 1663–1670.

13. Kobayashi, S.; Ishii, A.; Toyota, M. *Synlett* **2008**, *2008*, 1086–1090.

14. Sanseverino, J.; Chambron, J.-C.; Aubert, E.; Espinosa, E. *J. Org. Chem.* **2011**, *76*, 1914–1917.

15. Garcia, C.; Andraud, C.; Collet, A. *Supramol. Chem.* **1992**, *1*, 31–45.

16. Pohmakotr, M.; Kambutong, S.; Tuchinda, P.; Kuhakarn, C. *Tetrahedron* **2008**, *64*, 6315–6323.

17. (a) Feldman, K. S.; Fodor, M. D. *J. Org. Chem.* **2009**, *74*, 3449–3461. (b) Feldman, K. S.; Fodor, M. D. *J. Am. Chem. Soc.* **2008**, *130*, 14964–14965. (c) Feldman, K. S.; Fodor, M. D.; Skoumbourdis, A. P. *Synthesis* **2009**, 3162–3173.

18. (a) Lawrence, A. K.; Gademann, K. *Synthesis* **2008**, 331–351. (b) Smith, L. H. S.; Coote, S. C.; Sneddon, H. F.; Procter, D. J. *Angew. Chem. Int. Ed.* **2010**, *49*, 5832–5844.

19. Kobayashi, J. I.; Hirasawa, Y.; Yoshida, N.; Morita, H. *J. Org. Chem.* **2001**, *66*, 5901–5904.

20. Ishiuchi, K. I.; Kubota, T.; Hoshino, T.; Obara, Y.; Nakahata, N.; Kobayashi, J. I. *Bioorg. Med. Chem.* **2006**, *14*, 5995–6000.

21. Nishimura, T.; Unni, A. K.; Yokoshima, S.; Fukuyama, T. *J. Am. Chem. Soc.* **2010**, *133*, 418–419.

22. Shimizu, Y.; Shi, S.-L.; Usuda, H.; Kanai, M.; Shibasaki, M. *Tetrahedron* **2010**, *66*, 6569–6584.

23. (a) Seiple, I. B.; Su, S.; Young, I. S.; Nakamura, A.; Yamaguchi, J.; Jørgensen, L.; Rodriguez, R. A.; O'Malley, D. P.; Gaich, T.; Köck, M.; Baran, P. S. *J. Am. Chem. Soc.* **2011**, *133*, 14710–14726. (b) Seiple, I. B.; Su, S.; Young, I. S.; Lewis, C. A.; Yamaguchi, J.; Baran, P. S. *Angew. Chem. Int. Ed.* **2010**, *49*, 1095–1098. (c) Li, Q.; Hurley, P.; Ding, H.; Roberts, A. G.; Akella, R.; Harran, P. G. *J. Org. Chem.* **2009**, *74*, 5909–5919. (d) Zancanella, M. A.; Romo, D. *Org. Lett.* **2008**, *10*, 3685–3688. (e) Namba, K.; Kaihara, Y.; Yamamoto, H.; Imagawa, H.; Tanino, K.; Williams, R. M.; Nishizawa, M. *Chem. Eur. J.* **2009**, *15*, 6560–6563.

24. (a) Feldman, K. S.; Nuriye, A. Y. *Org. Lett.* **2010**, *12*, 4532–4535. (b) Feldman, K. S.; Nuriye, A. Y.; Li, J. *J. Org. Chem.* **2011**, *76*, 5042–5060.

25. Halder, S.; Satyam, A. *Tetrahedron Lett.* **2011**, *52*, 1179–1182.

26. (a) Tayu, M.; Higuchi, K.; Inaba, M.; Kawasaki, T. *Org. Biomol. Chem.* **2013**, *11*, 496–502. (b) Higuchi, K.; Tayu, M.; Kawasaki, T. *Chem. Commun.* **2011**, *47*, 6728–6730.

27. (a) Kobayashi, K.; Suzuki, T.; Horiuchi, M.; Shiroyama, Y.; Konishi, H. *Synthesis* **2011**, 2897–2906. (b) Kobayashi, K.; Kanbe, Y.; Horiuchi, M. *Synthesis* **2011**, 3429–3434.

28. Sánchez-Obregón, R.; Salgado, F.; Ortiz, B.; Díaz, E.; Yuste, F.; Walls, F.; García Ruano, J. L. *Tetrahedron* **2007**, *63*, 10521–10527.

29. (a) Kobatake, T.; Yoshida, S.; Yorimitsu, H.; Oshima, K. *Angew. Chem. Int. Ed.* **2010**, *49*, 2340–2343. (b) Iwasaki, M.; Fujino, D.; Wada, T.; Kondoh, A.; Yorimitsu, H.; Oshima, K. *Chem. Asian J.* **2011**, *6*, 3190–3194. (c) Murakami, K.; Imoto, J.; Matsubara, H.; Yoshida, S.; Yorimitsu, H.; Oshima, K. *Chem. Eur. J.* **2013**, *19*, 5625–5630.

30. (a) Huang, X.; Maulide, N. *J. Am. Chem. Soc.* **2011**, *133*, 8510–8513. (b) Huang, X.; Patil, M.; Farès, C.; Thiel, W.; Maulide, N. J. *Am. Chem. Soc.* **2013**, *135*, 7312–7323.

31. (a) Eberhart, A. J.; Procter, D. J. *Angew. Chem. Int. Ed.* **2013**, *52*, 4008–4011. (b) Eberhart, A. J.; Imbriglio, J. E.; Procter, D. J. *Org. Lett.* **2011**, *13*, 5882–5885.
32. (a) Kobatake, T.; Fujino, D.; Yoshida, S.; Yorimitsu, H.; Oshima, K. *J. Am. Chem. Soc.* **2010**, *132*, 11838–11840. (b) Ookubo, Y.; Wakamiya, A.; Yorimitsu, H.; Osuka, A. *Chem. Eur. J.* **2012**, *18*, 12690–12697.
33. Miller, M.; Tsang, W.; Merritt, A.; Procter, D. J. *Chem. Commun.* **2007**, 498–500.
34. Ovens, C.; Martin, N. G.; Procter, D. J. *Org. Lett.* **2008**, *10*, 1441–1444.
35. Ovens, C.; Vogel, J. C.; Martin, N. G.; Procter, D. J. *Chem. Commun.* **2009**, 3101–3103.
36. Miller, M.; Vogel, J. C.; Tsang, W.; Merrit, A.; Procter, D. J. *Org. Biomol. Chem.* **2009**, *7*, 589–597.
37. Coote, S. C.; Quenum, S.; Procter, D. J. *Org. Biomol. Chem.* **2011**, *9*, 5104–5108.
38. (a) Le Cesne, A. *Expert Rev. Anticancer Ther.* **2013**, *13*, s11–s19. (b) Reichardt, P. *Expert Rev. Anticancer Ther.* **2013**, *13*, s25–s30.
39. Smith, L. H. S.; Nguyen, T. T.; Sneddon, H. F.; Procter, D. J. *Chem. Commun.* **2011**, *47*, 10821–10823.
40. (a) Pellissier, H. *Chem. Rev.* **2012**, *113*, 442–524. (b) Hussain, M. M.; Walsh, P. J. *Acc. Chem. Res.* **2008**, *41*, 883–893. (c) Wasilke, J.-C.; Obrey, S. J.; Baker, R. T.; Bazan, G. C. *Chem. Rev.* **2005**, *105*, 1001–1020. (d) Volla, C. M. R.; Atodiresei, I.; Rueping, M. *Chem. Rev.* **2014**, *114*, 2390–2431.
41. Li, Z.; Li, H.; Guo, X.; Cao, L.; Yu, R.; Li, H.; Pan, S. *Org. Lett.* **2008**, *10*, 803–805.
42. Haimowitz, T.; Fitzgerald, M. E.; Winkler, J. D. *Tetrahedron Lett.* **2011**, *52*, 2162–2164.
43. Hugenberg, V.; Haufe, G. *J. Fluorine Chem.* **2012**, *143*, 238–262.
44. (a) Tosh, D. K.; Choi, W. J.; Kim, H. O.; Lee, Y.; Pal, S.; Hou, X.; Choi, J.; Choi, S.; Jeong, L. S. *J. Org. Chem.* **2008**, *73*, 4259–4262. (b) Taniike, H.; Inagaki, Y.; Matsuda, A.; Minakawa, N. *Tetrahedron* **2011**, *67*, 7977–7982. (c) Jayakanthan, K.; Johnston, B. D.; Pinto, B. M. *Carbohydr. Res.* **2008**, *343*, 1790–1800. (d) Jeong, L. S.; Choi, Y. N.; Tosh, D. K.; Choi, W. J.; Kim, H. O.; Choi, J. *Bioorg. Med. Chem.* **2008**, *16*, 9891–9897.
45. (a) Jeong, L. S.; Tosh, D. K.; Kim, H. O.; Wang, T.; Hou, X.; Yun, H. S.; Kwon, Y.; Lee, S. K.; Choi, J.; Zhao, L. X. *Org. Lett.* **2007**, *10*, 209–212. (b) Alexander, V.; Choi, W. J.; Chun, J.; Kim, H. O.; Jeon, J. H.; Tosh, D. K.; Lee, H. W.; Chandra, G.; Choi, J.; Jeong, L. S. *Org. Lett.* **2010**, *12*, 2242–2245.
46. Urabe, D.; Yamaguchi, H.; Someya, A.; Inoue, M. *Org. Lett.* **2012**, *14*, 3842–3845.

INDEX

Note: Page numbers in italics refer to figures and page numbers in bold refer to tables

Molecular Rearrangements in Organic Synthesis, First Edition. Edited by Christian M. Rojas.

Printed and bound by CPI Group (UK) Ltd, Croydon, CR0 4YY
16/09/2026
14963423-0007